DATE			

DESIGN OF FEEDBACK CONTROL SYSTEMS

Second Edition

Gene H. Hostetter
University of California, Irvine

Clement J. Savant, Jr.
California State University, Los Angeles

Raymond T. Stefani
California State University, Long Beach

Saunders College Publishing
A Division of Holt, Rinehart and Winston, Inc.

Fort Worth · Chicago · San Francisco · Philadelphia
Montreal · Toronto · London · Sydney · Tokyo

TO
GENE, DONNA, COLLEEN, AND KRISTEN
BARBARA
ANN, TED, AND RICK

Publisher   *Ted Buchholz*
Acquisitions Editor   *Deborah Moore*
Senior Project Manager   *Marc Sherman*
Production Manager   *Roger Kasunic*
Design Supervisor   *Judy Allan*

Library of Congress Cataloging-in-Publication Data

Hostetter, G. H. 1939–
    Design of feedback control systems.

    Includes bibliographies and index.
    1. Feedback control systems.   I. Savant, C. J.
II. Stefani, Raymond T.   III. Title.
TJ216.H63   1989   629.8′312   88-26362
ISBN 0-03-013492-7

Printed in the United States of America

012   039   98765432

Preface

In writing this book our intention has been to provide a clear, understandable, and comprehensive textbook, a sort of window into the world of control system science. We have tried to clarify and explain topics in a step-by-step manner. Our aim is to bring even the more challenging concepts well within the grasp of the average student. We want all students to share in the learning experience: from the average to the superlative, whether at a large university or small, and whether mainframe computers or only hand-held calculators are available.

We are pleased with the acceptance of the first edition of *Design of Feedback Control Systems*, and have drawn from that experience to clarify and reorganize the material into an improved second edition. We have greatly enhanced the design approach of the first edition with many more design examples and with a more complete method of control system design.

Each chapter begins with a preview which places that chapter's contents into perspective with the other chapters, and often into historical perspective as well. Throughout each chapter, the main principles are developed in sufficient detail so that the average reader can attain expertise. A multitude of drill problems and solutions is provided so that the reader can gauge the level of understanding, then repeat material where understanding is weak or proceed to the next section. The last few sections of each chapter provide comprehensive design examples that demonstrate the usefulness of that chapter's material. These examples span a wide spectrum of topics: devices such as a force-balance accelerometer and a phase-locked loop; biomedical applications such as an insulin delivery system and an artificial limb; and large-scale systems such as a high-performance jet fighter and a flexible spacecraft. Each chapter closes with a review as a starting point for exam preparation for students or as reference material in an industrial environment.

USE OF THIS TEXTBOOK

The text can be divided into five main areas: modeling, classical analysis, classical design, state variable methods, and digital control.

Modeling is covered in Chap. 1. Topics include: creating differential equations for common electromechanical devices, solving those equations, clarifying system structure using block diagrams, and obtaining the transfer function.

Classical analysis is covered in Chaps. 2, 3, 4, and 6 (through Sec. 6.7). Design concepts are spread throughout, but analysis is the main thrust. Topics include: analyzing the behavior of systems describable by first- or second-order linear differential equations, determining stability using Routh-Hurwitz, analyzing the significance and performance of tracking systems with power-of-time inputs, and using root locus and Bode/Nyquist methods to determine stability.

Classical design is covered in Chaps. 5 and 6 (Sec. 6.8 through the end). It is shown that design is an iterative process that requires establishing and evaluating alternatives. Design concepts are also spread throughout the chapters dealing primarily with classical analysis. A common system is used and compensators are selected using the root locus viewpoint in Chap. 5 and the Bode/Nyquist viewpoint in Chap. 6. The use of a common example permits comparison of the different compensators and of the two approaches.

State variable methods are covered in Chap. 7. Topics include: state variable representation, solving for the time response, controllability, observability, gain selection, steady state effects, observers, and quadratic optimal control.

Digital control is covered in Chap. 8. Topics include: sampling effects, state-variable representation, and stability analysis.

The five areas represent building blocks to construct a course. We have purposely included more material than a three-semester unit or a four-quarter unit course would normally cover. This extra material is intended to give the instructor flexibility in structuring a course to meet the needs of the program, the university, and the community served. We suggest that it is better to cover a smaller number of topics well rather than a larger number poorly.

For example, a basic-level control systems course could encompass classical and modern concepts by covering Chaps. 1 through 6 and Secs. 7.1 through 7.7 (including a review of the matrix material in Appendix A).

For those students who already have an understanding of linear system methods, including matrix methods, much of Chap. 1, parts of Chap. 7, and Appendix A can be omitted, leaving time to cover most of Chaps. 2 through 8.

A course emphasizing design (with appropriate design projects) could cover Secs. 6.7 through 6.16, 7.7 through 7.11, and 8.10. The possibilities are endless.

Given the nature of the technology, we agree with the emphasis placed upon design by the Accrediting Board for Engineering and Technology (ABET). We suggest that design be a focal point of course organization.

USE OF COMPUTERS

We encourage the use of computers from mainframe to PC to hand-held devices. Computational packages are readily available to carry out the root locus methods of Chapters 4 and 5, the frequency response methods of Chapter 6, and the state variable methods of Chapter 7. For example, some excellent PC packages include Turbo Pascal's scientific and graphics library, MATLAB, Ctrl-C, and Matrix-x. A package of PC routines written by Patrick Wang may be obtained from the publisher as a companion to this text. The computer program user manual is entitled *Computer Programs for Design of Feedback Control Systems*, 2nd Ed.

We also inject a word of caution. It is absolutely essential that the program user perform an analysis of the problem by hand to predict general trends and results. If the computer results are significantly different than expected, encoding errors (or lack of understanding) can be corrected. However, if the computer output is accepted with blind faith, then the user and the computer reverse their master/slave roles. A partnership should exist between the textbook, the computer, the instructor, and the student as the learning process unfolds.

CHANGES IN THE SECOND EDITION

For those who have used the first edition of this text, we wish to briefly review the many changes that we made to improve our educational product. Based on comments from faculty, students, and reviewers, we have introduced several hundred minor clarifications to better explain the material. A large number of new problems may be found at the end of each chapter. About 90 percent of the problems are either new or changed as compared with the first edition.

Chapter 1 of the second edition contains material from the first two chapters of the first edition. Block diagrams and signal flow graphs are now in Chap. 1 to provide a logical sequence in which we write the differential equations of important electromechanical systems, solve the equations using Laplace transforms, create a block diagram using Laplace-transform variables, and then obtain the transfer function. We now use an easier-to-understand presentation of spring-mass-damper translational and rotational systems and the resulting free-body diagrams. We have added such important devices as accelerometers and gyroscopes. The table of electromechanical devices contains a better format. A block diagram now helps to visualize the positioning servo dynamics. A force-balance accelerometer design example is added.

Chapter 2 of the second edition generally follows Chaps. 2 and 3 of the first edition, except for the deletion of block diagrams, signal flow graphs, and Mason's rule, which are now in Chap. 1, and the artificial limb example now in Chap. 4. Performance is thus carried from first order to second order and then to the stability of any order system using Routh-Hurwitz. The preview has been rewritten to put the material into better perspective, both historically and in terms of this text. The settling time curve is explained in greater detail. The present table 2.1 is improved, and polynomial factoring is now included.

Chapter 3 of the second edition generally follows Chap. 4 of the first edition. The preview and second section have been rewritten to better explain the significance of tracking systems and especially power-of-time inputs. The disturbance section now refers back to the positioning servo example.

Chapter 4 of the second edition generally follows Chap. 3 of the first edition, except that compensation is now in Chap. 5. The preview better explains the relevance of the root locus method and the proper use of computers. The various sketching rules (for example, locating the centroid) are proven in greater detail. Chap. 4 now includes the artificial limb design example. Two new design examples are added: a hydroelectric generating system and control of a flexible spacecraft.

Chapter 5 of the second edition is a greatly expanded version of the root locus design material from Chap. 5 of the first edition. Root locus design now occupies an entire chapter due to the relative importance of that topic. Each compensator design is applied to the same example system. By tabularizing the results, the relative merits of the compensators are clearly exhibited. The PID compensator was redesigned to provide improved performance as compared with the other compensators. A new example is included: design of a high-performance (open-loop unstable) aircraft using the root locus viewpoint. This chapter should help universities meet ABET design requirements.

Chapters 6 through 8 of both editions are similar in content, but improved in presentation. A number of clarifying statements have been added to Chap. 6 to better present frequency response methods. The material on sketching frequency domain magnitude and phase using the pole-zero plot has been moved earlier in the chapter to a more logical location. Bode plots then follow. The presentation of Nyquist plots has been reorganized and made clearer. Answers are provided for all Bode and Nyquist drill problems. Sections 6.7 through 6.13 have been added (design of compensators from the frequency domain viewpoint). The compensation material parallels the root locus coverage so the tabularized results can be compared. Section 6.14 contains a new example: design of a flexible spacecraft from the Bode viewpoint.

Besides adding clarifying statements throughout Chapter 7, we have added observers (Sec. 7.8) and quadratic optimal control (Sec. 7.9). Both topics are important state variable design methods. The magnetically elevated train material in Sec. 7.10 was revised to include more realistic values and to include the use of an observer.

From Chap. 8 we deleted the specifics of A/D and D/A circuitry, and we moved computer chronology to an earlier position. Our intention is to allow the instructor to make more efficient use of class time since this topic is probably covered later in the term.

Appendices A and B are new. Appendix A briefly reviews matrix algebra to serve as reference material for Chap. 7. Appendix B contains two algorithms. The first can be used to compute crossover frequencies and the second can be used to compute breakaway points. Each algorithm can be programmed on a hand-held computer.

ACKNOWLEDGMENTS

So many people have shaped our professional careers that a second text would be needed to list them all. We wish to recognize the faculty, staff, and students of the California State University, Long Beach, the California State University, Los Angeles, and the University of California, Irvine, for their help and support over the years. We wish to thank colleagues and reviewers from across the country and abroad for their advice. These include: B. Atabek, The Catholic University of America; R. Burns, DeVry Institute of Technology; R. Christiansen, Brigham Young University; S. K. Dalta, California Polytechnic State University; B. Egbert, Wichita State University; H. K. Eldeib, George Mason University; J. Feele, University of Idaho; J. Froya, Rose-Hulman Institute of Technology; W. Higgins, Arizona State University; P. Lewis, Michigan Technological University; and R. Molholland, University of Oklahoma. We gratefully thank those who prepared the manuscripts for both editions: Cynthia Klepadlo, Sarah Yap, and Hoang Khong. We thank Mohammed Zahzah for preparing the solutions manual that accompanies this text. We also wish to express our appreciation for the efforts of the editors and staff of Holt, Rinehart and Winston, Inc., and in particular Senior Project Manager Marc Sherman, who provided invaluable guidance and focus throughout the preparation of this revision. Finally, we thank the reader for supporting this text. We would appreciate your comments and suggestions as we move toward our goal of providing the best possible control systems textbook.

Gene H. Hostetter
Clement J. Savant, Jr.
Raymond T. Stefani
June 1988

Publisher's note: Shortly before this book was published, Gene H. Hostetter passed away after a long illness. The publisher and co-authors wish to express their sorrow over the loss of their esteemed colleague, and to recognize Dr. Hostetter's myriad achievements in electrical engineering.

SAUNDERS COLLEGE PUBLISHING
A Division of Holt, Rinehart and Winston, Inc.

SERIES IN ELECTRICAL ENGINEERING

M.E. Van Valkenburg, *Senior Consulting Editor*
Adel S. Sedra, *Series Editor/Electrical Engineering*
Michael R. Lightner, *Series Editor/Computer Engineering*

Contents

Chapter

1

Continuous-Time System Description

1.1 PREVIEW

The first conscious use of feedback control of a physical system by mankind lives in prehistory. Possibly it was a spillway in an irrigation network, where excess water was automatically drained. Development of a mathematical framework for the description, analysis, and design of control systems dates from the introduction of James Watt's flyball governor (1760), which was used to regulate the speed of steam engines, and the subsequent work by James Clerk Maxwell (ca. 1868) and others to improve the design and extend its applicability.

Since that era, the theory and practice of control system design advanced rapidly. Important new concepts and tools were developed in connection with telephone and radio communications in the 1920s and 1930s. Rather poorly performing electronic devices, including amplifiers and modulators, were dramatically improved by feedback. World War II further accelerated the development of classical control theory and practice. Heavy guns had to be rapidly and accurately positioned. Precise navigation and target tracking were increasingly important, and aircraft performance was improved greatly with incorporation of complex control systems to aid the pilot. Later, "automation" became a household word as industry began to depend more and more upon automatically controlled machinery.

Today, feedback control systems are pervasive in industry and in our everyday lives. They range from governmental regulation (such as that governing monetary policy) to automated and highly flexible manufacturing plants to sophisticated automobiles, household appliances, and entertainment systems. It is our purpose to learn to design feedback control systems for a wide variety of applications.

Control system designers find that block diagrams provide a particularly useful way to visualize the interconnections of system components, thus revealing the system structure. Successful design begins by creating a mathematical model of the system to be stabilized. Next, the contents of the blocks within a diagram must be identified. Finally, values must be selected for those parameters that are adjustable and sometimes additional components must be added to provide acceptable performance.

This chapter begins by defining basic control system terminology. Since design requires a model of each system of interest, the behaviors of many typical electrical, mechanical, and electromechanical systems are described. The resulting differential equations must be rendered into a form useful to the controls engineer. That goal can be accomplished by Laplace transforming each differential equation and then generating a relationship, the transmittance, between the input and output of each block of a control system. Those transmittances become the contents of the blocks in the system block diagram.

The block diagram can be reduced to just one input-output relationship, the system overall transfer function, by converting the block diagram into an equivalent form, the signal flow graph, and then applying a set of rules. Subsequent chapters will describe the design steps that follow, once the block diagram is defined and the transfer function is available.

All the chapters of this text conclude with examples that are intended to reinforce the key points of the chapter in an interesting and informative manner. Chapter 1 concludes with discussions of a positioning servo and a force-balance accelerometer.

Where the material in the first chapter involves subjects already known to the reader from previous experience, the text provides a coherent review. Such topics need not be covered during classroom lectures. The emphasis here is on using, rather than proving, results.

1.2 THE FEEDBACK CONCEPT AND MODELING

Control System Terminology

Control systems influence every facet of modern life. Automatic washers and dryers, microwave ovens, chemical process plants, navigation and guidance

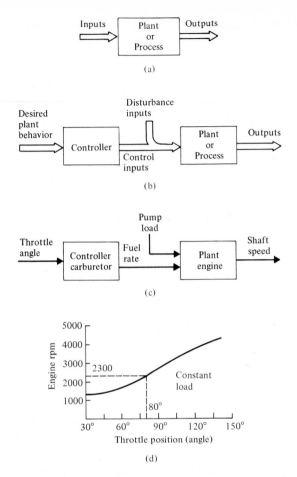

Figure 1.1 **(a)** A plant or process to be controlled. **(b)** An open-loop control system. **(c)** Example of an open-loop control system. **(d)** An engine speed versus throttle angle curve.

systems, space satellites, pollution control, mass transit, and economic regulation are a few examples. A control system is, in the broadest sense, any interconnection of components to provide a desired function.

The portion of a system which is to be controlled is called the *plant* or the *process*. It is affected by applied signals, called *inputs*, and produces signals of particular interest, called *outputs*, as indicated in Fig. 1.1a. The plant is fixed insofar as the control system designer is concerned. Whether the plant is an automobile engine, electrical generator, or nuclear reactor, it is the designer's job to ensure that the plant operates as required. Other components must be specially created and connected as a means to that end.

A *controller* may be used to produce a desired behavior of the plant, as shown in Fig. 1.1b. The controller generates plant input signals designed to produce desired outputs. Some of the plant inputs are accessible to the designer and some are generally not available. The inaccessible input signals are often disturbances to the plant. The double lines in the figure indicate that several signals of each type may be involved. This system is termed *open-loop* because the control inputs are not influenced by the plant outputs; that is, there is no *feedback* around the plant.

Such an open-loop control system has the advantage of simplicity, but its performance is highly dependent upon the properties of the plant, which may vary with time. The disturbances to the plant may also create unwanted response which it would be desirable to reduce.

As an example, suppose a gasoline engine is used to drive a large pump, as depicted in Fig. 1.1c. The carburetor and engine comprise a common type of control system wherein large-power output is controlled with a small-power input. The carburetor is the controller in this case, and the engine is the plant. The fuel rate is the control input, and the pump load is a disturbance signal. The desired plant output, a certain engine shaft speed, may be obtained by adjusting the throttle angle. The single lines in the figure indicate individual signals.

A representative plot of engine speed versus throttle angle is sketched in Fig. 1.1d. This "calibration curve" gives the engine speed for a given throttle setting, at constant load on the engine. To produce an engine speed of 2300 rpm, for example, set the throttle angle to 80°. If the engine should become untuned (a change in the plant) or if the load should change (a disturbance), the calibration curve would change, and an 80° throttle angle would no longer produce a 2300-rpm engine speed.

In applications such as automatic washing machines and variable-speed hand drills, maintaining an accurate calibration curve is of little importance, within bounds. In other applications, such as laboratory instrumentation systems, it suffices to calibrate the system, reestablishing knowledge of the input-output relation often enough to obtain the desired accuracy. In systems such as the automobile, the human operator is capable of adjusting to changes and distur-bances in the plant. In driving another's automobile for the first time, a new sense of "feel" must be established because no two automobiles produce exactly the same engine performance with the same accelerator setting.

The Feedback Concept

If the requirements of the system cannot be satisfied with an open-loop control system, a *closed-loop* or *feedback* system is desirable. A path (or loop) is provided from the output back to the controller. Some or all of the system outputs are measured and used by the controller, as indicated in Fig. 1.2a. The controller may then compare a desired plant output with the actual output and act to reduce the difference between the two.

Suppose that the system comprising a gasoline engine driving a pump is arranged in a closed-loop manner. One possible feedback control configuration is shown in Fig. 1.2b. A tachometer produces a voltage proportional to the engine shaft speed. The input voltage, which is proportional to the desired speed, is set

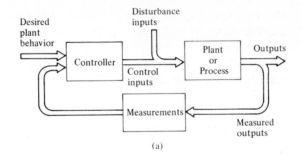

(a)

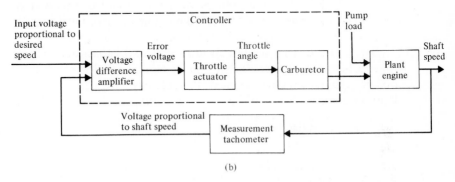

(b)

Figure 1.2 **(a)** Closed-loop or feedback control. **(b)** A closed-loop engine control system.

with a potentiometer. The tachometer voltage is subtracted from the input voltage, giving an error voltage which is proportional to the difference between the actual speed and the desired speed.

The error voltage is then amplified and used to position the throttle. The throttle actuator could be a reversible electric motor, geared to the throttle arm. When the engine shaft speed is equal to the desired speed (when the difference or *error* is zero), the throttle remains fixed. If a change in load or a change in the engine components should occur in the system, and the actual speed is no longer equal to the desired speed, the error voltage becomes nonzero, causing the throttle setting to change so that the actual speed approaches the desired speed. The controller here consists of the voltage difference amplifier, throttle actuator, and carburetor.

Some of the advantages which feedback control offer to the designer are

1. *Increased accuracy.* The closed-loop system may be designed to drive the error between desired and measured response to zero.
2. *Reduced sensitivity to changes in components.* As in the previous example, the system may be designed to seek zero error despite changes in the plant.
3. *Reduced effects of disturbances.* The effects of disturbances to the system may be greatly attenuated.

4. *Increased speed of response and bandwidth.* Feedback may be used to increase the range of frequencies over which a system will respond and to make it respond more desirably. A satellite booster rocket, for example, has aerodynamics resembling those of a giant broomstick. It may, with feedback, behave with beauty and grace.

Electrical Components

Electrical networks are governed by the two Kirchoff laws:

1. The algebraic sum of voltages around a closed loop equals zero.
2. The algebraic sum of currents flowing into a circuit node equals zero.

Network element models include resistors, capacitors, inductors, voltage sources, and current sources. The voltage-current relations for these are summarized in Table 1.1. One systematic method of network analysis consists of defining a loop current in each loop of a network, equating the algebraic sums of the voltages

Table 1.1 ELECTRICAL ELEMENT VOLTAGE-CURRENT RELATIONS

Resistor	Inductor	Capacitor
$v(t) = Ri(t)$	$v(t) = L \dfrac{di}{dt}$	$v(t) = \dfrac{1}{C} \displaystyle\int_{-\infty}^{t} i(t)dt$
$i(t) = \dfrac{1}{R} v(t)$	$i(t) = \dfrac{1}{L} \displaystyle\int_{-\infty}^{t} v(t)dt$	$i(t) = C \dfrac{dv}{dt}$

Voltage Source	Current Source
$v(t)$ a given function of time	$i(t)$ a given function of time
$v(t)$ expressed in terms of other network voltages or currents	$i(t)$ expressed in terms of other network voltages or currents

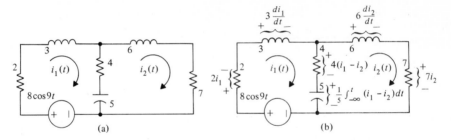

Figure 1.3 Writing simultaneous loop equations for an electrical network.
(a) Electrical network example for loop analysis. **(b)** Network with element voltages expressed in terms of the loop currents.

around each loop to zero. For example, in the network of Fig. 1.3, the loop currents $i_1(t)$ and $i_2(t)$ have been defined. In Fig. 1.3b the resistor, inductor, and capacitor voltages have been expressed in terms of the loop currents. The application of Kirchhoff's voltage law around the i_1 loop gives.

$$2i_1 + 3\frac{di_1}{dt} + 4(i_1 - i_2) + \frac{1}{5}\int_{-\infty}^{t}(i_1 - i_2)\,dt = 8\cos 9t$$

Similarly, around the i_2 loop,

$$6\frac{di_2}{dt} + 7i_2 - 4(i_1 - i_2) - \frac{1}{5}\int_{-\infty}^{t}(i_1 - i_2)\,dt = 0$$

Collecting terms, the following simultaneous integrodifferential equations in i_1 and i_2 result:

$$\begin{cases} 3\dfrac{di_1}{dt} + 6i_1 + \dfrac{1}{5}\int_{-\infty}^{t} i_1\,dt - 4i_2 - \dfrac{1}{5}\int_{-\infty}^{t} i_2\,dt = 8\cos 9t \\[3mm] -4i_1 - \dfrac{1}{5}\int_{-\infty}^{t} i_1\,dt + 6\dfrac{di_2}{dt} + 11i_2 + \dfrac{1}{5}\int_{-\infty}^{t} i_2\,dt = 0 \end{cases}$$

In the nodal method of network analysis, one node in the network is chosen as the reference node and voltages between the reference node and each other node are defined. Expressing the element currents in terms of node voltages and applying Kirchhoff's current law at each node except the reference node gives the same number of independent simultaneous integrodifferential equations as there are node voltages. For the network of Fig. 1.4a, the node voltages are labeled $v_1(t)$ and $v_2(t)$. In Fig. 1.4b, the branch currents are expressed in terms of these node voltages. Applying Kirchhoff's current law at node 1 gives

$$\frac{1}{2}\int_{-\infty}^{t} v_1\,dt + \frac{1}{3}v_1 + 4\frac{d}{dt}(v_1 - v_2) + \frac{1}{5}(v_1 - v_2) = 12$$

At node 2,

$$\frac{1}{6}v_2 - 4\frac{d}{dt}(v_1 - v_2) - \frac{1}{5}(v_1 - v_2) = -\sin t$$

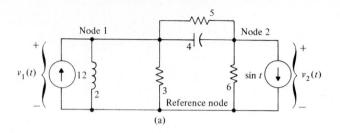

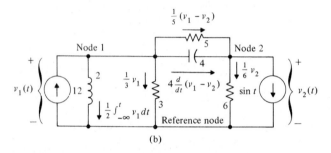

Figure 1.4 Writing simultaneous nodal equations for an electrical network.
(a) Electrical network example for nodal analysis.   **(b)** Network with element currents expressed in terms of the node voltages.

Collecting terms, the following two simultaneous integrodifferential equations in $v_1(t)$ and $v_2(t)$ result:

$$\begin{cases} 4\dfrac{dv_1}{dt} + \dfrac{8}{15}v_1 + \dfrac{1}{2}\displaystyle\int_{-\infty}^{t} v_1\,dt - 4\dfrac{dv_2}{dt} - \dfrac{1}{5}v_2 = 12 \\[3mm] -4\dfrac{dv_1}{dt} - \dfrac{1}{5}v_1 + 4\dfrac{dv_2}{dt} + \dfrac{11}{30}v_2 = -\sin t \end{cases}$$

Simple models for other common electrical and electronic devices are summarized in Table 1.2.

Translational Mechanical Components

The force and position relations for the translational mechanical mass, spring, and damper elements are given in Table 1.3. As in a free-body diagram, the forces shown are those applied to the element. A method of analysis for translational mechanical systems involving these elements is as follows:

1. **Define positions with directional senses for each mass in the system.**
2. **Draw free-body diagrams for each of the masses, expressing the forces on them in terms of mass positions.**

Table 1.2 SIMPLE MODELS OF SOME ELECTRICAL AND ELECTRONIC DEVICES

1. *Transformer* with core in linear region

Resistance of the coil wires can be included as additional resistors at each port.

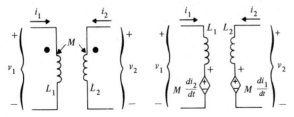

2. *Ideal transformer*

The ideal transformer models a transformer with perfect magnetic coupling, $M = \sqrt{L_1 L_2}$. For the ideal transformer,

$$v_2 = \frac{N_2}{N_1} v_1 \qquad i_2 = -\frac{N_1}{N_2} i_1$$

where N_1 and N_2 are the number of turns of the L_1 and L_2 coils, respectively.

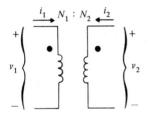

3. *Operational amplifier*

The idealized operational amplifier produces an output voltage that is proportional to the difference between two input voltages.

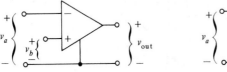 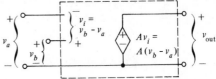

---

3. Write an equation for each mass, equating the algebraic sum of forces acting in the same direction to zero.

This procedure is applied to the system of Fig. 1.5a, where mass positions have been defined. In Fig. 1.5b, free-body diagrams for the two masses are shown. Equating forces for the first mass gives

$$4 \frac{d^2 x_1}{dt^2} + 7x_1 + 2(x_1 - x_2) + 6 \frac{d}{dt}(x_1 - x_2) = 0$$

Table 1.3   TRANSLATIONAL MECHANICAL NETWORK FORCE-POSITION RELATIONS

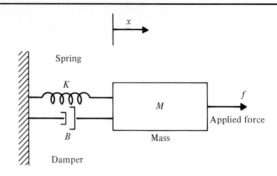

Spring-mass-damper system. At equilibrium (with no applied force) x is zero.

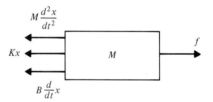

Free-body diagram.

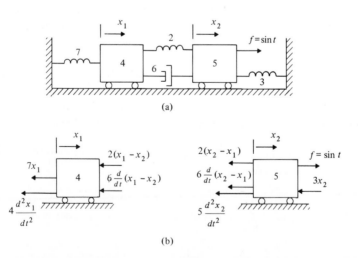

Figure 1.5 Writing simultaneous translational mechanical network equations. **(a)** A translational mechanical network. **(b)** Free-body diagrams for the masses.

Equating forces acting to the left on the second mass produces a negative sign on f because f acts to the right and because the effects of the inertia, spring, and damper act to retard changes in x while f acts to increase x.

$$5\frac{d^2 x_2}{dt^2} - f + 2(x_2 - x_1) + 6\frac{d}{dt}(x_2 - x_1) + 3x_2 = 0$$

Collecting terms, the two simultaneous differential equations, in x_1 and x_2, for the translational system are

$$\begin{cases} 4\frac{d^2 x_1}{dt^2} + 6\frac{dx_1}{dt} + 9x_1 - 6\frac{dx_2}{dt} - 2x_2 = 0 \\ -6\frac{dx_1}{dt} - 2x_1 + 5\frac{d^2 x_2}{dt^2} + 6\frac{dx_2}{dt} + 5x_2 = \sin t \end{cases}$$

Rotational Mechanical Components

Obtaining differential equations for angular motion is similar to that for translational motion. Torque-position relations for rotational elements are summarized in Table 1.4. The torques shown are those applied to the element. An analysis procedure is as follows:

1. **Define angular positions with directional senses for each rotational mass.**
2. **Draw free-body diagrams for each of the rotational masses, expressing each torque in terms of the angular positions of the masses.**
3. **Write an equation for each rotational mass, equating the algebraic sum of torques on it to zero.**

This procedure is applied to the rotational system of Fig. 1.6a. Free-body diagrams for this system are drawn in Fig. 1.6b. For the first rotational mass, equating torques give the equation.

$$3\frac{d^2 \theta_1}{dt^2} + 2\theta_1 + 5(\theta_1 - \theta_2) + 6\frac{d\theta_1}{dt} = 0$$

The applied torque, $\tau = 20$, to the second rotational mass is in the sense of increasing θ_2, so the torque equation involving it is as follows:

$$4\frac{d^2 \theta_2}{dt^2} + 5(\theta_2 - \theta_1) + 8\frac{d\theta_2}{dt} - 20 = 0$$

Collecting terms, the two simultaneous differential equations in θ_1 and θ_2 are

$$\begin{cases} 3\frac{d^2 \theta_1}{dt^2} + 6\frac{d\theta_1}{dt} + 7\theta_1 - 5\theta_2 = 0 \\ -5\theta_1 + 4\frac{d^2 \theta_2}{dt^2} + 8\frac{d\theta_1}{dt} + 5\theta_2 = 20 \end{cases}$$

Table 1.4   ROTATIONAL MECHANICAL ELEMENT TORQUE-ANGLE RELATIONS

Rotational Mass
(Inertia)

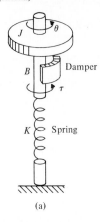

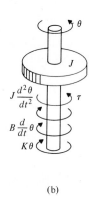

(a)   (b)

(a) Rotational spring-mass-damper system.   (b) Free-body diagram for the system
At equilibrium (with no applied torque)   of (a)
the angle θ is zero.

Gear Train

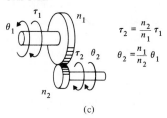

(c)

(c) Gear train

Gear Trains and Transformers

The mechanical network of Fig. 1.7a involves a gear train. It has rotational equations

$$
\begin{cases}
J_1 \dfrac{d^2\theta_1}{dt^2} + B_1 \dfrac{d\theta_1}{dt} + \tau_1 = \tau \\[2mm]
J_2 \dfrac{d^2\theta_2}{dt^2} + B_2 \dfrac{d\theta_2}{dt} - \tau_2 = 0
\end{cases}
\qquad\qquad \textbf{(1.1)}
$$

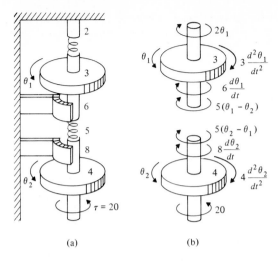

Figure 1.6 Writing simultaneous rotational mechanical network equations. **(a)** A rotational mechanical network. **(b)** Free-body diagram for the inertias.

as indicated in the free-body diagrams of Fig. 1.7b. The gear relations are

$$
\begin{cases}
\tau_2 = \dfrac{n_2}{n_1}\,\tau_1 \\[4mm]
\theta_2 = \dfrac{n_1}{n_2}\,\theta_1
\end{cases}
\tag{1.2}
$$

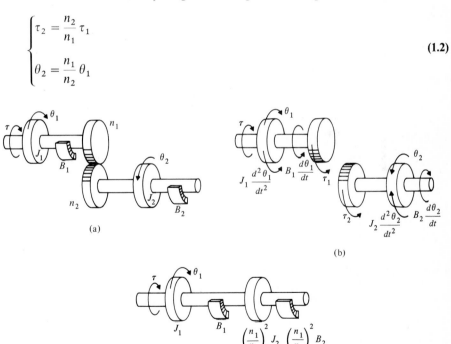

Figure 1.7 A mechanical network coupled by gears.

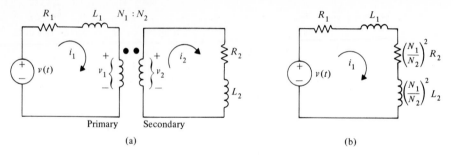

Figure 1.8 An electrical network involving a transformer.

Substituting the gear relations (1.2) into the second equation of (1.1) gives

$$\tau_1 = \left(\frac{n_1}{n_2}\right)^2 \left(J_2 \frac{d^2\theta_1}{dt^2} + B_2 \frac{d\theta_1}{dt}\right)$$

Eliminating τ_1 in the first equation of (1.1), there results

$$J_1 \frac{d^2\theta_1}{dt^2} + B_1 \frac{d\theta_1}{dt} + \left(\frac{n_1}{n_2}\right)^2 J_2 \frac{d^2\theta_1}{dt^2} + \left(\frac{n_1}{n_2}\right)^2 B_2 \frac{d\theta_1}{dt} = \tau$$

in which the gear load has been reflected to the left side of the gear train. An equivalent mechanical network is given in Fig. 1.7c.

The electrical equivalent of a gear train is a transformer. The electrical network of Fig. 1.8a involves an ideal transformer. The transformer windings are labeled "primary" and "secondary," with the primary winding closest to the voltage source $v(t)$. Simultaneous loop equations for the network are

$$\begin{cases} L_1 \dfrac{di_1}{dt} + R_1 i_1 + v_1 = v \\[3mm] L_2 \dfrac{di_2}{dt} + R_2 i_2 - v_2 = 0 \end{cases} \qquad \text{(1.3)}$$

where the ideal transformer voltages and currents are related by the turns ratio:

$$\begin{cases} v_2 = \dfrac{N_2}{N_1} v_1 \\[3mm] i_2 = \dfrac{N_1}{N_2} i_1 \end{cases} \qquad \text{(1.4)}$$

Substituting the transformer relations (1.4) into the second equation of (1.3) gives

$$v_1 = \left(\frac{N_1}{N_2}\right)^2 L_2 \frac{di_1}{dt} + \left(\frac{N_1}{N_2}\right)^2 R_2 i_1$$

Eliminating v_1 in the first equation of (1.1),

$$L_1 \frac{di_1}{dt} + R_1 i_1 + \left(\frac{N_1}{N_2}\right)^2 L_2 \frac{di_1}{dt} + \left(\frac{N_1}{N_2}\right)^2 R_2 i_1 = v$$

The secondary load, consisting of L_2 and R_2, is said to be reflected to the primary side of the transformer, through the square of the transformer turns ratio, as in the equivalent circuit of Fig. 1.8b.

This electrical network and the mechanical network of Fig. 1.7 are analogous to one another. That is, they are described by equations of the same form, with the following equivalent quantities:

$$i_1 \leftrightarrow \frac{d\theta_1}{dt}$$

$$i_2 \leftrightarrow \frac{d\theta_2}{dt}$$

$$v \leftrightarrow \tau$$

$$L_1 \leftrightarrow J_1$$

$$R_1 \leftrightarrow B_1$$

$$L_2 \leftrightarrow J_2$$

$$R_2 \leftrightarrow B_2$$

$$\frac{N_1}{N_2} \leftrightarrow \frac{n_1}{n_2}$$

Other analogies are possible if an alternate electrical circuit is used. For example, a circuit with a current source as a forcing function would provide an analogy between current and torque. All other analogies would change accordingly.

Electromechanical Components

Many electromechanical devices are encountered in engineering and scientific applications. Solenoids, actuators, motors, generators, gyroscopes, accelerometers, and loudspeakers are just a few of these. For many control systems it is necessary to deal with equations for a combination of electrical and mechanical components. Table 1.5 gives idealized equations for several common electromechanical devices. These devices operate as indicated over a suitable range of parameters and conditions.

A potentiometer contains a slider which moves along a resistance element as in part 1 of Table 1.5. The potentiometer has a voltage V applied across the entire resistance while a fraction of that voltage appears across the output. That fraction depends on the ratio of the angle subtended by the slider compared to the maximum angle.

Part 2 of Table 1.5 contains a model for a dc motor with a fixed field. An input voltage causes a current to flow in the armature of the dc motor. If that current creates a sufficiently large field to interact with the fixed field, the armature begins to turn. The armature turns because a torque is applied to the inertia and friction of

Table 1.5   SIMPLE MODELS FOR COMMON ELECTROMECHANICAL ELEMENTS

1. *Voltage-Driven Potentiometer*

   θ_{max} is the maximum angular position θ. For $\theta = \theta_{max}$, $R_1 = R$.

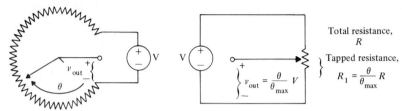

2. *DC Motor (Constant Field)*

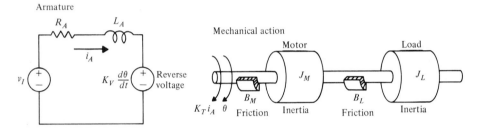

3. *DC Generator (Constant Field)*

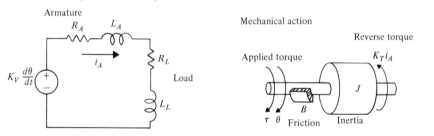

4. *Tachometer*

   The tachometer is a special case of a dc generator in which the field is replaced by a permanent magnet, which is equivalent to having a constant field current. It is normally used with a very small electrical load, so that i_a is nearly zero. Friction and inertia are typically made as small as is practical.

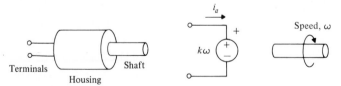

5. *Linear Actuator (Solenoid)*

   The constant k depends upon the amount of electromechanical coupling and is the same number in both relations when they are expressed in consistent units.

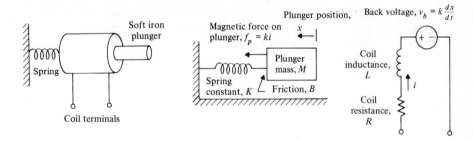

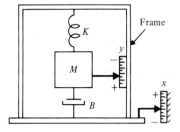

6. Seismic (Mechanical) Accelerometer

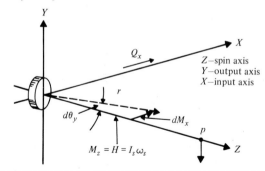

7. Gyroscope

the motor and to the inertia and friction of the load connected to the motor. That torque is proportional to the armature current. As the armature spins, a reverse voltage is induced so as to oppose the input voltage to the armature circuit.

The dc motor model is governed by

$$L_A \frac{di_A}{dt} + R_A i_A + K_V \frac{d\theta}{dt} = v_i$$

$$(J_M + J_L) \frac{d^2\theta}{dt^2} + (B_M + B_L) \frac{d\theta}{dt} - K_T i_A = 0$$

Special-purpose dc motors such as control motors can usually be analyzed using the above model and values specified by the manufacturer.

Part 3 of Table 1.5 contains a model for a dc generator with a fixed field. Some external means is employed to cause the generator to rotate (for example, hydrological forces or steam pressure). As the armature rotates in the fixed field, a voltage is induced across the armature coils. That voltage can be applied to a resistive-inductive load, for example. The field created by the armature current will create a reverse torque which opposes the applied torque. The dc generator model is governed by

$$(L_A + L_L)\frac{di_A}{dt} + (R_A + R_L)i_A - K_V\frac{d\theta}{dt} = 0$$

$$J\frac{d^2\theta}{dt^2} + B\frac{d\theta}{dt} + K_T i_A = T$$

As shown in part 4 of Table 1.5, a tachometer is a special dc generator whose output voltage is directly proportional to the tachometer's rate of angular rotation. Ideally, the armature supplies no current and the inertia is zero. A tachometer is usually connected to a rotating device so as to measure its rate of rotation without hindering the operation of the device being monitored.

The solenoid in part 5 converts electrical energy into linear motion. The soft iron plunger moves due to the magnetic force of attraction created by current flowing in an RL circuit (for example, in the ignition system of an automobile where the solenoid acts as a switch). The constant k that appears in two places has the same numerical value if consistent units are used. The k associated with force could have units of newtons/ampere. In the electrical circuit k has units of volts/(meters/sec). Voltage has units of joules/coulomb, the same as units of newton-meters/coulomb. Therefore,

$$\frac{\text{volts}}{\text{meters/sec}} = \frac{\text{newton-meters}}{\text{coulomb-meters/sec}}$$

is the same as units of newtons/(coulomb/sec) or newtons/ampere. The two values of k are the same.

A "seismic," or mechanical, accelerometer schematic is depicted in part 6 of Table 1.5. The instrument consists of a mass suspended from a frame by a spring. Damping is provided either mechanically or electrically, and a pickoff measures the mass position y with respect to the frame. In part 6, x is the displacement of the frame with respect to the body whose acceleration is to be measured. K is the spring constant of the suspension, and B is the viscous damping constant. Since y is measured with respect to the frame, the force on the mass due to the spring is $-Ky$ and the force due to the damper is $-B(dy/dt)$. The position of the mass is $y - x$. A free-body diagram can be used to obtain the differential equation of the mass with respect to the fixed reference.

$$M\frac{d^2(y - x)}{dt^2} + B\frac{dy}{dt} + Ky = 0$$

The output y of the accelerometer must be measured by some position sensor (for example, by a linear potentiometer). The position sensor provides a signal y_m where

$$\frac{y_m}{y} = K_A$$

The purpose of this device is to measure

$$a = \frac{d^2 x}{dt^2}$$

the acceleration of the frame with respect to the x-coordinate system. The dynamic behavior of the measured position is, therefore,

$$\frac{d^2 y_m}{dt^2} + \left(\frac{B}{M}\right)\frac{dy_m}{dt} + \left(\frac{K}{M}\right)y_m = K_A a$$

If a is a constant and if the mass achieves a fixed position, then y_m becomes proportional to the acceleration of the frame.

$$y_m = \left(K_A \frac{M}{K}\right)a$$

Accelerometers of this type fit in the category of spring-mass instruments. In these instruments the acceleration to be measured appears as a proportional force (or torque) which acts upon a small mass (or moment of inertia). This force is balanced against a spring, and the deflection of this spring becomes a measure of the force and therefore the acceleration. For static measurements, only a spring and a mass or inertia are necessary. Under dynamic conditions damping is required to dissipate vibrational energy. Accelerometers employ various forms of damping. They have an accuracy which depends directly upon the linearity of the spring. Hysteresis, nonlinearities, or nonsymmetrical properties of this spring will result in errors in the instrument. For this and other reasons which are related to open-loop computation, units of this type are generally less accurate and have a smaller range than those of the force-balance or "servo" type. Section 1.8 contains a design example of a force-balance accelerometer.

Part 7 of Table 1.5 shows a simple model for a gyroscope. A gyroscope consists of a wheel mounted on a shaft and arranged to be spun at high angular velocity. Frequently, the wheel is mounted in a system of gimbals that permits complete freedom of movement of all three axes.

The most useful characteristic of the gyroscope is its tendency to maintain its spin axis in a fixed direction in space. This phenomenon is best explained by a consideration of rotation dynamics.

A simple model is used to gain insight into the operation of the gyroscope. In this model the effects of moments of inertia of the wheel and gimbal system about axes other than the spin axis are neglected. The resulting equation will fail to show certain characteristics. For many purposes, however, the results are adequate. For this derivation the following nomenclature is needed:

$$M_x, M_y, M_z = \text{components of angular momentum about } x, y, \text{ and } z \text{ axes}$$
$$I_s = \text{moment of inertia of wheel about spin axis}$$
$$\omega_s = \text{angular velocity of wheel}$$
$$H = I_s \omega_s = \text{angular momentum of wheel}$$

These quantities are identified in part 7 of Table 1.5.

At time $t = 0$, suppose the torque Q_x is applied about the x axis by pressing down on the gyro housing at point p. Initially, $M_z = I_s \omega_s = H$, $M_x = M_y = 0$, and the angular momentum of spin lies along OZ and has a magnitude H. Since the rate of change of angular momentum of a system is equal to the applied torque, the following expression can be written:

$$Q_x = \frac{dM_x}{dt}$$

This is expressed in different form

$$dM_x = Q_x\, dt$$

If this term is added vectorially to the initial angular momentum, a new value is obtained. $M_z + dM_x$ is separated from the initial value by an angle $d\theta_y$ which is given by

$$d\theta_y = \frac{dM_x}{H} = \frac{Q_x\, dt}{H}$$

and is written as

$$\omega_y = \frac{d\theta_y}{dt} = \frac{Q_x}{H}$$

The gyro is thus rotating about the OY axis with a velocity ω_y. This is the fundamental gyroscopic law: A torque about any axis other than the spin axis produces a velocity about the axis which is orthogonal to the applied-torque axis. Because of this property the gyro is an important instrument for measuring torques ($Q_x = H\omega_y$) by measuring this orthogonal or precession velocity.

Alternatively, a large gyroscope can be used to obtain a stabilizing counter-torque. For applied torques about the x or y axis, the gyro supplies an equal countertorque which prevents motion in the direction of the applied torque as long as the gyro can precess. Once the precision angle θ has reached 90°, the gyroscope is in a state of "gimbal lock" and ceases to function as described. In the state of gimbal lock the OZ axis has precessed into the OX axis, about which the torque is being applied. With a torque applied about the OX, or spin, axis, the gyro ceases to produce a countertorque. Thus the gyro tends to rotate to align its spin axis in the direction of applied torque.

DRILL PROBLEMS

D1.1. List five examples of control systems that do not employ feedback. Draw illustrative diagrams for each.

D1.2. List five examples of control systems that employ feedback. Draw illustrative diagrams for each.

D1.3. Identify the input(s), output(s), and major parts of the following control systems. Which are open-loop and which are closed-loop?

(a) A heater with thermostat

(b) A toaster

(c) A human being reaching to touch an object

(d) A human being steering an automobile

(e) An electric generating station

D1.4. Draw diagrams similar to Fig. 1.2 for the following systems:

(a) Control of human skin temperature by sweating

(b) An automatic traffic light system at an intersection which varies the ratio of green light times in the two cross sections with the traffic ratio in the two directions

(c) A simple economic model which includes as inputs government tax level, private business investment, and consumer spending and has per capita income after taxes as the output

(d) Control of a nuclear reactor

(e) The teaching and learning process with feedback

D1.5. Write simultaneous integrodifferential loop equations for the following electrical networks in terms of the indicated loop currents:

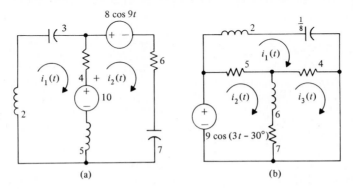

(a)   (b)

D1.6. Write simultaneous integrodifferential nodal equations for the following electrical networks in terms of the indicated node voltages. Node numbers are circled.

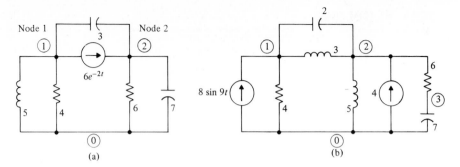

(a)   (b)

D1.7. Write simultaneous differential equations for the following translational mechanical networks in terms of the indicated mass positions. The mass positions are defined so that when all positions are zero, the spring forces are zero.

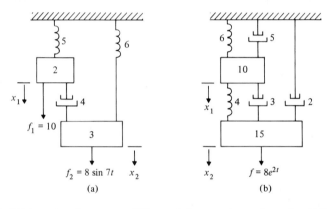

(a)   (b)

D1.8. Write simultaneous differential equations for the following rotational mechanical networks in terms of the indicated rotational mass angles. The angles are defined so that when all angles are zero, the spring torques are zero.

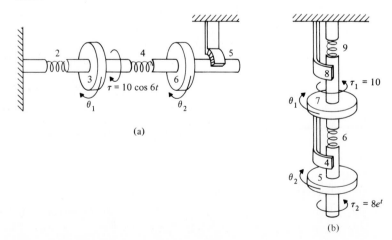

(a)

(b)

D1.9. Find $\theta(t)$. Mass and damping effects in this system are negligible.

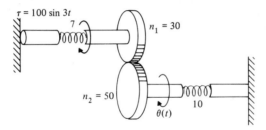

1.3 LAPLACE TRANSFORM

Definition and Properties

The expression of a system description in the language of mathematics is termed *modeling*. Modeling involves idealizations whereby the important aspects of a problem are isolated and the minor ones ignored, so that simplicity with sufficient accuracy is obtained. When the system models consist of linear, constant-coefficient integrodifferential equations, Laplace transform methods can be used to advantage. The Laplace transform of a function $f(t)$ is

$$\mathcal{L}[f(t)] = F(s) = \int_{0^-}^{\infty} f(t)e^{-st}\, dt$$

The inverse Laplace transform recovers the original function for $t \geq 0$ and gives zero for times prior to $t = 0$:

$$\mathcal{L}^{-1}[F(s)] = \frac{1}{2\pi j}\int_{\sigma - j\infty}^{\sigma + j\infty} F(s)e^{st}\, ds = \begin{cases} f(t), & t \geq 0 \\ 0, & t < 0 \end{cases} = f(t)u(t)$$

where $u(t)$ is the unit step function. When functions with discontinuities at $t = 0$ are involved, it is inconvenient to have $t = 0$ as a limit to the Laplace transform integral. The most useful and common definition begins the integration just before $t = 0$, at $t = 0^-$.

Table 1.6 is a collection of common functions $f(t)$ and their Laplace transforms, $F(s)$. As these Laplace transforms are unique, the table can also be used to find time functions from transforms. Table 1.7 gives important Laplace transform properties.

Solving Differential Equations

By Laplace-transforming linear constant-coefficient integrodifferential equations, one obtains linear algebraic equations, which can then be solved for the transforms

Table 1.6   SELECTED LAPLACE TRANSFORMS

$f(t)$	$F(s)$
$\delta(t)$ (unit impulse)	1
$u(t)$ (unit step)	$\dfrac{1}{s}$
$tu(t)$	$\dfrac{1}{s^2}$
$t^n u(t)$	$\dfrac{n!}{s^{n+1}}$
$e^{-at}u(t)$	$\dfrac{1}{s+a}$
$te^{-at}u(t)$	$\dfrac{1}{(s+a)^2}$
$t^n e^{-at}u(t)$	$\dfrac{n!}{(s+a)^{n+1}}$
$(\sin bt)u(t)$	$\dfrac{b}{s^2+b^2}$
$(\cos bt)u(t)$	$\dfrac{s}{s^2+b^2}$
$(t \sin bt)u(t)$	$\dfrac{2bs}{(s^2+b^2)^2}$
$(t \cos bt)u(t)$	$\dfrac{s^2-b^2}{(s^2+b^2)^2}$
$(e^{-at} \sin bt)u(t)$	$\dfrac{b}{(s+a)^2+b^2}$
$(e^{-at} \cos bt)u(t)$	$\dfrac{(s+a)}{(s+a)^2+b^2}$
$Ae^{-at} \cos(bt + \theta)u(t)$	$\dfrac{(A/2)\,e^{j\theta}}{s+a-jb} + \dfrac{(A/2)\,e^{-j\theta}}{s+a+jb} = \dfrac{\text{first-degree numerator}}{(s+a)^2+b^2}$

Table 1.7 FUNDAMENTAL LAPLACE TRANSFORM PROPERTIES

$\mathscr{L}[kf(t)] = kF(s)$, k a constant

$\mathscr{L}[f_1(t) + f_2(t)] = F_1(s) + F_2(s)$

$\mathscr{L}[f_1(t)f_2(t)]$ does *not* equal $F_1(s)F_2(s)$

$\mathscr{L}[f(t - T)] = e^{-sT}F(s)$, T a constant, provided $f(t)$ and $f(t - T)$ are both zero prior to $t = 0$

$\mathscr{L}[f(at)] = \dfrac{1}{a}F\left(\dfrac{s}{a}\right)$, a is a positive constant

$\mathscr{L}[e^{-at}f(t)] = F(s + a)$

$\mathscr{L}\left[\dfrac{df}{dt}\right] = sF(s) - f(0^-)$

$\mathscr{L}\left[\dfrac{d^2f}{dt^2}\right] = s^2F(s) - sf(0^-) - f'(0^-)$

$\mathscr{L}\left[\dfrac{d^nf}{dt^n}\right] = s^nF(s) - s^{n-1}f(0^-) - s^{n-2}f'(0^-) - \cdots - sf^{[n-2]}(0^-) - f^{[n-1]}(0^-)$

$\mathscr{L}\left[\displaystyle\int_{0^-}^t f(t)dt\right] = \dfrac{F(s)}{s}$

$\mathscr{L}\left[\displaystyle\int_{-\infty}^t f(t)dt\right] = \dfrac{F(s)}{s} + \dfrac{1}{s}\displaystyle\int_{-\infty}^{0^-} f(t)dt$

$\mathscr{L}[tf(t)] = -\dfrac{dF(s)}{ds}$

$\mathscr{L}[t^2f(t)] = \dfrac{d^2F(s)}{ds^2}$

$\mathscr{L}[t^nf(t)] = (-1)^n\dfrac{d^nF(s)}{ds^n}$

of the solutions. Initial conditions can be included when the equations are transformed. For example, transforming

$$\frac{d^2y}{dt^2} + 9\frac{dy}{dt} + 2y = 6e^{-4t}$$

with

$$y(0^-) = 2$$
$$y'(0^-) = -4$$

gives

$$s^2 Y(s) - 2s + 4 + 9[sY(s) - 2] + 2Y(s) = \frac{6}{s + 4}$$

$$Y(s) = \frac{6}{(s + 4)(s^2 + 9s + 2)} + \frac{2s + 14}{s^2 + 9s + 2}$$

The Laplace transform method is also easily applied to systems described by simultaneous integrodifferential equations. Transforming the simultaneous equations

$$\begin{cases} \dfrac{dy_1}{dt} + 2y_1 - \dfrac{3\,dy_2}{dt} = r(t) \\[2mm] y_1 + \dfrac{dy_2}{dt} + 4\displaystyle\int_0^t y_2\,dt = 0 \end{cases}$$

gives

$$\begin{cases} sY_1(s) - y_1(0^-) + 2Y_1(s) - 3sY_2(s) + 3y_2(0^-) = R(s) \\[2mm] Y_1(s) + sY_2(s) - y_2(0^-) + \dfrac{4}{s}\,Y_2(s) = 0 \end{cases}$$

or

$$\begin{cases} (s + 2)Y_1(s) - 3sY_2(s) = R(s) + y_1(0^-) - 3y_2(0^-) \\[2mm] Y_1(s) + \left(s + \dfrac{4}{s}\right)Y_2(s) = y_2(0^-) \end{cases}$$

which can be solved for $Y_1(s)$ and $Y_2(s)$, given the input $R(s)$ and the initial conditions $y_1(0^-)$ and $y_2(0^-)$.

Partial Fraction Expansion

Determining inverse Laplace transforms of rational functions involves expansion into partial fractions. If the numerator polynomial is of lower order than the denominator polynomial and if the denominator polynomial has no repeated roots, then the residues $K_1, K_2, K_3, \ldots$ can be found such that

$$Y(s) = \frac{\text{Numerator polynomial}}{(s + a)(s + b)(s + c)\cdots} = \frac{K_1}{s + a} + \frac{K_2}{s + b} + \frac{K_3}{s + c} + \cdots$$

The individual terms in the expansion represent exponential functions of time after $t = 0$:

$$y(t) = K_1 e^{-at} + K_2 e^{-bt} + K_3 e^{-ct} + \cdots \qquad t \geq 0$$

One way of finding residues is to recombine terms and solve the simultaneous linear algebraic equations that result from equating coefficients. The following is an example:

$$Y(s) = \frac{-3s^2 + 4}{s^3 + 5s^2 + 6s} = \frac{-3s^2 + 4}{s(s + 2)(s + 3)} = \frac{K_1}{s} + \frac{K_2}{s + 2} + \frac{K_3}{s + 3}$$

$$= \frac{(K_1 + K_2 + K_3)s^2 + (5K_1 + 3K_2 + 2K_3)s + 6K_1}{s(s + 2)(s + 3)}$$

$$\begin{cases} K_1 + K_2 + K_3 = -3 \\ 5K_1 + 3K_2 + 2K_3 = 0 \\ 6K_1 \qquad\qquad = 4 \end{cases} \qquad \begin{cases} K_1 = \dfrac{2}{3} \\ K_2 = 4 \\ K_3 = \dfrac{-23}{3} \end{cases}$$

$$Y(s) = \frac{\dfrac{2}{3}}{s} + \frac{4}{s + 2} + \frac{-\dfrac{23}{3}}{s + 3}$$

$$y(t) = \frac{2}{3} + 4e^{-2t} - \frac{23}{3}e^{-3t} \qquad t \geq 0$$

A faster method for finding a residue other than for a repeated root is to multiply each side of the expansion by the denominator term, then evaluate the result at the value of s which makes that denominator term zero:

$$\frac{(s + a)\text{ numerator}}{(s + a)(s + b)(s + c)\cdots} = \frac{(s + a)K_1}{s + a} + \frac{(s + a)K_2}{s + b} + \frac{(s + a)K_3}{s + c} + \cdots$$

$$\left.\frac{\text{Numerator}}{(s + b)(s + c)\cdots}\right|_{s = -a} = K_1$$

Applying this method to the previous example gives the residues quite easily:

$$K_1 = \left.\frac{-3s^2 + 4}{(s + 2)(s + 3)}\right|_{s=0} = \frac{2}{3}$$

$$K_2 = \left.\frac{-3s^2 + 4}{s(s + 3)}\right|_{s=-2} = 4$$

$$K_3 = \left.\frac{-3s^2 + 4}{s(s + 2)}\right|_{s=-3} = -\frac{23}{3}$$

If the numerator polynomial is not of lower order than the denominator polynomial, the denominator must be divided into the numerator until a remainder polynomial of lower order than the denominator is obtained, giving

$$\frac{\text{Numerator polynomial}}{(s + a)(s + b)(s + c)\cdots} = \frac{\text{dividend}}{\text{polynomial}} + \frac{\text{remainder polynomial}}{(s + a)(s + b)(s + c)\cdots}$$

For example,

$$Y(s) = \frac{3s^2 - 4s + 1}{s^2 + 5s + 6} = 3 + \frac{-19s - 17}{s^2 + 5s + 6}$$

$$= 3 + \frac{21}{s + 2} + \frac{-40}{s + 3}$$

$$y(t) = 3\delta(t) + 21e^{-2t} - 40e^{-3t} \qquad t \geq 0$$

A constant term in the Laplace transform corresponds to an impulsive time function.

If denominator roots are repeated, the corresponding terms in the partial function expansion are as follows:

$$\frac{\text{Numerator}}{(s + a)^n D(s)} = \frac{K_1}{s + a} + \frac{K_2}{(s + a)^2} + \cdots + \frac{K_n}{(s + a)^n} + \text{other}$$

For example,

$$Y(s) = \frac{3s - 4}{(s + 1)(s + 2)^2} = \frac{K_1}{s + 1} + \frac{K_2}{s + 2} + \frac{K_3}{(s + 2)^2}$$

$$= \left[\frac{(K_1 + K_2)s^2 + (4K_1 + 3K_2 + K_3)s + (4K_1 + 2K_2 + K_3)}{(s + 1)(s + 2)^2} \right]$$

$$\begin{cases} K_1 + K_2 = 0 \\ 4K_1 + 3K_2 + K_3 = 3 \\ 4K_1 + 2K_2 + K_3 = -4 \end{cases} \qquad \begin{cases} K_1 = -7 \\ K_2 = 7 \\ K_3 = 10 \end{cases}$$

$$Y(s) = \frac{-7}{s + 1} + \frac{7}{s + 2} + \frac{10}{(s + 2)^2}$$

$$y(t) = -7e^{-t} + 7e^{-2t} + 10te^{-2t} \qquad t \geq 0$$

The inverse Laplace transform of a repeated root is of the form

$$\mathcal{L}^{-1}\left[\frac{K_n}{(s + a)^n} \right] = \frac{K_n}{(n - 1)!} t^{n-1} e^{-at} u(t)$$

For a repeated root, the residue evaluation method works only for the highest-order repeated term in the expansion. For example, with

$$Y(s) = \frac{4s^2 - 1}{(s + 2)^3} = \frac{K_1}{s + 2} + \frac{K_2}{(s + 2)^2} + \frac{K_3}{(s + 2)^3}$$

evaluation gives

$$\frac{(4s^2 - 1)(s + 2)^3}{(s + 2)^3} = K_1(s + 2)^2 + K_2(s + 2) + K_3$$

$$K_3 = (s + 2)^3 Y(s)\big|_{s = -2} = (4s^2 - 1)\big|_{s = -2} = 15$$

Multiplying both sides of the expansion by $(s + 2)^2$ or $(s + 2)$ will leave $(s + 2)$ denominator factors, however, so K_1 and K_2 cannot be determined in this manner.

An alternate method to that of cross-multiplying and equating coefficients is to multiply both sides of the expansion by the repeated root term and differentiate with respect to s:

$$\frac{d}{ds}\left[\frac{(4s^2 - 1)(s + 2)^3}{(s + 2)^3}\right] = \frac{d}{ds}[K_1(s + 2)^2 + K_2(s + 2) + K_3]$$

$$8s = 2K_1(s + 2) + K_2 \tag{1.5}$$

Evaluating at $s = -2$ gives

$$8s|_{s = -2} = K_2 = -16$$

Differentiating (1.5) a second time with respect to s and evaluating,

$$\frac{d}{ds}(8s)\bigg|_{s = -2} = 2K_1$$

$$K_1 = 4$$

In general, the coefficients of the repeated root terms of a partial fraction expansion, repeated p times,

$$Y(s) = \frac{K_1}{s + a} + \frac{K_2}{(s + a)^2} + \cdots + \frac{K_p}{(s + a)^p} + \cdots \text{ terms for other different roots}$$

are given by

$$K_i = \frac{1}{(p - i)!}\frac{d^{p-i}}{ds^{p-i}}\{(s + a)^p Y(s)\}|_{s = -a} \qquad i = 1, 2, \ldots, p$$

For complex root terms, the corresponding residues are complex numbers that are complex conjugates of one another. It is thus only necessary to calculate one of the residues for a set of conjugate terms. For the transform

$$Y(s) = \frac{-s + 8}{s(s^2 + 2s + 10)} = \frac{-s + 8}{s(s + 1 - j3)(s + 1 + j3)}$$

$$= \frac{K_1}{s} + \frac{K_2}{s + 1 - j3} + \frac{K_3 = K_2^*}{s + 1 + j3}$$

then

$$K_1 = \frac{-s + 8}{s^2 + 2s + 10}\bigg|_{s = 0} = 0.8$$

$$K_2 = \frac{-s + 8}{s(s + 1 + j3)}\bigg|_{s = -1 + j3} = \frac{9 - j3}{(-1 + j3)(j6)} = 0.5e^{j143°}$$

so that

$$Y(s) = \frac{0.8}{s} + \frac{0.5e^{j143°}}{s + 1 - j3} + \frac{0.5e^{-j143°}}{s + 1 + j3}$$

Using the last entry in Table 1.6, the corresponding time function is

$$y(t) = 0.8 + e^{-t} \cos(3t + 143°) \qquad t \geq 0$$

Response Terms

The type of time function corresponding to each partial fraction expansion term for a Laplace-transformed signal depends upon the term's root location in the complex plane and upon whether or not the root is repeated. Table 1.8 shows representative time functions associated with various transform denominator root locations.

Table 1.8 LAPLACE TRANSFORM DENOMINATOR ROOT LOCATIONS AND CORRESPONDING TIME FUNCTIONS

Denominator Root Locations	Time Function

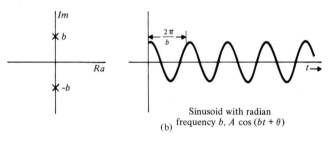

(a) Constant K

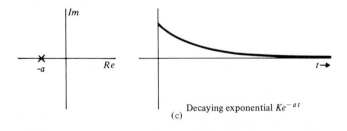

(b) Sinusoid with radian frequency b, $A \cos(bt + \theta)$

(c) Decaying exponential Ke^{-at}

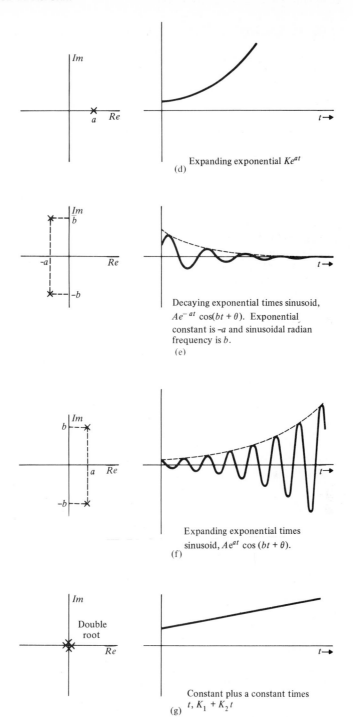

(d) Expanding exponential Ke^{at}

Decaying exponential times sinusoid, $Ae^{-at} \cos(bt + \theta)$. Exponential constant is $-a$ and sinusoidal radian frequency is b.

(e)

Expanding exponential times sinusoid, $Ae^{at} \cos(bt + \theta)$.

(f)

Constant plus a constant times t, $K_1 + K_2 t$

(g)

Table (*Continued*)

Table 1.8 (*Continued*)

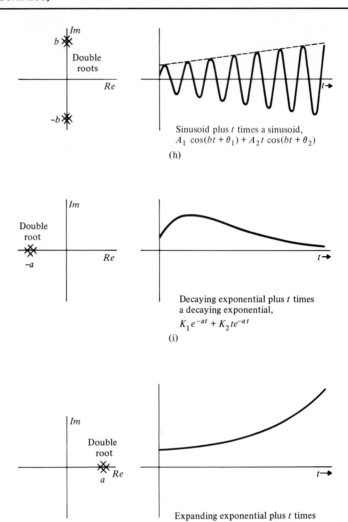

Sinusoid plus t times a sinusoid,
$A_1 \cos(bt + \theta_1) + A_2 t \cos(bt + \theta_2)$

(h)

Decaying exponential plus t times
a decaying exponential,
$K_1 e^{-at} + K_2 t e^{-at}$

(i)

Expanding exponential plus t times
an expanding exponential,
$K_1 e^{at} + K_2 t e^{at}$

(j)

DRILL PROBLEMS

D1.10. For the following Laplace-transformed signals, find $y(t)$ for $t \geq 0$:

(a) $Y(s) = \dfrac{s}{s + 2}$

Ans $\delta(t) - 2e^{-2t}$

(b) $Y(s) = \dfrac{3s - 5}{s^2 + 4s + 2}$

    Ans $5.39e^{-3.414t} - 2.39e^{-0.586t}$

(c) $Y(s) = \dfrac{3 - 6e^{-2s}}{(s + 2)(s + 3)}$

    Ans $3e^{-2t} - 3e^{-3t} - [6e^{-2(t-2)} - 6e^{-3(t-2)}]u(t - 2)$

(d) $Y(s) = \dfrac{10}{s^3 + 2s^2 + 5s}$

    Ans $2 + e^{-t}(-2\cos 2t - \sin 2t) = 2 + 2.24e^{-t}\cos(2t + 153°)$

(e) $Y(s) = \dfrac{4(s + 1)}{(s + 2)(s + 3)^2}$

    Ans $-4e^{-2t} + 4e^{-3t} + 8te^{-3t}$

D1.11. Use Laplace transform methods to solve the following differential equations for $t \geq 0$ with the indicated boundary conditions:

(a) $\dfrac{dy}{dt} + 4y = 6e^{2t}$

    $y(0^-) = 3$

    Ans $e^{2t} + 2e^{-4t}$

(b) $\dfrac{dy}{dt} + y = 3\cos 2t$

    $y(0^-) = 0$

    Ans $-\frac{3}{5}e^{-t} + \frac{3}{5}\cos 2t + \frac{6}{5}\sin 2t$
                   $= -\frac{3}{5}e^{-t} + 1.34\cos(2t - 63.4°)$

(c) $\dfrac{d^2y}{dt^2} + 7\dfrac{dy}{dt} + 12y = 10$

    $y(0^-) = 3$

    $y'(0^-) = 0$

    Ans $\frac{10}{12} - \frac{13}{2}e^{-4t} + \frac{26}{3}e^{-3t}$

(d) $\dfrac{d^2y}{dt^2} + 4\dfrac{dy}{dt} + 20y = 4$

    $y(0^-) = -2$

    $y'(0^-) = 0$

    Ans $\frac{1}{5} + e^{-2t}(-\frac{11}{5}\cos 4t - \frac{11}{10}\sin 4t)$
                   $= \frac{1}{5} + 2.46e^{-2t}\cos(4t + 153°)$

(e) $\dfrac{d^3 y}{dt^3} + 5\dfrac{d^2 y}{dt^2} + 6\dfrac{dy}{dt} = 0$

$y(0^-) = 3$

$y'(0^-) = -2$

$y''(0^-) = 7$

Ans $\dfrac{15}{6} + e^{-3t} - \dfrac{1}{2}e^{-2t}$

1.4 TRANSFER FUNCTIONS AND STABILITY

Transfer Functions

One of the most powerful tools of control system analysis and design is the transfer function representation, which is a generalization of the impedance concept in electrical and mechanical networks. For a single-input, single-output system with input $r(t)$ and output $y(t)$, the transfer function relating the output to the input is defined as

$$T(s) = \frac{Y(s)}{R(s)}\bigg|_{\text{when all initial conditions are zero}}$$

To find the transfer function, Laplace-transform the system equations, with zero initial conditions, and form the ratio of output transform to input transform.

Consider the system described by

$$\frac{d^2 y}{dt^2} + 6\frac{dy}{dt} + 8y = -\frac{dr}{dt} + 5r \tag{1.6}$$

Laplace-transforming with zero initial conditions,

$$s^2 Y(s) + 6s Y(s) + 8Y(s) = -sR(s) + 5R(s)$$

and

$$T(s) = \frac{Y(s)}{R(s)} = \frac{-s + 5}{s^2 + 6s + 8} \tag{1.7}$$

In systems described by linear, constant-coefficient integrodifferential equations, every Laplace-transformed signal is related to every other such signal by a transfer function. To avoid confusion, the term *transfer function* is reserved to describe the input-output relation and *transmittance* is used to denote the similar relation between a pair of signals other than the input and output.

Response Components

The system output when the initial conditions are all zero is termed the *zero-state* response component. Its Laplace transform is given simply by the product of the transfer function and the input transform. If the system of (1.6) and (1.7) has zero initial conditions and if the input is

$$r(t) = 7e^{-3t}$$

the system output is given by

$$Y_{\text{zero state}}(s) = T(s)R(s) = \frac{7(-s + 5)}{(s^2 + 6s + 8)(s + 3)}$$

If the system initial conditions are not zero, there is an additional output component present, the *zero-input* part of the response. The form of the zero-input response can be found from the transfer function denominator, provided that there have been no unwarranted cancellations of terms made between numerator and denominator of the transfer function. It is possible, although perhaps unlikely, that system parameters are just the right numbers so that a factor in the transfer function numerator cancels a denominator factor, causing the corresponding term in the system's zero-input response to be overlooked.

For a transfer function

$$T(s) = \frac{b_m s^m + b_{m-1} s^{m-1} + \cdots + b_1 s + b_0}{a_n s^n + a_{n-1} s^{n-1} + \cdots + a_1 s + a_0}$$

a differential equation relating the output $y(t)$ to the input $r(t)$ is

$$a_n \frac{d^n y}{dt^n} + a_{n-1} \frac{d^{n-1} y}{dt^{n-1}} + \cdots + a_1 \frac{dy}{dt} + a_0 y$$

$$= b_m \frac{d^m r}{dt^m} + b_{m-1} \frac{d^{m-1} r}{dt^{m-1}} + \cdots + b_1 \frac{dr}{dt} + b_0 r$$

Recall that $T(s)$ was computed by assuming that all initial conditions are zero. The above differential equation can be Laplace-transformed again so that initial condition terms are restored. Including the initial condition terms in the Laplace-transformed equations gives

$$a_n[s^n Y(s) - s^{n-1} y(0^-) - s^{n-2} y'(0^-) - \cdots]$$

$$+ a_{n-1}[s^{n-1} Y(s) - s^{n-2} y(0^-) - s^{n-3} y'(0^-) - \cdots] + \cdots + a_0 Y(s)$$

$$= b_m[s^m R(s) - s^{m-1} r(0^-) - s^{m-2} r'(0^-) - \cdots] + \cdots + b_0 R(s)$$

$$(a_n s^n + a_{n-1} s^{n-1} + \cdots + a_1 s + a_0) Y(s) = (b_m s^m + \cdots + b_0) R(s)$$

$$+ \text{polynomial in } s \text{ with coefficients dependent upon initial conditions}$$

$$Y(s) = T(s)R(s) + \frac{\text{polynomial in } s \text{ with coefficients dependent upon initial conditions}}{a_n s^n + a_{n-1} s^{n-1} + \cdots + a_1 s + a_0}$$

$$\underbrace{\qquad\qquad}_{\substack{\text{zero-state} \\ \text{component}}} \qquad \underbrace{\qquad\qquad\qquad\qquad}_{\substack{\text{zero-input} \\ \text{component}}}$$

The zero-input component transform has the same denominator polynomial as $T(s)$.

The transfer function denominator polynomial is the system *characteristic polynomial*, and the roots of the characteristic polynomial, that is, the solutions of the characteristic equation

$$a_n s^n + a_{n-1} s^{n-1} + \cdots + a_1 s + a_0 = 0$$

are known as the system's *characteristic roots* (or *poles*).

For the example system of (1.6), the zero-input response is given by

$$Y_{\text{zero input}}(s) = \frac{c_1 s + c_2}{s^2 + 6s + 8}$$

where c_1 and c_2 are constants which depend upon the specific initial conditions. The complete response for general initial conditions is of the form

$$Y(s) = Y_{\text{zero-state}}(s) + Y_{\text{zero input}}(s) = \frac{7(-s+5)}{(s^2 + 6s + 8)(s+3)} + \frac{c_1 s + c_2}{s^2 + 6s + 8} \qquad \textbf{(1.8)}$$

An alternative to the zero input/zero state decomposition of system response is to separate the response into *natural* and *forced* parts. The natural component consists of all the characteristic root terms in the partial fraction expansion for the response. The forced response component is the remainder of the response and is composed of the terms associated with the input transform. For example, for the system (1.6), (1.7), the characteristic roots are $s = -2$ and $s = -4$. The response (1.8) expands in partial fractions as

$$Y(s) = \underbrace{\frac{-7s + 35}{(s+2)(s+4)(s+3)}}_{\substack{\text{zero-state} \\ \text{component}}} + \underbrace{\frac{c_1 s + c_2}{(s+2)(s+4)}}_{\substack{\text{zero-input} \\ \text{component}}}$$

$$= \underbrace{\frac{-56}{s+3}}_{\substack{\text{forced} \\ \text{component}}} + \underbrace{\frac{K_1}{s+2} + \frac{K_2}{s+4}}_{\substack{\text{natural} \\ \text{component}}}$$

where the constants K_1 and K_2 depend upon the initial conditions. Both the zero-input and zero-state response components generally contribute to the natural response component.

Control system designers seldom explicitly calculate zero-input or natural components of system response because of their dependence upon the specific initial conditions, which are often unknown or of little concern. It is usually enough to know the natural response form and to know that the terms decay to zero.

If desired, however, system initial conditions can be considered to be system inputs, and transfer functions can be found which relate the outputs to the initial condition inputs. For a system described by

$$\frac{d^2y}{dt^2} + 3\frac{dy}{dt} + 2y = 4r(t)$$

Laplace-transforming with the initial conditions included gives

$$s^2Y(s) - sy(0^-) - y'(0^-) + 3[sY(s) - y(0^-)] + 2Y(s) = 4R(s)$$

$$Y(s) = \left[\frac{4}{s^2 + 3s + 2}\right]R(s) + \left[\frac{s+3}{s^2 + 3s + 2}\right]y(0^-) + \left[\frac{1}{s^2 + 3s + 2}\right]y'(0^-)$$

$$= T_1(s)R(s) + T_2(s)y(0^-) + T_3(s)y'(0^-)$$

Multiple Inputs and Outputs

If a system has several inputs $r_1(t), r_2(t), \ldots$ and/or several outputs $y_1(t), y_2(t), \ldots$, there is a transfer function which relates each one of the outputs to each one of the inputs, when all other inputs are zero:

$$T_{ij}(s) = \left.\frac{Y_i(s)}{R_j(s)}\right|_{\substack{\text{When all initial conditions are zero and} \\ \text{when all inputs except } R_j \text{ are zero}}}$$

```
           ⌐input number
           ⌐output number
```

In general, when the system initial conditions are zero, the outputs are given by

$$Y_1(s) = T_{11}(s)R_1(s) + T_{12}(s)R_2(s) + T_{13}(s)R_3(s) + \cdots$$

$$Y_2(s) = T_{21}(s)R_1(s) + T_{22}(s)R_2(s) + T_{23}(s)R_3(s) + \cdots$$

$$Y_3(s) = T_{31}(s)R_1(s) + T_{32}(s)R_2(s) + T_{33}(s)R_3(s) + \cdots$$
$$\vdots$$

Suppose that a two-input, two-output system is described by the following differential equations, where r_1 and r_2 are the inputs and y_1 and y_2 are the outputs:

$$\begin{cases} \dfrac{dy_1}{dt} + 2y_1 = r_1 + 5r_2 \\[3mm] y_1 + \dfrac{dy_2}{dt} + 3y_2 = 4r_2 + \dfrac{dr_2}{dt} \end{cases}$$

Laplace-transforming with zero initial conditions,

$$\begin{cases} (s+2)Y_1(s) & = R_1(s) + & 5R_2(s) \\ Y_1(s) + (s+3)Y_2(s) = & (s+4)R_2(s) \end{cases} \tag{1.9}$$

To find the transfer function which relates Y_1 to R_1, set R_2 to zero and solve for Y_1:

$$\begin{cases} (s+2)Y_1(s) & = R_1(s) \\ Y_1(s) + (s+3)Y_2(s) = 0 \end{cases} \qquad Y_1(s) = \frac{1}{s+2} R_1(s)$$

Then

$$T_{11} = \frac{Y_1(s)}{R_1(s)} = \frac{1}{s+2}$$

Solving, instead, for Y_2,

$$Y_2(s) = \frac{-1}{(s+2)(s+3)} R_1(s) \qquad T_{21}(s) = \frac{Y_2(s)}{R_1(s)} = \frac{-1}{(s+2)(s+3)}$$

Similarly, setting all inputs but R_2 to zero in (1.9),

$$\begin{cases} (s+2)Y_1(s) & = 5R_2(s) \\ Y_1(s) + (s+3)Y_2(s) = (s+4)R_2(s) \end{cases}$$

$$Y_1(s) = \frac{5}{s+2} R_2(s) \qquad T_{12}(s) = \frac{Y_1(s)}{R_2(s)} = \frac{5}{s+2}$$

And

$$Y_2(s) = \frac{s^2 + 6s + 3}{(s+2)(s+3)} R_2(s) \qquad T_{22}(s) = \frac{Y_2(s)}{R_2(s)} = \frac{s^2 + 6s + 3}{(s+2)(s+3)}$$

For zero initial conditions and inputs,

$$r_1(t) = 6 \sin 4t$$

$$r_2(t) = 10$$

the outputs are given by

$$Y_1(s) = T_{11}(s)R_1(s) + T_{12}(s)R_2(s) = \left(\frac{1}{s+2}\right)\left(\frac{24}{s^2+16}\right) + \left(\frac{5}{s+2}\right)\left(\frac{10}{s}\right)$$

and

$$Y_2(s) = T_{21}(s)R_1(s) + T_{22}(s)R_2(s)$$

$$= \frac{-1}{(s+2)(s+3)}\left(\frac{24}{s^2+16}\right) + \frac{s^2+6s+3}{(s+2)(s+3)}\left(\frac{10}{s}\right)$$

The natural part of the output y_1 will have an $\exp(-2t)$ contribution from T_{11} and another $\exp(-2t)$ contribution from T_{12} and so is of the form

$$y_{1\text{natural}} = K_1 e^{-2t}$$

The natural response in y_2 has $\exp(-2t)$ and $\exp(-3t)$ terms from each of T_{21} and T_{22}, so it is of the form

$$y_{2\text{natural}} = K_2 e^{-2t} + K_3 e^{-3t}$$

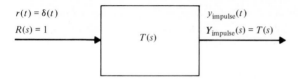

Figure 1.9 The transfer function of a single-input system is the Laplace transform of its unit impulse response.

Stability

The transfer function is also the Laplace transform of the system's unit impulse response, since if the input is a unit impulse,

$$r(t) = \delta(t) \qquad R(s) = 1$$

the output, with zero initial conditions, is

$$Y_{\text{impulse}}(s) = T(s)R(s) = T(s)$$

as indicated in Figure 1.9. A system is *input-output stable* if and only if its impulse response decays asymptotically to zero with time. As the transfer function is the Laplace transform of the unit impulse response, a system is stable if and only if all of its characteristic roots (poles) are to the left of the imaginary axis on the complex plane. For a stable system, the natural response component and the zero-input response component decay with time.

Referring to Table 1.8, characteristic roots in the right half of the complex plane (to the right of the imaginary axis) correspond to terms in the natural response that expand in time. Repeated roots along the imaginary axis also give expanding response. If *any* of the roots of the characteristic polynomial are in the right half-plane or are repeated and on the imaginary axis (including at the origin), one or more of the natural response terms expands with time, and the system is unstable. If *all* characteristic roots are in the left half-plane, the natural response terms all decay with time, and the system is stable.

For example, a system with transfer function

$$T(s) = \frac{-8s}{(s + 3)(s + 4)}$$

has characteristic roots $s = -3$ and $s = -4$, and thus is stable. The natural part of the system's response is of the form

$$y_{\text{natural}}(t) = K_1 e^{-3t} + K_2 e^{-4t}$$

which decays with time. A system with transfer function

$$T(s) = \frac{-s^2 + 8}{(s^2 + 2s + 5)(s - 4)}$$

has characteristic roots $s = -1 + j2$, $s = -1 - j2$, and $s = 4$. This system, because of the right half-plane root $s = 4$, is unstable. The natural part of the response is of the form

$$y_{\text{natural}}(t) = Ae^{-t}[\cos(2t + \theta)] + Ke^{4t}$$

and the e^{4t} term expands with time.

Characteristic roots on the imaginary axis, if not repeated, give an impulse response that neither expands nor decays with time. A system is *marginally stable* if it has no right half-plane or repeated imaginary axis roots, but there are non-repeated imaginary axis characteristic roots. If the system has one or more characteristic roots to the right of the imaginary axis or any repeated imaginary axis roots, its impulse response expands with time and the system is *unstable*.

A multiple-input and/or multiple-output system is input-output stable only if all denominator roots (poles) of *all* its transfer functions are in the left half of the complex plane.

DRILL PROBLEMS

D1.12. For systems with input $r(t)$ and output $y(t)$ that are described by the following equations, find the system transfer functions:

(a) $\dfrac{d^2y}{dt^2} + 3\dfrac{dy}{dt} + 7y = 6r$

    Ans $6/(s^2 + 3s + 7)$

(b) $\dfrac{d^3y}{dt^3} + 6\dfrac{d^2y}{dt^2} + 2\dfrac{dy}{dt} + 4y = -5\dfrac{d^2r}{dt^2} + 8\dfrac{dr}{dt}$

    Ans $(-5s^2 + 8s)/(s^3 + 6s^2 + 2s + 4)$

(c) $y(t) = 8r(t - 3)$

    Ans $8e^{-3s}$

(d) $\begin{cases} \dfrac{dx_1}{dt} = -3x_1 + x_2 + 4r \\[2mm] \dfrac{dx_2}{dt} = -2x_1 - r \\[2mm] y = x_1 - 2x_2 \end{cases}$

    Ans $(6s + 21)/(s^2 + 3s + 2)$

D1.13. For the following systems with outputs y and inputs r, find the transfer functions:

(a) $\dfrac{d^2 y}{dt^2} + 3\dfrac{dy}{dt} + 7y = 6r_1 + 5r_2 - 4\dfrac{dr_2}{dt}$

Ans $T_{11}(s) = 6/(s^2 + 3s + 7);$ $\qquad$ $T_{12}(s) = (-4s + 5)/(s^2 + 3s + 7)$

(b) $\begin{cases} \dfrac{d^2 y_1}{dt^2} + 6\dfrac{dy_1}{dt} + 2y_1 = \dfrac{dr}{dt} - 3r \\[2mm] \dfrac{dy_2}{dt} + 6y_2 = 4\dfrac{dr}{dt} \end{cases}$

Ans $T_{11}(s) = (s - 3)/(s^2 + 6s + 2);$ $\qquad$ $T_{21}(s) = 4s/(s + 6)$

(c) $\begin{cases} \dfrac{d^3 y_1}{dt^3} + 7\dfrac{d^2 y_1}{dt^2} + 6\dfrac{dy_1}{dt} + y_1 = \dfrac{d^2 r_1}{dt^2} + 3r_1 + r_2 \\[2mm] \dfrac{d^3 y_2}{dt^3} + 7\dfrac{d^2 y_2}{dt^2} + 6\dfrac{dy_2}{dt} + y_2 = 4\dfrac{dr_2}{dt} \end{cases}$

Ans $T_{11}(s) = (s^2 + 3)/(s^3 + 7s^2 + 6s + 1);$

$T_{12}(s) = 1/(s^3 + 7s^2 + 6s + 1);$ $\qquad$ $T_{21}(s) = 0;$

$T_{22}(s) = 4s/(s^3 + 7s^2 + 6s + 1)$

D1.14. Find the zero-state response for $t \geq 0$ of the systems with the following transfer functions and inputs:

(a) $T(s) = \dfrac{4}{s + 3}$

$r(t) = u(t)$

Ans $\frac{4}{3} - \frac{4}{3}e^{-3t}$

(b) $T(s) = \dfrac{3s}{s + 2}$

$r(t) = \delta(t)$

Ans $3\delta(t) - 6e^{-2t}$

(c) $T(s) = \dfrac{-5s}{s^2 + 4s + 3}$

$r(t) = 6u(t)e^{-2t}$

Ans $15e^{-t} - 60e^{-2t} + 45e^{-3t}$

(d) $T(s) = \dfrac{4}{s + 3}$

$r(t) = 3u(t) \cos 2t$

Ans $-\frac{36}{13}e^{-3t} + \frac{36}{13} \cos 2t + \frac{24}{13} \sin 2t$

$\qquad = -\frac{36}{13}e^{-3t} + 3.32 \cos(2t - 33.7°)$

D1.15. Find the complete response for $t \geq 0$ of each of the following systems with the indicated input and initial conditions. Identify the zero-state and zero-input response components and the forced and natural response components.

(a) $T(s) = \dfrac{4}{s + 3}$

$r(t) = u(t)$

$y(0^-) = -2$

Ans $\frac{4}{3} - \frac{10}{3}e^{-3t}, \frac{4}{3} - \frac{4}{3}e^{-3t}, -2e^{-3t}, \frac{4}{3}, -\frac{10}{3}e^{-3t}$

(b) $T(s) = \dfrac{10}{s + 4}$

$r(t) = \delta(t)$

$y(0^-) = 0$

Ans $10e^{-4t}, 10e^{-4t}, 0, 0, 10e^{-4t}$

(c) $T(s) = \dfrac{s - 5}{s^2 + 3s + 2}$

$r(t) = u(t)$

$y(0^-) = -3$

$y'(0^-) = 4$

Ans $-\frac{5}{2} + 4e^{-t} - \frac{9}{2}e^{-2t}, \quad -\frac{5}{2} - \frac{7}{2}e^{-2t} + 6e^{-t}, \quad -2e^{-t} - e^{-2t}, \quad -\frac{5}{2},$
$4e^{-t} - \frac{9}{2}e^{-2t}$

D1.16. Classify the systems with the following transfer functions as being stable, marginally stable or unstable:

(a) $T(s) = \dfrac{10(s - 3)}{(s + 1)^2(s + 3)}$

Ans stable

(b) $T(s) = \dfrac{s^2 - 4s}{(s^2 + 9)(s + 10)}$

Ans marginally stable

(c) $T(s) = \dfrac{10}{s^2 + 2s + 10}$

Ans stable

(d) $T(s) = \dfrac{10s + 1}{s(s^2 - 2s + 10)}$

Ans unstable

(e) $T_{11}(s) = \dfrac{s}{(s + 1)(s + 3)}$

$T_{12}(s) = \dfrac{3}{(s + 1)(s + 3)}$

$T_{21}(s) = \dfrac{-4s + 2}{(s + 1)(s + 3)}$

$T_{22}(s) = \dfrac{s^2}{(s + 1)(s + 3)(s - 5)}$

Ans unstable

(f) $T_{11}(s) = \dfrac{3(s - 2)^2}{(s^2 + s + 3)^2}$

$T_2(s) = \dfrac{9}{(s^2 + s + 3)^2}$

Ans stable

D1.17. A three-input, two-output system has the following transfer functions and inputs. Find the two outputs for $t \geq 0$ if all system initial conditions are zero:

$T_{11}(s) = \dfrac{3}{s + 2}$ $\qquad$ $T_{22}(s) = \dfrac{-6s + 4}{s^2 + 5s + 6}$

$T_{12}(s) = \dfrac{s}{s + 3}$ $\qquad$ $T_{23}(s) = \dfrac{s + 2}{s + 3}$

$T_{13}(s) = \dfrac{s + 7}{s^2 + 5s + 6}$ $\qquad$ $\begin{aligned} r_1(t) &= u(t) \\ r_2(t) &= \delta(t) \end{aligned}$

$T_{21}(s) = \dfrac{10}{(s + 2)(s + 3)}$ $\qquad$ $r_3(t) = 6u(t)e^{-2t}$

Ans $\delta(t) + \frac{3}{2} - \frac{51}{2}e^{-2t} + 21e^{-3t} + 30te^{-2t}, \qquad \frac{10}{6} + 11e^{-2t} - \frac{38}{3}e^{-3t}$

1.5 BLOCK DIAGRAMS

Block Diagram Elements

Block diagrams are used to describe the component parts of systems. They offer an alternative to dealing directly with equations. A *block* is used to indicate a proportional relationship between two Laplace-transformed signals. The proportionality function, or *transmittance*, relates incoming and outgoing signals and is indicated within the block. A *summer* is used to show additions and subtractions of signals. A summer can have any number of incoming signals, but only one outgoing signal. The algebraic signs to be used in the summation are indicated next to the arrowhead for each incoming signal. A *junction* (sometimes termed a "pickoff point") indicates that the same signal is to go several places. Examples of each of these elements are shown in Fig. 1.10.

For example, a system that satisfies the second-order linear differential equation

$$\frac{d^2y}{dt^2} + 4\frac{dy}{dt} + 13y = 4r$$

has the transmittance

$$T(s) = \frac{Y(s)}{R(s)}\bigg|_{\text{initial conditions}=0}$$

$$= \frac{4}{s^2 + 4s + 13}$$

so that Fig. 1.11 contains the transmittance in a block.

A temperature control system provides a more complicated example. If x represents the heat applied to some object, suppose the object's temperature y satisfies

$$\frac{dy}{dt} + by = bx$$

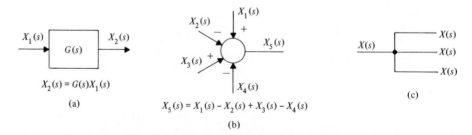

$$X_2(s) = G(s)X_1(s)$$

(a)

$$X_5(s) = X_1(s) - X_2(s) + X_3(s) - X_4(s)$$

(b)

(c)

Figure 1.10 Elements of block diagrams.

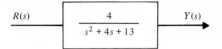

Figure 1.11 Transmittance of a second-order system.

creating the transmittance

$$T_1(s) = \left.\frac{Y(s)}{X(s)}\right|_{\text{initial conditions}=0}$$

$$= \frac{b}{s+b}$$

which becomes one of the blocks in Fig. 1.12. If the system compares the desired temperature r to the actual temperature y, the error e results, where

$$e = r - y$$

$$E(s) = R(s) - Y(s)$$

The difference is created in Fig. 1.12 by using a junction to produce the measurement of y for comparison with r and then using a summer with one negative sign to produce e.

Finally, suppose the oven operates upon e so as to modify the heat applied to the object, where

$$\frac{dx}{dt} + ax = ae$$

creating the transmittance

$$T_2(s) = \left.\frac{X(s)}{E(s)}\right|_{\text{initial conditions}=0}$$

$$= \frac{a}{s+a}$$

which completes Fig. 1.12. Figure 1.12 contains two blocks, one summer, and one junction.

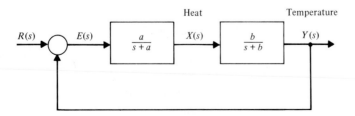

Figure 1.12 Block diagram of a temperature control system.

Block Diagram Reduction

Rearranging system block diagrams to affect simplification or special structures is termed *block diagram algebra*. Since the block diagrams represent Laplace-transformed system equations, manipulating a diagram is equivalent to algebraic manipulation of the original equations, but diagram manipulation is generally easier than dealing with the equations directly and also provides better physical insight about the structure of the system. For a single-input, single-output block diagram, *reduction* means simplifying the composite diagram to the point where it is a single block, displaying the transfer function relating the output to the input. In reducing a block diagram, it is helpful to proceed step by step, always maintaining the same overall relationship between input and output.

Some useful simplifications are the following. Two blocks in *cascade* (or *series*), with no additional connections between them, are equivalent so far as the incoming and outgoing signals are concerned to a single "product of transmittances" block, as indicated in Fig. 1.13a. Two blocks in *tandem* (or *parallel*), Fig. 1.13b, are equivalent to a single "sum of transmittances" block. This result is modified if there are other signs besides pluses on the summer; Fig. 1.13c shows one such example.

For two blocks in a *feedback* configuration, Fig. 1.14, two possible algebraic signs on the summer are considered. $G(s)$ is termed the *forward transmittance* and

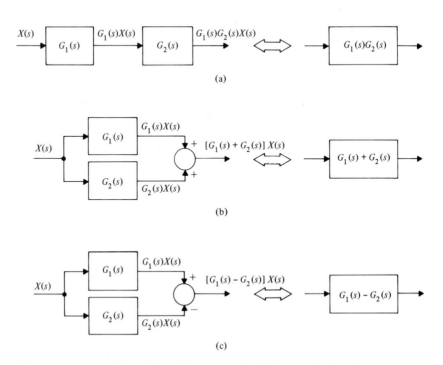

Figure 1.13 Equivalents of blocks **(a)** in cascade, **(b)** in tandem with positive summer signs, **(c)** in tandem with negative summer sign.

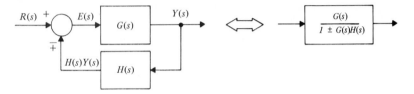

Figure 1.14 The feedback configuration.

$H(s)$ is the *feedback transmittance* in this arrangement. The relationship between the signals are

$$\begin{cases} Y(s) = G(s)E(s) \\ E(s) = R(s) \mp H(s)Y(s) \end{cases}$$

Solving for $Y(s)$ in terms of $R(s)$ by eliminating $E(s)$,

$$Y(s) = G(s)[R(s) \mp H(s)Y(s)] = G(s)R(s) \mp G(s)H(s)Y(s)$$

there results

$$T(s) = \frac{Y(s)}{R(s)} = \frac{G(s)}{1 \pm G(s)H(s)}$$

A negative sign on the feedback summation in Fig. 1.14 results in a positive algebraic sign in the denominator of $T(s)$; a plus sign on the summer gives a minus sign.

Other useful equivalences are given in Table 1.9. In Fig. 1.15, several of the equivalences are used to reduce a block diagram of a single-input, single-output system, to find its transfer function.

Multiple Inputs and Outputs

For a multiple-input, multiple-output system, block diagram reduction involves finding each of the system transfer functions. This is done by considering only one

Table 1.9 OTHER USEFUL BLOCK DIAGRAM EQUIVALENCES

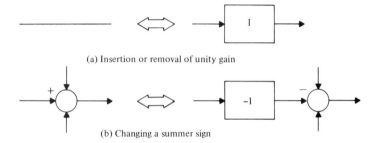

(a) Insertion or removal of unity gain

(b) Changing a summer sign

Table (*Continued*)

Table 1.9 (*Continued*)

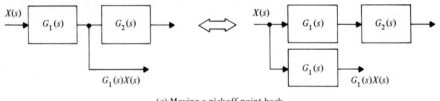

(c) Moving a pickoff point back

(d) Moving a pickoff point forward

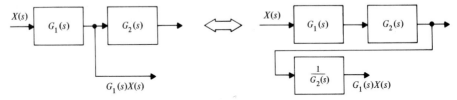

(e) Combining or expanding summations

(f) Combining or expanding junctions

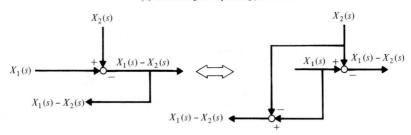

(g) Moving a pickoff point behind a summation

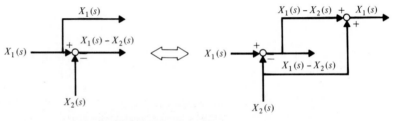

(h) Moving a pickoff point forward of a summation

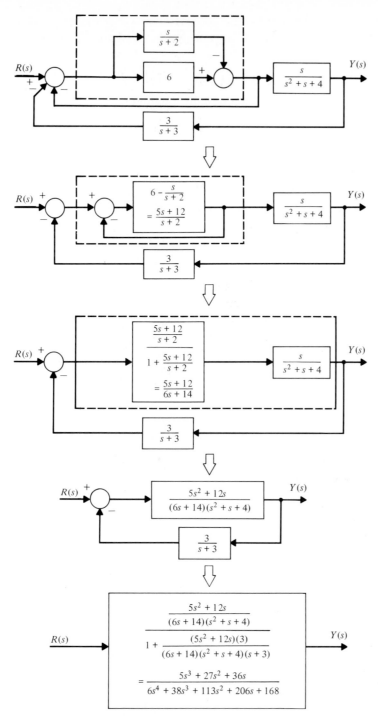

Figure 1.15 Example of block diagram reduction for a single-input, single-output system.

output at a time and setting all but one input to zero to determine the transfer function relating that output to that input. For example, see the four transfer functions for the two-input, two-output system found in Fig. 1.16.

The input and output signals in a two-input system are related as shown in the canonical block diagram of Fig. 1.17. Every block diagram is reducible to a similar equivalent form, where the transfer functions are placed in evidence.

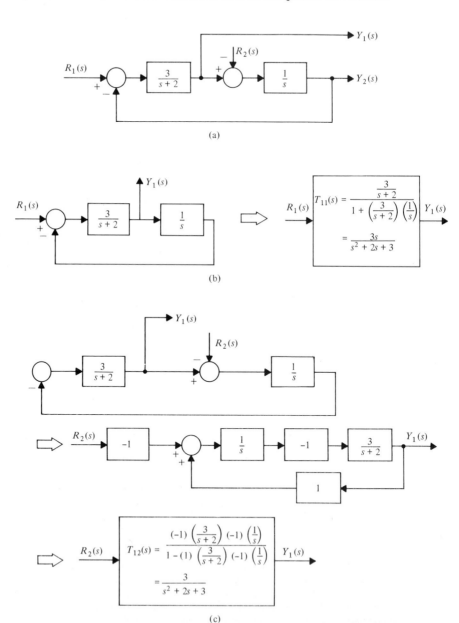

(a)

(b)

(c)

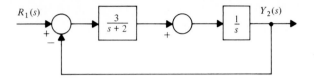

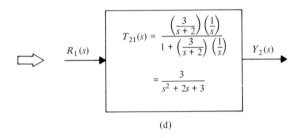

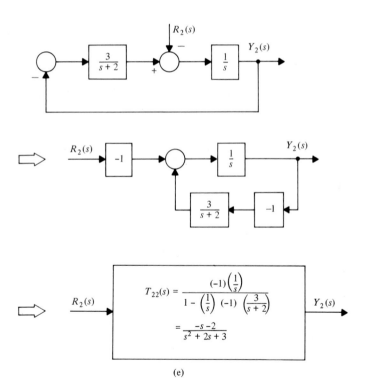

Figure 1.16 Example of multiple-input, multiple-output block diagram reduction.
(a) A two-input, two-output system. **(b)** Block diagram reduction to find $T_{11}(s)$. Input R_2 is set to zero and output Y_2 is ignored. **(c)** Reduction to find $T_{12}(s)$. R_1 is set to zero and Y_2 is ignored. **(d)** Reduction to find $T_{21}(s)$. R_2 is set to zero and Y_1 is ignored.
(e) Reduction to find $T_{22}(s)$. R_1 is set to zero and Y_1 is ignored.

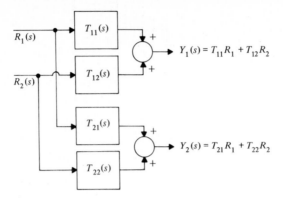

Figure 1.17 Canonical block diagram for a two-input, two-output system.

DRILL PROBLEMS

D1.18. Reduce the following block diagrams, obtaining the system transfer function.

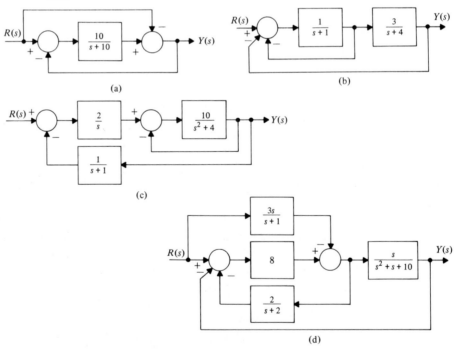

Ans (a) $-s/(s + 20)$; (b) $3/(s^2 + 6s + 11)$;
(c) $20(s + 1)/(s^4 + s^3 + 14s^2 + 14s + 20)$;
(d) $[-3s^2(s + 2) + 8s(s + 2)(s + 1)]/(s^4 + 28s^3 + 71s^2 + 224s + 180)$

D1.19. Find the six system transfer functions of the following system.

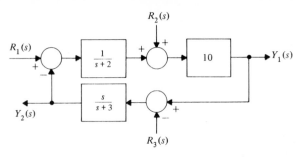

Ans $10(s + 3)/(s^2 + 15s + 6)$; $10(s + 2)(s + 3)/(s^2 + 15s + 6)$;
$10s/(s^2 + 15s + 6)$; $10s/(s^2 + 15s + 6)$;
$10s(s + 2)/(s^2 + 15s + 6)$; $-s(s + 2)/(s^2 + 15s + 6)$

D1.20. If all the system initial conditions are zero, find the Laplace transform of the output for the given system inputs.

$$r_1(t) = 3e^{-t}$$

$r_2(t) = 4u(t)$, where $u(t)$ is the unit step function

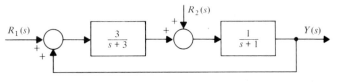

Ans $(4s^2 + 25s + 12)/[s^2(s + 4)(s + 1)]$

1.6 SIGNAL FLOW GRAPHS

Another way of depicting the relations between Laplace-transformed signals in a system is with a signal flow graph. Signal flow graphs have two elements, the *branch* and the *node*, as indicated in Fig. 1.18. The branch is equivalent to a block in the language of block diagrams. The node is equivalent to a summer, with all plus signs, followed by a junction. The signal at any node is the sum of the signals

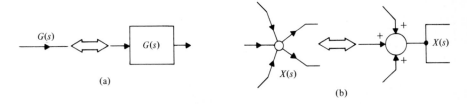

Figure 1.18 Elements of signal flow graphs.

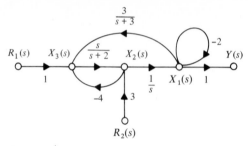

Figure 1.19 A signal flow graph.

coming into the node from branches. The signal entering a node from a branch is, in turn, the signal at the tail of the branch arrow times the branch transmittance. The signal flow graph of Fig. 1.19 represents the following Laplace-transformed equations:

$$
\begin{cases}
X_1(s) = \dfrac{1}{s} X_2(s) - 2X_1(s) \\[2ex]
X_2(s) = \dfrac{s}{s+2} X_3(s) + 3R_2(s) \\[2ex]
X_3(s) = \dfrac{3}{s+3} X_1(s) - 4X_2(s) + R_1(s) \\[2ex]
Y(s) = X_1(s)
\end{cases}
$$

It is customary to bring inputs into and outputs out of a signal flow graph through single branches, to set them off from the rest of the diagram.

The fact that a signal flow graph contains fewer elements than a block diagram results in a straightforward method, called *Mason's gain rule*, for determining the transfer function of a system directly from its signal flow graph without reduction. To explain the application of Mason's gain rule, some terms will first be defined and illustrated with the example signal flow graph of Fig. 1.20.

A *path* is any succession of branches, from input to output, in the direction of the arrows, which does not pass any node more than once, as indicated for the example in Fig. 1.20a. The *path gain* is the product of the transmittances of the branches comprising the path. For example,

$$
\begin{cases}
P_1 = \left(\dfrac{1}{s+1}\right)\left(\dfrac{1}{s^2+s}\right)(10)\left(\dfrac{1}{s}\right)\left(\dfrac{1}{s}\right) \\[2ex]
P_2 = \left(\dfrac{1}{s^2+4}\right)\left(\dfrac{8}{s+8}\right)\left(\dfrac{1}{s}\right)\left(\dfrac{1}{s}\right)
\end{cases}
$$

A *loop* is any closed succession of branches, in the direction of the arrows, which does not pass any node more than once, as in Fig. 1.20b. The *loop gain* is the

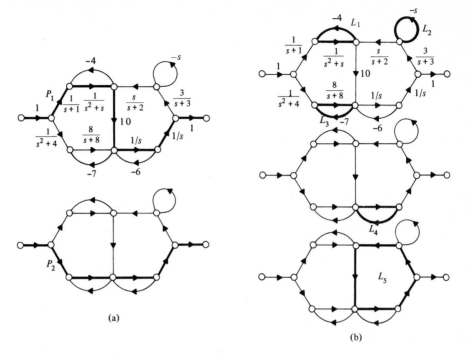

Figure 1.20 **(a)** Paths. **(b)** Loops in a Mason's gain rule example.

product of the transmittances of the branches comprising the loop. For the example,

$$\begin{cases} L_1 = \dfrac{-4}{s^2 + s} \\[2mm] L_2 = -s \\[2mm] L_3 = \dfrac{-56}{s + 8} \\[2mm] L_4 = \dfrac{-6}{s} \\[2mm] L_5 = (10)\left(\dfrac{1}{s}\right)\left(\dfrac{1}{s}\right)\left(\dfrac{3}{s+3}\right)\left(\dfrac{s}{s+2}\right) \end{cases}$$

Two loops are said to be *touching* if they have any node in common. Otherwise, they are nontouching. Similarly, a loop and a path are touching if they have any node in common.

The *determinant* of a signal flow graph is

$$\Delta = 1 - \text{(sum of all loop gains)} + \text{(sum of products of gains of all combina-}$$
tions of 2 nontouching loops) − (sum of products of gains of all
combinations of 3 nontouching loops) + ⋯

For the example,

$$
\begin{aligned}
\Delta = 1 - (L_1 + L_2 + L_3 + L_4 + L_5) + \\
(L_1 L_2 + L_1 L_3 + L_1 L_4 + L_2 L_3 + L_2 L_4) \\
- (L_1 L_2 L_3 + L_1 L_2 L_4)
\end{aligned}
$$

The *cofactor* of a path is the determinant of the signal flow graph formed by deleting all loops touching the path. In the example,

$$
\Delta_1 = 1 - L_2 = 1 + s
$$

$$
\Delta_2 = 1 - (L_1 + L_2) + (L_1 L_2) = 1 + \frac{4}{s^2 + s} + s + \frac{4s}{s^2 + s}
$$

Mason's gain rule is as follows: The transfer function of a system with single-input, single-output signal flow graph is

$$
T(s) = \frac{P_1 \Delta_1 + P_2 \Delta_2 + P_3 \Delta_3 + \cdots}{\Delta}
$$

For the example system, since there are two paths,

$$
T(s) = \frac{P_1 \Delta_1 + P_2 \Delta_2}{\Delta}
$$

where P_1, P_2, Δ, Δ_1, and Δ_2 are given above. The definitions related to Mason's gain rule are listed in Table 1.10.

Table 1.10   SIGNAL FLOW GRAPH DEFINITIONS

Path:	A succession of branches, from input to output, in the direction of the arrows, which does not pass any node more than once.
Path gain:	Product of the transmittances of the branches of the path. For the ith path, the path gain is denoted by P_i.
Loop:	A closed succession of branches, in the direction of the arrows, which does not pass any node more than once.
Loop gain:	Product of the transmittances of the branches of the loop.
Touching:	Loops with one or more nodes in common are termed touching. A loop and a path are touching if they have a common node.
Determinant:	The determinant of a signal flow graph is $\Delta = 1 -$ (sum of all loop gains) + (sum of products of gains of all combinations of 2 nontouching loops) $-$ (sum of products of gains of all combinations of 3 nontouching loops) $+ \cdots$
Cofactor:	The cofactor of the ith path, denoted by Δ_i, is the determinant of the signal flow graph formed by deleting all loops touching path i.
Mason's gain rule:	$T(s) = \dfrac{P_1 \Delta_1 + P_2 \Delta_2 + \cdots}{\Delta}$

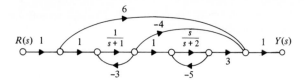

Figure 1.21 Another single-input, single-output signal flow graph.

A less involved example is shown in Fig. 1.21, for which

$$T(s) = \frac{P_1\Delta_1 + P_2\Delta_2 + P_3\Delta_3}{\Delta}$$

$$= \frac{6\left[1 + \dfrac{3}{s+1} + \dfrac{5s}{s+2} + \dfrac{15s}{(s+1)(s+2)}\right] + \left(\dfrac{-4}{s+1}\right)\left(1 + \dfrac{5s}{s+2}\right) + \left[\left(\dfrac{3}{s+1}\right)\left(\dfrac{s}{s+2}\right)\right]}{1 + \dfrac{3}{s+1} + \dfrac{5s}{s+2} + \dfrac{15s}{(s+1)(s+2)}}$$

$$= \frac{36s^2 + 135s + 40}{6s^2 + 26s + 8}$$

For multiple-input, multiple-output systems, Mason's gain rule is simply applied repeatedly for each different combination of output and input. The signal flow graph of Fig. 1.22a is of a system with two inputs and two outputs. The corresponding single-input, single-output signal flow graphs for calculating the

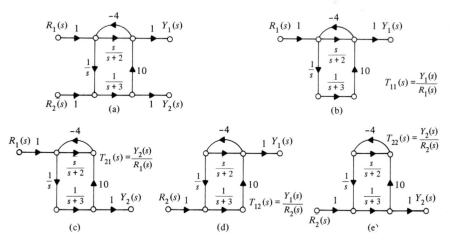

Figure 1.22 A multiple-input, multiple-output signal flow graph and its transfer functions.

four associated transfer functions are shown in Fig. 1.22b–e. Using Mason's gain rule, the system transfer functions are as follows:

$$T_{11}(s) = \frac{Y_1(s)}{R_1(s)} = \frac{\left(\dfrac{s}{s+2}\right)(1) + \dfrac{10}{s(s+3)}(1)}{1 - \left(\dfrac{-4s}{s+2}\right) - \left[\dfrac{-40}{s(s+3)}\right]} = \frac{s^3 + 3s^2 + 10s + 20}{5s^3 + 17s^2 + 46s + 80}$$

$$T_{21}(s) = \frac{Y_2(s)}{R_1(s)} = \frac{\left[\dfrac{1}{s(s+3)}\right](1)}{1 - \left(\dfrac{-4s}{s+2}\right) - \left[\dfrac{-40}{s(s+3)}\right]} = \frac{s+2}{5s^3 + 17s^2 + 46s + 80}$$

$$T_{12}(s) = \frac{Y_1(s)}{R_2(s)} = \frac{\left(\dfrac{10}{s+3}\right)(1)}{1 - \left(\dfrac{-4s}{s+2}\right) - \left[\dfrac{-40}{s(s+3)}\right]} = \frac{10s^2 + 20s}{5s^3 + 17s^2 + 46s + 80}$$

$$T_{22}(s) = \frac{Y_2(s)}{R_2(s)} = \frac{\left(\dfrac{1}{s+3}\right)\left[1 - \left(\dfrac{-4s}{s+2}\right)\right]}{1 - \left(\dfrac{-4s}{s+2}\right) - \left[\dfrac{-40}{s(s+3)}\right]} = \frac{5s^2 + 2s}{5s^3 + 17s^2 + 46s + 80}$$

DRILL PROBLEMS

D1.21. Write a set of simultaneous Laplace transformed equations for the signal flow graph:

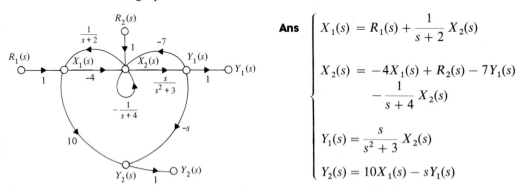

Ans
$$\begin{cases} X_1(s) = R_1(s) + \dfrac{1}{s+2}X_2(s) \\[2mm] X_2(s) = -4X_1(s) + R_2(s) - 7Y_1(s) \\[1mm] \qquad\quad - \dfrac{1}{s+4}X_2(s) \\[2mm] Y_1(s) = \dfrac{s}{s^2+3}X_2(s) \\[2mm] Y_2(s) = 10X_1(s) - sY_1(s) \end{cases}$$

D1.22. Use Mason's gain rule to find the transfer function of each system.

Ans (a) $(-12s^2 - 66s - 54)/(8s^2 + 17s + 3)$;
(b) $(56s^2 + 145s + 103)/(s+1)^2$;
(c) $(3s + 2)/(s^2 + 4s + 5)$;   (d) $(3s + 2)/(s^2 + 4s + 5)$

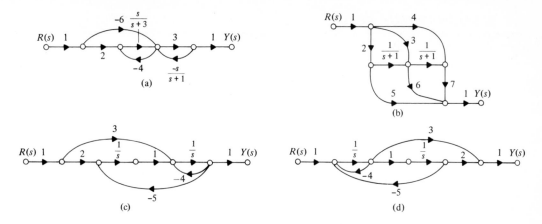

D1.23. Use Mason's gain rule to find the six transfer functions of the following system.

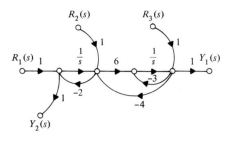

Ans $6/(s^2 + 29s + 6)$,
$6s/(s^2 + 29s + 6)$,
$s(s + 2)/(s^2 + 29s + 6)$,
$s(s + 27)/(s + 29s + 6)$,
$-s(2s + 6)/(s^2 + 29s + 6)$,
$8s^2/(s^2 + 29s + 6)$

D1.24. If all system initial conditions are zero, find the Laplace transforms of the outputs for the given system inputs.

$r_1(t) = 4 \sin t$

$r_2(t) = 3\delta(t)$, where $\delta(t)$ is the unit impulse

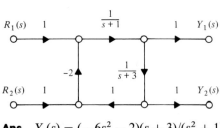

Ans $Y_1(s) = (-6s^2 - 2)(s + 3)/(s^2 + 1)(s^2 + 4s + 5)$
$Y_2(s) = (-6s^2 - 2)/(s^2 + 1)(s^2 + 4s + 5)$

1.7 A POSITIONING SERVO

A simple but practical feedback control system is diagramed in Fig. 1.23. It is a positioning system or *position servo* for a large microwave antenna. The antenna is modeled as a mass having a large moment of inertia, J. An output potentiometer measures the output shaft position, converting the position to a proportional voltage according to

$$v_o = K_p \theta$$

where θ is the output shaft angle in radians and v_o is the output potentiometer voltage; K_p, the constant of proportionality between shaft position and potentiometer voltage, is the total voltage V divided by the maximum rotation of the potentiometer:

$$K_p = \frac{V}{\theta_{max}} \quad \text{V/rad}$$

The input potentiometer slider position r is converted to a voltage with a potentiometer identical to the output potentiometer:

$$v_r = K_p r$$

The difference between the two potentiometer signals is then amplified with gain A_1,

$$v_1 = A_1(v_r - v_o) = A_1 K_p(r - \theta)$$

where v_1 is the error voltage output of the difference amplifier. This voltage is then further amplified with gain A_2 and is applied to the motor terminals,

$$v_2 = A_2 v_1 = A_1 A_2 K_p(r - \theta) \tag{1.10}$$

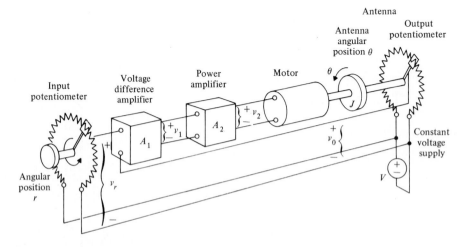

Figure 1.23 A position servo.

where v_2 is the motor voltage. The second amplifier is a power amplifier which is capable of supplying the electrical power necessary to drive the motor.

The motor chosen for this application has a negligible armature inductance. From Table 1.5, part 2, the motor produces a torque

$$\tau = K_T i_A$$

$$= \frac{K_T\left[v_2 - K_V\left(\dfrac{d\theta}{dt}\right)\right]}{R}$$

$$= k_1 v_2 - k_2 \frac{d\theta}{dt} = k_1 A_1 A_2 K_p(r - \theta) - k_2 \frac{d\theta}{dt}$$

Equating the applied torque on the rotational antenna mass to the inertial force, there results

$$\tau = J \frac{d^2\theta}{dt^2} = k_1 A_1 A_2 K_p(r - \theta) - k_2 \frac{d\theta}{dt} \tag{1.11}$$

Rearranging,

$$J \frac{d^2\theta}{dt^2} + k_2 \frac{d\theta}{dt} + k_1 A_1 A_2 K_p \theta = k_1 A_1 A_2 K_p r(t)$$

Some of the equation coefficients, and thus some of the system properties, can be selected by the designer by appropriately choosing the control components. Other parameters, such as the moment of inertia of the load, J, cannot be changed.

The transfer function relating the input position $R(s)$ to the output position $\theta(s)$ is given by

$$J s^2 \theta(s) + k_2 s \theta(s) + k_1 A_1 A_2 K_p \theta(s) = k_1 A_1 A_2 K_p R(s)$$

$$T_1(s) = \frac{\theta(s)}{R(s)} = \frac{k_1 A_1 A_2 K_p}{J s^2 + k_2 s + k_1 A_1 A_2 K_p}$$

If the antenna experiences an external torque, say from wind, in addition to the motor's torque, Eq. (1.11) becomes, instead,

$$\tau = J \frac{d^2\theta}{dt^2} - \tau_{\text{wind}} = k_1 A_1 A_2 K_p(r - \theta) - k_2 \frac{d\theta}{dt}$$

$$J \frac{d^2\theta}{dt^2} + k_2 \frac{d\theta}{dt} + k_1 A_1 A_2 K_p \theta = k_1 A_1 A_2 K_p r(t) + \tau_{\text{wind}}(t)$$

where τ_{wind} is the external torque in the sense of increasing θ. The transfer function relating $\theta(s)$ to $R(s)$ is unchanged, but now a disturbance input $\tau_{\text{wind}}(t)$ has been included in the model. The transfer function relating this input to θ is found by setting r to zero and Laplace-transforming with zero initial conditions:

$$Js^2\theta(s) + k_2 s\theta(s) + k_1 A_1 A_2 K_p \theta(s) = \mathcal{T}_{\text{wind}}(s)$$

$$T_2(s) = \frac{\theta(s)}{\mathcal{T}_{\text{wind}}(s)} = \frac{1}{Js^2 + k_2 s + k_1 A_1 A_2 K_p}$$

In general, with zero initial conditions,

$$\theta(s) = T_1(s)R(s) + T_2(s)\mathcal{T}_{\text{wind}}(s)$$

$$= \frac{k_1 A_1 A_2 K_p/J}{s^2 + (k_2/J)s + (k_1 A_1 A_2 K_p/J)} R(s)$$

$$+ \frac{1/J}{s^2 + (k_2/J)s + (k_1 A_1 A_2 K_p/J)} \mathcal{T}_{\text{wind}}(s)$$

The two transfer functions can also be derived using block diagram or signal flow graph methods. Figure 1.24 depicts the block diagram and signal flow graph for the above Laplace-transformed equations.

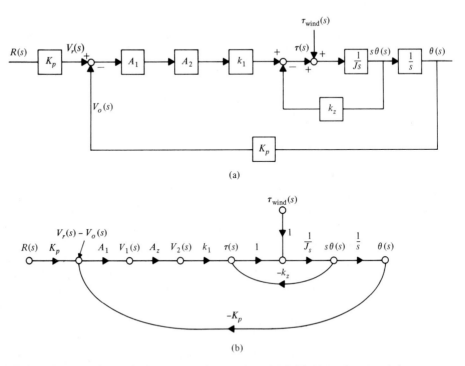

(a)

(b)

Figure 1.24 **(a)** Block diagram for the system of Fig. 1.23. **(b)** Signal flow graph for the system of Fig. 1.23.

By examining Fig. 1.24, two loops are apparent:

$$L_1 = \frac{-k_2}{Js}$$

$$L_2 = \frac{-k_1 A_1 A_2 K_p}{Js^2}$$

If path 1 denotes the path from $R(s)$ to $\theta(s)$, then

$$P_1 = \frac{k_1 A_1 A_2 K_p}{Js^2} \qquad \Delta_1 = 1$$

If path 2 denotes the path from the disturbance torque to $\theta(s)$, then

$$P_2 = \frac{1}{Js^2} \qquad \Delta_2 = 1$$

The two transfer functions follow easily by using Mason's gain rule:

$$T_1(s) = \frac{P_1 \Delta_1}{1 + L_1 + L_2}$$

$$T_2(s) = \frac{P_2 \Delta_2}{1 + L_1 + L_2}$$

The transfer functions are fairly simple here because loops 1 and 2 touch and because paths 1 and 2 touch both loops.

The block diagram (or signal flow graph) not only makes the interrelations among variables easy to visualize, but the transfer functions are also easy to derive when the paths and loops are clearly identified. This simple example illustrates the importance that block diagram depiction has to control system analysis. Design can then follow where adjustable values are selected to meet specifications.

1.8 A FORCE-BALANCE ACCELEROMETER

A force-balance accelerometer, shown in the schematic of Fig. 1.25 overcomes some of the disadvantages of the seismic instruments. In this system a mass is allowed to move along the acceleration-sensitive axis. The position of this mass is measured with a position pickoff. The output voltage from this position transducer is amplified with a high-gain amplifier whose output is a current. The current flows through the windings of a forcer which forces the mass back to its original null position. The forcer current, necessary to zero the position of this mass, is proportional to acceleration, and it is measured as a voltage across a resistor in series with the forcer coil. In this system high damping, which is independent of temperature, and good accuracy are obtained by means of appropriate equalization.

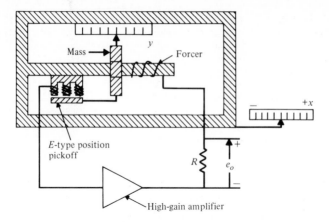

Figure 1.25 Schematic diagram of a force-balance accelerometer.

Analysis of the force-balance accelerometer of Fig. 1.25 is based upon the block diagram of Fig. 1.26. If the mass is free to move along the rod with essentially zero damping, the force on the mass is related to the displacement as follows:

$$F(s) = -Ms^2(Y(s) + X(s)) \tag{1.12}$$

or, rewritten,

$$Y(s) = -\frac{1}{Ms^2}(F(s) - Ms^2X(s)) = -\frac{1}{Ms^2}(F(s) - MA(s)) \tag{1.13}$$

s is the Laplace-transform operator, and zero initial conditions are assumed. Equation (1.13) is represented by the first block whose output is Y, which is converted to a voltage with a position pickoff (transfer function K_s) and equalized (G_e). Current from the amplifier passes through a forcer, with transfer function K_f in force units, per milliampere. This current forces the mass back to a null position. The output voltage is taken at the output of a final block with transfer function R. The force fed back cancels the input acceleration.

With an equalizer of the form

$$G_e(s) = \alpha s + 1 \tag{1.14}$$

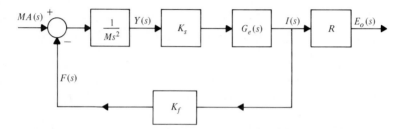

Figure 1.26 Block diagram of a force-balance accelerometer.

the system can be made stable. The damping is electrically controllable (as α is varied) and so is independent of temperature effects. The closed-loop system has the equation

$$\frac{E_o(s)}{MA(s)} = \frac{RK_sK(\alpha s + 1)/Ms^2}{[K_sK(\alpha s + 1)K_f/Ms^2] + 1}$$

which reduces to

$$\frac{E_o(s)}{A(s)} = \frac{RK_sK(\alpha s + 1)}{s^2 + (K_sKK_f\alpha/M)s + K_sKK_f/M}$$

Force-balance instruments, which are applicable to many types of measurement, have an advantage in that the mechanical system inaccuracies, such as spring hysteresis, are reduced. Accuracy of the force-balance system depends, to a large extent, on the accuracies with which one can build a forcer.

1.9 SUMMARY

Control systems generally involve a plant or process with inputs and outputs. Controllers are designed to generate control inputs to the plant which will produce a desired plant behavior. Feedback, if used, can result in improved accuracy and speed of response and reduced dependence upon specific components and less sensitivity to disturbances. Determining system equations is the starting point in most control system analysis. Relations for electrical, electronic, translational mechanical, rotational mechanical, and electromechanical components are summarized in Tables 1.1 through 1.5.

Laplace transforms and properties are listed in Tables 1.6 and 1.7. Partial fraction expansion is used to express a rational function as simple terms to which transform tables apply. A proper rational function ($m < n$) with distinct denominator roots expands in partial fractions as

$$Y(s) = \frac{b_ms^m + b_{m-1}s^{m-1} + \cdots + b_1s + b_0}{(s - s_1)(s - s_2)\cdots(s - s_n)} = \frac{K_1}{s - s_1} + \frac{K_2}{s - s_2} + \cdots + \frac{K_n}{s - s_n}$$

The residues of the expansion, $K_1, K_2, \ldots, K_n$, can be found by comparing coefficients or by evaluation:

$$K_i = [(s - s_i)Y(s)]_{s=s_i}$$

When there are repeated denominator roots (poles), the corresponding terms in the partial fraction expansion are

$$Y(s) = \frac{b_ms^m + b_{m-1}s^{m-1} + \cdots + b_1s + b_0}{(\text{nonrepeated root factors})(s - s_r)^i}$$

$$= \frac{\text{partial fraction terms}}{\text{for nonrepeated roots}} + \frac{K_1}{s - s_r} + \frac{K_2}{(s - s_r)^2} + \cdots + \frac{K_p}{(s - s_r)^p}$$

The constants $K_1, K_2, \ldots, K_i$ can be found by equating coefficients or with

$$K_i = \frac{1}{(p-i)!} \frac{d^{p-i}}{ds^{p-i}} [(s - s_r)^p Y(s)]_{s=s_r}$$

If the numerator order is not lower than the denominator order, the denominator is divided into the numerator for the number of steps required to yield a proper remainder, then the remainder is expanded in partial fractions.

The time responses corresponding to various partial fraction terms are shown in Table 1.8.

The transfer function of a single-input, single-output, linear, time-invariant system is the ratio of the Laplace transform of the output to the Laplace transform of the input, when all initial conditions are zero. Systems described by integrodifferential equations have transfer functions which can be expressed as the ratio of two polynomials in s.

System response is commonly considered to be composed of component parts in either of two ways:

1. Zero-state and zero-input
2. Forced and natural

The zero-state response component is the system response when all initial conditions are zero,

$$Y_{\text{zero-state}}(s) = T(s)R(s)$$

where $T(s)$ is the system transfer function and $R(s)$ is the Laplace transform of the input. The zero-input response component is the system response due to nonzero initial conditions when the input is zero. The forced component contains those terms, due to the inputs, which do not correspond to system characteristic roots and the characteristic root terms.

A system is stable if, for any initial conditions, all of the terms in its natural (and thus its zero-input) response decay with time. Such will be the case if and only if all system characteristic roots are in the left half of the complex plane. Nonrepeated characteristic roots on the imaginary axis correspond to natural response terms which neither decay nor expand in time. A system with all characteristic roots in the left half-plane except for one or more nonrepeated imaginary axis roots is termed marginally stable. If any characteristic roots are in the right half of the complex plane or are repeated on the imaginary axis, the system is unstable.

For multiple-input, multiple-output systems, there is a transfer function for each combination of input and output:

$$T_{ij}(s) = \left. \frac{Y_i(s)}{R_j(s)} \right|_{\substack{\text{all initial conditions and all inputs} \\ \text{except } r_j \text{ set to zero}}}$$

where the y_i terms are outputs and the r_j terms are inputs. For zero initial conditions, the outputs are given by

$$Y_1(s) = T_{11}(s)R_1(s) + T_{12}(s)R_2(s) + \cdots$$
$$Y_2(s) = T_{21}(s)R_1(s) + T_{22}(s)R_2(s) + \cdots$$
$$\vdots$$

Block diagrams of systems consist of blocks, summers, and junctions. They are used to graphically describe system components and their interconnections. To reduce a block diagram of a single-input, single-output system means to find, using block diagram equivalences, the single-block diagram containing the overall transfer function of the system. Basic equivalences are shown in Figs. 1.13 and 1.14 and Table 1.9. Multiple-input, multiple-output block diagrams are reduced by reducing a series of single-input, single-output diagrams, one for each different combination of input and output.

A signal flow graph for a system consists of nodes and branches. The signal at any node is the sum of the incoming signals through branches. A branch in a signal flow graph is equivalent to a block in a block diagram, and a node is equivalent to a summer, with all plus signs, followed by a junction. The transfer function of a single-input, single-output system is given in terms of its signal flow graph by Mason's gain rule,

$$T(s) = \frac{\sum_i P_i \Delta_i}{\Delta}$$

where the P_i terms are the path gains, Δ_i is the cofactor of the ith path, and Δ is the signal flow graph determinant, as defined in Table 1.10. Mason's gain rule is applied to find the transfer functions of multiple-input, multiple-output systems by considering one input and one output at a time.

The chapter concluded with two examples of systems to which the methods of Chapter 1 can be applied. The equations of motion for a pointing servo were developed. The transfer functions follow by manipulating the differential equations or by applying Mason's rule to either the signal flow graph or (with appropriate caution) to the block diagram. The block diagram clarifies the structure of the system while facilitating computation of the transfer function. An accelerometer was analyzed which employs a force-balance mechanism to generate acceleration measurements.

REFERENCES

The references given here and in the following chapters trace the history of topics presented in the text. While by no means comprehensive, they give a series of milestones in the development and understanding of these ideas. This, too, is our way of acknowledging those works from which we learned and to which we all owe a great deal.

Feedback

Black, H. S., "Inventing the Negative Feedback Amplifier," *IEEE Spectrum* (December 1977).

Blackman, R. B., "Effect of Feedback on Impedance." *Bell Syst. Tech. J.* 22 (October 1943).

Bode, H. W., "Feedback—The History of an Idea," in *Selected Papers on Mathematical Trends in Control Theory*. New York: Dover, 1964.

Fuller, A. T., "The Early Development of Control Theory." *Trans. ASME J. of Dynamic Systems, Measurement and Control* 96G (June 1976).

————, "The Early Development of Control Theory II." *Trans. ASME J. of Dynamic Systems, Measurement and Control* 98G (September 1976).

Maxwell, J. C., "On Governors." *Proc. Roy. Soc.* 16 (1868).

Mayr, O., "The Origins of Feedback Control." *Sci. Amer.* (October 1970).

Nyquist, H., "Regeneration Theory." *Bell Syst. Tech. J.* 11 (January 1932).

Wolf, A. *A History of Science, Technology and Philosophy in the Eighteenth Century*. New York: McGraw-Hill, 1939.

System Equations

Close, C. M., and Frederick, D. K., *Modeling and Analysis of Dynamic Systems*. Boston: Houghton Mifflin, 1978.

Cannon, R. H., Jr., *Dynamics of Physical Systems*. New York: McGraw-Hill, 1967.

Harmon, W. W., and Lytle, D. W., *Electrical and Mechanical Networks*. New York: McGraw-Hill, 1962.

Hostetter, G. H., *Engineering Network Analysis*. New York: Harper & Row, 1984.

Luenberger, D. G., *Introduction to Dynamic Systems*. New York: Wiley, 1979.

Perkins, W. R., and Cruz, J. B., Jr., *Engineering of Dynamic Systems*. New York: Wiley, 1969.

Van Valkenburg, M. E., *Network Analysis*. Englewood Cliffs, N.J.: Prentice-Hall, 1974.

Laplace Transformation, Transfer Functions, and Stability

Aseltine, J. A., *Transform Methods in Linear System Analysis*. New York: McGraw-Hill, 1958.

Gardner, M. F., and Barnes, J. L., *Transients in Linear Systems*. New York: Wiley, 1942.

LePage, W. R., *Complex Variables and the Laplace Transform for Engineers*. New York: McGraw-Hill, 1961.

Ley, B. J., Lutz, S. G., and Rehberg, C. F., *Linear Circuit Analysis*. New York: McGraw-Hill, 1959.

Savant, C. J., Jr., *Fundamentals of the Laplace Transformation*. New York: McGraw-Hill, 1962.

Signal Flow Graphs

Dertouzos, M. L., Athans, M., Spann, R. N., and Mason, S. J., *Systems, Networks and Computation*. New York: McGraw-Hill, 1973.

Mason, S. J., "Feedback Theory: Some Properties of Signal Flow Graphs." *Proc. IRE* 41 (September 1953).

————. "Feedback Theory: Further Properties of Signal Flow Graphs." *Proc. IRE* 44 (July 1956).

Automobile Control Systems

Jurgen, R. K., "Drivers Get More Options in 1983." *IEEE Spec.* (November 1982): 30–36.
_____. "Detroit Unveils Sophisticated Electronics." *IEEE Spec.* (October 1983): 33–39.
_____. "More Electronics in Detroit's 1985 Models." *IEEE Spec.* (October 1984): 54–60.
_____. "Detroit's 1987 Models: New Electronic Inroads." *IEEE Spec.* (October 1986): 68–72.

Generating and Controlling Power

Fischetti, M., and Zorpette, G., "Power and Energy." *IEEE Spec.* (January 1986): 65–45.
Gaushell, D. J., "Automating and Power Grid." *IEEE Spec.* (October 1985): 39–45.
Zorpette, G., "HVDC: Wheeling Lots of Power." *IEEE Spec.* (June 1985): 30–36.

PROBLEMS

1. Write simultaneous loop equations for the electrical networks of Fig. P1.1. Then Laplace-transform the equations, taking all initial conditions to be zero.

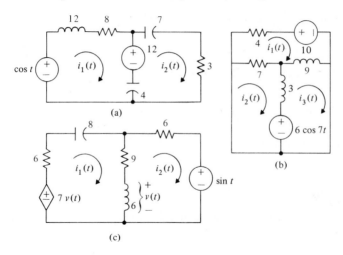

Figure P1.1

2. If all conditions of the networks of Fig. P1.2 are zero, find $v(t)$, $t \geq 0$.

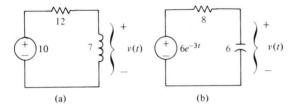

Figure P1.2

Ans (a) $v(t) = 10\, e^{-(12/7)t}u(t)$; (b) $v(t) = (0.041\, e^{-t/48} - 0.041\, e^{-3t})u(t)$

3. Write simultaneous nodal equations for the electrical networks of Fig. P1.3. Then Laplace-transform the equations, taking all initial conditions to be zero.

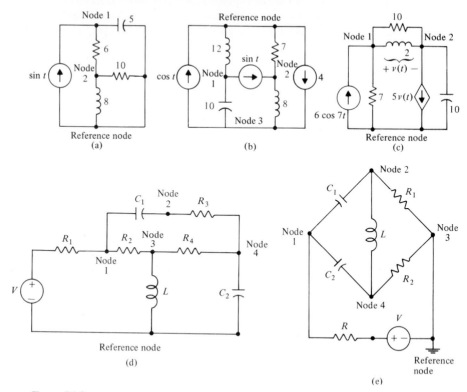

Figure P1.3

4. All initial conditions in the networks of Fig. P1.4 are zero. Find $I(s)$, then find $i(t)$ for $t \geq 0$.

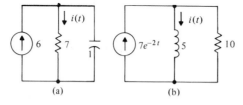

Figure P1.4

5. A useful model for a voltage amplifier is shown in Fig. P1.5a. The incoming voltage is amplified with gain A and produces an output voltage $Av_i(t)$. When feedback is added to an amplifier, as in Fig. P1.5b, desirable properties often result.

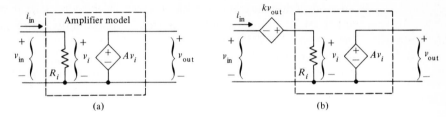

Figure P1.5

For this amplifier without feedback, the voltage gain is

$$G = \frac{v_{\text{out}}}{v_{\text{in}}} = A$$

Show that the voltage gain of the amplifier with feedback is

$$\frac{A}{1 - kA}$$

For positive values of kA less than unity, the gain is increased by the feedback, a circumstance which was exploited (as the "regenerative" receiver) in the early days of radio when high-gain amplifiers were very difficult to obtain otherwise. Unfortunately, all of the minor shortcomings of the basic amplifier are emphasized by positive feedback, and $kA \geq 1$ results in instability. For negative values of kA, the voltage gain is reduced, but in return performance in other respects is improved.

The amplifier without feedback has input resistance

$$R_{\text{in}} = \frac{v_{\text{in}}}{i_{\text{in}}} = R_i$$

Show that the input resistance of the amplifier with feedback is $R_i(1 - kA)$. For negative kA, the input resistance is increased.

6. Use the equivalent circuits of Prob. 1.5 to find $V(s)$ and $v(t)$, $t \geq 0$, for the electronic network of Fig. P1.6. Assume zero initial conditions.

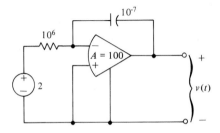

Figure P1.6

Ans $v(t) = (200e^{-10t/101} - 200)u(t)$

7. The connection diagram for a bipolar transistor feedback ac amplifier is given in Fig. P1.7a. For sufficiently small signals composed of midrange frequencies, the network of Fig. P1.7b is an accurate model. In terms of the various parameters, find the transfer function that relates $V_{out}(s)$ to $V_{in}(s)$. As only sources and resistors are involved in this model, the transfer function will be a constant.

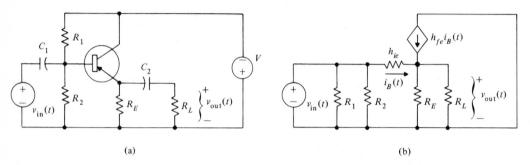

(a)   (b)

Figure P1.7

8. Write simultaneous differential equations for the translational mechanical networks of Fig. P1.8. Then Laplace-transform the equations, taking all initial conditions to be zero.

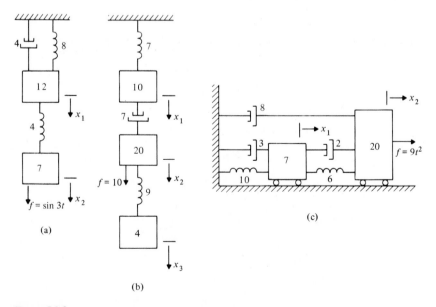

(a)   (b)   (c)

Figure P1.8

9. Find $X(s)$ and find $x(t)$, $t \geq 0$ for the networks of Fig. P1.9, for which all initial conditions are zero.

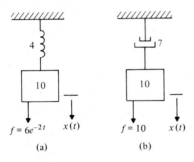

f = 6e^{-2t}   x(t)   f = 10   x(t)
        (a)   (b)

Figure P1.9

Ans (a) $(0.44 \cos(0.63t - 107°) - 0.13\, e^{-2t})u(t)$;
(b) $(1.43t + 2.04e^{-0.7t} - 2.04)u(t)$

10. Write simultaneous differential equations for the rotational mechanical networks of Fig. P1.10. Then Laplace-transform the equations, taking all initial conditions to be zero.

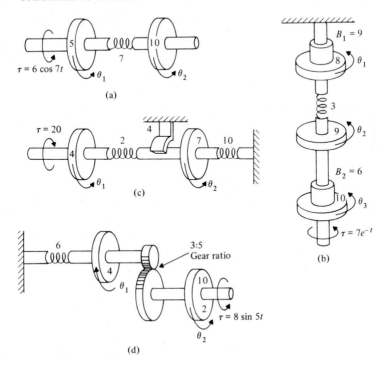

Figure P1.10

11. Find $\theta(s)$ and $\theta(t)$, $t \geq 0$, for the networks of Fig. P1.11, for which all initial conditions are zero.

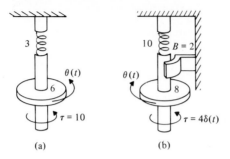

(a)   (b)

Figure P1.11

Ans   (a) $(3.33 - 3.33 \cos 0.707t)u(t)$;   (b) $(0.44e^{-1.25t} \sin 1.1t)u(t)$

12. Draw an electrical network that is analogous to the translational mechanical network of Fig. P1.12.

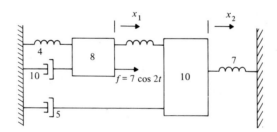

Figure P1.12

13. (a) Draw a translational mechanical network that is analogous to the electrical network of Fig. P1.13.

(b) Draw a rotational mechanical network that is analogous to the same electrical network (and to the translational mechanical network).

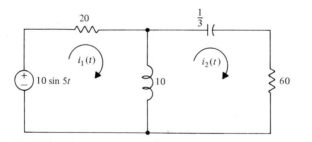

Figure P1.13

14. A linear actuator is driven by a sinusoidal voltage source with an internal resistance of R_1. The actuator plunger is connected to a load mass as shown in Fig. P1.14. When the load mass position x is zero, the spring force is zero. The solenoid's electromagnetic coupling constant is k. Find the differential equations describing $x(t)$.

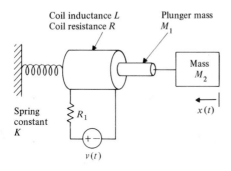

Figure P1.14

Ans $k\dfrac{dx}{dt} + \dfrac{di}{dt} + (R_1 + R_2)i = v(t); \quad (M_1 + M_2)\dfrac{d^2x}{dt^2} + kx - ki = 0$

15. Draw block diagrams to represent the following Laplace-transformed equations. The R signals are inputs, and the Y signals are outputs. This is a *synthesis* problem and has, in each case, many possible solutions.

(a)
$$\begin{cases} X_1(s) = \dfrac{3s}{s+1}\, R(s) + X_2(s) \\[2mm] X_2(s) = R(s) + 6X_1(s) \\[2mm] Y(s) = R(s) + \dfrac{10}{s^2 + 16}\, X_2(s) \end{cases}$$

(b)
$$\begin{cases} \left(\dfrac{10}{s+1}\right)X_1 - 6X_2 + 4R + \left(\dfrac{1}{s}\right)Y_1 = Y_2 \\[2mm] 8X_1 + \left(\dfrac{s}{s+1}\right)X_2 - \left(\dfrac{4s}{s+2}\right)R = 0 \\[2mm] \left(\dfrac{4}{s+2}\right)X_1 - \left(\dfrac{s}{s+2}\right)X_2 + \left(\dfrac{10}{s^2 + 4}\right)Y_1 = Y_2 \\[2mm] \left(\dfrac{4}{s+2}\right)X_1 + 10X_2 + \left(\dfrac{1}{s}\right)R + 10Y_2 = 0 \end{cases}$$

16. For the system equations of Prob. 15, draw instead representative signal flow graphs.

17. Using equivalences, reduce the block diagrams in Fig. P1.17 to single blocks or to a multiple-block canonical form, displaying the system transfer function(s).

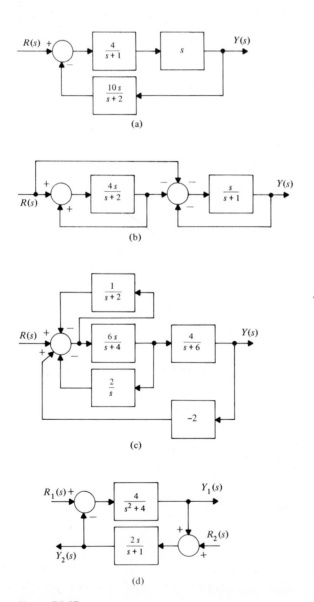

(a)

(b)

(c)

(d)

Figure P1.17

Ans (a) $T(s) = \dfrac{4s(s + 2)}{41s^2 + 3s + 2}$; (b) $T(s) = \dfrac{-s^2 - 2s}{-6s^2 + s + 2}$;

(c) $T(s) = \dfrac{24s(s + 2)}{s^3 + 73s^2 + 246s + 216}$

18. Find the transfer function(s) relating output(s) to input(s) for each of the signal flow graphs of Fig. P1.18, using Mason's gain rule.

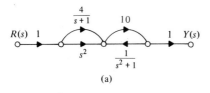

(a)

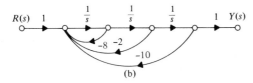

(b)

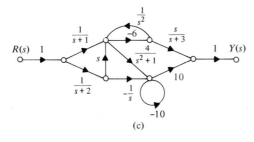

(c)

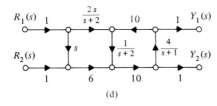

(d)

Figure P1.18

19. Write simultaneous integrodifferential equations for the electromechanical system of Fig. P1.19. Assume that the field current has reached a constant value so that the field is constant.

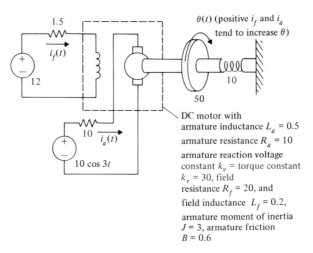

$\theta(t)$ (positive i_f and i_a tend to increase θ)

DC motor with
armature inductance $L_a = 0.5$
armature resistance $R_a = 10$
armature reaction voltage
constant k_v = torque constant
$k_\tau = 30$, field
resistance $R_f = 20$, and
field inductance $L_f = 0.2$,
armature moment of inertia
$J = 3$, armature friction
$B = 0.6$

Figure P1.19

20. In vibration studies, the human body is often modeled by springs, masses, and dampers. For the model of a seated body with applied force f, Fig. P1.20, find the system equations.

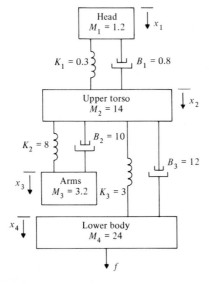

Figure P1.20

21. The mechanical device shown in Fig. P1.21 is an *accelerometer*. It is designed so that the position x_2 of the mass with respect to the case is approximately proportional to the case acceleration, d^2x_1/dt^2. Find the transmittance that relates x_2 to the case acceleration,

$$r(t) = \frac{d^2x_1}{dt^2}$$

in terms of K, M, and B, If M is 0.05, select K and B so the characteristic polynomial has both roots at -100.

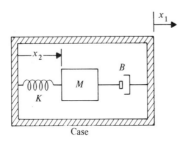

Figure P1.21

22. The position control system for a spacecraft platform is governed by the following approximate equations:

$$
\begin{cases}
\dfrac{d^2p}{dt^2} + \dfrac{dp}{dt} + 4p = \theta \\[2mm]
v_1 = r - p \\[2mm]
\dfrac{d\theta}{dt} = 0.4v_2 \\[2mm]
v_2 = 7v_1
\end{cases}
$$

The variables involved are as follows:

$r(t) =$ desired platform position (input)

$p(t) =$ actual platform position (output)

$v_1(t) =$ amplifier input voltage

$v_2(t) =$ amplifier output voltage

$\theta(t) =$ motor shaft position

Draw a block diagram of the system, identifying the component parts and their transmittances. Then determine the system transfer function.

23. A simplified block diagram of an aircraft roll control is given in Fig. P1.23. Find its transfer functions.

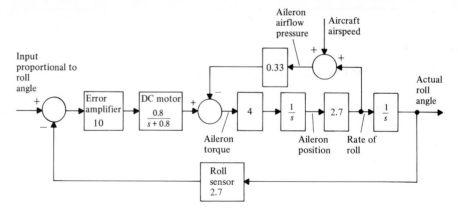

Figure P1.23

24. A system without an input that can be used to generate a sinusoidal output signal is shown in Fig. P1.24. When the differential equation describing $y(t)$ has a characteristic equation with roots on the imaginary axis of the complex plane, the system's zero-input response is sinusoidal. Find the value of the adjustable constant K, in terms of a, for which the system has a sinusoidal output. Also find the hertz frequency of the oscillations in terms of the constant a.

Achieving characteristic roots precisely on the imaginary axis is impossible for inexact K. If the characteristic roots are slightly to the right of the imaginary axis, the oscillation amplitude will increase exponentially. If they are to the left of the imaginary axis, the oscillations will decay exponentially in time. In practice, another control system can be used to slowly adjust K to maintain nearly constant sinusoidal amplitude.

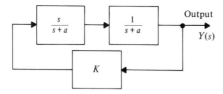

Figure P1.24

Ans $K = 2a$; frequency $= (a/2\pi)$ Hz

25. Write transfer functions for the electrical networks of Fig. P1.25. All networks are driven from zero source impedance and into infinite load impedance. Assume zero initial conditions.

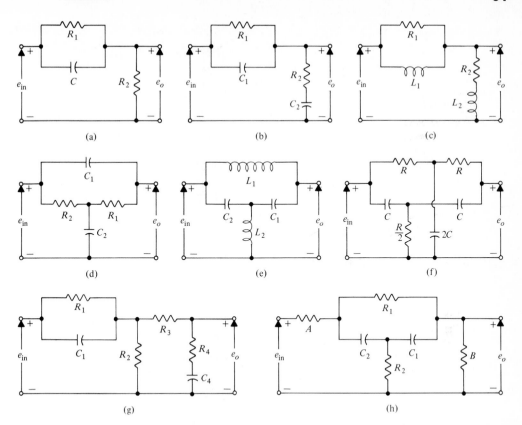

Figure P1.25

Ans (a) $\dfrac{E_o}{E_{\text{in}}} = \dfrac{s + 1/a\tau_a}{s + 1/\tau_a}$ where $a = 1 + \dfrac{R_1}{R_2}$, $\tau_a = \dfrac{R_1 R_2 C}{R_1 + R_2}$

(d) $\dfrac{E_o}{E_{\text{in}}} = \dfrac{s^2/\omega_0^2 + rns/\omega_0 + 1}{s^2/\omega_0 + ns/\omega_0 + 1}$

$\omega_0 = R_1 R_2 C_1 C_2 \qquad r = \dfrac{C_1(R_1 + R_2)}{C_1(R_1 + R_2) + C_2 R_2}$

$n = \dfrac{C_1(R_1 + R_2) + C_2 R_2}{\omega_0}$

(f) $\dfrac{E_o}{E_{\text{in}}} = \dfrac{\tau^2 s^2 + 1}{\tau^2 s^2 + 4\tau s + 1}$, $RC = \tau$

26. Write the Laplace-transformed differential equations for the mechanical system shown in Fig. P1.26; $v(t)$ is a velocity input driving function. Assume zero initial conditions. Find the transfer function.

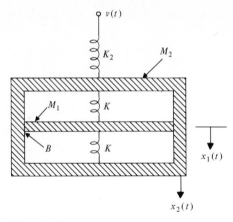

Figure P1.26

Ans $\dfrac{X_1(s)}{V(s)} = \dfrac{(K_2/s)(Bs + 2K)}{(M_1 s^2 + Bs + 2K)(M_2 s^2 + Bs + 2K + K_2) - (Bs + 2K)^2}$

27. Find the transfer function for the fluid coupling shown in Fig. P1.27.

$$\frac{\theta_3(s)}{\tau(s)} = G_2(s)$$

Assume zero initial conditions.

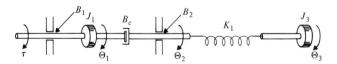

Figure P1.27

28. Find the transfer function $\theta_1(s)/T(s)$ for the system of Fig. P1.28.

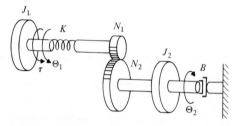

Figure P1.28

29. Find the transfer function $X(s)/F(s)$ for the system in Fig. P1.29.

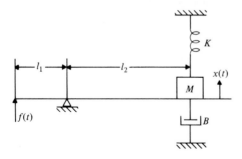

Figure P1.29

Ans $\dfrac{-l_1/l_2}{Ms^2 + Bs + K}$

30. Find the transfer function $G(s) = Y(s)/E(s)$ for the system in Fig. P1.30. Assume back-emf constant k_1 and force constant k_2 are equal.

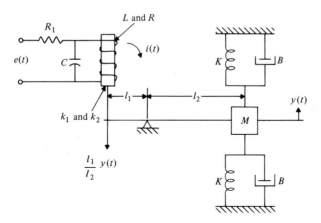

Figure P1.30

31. Find the transfer function E_o/Y of the system of Fig. P1.31 used to measure high-frequency displacements of a shake table. The potentiometer measures the difference in displacement between M_1 and M_2. The driving function is a position $y(t)$. Ignore the acceleration of gravity, and assume infinite input and zero output impedance of the amplifier. The maximum travel of the linear motion potentiometer is d. Assume zero initial conditions.

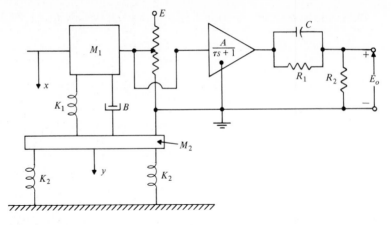

Figure P1.31

$$\textbf{Ans} \quad \frac{E_o(s)}{Y(s)} = \frac{+ \, AER_2 M_1 s^2 (sR_1 C + 1)}{d(R_1 + R_2)(M_1 s^2 + Bs + K_1)(\tau s + 1)\left(sC \dfrac{R_1 R_2}{R_1 + R_2} + 1\right)}$$

32. Analyze the system of Fig. P1.32 by finding

$$T_{11}(s), \qquad T_{12}(s), \qquad T_{21}(s), \quad \text{and} \quad T_{22}(s),$$

where

$$Y_1(s) = T_{11}(s)R_1(s) + T_{12}(s)R_2(s)$$
$$Y_2(s) = T_{21}(s)R_1(s) + T_{22}(s)R_2(s)$$

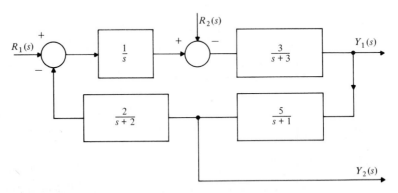

Figure P1.32

Continuous-Time System Response

2.1 PREVIEW

In order to design a control system for some practical application, a series of steps usually occurs. For example, using the methods of Chap. 1, we can examine a variety of electrical and mechanical systems and write a set of differential equations to describe their operation. In doing so, we draw upon techniques developed over several hundred years. In the late 1600s Newton and Leibnitz developed methods for writing and solving differential equations. These and other related topics are now called calculus. Over 100 years later, Laplace developed a transform to aid in solving those equations.

In the 1890s, about 100 years after Laplace, steam engines represented a challenge to controls designers in that steam pressure and various rotational velocities had to be controlled. The efforts of a mathematician (Hurwitz) and an engineer (Routh) resulted in a test of performance using the coefficients of the characteristic polynomial, a polynomial resulting from the same methods presented in Chap. 1. Now that we are living in a technology about 100 years after Routh and Hurwitz (some 300 years after Newton and Leibnitz) we inherit the legacy of the past in understanding the behavior of today's technology using their methods.

This chapter begins by recognizing that a characteristic polynomial can be factored into first- and second-order terms with real valued coefficients. If the behavior of first- and second-order systems are well understood, the behavior of higher-order systems follows as a combination of the first- and second-order building blocks.

Next, it is useful to understand whether or not an automobile or aircraft described by similar equations could give us a comfortable and safe ride. Certain definitions are presented which clarify the quality of performance in terms of a system's stability.

Even though computing capability has grown enormously since the time of Routh and Hurwitz, their method remains a valuable tool for determining a range of values for an unknown parameter so that the stability of the resulting closed-loop system is ensured.

Chapter 2 concludes with examples of two systems that illustrate the power of the analytical methods. A delivery system is selected so that a desired flow of insulin can be induced to flow in a diabetic's bloodstream. An application of a very different sort shows that the periodic vibration of turbine engines can cause excessive wing deflection in an aircraft.

2.2 RESPONSE OF FIRST-ORDER SYSTEMS

In a first-order system, the output $y(t)$ and input $r(t)$ are related by a differential equation of the form (for $m \leq 1$)

$$\frac{dy}{dt} + a_0 y = b_m \frac{d^m r}{dt^m} + b_0 r = f(t)$$

where the input terms, possibly involving derivatives of $r(t)$, form the driving function of the equation, $f(t)$. The corresponding transfer function is

$$T(s) = \frac{b_m s^m + b_0}{s + a_0}$$

A system is considered to be stable if the natural response decays to zero. The denominator of $T(s)$ is the characteristic polynomial whose roots must all be in the left half-plane to force all the natural response terms to decay to zero, and therefore to force the system to be stable. In the first-order case above, the system is stable if and only if a_0 is > 0.

The first-order system described by

$$\frac{dy}{dt} + a_0 y = b_0 r \tag{2.1}$$

where a_0 and b_0 are constants, has transfer function

$$T(s) = \frac{b_0}{s + a_0}$$

Laplace-transforming the input–output equation (2.1),

$$sY(s) - y(0^-) + a_0 Y(s) = b_0 R(s)$$

$$Y(s) = \underbrace{\frac{b_0}{s + a_0} R(s)}_{\substack{\text{zero-state} \\ \text{component}}} + \underbrace{\frac{y(0^-)}{s + a_0}}_{\substack{\text{zero-input} \\ \text{component}}}$$

For a step input signal

$$r(t) = Au(t) \qquad R(s) = \frac{A}{s}$$

and zero initial conditions,

$$Y(s) = T(s)R(s) = \frac{b_0 A}{s(s + a_0)} = \frac{(b_0 A/a_0)}{s} + \frac{(-b_0 A/a_0)}{s + a_0}$$

$$y(t) = \left(\frac{b_0 A}{a_0} - \frac{b_0 A}{a_0} e^{-a_0 t} \right) u(t)$$

which is sketched in Fig. 2.1. If the characteristic root $s = -a_0$ is negative (a_0 positive), the system is stable and the exponential natural (or "transient") response term decays with time, leaving the constant forced (or "steady state") term. For a positive characteristic root (negative a_0), the exponential term expands with time, and the system is unstable.

If, instead, the initial conditions are not zero,

$$Y(s) = T(s)R(s) + \frac{y(0^-)}{s + a_0} = \frac{b_0 A}{s(s + a_0)} + \frac{y(0^-)}{s + a_0}$$

$$= \frac{(b_0 A/a_0)}{s} + \frac{y(0^-) - (b_0 A/a_0)}{s + a_0}$$

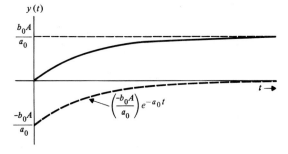

Figure 2.1 First-order system step response with zero initial conditions.

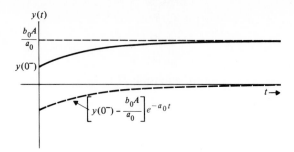

Figure 2.2 First-order system step response with nonzero initial conditions.

giving

$$y(t) = \left\{ \frac{b_0 A}{a_0} + \left[y(0^-) - \frac{b_0 A}{a_0} \right] e^{-a_0 t} \right\} u(t)$$

The amplitude of the exponential term is changed, as illustrated in Fig. 2.2.

The response to other inputs is calculated similarly, using the Laplace transform of the input signal. The natural exponential term in the response consists of contributions from both the zero-state and the zero-input parts of the response. In general, the natural and zero-input responses die out in time when the system is stable.

The *time constant* of a stable first-order system is

$$\tau = \frac{1}{a_0}$$

It is the time interval over which the exponential $K \exp(-a_0 t)$ decays by a factor of $e^{-1} = 0.37$. First-order system transfer functions (and similar terms in higher-order transfer functions) are sometimes placed in the form

$$T(s) = \frac{\text{numerator polynomial}}{s + a_0} = \frac{(\text{numerator polynomial})/a_0}{(1/a_0)s + 1}$$

$$= \frac{(\text{numerator polynomial})/a_0}{\tau s + 1}$$

to show the time constant explicitly.

For the first-order system

$$\frac{dy}{dt} + a_0 y = b_1 \frac{dr}{dt} + b_0 r$$

$$Y(s) = \underbrace{\frac{b_1 s + b_0}{s + a_0} R(s)}_{\text{zero-state component}} + \underbrace{\frac{y(0^-) - b_1 r(0^-)}{s + a_0}}_{\text{zero-input component}}$$

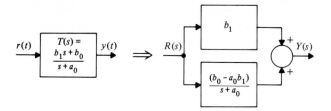

Figure 2.3 Tandem representation of a first-order system.

The transfer function relating $Y(s)$ to $R(s)$ is

$$T(s) = \frac{b_1 s + b_0}{s + a_0}$$

which can be expressed as

$$T(s) = b_1 + \frac{(b_0 - a_0 b_1)}{s + a_0}$$

Such a system can be considered to be composed of a gain b_1 in tandem with a first-order system with constant transfer function numerator, as indicated in Fig. 2.3. This system output thus contains a component proportional to the input, in addition to a component of the previous type.

Growth of bacteria and many other biological activities are commonly modeled by a first-order differential equation and have time responses with a single exponential term. Electrical RL and RC circuits and radioactive decay also exhibit first-order behavior.

DRILL PROBLEMS

D2.1. Find and also sketch the response of the systems with the following transfer functions, inputs, and initial conditions:

(a) $T(s) = \dfrac{3}{s + 3}$

$r(t) = 6u(t)$

$y(0^-) = 10$

Ans $6 + 4e^{-3t}$

(b) $T(s) = \dfrac{1}{s + 10}$

$r(t) = 3u(t) \cos 10t$

$y(0^-) = 0$

Ans $0.212 \cos(10t - 45°) - 0.15e^{-10t}$

(c) $T(s) = \dfrac{s}{s + 1000}$

$r(t) = 7u(t)$

$y(0^-) = 4$

Ans $11e^{-1000t}$

(d) $T(s) = \dfrac{20s}{s + 300}$

$r(t) = 8u(t) \sin 100t$

$y(0^-) = -10$

Ans $50.6 \cos(100t - 18.4°) - 58e^{-300t}$

(e) $T(s) = \dfrac{-4s + 20}{s + 300}$

$r(t) = 10u(t)$

$y(0^-) = 0$

Ans $0.67 - 40.67e^{-300t}$

D2.2. Find the time constants of the following systems:

(a) $T(s) = \dfrac{4s - 1}{3s + 2}$

Ans $\dfrac{3}{2}$ sec

(b) $\dfrac{dy}{dt} + 4y = -3\dfrac{dr}{dt}$

Ans $\dfrac{1}{4}$ sec

(c)

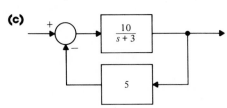

Ans $\dfrac{1}{53}$ sec

2.3 RESPONSE OF SECOND-ORDER SYSTEMS

Time Response

In a second-order system, the output $y(t)$ and the input $r(t)$ are related by a second-order linear differential equation in $y(t)$. If $y(t)$ is the dependent variable, up to $m = 2$ derivatives of $r(t)$ are permitted. If m exceeds 2, then $r(t)$ would become the dependent variable.

$$\frac{d^2y}{dt^2} + a_1 \frac{dy}{dt} + a_0 y = b_m \frac{d^m r}{dt^m} + \cdots + b_1 \frac{dr}{dt} + b_0 r$$

As mentioned above, stability implies that the natural response decays to zero. The denominator of $T(s)$ is the characteristic polynomial whose roots must all be in the left half plane to force all the natural response terms to decay to zero, and therefore to force the system to be stable. In the second-order case, both a_0 and a_1 must be >0 for stability.

$$T(s) = \frac{b_m s^m + \cdots + b_1 s + b_0}{s^2 + a_1 s + a_0}$$

and the system response is of the form

$$Y(s) = \underbrace{\frac{b_m s^m + \cdots + b_1 s + b_0}{s^2 + a_1 s + a_0} R(s)}_{\text{zero-state component}} + \underbrace{\frac{\left(\begin{array}{c} \text{first-degree numerator} \\ \text{polynomial dependent on} \\ \text{initial conditions} \end{array}\right)}{s^2 + a_1 s + a_0}}_{\text{zero-input component}}$$

The characteristic polynomial of a second-order system is

$$s^2 + a_1 s + a_0 = (s - s_1)(s - s_2)$$

with roots s_1 and s_2 given by the quadratic formula

$$s_1, s_2 = \frac{-a_1 \pm \sqrt{a_1^2 - 4a_0}}{2}$$

OVERDAMPED RESPONSE

If the characteristic roots s_1 and s_2 are real and distinct, the natural response of the system

$$\frac{d^2y}{dt^2} + a_1 \frac{dy}{dt} + a_0 y = b_1 \frac{dr}{dt} + b_0 r \tag{2.2}$$

is

$$Y_{\text{natural}}(s) = \frac{K_1}{s - s_1} + \frac{K_2}{s - s_2}$$

$$y_{\text{natural}}(t) = K_1 e^{s_1 t} + K_2 e^{s_2 t} \qquad t \geq 0$$

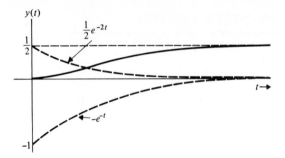

Figure 2.4 Step response of an overdamped second-order system.

which is the sum of two real exponential terms. Such a system is termed *overdamped* when s_1 and s_2 are both negative. For example, the overdamped second-order system with transfer function

$$T(s) = \frac{1}{s^2 + 3s + 2}$$

with unit step input

$$R(s) = \frac{1}{s}$$

and zero initial conditions has response given by

$$Y(s) = \frac{1}{s(s^2 + 3s + 2)} = \frac{\frac{1}{2}}{s} + \frac{-1}{s+1} + \frac{\frac{1}{2}}{s+2}$$

$$y(t) = [\tfrac{1}{2} - e^{-t} + \tfrac{1}{2} e^{-2t}]u(t)$$

This response is sketched in Fig. 2.4. Nonzero initial conditions result in different amplitudes for the two exponential natural terms.

CRITICALLY DAMPED RESPONSE
If the two characteristic roots are equal, the second-order system (2.2) has

$$Y_{\text{natural}}(s) = \frac{\text{numerator polynomial}}{s^2 + a_1 s + a_0} = \frac{\text{numerator polynomial}}{(s - s_1)^2}$$

$$= \frac{K_1}{s - s_1} + \frac{K_2}{(s - s_1)^2}$$

for which the corresponding time function after $t = 0$ is

$$y_{\text{natural}}(t) = K_1 e^{s_1 t} + K_2 t e^{s_1 t} \qquad t \geq 0$$

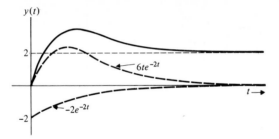

Figure 2.5   Step response of a critically damped second-order system.

Such a second-order system is said to be *critically damped* where s_1 is negative. The critically damped system with transfer function

$$T(s) = \frac{10s + 8}{s^2 + 4s + 4}$$

for example, has unit step response, with zero initial conditions, given by

$$Y(s) = \frac{10s + 8}{s(s^2 + 4s + 4)} = \frac{2}{s} + \frac{-2}{s + 2} + \frac{6}{(s + 2)^2}$$

for which

$$y(t) = [2 - 2e^{-2t} + 6te^{-2t}]u(t)$$

This response is sketched in Fig. 2.5. Other initial conditions result in different amplitudes for the $\exp(-2t)$ and $t\exp(-2t)$ natural terms, but the same character of response.

UNDERDAMPED RESPONSE
If the roots of the characteristic polynomial are complex numbers, they are complex conjugates of one another,

$$s_1, s_2 = -a \pm j\omega$$

and the natural response component is of the form

$$Y_{\text{natural}}(s) = \frac{\text{numerator polynomial}}{s^2 + a_1 s + a_0} = \frac{\text{numerator polynomial}}{(s + a - j\omega)(s + a + j\omega)}$$

$$= \frac{\text{numerator polynomial}}{(s + a)^2 + \omega^2}$$

corresponding to the time behavior

$$y_{\text{natural}}(t) = [Ae^{-at}\cos(\omega t + \theta)]u(t)$$

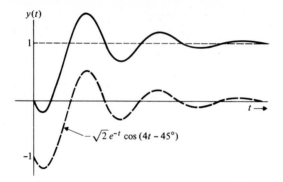

Figure 2.6 Step response of an underdamped second-order system.

This type of second-order system is termed *underdamped* where a is positive. For example, the underdamped system with transfer function

$$T(s) = \frac{-3s + 17}{s^2 + 2s + 17}$$

has unit step response

$$Y(s) = \frac{-3s + 17}{s(s^2 + 2s + 17)} = \frac{1}{s} + \frac{-(s + 5)}{(s + 1)^2 + (4)^2} = \frac{1}{s} + \frac{Me^{j\theta}}{s + 1 - j4} + \frac{Me^{-j\theta}}{s + 1 + j4}$$

Using the evaluation method to find M and θ, there results

$$Me^{j\theta} = \frac{-s - 5}{s + 1 + j4}\bigg|_{s = -1 + j4} = \frac{1 - j4 - 5}{j8} = -\frac{1}{2} + j\frac{1}{2} = \frac{\sqrt{2}}{2}e^{j135°}$$

From Table 1.6,

$$y(t) = [1 + \sqrt{2}e^{-t}\cos(4t + 135°)]u(t)$$
$$= [1 - \sqrt{2}e^{-t}\cos(4t - 45°)]u(t)$$

This response is sketched in Fig. 2.6. Other initial conditions give different constants A and θ.

Undamped Natural Frequency and Damping Ratio

Underdamped second-order systems have a natural response that is described by a radian frequency of oscillation ω and an exponential constant σ, that are found from the system's characteristic polynominal:

$$s^2 + a_1 s + a_0 = (s + \sigma)^2 + \omega^2$$

Alternatively, underdamped second-order system response is described by the *undamped natural frequency* ω_n and the *damping ratio* ζ:

$$s^2 + a_1 s + a_0 = s^2 + 2\zeta\omega_n s + \omega_n^2$$

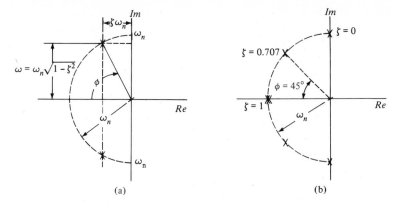

Figure 2.7 Underdamped natural frequency and damping ratio for a second-order response term. **(a)** Relation to characteristic root locations. **(b)** Damping ratios corresponding to various root locations.

The two sets of quantities are related by

$$\begin{cases} \sigma = \zeta\omega_n \\ \omega = \omega_n\sqrt{1 - \zeta^2} \end{cases}$$

For ζ between 0 and 1, the characteristic roots lie on a circle of radius ω_n about the origin in the left half of the complex plane, as shown in Fig. 2.7. For $\zeta = 0$, the roots are on the imaginary axis; and for $\zeta = 1$, both roots are on the negative real axis, repeated. The undamped natural frequency ω_n is the radian frequency at which the oscillations would occur if the damping ratio ζ were zero. If ζ were zero, the system natural response would have the form

$$Y_{\text{natural}}(s) = \frac{\text{numerator polynomial}}{s^2 + \omega_n^2}$$

$$y_{\text{natural}}(t) = [A \cos(\omega_n t + \theta)]u(t)$$

The damping ratio is related to the *damping angle* ϕ in Fig. 2.7 by

$$\zeta = \cos \phi$$

Consider the second-order system with transfer function

$$T(s) = \frac{100}{s^2 + 3s + 13}$$

The undamped natural frequency of the system is

$$\omega_n = \sqrt{13} = 3.6$$

and the damping ratio is given by

$$2\zeta\omega_n = 3 \qquad \zeta = \frac{3}{2\omega_n} = \frac{3}{7.2} = 0.42$$

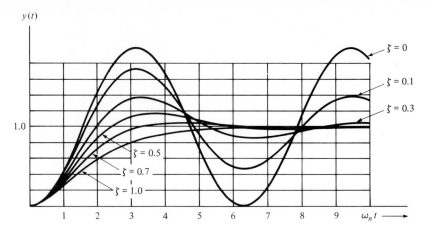

Figure 2.8 Normalized step response of a second-order system with constant transfer function numerator.

The true oscillation frequency is

$$\omega = \omega_n \sqrt{1 - \zeta^2} = 3.6(0.9) = 3.27$$

A set of normalized step response curves for underdamped second-order systems of the form

$$T(s) = \frac{\omega_n^2}{s^2 + 2\zeta\omega_n s + \omega_n^2}$$

is shown in Fig. 2.8. For $\zeta = 0$, the oscillations continue forever. Larger values of ζ give more rapid decay of the oscillations but a slower rise of the response. For $\zeta = 1$, the system is critically damped.

DRILL PROBLEMS

D2.3. Find and also sketch the response of the system with the following transfer functions, inputs, and initial conditions:

(a) $T(s) = \dfrac{s}{s^2 + 7s + 8}$

$r(t) = u(t)$

$y(0^-) = 7$

$y'(0^-) = -4$

Ans $8.62e^{-1.4t} - 1.62e^{-5.6t}$

(b) $T(s) = \dfrac{4}{s^2 + 4s + 4}$

$r(t) = 0$

$y(0^-) = -3$

$y'(0^-) = 2$

Ans $-3e^{-2t} - 4te^{-2t}$

(c) $T(s) = \dfrac{3}{s^2 + 0.5s + 4}$

$r(t) = 2u(t)$

$y(0^-) = 0$

$y'(0^-) = 0$

Ans $1.5 + 1.51e^{-0.25t} \cos(1.98t + 173°)$

(d) $T(s) = \dfrac{4s - 20}{s^2 + 4s + 29}$

$r(t) = 10\delta(t)$

$y(0^-) = 0$

$y'(0^-) = 6$

Ans $67.85e^{-2t} \cos(5t + 54°)$

(e) $T(s) = \dfrac{s^2}{s^2 + 2s + 17}$

$r(t) = 0$

$y(0^-) = 10$

$y'(0^-) = 0$

Ans $10.3e^{-t} \cos(4t - 14°)$

D2.4. Determine which of the following second-order systems are underdamped, which are critically damped, and which are overdamped:

(a) $T(s) = \dfrac{9s^2 + 3s + 10}{s^2 + 5s + 2}$

(b) $T(s) = \dfrac{s^2 - 2s}{s^2 + 6s + 9}$

(c) $T(s) = \dfrac{64}{3s^2 + 4s + 5}$

(d) $T(s) = \dfrac{19s - 20}{s^2 + s + 100}$

(e) $T(s) = \dfrac{s^2 + 2s + 100}{s^2 + 7s + 49}$

Ans (a) overdamped; (b) critically damped; (c) underdamped;
(d) underdamped; (e) underdamped

D2.5. For second-order systems with the following transfer functions, determine the undamped natural frequency, the damping ratio, and the oscillation frequency:

(a) $T(s) = \dfrac{100}{s^2 + s + 100}$

Ans   10, 0.05, 9.99

(b) $T(s) = \dfrac{3s - 49}{s^2 + 3s + 49}$

Ans   7, 0.214, 6.84

(c) $T(s) = \dfrac{s^2 + 9s}{s^2 + 4s + 10}$

Ans   3.16, 0.632, 2.45

(d) $T(s) = \dfrac{s^2 + 20}{s^2 + 2s + 20}$

Ans   4.47, 0.224, 4.36

(e) $T(s) = \dfrac{-3s + 0.7}{s^2 + 0.3s + 4}$

Ans   2, 0.075, 1.99

D2.6. Find the constant k for which the system with transfer function $T(s)$ has the given second-order response property.

(a) $T(s) = \dfrac{10}{s^2 + 40s + k}$

$\zeta = 0.7$

Ans   816

(b) $T(s) = \dfrac{ks + 6}{s^2 + ks + 49}$

$\omega = 4$

Ans 11.49

(c) $T(s) = \dfrac{20s}{3s^2 + 2s + k + 4}$

$\zeta = 0.1$

Ans 29.33

(d) $T(s) = \dfrac{s^2 - 6}{ks^2 + s + 6}$

$\omega_n = 2$

Ans 1.5

2.4 HIGHER-ORDER SYSTEM RESPONSE

Natural Response Terms

The natural response of third- and higher-order systems consists of a sum of terms, one term for each characteristic root. For each distinct real characteristic root, there is a real exponential term in the system natural response. For each pair of complex conjugate roots there is a pair of complex exponential terms which are better expressed as an exponential times a sinusoid. Repeated roots give additional terms involving powers of time times the exponential. For example, a system with transfer function

$$T(s) = \frac{-8s^2 + 5}{s^4 + 9s^3 + 37s^2 + 81s + 52} = \frac{-8s^2 + 5}{(s + 1)(s + 4)(s^2 + 4s + 13)}$$

has a natural response of the form

$$Y_{natural}(s) = \frac{K_1}{s + 1} + \frac{K_2}{s + 4} + \frac{K_3 s + K_4}{(s + 2)^2 + (3)^2}$$

or

$$y_{natural}(t) = K_1 e^{-t} + K_2 e^{-4t} + A e^{-2t} \cos(3t + \theta) \qquad t \geq 0$$

A system with transfer function

$$T(s) = \frac{-7s^3 + 6s^2 + 9s + 23}{(s + 1)(s + 3)(s^2 + 4s + 13)(s^2 + 8s + 17)}$$

has natural response of the form

$$Y_{\text{natural}}(s) = \frac{K_1}{s+1} + \frac{K_2}{s+3} + \frac{K_3}{s+2+j3}$$

$$+ \frac{K_3^*}{s+2-j3} + \frac{K_4}{s+4+j} + \frac{K_4^*}{s+4-j}$$

$$= \frac{K_1}{s+1} + \frac{K_2}{s+3} + \frac{(\text{first-order numerator})}{s^2+4s+13} + \frac{(\text{first-order numerator})}{s^2+8s+17}$$

$$y_{\text{natural}}(t) = K_1 e^{-t} + K_2 e^{-3t} + A_1 e^{-2t} \cos(3t + \theta_1)$$

$$+ A_2 e^{-4t} \cos(t + \theta_2) \qquad t \geq 0$$

One exponential term has a one-second time constant, and the time constant of the other exponential term is one-third second. The first damped sinusoid has undamped natural frequency and damping ratio.

$$\omega_n = \sqrt{13} = 3.6 \text{ rad/s} \qquad \zeta = \frac{4}{2\omega_n} = 0.555$$

The second damped sinusoid has

$$\omega_n' = \sqrt{17} = 4.12 \text{ rad/s} \qquad \zeta' = \frac{8}{2\omega_n'} = 0.97$$

Rise Time, Overshoot, and Settling Time

The quality of the performance of a stable system is commonly characterized by the *rise time*, *overshoot*, and *settling time* of its response to a step input. As indicated in Fig. 2.9, rise time is the interval of time required for the step response of a system to go from 10 to 90% of its final value. Overshoot is the percent difference between the maximum and the steady state values of the response. Settling time is the time required before the system response remains within $\pm 5\%$ of the final value. Other definitions of rise time (for example, 5 to 95%) and settling time ($\pm 10\%$) are possible, but will not be used in this text.

The real roots of most higher-order systems are often widely separated so that most terms in the response decay much more quickly than the others. When all but two roots of a transfer function exhibit rapid decay, the response appears to be second order, although the system may actually be third or higher order.

Rise time, overshoot, and settling time depend on the roots of the characteristic polynomial, the initial conditions, and the command. Considerable insight into those performance measures can be gained by considering only second-order underdamped systems with zero initial conditions and a unit step command. Other step commands would multiply the response values accordingly.

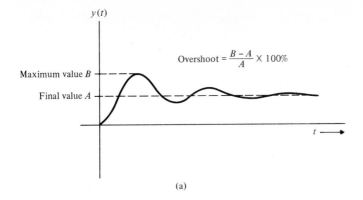

(a)

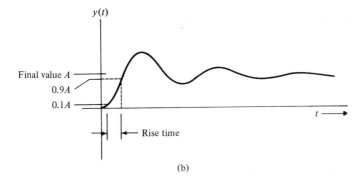

(b)

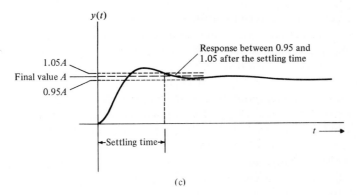

(c)

Figure 2.9 Step response specifications. **(a)** Overshoot. **(b)** Rise time. **(c)** Settling time.

For an underdamped second-order system with transfer function of the form

$$T(s) = \frac{\omega_n^2}{s^2 + 2\zeta\omega_n s + \omega_n^2}$$

the step response has Laplace transform

$$Y(s) = \frac{A}{s} T(s) = \frac{A\omega_n^2}{s(s^2 + 2\zeta\omega_n s + \omega_n^2)}$$

For a unit step command A is 1, and so

$$Y(s) = \frac{k_1}{s} + \frac{k_2}{s + \sigma - j\omega} + \frac{k_3}{s + \sigma + j\omega}$$

$$\sigma = \zeta\omega_n$$

$$\omega = \omega_n\sqrt{1 - \zeta^2}$$

$$k_1 = 1$$

$$k_2 = \frac{1}{2z} \left| \underline{90° + \tan^{-1}\frac{\omega}{\sigma}} \right.$$

$$z = \sqrt{1 - \zeta^2}$$

The time response is

$$y = 1 + \frac{1}{z} e^{-\zeta\omega_n t} \cos\left(\omega_n tz + \tan^{-1}\frac{z}{\zeta} + 90°\right)$$

The time response depends on the product of ω_n times t. It is common practice to normalize curves of this type by using $\omega_n t$ as the time axis. The value of time can always be computed later when the undamped natural frequency is known. For example, plots such as those in Fig. 2.9 can be created for each value of the damping ratio (with $\omega_n t$ being the horizontal axis).

The horizontal axis for Figs. 2.10a–c is the damping ratio. From Fig. 2.8, it is obvious that as damping ratio diminishes, the time response moves upward more quickly, so that rise time (Fig. 2.10a) diminishes as damping ratio diminishes assuming that undamped natural frequency remains constant. However, as damping ratio diminishes, the percent overshoot (Fig. 2.10b) increases. Rise time and percent overshoot move in opposite directions as a function of damping ratio. Settling time provides a compromising viewpoint for designing a second-order system, in that the speed of response measured by rise time is tempered with the undesirable feature of overshooting (or undershooting) the final value and then having to move back in the opposite direction.

In the following discussion, the product of undamped natural frequency times settling time will be called the normalized settling time. The product of undamped natural frequency times time will be called normalized time. The normalized settling time versus damping ratio curve of Fig. 2.10c can be understood by examining Fig. 2.10d, which has an exaggerated scale to facilitate this discussion. If

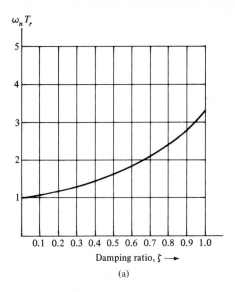

(a)

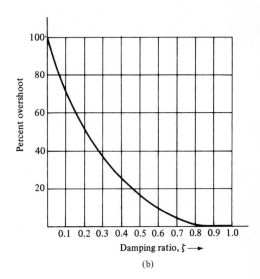

(b)

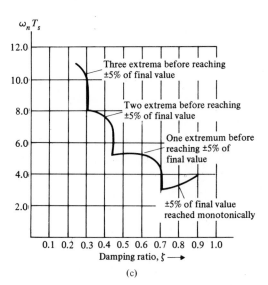

(c)

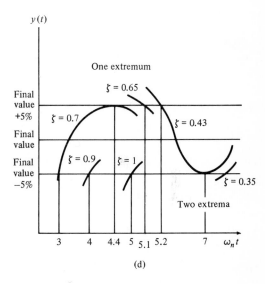

(d)

Figure 2.10 Step response performance of second-order systems with transfer functions having a constant numerator. **(a)** Rise time. **(b)** Overshoot. **(c)** Settling time. **(d)** Time response of $y(t)$ for various damping ratios (not to scale).

the damping ratio is 1.0, the lower limit is crossed when the normalized time is 5, but the upper limit is never crossed. As the damping ratio is reduced to just above 0.7, the normalized settling time diminishes to 4 and then to 3 seconds. Therefore, normalized settling time decreases as damping ratio decreases from 1 down to about 0.7 sec. When the damping ratio is about 0.7, the time response overshoots about 5%, reaching the upper limit after about 4.4 units of normalized time. Normalized settling time jumps, therefore, from 3 to 4.4. As the damping ratio drops from 0.7 down to about 0.43, the upper limit is entered after the derivative goes to zero once (one extremum), but the lower limit is not subsequently exceeded. The normalized settling time increases from 4.4 to about 5.3 as the damping ratio drops from 0.7 to 0.55. As the damping ratio drops from 0.55 to 0.43, the normalized settling time actually drops slightly to about 5.2. When the damping ratio is about 0.43, the lower limit is exceeded, so the normalized settling time jumps to about 7. For damping ratios from 0.43 down to about 0.3, it takes longer for the lower limit to be entered, so that normalized settling time increases while $y(t)$ traverses two time points where the derivative is zero (two extrema).

In general, the normalized settling time curve is comprised of an infinite number of segments, where, for each range of decreasing damping ratios, a plus or minus 5% limit is eventually exceeded, calling for one additional extremum (and another segment). Because the threshold is chosen to be plus or minus 5%, settling time is a minimum for a damping ratio of about 0.7. That value is often used in control system design. Chapters 5 and 6 will explore further uses of rise time and settling time in designing control systems.

DRILL PROBLEMS

D2.7. Find the form of the natural response of systems with the following transfer functions:

(a) $T(s) = \dfrac{100}{(s^2 + 4s + 4)(s^2 + 4s + 5)}$

    Ans $K_1 e^{-2t} + K_2 t e^{-2t} + K_3 e^{-2t} \cos(t + \theta)$

(b) $T(s) = \dfrac{3s - 12}{s^3 + 4s^2 + 13s}$

    Ans $K_1 + K_2 e^{-2t} \cos(3t + \theta)$

(c) $T(s) = \dfrac{s^2}{3(s + 3)^2(s^2 + 2s + 10)}$

    Ans $K_1 e^{-3t} + K_2 t e^{-3t} + K_3 e^{-t} \cos(3t + \theta)$

(d) $T(s) = \dfrac{5(s^2 + 2s + 1)(s^2 + 2s + 2)}{(s + 1)^2(s^2 + 4)(s + 8)}$

    Ans $K_1 e^{-t} + K_2 t e^{-t} + K_3 \cos(2t + \theta) + K_4 e^{-8t}$

(e) $T(s) = \dfrac{6s^3 - 4s^2 + 2s + 400}{(s^2 + s + 10)(s^2 + s + 20)}$

    Ans $K_1 e^{-0.5t} \cos(3.12t + \theta_1) + K_2 e^{-0.5t} \cos(4.44t + \theta_2)$

D2.8. Using the curves given in the text, find approximately the percent overshoot, rise time, and settling time of the following systems when driven by a step input:

(a) $T(s) = \dfrac{100}{s^2 + 4s + 100}$

Ans 54%, 0.12, 1.1

(b) $T(s) = \dfrac{49}{s^2 + 4s + 49}$

Ans 40%, 0.18, 1.5

(c) $T(s) = \dfrac{60}{2s^2 + 8s + 30}$

Ans 18%, 0.44, 1.29

(d) $T(s) = \dfrac{75}{s^2 + 3s + 20}$

Ans 32%, 0.34, 1.7

2.5 STABILITY TESTING

Coefficient Tests

For first- and second-order systems, stability is determined by inspection of the characteristic polynomial. A first- or second-order polynomial has all roots in the left half of the complex plane if and only if all polynomial coefficients have the same algebraic sign. For example,

$$3s^2 + s + 10$$

is the characteristic polynomial of a stable system, while

$$3s^2 + s - 10$$

represents an unstable system.

For higher-order polynomials, representing higher-order systems, the algebraic signs of the polynomial coefficients may or may not yield information as to stability. A polynomial with all roots in the left half-plane (LHP) has factors of the form

$$(s + a) \qquad a > 0 \qquad \text{(real axis root in the LHP)}$$

and

$$(s^2 + bs + c) \qquad b > 0 \text{ and } c > 0 \qquad \text{(two LHP roots, perhaps complex conjugate)}$$

When multiplied out, such a polynomial must have all coefficients of the same algebraic sign, all positive or all negative. No coefficient can be zero ("missing") in a system with LHP roots because there are no minus signs involved and thus no way for a coefficient to be canceled.

If imaginary axis roots exist in the polynomial, factors of the following forms can be present, in addition to the others:

(s) (root at the origin)

and

$(s^2 + a)$ $a > 0$ (complex conjugate roots on the imaginary axis)

With such roots present, all polynomial coefficients must be of the same algebraic sign, but some coefficients can be zero.

Right half-plane (RHP) roots involve factors of the form

$(s - a)$ $a > 0$ (real axis root in the RHP)

and

$(s^2 - as + b)$ $a > 0$ and $b > 0$ (two roots in the RHP, perhaps complex conjugate)

The presence of such factors may or may not cause differing algebraic signs of the coefficients and (by cancellation) zero coefficients.

Table 2.1 summarizes the information conveyed by these coefficient tests. For example, the polynomial

$$7s^6 + 5s^4 - 3s^3 - 2s^2 + s + 10$$

definitely has one or more RHP roots, indicated by the differing algebraic signs of the coefficients. Examination of the coefficient signs yields no information about root locations for the following polynomial:

$$8s^5 + 6s^4 + 3s^3 + 2s^2 + 7s + 10$$

The polynomial

$$s^6 + 3s^5 + 2s^4 + 8s^2 + 3s + 17$$

has imaginary axis roots or RHP roots or both, indicated by the missing s^3 term. Imaginary axis roots in the above polynomial, if they are present, are complex

Table 2.1 POLYNOMIAL COEFFICIENT TESTS

Properties of the Polynomial Coefficients	Conclusion about Roots from the Coefficient Test
Differing algebraic signs	At least one RHP root
Zero-valued coefficients	Imaginary axis or RHP roots or both
All of the same algebraic sign, none zero	No direct information

conjugate since if there were an imaginary axis root at $s = 0$, s would be a factor of the polynomial.

Routh-Hurwitz Testing

The Routh-Hurwitz test is a numerical procedure for determining the numbers of right half-plane (RHP) and imaginary axis (IA) roots of a polynomial.

An example of the Routh-Hurwitz test of a polynomial

$$p(s) = 2s^4 + 3s^3 + 5s^2 + 2s + 6$$

is as follows. First, the initial part of the array is formed. The powers of s are written to the left and the polynomial coefficients are alternated between the first and second rows, as shown. It is helpful to imagine the rows to continue to the right with entries of zeros.

$$
\begin{array}{c|ccc}
s^4 & 2 & 5 & 6 \\
s^3 & 3 & 2 \\
s^2 & & \\
s^1 & & \\
s^0 & & \\
\end{array}
$$

The array is completed by proceeding, row by row, calculating the elements of the next row. Each element calculated is derived from four elements in the above two rows, two of them at the left column and two in the column to the right of the element being calculated. In each case, the calculated element is the negative of the determinant of the four elements above, divided by the lower left element above. For the example, the first element of the s^2 row is

$$-\frac{\begin{vmatrix} 2 & 5 \\ 3 & 2 \end{vmatrix}}{3} = \frac{11}{3}$$

The second element of the s^2 row is

$$-\frac{\begin{vmatrix} 2 & 6 \\ 3 & 0 \end{vmatrix}}{3} = 6$$

The first element of the s^1 row is

$$-\frac{\begin{vmatrix} 3 & 2 \\ \frac{11}{3} & 6 \end{vmatrix}}{\frac{11}{3}} = -\frac{32}{11}$$

and so on.

s^4	2	5	6		s^4	2	5	6		s^4	2	5	6

$$
\begin{array}{c|ccc}
s^4 & 2 & 5 & 6 \\
s^3 & 3 & 2 \\
s^2 & \dfrac{11}{3} \\
s^1 \\
s^0
\end{array}
\qquad
\begin{array}{c|ccc}
s^4 & 2 & 5 & 6 \\
s^3 & 3 & 2 & 0 \\
s^2 & \dfrac{11}{3} & 6 \\
s^1 \\
s^0
\end{array}
\qquad
\begin{array}{c|ccc}
s^4 & 2 & 5 & 6 \\
s^3 & 3 & 2 \\
s^2 & \dfrac{11}{3} & 6 \\
s^1 & -\dfrac{32}{11} \\
s^0
\end{array}
$$

The completed Routh-Hurwitz array is shown below. The number of RHP roots of $p(s)$ is the number of algebraic sign changes in the elements of the left column of the array, proceeding from top to bottom. For this example, there are two sign changes in the left column, as indicated with the arrows; therefore $p(s)$ has two RHP roots:

$$
\begin{array}{c|ccc}
s^4 & 2 & 5 & 6 \\
s^3 & 3 & 2 \\
s^2 & \dfrac{11}{3} & 6 \\
s^1 & -\dfrac{32}{11} \\
s^0 & 6
\end{array}
$$

If $p(s)$ is the denominator polynomial of a system's transfer function, that system is unstable.

As another example, a system with transfer function

$$
T(s) = \frac{5s^2 - 7s + 2}{s^4 + 2s^3 + 3s^2 + 4s + 1}
$$

has characteristic polynomial

$$
s^4 + 2s^3 + 3s^2 + 4s + 1
$$

which has the Routh-Hurwitz array below. There are no algebraic sign changes in the left column, so the polynomial has no RHP roots.

$$
\begin{array}{c|ccc}
s^4 & 1 & 3 & 1 \\
s^3 & 2 & 4 \\
s^2 & 1 & 1 \\
s^1 & 2 \\
s^0 & 1
\end{array}
$$

As one gains practice in completing Routh-Hurwitz arrays, it is easier to form the negative of the determinant by simply evaluating the difference of element products in reverse order.

Each array has properties which serve as a partial check on its correct completion. The number of nonzero row entries is normally reduced by one every two rows, with just one nonzero element in the s^1 row and in the s^0 row. And the last coefficient of the polynomial appears periodically as the last nonzero entry in every other row.

Left-Column Zeros of the Array

It sometimes happens that the polynomial coefficients are such that a zero occurs in the left column of the array, so that the array cannot be completed. The situation where there is a zero at the left of a row, but the entire row does not consist of zeros, is termed a *left-column zero*. For example, the polynomial

$$p(s) = 3s^4 + 6s^3 + 2s^2 + 4s + 5 \tag{2.3}$$

has an array that begins as follows:

$$
\begin{array}{c|ccc}
s^4 & 3 & 2 & 5 \\
s^3 & 6 & 4 & \\
s^2 & 0 & 5 & \\
s^1 & & & \\
s^0 & & &
\end{array}
$$

The array cannot be completed in the usual way, because of the necessity to divide by zero.

When a left-column zero occurs, it is easiest to form a new polynomial with an additional known root, increasing its order but changing the coefficients so that a left-column zero does not occur. For example, adding an additional LHP root at $s = -1$ to the previous polynomial (2.3) gives

$$
\begin{aligned}
p'(s) &= (s + 1)(3s^4 + 6s^3 + 2s^2 + 4s + 5) \\
&= 3s^5 + 9s^4 + 8s^3 + 6s^2 + 9s + 5
\end{aligned}
$$

$$
\begin{array}{c|ccc}
s^5 & 3 & 8 & 9 \\
s^4 & 9 & 6 & 5 \\
s^3 & 6 & \dfrac{22}{3} & \\
s^2 & -5 & 5 & \\
s^1 & \dfrac{40}{3} & & \\
s^0 & 5 & &
\end{array}
$$

This polynomial has two **RHP** roots, so the original polynomial $p(s)$ has two **RHP** roots. In digital computer programming of the test, an LHP root at a random location is introduced whenever the array has a left-column entry very close to zero. If the new polynomial still has a left-column zero, or nearly so, a different random choice is made.

The left-column zero situation can also be resolved through realizing that it is the result of the polynomial coefficients being exactly certain numbers. If the coefficients were only slightly different, the left-column zero would instead be a small positive or negative number. Imagine that the polynomial coefficients have been altered very slightly, so that they are not the exact values which give the left-column zero; instead, they give some tiny nonzero number ε. The small number ε can be considered to be either positive or negative, but it is usually easiest to imagine it to be positive. For the original polynomial (2.3), replacing the left-column zero by ε and completing the array in terms of ε gives

$$
\begin{array}{c|ccc}
s^4 & 3 & 2 & 5 \\
s^3 & 6 & 4 & \\
s^2 & \varepsilon & 5 & \\
s^1 & \dfrac{4\varepsilon - 30}{\varepsilon} & & \\
s^0 & 5 & &
\end{array}
$$

In the limit as $\varepsilon \to 0$, using positive ε for convenience, the left-column entries become as shown below, since

$$
\lim_{\varepsilon \to 0} 4 - \frac{30}{\varepsilon} = -\infty
$$

$$
\begin{array}{c|c}
s^4 & 3 \\
s^3 & 6 \\
s^2 & 0 \\
s^1 & -\infty \\
s^0 & 5
\end{array}
$$

There are two algebraic sign changes in the left column as ε approaches zero in the example, and so this polynomial has two RHP roots.

Whether ε should be a small positive number or a small negative number depends on the system being tested. If the left-column values are all positive, prior to the left-column zero, then ε should be chosen to be a small positive number. A small negative number would then give a false sign change. In general, ε should be chosen so as to yield as few left-column sign changes as possible. Other resulting sign changes will then be correct.

Another example of a Routh-Hurwitz array with a left-column zero is shown below:

$$
\begin{array}{c|ccc}
s^5 & 1 & 2 & 1 \\
s^4 & 1 & 3 & 4 \\
s^3 & -1 & -3 & \\
s^2 & \varepsilon & 4 & \\
s^1 & \dfrac{-3\varepsilon + 4}{\varepsilon} & & \\
s^0 & 4 & &
\end{array}
\qquad \rightarrow \qquad
\begin{array}{c|c}
s^5 & 1 \\
s^4 & 1 \\
s^3 & -1 \\
s^2 & 0 \\
s^1 & +\infty \\
s^0 & 4
\end{array}
$$

Should ε instead be taken to be negative, the limit will give the same number of left-column sign changes (although not necessarily at the same locations in the array) provided that the polynomial has no imaginary axis roots. If imaginary axis roots are present, different perturbations of the polynomial coefficients can cause different numbers of RHP roots to be detected since a small polynomial coefficient change may move imaginary axis roots into either the LHP or the RHP, depending upon the nature of the perturbation.

When there are imaginary axis roots in the polynomial, a special circumstance termed *premature termination*, discussed in the next section, occurs. Except in this case, the ε method, with positive ε, can be safely used.

Premature Termination of the Array

The situation where an entire row of zeros occurs in the Routh-Hurwitz array is termed a *premature termination*. For example, the test of the polynomial

$$p(s) = s^5 + 2s^4 + 8s^3 + 11s^2 + 16s + 12$$

prematurely terminates at the s^1 row:

$$
\begin{array}{c|ccc}
s^5 & 1 & 8 & 16 \\
s^4 & 2 & 11 & 12 \\
s^3 & \frac{5}{2} & 10 & \\
s^2 & 3 & 12 & \\
s^1 & 0 & 0 & \\
s^0 & & &
\end{array}
$$

Premature termination occurs whenever there is an even or an odd polynomial divisor of the original polynomial. The coefficients of the even or odd divisor polynomial are those given in the row above the row of zeros:

$$p_{\text{divisor}}(s) = 3s^2 + 12 = 3(s^2 + 4) = 3(s + 2j)(s - 2j)$$

The tested polynomial thus has two imaginary axis roots, at $s = \pm 2j$:

$$
\begin{array}{r}
s^3 + 2s^2 + 4s\ + 3 \\[2pt]
\hline
\end{array}
$$

$$
s^2 + 4 \,\overline{\big)\, s^5 + 2s^4 + 8s^3 + 11s^2 + 16s + 12}
$$

$$
\begin{array}{l}
\quad s^5 \qquad\ \ + 4s^3 \\
\hline
\quad 2s^4 + 4s^3 + 11s^2 + 16s + 12 \\
\quad 2s^4 \qquad\ + 8s^2 \\
\hline
\qquad\quad 4s^3 + 3s^2 + 16s + 12 \\
\qquad\quad 4s^3 \qquad\ + 16s \\
\hline
\qquad\qquad\quad 3s^2 \qquad + 12 \\
\qquad\qquad\quad 3s^2 \qquad + 12 \\
\hline
\qquad\qquad\qquad\qquad\ 0
\end{array}
$$

To complete the array, the row of zeros is replaced by the coefficients of the derivative of the divisor polynomial:

$$
\frac{dp_{\text{divisor}}(s)}{ds} = 6s
$$

s^5	1	8	16
s^4	2	11	12
s^3	$\dfrac{5}{2}$	10	
s^2	3	12	
s^1	6		
s^0	12		

Even and odd polynomials have root locations that are symmetric about the imaginary axis. An odd polynomial always has s as a factor. Whether or not s is a factor is evident from a glance at the polynomial, and the remainder is an even polynomial. For any even polynomial, replacing s by $(-s)$ leaves the polynomial unchanged. Thus the roots of an even polynomial, besides occurring in conjugate pairs, are also symmetrical about the imaginary axis. There are three basic types of factors possible in an even polynomial. One type is

$$(s + ja)(s - ja) = (s^2 + a^2)$$

which consists of complex conjugate roots on the imaginary axis, as indicated in Fig. 2.11a. Another is

$$(s + a)(s - a) = (s^2 - a^2)$$

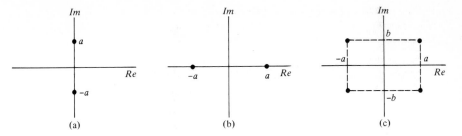

Figure 2.11 Even polynomial root locations.

which are symmetrical roots on the real axis, one in the LHP and one in the RHP, as in Fig. 2.11b. The third type of factor, Fig. 2.11c, involves complex roots in quadrature, one pair in the LHP and one pair in the RHP:

$$(s + a + jb)(s + a - jb)(s - a + jb)(s - a - jb)$$
$$= s^4 + 2(b^2 - a^2)s^2 + (a^2 + b^2)^2$$

The additional symmetry of even polynomial roots about the imaginary axis allows determination of the number of imaginary axis roots. Each RHP root of an even polynomial must be matched by just one corresponding LHP root. Thus if an even polynomial is of sixth order and is known to have just one RHP root, it has just one LHP root, and the remaining four roots must be on the imaginary axis. If an eighth-order polynomial has three RHP roots, it must have three LHP roots and two imaginary axis roots. For example, for the polynomial

$$s^6 + s^5 + 5s^4 + s^3 + 2s^2 - 2s - 8$$

the coefficient tests indicate the presence of at least one RHP root. The Routh-Hurwitz test begins as follows:

$$
\begin{array}{c|cccc}
s^6 & 1 & 5 & 2 & -8 \\
s^5 & 1 & 1 & -2 & \\
s^4 & 4 & 4 & -8 & \\
s^3 & 0 & 0 & &
\end{array}
$$

So

$$4s^4 + 4s^2 - 8$$

is a divisor of the original polynomial. Replacing the row of zeros by the coefficients of the derivative of the divisor polynomial,

$$\frac{d}{ds}(4s^4 + 4s^2 - 8) = 16s^3 + 8s$$

The completed array is as follows:

$$
\begin{array}{c|rrrr}
s^6 & 1 & 5 & 2 & -8 \\
s^5 & 1 & 1 & -2 & \\
\hline
s^4 & 4 & 4 & -8 & \\
s^3 & 16 & 8 & & \\
s^2 & 2 & -8 & & \\
s^1 & 72 & & & \\
s^0 & -8 & & &
\end{array}
$$

The test of the even polynomial divisor is that portion of the array below the dashed line. There is one left-column sign change in that portion, so the divisor has one RHP root, and thus one LHP root. The other two divisor roots must then be on the imaginary axis. When a polynomial has imaginary axis roots, the Routh-Hurwitz array will *definitely* contain a row of zeros, with all such roots contained within the divisor polynomial.

Since the entire array has one left-column sign change, the entire polynomial has one RHP root (in the even divisor). There are exactly two imaginary roots because if there are any, they must be in the divisor. The remaining roots must then be in the LHP. It is concluded that the example polynomial has the following numbers of the various types of roots:

$$
\text{RHP} = 1
$$
$$
\text{LHP} = 3
$$
$$
\text{IA} = 2
$$

This information can be found from the Routh-Hurwitz test for any polynomial.

Another array with a premature termination is the following:

$$
\begin{array}{c|rrr}
s^4 & 1 & 9 & 20 \\
s^3 & 6 & 24 & \\
\hline
s^2 & 5 & 20 & \\
s^1 & 10 & & \\
s^0 & 20 & &
\end{array}
$$

For this polynomial, there are no RHP roots. The second-order even polynomial divisor has no RHP roots and thus no LHP roots. Hence there must be two roots on the imaginary axis. The remaining two roots are not on the imaginary axis; all such roots must be in the divisor. They can only be in the LHP.

The polynomial

$$
s^5 + s^4 + 6s^3 + 6s^2 + 25s + 25
$$

prematurely terminates at the s^3 row:

$$
\begin{array}{c|rrr}
s^5 & 1 & 6 & 25 \\
\hline
s^4 & 1 & 6 & 25 \\
s^3 & 4 & 12 & \\
s^2 & 3 & 25 & \\
s^1 & -\dfrac{64}{3} & & \\
s^0 & 25 & &
\end{array}
$$

The fourth-order even polynomial divisor has two RHP roots, and so its remaining two roots must be in the LHP. There can thus be no imaginary axis roots. The entire polynomial has two RHP roots, and the remaining three roots must be in the LHP.

DRILL PROBLEMS

D2.9. What can be determined about the roots of the following polynomials from the coefficient tests?

(a) $-3s^4 + 2s^3 + s + 10$

Ans at least one RHP root

(b) $4s^4 + 3s^3 + 10s^2 + 8s + 1$

Ans nothing

(c) $s^5 + 4s^3 + 8$

Ans imaginary axis (IA) or RHP roots or both

(d) $s^6 + 6s^4 + 3s^2 + 10$

Ans IA or RHP roots or both

D2.10. How many roots of each of the following polynomials are in the right half of the complex plane?

(a) $s^3 + 2s^2 + 3s + 4$

Ans 0

(b) $s^4 - 6s^3 + 7s^2 + 2s + 4$

Ans 2

(c) $0.3s^4 + 1.1s^3 + 0.7s^2 + s + 2.1$

Ans 2

(d) $s^5 + s^4 + 2s^3 + 3s^2 + \frac{1}{2}$

Ans 4

(e) $2s^5 + s^4 + 2s^3 + 4s^2 + s + 6$

Ans 2

D2.11. The Routh-Hurwitz tests for the following polynomials might involve left-column zeros. For each polynomial, use the array to find the number of roots in the right half of the complex plane.

(a) $s^3 + 2s + 3$

Ans 2

(b) $3s^4 + 6s^3 + 2s^2 + 4s + 5$

Ans 2

(c) $2s^4 + 2s^3 + s^2 + s - 3$

Ans 1

(d) $s^5 + s^4 + 3s^3 + 2s^2 + 4s + 2$

Ans 2

D2.12. The Routh-Hurwitz tests for the following polynomials might involve premature terminations of the arrays. For each, complete the array and determine the number of roots in the right half of the complex plane.

(a) $s^4 + 8s^2 - 7$

Ans 1

(b) $s^4 + 2s^3 + 9s^2 + 4s + 14$

Ans 0

(c) $s^5 + s^3 + 2s$

Ans 2

(d) $s^5 + 3s^4 + 4s^3 + 7s^2 + 4s + 2$

Ans 0

D2.13. For each of the following polynomials, how many roots are in the LHP, how many are in the RHP, and how many are on the imaginary axis?

(a) $s^4 + 3s^2 + 4$

Ans 2RHP, 2LHP

(b) $s^4 + 2s^3 + 5s^2 - 4s - 14$

Ans 1RHP, 3LHP

(c) $s^5 + 2s^4 + 3s^3 + 6s^2 + 2s + 4$

Ans 1LHP, 4IA

(d) $3s^5 + 2s^3 + s$

Ans 2RHP, 2LHP, 1IA

(e) $2s^5 + 4s^4 + s^3 + 2s^2 + 3s + 6$

Ans 3LHP, 2RHP

D2.14. Are the following systems stable?

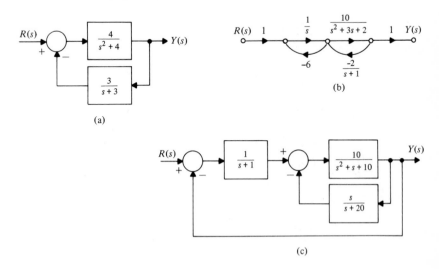

(a)

(b)

(c)

Ans (a) unstable; (b) stable; (c) stable

2.6 PARAMETER SHIFTING

Adjustable Systems

It is often desired to know for what range or ranges of an adjustable parameter K a system is stable. For example, suppose the transfer function of a system is, in terms of K,

$$T(s) = \frac{4}{s^2 + 2s + K}$$

This system is stable for all positive values of K because the roots of a quadratic are in the LHP if and only if all coefficients have the same algebraic sign. For $K = 0$,

$$T(s) = \frac{4}{s(s + 2)}$$

and the system is marginally stable because of the imaginary axis characteristic root at $s = 0$. For negative values of K, the transfer function has a characteristic root in the RHP and the system is unstable. The system with transfer function

$$T(s) = \frac{2s + K}{s^2 + (2 + K)s + 4}$$

is, similarly, stable for $K > -2$.

For systems with characteristic polynomials of higher order, the Routh-Hurwitz test is a useful tool for determining stability in terms of a constant but adjustable parameter. Suppose

$$s^4 + 2s^3 + 4s^2 + 2s + K$$

is the denominator polynomial of a system transfer function, in terms of an adjustable constant K. The Routh-Hurwitz test in terms of K is as follows:

s^4	1	4	K
s^3	2	2	
s^2	3	K	
s^1	$\dfrac{6 - 2K}{3}$		
s^0	K		

All of the left column array entries must be of the same algebraic sign if the polynomial is to have no RHP roots. Thus

$$\frac{6 - 2K}{3} > 0 \quad \text{and} \quad K > 0$$

or

$$0 < K < 3$$

for system stability. For $K = 3$, the array contains a row of zeros, $3s^2 + 3$ is a factor of the polynomial, and the system is marginally stable. For $K = 0$, s is a factor of the polynomial and, again, the system is marginally stable.

For the adjustable polynomial

$$s^4 + 2s^3 + 4s^2 + Ks + 6$$

the Routh-Hurwitz array is the following:

s^4	1	4	6
s^3	2	K	
s^2	$4 - \dfrac{K}{2}$	6	
s^1	$K - \dfrac{12}{4 - K/2}$		
s^0	6		

If the polynomial is to have all LHP roots,

$$4 - \frac{K}{2} > 0 \qquad K < 8$$

and

$$K - \frac{12}{4 - K/2} > 0$$

Making use of the requirement that $4 - K/2$ be positive, from the first inequality, the second inequality can be multiplied by the positive number $4 - K/2$. If the inequality were multiplied by a negative number, its sense would be reversed. Then

$$\left(4 - \frac{K}{2}\right)\left(K - \frac{12}{4 - K/2}\right) = -\tfrac{1}{2} K^2 + 4K - 12 > 0$$

The quadratic function

$$-\tfrac{1}{2} K^2 + 4K - 12$$

is negative for large negative K and is negative for large positive K. To determine if there are intermediate values of K for which the function is positive, its roots are found using the quadratic formula:

$$K = \frac{-4 \pm \sqrt{16 - 24}}{2(-\tfrac{1}{2})}$$

As the function's roots are complex, it is concluded that the inequalities cannot be satisfied for any (real) K. The original polynomial thus has RHP roots for all K.

Testing Relative Stability

A system with all characteristic roots in the LHP but with one or more roots only slightly to the left of the imaginary axis has a natural response that decays very slowly. The longer the distance from the imaginary axis to the nearest characteristic root of a stable system, the faster the slowest decaying term in the system's natural response dies out. The distance on the complex plane between the nearest characteristic root and the imaginary axis is termed the *relative stability* of the system. Normally, the relative stability concept is used only in connection with stable systems.

For the characteristic polynomial

$$s^4 + 14s^3 + 73s^2 + 168s + 144$$

a Routh-Hurwitz test shows all roots to be in the LHP:

s^4	1	73	144
s^3	14	168	
s^2	61	144	
s^1	$\dfrac{8232}{61}$		
s^0	144		

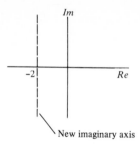

Figure 2.12 Axis shift on the complex plane.

Shifting the imaginary axis 2 units to the left, as indicated in Fig. 2.12, is accomplished by substituting $\sigma - 2$ for each s in the original polynomial:

$$(\sigma - 2)^4 + 14(\sigma - 2)^3 + 73(\sigma - 2)^2 + 168(\sigma - 2) + 144$$

$$= (\sigma^4 - 8\sigma^3 + 24\sigma^2 - 32\sigma + 16) + 14(\sigma^3 - 6\sigma^2 + 12\sigma - 8)$$

$$+ 73(\sigma^2 - 4\sigma + 4) + 168\sigma - 336 + 144$$

$$= \sigma^4 + 6\sigma^3 + 79\sigma^2 + 12\sigma + 4$$

A Routh-Hurwitz test on the polynomial with shifted axis is given below. There are no RHP roots of the shifted polynomial in σ, so the original polynomial has all roots to the left of $s = -2$. The system thus has a relative stability of at least 2 units:

$$
\begin{array}{c|ccc}
\sigma^4 & 1 & 79 & 4 \\
\sigma^3 & 6 & 12 & \\
\sigma^2 & 77 & 4 & \\
s^1 & \dfrac{900}{77} & & \\
\sigma^0 & 4 & &
\end{array}
$$

The characteristic polynomial

$$s^4 + 2s^3 + 3s^2 + s + 1$$

has all roots in the LHP, as is verified by a Routh-Hurwitz test:

$$
\begin{array}{c|ccc}
s^4 & 1 & 3 & 1 \\
s^3 & 2 & 1 & \\
s^2 & \dfrac{5}{2} & 1 & \\
s^1 & \dfrac{1}{5} & & \\
s^0 & 1 & &
\end{array}
$$

If it is desired to determine if all of the roots of this polynomial are to the left of $s = -1$ on the complex plane, $\sigma - 1$ is substituted for each s in the polynomial:

$$(\sigma - 1)^4 + 2(\sigma - 1)^3 + 3(\sigma - 1)^2 + (\sigma - 1) + 1 = \sigma^4 - 2\sigma^3 + 3\sigma^2 - 3\sigma + 2$$

The σ-polynomial is the s-polynomial with the imaginary axis shifted 1 unit to the left. The coefficient test shows the σ-polynomial to have RHP roots; therefore, there are roots of the s-polynomial to the right of $s = -1$. A Routh-Hurwitz test of the σ-polynomial is as follows:

$$
\begin{array}{c|ccc}
\sigma^4 & 1 & 3 & 2 \\[4pt]
\sigma^3 & -2 & -3 \\[4pt]
\sigma^2 & \dfrac{3}{2} & 2 \\[8pt]
\sigma^1 & -\dfrac{1}{3} \\[8pt]
\sigma^0 & 2
\end{array}
$$

The original polynomial has no RHP roots and four roots, to the right of $s = -1$.

Shifting the imaginary axis, instead, $\frac{1}{2}$ unit to the left on the complex plane gives

$$\left(\sigma - \frac{1}{2}\right)^4 + 2\left(\sigma - \frac{1}{2}\right)^3 + 3\left(\sigma - \frac{1}{2}\right)^2 + \left(\sigma - \frac{1}{2}\right) + 1$$

$$= (\sigma^4 - 2\sigma^3 + 1.5\sigma^2 - 0.5\sigma + 0.0625) + 2(\sigma^3 - 1.5\sigma^2 + 0.75\sigma - 0.125)$$

$$+ 3(\sigma^2 - \sigma + 0.25) + \sigma - 0.5 + 1$$

$$= \sigma^4 + 3\sigma^2 - \sigma + 1.0625$$

which has the following Routh-Hurwitz test:

$$
\begin{array}{c|cccc}
\sigma^4 & 1 & 3 & 1.0625 & 1 \\[6pt]
\sigma^3 & \varepsilon & -1 & & 0 \\[6pt]
\sigma^2 & \dfrac{3\varepsilon + 1}{\varepsilon} & 1.0625 & \rightarrow & +\infty \\[10pt]
\sigma^1 & -1 - \dfrac{1.0625\varepsilon^2}{3\varepsilon + 1} & & & -1 \\[10pt]
\sigma^0 & 1.0625 & & & 1.0625
\end{array}
$$

There are two roots to the right of $s = -\frac{1}{2}$. The original polynomial thus must have two roots between $s = -1$ and $s = -\frac{1}{2}$, and two roots between $s = -\frac{1}{2}$ and $s = 0$.

Although it is interesting to test for relative stability using the Routh-Hurwitz test, the methods that will be covered in Chap. 4 are easier to apply and give a broader view of the stability (or instability) of a closed-loop system.

Polynomial Factoring

In order to extract all the roots of a polynomial $p(s)$, a digital computer program must be able to extract real roots (requiring a one-dimensional search) and complex conjugate roots (requiring a two-dimensional search).

A complex number $s = x + jy$ is a root of a polynomial $p(s)$ only if both

$$\begin{cases} \text{Re}[p(s = x + jy)] = 0 \\ \text{Im}[p(s = x + jy)] = 0 \end{cases} \tag{2.4}$$

One popular root solving procedure is Bairstow's method. That method avoids complex arithmetic by seeking real-valued coefficients a and b so that

$$p_1(s) = s^2 + as + b$$

is a factor of $p(s)$. The quadratic formula can then be applied to determine whether a pair of complex conjugate roots or a pair of real roots are present. Bairstow's method selects successive values of a and b so that successive division of $p_1(s)$ into $p(s)$ results in a remainder that diminishes toward zero. As soon as the remainder is acceptably small, a lower-order polynomial is computed by dividing out $p_1(s)$. The process continues until all roots (including multiple ones) are extracted.

Another attractive factoring method involves repeated shifts of the imaginary axis and Routh-Hurwitz testing to locate the real parts of a polynomial's roots. The search for the real parts is one-dimensional, after which other one-dimensional searches can be done for the imaginary parts of the roots, using the divisor polynomial from the Routh-Hurwitz test.

DRILL PROBLEMS

D2.15. For what range(s), if any, of the adjustable constant K are all roots of the following polynomials in the left half of the complex plane?

(a) $s^3 + (2 + K)s^2 + (8 + K)s + 6$

　　Ans　$-1.12 < K$

(b) $2s^3 + (6 - 2K)s^2 + (4 + 3K)s + 10$

　　Ans　$-\frac{1}{3} < K < 2$

(c) $s^4 + (10 + K)s^3 + 9s + 11$

　　Ans　No value of K

(d) $s^4 + s^3 + 3s^2 + 2s + 4 + K$

　　Ans　$-4 < K < -2$

D2.16. Find the range(s) of positive, constant K, if any, for which the following systems are stable.

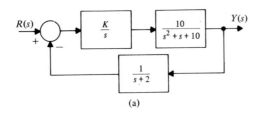

(a)

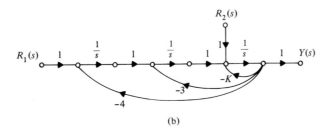

(b)

Ans (a) $0 < K < \frac{32}{9}$; (b) $\frac{4}{3} < K$

D2.17. Determine the number of units of relative stability of the systems with the following transfer functions:

(a) $T(s) = \dfrac{20}{(s + 2)^2(s^2 + 5s + 12)}$

Ans 2

(b) $T(s) = \dfrac{4(s + 2)(s^2 + 9)^2}{(s + 3)(s^2 + s + 8)^2}$

Ans $\frac{1}{2}$

(c) $T(s) = \dfrac{s^4}{(s + 1)(s^2 + 2s + 4)(s^2 + 2s + 7)}$

Ans 1

(d) $T(s) = \dfrac{-3s^3 + 3s^2 + 7}{s(s^2 + s + 10)(s^2 + 3s + 8)(s + 4)(s + 5)}$

Ans 0

D2.18. Find the number of units of relative stability of each of the following systems:

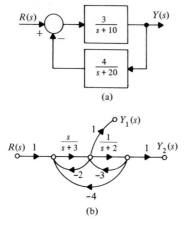

(a)

(b)

Ans (a) 11.4; (b) 1.21

2.7 AN INSULIN DELIVERY SYSTEM

Control system methods have been applied to the biomedical field to create an implanted insulin delivery system for diabetics. When food is eaten and digested, sugars, mainly glucose, are absorbed into the bloodstream. Normally, the pancreas secretes insulin into the bloodstream to metabolize the sugar. A diabetic's pancreas secretes insufficient insulin to metabolize blood sugar; blood sugar levels can thus become high enough to threaten damage to the body's organs.

One solution to this problem is for the diabetic to take one injection of insulin each day. In Fig. 2.13a, typical blood sugar and insulin concentration histories for a day are shown for a normal person. In Fig. 2.13b, the blood sugar and insulin concentration histories for a day are shown for a diabetic who takes one insulin injection in the morning. Notice that blood sugar is often higher than normal, but the sugar concentration is far less than would be the case if no insulin had been injected. A higher dose of insulin could be taken in the morning to counteract the low insulin residual after dinner, but blood sugar concentration might be driven unacceptably low in the morning (hypoglycemia), causing weakness, trembling, and possibly fainting. Three injections a day, one before each meal, are generally infeasible due to their damage to the veins and skin tissue.

One automatic control system of interest consists of a tiny insulin reservoir, control motor, and pump that is implanted in the body below the diaphragm. This electronic pancreas delivers insulin into the peritoneum using preprogrammed commands intended to establish insulin levels close to the levels of a normal individual. The pump runs at higher rates after meals than otherwise. The diabetic must time meals to complement the behavior of the implanted system, but

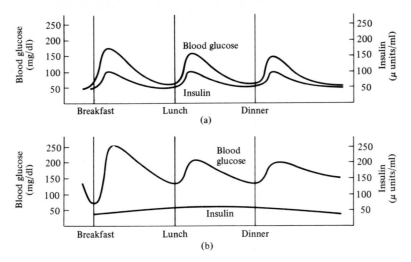

Figure 2.13 Typical blood sugar and insulin concentrations. **(a)** Normal person. **(b)** Diabetic with one daily insulin injection.

injections are required only every few weeks, to refill the insulin reservoir. Systems of this type have operated for many years without malfunction.

The methods of this chapter can be used to design such an insulin delivery system, which is the open-loop one shown in Fig. 2.14. A signal generator is programmed to drive the motor pump in such a way that the insulin delivery rate $i(t)$ approximates a desired delivery rate $i_D(t)$.

Figure 2.15a shows an approximate desired insulin rate $I_D(t)$ in cm³/sec for one-third day, beginning with a meal. Figure 2.15b shows a similar function,

$$i(t) = Ate^{-at}u(t)$$

which has the particularly simple Laplace transform

$$I(s) = \mathscr{L}[i(t)] = \frac{A}{(s + a)^2}.$$

A good approximation of $i_D(t)$ by $i(t)$ occurs when the constants A and a are selected so that $i(t)$ is maximum at $t = 3600$, as is $i_D(t)$, and so that the areas under the two curves are equal, with value 0.17 cm³:

$$\frac{di}{dt} = -aAte^{-at} + Ae^{-at} = A(1 - at)e^{-at}$$

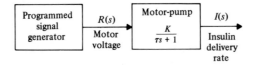

Figure 2.14 The implanted open-loop insulin delivery system.

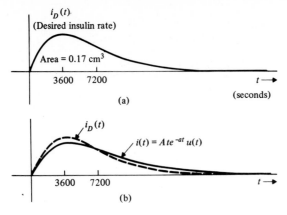

Figure 2.15 Approximating the required insulin rate with a time-weighted exponential. **(a)** Insulin rate required for a diabetic. **(b)** The function $i(t) = Ate^{-at}u(t)$.

For the maximum of $i(t)$ to occur at $t = 3600$,

$$\left.\frac{di}{dt}\right|_{t=3600} = A(1 - 3600a)e^{-3600a} = 0 \qquad a = \tfrac{1}{3600} = 2.78 \times 10^{-4}$$

The area under the $i(t)$ curve after $t = 0$ is

$$\int_0^\infty Ate^{-at}\,dt = A\left[-\frac{1}{a}te^{-at} - \frac{1}{a^2}e^{-at}\right]_0^\infty = \frac{A}{a^2}$$

Equating to the desired area of 0.17 cm³ gives

$$\frac{A}{a^2} = (3600)^2 A = 0.17 \qquad A = 1.31 \times 10^{-8}$$

If $i(t)$ is to be produced by the system input $r(t)$, one has

$$I(s) = \frac{A}{(s + a)^2} = \frac{K}{\tau s + 1} R(s)$$

giving the required programmed signal for each mealtime:

$$R(s) = \frac{A(\tau s + 1)/K}{(s + a)^2}$$

For a motor pump with

$$\tau = 5 \text{ sec} \qquad K = 2.3 \times 10^{-6} \text{ cm}^3/\text{volt-sec}$$

and with the delivery rate for which

$$a = 2.78 \times 10^{-4} \text{ sec}^{-1} \qquad A = 1.31 \times 10^{-8} \text{ cm}^3/\text{sec}^2$$

then

$$R(s) = \frac{(1.31 \times 10^{-8})(5s + 1)/2.3 \times 10^{-6}}{(s + 2.78 \times 10^{-4})^2}$$

$$= \frac{K_1}{(s + 2.78 \times 10^{-4})} + \frac{K_2}{(s + 2.78 \times 10^{-4})^2}$$

giving

$$K_1 = 28.5 \times 10^{-3} \qquad K_2 = 5.7 \times 10^{-3} - 2.78 \times 10^{-4} K_1 = 5.69 \times 10^{-3}$$

The programmed motor drive signal is thus to be

$$R(s) = \frac{28.5 \times 10^{-3}}{s + 2.78 \times 10^{-4}} + \frac{5.69 \times 10^{-3}}{(s + 2.78 \times 10^{-4})^2}$$

$$r(t) = [28.5e^{-(2.78 \times 10^{-4})t} + 5.69te^{-(2.78 \times 10^{-4})t}]10^{-3}u(t) \quad \text{volts}$$

which is sketched in Fig. 2.16a. Repetition, three times a day, of the motor drive signal will provide insulin delivery for periodic meals, as is shown in Fig. 2.16b.

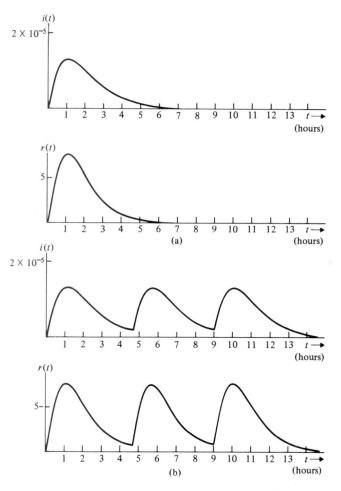

Figure 2.16 Response of the insulin delivery system. **(a)** Motor voltage and resulting insulin rate for one mealtime. **(b)** Repetitive motor voltage and insulin rate for three meals a day.

The insulin delivery system described in this section entered active use in 1981. A second-generation titanium insulin pump entered active service by the mid 1980s. The programmed rate of infusion, amount of insulin in the reservoir, and battery charge can be read automatically and transmitted over the telephone to the patient's doctor. The doctor can reprogram the pump over the same telephone line. As before, the reservoir is refilled by injection through the skin and into the reservoir.

A third-generation insulin pump entered development after 1985. The insulin pump operates in a closed-loop mode so that direct measurement of the blood glucose level causes the pump to track the patient's insulin requirements. This third-generation system frees the patient from some aspects of dietary regimen.

2.8 ANALYSIS OF AN AIRCRAFT WING

Students commonly perform an experiment in physics where a tuning fork is struck above an empty tube, resulting in a resonating tone. One can obtain the same effect by blowing into a bottle at just the right angle. What occurs is that the blowing or the tuning fork is at a natural frequency of the hollow glass chamber, causing a reinforcement or resonating effect. This effect can be catastrophic when dealing with certain mechanical systems. For example, in 1939 a bridge over Puget Sound at the Tacoma Narrows began to sway and twist due to a wind that whistled down the Narrows at just the right speed. In a mechanical sense the bridge was excited at a resonance, resulting in deflections that exceeded its structural limit and led to collapse. A similar effect must be considered in aircraft wing design. Structural failure of wings on certain turbine-driven jet aircraft in the 1960s was traced to a mechanical resonance excited by the jet turbine engines. The following problem illustrates the problem of mechanical resonance for an aircraft wing.

Figure 2.17a depicts an aircraft wing with an applied force $f(t)$ and resulting deflection $x(t)$ at the wingtip. We will ignore aerodynamic and accelerative forces and consider only a sinusoidal excitation of the wing caused by the turbine:

$$f(t) = D \cos \omega t$$

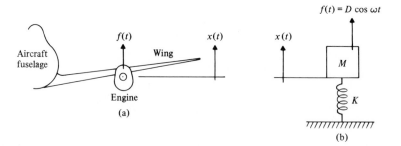

Figure 2.17   Deflection of an aircraft wing due to jet turbine vibrational force.   **(a)** The aircraft wing.   **(b)** Model of aircraft wing.

Figure 2.17b shows a simple mechanical model of the aircraft wing, a spring-mass system excited by the force $f(t)$. The differential equation which relates $f(t)$ to $x(t)$ is

$$M \frac{d^2 x}{dt^2} + Kx = f(t) = D \cos \omega t$$

Laplace-transforming,

$$Ms^2 X(s) - Msx(0^-) - Mx'(0^-) + KX(s) = \frac{Ds}{s^2 + \omega^2}$$

$$X(s) = \underbrace{\frac{Ds}{(s^2 + \omega^2)(Ms^2 + K)}}_{\substack{\text{zero-state response} \\ \text{component}}} + \underbrace{\frac{Msx(0^-) + Mx'(0^-)}{Ms^2 + K}}_{\substack{\text{zero-input response} \\ \text{component}}}$$

For zero initial conditions,

$$X(s) = \frac{(D/M)s}{(s^2 + \omega^2)(s^2 + K/M)} = \frac{K_1 s + K_2}{s^2 + \omega^2} + \frac{K_3 s + K_4}{s^2 + K/M}$$

$$= \frac{(K_1 + K_3)s^3 + (K_2 + K_4)s^2 + (K_1 K/M + K_3 \omega^2)s + (K_2 K/M + K_4 \omega^2)}{(s^2 + \omega^2)(s^2 + K/M)}$$

giving

$$
\begin{cases}
K_1 & + & K_3 & & = 0 \\
& K_2 & & + & K_4 = 0 \\
\dfrac{K}{M} K_1 & + & \omega^2 K_3 & & = \dfrac{D}{M} \\
& \dfrac{K}{M} K_2 & & + & \omega^2 K_4 = 0
\end{cases}
$$

These have solution

$$K_2 = K_4 = 0; \qquad K_1 = -K_3 = \frac{D/M}{(K/M) - \omega^2}$$

giving

$$X(s) = \frac{\dfrac{D/M}{(K/M) - \omega^2} s}{s^2 + \omega^2} + \frac{\dfrac{-D/M}{(K/M) - \omega^2} s}{s^2 + (K/M)}$$

$$x(t) = \frac{D/M}{(K/M) - \omega^2} \cos \omega t - \frac{D/M}{(K/M) - \omega^2} \cos \sqrt{\frac{K}{M}} t, \qquad t \geq 0$$

The following data for an aircraft wing where the frequency ω of the turbine and the natural frequency $\sqrt{K/M}$ are close to one another:

$M = 1000$ lb of mass

$D = 3000$ lb of force

$K = 400,800.4$ lb of force/ft of deflection

$\omega = 20$ rad/sec

Deflections exceeding 5 ft are considered to be beyond the wing's structural limit. With these numbers,

$$x(t) = 3.748(\cos 20t - \cos 20.02t) \qquad t \geq 0$$

Maximum deflection will obviously occur at times, assuming they exist, when

$$\cos 20t = 1 \qquad \text{and} \qquad \cos 20.02t = -1$$

These are times for which

$$20t = n2\pi \qquad n \text{ an integer}$$

and

$$20.02t = \pi + m2\pi \qquad m \text{ an integer}$$

one such solution being $n = m = 500$, $t = 50\pi$. The amount of maximum deflection will be

$$x_{max} = 3.748(2) = 7.496 \text{ ft}$$

To reduce the deflection amplitude, the wing can be stiffened to increase K to obtain a natural frequency $\sqrt{K/M}$ much larger than ω, giving a much smaller value for the amplitudes

$$K_1 = -K_3 = \frac{D/M}{(K/M) - \omega^2}$$

Stiffening the wing slightly to give

$$K = 441,000 \text{ lb of force/ft}$$

of deflection results instead in

$$K_1 = -K_3 = \frac{D/M}{(K/M) - \omega^2} = 0.073$$

$$x(t) = 0.073(\cos 20t - \cos 21t)$$

for which

$$x_{max} = 0.073(2) = 0.146 \text{ ft}$$

This is the solution that was used to prevent wing fatigue subsequent to the aircraft failures in the 1960s.

2.9 SUMMARY

A first-order system has transfer function

$$T(s) = \frac{b_m s^m + \cdots + b_1 s + b_0}{s + a_0}$$

It is stable for any $a_0 > 0$. First-order system response to an input $r(t)$ is of the form

$$Y(s) = T(s)R(s) + \frac{\begin{pmatrix} \text{constant numerator} \\ \text{dependent on initial} \\ \text{conditions} \end{pmatrix}}{s + a_0}$$

The natural component of the response is of the form

$$y_{\text{natural}}(t) = Ke^{-a_0 t}$$

and has time constant

$$\tau = \frac{1}{a_0}$$

A second-order system has transfer function

$$T(s) = \frac{b_m s^m + \cdots + b_1 s + b_0}{s^2 + a_1 s + a_0}$$

It is stable if and only if both a_1 and a_0 are positive. Second-order system response is of the form

$$Y(s) = T(s)R(s) + \frac{\begin{pmatrix} \text{first-degree numerator} \\ \text{polynomial dependent} \\ \text{on initial conditions} \end{pmatrix}}{s^2 + a_1 s + a_0}$$

If the characteristic polynomial

$$s^2 + a_1 s + a_0 = (s - s_1)(s - s_2)$$

has roots s_1 and s_2 that are real and distinct, the natural response component is *overdamped*:

$$y_{\text{natural}}(t) = K_1 e^{s_1 t} + K_2 e^{s_2 t}$$

If the characteristic roots are equal, $s_1 = s_2$, the natural response is *critically damped*:

$$y_{\text{natural}}(t) = K_1 e^{s_1 t} + K_2 t e^{s_1 t}$$

If the characteristic roots are complex numbers,

$$s_1, s_2 = \sigma \pm j\omega$$

the natural response component is *underdamped*:

$$y_{\text{natural}}(t) = K_1 e^{s_1 t} + K_2 e^{s_2 t} = A e^{-\sigma t} \cos(\omega t + \theta)$$

Underdamped second-order system natural response is also described by the undamped natural frequency ω_n and damping ratio ζ. In terms of these, the characteristic polynomial is

$$s^2 + 2\zeta\omega_n s + \omega_n^2$$

and

$$\sigma = \zeta\omega_n$$

$$\omega = \omega_n\sqrt{1 - \zeta^2}$$

For ζ between 0 and 1, the characteristic roots lie on a circle of radius ω_n about the origin in the left half of the complex plane. The *damping angle* ϕ is related to the damping ratio by

$$\zeta = \cos\phi$$

Higher-order systems have natural response that consists of a sum of terms, one term for each characteristic root. For each real, distinct characteristic root, there is a real exponential natural response term. Repeated characteristic roots give response terms involving powers of time times an exponential. For each pair of complex conjugate roots, there are a pair of complex exponential terms which combine into the exponential-times-sinusoid form.

One set of measures of a stable system's performance is its *rise time, overshoot,* and *setting time* to a step input. These were plotted for specific second-order systems with

$$T(s) = \frac{\omega_n^2}{s^2 + 2\zeta\omega_n s + \omega_n^2}$$

in Fig. 2.10. In Chap. 3, we consider tracking system performance in more detail.

A system is stable if and only if all of its characteristic roots are in the left half of the complex plane (LHP). First- and second-order polynomials have all roots in the LHP if and only if all polynomial coefficients have the same algebraic sign. Occasionally, the presence of RHP and imaginary roots of a high-order polynomial can be detected by an examination of the algebraic signs of the polynomial coefficients, as is summarized in Table 2.1.

The Routh-Hurwitz test provides a convenient, definitive method of ascertaining system stability in general. The number of algebraic sign changes in the left column of the Routh-Hurwitz array equals the number of RHP roots of the polynomial tested. A left-column zero situation is when there is a zero entry at the left of a row but there is at least one nonzero entry in the rest of the row. To complete an array with a left-column zero, replace the zero with ε and take the limit of the left-column entries as ε approaches zero. Alternatively, additional known root terms can be multiplied into the original polynomial to give a polynomial of higher order but without the left-column zero.

A premature termination of a Routh-Hurwitz array is when any row, through the s^0 row, consists solely of zeros. A premature termination indicates that the tested polynomial has an even or odd polynomial divisor. The coefficients of the divisor polynomial are given by the entries in the row above the row of zeros. To complete a prematurely terminated array, replace the row of zeros with the coefficients of the derivative with respect to s of the divisor polynomial and complete the array as usual. The completed array includes the test of the divisor polynomial which must have equal numbers of RHP and LHP roots. Any remaining roots of an even or odd divisor polynomial must be on the imaginary axis.

An adjustable polynomial has coefficients dependent upon a parameter K. The Routh-Hurwitz test can be performed in terms of K to determine the range(s) of K for which all polynomial roots are in the LHP. The relative stability of a stable system is the distance from the imaginary axis to the nearest characteristic root (pole). To shift the imaginary axis on the complex plane a units to the left, replace the complex variable by $\sigma - a$. The Routh-Hurwitz test can be used to determine the number of roots of a polynomial which are to the right of $s = -a$ by replacing each s in the polynomial by $\sigma - a$ and testing the shifted polynomial.

REFERENCES

System Response
DiStefano, J. J. III, Stubberud, A. R., and Williams, I. J., *Feedback and Control Systems (Schaum's Outline)*. New York: McGraw-Hill, 1967.
Lewis, L. J., Reynolds, D. K., Bergseth, F. R., and Alexandro, F. J., Jr., *Linear System Analysis*. New York: McGraw-Hill, 1969.

Routh-Hurwitz Testing
Clark, R. N., *Introduction to Automatic Control Systems*. New York: Wiley, 1962.
Fuller, A. T., ed., *Stability of Motion*. London: Taylor and Francis, and New York: Halsted Press, 1975.
Guillemin, E. A., *The Mathematics of Circuit Analysis*. New York: Wiley, 1949.
Hurwitz, A., "On the Conditions Under Which an Equation Has Only Roots with Negative Real Parts." *Math. Analen* 46 (1895).
Routh, E. J., *Stability of a Given State of Motion*. London: Macmillan, 1877.
_____. *Dynamics of a System of Rigid Bodies*. New York: Macmillan, 1892.
Savant, C. J., Jr., *Basic Feedback Control System Design*. New York: McGraw-Hill, 1958.
Truxal, J. C., *Control System Synthesis*. New York: McGraw-Hill, 1955.

Polynomial Factoring
Hostetter, G. H., "Using the Routh-Hurwitz Test to Determine the Numbers and Multiplicities of Real Roots of a Polynomial." *IEEE Trans. Circ. Syst.*, August 22, 1975.
_____, "On Polynomial Factorization Using Routh-Hurwitz Testing." *IEEE Trans. Ed.*, vol. 26 (1983).
Mastascusa, E. J., Rave, W. C., and Turner, B. M., "Polynomial Factorization Using the Routh Criterion." *Proc. IEEE* 59 (1971).

Insulin Delivery and Biomedical Applications

Albisser, A. M., "Review of Artificial Pancreas Research." *Arch. Int. Med.* 137 (May 1977): 639–649.

Blackshear, P. J., "Implantable Drug-Delivery Systems." *Sci. Amer.*, December 1979, pp. 66–73.

Horgan, J., "Medical Electronics." *IEEE Spectrum* (January 1985): 89–94.

Spencer, W. J., "For Diabetics: An Electronic Pancreas." *IEEE Spec.* 15, no. 6 (June 1978): 38–42.

Aircraft Dynamics

Etkin, B. B., *Dynamics of Atmospheric Flight.* New York: Wiley, 1972.

Kolk, R. W., *Modern Flight Dynamics.* Englewood Cliffs, N.J.: Prentice-Hall, 1961.

PROBLEMS

1. Find and sketch the response $y(t)$, $t \geq 0$, of the first-order systems with the following transfer functions, inputs $r(t)$, and initial conditions. Indicate the natural part of the response and find its time constant.

(a) $T(s) = \dfrac{8}{s + 8}$

$r(t) = \delta(t)$, an impulse

$y(0^-) = -4$

(b) $T(s) = \dfrac{-4s}{3s + 10}$

$r(t) = u(t)$, a step

$y(0^-) = 5$

(c) $T(s) = \dfrac{4s + 1}{2s + 8}$

$r(t) = u(t)$, a step

$y(0^-) = 3$

(d) $T(s) = \dfrac{3}{s + 10}$

$r(t) = 2 \cos 4t$

$y(0^-) = 8$

(e) $T(s) = \dfrac{4}{2s + 1}$

$r(t) = 6e^{-(1/2)t}$

$y(0^-) = 8$

2. When the heater for an industrial controlled-temperature chamber was turned off, the measured temperature decayed as shown in the Table. What is the chamber's time constant? What is the temperature outside the oven?

OVEN TEMPERATURE DECAY DATA

Time	Temperature (°C)
14:23:10	120
14:31:00	108
14:39:30	92
14:48:35	80.5
15:05:00	63
15:34:00	40.5

Ans $\tau = 27$ sec; temp. outside $= 40°$

3. Find and sketch the response $y(t)$, $t \geq 0$, of the second-order systems with the following transfer functions, inputs $r(t)$, and zero initial conditions. Indicate the natural part of the response.

(a) $T(s) = \dfrac{8}{s^2 + 4s + 10}$

$r(t) = \delta(t)$, an impulse

(b) $T(s) = \dfrac{s - 2}{s^2 + 6s + 9}$

$r(t) = u(t)$, a step

(c) $T(s) = \dfrac{s + 1}{s^2 + 4s + 1}$

$r(t) = u(t)$, a step

(d) $T(s) = \dfrac{10}{s^2 + s + 10}$

$r(t) = 2t$

(e) $T(s) = \dfrac{6s}{s^2 + 4s + 13}$

$r(t) = 4 \sin 2t$

4. For the underdamped systems with the following transfer functions, find the undamped natural frequency ω_n, the damping ratio ζ, the exponential constant $a = \zeta\omega_n$, and the oscillation frequency ω:

(a) $T(s) = \dfrac{-s^2}{s^2 + 4s + 25}$

(b) $T(s) = \dfrac{s^2 + 6s + 10}{4s^2 + 2s + 50}$

(c) $T(s) = \dfrac{20}{3s^2 + s + 10}$

(d) $T(s) = \dfrac{s - 2}{s^2 + 6}$

(e) $T(s) = \dfrac{1}{2s^2 + 6s + 20}$

Ans (a) 5, 0.4, 2, 4.58   (c) 1.82, 0.09, 0.16, 1.8

5. Find a controller transmittance $G(s)$ such that the overall system of Fig. P2.5 is second-order and critically damped. A solution to this problem is not unique.

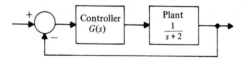

Figure P2.5

6. For the system of Fig. P2.6, find the constant value of gain, K, for which the damping ratio of the overall system is 0.7. For this value of K, what is the system's undamped natural frequency?

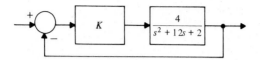

Figure P2.6

Ans 17.8, 8.57

7. For the system of Fig. P2.7, find a gain K for which the natural component of the output decays at least as rapidly as $\exp(-10t)$. A solution to this problem is not unique.

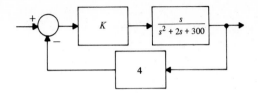

Figure P2.7

8. Draw "block" diagrams to represent the transfer function

$$T(s) = \frac{10(s-1)(s+4)}{(s+2)(s+3)(s+6)}$$

as follows:

(a) In terms of cascaded (end-to-end connection) first-order blocks.

(b) In terms of tandem (same input, outputs summed) first-order blocks.

(c) As a nontrivial combination of cascaded and tandem first-order blocks.

9. Identify the type of natural response (overdamped, critically damped, or underdamped) associated with each of the following characteristic polynomials:

(a) $s^2 + 8s + 8$

(b) $s^2 + s + 4$

(c) $2s^2 + 9s + 3$

(d) $4s^2 + s + 10$

(e) $s^2 + 6s + 9$

   Ans (a) overdamped; (c) overdamped; (e) critically damped

10. Use the Routh-Hurwitz test to show that all roots of a cubic,

$$s^3 + a_2 s^2 + a_1 s + a_0$$

are in the LHP if and only if a_2, a_1, and a_0 are positive and

$$a_1 a_2 > a_0$$

11. Determine a range of values for K, if a range exists, for which each of the following is stable.

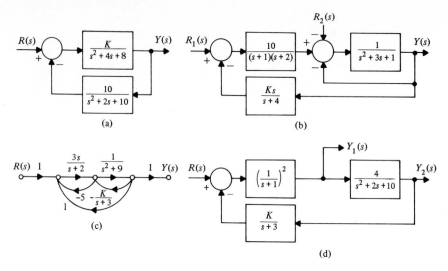

(a)

(b)

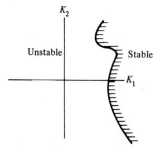

(c)

(d)

Figure P2.11

Ans (a) $-24 < K < 22.2$; (c) $-27 < K < 0.45$

12. A plot such as the example sketch of Fig. P2.12 shows the range of values of two parameters K_1 and K_2 for which a system is stable. It is called a *stability boundary diagram*. Draw such a diagram for a system with characteristic equation

$$s^2 + (2 + 0.5K_1)s + (K_1 + K_2) = 0$$

Figure P2.12

13. For what range(s) of the adjustable parameter K do the following polynomials have all roots in the LHP?

(a) $Ks^3 + 2s^2 + 7s + 10$

(b) $s^3 + Ks^2 + 7s + 10$

(c) $s^3 + 2s^2 + Ks + K$

14. Find the range(s) of the adjustable parameter $K > 0$ for which the systems of Fig. P2.14 are stable.

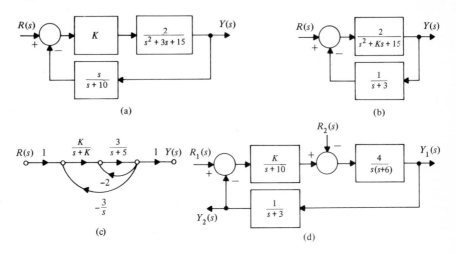

(a)

(b)

(c)

(d)

Figure P2.14

15. Determine if the system of Fig. P2.15 has a relative stability in excess of 3 units.

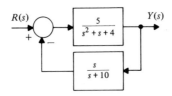

Figure P2.15

Ans No.

16. For what range(s) of the adjustable constant $K > 0$, if any, is the relative stability of the system of Fig. P2.16 greater than 2 units?

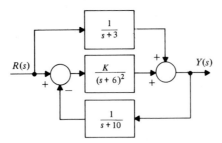

Figure P2.16

17. The required insulin delivery rate varies considerably from person to person because of differences in body chemistry. For the insulin delivery system (Sec. 2.7), suppose that a peak delivery of insulin must occur at 2 hours (7200 sec) instead of at 1 hour. Let the total insulin volume to be delivered for each meal be 0.35 cm^3/sec instead of 0.17 cm^3/sec. Find a single cycle of the required motor drive signal.

18. For the aircraft wing problem (Sec. 2.8), instead of stiffening the wing with the mass remaining constant, suppose the stiffness remains constant but the mass M of the wing varies. K_1 represents half the maximum wing deflection. Plot K_1 versus M for M between 990 and 1000 lb of mass.

19. For the original M, D, and K of the aircraft wing problem (Sec. 2.8), plot K_1 versus ω for values of ω from 19.8 to 20 rad/sec.

20. A savings account with an initial deposit of $1000 and 0.7% interest per month has a balance during the nth month of

$$b(n) = \$1000(1.007)^n, \qquad n = 0, 1, 2, \ldots$$

This discrete function of the integer values n can be modeled by a continuous exponential function

$$f(t) = Ae^{at}$$

which has the same values as $b(n)$ for integer values of t, as indicated in Fig. P2.20. For convenience, t can be measured in months. For this model, find the constants A and a.

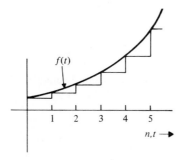

Figure P2.20

Ans $A = 1000$, $a = 0.0069$

21. Power steering for an automobile is a feedback system which can be modeled as in Fig. P2.21. For a unit step input $A(s)$, find the values of K_1 and K_2, if possible, for which the response $w(t)$ is critically damped and has a forced response of 0.4 unit. Repeat for a damping ratio of 0.707 and a forced response of 0.23 unit.

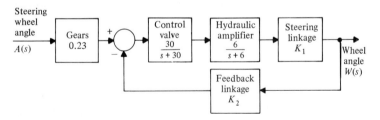

Figure P2.21

22. A feedback control system which is designed to maintain constant torque on a rotary shaft is modeled in Fig. P2.22. The torque sensor monitors strain on a section of the shaft, which is nearly proportional to the applied shaft torque.

    For a step input of desired torque, choose the constants k_1 and k_2 in the controller, if possible, so that the system response is oscillatory with a damping ratio of 0.7 and an undamped natural frequency of 6 rad/sec.

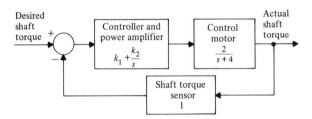

Figure P2.22

Ans $k_1 = 2.2, k_2 = 18$

23. A simple model for the roll stabilizer on a large ship is given in Fig. P2.23. Find the two system transfer functions in terms of the fin actuator gain K, and determine the range of $K > 0$ for which the system is stable. Then, for $K = 1$, use the Routh-Hurwitz test to determine the relative stability of the system to within 0.5 unit.

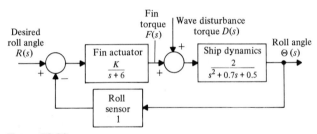

Figure P2.23

24. When paper is rolled in a paper mill, it is extremely important to maintain a specific tension as the paper is wound. A model for the control of this process is given in Fig. P2.24 in terms of various constants. On the basis of your imagination of the behavior of the components involved, choose nonzero values for K_1, K_2, a_1, a_2, a_3, and a_4, and determine if your system model is stable.

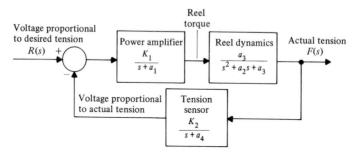

Figure P2.24

25. A motor shaft velocity control system model is given in Fig. P2.25. For what range of $K > 0$ is this system's relative stability greater than 50 units?

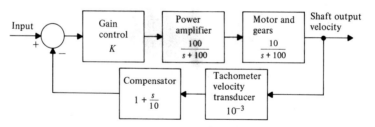

Figure P2.25

Ans   No $K > 0$ will work.

26. A motor shaft position control system is modeled in Fig. P2.26. Find the relative stability of the system as a function of the positive gain K.

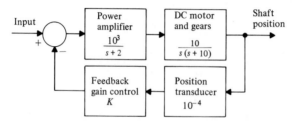

Figure P2.26

27. The following is the forward path transmittance for a unity feedback system. For what K is the closed-loop system marginally stable? At that K, where is the imaginary axis root?

$$G(s) = \frac{K(s + 8)}{s(s + 1)(s^2 + 2s + 1)}$$

28. The block diagram for a radar antenna tracking system is shown in Fig. P2.28. K_p is 3.18. Set A so the damping ratio of the complex conjugate closed-loop roots is 0.1.

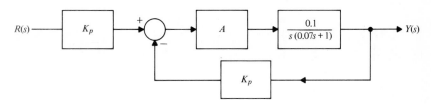

Figure P2.28

Ans $A = 1122$

Chapter
3

Tracking System Performance

3.1 PREVIEW

Two aspects of performance are often considered when a control system is designed: the transient performance and the steady state performance. For example, consider an automobile that is being driven along some level, smooth highway. The driver pushes down on the accelerator pedal a few more inches and holds the pedal stationary. The automobile then slowly gathers speed. The time history of speed would be the transient performance. A high-performance car would be designed to be far more responsive than a car driven for transportation to work. When the automobile speedometer reads a constant velocity, steady state has been achieved. The final (steady state) velocity depends on how aerodynamic the car is, road and wind conditions, and the efficiency of the engine, to name just a few factors.

As another example, consider a radar surveillance operator trying to detect enemy aircraft. The radar dish should quickly and smoothly move to sweep a suspicious portion of the sky (the transient response) and eventually achieve a desired sweep rate (the steady state result). Sometimes the steady state result is a constant (the automobile velocity) and sometimes a signal that increases linearly with time (the angular position of the radar dish).

In this chapter we analyze the tendency of a closed-loop system to follow a desired command. Chapters 1 and 2 were dedicated to defining the differential equations, the transfer function, and the stability of a system. Stability is defined in terms of the natural response, so stability is a property of the transient aspect of performance.

In Chap. 3 emphasis shifts to steady state performance, in particular to the tendency of the system to follow a desired command. Special attention is focused on commands that are powers of time, since most commands can be approximated by the sum of a step, ramp, etc. As system parameters change, the system transfer function may also change. The study of that cause-effect relationship, sensitivity, is also covered. The chapter ends with two design examples. Accurate velocity and positioning control is designed for an electric rail transportation system. Another example considers the aquisition characteristics of a phase-locked loop for a citizen's band radio receiver.

3.2 ANALYZING TRACKING SYSTEMS

Importance of Tracking Systems

A control system is often part of a larger device containing a guidance system that creates the commands that the control system acts upon as in Fig. 3.1a. The driver of an automobile is actually a sort of guidance system. The driver monitors environmental conditions and the destination. The driver issues commands to the automobile by depressing the pedals and by using the steering wheel. The driver expects prompt, smooth activation of the ever-changing commands as conditions change. A missile has an electronic guidance system to monitor the target motion and to compute corrective commands for the missile so as to intercept the target. When Armstrong and Aldrin landed on the moon, a guidance system caused the spacecraft to automatically fly near enough to the final destination for the crew to take over the terminal phases of the descent.

From the viewpoint of the guidance system the control system should quickly respond to the commands. From the viewpoint of what should be a very responsive control system, the guidance commands should appear to be fairly slowly varying functions of time. In fact, a command like that in Fig. 3.1b can be approximated (as far as the control system is concerned) by the sum of steps, ramps, etc.

In general, the input $r(t)$ can be written as a power series in terms of powers of t. For example

$$r(t) = r(a) + \frac{dr}{dt}\bigg|_{t=a}(t-a) + \frac{(d^2r/dt^2)|_{t=a}}{2!}(t-a)^2 + \frac{(d^3r/dt^3)|_{t=a}}{3!}(t-a)^3 + \cdots$$

$$= A_0 + A_1 t + A_2 t^2 + A_3 t^3 + \cdots$$

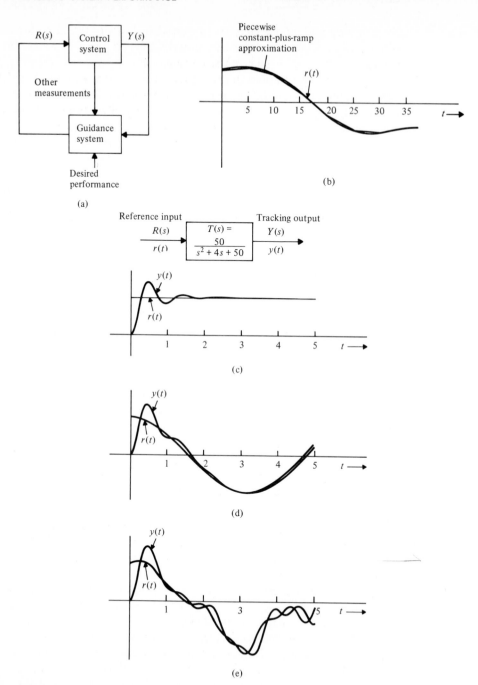

Figure 3.1   Performance of a tracking system.   **(a)** Guidance and control.
(b) Approximating a function with powers of time.   **(c)** Unit step reference input.
(d) Arbitrary slowly varying reference input.   **(e)** Arbitrary more quickly varying reference
input.

The viewpoint of this chapter is to examine how a control system responds to commands, in particular those that are powers of t, because more complicated commands are expressible in terms of powers of t.

A tracking system is a control system that creates an output which tracks (follows) the input to some level of tolerance. Figures 3.1c–e show how a typical tracking system responds to several reference inputs. The system shown is the elevation control system for a shipboard satellite dish antenna. The overall transfer function model is:

$$T(s) = \frac{50}{s^2 + 4s + 50}$$

The unit step response of this tracking system is shown in Fig. 3.1c. It has Laplace transform

$$Y(s) = T(s)\left(\frac{1}{s}\right) = \frac{50}{s(s^2 + 4s + 50)}$$

$$= \underbrace{\frac{1}{s}}_{\substack{\text{forced} \\ \text{component}}} + \underbrace{\frac{-s - 4}{s^2 + 4s + 50}}_{\substack{\text{natural} \\ \text{component}}}$$

and is, as a function of time after $t = 0$,

$$y(t) = \underbrace{1}_{\substack{\text{forced} \\ \text{component}}} + \underbrace{1.04e^{-2t}\cos(6.78 + 163.6°)}_{\substack{\text{natural} \\ \text{component}}}$$

The natural component of the response dies out, with a time constant of $\frac{1}{2}$ second, leaving a forced response component which equals the constant unit reference input.

The tracking of a relatively slowly varying but otherwise arbitrary reference input is shown in Fig. 3.1d. The output does track the reference input pretty well. When the reference input for this system varies more quickly, Fig. 3.1e, the performance is poorer.

A good tracking system has a natural response that decays rapidly and without excessive fluctuations, leaving the forced component of the response. The natural response component depends on the system initial conditions but is otherwise not affected by the specific reference input. The forced response component, which is the tracking system response after the natural component has decayed, should adequately track reference inputs of the class to be encountered. For example, in a steel rolling mill control system where step changes in the desired steel thickness are to be made, fast, smooth decay of the natural response and accuracy of the forced response for constant inputs is of primary importance. In a high-perfor-mance terrain-following aircraft altitude control, not only must the natural response component decay quickly, but the forced response should accurately track any reference input within the plane's safe physical limits.

The analysis and design of tracking systems can be separated into two parts:

1. Locating the characteristic roots (poles) of the transfer function. These determine the character of the system's natural response component. Normally, it is desired that the natural response decays rapidly, and that any oscillatory terms be well damped.
2. Tracking of the reference input by the forced response of the system for those kinds of inputs to be encountered in practice.

In the remainder of this section we consider the natural response component and in subsequent sections, we analyze tracking system forced response. Additionally, the designer is concerned with

3. Performance when the plant model is inaccurate.
4. Tracking system response due to unwanted, inaccessible disturbance inputs.

Natural Response, Relative Stability and Damping

The relative stability of a system is the distance into the left half of the complex plane from the imaginary axis to the nearest characteristic root or roots. For example, the system with the characteristic roots (or poles) shown in Fig. 3.2a has a relative stability of 2 units. The slowest-decaying term in this system's natural response component decays as $\exp(-2t)$. For the natural response to decay at least as quickly as $\exp(-\sigma t)$, a system must have a relative stability of at least σ units. That is, its characteristic roots (poles) must be on or to the left of the line $\text{Re}(s) = -\sigma$ as indicated in Fig. 3.2b.

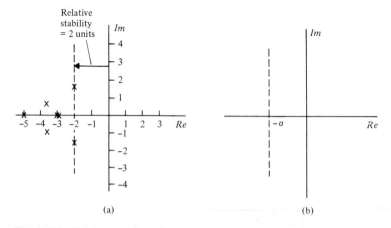

(a)   (b)

Figure 3.2 Relative stability. **(a)** An example. **(b)** Region of greater relative stability than σ.

A pair of complex conjugate characteristic roots

$$s_1, s_2 = -a \pm jb$$

gives rise to a damped oscillatory term in the natural response component of the form

$$y_i(t) = Ae^{-at} \cos(bt + \theta)$$

where the constants A and θ depend on the initial conditions. The damping ratio of such a term is (Sec. 2.3)

$$\zeta = \cos \phi$$

where ϕ is the damping angle, as shown in Fig. 3.3a. A low damping ratio, which occurs for ϕ near 90°, is undesirable for most tracking systems because it means the output will likely exhibit large fluctuations as its natural response is decaying. It is

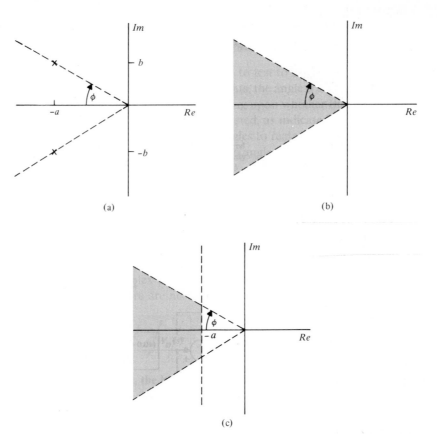

Figure 3.3 Damping ratio for response terms corresponding to complex conjugate transfer function characteristic root (pole) pairs. **(a)** Damping angle. **(b)** Region of more than a certain amount of damping. **(c)** Region of greater relative stability than σ and more than a certain amount of damping.

therefore usually important that the characteristic roots (poles) of a tracking system be within a region of more than a certain amount of damping as shown in Fig. 3.3b. Often, the choice of maximum damping angle $\phi = 45°$ is made, corresponding to a minimum damping ratio of 0.707.

When the two requirements of relative stability and damping ratio are combined, the result is Fig. 3.3c. The characteristic roots (poles) of a tracking system should be to the left of the line $\text{Re}(s) = -\sigma$, where σ is the minimum desired relative stability. The natural component of the system's response then decays at least as quickly as $\exp(-\sigma t)$, leaving the forced response component. The characteristic roots (poles) should be within a maximum damping angle ϕ so that oscillatory terms in the natural component of the response are damped sufficiently quickly.

DRILL PROBLEMS

D3.1. Determine the relative stability of systems with the following transfer functions. Also determine the damping ratios of any complex conjugate pairs of characteristic roots.

(a) $T(s) = \dfrac{5s^2 - 1}{(s + 3)^2(s + 4)(s + 6)}$

Ans 3

(b) $T(s) = \dfrac{4s^2 - 3s + 1}{(s + 2)(s^2 + 3s + 10)^2}$

Ans 1.5, 0.47

(c) $T(s) = \dfrac{24}{(s^2 + s + 4)(s^2 + 2s + 6)}$

Ans $\frac{1}{2}$, 0.25, 0.41

(d) $T(s) = \dfrac{(s + 4)^2(s - 2)}{(s + 2)(s^2 + 5s + 20)(s + 3)}$

Ans 2, 0.56

3.3 FORCED RESPONSE

Steady State Error

The error between the output and input of a system is given by

$$E(s) = R(s) - Y(s) = R(s) - T(s)R(s) = [1 - T(s)]R(s) = T_E(s)R(s)$$

The error between the input and the output of the system has the transmittance

$$T_E(s) = 1 - T(s) = \left.\frac{E(s)}{R(s)}\right|_{\substack{\text{zero initial} \\ \text{conditions}}}$$

which relates the error and input transforms. The poles of $T_E(s)$ are the same as the poles of $T(s)$. Like all system signals, the error signal is generally composed of natural and forced parts. The natural part is composed of terms, one for each pole of $T_E(s)$, or $T(s)$, with amplitudes dependent on initial conditions. For a system with relative stability σ, this natural response decays to zero at least as fast as $\exp(-\sigma t)$.

The forced part of the error signal is

$$e_{\text{forced}}(t) = r(t) - y_{\text{forced}}(t)$$

If the system output tracks a reference input $r(t)$ well, then $e_{\text{forced}}(t)$ will be small. For perfect tracking,

$$y_{\text{forced}}(t) = r(t)$$

and

$$e_{\text{forced}}(t) = 0$$

Initial and Final Values

The *initial value* of a function of time $y(t)$ is related to the function's Laplace transform by

$$y(0) = \lim_{s \to \infty} [sY(s)]$$

For example, the initial value of the function with Laplace transform

$$Y(s) = \frac{-4s^4 + 3s^3 + s^2 - s + 1}{3s^5 - 2s^4 + s^3 - s + 10}$$

is

$$y(0) = \lim_{s \to \infty} [sY(s)] = -\tfrac{4}{3}$$

In general, the values $y(0^-)$, $y(0)$, and $y(0^+)$ for a Laplace-transformable function can differ. If there is an impulse in $y(t)$ at $t = 0$, then $y(0)$ will be infinite.

For functions with rational transforms, the initial-value theorem is especially easy to visualize. The partial fraction expansion of a rational function can have the representative types of terms in Table 3.1. For each, and for other terms as well, multiplying by s and taking the limit as s goes to infinity gives the correct contribution to $y(0)$.

The *final value* (or *steady state value*) of a function $y(t)$ is

$$\lim_{t \to \infty} y(t)$$

Table 3.1   APPLICATION OF THE INITIAL-VALUE AND FINAL-VALUE THEOREMS TO REPRESENTATIVE LAPLACE TRANSFORM TERMS

Transform $Y(s)$	Time function $y(t), t \geq 0$	$\lim_{s \to \infty} sY(s)$	$\lim_{s \to 0} sY(s)$
A	$A\delta(t)$	∞	0
$\dfrac{A}{s}$	A	A	A
$\dfrac{A}{s^2}$	At	0	∞
$\dfrac{A}{s+a}$	Ae^{-at}	A	0
$\dfrac{Ab}{s^2 + b^2}$	$A \sin bt$	0	0
$\dfrac{Ab}{(s+a)^2 + b^2}$	$Ae^{-at} \sin bt$	0	0
$\dfrac{As}{s^2 + b^2}$	$A \cos bt$	A	0
$\dfrac{A(s+a)}{(s+a)^2 + b^2}$	$Ae^{-at} \cos bt$	A	0

If it exists and is finite, the final value is related to the function's Laplace transform by what is called the final-value theorem:

$$\lim_{t \to \infty} y(t) = \lim_{s \to 0} [sY(s)]$$

For a final value of $y(t)$ to exist, all denominator roots of $Y(s)$ must be in the LHP except possibly for one root at $s = 0$. It is the root at $s = 0$, corresponding to a constant term in $y(t)$ after $t = 0$, which then contributes a nonzero final value. Application of this result is simply a calculation of the residue of a K/s term in the partial fraction expansion of $Y(s)$. The function

$$Y(s) = \frac{-4s^3 - s^2 + 7s + 3}{s^3 + 9s^2 + 2s}$$

for example, has all denominator roots in the LHP except for the single $s = 0$ root. The final value of $y(t)$ is

$$\lim_{t \to \infty} y(t) = \lim_{s \to 0} [sY(s)] = \frac{3}{2}$$

Unfortunately, the final-value theorem gives answers even when a final value does not exist, as is demonstrated by the entries in Table 3.1. The transform

$$Y(s) = \frac{-3s^2 + 4}{s^4 + 5s^3 + 8s^2 + 4s} = \frac{-3s^2 + 4}{s(s+1)(s+2)^2}$$

$$= \frac{K_1}{s} + \frac{K_2}{s+1} + \frac{K_3}{s+2} + \frac{K_4}{(s+2)^2}$$

represents a time function of the form

$$y(t) = [K_1 + K_2 e^{-t} + K_3 e^{-2t} + K_4 t e^{-2t}]u(t)$$

which has the final value K_1. The final value is correctly given by

$$\lim_{t\to\infty} y(t) = \lim_{s\to 0} [sY(s)] = \frac{4}{4} = 1 = K_1$$

The Laplace transform

$$Y(s) = \frac{s - 6}{s^4 + s^3 + 3s^2 - 5s} = \frac{s - 6}{s(s-1)[(s+1)^2 + 4]}$$

$$= \frac{K_1}{s} + \frac{K_2}{s-1} + \frac{K_3 s + K_4}{(s+1)^2 + (2)^2}$$

is of a time function of the form

$$y(t) = [K_1 + K_2 e^t + A e^{-t} \cos(2t + \theta)]u(t)$$

which does not have a finite final value because of the e^t term. However, application of the final value theorem gives

$$\lim_{s\to 0} [sY(s)] = \tfrac{6}{5}$$

which is incorrect.

For the Laplace transform

$$Y(s) = \frac{6s + 7}{s(s^3 + s^2 + 2s + 3)}$$

the limit is

$$\lim_{s\to 0} [sY(s)] = \tfrac{7}{3}$$

However, the final value is infinite since $Y(s)$ has denominator roots in the RHP due to the third order polynomial factor.

$$
\begin{array}{c|cc}
s^3 & 1 & 2 \\
s^2 & 1 & 3 \\
s^1 & -1 & \\
s^0 & 3 &
\end{array}
$$

Steady State Errors to Power-of-Time Inputs

The standard ith degree power-of-time inputs have Laplace transforms

$$R_i(s) = \frac{1}{s^{i+1}}$$

The corresponding time functions are the unit step,

$$R_0(s) = \frac{1}{s} \qquad r(t) = u(t)$$

the unit ramp,

$$R_1(s) = \frac{1}{s^2} \qquad r(t) = tu(t)$$

one-half the unit parabola,

$$R_2(s) = \frac{1}{s^3} \qquad r(t) = \tfrac{1}{2}t^2 u(t)$$

and so on. If the input to a tracking system is a power-of-time input, the error signal is given by

$$E_i(s) = T_E(s)R_i(s) = \frac{1}{s^{i+1}} T_E(s)$$

As a function of time, the error will consist of natural response terms, one for each pole of $T_E(s)$, and a forced response consisting of power-of-time terms, through the ith degree term. For example, a system with transfer function

$$T(s) = \frac{6}{s^2 + 3s + 2}$$

has error transmittance

$$T_E(s) = 1 - T(s) = \frac{s^2 + 3s - 4}{s^2 + 3s + 2}$$

Its error to the standard ramp input

$$R(s) = \frac{1}{s^2} \qquad r(t) = tu(t)$$

is

$$E(s) = T_E(s)R(s) = \frac{s^2 + 3s - 4}{s^2(s^2 + 3s + 2)}$$

$$= \underbrace{\frac{\tfrac{9}{2}}{s} + \frac{-2}{s^2}}_{\substack{\text{forced response} \\ \text{component}}} + \underbrace{\frac{-6}{s+1} + \frac{\tfrac{3}{2}}{s+2}}_{\substack{\text{natural response} \\ \text{component}}}$$

or

$$e(t) = \underbrace{\tfrac{9}{2} - 2t}_{\substack{\text{forced response} \\ \text{component}}} \quad \underbrace{-6e^{-t} + \tfrac{3}{2}e^{-2t}}_{\substack{\text{natural response} \\ \text{component}}}$$

after $t = 0$.

For a stable system driven by a power-of-time input, the forced component of the error can do only one of three things:

1. The forced error can be zero, meaning that after the natural response has died out, the error is zero and the tracking system output equals the power-of-time reference input.
2. The forced error can be a constant so that after the natural response decays to zero, the error is constant and the tracking system output and the reference input differ by a constant.
3. The forced error can involve a nonzero term proportional to t or a higher power of t, in which case the error grows without bound.

These three situations are easily distinguished, without calculating $e(t)$, by applying the final value theorem to $E(s)$. For a stable system, if the final value of $e(t)$ is zero, then the situation must be that of (1), ideal tracking of the input:

$$y_{\text{forced}}(t) = r(t)$$

The final-value theorem does not indicate how quickly the final value is achieved but it does indicate what the final value will be. If the final value of $e(t)$ is a finite, nonzero constant, the situation is that of (2), where $y_{\text{forced}}(t)$ and $r(t)$ differ by that constant. If $E(s)$ has more than a single pole at $s = 0$ (the final value theorem does not apply then), the final value of $e(t)$ is infinite, and we have situation (3).

For a system with transfer function

$$T(s) = \frac{10}{s^2 + 3s + 10}$$

the error to a unit step input is given by

$$E(s) = [1 - T(s)]R(s) = \frac{s^2 + 3s}{s(s^2 + 3s + 10)}$$

Its final value is

$$\lim_{t \to \infty} e(t) = \lim_{s \to 0} sE(s) = \lim_{s \to 0} \frac{s^2 + 3s}{s^2 + 3s + 10} = 0$$

so after this system's natural response decays to zero, it tracks any constant input with zero error. The error of this system to a unit ramp input is given by

$$E(s) = [1 - T(s)]R(s) = \frac{s^2 + 3s}{s^2(s^2 + 3s + 10)}$$

and has final value

$$\lim_{t \to \infty} e(t) = \lim_{s \to 0} sE(s) = \lim_{s \to 0} \frac{s + 3}{s^2 + 3s + 10} = \frac{3}{10}$$

After the system's natural response has died out, the output tracks any constant input with an error of $\frac{3}{10}$. For a standard parabolic input

$$R(s) = \frac{1}{s^3} \qquad r(t) = \tfrac{1}{2}t^2 u(t)$$

the error is given by

$$E(s) = [1 - T(s)]R(s) = \frac{s^2 + 3s}{s^3(s^2 + 3s + 10)}$$

and

$$\lim_{t \to \infty} e(t) = \infty$$

DRILL PROBLEMS

D3.2. For each of the following Laplace-transformed signals, find the initial value $y(0)$.

(a) $Y(s) = \dfrac{4s - 1}{s^2 + 3s}$

    Ans 4

(b) $Y(s) = \dfrac{5}{s^3 + 2s^2 + 11}$

    Ans 0

(c) $Y(s) = \dfrac{6e^{-2s}}{s + 4}$

    Ans 0

(d) $Y(s) = \dfrac{30s}{s^3 + 2s^2 + 11s + 3}$

    Ans 0

D3.3. For each of the following Laplace-transformed signals, find the final value if it exists. If a final value of the signal does not exist, so state.

(a) $Y(s) = \dfrac{6s^3 - 3s^2 + 2s - 4}{s^4 + 3s^3 + 2s^2 + s + 10}$

Ans does not exist

(b) $Y(s) = \dfrac{s^2 - 2s + 10}{(s^2 + 2s + 10)(s^2 + 3s)}$

Ans $\frac{1}{3}$

(c) $Y(s) = 7e^{-3s}$

Ans 0

(d) $Y(s) = \dfrac{6s^2 + 5}{s^3 + 2s^2 + 11s}$

Ans $\frac{5}{11}$

(e) $Y(s) = \dfrac{30}{s^3 + 2s^2 + 11s + 3}$

Ans 0

D3.4. For each of the following systems, the input $R(s)$ is a unit step. Find the steady state value of the output signal $Y(s)$ if it exists.

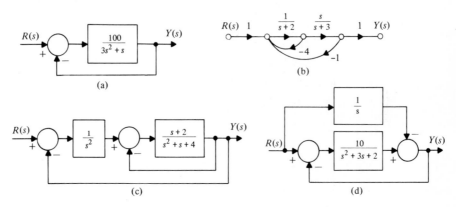

(a)

(b)

(c)

(d)

Ans (a) 1; (b) 0; (c) 1; (d) does not exist

D3.5. For systems with the following transfer functions, find steady state errors, if they exist, between output and input for unit step, ramp, and parabolic inputs:

(a) $T(s) = \dfrac{-3s^2 + 5}{s^3 + 3s^2 + 2s + 5}$

Ans $0, \frac{2}{5}, \infty$

(b) $T(s) = \dfrac{2s^2 + 2s + 5}{s^3 + 2s^2 + 2s + 5}$

Ans   infinite errors

(c) $T(s) = \dfrac{3s^2 + 2s + 10}{(s + 2)^2(s^2 + 2s + 10)}$

Ans $\frac{3}{4}$, ∞, ∞

(d) $T(s) = \dfrac{2s - 1}{2s^4 + 4s^3 + 4s^2 + 2s + 1}$

Ans   2, ∞, ∞

3.4 POWER-OF-TIME ERROR PERFORMANCE

System Type Number

Steady state error to power-of-time inputs is intimately related to the number of factors of s in the numerator of the error transmittance $T_E(s)$, the *type number* of the system. If $T_E(s)$ has no factor of s in its numerator, the type number is 0 and the steady state error to a step input

$$R(s) = \frac{A}{s}$$

is

$$\lim_{t \to \infty} e_{\text{step}}(t) = \lim_{s \to 0} s T_E(s) R(s) = \lim_{s \to 0} s T_E(s) \frac{A}{s} = A T_E(0)$$

which is finite. Figure 3.4 illustrates how a system can respond to give a constant steady state error. For no factors of s in the numerator of $T_E(s)$, higher power-of-t inputs give infinite steady state error. For a ramp input,

$$R(s) = \frac{A}{s^2}$$

the error

$$E(s) = T_E(s) \frac{A}{s^2}$$

has a repeated denominator root at $s = 0$, indicative of a ramp term in $e(t)$. The final-value theorem does not apply in this case because $e(t)$ does not approach a final value. Taking the limit

$$\lim_{s \to 0} s T_E(s) \frac{A}{s^2} = \infty$$

does give the correct answer, though.

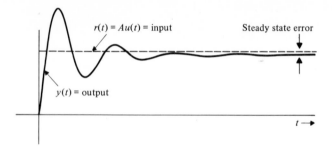

Figure 3.4 Type 0 system step response.

If the system is stable and $T_E(s)$ has one factor of s in the numerator, the steady state error to a step is zero:

$$\lim_{t \to \infty} e_{\text{step}}(t) = \lim_{s \to 0} sT_E(s)\frac{A}{s} = AT_E(0) = 0$$

To a ramp, the error is constant:

$$\lim_{t \to \infty} e_{\text{ramp}}(t) = \lim_{s \to 0} sT_E(s)\frac{A}{s^2} = A \lim_{s \to 0} \frac{T_E(s)}{s}$$

The constant steady state error results by cancellation with the s factor in the numerator of $T_E(s)$. Figure 3.5 illustrates how a system can respond to a ramp input to give a finite steady state error. For higher power-of-t inputs, the error of such a system is infinite, since

$$E(s) = T_E(s)\frac{A}{s^n}$$

has a repeated $s = 0$ denominator root.

A factor of s^2 in the numerator of $T_E(s)$ means a type 2 system. For a type 2 system, there is zero steady state error to a step, zero steady state error to a ramp,

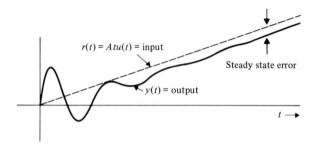

Figure 3.5 Type 1 system ramp response.

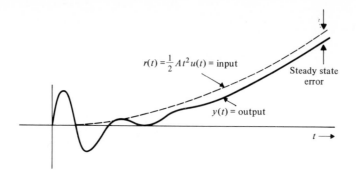

Figure 3.6 Type 2 system parabolic response.

finite error to a parabola (Fig. 3.6), and ever-increasing error for higher power-of-t inputs. The behavior of various system types for power-of-time inputs is summarized in Table 3.2.

Unity Feedback Systems

When a control system has unity feedback, the input r and output y are compared directly, as in Fig. 3.7. The error signal drives the forward transmittance $G(s)$. System type is determined by the number of $s = 0$ numerator roots of

$$T_E(s) = 1 - T(s) = 1 - \frac{G(s)}{1 + G(s)} = \frac{1}{1 + G(s)}$$

Table 3.2 STEADY STATE ERRORS

System Type (Number of $s = 0$ numerator roots of the error transmittance T_E)	Steady State Error to Step Input $r(t) = Au(t)$ $R(s) = A/s$	Steady State Error to Ramp Input $r(t) = Atu(t)$ $R(s) = A/s^2$	Steady State Error to Parabolic Input $r(t) = \frac{1}{2}At^2u(t)$ $R(s) = A/s^3$	Steady State Error to Input $r(t) = \frac{1}{6}At^3u(t)$ $R(s) = A/s^4$
0	$AT_E(0)$	∞	∞	∞
1	0	$A\lim\limits_{s\to 0}\left[\dfrac{T_E(s)}{s}\right]$	∞	∞
2	0	0	$A\lim\limits_{s\to 0}\left[\dfrac{T_E(s)}{s^2}\right]$	∞
3	0	0	0	$A\lim\limits_{s\to 0}\left[\dfrac{T_E(s)}{s^3}\right]$
$\vdots$				

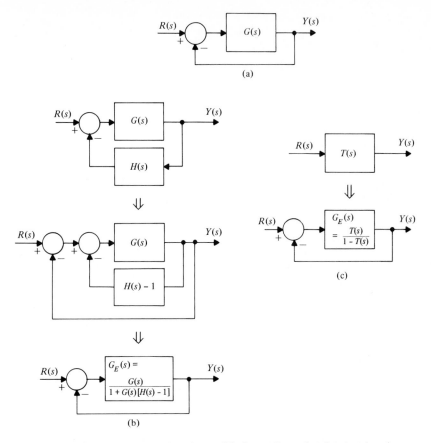

Figure 3.7 **(a)** A unity feedback system. **(b)** Converting a feedback system to an equivalent unity feedback system. **(c)** General conversion of a system to an equivalent unity feedback system.

For a unity feedback system, the system type can be determined by the number of $s = 0$ denominator roots of the transmittance $G(s)$. If $G(s)$ consists of the ratio of polynomials

$$G(s) = \frac{p(s)}{q(s)}$$

then

$$T_E(s) = \frac{1}{1 + G(s)} = \frac{q(s)}{p(s) + q(s)}$$

Hence it is seen that adding factors of s to $q(s)$ raises the system type number, so long as the resulting system is stable (and the system is unity feedback).

A system that does not have unity feedback but does have a transfer function $T(s)$ possesses an equivalent unity feedback transmittance $G_E(s)$ as in Fig. 3.7c. For

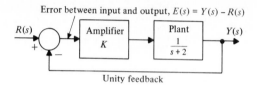

Error between input and output, $E(s) = Y(s) - R(s)$

Figure 3.8 A unity feedback system.

example, any system of the form of Fig. 3.7b has the indicated $G_E(s)$. In general, a unity feedback system [for which $G_E(s)$ is just $G(s)$], should be treated by the methods that follow. It will usually be easier to treat a non-unity feedback system using the $T_E(s)$ approach.

Consider the unity feedback system of Fig. 3.8. The difference between the input and the output is formed, amplified, and applied to the plant input in such a way as to reduce this difference. The system transfer function is

$$T(s) = \frac{Y(s)}{R(s)} = \frac{K/(s+2)}{1 + K/(s+2)} = \frac{K}{s + 2 + K}$$

For a step input,

$$R(s) = \frac{A}{s} \qquad Y(s) = T(s)R(s) = T(s)\frac{A}{s}$$

and the error between input and output is

$$E(s) = R(s) - Y(s) = \frac{A}{s}[1 - T(s)] = \frac{A}{s}\left(\frac{s+2}{s+2+K}\right)$$

The error reaches a steady state for any positive K (in fact, for any K larger than -2) and is

$$\lim_{t \to \infty} e(t) = \lim_{s \to 0} sE(s) = \lim_{s \to 0} A\left(\frac{s+2}{s+2+K}\right) = \frac{2A}{2+K}$$

which can be made arbitrarily small in this case by choosing a sufficiently large amplifier gain K.

For a ramp, parabolic, or higher power-of-t input, the error for the example system becomes infinite. For a ramp input

$$R(s) = \frac{A}{s^2}$$

we have

$$E(s) = \left(\frac{A}{s^2}\right)\left(\frac{s+2}{s+2+K}\right) = \frac{K_1}{s} + \frac{K_2}{s^2} + \frac{K_3}{s+2+K}$$

The term K_2/s^2 in the partial fraction expansion of $E(s)$ corresponds to the time function $K_2 t$, which grows without bound.

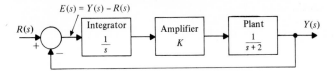

$E(s) = Y(s) - R(s)$

Figure 3.9 A tracking system with error integration.

A simple method of obtaining zero steady state error for a step input is to drive the plant with a signal proportional to the integral of the error. This places an $s = 0$ pole in the system's forward transmittance, raising the type number. The addition of an integrator to the previous example system is shown in Fig. 3.9. For this system,

$$T(s) = \frac{K/s(s+2)}{1 + K/s(s+2)} = \frac{K}{s^2 + 2s + K}$$

For a step input

$$R(s) = \frac{A}{s}$$

the new error signal is given by

$$E(s) = R(s) - Y(s) = \frac{A}{s}[1 - T(s)] = \frac{A}{s}\left(\frac{s^2 + 2s}{s^2 + 2s + K}\right)$$

The steady state error for any positive K is

$$\lim_{t \to \infty} e(t) = \lim_{s \to 0} sE(s) = 0$$

This result occurs because of the factor of s in the numerator of $E(s)$. Without the integrator, the numerator of $E(s)$ contains no such factor and a finite steady state error results for a step input.

For a ramp input

$$R(s) = \frac{A}{s^2}$$

the steady state error is

$$\lim_{t \to \infty} e(t) = \lim_{s \to 0} \frac{A(s+2)}{s^2 + 2s + K} = \frac{2A}{K}$$

Unity Feedback Error Coefficients

The *steady state error coefficients* of a unity feedback system are

$$\kappa_i = \lim_{s \to 0} s^i G(s)$$

Since the error transmittance is

$$T_E(s) = 1 - T(s) = 1 - \frac{G(s)}{1 + G(s)} = \frac{1}{1 + G(s)}$$

the error to a power-of-t input

$$R(s) = \frac{A}{s^i}$$

is

$$E(s) = T_E(s)R(s) = \frac{A}{s^i[1 + G(s)]}$$

When the limits exist and are finite,

$$\left(\begin{matrix}\text{Steady state error} \\ \text{to input } A/s^i\end{matrix}\right) = \lim_{s \to 0} sE(s) = \lim_{s \to 0} \frac{A}{s^{i-1}[1 + G(s)]}$$

For a step input, $i = 1$,

$$\left(\begin{matrix}\text{Steady state error} \\ \text{to input } A/s\end{matrix}\right) = \lim_{s \to 0} \frac{A}{1 + G(s)} = \frac{A}{1 + \kappa_0}$$

For a ramp input, $i = 2$,

$$\left(\begin{matrix}\text{Steady state error} \\ \text{to input } A/s^2\end{matrix}\right) = \lim_{s \to 0} \frac{A}{s[1 + G(s)]} = \lim_{s \to 0} \frac{A}{sG(s)} = \frac{A}{\kappa_1}$$

For higher power-of-t inputs,

$$\left(\begin{matrix}\text{Steady state error} \\ \text{to input } A/s^i\end{matrix}\right) = \lim_{s \to 0} \frac{A}{s^{i-1}[1 + G(s)]} = \lim_{s \to 0} \frac{A}{s^{i-1}G(s)}$$

$$= \frac{A}{\kappa_{i-1}} \qquad i = 2, 3, 4, \ldots$$

These relations are summarized in Table 3.3. For the previous unity feedback type 0 system of Fig. 3.8,

$$G(s) = \frac{K}{s + 2}$$

$$\kappa_0 = \lim_{s \to 0} G(s) = \frac{K}{2}$$

giving

$$\left(\begin{matrix}\text{Steady state error} \\ \text{to input } A/s\end{matrix}\right) = \frac{A}{1 + \kappa_0} = \frac{2A}{2 + K}$$

Table 3.3 STEADY STATE ERROR OF UNITY FEEDBACK SYSTEMS IN TERMS OF ERROR COEFFICIENTS

System Type	Steady State Error to Step Input $R(s) = A/s$	Steady State Error to Ramp Input $R(s) = A/s^2$	Steady State Error to Parabolic Input $R(s) = A/s^3$	Steady State Error to Input $R(s) = A/s^4$
0	$\dfrac{A}{1 + \kappa_0}$	∞	∞	∞
1	0	$\dfrac{A}{\kappa_1}$	∞	∞
2	0	0	$\dfrac{A}{\kappa_2}$	∞
3	0	0	0	$\dfrac{A}{\kappa_3}$
⋮				

as was found earlier. When an integrator is added to this system, as in Fig. 3.9, to make it type 1,

$$\kappa_0 = \lim_{s \to 0} G(s) = \lim_{s \to 0} \frac{K}{s(s + 2)} = \infty$$

and the step error is

$$\left(\begin{array}{l}\text{Steady state error} \\ \text{to input } A/s\end{array}\right) = \frac{A}{1 + \kappa_0} = 0$$

For a ramp input

$$\kappa_1 = \lim_{s \to 0} \frac{K}{s + 2} = \frac{K}{2}$$

and

$$\left(\begin{array}{l}\text{Steady state error} \\ \text{to input } A/s^2\end{array}\right) = \frac{A}{\kappa_1} = \frac{2A}{K}$$

DRILL PROBLEMS

D3.6. Find the output-input error transmittance of each of the following systems. Then determine the type number of each system and, if the response reaches steady state, steady state errors to unit step, and to unit ramp inputs.

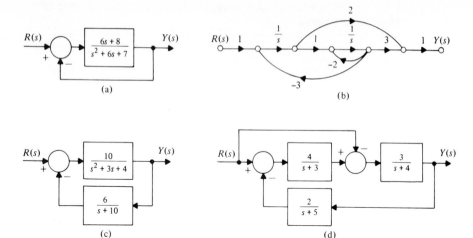

(a)

(b)

(c)

(d)

Ans $0, \frac{7}{15}, \infty; 0, -\frac{4}{3}, \infty; 1, 0, \frac{6}{25}; 0, \frac{23}{28}, \infty$

D3.7. For unity feedback systems with the following forward transmittances, determine each system type number and, if the response reaches steady state, find steady state output-input errors to unit step and to unit ramp inputs.

(a) $G(s) = \dfrac{10}{s^3 + 8s^2 + 2s}$

Ans $1, 0, \frac{1}{5}$

(b) $G(s) = \dfrac{1}{s^3 + 2s^2 + s + 3}$

Ans 0, unstable

(c) $G(s) = \dfrac{4(s + 1)}{(s + 2)(s + 3)(s^2 + s + 10)}$

Ans $0, \frac{15}{16}, \infty$

(d) $G(s) = \dfrac{2s + 1}{2s^4 + 4s^3 + 4s}$

Ans 1, unstable

D3.8. Find the error coefficients for each of the following systems. Then, if the responses reach steady state, find the steady state output-input errors to unit step and to unit ramp inputs.

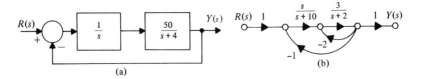

(a)

(b)

Ans $\kappa_0 = \infty$, $\kappa_1 = 12.5$, $\kappa_2 = 0, 0, 0.08$; $\kappa_0 = \frac{3}{80}$, $\kappa_1 = 0, \frac{80}{83}, \infty$

3.5 PERFORMANCE INDICES AND OPTIMAL SYSTEMS

It generally happens that a system design problem reaches the point where one or more parameters are to be selected to give the best performance. If a measure or index of performance can be expressed mathematically, the problem can be solved for the best choice of the adjustable parameters. The resulting system is termed *optimal* with respect to the selection criteria. Optimization generally proceeds as in Fig. 3.10. A given system provides a value for some given performance index. The adjustable parameters are modified so as to maximize (or minimize, depending on the application) the performance index. The term optimization pertains to both maximization and minimization. Optimization can occur before the system begins operation (off-line optimization) or the parameters may vary as the system operates (on-line optimization).

Selection of an appropriate performance index is as much a part of the design process as fabricating the final system. An optimal value for an inappropriate performance measure may result in poor performance, by other standards. Imagine that a system minimizes rather than maximizes profit, for example.

A commonly used performance index is the integral of the square of the error to a step input

$$I_S = \int_0^\infty e_{\text{step}}^2(t)\, dt$$

If the step error is expressed as a function of the adjustable parameters, the index can be minimized with respect to the parameters, yielding the optimal parameter values.

Consider the adjustable system of Fig. 3.11a, for which it is desired to choose the parameter k to give minimum integral square error to a step input. The system

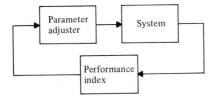

Figure 3.10 Optimal parameter selection.

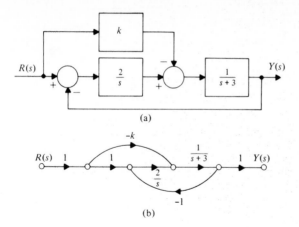

Figure 3.11 An adjustable feedback system. **(a)** System block diagram. **(b)** System signal flow graph.

transfer function, using Mason's gain rule on the system signal flow graph in Fig. 3.11b, is

$$T(s) = \frac{2/s(s+3) - k/(s+3)}{1 + 2/s(s+3)} = \frac{2 - ks}{s^2 + 3s + 2}$$

The error transmittance is

$$T_E(s) = 1 - T(s) = \frac{s^2 + (3+k)s}{s^2 + 3s + 2}$$

and the error to a step input is given by

$$E(s) = \frac{A}{s} T_E(s) = A \frac{s + 3 + k}{s^2 + 3s + 2} = A \left(\frac{k+2}{s+1} + \frac{-k-1}{s+2} \right)$$

As a function of time, the error signal is

$$e(t) = \mathcal{L}^{-1}[E(s)] = A[(k+2)e^{-t} - (k+1)e^{-2t}]u(t)$$

The square of the error is

$$e^2(t) = A^2[(k+2)^2 e^{-2t} - 2(k+1)(k+2)e^{-3t} + (k+1)^2 e^{-4t}]u(t)$$

and the integral square error, in terms of k, is

$$I_s(k) = \int_0^\infty e^2(t)dt = A^2 \left[(k^2 + 4k + 4)\left(\frac{e^{-2t}}{-2}\right) - 2(k^2 + 3k + 2)\left(\frac{e^{-3t}}{-3}\right) \right.$$

$$\left. + (k^2 + 2k + 1)\left(\frac{e^{-4t}}{-4}\right) \right]_0^\infty$$

$$= A^2[(k^2 + 4k + 4)(\tfrac{1}{2}) - 2(k^2 + 3k + 2)(\tfrac{1}{3}) + (k^2 + 2k + 1)(\tfrac{1}{4})]$$

$$= \frac{A^2}{12} [k^2 + 6k + 11]$$

As a function of k, $I_S(k)$ is a parabola, with minimum given by

$$\frac{dI_S}{dk} = \frac{A^2}{12}[2k + 6] = 0 \qquad k = -3$$

Thus the optimal system, in the sense of minimum integral square error to a step with all parameters but k fixed, is the one with this value of k.

Determination of the "best" damping ratio for a second-order system offers another example of the use of performance indices. Consider the stable second-order system with complex conjugate characteristic roots, with transfer function of the form

$$T(s) = \frac{\omega_n^2}{s^2 + 2\zeta\omega_n s + \omega_n^2} \qquad\qquad (3.1)$$

For a unit step input, the error between output and input is given by

$$E(s) = \frac{1}{s}[1 - T(s)] = \frac{1}{s}\left[\frac{s^2 + 2\zeta\omega_n s}{s^2 + 2\zeta\omega_n s + \omega_n^2}\right]$$

$$= \frac{s + 2\zeta\omega_n}{s^2 + 2\zeta\omega_n s + \omega_n^2} = \frac{s + 2\zeta\omega_n}{(s + \zeta\omega_n)^2 + (\omega_n\sqrt{1 - \zeta^2})^2}$$

$$= \frac{s + \zeta\omega_n}{(s + \zeta\omega_n)^2 + (\omega_n\sqrt{1 - \zeta^2})^2} + \left(\frac{\zeta}{\sqrt{1 - \zeta^2}}\right)\frac{\omega_n\sqrt{1 - \zeta^2}}{(s + \zeta\omega_n)^2 + (\omega_n\sqrt{1 - \zeta^2})^2}$$

$$e(t) = \mathcal{L}^{-1}[E(s)]$$

$$= e^{-\zeta\omega_n t}\left[\cos(\omega_n\sqrt{1 - \zeta^2}t) + \frac{\zeta}{\sqrt{1 - \zeta^2}}\sin(\omega_n\sqrt{1 - \zeta^2}t)\right]u(t)$$

The integral of the square of the error to a step input is as follows:

$$I_S = \int_0^\infty e^2(t)\, dt$$

$$= \int_0^\infty e^{-2\zeta\omega_n t}\left[\cos(\omega_n\sqrt{1 - \zeta^2}t) + \frac{\zeta}{\sqrt{1 - \zeta^2}}\sin(\omega_n\sqrt{1 - \zeta^2}t)\right]^2 dt$$

Letting

$$t' = \omega_n t \qquad dt' = \omega_n\, dt$$

$$I_S = \frac{1}{\omega_n}\int_0^\infty e^{-2\zeta t'}\left[\cos(\sqrt{1 - \zeta^2}t') + \frac{\zeta}{\sqrt{1 - \zeta^2}}\sin(\sqrt{1 - \zeta^2}t')\right]^2 dt'$$

A computer-generated plot of $\omega_n I_S(\zeta)$ is given as Fig. 3.12. Minimal mean square error to a step input for the system (3.1) occurs for $\zeta = 0.5$.

Other useful performance indices can involve errors for other test signals, such as a ramp input, and include the integral of the magnitude of the error,

$$I_M = \int_0^\infty |e(t)|\, dt$$

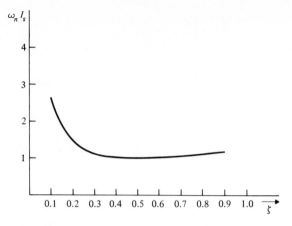

Figure 3.12   Integral square error performance measure for a certain second-order system with adjustable damping ratio.

and integrals where the square and magnitude are weighted with powers of time t to emphasize the behavior at large t:

$$I_{TS} = \int_0^\infty t e^2(t)\, dt \qquad I_{TM} = \int_0^\infty t|e(t)|\, dt$$

Figure 3.13 shows the performance measures I_M, I_{TS}, and I_{TM} for the second-order system (3.1). Clearly, the optimum value of ζ depends upon the definition of "goodness," the performance measure used. Minimum integral magnitude error for the example system occurs for $\zeta = 0.67$. Minimum I_{TS} occurs for $\zeta = 0.6$, and minimum I_{TM} for $\zeta = 0.7$.

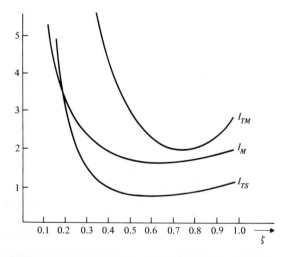

Figure 3.13   Other performance measures for a certain second-order system with adjustable damping ratio.

In some situations, such as in the control of a spacecraft for minimal fuel use, the performance index is clearly given by the design objectives. In others, an index, if used, must be chosen somewhat arbitrarily. In the latter case, choosing desired response characteristics such as rise time, overshoot, and steady state error directly may be much more sensible than choosing a performance index.

DRILL PROBLEMS

D3.9. Use Figs. 3.13 and 3.14 to determine the optimum choice of the parameter k for the given transfer function with the indicated step response performance measure.

(a) $T(s) = \dfrac{100}{s^2 + ks + 100}$

Integral square error I_S

Ans 10

(b) $T(s) = \dfrac{k}{2s^2 + s + k}$

Integral magnitude error I_M

Ans 0.278

(c) $T(s) = \dfrac{k}{s^2 + ks + k}$

Integral time-weighted square error I_{TS}

Ans 1.44

(d) $T(s) = \dfrac{10k}{s^2 + 8s + 10k}$

Integral time-weighted magnitude error I_{TM}

Ans 3.26

D3.10. Find the optimum choice of the parameter k, with a minimum square error step response performance measure, for a system with output-input error transmittance

$$T_E(s) = \frac{ks^2 + (1 - k)s}{s^2 + 3s + 2}$$

Ans $\frac{1}{3}$

3.6 SYSTEM SENSITIVITY

Calculating the Effects of Changes in Parameters

One of the major advantages of feedback is that it can be used to make the response of a system relatively independent of certain types of changes or inaccuracies in the plant model. For example, in the system of Fig. 3.14a, the nominal system transfer function is

$$T(s) = \frac{400/(s+2)}{1+400/(s+2)} = \frac{400}{s+402}$$

Now suppose one of the plant parameters changes or is wrongly modeled, as in 3.14b. For $k_1 = 1$, the plant is the nominal one; other values of k_1 correspond to perturbations from the nominal plant. In terms of k_1,

$$T(s) = \frac{400/(s+2k_1)}{1+400/(s+2k_1)} = \frac{400}{s+400+2k_1}$$

and it is seen that even 50% changes in the parameter, ranging from $k_1 = \frac{1}{2}$ to $k_1 = \frac{3}{2}$, result in a relatively minor change in $T(s)$. Even negative values of k_1 (say $k_1 = -1$), for which the plant is unstable, give much the same, stable, overall transfer function $T(s)$.

The system's steady state error to a unit step input is, in terms of k_1,

$$\lim_{s \to 0} s\left(\frac{1}{s}\right)[1 - T(s)] = \lim_{s \to 0} \frac{s+2k_1}{s+400+2k_1} = \frac{2k_1}{400+2k_1}$$

which is dominated by the factor of 400 and is nearly proportional to k_1.

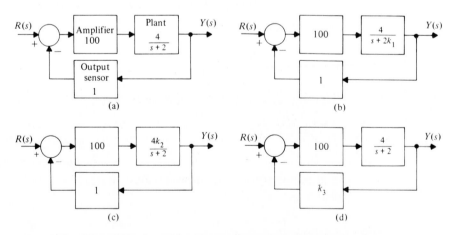

Figure 3.14 Determining the effects of changes in the parameters of a feedback system. **(a)** The nominal system. **(b)** Perturbation of one plant parameter. **(c)** Perturbation of another plant parameter. **(d)** Perturbation of the sensor gain.

For the parameter perturbed by k_2 in Fig. 3.14c,

$$T(s) = \frac{400k_2/(s + 2)}{1 + 400k_2/(s + 2)} = \frac{400k_2}{s + 400k_2 + 2}$$

For this parameter, the system's steady state error (between output and input) to a unit step input is

$$\lim_{s \to 0} s\left(\frac{1}{s}\right)[1 - T(s)] = \lim_{s \to 0} \frac{s + 2}{s + 400k_2 + 2} = \frac{2}{400k_2 + 2}$$

This steady state error is dominated by the factor of $400k_2$ for moderate changes in k_2 from the nominal $k_2 = 1$ and is nearly inversely proportional to k_2. Changes from the nominal amplifier gain of 400 will produce the same effects on $T(s)$ and its step response.

If the sensor gain is perturbed, as in Fig. 3.14d,

$$T(s) = \frac{400/(s + 2)}{1 + 400k_3/(s + 2)} = \frac{400}{s + 400k_3 + 2}$$

The steady state error to a unit step input is

$$\lim_{s \to 0} s\left(\frac{1}{s}\right)[1 - T(s)] = \frac{s + 400(k_3 - 1) + 2}{s + 400k_3 + 2} = \frac{400(k_3 - 1) + 2}{400k_3 + 2}$$

which can become quite large in comparison to the previous expressions for comparable percent parameter changes. In this case, the result is expected, since an error by the sensor in the perceived plant output is indistinguishable by the rest of the system from an actual output error.

Sensitivity Functions

In general, the *sensitivity* of a single-input, single-output system transfer function to changes in a specific parameter a is defined as

$$S_a = \lim_{\Delta a \to 0} \frac{\Delta T/T}{\Delta a/a} = \lim_{\Delta a \to 0} \frac{a}{T} \frac{\Delta T}{\Delta a} = \frac{a}{T} \frac{\partial T}{\partial a}$$

It is the limiting ratio of the fractional change in the transfer function to the fractional change in the parameter.

For example, the feedback system with constant "block" transmittances in Fig. 3.15 might represent a feedback amplifier over a range of frequencies. The transfer function of this system is the constant

$$T = \frac{G}{1 + GH} = \frac{10}{1 + \frac{10}{3}} = \frac{30}{13}$$

The sensitivity of T to changes in G is

$$S_G = \frac{G}{T} \frac{\partial T}{\partial G} = \frac{1}{1 + GH}$$

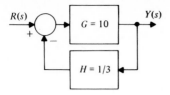

Figure 3.15 Finding sensitivity of a feedback system with constant transmittances.

which for $G = 10$, $H = \frac{1}{3}$ is

$$S_G = \frac{1}{1 + \frac{10}{3}} = \frac{3}{13}$$

This is to say that, with the feedback, the transfer function changes only $\frac{3}{13}$ as much with small changes in G as it would without feedback.

The sensitivity of T to changes in H is

$$S_H = \frac{H}{T}\frac{\partial T}{\partial H} = \frac{-GH}{1 + GH}$$

which for $G = 10$, $H = \frac{1}{3}$ is

$$S_H = \frac{-\frac{10}{3}}{1 + \frac{10}{3}} = -\frac{10}{13}$$

The transfer function is affected by changes in H much more than by changes in G. The minus sign in S_H indicates that T decreases with an increase in H.

For the feedback system of Fig. 3.16, suppose that the plant parameter a is nominally $a = 2$ but is subject to small changes about the nominal value. The transfer function of the system is

$$T(s) = \frac{1/(s + a)}{1 + K/(s + a)} = \frac{1}{s + a + K}$$

The sensitivity of $T(s)$ to changes in a is given by

$$S_a = \frac{a}{T}\frac{\partial T}{\partial a} = a(s + a + K)\frac{-1}{(s + a + K)^2} = \frac{-a}{s + a + K}$$

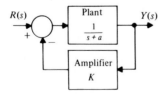

Figure 3.16 Finding the sensitivity of a feedback system to changes in a plant parameter.

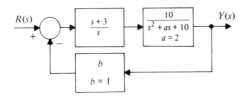

Figure 3.17 Calculating sensitivities to parameter changes for another system.

Sensitivities are generally functions of the complex variable s. For $a = 2$, the sensitivity is

$$S_a = \frac{-2}{s + 2 + K}$$

which can, for any s, be reduced by making K sufficiently large. Without control, $K = 0$ and S_a becomes fixed as a function of s. Without control S_a has a value that becomes infinite for $s = -2$. In general, control provides a mechanism for reducing sensitivity to parameter variations in much the same way that control can reduce the influence of disturbances.

As a more involved example, consider the precision positioning system of Fig. 3.17, for which the transfer function, as a function of the parameter a, is

$$T(s, a) = \frac{\left(\dfrac{s + 3}{s}\right)\left(\dfrac{10}{s^2 + as + 10}\right)}{1 + \left(\dfrac{s + 3}{s}\right)\left(\dfrac{10}{s^2 + as + 10}\right)} = \frac{10s + 30}{s^3 + as^2 + 20s + 30}$$

$$\frac{\partial T}{\partial a} = \frac{-s^2(10s + 30)}{(s^3 + as^2 + 20s + 30)^2}$$

and

$$S_a = \frac{a}{T}\frac{\partial T}{\partial a} = \left[\frac{a}{\left(\dfrac{10s + 30}{s^3 + as^2 + 20s + 30}\right)}\right]\left[\frac{-s^2(10s + 30)}{(s^3 + as^2 + 20s + 30)^2}\right]$$

$$= \frac{-as^2}{s^3 + as^2 + 20s + 30}$$

For the nominal value of $a = 2$,

$$S_a = \frac{-2s^2}{s^3 + 2s^2 + 20s + 30}$$

For small changes in a about the nominal value of $a = 2$,

$$S_a \cong \frac{a}{T}\frac{\Delta T}{\Delta a}$$

$$\frac{\Delta T(s)}{T(s)} \cong \frac{\Delta a}{a} S_a = \frac{\Delta a}{a}\frac{-2s}{s^3 + 2s^2 + 20s + 30} \tag{3.2}$$

For a specific value of s, (3.2) relates fractional changes in the transfer function to fractional small changes in the parameter. In calculating the response to a step input to the system, for example, the transfer function is evaluated at $s = 0$ to obtain the residue corresponding to the forced response. Equation (3.2) with $s = 0$ gives

$$\frac{\Delta T(0)}{T(0)} \cong 0 \qquad \Delta T(0) \cong 0$$

for small changes Δa.

For the parameter b in the feedback path of the system of Fig. 3.17,

$$T(s, b) = \frac{\left(\dfrac{s + 3}{s}\right)\left(\dfrac{10}{s^2 + 2s + 10}\right)}{1 + b\left(\dfrac{s + 3}{s}\right)\left(\dfrac{10}{s^2 + 2s + 10}\right)} = \frac{10s + 30}{s^3 + 2s^2 + (10 + 10b)s + 30b}$$

so that

$$\frac{\partial T}{\partial b} = \frac{-(10s + 30)(10s + 30)}{(s^3 + 2s^2 + 10s + 10bs + 30b)^2}$$

For the nominal $b = 1$,

$$\frac{\partial T}{\partial b} = \frac{-(10s + 30)(10s + 30)}{(s^3 + 2s^2 + 20s + 30)^2}$$

and

$$S_b = \frac{b}{T}\frac{\partial T}{\partial b} = \left(\frac{1}{\left(\dfrac{10s + 30}{s^3 + 2s^2 + 20s + 30}\right)}\right)\left[\frac{-(10s + 30)(10s + 30)}{(s^3 + 2s^2 + 20s + 30)^2}\right]$$

$$= \frac{-10s - 30}{s^3 + 2s^2 + 20s + 30}$$

For small changes Δb,

$$\frac{\Delta T(s)}{T(s)} \cong \frac{\Delta b}{b}S_b = \left(\frac{\Delta b}{b}\right)\frac{-10s - 30}{s^3 + 2s^2 + 20s + 30}$$

At $s = 0$, for calculating the forced response to a step input,

$$\frac{\Delta T(0)}{T(0)} \cong \left(\frac{\Delta b}{b}\right)(-1)$$

Changes in T are proportional (in the opposite sense) to changes in b.

Sensitivity to Disturbance Signals

Another major advantage of feedback is that it can be used to reduce the effects of disturbance inputs upon system response. For example, for the thermal control system of Fig. 3.18a, a disturbance signal $D(s)$ affects the plant but is not accessible to the designer. The transfer function relating $Y(s)$ to $D(s)$ is

$$T_D(s) = \frac{1}{s+2}$$

For a unit step disturbance input, the final value of the output due to the disturbance is given by

$$Y(s) = \frac{1}{s}\frac{1}{s+2}$$

$$\lim_{t\to\infty} y(t) = \lim_{s\to 0} sY(s) = \frac{1}{2}$$

If the plant is driven in the feedback arrangement of Fig. 3.18b, the transfer function relating $Y(s)$ to $D(s)$ becomes, instead,

$$T_D(s) = \frac{1/(s+2)}{1-[-K/(s+2)]} = \frac{1}{s+2+K}$$

For a unit step disturbance input to the feedback system, the resulting steady state output is given by

$$Y(s) = \frac{1}{s}\left(\frac{1}{s+2+K}\right)$$

$$\lim_{t\to\infty} y(t) = \lim_{s\to 0} sY(s) = \frac{1}{2+K}$$

which can be made arbitrarily small by making K sufficiently large.

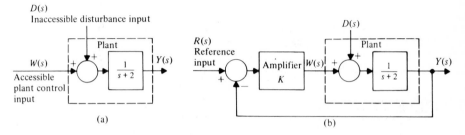

(a)

(b)

Figure 3.18 Disturbance input to a system with and without feedback. **(a)** Plant with control and disturbance inputs. **(b)** Plant with feedback.

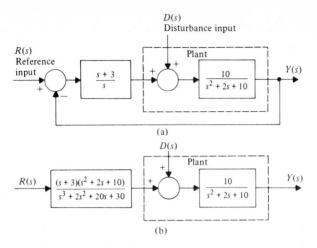

Figure 3.19   Reducing the effects of disturbance signals with feedback.   **(a)** A feedback control system.   **(b)** An open-loop system with the same relation between $Y(s)$ and $R(s)$.

Another example system is shown in Fig. 3.19a. The disturbance signal $D(s)$ represents an unwanted, largely unknown effect upon the plant. The two system transfer functions are

$$T_R(s) = \frac{Y(s)}{R(s)} = \frac{\dfrac{10(s+3)}{s(s^2+2s+10)}}{1 + \dfrac{10(s+3)}{s(s^2+2s+10)}} = \frac{10(s+3)}{s^3+2s^2+20s+30}$$

$$T_D(s) = \frac{Y(s)}{D(s)} = \frac{\dfrac{10}{s^2+2s+10}}{1 + \dfrac{10(s+3)}{s(s^2+2s+10)}} = \frac{10s}{s^3+2s^2+20s+30}$$

The system is stable, as is easily determined from a Routh-Hurwitz test:

$$
\begin{array}{c|cc}
s^3 & 1 & 20 \\
s^2 & 2 & 30 \\
s^1 & 5 & \\
s^0 & 30 &
\end{array}
$$

A unit step disturbance to this system will produce zero contribution to the steady state output since

$$\lim_{s \to 0} s\left(\frac{1}{s}\right) T_D(s) = \lim_{s \to 0} \frac{10s}{s^3+2s^2+20s+30} = 0$$

Now consider the open-loop (nonfeedback) system of Fig. 3.19b. It has the same transfer function relating $Y(s)$ and $R(s)$ as does the feedback system. For this system, however, the relationship between the output and the disturbance is not modified by feedback:

$$T_D(s) = \frac{Y(s)}{D(s)} = \frac{10}{s^2 + 2s + 10}$$

A unit step disturbance of the open-loop system will produce a unit contribution to the steady state output:

$$\lim_{s \to 0} s\left(\frac{1}{s}\right) T_D(s) = \lim_{s \to 0} \frac{10}{s^2 + 2s + 10} = 1$$

It is difficult to generalize about methods for improving disturbance rejection with feedback because most practical situations involve many specific structural constraints. For example, it may or may not be acceptable to sense (provide as an output) a certain plant signal or to supply a control signal to a certain part of the plant.

Section 1.7 contains a positioning servo for a microwave antenna. A control motor having a linear torque-speed curve rotates the antenna. The wind provides a disturbance torque that is beyond the control of a designer; however, the effects of the wind can be rendered acceptable. The wind torque acts to rotate the antenna in the same direction as for a positive input. The response due to a unit step input is

$$\lim_{t \to \infty} \theta(t) = \lim_{s \to 0} s\theta(s)$$

$$= \lim_{s \to 0} \frac{sT_1(s)}{s} = T_1(0)$$

$$= 1$$

The response due to a unit step disturbance torque is

$$\lim_{t \to \infty} \theta(t) = \lim_{s \to 0} s\theta(s)$$

$$= \lim_{s \to 0} \frac{sT_2(s)}{s} = T_2(0)$$

$$= \frac{1}{k_1 A_1 A_2 K_p}$$

This analysis demonstrates that the response due to the wind torque (as compared to the response due to the input) can be reduced by selecting large gain values (subject to stability considerations).

DRILL PROBLEMS

D3.11. For the feedback system with constant "block" transmittances, find the sensitivity of each of the four transfer functions to small changes in G_1 and to small changes in G_2.

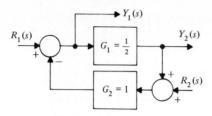

Ans sensitivity of T_{11} to G_1 is $-\frac{1}{3}$

D3.12. Find the sensitivities of the transfer functions of the following systems to small changes in k_1, k_2, and k_3 about the given nominal values:

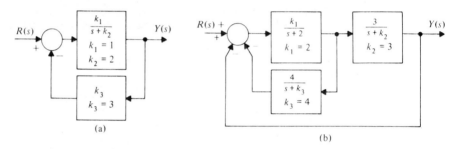

(a)   (b)

Ans (a) $(s + 2)/(s + 5)$, $-2/(s + 5)$, $-3/(s + 5)$;
(b) $(s^3 + 9s^2 + 26s + 24)/(s^3 + 9s^2 + 4s + 72)$,
$-3(s^2 + 6s + 16)/(s^3 + 9s^2 + 4s + 72)$,
$(3s^3 + 27s^2 + 128s + 240)/(s + 4)(s^3 + 9s^2 + 4s + 72)$

D3.13. For each of the following systems, find the steady state error to a unit step input $R(s)$ and the steady state error to a unit step disturbance input $D(s)$.

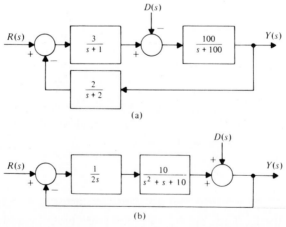

(a)

(b)

Ans $\frac{1}{4}, \frac{5}{4}; 0, 1$

3.7 AN ELECTRIC RAIL TRANSPORTATION SYSTEM

The control of transportation systems is an interesting design area. Many European trains are electric, so control of electric traction motors is common in that area of the world. Desired speed inputs to the system are made by the operator, with an override occurring in case of emergency. A similar system is used for the 100 mi/hr Japanese Kyoto-to-Tokyo train. In San Francisco, the BART (Bay Area Rapid Transit) system is designed to automatically vary the speed as conditions warrant, without human intervention.

Manned aircraft and space flight are further examples of transportation systems where control inputs to the vehicle are generated by an autopilot. The same methodology can be applied to bus, car, and passenger train operation to improve performance. In this example, a velocity and position control system for a rail vehicle is examined. It is similar to systems employed for passenger trains in Germany and Switzerland.

Figure 3.20a shows the relation between motor drive and velocity for an electric rail car. A unit step input $D(s)$ will produce a steady state car velocity given by

$$\lim_{s \to 0} s \left(\frac{1}{s} \right) \left(\frac{15}{s + 0.1} \right) = 150 \text{ ft/sec}$$

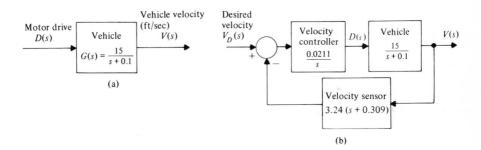

(a)

(b)

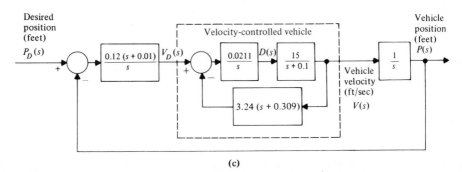

(c)

Figure 3.20 Controlling a transportation vehicle. **(a)** Vehicle model. **(b)** Vehicle with velocity control feedback loop. **(c)** Vehicle with velocity and position control.

with a time constant of 10 sec. In perhaps more familiar terms, in 10 sec, the car will accelerate to

$$(150)(0.63) = 94.5 \text{ ft/sec}$$

$$(94.5)(3600/5280) = 64 \text{ mi/hr}$$

$$(94.5)(3600/3281) = 103.7 \text{ km/hr}$$

Automobile racing enthusiasts would characterize this vehicle by saying that it can accelerate from 0 to 60 mi/hr in 8.8 sec or from 0 to 100 km/hr in 9.3 sec.

In Fig. 3.20b, the vehicle is shown as part of a feedback system for velocity control, where V_D is the desired velocity input. This system has transfer function

$$T_V(s) = \frac{\left(\dfrac{0.0211}{s}\right)\left(\dfrac{15}{s + 0.1}\right)}{1 + \left(\dfrac{0.0211}{s}\right)\left(\dfrac{15}{s + 0.1}\right)(3.24)(s + 0.309)} = \frac{0.317}{s^2 + 1.125s + 0.317}$$

The error between desired and actual velocity is

$$E_V(s) = V_D(s) - V(s) = V_D(s) - T_V(s)V_D(s) = [1 - T_V(s)]V_D(s)$$

The error signal in this case is not the summer signal because this system does not have unity feedback. The error transmittance is

$$T_{VE}(s) = 1 - T_V(s) = \frac{s(s + 1.125)}{s^2 + 1.125s + 0.317}$$

As the system is type 1, its steady state error to a step input will be zero:

$$\lim_{s \to 0} s\left(\frac{1}{s}\right)\left[\frac{s(s + 1.125)}{s^2 + 1.125s + 0.317}\right] = 0$$

Its normalized steady state error to a ramp input is

$$\lim_{s \to 0} s\left(\frac{1}{s^2}\right)\left[\frac{s(s + 1.125)}{s^2 + 1.125s + 0.317}\right] = \frac{1.125}{0.317} = 3.55 \text{ ft/sec}$$

The velocity control feedback system has the characteristic equation

$$s^2 + 1.125s + 0.317 = 0$$

and repeated characteristic roots s_1, $s_2 = -0.56$. Its relative stability is 0.56 unit, and its natural response dies out as fast as $\exp(-0.5t)$ compared to the natural response, $\exp(-0.1t)$, of the vehicle alone, Fig. 3.20a. To achieve control of the vehicle position, a second feedback loop has been added to the system in Fig. 3.20c. The transfer function of the complete system is

$$T(s) = \cfrac{\left[\dfrac{0.12(s + 0.01)}{s}\right]\left(\dfrac{0.317}{s^2 + 1.125s + 0.317}\right)\left(\dfrac{1}{s}\right)}{1 + \left[\dfrac{0.12(s + 0.01)}{s}\right]\left(\dfrac{0.317}{s^2 + 1.125s + 0.317}\right)\left(\dfrac{1}{s}\right)}$$

$$= \frac{0.038s + 0.00038}{s^4 + 1.125s^3 + 0.317s^2 + 0.038s + 0.00038}$$

The system is stable, as a Routh-Hurwitz test easily shows:

$$
\begin{array}{c|ccc}
s^4 & 1 & 0.317 & 0.00038 \\
s^3 & 1.125 & 0.038 \\
s^2 & 0.283 & 0.00038 \\
s^1 & 0.036 \\
s^0 & 0.00038
\end{array}
$$

The error between desired and actual position is

$$E(s) = P_D(s) - P(s) = [1 - T(s)]P_D(s)$$

giving the following error transmittance:

$$T_E(s) = 1 - T(s) = \frac{s^2(s^2 + 1.125s + 0.317)}{s^4 + 1.125s^3 + 0.317s^2 + 0.038s + 0.00038}$$

The system is type 2; it exhibits zero steady state error to a step and to a ramp input $P_D(s)$. A ramp desired position constitutes a step in desired velocity. In the steady state, this system will thus approach zero error in both position and velocity.

3.8 PHASE-LOCKED LOOP FOR A CB RECEIVER

The phase-locked loop (PLL) is an important component of many telecommunications systems. It is used to demodulate the stereo channel in FM broadcast receivers, to detect and maintain the color subcarrier in color television receivers, to generate precise frequencies in citizen band (CB) receivers, and in many other applications.

Figure 3.21a shows a pictorial diagram of the operation of a superhetrodyne receiver for the citizen band. The signal from the receiver's antenna and a sinusoidal mixer signal generated in the receiver are mixed in such a way as to produce sums and differences of the antenna signal frequencies with the sinusoidal mixing frequency signal. Those difference frequencies which are centered at 10.7 MHz are passed by the bandpass filter and detected. By changing the mixing signal frequency, different incoming channels are translated to the 10.7-MHz passband to be detected.

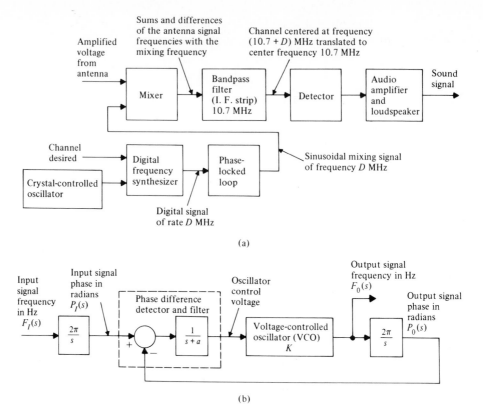

Figure 3.21 Phase-locked loop for a citizen band receiver. **(a)** Description of a CB receiver. **(b)** Phase-locked loop model.

From a highly stable crystal-controlled oscillator, logic circuits produce a digital waveform at the proper mixing frequency for the desired channel to be detected. The mixer requires a smooth, sinusoidal mixing signal, and it is the purpose of the phase-locked loop to "lock" the frequency of a voltage-controlled oscillator to the digital rate. In some receivers, a PLL is also used to aid in the frequency synthesis, but that situation will not be considered here.

A phase-locked loop model is given in Fig. 3.21b. The phase of a sinusoidal signal is proportional to the integral of its frequency. The difference in phase between the input and output signals is sensed and used to control the frequency of an oscillator in such a way that difference in phase (and thus in frequency) is reduced. Commercial phase-locked loops may differ from this arrangement in that the phase difference detector is nonlinear and capable of sensing phase differences only as large as π or 2π rad. Nonetheless, this simple model adequately predicts PLL performance for most applications. The system block diagram of Fig. 3.21b is unusual, compared with previous examples, because it relates frequencies and phases of signals, not the signals themselves.

The transfer function of the system that relates output frequency to input frequency is

$$T(s) = \frac{F_O(s)}{F_I(s)} = \frac{\dfrac{2\pi}{s}\left(\dfrac{1}{s+a}\right)(K)}{1 + \left(\dfrac{1}{s+a}\right)(K)\left(\dfrac{2\pi}{s}\right)} = \frac{2\pi K}{s^2 + as + 2\pi K}$$

The error between input and output frequency is given by

$$E(s) = F_I(s) - F_O(s) = [1 - T(s)]F_I(s)$$

and has transmittance

$$T_E(s) = 1 - T(s) = \frac{s^2 + as}{s^2 + as + 2\pi K}$$

The system is type 1 and has zero steady state error to a step input frequency change.

The choice of filter time constant $(1/a)$ and voltage-controlled oscillator gain K is based upon the desired characteristics of the system when responding to a change in desired frequency. A reasonable requirement would be for the minimum mean square error damping ratio of 0.5 and a settling time of 0.1 sec.

The system transfer function

$$T(s) = \frac{2\pi K}{s^2 + as + 2\pi K} = \frac{\omega_n^2}{s^2 + 2\zeta\omega_n s + \omega_n^2}$$

is of the form analyzed in detail in Sec. 2.3, with

$$\omega_n = \sqrt{2\pi K} \qquad \text{and} \qquad \zeta = \frac{a}{2\omega_n} = \frac{a}{2\sqrt{2\pi k}} \tag{3.3}$$

Using Fig. 2.10c, which shows normalized settling time as a function of damping ratio, for a 0.5 damping ratio,

$$\omega_n T_S = 5.3 \tag{3.4}$$

For a settling time $T_S = 0.1$, Eq. (3.4) gives

$$\omega_n = \frac{5.3}{0.1} = 53$$

Using (3.3),

$$K = \frac{\omega_n^2}{2\pi} = \frac{(53)^2}{6.28} = 447$$

and with $\zeta = 0.5$,

$$a = 2\omega_n\zeta = 2(53)(0.5) = 53$$

This preliminary design is now examined for "worst case" behavior in changing from channel to channel. The lowest CB carrier frequency is 26.97 MHz, and the highest is 27.26 MHz, corresponding respectively to PLL input frequencies of

$$26.97 - 10.7 = 16.27 \text{ MHz} \quad \text{and} \quad 27.26 - 10.7 = 16.56 \text{ MHz}$$

The largest change in PLL input frequency between stations will be from 16.27 to 16.56 MHz, a step change of 290,000 Hz. By design, the frequency settles to within 5%,

$$(0.05)(290,000) = 14,500 \text{ Hz}$$

in 0.1 sec, but it will take about 3 times that long for the frequency to settle to below about 50 Hz, which is necessary for intelligible reception. A 0.3-sec maximum time to change stations is probably quite acceptable.

When the device is first turned on, however, it is possible that it will have to respond to a step change of up to 16.56 MHz. Although the behavior of a PLL for such a change in input frequency will likely be nonlinear at first, the linear model will be used to predict an approximate initial length of time until the PLL output frequency has settled to within about 50 Hz. In the first 0.1 sec, the response will settle to

$$(0.05)(16,560,000) = 828,000 \text{ Hz}$$

As the envelope of the second-order oscillatory natural behavior is exponential, the response will settle to within 5% of this value in the next 0.1 sec:

$$(0.05)(828,000) = 41,400 \text{ Hz}$$

Continuing

$$(0.05)(41,400) = 2070 \text{ Hz}$$

$$(0.05)(2070) = 103.5 \text{ Hz}$$

so an acceptable "worst case" initial tuning is predicted to be within about 0.5 sec.

This system is type 1 insofar as a step change in frequency is concerned, so that zero steady-state frequency error results. The transfer function relating $P_i(s)$ and $P_o(s)$ is also type 1. However, a constant frequency implies that phase (the integral of frequency) is a ramp so that some steady state phase error occurs. That error can be controlled by raising K or by increasing system type to 2.

3.9 SUMMARY

The concern of this chapter is the analysis of tracking system performance. In subsequent chapters, these analysis tools will be brought to bear on tracking system design. The analysis (and design) of tracking systems can be separated into two parts:

1. Location of the system poles (characteristic roots) to give natural responses that decay sufficiently rapidly and have acceptable damping ratios for oscillatory terms. It is usually desirable that the pole locations be within a region on the complex plane of the shape shown in Figure 3.3c. The amplitudes of the natural response terms depend only on the system's initial conditions, so the designer can normally only choose the manner in which the natural response decays to zero.

2. A forced response component that tracks the reference input well for the class of inputs to be encountered. This can be characterized by the system's steady state errors to power-of-time inputs.

For any function $y(t)$, the initial value is related to its Laplace transform by

$$y(0) = \lim_{s \to \infty} sY(s)$$

If $y(t)$ has a final value, it is given by

$$\lim_{t \to \infty} y(t) = \lim_{s \to 0} sY(s)$$

If the function $y(t)$ does not have a final value, the limit can give a misleading result.

Tracking systems are designed so that the forced component of the system output, as nearly as possible, equals the input. An important measure of performance in this regard is the steady state error between input and output for step, ramp, parabolic, and other power-of-time input signals. When the error has a final value, the steady state error to an input

$$r_i(t) = \frac{1}{i!} t^i u(t) \qquad R_i(s) = \frac{1}{s^{i+1}}$$

is given by the final-value theorem as

$$\lim_{s \to 0} \frac{1}{s^i} [1 - T(s)]$$

System type number is the number of $s = 0$ numerator roots in the error transmittance

$$T_E(s) = 1 - T(s)$$

Type number determines the steady state power-of-time error properties of a system, as summarized in Table 3.2. A system that has finite steady state error for the ith power-of-t input will have zero steady state error for lower powers of t and infinite steady state error for higher powers of t.

Unity feedback systems have error transmittance

$$T_E(s) = 1 - T(s) = \frac{1}{1 + G(s)}$$

Their system type number, the number of $s = 0$ numerator roots in $T_E(s)$, is also the number of $s = 0$ denominator roots of $G(s)$. Unity feedback system error coefficients are defined as

$$\kappa_i = \lim_{s \to 0} s^i G_E(s)$$

and these are related to steady state errors in Table 3.3.

If the quality of performance of a system can be expressed with a performance measure I in terms of the adjustable system parameters, they can be selected by the mathematical process of finding the set of parameters that maximize or minimize I. A common performance index for a tracking system is the integral square error to a step input:

$$I_S = \int_0^\infty e_{\text{step}}^2(t)\, dt$$

The mathematics of finding extrema of functions I of many variables is formidable in most cases, and results can be quite dependent upon the specific performance measure used.

Feedback can be used to make system response relatively independent of inaccuracies in some of the system's parameters. A feedback system can thus be designed to perform well even when the controlled plant parameters are not known accurately or when they drift with time. In general, the sensitivity of a transfer function T to a change in a parameter a is

$$S_a = \lim_{\Delta a \to 0} \frac{\Delta T/T}{\Delta a/a} = \frac{a}{T} \frac{\partial T}{\partial a}$$

It is the fractional rate of change of T with fractional change in a. Sensitivities generally depend upon the complex variable s.

Feedback is also used to reduce the effects of disturbance signals upon a system's response. A tracking feedback arrangement that compares the output with a desired reference signal and works to drive the error between the two to zero tends to lessen disturbance effects.

The velocity and position control system for a rail transportation system illustrated how one can use multiple feedback loops in tracking system design. In this example, an inner loop effected velocity control of the electric train. Then, an outer feedback loop involving error integration was used to achieve position control. Finally, step response was seen to be a highly useful and easily visualized test signal to use in analyzing the performance of a phase-locked loop for a citizen band receiver.

REFERENCES

Pole Placement and Steady State Error
James, H. M., Nichols, N. B., and Philips, R. S., *Theory of Servomechanisms* (MIT Radiation Laboratory Series, vol. 25). New York: McGraw-Hill, 1947.

DiStefano, J. J. III, Stubberud, A. R., and Williams, I. J., *Feedback and Control Systems* (*Schaum's Outline*). New York: McGraw-Hill, 1967).

Chestnut, H., and Mayer, R. W., *Servomechanisms and Regulating System Design*, vol. 1. New York: Wiley, 1959.

Savant, C. J. Jr., *Basic Feedback Control System Design.* New York: McGraw-Hill, 1958.

Truxal, J. C., *Control System Synthesis.* New York: McGraw-Hill, 1955.

Sensitivity

Cruz, J. B. Jr., *Feedback Systems.* New York: McGraw-Hill, 1972.

_____, ed. *System Sensitivity Analysis.* Stroudsburg, Pa: Dowden, 1973.

Horowitz, I. M., *Synthesis of Feedback Systems.* New York: Academic Press, 1963.

Kreindler, E., "On the Definition and Application of the Sensitivity Function." *J. Franklin Inst.* 285 (January 1968).

Tomovic, R., *Sensitivity Analysis of Dynamic Systems.* New York: McGraw-Hill, 1963.

Disturbance Rejection

Friedland, B., *Control System Design: An Introduction to State Space Methods.* New York: McGraw-Hill, 1986.

Hostetter, G. H., *Digital Control System Design.* New York: Holt, Rinehart and Winston, 1987.

Optimal Control

Athans, M., "The Status of Optimal Control Theory and Applications for Deterministic Systems." *IEEE Trans. Auto. Contr.*, July 1966.

Dorf, R. C., *Time-Domain Analysis and Design of Control Systems.* Reading, Mass.: Addison-Wesley, 1965.

McCausland, I., *Introduction to Optimal Control.* New York: Wiley, 1969.

Sage, A. P., and White, C. C. III., *Optimum Systems Control.* Englewood Cliffs, N.J.: Prentice-Hall, 1977.

Schultz, D. G., and Melsa, J. L., *State Functions and Linear Control Systems.* New York: McGraw-Hill, 1967.

Transportation Systems

Friedlander, G. D., "Electronics and Swiss Railways," *IEEE Spec.* 11, no. 9 (September 1974): 68–75.

Kaplan, G., "Microprocessors Monitor Rail Car Systems; Software Optimizes Headway in Miami's People Mover." *IEEE Spec.* (January 1987): 59–61.

_____, "Rail Transportation." *IEEE Spec.* (January 1984): 82–85.

Stefani, R. T., "Design and Simulation of an Automobile Guidance and Control System." *Computers Ed.* (*COED*) *Trans. Amer. Soc. Eng. Ed.* (*ASEE*), January 1978.

PROBLEMS

1. Find the relative stability of the feedback systems of Fig. P3.1. If there are complex conjugate pairs of characteristic roots, also find the damping ratios.

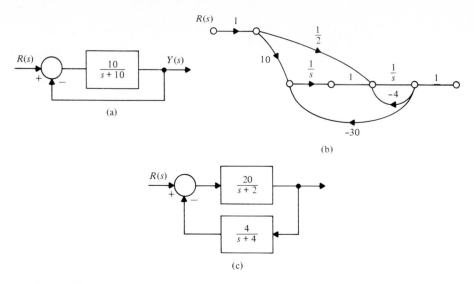

(a)

(b)

(c)

Figure P3.1

2. If possible, find a value of the adjustable constant k so that the system of Fig. P3.2 has a relative stability of at least 2 units and a minimum damping ratio of 0.5.

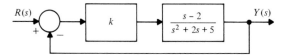

Figure P3.2

3. Find the initial values $y(0)$ of the signals with the following Laplace transforms.

(a) $Y(s) = \dfrac{-5s^3 + 3s^2 + 2}{s^3 + 4s^2 + 8s + 12}$

(b) $Y(s) = \dfrac{6s + 6se^{-4s}}{s^2 + 8s + 2}$

(c) $Y(s) = \dfrac{4s^2 - 3s + 4}{11s^4 + 4s^3 + 2s^2 + 10s}$

(d) $Y(s) = \dfrac{9s^2 + 2}{s^5 + 9s^4 - 4s^3}$

Ans (a) ∞; (c) 0

4. Determine if signals with the following Laplace transforms have a final value. If a finite final value exists, use the final-value theorem to find it.

(a) $Y(s) = \dfrac{-3s^2 + 2s + 4}{s^3 + 7s^2 + 8s + 6}$

(b) $Y(s) = \dfrac{6s^2 + 8s + 2}{s^4 + 2s^3 + 4s^2 + 9s + 3}$

(c) $Y(s) = \dfrac{4s^{-2s}}{3s^4 + 6s^3 + 2s^2 + 2s + 12}$

(d) $Y(s) = \dfrac{10}{s^5 + 3s^4 + 7s^3 + 4s^2 + 4s}$

Ans (a) exists, 0; (d) exists, $\frac{10}{4}$

5. For the systems of Fig. P3.5, find the steady state error for a unit step input. The error signal is $r(t) - y(t)$.

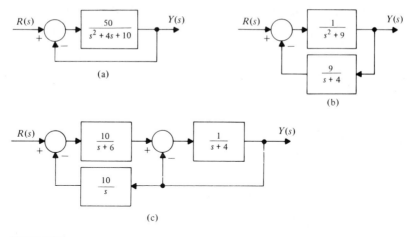

(a)

(b)

(c)

Figure P3.5

6. Find the type number and steady state errors with unit step, ramp, and parabolic inputs for the tracking systems with the following transfer functions:

(a) $T(s) = \dfrac{50}{s^2 + 4s + 50}$

(b) $T(s) = \dfrac{4s + 50}{s^2 + 4s + 50}$

(c) $T(s) = \dfrac{5}{s^2 + 4s + 50}$

(d) $T(s) = \dfrac{s + 10}{s^3 + 4s^2 + s + 10}$

(e) $T(s) = \dfrac{2s^2 + s + 4}{s^4 + 3s^3 + 10s^2 + s + 4}$

Ans (a) 1, 0, $\frac{4}{50}$, ∞; (c) 0, 0.9, ∞, ∞; (e) 2, unstable

---

7. The forward transmittances of unity feedback tracking systems are given below. For each, find the type number of the system, the steady state error coefficients κ_0, κ_1, and κ_2, and the steady state errors to unit step, ramp, and parabolic inputs.

(a) $G_E(s) = \dfrac{-3s + 10}{s^4 + 3s^3 + 2s^2 + 4s}$

(b) $G_E(s) = \dfrac{10}{s^3 + 2s^2 + 10s}$

(c) $G_E(s) = \dfrac{6s^2 - 3s + 10}{s^4}$

(d) $G_E(s) = \dfrac{3s + 4}{s^4 + 8s^3}$

(e) $G_E(s) = \dfrac{4s^2 + 9s + 2}{s^4 + 8s^3}$

Ans (b) 1, ∞, 1, 0, 0, 1, ∞; (d) 4, unstable

---

8. Choose a transmittance $G(s)$ for the block so that the overall system of Fig. P3.8 is type 2. Many different choices for $G(s)$ are possible.

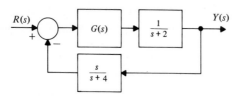

Figure P3.8

9. Find the sensitivities of the following transfer functions to small changes in the parameter a about the given nominal values:

(a) $T(s) = \dfrac{4a}{s + a};\quad a = 3$

(b) $T(s) = \dfrac{4}{s^2 + as + 4};\quad a = 2$

(c) $T(s) = \dfrac{as + 10}{s^2 + as + 10};\quad a = 4$

Ans (a) $s/(s + 3)$; (b) $-2s/(s^2 + 2s + 2)$

10. It is sometimes possible to eliminate the effects of an inaccessible disturbance upon the output of a feedback system entirely. When this is done, the disturbance is said to be *decoupled* from the output. For the system of Fig. P3.10, find a "block" transmittance $G(s)$, if possible, for which $D(s)$ is decoupled from $Y(s)$.

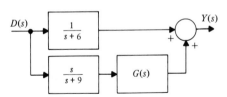

Figure P3.10

11. For the system of Fig. P3.11, find the value of k, if it exists, such that for a step input $r(t)$, the integral square error between $y(t)$ and $r(t)$ is minimum.

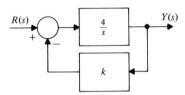

Figure P3.11

12. For signals with the following Laplace transforms, find values of the adjustable constants k, if possible, for which

$$I_S = \int_0^\infty e^2(t)\, dt$$

is smallest:

(a) $E(s) = \dfrac{ks + 2}{s^2 + 7s + 10}$

(b) $E(s) = \dfrac{-6s + k}{s^2 + 5s + 6}$

Ans   (a) $k = 0.73$

13. Tracking systems such as the one in Fig. P3.13, in which the reference input is zero (since it is missing), are called *regulators*. It is desired to keep the output as near to zero as is possible in the system, even when there are disturbances present. For what range of the adjustable constant K, if any, is the steady state value of $y(t)$ less than 0.1 when $d(t)$ is a unit step signal?

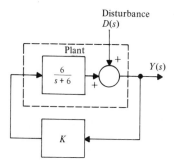

Figure P3.13

14. For the electric rail system of Sec. 3.7, suppose the vehicle transmittance is instead

$$G(s) = \frac{10}{s + 0.4}$$

Find the steady state position error for a 10,000-ft step change in desired position.

Ans   0

15. For the electric rail transportation system of Sec. 3.7 with the parameter values given in the text, suppose that an electrical malfunction causes a unit step to be added to $V_D(s)$ before the signal is applied to the velocity-controlled vehicle subsystem. Find the resulting steady state error in vehicle position.

16. For the phase-locked loop of Sec. 3.8, find a and K such that the system's natural response is critically damped and has suitable speed of response for the CB receiver application.

Ans $K = 168$, $a = 65$, rise time $= 0.1$ sec

17. For the citizen band receiver of Sec. 3.8, with the values of K and a chosen in the text, suppose a listener is initially tuned to the station at 26.97 MHz. The listener suddenly tunes to the station at 27.26 MHz and remains tuned for 1 sec. Curious about the end of a message the listener had been tuned to, the receiver is suddenly returned to the station at 26.97 MHz. Sketch $f_0(t)$.

18. A power plant frequency control system has the block diagram given in Fig. P3.18. Find, in terms of K, relative stability, steady state errors due to power-of-t reference inputs, and steady state error due to a unit step change in load torque. Choose, on the basis of your feeling of how a large power generator should be controlled, a value for K. For this value of K, find the form of the system's natural response.

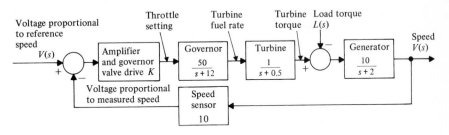

Figure P3.18

19. A simplified block diagram for a chemical process control system is shown in Fig. P3.19. The controller is called a **PID** (or *three-term*) type, because it develops a signal which is a linear combination of terms that are proportional to, the derivative of, and the integral of the incoming signal $f(t)$. If possible, choose values for k_1, k_2, and k_3 such that the resulting system has a relative stability of at least 5 units and zero steady state error to a step input.

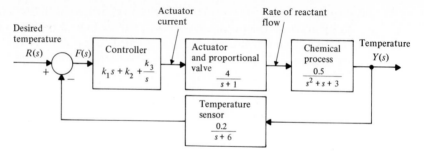

Figure P3.19

20. Figure P3.20 shows a simplified model of a submarine depth control when the submarine is submerged. The ship-settling dynamics transmittance is

$$G(s) = \frac{10^4}{s^2 + 3 \times 10^3 s + 14 \times 10^6}$$

(a) Carefully explain the meaning of the "block" with transmittance $(1000/s)$. Is this a component of the system in the same sense as an amplifier or a motor might be?

(b) Determine if the system is stable.

(c) Find the steady state change in actual depth due to a unit step change in desired depth.

(d) A sonar beacon is released by the ship, causing a unit step change in the weight of the vessel. Find the steady state change in actual depth.

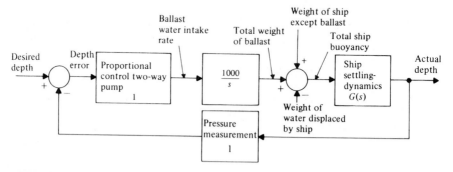

Figure P3.20

Ans (b) Stable; (c) 1; (d) 0

21. A tape loop positioning system for a digital computer tape drive is diagramed in Fig. P3.21. Find values of the constants $k_1, k_2,$ and $k_3,$ if possible, so that this system has a relative stability in excess of 10 units and a steady state error no greater than 5%. The motor time constant k_3 cannot be less than 0.2.

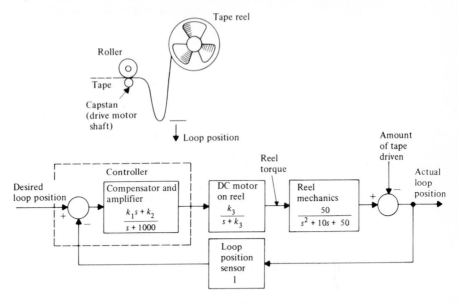

Figure P3.21

22. The table below gives approximate measurements of the output $y(t)$ of a system at various times t when the system input is a step,

$$r(t) = 8.3u(t)$$

(a) Using the data, find an approximate first-order linear, time-invariant system model, specifying its transfer function.

(b) Repeat, but find an approximate second-order linear, time-invariant system model. Specify the approximate system transfer function.

AN INPUT-OUTPUT RECORD FOR A SYSTEM

Time in seconds, t	Output value $y(t)$
0.0	0.0
0.15	1.1
0.3	2.2
0.45	2.8
0.6	3.3
0.75	3.5
0.9	3.5
1.05	3.4
1.2	3.3
1.35	3.3

23. The forward transmittances of unity feedback tracking systems are given below. For each, find the type number of the system, the steady state error coefficients κ_0, κ_1, and κ_2, and the steady state errors to unit step, ramp, and parabolic inputs.

(a) $G_E(s) = \dfrac{K}{(s + 10)(s + 100)}$

(b) $G_E(s) = \dfrac{K}{s(s + 10)(s + 100)}$

   Ans $1, \infty, 10^{-3} K, 0, 0, 10^3/K, \infty$

(c) $G_E(s) = \dfrac{Ks}{(s + 1)(s + 10)(s + 100)}$

   Ans $0, 10^{-3} K, 0, 0, 1/(1 + 10^{-3}K), \infty, \infty$

(d) $G_E(s) = \dfrac{K}{s^2(s^2 + 2\zeta\omega_n s + \omega_n^2)}$

(e) $G_E(s) = \dfrac{100}{(s^2 + 2\zeta\omega_n s + \omega_n^2)}$

(f) $G_E(s) = \dfrac{100(s + 10)(s + 50)}{s^3(s^2 + 2\zeta\omega_n s + \omega_n^2)}$

   Ans $3, \infty, \infty, \infty,$ (unstable)

24. Compute the step, ramp and parabolic error coefficients for the non-unity feedback system of Fig. P3.24.

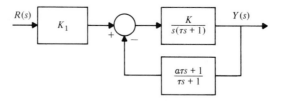

Figure P3.24

25. The following questions pertain to the system of Fig. P3.25. A control motor with a linear torque-speed curve rotates a missile launcher while wind gusts provide a disturbance torque thwarting accurate targeting.

(a) When the disturbance torque is zero, compute the output if the input $r(t)$ is a step and if the input $r(t)$ is a ramp.

(b) If the disturbance torque $\tau_D(t)$ is a constant T_0, compute the steady state output due to that disturbance.

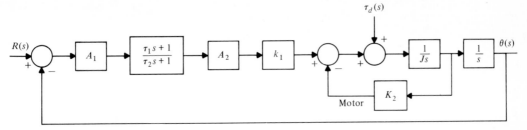

Figure P3.25

Ans $T_0/A_1 A_2 k_1$

26. Repeat Prob. 25b if the disturbance torque is bt.

27. Repeat Prob. 26 where A_2 becomes K/s.

   Ans 0

28. Repeat Prob. 26 where A_2 becomes Ks.

   Ans infinity

29. For the open-loop plant for a unity feedback system

$$G(s) \frac{K(s + s_1)}{s^2(s + s_2)(s + s_4)}$$

   (a) State the type of system.

   (b) Find κ_0, κ_1, and κ_2.

   Ans $\kappa_0 = \infty = \kappa_1$, $\kappa_2 = K s_1/s_2 s_4$

30. In a position servo, addition of "rate feedback" increases the damping. The block diagram for the system is shown in Fig. P3.30.

   (a) Derive the damping ratio and resonant frequency and show how they depend on τ.

   (b) Calculate the steady state error in response to a unit ramp function input.

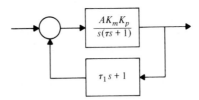

Figure P3.30

Root Locus Analysis

4.1 PREVIEW

In Chap. 1, we learned that differential equations could be written for electrome-chanical devices often associated with control system applications. The equations can be solved and the response can be divided into forced and natural components. Chapter 2 established definitions related to the natural response, especially definitions of types of stability. Stability is generally considered to be a property of the natural response, since the natural response is due to the system and not due to outside influences by which the control input evolves. The other response compo-nent was considered in Chap. 3: the steady state response; that is, the tendency of a device to follow (track) a command.

In Chap. 4, we return to stability, a property of the natural response, but we provide a much broader and more useful measure of that stability (as compared with Chap. 3). The Routh-Hurwitz approach provides a generally yes or no answer to the stability question, as would be expected of a method predating modern digital computation (the method having been developed in the 1890s). Although a range of values emerge for a variable gain (so that the closed-loop system is stable), there is no real advice as to which of these gain values are preferable.

A logical approach to determining stability is to extract the roots of the characteristic polynomial as the adjustable gain varies (and indeed computational packages are readily available to do just that). Faced with poor root solving capabilities in his era, Walter Evans developed a set of rules in the mid-1940s by which the path traced by the closed-loop characteristic equation roots could be sketched to reasonable accuracy as the gain varies. This plot is referred to as a root locus.

At this point in our discussion, it might seem logical to let today's computers do all the work, and to skip the sketching rules altogether. Suppose a beginning trigonometry student reaches a comparable conclusion since trig functions are readily available on hand-held calculators. That student might attempt to key in the cosine of 60° and write down the computer display of 1.732. A nearby student who had studied trig would conclude that the tangent of 60° was obtained instead. The student who owns the calculator would not otherwise recognize that an error had been made.

By all means, computational packages should be used to remove the drudgery of repetitive computation and plotting, but the user must be able to perceive obviously erroneous results from those generally expected when using the sketching rules to follow. If a computer is used without being able to critically evaluate the numerical output, in a very real sense, the student and the computer change roles in terms of which is the master and which is the slave, blindly following orders without question. With this caution in mind, we refer the user to computational packages like those mentioned in the preface or to a PC-compatible package available from the publisher to accompany this text.

Root locus methods are so widely used in industry that two chapters are devoted to the topic. Chapter 4 is devoted to a basic understanding of root locus principles, while Chap. 5 deals with root locus compensation, a design application of the method.

The versatility of the root locus method is demonstrated by the four examples that terminate Chap. 4. A system is designed so as to track a light source. A control system is developed in which minute electrical muscle impulses are amplified to drive an artificial limb. A hydroelectric generating system is designed by modeling the system using transmittances and then using root locus techniques to stabilize the voltage-controlled response. A root locus analysis of a flexible spacecraft demonstrates that omitting flexible dynamics could cause the designer to overlook a potential cause of instability.

4.2 POLE-ZERO PLOTS

Poles and Zeros

The *zeros* of a function are the values of the variable for which the function is zero. The *poles* of a function are the values of the variable for which the function is infinite, for which its inverse is zero. For a rational function, the zeros are the roots of the numerator polynomial and the poles are the roots of the denominator polynomial. The function

$$F(s) = \frac{-3s^3 + 6s^2 - 3s + 6}{2s^4 + 12s^3 + 36s^2 + 8s} = \frac{-3(s-2)(s+j)(s-j)}{2s(s+4)(s+1+3j)(s+1-3j)}$$

has zeros at $s = 2$ and $s = \pm j$. Its poles are at $s = 0$, $s = -4$, and $s = -1 \pm 3j$. In general, a rational function can be placed in the factored form

$$F(s) = \frac{b_m s^m + b_{m-1}s^{m-1} + \cdots + b_1 s + b_0}{a_n s^n + a_{n-1}s^{n-1} + \cdots + a_1 s + a_0} = \frac{k(s-z_1)(s-z_2)\cdots(s-z_m)}{(s-p_1)(s-p_2)\cdots(s-p_n)}$$

where its zeros, $z_1, z_2, \ldots, z_m$ and its poles, $p_1, p_2, \ldots, p_n$ are in evidence. The constant

$$k = \frac{b_m}{a_n}$$

is the *multiplying constant* of the function.

When the poles and zeros of a function are plotted on the complex plane, the result is a *pole-zero plot*, from which important properties of the function can be visualized. The zero locations are indicated by O on the plot and pole locations are indicated by X. Figure 4.1 shows pole-zero plots for the following functions:

$$F_1(s) = \frac{4s + 5}{s^3 + 4s^2 + 13s} = \frac{4(s + \frac{5}{4})}{s(s + 2 + 3j)(s + 2 - 3j)}$$

$$F_2(s) = \frac{-s^2 - 4}{2s^2 + 14s + 24} = \frac{-\frac{1}{2}(s + 2j)(s - 2j)}{(s + 3)(s + 4)}$$

We use the notation of Clark (1962) and others whereby the multiplying constant of a rational function is placed in a box at the right of the pole-zero plot. The rational function is then entirely given by the plot. This arrangement is by no means common in practice, but the alternatives of separately accounting for the multiplying constant or dealing only with functions with unity multiplying constant are apt to result in unnecessary complication, confusion, and error.

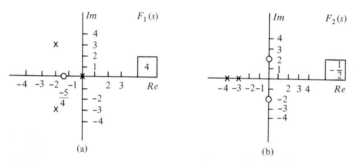

(a)   (b)

Figure 4.1 Pole-zero plots.

Graphical Evaluation

A rational function

$$F(s) = \frac{k(s - z_1)(s - z_2) \cdots (s - z_m)}{(s - p_1)(s - p_2) \cdots (s - p_n)}$$

when evaluated at a specific value of the variable, $s = s_0$, is

$$F(s_0) = \frac{k(s_0 - z_1)(s_0 - z_2) \ldots (s_0 - z_m)}{(s_0 - p_1)(s_0 - p_2) \ldots (s_0 - p_n)}$$

On a pole-zero plot, suppose a directed line segment is drawn from the position of a pole, say p_1, to the value s_0 at which the function is to be evaluated. The segment has length $|s_0 - p_1|$ and makes the angle $\angle(s_0 - p_1)$ with the real axis, as indicated in Fig. 4.2. Thus,

$$F(s_0) = \frac{k(|s_0 - z_1|e^{j\angle(s_0 - z_1)})(|s_0 - z_2|e^{j\angle(s_0 - z_2)}) \cdots}{(|s_0 - p_1|e^{j\angle(s_0 - p_1)})(|s_0 - p_2|e^{j\angle(s_0 - p_2)}) \cdots}$$

$$|F(s_0)| = \frac{|k| \begin{pmatrix} \text{product of the lengths of the directed} \\ \text{line segments from the zeros to } s_0 \end{pmatrix}}{\begin{array}{l} \text{product of the lengths of the directed} \\ \text{line segments from the poles to } s_0 \end{array}}$$

$$\angle F(s_0) = \begin{bmatrix} \text{(sum of the angles of the directed line segments} \\ \text{from the zeros to } s_0) - \text{(sum of the pole angles)} \\ + 180° \text{ if } k \text{ is negative} \end{bmatrix}$$

Should k be positive, the 180° is not added to the angle, and $|k| = k$.

For example, for $F(s)$ with the pole-zero plot of Fig. 4.3, a graphical evaluation at $s_0 = -1 + j3$ gives

$$|F(s = -1 + j3)| = \frac{6(5)}{(3)(3)(5.4)(2.2)} = 0.28$$

$$\angle F(s = -1 + j3) = 143° - 90° - 90° - 68° - 27° + 180° = 48°$$

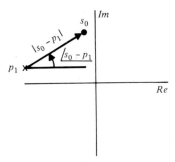

Figure 4.2 Evaluating a rational function at a point $s = s_0$.

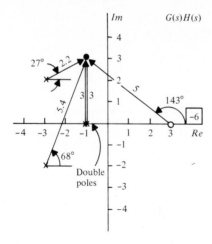

Figure 4.3   Graphical evaluation of a rational function.

DRILL PROBLEMS

D4.1. Draw pole-zero plots for the following functions. Include the multiplying constant with the plot.

(a) $F(s) = \dfrac{3s - 1}{s^2 + 2s}$

> **Ans**   zero at $s = \frac{1}{3}$; poles at $s = 0$ and $s = -2$; multiplying constant 3

(b) $F(s) = \dfrac{9s^2 + 1}{(s^2 + 8s + 17)^2}$

> **Ans**   zeros at $s = \pm j\frac{1}{3}$; repeated poles at $s = -4 \pm j$; multiplying constant 9

(c) $F(s) = \dfrac{-2s^2 + 6s + 3}{(s^2 + 3s + 8)(s^2 + 6s + 15)}$

> **Ans**   zeros at $s = 3.44$ and $s = -0.44$; poles at $-1.5 \pm j2.4$ and $-3 \pm j2.45$; multiplying constant -2

(d) $F(s) = \dfrac{(3s + 1)(2s + 1)}{(4s + 1)(7s + 1)^2}$

> **Ans**   zeros at $s = -\frac{1}{3}$ and $s = -\frac{1}{2}$; pole at $s = -\frac{1}{4}$ and repeated pole at $s = -\frac{1}{7}$; multiplying constant 3/98

D4.2. Find the rational functions represented by the following pole-zero plots.

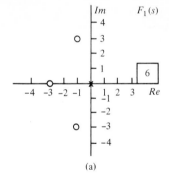

(a)

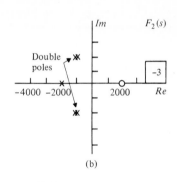

(b)

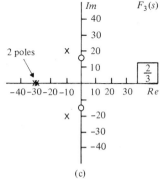

(c)

Ans (a) $6(s + 3)(s^2 + 2s + 10)/s$; (b) $-3(s - 2000)/(s + 2000)(s^2 + 2000s + 5 \times 10^6)^2$; (c) $\frac{2}{3}(s^2 + 225)/(s + 30)^2(s^2 + 20s + 500)$

D4.3. Carefully sketch pole-zero plots for the following functions, then graphically evaluate the functions at the indicated value of the variable s.

(a) $F(s) = \dfrac{10(s - 2)}{(s + 1)(s + 2)}$

$s = j3$

Ans $3.2e^{-j5°}$

(b) $F(s) = \dfrac{4s^2 + 32}{(s^2 + 8s + 20)(s + 2)}$

$s = 2 + j$

Ans $0.3e^{-j11°}$

(c) $F(s) = \dfrac{4(s^2 - 4s + 5)^2}{(s + 3)^2(s^2 + 6s + 10)}$

$s = j3$

Ans $2.0e^{-j34°}$

4.3 ROOT LOCUS FOR FEEDBACK SYSTEMS

Angle Criterion

A *root locus plot* is a drawing of the loci of the poles of a rational function as some system parameter is varied. The basic root locus problem applies directly to the simple feedback system of Fig. 4.4, for which the transfer function is

$$T(s) = \frac{KG(s)}{1 + KG(s)H(s)}$$

where the constant gain K is the parameter of interest. The poles of the transfer function are the roots of

$$1 + KG(s)H(s) = 0$$

which depend upon the parameter K. The product of the forward transmittance $KG(s)$ and the feedback transmittance $H(s)$ is termed the open-loop transmittance (or gain) of the system. The poles and zeros of $G(s)H(s)$ are called the open-loop poles and zeros, while the poles and zeros of $T(s)$ are closed-loop poles and zeros. The open-loop transfer function can also be visualized as being the transfer function $Y_1(s)/R(s)$ when the loop in Fig. 4.4 is broken at the point denoted by x.

We consider now K in the range $0 \le K < \infty$. If

$$1 + KG(s)H(s) = 0$$

then

$$G(s)H(s) = -\frac{1}{K}$$

and for positive K, this means that a point s which is a pole of $T(s)$ makes

$$|G(s)H(s)| = \frac{1}{K}$$

and

$$\angle G(s)H(s) = \text{odd multiple of } 180°$$

Suppose that there is a point s for which the second of these conditions is satisfied. Then whatever the magnitude of $G(s)H(s)$ for this value of s, there is a correspond-

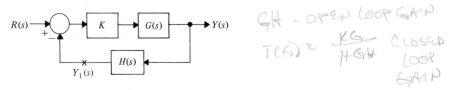

Figure 4.4   A simple feedback system with adjustable gain K.

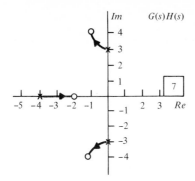

Figure 4.5 A root locus plot.

ing value of K. Thus any point s for which $\angle G(s)H(s) = 180°$ is a point of the root locus, for some positive value of K.

Root Locus Properties

Figure 4.5 shows an example root locus plot for a feedback system with open-loop transmittance

$$KG(s)H(s) = \frac{K(s + 2)(s^2 + 2s + 17)}{(s + 4)(s^2 + 9)}$$

It consists of a pole-zero plot for $G(s)H(s)$, which is generally easy to construct because $G(s)$ and $H(s)$, being components of the system, are usually known in factored or partially factored form. Superimposed upon the pole-zero plot for $G(s)H(s)$ are the curves that are the loci of the poles of $T(s)$ as K varies from zero to infinity. The locus segments are symmetrical about the real axis, and the sense of increasing K usually is indicated on each segment. As $K \to 0, |G(s)H(s)| \to \infty$, and as $K \to \infty, |G(s)H(s)| \to 0$. This is to say that the poles of $T(s)$ are near the poles of $G(s)H(s)$ for small K and are near the zeros of $G(s)H(s)$ for large K. The loci begin on the poles of GH and end on the zeros of GH.

To determine if a given point s_0 is a point on the root locus for some value of K between zero and $+\infty$, it is only necessary to determine whether or not the angle of $G(s)H(s)$ is $180°$. This determination is easily made graphically, using directed line segments:

$$\angle G(s_0)H(s_0) = \text{sum of zero angles to } s_0$$

$$- \text{ sum of pole angles to } s_0$$

$$+ 180° \text{ if the multiplying constant is negative}$$

For the GH product with pole-zero plot given in Fig. 4.6a, the angle of $G(s)H(s)$ for the indicated value of s is (approximately)

$$63° - (56° + 30° + 27°) = -50°$$

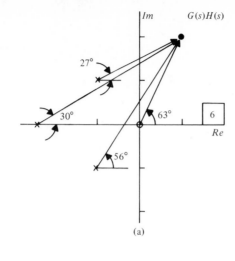

(a)

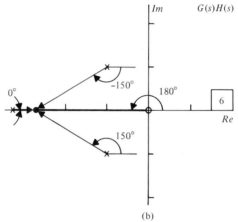

(b)

Figure 4.6   Testing the angle of $G(s)H(s)$ to determine if a point is on the root locus.

Thus the indicated point is not on the root locus. For the point tested in Fig. 4.6b.

$$180° - (0° + 150° - 150°) = 180°$$

Hence that point is on the root locus. There is a pole of $T(s)$ there for some positive value of K.

DRILL PROBLEM

D4.4. Graphically find the angle of $G(s)H(s)$ at each of the indicated points.

   Ans   (a) $180°$;   (b) $256°$;   (c) $-117°$

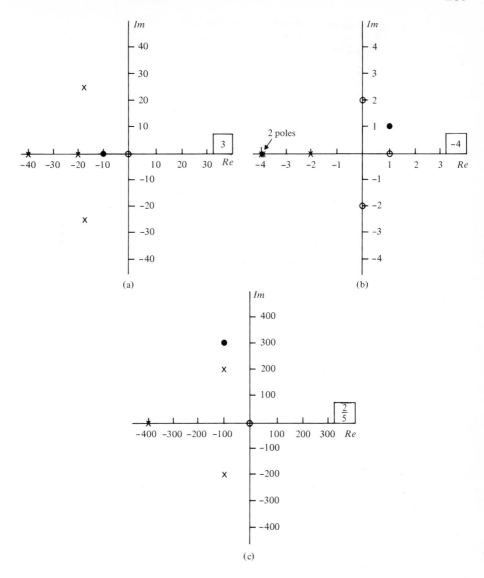

(a)

(b)

(c)

4.4 ROOT LOCUS CONSTRUCTION

Construction Principles

At first, we consider only the most common case where the parameter K of interest is nonnegative and where the multiplying constant of the $G(s)H(s)$ product is positive. There are several simple rules that allow approximate root locus sketches to be made easily and rapidly. These are listed in Table 4.1. If greater accuracy is

Table 4.1 BASIC ROOT LOCUS PRINCIPLES

1. The branches of the locus are continuous curves that start at each of the n poles of GH, for $K = 0$. As K approaches $+\infty$, the locus branches approach the m zeros of GH. Locus branches for excess poles extend infinitely far from the origin; for excess zeros, locus segments extend from infinity.
2. The locus includes all points along the real axis to the left of an odd number of poles plus zeros of GH.
3. As K approaches $+\infty$, the branches of the locus become asymptotic to straight lines with angles

$$\theta = \frac{180° + i360°}{n - m}$$

for $i = 0, \pm 1, \pm 2, \ldots$ until all $n - m$ or $m - n$ angles are obtained, where n is the number of poles and m is the number of zeros of GH.
4. The starting point of the asymptotes, the centroid of the pole-zero plot, is on the real axis at

$$\sigma = \frac{\Sigma \text{ pole values of } GH - \Sigma \text{ zero values of } GH}{n - m}$$

5. Loci leave the real axis at a gain K that is the maximum K in that region of the real axis. Loci enter the real axis at the minimum value of K in that region of the real axis. These points are termed breakaway points and entry points, respectively. A pair of locus segments leave or enter the real axis at angles of $\pm 90°$.
6. The angle of departure ϕ of a locus branch from a complex pole is given by

(sum of the other GH pole angles to the pole under consideration) $+ \phi -$ (sum of the GH zero angles to the pole) $= 180°$

The angle of approach ϕ' of a locus branch to a complex zero is given by

(sum of the GH pole angles to the zero under consideration) $-$ (sum of the other GH zero angles to the zero) $- \phi' = 180°$

Multiple angles of departure and approach at repeated complex poles of GH are found similarly, using multiple contributions of ϕ or ϕ' and equating to $180° \pm i360°$.

---

needed in a region of the complex plane, points near the approximate loci can be tested to determine their locations more accurately.

Because most transmittances encountered in practice have more poles than zeros, and because $180°$ and $-180°$ are the same angle, the negative of the angle of the GH product is usually calculated, and the angle criterion becomes

Sum of pole angles to $s_0 -$ sum of zero angles to $s_0 = - \angle G(s_0)H(s_0) = 180°$

We will use this "reverse angle" evaluation in the development to follow.

Rule 1. Loci Branches

The branches of the locus are continuous curves that start at each of the n poles of GH, for $K = 0$. As K approaches $+\infty$, the locus branches approach the m zeros of

GH. Locus branches for excess poles extend infinitely far from the origin; for excess zeros, locus segments extend from infinity.

If $G(s)H(s)$ has more poles than zeros, some of the segments of the root locus, which start on poles (for $K = 0$) do not have a zero to end upon (for $K \to \infty$). These segments of the locus extend from the poles, infinitely far from the origin of the complex plane. It is said that these loci extend "to infinity." If $G(s)H(s)$ has more poles n than zeros m, m segments of the locus extend from a pole to a zero, and $n - m$ excess segments each start at a pole and extend infinitely far from the origin. Segments never extend from a pole to infinity and then back from infinity to a zero. If $G(s)H(s)$ has more zeros than poles, the situation is similar, with n segments extending from a pole to a zero and $m - n$ excess segments coming from infinity to a zero. Some examples of root locus plots are given in Table 4.2.

Table 4.2 SOME ROOT LOCUS PLOTS

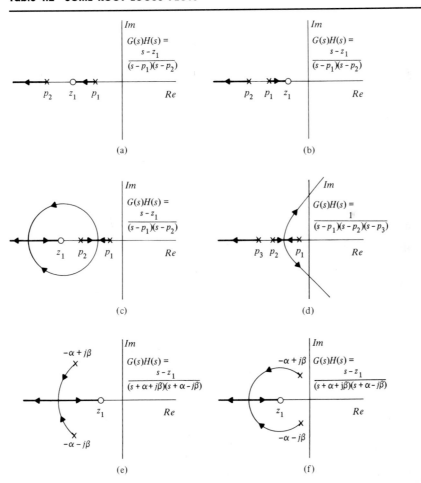

$$\text{(a)} \quad G(s)H(s) = \frac{s - z_1}{(s - p_1)(s - p_2)}$$

$$\text{(b)} \quad G(s)H(s) = \frac{s - z_1}{(s - p_1)(s - p_2)}$$

$$\text{(c)} \quad G(s)H(s) = \frac{s - z_1}{(s - p_1)(s - p_2)}$$

$$\text{(d)} \quad G(s)H(s) = \frac{1}{(s - p_1)(s - p_2)(s - p_3)}$$

$$\text{(e)} \quad G(s)H(s) = \frac{s - z_1}{(s + \alpha + j\beta)(s + \alpha - j\beta)}$$

$$\text{(f)} \quad G(s)H(s) = \frac{s - z_1}{(s + \alpha + j\beta)(s + \alpha - j\beta)}$$

Table (*Continued*)

Table 4.2 (*Continued*)

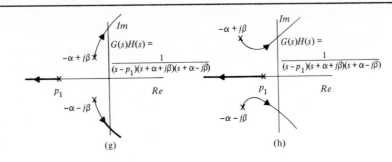

$$G(s)H(s) = \frac{1}{(s - p_1)(s + \alpha + j\beta)(s + \alpha - j\beta)}$$

(g)

$$G(s)H(s) = \frac{1}{(s - p_1)(s + \alpha + j\beta)(s + \alpha - j\beta)}$$

(h)

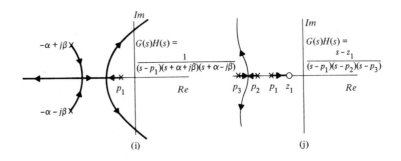

$$G(s)H(s) = \frac{1}{(s - p_1)(s + \alpha + j\beta)(s + \alpha - j\beta)}$$

(i)

$$G(s)H(s) = \frac{s - z_1}{(s - p_1)(s - p_2)(s - p_3)}$$

(j)

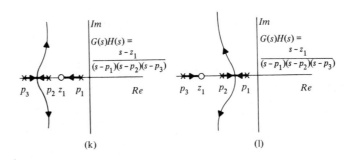

$$G(s)H(s) = \frac{s - z_1}{(s - p_1)(s - p_2)(s - p_3)}$$

(k)

$$G(s)H(s) = \frac{s - z_1}{(s - p_1)(s - p_2)(s - p_3)}$$

(l)

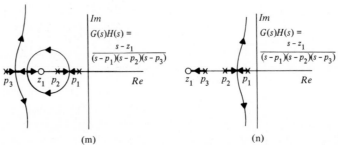

$$G(s)H(s) = \frac{s - z_1}{(s - p_1)(s - p_2)(s - p_3)}$$

(m)

$$G(s)H(s) = \frac{s - z_1}{(s - p_1)(s - p_2)(s - p_3)}$$

(n)

Table (*Continued*)

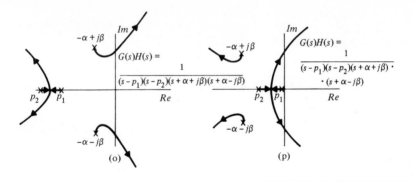

Rule 2. Real Axis Segments

The locus includes all points along the real axis to the left of an odd number of poles plus zeros of *GH*.

The easiest points on the complex plane to test to see if they are on the root locus are points on the real axis. For these points, the angle contribution of each real axis pole or zero is either $0°$ or $180°$, depending upon whether the root is to the right or to the left of the real axis point being tested, as indicated in Figs. 4.7a, b. A set of complex conjugate roots contributes angles to real axis points which are negatives of one another, so the net contribution to angle of a complex set of roots is zero, as indicated in Fig. 4.7c.

A point on the real axis is thus on the root locus if and only if it is to the left of an odd number of roots (poles and zeros), so that the angle of *GH* at that point is an odd multiple of $180°$. The real axis root locus segments of several systems are sketched in Fig. 4.8. In Fig. 4.8a, the loci are entirely along the real axis. In Fig. 4.8b, one segment extends from the double pole to the zero at $s = 0$, and one extends from the double pole to infinity. Two other root locus segments will extend from the complex poles; the sketching of these segments will be discussed later. In Fig. 4.8c, there are no real axis locus segments.

Rule 3. Asymptotic Angles

As *K* approaches $+\infty$, the branches of the locus become asymptotic to straight lines with angles

$$\theta = \frac{180° + i360°}{n - m}$$

for $i = 0,\ \pm 1,\ \pm 2, \ldots$ until all $(n - m)$ or $(m - n)$ angles are obtained; *n* is the number of poles and *m* is the number of zeros of *GH*.

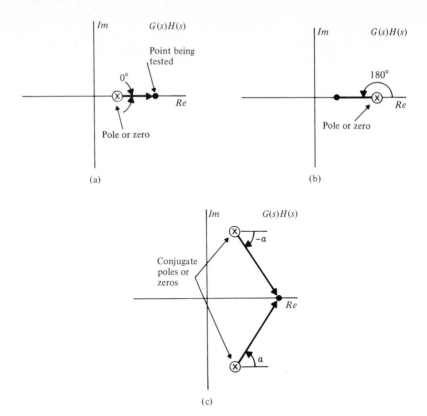

Figure 4.7 Testing points on the real axis.

At points on the complex plane very far from all of the poles and zeros of $G(s)H(s)$, the angle of GH is virtually

(Number of poles − number of zeros)θ

where θ is the angle of the point itself, as indicated in Fig. 4.9. At large distances from the cluster of poles and zeros, root loci extending to or from infinity approach straight-line asymptotes at angles given by

(Number of poles − number of zeros)$\theta = 180° \pm i360°$

It is important to include the multiples of 360° in this formulation because it is from the term that multiple solutions, where there is more than one asymptotic angle, arise.

If, for example, the GH product has two zeros and six poles, the asymptotic angles are given by

$$(6-2)\theta = 180° \pm i360° \qquad \theta = \frac{180° \pm i360°}{4}$$

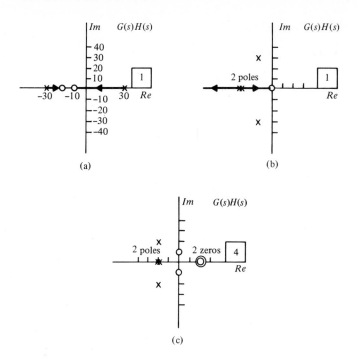

(a)

(b)

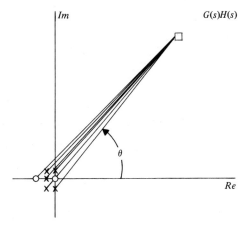

(c)

Figure 4.8 Example of real axis root locus segments.

Substituting various integer values of i, say 0, 1, 2, 3, ..., the four different angles

$$\theta = +45°, +135°, 225°, 315°$$
$$= \pm45°, \pm135°$$

result. Substitutions of additional integers simply give repetitions of the same angles. There are $n - m = 4$ different asymptotic angles in this example.

Figure 4.9 Angle contributions far from the poles and zeros of *GH.*

Rule 4. Centroid of the Asymptotes

The starting point of the asymptotes, the centroid of the pole-zero plot, is on the real axis at

$$\sigma = \frac{\Sigma \text{ pole values of } GH - \Sigma \text{ zero values of } GH}{n - m}$$

The centroid is a sort of reverse vanishing point in that, as the closed-loop poles move to some distance away from the real axis, an observer looking backward would see the centroid as an apparent point of departure from the real axis. When $n - m$ is either zero or 1, the centroid is not used. For those values of $n - m$, closed-loop poles do not move off the real axis on the way toward infinity. When $n - m$ is zero, all the open-loop poles have an open-loop zero to approach as the gain rises toward infinity. When $n - m$ is 1, one closed-loop pole moves toward infinity along the negative real axis (moving at an asymptote angle of $180°$).

For large gain values, those s values on the root locus are approximately given by

$$s = \sigma + m(K)$$

where $m(K)$ is a large magnitude that grows with K. Since

$$s - \sigma = m(K)$$

each s in the equation

$$1 + KG(s)H(s) = 1 + \frac{KN(s)}{D(s)}$$

causes that equation to be

$$1 + KG(s)H(s) = 1 + \frac{K}{(s - \sigma)^{n-m}}$$

Because s is large, the polynomials $D(s)$ and $N(s)$ can be approximated by the first two terms

$$D(s) \approx s^n + a_{n-1}s^{n-1}$$

$$N(s) \approx s^m + b_{m-1}s^{m-1}$$

Using long division,

$$\frac{D(s)}{N(s)} \approx s^{n-m} + (a_{n-1} - b_{m-1})s^{n-m-1}$$

Using the binomial expansion,

$$(s - \sigma)^{n-m} \approx s^{n-m} + (n - m)\sigma s^{n-m-1}$$

The two expressions can be equated

$$\sigma = \frac{a_{n-1} - b_{m-1}}{n - m}$$

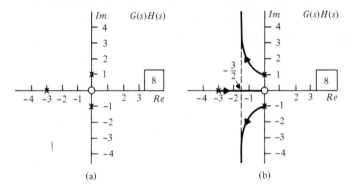

Figure 4.10 Root locus construction with asymptotes and centroid.

The coefficient for the second highest power of s is always the sum of the roots of that polynomial; therefore, the numerator of the expression for σ is the sum of the pole values of GH (the sum of the roots of $D(s)$) minus the sum of the zero values for GH [the sum of the roots of $N(s)$].

Consider the system with pole-zero plot given in Fig. 4.10a. A real axis segment of the locus extends from the real axis pole to the zero at $s = 0$. The other two locus segments extend from the imaginary axis poles to infinity. Their asymptotic angles are given by

$$(3 - 1)\theta = 180° \pm i360° \qquad \theta = 90°, -90°$$

The centroid of the asymptotes is

$$\sigma = \frac{(0 + j + 0 - j - 3) - 0}{3 - 1} = -\frac{3}{2}$$

The imaginary part contributions of conjugate sets of roots always cancel one another, so only the real parts of the root locations need to be included in the centroid calculation. The complete root locus is shown in Fig. 4.10b.

DRILL PROBLEMS

D4.5. Sketch root locus plots, for an adjustable constant K between 0 and $+\infty$, for systems with the following GH products. Find the asymptotic angles and centroid if applicable.

(a) $G(s)H(s) = \dfrac{3s}{(s + 2)(s^2 + 6s + 18)}$

   Ans $\pm 90°;\ -4$

(b) $G(s)H(s) = \dfrac{10}{(s + 6)(s^2 + 8s + 41)^2}$

Ans $\pm 36°, \pm 108°, -180°; -4.4$

(c) $G(s)H(s) = \dfrac{1}{(s^2 + 8s + 41)(s^2 + 2s + 5)}$

Ans $\pm 45°, \pm 135°; -2.5$

(d) $G(s)H(s) = \dfrac{7}{(s + 1)(s^2 + 10s + 26)}$

Ans $\pm 60°, -180°; -\frac{11}{3}$

(e) $G(s)H(s) = \dfrac{2(s^2 + 4s + 3)}{(s + 2)^2(s + 4)}$

Ans $-180°$; centroid not applicable

4.5 MORE ABOUT ROOT LOCUS

Root Locus Calibration

The values of the adjustable constant K corresponding to various points on a root locus can be found by applying the relation

$$|G(s)H(s)| = \dfrac{1}{|K|}$$

for points on the locus. The magnitude $|G(s)H(s)|$ can be found graphically for a point on the locus using

$$|G(s)H(s)| = \dfrac{\left(\begin{array}{c}\text{magnitude of the}\\\text{multiplying constant}\end{array}\right)\left(\begin{array}{c}\text{product of}\\\text{zero distances}\end{array}\right)}{\text{product of pole distances}}$$

For the system of Fig. 4.11, for example, the value of K corresponding to the imaginary axis points on the locus, where $T(s)$ is marginally stable, is given approximately by

$$|G(s)H(s)| = \dfrac{10}{(3.3)(4.1)(5.0)} = \left|\dfrac{1}{K}\right| \qquad K = 6.8$$

Thus for K greater than about this value, the overall system will be unstable. If the root locus is only approximate, this solution for K is approximate, too. More accuracy can easily be obtained, if needed, by testing the angles of some points near the approximate locus and refining the solution for K.

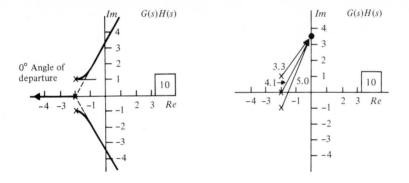

Figure 4.11 Calibrating a point on the root locus.

Rule 5. Breakaway and Entry Points

Loci leave the real axis at a gain K that is the maximum K in that region of the real axis. Loci enter the real axis at the minimum value of K in that region of the real axis. These points are termed breakaway points and entry points respectively. A pair of locus segments leave or enter the real axis at angles of $\pm 90°$.

There are at least two distinctly different ways to compute breakaway and entry points. First, the gain K as a function of s can be differentiated with respect to s, the result set equal to zero, and the resulting polynomial solved to obtain the answer. If a fourth-order polynomial is differentiated, a third-order polynomial must be root solved. As the order of the system rises, the differentiation approach becomes increasingly more difficult. However, Appendix B contains an algorithm that automatically searches for a value of s that maximizes K (breakaway point) or minimizes K (entry point). The algorithm can be programmed on anything from a mainframe computer to a hand-held calculator. The algorithm requires an initial value of s somewhere near to the eventual answer. The pole-zero plot normally provides a reasonably accurate initial trial point.

The second (and more obvious) method is to try various values of s until a value results in the largest (or smallest) K to a desired accuracy. In the examples that follow, this trial-and-error approach is used. More adept programmers are encouraged to implement Appendix B.

For the system with $G(s)H(s)$ as in Fig. 4.12, there is a real axis root locus segment, but it is not complete because it does not extend from an open-loop pole to an open-loop zero. For this system, there are asymptotes with angles

$$\theta = \frac{180° + i360°}{2} = 90°, -90°$$

and centroid

$$\sigma = \frac{-3 - 1}{2} = -2$$

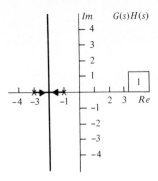

Figure 4.12 Breakaway of loci from the real axis.

As K varied from zero to infinity, the poles of the overall transfer function begin as the poles of GH. For larger K, the overall transfer function $T(s)$ has two real axis poles which become closer and closer together with increasing K. At some gain K there are two repeated poles in $T(s)$ at $s = -2$. For still larger K, the poles are complex conjugates with larger and larger imaginary parts. In this simple case, the locus breaks away from the real axis at the centroid, $s = -2$, as can be verified by examining the angle of $G(s)H(s)$ at points slightly above and below $s = -2$.

In more involved systems, a breakaway of two root locus segments from the real axis is also at 90° angles, but the breakaway point is not necessarily midway between the real axis GH roots. For a segment extending from poles, as in the example of Fig. 4.13a, the real axis point at which the loci breakaway will correspond to the largest value of K for locus points on that segment. Calculations of K for several real axis test points between 0 and -5 are summarized in Table 4.3. From these, it is seen that the largest value of K on that real axis segment is in the vicinity of $s = -2.9$, which must be the location of the breakaway point to an accuracy of 0.1 unit. Since the breakaway point has one digit to the left of the decimal point, the breakaway point is computed to 2 significant digits.

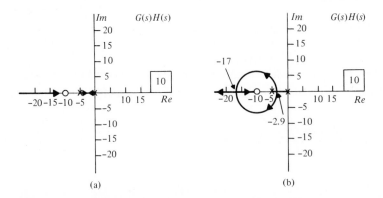

Figure 4.13 Calculating breakaway and entry points.

Table 4.3 BREAKAWAY POINT CALCULATIONS FOR THE EXAMPLE OF FIGURE 4.13

s_0	$|G(s_0)H(s_0)| = 1/K$	K
-2	$\dfrac{10(8)}{(3)(2)}$	0.075
-2.5	$\dfrac{10(7.5)}{(2.5)(2.5)}$	0.083
-2.8	$\dfrac{10(7.2)}{(2.2)(2.8)}$	0.0855
-2.9	$\dfrac{10(7.1)}{(2.1)(2.9)}$	0.0858
-3	$\dfrac{10(7)}{(2)(3)}$	0.0857
-3.5	$\dfrac{10(6.5)}{(1.5)(3.5)}$	0.081
-4	$\dfrac{10(6)}{(1)(4)}$	0.067

In the example, Fig. 4.13b, there is also a point to the left of $s = -10$ on the real axis where the loci enter the real axis. The real axis entry point of the loci will correspond to the smallest value of K on that segment of the axis, since for larger K, the loci approach the zero and $-\infty$. Calculations of K for real axis test points to the left of $s = -10$ are given in Table 4.4, and it is seen that the entry point of the locus is at $s = -17$ to an accuracy of 0.5 units. Since the entry point has 2 digits to the left of the decimal point, an accuracy of 0.1 unit means 3 significant digits while an accuracy of 1 unit means 2 significant digits. An accuracy of 0.5 units can be thought of as $2\frac{1}{2}$ significant digits.

Rule 6. Angles of Departure and Approach

The angle of departure ϕ of a locus branch from a complex pole is given by

(Sum of the other GH pole angles to the pole under consideration)
$+ \phi -$ (sum of the GH zero angles to the pole)
$= 180°$

The angle of approach ϕ' of a locus branch to a complex zero is given by

(Sum of the GH pole angles to the zero under consideration)
$-$ (sum of the other GH zero angles to the zero) $- \phi'$
$= 180°$

Table 4.4 ENTRY POINT CALCULATIONS FOR THE EXAMPLE OF FIGURE 4.13

s_0	$\|G(s_0)H(s_0)\| = 1/K$	K
-15	$\dfrac{10(5)}{(10)(15)}$	3.00
-16	$\dfrac{10(6)}{(11)(16)}$	2.93
-17	$\dfrac{10(7)}{(12)(17)}$	2.91
-17.5	$\dfrac{10(7.5)}{(12.5)(17.5)}$	2.917
-18	$\dfrac{10(8)}{(13)(18)}$	2.925
-20	$\dfrac{10(10)}{(15)(20)}$	3.00
-25	$\dfrac{10(15)}{(20)(25)}$	3.33

Multiple angles of departure and approach at repeated complex roots of *GH* are found similarly, using multiple contributions of ϕ or ϕ' and equating to $180° \pm i360°$.

For a set of complex conjugate poles of $G(s)H(s)$, the angle at which the root locus leaves one of the poles is found by considering a point on the locus branch very close to the pole, as in the example of Fig. 4.14. Since the point is very close to the pole under consideration, the angles to the point are virtually the angles to the pole itself. Solving

(Sum of other pole angles to the pole under consideration)
$\quad + \phi -$ (sum of zero angles to the pole)
$= 180°$

will give the angle of departure, ϕ. Of course, the angle of departure from the lower pole of the conjugate set is the negative of the upper pole's angle of departure. Angles of approach to complex zeros are found similarly:

(Sum of pole angles to the zero under consideration)
$\quad -$ (sum of other zero angles to the zero) $- \phi'$
$= 180°$

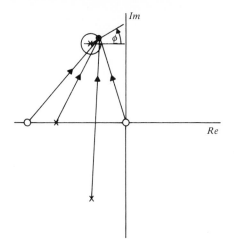

Figure 4.14 Finding an angle of departure.

Consider the system with *GH* product given in Fig. 4.15a. For a point on the root locus near the top complex pole, approximately,

$$90° + 50° + \phi - 140° = 180° \qquad \phi = 180°$$

where ϕ is the angle of departure of the locus from the pole. There is a complete real axis segment of the locus between the real axis pole and zero. The other two locus segments extend from the complex poles to infinity, with asymptotic angles given by

$$(3 - 1)\theta = 180° \pm i360° \qquad \theta = \pm 90°$$

and centroid

$$\sigma = \frac{(-4 - \frac{3}{2} - \frac{3}{2}) - 2}{3 - 1} = -\frac{9}{2}$$

A completed root locus sketch is given in Fig. 4.15b.

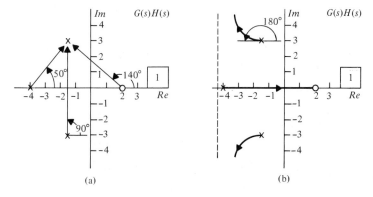

(a)   (b)

Figure 4.15 Root locus construction using angles of departure.

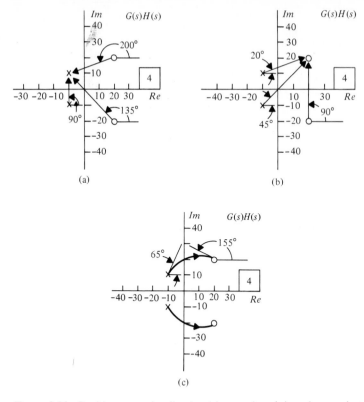

Figure 4.16 Root locus construction involving angles of departure and angles of approach.

Consider the system of Fig. 4.16a. For a point on the root locus near the top complex pole, approximately

$$90° + \phi - 200° - 135° = 180° \qquad \phi = 65°$$

where ϕ is the angle of departure of the locus from the top pole. For a point near the top complex zero, Fig. 4.16(b), the angle ϕ' of arrival of the locus is given approximately by

$$20° + 45° - 90° - \phi' = 180°; \qquad \phi' = 155°$$

A complete root locus sketch is shown in Fig. 4.16c.

At repeated complex roots of GH, more than one locus segment begins or ends at the root location, so more than one angle of departure or approach will be found, a different angle for each locus segment. For the system of Fig. 4.17a, the angle contribution to a point near the top double pole is given approximately by

$$90° + 90° + 108° + \phi + \phi - 124° = 180° \pm i360° \qquad \phi = 8°, 188°$$

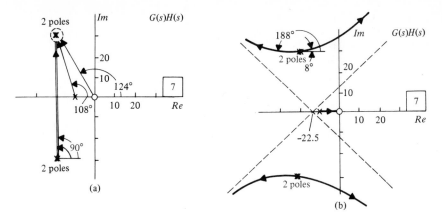

Figure 4.17 Angles of departure from multiple poles.

Inclusion of the multiples of 360° is important here because it is this term that gives the multiple solutions. It is easy to see that multiple departure (or arrival) angles will be evenly spaced around the multiple roots. To complete the root locus sketch as in Fig. 4.17b, the real axis locus segment is drawn and the asymptotic angles and centroid are found:

$$(5-1)\theta = 180° \pm i360° \qquad \theta = \pm 45°, \pm 135°$$

$$\sigma = \frac{-20-20-20-20-10}{5-1} = -22.5$$

Another Example

As an example of application of the six root locus construction rules, consider the open-loop transfer function

$$KG(s)H(s) = \frac{K}{s(s+3)(s^2+6s+64)}$$

A step-by-step procedure is as follows:

Locate the open-loop poles and zeros and plot them (rule 1). There are no zeros of GH. The poles of GH are located at 0, -3, $-3+j7.4$, and $-3-j7.4$.

Locate real axis portions of the locus (rule 2). The real axis segment between $s=0$ and $s=-3$ is on the root locus. The root locus diagram so far is given in Fig. 4.18(a).

Determine the angles of the asymptotes (rule 3). The asymptotic angles are given by

$$\theta = \frac{180° \pm i360°}{4-0} = 45°, 135°, -135°, -45°$$

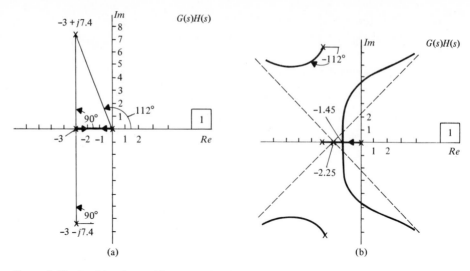

Figure 4.18 Applying the root locus construction rules.

Determine the centroid of the asymptotes (rule 4).

$$\sigma = \frac{0 - 3 - 3 + j7.4 - 3 - j7.4}{4 - 0} = \frac{-9}{4} = -2.25$$

Find the real axis breakaway point (rule 5). The values of K corresponding to various real axis points on the locus between $s = 0$ and $s = -3$ are found using

$$K = \left| \frac{1}{G(s)H(s)} \right| = |s(s + 3)(s^2 + 6s + 64)|$$

With the aid of a pocket calculator, very accurate results will be obtained

Value of s	Value of K
-1.3	127.93
-1.4	128.93
-1.45	129.01
-1.5	128.81
-1.6	127.59

A value of -1.45 is obtained quickly. The real axis breakaways are at $\pm 90°$.

Determine the angle of departure from the top pole (rule 6). Using the construction of Fig. 4.18a,

$$112° + 90° + 90° + \phi = 180°$$

$$\phi = -112°$$

The completed root locus diagram is given in Fig. 4.18b. Of course, in many situations, only certain of the construction rules apply.

Computer-Aided Root Locus

To the designer, the usefulness of root locus is primarily the ability to quickly visualize how changing a parameter changes the feedback system's pole locations. With very little effort the character of the root locus can be determined, and a decision made as to whether or not to examine a potential design further. Digital computer-generated data is most often used to refine a rough root locus sketch, as final checks on a completed design, and in studies of system sensitivity.

Figure 4.19 shows an example of a computer-generated root locus plot for a system with

$$G(s)H(s) = \frac{s}{(s + 2 + j3)^2(s + 2 - j3)^2(s + 6)}$$

As is common with such programs, the feedback system characteristic polynomial,

$$s^5 + 14s^4 + 90s^3 + 356s^2 + (793 + K)s + 1014$$

is factored for various values of K and the factors plotted on the complex plane. In the example of Fig. 4.19, it is left to the user to draw the arrows. As more points are added, the resulting smooth curves become increasingly more accurate.

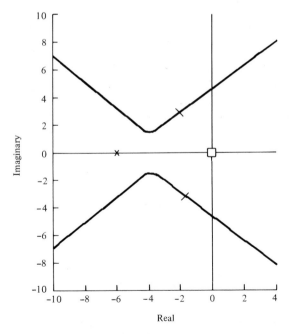

Figure 4.19 Example of a computer-generated root locus plot.

DRILL PROBLEMS

D4.6. Sketch root locus plots for the following systems. Find asymptotic angles, centroid, approximate breakaway points, angles of departure, and angles of approach where applicable.

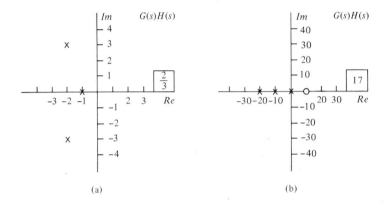

(a)   (b)

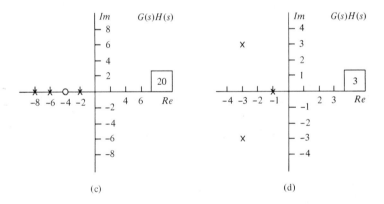

(c)   (d)

Ans (a) $\pm 60°$, $-180°$; $-\frac{5}{3}$; none; $\mp 18°$; none; (b) $\pm 90°$; -20; -15.3; none; none; (c) $\pm 90°$; -6; -6.87; none; none; (d) $\pm 60°$, $-180°$; $-\frac{7}{3}$; none; ∓ 33.7; none

D4.7. For systems with the following root locus plots, using graphical evaluation, find the value of the adjustable constant K for which the overall transfer function has a pole at the location indicated by the dot.

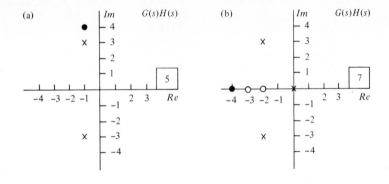

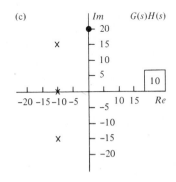

Ans (a) $K = 1.4$; (b) $K = \frac{26}{7}$, (c) $K = 910$

4.6 ROOT LOCUS FOR OTHER SYSTEMS

Systems with Other Forms

Problems other than the basic one, which has been considered exclusively here to this point, can be cast in the root locus form,

$$1 + K\,\frac{a(s)}{b(s)} = 0$$

or

$$b(s) + Ka(s) = 0$$

where $a(s)$ and $b(s)$ are known polynomials. For example, the adjustable system of Fig. 4.20a, where K is adjustable, has an overall transfer function

$$T(s) = \frac{\dfrac{s}{s+2}}{1 + \dfrac{s}{s+2}\dfrac{K}{s+K}} = \frac{s(s+K)}{s^2 + 2s + K(2s+2)} = \frac{\dfrac{s(s+K)}{s^2 + 2s}}{1 + K\dfrac{2s+2}{s^2 + 2s}}$$

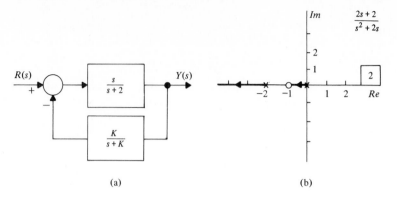

(a)   (b)

Figure 4.20 Root locus for a system with an adjustable time constant.

Here the equivalent GH product can be taken to be

$$G(s)H(s) = \frac{2(s+1)}{s^2 + 2s}$$

The root locus for K ranging from zero to infinity is shown in Fig. 4.20b. For the unity feedback system with adjustable damping in Fig. 4.21a,

$$T(s) = \frac{\dfrac{2s-4}{s^2 + 6\zeta s + 9}}{1 + \dfrac{2s-4}{s^2 + 6\zeta s + 9}} = \frac{2s-4}{s^2 + 2s + 5 + 6\zeta s} = \frac{\dfrac{2s-4}{s^2 + 2s + 5}}{1 + \zeta \dfrac{6s}{s^s + 2s + 5}}$$

The equivalent GH product can be taken to be

$$G(s)H(s) = \frac{6s}{s^2 + 2s + 5}$$

The root locus for ζ in the range from zero to infinity is shown in Fig. 4.21b.

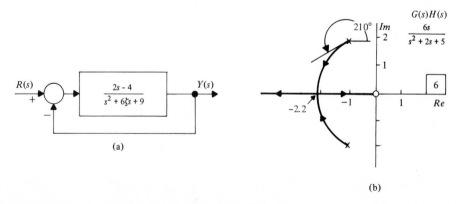

(a)   (b)

Figure 4.21 Root locus of a system with adjustable damping.

Negative Parameter Ranges

When it is desired to determine the root locus for a parameter K that ranges from zero to minus infinity, the applicable relations are

$$\begin{cases} \angle G(s)H(s) = 0° \pm i360° \\ |G(s)H(s)| = \dfrac{1}{|K|} \end{cases}$$

Points on the root locus are values of s for which the angle of the GH product is $0°$.

Construction of the root locus is similar to the $180°$ angle procedure, with real axis branches being anywhere *not* to the left of an odd number of roots, asymptotic angles given by

$$(\text{Number of poles} - \text{number of zeros})\theta = 0° \pm i360°$$

and with similar expressions for angles of departure and approach. Locus segments begin on the poles of $G(s)H(s)$ (for $K \to 0$) and approach the zeros of $G(s)H(s)$ (for $K \to -\infty$), just as with the ordinary locus.

For example, the system with GH product given in Fig. 4.22a has the real axis loci shown for negative K. The asymptotic angles are given by

$$(4 - 1)\theta = 0° \pm i360°$$

$$\theta = 0°, 120°, -120°$$

The centroid is

$$\sigma = \frac{-1 + j2 - 1 - j2 + 2 - (-3)}{4 - 1} = 1$$

Angle of departure ϕ from the top pole is given by

$$90° + 146.3° + 116.6° + \phi - 45° = 0°$$

$$\phi = -307.9° = 52.1°$$

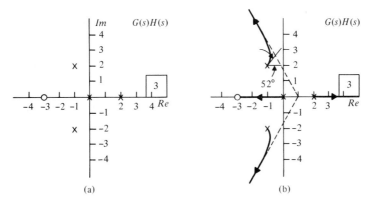

Figure 4.22 Root locus construction for negative K.

The completed root locus plot is shown in Fig. 4.22b.

The same $0°$ locus situation occurs, too, for positive K if the multiplying constant of $G(s)H(s)$ is negative. Then, although

$$\angle G(s)H(s) = 180°$$

is required, the angle contributions from the poles and zeros of GH must total $0°$ since $180°$ is contributed to the angle by the negative multiplying constant.

DRILL PROBLEMS

D4.8. Develop root locus plots for the following systems, for K ranging from 0 to $+\infty$. Find asymptotic angles, centroid, approximate breakaway points, angles of departure, and angles of approach where applicable.

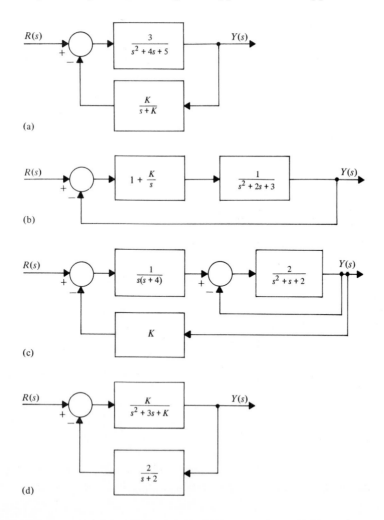

(a)

(b)

(c)

(d)

Ans (a) $G(s)H(s) = (s + 2 + j2)(s + 2 - j2)/s(s + 2 + j1)(s + 2 - j1)$;
$180°$; none; none; $\mp 63.4°$; $\pm 45°$

(b) $G(s)H(s) = 1/s(s^2 + 2s + 4)$; $\pm 60°$, $180°$; $-2/3$; none; $\mp 30°$; none

(c) $G(s)H(s) = 2/s(s + 4)(s^2 + s + 4)$; $\pm 45°$, $\pm 135°$; -1.25; -2.8;
$\mp 43.4°$; none

(d) $G(s)H(s) = (s + 4)/s(s^2 + 5s + 6)$; $\pm 90°$; $-\frac{1}{2}$; -0.9; none; none

D4.9. Develop root locus plots for systems with the following overall transfer function, for the indicated range of the adjustable parameter K:

(a) $T(s) = \dfrac{6Ks + 7}{s^3 + Ks^2 + (2K + 9)s + K}$

$0 < K < \infty$

Ans $G(s)H(s) = (s + 1)^2/s(s^2 + 9)$

(b) $T(s) = \dfrac{10}{(1 - K)s^2 + 3s + 2}$

$0 < K < \infty$

Ans $G(s)H(s) = -[s^2/(s + 1)(s + 2)]$

(c) $T(s) = \dfrac{K\dfrac{s}{s + 4}}{1 + K\dfrac{s(s - 3)}{(s + 2)^2(s + 4)}}$

$-\infty < K < 0$

Ans $G(s)H(s) = s(s - 3)/(s + 2)^2(s + 4)$

(d) $T(s) = \dfrac{6s^2 + Ks + 2}{s^3 + 2s^2 + (5 + 2K)s + 2K}$

$-\infty < K < \infty$

Ans $G(s)H(s) = 2(s + 1)/s(s^2 + 2s + 5)$

4.7 A LIGHT SOURCE TRACKING SYSTEM

This example system is designed to follow, in one dimension, a moving light source. As pictured in Fig. 4.23a, when equal light intensities are detected by the two photodiodes, the electrical bridge is balanced, and zero voltage is applied to the drive motor. When one photodiode receives more light than the other, the bridge is unbalanced, and a nonzero voltage is amplified and applied to the drive motor

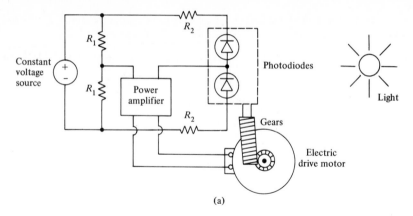

(a)

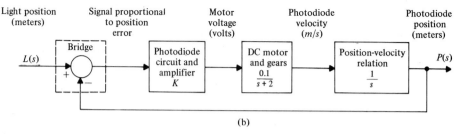

(b)

Figure 4.23 Light source tracking system. **(a)** Physical arrangement. **(b)** Block diagram model.

which then moves the photodiodes toward the equal-light-intensity position. Similar systems are used for precision machine tool alignment, where the light is reflected from a calibrated scale or transmitted through a tiny hole in the tool or the work. Variations of this system are used to track the sun or another star in navigation systems, to follow aircraft in collision avoidance systems, and to track the recording path on optical videodiscs.

For small signals, a block diagram of the system is shown in Fig. 4.23b. The system transfer function is, in terms of the gain constant K,

$$T(s) = \frac{\dfrac{0.1K}{s(s + 2)}}{1 + \dfrac{0.1K}{s(s + 2)}} = \frac{0.1K}{s^2 + 2s + 0.1K}$$

which is stable for all $K > 0$. A relative stability of two units for a system means that the natural component of system response decays with time as $\exp(-2t)$, that is, with a $\frac{1}{2}$-sec time constant. This degree of stability cannot be achieved with the system. This is evident from the root locus plot for Fig. 4.24a, where it is seen that the system's relative stability (the distance from the imaginary axis to the nearest pole) is always equal to one unit. System response to a unit step change in

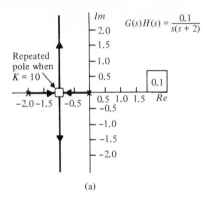

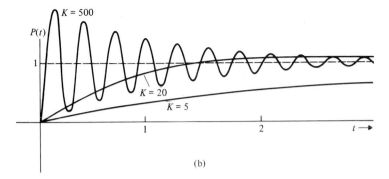

(b)

Figure 4.24 Root locus and typical responses for the original light source tracking system.

light position, for various representative values of K, are shown in Fig. 4.24b. For each, the settling time is relatively long, in consequence of the small degree of relative stability.

The performance of this system can be improved substantially by the addition of velocity feedback as well as the position feedback. A tachometer coupled to the drive motor shaft will produce a voltage nearly proportional to the motor speed, which in turn is proportional to the photodiode velocity. Adding a fraction of this voltage to the bridge voltage (which is amplified to drive the motor) results in the block diagram of Fig. 4.25a. Using Mason's gain rule on the system's signal flow graph in Fig. 4.25b,

$$T'(s) = \frac{\dfrac{0.1K}{s+2}\dfrac{1}{s}}{1 - (-4K')\dfrac{0.1K}{s+2} - (-1)\dfrac{0.1K}{s+2}\dfrac{1}{s}} = \frac{0.1K}{s^2 + (2 + 0.4KK')s + 0.1K}$$

Here it is seen that both coefficients of the characteristic polynomial can be chosen at will by the designer by selecting appropriate values of K and K'.

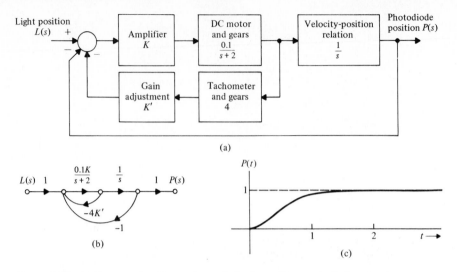

Figure 4.25 Light source tracking system with both position and rate feedback.
(a) Block diagram. **(b)** Signal flow graph. **(c)** Unit step response with $K = 250$,
$K' = 0.08$.

For example, a characteristic polynomial

$$(s + 5)^2 = s^2 + 10s + 25 = s^2 + (2 + 0.4KK')s + 0.1K$$

is achieved with

$$K = 250 \qquad K' = 0.08$$

With these values of K and K', the system's step response is critically damped and
has a relative stability of 5 units. Its step response is shown in Fig. 4.25c.

4.8 AN ARTIFICIAL LIMB

Development of prosthetics has paralleled wars and natural calamities. Until
recently, all but a very few artificial limbs were nonactive devices connected to what
remained of the limb. Lower arm devices required the user to have an elbow joint,
and lower leg devices required a usable knee joint. The "Boston Arm" was an early
active artificial limb that used a torque motor and velocity feedback. The
microprocessor revolution of the 1980s spawned more advanced artifical limbs
capable of greater dexterity and more precise operation. Advanced research
included creation of a system that bypasses a damaged spinal cord so that muscle-
electric signals from the upper body cause a balance-control system to facilitate
movement of the legs and locomotion for an otherwise paralyzed patient. The
example that follows exhibits typical design techniques for an artificial limb
although no actual working system is intended.

The human body has a time delay between decisions of the brain and reception
of signals by the muscles, so rational limits must be placed upon the capabilities of

an automatic artificial limb. If the device is too strong and sensitive, it could operate too rapidly for the brain and body to control, making the human-machine team a bionic pretzel. On the other hand, too little sensitivity could result in a bionic statue. Between the two extremes lie useful designs.

Figure 4.26a shows a block diagram for a bionic arm in a closed-loop system with the body. For simplicity, motion is considered in one dimension only. The brain monitors the desired position and the sensed position, generating an error signal to the nervous system. Special sensors pick up the electric muscle impulses (myoelectric signals), and an amplifier produces a voltage that drives a dc control motor. The motor circuit involves tachometer feedback, as shown. The output of the motor circuit is the velocity of the limb in one dimension which, when integrated, is the limb position.

When a control system is to be designed for a complicated system such as this, some course of action must be planned. In this case, a simplified block diagram will be created and the control will evolve by stabilizing the loops, starting with the inner loops, and then progressing outward until all parameters are selected.

A simplified model for the system is given in Fig. 4.26b. The action of the brain is approximated by the transmittance

$$G_B(s) = 1 + \frac{0.1}{s}$$

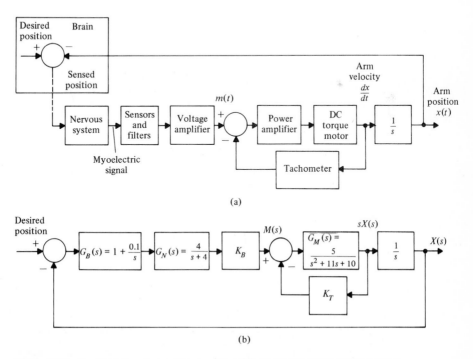

(a)

(b)

Figure 4.26 Prosthetic control system model. **(a)** Prosthetic and human control system model. **(b)** Simplified model for the system.

which involves consideration of both the position error and its integral. The nervous system is modeled by the first-order system with transmittance

$$G_N(s) = \frac{1/T}{s + 1/T} = \frac{4}{s + 4}$$

where $T = \frac{1}{4}$, the time constant, is approximately the neuromuscular delay time. The myoelectric signal is sensed and amplified with a gain K_B to form the amplified voltage $m(t)$.

The power amplifier, control motor, and mechanical load have a second-order transmittance, relating motor control voltage $m(t)$ to arm velocity, modeled as

$$G_M(s) = \frac{5}{s^2 + 11s + 10} = \frac{5}{(s + 1)(s + 10)}$$

This block exhibits time constants of 1 sec, associated with mechanical inertia, and 1/10 sec due primarily to the motor itself. The step response of $G_M(s)$ is sketched in Fig. 4.27a. The tachometer, with constant transmittance K_T, provides feedback for the motor and arm, giving the following overall transmittance of those components:

$$G_T(s) = \frac{G_M(s)}{1 + K_T G_M(s)} = \frac{\dfrac{5}{s^2 + 11s + 10}}{1 + \dfrac{5K_T}{s^2 + 11s + 10}} = \frac{5}{s^2 + 11s + (10 + 5K_T)}$$

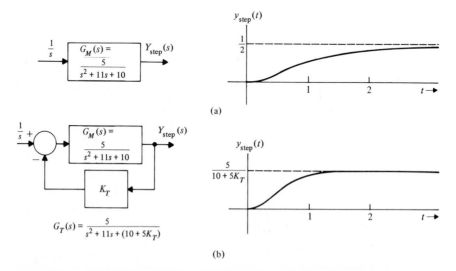

(a)

(b)

Figure 4.27 Improving the motor-arm performance with tachometer feedback and critical damping. **(a)** Step response of the motor and arm without tachometer feedback. **(b)** Step response with the tachometer feedback.

$$G_M(s) = \frac{5}{s^2 + 11s + 10}$$

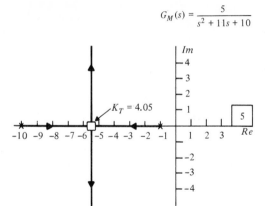

Figure 4.28 Root locus of the motor-arm subsystem.

A root locus plot for this motor-arm subsystem, $G_T(s)$, in terms of the adjustable gain K_T is shown in Fig. 4.28. The transmittance $G_T(s)$ has critical damping when its denominator polynomial is

$$s^2 + 11s + (10 + 5K_T) = (s + 5.5)^2 = s^2 + 11s + 30.25$$

which is obtained for

$$10 + 5K_T = 30.25 \qquad K_T = 4.05$$

For this value of K_T, the subsystem has a maximum relative stability of 5.5 units; larger values of K_T reduce the damping but do not increase relative stability. This local feedback makes the subsystem act as if it were a better motor, with faster response. It is best not to overdo a good thing, however. Huge improvements in response speed, when they are possible, generally require that the device being controlled be driven with very large input signals, as is the case if one wants a motor to quickly speed up. Very large inputs can cause damage and usually will drive the device into a nonlinear region of operation.

The entire system is redrawn in Fig. 4.29a for the tachometer gain $K_T = 4.05$ which gives critical damping of the motor-arm block. The transfer function of the entire system is

$$T(s) = \frac{G_B(s)G_N(s)K_BG_T(s)\left(\dfrac{1}{s}\right)}{1 + G_B(s)G_N(s)K_BG_T(s)\left(\dfrac{1}{s}\right)} = \frac{\dfrac{20(s + 0.1)K_B}{s^2(s + 4)(s^2 + 11s + 30.25)}}{1 + \dfrac{20(s + 0.1)K_B}{s^2(s + 4)(s^2 + 11s + 30.25)}}$$

$$= \frac{20(s + 0.1)K_B}{s^5 + 15s^4 + 74.25s^3 + 121s^2 + 20K_Bs + 2K_B}$$

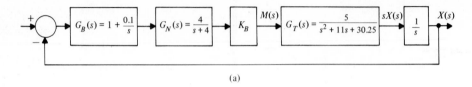

(a)

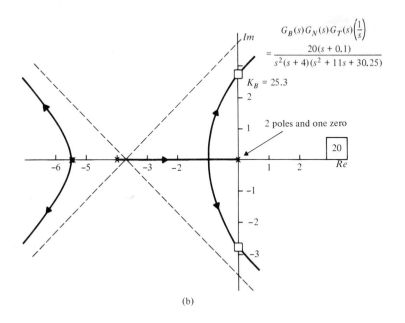

(b)

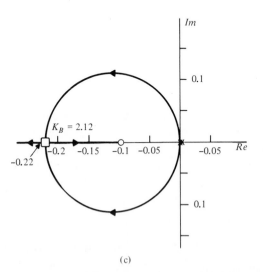

(c)

Figure 4.29 Root locus for the complete system. **(a)** Block diagram. **(b)** The big picture. **(c)** Closeup of the region near the origin.

A root locus plot in terms of K_B is shown in Fig. 4.29b. There are two poles at the origin and a zero at $s = -0.1$ in the open-loop transmittance $G_B(s)G_N(s)G_T(s)(1/s)$. However, except very near to these roots, the loci are virtually the same as if a single pole were in that region.

A Routh-Hurwitz test for the stability of $T(s)$ is as follows:

$$
\begin{array}{c|ccc}
s^5 & 1 & 74 & 20K_B \\[4pt]
s^4 & 15 & 121 & 2K_B \\[4pt]
s^3 & 65.9 & 19.86K_B & \\[4pt]
s^2 & 121 - 4.52K_B & 2K_B & \\[4pt]
s^1 & \dfrac{2271K_B - 89.76K_B^2}{121 - 4.52K_B} & & \\[8pt]
s^0 & 2K_B & &
\end{array}
$$

For stability,

$$
\begin{cases}
121 - 4.52K_B > 0 \\
K_B(2271 - 89.76K_B) > 0 \\
2K_B > 0
\end{cases}
$$

or

$$
\begin{cases}
K_B < 26.8 \\
K_B < 25.3 \qquad \text{for positive } K_B \\
K_B > 0
\end{cases}
$$

Thus the system is stable for

$$
0 < K_B < 25.3
$$

and the root locus must cross the imaginary axis at $K_B = 25.3$.

The region of the root locus plot near the origin of the complex plane is magnified in Fig. 4.29c. The approximate entry point of the loci to the real axis is found by computing the minimum K_B for which there are real axis roots:

Value of s	Value of K_B
-0.18	2.189
-0.19	2.155
-0.20	2.135
-0.21	2.126
-0.22	2.125
-0.23	2.130

The maximum relative stability of the closed-loop system occurs when $K_B = 2.12$, so this choice will be made. The overall system transfer function is then

$$T(s) = \frac{42.4(s + 0.1)}{s^5 + 15s^4 + 74.25s^3 + 121s^2 + 42.4s + 4.24}$$

Now, with $K_B = 2.12$, suppose we investigate the root locus of the overall system when K_T is again varied. It might be that this gain could vary significantly with conditions such as time and temperature, being dependent as it is on the tachometer field magnetization. For this K_T variable, the system transfer function is

$$T(s) = \frac{K_B G_B(s)G_N(s)G_T(s)(1/s)}{1 + K_B G_B(s)G_N(s)G_T(s)(1/s)}$$

$$= \frac{42.4(s + 0.1)}{s^5 + 15s^4 + 54s^3 + 40s^2 + 42.4s + 4.24 + K_T(5s^3 + 20s^2)}$$

$$= \frac{\text{numerator}}{1 + K_T \dfrac{5s^3 + 20s^2}{s^5 + 15s^4 + 54s^3 + 40s^2 + 42.4s + 4.24}}$$

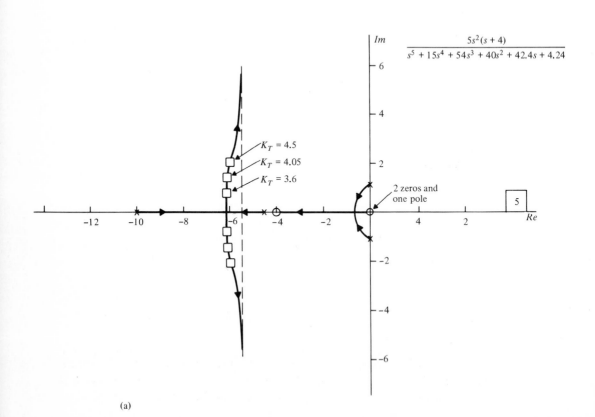

$$\frac{5s^2(s + 4)}{s^5 + 15s^4 + 54s^3 + 40s^2 + 42.4s + 4.24}$$

(a)

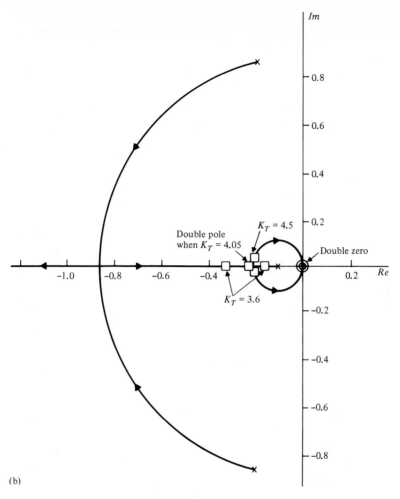

(b)

Figure 4.30   Root locus for $K_B = 2.12$ and K_T variable.

To do a complete root locus plot, the following polynomial must be factored

$$s^5 + 15s^4 + 54s^3 + 40s^2 + 42.4s + 4.24$$

$$= (s + 0.11)(s + 4.5)(s + 9.9)(s + 0.24 + j0.90)(s + 0.24 - j0.90)$$

and this has been done (approximately) with a digital computer factoring routine.

The root locus, on a relatively large scale, is shown in Fig. 4.30a. At this scale, the two "open-loop" zeros at the origin and the pole at $s = -0.11$ are merged into a single zero.

The region of the plot near the origin is expanded in Fig. 4.30b. Closed-loop pole locations for the design value of $K_T = 4.05$ and for values of K_T about 10% higher and lower than the design value are shown. These indicate the degree of sensitivity of the feedback system's pole locations to changes in the gain K_T.

4.9 A HYDROELECTRIC GENERATING SYSTEM

Moving water has long been a useful source of energy. Water-driven grinders and pounders were extensively used in Northern Europe before the invention of steam-driven machines. Indeed, the industrial revolution required large amounts of water so Northern Europe benefited from that resource.

Today Switzerland and Norway generate most of their electrical energy using the hydroelectric method. Both countries receive extensive snowfall and both countries have sharp variations in altitude. Switzerland uses much of its generated electrical energy to drive an electrified train system. The relatively cheap source of energy in Norway allows that country to provide nearly all-electric homes and to encourage heavy industry (aluminum manufacturing for example).

In the 1980s, Canada began a massive hydroelectric project to provide some measure of energy independence. In Canada's case, a vast watershed area with a small altitude difference is exploited whereas Norway and Switzerland exploit smaller watersheds with greater altitude differences.

Figure 4.31 shows a typical hydroelectric system. A reservoir impounds water at some height h above the turbine inlet. If the pipe from the reservoir to the inlet is full and flow is not restricted, the relative incompressibility of water suggests that Bernoulli's equation for fluid flow may be used.

$$\tfrac{1}{2}dv^2 + p + dgz = \text{constant}$$

where d = density of water, lb/ft^3

v = velocity of water, ft/sec

p = static pressure, lb/ft^2

z = height above the inlet, ft

g = acceleration due to gravity, ft/sec^2

The first term in Bernoulli's equation is called dynamic pressure. The second term is the static (barometric) pressure, and the third term is due to the pressure head (height above the inlet).

The velocity of water inflow to the turbine blades can be computed by using Bernoulli's equation at the reservoir surface (where $v = 0$ and $z = h$) and at the turbine inlet (where $z = 0$). Assuming that the barometric pressures are nearly equal,

$$\tfrac{1}{2}dv^2 + p = p + dgh$$

$$v = \sqrt{2gh}$$

The velocity inflow depends only on g and h (assuming ideal behavior).

Figure 4.32 shows a Kaplan-type turbine with variable pitch blade angles. If the blades are pitched so the water flows parallel to the blades, the effective angle is zero and no torque is imparted to the turbines. As blade pitch increases, the

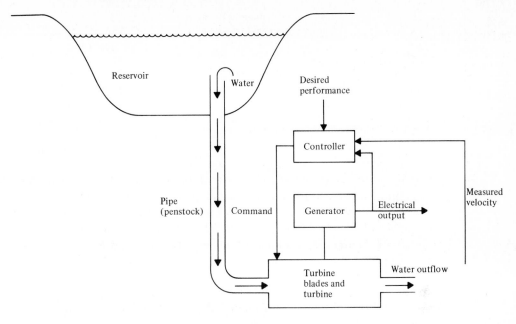

Figure 4.31 Hydroelectric generation system.

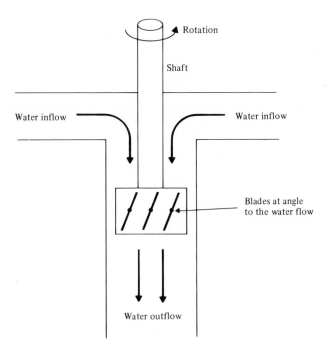

Figure 4.32 Kaplan-type adjustable pitch turbine blades.

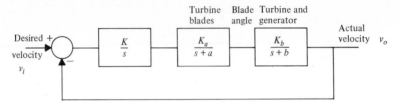

Figure 4.33 Linearized velocity control loop.

incoming water is deflected. The loss of momentum of the water goes to overcoming friction and to rotating the inertial mass consisting of the turbine shaft connected to an electrical generator. In Fig. 4.31, the generator output and other environmental conditions cause a controller to vary the blade pitch.

Figure 4.33 contains a simplified (and linearized) model for a velocity control loop. The ultimate goal is to provide a steady flow of electricity. One variable to be controlled is the turbine velocity, hence Fig. 4.33.

Suppose the distance from the turbine blades to the reservoir surface is 900 ft. The acceleration due to gravity is about 32 ft per second. The ideal water velocity at the turbine inlet is

$$v = \sqrt{2 \times 32 \times 900} = 240 \text{ (ft/sec)}$$

A first-order actuator is assumed. That actuator changes the turbine blade pitch. As a result of the inlet flow of velocity v, the turbine parameter K_a causes the turbine to rotate, given the blade pitch angle. The turbine dynamics are also assumed to be first order.

An engineer could obtain approximate time constants by testing the system. Suppose the turbine is blocked from rotation. If a command to the turbine blades results in a 15-sec time constant, then $1/a$ is 15. If a constant velocity error signal equivalent to 150 ft/sec causes a steady state 15° deflection

$$\lim_{s \to 0} s\left(\frac{150}{s}\right)\left(\frac{K_a}{s+a}\right) = \frac{150\,K_a}{a} = 15$$

$$K_a = \frac{15a}{150} = \frac{1}{150} \text{ [degrees/(ft)/(sec)]}$$

Suppose the blades remain locked at 15° but the turbine is freed. If the turbine moves to a steady state velocity of 150 ft/sec with a time constant of 2 min, then K_b is $\frac{1}{120}$ and

$$\lim_{s \to 0} s\, \frac{15}{s}\, \frac{K_b}{s+b} = \frac{15K_b}{b} = 150$$

$$K_b = \frac{150b}{15} = \frac{1}{12} \text{ [ft/(sec)(degree)]}$$

The root locus for variable K follows from

$$KG(s)H(s) = \frac{K(\frac{1}{12})(\frac{1}{150})}{s(s + \frac{1}{15})(s + \frac{1}{120})}$$

$$= \frac{960K}{(120s)(120s + 8)(120s + 1)}$$

$$= \frac{960K}{s'(s' + 8)(s' + 1)}$$

The product $KG(s)H(s)$ becomes more convenient if the low-valued coefficients are replaced by defining s' to be $120s$. The root locus (as a function of s') appears in Fig. 4.34. The turbine would encounter fewer fluid flow abnormalities if the velocity does not oscillate about a desired point; hence, the dominant roots should not be underdamped. K could be selected to cause the dominant roots to be located at the breakaway point. For any s' on the root locus

$$K = \frac{|s'(s' + 8)(s' + 1)}{960}$$

The breakaway point (and therefore K) could be computed by trial and error; but here, K is only a third-order function of s'. The value of s' for maximum K can be computed by solving

$$\frac{dK}{ds'} = 0 = \frac{3(s')^2 + 18(s') + 8}{960}$$

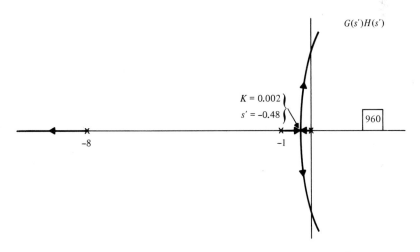

Figure 4.34 Root locus for the system in Fig. 4.33 with a change of variables to s'.

The two solutions to the quadratic are -0.483 and -5.52. The first solution lies between 0 and -1. Since the breakaway point s' equals -0.483, then K becomes 0.002. At the breakaway point

$$s = \frac{s'}{120} = -0.004$$

The dominant time constant under control is now 250 sec.

The root locus appears to show how closed-loop poles change as K deviates from the design value of 0.002. Actually, the root locus is valid in a narrow region around the design values. Fluid flow conditions (when the system is more heavily or less heavily driven) could cause the actual behavior to differ significantly from this linearized model. It would not be possible to command a velocity to exceed the stream velocity of 240 ft/sec; hence steady state accuracy is dependent on the command.

4.10 CONTROL OF A FLEXIBLE SPACECRAFT

When observing an aircraft, a person sees an apparently smooth motion through the sky. If the same person observes a satellite through a telescope, the satellite seems to follow a smooth trajectory. Actually, each spacecraft has a much more complicated motion, which can influence the design of an automatic control system.

Very large satellites often extend solar panels to generate and store electrical energy. As a result, the satellite is very flexible. When a thruster fires to cause the satellite to change postion, the entire vehicle moves (rigid motion) but parts of the vehicle will bend and oscillate (flexible motion) in the same way that a taut guitar string oscillates when plucked.

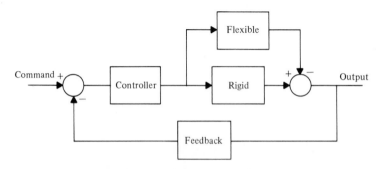

Figure 4.35 Block diagram of a flexible spacecraft.

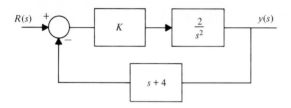

Figure 4.36 Rigid behavior example.

Figure 4.35 shows the block diagram of a controlled flexible spacecraft. Often the controller output causes the rigid and flexible behaviors to move in opposite directions. Ignoring the flexible behavior can be a tragic mistake, especially for a manned spacecraft or airplane.

As a simple example of designing a control system that considers only rigid behavior, Fig. 4.36 causes

$$KG(s)H(s) = \frac{K2(s+4)}{s^2}$$

This system uses rate and position feedback to stabilize the system. The root locus of Fig. 4.37 seems to indicate that the system is stable for all positive values of K.

Suppose the spacecraft is actually quite flexible and (including at least one mode of vibration) Fig. 4.38 results. Actually, there may be many higher-order vibration modes (resonances as in a musical instrument). From Fig. 4.38, the two forward path transmittances must be combined.

$$G(s) = \frac{2}{s^2} - \frac{1}{s^2+s+1} = \frac{s^2+2s+2}{s^2(s^2+s+1)}$$

$$KG(s)H(s) = \frac{K(s+1+j1)(s+1-j1)(s+4)}{s^2(s+0.5+j0.866)(s+0.5-j866)}$$

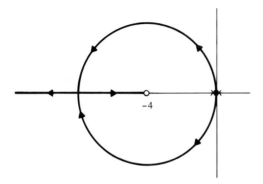

Figure 4.37 Root locus for Fig. 4.36.

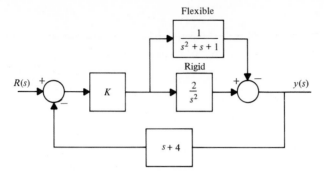

Figure 4.38 Flexible and rigid behavior example.

Figure 4.39 shows the root locus where the one flexible mode is included. It now appears that the system can become unstable for some positive values of K, a conclusion that would not result if the flexible mode is ignored.

A designer can select values of K in Fig. 4.39 to ensure stability. It is also possible to redesign the spacecraft so the vehicle is more rigid. The poles and zeros due to the flexible mode (poles and zeros will always appear in pairs) can be moved away from the vertical axis so as to render the system stable over a larger gain range (perhaps for all positive K).

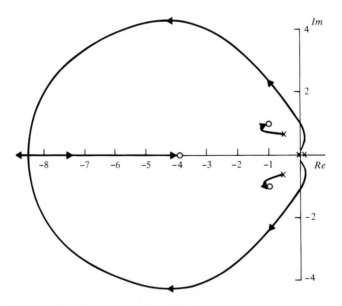

Figure 4.39 Root locus for Fig. 4.38.

4.11 SUMMARY

A pole-zero plot of a rational function consists of indications on the complex plane of the locations of the function's poles and zeros. The addition of the multiplying constant of the function to the diagram completely specifies the function.

To graphically evaluate a rational function $F(s)$ at a particular value $s = s_0$, directed line segments are drawn from each zero and each pole of $F(s)$ to the point s_0. The magnitude of the evaluated function is the product of the magnitude of the function's multiplying constant and the lengths of the directed line segments from the zeros, divided by the product of the lengths of the directed line segments from the poles. The angle of the evaluated function is the sum of the angles of the directed line segments from the zeros minus the sum of the angles of the directed line segments from the poles. An additional $180°$ must be added or subtracted from the angle of the function if the multiplying constant is negative.

A root locus plot is a pole-zero plot of the rational open-loop system transmittance $G(s)H(s)$, upon which is superimposed the locus of the roots

$$1 + KG(s)H(s) = 0$$

as K is varied from zero to $+\infty$. The root locus is symmetrical about the real axis. A value of s for which the angle of $G(s)H(s)$ is $180°$ (plus or minus any multiple of $360°$) is a point on the root locus, corresponding to some value of K. The six rules for root locus plot construction are summarized in Table 4.1, and Table 4.2 gives many root locus examples. The value of the constant K corresponding to a specific point on the locus is found using

$$|G(s)H(s)| = \frac{1}{|K|} = \frac{\left(\begin{array}{c}\text{magnitude of the} \\ \text{multiplying constant}\end{array}\right)\left(\begin{array}{c}\text{product of} \\ \text{zero distances}\end{array}\right)}{\text{product of pole distances}}$$

It was shown how root locus methods can be applied to other systems besides those of the standard form and that it is straightforward to extend the technique to systems with negative adjustable parameters.

A light source tracking system example illustrated some of the power of root locus for design. A control system for an artificial limb involved the design of subsystems as part of an overall system design. A velocity control system for a hydroelectric generating system was analyzed. A root locus analysis of a flexible spacecraft demonstrated that omitting flexible dynamics can cause a designer to overlook a potential cause of instability. In the next chapter, we apply root locus methods to the systematic design of feedback tracking control systems.

REFERENCES

Root Locus
Clark, R. N., *Introduction to Automatic Control Systems.* New York: Wiley, 1962.
Evans, W. R., "Graphical Analysis of Control Systems." *Trans. AIEE* 67 (1948): 547–551.

————, "Control System Synthesis by the Root Locus Method." *Trans AIEE* 69 (1950): 67–69.

————, *Control System Dynamics*, New York: McGraw-Hill, 1954.

Truxal, J. G. *Control System Synthesis*, New York: McGraw-Hill, 1955.

Tracking Systems

Ahrendt, W. R., and Savant, C. J. Jr., *Servomechanism Practice*. New York: McGraw-Hill, 1960.

Bode, H. W., *Network Analysis and Feedback Amplifier Design*. Princeton, N. J.: Van Nostrand, 1945.

Cannon, R. H. Jr., *Dynamics of Physical Systems*. New York: McGraw-Hill, 1967.

Chestnut, H., and Mayer, R. W., *Servomechanisms and Regulating System Design*, vol. 1. New York: Wiley, 1959.

Newton, G. C.; Gould, L. A.; and Kaiser, J. F., *Analytical Design of Linear Feedback Controls*. New York: Wiley, 1957.

Savant, C. J. Jr., *Control System Design*. New York: McGraw-Hill, 1964.

Prosthetics

Allan, R., "Electronics Aids the Disabled." *IEEE Spec.* 13, no. 11 (November 1976): 36–40.

"Brain Controls Use of Artificial Arm." *Prod. Eng.*, October 7, 1968, p. 23.

"Novel Artificial Arm Gives Wearer Six Degrees of Freedom." *Prod. Eng.*, October 20, 1969, p. 15.

Horgan, J., "Medical Electronics." *IEEE Spec.* (January 1984): 90–93.

Raibert, M. A., and Sutherland, I. E., "Machines That Walk." *Sci. Am.* (January 1983): 44–53.

Rauch, H., ed., "Dextrous Robotic Hand." *IEEE Control Systems Magazine.* (December 1986.)

Hydraulic Turbines

Housner, G. W., and Hudson, D. E., *Applied Mechanics: Dynamics*. New York: D. Van Nostrand, 1959, p. 277.

Little and Ives Complete Book of Science Illustrated, New York: D. Van Nostrand, 1963, pp. 793, 974, 975.

Yule, J. D., *Concise Encyclopedia of Science and Technology*. New York: Crescent, 1985, pp. 293–295.

Hydroelectric Power

Fischetti, M. A., "Quebec Hydro: La Grande Tour." *IEEE Spec.* (October 1986): 30–36.

Kohl, L., "Quebec's Northern Dynamo." *National Geographic* (March 1982): 406–418.

Moraes, J. and Salatko, V. F., "Coming: 12600 Megawatts at Itaipu Island." *IEEE Spec.* (August 1983): 46–51.

Control of Flexible Spacecraft

Larson, V., and Likins, P. W., "Optimal Estimation and Control of Elastic Spacecraft," in *Advances in Control and Dynamic Systems*. New York: Academic Press, 1977.

Martin, G. D., and Bryson, A. E., "Attitude Control of a Flexible Spacecraft." *Journal of Guidance and Control* (January–February 1980): 37–41.

PROBLEMS

1. Draw pole-zero plots for the following rational functions:

(a) $F(s) = \dfrac{-3s^2 + 4}{(8s + 1)(s + 7)^2}$

(b) $F(s) = \dfrac{8(s^2 - 4s + 10)(1 + 4s)}{s^3(s^2 + 4s + 10)}$

(c) $F(s) = \dfrac{100s(s^2 - 4s + 10)^2}{(s + 6)^2(s^2 + 4s + 10)^2}$

(d) $F(s) = \dfrac{3s^3 + 9s}{(s^2 + 2s + 20)(s^2 + 2s + 8)}$

2. For each of the following GH products, construct root locus sketches. Find the range of the positive adjustable gain K in

$$T(s) = \frac{KG(s)}{1 + KG(s)H(s)}$$

for which each overall system is stable.

(a) $G(s)H(s) = \dfrac{1}{s(4s + 1)}$

(b) $G(s)H(s) = \dfrac{3s + 1}{s(4s + 1)}$

(c) $G(s)H(s) = \dfrac{s + 1}{s(4s + 1)}$

(d) $G(s)H(s) = \dfrac{1}{s(s + 1)(4s + 1)}$

(e) $G(s)H(s) = \dfrac{1}{s^2 + 10s + 40}$

(f) $G(s)H(s) = \dfrac{s + 4}{s^2 + s + 6}$

(g) $G(s)H(s) = \dfrac{s}{(s + 2)(s + 10)}$

(h) $G(s)H(s) = \dfrac{1}{s(s + 2)(s^2 + 2s + 50)}$

Ans

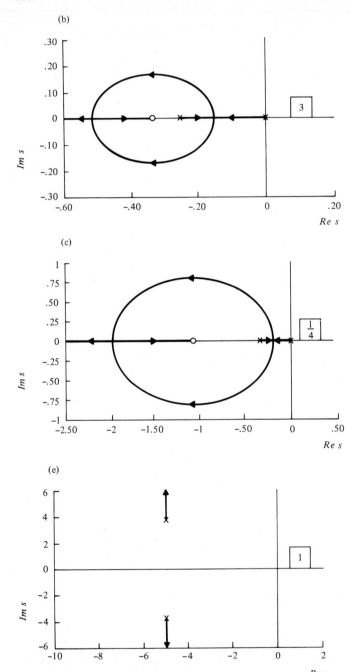

Figure P4.2

3. Sketch root locus plots for the systems of Fig. P4.3, for K between zero and $+\infty$. Find asymptotes, centroid, angles of departure, angles of approach, and approximate breakaway points where applicable.

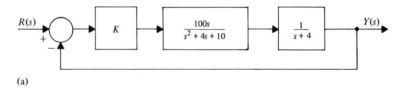

(a)

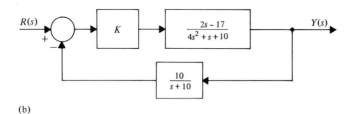

(b)

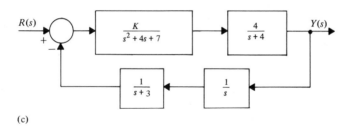

(c)

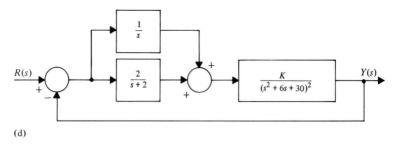

(d)

Figure P4.3

4. Sketch root locus plots for the systems with pole-zero plots given in Fig. P4.4. Find asymptotes, centroid, angles of departure, angles of approach, and approximate breakaway points where applicable.

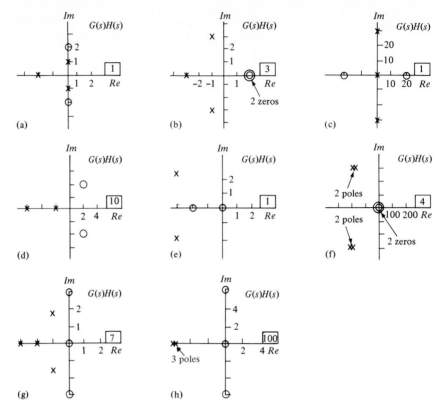

(a)   (b)   (c)

(d)   (e)   (f)

(g)   (h)

Figure P4.4

Ans

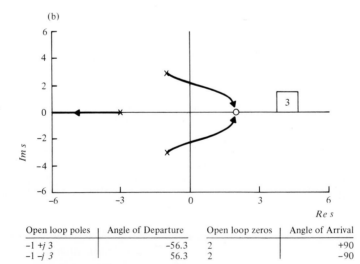

(b)

Open loop poles	Angle of Departure	Open loop zeros	Angle of Arrival
$-1 + j\,3$	-56.3	2	$+90$
$-1 - j\,3$	56.3	2	-90

Center of gravity = −9
Asymptotic angles :−180

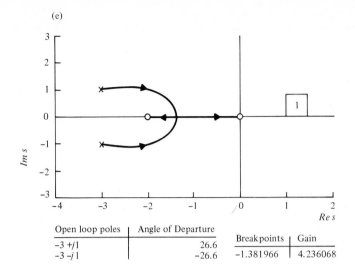

(e)

Open loop poles	Angle of Departure
−3 +j1	26.6
−3 −j1	−26.6

Breakpoints	Gain
−1.381966	4.236068

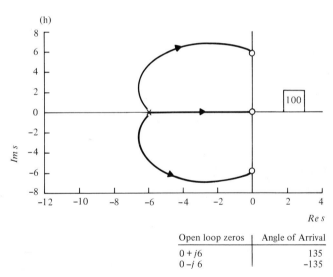

(h)

Open loop zeros	Angle of Arrival
0 + j6	135
0 −j 6	−135

5. For the systems with the following GH products, use graphical root locus methods to find an approximate value of adjustable gain K for which (if possible) there is a complex conjugate set of roots of the overall system with damping ratio $\zeta = 0.2$.

(a) $\dfrac{1}{s(s + 1)(0.2s + 1)}$

(b) $\dfrac{100(s + 0.2)}{s(s + 1)(0.2s + 1)(s + 20)}$

(c) $\dfrac{100(s + 2)}{s(s + 1)(0.2s + 1)(s + 200)}$

(d) $\dfrac{100(s + 5)}{s(s + 1)(0.2s + 1)(s + 500)}$

Ans

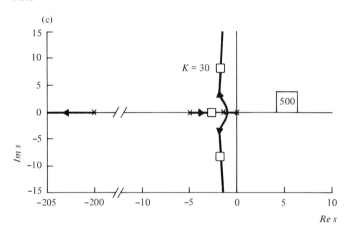

Figure P4.5

6. On the complex plane, sketch the locus of the roots of the following functions, as the parameter K is varied from zero to $+\infty$:

(a) $T(s) = \dfrac{6}{s^2 + K}$

(b) $T(s) = \dfrac{s^2 + 7s - 2}{Ks^2 + 7s + 2}$

(c) $T(s) = \dfrac{100}{s^2 + Ks + 8}$

(d) $T(s) = \dfrac{-10}{s^2 + (K - 2)s + 7 + K}$

7. Sketch the root locus, for K from zero to $+\infty$, for a system with

$$G(s)H(s) = -\dfrac{s + 4}{(s + 1)^2(s^2 + 4s + 9)}$$

Note the negative algebraic sign.

8. Sketch the root locus for K in the range from zero to *minus* infinity for a system with

$$G(s)H(s) = \frac{2(s-4)^2}{(s+5)(s^2+4s+10)}$$

Using this root locus, approximately locate the poles of

$$T(s) = \frac{G(s)}{1 + KG(s)H(s)}$$

for $K = -2$.

Ans

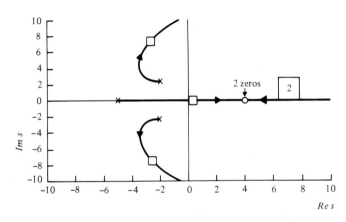

Figure P4.8

9. On the complex plane, sketch the locus of the poles of the function

$$T(s) = \frac{-10s + 3}{4s^2 + 2s + Ks - 7K}$$

as the parameter K is varied through the range $-\infty$ to $+\infty$.

10. For the light source tracking system (Sec. 4.7), suppose that an electrical compensator network is added instead of the tachometer, as in Fig. P4.10. Find gain constants K_1 and K_2, if possible, such that the system has a relative stability of at least 5 units.

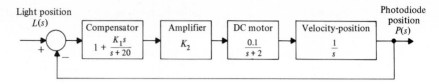

Figure P4.10

11. Sketch the root locus for

$$G(s)H(s) = \frac{1}{s(s + 1)(s + 2)(s + 10)}$$

For $K = 200$, are there any characteristic equation roots in the right half-plane?

Ans Yes

12. For the artificial limb system (Sec. 4.8) with $T = \frac{1}{4}$ sec, suppose that more tachometer feedback is used so that $K_T = 10$. Use root locus methods to find an acceptable value of K_B. Using graphical methods, find, approximately, the maximum value of K_B for which the overall system is stable.

13. A simple block diagram of a control rod positioning system for a nuclear power plant is given in Fig. P4.13. Sketch a root locus plot for the system. Graphically test points along the imaginary axis to accurately determine where the loci cross into the RHP. Then determine graphically the value of K for which the system is marginally stable.

For a system of relative low order such as this one, the same results are perhaps more easily obtained by Routh-Hurwitz testing in terms of K. With greater system complexity, a Routh-Hurwitz test becomes hopelessly involved, but root locus methods remain very suitable.

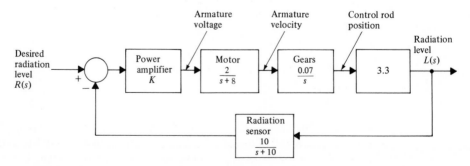

Figure P4.13

Ans

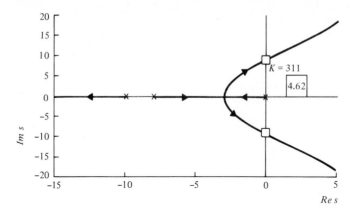

14. A linearized inventory control system model is given in Fig. P4.14. Based upon the difference between desired and actual inventory levels, management sets the production quota which in turn determines the production rate. Using graphical root locus methods, determine the production time constant

$$\tau = \frac{1}{K}$$

in the range

$$1 < \tau < 10$$

which gives greatest relative stability of the system.

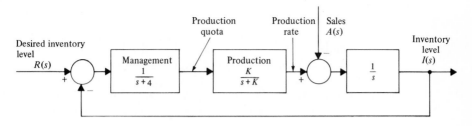

Figure P4.14

15. A simplified model of an automobile pollution control system is shown in Fig. P4.15. Sketch the system root locus for $K > 0$, paying particular attention to the manner in which the loci leave the repeated poles. Use the root locus to determine, approximately, the value of K, if any, for which the system's natural response will decay at least as rapidly as $\exp(-0.4t)$.

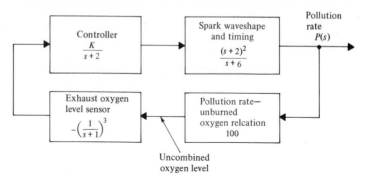

Figure P4.15

Ans

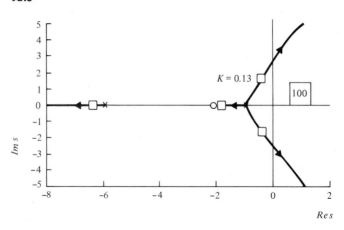

16. A linearized model of a frequency-locked loop is given in Fig. P4.16. The frequency difference between an incoming sinusoidal signal and an oscillator is sensed and integrated to produce a voltage which drives the voltage-controlled oscillator. The oscillator's frequency is proportional to the control voltage. When there is zero frequency difference, the integrator input is zero, its output is constant, and the oscillator frequency equals the incoming frequency. For incoming signals which are distorted and corrupted by noise, the frequency-locked loop produces a nearly pure sinusoid, locked in frequency to the incoming signal. The effect is as if the incoming signal had been processed by a very sophisticated filter which had removed virtually all of its noise and distortion. Use root locus procedures to select a value of positive K such that the damping ratio of the dominant roots is 0.2.

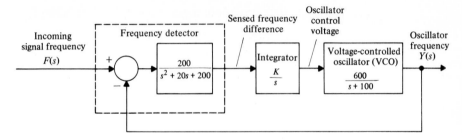

Figure P4.16

17. A position servosystem model is given in Fig. P4.17. The compensator is an imperfect attempt to integrate the sensed velocity to obtain position feedback. Use root locus procedures to find positive values of the adjustable parameter K for which the system which relates shaft velocity to the input is stable.

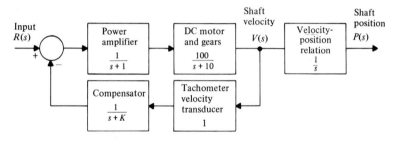

Figure P4.17

18. Find the root locus diagrams for

(a) $G(s)H(s) = \dfrac{1}{s(\tau_1 s + 1)(\tau_2^2 s^2 + \tau_2 s + 1)(\tau_3 s + 1)}$

(b) $G(s)H(s) = \dfrac{1}{s(\tau_1 s + 1)(\tau_2 s + 1)^2(\tau_3 s + 1)}$

where $\tau_1 = 2\tau_2$ and $\tau_1 = 3\tau_3$. Determine accurately the initial directions of the motions of the roots, the asymptotic directions, and the centroid.

19. For the system with a GH product of the form

$$G(s)H(s) = \frac{(s + \alpha)(s + \alpha/2)}{s^2(s + 10)^2}$$

sketch on one diagram a root locus for each of the following values of α: 1, 4, 6.67, 9, 12.

20. Determine the initial angles of departure from the complex poles and sketch the root-locus diagrams for the following GH products:

(a) $\dfrac{1 + \tau_1 s}{\tau_2^2 s^2 + 2\zeta\tau_2 s + 1}$ $\dfrac{\zeta}{\tau_2} < \dfrac{1}{\tau_1}$

(b) $\dfrac{1 + \tau_1 s}{\tau_2^2 s^2 + 2\zeta\tau_2 s + 1}$ $\dfrac{\zeta}{\tau_2} > \dfrac{1}{\tau_1}$

21. A simple position servo is expressed in block diagram form as shown in Fig. P4.21.

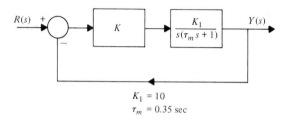

$K_1 = 10$
$\tau_m = 0.35$ sec

Figure P4.21

(a) Is the system stable for all values of gain?

(b) Upon construction of the above servo, it is found to have zero damping for a gain $K = 30$. The instability is due to parasitic time lags in the synchros, amplifier, and motor. Approximate lumping these effects together using an additional time lag in the forward transfer function, i.e., by a factor $1/(1 + \tau_a s)$. What is the value of τ_a?

Ans $\tau_a = 0.00336$

22. Sketch root locus diagrams for the following systems:

(a) $GH = \dfrac{1}{s^4 + 16}$

(b) $GH = \dfrac{1}{s^4 - 16}$

(c) $GH = \dfrac{1}{(s^2 + 4)(s^2 - 1)}$

(d) $GH = \dfrac{1}{(s^2 + 1)(s^2 - 4)}$

Ans

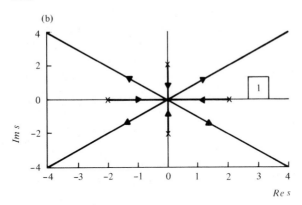

(b)

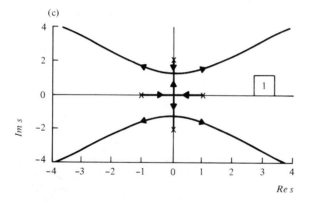

(c)

Figure P4.22

23. Sketch the root locus for the system with the following GH product. Determine all angles of departure and arrival, asymptote angles, and the centroid.

$$G(s)H(s) = \frac{[(s + 2)^2 + 4]^2}{s^3(s + 10)^2}$$

24. For Prob. 23, find the value of the gain K so the dominant closed-loop roots have a damping ratio of 0.1.

25. For the system of Fig. P4.25 determine the gain K of that the dominant closed-loop roots have a damping ratio of 0.2.

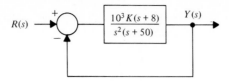

Figure P4.25

26. Find the gain K so that the system with the following GH product has dominant closed-loop roots with a damping ratio of 0.2.

$$G(s)H(s) = \frac{K(s+6)}{s(s^2+4s+13)}$$

Ans $K = 6$

27. For the system of Fig. P4.27:

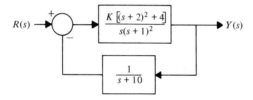

Figure P4.27

(a) Sketch the root locus and compute all angles of arrival, angles of departure, break-in points, entry points, and the centroid.

(b) Determine the gain K so the system has closed-loop dominant roots with a damping ratio of 0.4.

Chapter

5

Root Locus Design

5.1 PREVIEW

Engineering practice in general and control system techniques in particular may be divided into two related areas: analysis and design. Analysis answers the question: "How does it work?" Design responds to the demand: "Make it work!" Most of the material in the previous chapters provides the control system engineer with analytical tools. For example, given a differential equation, the Laplace transform may be used to generate the time response from which the time constant of a first-order system or the damping ratio and undamped natural frequency of an underdamped second order system may be obtained. The root locus demonstrates the effect that a variation in some gain would have upon the closed-loop characteristic equation roots. These are analytical tools but each can be used to evaluate a design.

Chapter 5 begins by showing that certain closed-loop roots create terms that dominate the time response so that an approximate relationship exists between the roots of the characteristic polynomial and such characteristics as rise time and settling time.

If the selection of one variable (perhaps gain) causes the design requirements to be met, design is accomplished by a very simple unity feedback control system. That system will be referred to as uncompensated. If that design cannot meet specifications, additional components must be added. In this chapter, forward path and feedback path compensators will be considered.

The designer must vary the available component parameters until the design requirements are met. The design process results in the iterative scheme: select, analyze, change the values, analyze, change the values again, analyze, until proper performance results.

Typical design specifications include steady state effects (where does the system go?) and stability effects (how does it get there?). This chapter focuses on the error coefficients as a measure of steady state accuracy and the location of dominant roots (nearest the vertical axis) to measure stability. From these dominant roots, the designer can compute values for the rise time and settling time caused by those dominant roots. Alternatively, the designer could focus on rise time and settling time as a measure of stability, although that will not be done here.

In the root loci that follow, the symbol "X" denotes an open-loop pole, the symbol "O" denotes an open-loop zero, while a box denotes a closed-loop transfer function pole for some gain value. Often the boxes denote a set of closed-loop poles that meet performance specifications.

5.2 RELATIONSHIP BETWEEN ROOT LOCUS AND TIME DOMAIN

A control system is usually designed using either the root locus, the time response, or the Bode/Nyquist plot (frequency domain). In the root locus, the system gain varies. For the time response, time is varied. In the Bode/Nyquist plot, frequency varies. Although the three approaches portray three different views of a control system, relationships exist among these methods. Chapter 6 will discuss Bode/Nyquist methods. In order to design a control system using root locus methods (the topic of this chapter), the relationship between the root locus and the resulting time response should be understood.

Figure 5.1 shows a second-order control system. A higher-order system will be far more complex. However, if two dominant roots exist that are much closer to the vertical axis than any other, Fig. 5.1 approximately describes that higher-order system also.

The closed-loop transfer function for Fig. 5.1 has the standard form including the damping ratio ζ and the undamped natural frequency ω_n.

$$T(s) = \frac{Y(s)}{R(s)} = \frac{\omega_n^2}{s^2 + 2\zeta\omega_n s + \omega_n^2}$$

If the damping ratio is between zero and one, complex conjugate poles of $T(s)$ result.

$$s^2 + 2\zeta\omega_n s + \omega_n^2 = (s + \sigma + j\omega)(s + \sigma - j\omega)$$
$$= s^2 + 2\sigma s + \sigma^2 + \omega^2$$

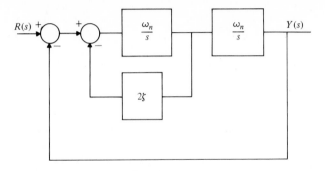

Figure 5.1 Second-order system.

Figure 5.2 shows the complex conjugate closed-loop poles for $T(s)$ denoted by a box. From the figure and the characteristic polynomial, the damping ratio and the angle are related.

$$\zeta = \frac{\sigma}{\omega_n} = \cos \phi$$

Notice that ω_n is the length of the vector from the origin to the complex conjugate poles. Figure 5.3 shows loci of constant ζ with increasing ω_n (straight lines projecting outward) and loci of constant ω_n with increasing ζ (semicircular arcs).

Figures 5.4 and 5.5 were discussed previously in Chap. 2. Figure 5.4 contains $\omega_n T_r$ versus damping ratio. Figure 5.5 contains $\omega_n T_s$ versus damping ratio where T_r refers to rise time, T_s refers to settling time, and ω_n is the undamped natural frequency. The effect of constant damping ratio and variable undamped natural frequency can be evaluated by recognizing that each value of damping ratio

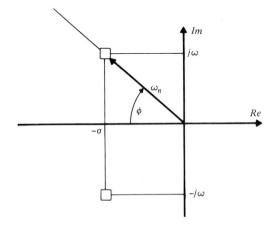

Figure 5.2 Complex conjugate poles.

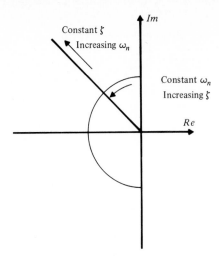

Figure 5.3   Changes in damping ratio and undamped natural frequency.

establishes a point k_1 on the curve of Fig. 5.4 and a point k_2 on the curve of Fig. 5.5. Then

$$T_r = \frac{k_1}{\omega_n}$$

$$T_s = \frac{k_2}{\omega_n}$$

therefore rise time decreases as ω_n increases (the closed-loop pole moves outward along a line of constant damping ratio). Similarly, setting time diminishes.

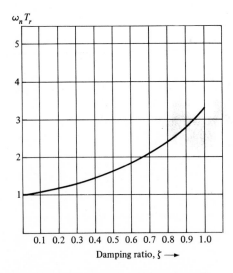

Figure 5.4   Rise time versus damping ratio.

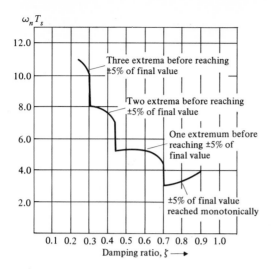

Figure 5.5 Settling time versus damping ratio.

If ω_n remains fixed while ζ increases, Figs. 5.4 and 5.5 portray the relationship between rise time and settling time, respectively, versus ζ. For increased damping ratio, rise time increases. As ζ rises from 0 to 0.7, settling time diminishes to a minimum. As ζ rises from 0.7 to 1, settling time increases.

If the designer creates a system in which a closed-loop pole moves generally outward along a line of constant damping ratio, the resulting increase in the undamped natural frequency causes a decrease in both settling time and rise time. Similarly, if some design change causes the closed-loop pole to move in a semicircular path, the damping ratio alone would change, and therefore settling time can be tuned to a minimum value.

As several roots coalesce, the influence of the root locus upon transient performance becomes harder to predict; however, the actual time response can be generated by using analog or digital simulation. The remainder of this chapter focuses upon modifying the location of the dominant roots. Steady state performance will be shaped by modifying the error coefficients.

DRILL PROBLEMS

D5.1. For the following characteristic polynomial indicate the damping ratio and undamped natural frequency. Using Figs. 5.4 and 5.5, estimate the rise time and settling time.

$$s^2 + 2s + 4$$

Ans 0.5, 2, 0.85, 2.6

D5.2. Repeat D5.1 for

$$s^2 + 3.2s + 4$$

Ans 0.8, 2, 0.85, 1.3

D5.3. Repeat D5.1 for

$$s^2 + 4s + 16$$

Ans 0.5, 4, 0.425, 1.3

D5.4. For a second-order system it is found that the rise time is 0.5 sec and that the settling time is 2.6 sec. Use Figs. 5.4 and 5.5 to estimate the damping ratio and undamped natural frequency.

Ans 0.4, 3

D5.5. Repeat D5.4 for a rise time of 0.3 sec and a settling time of 1.56 sec.

Ans 0.4, 5

D5.6. For the closed-loop system below, what are the rise time and settling time for $K = 4$?

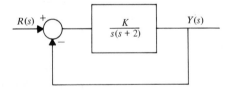

Ans 0.85, 2.5

D5.7. Repeat D5.6 for $K = 16$.

Ans 0.3, 2.8

D5.8. What range of values for K causes the rise time to be less than 0.8 sec for the system of D5.6?

Ans $K > 4$

D5.9. What value for K causes the minimum settling time for the system of D5.6?

Ans $K = 2$

5.3 COMPENSATION

The simple tracking control system configuration of Fig. 5.6 is created by providing unity feedback around the plant $G_p(s)$ and adding an error signal amplifier with gain K. The amplifier gain is adjusted to provide acceptable performance, if possible. Typically, the designer is interested in the closed-loop system's relative stability and its steady state error performance. Other considerations such as step response overshoot may also be of concern.

If adequate performance cannot be obtained with output feedback alone, additional transmittances termed *compensators* may be added to the system. Figure 5.7 shows several simple compensator configurations for single-input, single-output plants. In Fig. 5.7a, the compensator $G_c(s)$ is inserted into the system's forward path. In Fig. 5.7b, the compensator $H_c(s)$ is placed in the feedback path around the plant, forming a *feedback* compensator. A system involving both feedback and cascade compensation is given in Fig. 5.7c.

Tables 5.1 and 5.2 contain typical cascade and feedback compensators and the effect of each compensator on relative stability and on steady state error. A compensator influences the root locus of a system due to the poles and zeros of that compensator. The gain K with a compensator included may differ from the value used where only the gain appears in the forward path. The system of Fig. 5.6 will be referred to as being uncompensated in that only a gain is used to stabilize the system.

A cascade integral compensator adds a pole at the origin. System type number increases by 1. If the resulting system is stable, steady state error for a step into a previously type 0 system and steady state error for a ramp into a previously type one system disappear. However, the difference between the number of open-loop poles n and the number of open-loop zeros m increases by 1. Because $n - m$ increases, asymptote angles are decreased (pointing more into the right half-plane). The resulting system plus compensator has diminished relative stability due to that increase of $n - m$. Usually, integral compensation is avoided.

Cascade integral plus proportional compensation provides the benefit of cascade integral compensation. System type number increases by 1, thus eliminating steady state error. Because a new pole and zero are added, $n - m$ and the

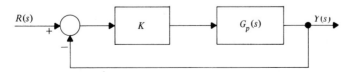

Figure 5.6 Unity feedback control system.

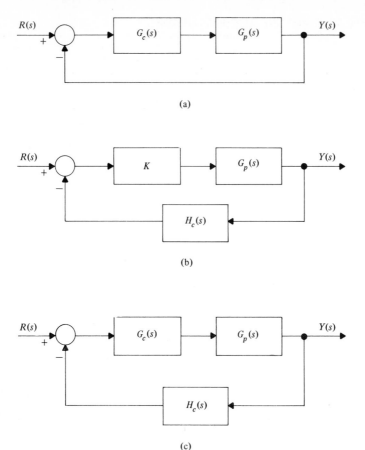

Figure 5.7 Compensator configurations. **(a)** Cascade compensated system. **(b)** Feedback compensated system. **(c)** System with feedback and cascade compensation.

asymptote angles are unchanged. The centroid moves to the right, however, so that the closed-loop poles can also move somewhat to the right, reducing relative stability. Attention focuses on the dominant roots of the uncompensated system of Fig. 5.6. The value of a for the compensator should cause the centroid to move only slightly to the right as compared with the location of the dominant roots (uncompensated). The value of K with the compensator can be equal to the uncompensated value. The integral plus proportional compensator requires an active device (perhaps an operational amplifier with capacitive feedback).

A cascade lag compensator adds a pole and a zero where the pole is not located at the origin as with the integral plus proportional case. Type number remains the same; however, the value of a/b can cause the error coefficient to increase, especially where the value of K is not changed from the uncompensated case. As with the integral plus proportional compensator, the centroid moves to the right.

Table 5.1 COMMON TYPES OF COMPENSATORS

Compensator	Transmittance	Typical Effect on Steady State Errors	Typical Effect on Relative Stability
Cascade integral	$G_c(s) = \dfrac{K}{s}$	Greatly improved	Greatly reduced
Cascade intergral plus proportional	$G_c(s) = \dfrac{K(s + a)}{s}$	Greatly improved	Reduced
Cascade lag	$G_c(s) = \dfrac{K(s + a)}{s + b}$ $b < a$	Improved	Reduced
Cascade lead	$G_c(s) = \dfrac{K(s + a)}{(s + b)}$ $a < b$	Somewhat improved or somewhat worse	Increased
Cascade lag-lead	$G_c(s) = K\left(\dfrac{s + a}{s + b}\right)_{\text{lag}}$ $\times \left(\dfrac{s + a}{s + b}\right)_{\text{lead}}$	Improved	Increased
Feedback rate	$H_c(s) = 1 + As$	Somewhat improved or somewhat worse	Increased
Proportional integral derivative	$G_c(s) = \dfrac{K(s^2 + as + b)}{s}$	Greatly improved	Increased

The value of a may be chosen as with the integral plus proportional compensator. Here a/b is chosen to change the error coefficient. A passive RC circuit may be used to create the lag compensator.

A cascade lead compensator may also be created by a passive RC circuit. The zero is closer to the origin than the pole. The centroid and the dominant roots move to the left. The error coefficient might increase or decrease depending on the gain used with the compensator. The value of K must be at least b/a times the uncompensated value of K to maintain steady state accuracy. A multivariable root locus demonstrates that a should be almost equal to 2 times the damping ratio times the undamped natural frequency for the uncompensated dominant poles.

A cascade lag-lead compensator combines the attributes of the lag and lead compensator. The error coefficient can be increased (as with a lag compensator) and the dominant roots can be moved to the left (as with a lead compensator). Where the ratio of a/b for the lag compensator equals the ratio b/a for the lead compensator, the error coefficient becomes directly proportional to K.

Table 5.2 COMMON TYPES OF COMPENSATORS

Compensator	Change in $n - m$	Change in Centroid if $n - m$ Is Constant	Change in Type Number	$\dfrac{K_i(\text{comp})}{K_i(\text{uncomp})}$
Cascade integral	$+1$		$+1$	∞
Cascade integral plus proportional	0	$\dfrac{a}{n-m}$	$+1$	∞
Cascade lag $(a > b)$	0	$\dfrac{a - b}{n - m}$	0	$\dfrac{K(\text{comp})}{K(\text{uncomp})}\left(\dfrac{a}{b}\right)$
Cascade lead $(b > a)$	0	$\dfrac{-(b - a)}{n - m}$	0	$\dfrac{K(\text{comp})}{K(\text{uncomp})}\left(\dfrac{a}{b}\right)$
Cascade lag-lead	0	$\dfrac{(a - b)_{\text{lag}} - (b - a)_{\text{lead}}}{n - m}$	0	$\dfrac{K(\text{comp})}{K(\text{uncomp})}\left(\dfrac{a}{b}\right)_{\text{lag}}\left(\dfrac{a}{b}\right)_{\text{lead}}$
Feedback rate	-1		0	$\dfrac{K(\text{comp})}{K(\text{comp}) + AK(\text{comp})K_1(\text{uncomp})}$
Proportional integral derivative	-1		$+1$	∞

A cascade proportional-integral-derivative (PID) compensator adds two zeros and a pole. The pole is at the origin. System type number increases by one as with the cascade-integral plus proportional compensator. Steady state error diminishes accordingly. The value of $n - m$ diminishes by one which increases the asymptote angles. Since those angles point less directly toward the right half-plane, the result is a more stable system (dominant roots move to the left). The two zeros may be selected to attract dominant roots toward favorable locations to the left of their uncompensated sites. The new zeros will cause the centroid to move to the right; hence the zeros must not be too far left.

A feedback rate compensator causes the rate of change of the output to be fed back along with the output. These two signals combine to add a new zero to the root locus. Since $n - m$ decreases by 1, the system becomes more stable. System type number is unchanged; hence, rate feedback compensation (like lead compensation) is primarily intended to improve stability. If K with rate feedback compensation exceeds the uncompensated value of K, the steady state error can be reduced.

Each of the seven compensators from Tables 5.1 and 5.2 will be applied to the same problem. The design of each compensator will be considered in detail. The resulting designs are summarized in Table 5.3.

Table 5.3 SUMMARY OF DESIGNS

Compensator	Transmittance	Ramp Error Coefficient	Closed-Loop Zeros	Closed-Loop Poles	Dominant Roots ω_n	Dominant Roots ζ
Uncompensated	$K = 3$	0.6	None	$-0.767 \pm j0.793,$ -2.47	1.10	0.70
Cascade integral	$G_c(s) = \dfrac{K}{s}$	∞		Unstable		
Cascade integral plus proportional	$G_c(s) = \dfrac{3(s + 0.1)}{s}$	∞	-0.1	$-0.123, -0.724 \pm j0.694,$ -2.43	1.00	0.72
Cascade lag	$G_c(s) = \dfrac{3(s + 0.1)}{s + 0.01}$	6	-0.1	$-0.120, -0.729 \pm j0.705,$ -2.43	1.01	0.72
Cascade lead	$G_c(s) = \dfrac{60(s + 1.6)}{s + 16}$	1.2	-1.6	$-1.11, -1.32 \pm j1.9,$ -16.3	2.31 1.11	0.57 1.00
Cascade lag-lead	$G_c(s) = \dfrac{60(s + 0.1)}{s + 0.01}$ $\times \dfrac{s + 1.6}{s + 16}$	12	$-0.1, -1.6$	$-0.109, -1.06$ $-1.29 \pm j1.86,$ -16.3	2.26 1.06	0.57 1.00
Feedback rate	$H_c(s) = \frac{1}{4}(s + 4)$ $K = 6$	0.92	None	$-0.845 \pm j1.37,$ -2.31	1.61	0.52
PID	$G_c(s) = \dfrac{4(s + 1.25)^2}{s}$	∞	$-1.5, -1.5$	$-1 \pm j1.22,$ $-1 \pm j1.22$	1.58	0.63

5.4 THE UNCOMPENSATED SYSTEM WITH FEEDBACK

In the following sections of this chapter, the use of each of these compensators in systems with an example plant transmittance

$$G_p(s) = \frac{1}{s(s + 2 + j)(s + 2 - j)} = \frac{1}{s(s^2 + 4s + 5)}$$

will be illustrated, along with design strategies and methods. For the simple feedback configuration of Fig. 5.6, the root locus in terms of the adjustable error signal gain K is given in Fig. 5.8. The asymptotic angles are $\pm 60°$ and $180°$, and the centroid of the asymptotes is at $s = -\frac{4}{3}$. If $K = 3$, the closed-loop poles are at $s = -0.767 \pm j0.793$ and -2.47, as indicated.

The ramp error coefficient of the system is

$$K_1 = \lim_{s \to 0} sKG_p(s) = \frac{K}{5} = 0.6$$

For a unit ramp input, the steady state error between output and input is thus given by

$$\text{Steady state error to a ramp input} = \frac{1}{K_1} = 1.67$$

If this steady state error is too large, K could be increased; however, the closed-loop poles would then be further to the right on the complex plane, giving less relative stability. If greater relative stability is required, the gain K could be reduced, but the steady state ramp error would increase.

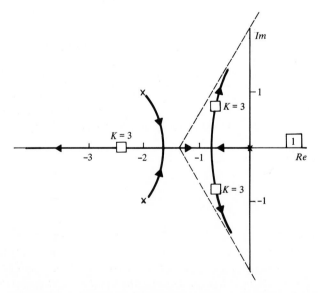

Figure 5.8 Root locus for the example uncompensated feedback system.

DRILL PROBLEM

D5.10. An uncompensated feedback system of the form of Fig. 5.6 has

$$G_p(s) = \frac{1}{(s + 4)^2}$$

Select K so that the closed-loop poles are at $-4 \pm j4$. What is K_0, the step error coefficient? Where are the closed-loop zeros?

Ans $K = 16$; $K_0 = 1$; no closed-loop zeros

5.5 CASCADE INTEGRAL COMPENSATION

For the cascade compensation configuration of Fig. 5.7, the open-loop system transmittance is

$$G(s) = G_c(s)G_p(s)$$

Assuming that the overall system is stable, the steady state error coefficients are

$$K_i = \lim_{s \to 0} s^i G(s) = \lim_{s \to 0} s^i G_c(s)G_p(s)$$

Steady state performance may be improved by adding one or more poles, in the compensator, at or near $s = 0$. If a pole is added at $s = 0$, the system type number is increased by 1. Adding a pole very close to $s = 0$ for a type i system does not change the system type number but increases K_i and thus reduces the steady state error to a t^i input.

The addition of open-loop poles at or near $s = 0$, however, tends to reduce the relative stability of the closed-loop system or to make it unstable. Thus, open-loop zeros are often also placed in the compensator transmittance to "draw away" the root locus into the LHP. The usual design strategy is then to place compensator poles near $s = 0$ to improve steady state performance and to place compensator zeros so that the system root locus gives sufficient relative stability for a suitable value of the compensator multiplying constant K.

The cascade integral compensator

$$G_c(s) = \frac{K}{s}$$

adds a single pole at $s = 0$ to the system's open-loop transmittance. It thus raises the system type number, improving steady state error performance provided that the resulting compensated system is stable. The addition of an open-loop pole at $s = 0$ tends to reduce the relative stability because the additional pole in the open-loop transmittance means that the locus will have one more asymptotes. The

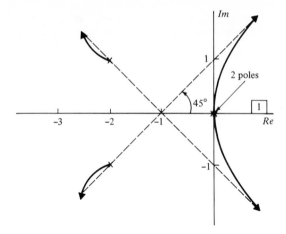

Figure 5.9 Root locus of the example system with cascade integral compensation.

asymptotic angles of the root locus being evenly spaced, this increases the likelihood of closed-loop poles being near or in the RHP.

For the example plant $G_p(s)$, the compensated system transfer function is

$$T(s) = \frac{\dfrac{K}{s}\left[\dfrac{1}{s(s^2 + 4s + 5)}\right]}{1 + \dfrac{K}{s}\left[\dfrac{1}{s(s^2 + 4s + 5)}\right]} = \frac{K}{s^2(s^2 + 4s + 5) + K}$$

A root locus plot, in terms of the adjustable constant K, is given in Fig. 5.9.

Unfortunately, this compensated system is unstable for all positive K. If a system is stable, the final value theorem is applicable and the ramp error coefficient may be used to predict the steady state error for a ramp input (for example, the steady state error for a unit ramp input is $1/K_1$). For this unstable system, the response is unbounded so that no steady state error exists and the ramp error coefficient is meaningless. It would be

$$K_1 = \lim_{s \to 0} sG_c(s)G_p(s)$$

$$= \lim_{s \to 0} \frac{K}{s(s^2 + 4s + 5)} = \infty$$

indicating zero steady state error to a ramp input because the compensator has raised the system type from type 1 to type 2. However, the detrimental aspect of this compensator, that of reducing the system's relative stability, makes this approach unusable for the example plant.

5.6 CASCADE INTEGRAL PLUS PROPORTIONAL COMPENSATION

The cascade integral plus proportional compensator has transmittance of the form

$$G_c(s) = A_p + \frac{A_i}{s} = \frac{K(s+a)}{s}$$

A pole at $s = 0$ is added to the open-loop system transmittance $G_c(s)G_p(s)$ while a zero is added at $s = -a$.

System type increases by 1 because of the added open-loop pole at the origin. Hence, for a stable design, steady state error performance is improved. This compensator, having both a pole and a zero, does not change the difference between the number of open-loop poles and open-loop zeros. Thus the compensated system has the same root locus asymptotic angles as does the uncompensated one, typically meaning a greater relative stability than with integral compensation.

The centroid of the asymptotes does change, however. For the compensated system with open-loop transmittance $G_c(s)G_p(s)$,

$$\sigma = \frac{\Sigma \text{ poles of } G_c G_p - \Sigma \text{ zeros of } G_c G_p}{\text{number of poles of } G_c G_p - \text{number of zeros of } G_c G_p}$$

As the compensator contributes a pole at $s = 0$ and a zero at $s = -a$,

$$\sigma = \frac{\Sigma \text{ poles of } G_p - \Sigma \text{ zeros of } G_p + a}{\text{number of poles of } G_p - \text{number of zeros of } G_p}$$

For a positive constant a (an LHP compensator zero), σ is moved to the right, tending to reduce relative stability from that of the uncompensated feedback system by an amount proportional to a.

For the cascade integral plus proportional compensator, there are two adjustable parameters, K and a, and the designer may alternate between adjusting one parameter then the other until a satisfactory design results. A reasonable starting point for such an iterative design procedure is with $a = 0$, for which the compensator transmittance is

$$G_c(s) = \frac{K(s+0)}{s} = K$$

which is that for the uncompensated feedback system. The previous root locus plot of Fig. 5.8 then shows the effect of adjustable K when $a = 0$. Choosing $K = 3$, as was done for the uncompensated system, this compensator transmittance is, in terms of a,

$$G_c(s) = \frac{3(s+a)}{s}$$

giving an overall system transfer function

$$T(s) = \frac{G_c(s)G_p(s)}{1 + G_c(s)G_p(s)}$$

$$= \frac{\dfrac{3(s + a)}{s}\left[\dfrac{1}{s(s^2 + 4s + 5)}\right]}{1 + \dfrac{3(s + a)}{s}\left[\dfrac{1}{s(s^2 + 4s + 5)}\right]}$$

$$= \frac{3(s + a)}{s^2(s^2 + 4s + 5) + 3s + 3a}$$

$$= \frac{\text{numerator}}{1 + \dfrac{3a}{s^2(s^2 + 4s + 5) + 3s}}$$

$$= \frac{\text{numerator}}{1 + \dfrac{3a}{s(s^3 + 4s^2 + 5s + 3)}}$$

$$= \frac{\text{numerator}}{1 + \dfrac{3a}{s(s + 0.767 + j0.793)(s + 0.767 - j0.793)(s + 2.47)}}$$

With K equal to the uncompensated value of 3 and a adjustable, two of the open-loop poles are the dominant closed-loop poles of the uncompensated system (the poles at $-0.767 \pm j0.793$). As a increases in the root locus of Fig. 5.10a, those two dominant roots migrate into the right half-plane so the root locus for variable a suggests that a should be small. A value of $a = 0.1$ provides a set of roots that have not migrated too far to the right, while moving the root at the origin to the left.

The same result can be obtained by selecting a so the centroid moves to the right by a small amount compared with the undamped natural frequency of the uncompensated dominant roots. For those dominant roots

$$s^2 + 2\zeta\omega_n s + \omega_n^2 = (s + 0.767 + j0.793)(s + 0.767 - j0.793)$$

$$= s^2 + 1.534s + 1.217$$

$$\omega_n = \sqrt{1.217} = 1.10 \qquad \zeta = \frac{1.534}{2 \times 1.10} = 0.7$$

Compared with the uncompensated case the centroid moves to the right by

$$\frac{a}{n - m}$$

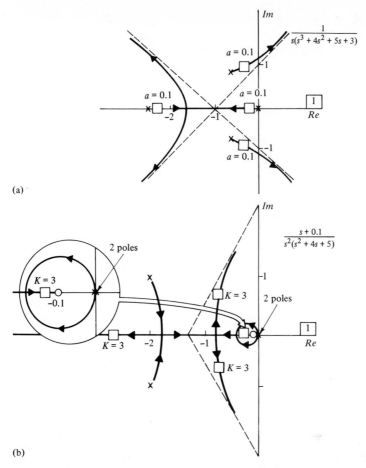

Figure 5.10 Root loci of the example system with cascade integral plus proportional compensation. **(a)** $K = 3$, a variable. **(b)** K variable, $a = 0.1$.

If a is selected to be

$$a = 0.1\omega_n$$

then the centroid should disturb the dominant roots by an acceptably small amount while the system type number increases by 1. Therefore

$$a = 0.1 \times 1.10 \simeq 0.1$$

is a logical choice which substantiates the root locus of Fig. 5.10a.

With $a = 0.1$, the root locus of Fig. 5.10b follows. With $K = 3$ the iterative design procedure terminates.

$$G_c(s) = \frac{3(s + 0.1)}{s}$$

and

$$T(s) = \frac{G_c(s)G_p(s)}{1 + G_c(s)G_p(s)} = \frac{\dfrac{3(s + 0.1)}{s^2(s^2 + 4s + 5)}}{1 + \dfrac{3(s + 0.1)}{s^2(s^2 + 4s + 5)}}$$

$$= \frac{3(s + 0.1)}{s^4 + 4s^3 + 5s^2 + 3s + 0.3}$$

The compensated closed-loop system poles are at $s = -0.123$, $-0.724 \pm j0.694$, and -2.43. Because of the compensator's added pole at the origin, the steady state error to a ramp input is zero.

The closed-loop pole at -0.123 is nearly canceled by the closed-loop zero at -0.1. The dominant roots are at $-0.724 \pm j0.694$. This near cancellation is visible from a partial fraction expansion of $T(s)$

$$T(s) = -\frac{0.0357}{s + 0.123} + \frac{0.895}{s + 2.43} + \frac{1.19e^{-j111.2°}}{s + 0.724 - j0.694} + \frac{1.19e^{j111.2°}}{s + 0.724 + j0.694}$$

A natural response due to some forcing function [for example, $\delta(t)$] would contain a very small amplitude response due to the pole at -0.123 (the natural response is therefore dominated by the complex conjugate terms).

DRILL PROBLEMS

D5.11. Design cascade integral plus proportional compensators for systems with the following plant transmittances. Select compensator gains K using root locus plots. For the values of K selected, find the steady state error coefficients and the relative stability of each system.

(a) $G_p(s) = \dfrac{1}{(s + 4)^2}$

(b) $G_p(s) = \dfrac{s + 1}{s(s + 4)}$

(c) $G_p(s) = \dfrac{s + 1}{s^2 + 4s + 5}$

D5.12. Sketch root locus plots, in terms of the compensator multiplying constant K, for cascade integral plus proportional compensated systems with the plant transmittances of the previous problem. If possible, choose the compensator zero location and K, by trial and error, to achieve better relative stability and steady state error performance than with only integral compensation.

5.7 CASCADE LAG COMPENSATION

A cascade lag compensator has transmittance of the form

$$G_c(s) = \frac{K(s + a)}{s + b}$$

where K, a, and b are positive constants and $a > b$ so that the compensator zero is to the left of the compensator pole on the complex plane. The proportional plus integral compensator is a special case of lag compensation, for which the constant b is zero. The pure integration operation required when $b = 0$ is often difficult to achieve in practice, so this more general form is of considerable practical interest.

For nonzero b, this compensator does not increase the system type number. However, steady state error performance can be improved over that of the uncompensated feedback system.

The error coefficient for an uncompensated system is

$$K_i = K \lim_{s \to 0} s^i G_p(s)$$

For a lag compensated system

$$K_i = \lim_{s \to 0} s^i G_c(s) G_p(s)$$

$$= K\left(\frac{a}{b}\right) \lim_{s \to 0} s^i G_p(s)$$

The ratio of the error coefficient of the uncompensated system to the error coefficient for the cascade lag compensated system is

$$\frac{K_i \text{ (compensated)}}{K_i \text{ (uncompensated)}} = \frac{K \text{ (compensated)}}{K \text{ (uncompensated)}} \left(\frac{a}{b}\right)$$

The centroid for a cascade lag compensated system moves to the right by

$$\frac{a - b}{n - m}$$

One design approach is to select the ratio a/b to equal the factor by which the error coefficient is to be increased. If that factor is d, then

$$b = \frac{a}{d} \qquad d > 1$$

$$K \text{ (compensated)} = K \text{ (uncompensated)}$$

The error coefficient increases by the factor d.

$$K_i \text{ (comp)} = K_i \text{ (uncomp)} \, d$$

If d approaches infinity, the cascade lag compensator becomes cascade integral plus proportional; however, a large value of d creates a problem in fabricating a reliable *RC* circuit with a zero and a pole widely separated. In practice, the value of *d* is chosen around 10.

The centroid moves to the right by

$$\frac{a(1 - 1/d)}{n - m}$$

The value of *a* could be selected by a root locus approach (for fixed *K* and *d*) or by selecting

$$a = 0.1\omega_n$$

Where ω_n is the undamped natural frequency of the dominant uncompensated roots and where the centroid moves a small amount compared to ω_n.

For the example problem, the ramp error coefficient for the uncompensated case is 0.6 while the dominant roots are at $-0.767 \pm j0.793$. The ramp error coefficient can be multiplied by 10 with a modest motion to the right of the dominant roots if

$$d = 10$$

$$K_1 = 6$$

$$\omega_n = \sqrt{(0.767)^2 + (0.793)^2} = 1.10$$

$$a = 0.1 \times 1.10 \approx 0.1$$

$$G_c(s) = \frac{3(s + 0.1)}{s + 0.01}$$

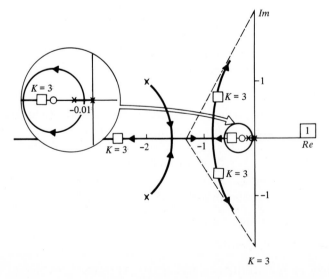

Figure 5.11 Root locus for the example system with cascade lag compensation.

The root locus appears in Figure 5.11. See Table 5.3. The closed-loop pole at -0.12 is nearly cancelled by the closed-loop zero at -0.10. The transient response is dominated by the closed-loop poles $-0.729 \pm j0.705$. The ramp error coefficient rises to 6.

5.8 CASCADE LEAD COMPENSATION

A cascade lead compensator has the same form as a cascade lag compensator

$$G_c(s) = \frac{K(s + a)}{s + b}$$

except that $b > a$. The compensator pole is to the left of the zero. The centroid moves to the left by

$$\frac{-(b - a)}{n - m}$$

Since $n - m$ remains the same, the asymptote angles are unchanged. The root locus (for fixed K) moves somewhat to the left. The lead compensator primarily influences system stability. The performance of the cascade lead compensator is more easily evaluated by replacing b with ad, where $d > 1$. The ith error coefficient of the compensated system is

$$K_i = \lim_{s \to 0} s^i G_c(s) G_p(s)$$

$$= \frac{K}{d} \lim_{s \to 0} s^i G_p(s)$$

The ratio of the error coefficient for the uncompensated system to the error coefficient for the cascade lead compensated system is

$$\frac{K_i \,(\text{compensated})}{K_i \,(\text{uncompensated})} = \frac{K \,(\text{compensated})/d}{K \,(\text{uncompensated})}$$

The value of K for the compensated system must be at least d times the value of K for the uncompensated system in order for the error coefficient to remain the same.

Three values must be selected: K, a, and d. A designer could select a value for K and d and then obtain a root locus for variable a. With a selected, other values of K and d could be tried until the design is acceptable.

It is possible to produce some general results by examining Fig. 5.12a. Let ω_n represent the undamped natural frequency of the uncompensated dominant roots. Similarly, ζ is the damping ratio for those dominant roots.

If $K = 1$ and $a = 0$, the transfer function is in standard form and the uncompensated case is unchanged.

$$\frac{Y(s)}{R(s)} = \frac{\omega_n^2}{s^2 + 2\zeta\omega_n s + \omega_n^2}$$

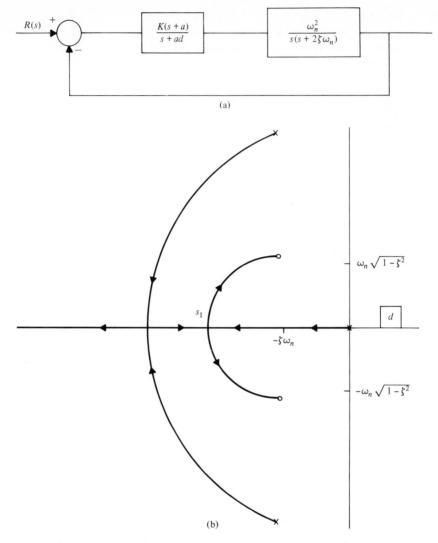

Figure 5.12 (a) Lead compensation of the uncompensated dominant roots. **(b)** Root locus for variable a.

The following analysis assumes the other uncompensated closed-loop poles are far removed from the dominant poles and that no zeros exist. The actual system could appear in Fig. 5.12, but the goal is to understand the general effect of lead compensation upon the dominant roots.

The closed-loop transfer function including K, a, and d is

$$\frac{Y(s)}{R(s)} = \frac{K(s + a)\omega_n^2}{s^3 + s^2(2\zeta\omega_n + ad) + s(ad2\zeta\omega_n + K\omega_n^2) + Ka\omega_n^2}$$

If K and d are fixed, the root locus for adjustable a follows by factoring.

$$\frac{Y(s)}{R(s)} = \frac{\text{numerator}}{1 + \dfrac{ad[s^2 + 2\zeta\omega_n s + (K/d)\omega_n^2]}{s(s^2 + 2\zeta\omega_n s + K\omega_n^2)}}$$

For the uncompensated case ($K = 1$, $a = 0$),

$$K_1 = \frac{\omega_n}{2\zeta}$$

For the compensated case

$$K_1 = \left(\frac{K}{d}\right)(\omega_n/2\zeta)$$

Suppose a designer chooses to adjust K to maintain the same error coefficient with the compensator as with the uncompensated system. Then $K = d$ and the root locus for adjustable a requires plotting

$$KG(s)H(s) = \frac{ad(s^2 + 2\zeta\omega_n s + \omega_n^2)}{s(s^2 + 2\zeta\omega_n s + d\omega_n^2)}$$

The open-loop zeros are the same as the uncompensated closed-loop poles. A wise choice of a should cause the resulting dominant closed-loop poles to be located to the left of the uncompensated closed-loop poles (the open-loop zeros for adjustable a). Figure 5.12b shows a typical root locus with $d = 10$.

The best choice for a causes a pair of closed-loop poles to be located at s_1, in the sense that a lower value of a creates one real root (closer to the origin) with a longer time constant while a larger value of a creates a complex conjugate pair whose real part also exhibits a longer time constant. Table 5.4 shows the values of a for various values of damping ratio and the resulting location of s_1.

In general, the best choice of a is

$$a \simeq 2\zeta\omega_n$$

Table 5.4 OPTIMAL LOCATION FOR a ($d = 10$)

ζ	a/ω_n	s_1/ω_n
0.3	0.7	-1.1
0.4	0.85	-1.1
0.5	1.01	-1.0
0.6	1.225	-1.0
0.7	1.6	-1.0

where a double pair of dominant closed-loop poles results at $-\omega_n$ with unity damping. An actual system may behave somewhat differently; hence the value of a, K, and d are only starting points for an iterative design.

Returning to the example problem, the uncompensated dominant roots provide

$$\omega_n = 1.11$$

$$\zeta = 0.7$$

The above analysis suggests (if $d = 10$) that

$$a = 2(0.7)(1.11) \simeq 1.6$$

Suppose the gain must double the ramp error coefficient. Then

$$K \text{ (compensated)} = 2dK \text{ (uncompensated)}$$

$$K = 2 \times 10 \times 3 = 60$$

$$G_c(s) = \frac{60(s + 1.6)}{s + 16}$$

The root locus is shown in Fig. 5.13. The closed-loop poles at $-1.32 \pm j1.9$ have indeed been moved to the left of their uncompensated locations. The closed-loop pole at -1.11 would also dominate transient response. The remaining closed-loop

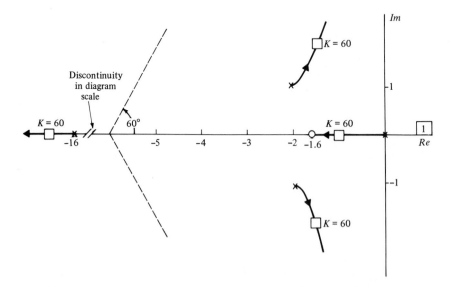

Figure 5.13   Root locus for the example system with cascade lead compensation.

pole is at -16.3. The ramp error coefficient is doubled to 1.2. The corresponding overall system transfer function is

$$T(s) = \cfrac{\cfrac{60(s + 1.6)}{s + 16}\left[\cfrac{1}{s(s^2 + 4s + 5)}\right]}{1 + \cfrac{60(s + 1.6)}{s + 16}\left[\cfrac{1}{s(s^2 + 4s + 5)}\right]}$$

$$= \frac{60(s + 1.6)}{s^4 + 20s^3 + 69s^2 + 140s + 96}$$

DRILL PROBLEM

D5.13. Sketch root locus plots, in terms of the compensator multiplying constant K, for cascade lead compensated systems with the following plant transmittances. If possible, choose the compensator pole and zero locations and K, by trial and error, to achieve better relative stability and better steady state error performance than with the uncompensated feedback system.

(a) $G_p(s) = \dfrac{1}{(s + 4)^2}$

(b) $G_p(s) = \dfrac{1}{s(s + 4)}$

(c) $G_p(s) = \dfrac{s + 1}{s(s + 4)}$

5.9 CASCADE LAG-LEAD COMPENSATION

A cascade lag compensator improves steady state accuracy whereas a cascade lead compensator improves relative stability. The best attributes of both compensators can be combined where

$$G_c(s) = K\left[\frac{s + a}{s + b}\right]_{\text{lag}}\left[\frac{s + a}{s + b}\right]_{\text{lead}}$$

The design is simplified if the pole-zero ratios are set by

$$\left[\frac{b}{a}\right]_{\text{lead}} = \left[\frac{a}{b}\right]_{\text{lag}} = d$$

The ratio of error coefficients becomes

$$\frac{K_i\,(\text{compensated})}{K_i\,(\text{uncompensated})} = \frac{K\,(\text{compensated})}{K\,(\text{uncompensated})}$$

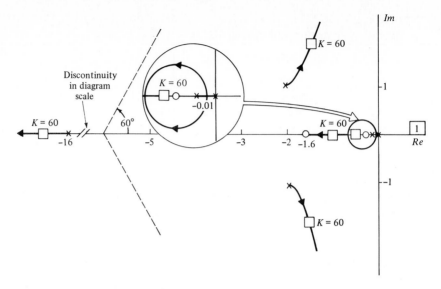

Figure 5.14   Root locus for the example system with cascade lag-lead compensation.

A multivariable approach can be employed where the variables are K, a_{lag}, a_{lead}, and d. A starting point could be (using the uncompensated dominant undamped natural frequency)

$$a_{lag} \simeq 0.1\omega_n$$

$$a_{lead} \simeq 2\zeta\omega_n$$

Returning to the example problem, the ramp error coefficient can be multiplied by a factor of 10 as compared to the lead case and by a factor of 20 as compared to the uncompensated case by combining the lag and lead compensators with $K = 60$:

$$G_c(s) = \frac{60(s + 0.1)(s + 1.6)}{(s + 0.01)(s + 10)}$$

The root locus appears in Fig. 5.14. The closed-loop poles and ramp error coefficient are in Table 5.3.

DRILL PROBLEM

D5.14. Sketch root locus plots, in terms of the compensator multiplying constant K, for cascade lag-lead compensated systems with the following plant transmittances. If possible, choose compensator pole and zero locations and K, by trial and error, to achieve better relative stability and better steady state error performance than with the uncompensated feedback system.

(a) $G_p(s) = \dfrac{1}{(s+4)^2}$

(b) $G_p(s) = \dfrac{1}{s(s+2)}$

(c) $G_p(s) = \dfrac{s^2 + 4s + 5}{s^2(s+4)}$

5.10 FEEDBACK RATE COMPENSATION

A feedback compensated system, Fig. 5.7b, with feedback rate compensation has

$$H_c(s) = 1 + As$$

The feedback is proportional to the output derivative (rate of change) in addition to unity output feedback.

An equivalent forward path transmittance for unity feedback would be

$$G_E(s) = \frac{KG_p(s)}{1 + AKsG_p(s)}$$

The ith error coefficient is

$$K_i = \lim_{s \to 0} s^i G_E(s)$$

Suppose the above K is referred to as $K(\text{comp})$ and suppose K_i is multiplied and divided by the uncompensated value of K, $K(\text{uncomp})$.

$$K_i(\text{comp}) = \lim_{s \to 0} \frac{K(\text{comp}) s^i K(\text{uncomp}) G_p(s)}{K(\text{uncomp}) + AK(\text{comp}) s K(\text{uncomp}) G_p(s)}$$

The compensated value of K_i may be computed in terms of error coefficients for the uncompensated system

$$K_i(\text{comp}) = \frac{K(\text{comp}) K_i(\text{uncomp})}{K(\text{uncomp}) + AK(\text{comp}) K_1(\text{uncomp})}$$

$$\frac{K_i(\text{comp})}{K_i(\text{uncomp})} = \frac{K(\text{comp})}{K(\text{uncomp}) + AK(\text{comp}) K_1(\text{uncomp})}$$

The ith error coefficient could be increased or decreased, depending on the system.

The designer must select K and A. The zero added by the rate feedback device is at $-1/A$ since

$$H_c(s) = A\left(s + \frac{1}{A}\right)$$

The value of $n - m$ decreases by 1 because one open-loop zero is added. Asymptote angles increase, generally improving stability. The centroid may move to the right due to the last term of

$$\sigma = \frac{\sum \text{open-loop poles of } G_p(s) - \sum \text{open-loop zeros of } G_p(s)}{n - m} + \frac{\frac{1}{A}}{n - m}$$

For the example system

$$\sigma = \frac{-2 - 2}{2} + \frac{1/A}{2} = -2 + \frac{1}{2A}$$

Since the open-loop zero at $-1/A$ is not a closed-loop zero, a designer should not place the compensator zero close to the origin (placing a closed-loop pole there) or too far from the origin (moving the centroid into the right half-plane). The centroid can be placed at the origin by selecting

$$0 = -2 + \frac{1}{2A}$$

$$\frac{1}{A} = 4$$

$$A = \tfrac{1}{4}$$

The root locus is shown in Fig. 5.15. A value of $K = 6$ seems to provide good dominant roots (compared to the uncompensated case). Reducing K would reduce

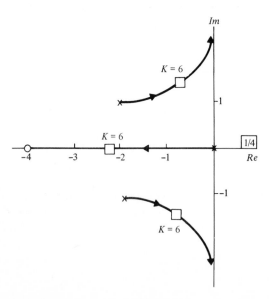

Figure 5.15 Root locus of the example system with feedback rate compensation.

the error coefficient and move the real root to the right. The ramp error coefficient (compensated) becomes (with $K = 6$)

$$K_1 = \frac{(6)(0.6)}{3 + (\frac{1}{4})(6)(0.6)} = \frac{3.6}{3.9} = 0.92$$

The dominant closed-loop poles are at $-0.845 \pm j1.37$. The remaining closed-loop pole is at -2.31.

DRILL PROBLEMS

D5.15. Design feedback rate compensators for systems with each of the following plant transmittances. If possible, choose the compensator zero location and the error signal gain K, by trial and error, to achieve better relative stability and better steady state error performance than with the uncompensated feedback system.

(a) $G_p(s) = \dfrac{1}{(s + 4)^2}$

(b) $G_p(s) = \dfrac{1}{s(s + 4)}$

(c) $G_p(s) = \dfrac{1}{s(s + 4)^2}$

D5.16. Use root locus methods to design compensators for systems with each of the following plant transmittances. Each compensator should involve a combination of both rate feedback and integral cascade compensation. For each design, find the steady state error coefficients and the amount of relative stability.

(a) $G_p(s) = \dfrac{1}{s(s + 4)^2}$

(b) $G_p(s) = \dfrac{s + 1}{(s + 4)^2}$

(c) $G_p(s) = \dfrac{s + 4}{s(s^2 + 2s + 5)}$

5.11 PROPORTIONAL-INTEGRAL-DERIVATIVE COMPENSATION

A cascade proportional-integral-derivative (PID) compensator has a transmittance of the form

$$G_c(s) = A_p + \frac{A_i}{s} + A_d s = \frac{K(s^2 + as + b)}{s}$$

Controllers of this type are widely used in process control applications in industry.

Provided that the resulting feedback system is stable, the added open-loop pole at the origin increases the system type number by 1. Relative stability is apt to be improved because the net addition of a zero will result in one less root locus asymptote.

The two zeros created by the PID compensator may be designated s_1 and s_2. The numerator of the PID compensator is

$$K(s^2 + as + b) = K(s - s_1)(s - s_2)$$
$$= K[s^2 - (s_1 + s_2)s + s_1 s_2]$$

Notice that the negative sum of the two compensator zeros is equal to a. That fact can be used to compute the centroid when the PID compensator is added to an uncompensated system as in the example where

$$\sigma = \frac{-2 + j1 - 2 - j1 - 0 - 0 - (s_1 + s_2)}{2}$$

$$= \frac{-4 - (s_1 + s_2)}{2} = -2 + \frac{a}{2}$$

If both zeros of the PID compensator are placed at -2, the centroid becomes zero. The asymptote angles are $\pm 90°$, so the root locus of Fig. 5.16a results. As the gain increases, the pair of open-loop poles at the origin move into the left half-plane and then move up the vertical axis. The open-loop complex conjugate poles move toward the double PID zeros. Because the dominant roots are always relatively close to the vertical axis, the design of Figure 5.16a would be unacceptable.

If both PID zeros remain equal and move to -1, the centroid (carrying the asymptotes with it) moves also to -1. The root locus of Fig. 5.16b reveals that the open-loop poles at the origin now move toward the double PID zeros while the complex conjugate open-loop poles now move toward the asymptotes. The sets of open-loop poles have switched destinations.

When both PID zeros are placed at -1.25, the centroid becomes -0.75. The root locus of Fig. 5.16c is very interesting because the pairs of closed-loop poles come together. This indicates that when the double PID zeros are to the right of -1.25, the open-loop poles at the origin are attracted to the double PID zeros. In the opposite case (when the double PID zeros are to the left of -1.25), the open-loop poles at the origin move toward the vertical asymptotes. With the double PID zeros being located at -1.25, a gain of $K = 4$ locates double closed-loop poles at $-1 \pm j1.22$. This choice of double PID zeros provides a good value for the undamped natural frequency (1.58) and a damping ratio of 0.63. Rise time and settling time would be lower than for the uncompensated case, so the design process can end here.

With $K = 4$ the transfer function is

$$T(s) = \frac{G_p(s)G_c(s)}{1 + G_p(s)G_c(s)}$$

$$= \frac{4(s + 1.25)^2}{s^4 + 4s^3 + 9s^2 + 10s + 6.25}$$

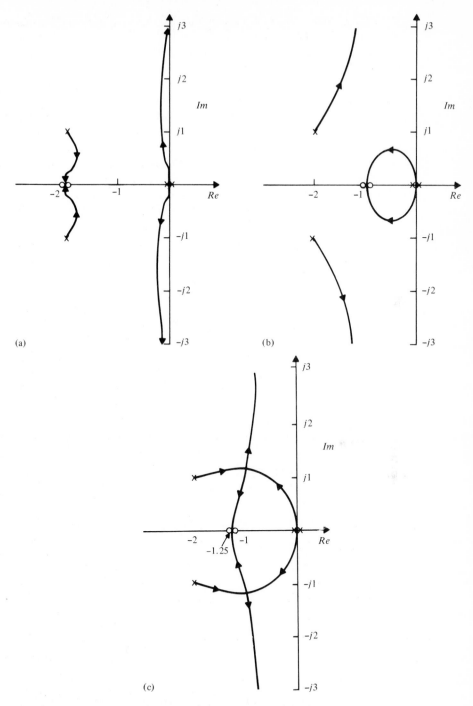

Figure 5.16 Root loci for PID compensation. **(a)** PID zeros at -2. **(b)** PID zeros at -1. **(c)** PID zeros at -1.25.

Many other choices exist where the PID zeros are not equal. For example, the zeros could be placed at -0.1 and -4 which would combine the cascade integral plus proportional and rate feedback compensators designed previously. Since the double PID zero design with the zeros at -1.25 provides a higher dominant damping ratio than for that rate feedback design, the design of Figure 5.16a will be chosen (for $K = 4$) and these results are contained in Table 5.3.

Because of the number of choices that are offered, PID design is a very popular procedure.

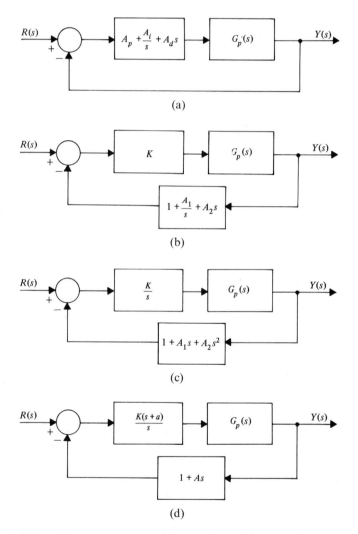

Figure 5.17 Some PID compensation arrangements. **(a)** Cascade compensation. **(b)** Feedback compensation. **(c)** A combination of cascade and feedback compensation. **(d)** Another combination of cascade and feedback compensation.

Figure 5.17 shows a number of different feedback system configurations for which the open-loop system transmittance contains the same PID compensator terms. Each of these systems shares the same root locus plot, but their steady state error performances and other characteristics differ somewhat. A designer selects a system configuration based upon the ease of developing and generating needed signals and the required performance characteristics.

DRILL PROBLEM

D5.17. Use root locus methods to design cascade PID compensators for systems with the following plant transmittances. For each design, find the steady state error coefficients and the amount of relative stability of the closed-loop system.

(a) $G_p(s) = \dfrac{1}{(s + 4)^2}$

(b) $G_p(s) = \dfrac{s + 1}{s(s^2 + 4s + 5)}$

(c) $G_p(s) = \dfrac{s^2 + 2s + 5}{s(s + 4)}$

5.12 AN UNSTABLE HIGH-PERFORMANCE AIRCRAFT

Control system engineers may find it interesting that the design of high-performance aircraft has come full circle since the Wright Brothers. Far from being experimental tinkerers, the Wright Brothers performed careful analysis and wind tunnel testing prior to their first successful powered flight with a human at the controls.

The German Otto Lilienthal carried out extensive experimentation using kites and human-controlled gliders from 1891 until his death in a stall-induced crash in 1899. The Wright Brothers found errors in Lilienthal's aerodynamic data. Determined to design a very responsive aircraft, one in which a potential stall could be quickly eradicated, they purposely designed the famous Flyer of 1903 to be open-loop unstable. Some historians suggest that their experience building bicycles [which must be stabilized by constant (often unrealized) weight shifting by the rider] led naturally to creation of an aircraft that required weight shifting by the pilot to provide stability.

Subsequent designers learned enough about control surfaces to design aircraft that were responsive enough to avoid stalls but were stable (forgiving) enough to

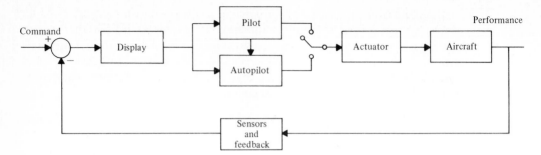

Figure 5.18   An aircraft control system.

be more easily flown. From 1903 through the mid 1980s, open-loop stable operation was commonplace for passenger aircraft, fighter aircraft, and satellites. In order to provide greater responsiveness in combat, the F-16 was designed (in the mid 1970s) to be open-loop unstable, thus returning to the 80-year-old Wright Brothers' scheme. Unlike the Wright Flyer, the pilot cannot actually control the F-16, so that an autopilot must be used constantly. In the remainder of this section, design of piloted aircraft will be considered from the viewpoint of the root locus principles in this chapter.

Figure 5.18 shows a typical aircraft control system. The actuator consists of those electrical, mechanical, and hydraulic devices that move the flaps, elevators, fuel flow controllers, and other devices that cause the aircraft to vary its flight. Sensors provide information on velocity, heading, rate of rotation, and other flight data. This information is combined with the desired flight characteristics (commands) on displays visible to the pilot or electronically available to the autopilot. The autopilot should be able to fly the aircraft on a heading and under conditions set by the pilot. The command often consists of a predetermined heading or precoded warning strategies that are displayed as mentioned above. In combat the command might represent a radar-generated image of an enemy aircraft or of a ground-based target.

Figure 5.19 shows a simple version of Fig. 5.18. Flight would be in one coordinate plane, where a coordinate plane refers to that part of our three-dimensional world that is modeled. Often design focuses on a forward-moving aircraft that moves somewhat up or down without moving right or left and without rolling (rotating the wingtips). Such a study is called pitch plane design. Figure 5.19 is typical of pitch plane dynamics.

The aircraft is modeled with a second-order transfer function having underdamped poles. The actuator may be approximated by a first order $G_1(S)$. Where position, rate, and acceleration feedback occurs, $H(S)$ would contain two zeros. The pilot is assumed to provide integral plus proportional compensation. The pole at $-g$ represents time delay in the pilot's audio-visual-brain-neuromuscular system. The display is assumed to be the difference between the command and the output of $H(S)$.

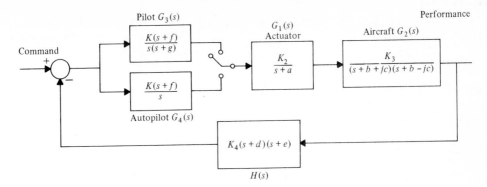

Figure 5.19 Aircraft control system dynamics.

It is assumed that K is adjustable by the pilot or by the autopilot but that K_2, K_3, and K_4 are fixed. For a piloted aircraft the root locus (for adjustable K) results from

$$KG(S)H(S) = K\left[\frac{s+f}{s(s+g)}\,G_1(S)G_2(S)\right]H(S)$$

For an autopiloted aircraft

$$KG(S)H(S) = K\left[\frac{s+f}{s}\,G_1(S)G_2(S)\right]H(S)$$

The two systems differ because the autopiloted system does not have the delay represented by the term with a pole at $-g$.

Figure 5.20 contains the root locus for a pilot-controlled stable aircraft. The system has a closed-loop zero located at $-f$ since $-f$ is a forward-path zero. The

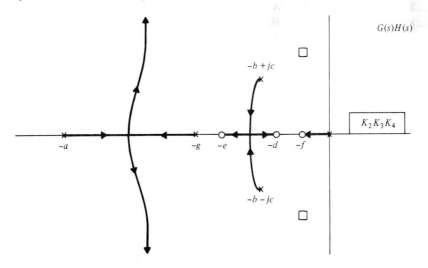

Figure 5.20 Pilot-controlled stable aircraft.

remaining coefficients show typical relative locationships. The closed-loop pole near $-f$ is nearly canceled by the closed-loop zero; hence the dominant poles (those most influencing the transient response) are the next two closed-loop poles further from the vertical axis (those leaving $-b \pm jc$). The pilot closes the loop by reacting to various cues and by moving the control stick accordingly. An inactive pilot opens the loop creating about the same action as setting K to zero. The aircraft then operates due to its natural response, that is, due to values of a, b, c, g, K_2, K_3, and K_4. Inattentiveness by the pilot does not result in instability. First flown in late 1935, the DC-3 was particularly stable (forgiving). Many of these craft are still in service.

If the autopilot controls the system as in Fig. 5.21, any selected value for K determines a stable response. The pole at $-g$ is eliminated hence large values for K do not create a lightly damped pole pair to the left of $-g$ as in Fig. 5.20.

Suppose the enemy has a plane with less rise time (more responsive performance) than the aircraft just discussed. To counteract this enemy, the closed-loop dominant poles should be moved to a point identified by the box in Figs. 5.20 and 5.21. Clearly no value of K can create this more rapid response.

Recall from Chap. 2 that a second-order characteristic polynomial can be written

$$s^2 + 2\zeta\omega_n s + \omega_n^2 = (s + \sigma + j\omega)(s + \sigma - j\omega)$$

The damping ratio is

$$\zeta = \frac{\sigma}{\omega_n} = \frac{\sigma}{\sqrt{\sigma^2 + \omega^2}}$$

For dominant complex conjugate roots at $-\sigma \pm j\omega$, if ω_n is held constant, a reduced damping ratio implies a lower value for ζ and a resulting reduction in rise time. In Figs. 5.20 and 5.21, the desired dominant pole locations imply less rise time and (by that definition of performance) better response.

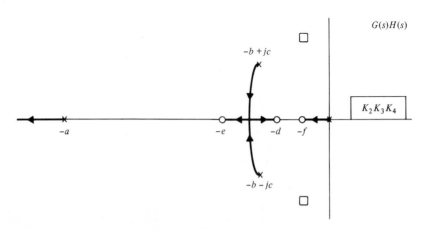

Figure 5.21 Autopilot-controlled stable aircraft.

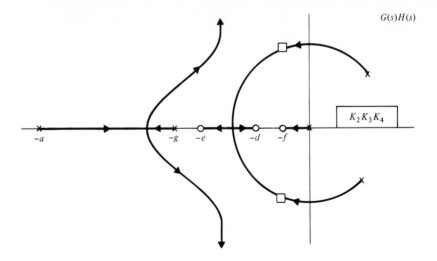

Figure 5.22 Pilot-controlled unstable aircraft.

In Fig. 5.22 the aircraft has been redesigned and is open-loop unstable [the poles of $G_2(S)$ are in the right half-plane]. With proper attention, the pilot can generate a value of K that provides the desired performance. The pilot must always maintain active control (as a rider does on a bicycle); otherwise any disturbance causes the system to go out of control. Pilot error or fatigue renders the design of Fig. 5.22 unreliable. Only the autopilot-controlled system of Fig. 5.23 provides a safe and reliable performance.

The pilot of an aircraft like the F-16 uses a standard-appearing stick, rudder, and other controls. However, the pilot is providing information to the autopilot

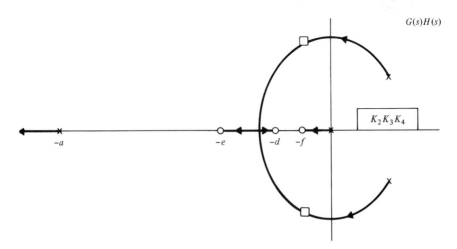

Figure 5.23 Autopilot-controlled unstable aircraft.

(Fig. 5.18) which actually moves all control surfaces and stabilizes the craft against disturbances which would otherwise grow so large as to compromise the survivability of the aircraft.

The partnership of autopilot, pilot, and open-loop unstable aircraft creates a closed-loop system with the desired dominant, combat-competitive behavior.

5.13 SUMMARY

In feedback tracking systems, compensators are used to improve relative stability, steady state error performance, and perhaps other characteristics such as damping ratio and overshoot. The root locus method is an especially powerful tool for compensator design because the effects of additional open-loop compensator poles and zeros upon the closed-loop pole locations are easily and quickly visualized.

A unity feedback system with a cascade compensator has type number equal to the number of $s = 0$ poles in the forward transmittance $G(s)$. Provided the resulting design has sufficient relative stability for the application at hand, the system type number may be increased by introducing additional $s = 0$ poles in the compensator. In cascade integral compensation, the compensator is an integrator having a single $s = 0$ pole.

Unfortunately, the addition of $s = 0$ poles to a system's open-loop transmittance tends to greatly reduce closed-loop relative stability, often resulting in unstable systems. Additional compensator zeros, placed to draw the root loci to the left on the complex plane, are then desirable. The cascade integral plus proportional compensator introduces a pole at $s = 0$ and one zero, while the PID compensator has an $s = 0$ pole and two zeros.

Accurate integrators are expensive or difficult to build for many applications, but steady state system performance can still be improved by placing compensator poles near $s = 0$ in the LHP. The system type number is not increased, but steady state error may be thereby reduced. Cascade lag compensation is often used for this purpose. Lead compensation, on the other hand, typically increases relative stability with little or no improvement in steady state error.

Placing compensation elements in the feedback path gives the designer additional flexibility. Relative stability depends only on loop transmittance, while steady state error performance depends upon how the compensation is distributed between the forward and feedback transmittances.

Increased compensator complexity will generally give improved system performance, although ultimately one should expect a diminishing return because of unavoidable inaccuracies.

REFERENCES

Compensator Design

D'Azzo, J. J., and Houpis, C. H., *Linear Control System Analysis and Design.* New York: McGraw-Hill, 1975.

Dorf, R. C., *Modern Control Systems.* Reading, Mass.: Addison-Wesley, 1967.

Fallside, F., *Control System Design by Pole-Zero Assignment.* New York: Academic Press, 1977.

Garnell, P., and East, D. J., *Guided Weapon Control Systems.* New York: Pergamon Press, 1977.

Kuo, B. C., *Automatic Control Systems.* Englewood Cliffs, N.J.: Prentice-Hall, 1967.

Truxal, J. G., *Control System Synthesis.* New York: McGraw-Hill, 1955.

Aircraft Design

Chant, C., *Aviation: An Illustrated History.* Seacaucus, N.J.: Chartwell Books, 1978.

Culic, F. E. C., "The Origins of the First Powered, Man-Carrying Airplane." *Scientific American* (July 1979): 86–100.

Encyclopedia of Transportation. Chicago: Rand McNally, 1976.

Markowski, M. A. "Ultralight Airplanes." *Scientific American* (July 1982): 62–68.

Reed, F., "The Electric Jet." *Air and Space* (December 1986/January 1987): 42–48.

PROBLEMS

1. Each of the following systems consists of the plant driven by an error signal amplifier with gain K, as in Fig. 5.6. For each of the following plant transmittances and amplifier gains, use root locus methods to find, approximately, the closed-loop system poles. Then find the step error coefficient K_0.

(a) $G_p(s) = \dfrac{s + 4}{s + 10}$

   $K = 10$

(b) $G_p(s) = \dfrac{s + 10}{s + 4}$

   $K = 10$

   Ans $-9.33; 20$

(c) $G_p(s) = \dfrac{s + 10}{(s + 2)^2}$

   $K = 10$

(d) $G_p(s) = \dfrac{1}{(s + 10)(s + 4)}$

   $K = 58$

   Ans $-7 \pm j7; 1.45$

2. Design cascade integral compensators for unity feedback systems with the following plant transmittances. Use root locus plots to select compensator gains K so that the damping ratio of the dominant roots is 0.7, if possible. For each design, find the steady state error coefficients and the dominant roots.

(a) $G_p(s) = \dfrac{6}{(s+4)^2}$

(b) $G_p(s) = \dfrac{s+1}{(s+4)^3}$

(c) $G_p(s) = \dfrac{s+9}{(s^2+4s+40)}$

(d) $G_p(s) = \dfrac{2s^2+10s+13}{(s+4)(s^2+3s+10)}$

Ans (b) $K = 29$; (d) not possible

3. Design cascade integral plus proportional compensators for unity feedback systems with the following plant transmittances. Use root locus plots to select the compensator gain K where the compensator zero is at -10 and the effective damping ratio is 0.7. For each design find the steady state error coefficients and the dominant roots.

(a) $G_p(s) = \dfrac{10}{s^2+4s+10}$

(b) $G_p(s) = \dfrac{s+4}{s^2+10}$

(c) $G_p(s) = \dfrac{2s+20}{s^2+9}$

(d) $G_p(s) = \dfrac{6}{(s+5)(s^2+4s+10)}$

4. A plant to be controlled has transmittance

$$G_p(s) = \dfrac{s+8}{s(s^2+4s+40)}$$

Use the root locus methods to select the gain K for each of the following compensated systems. The effective damping ratios must be at least 0.5.

(a) Uncompensated (unity) feedback.

(b) Cascade lag compensated with the compensator $3K(s+0.1)/(s+0.01)$.

(c) Rate feedback compensated with the compensator $\frac{1}{8}(s + 8)$.

For each design, find the steady state errors to step, ramp, and parabolic inputs and the dominant roots.

5. Use root locus methods to select the gain K for a PID compensator and plant

$$G_p(s) = \frac{1}{(s + 4)(s^2 + 10s + 50)}$$

$$G_c(s) = K\frac{(s + 1.25)^2}{s}$$

The system should have a relative stability of at least 0.86. Determine the steady state error to a ramp input.

Ans $K = 750$, $e_{\text{ramp}}(\infty) = 0.17A$

6. A fine positioning system for an elevator when it is in the vicinity of the correct position is modeled in Fig. P5.6. Use root locus methods to locate the overall system poles as a function of $K_1 > 0$, when $K_2 = 0$. Then, with a specific choice of K_1, use root locus methods to determine the overall system poles as a function of $K_2 > 0$. Using several iterations of choosing K_1, then choosing K_2, find values of K_1 and K_2, if possible, for which the system has a pair of complex poles with damping ratio approximately 0.7

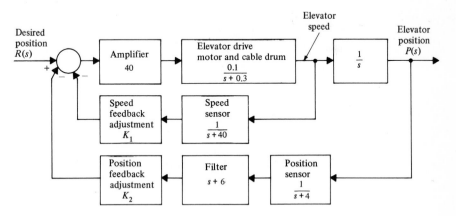

Figure P5.6

7. A linearized inventory control system model is given in Fig. P5.7. Based upon the difference between desired and actual inventory levels, management sets the production quota which in turn determines the production rate. Using graphical root locus methods, determine the production time constant

$$\tau = \frac{1}{K}$$

in the range

$$1 < \tau < 10$$

which gives greatest relative stability of the system.

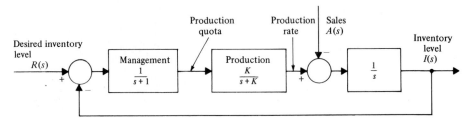

Figure P5.7

Ans $K = 1$

8. The pitch control system for high-altitude aircraft is modeled in Fig. P5.8. Find the value of the adjustable constant $K > 0$ for which the potentially complex pole pair of the closed-loop system has critical damping.

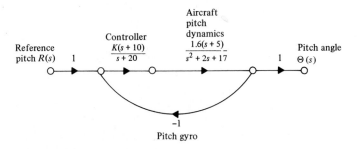

Figure P5.8

Ans $K = 70.5$

Note: For most of the remaining problems, no answers are shown. These problems are purposely open-ended to allow the instructor or the student to exercise individual judgment in selecting design objectives and the method of solution. For example, these problems can serve as class projects. Differing objectives yield differing solutions hence no unique solution exists.

9. Design constant-gain error feedback compensators for plants with the following transmittances. Use your own judgment as to the best trade-off between steady state errors and pole placement.

(a) $G_p(s) = \dfrac{2}{s+3}$

(b) $G_p(s) = \dfrac{3s}{s^2+4}$

(c) $G_p(s) = \dfrac{5}{s^2+4s+5}$

(d) $G_p(s) = \dfrac{1}{s^3+6s^2+9s}$

(e) $G_p(s) = \dfrac{s+1}{s(s^2+6s+13)}$

10. It is desired that, for a plant with transmittance

$$G_p(s) = \frac{s+4}{(s+1)(s+2+j)(s+2-j)}$$

the steady state error to a step input be no more than 2%. If possible, design a stable constant-gain error feedback control of the plant with this property. What is the relative stability of your design? If there are complex conjugate closed-loop poles, what is the associated damping ratio?

11. The control of a medical patient's heart rate by an implanted pacemaker with a heart rate sensor is modeled as shown. Use a root locus diagram to design the pacemaker gain K for response with critical damping. In terms of K, what is the steady state error to an input

$$r(t) = (80 + 0.2t)u(t)$$

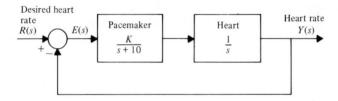

Figure P5.11

Ans   25; 2/K

12. Design cascade integral compensators for unity feedback systems with the following plant transmittances. Use root locus plots to select the integrator gain K. For each design, find the steady state errors to unit step and ramp inputs, the amount of relative stability, and the damping ratios of any complex closed-loop poles.

(a) $G_p(s) = \dfrac{9}{(s + 3)^2}$

(b) $G_p(s) = \dfrac{s + 2}{(s + 3)^3}$

(c) $G_p(s) = \dfrac{s + 5}{(s^2 + 4s + 20)}$

(d) $G_p(s) = \dfrac{s^2 + 4s + 13}{(s + 3)(s^2 + 2s + 10)}$

13. Design cascade integral-plus-proportional compensators for unity feedback systems with the following plant transmittances. Use root locus plots to help select the compensator parameters. For each design, find steady state errors to unit step and ramp inputs, the amount of relative stability, and the damping ratios of any complex closed-loop poles.

(a) $G_p(s) = \dfrac{10}{s^2 + 7s + 10}$

(b) $G_p(s) = \dfrac{s + 2}{(s^2 + 4s + 3)s}$

(c) $G_p(s) = \dfrac{s + 4}{s^2 + 5s + 6}$

(d) $G_p(s) = \dfrac{6s + 30}{s^2 + 4s + 13}$

14. Design cascade lag compensators for unity feedback systems with the following plant transmittances. Use root locus plots to help select the compensator parameters, emphasizing small steady state power-of-time errors. Find the steady state errors to unit step and ramp inputs, the amount of relative stability, and the damping ratios of any complex closed-loop poles.

(a) $G_p(s) = \dfrac{3s + 1}{s(s^2 + 4s + 13)}$

(b) $G_p(s) = \dfrac{3}{s^2 + 10s}$

(c) $G_p(s) = \dfrac{s + 3}{(s + 2)(s + 4)(s + 6)}$

15. Design cascade lead compensators for unity feedback systems with the following plant transmittances. Use root locus plots to help select the compensator parameters, emphasizing pole placement. Find the steady state errors to unit step and ramp inputs, the amount of relative stability, and the damping ratios of any complex closed-loop poles.

(a) $G_p(s) = \dfrac{1}{s^2}$

(b) $G_p(s) = \dfrac{10}{s^2 + 10s}$

(c) $G_p(s) = \dfrac{s + 2}{s^2 + 2s + 10}$

(d) $G_p(s) = \dfrac{10}{s(s^2 + 2s + 10)}$

16. Design cascade lag-lead compensators for unity feedback systems with the following plant transmittances. Use root locus plots to help select the compensator parameters. Find the steady state errors to step and ramp inputs, the amount of relative stability, and the damping ratios of any complex closed-loop poles.

(a) $G_p(s) = \dfrac{1}{s(s + 10)^2}$

(b) $G_p(s) = \dfrac{s + 1}{(s + 10)^2}$

(c) $G_p(s) = \dfrac{s + 3}{s^2(s + 10)}$

(d) $G_p(s) = \dfrac{s + 4}{s(s^2 + 2s + 5)}$

17. Design feedback rate compensators for systems with the following plant transmittances. Use root locus plots to help select the compensator parameters. Find the steady state errors to step and ramp inputs, the amount of relative stability, and the damping ratios of any complex closed-loop poles.

(a) $G_p(s) = \dfrac{1}{s(s + 10)^2}$

(b) $G_p(s) = \dfrac{1}{s^2(s + 10)}$

(c) $G_p(s) = \dfrac{1}{s^2(s + 10)^2}$

18. Design cascade PID compensators for unity feedback systems with the following plant transmittances. Use root locus plots to help select the compensator parameters. Find the steady state errors to step and ramp inputs, the amount of relative stability, and the damping ratios of any complex closed-loop poles.

(a) $G_p(s) = \dfrac{1}{s(s + 4)^2}$

(b) $G_p(s) = \dfrac{s + 1}{s^2(s + 4)^2}$

(c) $G_p(s) = \dfrac{s + 1}{s(s^2 + 4s + 5)}$

(d) $G_p(s) = \dfrac{s^2 + 2s + 5}{s(s + 4)}$

19. A plant to be controlled has transmittance

$$G_p(s) = \frac{s + 5}{s(s^2 + 4s + 10)}$$

Use root locus methods to design tracking systems of each of the following types:

(a) Constant-gain error feedback

(b) Cascade lag compensated

(c) Rate feedback with the compensator zero at $s = -8$

(d) Cascade lead compensation with the compensator pole located 10 times further in the LHP than the compensator zero

For each design, find the steady state error to unit step, ramp, and parabolic inputs, the amount of relative stability, and the damping ratios of any complex closed-loop poles.

20. For a plant with transmittance

$$G_p(s) = \frac{1}{(s + 1)(s^2 + 4s + 5)}$$

it is desired that the steady state error to a step input be no more than 5%, and with this requirement that the relative stability be maximum (or nearly so), providing no closed-loop poles have damping ratios greater than 0.7. If possible, design a controller with cascade lag compensation having these properties.

21. An electrohydraulic positioning mechanism is shown schematically in Fig. P5.21a. The block diagram of a linearized model of this system is shown in Fig. P5.21b. Use root locus methods to choose the gain K for a damping ratio of 0.7 for the dominant roots. For this value of K, what is the steady state error to a unit ramp input.

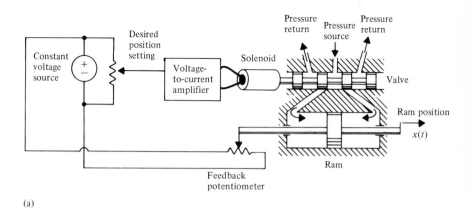

(a)

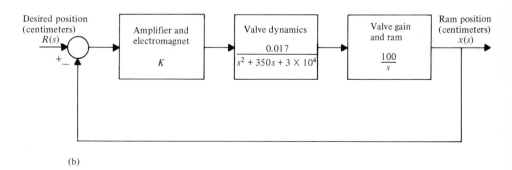

(b)

Figure P5.21

22. A large air temperature control system is arranged as shown in Fig. P5.22a. A block diagram of its model is given in Fig. P5.22b. Choose the gain K for minimum (or nearly so) steady state error, provided that the relative stability is at least 0.75 unit and the damping ratio of any complex poles is greater than 0.707.

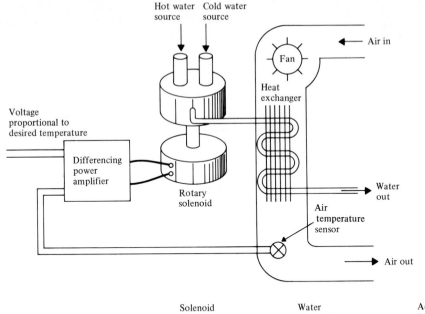

(a)

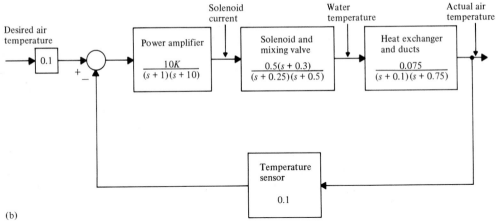

(b)

Figure P5.22

23. The vertical axis control of an "active suspension" for a cargo truck is modeled by the relatively "stiff" mechanical suspension transmittance

$$G_m(s) = \frac{13}{s^2 + 4s + 13}$$

with input provided by the electromechanical driver having transmittance (Fig. P5.23)

$$G_d(s) = \frac{1}{(s + 5)(s + 10)}$$

To improve the truck's vertical tracking of roads, the height of the axle from the road is sensed and an electrically driven hydraulic actuator is used to give smoother response to bumps in the roadway. Use a PID compensator with $a = 4$ and $b = 13$. Select K so that the effective damping ratio is 0.7.

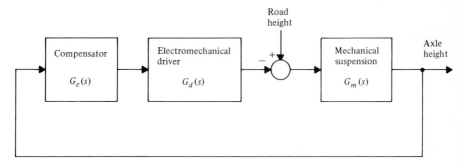

Figure P5.23

24. Without a feedback control system, most helicopters are unstable. For helicopter dynamics having unstable transmittance

$$G_p(s) = \frac{10(s + 0.05)}{(s + 0.5)[(s - 0.2)^2 + (0.4)^2]}$$

relating pitch angle to blade control, Fig. P5.24, design a feedback compensator

$$\frac{K(s + a)}{s + b}$$

that gives an effective damping ratio of about 0.7. For your design, find the steady state error to a unit step disturbance such as might result from a constant wind.

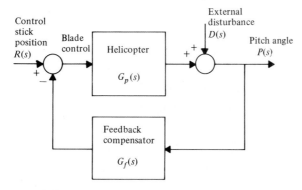

Figure P5.24

25. Use cascade lead compensation to stabilize the unstable plant with transmittance

$$G_p(s) = \frac{1}{s(s-1)}$$

using negative feedback.

26. Design feedback rate compensation for plants with the following transmittances. Use your own judgment as to the best tradeoff between steady state error performance and pole placement (effective damping ratio).

(a) $G_p(s) = \dfrac{4(s+2)}{(s+3)^2(s^2+2s+5)}$

(b) $G_p(s) = \dfrac{s-1}{s(s^2+2s+5)}$

(c) $G_p(s) = \dfrac{10}{s^2(s^2+4s+5)}$

(d) $G_p(s) = \dfrac{25}{(s^2+4s+5)^2}$

27. Design controllers with cascade lag compensation for plants with the following transmittances. Emphasize steady state power-of-time error performance over pole placement.

(a) $G_p(s) = \dfrac{1}{(s+4)^2}$

(b) $G_p(s) = \dfrac{s+1}{(s+3)^2}$

(c) $G_p(s) = \dfrac{1}{s(s+2)}$

(d) $G_p(s) = \dfrac{s^2+4s+5}{s^2(s+4)^2}$

28. Design controllers with cascade integral compensation for plants with the following transmittances. Use your own judgment as to the best trade-off between steady state errors and pole placement (effective damping ratio).

(a) $G_p(s) = \dfrac{3}{s+2}$

(b) $G_p(s) = \dfrac{1}{s^2 + 10}$

(c) $G_p(s) = \dfrac{s}{s^2 + 4s + 3}$

(d) $G_p(s) = \dfrac{s + 5}{(s + 3)^3}$

29. For a plant with transmittance

$$G_p(s) = \frac{(s + 10)^2}{(s + 4)^2}$$

it is desired that the steady state error to a ramp input be no more than 10%. If possible, design a stable controller with integral compensation with this property. What is the relative stability of your design?

30. Design controllers with cascade integral-plus-proportional compensation for plants with the following transmittances. Use your own judgment as to the best trade-off between steady state errors and pole placement.

(a) $G_p(s) = \dfrac{1}{s + 2}$

(b) $G_p(s) = \dfrac{s + 1}{s + 3}$

(c) $G_p(s) = \dfrac{s}{s^2 + 16}$

(d) $G_p(s) = \dfrac{s + 4}{(s + 3)^3}$

31. It is desired that, for a plant with transmittance

$$G_p(s) = \frac{3}{(s + 2 + j)(s + 2 - j)}$$

the steady state error to a ramp input be no more than 10% and that the relative stability be at least 1 unit. If possible, design a cascade integral-plus-proportional controller with these properties.

Chapter
6

Frequency Response Methods

6.1 PREVIEW

Whenever a radio or TV is tuned to a particular station, some circuit's performance varies so as to tune in the desired signal and reject undesired signals. It is logical that a key to designing an effective communication system is to understand the behavior of that system as a function of the incoming frequencies.

Research by Nyquist in the 1930s led to a book by Bode in the 1940s in which the same frequency response methods used to understand electrical circuits were used to understand the stability of control systems. Once a system's performance is understood, remedies can be created to reduce unwanted effects.

The technology of the 1930s was based on analysis and design using slide rules and simple (by current standards) mechanical computational aids. Frequency response computations are fairly easy to perform so these methods were applied with vigor.

Three viewpoints are generally taken when designing and analyzing control system behavior. First, behavior as a function of s is evaluated using Routh-Hurwitz methods (dating to the 1890s when steam engines used mechanical governors) and root locus methods (dating to Walter Evans in the late 1940s). Second, time domain behavior can be determined by inverse Laplace-transforming or by applying techniques to be developed in Chaps. 7 and 8. Third, behavior as a function of frequency requires use of $s = j\omega$; this approach is the topic of this chapter.

The three approaches are not mutually exclusive. In Chap. 5 it was pointed out that dominant pole locations in the s-plane imply time domain response in terms of undamped natural frequency and damping ratio. Later in Chap. 6 we will show that frequency domain figures-of-merit are also related to dominant pole locations (and therefore to the time domain).

Root locus, time domain, and frequency domain methods provide a comprehensive viewpoint of the strengths and weaknesses of a system. All three methods should be applied to fully understand (and improve) system performance.

As was mentioned in the preface and in the preview for Chap. 4, a number of computational packages are available, including one from the publisher to accompany this text. However, we again stress the importance of the user backing up and double-checking computer results with manual sketches and numerical values. In many cases, very accurate frequency response data and sketches can be arrived at without use of a PC or other large-scale computational device.

6.2 FREQUENCY RESPONSE

Forced Sinusoidal Response

In the development to follow, a general symbol for a transmittance, $F(s)$, will be used. In later sections of this chapter, $F(s)$ is taken to represent the transmittance of a system component, the overall system transfer function $T(s)$, or the loop transmittance $G(s)H(s)$ of a feedback system, depending upon the application at hand.

The response of a system with transmittance $F(s)$ to a sinusoidal input signal, Fig. 6.1a,

$$r(t) = B \cos(\omega t + \beta)$$

generally consists of both forced and natural components. The forced part of the response is also sinusoidal, with the same frequency ω as the input. Generally, the amplitude and the phase angle of the forced sinusoidal response are different from those of the input and they depend upon the input frequency:

$$y(t) = y_{\text{natural}}(t) + C \cos(\omega t + \gamma)$$

One way of calculating the forced sinusoidal response of a system is to solve, instead, the related phasor problem of Fig. 6.1b, in which the forced sinusoidal

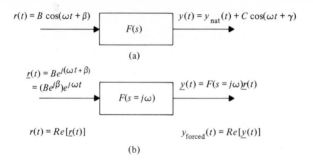

(a)

(b)

Figure 6.1 Response of a transmittance to a sinusoidal input. **(a)** Form of the solution. **(b)** Phasor solution for forced sinusoidal response.

signals are replaced by complex exponentials, the real parts of which are the actual signals of interest:

$$\underline{r}(t) = Be^{j(\omega t + \beta)} \qquad Re[\underline{r}(t)] = B \cos(\omega t + \beta)$$

$$\underline{y}(t) = Ce^{j(\omega t + \gamma)} \qquad Re[\underline{y}(t)] = C \cos(\omega t + \gamma)$$

Then

$$\frac{\underline{y}(t)}{\underline{r}(t)} = \frac{Ce^{j(\omega t + \gamma)}}{Be^{j(\omega t + \beta)}} = \frac{C}{B} e^{j(\gamma - \beta)} = F(s = j\omega)$$

$$y_{\text{forced}}(t) = C \cos(\omega t + \gamma) = Re[\underline{y}(t)] = A(\omega)B \cos[\omega t + \beta + \Phi(\omega)]$$

The magnitude of the transmittance when evaluated at $s = j\omega$ is the ratio of output amplitude to input amplitude:

$$A(\omega) = |F(s = j\omega)| = \frac{C}{B}$$

The angle of the transmittance, for $s = j\omega$, is the difference in phase angles between output and input:

$$\Phi(\omega) = \angle F(s = j\omega) = \gamma - \beta$$

As a numerical example, consider the transmittance

$$F(s) = \frac{6}{s + 4}$$

and the input signal

$$r(t) = 3 \cos(7t + 20°)$$

as in Fig. 6.2. For $s = j7$, the transmittance is

$$F(s = j7) = \frac{6}{j7 + 4} = 0.74e^{-j60°}$$

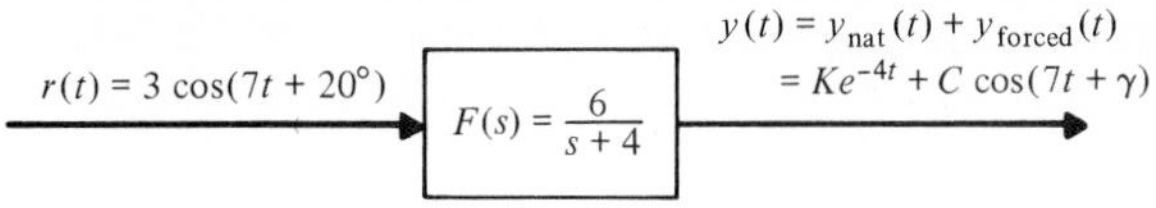

$$r(t) = 3 \cos(7t + 20°) \quad \boxed{F(s) = \frac{6}{s + 4}} \quad \begin{array}{l} y(t) = y_{\text{nat}}(t) + y_{\text{forced}}(t) \\ = Ke^{-4t} + C \cos(7t + \gamma) \end{array}$$

Figure 6.2 Finding forced sinusoidal response.

The forced sinusoidal output has amplitude and phase angle

$$C = (0.74)(3) = 2.22 \qquad \gamma = 20° + (-60°) = -40°$$

giving

$$y_{\text{forced}}(t) = 2.22 \cos(7t - 40°)$$

Frequency Response Measurement

When a transmittance $F(s)$ is driven by a sinusoidal input signal, the ratio of the amplitude of the forced output to the amplitude of the input is

$$A(\omega) = \frac{\text{amplitude of sinusoidal output}}{\text{amplitude of sinusoidal input}} = |F(s = j\omega)|$$

which is a function of the radian frequency of the sinusoid, ω. The difference in the phase angles of the output and the input is

$$\Phi(\omega) = \text{phase angle of sinusoidal output} - \text{phase angle of sinusoidal input}$$

$$= \angle F(s = j\omega)$$

To measure the frequency response of a stable component or system at some frequency, apply a sinusoidal input of that frequency. Choose any convenient amplitude which is not so large as to overload the system, yet not so small that the system signals are masked by noise. Wait until the system natural behavior, arising from the connection of the input, dies out. Measure the amplitudes of the input and output sinusoids and form the ratio

$$A = \frac{\text{output sinusoid amplitude}}{\text{input sinusoid amplitude}}$$

The phase shift is the difference in phase between the forced-output sinusoid and the sinusoidal input signal. If the two signals can be plotted side by side, the phase shift is found by comparing the plots. Stroboscopic light techniques are especially useful for comparing the phases of mechanical elements in rapid motion. If the system signals are electrical or can be monitored by electrical transducers, the elliptical pattern formed by plotting the input versus the output signal on an oscilloscope or instruments known as phase meters or vector voltmeters can be used to display the amount of phase shift.

With a number of such measurements of amplitude ratio and phase shift at various frequencies, curves for $A(\omega)$ and $\phi(\omega)$ can be sketched.

Response at Low and High Frequencies

There are some considerations that make the determination of amplitude and phase shift curves from a limited number of measurements more accurate. If the transmittance is a rational function

$$F(s) = \frac{b_m s^m + b_{m-1} s^{m-1} + \cdots + b_1 s + b_0}{s^n + a_{n-1} s^{n-1} + \cdots + a_1 s + a_0}$$

then

$$F(j\omega) = \frac{b_m (j\omega)^m + b_{m-1}(j\omega)^{m-1} + \cdots + b_1(j\omega) + b_0}{(j\omega)^n + a_{n-1}(j\omega)^{n-1} + \cdots + a_1(j\omega) + a_0}$$

For large values of ω, all but the highest powers of ω in the numerator and denominator can be ignored:

$$F(j\omega) \cong \frac{b_m (j\omega)^m}{(j\omega)^n} = b_m j^{m-n} \omega^{m-n}$$

The amplitude curve approaches a power of ω, and the phase curve approaches a multiple of $90°$.

For small ω, all but the lowest powers of ω in the transmittance numerator and denominator are negligible. If a_0 and b_0 are nonzero,

$$F(j\omega) \cong \frac{b_1(j\omega) + b_0}{a_1(j\omega) + a_0} \cong \frac{b_0}{a_0}$$

In general, at low frequencies, a rational transmittance's amplitude curve approaches a power of ω or a constant and the phase curve approaches a multiple of $90°$.

For example, consider

$$F(s) = \frac{6s^3 + 2s^2 + 3s}{s^5 + 4s^4 + 2s^3 + s^2 + s + 10}$$

At high frequencies,

$$F(j\omega) = \frac{6(j\omega)^3}{(j\omega)^5} = -6\omega^{-2}$$

$$A(\omega) = |F(j\omega)| \cong \frac{6}{\omega^2} = 6\omega^{-2} \qquad \Phi(\omega) = \angle F(j\omega) \cong 180°$$

At low frequencies,

$$F(j\omega) \cong \frac{3(j\omega)}{10}$$

$$A(\omega) = |F(j\omega)| \cong \tfrac{3}{10}\omega \qquad \Phi(\omega) = \angle F(j\omega) \cong 90°$$

Once the experimenter is certain that measurements are being made in the small-ω or large-ω region, a few measurements will suffice for the entire region. It is most useful to make measurements at more closely spaced frequencies where the largest changes in amplitude or phase occur.

DRILL PROBLEMS

D6.1. Find the forced sinusoidal response of each of the following transmittances to the indicated input signals:

(a) $F(s) = \dfrac{s}{s + 3}$

$r(t) = 7 \cos(3t - 40°)$

Ans $(7/\sqrt{2}) \cos(3t + 5°)$

(b) $F(s) = \dfrac{4}{s + 2}$

$r(t) = 6 \cos(5t + 30°)$

Ans $4.46 \cos(5t - 38°)$

(c) $F(s) = \dfrac{10}{s^2 + 3s + 10}$

$r(t) = 8 \cos(2t + 70°)$

Ans $(40/3\sqrt{2}) \cos(2t + 25°)$

D6.2. For transmittances with the following amplitude ratios and phase shift functions, find the forced sinusoidal response to the given input signal:

(a) $A(\omega) = \dfrac{1}{\sqrt{\omega^2 + 100}}$

$\Phi(\omega) = \tan^{-1} \dfrac{\omega}{10}$

$r(t) = 7 \cos(6t + 80°)$

Ans $0.6 \cos(6t + 111°)$

(b) $A(\omega) = \dfrac{4}{\sqrt{\omega^2 + 4}}$

$\Phi(\omega) = 90° - \tan^{-1} \dfrac{\omega}{2}$

$r(t) = 3 \cos 2t$

Ans $(3\sqrt{2}) \cos(2t + 45°)$

(c) $A(\omega) = 4$

$\Phi(\omega) = -3\omega \text{ (rad)}$

$r(t) = 10 \cos(5t - 30°)$

Ans $40 \cos(5t - 170°)$

Graphical Frequency Response Methods

Given the pole-zero plot for a transmittance, its evaluation for various values of $s = j\omega$ can be done graphically by drawing sets of directed line segments to points of evaluation on the imaginary axis. With a little practice, the amplitude and phase shift curves can be roughly sketched from just a few evaluations.

For example, the transmittance with the pole-zero plot given in Fig. 6.3a is

$$F(s) = \frac{s + 3}{(s + 1 + j3)(s + 1 - j3)}$$

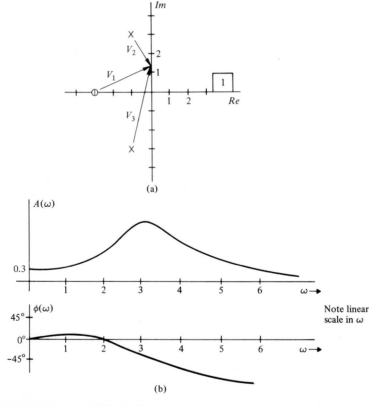

(a)

(b)

Figure 6.3 Sketching frequency response from a pole-zero plot. **(a)** Pole-zero plot. **(b)** Approximate frequency response.

where the directed line segments are defined by vectors v_1, v_2, and v_3. For s restricted to the imaginary axis

$$F(s = j\omega) = \frac{|v_1|(\underline{/v_1} - \underline{/v_2} - \underline{/v_3})}{|v_2||v_3|}$$

As $s = j\omega$ moves up the vertical axis, the three vectors change magnitude and direction so as to form the magnitude and phase versus linear frequency curves in Fig. 6.3b. This figure helps to visualize the significance of the pole-zero plot, although we shall soon see that most magnitude and phase curves are plotted versus the logarithm of frequency, rather than versus linear frequency.

The amplitude curve for a transmittance can be easily visualized from the pole-zero plot by imagining a very flexible rubber sheet suspended over the complex plane. The sheet is poked up by thin rods at each pole location and tacked down to the plane at each zero location. The height of the rubber sheet will represent $|F(s)|$ for each value of s. The frequency response amplitude is a cross section of the sheet displacement along the imaginary axis, as indicated in Fig. 6.4.

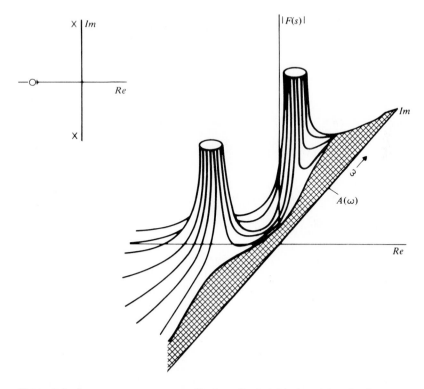

Figure 6.4 Frequency response amplitude as the height of a rubber sheet.

To use the rubber sheet analogy effectively, the height of the sheet at large distances from the origin, that is, as $|s| \to \infty$, must be determined. If

$$F(s) = \frac{b_m s^m + b_{m-1} s^{m-1} + \cdots + b_1 s + b_0}{a_n s^n + a_{n-1} s^{n-1} + \cdots + a_1 s + a_0}$$

$$\lim_{|s| \to \infty} |F(s)| = \left| \frac{b_m}{a_n} \right| \frac{|s|^m}{|s|^n}$$

If the number of poles of $F(s)$ is greater than its number of zeros,

$$\lim_{|s| \to \infty} |F(s)| = 0$$

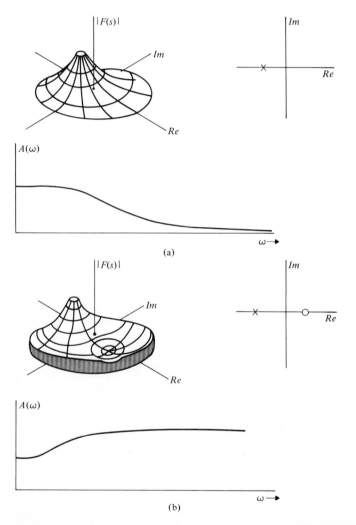

(a)

(b)

Figure 6.5 Using the rubber sheet analogy to visualize frequency response amplitude ratios. **(a)** More poles than zeros. **(b)** Equal number of poles and zeros.

and the rubber sheet is visualized as being tacked down to the complex plane at large distances from the origin, as in Fig. 6.5a.

If $F(s)$ has an equal number of poles and zeros,

$$\lim_{|s| \to \infty} |F(s)| = \left| \frac{b_m}{a_n} \right|$$

and the rubber sheet approaches a fixed height above the complex plane, as in Fig. 6.5b. If there are more zeros than poles in $F(s)$,

$$\lim_{|s| \to \infty} |F(s)| = \infty$$

This case is unusual in practical applications because it means that the amplitude ratio of the transmittance increases with frequency without bound.

Imaginary axis zeros and poles give zero or infinite amplitude for the values of ω where they occur. The phase shift is discontinuous at these values of ω, but limiting values of the phase shift are found by considering imaginary axis points slightly below and slightly above the zero or pole. An illustrative example is given in Fig. 6.6. Complex conjugate poles on the imaginary axis mean that the corresponding system natural behavior is sinusoidal, neither decaying nor expanding with time. If such a system is *driven* with a sinusoidal signal of the same frequency as this natural behavior, the output becomes larger and larger. Of course, it cannot really become infinite in a practical system, since nonlinearities will

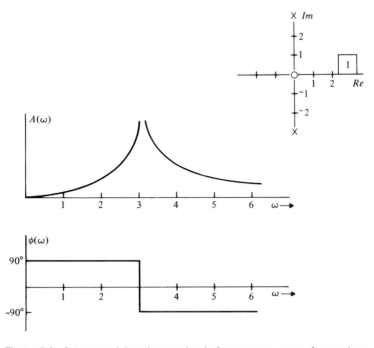

Figure 6.6 Pole-zero plot and approximate frequency response for a system with imaginary axis roots.

eventually be reached which limit the output. Possibly, the system will be destroyed or it will not be possible to continue to supply the ever-increasing energy needed to maintain the input signal. A linear, time-invariant model does not take these practical constraints into account.

Complex conjugate poles in the LHP near the imaginary axis represent slowly decaying oscillatory natural behavior. Driving such a system with a sinusoidal signal of the same or nearly the same frequency as the natural behavior oscillations results in a relatively large, but not infinite, forced output. That is, there is typically a corresponding peak in the system's frequency response, termed a *resonance peak*.

DRILL PROBLEM

D6.3. Sketch approximate frequency response curves (both amplitude and phase shift) for functions with the following pole-zero plots:

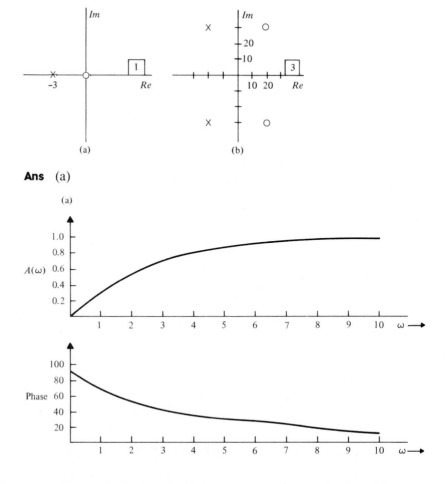

Ans (a)

Ans (b)

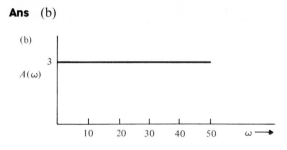

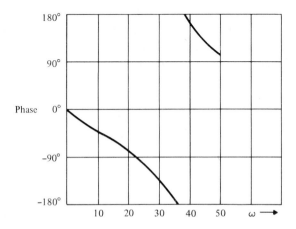

6.3 BODE PLOTS

Amplitude Plots in Decibels

The magnitude of the product of complex numbers is the product of the individual magnitudes, and the angle of a product is the sum of the individual angles. A transmittance that is the product of several simple terms thus has an amplitude curve that is the product of the amplitude curves for the individual terms. The overall phase shift curve is the sum of the individual phase shift curves.

If, instead of dealing with the amplitude curves directly, their logarithms are used, multiplication of individual amplitude curves is, in terms of logarithms, addition. Commonly, *decibels* (abbreviated dB) are used for the description of frequency response amplitude ratios:

$$\text{dB} = 20 \log_{10} A(\omega) = 20 \log_{10} |F(s = j\omega)|$$

If

$$F(s) = F_1(s)F_2(s)F_3(s)\dots$$

then

$$F \text{ (in dB)} = F_1 \text{ (in dB)} + F_2 \text{ (in dB)} + F_3 \text{ (in dB)} + \cdots$$

There is little reason for this choice of common logarithm units; the choice dates from Alexander Graham Bell's study of the response of the human ear. In electromagnetics and some other applications, the natural logarithm is used, and the definition does not appear to be so contrived. The units then are *nepers*.

For rational functions $F(s)$, it is only necessary to be able to plot amplitude and phase shift for the following types of terms:

Constants
Poles and zeros at the origin of the complex plane
Real axis poles and zeros
Complex conjugate pairs of poles and zeros

A rational function can be factored into terms of these types, and the individual dB and phase shift curves plotted. The complete dB curve is then the sum of the component dB curves, and the complete phase shift curve is the sum of the individual phase shift curves.

It is usually most convenient to plot dB and phase shift curves with a logarithmic scale for ω, in which event they are termed *Bode plots*, after Hendrik W. Bode. Table 6.1 shows Bode plots for constant and power-of-s transmittances.

Table 6.1a contains dB values for several values of amplitude. Commonly used dB values are 40 dB ($A = 100$), 20 dB ($A = 10$), 6 dB ($A \cong 2$), 3 dB ($A \cong \sqrt{2}$), 0 dB ($A = 1$), -20 dB ($A = 0.1$), and -40 dB ($A = 0.01$). Inexpensive pocket-sized calculators have replaced the now antiquated slide rule which the authors used in their student days to compute dB and phase values. The student should attempt to gain a perception of dB and phase trends so as to catch obvious programming errors. A partnership must exist between the computer user and the computer. The user must be able to double check computer output by (what used to be called) back-of-the-envelope analysis.

A positive constant k has constant amplitude in dB

$$20 \log_{10} k$$

and angle $0°$. A negative constant, $-k$, has dB amplitude

$$20 \log_{10}|k|$$

and angle $180°$. For example,

$$F = 100$$

Table 6.1 LOGARITHMIC FREQUENCY RESPONSE (BODE) PLOTS FOR SOME BASIC TERMS

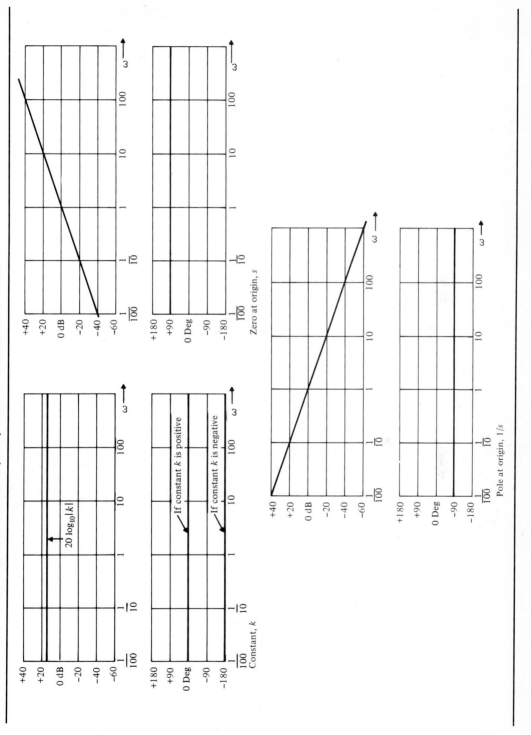

Table 6.1a   DECIBELS

Intensity Ratio A	dB $= 20 \log_{10} A$	Intensity Ratio A	dB $= 20 \log_{10} A$
1.000	0.0	1.000	-0.0
1.012	0.1	0.989	-0.1
1.023	0.2	0.977	-0.2
1.035	0.3	0.966	-0.3
1.047	0.4	0.955	-0.4
1.059	0.5	0.944	-0.5
1.072	0.6	0.933	-0.6
1.084	0.7	0.923	-0.7
1.096	0.8	0.912	-0.8
1.109	0.9	0.902	-0.9
1.122	1.0	0.891	-1.0
1.189	1.5	0.841	-1.5
1.259	2.0	0.794	-2.0
1.334	2.5	0.750	-2.5
1.413	3.0	0.703	-3.0
1.496	3.5	0.668	-3.5
1.585	4.0	0.631	-4.0
1.679	4.5	0.596	-4.5
1.778	5.0	0.562	-5.0
1.884	5.5	0.531	-5.5
1.995	6.0	0.501	-6.0
2.113	6.5	0.473	-6.5
2.239	7.0	0.447	-7.0
2.371	7.5	0.422	-7.5
2.512	8.0	0.398	-8.0
2.661	8.5	0.376	-8.5
2.818	9.0	0.355	-9.0
2.985	9.5	0.335	-9.5
3.162	10	0.316	-10
3.548	11	0.282	-11
3.981	12	0.251	-12
4.467	13	0.224	-13
5.012	14	0.200	-14
5.62	15	0.178	-15
6.31	16	0.159	-16
7.03	17	0.141	-17
7.94	18	0.126	-18
8.91	19	0.112	-19
10^1	20	10^{-1}	-20
3.16×10^1	30	3.16×10^{-2}	-30
10^2	40	10^{-2}	-40
3.16×10^2	50	3.16×10^{-3}	-50
10^3	60	10^{-3}	-60
3.16×10^3	70	3.16×10^{-4}	-70
10^4	80	10^{-4}	-80
3.16×10^4	90	3.16×10^{-5}	-90
10^5	100	10^{-5}	-100
3.16×10^5	110	3.16×10^{-6}	-110
10^6	120	10^{-6}	-120

has constant amplitude ratio and constant phase shift

$$dB = 20 \log_{10}(100) = 40 \ dB \qquad \Phi = 0°$$

as plotted in Fig. 6.7a. The transmittance

$$F = -\tfrac{1}{10}$$

has the curves of Fig. 6.7b.

For the transmittance

$$F(s) = s$$

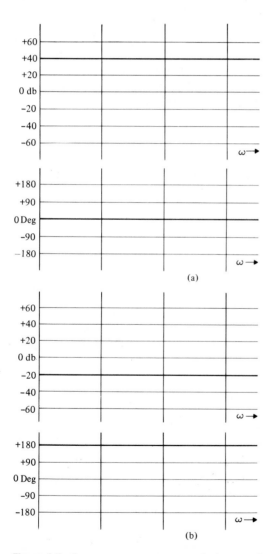

(a)

(b)

Figure 6.7 Frequency response curves for two constant transmittances.
(a) $F = 100$. **(b)** $F = -\tfrac{1}{10}$.

the curves are

$$A(\omega) = |F(s = j\omega)| = \omega \qquad dB = 20 \log_{10} A(\omega) = 20 \log_{10} \omega$$

and

$$\Phi(\omega) = F(s = j\omega) = 90°$$

Since dB is plotted versus the logarithm of ω, the Bode plot is a straight line with slope 20 dB per decade of ω, passing through 0 dB at $\omega = 1$, as shown in Table 6.1. The curves for

$$F(s) = s^2$$

shown in Fig. 6.8a are just the sums of two $F(s) = s$ curves, that is, double the plots for s. In fact, the nth power of any transmittance has curves which are n times the plots for the original transmittance. The Bode plots for

$$F(s) = \frac{1}{s} = s^{-1}$$

(Figure 6.8b) are the negatives of the dB and phase angle plots for $F(s) = s$.

Real Axis Roots

Consider a transmittance that is a left half-plane (LHP) zero, of the form

$$F(s) = \frac{s + a}{a} = 1 + \frac{s}{a}$$

where a is a positive number. Instead of considering a single zero term, $s + a$, terms are put in this form because the relations are simplest when

$$F(s = j\omega) = 1 + j\frac{\omega}{a}$$

For this term,

$$|F(s = j\omega)| = \sqrt{1 + \left(\frac{\omega}{a}\right)^2} \qquad \angle F(s = j\omega) = \tan^{-1}\left(\frac{\omega}{a}\right)$$

and

$$dB = 20 \log_{10}|F(s = j\omega)| = 10 \log_{10}\left[1 + \left(\frac{\omega}{a}\right)^2\right]$$

For $\omega \ll a$,

$$F(s = j\omega) \cong 1 \qquad dB \cong 10 \log_{10} 1 = 0 \qquad \angle F(s = j\omega) \cong \tan^{-1} 0 = 0$$

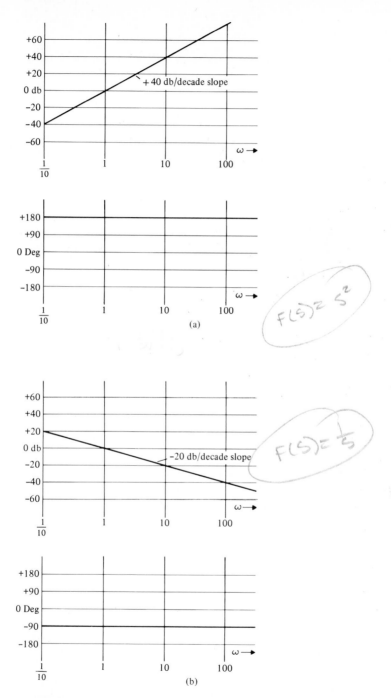

Figure 6.8 Bode plots for transmittances having poles and zeros at the origin of the complex plane. **(a)** Frequency response plots for $F(s) = s^2$. **(b)** Frequency response plots for $F(s) = 1/s$.

For $\omega \gg a$,

$$F(s = j\omega) \cong j\frac{\omega}{a} \qquad \text{dB} \cong 20 \log_{10} \frac{\omega}{a} = 20 \log_{10} \omega - 20 \log_{10} a$$

$$\angle F(s = j\omega) \cong 90°$$

The actual and the approximate curves are shown in Fig. 6.9.

The dB curve can be approximated quite nicely by a curve along 0 dB up to $\omega = a$, termed the *break frequency* (or corner frequency) then a line sloping upward beyond $\omega = a$ with a slope of $+20$ dB/decade of ω. The maximum error, using this approximation, will be at the break frequency, where the actual value of the amplitude is

$$|F(ja)| = 20 \log_{10} \sqrt{2} = 3.01 \text{ dB}$$

In practice, one sketches the approximate curve and, if greater accuracy is desired, modifies it slightly so that it is smooth and goes through 3 dB at the break frequency.

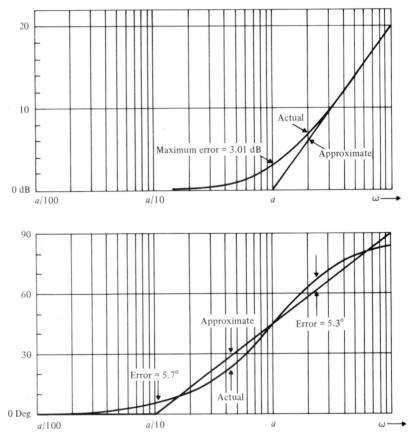

Figure 6.9 Bode plot for $F(s) = (s + a)/a$.

The three-segment phase shift approximation indicated in Fig. 6.9 is accurate to about 6°. In the approximation, the angle is 0° up to one-tenth the break frequency, rises 45° per decade through 45° at the break frequency, and continues at 45° per decade slope up to ten times the break frequency. Beyond 10 times the break frequency, the approximate angle is 90°. At the break frequency a, the actual and approximate curves are equal, since

$$\angle F(ja) = \tan^{-1}\left(\frac{1}{1}\right) = 45°$$

As a numerical example, the transmittance

$$F(s) = \frac{s + 10}{10}$$

has the specific Bode curves of Fig. 6.10. Approximately, the dB curve is along 0 dB up to the break frequency, $\omega = 10$, then the dB rises 20 dB per decade of ω. The

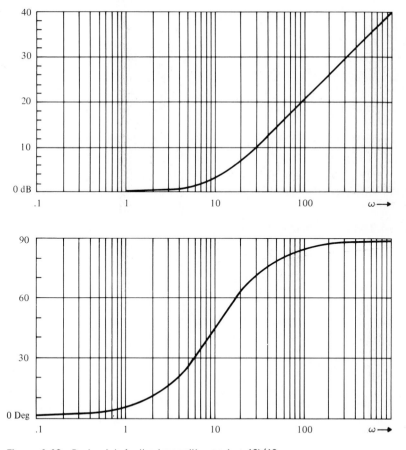

Figure 6.10 Bode plots for the transmittance $(s + 10)/10$.

phase shift curve is approximately $0°$ up to one-tenth the break frequency, and $90°$ beyond 10 times the break frequency. In between, it is approximately a straight line with slope $45°$ per decade through $45°$ at $\omega = 10$.

Bode plots for real axis LHP poles and for RHP (right half-plane) zeros and poles are given in Table 6.2. The Bode frequency response plots for a real axis LHP pole term,

$$F(s) = \frac{a}{s + a} = \frac{1}{1 + (s/a)}$$

have dB and phase shift curves that are negatives of the curves for a corresponding LHP zero. The dB curve is

$$dB = 20 \log_{10} |F(s = j\omega)| = 20 \log_{10} \left| \frac{1}{1 + j(\omega/a)} \right| = -20 \log_{10} \left| 1 + j\left(\frac{\omega}{a}\right) \right|$$

which is the negative of that for a zero. The phase shift is, similarly,

$$\angle F(s = j\omega) = \left| \frac{1}{1 + j(\omega/a)} \right. = -\left[1 + j\left(\frac{\omega}{a}\right) \right]$$

RHP zeros and poles differ from their LHP counterparts by the algebraic sign of the phase shift; amplitude ratios are the same as for LHP roots of the same type.

For example, the transmittance

$$F(s) = \frac{-10}{s - 10}$$

an RHP pole term, has a dB curve that is the negative of that for the zero term $(s + 10)/10$ and a phase shift that is the same as that for the zero, as shown in Fig. 6.11. Although this transmittance, having a RHP pole, is unstable, it could be part of an overall system of interest that is stable.

Products of Transmittance Terms

Consider plotting Bode frequency response curves for

$$F(s) = \frac{50(s + 2)}{s(s + 10)}$$

First, $F(s)$ is decomposed into factors for which the frequency response is known and can easily be sketched:

$$F(s) = 10\left(\frac{1}{s}\right)\left(\frac{s + 2}{2}\right)\left(\frac{10}{s + 10}\right)$$

the individual terms have Bode plots represented by the straight line approximations in Fig. 6.12a. The sums of these curves, shown in Fig. 6.12b, are the plots for $F(s)$. Rounding the corners of the straight line approximation by 3 dB at $\omega = 2$ and $\omega = 10$ further improves the approximation.

Table 6.2   LOGARITHMIC FREQUENCY RESPONSE (BODE) PLOTS FOR REAL AXIS ROOT TERMS

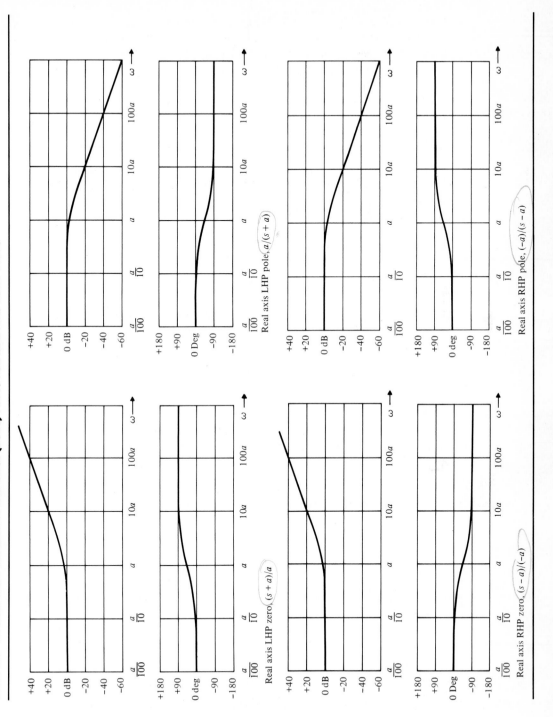

339

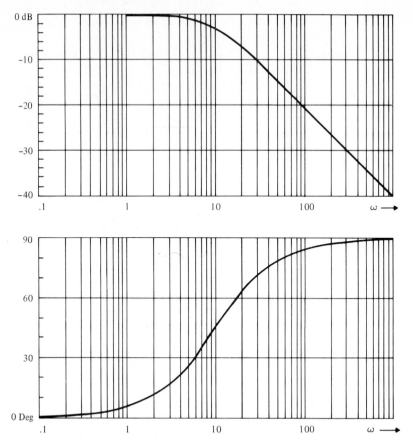

Figure 6.11 Bode plots for the transmittance $-10(s - 10)$.

To be complete, the frequency response curves should be drawn for a range of frequency which includes all the detail of the response. A minimum range of frequency for the above example is from below 1 rad/sec to above 100 rad/sec. Outside that range, the curves continue as straight lines. With some experience and practice, frequency response curves for transfer functions with real poles and zeros can be sketched by considering the various changes in slope that occur for each break frequency.

Since the corner frequencies seldom are a decade apart, it is necessary to obtain an equation for each straight-line segment so as to compute the dB and phase values at the ends. It could also be useful to quickly approximate dB and phase values somewhere along a curve, rather than to compute the exact value.

A straight line is often written using y as the vertical-axis variable where x_0, y_0 is a beginning point for the straight line and m is the slope.

$$y = y_0 + m(x - x_0)$$

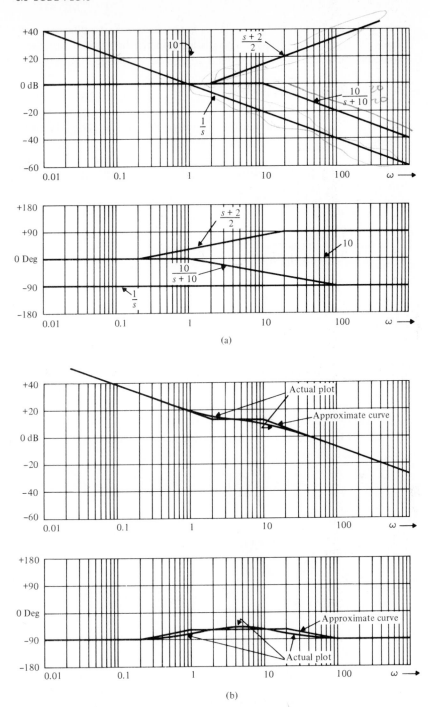

Figure 6.12 Example of Bode plot construction. **(a)** Bode plots of individual terms. **(b)** Complete plots, consisting of sums of terms.

For a dB plot, the slope is in units of dB per decade. For a phase plot, the slope is in units of degrees per decade. The distance in the horizontal (frequency) direction must be measured in decades. That can be accomplished (where $x = \omega$) by

$$\text{decades of } (\omega - \omega_0) = \log_{10} \frac{\omega}{\omega_0}$$

When $\omega = \omega_0$, there is no decade of change. When $\omega = 10\omega_0$, the equation yields 1 decade.

Suppose a straight-line equation is needed for the Bode magnitude plot of Fig. 6.12 for frequencies between 1 and 2 rad/sec. At 1 rad/sec, the amplitude is 20 dB and the slope is -20 dB/decade.

$$y_0 = 20 \text{ dB}$$

$$\omega_0 = 1$$

$$m = -20 \text{ dB/decade}$$

$$y = 20 - 20 \log_{10} \left(\frac{\omega}{1} \right)$$

When $\omega = 2$, then

$$y = 20 - 20 \log_{10} 2 = 14 \text{ dB}$$

For frequencies above 10 rad/sec, the approximate curve is

$$y = 14 - 20 \log_{10} \left(\frac{\omega}{10} \right)$$

For the phase curve where frequencies lie between 0.2 and 1 rad/sec, the approximate phase is

$$y = -90° + 45 \log_{10} \left(\frac{\omega}{0.2} \right)$$

At 1 rad/sec,

$$y = -90° + 45° \log_{10} \left(\frac{1}{0.2} \right) = -58.5°$$

For frequencies above 20 rad/sec,

$$y = -58.5° - 45° \log_{10} \left(\frac{\omega}{20} \right)$$

The frequency at which zero dB occurs can be approximated by

$$0 = 14 - 20 \log_{10} \frac{\omega}{10}$$

$$\log_{10} \left(\frac{\omega}{10} \right) = \left(\frac{14}{20} \right) = 0.7$$

$$\omega = 10 \times 10^{0.7} = 50 \text{ rad/sec}$$

Appendix B contains a computer-aided procedure by which the exact (not the asymptotic) frequency can be computed for a given dB or phase value. Of course a trial-and-error procedure could also be used.

Complex Roots

Bode frequency response curves for complex conjugate pole or zero terms are plotted by hand, if desired, from straight-line asymptotes similar to simple real axis poles and zeros. Factor a set of complex conjugate poles or zeros to the form

$$F_1(s) = \frac{\omega_n^2}{s^2 + 2\zeta\omega_n s + \omega_n^2}$$

for a set of complex conjugate poles, or

$$F_2(s) = \frac{s^2 + 2\zeta\omega_n s + \omega_n^2}{\omega_n^2}$$

for a set of complex conjugate zeros.

First, we consider a complex conjugate set of poles $F_1(s)$. For small values of ω compared to ω_n,

$$F_1(s = j\omega) \cong 1$$

For large values of ω compared to ω_n,

$$F_1(s = j\omega) \cong -\frac{\omega_n^2}{\omega^2}$$

which is a dB curve with slope -40 dB/decade and a phase shift of $180°$. At $\omega = \omega_n$,

$$F_1(j\omega_n) = \frac{1}{j2\zeta}$$

for which the dB curve has value

$$dB = 20 \log_{10} \frac{1}{2\zeta}$$

and the phase shift is $-90°$.

Plots of amplitude and phase shift for various damping ratios for complex conjugate pole terms are given in Fig. 6.13. For $\zeta \geq 1$, the poles are real, not complex, and so can be handled by the previous real axis pole methods. Figure 6.13c shows the true dB and phase values (for various actual damping ratios) minus the approximate curve (the asymptotic curve plotted with a damping ratio of 1). With a bit of practice a designer can sketch the asymptotic curve for a damping ratio of 1 and then add the correctional curve (from Fig. 6.13c) for the actual damping ratio. Curves of reasonable accuracy can be derived in a fairly short time.

For damping ratios between 0.5 and 1, the true dB curve lies generally below the asymptotic curve; so the correctional dB curve is generally negative. For damping ratios between 0 and 0.5, the true curve lies generally above the correctional curve; so the correctional dB values are generally positive. The true phase curve forms an S-shaped curve with the asymptotic curve; therefore, the correctional phase curves are of both signs with magnitudes depending on the actual damping ratio.

As an example, consider the transmittance

$$F(s) = \frac{1000}{s^2 + 2s + 100} = (10)\left(\frac{100}{s^2 + 2s + 100}\right)$$

For the complex conjugate pole term

$$\omega_n = \sqrt{100} = 10$$

$$2\zeta\omega_n = 20\zeta = 2 \qquad \zeta = \frac{1}{10}$$

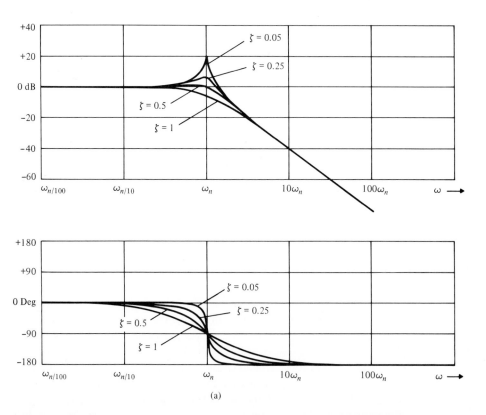

Figure 6.13 Bode plots for second-order terms. **(a)** Normalized curves for several values of the damping ratio.

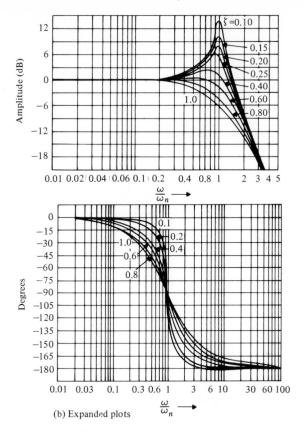

(b) Expanded plots

Figure 6.13 (b) Expanded plots.

First the straight-line approximations shown in Fig. 6.14 are drawn. The conjugate pair of roots are first approximated as if they were critically damped, of the form

$$\frac{100}{s^2 + 20s + 100} = \left(\frac{10}{s + 10}\right)^2$$

The dB approximation is 0 dB (+20 dB for the constant factor of 10) to the break frequency of $\omega = 10$, then -40 dB/decade slope thereafter. The angle approximation is $0°$ to one-tenth the break frequency, then $-90°$/decade slope, passing through $-90°$ at the break frequency and continuing until 10 times the break frequency.

The corrections of Fig. 6.13, for $\zeta = \frac{1}{10}$, are then applied to the approximate curves to give the final result in Fig. 6.14. Transferring several points from each curve and then drawing a smooth curve through the result will usually suffice. As the error of the straight-line approximation is sizable, it is important that the corrections be made.

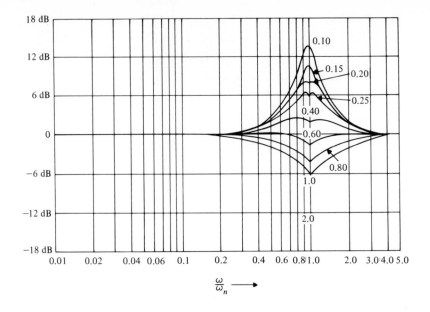

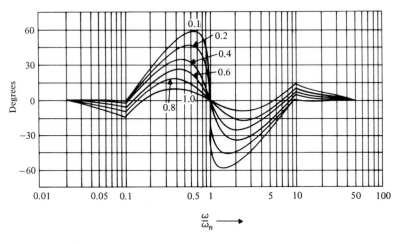

(c) Correctional curves (true-asymptotic)

Figure 6.13 (c) Correctional curves (true-asymptotic).

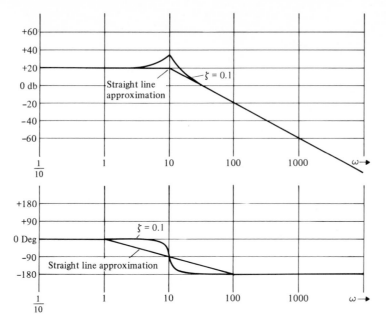

Figure 6.14 Frequency response plots for $F(s) = 1000/(s^2 + 2s + 100)$.

Table 6.3 shows Bode plots for complex conjugate LHP and RHP pole and zero pairs. Conjugate LHP zeros of the form

$$F(s) = \frac{s^2 + 2\zeta\omega_n s + \omega_n^2}{\omega_n^2}$$

have Bode plots that are the negatives of the curves for the corresponding conjugate poles. Conjugate RHP poles and zeros,

$$F(s) = \frac{\omega_n^2}{s^2 - 2\zeta\omega_n s + \omega_n^2}$$

and

$$F(s) = \frac{s^2 - 2\zeta\omega_n s + \omega_n^2}{\omega_n^2}$$

differ from their LHP counterparts in algebraic sign of the phase shift; the dB amplitude ratios are the same as for LHP roots of the same type.

As another example, consider the transmittance

$$F(s) = \frac{s^2 + s + 8}{s^2}$$

Decomposing,

$$F(s) = (8)\left(\frac{s^2 + s + 8}{8}\right)\left(\frac{1}{s}\right)^2$$

Table 6.3 BODE PLOTS FOR COMPLEX CONJUGATE POLE AND ZERO PAIRS

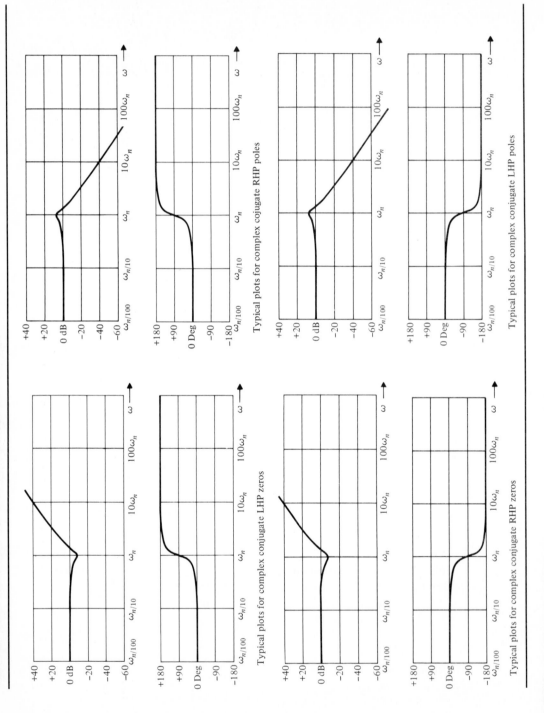

Typical plots for complex conjugate LHP zeros

Typical plots for complex cojugate RHP poles

Typical plots for complex conjugate LHP zeros

Typical plots for complex conjugate RHP zeros

Typical plots for complex conjugate LHP poles

348

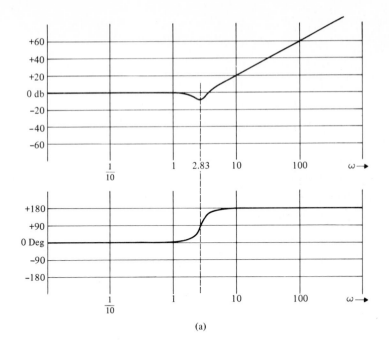

(a)

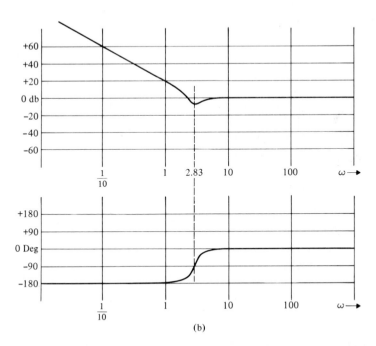

(b)

Figure 6.15 Bode plots of a transmittance involving complex roots. **(a)** Approximate frequency response curves for $F(s) = (s^2 + s + 8)/8$. **(b)** Approximate frequency response curves for $F(s) = (s^2 + s + 8)/s^2$.

The set of complex conjugate zeros has curves that are the negatives of the given curves for a pole, with

$$\omega_n = \sqrt{8} = 2\sqrt{2} = 2.83$$

$$2\zeta\omega_n = 5.66\zeta = 1 \qquad \zeta = 0.177$$

from which the curves of Fig. 6.15 are sketched.

DRILL PROBLEM

D6.4. Sketch frequency response curves (both amplitude in dB and phase shift) for the following transmittances:

(a) $F(s) = \dfrac{1}{s + 1000}$

Ans

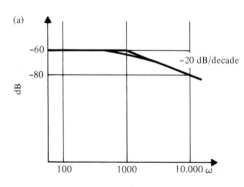

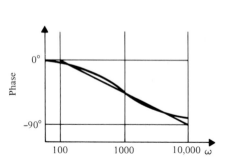

(b) $F(s) = \dfrac{1}{(s + 10)^3}$

Ans

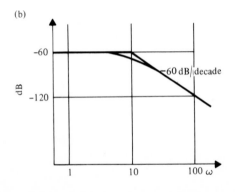

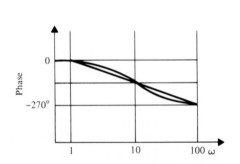

(c) $F(s) = \dfrac{s - 10}{s + 10}$

Ans

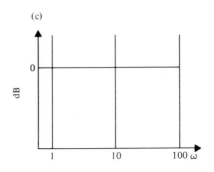

(c)

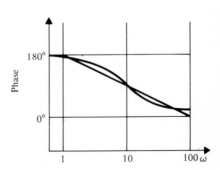

(d) $F(s) = \dfrac{10}{s^2 + s + 4}$

Ans

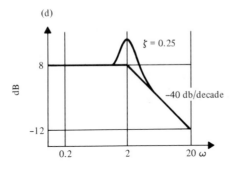

(d)

$\zeta = 0.25$

-40 db/decade

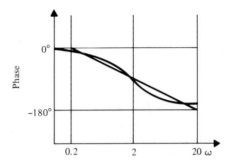

(e) $F(s) = \dfrac{s^2 - 4s + 30}{(s + 10)^2}$

Ans

(e)

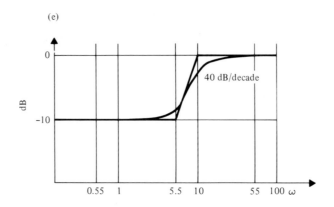

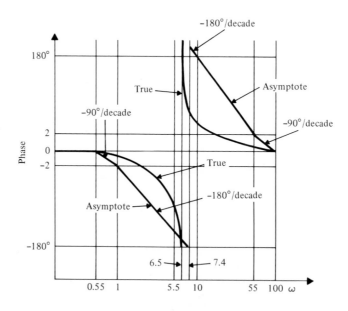

(f) $F(s) = \dfrac{s^2 + 2s + 100}{s^2 + 10s + 100}$

Ans

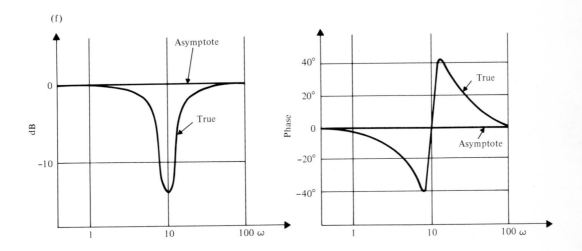

(g) $F(s) = \dfrac{s}{s^2 + 20s + 100}$

Ans

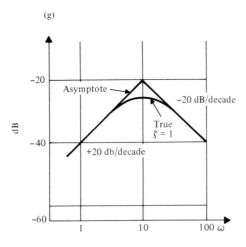
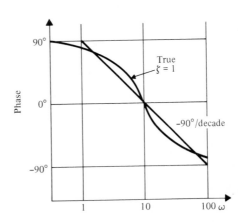

6.4 USING EXPERIMENTAL DATA

Finding Models

One of the most important and powerful uses of the frequency response method of system design is in determining component transmittances. For many practical system components, such as pneumatic valves and airframes, analytic expressions for transmittances are difficult to obtain from theory. If a frequency response test can be performed, however, the transmittance can be determined experimentally.

Pneumatic valves, for example, do not readily lend themselves to analytic determination of their transfer functions. Very often, however, a frequency test can be run on these components and the dB gain and phase shift versus logarithmic frequency can be plotted. Then design can be facilitated by the use of approximate transfer functions determined from the experimentally obtained frequency plots.

Consider an example transmittance for which the experimental amplitude and phase characteristics are as tabulated in Table 6.4. It is desired to obtain a transfer function which approximates these characteristics. If the characteristics are plotted, as in Fig. 6.16, a series of straight-line asymptotes can be fitted to these data for both amplitude and phase. By use of the slopes and corresponding break frequencies, a transfer function is obtained. For the example given, an approximate transmittance is

$$F(s) = 16\left(\frac{15}{s + 15}\right)\left(\frac{150}{s + 150}\right) = \frac{16.0}{(0.07s + 1)(0.007s + 1)}$$

Often the phase versus logarithmic frequency, as calculated from the approximate transfer function, will not completely agree with the corresponding experi-

Table 6.4 EXPERIMENTAL FREQUENCY RESPONSE DATA

f	ω	Gain (dB)	Phase Shift (deg)
60	377	−7.75	−155
50	314	−4.3	−150
40	251	−0.2	−145
35	219	0.75	−140
25	157	5.16	−135
20	126	7.97	−120
16	100	10.5	−110
10	63	15.0	−100
7	44	16.9	−85
2.5	16	20.4	−45
1.3	8	21.6	−30
0.22	1.38	24.0	−5
0.16	1.0	24.1	0

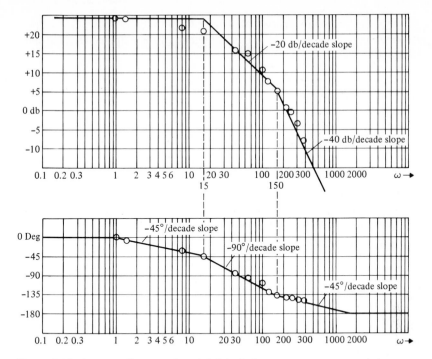

Figure 6.16 Incorporating experimental data for frequency response analysis.

mental curve. The problem of obtaining the best match for both amplitude and phase curves is simplified if linear asymptotes are used for both amplitude and phase curves.

Use of this approximated transfer function, in conjunction with the remaining analytically obtainable transfer functions, permits the engineer to analyze the system.

Irrational Transmittances

Another advantage of frequency response methods is that it is not necessary to restrict the type of transfer function to rational polynomials. Frequency response methods can be brought to bear on such irrational transfer functions as

$$F(s) = \sqrt{s}$$

and

$$F(s) = \cos s$$

An irrational transmittance of considerable practical importance is of the form

$$F(s) = e^{-\tau s}$$

where τ is a positive constant. This transmittance represents the time delay of the incoming signal by τ sec. In the language of the Laplace transformation,

$$\mathcal{L}^{-1}[Y(s)e^{-\tau s}] = y(t - \tau)$$

One type of time delay scheme involves recording the input signal on magnetic tape. A time delay is obtained as the tape moves from the record to the playback head. This arrangement is often used on broadcast interview programs to allow censorship of the program before it is aired. Other simple time delay systems are transmission lines, digital shift registers, conveyor belts, and audio reverberation generators.

The transmittance for a system that is only delayed in time emerges with no change in amplitude versus frequency, but it does undergo a change in phase. The higher the frequency, the greater the phase shift for the same time delay. For

$$F(s) = e^{-s\tau} \qquad F(j\omega) = e^{j\omega\tau}$$

and

$$A(\omega) = |F(j\omega)| \qquad \Phi(\omega) = \angle F(j\omega) = -\omega\tau \qquad \text{rad}$$

The frequency response of the time delay system

$$F(s) = e^{-(1/2)s}$$

is plotted on a linear scale of frequency in Fig. 6.17. On a logarithmic frequency scale, the phase shift curve is more and more compressed for larger values of ω.

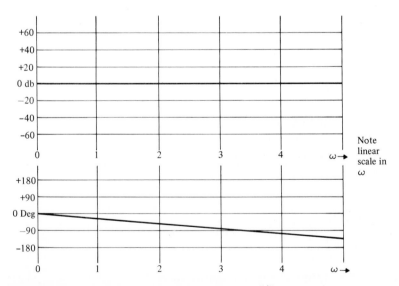

Figure 6.17 Frequency response plots for $F(s) = e^{-(1/2)s}$.

DRILL PROBLEMS

D6.5. Determine an approximate transmittance from the following experimental frequency response data:

ω	dB	Phase
0.1	-20	0
0.5	-21	0
1.0	-21	$-9°$
2.0	-22	$-54°$
3.0	-24	$-90°$
5.0	-28	$-135°$
10.0	-40	$-170°$
30.0	-60	$-178°$
100.0	-84	$-180°$

Ans $F(s) = \dfrac{0.1}{(s/3 + 1)^2}$

D6.6. Find and sketch the frequency response (both amplitude and phase shift) for the following irrational functions:

(a) $F(s) = 6e^{-0.2s}$

 Ans $6, -0.2\omega$

(b) $F(s) = \dfrac{e^{-4s}}{s}$

 Ans $1/\omega, -4\omega - \pi/2$

(c) $F(s) = \sqrt{s}$

 Ans $\sqrt{\omega}, 45°$ or $225°$

6.5 GAIN AND PHASE MARGINS

Feedback System Stability

Many practical systems are of the simple feedback type of Fig. 6.18. The overall transfer function that relates the output $Y(s)$ to the input $R(s)$ is

$$T(s) = \frac{G(s)}{1 + G(s)H(s)} \tag{6.1}$$

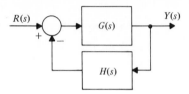

Figure 6.18 A feedback system.

If the factors of $G(s)H(s)$ are known, frequency plots of GH are relatively easy to construct, and they are of help in determining the properties of the overall system. Or the frequency response plots for GH might have been found experimentally.

Whether or not the overall system is stable can be determined in the following way: Suppose the transmittance $G(s)H(s)$, is replaced by $KG(s)H(s)$ where K is a constant. Then

$$T'(s) = \frac{G(s)}{1 + KG(s)H(s)} \tag{6.2}$$

This related system has a transfer function in the form of a root locus problem with adjustable K. When $K = 1$, the new system (6.2) is the actual system of interest (6.1). For $K = 0$, the poles of $T'(s)$ are the poles of $G(s)H(s)$. As $K \to \infty$, the loci extend to the zeros of $G(s)H(s)$ in the usual way, but we are here only interested in the part of the locus from $K \cong 0$ to $K = 1$.

If the open-loop transmittance $G(s)H(s)$ is stable and if $T(s)$ is unstable, then root loci extending from the LHP poles of GH must cross into the RHP for some value of K between 0 and 1. In particular, a locus segment must cross the imaginary axis. There must be some intermediate value of K for which $T'(s)$ has a pole or poles on the imaginary axis, at $s = j\omega$:

$$|KG(s = j\omega)H(s = j\omega)| = 1 \quad \text{and} \quad \angle KG(s = j\omega)H(s = j\omega) = 180°$$

A positive constant K only affects the amplitude ratio, so $T'(s)$ has poles on the imaginary axis for K between 0 and 1 if and only if

$$|G(s = j\omega)H(s = j\omega)| = \frac{1}{K} > 1 \quad \text{and} \quad \angle G(s = j\omega)H(s = j\omega) = 180°$$

That is, for an imaginary axis locus crossing at $s = j\omega$, the magnitude of the GH product must be greater than unity (0 dB) at a frequency ω where the angle of the GH product is 180°.

For example, the frequency response plots for

$$G(s)H(s) = \frac{100}{(s + 1)^3}$$

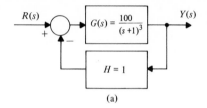

(a)

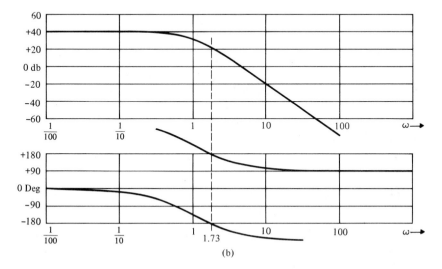

(b)

Figure 6.19 Feedback system example.

in the system of Fig. 6.19a are given in Fig. 6.19b. To determine whether the overall system is stable, one imagines the related system

$$T'(s) = \frac{G(s)}{1 + KG(s)H(s)} = \frac{100}{(s + 1)^3 + 100K}$$

For $K = 0$, $T'(s)$ is stable, its three poles being at $s = -1$. Various values of K between 0 and 1 do not affect the phase shift curve for $KG(s)H(s)$ but result in an amplitude curve with the same shape as that for $G(s)H(s)$, but at various lower levels, as indicated in Fig. 6.20.

The frequency at which the phase of GH is $-180°$ in Fig. 6.19 can be found easily for this example. Because of the third-order denominator term:

$$\underline{/G(s = j\omega)H(s = j\omega)} = -3 \tan^{-1}(\omega)$$

$$= -180°$$

$$\tan^{-1}(\omega) = 60°$$

$$\omega = \tan(60°) = 1.73 \text{ rad/sec}$$

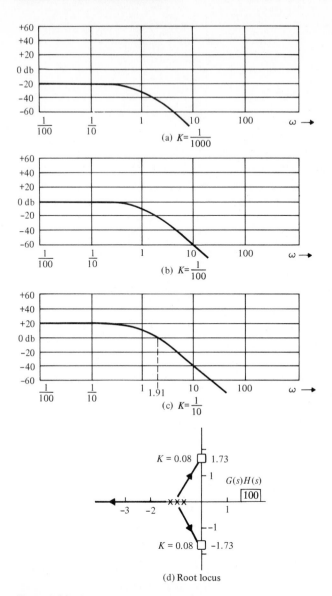

Figure 6.20 Bode amplitude curves for specific values K and the root locus for variable K. **(a)** $K = \frac{1}{1000}$. **(b)** $K = \frac{1}{100}$. **(c)** $K = \frac{1}{10}$. **(d)** Root locus.

When the transmittance GH is more complicated, the frequency for a phase of $-180°$ can be found by trial and error or by the computer-aided procedure of Appendix B.

A quick look at Fig. 6.20a (for $K = \frac{1}{1000}$) and Fig. 6.20b (for $K = \frac{1}{100}$) indicates that 0 dB occurs at a frequency well below 1.73 rad/sec where the phase is $-180°$.

When $K = \frac{1}{10}$, the actual dB values can be easily found

$$dB[A(\omega)] = 40 + 20 \log_{10} K - 60 \log_{10} \sqrt{1 + \omega^2}$$

Where 0 dB occurs,

$$0 = 20 - 30 \log_{10}(1 + \omega^2)$$
$$\log_{10}(1 + \omega^2) = \tfrac{2}{3}$$
$$\omega = \sqrt{10^{2/3} - 1} = 1.91$$

which is slightly above 1.73 rad/sec. Zero dB occurs for

$$20 \log_{10} K = 30 \log_{10}(1 + 1.73^2) - 40$$
$$K = 0.08$$

For this value of K,

$$KG(s = j1.73)H(s = j1.73) = -1$$

and $T'(s)$ has a set of complex conjugate poles at $s = \pm j1.73$. As K is varied from zero, where all the poles of $T'(s)$ are in the LHP, to unity, where $T'(s) = T(s)$, there is an intermediate value of K for which $T'(s)$ has imaginary axis poles. This means that the loci of two of the poles of $T'(s)$ must extend from the LHP, across the imaginary axis (at $\omega = \pm 1.73$) to the RHP as K goes from zero to 1. Each of the two root locus branches crosses only once; so it is concluded that $T(s)$ has RHP poles and the system is unstable.

The root locus for this system is shown in Fig. 6.20d. The vertical axis is crossed as predicted. The Bode plot provides a convenient way to determine stability and also to determine the location of vertical axis root crossings without actually obtaining the root locus. Historically, Bode methods preceded root locus methods, partially due to ease of computation.

There is the possibility that a segment of the root locus lies along the real axis, crossing the imaginary axis at the origin. This situation is easy to overlook if the frequency response plots are made versus log frequency because the response at $\omega = 0$ is never explicitly plotted. If

$$G(s)H(s) = \frac{s - 10}{s + 1}$$

the loop transmittance frequency response is as plotted in Fig. 6.21. The phase shift is 180° at $\omega = 0$. (It is nearly the asymptotic value below $\omega = 1/10$, but reaches precisely 180° only at $\omega = 0$.) As K is increased from zero to unity, the amplitude curve for $KG(s)H(s)$ has various positions below that of the amplitude curve for $G(s)H(s)$ in Fig. 6.21.

At the value $K = 1/10$, the amplitude of KGH is at 0 dB at $\omega = 0$, so there is a pole of $T'(s)$ at $s = j0$ for $K = 1/10$. As K is further increased to unity, there are no other values of ω for which

$$KG(s = j\omega)H(s = j\omega) = -1$$

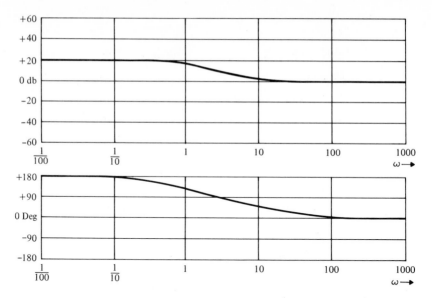

Figure 6.21   *GH* product for a system with an imaginary axis locus crossing at $\omega = 0$.

A root locus segment for $T'(s)$ must extend from the LHP, crossing the imaginary axis at $\omega = 0$, into the RHP. The original transfer function $T(s)$ of this feedback system is thus unstable.

Consider the feedback system with irrational transmittance

$$G(s)H(s) = \frac{100e^{-0.07s}}{s + \frac{1}{10}}$$

The frequency response for *GH* is given in Fig. 6.22. It is concluded that the overall system is unstable because the amplitude is larger than 0 dB when the phase is 180° and because the open-loop transmitance has only LHP poles.

Finding Gain Margin and Phase Margin

Suppose it is known that a certain design results in a stable feedback system. The designed *GH* product will only be approximated in practice, owing to component tolerances. How much change in amplitude and in phase of the *GH* product can be tolerated before the overall system becomes unstable? The additional amplitude of the *GH* product and the additional phase angle of the product which result in imaginary axis roots of $T(s)$ are measures of the allowable tolerances in $G(s)H(s)$ for overall system stability.

A *phase crossover frequency* is any frequency at which the phase shift of the *GH* product is $\pm 180°$. The *gain margin* of a feedback system is the additional dB amplitude necessary to make the amplitude of *GH* unity at a phase crossover frequency. If the phase shift of *GH* crosses $\pm 180°$ at more than one frequency, the

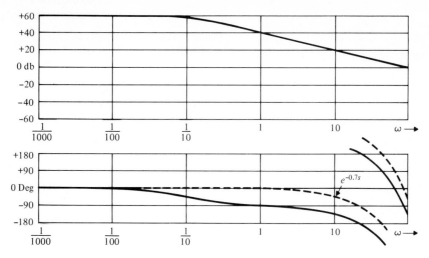

Figure 6.22 An irrational loop transmittance.

gain margin is the smallest value of the possibilities. For example, frequency response plots for

$$G(s)H(s) = \frac{s^3}{(s + 1)^3}$$

are given in Fig. 6.23. At the phase crossover frequency the amplitude curve is at -18 dB, giving an 18-dB gain margin. The loop gain could be increased by 18 dB before instability would result.

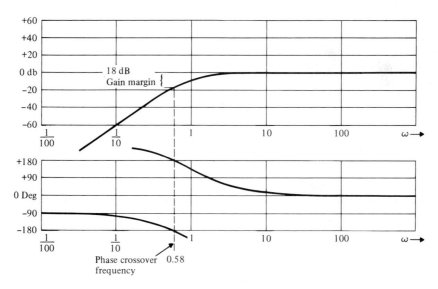

Figure 6.23 Gain margin of a simple feedback system. Frequency response plotted is for $G(s)H(s)$.

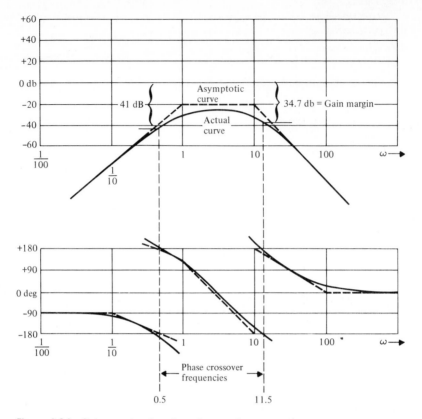

Figure 6.24 Gain margin when there is more than one phase crossover frequency.

For the system with

$$G(s)H(s) = \frac{1000s^3}{(s+1)^3(s+10)^4}$$

there are two phase crossover frequencies, as shown in Fig. 6.24. The gain margin is the smaller of the two candidates. If there is no phase crossover frequency, the gain margin can be said to be infinite.

A *gain crossover frequency* is any frequency at which the amplitude ratio for *GH* is 0 dB. The *phase margin* is the additional *negative* phase shift necessary to make the phase of *GH* equal to ±180° at a gain crossover frequency. The phase margin for an example feedback system with

$$G(s)H(s) = \frac{s^2}{s+1}$$

is indicated in Fig. 6.25.

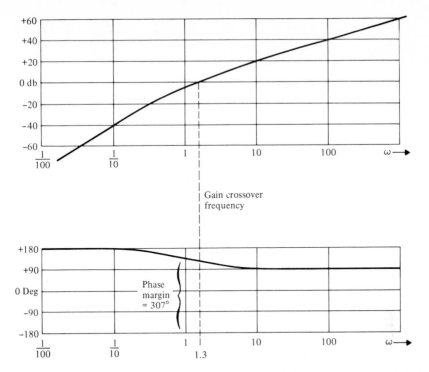

Figure 6.25 Phase margin for a simple feedback system. Frequency response plotted is for $G(s)H(s)$.

If there is more than one gain crossover frequency, there is more than a single phase margin. For

$$G(s)H(s) = \frac{10s}{(s + 1)^2}$$

the two phase margins are shown in Fig. 6.26. A decrease of $101°$ or an increase of $360° - 258° = 102°$ in the loop transmittance phase shift results in instability.

The design of a closed-loop system can be accomplished by using experimental data directly, without approximating the transmittances of complicated components from their experimental data. The experimental data are plotted and the frequency responses of analytically known components are added directly to the experimental data. The resulting combined loop transmittance is used in the conventional fashion to determine stability and gain and phase margins.

As an example of this powerful technique, consider the yaw control system for a ground-effect vehicle which is modeled in the block diagram of Fig. 6.27a. Experimental data for the vehicle are used in part to construct a frequency response plot for the loop gain $G_1(s)G_2(s)$. The experimental data are not approximated with asymptotes or an equation here. Instead, the experimental data for $G_2(s)$ are

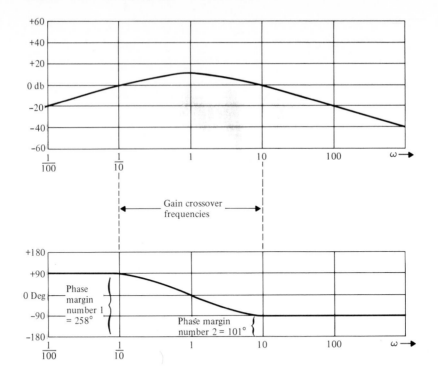

Figure 6.26   Phase margins when there are two gain crossover frequencies.

plotted and the analytical response for $G_1(s)$ is added in Fig. 6.27b. The system of Figure 6.27 has significant gain and phase margin; hence, the system is stable.

Before going on to more topics and applications, we will gather together the definitions from this section.

> **Phase crossover frequency (rad/sec)** = frequency at which the phase of GH is $-180°$
>
> **Gain margin (dB)** = $-$(dB of GH measured at the phase crossover frequency)
>
> **Gain crossover frequency (rad/sec)** = frequency at which the magnitude of GH is 0 dB
>
> **Phase margin (degrees)** = $180°$ + phase of GH measured at the gain crossover frequency (count first quadrant angles as positive and other quadrant angles as negative)

The phase margin in Fig. 6.25 is computed from

Phase margin = $180° + 127° = 307°$

The two phase margins in Fig. 6.26 follow from

First phase margin = $180° + 78° = 258°$

Second phase margin = $180° - 79° = 101°$

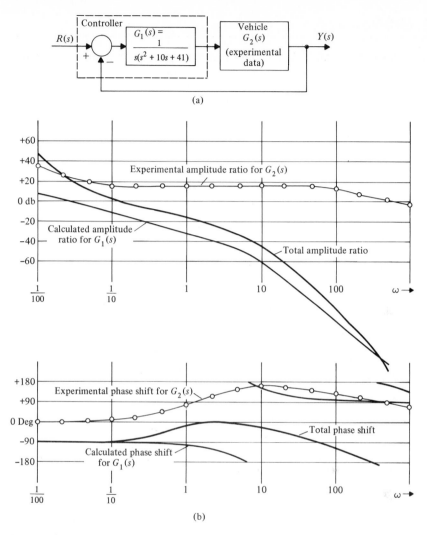

Figure 6.27 Incorporating experimental data into frequency response plots. **(a)** Model of yaw system. **(b)** Frequency response for the transmittance $G_1(s)G_2(s)$.

The phase crossover frequency and the gain crossover frequency may be found in at least three ways

1. Estimate them using the straight-line asymptotic approximations.
2. Apply trial and error to GH.
3. Apply the computer-aided method from Appendix B. This method can be programmed in high-level languages like FORTRAN and BASIC or even by key-stroke programming a pocket-sized calculator.

DRILL PROBLEMS

D6.7. Find gain margins and phase margins (if they exist) for feedback systems with the following loop transmittances:

(a) $G(s)H(s) = \dfrac{2000}{(s + 2)(s + 7)(s + 16)}$

    Ans 5.4 dB, 20°

(b) $G(s)H(s) = \dfrac{20}{s(s^2 + 7s + 140)}$

    Ans 33.8 dB, 89°

(c) $G(s)H(s) = \dfrac{-s}{(s + 100)^3}$

    Note the negative algebraic sign.

    Ans 88.5 dB, infinite phase margin

(d) $G(s)H(s) = \dfrac{e^{-0.1s}}{s}$

    Ans 23.9 dB, 84°

D6.8. For the following systems, use frequency response methods to determine the range of the positive constant K for which the following systems are stable.

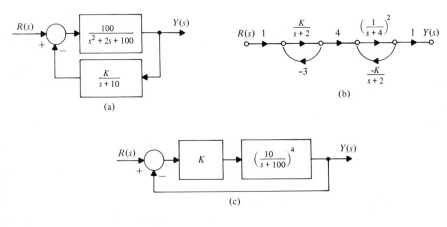

(a)   (b)   (c)

Ans (a) $K < 4.4$;   (b) $K < 288$;   (c) $K < 40{,}000$

6.6 NYQUIST METHODS

Nyquist's work in the 1930s led to a procedure that can determine whether or not a system is stable by using the frequency response of the open-loop transmittance $G(s)H(s)$. Since the frequency response could be calculated fairly easily (given the slide rule-based technology of that era) Nyquist used calculus to prove that the number of right half-plane poles could be found for an electrical network (radio and TV were under active development then) or for a control system (to which the method is now primarily used).

The Nyquist procedure can be understood by dividing the discussion into two parts: generating the plot and interpreting the plot. The same division was helpful when Bode plots were generated and then the gain and phase margins were interpreted. First, then, a Nyquist (in fact a polar) plot must be generated. The Nyquist plot approach is similar to casting an infinitely large net over the entire right half-plane and then hauling in the net. The controls engineer examines the contents of the net to see if any RHP poles are ensnared. To accomplish the counting of those poles the boundary of the RHP in Fig. 6.28 is mapped into another shape using the open-loop transmittance $G(s)H(s)$ for all values of s along the boundary of the RHP. The result is plotted using a polar plot in the complex plane.

Generating the Nyquist Plot

For a Nyquist plot, a closed contour is mapped to the GH-plane. The Nyquist contour includes the negative imaginary axis, the positive imaginary axis, and a half-circle of arbitrarily large radius in the RHP of the s plane, as shown in Fig.

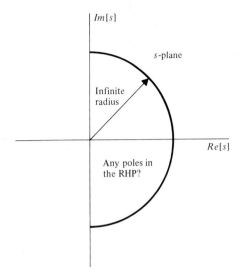

Figure 6.28 Boundary of the right half-plane.

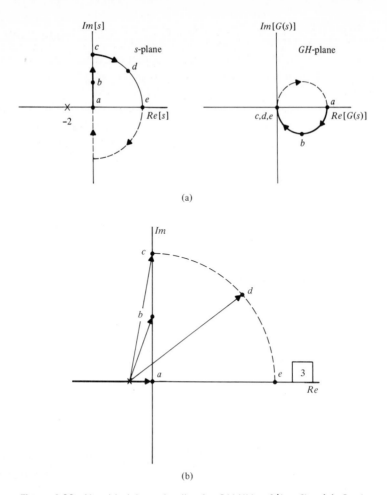

Figure 6.29   Nyquist plot construction for $G(s)H(s) = 3/(s + 2)$.   **(a)** Contour mapping.   **(b)** Magnitude and angle calculations with directed line segments. Points on the s-plane are not to scale.

6.29a. A polar plot of $G(s)H(s)$ for values of s along the positive imaginary axis and the upper part of the arbitrarily large circular path is shown as a solid curve on the GH plane in that figure. The magnitude and angle of GH for selected points on the s-plane contour were calculated graphically, using directed line segments in Fig. 6.29b. The dashed mirror image of this top portion of the closed Nyquist contour in the s plane produces the dashed mirror image GH-plane curve, as in the figure, because changing the imaginary part of s maintains the magnitude of $G(s)H(s)$ but changes the sign of the phase angle. A point a in the s plane is mapped to $G(a)H(a)$ in the $G(s)H(s)$ plane where that mapped point is labeled a.

One sketches a Nyquist plot by calculating, graphically, or otherwise, the magnitude and angle of $G(s)H(s)$ for several representative points s along the upper portion of the semicircular closed contour in the s plane. Then that portion of the plot is reflected across the real axis in the GH plane to form the closed curve which is the Nyquist plot. It is helpful to indicate with arrows the sense of traversal of the Nyquist plot for clockwise traversal of the contour in the s plane.

If $G(s)H(s)$ has a root on the imaginary axis, as in the example of Fig. 6.30, the path in the s plane is modified to make circular detours of arbitrarily small radius into the RHP around the roots. Figure 6.31 shows another example Nyquist plot in which small detours are made around a complex conjugate set of poles. In this case, the Nyquist plot in the GH plane extends from point a to point b, through c to

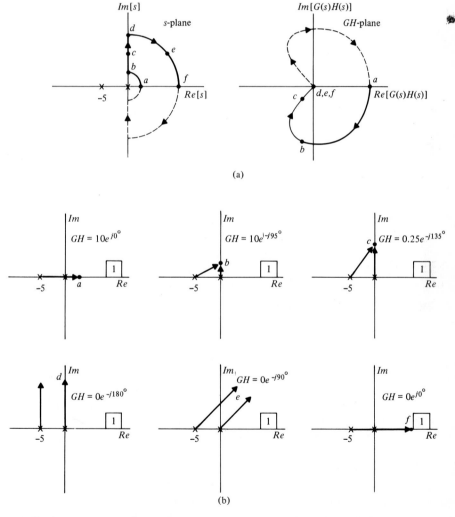

(a)

(b)

Figure 6.30 Nyquist plot construction for $G(s)H(s) = 1/s(s + 5)$. **(a)** Contour mapping. **(b)** Pole-zero calculations.

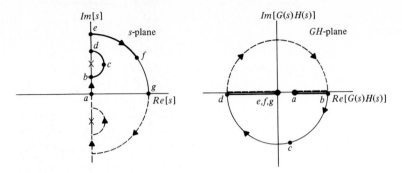

Figure 6.31 Nyquist plot construction for $G(s)H(s) = 4/(s^2 + 4)$. The curves are not precisely to scale.

points d, e, and f, just below the real axis. The corresponding mirror-image portions of the plot are above the real axis, as shown.

Other examples of Nyquist plots are given in Table 6.5. The closed path in the s plane always results in a closed curve in the GH plane. It should perhaps be emphasized that very few people begin with a natural talent for rapidly sketching Nyquist plots. A careful step-by-step procedure, involving the calculation or estimation of $G(s)H(s)$ for a number of points s, is generally the best approach.

Interpreting the Nyquist Plot

Nyquist showed that a relationship exists between the way the Nyquist plot of the open-loop transmittance $G(s)H(s)$ encircles the -1 point and the number of RHP poles of the closed-loop system. An important value is the number of clockwise (CW) encirclements of the -1 point. In Fig. 6.32 that value can be established by looking at a segment of the Nyquist plot in the fourth quadrant of the GH plane. A vector can be drawn outward from the -1 point. If the vector is crossed by the Nyquist plot twice in a CW direction and once in a counterclockwise (CCW) direction, the result is interpreted as one CW encirclement. CCW encirclements are considered to be negative.

Nyquist proved that the number of RHP poles of the closed-loop transfer function $T(s)$,

$$T(s) = \frac{G(s)}{1 + G(s)H(s)}$$

is given by

$$\begin{pmatrix} \text{Number of CW} \\ \text{encirclements of} \\ \text{the } -1 \text{ point on} \\ \text{the complex plane} \end{pmatrix} = \begin{pmatrix} \text{number of RHP} \\ \text{poles of } T(s) \end{pmatrix} - \begin{pmatrix} \text{number of RHP} \\ \text{poles of } G(s)H(s) \end{pmatrix}$$

Table 6.5 SOME OTHER NYQUIST PLOT EXAMPLES

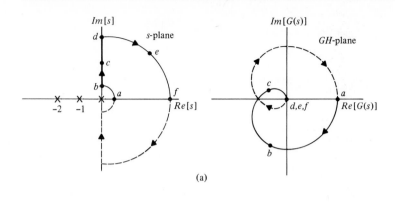

(a)

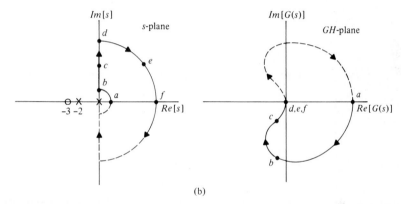

(b)

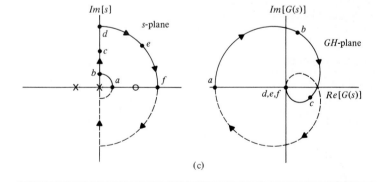

(c)

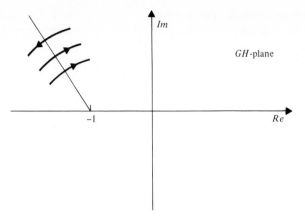

Figure 6.32 Segment of a Nyquist plot used to interpret the number of CW encirclements of −1. (Here there is one CW encirclement.)

Or

$$\begin{pmatrix} \text{Number of RHP} \\ \text{poles of } T(s) \end{pmatrix} = \begin{pmatrix} \text{number of CW} \\ \text{encirclements of} \\ \text{the } -1 \text{ point on} \\ \text{the } GH \text{ plane} \end{pmatrix} + \begin{pmatrix} \text{number of RHP} \\ \text{poles of } G(s)H(s) \end{pmatrix}$$

For example, a system with

$$G(s)H(s) = \frac{2(s + 3)}{(s + 2)^2(s - 1)}$$

has the Nyquist plot given in Fig. 6.33. $G(s)H(s)$ has one RHP pole, but the Nyquist plot circles the −1 point once in a CCW sense, so

$$\begin{pmatrix} \text{Number of RHP} \\ \text{poles of } T(s) \end{pmatrix} = (-1) + (1) = 0$$

and the overall system is stable.

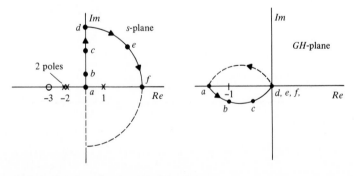

Figure 6.33 Applying the Nyquist criterion.

Table 6.6 shows a number of Nyquist plots for systems with rational transmittances. By counting encirclements of the -1 point, it is possible to conclude whether or not each system is stable. If a system is unstable, the number of RHP closed-loop poles can easily be computed. The results are:

a, b, c, d always stable
e, f stable or unstable with two closed-loop RHP poles
g always stable
h always unstable with one closed-loop RHP pole
i always unstable with two closed-loop RHP poles
j always stable

Several other Nyquist plots are shown in Fig. 6.34 for systems of the form

$$G(s)H(s) = \frac{Ke^{-s}}{s}$$

Table 6.6   A COLLECTION OF NYQUIST PLOTS

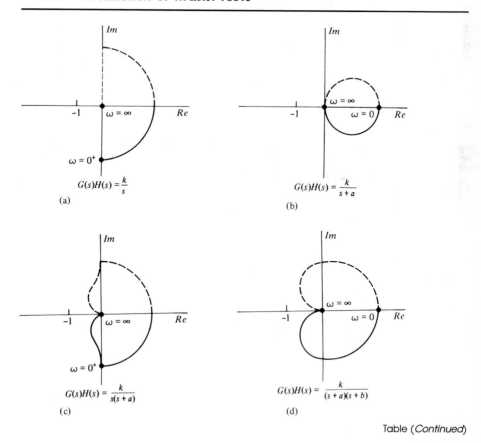

$$G(s)H(s) = \frac{k}{s}$$
(a)

$$G(s)H(s) = \frac{k}{s+a}$$
(b)

$$G(s)H(s) = \frac{k}{s(s+a)}$$
(c)

$$G(s)H(s) = \frac{k}{(s+a)(s+b)}$$
(d)

Table (*Continued*)

Table 6.6 (Continued)

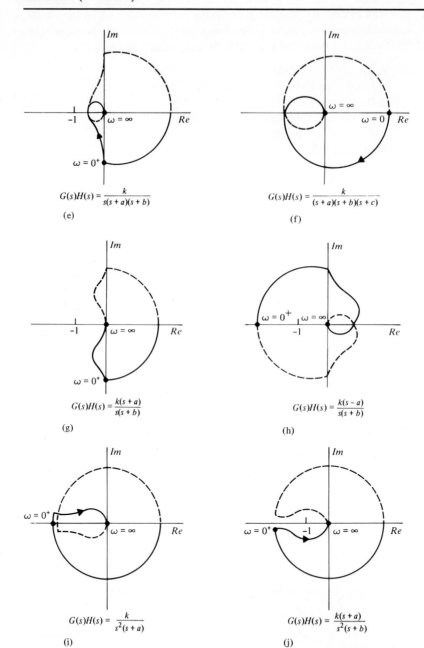

$$G(s)H(s) = \frac{k}{s(s+a)(s+b)}$$

(e)

$$G(s)H(s) = \frac{k}{(s+a)(s+b)(s+c)}$$

(f)

$$G(s)H(s) = \frac{k(s+a)}{s(s+b)}$$

(g)

$$G(s)H(s) = \frac{k(s-a)}{s(s+b)}$$

(h)

$$G(s)H(s) = \frac{k}{s^2(s+a)}$$

(i)

$$G(s)H(s) = \frac{k(s+a)}{s^2(s+b)}$$

(j)

for various values of the positive constant K. For sufficiently large K, this system's Nyquist plot circles the -1 point of the GH plane several times, as in Fig. 6.34a, indicating the presence of several RHP poles in the overall system transfer function $T(s)$. For a smaller value of K, the Nyquist plot is given in Fig. 6.34b with a single CW encirclement of the -1 point, indicating one RHP pole in $T(s)$. For $K = \pi/2$, Fig. 6.34c, the Nyquist curve passes through the -1 point and $T(s)$ has imaginary axis poles. For smaller K, the Nyquist plot is as in Fig. 6.34d and $T(s)$ is stable.

The gain and phase margins of simple feedback systems can be easily determined from their Nyquist plots. The negative real axis of the GH plane

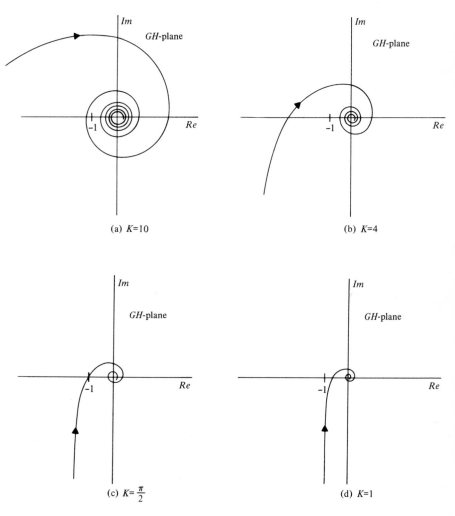

(a) $K=10$

(b) $K=4$

(c) $K=\dfrac{\pi}{2}$

(d) $K=1$

Figure 6.34 Nyquist plots for systems involving a time delay. For readability only the positive frequency part of the plot is drawn. **(a)** $K = 10$. **(b)** $K = 4$. **(c)** $K = \pi/2$. **(d)** $K = 1$.

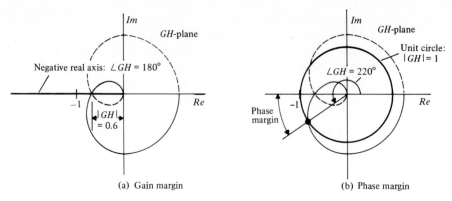

(a) Gain margin   (b) Phase margin

Figure 6.35   Gain and phase margins from the Nyquist plot. **(a)** Gain margin. **(b)** Phase margin.

represents an angle of 180°, so any crossing of that line by the polar frequency response portion of the Nyquist curve is at a phase crossover frequency. As indicated in Fig. 6.35a, the magnitude of the loop gain is the polar distance to the crossing, 0.6 in the example. In decibels, the gain margin is

$$20 \log_{10}\left(\frac{1}{0.6}\right) = 4.44 \text{ dB}$$

The unit circle on the GH plane represents unit magnitude, so any crossing of that circle by the Nyquist polar frequency response is at a gain crossover frequency. For the example of Fig. 6.35b, the angle of GH at the gain crossover frequency is 220° or $-140°$, corresponding to a phase margin of 40°.

DRILL PROBLEMS

D6.9.   Sketch Nyquist plots for feedback systems with the following loop transmittances, then use the plots to determine whether or not each system is stable:

(a) $G(s)H(s) = \dfrac{s}{s+4}$

Ans   stable

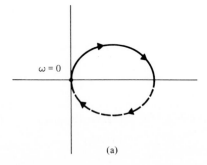

(a)

(b) $G(s)H(s) = \dfrac{10}{(s + 2)(s + 6)}$

Ans stable

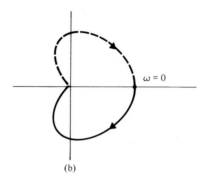

(b)

(c) $G(s)H(s) = \dfrac{s^2}{s^2 + 2s + 10}$

Ans stable

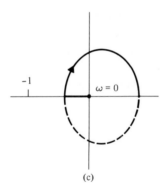

(c)

(d) $G(s)H(s) = \dfrac{2}{s^2(s + 3)}$

Ans unstable

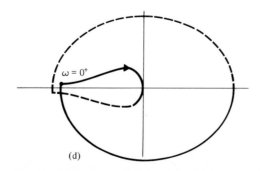

(d)

D6.10. For the following feedback systems, sketch Nyquist plots and use them to determine whether or not the system is stable.

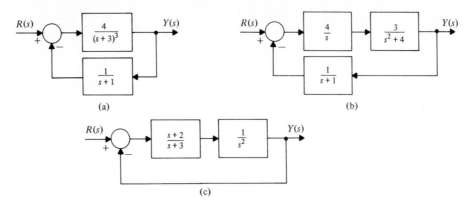

(a)

(b)

(c)

Ans (a) stable;   (b) unstable;   (c) stable

6.7 RELATION BETWEEN ROOT LOCUS, TIME DOMAIN, AND FREQUENCY DOMAIN

A control system is usually designed using either the root locus, the time response, or the Bode/Nyquist plot (frequency domain). In the root locus, the system gain varies. For the time response, time is varied. In the Bode/Nyquist plot, frequency varies. Although the three approaches portray three different views of a control system, relationships exist among these methods.

Figure 6.36 shows a second-order control system. A higher-order system will be far more complex. However, if two dominant roots exist that are much closer to the vertical axis than any other, Fig. 6.36 approximately describes that higher-order system also.

The closed-loop transfer function for Fig. 6.36 has the standard form including the damping ratio ζ and the undamped natural frequency ω_n.

$$T(s) = \frac{Y(s)}{R(s)} = \frac{\omega_n^2}{s^2 + 2\zeta\omega_n s + \omega_n^2}$$

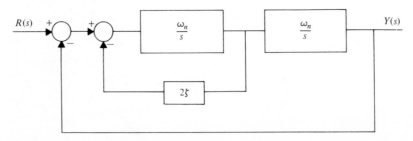

Figure 6.36   Second-order system.

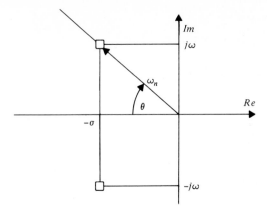

Figure 6.37 Complex conjugate poles.

If the damping ratio is between zero and 1, complex conjugate poles of $T(s)$ result.

$$s^2 + 2\zeta\omega_n s + \omega_n^2 = (s + \sigma + j\omega)(s + \sigma - j\omega)$$
$$= s^2 + 2\sigma s + \sigma^2 + \omega^2$$

Figure 6.37 shows the complex conjugate closed-loop poles for $T(s)$ denoted by a box. From the figure and the characteristic polynomial, the damping ratio and the angle are related.

$$\zeta = \frac{\sigma}{\omega_n} = \cos\phi$$

Notice that ω_n is the length of the vector from the origin to the complex conjugate poles. Figure 6.38 shows loci of constant ζ with increasing ω_n (straight lines projecting outward) and loci of constant ω_n with increasing ζ (semicircular arcs).

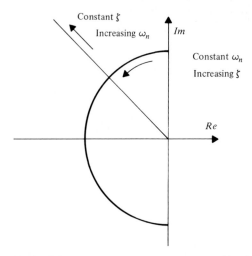

Figure 6.38 Changes in damping ratio and undamped natural frequency.

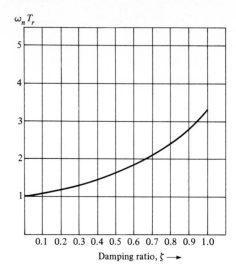

Figure 6.39 Rise time versus damping ratio.

Figures 6.39 and 6.40 were discussed in Section 5.2. It was pointed out that when damping ratio is constant, both rise time and settling time decrease as undamped natural frequency increases. For fixed undamped natural frequency, it was stated that increased damping ratio causes rise time to increase. Settling time is a minimum for a damping ratio of 0.7.

The above trends may be related to the frequency domain.

Figure 6.36 may be redrawn as in Fig. 6.41 where

$$G(j\omega) = \frac{\omega_n^2}{j\omega(j\omega + 2\zeta\omega_n)}$$

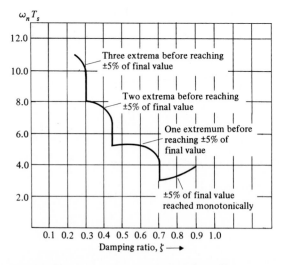

Figure 6.40 Settling time versus damping ratio.

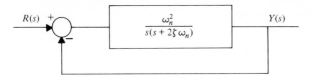

Figure 6.41 Second-order system redrawn.

The gain crossover frequency requires

$$|G(j\omega_\phi)| = 1$$

The phase margin is then

$$\text{PM} = 180° + \text{phase of } G(j\omega_\phi)$$

From the magnitude condition

$$\omega_n^2 = \omega_\phi\sqrt{\omega_\phi^2 + (2\zeta\omega_n)^2}$$

$$\omega_\phi^4 + 4\zeta^2\omega_n^2\omega_\phi^2 - \omega_n^4 = 0$$

The roots of the last expression follow by applying the quadratic formula in terms of ω_ϕ^2:

$$\omega_\phi^2 = \omega_n^2(-2\zeta^2 \pm \sqrt{4\zeta^4 + 1})$$

For ω_ϕ to be real-valued, the positive root must be used so that

$$\omega_\phi = \omega_n k$$

$$k = \sqrt{\sqrt{4\zeta^4 + 1} - 2\zeta^2}$$

Figure 6.42 shows a plot of k versus ζ. For most values of ζ, k is near unity so that the gain crossover frequency and the undamped natural frequency are closely related.

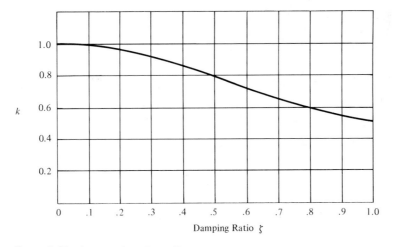

Figure 6.42 k versus damping ratio.

The phase margin becomes

$$PM = 180° - 90° - \tan^{-1}\left(\frac{k}{2\zeta}\right)$$

$$= 90° - \tan^{-1}\left(\frac{k}{2\zeta}\right)$$

$$= \tan^{-1}\left(\frac{2\zeta}{k}\right)$$

The last result is due to a trigonometric identity. Figure 6.43 shows phase margin in degrees versus ζ. Phase margin is directly proportional to damping ratio. For damping ratios less than 0.5, the phase margin (in degrees) is about 100 times the damping ratio. The slope of phase margin versus damping ratio diminishes as damping ratio varies from 0.5 to 1.

To summarize, for most values of damping ratio the gain crossover frequency and the undamped natural frequency are nearly equal. The phase margin is directly (and almost linearly) related to the damping ratio. Therefore, when a frequency domain design holds phase margin constant while increasing the gain crossover frequency, the resulting rise time and settling time would diminish in the time domain and the root locus would move outward along a line of constant damping ratio as in Fig. 6.38.

Good controls design involves examination of all three evaluative tools: root locus, time domain, and frequency domain. The above discussion demonstrates (approximately) how frequency domain design influences the other two portraits.

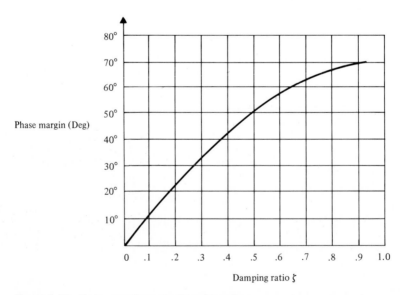

Figure 6.43 Phase margin versus damping ratio.

DRILL PROBLEM

D6.11. For each of the following unity feedback systems, a forward path transmittance is given. Compute the gain crossover frequency and the phase margin. Use Figs. 6.42 and 6.43 to approximate the damping ratio and the undamped natural frequency for the dominant closed-loop roots. Compare with the exact values.

(a) $G(s) = \dfrac{9}{s(s + 3)}$

  Ans 2.36 rad/sec, 51.8°, 0.5, 3 rad/sec, 0.5, 3 rad/sec

(b) $G(s) = \dfrac{100}{s(s + 3)(s + 9)}$

  Ans 2.66 rad/sec, 31.9°, 0.28, 2.89 rad/sec, 0.27, 3.11 rad/sec

(c) $G(s) = \dfrac{30(s + 20)}{s(s + 3)(s + 9)}$

  Ans 7.15 rad/sec, 4°, 0.04, 7.2 rad/sec, 0.03, 7.20 rad/sec

6.8 COMPENSATION USING BODE PLOTS

Figure 6.44 appeared in Chap. 5, where the influence of compensators upon the root locus was considered. The same compensators may be analyzed using the Bode plot concepts of this chapter. A designer may decide to adjust a gain K so as to provide an acceptable error coefficient, gain crossover frequency, phase margin, phase crossover frequency and gain margin. The resulting control system of Fig. 6.44a is called uncompensated in that only a gain K and a unity feedback structure are used.

If the uncompensated system cannot be designed to meet specifications, a cascade compensator (Fig. 6.44b) or a feedback compensator (Fig. 6.44c) may be selected. Table 6.7 contains a representative sample of the types of compensators often employed. These same compensators were considered in Chap. 5 as each affected root locus behavior. Now interest is focused on Bode plot behavior. Of course, the influence on steady state accuracy should always be considered (see Chap. 5 for the steady state effects of these compensators).

A cascade integral-plus-proportional compensator or a cascade lag compensator provides a negative phase angle where s is replaced by $j\omega$. Since the phase angle of the compensated system becomes more negative, phase margin tends to be reduced for fixed gain crossover frequency or gain crossover frequency tends to be reduced for fixed phase margin. These compensators are primarily intended for improvement of steady state accuracy.

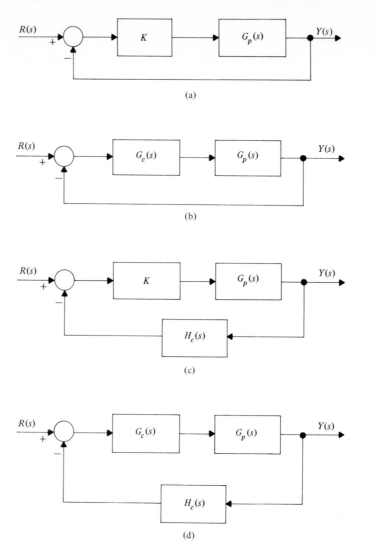

Figure 6.44 Compensator configurations. **(a)** Uncompensated unity feedback system. **(b)** Cascade compensated system. **(c)** Feedback compensated system. **(d)** Feedback and cascade compensated system.

The cascade lead compensator provides a positive phase angle. Therefore phase margin may be increased for fixed gain crossover frequency or gain crossover frequency may be increased for fixed phase margin.

The cascade lag-lead compensator combines the attributes of the lag and lead compensators so that improvements in steady state accuracy and in stability (through the gain crossover frequency and the phase margin) are possible.

Table 6.7 COMMON TYPES OF COMPENSATORS

Compensator	Transmittance	Typical Effect on Steady State Errors	Typical Effect on Phase Margin and Gain Crossover Frequency
Cascade integral-plus-proportional	$G_c(s) = \dfrac{K(s + a)}{s}$	Greatly improved	Reduced
Cascade lag	$G_c(s) = \dfrac{K(s + a)}{s + b}$ $b < a$	Improved	Reduced
Cascade lead	$G_c(s) = \dfrac{K(s + a)}{s + b}$ $a < b$	Somewhat improved or somewhat worse	Increased
Cascade lag-lead	$G_c(s) = K\left(\dfrac{s + a}{s + b}\right)_{\text{lag}}$ $\times \left(\dfrac{s + a}{s + b}\right)_{\text{lead}}$	Improved	Increased
Feedback rate	$H_c(s) = 1 + As$	Somewhat improved or somewhat worse	Increased

The feedback rate compensator provides a positive phase angle. The same benefits as with lead compensation results. Parameter selection follows by recognizing that the rate feedback compensator $H_c(s)$ represents the combination of two loops, one containing K and one containing both K and A.

6.9 UNCOMPENSATED SYSTEM

The effects of the compensators of Table 6.7 will be evaluated using the same uncompensated system as in Chap. 5. See Table 6.8. That is

$$G_p(s) = \frac{1}{s(s^2 + 4s + 5)}$$

With K equal to 3, the ramp error coefficient is 0.6. Figure 6.45 shows the Bode plot for the uncompensated system. The gain crossover frequency is at 0.576 rad/sec. The phase margin is 63.7°. Compensation may be applied.

Table 6.8 SUMMARY OF DESIGNS

Compensator	Transmittance	Ramp Error Coefficient	Gain Crossover Frequency	Phase Margin
Uncompensated	$K = 3$	0.6	0.58	63.7°
Cascade integral-plus-proportional	$G_c(s) = \dfrac{3(s + 0.0576)}{s}$	∞	0.58	57.8°
Cascade lag	$G_c(s) = \dfrac{3(s + 0.0576)}{s + 0.00576}$	6	0.58	58.5°
Cascade lead	$G_c(s) = \dfrac{60(s + 0.576)}{s + 5.76}$	1.2	2.45	47.9°
Cascade lag-lead	$G_c(s) = \dfrac{60(s + 0.0576)}{s + 0.00576}$ $\times \dfrac{s + 0.576}{s + 5.76}$	12	2.45	46.7°
Rate feedback	$H_c(s) = \frac{1}{2}(s + 2)$ $K = 10$	1.0	1.81	55.5°

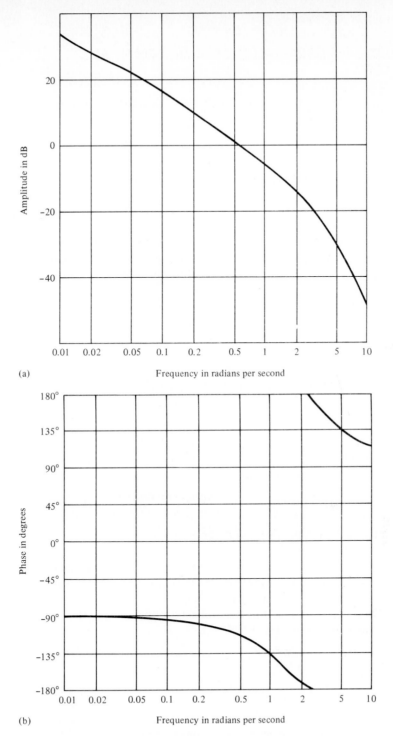

(a)

Amplitude in dB

Frequency in radians per second

(b)

Phase in degrees

Frequency in radians per second

Figure 6.45 Uncompensated system Bode plot. **(a)** Amplitude versus frequency. **(b)** Phase versus frequency.

6.10 CASCADE INTEGRAL PLUS PROPORTIONAL AND CASCADE LAG

The Bode plot for the cascade integral-plus-proportional compensator follows by replacing s with $j\omega$:

$$G_c(j\omega) = \frac{K(1 + j\omega/a)}{j\omega/a}$$

Similarly, the cascade lag compensator is as follows, where

$$G_c(j\omega) = \frac{Kd(1 + j\omega/a)}{1 + j\omega\, d/a}$$

As d rises toward infinity, the two compensators become equal (although the RC elements of the lag compensator would be very difficult to accurately realize).

Figures 6.46 and 6.47 contain the Bode plots for these two compensators. For convenience, the frequency axis is normalized by using ω/a. Also, K is set equal to 1. It would be a simple matter to include other values of K because all dB values would be increased by the dB values of K.

Each compensator provides negative phase angles. In fact, the term "lag" compensator results because negative phase angles provide a steady state sinusoidal output that "lags" behind a sinusoidal input of the same frequency.

Because each compensator creates a negative phase angle, the phase margin would be decreased for the combination of compensator and plant if K is adjusted so the gain crossover frequency remains unchanged. When the compensator is added:

Phase of $G_c(j\omega_\phi)G_p(j\omega_\phi)$ = phase of $G_c(j\omega_\phi)$ + phase of $G_p(j\omega_\phi)$

Phase margin (compensated) = $180°$ + phase of $G_p(j\omega_\phi)$ + phase of $G_c(j\omega_\phi)$

$$= \text{phase margin (uncompensated)}$$
$$+ \text{ phase of } G_c(j\omega_\phi)$$

If the phase margin must be maintained, then ω_ϕ must be reduced from the uncompensated value so the phase angle of the plant is increased by the same amount as the negative phase angle of the compensator. In either case, the stability of the system is worse. However, these compensators are primarily intended to improve steady state accuracy (see Chap. 5).

One way to reduce the destabilizing influence that the compensators have upon the closed-loop system is to select

$$a = \frac{\omega_\phi(\text{uncompensated})}{10}$$

The integral plus proportional compensator then has a phase angle at ω_ϕ of

$$\tan^{-1} 10 - 90° = -5.7°$$

If $d = 10$, the lag compensator has a phase angle at ω_ϕ of

$$\tan^{-1} 10 - \tan^{-1} 100 = -5.1°$$

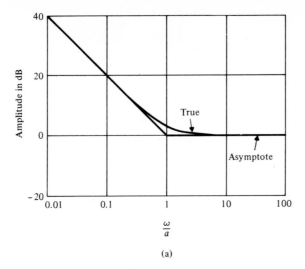

(a)

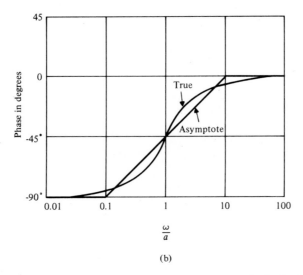

(b)

Figure 6.46   Integral plus proportional compensator Bode plot.

$$G'_c(j\omega) = \frac{1 + j\omega/a}{j\omega/a}$$

(a) Amplitude versus frequency.   **(b)** Phase versus frequency.

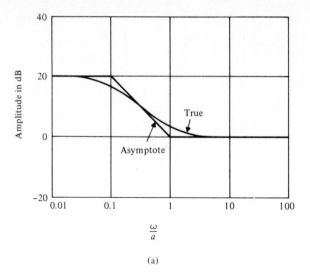

(a)

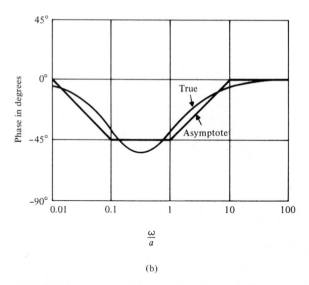

(b)

Figure 6.47 Lag compensator Bode plot ($d = 10$).

$$G_c'(j\omega) = \frac{10(1 + j\omega/a)}{1 + j10\omega/a}$$

(a) Amplitude versus frequency. **(b)** Phase versus frequency.

If the gain crossover frequency remains unchanged, phase margin is reduced by less than 6°.

For the example system

$$a = \frac{0.576}{10} = 0.0576$$

The resulting integral plus proportional compensator is

$$G_c(s) = \frac{K(s + 0.0576)}{s}$$

A value for K must somehow be chosen even though the Bode plot cannot be plotted without knowing K. One way out of that problem is to factor out K from the compensator

$$G_c(s) = KG_c'(s)$$

The compensated system

$$G_c(s)G_p(s) = KG_c'(s)G_p(s)$$

can be analyzed from the Bode plot of

$$G_c'(s)G_p(s)$$

Since K is usually positive, the phase with or without K is usually the same. If the designer selects a new gain crossover frequency (ω_{ϕ_1}) so as to create some phase margin, then K must be

$$K = \frac{1}{|G_c'(j\omega_{\phi_1})G_p(j\omega_{\phi_1})|}$$

If the designer selects K for some other reason (perhaps to create an error coefficient), then the new gain crossover frequency becomes such that

$$|G_c'(j\omega_{\phi_1})G_p(j\omega_{\phi_1})| = \frac{1}{K}$$

Figure 6.48 shows the integral-plus-proportional compensated system Bode plot with K factored out. If K is chosen to remain at the uncompensated value of 3, the gain crossover frequency (where the magnitude is $\frac{1}{3}$ or -9.5 dB) becomes 0.58 rad/sec with a phase margin of 57.8°. The phase margin is 6° less than for the uncompensated system.

For the lag compensator, suppose d is chosen to be 10 (perhaps the error coefficient is to be increased by a factor of 10). Then

$$G_c(s) = \frac{K(s + 0.0576)}{(s + 0.00576)}$$

Figure 6.49 shows the Bode plot of $G_c'(s)G_p(s)$. If K is chosen to be 3, the ramp error coefficient becomes 6. The gain crossover frequency (where the magnitude is $\frac{1}{3}$ or -9.5 dB) is 0.58 rad/sec with a phase margin of 58.5°. The phase margin is 5° less than for the uncompensated system.

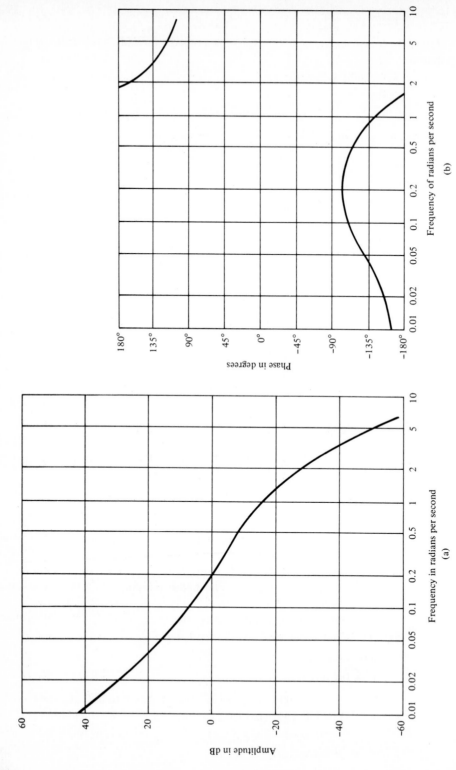

Figure 6.48 Bode plot of $G_c'(j\omega)G_p(j\omega)$ for integral plus proportional compensation. **(a)** Amplitude versus frequency. **(b)** Phase versus frequency.

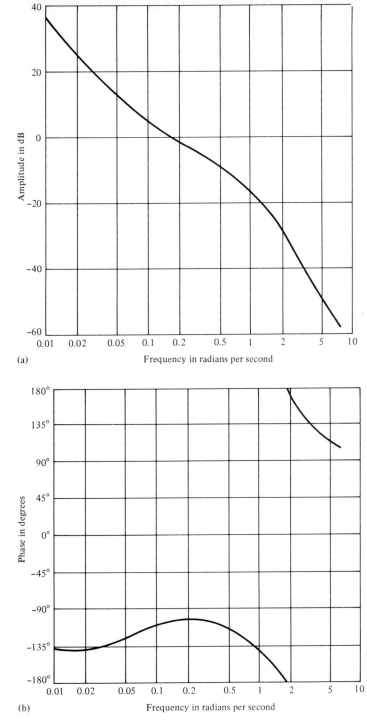

Figure 6.49 Bode plot of $G'_c(j\omega)G_p(j\omega)$ for lag compensation. **(a)** Amplitude versus frequency. **(b)** Phase versus frequency.

6.11 CASCADE LEAD COMPENSATION

A cascade lead compensator has the following form if d is defined to be b/a:

$$G_c(j\omega) = \frac{K}{d} \frac{(1 + j\omega/a)}{(1 + j\omega/(ad))}$$

Figure 6.50 shows the Bode plot of a lead compensator where K equals 1 and where the frequency axis is normalized to ω/a.

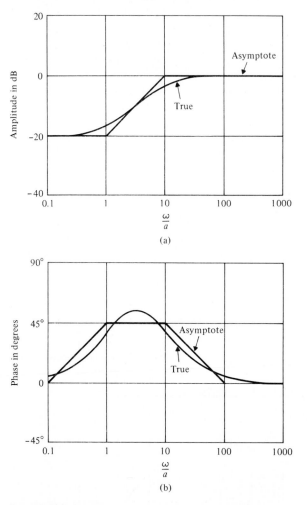

Figure 6.50   Lead compensator Bode plot ($d = 10$).

$$G_c'(j\omega) = \frac{0.1(1 + j\omega/a)}{1 + j0.1\omega/a}$$

(a) Amplitude versus frequency.   **(b)** Phase versus frequency.

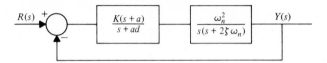

Figure 6.51 Lead compensation of the uncompensated dominant roots.

The phase angle is always positive which is why this compensator is called a "lead" compensator. A sinusoidal steady state output would "lead" a sinusoidal input of the same frequency. Since the phase angle is always positive, the phase margin is increased in the compensated case if the gain crossover frequency is unchanged. That fact follows from the same analysis as when lag compensators were discussed. If the same phase margin is needed, then the gain crossover frequency would increase.

A designer must select values for K, a, and d. The system of Fig. 6.51 is useful for providing some general clues as to Bode design procedure in the same way that the system of Fig. 6.51 provided guidance for root locus design. Suppose a damping ratio of 0.7 occurs for the uncompensated system dominant roots. Figure 6.42 suggests (for the uncompensated case) that

$$\frac{\omega_\phi}{\omega_n} = 0.65$$

The ramp error coefficient with the lead compensator included remains the same if K equals d. The Bode plot of Fig. 6.52 shows $G_c(s)G_p(s)$ where $a/\omega_n = 0.63$, $K = d = 10$, and where the frequency axis is normalized by using ω/ω_n. The compensated normalized gain crossover frequency and phase margin are

$$\frac{\omega_{\phi_1}}{\omega_n} = 1.12$$

Phase margin $= 101.9°$

If another value of a is chosen, a lower phase margin results. By trial and error, the best value of a (most phase margin) may be chosen for various damping ratios. The results are in Table 6.9. In general the best choice of a is approximately

$$a \cong \omega_\phi(\text{uncompensated})$$

For the example system

$$a \cong 0.576$$

This is only a starting point for what is often an iterative design procedure. The complexity of a given system or other criteria could conspire to require an adjustment to this value. If d equals 10, the lead compensator is

$$G_c(s) = \frac{K(s + 0.576)}{s + 5.76}$$

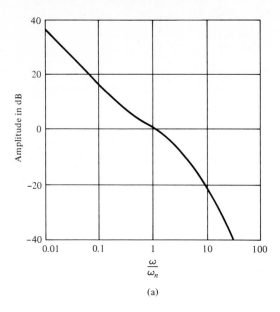

(a)

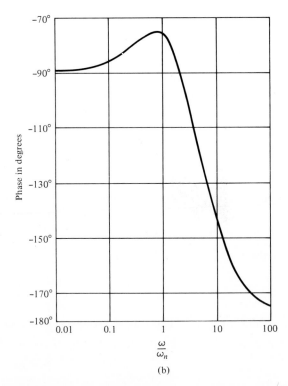

(b)

Figure 6.52   Bode plot for the system in Fig. 6.51 where the damping ratio = 0.7; $a/\omega_n = 0.63$; $K = d = 10$.   **(a)** Amplitude versus frequency.   **(b)** Phase versus frequency.

Table 6.9   LEAD COMPENSATION (BEST PHASE MARGIN)

Damping Ratio	Uncompensated System	Compensated System			$\dfrac{a}{\omega_\phi\text{(uncomp)}}$
	ω_ϕ/ω_n	a/ω_n	ω_ϕ/ω_n	Phase Margin	
0.3	0.91	0.65	1.52	75.3°	0.71
0.4	0.85	0.67	1.41	82.3°	0.78
0.5	0.79	0.67	1.31	89.2°	0.85
0.6	0.72	0.66	1.20	95.8°	0.92
0.7	0.65	0.63	1.12	101.9°	0.97

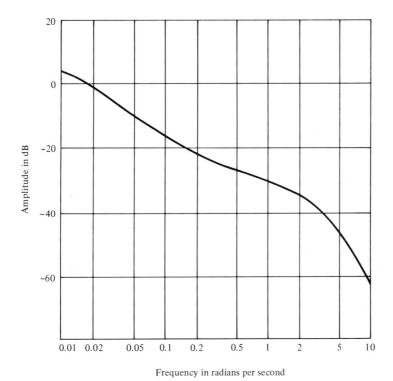

Frequency in radians per second

(a)

Figure 6.53   Bode plot of $G_c'(j\omega)G_p(j\omega)$ for lead compensation.   **(a)** Amplitude versus frequency.

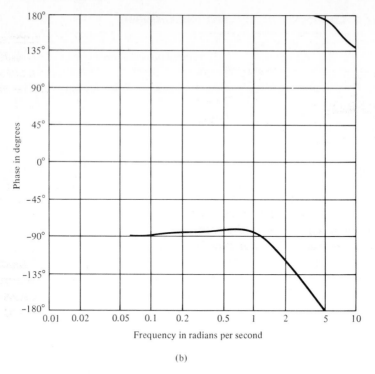

(b)

Figure 6.53 Bode plot of $G_c'(j\omega)G_p(j\omega)$ for lead compensation. **(b)** Phase versus frequency.

The Bode plot for $G_c'(s)G_p(s)$ is shown in Fig. 6.53. The ramp error coefficient can be doubled using $K = 60$. The gain crossover frequency (magnitude equals $\frac{1}{60}$ or -36 dB) is 2.45 rad/sec with a phase margin of 47.9°. The gain crossover frequency is about four times the uncompensated values while the phase margin is about 15° less. The phase margin could be increased by decreasing the gain crossover frequency.

6.12 LAG-LEAD COMPENSATION

The lag and lead compensators can be combined with a common value of d being used. With d equal to 10 the result here is

$$G_c(s) = \frac{K(s + 0.0576)(s + 0.576)}{(s + 0.00576)(s + 5.76)}$$

Figure 6.54 shows the Bode plot for $G_c'(s)G_p(s)$. The ramp error coefficient can be increased by a factor of 10 (compared to the lead case) and by a factor of 20 (compared to the uncompensated case) if K equals 60. The gain crossover

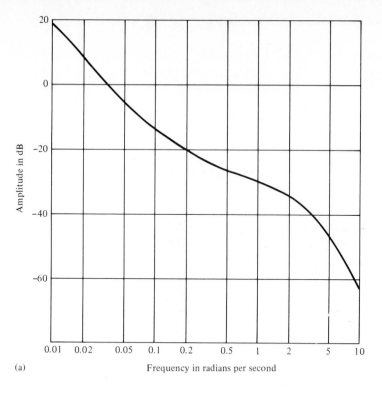

(a) Frequency in radians per second

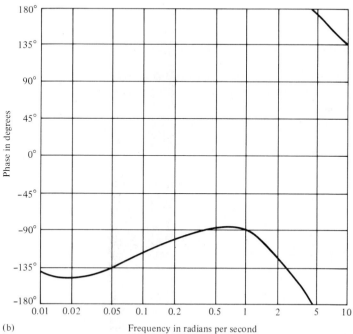

(b) Frequency in radians per second

Figure 6.54 Bode plot of $G_c'(j\omega)G_p(j\omega)$ for lag-lead compensation. **(a)** Amplitude versus frequency. **(b)** Phase versus frequency.

frequency occurs where the magitude is $\frac{1}{60}$ or -36 dB (2.45 rad/sec with a phase margin of 46.7°). The gain crossover frequency and phase margin are essentially the same as for the lead compensator; however, the ramp error coefficient is 10 times higher.

6.13 RATE FEEDBACK

The typical rate feedback system in Fig. 6.55 contains two feedback signals. The outer loop contains unity feedback while the inner loop provides rate of change of the output. Two parameters must be chosen: K and A.

One procedure with Bode methods is to stabilize the inner loop and then the entire system. The inner loop contains

$$KAsG_p(s)$$

the entire system then contains

$$KA\left(s + \frac{1}{A}\right)G_p(s)$$

For the example system, analysis and then design of the inner loop requires a close look at

$$\frac{KA}{s^2 + 4s + 5}$$

where one factor of s cancels. Figure 6.56 shows the Bode plot for the inner loop. Stability is ensured if

$$\frac{KA}{5} = 1$$

$$KA = 5$$

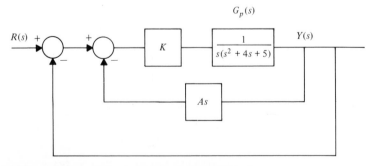

Figure 6.55 Rate feedback compensated system.

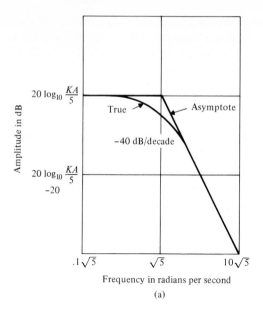

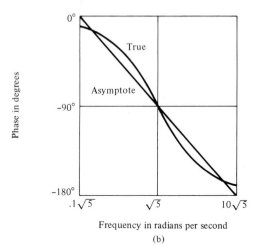

Figure 6.56 Bode plot for the inner loop of the system in Fig. 6.55.

$$\frac{KA}{s^2 + 4s + 5}$$

(a) Amplitude versus frequency. **(b)** Phase versus frequency.

because the magnitude can never exceed unity. For this inner loop, there is no gain crossover frequency. The magnitude when the phase is $-180°$ is negative infinity so the gain margin is infinite.

For the entire system, the designer must evaluate

$$\frac{KA(s + 1/A)}{s(s^2 + 4s + 5)}$$

where KA equals 5. Once $1/A$ is selected, the gain crossover frequency and the phase margin become fixed. The ramp error coefficient follows from Chap. 5. The computation of K_1 for the compensated system may operate upon compensated and uncompensated values.

$$K_1(\text{comp}) = \frac{K\,(\text{comp})\,K_1\,(\text{uncomp})}{K\,(\text{uncomp}) + AK\,(\text{comp})\,K_1\,(\text{uncomp})}$$

$$= \frac{(5/A)0.6}{3 + 5(0.6)}$$

$$= \frac{0.5}{A}$$

The last result can also be obtained for the compensated system using only compensated values

$$G_E(s) = \frac{KG_p(s)}{1 + AsKG_p(s)}$$

$$K_1\,(\text{comp}) = \lim_{s \to 0} sG_E(s)$$

Table 6.10 contains the gain crossover frequency, phase margin, and ramp error coefficient for several values of $1/A$.

As $1/A$ increases, the open-loop zero due to rate feedback moves to the left (higher break frequency). The ramp error coefficient and the gain crossover frequency increase but the phase margin decreases. If $1/A$ is 2, then K is 10, the gain crossover frequency is 1.81, the ramp error coefficient is 1, and the phase margin is 55.5°. As compared to the uncompensated case, the gain crossover frequency is

Table 6.10 RATE FEEDBACK DESIGN

$1/A$	K	Gain Crossover Frequency	Phase Margin	Ramp Error Coefficient
1	5	1.34	84.2°	0.5
2	10	1.81	55.5°	1.0
3	15	2.15	38.2°	1.5
4	20	2.41	26.2°	2.0
5	25	2.64	17.3°	2.5
6	30	2.83	10.4°	3.0

tripled. The ramp error coefficient is increased by more than 50%. The phase margin is reduced by 8°. That design would appear to provide a good balance for the trade-offs involved. Other considerations could favor other values of $1/A$, or the value of KA might have to be adjusted.

DRILL PROBLEMS

D6.12. Sketch the Bode plots for each of the following systems. From $KG_p(s)$ determine the gain crossover frequency and phase margin

(a) $KG_p(s) = \dfrac{32}{(s+4)^2}$

   Ans 4.00 rad/sec, 90°

(b) $KG_p(s) = \dfrac{16}{s(s+4)^2}$

   Ans 0.95 rad/sec, 63.4°

(c) $KG_p(s) = \dfrac{16(s+1)}{s^2(s+4)}$

   Ans 3.25 rad/sec, 33.8°

D6.13. For each of the uncompensated systems in D6.12 select a cascade integral-plus-proportional compensator to improve steady state accuracy while maintaining about the same gain crossover frequency and phase margin.

D6.14. For each of the uncompensated systems in D6.12 select a cascade lead compensator to improve steady state accuracy, gain crossover frequency and phase margin.

D6.15. For each of the uncompensated systems in D6.12 select a cascade lag-lead compensator to improve steady state accuracy, gain crossover frequency and phase margin.

D6.16. For each of the uncompensated systems in D6.12 select a rate feedback compensator to improve steady state accuracy, gain crossover frequency and phase margin.

6.14 FREQUENCY RESPONSE OF A FLEXIBLE SPACECRAFT

Chapter 4 contains a discussion of controlling a flexible spacecraft. Very large satellites extend solar panels to generate and store electrical energy. As a result, the satellite is very flexible. When a thruster fires to cause the satellite to change

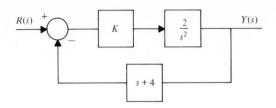

Figure 6.57 Rigid behavior.

position, the entire vehicle moves (rigid motion) about some center of gravity but individual parts of the vehicle bend and oscillate about local null points (flexible motion). Most aircraft and satellites have both rigid and flexible behaviors.

Figure 6.57 shows a hypothetical spacecraft where rigid behavior is included and flexible behavior is ignored. Figure 6.58 shows the spacecraft with flexible behavior included. The root locus for these systems concluded (in Chap. 4) that the flexible behavior has a strong destabilizing influence which must be included. Otherwise, the root locus would appear overly optimistic as a source of stability analysis. Frequency domain analysis is equally revealing.

From Fig. 6.57 (rigid behavior) a Bode plot must be constructed for

$$\frac{2K(s+4)}{s^2}$$

Suppose K is chosen to force the gain crossover frequency to be 4. Then

$$K = \frac{\omega_\phi^2}{2\sqrt{16 + \omega_\phi^2}} = \frac{16}{2\sqrt{32}} = \sqrt{2} = 1.414$$

The phase margin is

$$PM = 180° + \tan^{-1}(\omega_\phi/4) - 180° = 45°$$

The Bode plot appears in Fig. 6.59.

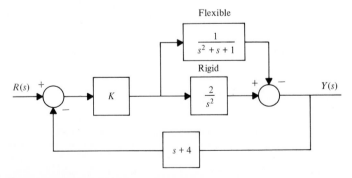

Figure 6.58 Flexible and rigid behavior.

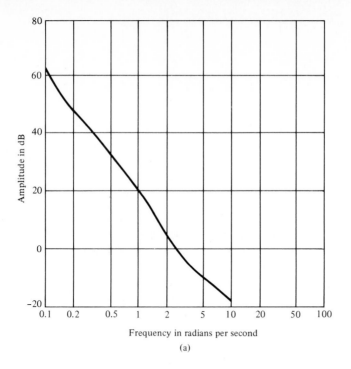

(a)

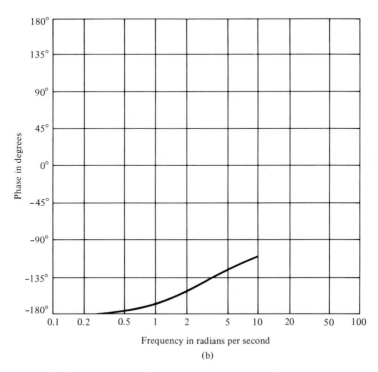

(b)

Figure 6.59 Bode plot for the system in Fig. 6.57. **(a)** Amplitude versus frequency. **(b)** Phase versus frequency.

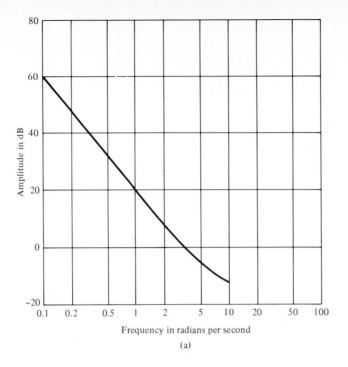

(a)

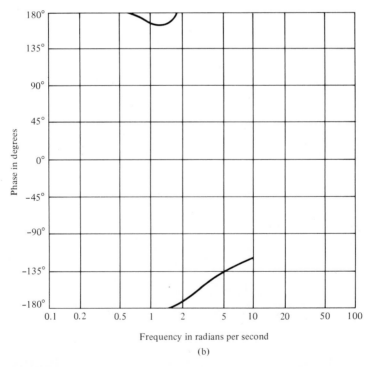

(b)

Figure 6.60 Bode plot for the system in Fig. 6.58. **(a)** Amplitude versus frequency. **(b)** Phase versus frequency.

From Fig. 6.58 (including flexible behavior) a Bode plot must be constructed for

$$\frac{1.414(s^2 + 2s + 2)(s + 4)}{s^2(s^2 + s + 1)}$$

The Bode plot appears in Fig. 6.60. The gain crossover frequency is 2.75 (not 4 as would result ignoring flexible behavior). The negative phase created by flexible behavior lowers the actual phase margin to 12.5°. The same conclusion results as before: flexible behavior must be included to obtain a true picture of stability.

6.15 COMPENSATING OPERATIONAL AMPLIFIERS

Single-chip integrated circuit operational amplifiers (op amps) are highly useful components for such applications as the synthesis of feedback system controllers. The type 101 and type 741 op amps have been widely used in industry and are, in a sense, standards to which other op amps are compared. Many other type numbers have been marketed, each improved or optimized for one application or another. And, multiple operational amplifiers are now routinely integrated on semiconductor chips with other electronic devices. Because of their wide acceptance and application, the discussion to follow will concern the 101 and 741 types, although the methods and results apply to most other types as well.

When an op amp is connected as shown in Fig. 6.61a, it forms a noninverting amplifier of the input voltage V_{in}, with gain fixed primarily by the external resistors R_F and R_A. To a first approximation, the operational amplifier itself is a high-gain voltage-differencing amplifier with a model given in Fig. 6.61b. For this circuit,

$$\begin{cases} V_a = \dfrac{R_A}{R_A + R_F} V_{out} \\ V_b = V_{in} \\ V_{out} = A(V_b - V_a) \end{cases}$$

These relations are represented as a feedback system in Fig. 6.61c.

The transfer function of the noninverting op amp configuration is, applying the feedback relation to Fig. 6.61c,

$$T = \frac{A}{1 + AR_A/(R_A + R_F)} = \frac{R_A + R_F}{R_A + (R_A + R_F)/A}$$

For sufficiently large open-loop op amp gain A,

$$T \cong \frac{R_A + R_F}{R_A} = 1 + \frac{R_F}{R_A}$$

and

$$V_{out} \cong \left(1 + \frac{R_F}{R_A}\right) V_{in}$$

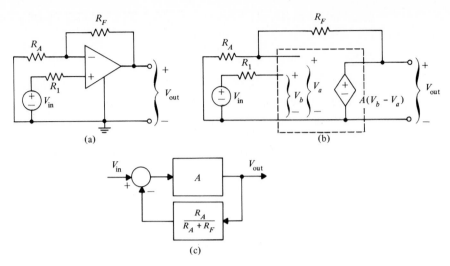

Figure 6.61   Analysis of an operational amplifier circuit.   **(a)** Noninverting amplifier connection of an op amp.   **(b)** A simple equivalent circuit.   **(c)** The amplifier as a feedback system.

Typical general-purpose operational amplifiers have low-frequency open-loop gains A on the order of 10^5, so the approximation is a good one at frequencies where A is large.

In a similar manner, approximate behavior of other operational amplifier configurations such as the ones listed in Table 6.11 can be calculated.*

An operational amplifier's transmittance A is actually a function of s, and typical frequency response plots of $A(s)$ are shown in Fig. 6.62a. While the transmittance is large and nearly constant at low frequencies, there are three significant high-frequency breaks corresponding to three real poles of $A(s)$. One of these poles is associated with the transistor input circuitry, one with the transistor gains themselves, and one is due to capacitive effects at the output.

The loop transmittance of the feedback connection of Fig. 6.61 is

$$G(s)H(s) = A(s)\,\frac{R_A}{R_A + R_F}$$

which simply involves a downward shift of the amplitude curve for $A(s)$ by the dB represented by the factor $R_A/(R_A + R_F)$, as in Fig. 6.62b. As the feedback factor

$$K = \frac{R_A}{R_A + R_F}$$

---

* The operational amplifier symbol used here includes both the amplifier itself and its power supply; hence the explicit ground connection. Electronic designers commonly omit drawing the ground connection by considering the power supply to be externally connected and not of interest.

Table 6.11   SOME COMMON HIGH-GAIN OPERATIONAL AMPLIFIER CONNECTIONS

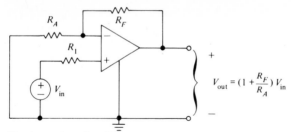

$$V_{out} = (1 + \frac{R_F}{R_A}) V_{in}$$

a. *Noninverting Amplifier*

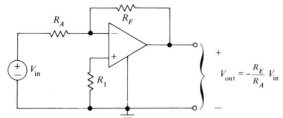

$$V_{out} = -\frac{R_F}{R_A} V_{in}$$

b. *Inverting Amplifier*

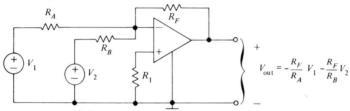

$$V_{out} = -\frac{R_F}{R_A} V_1 - \frac{R_F}{R_B} V_2$$

c. *Summing Amplifier*

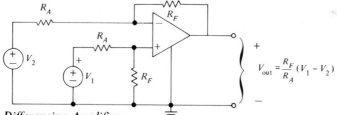

$$V_{out} = \frac{R_F}{R_A} (V_1 - V_2)$$

d. *Differencing Amplifier*

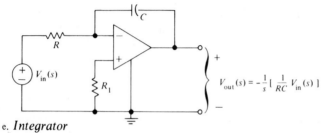

$$V_{out}(s) = -\frac{1}{s} [\frac{1}{RC} V_{in}(s)]$$

e. *Integrator*

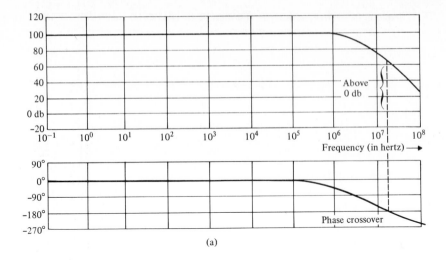

(a)

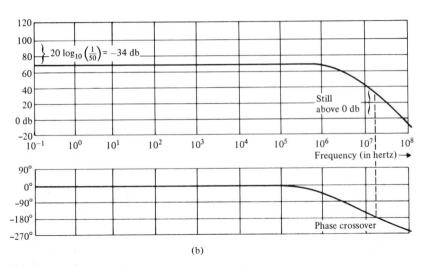

(b)

Figure 6.62 Operational amplifier frequency response. **(a)** Typical op amp transmittance $A(s)$. **(b)** Loop transmittance $A(s)R_A/(R_A + R_F)$ for a feedback factor $R_A/(R_A + R_F) = \frac{1}{50}$.

is adjusted from a small value toward unity, there will be an amount of feedback for which the loop gain amplitude ratio is unity at the same frequency for which the phase shift is 180°. Hence for feedback factors greater than this amount, the op amp circuit will be unstable.

For general-purpose applications, it is desirable that the operational amplifier be stable for all feedback factors K from zero to unity. The worst case of smallest gain margin is for $K = 1$, corresponding to $R_F = 0$. For $K = 1$, the noninverting configuration has unity overall gain at low frequencies. One approach to stabilization is to insert, with an *RC* network, an additional pole in *A(s)* which will reduce the loop transmittance amplitude ratio to below 0 dB at the phase crossover frequency, as shown in Fig. 6.63a. To do so typically requires that the added break frequency be at a few hertz, which greatly reduces the high-frequency performance

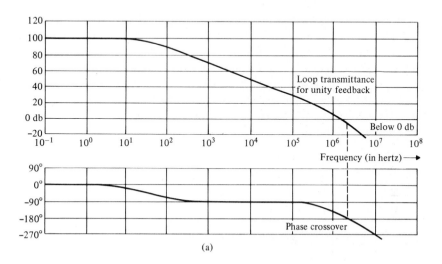

(a)

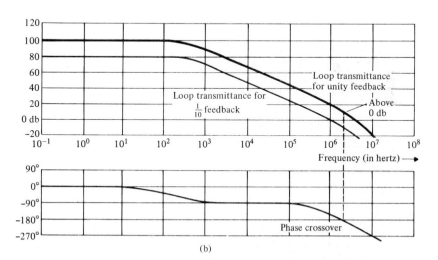

(b)

Figure 6.63 Open-loop frequency response of compensated operational amplifiers. **(a)** Open-loop frequency with an additional pole at a sufficiently low break frequency. The closed-loop system is stable for any feedback ratio of unity or less. **(b)** An additional pole with a higher break frequency. The closed-loop system is stable only for a sufficiently small feedback ratio.

of the amplifier. The type 741 op amp contains such an internal RC network-and is stable for all feedback factors including the worst case of unity feedback.

For feedback gain factors less than unity, the break frequency of the added pole can be increased as shown in Fig. 6.63b while still keeping the loop transmittance amplitude ratio below 0 dB at the phase crossover frequency. Stabilization of the type 101 op amp is done with an RC network for which the capacitor is connected externally. A 30-pF external capacitor will place the added break frequency in the same position as for the 741 op amp, making it unity-gain stable. A 3-pF capacitor will place the break frequency 10 times higher, giving better high-frequency response, but stability only for feedback factors of $\frac{1}{10}$ or less, corresponding to overall low-frequency gains of 10 or greater.

In more complicated compensation arrangements, both added zeros and poles are used to give sufficient gain margin while keeping the open-loop amplitude ratio large over a suitable range of frequency.

6.16 SUMMARY

When a transmittance $F(s)$ is driven with a sinusoidal signal, its forced response is also sinusoidal and of the same frequency. The ratio of sinusoidal output amplitude to input amplitude is the amplitude ratio

$$A(\omega) = |F(s = j\omega)|$$

as a function of radian frequency ω. The phase shift between input and output is

$$\Phi(\omega) = \angle F(s = j\omega)$$

At very small and at very large frequencies, the amplitude ratio of a rational transmittance is proportional to an integer power of ω and the phase shift is a multiple of 90°.

Frequency response curves are sketched from the pole-zero plot of a transmittance by considering directed line segments from the poles and zeros to various points on the imaginary axis. The frequency response amplitude curve may be visualized as the height along the imaginary axis of a rubber sheet laid over the complex plane, pushed up by poles and tacked down by zeros.

Bode plots are frequency response curves in a format which is especially convenient for rational transmittances. The amplitude ratio is plotted in decibels,

$$dB = 20 \log_{10} A(\omega)$$

and both amplitude ratio and phase shift are plotted on a logarithmic frequency scale. The frequency response contributions of real axis pole and zero terms are approximated well with straight-line segments, while the frequency response for complex conjugate pairs of roots may be constructed from standard, normalized curves.

A process for plotting dB and phase shift curves for complicated transmittances is the following:

1. Decompose the transmittance into simple factors.
2. Plot the dB and phase shift curves for each factor.
3. Add the individual dB curves to obtain the overall dB curve.
4. Add the individual phase shift curves to obtain the overall phase shift. Multiples of 360° may be added to or subtracted from the phase shift to keep that curve within a convenient range of angle.

Frequency response methods apply also to systems with irrational transmittances. One such system, of considerable practical importance, is the time delay

$$F(s) = e^{-st}$$

where τ, the delay time, is a constant. For $F(s)$, the frequency response amplitude ratio is a uniform 0 dB. The phase shift is proportional to frequency:

$$\Phi(\omega) = -\tau\omega$$

A major advantage of frequency response methods, in addition to their applicability to systems with irrational transmittances, is that experimentally derived data is easily incorporated. Approximate transmittances may be determined from experimental data, or the data may be used directly in analysis and design.

If a simple feedback system has a loop transmittance $G(s)H(s)$ which is stable, and there are no frequencies (including $\omega = 0$) for which the frequency response for $KG(s)H(s)$ passes simultaneously through 0 dB and 180° for any K between 0 and 1, the overall system is stable. If there is a single such 0-dB, 180° frequency, the overall system is unstable.

For a stable system, a phase crossover frequency is any frequency at which the phase shift curve for $G(s)H(s)$ crosses 180°. The gain margin of a simple feedback system is the smallest additional amplitude of $G(s)H(s)$ necessary to give unity amplitude at a phase crossover frequency. A gain crossover frequency of a simple feedback system is any frequency at which the amplitude ratio for $G(s)H(s)$ crosses 0 dB. The phase margin of a stable system is the additional phase lag necessary to give 180° phase shift at a gain crossover frequency.

A Nyquist plot consists of a curve on the complex plane representing the frequency response of the loop transmittance $G(s)H(s)$ of a simple feedback system. The Nyquist curve is a mapping of $G(s)H(s)$ for values of s along the closed curve from $-j\infty$ to $+j\infty$, then back to $-j\infty$ at large radius from the origin in the RHP. Small-radius RHP detours are made about any imaginary axis roots of $G(s)H(s)$. The number of RHP poles of the feedback system is equal to the algebraic number of clockwise encirclements of the point $s = -1$ by the Nyquist curve plus the number of RHP poles of $G(s)H(s)$.

Frequency domain figures of merit such as gain crossover frequency and phase margin can be related to the undamped natural frequency and the damping ratio of the dominant roots of the characteristic polynomial, and therefore to rise time and settling time. The root locus, Bode plot, and time response are therefore related in a way that influences design.

A designer may vary the gain of a unity feedback system to provide an acceptable error coefficient, gain crossover frequency, phase margin, phase cross-over frequency, and gain margin. If that uncompensated system cannot meet specifications, compensators may be added to the forward path and/or the feedback path.

Cascade integral-plus-proportional and cascade lag compensators provide a negative phase angle, thus reducing phase margin for a fixed gain crossover frequency. These compensators are primarily intended for improvement in steady state accuracy. Cascade lead and rate feedback compensators provide a positive phase angle, thus increasing phase margin for a fixed gain crossover frequency. These compensators are primarily intended for improvement in stability. A cascade lag-lead compensator combines the steady state improvement of lag compensation with the stability improvement of lead compensation.

Flexible spacecraft design and operational amplifier stabilization are diverse examples to which Bode/Nyquist methods may be successfully applied.

REFERENCES

Frequency Response Methods

Bode, H. W., *Network Analysis and Feedback Amplifier Design.* Princeton, N.J.: Van Nostrand, 1945.

James, H. M., Nichols, N. B., and Phillips, R. S., *Theory of Servomechanisms.* New York: McGraw-Hill, 1947.

Nyquist, H., "Regeneration Theory." *Bell Syst. Tech. J.* (January 1932): 126–147.

Savant, C. J., Jr., *Control System Design.* New York: McGraw-Hill, 1964.

Operational Amplifiers

Giles, J. N., *Fairchild Semiconductor Linear Integrated Circuits Applications Handbook.* Mountain View, Calif.: Fairchild Semiconductor Company, 1967.

Huelsman, L. P., *Theory and Design of Active R. C. Networks.* New York: McGraw-Hill, 1966.

Stout, D. F., and Kaufman, M., *Handbook of Operational Amplifiers.* 2d ed. London: Butterworths, 1979.

Tobey, G. E., Graeme, J. G., and Huelsman, L. P., *Operational Amplifiers—Design and Applications.* New York: McGraw-Hill, 1971.

PROBLEMS

1. For the following transmittances, find the forced sinusoidal response to the indicated input signals:

(a) $F(s) = \dfrac{1}{s + 8}$

$r(t) = 5 \cos 5t$

(b) $F(s) = \dfrac{s^2}{s+4}$

$r(t) = 100 \cos(4t + 40°)$

(c) $F(s) = \dfrac{10}{s^2 + 2s + 10}$

$r(t) = 4 \cos(5t - 70°)$

(d) $F(s) = \dfrac{s}{(s+4)(s+8)}$

$r(t) = 10 \cos(4t + 120°)$

(e) $F(s) = \dfrac{10}{(s+6)^2}$

$r(t) = 5 \sin 6t$

2. Sketch approximate frequency response curves (both amplitude ratio and phase shift) for functions with the pole-zero plots of Fig. P6.2.

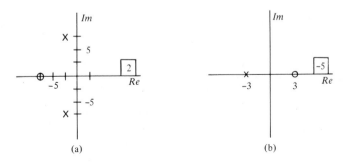

(a)   (b)

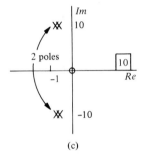   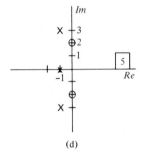

(c)   (d)

Figure P6-2

Ans (c)

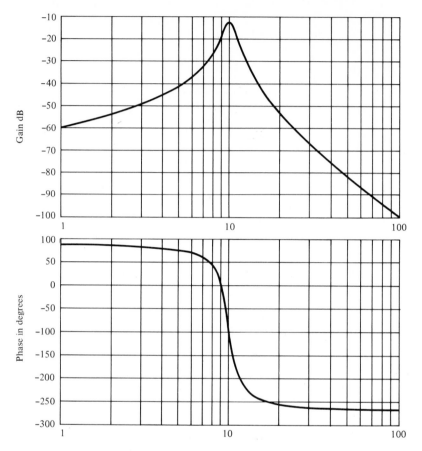

Frequency in radians per second

3. Draw Bode plots (both dB and phase shift) for the following transmittances:

(a) $F(s) = \dfrac{100}{s^2}$

(b) $F(s) = \dfrac{s}{s+6}$

(c) $F(s) = \dfrac{10(s + 6)}{s + 4}$

Ans (c)

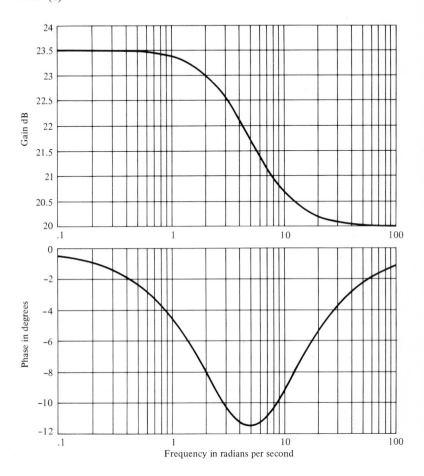

(d) $F(s) = \dfrac{-10}{s(s + 10)}$

(e) $F(s) = \dfrac{s + 1}{s(s + 10)}$

(f) $F(s) = \dfrac{1}{s^2 + 2s + 20}$

Ans (f)

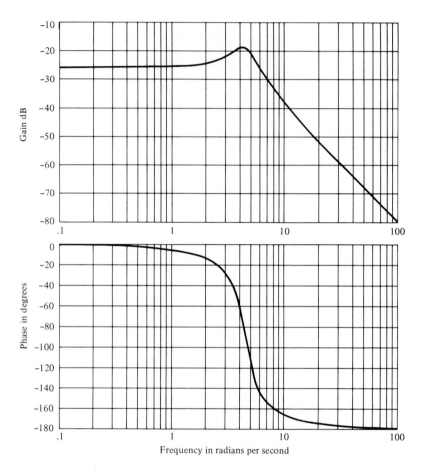

(g) $F(s) = \dfrac{1}{3s(s^2 + s + 8)}$

(h) $F(s) = \dfrac{s^2 - s + 4}{s^2 + s + 4}$

(i) $F(s) = \dfrac{s + 10}{s^2 + 10s + 100}$

Ans (i)

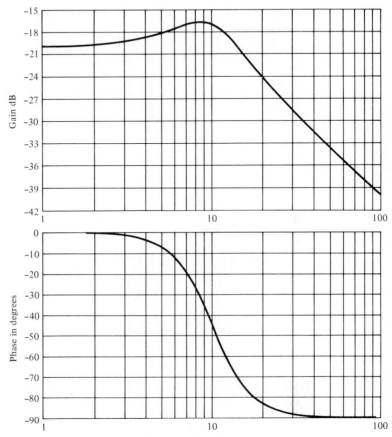

(j) $F(s) = \dfrac{4s - 1}{(s^2 + 2s + 4)^2}$

4. Draw Bode plots for the transmittances

$$F(s) = \dfrac{E_o(s)}{E_i(s)}\bigg|_{\substack{\text{zero initial}\\\text{conditions}}}$$

of the electrical networks of Fig. P6.4.

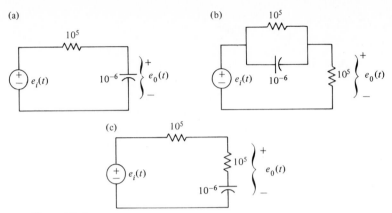

(a)

(b)

(c)

Figure P6-4

Ans (c)

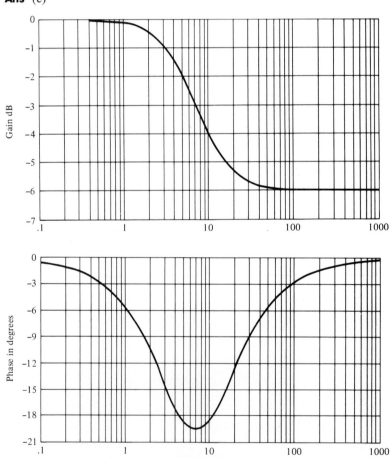

5. Draw Bode plots for the indicated transmittances

$$T(s) = \left.\frac{X(s)}{F(s)}\right|_{\substack{\text{zero initial} \\ \text{conditions}}}$$

of each of the translational mechanical networks of Fig. P6.5.

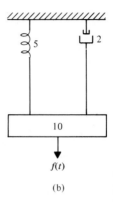

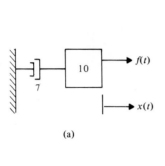

(a)

(b)

Figure P6-5

6. Draw Bode plots for the indicated transmittances

$$F(s) = \left.\frac{\Theta(s)}{\tau(s)}\right|_{\substack{\text{zero initial} \\ \text{conditions}}}$$

of each of the rotational mechanical networks of Fig. P6.6.

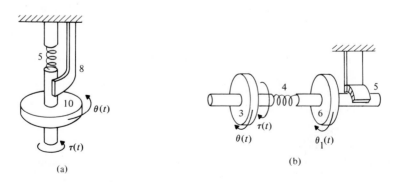

(a)

(b)

Figure P6-6

Ans (a)

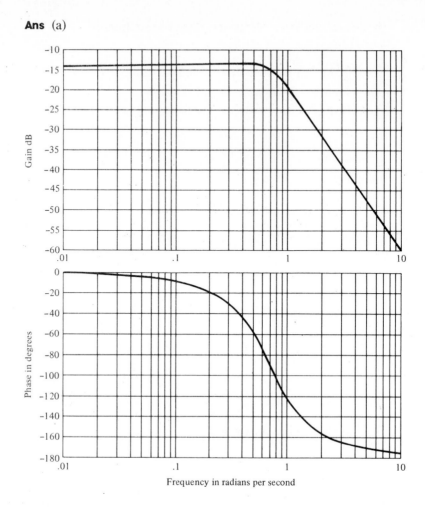

7. Find a transmittance that has the approximate dB curve as shown in Fig. P6.7. A *minimum-phase* transmittance has all poles and zeros in the left half of the complex plane. Find a minimum-phase solution.

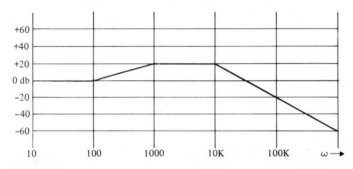

Figure P6-7

8. Find a stable transmittance that has the approximate phase shift curve of Fig. P6.8.

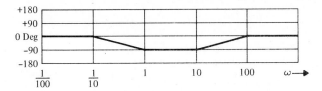

Figure P6-8

9. Experimental frequency response data for two transmittances $G(s)$ are given below. Find an approximate analytical expression for each of these $G(s)$ functions.

(a)

Radian Frequency ω	$G(s = j\omega)$
0.1	$0 - j20$
0.2	$-1.3 - j10$
0.4	$-0.8 - j5$
1.0	$-0.6 - j1.7$
2.0	$-0.4 - j0.6$
4.0	$-0.2 - j0.2$
8.0	$-0.05 - j0.008$
16.0	$-0.0022 + j0.0056$

(b)

Frequency f (Hz)	Amplitude ratio $\|G\|$ (dB)	Phase shift $\angle G$(deg)
120	−7.8	−165
100	−4.3	−160
80	−0.2	−160
70	0.75	−150
50	5.2	−140
40	8.0	−130
32	10.5	−125
20	15.0	−90
14	16.9	−55
5	20.4	−40
2.5	22.0	−30
0.5	24.0	−10
0	24.0	0

Ans (b) $G(s) \approx 158{,}000/(s + 100)^2$

10. Find gain margins and phase margins (if they exist) for feedback systems with the following loop transmittances:

(a) $G(s)H(s) = \dfrac{100}{(s+7)^3}$

(b) $G(s)H(s) = \dfrac{100}{s(s+10)^2}$

(c) $G(s)H(s) = \dfrac{s}{(s+4)^3}$

(d) $G(s)H(s) = \dfrac{10^7(s+1)^2(s+10)^2}{(s+6)^4}$

(e) $G(s)H(s) = \dfrac{1}{s(s+10)^2}$

> **Ans** (a) gain margin = 30 dB; (b) gain margin = 26 dB,
> phase margin = 80°

11. For the system of Fig. P6.11, use frequency response methods to determine values of $K > 0$, if any, that result in marginal stability of the overall system.

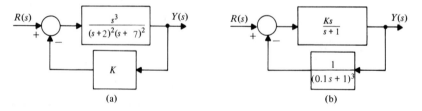

(a)   (b)

Figure P6-11

12. Use frequency response methods to find the ranges of the positive constant K (if any) for which each of the systems of Fig. P6.12 is stable.

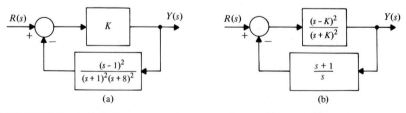

(a)   (b)

Figure P6-12 (a), (b)

13. Construct Bode plots for feedback systems with the following loop transmittances, and then use the plots to determine whether or not each system is stable.

(a) $G(s)H(s) = \dfrac{s+3}{(s+1)(s+7)}$

(b) $G(s)H(s) = \dfrac{10}{s(s+3)(s+7)}$

(c) $G(s)H(s) = \dfrac{6}{(s^2+1)(s+7)}$

(d) $G(s)H(s) = \dfrac{s^2+5}{s^2(s^2+4s+4)}$

14. Repeat Prob. 13 by sketching Nyquist plots for each of the *GH* functions. Determine whether or not each system is stable using the Nyquist plots.

Ans

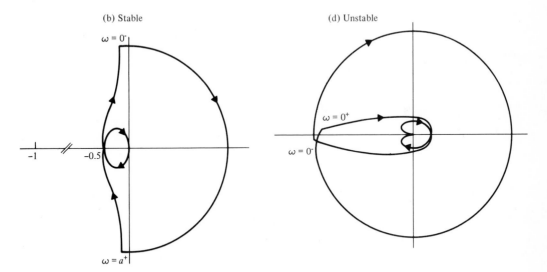

Figure P6-14

15. For the feedback systems of Fig. P6.15, sketch Nyquist plots and use them to determine whether or not each system is stable.

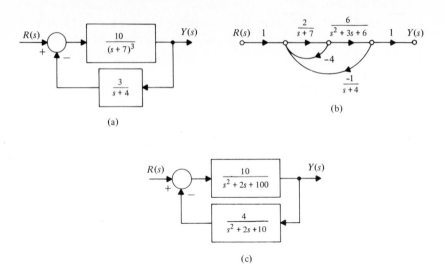

(a)

(b)

(c)

Figure P6-15

16. Use frequency response methods to determine the range of positive values of the constant a for which the system of Fig. P6.16 is stable.

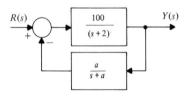

Figure P6-16

17. Fig. P6.17 shows a *Nichols chart*, which shows the graphical conversion of the frequency response of the open-loop transmittance $G(s = j\omega)$ of a unity feedback system to its closed-loop frequency response,

$$T(s = j\omega) = \frac{G(s = j\omega)}{1 + G(s = j\omega)}$$

This form of the chart deals with the amplitude ratio rather than dB. Write and test a digital computer or pocket calculator program to perform this computation. Given the dB and phase of G, your program should produce the dB and phase of T.

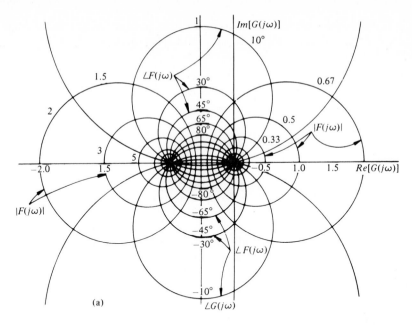

Figure P6-17

18. Carefully describe a method for measuring the frequency response of a control system component with an *unstable* transmittance by connecting it as part of a stable feedback system.

19. For the system of Fig. P6.19, how large can the constant K be for each positive integer value of n if the overall system is to be stable?

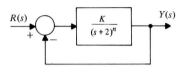

Figure P6-19

Ans $0 < K < \dfrac{2^n}{\cos^n\left(\dfrac{180°}{n}\right)}$

20. Find first-, second-, and third-order rational transmittances that approximate the desired "low-pass" amplitude ratio given in Fig. P6.20.

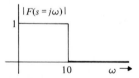

Figure P6-20

21. For a type 741 operational amplifier with open-loop frequency response given by Fig. 6.63a, find and sketch the frequency response of the closed-loop noninverting configuration with feedback factor

$$\frac{R_A}{R_A + R_F} = \frac{1}{100}$$

22. Use a type 741 operational amplifier in the circuit of Table 6.11b, with $R_F = 90$ kΩ and $R_A = 10$ kΩ. The transmittance of the 741 op-amp can be approximated by

$$G(s) = \frac{G_0}{1 + s/\omega_0}$$

where $G_0 = 10^5$ and $\omega_0 = 20\pi$ rad/sec. Determine the closed-loop voltage gain

$$A_v = \frac{V_{out}(s)}{V_{in}(s)}$$

23. Sketch the closed-loop frequency response for the op-amp system of Prob. 22.

24. Sketch the frequency response for the integrator of Table 6.11e with $R = 100$ kΩ, $C = 0.1$ μF, $G_0 = 10^5$, $\omega_0 = 20\pi$ rad/sec. Use a 741 op-amp with the transmittance of Prob. 22.

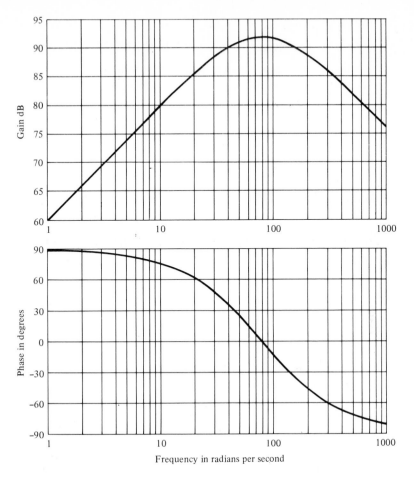

Figure P6-24

25. Two improved operational amplifier compensation arrangements are termed "two-pole" and "feedforward." In two-pole compensation, two additional poles are placed in the open-loop transmittance $A(s)$, allowing higher break frequencies than the single-pole arrangement, thus giving high gain over a wider range of frequency. For feedforward compensation, a pole and a zero are added to $A(s)$, with the zero placed to cancel or nearly cancel the pole in $A(s)$, with the lowest break frequency.

For an operational amplifier similar to the type 101 with

$$A(s) = \frac{10^{27}}{(s + 10^7)(s + 3 \times 10^7)(s + 10^8)}$$

design

(a) two-pole compensation and (b) feedforward compensation

such that the first open-loop break frequency is relatively high but the noninverting amplifier configuration is unity gain stable.

For each design, specify the compensator pole and zero locations and the frequency at which the open-loop transmittance is $\frac{1}{1000}$ its value of 10^5 at low frequencies.

26. An aircraft heading control system is diagrammed in Fig. P6.26. Use frequency response methods to determine a suitable value of "pilot gain" K.

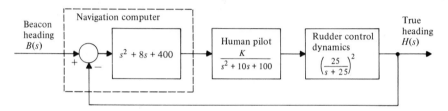

Figure P6-26

27. The following contains experimental data for the transmittance $G(s)$ of the hydraulic actuator in an aircraft control system.

Freq (rad)	Gain (dB)	Phase Shift (degrees)
0.01	20	−90
0.1	−2	−94
0.4	−14	−104
1.0	−24	−117
4.0	−47	−155
10.0	−63	−166
100.0	−100	−179

Use a cascade integral-plus-proportional compensator

$$\frac{K(s + 1)}{s}$$

to create the system of Fig. P6.27. Select the gain K so the phase margin is $45°$.

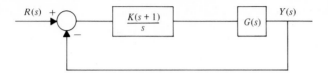

Figure P6-27

28. The position control system of Fig. P6.28 consists (from left to right across the forward path) of an adjustable electronic amplifier with transmittance A, a magnetic amplifier $G(s)$, whose frequency response data appears in the table of Fig. P6.9b, and a motor with the indicated second-order dynamics. Determine a value for A so the phase margin is 60°.

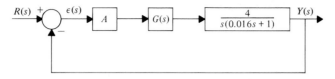

Figure P6-28

29. Sketch the Bode plot for a system with the following forward path transmittance. Select a forward path gain K so the system has a phase margin of 70°.

$$G_p(s) = \frac{1}{s(s^2 + 10s + 100)}$$

Ans $K = 291.5$

30. A block diagram model of a chemical process temperature control system is given in Fig. P6.30. Because the temperature sensor is located downstream in the fluid path from the heater, an appreciable time delay is involved. Find the maximum delay time K for which this system is stable.

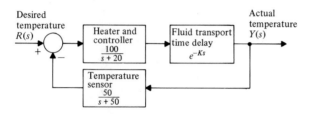

Figure P6-30

31. Apply cascade integral plus proportional compensation to:

$$G_p(s) = \frac{1000}{(s + 10)(s + 100)} \qquad K = 100$$

so the gain crossover frequency and phase margin remain near the uncompensated values.

32. Apply cascade lead compensation to:

$$G_p(s) = \frac{400}{(s + 20)^2} \qquad K = 100$$

so the gain crossover frequency, phase margin, and step error coefficients increase as compared to the uncompensated values.

33. Apply Bode rate feedback methods to Prob. 10 in Chap. 5 to improve the phase margin.

34. Repeat Prob. 13a, b in Chap. 5 using Bode methods.

Chapter 7

State Variable Methods for Continuous Time Systems

7.1 PREVIEW

Control system analysis and design methods are often divided into classical and modern categories. Unlike classical music (dating back hundreds of years), classical control theory deals with techniques in use before about 1950, while modern control theory relates to techniques applied after 1950.

Classical control embodies such methods as Routh-Hurwitz, root locus, and Bode/Nyquist. What these methods have in common is a use of transfer functions in the complex frequency (*s*) domain, the feedback of a small number of measurements, compensation to improve performance, and simplifying assumptions to approximate the time response. For example, the location of dominant closed-loop poles can be used to approximate rise time and settling time.

Modern control theory relies heavily on vector and matrix math to analyze system behavior with a heavy reliance on digital computers to carry out solutions (for example, a computer program can easily generate a closed-loop transfer function for a highly interconnected system when the use of Mason's gain rule would be quite difficult to apply manually). Many more signals are fed back as compared with classical controls. Steady state accuracy can be modified without adding compensators. Fundamental questions can be asked; for example, will the structure of a system thwart attempts at controlling the system?

Most of the mathematics associated with the so-called modern approach were developed well before 1950. In fact, a good deal was known about matrix-type operations in the era of Laplace and Fourier, who have provided us with techniques used in classical control theory. The rapid proliferation of low-cost, high-storage digital computers provided the tools which are now making matrix-based methods increasingly more useful. Since circuits are now generated with extremely high density, much more complicated control strategies become easily implementable with a microprocessor-based realization.

Appendix A is provided as a brief review of matrix algebra. That subject should be well understood in order to appreciate the material that follows.

7.2 SIMULATION DIAGRAMS

Up to this point in this textbook, all control systems have been shown using transmittances as functions of the complex frequency variable s. That approach is often called classical as compared with the so-called modern approach using time domain (differential) equations. A fundamental apparatus needed to describe a control system in the time domain is the use of state variables. In general, a system that can be described by an n-order linear differential equation can be defined by creating n state variables. For example, a system whose transfer function has a second-order denominator would require two state variables because if

$$\frac{Y(s)}{R(s)} = \frac{3(s + 1)}{s^2 + 2s + 4}$$

that system can be described by

$$\frac{d^2 y}{dt^2} + 2\frac{dy}{dt} + 4y = 3\frac{dr}{dt} + 3r$$

which is a second-order linear differential equation in y. By obtaining the transfer function, the system order is determined from the denominator. That order identifies the number of state variables that are needed. The problem, then, is to determine those state variables so the n choices are independent of each other.

Figure 7.1 shows the three common methods for selecting state variables. If the system transfer function (but not the actual structure of the system) is available, then Fig. 7.1a shows that phase or dual phase variables can be used to create a linkage, called a simulation diagram, between the classical and the modern. Each integrator output (as we will soon see) is defined as a state variable. If the actual system structure is known in block diagram form (for example, when a closed-loop system includes compensators), the resulting simulation diagram will follow the

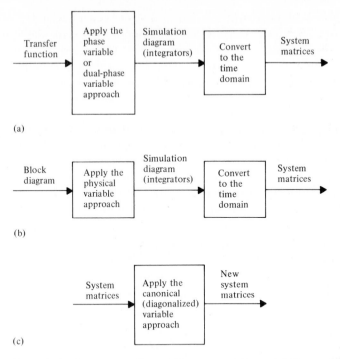

Figure 7.1 Creating a state variable representation. **(a)** Phase and dual phase variables. **(b)** Physical variables. **(c)** Canonical (diagonalized) variables.

system structure providing integrator outputs that have physical significance. These physical variables might include voltage, current, velocity, and position (for example Fig. 7.1b).

In either case mentioned, the ultimate result is a set of matrices. Finally, if the matrices are available (Fig. 7.1c), certain operations can be performed on them to create new system matrices of a particularly simple form (providing what are called a set of canonical state variables).

Physical variables are most closely related to real-world recognizable quantities while canonical variables are least related to real-world quantities and are the most theoretical. The phase and dual phase variables lie between the extremes of practical and theoretical.

Phase-Variable Form

An important problem in control system design is the synthesis of specific transmittances through the interconnection of simple components, as is needed for many of the controllers (or compensators) of the previous chapters. Synthesis is important also in the simulation of systems, where system behavior is predicted from a model governed by equivalent equations. Above all, the viewpoint of synthesis leads to fundamental techniques for system description, analysis, and design. These methods are systematic, compact, and suitable for computer analysis. They are also extendable to nonlinear and time-varying systems.

A basic component for synthesis is the integrator, a block or branch having transmittance $1/s$. A block diagram or signal flow graph composed only of constant transmittances and integrators is termed a *simulation diagram*. The order of such a system is simply the number of integrators present. Signal flow graphs are especially convenient for representing simulation diagrams because in many cases, system transfer functions are evident by inspection, using Mason's gain rule.

A transfer function that is the ratio of two polynomials in s is termed rational. If the numerator degree is less than the denominator degree, the transfer function is said to be *proper*. Any proper rational transfer function may be synthesized with a simulation diagram, that is, using only integration, multiplication by a constant, and summation operations. One very useful synthesis arrangement, known as the *phase-variable* form, is described below. The development, which is in terms of a specific numerical example for clarity, is applicable to any proper rational transfer function.

For the transfer function

$$T(s) = \frac{-5s^2 + 4s - 12}{s^3 + 6s^2 + s + 3} = \frac{-\dfrac{5}{s} + \dfrac{4}{s^2} + \dfrac{-12}{s^3}}{1 + \dfrac{6}{s} + \dfrac{1}{s^2} + \dfrac{3}{s^3}} = \frac{P_1 + P_2 + P_3}{1 - L_1 - L_2 - L_3}$$

dividing the numerator and denominator by the highest power of s term in the denominator places a 1 in the denominator and results in other numerator and denominator terms which are inverse powers of s, representing multiple integrations. In this form the transfer function may be interpreted as a Mason's gain rule expression. The numerator terms

$$\frac{-5}{s} + \frac{4}{s^2} + \frac{-12}{s^3}$$

are each taken to be paths through integrators and the paths are intermingled as in Fig. 7.2a so as to require a minimum number of integrators—in this case, three. The denominator terms

$$\frac{6}{s} + \frac{1}{s^2} + \frac{3}{s^3}$$

are taken to be loop gains. By placing each of these loops through the node to which $R(s)$ couples, all loops touch one another, so no product of loop gain terms is involved. All the loops touch each of the paths, so each path cofactor is unity.

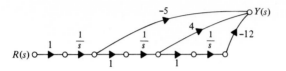

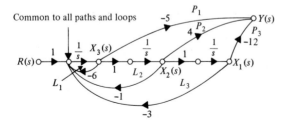

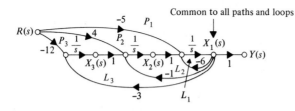

Figure 7.2 Phase-variable synthesis of a single-input, single-output system. **(a)** Paths in the simulation diagram. **(b)** Complete simulation diagram. **(c)** Synthesizing a transfer function in the dual phase-variable form.

In Fig. 7.2b, each integrator output signal has been labeled. These signals are termed the *state variables* of the system. This realization of the example transfer function is then described by the following Laplace-transformed equations:

$$X_1(s) = \frac{1}{s} X_2(s)$$

$$X_2(s) = \frac{1}{s} X_3(s)$$

$$X_3(s) = \frac{1}{s}[-3X_1(s) - X_2(s) - 6X_3(s) + R(s)]$$

$$Y(s) = -12X_1(s) + 4X_2(s) - 5X_3(s)$$

or

$$sX_1(s) = X_2(s)$$
$$sX_2(s) = X_3(s)$$
$$sX_3(s) = -3X_1(s) - X_2(s) - 6X_3(s) + R(s)$$
$$Y(s) = -12X_1 + 4X_2(s) - 5X_3(s)$$

As functions of time, the signals satisfy

$$\frac{dx_1}{dt} = x_2(t)$$

$$\frac{dx_2}{dt} = x_3(t)$$

$$\frac{dx_3}{dt} = -3x_1(t) - x_2(t) - 6x_3(t) + r(t)$$

$$y(t) = -12x_1(t) + 4x_2(t) - 5x_3(t)$$

which is a set of coupled first-order differential equations.

Dual Phase-Variable Form

Another especially convenient way to synthesize a transfer function with integrators is to arrange the signal flow graph so that all of the paths and all of the loops touch an output node. For the previous transfer function

$$T(s) = \frac{-5s^2 + 4s - 12}{s^3 + 6s^2 + s + 3}$$

$$= \frac{-\dfrac{5}{s} + \dfrac{4}{s^2} - \dfrac{12}{s^3}}{1 + \dfrac{6}{s} + \dfrac{1}{s^2} + \dfrac{3}{s^3}}$$

$$= \frac{P_1 + P_2 + P_3}{1 - L_1 - L_2 - L_3}$$

for example, the diagram of Fig. 7.2c shows this *dual phase-variable* arrangement. The output signal is derived from a single node, while the input signal is coupled to each integrator.

The Laplace transform relations describing this system are, in terms of the indicated state variables,

$$sX_1(s) = -6X_1(s) + X_2(s) - 5R(s)$$

$$sX_2(s) = -X_1(s) + X_3(s) + 4R(s)$$

$$sX_3(s) = -3X_1 - 12R(s)$$

$$Y(s) = X_1(s)$$

As functions of time, the signals satisfy

$$\frac{dx_1}{dt} = -6x_1(t) + x_2(t) - 5r(t)$$

$$\frac{dx_2}{dt} = -x_1(t) + x_3(t) + 4r(t)$$

$$\frac{dx_3}{dt} = -3x_1(t) - 12r(t)$$

$$y(t) = x_1(t)$$

Multiple Outputs and Inputs

Additional system outputs may be easily derived from the phase-variable arrangement. For example, the single-input, two-output system of Fig. 7.3a has the following transfer functions:

$$T_{11}(s) = \frac{Y_1(s)}{R(s)}\bigg|_{\substack{\text{initial} \\ \text{conditions} = 0}} = \frac{-5/s + 4/s^2 + (-12/s^3)}{1 + 6/s + 1/s^2 + 3/s^3}$$

$$= \frac{-5s^2 + 4s - 12}{s^3 + 6s^2 + s + 3}$$

$$T_{21}(s) = \frac{Y_2(s)}{R(s)}\bigg|_{\substack{\text{initial} \\ \text{conditions} = 0}} = \frac{3/s + 1/s^2 + (-6/s^3)}{1 + 6/s + 1/s^2 + 3/s^3}$$

$$= \frac{3s^2 + s - 6}{s^3 + 6s^2 + s + 3}$$

Additional inputs are easily added to the dual phase-variable arrangement. The example two-input, single-output system of Fig. 7.3b has the following transfer functions:

$$T_{11}(s) = \frac{Y(s)}{R_1(s)}\bigg|_{\substack{\text{initial} \\ \text{conditions} \\ \text{and } R_2 = 0}} = \frac{-5/s + 4/s^2 - 12/s^3}{1 + 6/s + 1/s^2 + 3/s^3}$$

$$= \frac{-5s^2 + 4s - 12}{s^3 + 6s^2 + s + 3}$$

$$T_{12}(s) = \frac{Y(s)}{R_2(s)}\bigg|_{\substack{\text{initial} \\ \text{conditions} \\ \text{and } R_1 = 0}} = \frac{10/s + 7/s^2 - 8/s^3}{1 + 6/s + 1/s^2 + 3/s^3}$$

$$= \frac{10s^2 + 7s - 8}{s^3 + 6s^2 + s + 3}$$

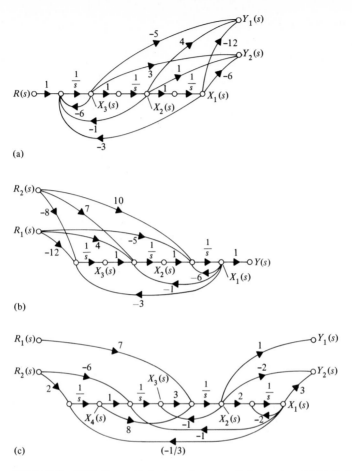

Figure 7.3 Achieving multiple outputs and multiple inputs. **(a)** Multiple outputs from the phase-variable arrangement. **(b)** Multiple inputs with the dual phase-variable arrangement. **(c)** System with both multiple inputs and multiple outputs.

The system described by the simulation diagram of Fig. 7.3c is neither in phase-variable nor dual phase-variable form. Its two inputs and two outputs are governed by the following Laplace-transformed equations:

$$sX_1(s) = -2X_1(s) + 2X_2(s)$$

$$sX_2(s) = -3X_2(s) + 3X_3(s) + 8X_4(s) + 7R_1(s)$$

$$sX_3(s) = -X_1(s) + X_4(s) - 6R_2(s)$$

$$sX_4(s) = -\tfrac{1}{3}X_1(s) + 2R_2(s)$$

$$Y_1(s) = X_2(s)$$

$$Y_2(s) = 3X_1(s) - 2X_2(s)$$

Applying Mason's gain rule to the simulation diagram, the four transfer functions describing this system are:

$$T_{11}(s) = \frac{Y_1(s)}{R_1(s)}\bigg|_{\substack{\text{initial} \\ \text{conditions} \\ \text{and } R_2 = 0}}$$

$$= \frac{(7/s)(1 + 2/s)}{1 - [-2/s - 3/s^2 - 6/s^3 - 2/s^4 - (16/3)/s^3 - 2/s^4] + (-2/s)(-3/s)}$$

$$= \frac{7s^3 + 14s^2}{s^4 + 5s^3 + \frac{34}{3}s^2 + 6s + 2}$$

$$T_{12}(s) = \frac{Y_1(s)}{R_2(s)}\bigg|_{\substack{\text{initial} \\ \text{conditions} \\ \text{and } R_1 = 0}}$$

$$= \frac{(-2/s^2)(1 + 2/s) + (6/s^3)(1 + 2/s)}{1 - [-2/s - 3/s^2 - 6/s^3 - (16/3)/s^3 - 2/s^4] + (-2/s)(-3/s)}$$

$$= \frac{-2s^2 + 2s + 12}{s^4 + 5s^3 + \frac{34}{3}s^2 + 6s + 2}$$

$$T_{21}(s) = \frac{Y_2(s)}{R_1(s)}\bigg|_{\substack{\text{initial} \\ \text{conditions} \\ \text{and } R_2 = 0}}$$

$$= \frac{(-14/s)(1 + 2/s) + 42/s^2}{1 - [-2/s - 3/s^2 - 6/s^3 - (16/3)/s^3 - 2/s^4] + (-2/s)(-3/s)}$$

$$= \frac{-14s^3 + 14s^2}{s^4 + 5s^3 + \frac{34}{3}s^2 + 6s + 2}$$

$$T_{22}(s) = \frac{Y_2(s)}{R_2(s)}\bigg|_{\substack{\text{initial} \\ \text{conditions} \\ \text{and } R_1 = 0}}$$

$$= \frac{(4/s^2)(1 + 2/s) - 12/s^3(1 + 2/s) - 12/s^3 + 36/s^4}{1 - [-2/s - 3/s^2 - 6/s^3 - (16/3)/s^3 - 2/s^4] + (-2/s)(-3/s)}$$

$$= \frac{4s^2 - 16s + 12}{s^4 + 5s^3 + \frac{34}{3}s + 2}$$

DRILL PROBLEM

D7.1 Draw simulation diagrams in either the phase-variable or the dual phase-variable form for systems with the following transfer functions:

(a) $T(s) = \dfrac{-4s + 3}{s^2 + 6s + 2}$

(b) $T(s) = \dfrac{-s^2 + 5s + 9}{3s^3 + 2s^2 + 4s + 1}$

(c) $T_{11}(s) = \dfrac{0.4s^2 + 1.4s + 0.8}{s^3 + 0.3s^2 + 1.7s + 0.2}$

$T_{12}(s) = \dfrac{-0.5s^2 + 0.7s - 1.9}{s^3 + 0.3s^2 + 1.7s + 0.2}$

(d) $T_{11}(s) = \dfrac{4s^2 - 1}{s^3 + 6s^2 + 2s + 5}$

$T_{21}(s) = \dfrac{3s + 6}{s^3 + 6s^2 + 2s + 5}$

7.3 STATE REPRESENTATION OF SYSTEMS

State Variable Equations

Phase-variable form is particularly convenient for the synthesis of single- and multiple-output systems, while in dual phase-variable form, single- and multiple-input systems are easily arranged. There are a whole spectrum of other ways of connecting integrators to achieve systems with desired transfer functions, including systems with both multiple inputs and multiple outputs. Moreover, the representation of systems in terms of integrators is useful not only for transfer function synthesis, but for the description of systems of all kinds, particularly those that are very complicated, for which a standard, compact notation is especially helpful.

A general state variable description of an nth-order system involves n integrators, the outputs of which are the state variables. The inputs of each of the integrators are driven with a linear combination of the state signals and the inputs:

$$sX_1(s) = a_{11}X_1(s) + a_{12}X_2(s) + \cdots + a_{1n}X_n(s) + b_{11}R_1(s) + \cdots + b_{1i}R_i(s)$$

$$sX_2(s) = a_{21}X_1(s) + a_{22}X_2(s) + \cdots + a_{2n}X_n(s) + b_{21}R_1(s) + \cdots + b_{2i}R_i(s)$$

$$\vdots$$

$$sX_n(s) = a_{n1}X_1(s) + a_{n2}X_2(s) + \cdots + a_{nn}X_n(s) + b_{n1}R_1(s) + \cdots + b_{ni}R_i(s)$$

$$(7.1)$$

In the time domain, these are a set of n first-order differential equations in the n state variables and the inputs:

$$\frac{dx_1}{dt} = a_{11}x_1 + a_{12}x_2 + \cdots + a_{1n}x_n + b_{11}r_1 + \cdots + b_{1i}r_i$$

$$\frac{dx_2}{dt} = a_{21}x_1 + a_{22}x_2 + \cdots + a_{2n}x_n + b_{21}r_1 + \cdots + b_{2i}r_i$$

$$\vdots$$

$$\frac{dx_n}{dt} = a_{n1}x_1 + a_{n2}x_2 + \cdots + a_{nn}x_n + b_{n1}r_1 + \cdots + b_{ni}r_i$$

These state equations are compactly written in matrix notation as

$$\frac{d}{dt}\begin{bmatrix} x_1 \\ x_2 \\ \vdots \\ x_n \end{bmatrix} = \begin{bmatrix} \dot{x}_1 \\ \dot{x}_2 \\ \vdots \\ \dot{x}_n \end{bmatrix} = \begin{bmatrix} a_{11} & a_{12} & \cdots & a_{1n} \\ a_{21} & a_{22} & \cdots & a_{2n} \\ \vdots & & & \\ a_{n1} & a_{n2} & \cdots & a_{nn} \end{bmatrix}\begin{bmatrix} x_1 \\ x_2 \\ \vdots \\ x_n \end{bmatrix}$$

$$+ \begin{bmatrix} b_{11} & \cdots & b_1 \\ b_{21} & \cdots & b_2 \\ \vdots & & \\ b_{n1} & \cdots & b_{ni} \end{bmatrix}\begin{bmatrix} r_1 \\ r_2 \\ \vdots \\ r_i \end{bmatrix}$$

or

$$\frac{d\mathbf{x}}{dt} = \dot{\mathbf{x}} = \mathbf{A}\mathbf{x} + \mathbf{B}\mathbf{r}$$

The column matrix of state variables

$$\mathbf{x} = \begin{bmatrix} x_1 \\ x_2 \\ \vdots \\ x_n \end{bmatrix}$$

is called the *state vector*. The inputs are arranged to form the *input vector*:

$$\mathbf{r} = \begin{bmatrix} r_1 \\ \vdots \\ r_i \end{bmatrix}$$

The system outputs are similarly arranged in an *output vector*,

$$\mathbf{y} = \begin{bmatrix} y_1 \\ \vdots \\ y_m \end{bmatrix}$$

related linearly to the state variables through the output equations:

$$\begin{cases} y_1 = c_{11}x_1 + c_{12}x_2 + \cdots + c_{1n}x_n \\ \vdots \\ y_m = c_{m1}x_1 + c_{m2}x_2 + \cdots + c_{mn}x_n \end{cases}$$

or

$$\begin{bmatrix} y_1 \\ \vdots \\ y_m \end{bmatrix} = \begin{bmatrix} c_{11} & c_{12} & \cdots & c_{1n} \\ \vdots & & & \\ c_{m1} & c_{m2} & \cdots & c_{mn} \end{bmatrix}\begin{bmatrix} x_1 \\ \vdots \\ x_n \end{bmatrix}$$

or

$$\mathbf{y} = \mathbf{C}\mathbf{x}$$

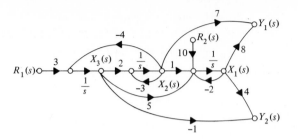

Figure 7.4 Simulation diagram for a certain two-input, two-output system.

The state equations describe how the system state vector evolves in time. One may imagine the tip of the vector tracing a curve, the *state trajectory*, in an n-dimensional space. The output equations describe how the output signals are related to the state.

Consider the two-input, two-output system of the simulation diagram of Fig. 7.4. The indicated state variables are governed by

$$\begin{cases} \dot{x}_1 = -2x_1 + x_2 + 5x_3 + 10r_2 \\ \dot{x}_2 = -3x_2 + 2x_3 \\ \dot{x}_3 = -4x_2 + 3r_1 \end{cases}$$

In matrix form, these state equations are

$$\begin{bmatrix} \dot{x}_1 \\ \dot{x}_2 \\ \dot{x}_3 \end{bmatrix} = \begin{bmatrix} -2 & 1 & 5 \\ 0 & -3 & 2 \\ 0 & -4 & 0 \end{bmatrix} \begin{bmatrix} x_1 \\ x_2 \\ x_3 \end{bmatrix} + \begin{bmatrix} 0 & 10 \\ 0 & 0 \\ 3 & 0 \end{bmatrix} \begin{bmatrix} r_1 \\ r_2 \end{bmatrix}$$

The outputs of this system are related to the state variables by

$$\begin{cases} y_1 = 8x_1 + 7x_2 \\ y_2 = 4x_1 - x_3 \end{cases}$$

In matrix form, these output equations are

$$\begin{bmatrix} y_1 \\ y_2 \end{bmatrix} = \begin{bmatrix} 8 & 7 & 0 \\ 4 & 0 & -1 \end{bmatrix} \begin{bmatrix} x_1 \\ x_2 \\ x_3 \end{bmatrix}$$

For systems described by linear constant-coefficient integrodifferential equations, the state variable arrangement is simply a standard form for the equations describing a system. Instead of dealing with a mixed collection of simultaneous system equations, some of first order, some of second order, some involving running integrals, and so on, additional manipulation of the original equations is done to place them in the standard form. The advantages of a standard form are that systematic methods may be easily brought to bear upon very involved problems and that a degree of unification results.

DRILL PROBLEMS

D7.2. Write state variable equations in matrix form for the systems described by the following simulation diagrams.

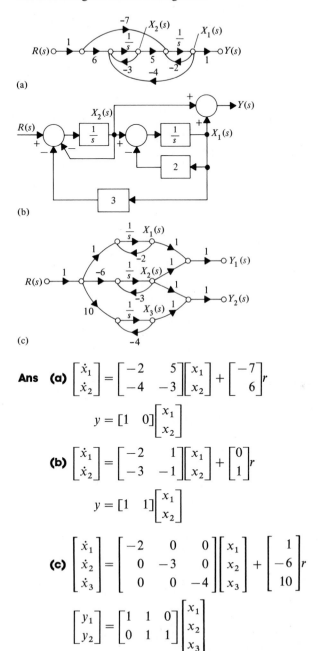

(a)

(b)

(c)

Ans (a)
$$\begin{bmatrix} \dot{x}_1 \\ \dot{x}_2 \end{bmatrix} = \begin{bmatrix} -2 & 5 \\ -4 & -3 \end{bmatrix} \begin{bmatrix} x_1 \\ x_2 \end{bmatrix} + \begin{bmatrix} -7 \\ 6 \end{bmatrix} r$$

$$y = \begin{bmatrix} 1 & 0 \end{bmatrix} \begin{bmatrix} x_1 \\ x_2 \end{bmatrix}$$

(b)
$$\begin{bmatrix} \dot{x}_1 \\ \dot{x}_2 \end{bmatrix} = \begin{bmatrix} -2 & 1 \\ -3 & -1 \end{bmatrix} \begin{bmatrix} x_1 \\ x_2 \end{bmatrix} + \begin{bmatrix} 0 \\ 1 \end{bmatrix} r$$

$$y = \begin{bmatrix} 1 & 1 \end{bmatrix} \begin{bmatrix} x_1 \\ x_2 \end{bmatrix}$$

(c)
$$\begin{bmatrix} \dot{x}_1 \\ \dot{x}_2 \\ \dot{x}_3 \end{bmatrix} = \begin{bmatrix} -2 & 0 & 0 \\ 0 & -3 & 0 \\ 0 & 0 & -4 \end{bmatrix} \begin{bmatrix} x_1 \\ x_2 \\ x_3 \end{bmatrix} + \begin{bmatrix} 1 \\ -6 \\ 10 \end{bmatrix} r$$

$$\begin{bmatrix} y_1 \\ y_2 \end{bmatrix} = \begin{bmatrix} 1 & 1 & 0 \\ 0 & 1 & 1 \end{bmatrix} \begin{bmatrix} x_1 \\ x_2 \\ x_3 \end{bmatrix}$$

D7.3. Draw simulation diagrams to represent the following systems:

(a)
$$\begin{bmatrix} \dot{x}_1 \\ \dot{x}_2 \end{bmatrix} = \begin{bmatrix} -2 & -3 \\ 4 & 1 \end{bmatrix} \begin{bmatrix} x_1 \\ x_2 \end{bmatrix} + \begin{bmatrix} 5 \\ -6 \end{bmatrix} r$$

$$y = \begin{bmatrix} 7 & -8 \end{bmatrix} \begin{bmatrix} x_1 \\ x_2 \end{bmatrix}$$

(b)
$$\begin{bmatrix} \dot{x}_1 \\ \dot{x}_2 \\ \dot{x}_3 \end{bmatrix} = \begin{bmatrix} 0 & 10 & 3 \\ 5 & 8 & 0 \\ -2 & -7 & -3 \end{bmatrix} \begin{bmatrix} x_1 \\ x_2 \\ x_3 \end{bmatrix} + \begin{bmatrix} 0 & 1 \\ 2 & 0 \\ -1 & 3 \end{bmatrix} \begin{bmatrix} r_1 \\ r_2 \end{bmatrix}$$

$$y = \begin{bmatrix} 0 & 4 & -3 \end{bmatrix} \begin{bmatrix} x_1 \\ x_2 \\ x_3 \end{bmatrix}$$

(c)
$$\begin{bmatrix} \dot{x}_1 \\ \dot{x}_2 \\ \dot{x}_3 \end{bmatrix} = \begin{bmatrix} -2 & 0 & -6 \\ 3 & 5 & 0 \\ -4 & 0 & 7 \end{bmatrix} \begin{bmatrix} x_1 \\ x_2 \\ x_3 \end{bmatrix} + \begin{bmatrix} 8 \\ -2 \\ 0 \end{bmatrix} r$$

$$\begin{bmatrix} y_1 \\ y_2 \end{bmatrix} = \begin{bmatrix} 0 & -1 & 1 \\ -1 & 1 & 0 \end{bmatrix} \begin{bmatrix} x_1 \\ x_2 \\ x_3 \end{bmatrix}$$

Transfer Functions

The transfer functions of a system represented in state variable form may be found by Laplace-transforming the state equations with zero initial conditions. In general, these are the equations in (7.1). Collecting the terms involving $\mathbf{X}(s)$, there results

$$\begin{bmatrix} (s - a_{11}) & -a_{12} & \cdots & -a_{1n} \\ -a_{21} & (s - a_{22}) & \cdots & -a_{2n} \\ \vdots & & & \\ -a_{n1} & -a_{n2} & \cdots & (s - a_{nn}) \end{bmatrix} \begin{bmatrix} X_1(s) \\ X_2(s) \\ \vdots \\ X_n(s) \end{bmatrix}$$

$$= \begin{bmatrix} b_{11} & b_{12} & \cdots & b_{1i} \\ b_{21} & b_{22} & \cdots & b_{2i} \\ \vdots & & & \\ b_{n1} & b_{n2} & \cdots & b_{ni} \end{bmatrix} \begin{bmatrix} R_1(s) \\ R_2(s) \\ \vdots \\ R_i(s) \end{bmatrix}$$

or

$$[s\mathbf{I} - \mathbf{A}]\mathbf{X}(s) = \mathbf{B}\mathbf{R}(s)$$

where $\mathbf{I}$ is the $n \times n$ identity matrix

$$\mathbf{I} = \begin{bmatrix} 1 & 0 & \cdots & 0 & 0 \\ 0 & 1 & \cdots & 0 & 0 \\ \vdots & & & & \\ 0 & 0 & \cdots & 0 & 1 \end{bmatrix}$$

Solving for the Laplace transform of the state vector,

$$\mathbf{X}(s) = [s\mathbf{I} - \mathbf{A}]^{-1}\mathbf{B}R(s)$$

The output and state vectors are related by

$$\begin{bmatrix} Y_1(s) \\ Y_2(s) \\ \vdots \\ Y_m(s) \end{bmatrix} = \begin{bmatrix} c_{11} & c_{12} & \cdots & c_{1n} \\ c_{21} & c_{22} & \cdots & c_{2n} \\ \vdots & & & \\ c_{m1} & c_{m2} & \cdots & c_{mn} \end{bmatrix} \begin{bmatrix} X_1(s) \\ X_2(s) \\ \vdots \\ X_n(s) \end{bmatrix}$$

or

$$\mathbf{Y}(s) = \mathbf{C}\mathbf{X}(s) = \{\mathbf{C}[s\mathbf{I} - \mathbf{A}]^{-1}\mathbf{B}\}R(s)$$

The $m \times i$ matrix in braces $\{\ \}$ above consists of the input-output transfer functions of the system, arranged as a matrix:

$$C[s\mathbf{I} - \mathbf{A}]^{-1}\mathbf{B} = \begin{bmatrix} T_{11}(s) & T_{12}(s) & \cdots & T_{1i}(s) \\ T_{21}(s) & T_{22}(s) & \cdots & T_{2i}(s) \\ \vdots & & & \\ T_{m1}(s) & T_{m2}(s) & \cdots & T_{mi}(s) \end{bmatrix}$$

For example, a single-input, single-output system with state equations

$$\begin{bmatrix} \dot{x}_1 \\ \dot{x}_2 \end{bmatrix} = \begin{bmatrix} -3 & 1 \\ -2 & 0 \end{bmatrix} \begin{bmatrix} x_1 \\ x_2 \end{bmatrix} + \begin{bmatrix} 4 \\ -5 \end{bmatrix} r$$

$$y = \begin{bmatrix} 1 & -1 \end{bmatrix} \begin{bmatrix} x_1 \\ x_2 \end{bmatrix}$$

has transfer function given by

$$T(s) = \begin{bmatrix} 1 & -1 \end{bmatrix} \begin{bmatrix} s+3 & -1 \\ 2 & s \end{bmatrix}^{-1} \begin{bmatrix} 4 \\ -5 \end{bmatrix}$$

$$= \begin{bmatrix} 1 & -1 \end{bmatrix} \frac{\begin{bmatrix} s & 1 \\ -2 & s+3 \end{bmatrix} \begin{bmatrix} 4 \\ -5 \end{bmatrix}}{s^2 + 3s + 2}$$

$$= \frac{\begin{bmatrix} 1 & -1 \end{bmatrix} \begin{bmatrix} (4s - 5) \\ (-5s - 23) \end{bmatrix}}{s^2 + 3s + 2}$$

$$= \frac{9s + 18}{s^2 + 3s + 2}$$

The two-input, two-output system

$$\begin{bmatrix} \dot{x}_1 \\ \dot{x}_2 \end{bmatrix} = \begin{bmatrix} -3 & 1 \\ -2 & 0 \end{bmatrix}\begin{bmatrix} x_1 \\ x_2 \end{bmatrix} + \begin{bmatrix} 4 & 6 \\ -5 & 0 \end{bmatrix}\begin{bmatrix} r_1 \\ r_2 \end{bmatrix}$$

$$\begin{bmatrix} y_1 \\ y_2 \end{bmatrix} = \begin{bmatrix} 1 & -1 \\ 8 & 1 \end{bmatrix}\begin{bmatrix} x_1 \\ x_2 \end{bmatrix}$$

is described by the transfer function matrix given by

$$\mathbf{T}(s) = \begin{bmatrix} 1 & -1 \\ 8 & 1 \end{bmatrix}\begin{bmatrix} s+3 & -1 \\ 2 & s \end{bmatrix}^{-1}\begin{bmatrix} 4 & 6 \\ -5 & 0 \end{bmatrix}$$

$$= \begin{bmatrix} 1 & -1 \\ 8 & 1 \end{bmatrix}\dfrac{\begin{bmatrix} s & 1 \\ -2 & s+3 \end{bmatrix}}{s^2 + 3s + 2}\begin{bmatrix} 4 & 6 \\ -5 & 0 \end{bmatrix}$$

$$= \dfrac{\begin{bmatrix} 1 & -1 \\ 8 & 1 \end{bmatrix}\begin{bmatrix} (4s-5) & 6s \\ (-5s-23) & -12 \end{bmatrix}}{s^2 + 3s + 2}$$

$$= \begin{bmatrix} \dfrac{9s+18}{s^2+3s+2} & \dfrac{6s+12}{s^2+3s+2} \\[2ex] \dfrac{27s-63}{s^2+3s+2} & \dfrac{48s-12}{s^2+3s+2} \end{bmatrix} = \begin{bmatrix} T_{11}(s) & T_{12}(s) \\ T_{21}(s) & T_{22}(s) \end{bmatrix}$$

where

$$T_{11}(s) = \dfrac{9s+18}{s^2+3s+2} = \dfrac{Y_1(s)}{R_1(s)}\bigg|_{\substack{\text{initial} \\ \text{conditions} \\ \text{and } R_2 = 0}}$$

$$T_{12}(s) = \dfrac{6s+12}{s^2+3s+2} = \dfrac{Y_1(s)}{R_2(s)}\bigg|_{\substack{\text{initial} \\ \text{conditions} \\ \text{and } R_1 = 0}}$$

$$T_{21}(s) = \dfrac{27s-63}{s^2+3s+2} = \dfrac{Y_2(s)}{R_1(s)}\bigg|_{\substack{\text{initial} \\ \text{conditions} \\ \text{and } R_2 = 0}}$$

$$T_{22}(s) = \dfrac{48s-12}{s^2+3s+2} = \dfrac{Y_2(s)}{R_2(s)}\bigg|_{\substack{\text{initial} \\ \text{conditions} \\ \text{and } R_1 = 0}}$$

All of the transfer functions of a system share the denominator polynomial

$$|s\mathbf{I} - \mathbf{A}|$$

where $\mathbf{A}$ is the state coupling matrix for the system, since

$$[s\mathbf{I} - \mathbf{A}]^{-1} = \dfrac{\text{adjoint } [s\mathbf{I} - \mathbf{A}]}{|s\mathbf{I} - \mathbf{A}|}$$

The nth-degree polynomial

$$|s\mathbf{I} - \mathbf{A}| = 0$$

is termed the characteristic polynomial of an $n \times n$ matrix $\mathbf{A}$ and the n roots of that polynomial are the characteristic roots of the matrix. A system is stable if and only if the characteristic roots of the state coupling matrix are all in the left half of the complex plane.

DRILL PROBLEM

D7.4. Find the transfer function matrices of the following systems:

(a)
$$\begin{bmatrix} \dot{x}_1 \\ \dot{x}_2 \end{bmatrix} = \begin{bmatrix} -2 & 3 \\ -1 & -1 \end{bmatrix} \begin{bmatrix} x_1 \\ x_2 \end{bmatrix} + \begin{bmatrix} 4 & 0 \\ -5 & 6 \end{bmatrix} \begin{bmatrix} r_1 \\ r_2 \end{bmatrix}$$

$$y = \begin{bmatrix} 7 & 8 \end{bmatrix} \begin{bmatrix} x_1 \\ x_2 \end{bmatrix}$$

Ans
$$\begin{bmatrix} \dfrac{-12s - 189}{s^2 + 3s + 5} & + & \dfrac{48s + 222}{s^2 + 3s + 5} \end{bmatrix}$$

(b)
$$\begin{bmatrix} \dot{x}_1 \\ \dot{x}_2 \end{bmatrix} = \begin{bmatrix} -3 & 4 \\ -2 & 0 \end{bmatrix} \begin{bmatrix} x_1 \\ x_2 \end{bmatrix} + \begin{bmatrix} 2 \\ 1 \end{bmatrix} r$$

$$\begin{bmatrix} y_1 \\ y_2 \end{bmatrix} = \begin{bmatrix} -4 & 6 \\ 5 & -1 \end{bmatrix} \begin{bmatrix} x_1 \\ x_2 \end{bmatrix}$$

Ans
$$\begin{bmatrix} \dfrac{-2s - 22}{s^2 + 3s + 8} \\[2ex] \dfrac{9s + 21}{s^2 + 3s + 8} \end{bmatrix}$$

(c)
$$\begin{bmatrix} \dot{x}_1 \\ \dot{x}_2 \\ \dot{x}_3 \end{bmatrix} = \begin{bmatrix} -4 & 0 & 2 \\ -1 & -1 & 0 \\ 3 & 0 & -3 \end{bmatrix} \begin{bmatrix} x_1 \\ x_2 \\ x_3 \end{bmatrix} + \begin{bmatrix} -3 & -2 \\ 4 & 1 \\ 0 & 0 \end{bmatrix} \begin{bmatrix} r_1 \\ r_2 \end{bmatrix}$$

$$\begin{bmatrix} y_1 \\ y_2 \end{bmatrix} = \begin{bmatrix} 1 & 1 & 1 \\ -1 & 0 & 1 \end{bmatrix} \begin{bmatrix} x_1 \\ x_2 \\ x_3 \end{bmatrix}$$

Ans
$$\begin{bmatrix} \dfrac{s^2 + 10s + 15}{(s + 1)(s^2 + 7s + 6)} & \dfrac{-s(s + 5)}{(s + 1)(s^2 + 7s + 6)} \\[3ex] \dfrac{3s}{s^2 + 7s + 6} & \dfrac{2s}{s^2 + 7s + 6} \end{bmatrix}$$

Change of State Variables

A nonsingular transformation of state variables results in a new system state representation with the same relation between inputs and outputs. A new set of n variables may be derived from the original n state variables through a constant transformation of the form

$$x_1' = p_{11}x_1 + p_{12}x_2 + \cdots + p_{1n}x_n$$

$$x_2' = p_{21}x_1 + p_{22}x_2 + \cdots + p_{2n}x_n$$

$$\vdots$$

$$x_n' = p_{n1}x_1 + p_{n2}x_2 + \cdots + q_{nn}x_n$$

If the transformation is nonsingular, the original variables may be recovered from the new ones through the inverse transformation

$$x_1 = q_{11}x_1' + q_{12}x_2' + \cdots + q_{1n}x_n'$$

$$x_2 = q_{21}x_1' + q_{22}x_2' + \cdots + q_{2n}x_n'$$

$$\vdots$$

$$x_n = q_{n1}x_1' + q_{n2}x_2' + \cdots + q_{nn}x_n'$$

In matrix notation,

$$
\begin{bmatrix} x_1' \\ x_2' \\ \vdots \\ x_n' \end{bmatrix}
=
\begin{bmatrix} p_{11} & p_{12} & \cdots & p_{1n} \\ p_{21} & p_{22} & \cdots & p_{2n} \\ \vdots & & & \\ p_{n1} & p_{n2} & \cdots & p_{nn} \end{bmatrix}
\begin{bmatrix} x_1 \\ x_2 \\ \vdots \\ x_n \end{bmatrix}
$$

or

$$\mathbf{x}' = \mathbf{P}\mathbf{x}$$

and

$$
\begin{bmatrix} x_1 \\ x_2 \\ \vdots \\ x_n \end{bmatrix}
=
\begin{bmatrix} q_{11} & q_{12} & \cdots & q_{1n} \\ q_{21} & q_{22} & \cdots & q_{2n} \\ \vdots & & & \\ q_{n1} & q_{n2} & \cdots & q_{nn} \end{bmatrix}
\begin{bmatrix} x_1' \\ x_2' \\ \vdots \\ x_n' \end{bmatrix}
$$

or

$$\mathbf{x} = \mathbf{Q}\mathbf{x}' = \mathbf{P}^{-1}\mathbf{x}'$$

A nonsingular change of state variables

$$\mathbf{x}' = \mathbf{P}\mathbf{x} \qquad \mathbf{x} = \mathbf{P}^{-1}\mathbf{x}'$$

in a state representation for a system

$$\dot{\mathbf{x}} = \mathbf{A}\mathbf{x} + \mathbf{B}\mathbf{r}$$

$$\mathbf{y} = \mathbf{C}\mathbf{x}$$

gives a representation of the same form, in terms of an alternative set of state variables:

$$\mathbf{P}^{-1}\dot{\mathbf{x}}' = \mathbf{AP}^{-1}\mathbf{x}' + \mathbf{Br}$$
$$\mathbf{y} = \mathbf{CP}^{-1}\mathbf{x}'$$
$$\dot{\mathbf{x}}' = (\mathbf{PAP}^{-1})\mathbf{x}' + (\mathbf{PB})\mathbf{r}$$
$$\mathbf{y} = (\mathbf{CP}^{-1})\mathbf{x}'$$

Defining

$$\mathbf{A}' = \mathbf{PAP}^{-1}$$
$$\mathbf{B}' = \mathbf{PB}$$
$$\mathbf{C}' = \mathbf{CP}^{-1}$$

the equations in terms of the new state variables $\mathbf{x}'$ are of the same form as the original equations:

$$\dot{\mathbf{x}}' = \mathbf{A}'\mathbf{x}' + \mathbf{B}'\mathbf{r}$$
$$\mathbf{y} = \mathbf{C}'\mathbf{x}'$$

The input-output relations of a system are unchanged by a nonsingular change of state variables; it is only the internal description, in terms of its state, that is changed. The matrix of system transfer functions for the prime system is given by

$$\mathbf{T}'(s) = \mathbf{C}'(s\mathbf{I} - \mathbf{A}')^{-1}\mathbf{B}'$$

Substituting the original system matrices, it is seen that the transfer functions are identical:

$$\begin{aligned}
\mathbf{T}'(s) &= \mathbf{CP}^{-1}(s\mathbf{I} - \mathbf{PAP}^{-1})^{-1}\mathbf{PB} \\
&= \mathbf{CP}^{-1}(s\mathbf{PP}^{-1} - \mathbf{PAP}^{-1})^{-1}\mathbf{PB} \\
&= \mathbf{CP}^{-1}[\mathbf{P}(s\mathbf{I} - \mathbf{A})\mathbf{P}^{-1}]^{-1}\mathbf{PB} \\
&= \mathbf{CP}^{-1}\mathbf{P}(s\mathbf{I} - \mathbf{A})^{-1}\mathbf{P}^{-1}\mathbf{PB} \\
&= \mathbf{C}(s\mathbf{I} - \mathbf{A})^{-1}\mathbf{B}
\end{aligned}$$

which is the same transfer function matrix as for the system described in terms of the original state variables. The matrix relations

$$\mathbf{PP}^{-1} = \mathbf{P}^{-1}\mathbf{P} = \mathbf{I}$$

and, for square matrices $\mathbf{X}$, $\mathbf{Y}$, and $\mathbf{Z}$,

$$(\mathbf{XYZ})^{-1} = \mathbf{Z}^{-1}\mathbf{Y}^{-1}\mathbf{X}^{-1}$$

are used in this derivation.

As a numerical example, consider the system

$$\begin{bmatrix} \dot{x}_1 \\ \dot{x}_2 \\ \dot{x}_3 \end{bmatrix} = \begin{bmatrix} -2 & 1 & 1 \\ -3 & 0 & 1 \\ 0 & 0 & 0 \end{bmatrix} \begin{bmatrix} x_1 \\ x_2 \\ x_3 \end{bmatrix} + \begin{bmatrix} 1 & 2 \\ 0 & 0 \\ -1 & 0 \end{bmatrix} \begin{bmatrix} r_1 \\ r_2 \end{bmatrix} = \mathbf{Ax} + \mathbf{Br}$$

$$\begin{bmatrix} y_1 \\ y_2 \end{bmatrix} = \begin{bmatrix} 1 & 0 & 1 \\ 0 & 0 & -2 \end{bmatrix} \begin{bmatrix} x_1 \\ x_2 \\ x_3 \end{bmatrix} = \mathbf{Cx}$$

The nonsingular transformation of state variables

$$\begin{bmatrix} x'_1 \\ x'_2 \\ x'_3 \end{bmatrix} = \begin{bmatrix} 1 & 0 & 1 \\ 0 & 1 & 2 \\ 0 & 0 & 4 \end{bmatrix} \begin{bmatrix} x_1 \\ x_2 \\ x_3 \end{bmatrix} = \mathbf{Px}$$

$$\begin{bmatrix} x_1 \\ x_2 \\ x_3 \end{bmatrix} = \begin{bmatrix} 1 & 0 & -\frac{1}{4} \\ 0 & 1 & -\frac{1}{2} \\ 0 & 0 & \frac{1}{4} \end{bmatrix} \begin{bmatrix} x'_1 \\ x'_2 \\ x'_3 \end{bmatrix} = \mathbf{P}^{-1}\mathbf{x}'$$

gives a new state representation as follows:

$$\mathbf{A}' = \mathbf{PAP}^{-1} = \begin{bmatrix} 1 & 0 & 1 \\ 0 & 1 & 2 \\ 0 & 0 & 4 \end{bmatrix} \begin{bmatrix} -2 & 1 & 1 \\ -3 & 0 & 1 \\ 0 & 0 & 0 \end{bmatrix} \begin{bmatrix} 1 & 0 & -\frac{1}{4} \\ 0 & 1 & -\frac{1}{2} \\ 0 & 0 & \frac{1}{4} \end{bmatrix}$$

$$= \begin{bmatrix} -2 & 1 & 1 \\ -3 & 0 & 1 \\ 0 & 0 & 0 \end{bmatrix} \begin{bmatrix} 1 & 0 & -\frac{1}{4} \\ 0 & 1 & -\frac{1}{2} \\ 0 & 0 & \frac{1}{4} \end{bmatrix} = \begin{bmatrix} -2 & 1 & \frac{1}{4} \\ -3 & 0 & 1 \\ 0 & 0 & 0 \end{bmatrix}$$

$$\mathbf{B}' = \mathbf{PB} = \begin{bmatrix} 1 & 0 & 1 \\ 0 & 1 & 2 \\ 0 & 0 & 4 \end{bmatrix} \begin{bmatrix} 1 & 2 \\ 0 & 0 \\ -1 & 0 \end{bmatrix} = \begin{bmatrix} 0 & 2 \\ -2 & 0 \\ -4 & 0 \end{bmatrix}$$

$$\mathbf{C}' = \mathbf{CP}^{-1} = \begin{bmatrix} 1 & 0 & 1 \\ 0 & 0 & -2 \end{bmatrix} \begin{bmatrix} 1 & 0 & -\frac{1}{4} \\ 0 & 1 & -\frac{1}{2} \\ 0 & 0 & \frac{1}{4} \end{bmatrix} = \begin{bmatrix} 1 & 0 & 0 \\ 0 & 0 & -\frac{1}{2} \end{bmatrix}$$

Simulation diagrams for the original and the transformed systems are given in Fig. 7.5. The two systems are indistinguishable from one another so far as input and output signals are concerned. In general, every system has an infinite number of different state representations, each involving a different choice of state variables. There is the possibility, then, of finding especially simple state variable system representations.

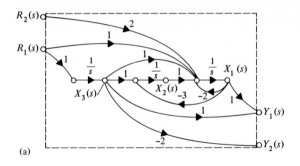

(a)

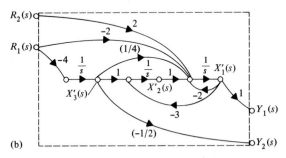

(b)

Figure 7.5 Simulation diagrams for two different representations of a system. **(a)** The system $\dot{\mathbf{x}} = \mathbf{Ax} + \mathbf{Br}$, $\mathbf{y} = \mathbf{Cx}$. **(b)** The system $\dot{\mathbf{x}}' = \mathbf{A'x}' + \mathbf{B'r}$, $\mathbf{y} = \mathbf{C'x}'$.

DRILL PROBLEM

D7.5. Make the indicated change of state variables, finding the new set of state and output equations in terms of $\mathbf{x}'$.

(a) $\begin{bmatrix} \dot{x}_1 \\ \dot{x}_2 \end{bmatrix} = \begin{bmatrix} -2 & 1 \\ -3 & 0 \end{bmatrix} \begin{bmatrix} x_1 \\ x_2 \end{bmatrix} + \begin{bmatrix} 4 \\ 5 \end{bmatrix} r$

$y = \begin{bmatrix} 1 & 0 \end{bmatrix} \begin{bmatrix} x_1 \\ x_2 \end{bmatrix}$

$\begin{bmatrix} x_1' \\ x_2' \end{bmatrix} = \begin{bmatrix} 2 & 1 \\ 4 & 3 \end{bmatrix} \begin{bmatrix} x_1 \\ x_2 \end{bmatrix}$

Ans $\begin{bmatrix} \dot{x}_1' \\ \dot{x}_2' \end{bmatrix} = \begin{bmatrix} -\frac{29}{2} & \frac{11}{2} \\ -\frac{67}{2} & \frac{25}{2} \end{bmatrix} \begin{bmatrix} x_1' \\ x_2' \end{bmatrix} + \begin{bmatrix} 13 \\ 31 \end{bmatrix} r$

$y = \begin{bmatrix} \frac{3}{2} & -\frac{1}{2} \end{bmatrix} \begin{bmatrix} x_1' \\ x_2' \end{bmatrix}$

(b) $\begin{bmatrix} \dot{x}_1 \\ \dot{x}_2 \end{bmatrix} = \begin{bmatrix} 2 & -6 \\ 12 & 16 \end{bmatrix} \begin{bmatrix} x_1 \\ x_2 \end{bmatrix} + \begin{bmatrix} 0 & -1 \\ 2 & 1 \end{bmatrix} \begin{bmatrix} r_1 \\ r_2 \end{bmatrix}$

$y = \begin{bmatrix} 1 & 1 \end{bmatrix} \begin{bmatrix} x_1 \\ x_2 \end{bmatrix}$

$\begin{bmatrix} x_1' \\ x_2' \end{bmatrix} = \begin{bmatrix} 1 & 2 \\ 1 & 4 \end{bmatrix} \begin{bmatrix} x_1 \\ x_2 \end{bmatrix}$

Ans $\begin{bmatrix} \dot{x}_1' \\ \dot{x}_2' \end{bmatrix} = \begin{bmatrix} 39 & -13 \\ 71 & -21 \end{bmatrix} \begin{bmatrix} x_1' \\ x_2' \end{bmatrix} + \begin{bmatrix} 4 & 1 \\ 8 & 3 \end{bmatrix} \begin{bmatrix} r_1 \\ r_2 \end{bmatrix}$

$y = \begin{bmatrix} \frac{3}{2} & -\frac{1}{2} \end{bmatrix} \begin{bmatrix} x_1' \\ x_2' \end{bmatrix}$

(c) $\begin{bmatrix} \dot{x}_1 \\ \dot{x}_2 \\ \dot{x}_3 \end{bmatrix} = \begin{bmatrix} -2 & 1 & 1 \\ -3 & 0 & 0 \\ 0 & 0 & 0 \end{bmatrix} \begin{bmatrix} x_1 \\ x_2 \\ x_3 \end{bmatrix} + \begin{bmatrix} 1 \\ 1 \\ 1 \end{bmatrix} r$

$\begin{bmatrix} y_1 \\ y_2 \end{bmatrix} = \begin{bmatrix} 2 & -2 & 1 \\ 0 & -1 & 1 \end{bmatrix} \begin{bmatrix} x_1 \\ x_2 \\ x_3 \end{bmatrix}$

$\begin{bmatrix} x_1' \\ x_2' \\ x_3' \end{bmatrix} = \begin{bmatrix} 1 & 0 & 1 \\ 0 & 1 & 3 \\ 0 & 0 & 4 \end{bmatrix} \begin{bmatrix} x_1 \\ x_2 \\ x_3 \end{bmatrix}$

Ans $\begin{bmatrix} \dot{x}_1 \\ \dot{x}_2 \\ \dot{x}_3 \end{bmatrix} = \begin{bmatrix} -2 & 1 & 0 \\ -3 & 0 & \frac{3}{4} \\ 0 & 0 & 0 \end{bmatrix} \begin{bmatrix} x_1 \\ x_2 \\ x_3 \end{bmatrix} + \begin{bmatrix} 2 \\ 4 \\ 4 \end{bmatrix} r$

$y = \begin{bmatrix} 2 & -2 & \frac{5}{4} \\ 0 & -1 & 1 \end{bmatrix} \begin{bmatrix} x_1' \\ x_2' \\ x_3' \end{bmatrix}$

7.4 DECOUPLING STATE EQUATIONS

As shown in Fig. 7.1a, a simulation diagram results in a set of system matrices when a transfer function is properly decomposed. Phase variables or dual phase variables result in a recognizable form for the **A** matrix. In this section, a properly chosen change of variables can create a set of **A**, **B**, and **C** matrices where the **A** matrix has a very simple (diagonal) form. The state variables (Fig. 7.1c) are often called canonical when the **A** matrix becomes diagonal.

Diagonal Forms for the Equations

When a nonsingular change of state variables in a system representation is made,

$$\mathbf{x}' = \mathbf{P}\mathbf{x} \qquad \mathbf{x} = \mathbf{P}^{-1}\mathbf{x}'$$

the new state coupling matrix $\mathbf{A}'$ is related to the original one $\mathbf{A}$ by

$$\mathbf{A}' = \mathbf{PAP}^{-1}$$

Such an operation on a matrix is termed a *similarity transformation*. One of the most important results of matrix algebra is that, provided that a square matrix $\mathbf{A}$ has no repeated characteristic roots, a similarity transformation $\mathbf{P}$ may be found for which

$$\mathbf{A}' = \mathbf{PAP}^{-1}$$

is diagonal, with the characteristic roots as the diagonal elements.

A similarity transformation which diagonalizes $\mathbf{A}$ can be created using a set of eigenvectors, one for each eigenvalue. The German word "eigen" means characteristic. These eigenvectors are not unique. Each eigenvector can be multiplied by a constant that works just as well. The following procedure can be used to get $\mathbf{P}$.

1. Find the eigenvalues s_i where

$$|s\mathbf{I} - \mathbf{A}| = 0$$

2. Find an eigenvector $\mathbf{x}_i$ for each s_i

$$[s_i\mathbf{I} - \mathbf{A}]\mathbf{x}_i = 0$$

3. Let $\mathbf{P}^{-1}$ be a matrix consisting of the eigenvectors

$$\mathbf{P}^{-1} = [\mathbf{x}_1 \quad : \quad \mathbf{x}_2 \quad : \quad \dots \quad : \quad \mathbf{x}_n]$$

$$\mathbf{PAP}^{-1} = \begin{bmatrix} s_1 & 0 & 0 & \cdot \\ 0 & s_2 & 0 & \cdot \\ \text{etc.} & & & \end{bmatrix}$$

For example, for the system

$$\begin{bmatrix} \dot{x}_1 \\ \dot{x}_2 \\ \dot{x}_3 \end{bmatrix} = \begin{bmatrix} -1 & -2 & 0 \\ 1 & 2 & 0 \\ -2 & -1 & -3 \end{bmatrix} \begin{bmatrix} x_1 \\ x_2 \\ x_3 \end{bmatrix} + \begin{bmatrix} 1 \\ 0 \\ 0 \end{bmatrix} r = \mathbf{Ax} + \mathbf{b}r$$

$$\begin{bmatrix} y_1 \\ y_2 \end{bmatrix} = \begin{bmatrix} 1 & 0 & 1 \\ 1 & -1 & 0 \end{bmatrix} \begin{bmatrix} x_1 \\ x_2 \\ x_3 \end{bmatrix} = \mathbf{Cx}$$

The characteristic polynomial is

$$\begin{vmatrix} s+1 & 2 & 0 \\ -1 & s-2 & 0 \\ 2 & 1 & s+3 \end{vmatrix} = s(s-1)(s+3)$$

The eigenvalues can be selected in any order. If

$$s_1 = 0 \qquad s_2 = 1 \qquad s_3 = -3$$

Then the first eigenvector is found by solving

$$\begin{bmatrix} s_1 + 1 & 2 & 0 \\ -1 & s_1 - 2 & 0 \\ 2 & 1 & s_1 + 3 \end{bmatrix} \begin{bmatrix} x_{11} \\ x_{21} \\ x_{31} \end{bmatrix} = \begin{bmatrix} 0 \\ 0 \\ 0 \end{bmatrix}$$

$$\mathbf{x}_1 = \begin{bmatrix} x_{11} \\ x_{21} \\ x_{31} \end{bmatrix}$$

With the first eigenvalue being zero

$$x_{11} + 2x_{21} \qquad = 0$$

$$-x_{11} - 2x_{21} \qquad = 0$$

$$2x_{11} + x_{21} + 3x_{31} = 0$$

The first two equations are equivalent, so an infinite number of solutions exist. It is possible to select x_{21} arbitrarily (say -1). Then x_{11} is 2 and x_{31} is -1. By proceeding in a similar way, three eigenvectors result:

$$\mathbf{x}_1 = \begin{bmatrix} 2 \\ -1 \\ -1 \end{bmatrix} \qquad \mathbf{x}_2 = \begin{bmatrix} 4 \\ -4 \\ -1 \end{bmatrix} \qquad \mathbf{x}_3 = \begin{bmatrix} 0 \\ 0 \\ 1 \end{bmatrix}$$

Any nonzero multiple of any eigenvector also works. Collecting these eigenvectors into $\mathbf{P}^{-1}$, the transformation

$$\mathbf{P} = \begin{bmatrix} 1 & 1 & 0 \\ -\frac{1}{4} & -\frac{1}{2} & 0 \\ \frac{3}{4} & \frac{1}{2} & 1 \end{bmatrix} \qquad \mathbf{P}^{-1} = \begin{bmatrix} 2 & 4 & 0 \\ -1 & -4 & 0 \\ -1 & -1 & 1 \end{bmatrix}$$

gives a state variable representation for which the state coupling matrix is diagonal:

$$\mathbf{A}' = \mathbf{PAP}^{-1} = \begin{bmatrix} 1 & 1 & 0 \\ -\frac{1}{4} & -\frac{1}{2} & 0 \\ \frac{3}{4} & \frac{1}{2} & 1 \end{bmatrix} \begin{bmatrix} -1 & -2 & 0 \\ 1 & 2 & 0 \\ -2 & -1 & -3 \end{bmatrix} \begin{bmatrix} 2 & 4 & 0 \\ -1 & -4 & 0 \\ -1 & -1 & 1 \end{bmatrix}$$

$$= \begin{bmatrix} 1 & 1 & 0 \\ -\frac{1}{4} & -\frac{1}{2} & 0 \\ \frac{3}{4} & \frac{1}{2} & 1 \end{bmatrix} \begin{bmatrix} 0 & 4 & 0 \\ 0 & -4 & 0 \\ 0 & -1 & -3 \end{bmatrix} = \begin{bmatrix} 0 & 0 & 0 \\ 0 & 1 & 0 \\ 0 & 0 & -3 \end{bmatrix}$$

$$\mathbf{B}' = \mathbf{PB} = \begin{bmatrix} 1 & 1 & 0 \\ -\frac{1}{4} & -\frac{1}{2} & 0 \\ \frac{3}{4} & \frac{1}{2} & 1 \end{bmatrix} \begin{bmatrix} 1 \\ 0 \\ 0 \end{bmatrix} = \begin{bmatrix} 1 \\ -\frac{1}{4} \\ \frac{3}{4} \end{bmatrix}$$

$$\mathbf{C}' = \mathbf{CP}^{-1} = \begin{bmatrix} 1 & 0 & 1 \\ 1 & -1 & 0 \end{bmatrix} \begin{bmatrix} 2 & 4 & 0 \\ -1 & -4 & 0 \\ -1 & -1 & 1 \end{bmatrix} = \begin{bmatrix} 1 & 3 & 1 \\ 3 & 8 & 0 \end{bmatrix}$$

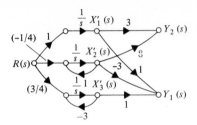

Figure 7.6 Simulation diagram for the diagonalized example system.

The system described by the primed state variables,

$$
\begin{bmatrix} \dot{x}'_1 \\ \dot{x}'_2 \\ \dot{x}'_3 \end{bmatrix} = \begin{bmatrix} 0 & 0 & 0 \\ 0 & 1 & 0 \\ 0 & 0 & -3 \end{bmatrix} \begin{bmatrix} x'_1 \\ x'_2 \\ x'_3 \end{bmatrix} + \begin{bmatrix} 1 \\ -\frac{1}{4} \\ \frac{3}{4} \end{bmatrix} r
$$

$$
\begin{bmatrix} y_1 \\ y_2 \end{bmatrix} = \begin{bmatrix} 1 & -3 & 1 \\ 3 & 8 & 0 \end{bmatrix} \begin{bmatrix} x'_1 \\ x'_2 \\ x'_3 \end{bmatrix}
$$

has the same relation between r and $\mathbf{y}$. Because the state coupling matrix is diagonal, however, the state equations are decoupled from one another. The system is represented in the form of three separate first-order systems, as in the simulation diagram of Fig. 7.6.

Finding a transformation matrix that diagonalizes a square matrix $\mathbf{A}$ with distinct characteristic roots is a fundamental technique of linear algebra. It is termed the *characteristic value problem* and is discussed in detail in most texts on linear algebra, including those cited in the references at the end of this chapter.

Diagonalization Using Partial Fraction Expansion

Another method of determining a diagonal form for a system involves partial fraction expansion. For a single-input, single-output system such as

$$
\begin{bmatrix} \dot{x}_1 \\ \dot{x}_2 \\ \dot{x}_3 \end{bmatrix} = \begin{bmatrix} -1 & -2 & 0 \\ 1 & 2 & 0 \\ -2 & -1 & -3 \end{bmatrix} \begin{bmatrix} x_1 \\ x_2 \\ x_3 \end{bmatrix} + \begin{bmatrix} 1 \\ 0 \\ 0 \end{bmatrix} r
$$

$$
y = \begin{bmatrix} 1 & 0 & 1 \end{bmatrix} \begin{bmatrix} x_1 \\ x_2 \\ x_3 \end{bmatrix}
$$

the transfer function is

$$T(s) = \mathbf{C}(s\mathbf{I} - \mathbf{A})^{-1}\mathbf{B}$$

$$= \begin{bmatrix} 1 & 0 & 1 \end{bmatrix} \begin{bmatrix} s+1 & 2 & 0 \\ -1 & s-2 & 0 \\ 2 & 1 & s+3 \end{bmatrix}^{-1} \begin{bmatrix} 1 \\ 0 \\ 0 \end{bmatrix}$$

$$= \begin{bmatrix} 1 & 0 & 1 \end{bmatrix} \frac{\begin{bmatrix} s^2+s-6 & -2s-6 & 0 \\ s+3 & s^2+4s+3 & 0 \\ -2s+3 & -s+3 & s^2-s \end{bmatrix}}{s^3+2s^2-3s} \begin{bmatrix} 1 \\ 0 \\ 0 \end{bmatrix}$$

$$= \frac{\begin{bmatrix} 1 & 0 & 1 \end{bmatrix} \begin{bmatrix} (s^2+s-6) \\ (s+3) \\ (-2s+3) \end{bmatrix}}{s^3+2s^2-3s} = \frac{s^2-s-3}{s^3+2s^2-3s}$$

Expanding this transfer function in partial fractions, there results

$$T(s) = \frac{s^2-s-3}{s(s-1)(s+3)} = \frac{1}{s} + \frac{-\frac{3}{4}}{s-1} + \frac{\frac{3}{4}}{s+3}$$

which may be considered as the tandem (or parallel) connection of first-order systems shown in Fig. 7.7a. Each of these first-order subsystems is drawn in state variable form in Fig. 7.7b, where the three integrator output signals are labeled as state variables. The state variable equations for this alternate system representation, which has the same transfer function as the original system, are

$$\dot{x}_1' = r$$

$$\dot{x}_2' = x_2' + r$$

$$\dot{x}_3' = -3x_3' + r$$

$$y = x_1' - \tfrac{3}{4}x_2' + \tfrac{3}{4}x_3'$$

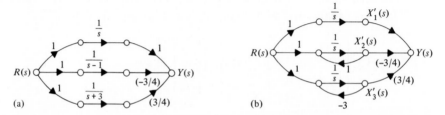

Figure 7.7 Diagonalizing a single-input, single-output system. **(a)** Tandem first-order subsystems from the partial fraction expansion of the transfer function. **(b)** Subsystems in simulation diagram form.

or

$$
\begin{bmatrix} \dot{x}'_1 \\ \dot{x}'_2 \\ \dot{x}'_3 \end{bmatrix} = \begin{bmatrix} 0 & 0 & 0 \\ 0 & 1 & 0 \\ 0 & 0 & -3 \end{bmatrix} \begin{bmatrix} x'_1 \\ x'_2 \\ x'_3 \end{bmatrix} + \begin{bmatrix} 1 \\ 1 \\ 1 \end{bmatrix} r
$$

$$
y = \begin{bmatrix} 1 & -\frac{3}{4} & \frac{3}{4} \end{bmatrix} \begin{bmatrix} x'_1 \\ x'_2 \\ x'_3 \end{bmatrix}
$$

which is diagonal.

Additional outputs of the same system may be added by adding further couplings of the state variables. The related single-input, two-output system of Fig. 7.8a has the following transfer functions:

$$
T_{11}(s) = \frac{Y_1(s)}{R(s)}\bigg|_{\substack{\text{initial} \\ \text{conditions}=0}} = \frac{1}{s} + \frac{-\frac{3}{4}}{s-1} + \frac{\frac{3}{4}}{s+3}
$$

$$
= \frac{s^2 - s - 3}{s(s-1)(s+3)}
$$

$$
T_{21}(s) = \frac{Y_2(s)}{R(s)}\bigg|_{\substack{\text{initial} \\ \text{conditions}=0}} = \frac{3}{s} + \frac{-2}{s-1} + \frac{\frac{3}{4}}{s+3}
$$

$$
= \frac{\frac{7}{4}s^2 - \frac{3}{4}s - 9}{s(s-1)(s+3)}
$$

An alternative diagonalized form for the original single-input, single-output system is given in Fig. 7.8b. In the new arrangement, the gains of 1, $-\frac{3}{4}$, and $\frac{3}{4}$ are placed on the input end of the diagram instead of the output end. The state variable equations for the modified system are slightly different,

$$
\begin{bmatrix} \dot{x}''_1 \\ \dot{x}''_2 \\ \dot{x}''_3 \end{bmatrix} = \begin{bmatrix} 0 & 0 & 0 \\ 0 & 1 & 0 \\ 0 & 0 & -3 \end{bmatrix} \begin{bmatrix} x''_1 \\ x''_2 \\ x''_3 \end{bmatrix} + \begin{bmatrix} 1 \\ -\frac{3}{4} \\ \frac{3}{4} \end{bmatrix} r
$$

$$
y = \begin{bmatrix} 1 & 1 & 1 \end{bmatrix} \begin{bmatrix} x''_1 \\ x''_2 \\ x''_3 \end{bmatrix}
$$

although the transfer function for the system remains the same:

$$
T(s) = \frac{Y(s)}{R(s)}\bigg|_{\substack{\text{initial} \\ \text{conditions}=0}} = \frac{1}{s} + \frac{-\frac{3}{4}}{s-1} + \frac{\frac{3}{4}}{s+3}
$$

$$
= \frac{s^2 - s - 3}{s(s-1)(s+3)}
$$

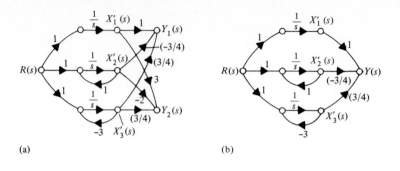

(a)   (b)

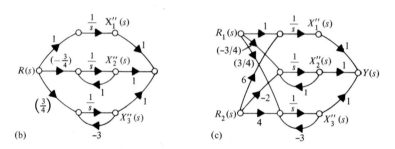

(b)   (c)

Figure 7.8 Diagonalizing multiple-output and multiple-input systems.   **(a)** Diagonalized one-input, multiple-output system.   **(b)** Alternative diagonal forms for the single-input, single-output example system.   **(c)** Diagonalized multiple-input, one-output system.

Different input coupling gains will give different transfer function numerator polynomials. The numerator polynomials for multiple-input, single-output systems may then be chosen at will by this method. The two-input, one-output system of Fig. 7.8c has the following two transfer functions:

$$T_{11}(s) = \left.\frac{Y(s)}{R_1(s)}\right|_{\substack{\text{initial} \\ \text{conditions} \\ \text{and } R_2 = 0}} = \frac{1}{s} + \frac{-\frac{3}{4}}{s-1} + \frac{\frac{3}{4}}{s+3}$$

$$= \frac{s^2 - s - 3}{s(s-1)(s+3)}$$

$$T_{12}(s) = \left.\frac{Y(s)}{R_2(s)}\right|_{\substack{\text{initial} \\ \text{conditions} \\ \text{and } R_1 = 0}} = \frac{6}{s} + \frac{-2}{s-1} + \frac{4}{s+3}$$

$$= \frac{8s^2 + 2s - 18}{s(s-1)(s+3)}$$

For the case of both multiple inputs and multiple outputs, both the input coupling gains and the output coupling gains must be selected. Two inputs and two

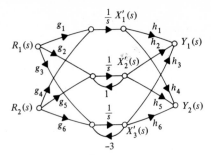

Figure 7.9 A diagonalized two-input, two-output system.

outputs in the example system, as in Fig. 7.9, would result in four transfer functions, of the form

$$T_{11}(s) = \frac{Y_1(s)}{R_1(s)}\Big|_{\substack{\text{initial}\\ \text{conditions}\\ \text{and } R_2 = 0}} = \frac{k_1 s^2 + k_2 s + k_3}{s(s-1)(s+3)}$$

$$T_{12}(s) = \frac{Y_1(s)}{R_2(s)}\Big|_{\substack{\text{initial}\\ \text{conditions}\\ \text{and } R_1 = 0}} = \frac{k_4 s^2 + k_5 s + k_6}{s(s-1)(s+3)}$$

$$T_{21}(s) = \frac{Y_2(s)}{R_1(s)}\Big|_{\substack{\text{initial}\\ \text{conditions}\\ \text{and } R_2 = 0}} = \frac{k_7 s^2 + k_8 s + k_9}{s(s-1)(s+3)}$$

$$T_{22}(s) = \frac{Y_2(s)}{R_2(s)}\Big|_{\substack{\text{initial}\\ \text{conditions}\\ \text{and } R_1 = 0}} = \frac{k_{10} s^2 + k_{11} s + k_{12}}{s(s-1)(s+3)}$$

where the 12 k terms are determined by the 12 gains, $g_1, \ldots, g_6$ and $h_1, \ldots, h_6$.

DRILL PROBLEMS

D7.6. Use the partial fraction method to find diagonal state equations for single-input, single-output systems with the following transfer functions:

(a) $T(s) = \dfrac{-5s + 7}{s^2 + 7s + 12}$

$$\textbf{Ans} \quad \begin{bmatrix} \dot{x}_1 \\ \dot{x}_2 \end{bmatrix} = \begin{bmatrix} -3 & 0 \\ 0 & -4 \end{bmatrix} \begin{bmatrix} x_1 \\ x_2 \end{bmatrix} + \begin{bmatrix} 22 \\ -27 \end{bmatrix} r$$

$$y = \begin{bmatrix} 1 & 1 \end{bmatrix} \begin{bmatrix} x_1 \\ x_2 \end{bmatrix}$$

(b) $T(s) = \dfrac{3s^2 - 2}{(s + 1)(s + 4)(s + 10)}$

Ans $\begin{bmatrix} \dot{x}_1 \\ \dot{x}_2 \\ \dot{x}_3 \end{bmatrix} = \begin{bmatrix} -1 & 0 & 0 \\ 0 & -4 & 0 \\ 0 & 0 & -10 \end{bmatrix} \begin{bmatrix} x_1 \\ x_2 \\ x_3 \end{bmatrix} + \begin{bmatrix} \frac{1}{27} \\ -\frac{23}{9} \\ \frac{149}{27} \end{bmatrix} r$

$y = \begin{bmatrix} 1 & 1 & 1 \end{bmatrix} \begin{bmatrix} x_1 \\ x_2 \\ x_3 \end{bmatrix}$

(c) $T(s) = \dfrac{4}{s^3 + 3s^2 + 2s}$

Ans $\begin{bmatrix} \dot{x}_1 \\ \dot{x}_2 \\ \dot{x}_3 \end{bmatrix} = \begin{bmatrix} 0 & 0 & 0 \\ 0 & -1 & 0 \\ 0 & 0 & -2 \end{bmatrix} \begin{bmatrix} x_1 \\ x_2 \\ x_3 \end{bmatrix} + \begin{bmatrix} 1 \\ 1 \\ 1 \end{bmatrix} r$

$y = \begin{bmatrix} 2 & -4 & 2 \end{bmatrix} \begin{bmatrix} x_1 \\ x_2 \\ x_3 \end{bmatrix}$

D7.7. The following systems have real characteristic roots. Find alternative diagonal state equations, and then draw a simulation diagram for the new equations.

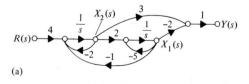

(a)

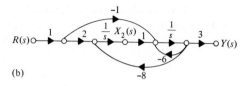

(b)

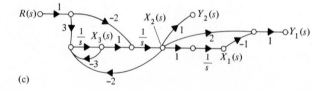

(c)

Ans (a) $\begin{bmatrix} \dot{x}'_1 \\ \dot{x}'_2 \end{bmatrix} = \begin{bmatrix} -3 & 0 \\ 0 & -4 \end{bmatrix} \begin{bmatrix} x'_1 \\ x'_2 \end{bmatrix} + \begin{bmatrix} 8 \\ 4 \end{bmatrix} r$

$y = \begin{bmatrix} 1 & 1 \end{bmatrix} \begin{bmatrix} x'_1 \\ x'_2 \end{bmatrix}$

(b) $\begin{bmatrix} \dot{x}'_1 \\ \dot{x}'_2 \end{bmatrix} = \begin{bmatrix} -2 & 0 \\ 0 & -4 \end{bmatrix} \begin{bmatrix} x'_1 \\ x'_2 \end{bmatrix} + \begin{bmatrix} 1 \\ 1 \end{bmatrix} r$

$y = \begin{bmatrix} 6 & -9 \end{bmatrix} \begin{bmatrix} x'_1 \\ x'_2 \end{bmatrix}$

(c) $\begin{bmatrix} \dot{x}'_1 \\ \dot{x}'_2 \\ \dot{x}'_3 \end{bmatrix} = \begin{bmatrix} 0 & 0 & 0 \\ 0 & -1 & 0 \\ 0 & 0 & -2 \end{bmatrix} \begin{bmatrix} x'_1 \\ x'_2 \\ x'_3 \end{bmatrix} + \begin{bmatrix} 1 \\ 1 \\ 1 \end{bmatrix} r$

$y = \begin{bmatrix} \frac{3}{2} & -3 & -\frac{5}{2} \\ 0 & -1 & -1 \end{bmatrix} \begin{bmatrix} x'_1 \\ x'_2 \\ x'_3 \end{bmatrix}$

Complex Conjugate Characteristic Roots

In general, diagonalized state equations for systems with complex characteristic roots involve state equations with complex coefficients. For example, the single-input, single-output system with transfer function

$$T(s) = \frac{6s^2 + 26s + 8}{(s + 2)(s^2 + 2s + 10)} = \frac{-2}{s + 2} + \frac{4 + j}{s + 1 + j3} + \frac{4 - j}{s + 1 - j3}$$

may be represented in terms of state variables as in the simulation diagram of Fig. 7.10a. The gains associated with the complex characteristic roots are generally complex numbers. The state equations, in terms of the indicated state variables, are given by

$$sX_1(s) = -2X_1(s) + R(s)$$

$$sX_2(s) = (-1 + j3)X_2(s) + R(s)$$

$$sX_3(s) = (-1 - j3)X_3(s) + R(s)$$

$$Y(s) = -2X_1(s) + (4 + j)X_2(s) + (4 - j)X_3(s)$$

or

$$\begin{bmatrix} \dot{x}_1 \\ \dot{x}_2 \\ \dot{x}_3 \end{bmatrix} = \begin{bmatrix} -2 & 0 & 0 \\ 0 & -1 + j3 & 0 \\ 0 & 0 & -1 - j3 \end{bmatrix} \begin{bmatrix} x_1 \\ x_2 \\ x_3 \end{bmatrix} + \begin{bmatrix} 1 \\ 1 \\ 1 \end{bmatrix} r$$

$$y = \begin{bmatrix} -2 & 4 + j & 4 - j \end{bmatrix} \begin{bmatrix} x_1 \\ x_2 \\ x_3 \end{bmatrix}$$

Although the individual physical components of this representation cannot be assembled, involving complex numbers as they do, the mathematical relationships are valid. To build such a system, or to represent it in a convenient form which does not involve complex numbers, the two complex conjugate component parts may be combined just as one commonly combines the corresponding conjugate partial fraction terms:

$$\frac{4 + j}{s + 1 + j3} + \frac{4 - j}{s + 1 - j3} = \frac{8s + 14}{s^2 + 2s + 10}$$

This portion of the system may be represented in phase-variable form, giving the real-number simulation diagram of Fig. 7.10b. The state equations for this alternative arrangement are given by

$$sX'_1(s) = -2X'_1(s) + R(s)$$

$$sX'_2(s) = X'_3(s)$$

$$sX'_3(s) = -10X'_2(s) - 2X'_3(s) + R(s)$$

$$Y(s) = -2X'_1(s) + 14X'_2(s) + 8X'_3(s)$$

or

$$\begin{bmatrix} \dot{x}'_1 \\ \dot{x}'_2 \\ \dot{x}'_3 \end{bmatrix} = \begin{bmatrix} -2 & 0 & 0 \\ 0 & 0 & 1 \\ 0 & -10 & -2 \end{bmatrix} \begin{bmatrix} x'_1 \\ x'_2 \\ x'_3 \end{bmatrix} + \begin{bmatrix} 1 \\ 0 \\ 1 \end{bmatrix} r$$

$$y = \begin{bmatrix} -2 & 14 & 8 \end{bmatrix} \begin{bmatrix} x'_1 \\ x'_2 \\ x'_3 \end{bmatrix}$$

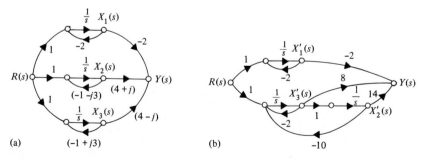

(a)   (b)

Figure 7.10 A system with complex characteristic roots. **(a)** Diagonalized system. **(b)** Alternative form for the diagonalized system, where the complex conjugate root terms have been combined and placed in phase-variable form.

It is thus possible to represent systems with one or more pairs of complex conjugate characteristic roots with diagonalized state equations involving complex numbers or in block diagonal form involving real numbers. The state equations

$$
\begin{bmatrix} \dot{x}_1 \\ \dot{x}_2 \\ \dot{x}_3 \\ \dot{x}_4 \\ \dot{x}_5 \\ \dot{x}_6 \end{bmatrix} =
\begin{bmatrix}
3 & 0 & 0 & 0 & 0 & 0 \\
0 & -4 & 0 & 0 & 0 & 0 \\
0 & 0 & 0 & 1 & 0 & 0 \\
0 & 0 & -17 & -2 & 0 & 0 \\
0 & 0 & 0 & 0 & 0 & 1 \\
0 & 0 & 0 & 0 & -10 & 3
\end{bmatrix}
\begin{bmatrix} x_1 \\ x_2 \\ x_3 \\ x_4 \\ x_5 \\ x_6 \end{bmatrix} +
\begin{bmatrix} 1 \\ 1 \\ 0 \\ 1 \\ 0 \\ 1 \end{bmatrix} r
$$

$$
y = \begin{bmatrix} 6 & -8 & 1 & -5 & 0 & 7 \end{bmatrix}
\begin{bmatrix} x_1 \\ x_2 \\ x_3 \\ x_4 \\ x_5 \\ x_6 \end{bmatrix}
$$

for example, which are in block diagonal form, represent a system with transfer function

$$
T(s) = \frac{6}{s-3} + \frac{-8}{s+4} + \frac{-5s+1}{s_2 + 2s + 17} + \frac{7s}{s^2 - 3s + 10}
$$

DRILL PROBLEM

D7.8. The following transfer functions for single-input, single-output systems involve complex characteristic roots. Find diagonal state equations for these systems. Then find an alternative block diagonal representation which does not involve complex numbers.

(a) $T(s) = \dfrac{10}{s^3 + 2s + 5s}$

Ans
$$
\begin{bmatrix} \dot{x}_1 \\ \dot{x}_2 \\ \dot{x}_3 \end{bmatrix} =
\begin{bmatrix} 0 & 0 & 0 \\ 0 & 0 & 1 \\ 0 & -5 & -2 \end{bmatrix}
\begin{bmatrix} x_1 \\ x_2 \\ x_3 \end{bmatrix} +
\begin{bmatrix} 2 \\ 0 \\ 1 \end{bmatrix} r
$$

$$
y = \begin{bmatrix} 1 & -4 & -2 \end{bmatrix}
\begin{bmatrix} x_1 \\ x_2 \\ x_3 \end{bmatrix}
$$

(b) $T(s) = \dfrac{3s^2 - 1}{(s^2 + 4)(s^2 + 4s + 5)}$

Ans
$$\begin{bmatrix} \dot{x}_1 \\ \dot{x}_2 \\ \dot{x}_3 \\ \dot{x}_4 \end{bmatrix} = \begin{bmatrix} 0 & 1 & 0 & 0 \\ -4 & 0 & 0 & 0 \\ 0 & 0 & 0 & 1 \\ 0 & 0 & -5 & -4 \end{bmatrix} \begin{bmatrix} x_1 \\ x_2 \\ x_3 \\ x_4 \end{bmatrix} + \begin{bmatrix} 0 \\ 1 \\ 0 \\ 1 \end{bmatrix} r$$

$$y = \begin{bmatrix} -\frac{1}{5} & \frac{4}{5} & 0 & -\frac{4}{5} \end{bmatrix} \begin{bmatrix} x_1 \\ x_2 \\ x_3 \\ x_4 \end{bmatrix}$$

(c) $T(s) = \dfrac{s^2 - 4s + 10}{(s + 2)(s^2 + 6s + 13)}$

Ans
$$\begin{bmatrix} \dot{x}_1 \\ \dot{x}_2 \\ \dot{x}_3 \end{bmatrix} = \begin{bmatrix} -2 & 0 & 0 \\ 0 & 0 & 1 \\ 0 & -13 & -6 \end{bmatrix} \begin{bmatrix} x_1 \\ x_2 \\ x_3 \end{bmatrix} + \begin{bmatrix} 1 \\ 0 \\ 1 \end{bmatrix} r$$

$$y = \begin{bmatrix} \frac{22}{5} & -\frac{118}{5} & -\frac{17}{5} \end{bmatrix} \begin{bmatrix} x_1 \\ x_2 \\ x_3 \end{bmatrix}$$

Repeated Characteristic Roots

The state equations for a system with repeated characteristic roots cannot be diagonalized. A block diagonal form, termed a *Jordan canonical form*, is commonly used when a simple representation is desired. For example, the single-input, single-output system with transfer function

$$T(s) = \frac{10s^2 + 51s + 56}{(s + 4)(s + 2)^2} = \frac{3}{s + 4} + \frac{-6}{s + 2} + \frac{7}{(s + 2)^2}$$

may be represented as in Fig. 7.11a. A simplification results when the $1/(s + 2)$ transmittance is used in common by two paths, as shown in Fig. 7.11b. A corresponding state variable representation is given in Fig. 7.11c.

The state equations are given by

$$sX_1(s) = -4X_1(s) + R(s)$$

$$sX_2(s) = -2X_2(s) + X_3(s)$$

$$sX_3(s) = -2X_3(s) + R(s)$$

$$Y(s) = 3X_1(s) + 7X_2(s) - 6X_3(s)$$

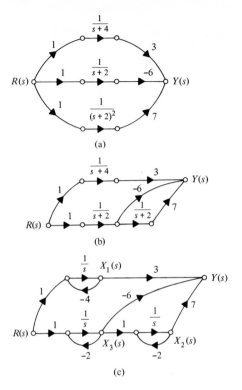

Figure 7.11 State equations for a system with repeated characteristic roots.
(a) Diagram showing each partial fraction term. **(b)** Diagram with common signal path through a repeated transmittance. **(c)** Diagram showing state variables.

or

$$
\begin{bmatrix} \dot{x}_1 \\ \dot{x}_2 \\ \dot{x}_3 \end{bmatrix} =
\begin{bmatrix} -4 & 0 & 0 \\ 0 & -2 & 1 \\ 0 & 0 & -2 \end{bmatrix}
\begin{bmatrix} x_1 \\ x_2 \\ x_3 \end{bmatrix} +
\begin{bmatrix} 1 \\ 0 \\ 1 \end{bmatrix} r
$$

$$
y = \begin{bmatrix} 3 & 7 & -6 \end{bmatrix}
\begin{bmatrix} x_1 \\ x_2 \\ x_3 \end{bmatrix}
$$

For three repetitions of a characteristic root, the corresponding transfer function partial fraction terms are

$$
\frac{k_1}{s+a} + \frac{k_2}{(s+a)^2} + \frac{k_3}{(s+a)^3}
$$

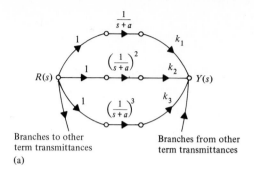

Branches to other
term transmittances

Branches from other
term transmittances

(a)

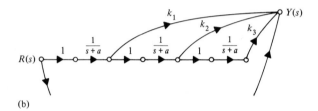

(b)

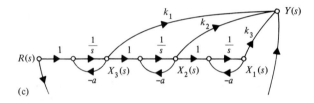

(c)

Figure 7.12   State variables for repeated roots.   **(a)** Diagram showing each partial
fraction term.   **(b)** Diagram using common signal paths.   **(c)** Diagram showing state
variables.

and the state variables may be defined as in Fig. 7.12. The resulting Jordan block
has the following structure:

$$
\begin{bmatrix} \dot{x}_1 \\ \dot{x}_2 \\ \dot{x}_3 \\ \vdots \end{bmatrix} =
\begin{bmatrix}
-a & 1 & 0 & 0 & 0 & \cdots \\
0 & -a & 1 & 0 & 0 & \cdots \\
0 & 0 & -a & 0 & 0 & \cdots \\
0 & 0 & 0 & & & \\
\vdots & \vdots & \vdots & \vdots & \vdots &
\end{bmatrix}
\begin{bmatrix} x_1 \\ x_2 \\ x_3 \\ \vdots \end{bmatrix} +
\begin{bmatrix} 0 \\ 0 \\ 1 \\ \vdots \end{bmatrix} r
$$

$$
y = \begin{bmatrix} k_3 & k_2 & k_1 & \cdots \end{bmatrix}
\begin{bmatrix} x_1 \\ x_2 \\ x_3 \\ \vdots \end{bmatrix}
$$

The state variable equations

$$\begin{bmatrix} \dot{x}_1 \\ \dot{x}_2 \\ \dot{x}_3 \\ \dot{x}_4 \\ \dot{x}_5 \\ \dot{x}_6 \end{bmatrix} = \begin{bmatrix} -2 & 1 & 0 & 0 & 0 & 0 \\ 0 & -2 & 0 & 0 & 0 & 0 \\ 0 & 0 & -3 & 0 & 0 & 0 \\ 0 & 0 & 0 & 4 & 1 & 0 \\ 0 & 0 & 0 & 0 & 4 & 1 \\ 0 & 0 & 0 & 0 & 0 & 4 \end{bmatrix} \begin{bmatrix} x_1 \\ x_2 \\ x_3 \\ x_4 \\ x_5 \\ x_6 \end{bmatrix} + \begin{bmatrix} 0 \\ 1 \\ 1 \\ 0 \\ 0 \\ 1 \end{bmatrix} r$$

$$y = [4 \quad -5 \quad 6 \quad 7 \quad -8 \quad 9] \begin{bmatrix} x_1 \\ x_2 \\ x_3 \\ x_4 \\ x_5 \\ x_6 \end{bmatrix}$$

for example, are in Jordan canonical form. They represent a system with transfer function

$$T(s) = \frac{-5}{s+2} + \frac{4}{(s+2)^2} + \frac{6}{s+3} + \frac{9}{s-4} + \frac{-8}{(s-4)^2} + \frac{7}{(s-4)^3}$$

DRILL PROBLEM

D7.9. The following systems have repeated characteristic roots. Find an alternate set of state equations in Jordan canonical form.

(a) $\begin{bmatrix} \dot{x}_1 \\ \dot{x}_2 \end{bmatrix} = \begin{bmatrix} 2 & 9 \\ -1 & -4 \end{bmatrix} \begin{bmatrix} x_1 \\ x_2 \end{bmatrix} + \begin{bmatrix} 4 \\ -3 \end{bmatrix} r$

$y = [2 \quad -6] \begin{bmatrix} x_1 \\ x_2 \end{bmatrix}$

Ans $\begin{bmatrix} \dot{x}'_1 \\ \dot{x}'_2 \end{bmatrix} = \begin{bmatrix} -1 & 1 \\ 0 & -1 \end{bmatrix} \begin{bmatrix} x'_1 \\ x'_2 \end{bmatrix} + \begin{bmatrix} 0 \\ 1 \end{bmatrix} r$

$y = [-60 \quad 26] \begin{bmatrix} x'_1 \\ x'_2 \end{bmatrix}$

(b) $\begin{bmatrix} \dot{x}_1 \\ \dot{x}_2 \\ \dot{x}_3 \end{bmatrix} = \begin{bmatrix} -6 & 1 & 0 \\ -9 & 0 & 1 \\ 0 & 0 & 0 \end{bmatrix} \begin{bmatrix} x_1 \\ x_1 \\ x_3 \end{bmatrix} + \begin{bmatrix} 1 \\ 2 \\ -1 \end{bmatrix} r$

$y = [3 \quad -2 \quad 1] \begin{bmatrix} x_1 \\ x_2 \\ x_3 \end{bmatrix}$

$$\textbf{Ans} \quad \begin{bmatrix} \dot{x}'_1 \\ \dot{x}'_2 \\ \dot{x}'_3 \end{bmatrix} = \begin{bmatrix} -3 & 1 & 0 \\ 0 & -3 & 1 \\ 0 & 0 & -3 \end{bmatrix} \begin{bmatrix} x'_1 \\ x'_2 \\ x'_3 \end{bmatrix} + \begin{bmatrix} 0 \\ 0 \\ 1 \end{bmatrix} r$$

$$y = \begin{bmatrix} 2 & -2 & 0 \end{bmatrix} \begin{bmatrix} x'_1 \\ x'_2 \\ x'_3 \end{bmatrix}$$

7.5 CONTROLLABILITY AND OBSERVABILITY

In the previous chapters, a control system was operated upon to provide acceptable phase margin, rise time, or other figures of merit. Perhaps the system is constructed in a way that conspires to thwart efforts at improving performance. By using state variable methods it is possible to answer fundamental questions about the ability of the control system designer to affect meaningful improvement in performance and to generate needed sensor measurements. The terms controllability and observability address those needs respectively.

A system is completely controllable if the system state $x(t_f)$ at time t_f can be forced to take on any desired value by applying a control input $r(t)$ over a period of time from t_0 until t_f. The definition does not restrict the choice of $\mathbf{r}(t)$. The idea is that it is possible to move the system state to any desired destination. Perhaps the system is (or is not) constructed in a way that allows control to take place. A test for controllability can easily be constructed.

A system is completely observable if any initial state vector $\mathbf{x}(t_0)$ can be reconstructed by examining the system output $\mathbf{y}(t)$ over some period of time from t_0 until t_f. There are no restrictions placed on the output. The definition indicates that any earlier value of the state vector is determinable by watching the output $\mathbf{y}(t)$. An automobile would be considered completely observable if, by monitoring speedometer (for speed), odometer (for distance), and steering wheel position (for turning), it is possible to determine where the car was parked before being driven.

For certain kinds of systems (with diagonal **A** matrices), the tests for controllability and observability are easy to apply. For a nondiagonal system, a test can also be constructed. For a system that is completely controllable, methods will be developed by which an appropriate control can be derived. Similarly, for a system that is completely observable, an observer will be designed to carry out that task of state reconstruction.

Figure 7.13 shows that controllability is tested assuming a zero-state response and that observability is tested assuming a zero-input response. The tests provide a worst-case scenario where the initial condition does not necessarily aid in control and an input does not necessarily aid in reconstruction of an earlier state. A system that passes the controllability test is usually applied in an environment that has a nonzero initial condition. Similarly, a system that passes the observability test (observers will be considered shortly) is usually applied in an environment that includes an input and control.

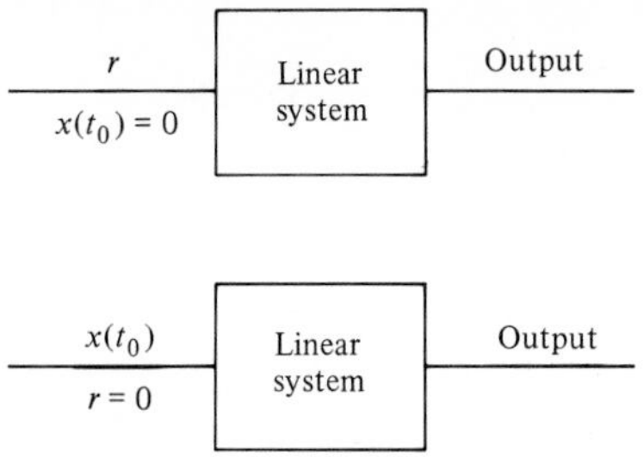

Figure 7.13 Significance of controllability and observability tests. **(a)** Test for controllability. **(b)** Test for observability.

Uncontrollable and Unobservable Modes

When a system is placed in diagonal form, its state equations are decoupled from one another. The natural response terms in each of the individual equation solutions are called *modes*. The natural component of each system output consists of a linear combination of the system modes. For example, the diagonal system

$$
\begin{bmatrix} \dot{x}_1 \\ \dot{x}_2 \\ \dot{x}_3 \end{bmatrix} = \begin{bmatrix} 8 & 0 & 0 \\ 0 & -4 & 0 \\ 0 & 0 & -10 \end{bmatrix} \begin{bmatrix} x_1 \\ x_2 \\ x_3 \end{bmatrix} + \begin{bmatrix} 1 \\ -2 \\ 3 \end{bmatrix} r
$$

$$
y = \begin{bmatrix} 4 & -5 & -6 \end{bmatrix} \begin{bmatrix} x_1 \\ x_2 \\ x_3 \end{bmatrix}
$$

is described by the decoupled differential equations

$$
\dot{x}_1 = 8x_1 + r
$$

$$
\dot{x}_2 = -4x_2 - 2r
$$

$$
\dot{x}_3 = -10x_3 + 3r
$$

The state variables have natural response components of the following form:

$$
x_{1,\,\text{natural}}(t) = K_1 e^{8t}
$$

$$
x_{2,\,\text{natural}}(t) = K_2 e^{-4t}
$$

$$
x_{3,\,\text{natural}}(t) = K_3 e^{-10t}
$$

If, for a *diagonalized* system, there is no path from any input to one of the decoupled equations, as in Fig. 7.14a, the corresponding mode is termed *uncontrollable*. That portion of the system is not affected by any input. If it has no uncontrollable modes, a system is said to be completely controllable.

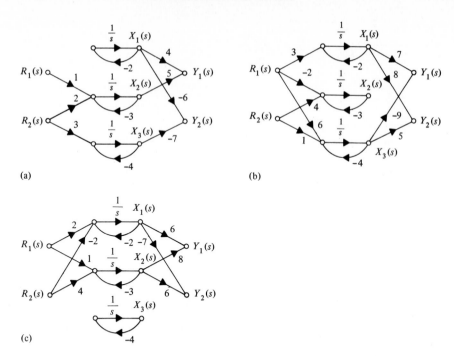

Figure 7.14 Systems with uncontrollable and unobservable modes. **(a)** Systems where the e^{-2t} mode is uncontrollable. **(b)** System where the e^{-3t} mode is unobservable. **(c)** System where the e^{-4t} mode is uncontrollable and unobservable.

For a *diagonalized* system, if one of the state variables of the decoupled equations is not added to any of the system outputs, as in Fig. 7.14b, the corresponding mode is termed *unobservable*. The response of that first-order differential equation is not visible from any output. A system with no unobservable modes is said to be completely observable. Modes may be both uncontrollable and unobservable, as in the example of Fig. 7.14c.

The system with diagonalized state equations

$$
\begin{bmatrix} \dot{x}_1 \\ \dot{x}_2 \\ \dot{x}_3 \end{bmatrix} = \begin{bmatrix} 2 & 0 & 0 \\ 0 & -3 & 0 \\ 0 & 0 & -4 \end{bmatrix} \begin{bmatrix} x_1 \\ x_2 \\ x_3 \end{bmatrix} + \begin{bmatrix} 1 & -2 \\ 0 & 0 \\ 0 & 0 \end{bmatrix} \begin{bmatrix} r_1 \\ r_2 \end{bmatrix}
$$

is not completely controllable. For this system, both the e^{-3t} and the e^{-4t} modes are not coupled to any input. The diagonal system

$$
\begin{bmatrix} \dot{x}_1 \\ \dot{x}_2 \\ \dot{x}_3 \end{bmatrix} = \begin{bmatrix} 3+j & 0 & 0 \\ 0 & 3-j & 0 \\ 0 & 0 & 6 \end{bmatrix} \begin{bmatrix} x_1 \\ x_2 \\ x_3 \end{bmatrix} + \begin{bmatrix} -1 & 2j \\ -1 & -2j \\ 0 & 4 \end{bmatrix} \begin{bmatrix} r_1 \\ r_2 \end{bmatrix}
$$

is completely controllable since the rows of its input coupling matrix are each nonzero.

The system with diagonalized state equations

$$
\begin{bmatrix} \dot{x}_1 \\ \dot{x}_2 \\ \dot{x}_3 \end{bmatrix} = \begin{bmatrix} -2 & 0 & 0 \\ 0 & 1 & 0 \\ 0 & 0 & 4 \end{bmatrix} \begin{bmatrix} x_1 \\ x_2 \\ x_3 \end{bmatrix} + \begin{bmatrix} 1 \\ 0 \\ 0 \end{bmatrix} r
$$

$$
\begin{bmatrix} y_1 \\ y_2 \end{bmatrix} = \begin{bmatrix} 1 & 0 & 3 \\ 0 & -2 & 3 \end{bmatrix} \begin{bmatrix} x_1 \\ x_2 \\ x_3 \end{bmatrix}
$$

is completely observable since there are no columns of zeros in its output coupling matrix. The diagonalized system

$$
\begin{bmatrix} \dot{x}_1 \\ \dot{x}_2 \end{bmatrix} = \begin{bmatrix} -3 & 0 \\ 0 & -10 \end{bmatrix} \begin{bmatrix} x_1 \\ x_2 \end{bmatrix} + \begin{bmatrix} 1 & 2 \\ 0 & 4 \end{bmatrix} \begin{bmatrix} r_1 \\ r_2 \end{bmatrix}
$$

$$
y = \begin{bmatrix} 1 & 0 \end{bmatrix} \begin{bmatrix} x_1 \\ x_2 \end{bmatrix}
$$

is not completely observable since the e^{-10t} mode does not couple to the system output.

The system with diagonalized state equations

$$
\begin{bmatrix} \dot{x}_1 \\ \dot{x}_2 \\ \dot{x}_3 \end{bmatrix} = \begin{bmatrix} -3 & 0 & 0 \\ 0 & 2 & 0 \\ 0 & 0 & 4 \end{bmatrix} \begin{bmatrix} x_1 \\ x_2 \\ x_3 \end{bmatrix} + \begin{bmatrix} -1 \\ 0 \\ 0 \end{bmatrix} r
$$

$$
y = \begin{bmatrix} 0 & 6 & 0 \end{bmatrix} \begin{bmatrix} x_1 \\ x_2 \\ x_3 \end{bmatrix}
$$

is neither completely controllable nor completely observable. The e^{-3t} mode is controllable but not observable; the e^{2t} mode is observable but not controllable; while the e^{4t} mode is both uncontrollable and unobservable.

For a system which is not diagonal, the presence or absence of complete controllability or complete observability is not obvious. A row of zeros in the input coupling matrix does *not* indicate lack of complete controllability; when diagonalized, such a system may or may not have a row of zeros in the transformed input coupling matrix. Similarly, a column of zeros in the output coupling matrix of a nondiagonal system does not necessarily indicate absence of complete observability.

For systems with repeated characteristic roots, which cannot be diagonalized, controllability and observability may be determined by examining the system matrices in Jordan form.

The Controllability Matrix

Fortunately, there is a much simpler method of determining system controllability than diagonalization. It can be shown that an nth order system, with or without repeated characteristic roots,

$$\dot{x} = Ax + Br$$

is completely controllable if and only if its controllability matrix

$$M_c = [B \mid AB \mid \cdots \mid A^{n-1}B]$$

is of full rank. The controllability matrix consists of the columns of B followed by the columns of AB, and so on.

For the system

$$\begin{bmatrix} \dot{x}_1 \\ \dot{x}_2 \\ \dot{x}_3 \end{bmatrix} = \begin{bmatrix} -2 & 1 & 2 \\ 4 & 0 & 3 \\ 1 & -1 & 0 \end{bmatrix} \begin{bmatrix} x_1 \\ x_2 \\ x_3 \end{bmatrix} + \begin{bmatrix} 0 & 4 \\ -5 & 0 \\ 0 & 0 \end{bmatrix} \begin{bmatrix} r_1 \\ r_2 \end{bmatrix}$$

A is 3×3 so

$$M_c = [B \mid AB \mid A^2B]$$

Using

$$AB = \begin{bmatrix} -2 & 1 & 2 \\ 4 & 0 & 3 \\ 1 & -1 & 0 \end{bmatrix} \begin{bmatrix} 0 & 4 \\ -5 & 0 \\ 0 & 0 \end{bmatrix} = \begin{bmatrix} -5 & -8 \\ 0 & 16 \\ 5 & 4 \end{bmatrix}$$

$$A^2B = A(AB) = \begin{bmatrix} -2 & 1 & 2 \\ 4 & 0 & 3 \\ 1 & -1 & 0 \end{bmatrix} \begin{bmatrix} -5 & -8 \\ 0 & 16 \\ 5 & 4 \end{bmatrix} = \begin{bmatrix} 20 & 40 \\ -5 & -20 \\ -5 & -24 \end{bmatrix}$$

then

$$M_c = \begin{bmatrix} 0 & 4 & -5 & -8 & 20 & 40 \\ -5 & 0 & 0 & 16 & -5 & -20 \\ 0 & 0 & 5 & 4 & -5 & -24 \end{bmatrix}$$

To be of full rank, the controllability matrix must have three linearly independent columns, which it does, since

$$\begin{vmatrix} 0 & 4 & -5 \\ -5 & 0 & 0 \\ 0 & 0 & 5 \end{vmatrix} \ne 0$$

The system

$$\begin{bmatrix} \dot{x}_1 \\ \dot{x}_2 \end{bmatrix} = \begin{bmatrix} 2 & 3 \\ 6 & -1 \end{bmatrix} \begin{bmatrix} x_1 \\ x_2 \end{bmatrix} + \begin{bmatrix} 1 \\ -2 \end{bmatrix} r$$

has controllability matrix

$$\mathbf{M}_c = [\mathbf{B} \vdots \mathbf{AB}]$$

where

$$\mathbf{AB} = \begin{bmatrix} 2 & 3 \\ 6 & -1 \end{bmatrix} \begin{bmatrix} 1 \\ -2 \end{bmatrix} = \begin{bmatrix} -4 \\ 8 \end{bmatrix}$$

Thus

$$\mathbf{M}_c = \begin{bmatrix} 1 & -4 \\ -2 & 8 \end{bmatrix}$$

which is not of rank 2 since

$$\begin{vmatrix} 1 & -4 \\ -2 & 8 \end{vmatrix} = 0$$

This system is not completely controllable.

The rank test with the controllability matrix does not indicate which mode is uncontrollable, but it is far simpler to apply the test than to diagonalize the state equations.

The Observability Matrix

To determine whether or not a nondiagonalized nth-order system is completely observable, its observability matrix

$$\mathbf{M}_o = \begin{bmatrix} \mathbf{C} \\ \hline \mathbf{CA} \\ \hline \vdots \\ \hline \mathbf{CA}^{n-1} \end{bmatrix}$$

may be formed. The system is completely observable if and only if the observability matrix is of full rank, that is, if $\mathbf{M}_o$ has n linearly independent rows.

The system

$$\begin{bmatrix} \dot{x}_1 \\ \dot{x}_2 \\ \dot{x}_3 \end{bmatrix} = \begin{bmatrix} 2 & 1 & 0 \\ -3 & 0 & 1 \\ 4 & 0 & 0 \end{bmatrix} \begin{bmatrix} x_1 \\ x_2 \\ x_3 \end{bmatrix} = \begin{bmatrix} 1 \\ 1 \\ 1 \end{bmatrix} r$$

$$y = \begin{bmatrix} 0 & 0 & 1 \end{bmatrix} \begin{bmatrix} x_1 \\ x_2 \\ x_3 \end{bmatrix}$$

for example, is completely observable:

$$\mathbf{CA} = [0 \quad 0 \quad 1]\begin{bmatrix} 2 & 1 & 0 \\ -3 & 0 & 1 \\ 4 & 0 & 0 \end{bmatrix} = [4 \quad 0 \quad 0]$$

$$\mathbf{CA}^2 = (\mathbf{CA})\mathbf{A} = [4 \quad 0 \quad 0]\begin{bmatrix} 2 & 1 & 0 \\ -3 & 0 & 1 \\ 4 & 0 & 0 \end{bmatrix}$$

$$= [8 \quad 4 \quad 0]$$

$$\mathbf{M}_o = \begin{bmatrix} 0 & 0 & 1 \\ 4 & 0 & 0 \\ 8 & 4 & 0 \end{bmatrix}$$

As another example, the system

$$\begin{bmatrix} \dot{x}_1 \\ \dot{x}_2 \end{bmatrix} = \begin{bmatrix} 1 & 0 \\ 1 & 1 \end{bmatrix}\begin{bmatrix} x_1 \\ x_2 \end{bmatrix} + \begin{bmatrix} 1 \\ 1 \end{bmatrix}r$$

$$\begin{bmatrix} y_1 \\ y_2 \end{bmatrix} = \begin{bmatrix} 1 & -1 \\ -2 & 2 \end{bmatrix}\begin{bmatrix} x_1 \\ x_2 \end{bmatrix}$$

$$\mathbf{M}_o = \begin{bmatrix} 1 & -1 \\ -2 & 2 \\ 0 & -1 \\ 0 & 2 \end{bmatrix}$$

The observability matrix has two linearly independent rows (1 and 3).

It is a very special case when a system is not completely observable or not completely controllable. The parameters describing the system must be "just right" for such situations to occur. Nevertheless, the special cases do occur frequently enough in practice to warrant careful consideration.

DRILL PROBLEM

D7.10 Use controllability and observability matrices to determine whether the following systems are completely controllable and whether these systems are completely observable.

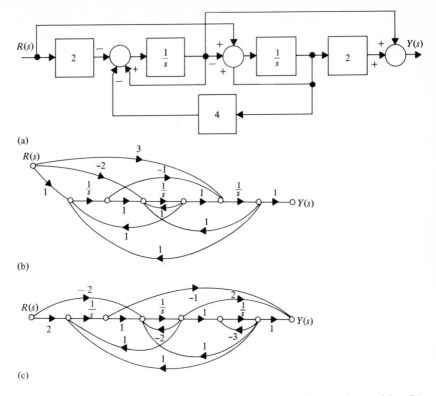

(a)

(b)

(c)

Ans (a) not completely controllable and not completely observable; (b) completely controllable and completely observable; (c) completely controllable but not completely observable

7.6 TIME RESPONSE FROM STATE EQUATIONS

Laplace Transform Solution

One method of calculating the state of a system as a function of time is to Laplace-transform the equations, solve for the transform of the signals of interest, then invert the transforms. The system outputs, being linear combinations of the state signals, are easily found from the state.

For example, consider the system

$$\begin{bmatrix} \dot{x}_1 \\ \dot{x}_2 \end{bmatrix} = \begin{bmatrix} -6 & 1 \\ -5 & 0 \end{bmatrix} \begin{bmatrix} x_1 \\ x_2 \end{bmatrix} + \begin{bmatrix} 0 \\ 1 \end{bmatrix} r$$

$$y = \begin{bmatrix} 1 & -2 \end{bmatrix} \begin{bmatrix} x_1 \\ x_2 \end{bmatrix}$$

with initial state

$$\begin{bmatrix} x_1(0^-) \\ x_2(0^-) \end{bmatrix} = \begin{bmatrix} -3 \\ 1 \end{bmatrix}$$

and input

$$r(t) = 7u(t)$$

where $u(t)$ is the unit step function.

The Laplace-transformed state equations are as follows:

$$\begin{cases} sX_1(s) + 3 = -6X_1(s) + X_2(s) \\ sX_2(s) - 1 = -5X_1(s) + \dfrac{7}{s} \end{cases}$$

$$\begin{cases} (s + 6)X_1(s) - X_2(s) = -3 \\ 5X_1(s) + sX_2(s) = 1 + \dfrac{7}{s} = \dfrac{s + 7}{s} \end{cases}$$

$$X_1(s) = \frac{\begin{vmatrix} -3 & -1 \\ \dfrac{s+7}{s} & s \end{vmatrix}}{\begin{vmatrix} s+6 & -1 \\ 5 & s \end{vmatrix}} = \frac{-3s + (s+7)/s}{s^2 + 6s + 5}$$

$$= \frac{-3s^2 + s + 7}{s(s+1)(s+5)} = \frac{\frac{7}{5}}{s} + \frac{-\frac{3}{4}}{s+1} + \frac{-\frac{73}{20}}{s+5}$$

$$x_1(t) = \tfrac{7}{5} - \tfrac{3}{4}e^{-t} - \tfrac{73}{20}e^{-5t} \qquad t \ge 0$$

Similarly,

$$X_2(s) = \frac{\begin{vmatrix} s+6 & -3 \\ 5 & \dfrac{s+7}{s} \end{vmatrix}}{s^2 + 6s + 5} = \frac{(s^2 + 13s + 42)/s + 15}{s^2 + 6s + 5}$$

$$= \frac{s^2 + 28s + 42}{s(s+1)(s+5)} = \frac{\frac{42}{5}}{s} + \frac{-\frac{15}{4}}{s+1} + \frac{-\frac{73}{20}}{s+5}$$

$$x_2(t) = \tfrac{42}{5} - \tfrac{15}{4}e^{-t} - \tfrac{73}{20}e^{-5t} \qquad t \ge 0$$

The system output is then

$$y(t) = x_1 - 2x_2 = -\tfrac{77}{5} + (+\tfrac{27}{4})e^{-t} + \tfrac{73}{20}e^{-5t}$$

Time-Domain Response of First-Order Systems

In many situations, it is advantageous to have an expression for the solution of a set of state equations as functions of time rather than in terms of Laplace transforms. For a first-order state variable system,

$$\frac{dx}{dt} = ax + br$$

$$sX(s) - x(0^-) = aX(s) + bR(s)$$

$$X(s) = \frac{x(0^-)}{s - a} + bR(s)\frac{1}{s - a}$$

$$x(t) = \mathcal{L}^{-1}\left\{\frac{x(0^-)}{s - a} + bR(s)\frac{1}{s - a}\right\}$$

$$= e^{at}x(0^-) + \text{convolution } [br(t), e^{at}]$$

$$= e^{at}x(0^-) + \int_{0^-}^{t} e^{a(t - \tau)}br(\tau)\, d\tau$$

the inverse transform of a product of Laplace transforms being the convolution of the corresponding time functions.

An alternate derivation of this result is as follows. Multiplying the first-order equation

$$\frac{dx}{dt} = ax + br$$

by the integrating factor e^{-at} gives

$$e^{-at}\frac{dx}{dt} - axe^{-at} = e^{-at}br(t)$$

The left side of this equation is seen to be the derivative of a product:

$$\frac{d}{dt}\left[e^{-at}x(t)\right] = e^{-at}br(t)$$

Integrating both sides, there results

$$e^{-at}x(t) = \int e^{-at}br(t)\, dt + C$$

where C is an arbitrary constant of integration.

If the integration is begun at time $t = 0^-$,

$$e^{-at}x(t) = \int_{0^-}^{t} e^{-a\tau}br(\tau)\, d\tau + C \qquad t \geq 0$$

where the variable of integration has been written as τ to avoid confusion with the integral's upper limit t. Substituting $t = 0$, the constant of integration is seen to be

$$C = x(0^-)$$

giving

$$x(t) = e^{at} \int_{0^-}^{t} e^{-a\tau} br(\tau)\, d\tau + e^{at}x(0^-)$$

$$= e^{at}x(0^-) + \int_{0^-}^{t} e^{a(t-\tau)} br(\tau)\, d\tau \qquad t \geq 0$$

The integral is the convolution of the function e^{at} and the input $r(t)$.
As a numerical example, consider the first-order system

$$\dot{x} = -2x + 3r$$

$$y = 4x$$

The general solution for $x(t)$ is

$$x(t) = e^{-2t}x(0^-) + \int_{0^-}^{t} 3e^{-2(t-\tau)} r(\tau)\, d\tau$$

If

$$x(0^-) = 10 \quad \text{and} \quad r = 5$$

then

$$x(t) = 10e^{-2t} + \int_{0^-}^{t} 15e^{-2(t-\tau)}\, d\tau$$

$$= 10e^{-2t} + 15e^{-2t} \left. \frac{e^{2\tau}}{2} \right|_{0^-}^{t}$$

$$= 10e^{-2t} + 15e^{-2t} \frac{e^{-2t} - 1}{2}$$

$$= \tfrac{5}{2}e^{-2t} + \tfrac{15}{2} \qquad t \geq 0$$

and the system output is

$$y(t) = 4x(t) = 10e^{-2t} + 30 \qquad t \geq 0$$

Time-Domain Response of Higher-Order Systems

In general, a state variable system

$$\dot{x} = Ax + Br$$

has state response given by

$$sX(s) - x(0^-) = AX(s) + BR(s)$$

$$[sI - A]X(s) = x(0^-) + BR(s)$$

$$X(s) = [sI - A]^{-1}x(0^-) + [sI - A]^{-1}BR(s)$$

Denoting the *state transition matrix* by

$$\Phi(t) = \mathscr{L}^{-1}\{[sI - A]^{-1}\}$$

then

$$x(t) = \Phi(t)x(0^-) + \text{convolution } [Br(t), \Phi(t)]$$

$$= \Phi(t)x(0^-) + \int_{0^-}^{t} \Phi(t - \tau)Br(\tau)\, d\tau$$

For example, for the system

$$\begin{bmatrix} \dot{x}_1 \\ \dot{x}_2 \end{bmatrix} = \begin{bmatrix} -3 & 1 \\ -2 & 0 \end{bmatrix}\begin{bmatrix} x_1 \\ x_2 \end{bmatrix} + \begin{bmatrix} 2 \\ -1 \end{bmatrix} r$$

the state transition matrix is given by

$$\Phi(t) = \mathscr{L}^{-1}\{[sI - A]^{-1}\}$$

$$= \mathscr{L}^{-1}\left\{\begin{bmatrix} s + 3 & -1 \\ 2 & s \end{bmatrix}^{-1}\right\}$$

$$= \mathscr{L}^{-1}\begin{bmatrix} \dfrac{s}{s^2 + 3s + 2} & \dfrac{1}{s^2 + 3s + 2} \\ \dfrac{-2}{s^2 + 3s + 2} & \dfrac{s + 3}{s^2 + 3s + 2} \end{bmatrix}$$

$$= \mathscr{L}^{-1}\begin{bmatrix} \dfrac{-1}{s + 1} + \dfrac{2}{2 + 2} & \dfrac{1}{s + 1} + \dfrac{-1}{s + 2} \\ \dfrac{-2}{s + 1} + \dfrac{2}{s + 2} & \dfrac{2}{s + 1} + \dfrac{-1}{s + 2} \end{bmatrix}$$

$$= \begin{bmatrix} -e^{-t} + 2e^{-2t} & e^{-t} - e^{-2t} \\ -2e^{-t} + 2e^{-2t} & 2e^{-t} - e^{-2t} \end{bmatrix}$$

The system state is, in terms of initial conditions and the inputs,

$$\begin{bmatrix} x_1(t) \\ x_2(t) \end{bmatrix} = \begin{bmatrix} (-e^{-t} + 2e^{-2t}) & (e^{-t} - e^{-2t}) \\ (-2e^{-t} + 2e^{-2t}) & (2e^{-t} - e^{-2t}) \end{bmatrix}\begin{bmatrix} x_1(0^-) \\ x_2(0^-) \end{bmatrix}$$

$$+ \int_{0^-}^{t} \begin{bmatrix} -e^{-(t-\tau)} + 2e^{-2(t-\tau)} & e^{-(t-\tau)} - e^{-2(t-\tau)} \\ -2e^{-(t-\tau)} + 2e^{-2(t-\tau)} & 2e^{-(t-\tau)} - e^{-2(t-\tau)} \end{bmatrix}\begin{bmatrix} 2 \\ -1 \end{bmatrix} r(\tau)\, d\tau$$

DRILL PROBLEMS

D7.11. Use Laplace transform methods to find the outputs of the following systems for $t \geq 0$ with the given inputs and initial conditions:

(a) $\dot{x} = -2x + r(t)$
$y = 10x$
$x(0^-) = 3$
$r(t) = 4e^{5t}$

Ans $\frac{170}{7}e^{-2t} + \frac{40}{7}e^{5t}$

(b) $\begin{bmatrix} \dot{x}_1 \\ \dot{x}_2 \end{bmatrix} = \begin{bmatrix} 0 & 1 \\ -12 & -7 \end{bmatrix} \begin{bmatrix} x_1 \\ x_2 \end{bmatrix} + \begin{bmatrix} 1 \\ 1 \end{bmatrix} r$

$y = \begin{bmatrix} 1 & -1 \end{bmatrix} \begin{bmatrix} x_1 \\ x_2 \end{bmatrix}$

$\begin{bmatrix} x_1(0^-) \\ x_2(0^-) \end{bmatrix} = \begin{bmatrix} 10 \\ 0 \end{bmatrix}$

$r(t) = u(t)$, where $u(t)$ is the unit step function.

Ans $Y(s) = (10s^2 + 190s + 20)/s(s + 3)(s + 4)$
$y(t) = \frac{5}{3} + \frac{460}{3}e^{-3t} - 145e^{-4t}$ $t \geq 0$

(c) $\begin{bmatrix} \dot{x}_1 \\ \dot{x}_2 \\ \dot{x}_3 \end{bmatrix} = \begin{bmatrix} -5 & 1 & 0 \\ -6 & 0 & 1 \\ 0 & 0 & 0 \end{bmatrix} \begin{bmatrix} x_1 \\ x_2 \\ x_3 \end{bmatrix} + \begin{bmatrix} 0 \\ 0 \\ 1 \end{bmatrix} r$

$y = \begin{bmatrix} 1 & 0 & 0 \end{bmatrix} \begin{bmatrix} x_1 \\ x_2 \\ x_3 \end{bmatrix}$

$\mathbf{x}(0) = \mathbf{0}$
$r(t) = \delta(t)$, where $\delta(t)$ is the unit inpulse.

Ans $\frac{1}{6} + \frac{1}{3}e^{-3t} - \frac{1}{2}e^{-2t}$ $t \geq 0$

D7.12. Calculate state transition matrices for systems with the following state coupling matrices **A**, using

$$\Phi(t) = \mathscr{L}^{-1}\{[s\mathbf{I} - \mathbf{A}]^{-1}\}:$$

(a) $\begin{bmatrix} -9 & 1 \\ -14 & 0 \end{bmatrix}$

Ans $\begin{bmatrix} -\frac{2}{5}e^{-2t} + \frac{7}{5}e^{-7t} & \frac{1}{5}e^{-2t} - \frac{1}{5}e^{-7t} \\ -\frac{14}{5}e^{-2t} + \frac{14}{5}e^{-7t} & \frac{7}{5}e^{-2t} - \frac{2}{5}e^{-7t} \end{bmatrix}$

(b) $\begin{bmatrix} 1 & -1 \\ 2 & -4 \end{bmatrix}$

Ans $\begin{bmatrix} 1.11e^{0.56t} - 0.11e^{-3.56t} & -0.24e^{0.56t} + 0.24e^{-3.56t} \\ 0.48e^{0.56t} - 0.48e^{-3.56t} & -0.11e^{0.56t} + 1.11e^{-3.56t} \end{bmatrix}$

System Response Computation

One advantage of placing system equations in a state variable form is that it is well suited to digital computer calculations. Computers are not particularly efficient at equation manipulation, Laplace transformation, and the like, but they excel at such repetitive tasks as matrix addition and multiplication. The capability of simulating a system, that is, investigating and testing its performance by modeling, is important to the designer, particularly for the common situation in which the plant is expensive and the design has to be correct when it is first installed.

The state transition matrix can be approximated by an $m + 1$ term Taylor series

$$\mathbf{\Phi}(\Delta t) = \mathbf{I} + \mathbf{A}(\Delta t) + \frac{1}{2} \mathbf{A}^2(\Delta t)^2 + \cdots + \left(\frac{1}{m!}\right)\mathbf{A}^m(\Delta t)^m$$

The convolution integral depends on the state transition matrix and on the input, both of which can be functions of time. However, if Δt is a very short time, then $\mathbf{r}(t)$ can be removed from the integral so that

Convolution integral $= \mathbf{D}(\Delta t)\mathbf{B}r(t)$

$$\mathbf{D}(\Delta t) = \mathbf{I} \, \Delta t + \frac{1}{2} \mathbf{A}(\Delta t)^2 + \left(\frac{1}{3!}\right)\mathbf{A}^2(\Delta t)^3$$

$$+ \cdots + \left(\frac{1}{(m + 1)!}\right)\mathbf{A}^m(\Delta t)^{m+1}$$

For sufficiently small time increments Δt, once can start with the initial state $\mathbf{x}(0)$ and calculate $\mathbf{x}(\Delta t)$ as follows:

$$\mathbf{x}(\Delta t) \cong (\mathbf{I} + \mathbf{A} \, \Delta t)\mathbf{x}(0) + (\mathbf{B} \, \Delta t)\mathbf{r}(0)$$

then $\mathbf{x}(2\Delta t)$ may be calculated from $\mathbf{x}(\Delta t)$,

$$\mathbf{x}(2\Delta t) \cong (\mathbf{I} + \mathbf{A} \, \Delta t)\mathbf{x}(\Delta t) + (\mathbf{B} \, \Delta t)\mathbf{r}(\Delta t)$$

and so on, obtaining approximate solutions for the state,

$$\mathbf{x}\{(k + 1)\Delta t\} \cong (\mathbf{I} + \mathbf{A} \, \Delta t)\mathbf{x}(k \, \Delta t) + (\mathbf{B} \, \Delta t)\mathbf{r}(k \, \Delta t)$$

For example, the response of the first-order system

$$\dot{x} = -2x + r$$

$$y = x$$

with

$$x(0^-) = 10$$

$$r(t) = 3 \sin t$$

is approximated by

$$x\{(k + 1)\Delta t\} \cong (1 - 2\Delta t)x(k\,\Delta t) + 3\Delta t \sin(k\,\Delta t)$$

with

$$x(0 \cdot \Delta t) = 10$$

Representative computer-generated plots of $x(t)$ are given in Fig. 7.15 for various choices of Δt. For a sufficiently small time increment Δt, the approximate response is very nearly the actual system response.

Another example system is the following:

$$\begin{bmatrix} \dot{x}_1 \\ \dot{x}_2 \end{bmatrix} = \begin{bmatrix} -2 & 1 \\ -3 & 0 \end{bmatrix}\begin{bmatrix} x_1 \\ x_2 \end{bmatrix} + \begin{bmatrix} 2 \\ -1 \end{bmatrix}r$$

$$y = \begin{bmatrix} 1 & -\frac{1}{2} \end{bmatrix}\begin{bmatrix} x_1 \\ x_2 \end{bmatrix}$$

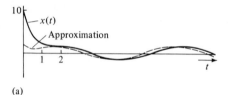

(a)

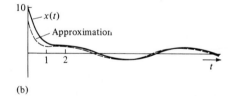

(b)

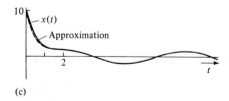

(c)

Figure 7.15 Computer-generated response plots for a first-order system. **(a)** Step size $\Delta t = 0.4$. **(b)** Step size $\Delta t = 0.2$. **(c)** Step size $\Delta t = 0.05$.

with

$$\begin{bmatrix} x_1(0^-) \\ x_2(0^-) \end{bmatrix} = \begin{bmatrix} -4 \\ 5 \end{bmatrix}$$

$$r(t) = \cos 0.25t$$

is approximated by

$$\begin{bmatrix} x_1\{(k+1)\Delta t\} \\ x_2\{(k+1)\Delta t\} \end{bmatrix} = \begin{bmatrix} 1 - 2\Delta t & \Delta t \\ -3\Delta t & 1 \end{bmatrix} \begin{bmatrix} x_1(k\,\Delta t) \\ x_2(k\,\Delta t) \end{bmatrix} + \begin{bmatrix} 2 \\ -1 \end{bmatrix} \Delta t \cos(0.25k\,\Delta t)$$

$$y\{(k+1)\Delta t\} = \begin{bmatrix} 1 & -\frac{1}{2} \end{bmatrix} \begin{bmatrix} x_1\{(k+1)\Delta t\} \\ x_2\{(k+1)\Delta t\} \end{bmatrix}$$

or

$$\begin{cases} x_1\{(k+1)\Delta t\} = (1 - 2\,\Delta t)x_1(k\,\Delta t) + \Delta t x_2(k\,\Delta t) + 2\,\Delta t \cos(0.25k\,\Delta t) \\ x_2\{(k+1)\Delta t\} = -3\Delta t x_1(k\,\Delta t) + x_2(k\,\Delta t) - \Delta t \cos(0.25k\,\Delta t) \\ y\{(k+1)\Delta t\} = x_1\{(k+1)\Delta t\} - \frac{1}{2}x_2\{(k+1)\Delta t\} \end{cases}$$

with

$$\begin{cases} x_1(0 \cdot \Delta t) = -4 \\ x_2(0 \cdot \Delta t) = 5 \end{cases}$$

Computer-generated response plots for this system are given in Fig. 7.16, where $\Delta t = 0.05$.

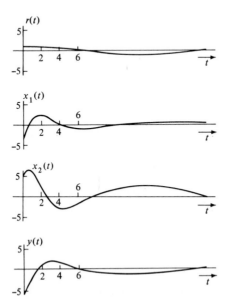

Figure 7.16 Computer-generated response plots for a second-order system.

Improved accuracy and reduced computation time may result from using more involved approximations—for example, matrix power series, predictor correctors, or Runge-Kutta methods.

DRILL PROBLEMS

D7.13. For the following systems, develop discrete-time approximation equations using the indicated time steps Δt.

(a) $\begin{bmatrix} \dot{x}_1 \\ \dot{x}_2 \end{bmatrix} = \begin{bmatrix} 3 & 1 \\ 2 & -1 \end{bmatrix} \begin{bmatrix} x_1 \\ x_2 \end{bmatrix} + \begin{bmatrix} 1 \\ 4 \end{bmatrix} r$

$y = [-3 \quad -1] \begin{bmatrix} x_1 \\ x_2 \end{bmatrix}$

$\Delta t = 0.2$

Ans $\mathbf{x}[(k+1)\Delta t] = \begin{bmatrix} 1.6 & 0.2 \\ 0.4 & 0.8 \end{bmatrix} \mathbf{x}(k\,\Delta t) + \begin{bmatrix} 0.2 \\ 0.8 \end{bmatrix} r(k\,\Delta t)$

$y = [-3 \quad -1] \mathbf{x}(k\,\Delta t)$

(b) $\begin{bmatrix} \dot{x}_1 \\ \dot{x}_2 \\ \dot{x}_3 \end{bmatrix} = \begin{bmatrix} 1 & 2 & 3 \\ 7 & -2 & -3 \\ 6 & 0 & 4 \end{bmatrix} \begin{bmatrix} x_1 \\ x_2 \\ x_3 \end{bmatrix} + \begin{bmatrix} 1 & -2 \\ -1 & 3 \\ 0 & 4 \end{bmatrix} \begin{bmatrix} r_1 \\ r_2 \end{bmatrix}$

$y = [5 \quad -2 \quad 1] \begin{bmatrix} x_1 \\ x_2 \\ x_3 \end{bmatrix}$

$\Delta t = 0.01$

Ans $\mathbf{x}[(k+1)\Delta t] = \begin{bmatrix} 1.01 & 0.02 & 0.03 \\ 0.07 & 0.98 & -0.03 \\ 0.06 & 0 & 1.04 \end{bmatrix} \mathbf{x}(k\,\Delta t)$

$+ \begin{bmatrix} 0.01 & -0.02 \\ -0.01 & 0.03 \\ 0 & 0.04 \end{bmatrix} \mathbf{r}(k\,\Delta t)$

$y = [5 \quad -2 \quad 1] \mathbf{x}(k\,\Delta t)$

D7.14. For the set of state equations

$\begin{bmatrix} \dot{x}_1 \\ \dot{x}_2 \end{bmatrix} = \begin{bmatrix} -2 & 1 \\ -3 & 0 \end{bmatrix} \begin{bmatrix} x_1 \\ x_2 \end{bmatrix} + \begin{bmatrix} 1 \\ 4 \end{bmatrix} r$

$y = [1 \quad -1] \begin{bmatrix} x_1 \\ x_2 \end{bmatrix}$

a discrete-time approximation is

$$\begin{bmatrix} x_1\{(k+1)\Delta t\} \\ x_2\{(k+1)\Delta t\} \end{bmatrix} = \begin{bmatrix} (1-2\Delta t) & \Delta t \\ -3\Delta t & 1 \end{bmatrix} \begin{bmatrix} x_1(k\,\Delta t) \\ x_2(k\,\Delta t) \end{bmatrix} + \begin{bmatrix} \Delta t \\ 4\Delta t \end{bmatrix} r(k\,\Delta t)$$

$$y(k\,\Delta t) = \begin{bmatrix} 1 & -1 \end{bmatrix} \begin{bmatrix} x_1(k\,\Delta t) \\ x_2(k\,\Delta t) \end{bmatrix}$$

If

$$\begin{bmatrix} x_1(0^-) \\ x_2(0^-) \end{bmatrix} = \begin{bmatrix} 10 \\ 0 \end{bmatrix} \qquad \text{and} \qquad r(t) = 2u(t)$$

where $u(t)$ is the unit step function, calculate approximate values for $\mathbf{x}(\Delta t)$, $\mathbf{x}(2\,\Delta t)$, and $\mathbf{x}(3\,\Delta t)$ for the following:

(a) $\Delta t = 0.2$

Ans $\begin{bmatrix} 6.4 \\ -4.4 \end{bmatrix}, \begin{bmatrix} 3.36 \\ -6.64 \end{bmatrix}, \begin{bmatrix} 1.09 \\ -7.06 \end{bmatrix}$

(b) $\Delta t = 0.1$

Ans $\begin{bmatrix} 8.2 \\ -2.2 \end{bmatrix}, \begin{bmatrix} 6.54 \\ -3.86 \end{bmatrix}, \begin{bmatrix} 6.37 \\ -5.02 \end{bmatrix}$

(c) $\Delta t = 0.02$

Ans $\begin{bmatrix} 9.84 \\ -0.44 \end{bmatrix}, \begin{bmatrix} 9.67 \\ -0.87 \end{bmatrix}, \begin{bmatrix} 9.503 \\ -1.29 \end{bmatrix}$

7.7 STATE FEEDBACK AND STEADY STATE RESPONSE

Figure 7.1 contained three ways that state variables could be organized to describe the dynamics of a system. Phase variables and canonical (diagonal) variables have already been described. Now, when an actual system structure is present (containing the plant, gains, and compensators), physically relevant quantities can be used to create a set of physical variables. As we shall see, each transmittance can be decomposed into a separate set of phase variables. All the phase variables can then be gathered together to describe the composite interconnected system.

System Description

The structure (blocks) of a system can be maintained when converting to state variables, rather than compute the transfer function and then use phase variables, which would lose all structural information about the system. Each block, instead, can be decomposed into an equivalent set of integrators. Systematic steps for converting a system block diagram to a set of state equations is illustrated with an example system in Fig. 7.17. The block diagram of Fig. 7.17a is converted to an

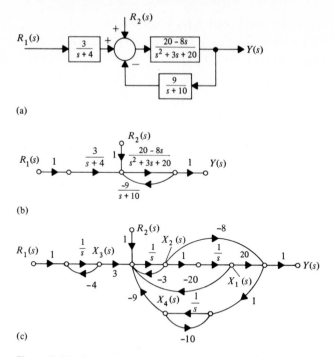

Figure 7.17 Constructing a simulation diagram. **(a)** Block diagram of the system. **(b)** Equivalent signal flow graph. **(c)** Signal for graph expanded into a simulation diagram.

equivalent signal flow graph in Fig. 7.17b. In Fig. 7.17c, simulation diagrams replace the individual transmittances. In each of these, the phase-variable form is used because it is simple and easily related to the coefficients of the transmittances. Care must be taken to preserve the signal relationships. For instance, the unit transmittance in the lower feedback loop is necessary; if it were removed, the $X_4(s)$ signal would couple through the -10 transmittance into $Y(s)$.

State equations for the system are then written by labeling each integrator output as a state variable and equating the sum of the input signals to each integrator to the derivative of the appropriate state variable:

$$\dot{x}_1 = x_2$$

$$\dot{x}_2 = -20x_1 - 3x_2 + 3x_3 - 9x_4 + r_2$$

$$\dot{x}_3 = 4x_3 + r_1$$

$$\dot{x}_4 = -10x_4 + y = 20x_1 - 8x_2 - 10x_4$$

$$y = 20x_1 - 8x_2$$

Occasionally, it is necessary to substitute, in terms of the inputs and state variables, for an intermediate signal such as y in the $\dot{x}_4$ equation above. In vector-matrix notation, these equations are as follows:

$$\begin{bmatrix} \dot{x}_1 \\ \dot{x}_2 \\ \dot{x}_3 \\ \dot{x}_4 \end{bmatrix} = \begin{bmatrix} 0 & 1 & 0 & 0 \\ -20 & -3 & 3 & -9 \\ 0 & 0 & -4 & 0 \\ 20 & -8 & 0 & -10 \end{bmatrix} \begin{bmatrix} x_1 \\ x_2 \\ x_3 \\ x_4 \end{bmatrix} + \begin{bmatrix} 0 & 0 \\ 0 & 1 \\ 1 & 0 \\ 0 & 0 \end{bmatrix} \begin{bmatrix} r_1 \\ r_2 \end{bmatrix}$$

$$y = \begin{bmatrix} 20 & -8 & 0 & 0 \end{bmatrix} \begin{bmatrix} x_1 \\ x_2 \\ x_3 \\ x_4 \end{bmatrix}$$

As another example, consider the system with the signal flow graph of Fig. 7.18a. The transmittance $G(s)$ does not have numerator degree less than the denominator degree, so an individual simulation diagram for it cannot be drawn. However, dividing denominator into numerator for one step gives

$$G(s) = 2 + \frac{-7}{s + 4}$$

which is represented in the equivalent signal flow graph of Fig. 7.18b.

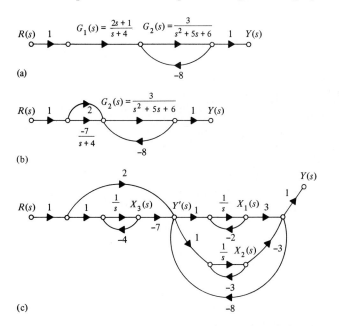

(a)

(b)

(c)

Figure 7.18 Another example of simulation diagram construction. **(a)** Signal flow graph of the system. **(b)** Improper rational transmittance expanded. **(c)** Simulation diagram.

For variety, the transmittance $G_2(s)$ is expanded in the diagonal form corresponding to

$$G_2(s) = \frac{3}{s+2} + \frac{-3}{s+3}$$

in the simulation diagram of Fig. 7.18c for which

$$\dot{x}_1 = -2x_1 + y'$$

$$\dot{x}_2 = -3x_2 + y'$$

$$\dot{x}_3 = -4x_3 + r$$

$$y = 3x_1 - 3x_2$$

Substituting for the intermediate signal

$$y' = 2r - 7x_3 - 8y = 2r - 7x_3 - 24x_1 + 24x_2$$

the following state equations result:

$$\begin{bmatrix} \dot{x}_1 \\ \dot{x}_2 \\ \dot{x}_3 \end{bmatrix} = \begin{bmatrix} -26 & 24 & -7 \\ -24 & 21 & -7 \\ 0 & 0 & -4 \end{bmatrix} \begin{bmatrix} x_1 \\ x_2 \\ x_3 \end{bmatrix} + \begin{bmatrix} 2 \\ 2 \\ 1 \end{bmatrix} r$$

$$y = \begin{bmatrix} 3 & -3 & 0 \end{bmatrix} \begin{bmatrix} x_1 \\ x_2 \\ x_3 \end{bmatrix}$$

State Feedback

A compact diagram of the state variable description of a system is given in Fig. 7.19a. The bold arrows represent signal vectors rather than individual signals, and the blocks show matrix operations. In this diagram,

$$s\mathbf{X}(s) = \mathbf{A}\mathbf{X}(s) + \mathbf{B}\mathbf{R}(s)$$

$$\mathbf{Y}(s) = \mathbf{C}\mathbf{X}(s)$$

In the time domain,

$$\frac{d\mathbf{x}}{dt} = \mathbf{A}\mathbf{x} + \mathbf{B}\mathbf{r}$$

$$\mathbf{y} = \mathbf{C}\mathbf{x}$$

An alternative diagram in terms of functions of time is given in Fig. 7.19b. When a change of state variables is made,

$$\mathbf{x}' = \mathbf{P}\mathbf{x} \qquad \mathbf{x} = \mathbf{P}^{-1}\mathbf{x}'$$

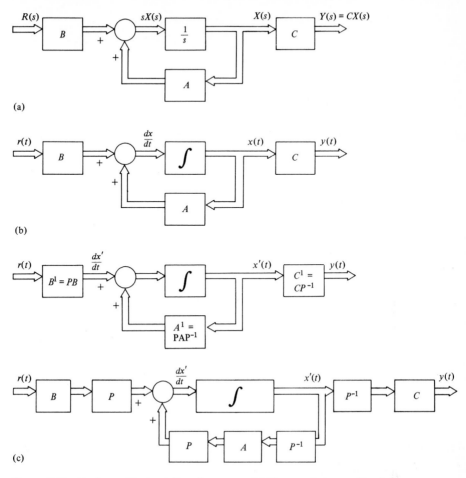

Figure 7.19 Vector-matrix simulation diagrams. **(a)** Diagram in terms of Laplace transforms. **(b)** Diagram in terms of functions of time. **(c)** Diagram with change of state variables.

the state equations become

$$\frac{d\mathbf{x}'}{dt} = (\mathbf{PAP}^{-1})\mathbf{x}' + (\mathbf{PB})\mathbf{r} = \mathbf{A}'\mathbf{x}' + \mathbf{B}'\mathbf{r}$$

$$\mathbf{y} = (\mathbf{CP}^{-1})\mathbf{x}' = \mathbf{C}'\mathbf{x}'$$

Fig. 7.19c shows the effect of a change of variables upon the simulation diagram.

For feedback system control an important design strategy is to sense the state variable signals and feed them back to the input through appropriate gains. Provided that the system is completely controllable, state feedback may be used to place the system's characteristic roots, the poles of its transfer functions, at any

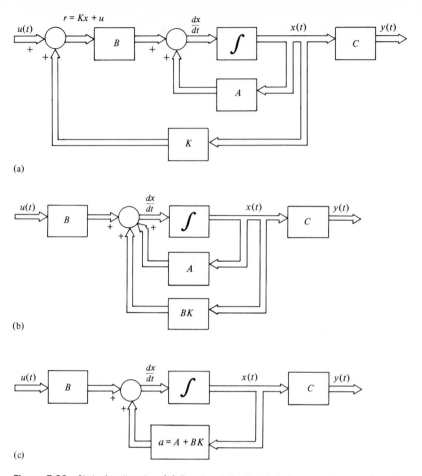

Figure 7.20 State feedback. **(a)** Feedback for the state in a tracking configuration, **r** = kx + **u**. **(b)** Simulation diagram rearranged to show feedback system state coupling matrix, $\mathscr{A}$ = **A** + **BK**.

desired locations. State feedback is illustrated in Fig. 7.20a. The open-loop system input is driven with a linear combination of the state signals and a reference input in a tracking configuration:

$$\mathbf{r} = \mathbf{Kx} + \mathbf{u}$$

Figs. 7.20b and c show the simulation diagram rearranged so that it is evident that the state coupling matrix of the feedback system is

$$\mathscr{A} = \mathbf{A} + \mathbf{BK}$$

As a numerical example of placing the system poles as desired with state feedback, consider the following single-input, single-output system which is described in phase-variable form:

$$\begin{bmatrix} \dot{x}_1 \\ \dot{x}_2 \\ \dot{x}_3 \end{bmatrix} = \begin{bmatrix} 0 & 1 & 0 \\ 0 & 0 & 1 \\ -5 & -7 & -3 \end{bmatrix} \begin{bmatrix} x_1 \\ x_2 \\ x_3 \end{bmatrix} + \begin{bmatrix} 0 \\ 0 \\ 1 \end{bmatrix} r(t)$$

$$y = \begin{bmatrix} -2 & 4 & 3 \end{bmatrix} \begin{bmatrix} x_1 \\ x_2 \\ x_3 \end{bmatrix}$$

This system is shown in the simulation diagram of Fig. 7.21a and, using Mason's gain rule, has transfer function

$$T(s) = \frac{3/s + 4/s^2 + -2/s^3}{1 + 3/s + 7/s^2 + 5/s^3}$$

$$= \frac{3s^2 + 4s - 2}{s^3 + 3s^2 + 7s + 5}$$

Since the characteristic equation factors as

$$s^3 + 3s^2 + 7s + 5 = (s + 1 + j2)(s + 1 - j2)(s + 1)$$

its poles are at $s = -1 - j2, -1 + j2$, and -1.

With state feedback,

$$r(t) = k_1 x_1 + k_2 x_2 + k_3 x_3 + u(t)$$

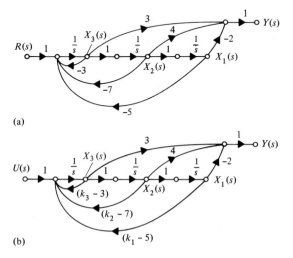

(a)

(b)

Figure 7.21 State feedback example. **(a)** Open-loop system. **(b)** System with state feedback.

the state equations are of the form

$$
\begin{bmatrix} \dot{x}_1 \\ \dot{x}_2 \\ \dot{x}_3 \end{bmatrix} = \begin{bmatrix} 0 & 1 & 0 \\ 0 & 0 & 1 \\ k_1 - 5 & k_2 - 7 & k_3 - 3 \end{bmatrix} \begin{bmatrix} x_1 \\ x_2 \\ x_3 \end{bmatrix} + \begin{bmatrix} 0 \\ 0 \\ 1 \end{bmatrix} u(t)
$$

$$
y = \begin{bmatrix} -2 & 4 & 3 \end{bmatrix} \begin{bmatrix} x_1 \\ x_2 \\ x_3 \end{bmatrix}
$$

as diagramed in Fig. 7.21b. The feedback system has transfer function, in terms of the feedback gain constants k,

$$
T(s) = \frac{3/s + 4/s^2 - 2/s^3}{1 + (3 - k_3)/s + (7 - k_2)/s^2 + (5 - k_1)/s^3}
$$

$$
= \frac{3s^2 + 4s - 2}{s^3 + (3 - k_3)s^2 + (7 - k_2)s + (5 - k_1)}
$$

The coefficients of the characteristic equation may be chosen at will, by appropriately selecting k_1, k_2, and k_3. If, for instance, it is desired that the system poles be located at $s = -4$, -4, and -5, the characteristic polynomial should then be

$$
(s + 4)(s + 4)(s + 5) = s^3 + 13s^2 + 56s + 80
$$

$$
= s^3 + (3 - k_3)s^2 + (7 - k_2)s + (5 - k_1)
$$

which will be the case for

$$
k_1 = -75
$$

$$
k_2 = -49
$$

$$
k_3 = -10
$$

It is true in general that appropriate state feedback will place the poles of any completely controllable system arbitrarily. For a single-input system, the phase-variable form of the state equations is especially convenient because the required values of the feedback gain constants may be determined by inspection, as was the case with the above example. In a multiple-input system, additional design freedom exists. For any completely observable system, when all of the state variables cannot be sensed directly, and fed back for control, they may be estimated (at the expense of increased system order) and the estimates fed back to achieve pole placement. State estimators for this purpose are termed *observers*.

Steady State Response to Power-of-Time Inputs

The steady state response to step or constant inputs may, as always, be found using system transfer functions and the final-value theorem. A generally easier procedure in terms of state equations

$$
\dot{\mathbf{x}} = \mathbf{A}\mathbf{x} + \mathbf{B}\mathbf{r}
$$

$$
\mathbf{y} = \mathbf{C}\mathbf{x}
$$

is the following: Assuming the system is stable, all steady state signals due to constant inputs will be constant, and

$$\dot{\mathbf{x}} = \mathbf{0}$$

giving

$$\mathbf{0} = \mathbf{Ax} + \mathbf{Br}$$

$$\mathbf{x} = -\mathbf{A}^{-1}\mathbf{Br}$$

$$\mathbf{y} = -\mathbf{CA}^{-1}\mathbf{Br}$$

The existence of $\mathbf{A}^{-1}$ is guaranteed for a stable system since

$$|\mathbf{A}| = 0$$

if and only if $s = 0$ is a characteristic root of $\mathbf{A}$, that is, if the system has a pole at $s = 0$.

Consider the system

$$\begin{bmatrix} \dot{x}_1 \\ \dot{x}_2 \\ \dot{x}_3 \end{bmatrix} = \begin{bmatrix} -2 & 1 & 0 \\ -3 & 0 & 1 \\ -4 & 0 & 0 \end{bmatrix} \begin{bmatrix} x_1 \\ x_2 \\ x_3 \end{bmatrix} + \begin{bmatrix} -1 \\ 5 \\ 0 \end{bmatrix} r(t)$$

$$\begin{bmatrix} y_1 \\ y_2 \end{bmatrix} = \begin{bmatrix} 1 & 0 & -1 \\ 2 & -2 & 0 \end{bmatrix} \begin{bmatrix} x_1 \\ x_2 \\ x_3 \end{bmatrix}$$

which is stable. For a unit step input $r(t)$, the steady state behavior of the system is governed by

$$\begin{bmatrix} 0 \\ 0 \\ 0 \end{bmatrix} = \begin{bmatrix} -2 & 1 & 0 \\ -3 & 0 & 1 \\ -4 & 0 & 0 \end{bmatrix} \begin{bmatrix} x_1 \\ x_2 \\ x_3 \end{bmatrix} + \begin{bmatrix} -1 \\ 5 \\ 0 \end{bmatrix} r(t)$$

$$\begin{bmatrix} x_1 \\ x_2 \\ x_3 \end{bmatrix} = -\begin{bmatrix} -2 & 1 & 0 \\ -3 & 0 & 1 \\ -4 & 0 & 0 \end{bmatrix}^{-1} \begin{bmatrix} -1 \\ 5 \\ 0 \end{bmatrix}$$

$$= -\begin{bmatrix} 0 & 0 & -\frac{1}{4} \\ 1 & 0 & -\frac{1}{2} \\ 0 & 1 & -\frac{3}{4} \end{bmatrix} \begin{bmatrix} -1 \\ 5 \\ 0 \end{bmatrix} = \begin{bmatrix} 0 \\ 1 \\ -5 \end{bmatrix}$$

The steady state values of the outputs are

$$\begin{bmatrix} y_1 \\ y_2 \end{bmatrix} = \begin{bmatrix} 1 & 0 & -1 \\ 2 & -2 & 0 \end{bmatrix} \begin{bmatrix} 0 \\ 1 \\ -5 \end{bmatrix} = \begin{bmatrix} 5 \\ -2 \end{bmatrix}$$

This method may be applied to ramp and higher power-of-t inputs if desired. For a unit ramp input to the example system,

$$r(t) = t$$

the system state will consist of constant and ramp components,

$$x_1(t) = \alpha_1 t + \beta_1$$
$$x_2(t) = \alpha_2 t + \beta_2$$
$$x_3(t) = \alpha_3 t + \beta_3$$

where the α and β terms are constants. Then

$$
\begin{bmatrix} \dot{x}_1 \\ \dot{x}_2 \\ \dot{x}_3 \end{bmatrix}
=
\begin{bmatrix} \alpha_1 \\ \alpha_2 \\ \alpha_3 \end{bmatrix}
=
\begin{bmatrix} -2 & 1 & 0 \\ -3 & 0 & 1 \\ -4 & 0 & 0 \end{bmatrix}
\begin{bmatrix} \alpha_1 t + \beta_1 \\ \alpha_2 t + \beta_2 \\ \alpha_3 t + \beta_3 \end{bmatrix}
+
\begin{bmatrix} -1 \\ 5 \\ 0 \end{bmatrix} t
$$

Equating coefficients of the powers of t, the α terms are governed by

$$
\begin{bmatrix} 0 \\ 0 \\ 0 \end{bmatrix}
=
\begin{bmatrix} -2 & 1 & 0 \\ -3 & 0 & 1 \\ -4 & 0 & 0 \end{bmatrix}
\begin{bmatrix} \alpha_1 \\ \alpha_2 \\ \alpha_3 \end{bmatrix}
+
\begin{bmatrix} -1 \\ 5 \\ 0 \end{bmatrix}
$$

$$
\begin{bmatrix} \alpha_1 \\ \alpha_2 \\ \alpha_3 \end{bmatrix}
=
-\begin{bmatrix} -2 & 1 & 0 \\ -3 & 0 & 1 \\ -4 & 0 & 0 \end{bmatrix}^{-1}
\begin{bmatrix} -1 \\ 5 \\ 0 \end{bmatrix}
=
\begin{bmatrix} 0 \\ 1 \\ -5 \end{bmatrix}
$$

which are always identical to the constant steady state response of the state to a unit step input. The β terms satisfy

$$
\begin{bmatrix} \alpha_1 \\ \alpha_2 \\ \alpha_3 \end{bmatrix}
=
\begin{bmatrix} -2 & 1 & 0 \\ -3 & 0 & 1 \\ -4 & 0 & 0 \end{bmatrix}
\begin{bmatrix} \beta_1 \\ \beta_2 \\ \beta_3 \end{bmatrix}
$$

$$
\begin{bmatrix} \beta_1 \\ \beta_2 \\ \beta_3 \end{bmatrix}
=
\begin{bmatrix} -2 & 1 & 0 \\ -3 & 0 & 1 \\ -4 & 0 & 0 \end{bmatrix}^{-1}
\begin{bmatrix} \alpha_1 \\ \alpha_2 \\ \alpha_3 \end{bmatrix}
=
\begin{bmatrix} 0 & 0 & -\frac{1}{4} \\ 1 & 0 & -\frac{1}{2} \\ 0 & 1 & -\frac{3}{4} \end{bmatrix}
\begin{bmatrix} 0 \\ 1 \\ -5 \end{bmatrix}
$$

$$
=
\begin{bmatrix} \frac{5}{4} \\ \frac{5}{2} \\ \frac{19}{4} \end{bmatrix}
$$

and the steady state outputs are

$$
\begin{bmatrix} y_1 \\ y_2 \end{bmatrix}
=
\begin{bmatrix} 1 & 0 & -1 \\ 2 & -2 & 0 \end{bmatrix}
\begin{bmatrix} t + \frac{5}{4} \\ \frac{5}{2} \\ -5t + \frac{19}{4} \end{bmatrix}
=
\begin{bmatrix} (5t - \frac{7}{2}) \\ (-2t - \frac{5}{2}) \end{bmatrix}
$$

DRILL PROBLEMS

D7.15. Find state variable equations for systems with the following signal flow graphs.

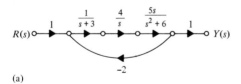

(a)

$R_1(s)$ ... $Y_1(s)$

$R_2(s)$ ... $Y_2(s)$

(b)

Ans (a) One possibility is the following:

$$\begin{bmatrix} \dot{x}_1 \\ \dot{x}_2 \\ \dot{x}_3 \\ \dot{x}_4 \end{bmatrix} = \begin{bmatrix} 0 & 1 & 0 & 0 \\ -6 & 0 & 1 & 0 \\ 0 & 0 & 0 & 4 \\ 0 & -10 & 0 & -3 \end{bmatrix} \begin{bmatrix} x_1 \\ x_2 \\ x_3 \\ x_4 \end{bmatrix} + \begin{bmatrix} 0 \\ 0 \\ 0 \\ 1 \end{bmatrix} r$$

$$y = \begin{bmatrix} 0 & 5 & 0 & 0 \end{bmatrix} \begin{bmatrix} x_1 \\ x_2 \\ x_3 \\ x_4 \end{bmatrix}$$

(b) One possibility is the following:

$$\begin{bmatrix} \dot{x}_1 \\ \dot{x}_2 \\ \dot{x}_3 \\ \dot{x}_4 \end{bmatrix} = \begin{bmatrix} -2 & 0 & 0 & 0 \\ 3 & -1 & 0 & 0 \\ 0 & 0 & 0 & 1 \\ 0 & 0 & -4 & 0 \end{bmatrix} \begin{bmatrix} x_1 \\ x_2 \\ x_3 \\ x_4 \end{bmatrix} = \begin{bmatrix} 1 & -6 \\ 0 & 0 \\ 0 & 0 \\ 0 & 10 \end{bmatrix} \begin{bmatrix} r_1 \\ r_2 \end{bmatrix}$$

$$\begin{bmatrix} y_1 \\ y_2 \end{bmatrix} = \begin{bmatrix} 3 & 0 & 0 & 0 \\ 0 & 1 & 1 & 0 \end{bmatrix} \begin{bmatrix} x_1 \\ x_2 \\ x_3 \\ x_4 \end{bmatrix}$$

D7.16. For the state feedback systems described by the following equations, choose the feedback gain constants k_i to place the closed-loop system poles at the indicated locations:

(a)
$$\begin{bmatrix} \dot{x}_1 \\ \dot{x}_2 \\ \dot{x}_3 \end{bmatrix} = \begin{bmatrix} 0 & 1 & 0 \\ 0 & 0 & 1 \\ -3 & -6 & -7 \end{bmatrix} \begin{bmatrix} x_1 \\ x_2 \\ x_3 \end{bmatrix} + \begin{bmatrix} 0 \\ 0 \\ 1 \end{bmatrix} r$$

$$r = \begin{bmatrix} k_1 & k_2 & k_3 \end{bmatrix} \begin{bmatrix} x_1 \\ x_2 \\ x_3 \end{bmatrix} + u$$

$$y = \begin{bmatrix} 2 & 0 & -1 \end{bmatrix} \begin{bmatrix} x_1 \\ x_2 \\ x_3 \end{bmatrix}$$

Closed-loop poles at $s = -3, -4,$ and -5

Ans $k_1 = -57, k_2 = -41, k_3 = -5$

(b)
$$\begin{bmatrix} \dot{x}_1 \\ \dot{x}_2 \\ \dot{x}_3 \end{bmatrix} = \begin{bmatrix} -2 & 1 & 0 \\ 4 & 0 & 1 \\ 0 & 0 & 0 \end{bmatrix} \begin{bmatrix} x_1 \\ x_2 \\ x_3 \end{bmatrix} + \begin{bmatrix} 0 \\ 0 \\ 1 \end{bmatrix} r$$

$$r = \begin{bmatrix} k_1 & k_2 & k_3 \end{bmatrix} \begin{bmatrix} x_1 \\ x_2 \\ x_3 \end{bmatrix} + 9$$

$$y = \begin{bmatrix} 2 & 0 & -1 \\ 1 & 1 & 0 \end{bmatrix} \begin{bmatrix} x_1 \\ x_2 \\ x_3 \end{bmatrix}$$

Closed-loop poles at $s = -3 \pm j3, -3$

Ans $k_1 = -30, k_2 = -26, k_3 = -7$

D7.17. For a unit step input, find the steady state output, if it exists, of each of the following systems:

(a)
$$\begin{bmatrix} \dot{x}_1 \\ \dot{x}_2 \end{bmatrix} = \begin{bmatrix} -2 & 2 \\ -3 & 0 \end{bmatrix} \begin{bmatrix} x_1 \\ x_2 \end{bmatrix} + \begin{bmatrix} 1 \\ 4 \end{bmatrix} r$$

$$y = \begin{bmatrix} -1 & 7 \end{bmatrix} \begin{bmatrix} x_1 \\ x_2 \end{bmatrix}$$

Ans $\frac{27}{6}$

$$
\textbf{(b)} \quad \begin{bmatrix} \dot{x}_1 \\ \dot{x}_2 \\ \dot{x}_3 \end{bmatrix} = \begin{bmatrix} -1 & 1 & 0 \\ 2 & 2 & 0 \\ 3 & -1 & 4 \end{bmatrix} \begin{bmatrix} x_1 \\ x_2 \\ x_3 \end{bmatrix} + \begin{bmatrix} 1 \\ 0 \\ -1 \end{bmatrix} r
$$

$$
y = \begin{bmatrix} 3 & -2 & 1 \end{bmatrix} \begin{bmatrix} x_1 \\ x_2 \\ x_3 \end{bmatrix}
$$

Ans system unstable

7.8 OBSERVER DESIGN

In order to fully implement the advantages of state variable feedback, all states should be fed back. For a fifth-order system, all five states should be eligible for state variable feedback. Perhaps two of those states are easily sensed, but the other three defy available off-the-shelf sensor measurement. The three hard-to-measure states must be estimated to an acceptable degree of accuracy by some scheme.

Control and estimation steps are separable for linear systems. Control gains are computed under the assumption that the measurements are available. The designer then creates an estimator intended to provide the states for purposes of control. Control will be considered later in this section. An observer is a special kind of estimator. The definition of observability provides a first step in understanding the structure of an observer.

A system is completely observable if any initial condition $\mathbf{x}(t_0)$ can be reconstructed from the system output $\mathbf{y}(t)$ over a period of time from t_0 until t_f. The test implies that $\mathbf{x}(t_0)$ can be constructed by some system as in Fig. 7.22. The observability test (if passed) is mathematically equivalent to requiring that the system output $\mathbf{y}$ and $n-1$ derivatives of $\mathbf{y}$ be linearly independent. Since the test assumes no input,

$$
\mathbf{y} = \mathbf{C}\mathbf{x}
$$

$$
\dot{\mathbf{y}} = \mathbf{C}\dot{\mathbf{x}} = \mathbf{C}\mathbf{A}\mathbf{x}
$$

$$
\ddot{\mathbf{y}} = \mathbf{C}\mathbf{A}\dot{\mathbf{x}} = \mathbf{C}\mathbf{A}^2\mathbf{x}
$$

$$
\mathbf{y}^{(n-1)} = \mathbf{C}\mathbf{A}^{(n-1)}\mathbf{x}
$$

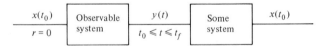

Figure 7.22 State reconstruction for an observable system.

Therefore, collecting $\mathbf{y}$ and its $n - 1$ derivatives,

$$
\begin{bmatrix} \mathbf{y} \\ \dot{\mathbf{y}} \\ \vdots \\ \mathbf{y}^{(n-1)} \end{bmatrix} = \begin{bmatrix} \mathbf{C} \\ \mathbf{CA} \\ \vdots \\ \mathbf{CA}^{(n-1)} \end{bmatrix} \mathbf{x} = \mathbf{M}_0 \mathbf{x}
$$

Mathematically, this means that the system output contains n linearly independent pieces of information from which it is possible to reconstruct the n pieces of information contained in $\mathbf{x}(t_0)$.

In general, the observability test is a worst case scenario, in that the input $\mathbf{r}(t)$ is considered to be absent. Certainly a system that passes the observability test should be reconstructible if the history of $\mathbf{r}(t)$ is also known. An observer (Fig. 7.23) is a linear system operating on both $\mathbf{y}$ and $\mathbf{r}$ in an effort to estimate some (or all) of the elements of $\mathbf{x}$. Since the observer has states $\mathbf{z}$ and operates (linearly) on $\mathbf{y}$ and $\mathbf{r}$, the matrices $\mathbf{D}$, $\mathbf{F}$, and $\mathbf{G}$ define the dynamics of $\mathbf{z}$:

$$
\dot{\mathbf{z}} = \mathbf{Dz} + \mathbf{Fy} + \mathbf{Gr}
$$

If the observer is properly designed, the states $\mathbf{z}$ should approach some linear combination of the system states ($\mathbf{Tx}$) as time approaches infinity. Somehow the matrices $\mathbf{D}$, $\mathbf{F}$, $\mathbf{G}$, and $\mathbf{T}$ must be computed by knowing $\mathbf{A}$, $\mathbf{B}$, and $\mathbf{C}$ and by requiring

$$
\mathbf{z} \to \mathbf{Tx} \quad \text{as} \quad t \to \infty
$$

The observer design proceeds by defining an error $\mathbf{e}$ between $\mathbf{z}$ and $\mathbf{Tx}$. Somehow, that error must disappear.

$$
\mathbf{e} = \mathbf{z} - \mathbf{Tx}
$$

$$
\dot{\mathbf{e}} = \dot{\mathbf{z}} - \mathbf{T\dot{x}}
$$

$$
= \mathbf{Dz} + \mathbf{Fy} + \mathbf{Gr} - \mathbf{T}[\mathbf{Ax} + \mathbf{Br}]
$$

$$
= \mathbf{Dz} + [\mathbf{FC} - \mathbf{TA}]\mathbf{x} + [\mathbf{G} - \mathbf{TB}]\mathbf{r}
$$

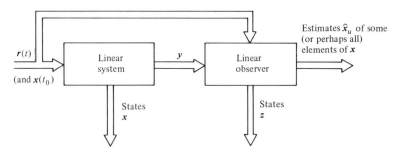

Figure 7.23   Vector block diagram of a system and its observer.

The rate of change of the error becomes independent of **r** if

$$\mathbf{G} = \mathbf{TB}$$

Suppose a requirement is imposed that

$$\mathbf{FC} - \mathbf{TA} = -\mathbf{DT}$$

The dynamics of **e** become

$$\dot{\mathbf{e}} = \mathbf{Dz} - \mathbf{DTx}$$
$$= \mathbf{D}[\mathbf{z} - \mathbf{Tx}]$$
$$= \mathbf{De}$$

By selecting the eigenvalues of **D** to be in the left half-plane, the error between **z** and **Tx** diminishes as time goes on.

Two kinds of observers are commonly used. An identity observer is so called because **T** is the identity matrix. For a fifth-order system where two measurements are already available and three are needed, the identity observer estimates all five states. As a sensor, the identity observer is redundant, but particularly easy to design. A reduced-order observer of minimum order only estimates those states not easily measurable by a sensor. For a fifth-order system with two measurable states, that reduced-order observer is of third order (the minimum order needed).

The two equations to be solved are

$$\mathbf{TA} - \mathbf{DT} = \mathbf{FC}$$
$$\mathbf{G} = \mathbf{TB}$$

For example, a fifth-order system might have states consisting of position, velocity, acceleration, the first derivative of acceleration (sometimes called the jerk derivative) and the second derivative of acceleration. Position and velocity can usually be measured (for example, an automobile odometer measures position and a speedometer measures velocity). Acceleration and its two derivatives are usually much harder to measure. If an identity observer is used, also five states would be estimated (including position and velocity which are already known). A reduced-order observer of minimum order would only estimate acceleration and its two derivatives.

Identity Observer

For the identity observer (where **T** = **I**) the equations simplify

$$\mathbf{A} - \mathbf{D} = \mathbf{FC}$$
$$\mathbf{G} = \mathbf{B}$$

The first equation means that

$$D = A - FC$$

The identity observer has dynamics

$$\dot{z} = [A - FC]z + Fy + Br$$
$$= Az + Br + F(y - Cz)$$

This observer reconstructs (estimates) the entire **x** vector (part of which is already measured). Using the hat (^) symbol to mean estimation

$$\hat{x}_u = z = \hat{x}$$
$$Cz = C\hat{x} = \hat{y}$$
$$\dot{z} = Az + Br + F(y - \hat{y})$$

The identity observer (in state-space form) appears in Fig. 7.24. The double lines imply vector quantities. The dynamic behavior of the identity observer includes two parts: one part is a duplicate of the system dynamics (using **A** and **B**) while the second part is an adjustive term depending on the difference between **y** and an estimate of **y**. As the estimate becomes correct, the adjustive term disappears and the observer becomes a duplicate of the system. If the observer is properly designed, the error between the true and estimated states should decay more quickly than the system responds. In that way the estimated states approach the desired states soon enough that the estimates are useful for purposes of control.

The design strategy for an identity observer follows:

1. Select the eigenvalues of **D** well to the left of those of **A** so the error diminishes quickly compared with the uncontrolled (and eventually the controlled) system.

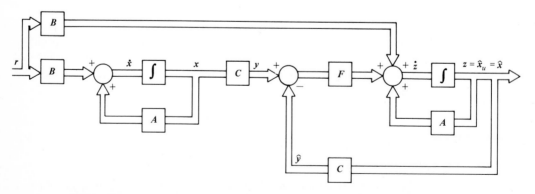

Figure 7.24   Vector block diagram of a system with an identity observer.

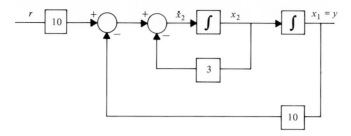

Figure 7.25 A second-order system where x_1 is measured.

2. Select **F** so that

$$|s\mathbf{I} - (\mathbf{A} - \mathbf{FC})| = |s\mathbf{I} - \mathbf{D}|$$

Notice that **D** is not needed, only the characteristic polynomial of **D**.

Figure 7.25 shows a second-order system to which an identity observer could be applied. Suppose x_1 is measured, but x_2 must be estimated. An identity observer would estimate x_1 (redundantly) while also providing the needed estimate of x_2. The observer would be second order. The system dynamics are

$$\begin{bmatrix} \dot{x}_1 \\ \dot{x}_2 \end{bmatrix} = \begin{bmatrix} 0 & 1 \\ -10 & -3 \end{bmatrix} \begin{bmatrix} x_1 \\ x_2 \end{bmatrix} + \begin{bmatrix} 0 \\ 10 \end{bmatrix} r$$

$$\mathbf{y} = \begin{bmatrix} 1 & 0 \end{bmatrix} \begin{bmatrix} x_1 \\ x_2 \end{bmatrix}$$

For an identity observer

$$\mathbf{z} = \begin{bmatrix} \hat{x}_1 \\ \hat{x}_2 \end{bmatrix}$$

The eigenvalues of **A** follow from

$$|s\mathbf{I} - \mathbf{A}| = s^2 + 3s + 10$$
$$= (s + 1.5 + j2.78)(s + 1.5 - j2.78)$$

The two design steps may be completed.

1. Since the observer eigenvalues should be well to the left of the eigenvalues of **A**, one possibility is to place both eigenvalues at -20.

2. Since $\mathbf{D} = \mathbf{A} - \mathbf{FC}$ and since there is one measurement $\mathbf{D}$ is 2×2; $\mathbf{F}$ is 2×1; $\mathbf{C}$ is 1×2.

$$\mathbf{A} - \mathbf{FC} = \begin{bmatrix} 0 & 1 \\ -10 & -3 \end{bmatrix} - \begin{bmatrix} f_{11} \\ f_{21} \end{bmatrix} [1 \quad 0]$$

$$= \begin{bmatrix} -f_{11} & 1 \\ -10 - f_{21} & -3 \end{bmatrix}$$

$$|s\mathbf{I} - (\mathbf{A} - \mathbf{FC})| = s^2 + s(3 + f_{11}) + 3f_{11} + 10 + f_{21}$$

$$= (s + 20)^2$$

$$= s^2 + 40s + 400$$

$$3 + f_{11} = 40 \quad \text{so} \quad f_{11} = 37$$

$$3f_{11} + 10 + f_{21} = 400 \quad \text{so} \quad f_{21} = 400 - 3(37) - 10$$

$$= 279$$

$$\mathbf{F} = \begin{bmatrix} 37 \\ 279 \end{bmatrix}$$

Now that $\mathbf{F}$ is available, the observer dynamics follow from

$$\dot{\mathbf{z}} = \mathbf{Az} + \mathbf{Br} + \mathbf{F}(\mathbf{y} - \hat{\mathbf{y}})$$

$$\mathbf{y} = x_1$$

$$\hat{\mathbf{y}} = z_1$$

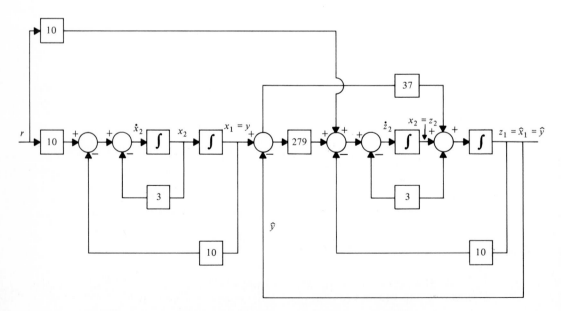

Figure 7.26 The second-order system of Fig. 7.25 with an identity observer.

so that

$$\dot{z}_1 = z_2 + 37(x_1 - z_1)$$

$$\dot{z}_2 = -10z_1 - 3z_2 + 10r + 279(x_1 - z_1)$$

The identity observer is connected to the system of Fig. 7.25 in Fig. 7.26.

DRILL PROBLEM

D7.18. For each of the following systems, x_1 is measured while x_2 must be estimated by an identity observer (which will also estimate x_1). Select **F** so both eigenvalues are as required. Write the equations for the observer and create a block diagram or signal flow graph for the interconnected system and observer.

(a) $A = \begin{bmatrix} -2 & -4 \\ 1 & -4 \end{bmatrix}$ $\qquad B = \begin{bmatrix} 2 \\ 0 \end{bmatrix}$

$C = \begin{bmatrix} 1 & 0 \end{bmatrix}$

Both eigenvalues of **D** should be -50.

Ans $F = \begin{bmatrix} 94 \\ -528 \end{bmatrix}$

$$\dot{z}_1 = -2z_1 - 4z_2 + 2r + 94(x_1 - z_1)$$

$$\dot{z}_2 = z_1 - 4z_2 - 528(x_1 - z_1)$$

(b) $A = \begin{bmatrix} -4 & -4 \\ 1 & -2 \end{bmatrix}$ $\qquad B = \begin{bmatrix} 0 \\ 2 \end{bmatrix}$

$C = \begin{bmatrix} 1 & 0 \end{bmatrix}$

Both eigenvalues of **D** should be -10.

Ans $F = \begin{bmatrix} 14 \\ -15 \end{bmatrix}$

$$\dot{z}_1 = -4z_1 - 4z_2 + 14(x_1 - z_1)$$

$$\dot{z}_2 = z_1 - 2z_2 + 2r - 15(x_1 - z_1)$$

Reduced-Order Observer

A reduced-order observer estimates only those states not measurable by a sensor. To carry out the design task, it is desirable to define the system dynamics where the **x** vector is organized using the sensor measurements (**y**) followed by the remaining states which are otherwise unavailable ($\mathbf{x}_u$). Even if the system is not organized that way at first, a change of variables can be used.

$$\mathbf{x} = \begin{bmatrix} \mathbf{y} \\ \mathbf{x}_u \end{bmatrix}$$

This decomposition leads to the definition of the observer states by

$$\mathbf{z} = [\mathbf{T}_1 \quad \mathbf{T}_2] \begin{bmatrix} \mathbf{y} \\ \hat{\mathbf{x}}_u \end{bmatrix}$$

It is common practice to select

$$\mathbf{T} = [-\mathbf{L} \quad \mathbf{I}]$$

Therefore,

$$\hat{\mathbf{x}}_u = \mathbf{z} + \mathbf{L}\mathbf{y}$$

The equation

$$\mathbf{G} = \mathbf{T}\mathbf{B}$$

is easily solved when $\mathbf{T}$ is known (since $\mathbf{B}$ is also known). In order to solve $\mathbf{T}$,

$$\mathbf{T}\mathbf{A} - \mathbf{D}\mathbf{T} = \mathbf{F}\mathbf{C}$$

the following conditions must hold (in which case $\mathbf{T}$ is uniquely solvable):

1. $\mathbf{A}$, $\mathbf{C}$ must be observable
2. $\mathbf{D}$, $\mathbf{F}$ must be controllable
3. The eigenvalues of $\mathbf{A}$ and $\mathbf{D}$ must be different.

The above three conditions cause the matrices $\mathbf{A}$, $\mathbf{D}$, $\mathbf{F}$, $\mathbf{C}$ to create a set of linearly independent equations in the elements of $\mathbf{T}$. Because of his work in proving the above, these observers are often called Luenberger observers. Item 2 causes $\mathbf{D}$ to replace $\mathbf{A}$ and $\mathbf{F}$ to replace $\mathbf{B}$ in the controllability test. In other words, the system must be observable, the observer must be controllable, and the error dynamics must differ from the system dynamics.

Some flexibility exists in selecting $\mathbf{F}$ (where $\mathbf{D}$, $\mathbf{F}$ must be controllable). The required equation is easier to understand if the component matrices are partitioned.

$$\mathbf{T}\mathbf{A} - \mathbf{D}\mathbf{T} = \mathbf{F}\mathbf{C}$$

$$[-\mathbf{L} \quad \mathbf{I}] \begin{bmatrix} \mathbf{A}_{11} & \mathbf{A}_{12} \\ \mathbf{A}_{21} & \mathbf{A}_{22} \end{bmatrix} - \mathbf{D}[-\mathbf{L} \quad \mathbf{I}] = \mathbf{F}[\mathbf{I}_1 \quad \mathbf{0}]$$

$\mathbf{A}_{11}$ is $m \times m$; $\mathbf{L}$ is $(n - m) \times m$
$\mathbf{A}_{12}$ is $m \times (n - m)$; $\mathbf{D}$ is $(n - m) \times (n - m)$
$\mathbf{A}_{21}$ is $(n - m) \times m$; $\mathbf{I}_1$ is $(m \times m)$
$\mathbf{A}_{22}$ is $(n - m) \times (n - m)$; $\mathbf{I}$ is $(n - m) \times (n - m)$
n = system order
m = order of $\mathbf{y}$ (measurements)

A design strategy for the reduced-order observer is:

1. Select **D** to be a matrix whose eigenvalues are well to the left (perhaps by a factor of 10 or more) from those of **A** (or of any desired closed-loop behavior).
2. Solve
   (a) $-\mathbf{L}\mathbf{A}_{11} + \mathbf{A}_{21} + \mathbf{D}\mathbf{L} = \mathbf{F}$
   (b) $-\mathbf{L}\mathbf{A}_{12} + \mathbf{A}_{22} - \mathbf{D} = \mathbf{0}$
   Select elements of **F** to cause as many elements as possible of **L** to become zero. Be sure that **D, F** is controllable.
3. Solve

$$\mathbf{G} = \mathbf{T}\mathbf{B}$$

In step 2, at most $m - 1$ columns of **L** can be zeroed out so that the design becomes independent (and therefore insensitive) to those elements of $\mathbf{A}_{11}$, $\mathbf{A}_{12}$, and **B** that are multiplied by **L**. Sometimes parameters located in the matrices are not known exactly. Although that case is not treated in this book, the possibility exists and good observer design can minimize the effect of uncertainty.

Once the matrices **D, F, G, L** are computed, the observer is ready to be fabricated. The observer dynamics are

$$\dot{\mathbf{z}} = \mathbf{D}\mathbf{z} + \mathbf{F}\mathbf{y} + \mathbf{G}\mathbf{r}$$

$$\hat{\mathbf{x}}_u = \mathbf{z} + \mathbf{L}\mathbf{y}$$

Figure 7.27 contains a vector block diagram showing the linear system and reduced-order observer dynamics. The observer estimates $\mathbf{x}_u$.

In Chaps. 5 and 6 compensators were applied to an uncompensated system where

$$KG_p(s) = \frac{3}{s(s^2 + 4s + 5)}$$

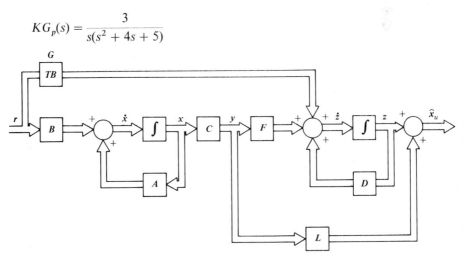

Figure 7.27 Vector block diagram of a state variable system with a reduced-order observer.

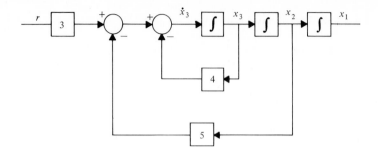

Figure 7.28   Third-order system where x_1 and x_2 are measurable by a sensor.

Figure 7.28 shows a state-space version of that system where

$$\frac{X_1(s)}{R(s)} = \frac{3}{s(s^2 + 4s + 5)}$$

Suppose the states marked x_1 and x_2 are measured but the state x_3 is to be reconstructed using a reduced-order observer. Here $n = 3$ and $m = 2$ so a first-order observer is called for. The system dynamics are

$$\begin{bmatrix} \dot{x}_1 \\ \dot{x}_2 \\ \dot{x}_3 \end{bmatrix} = \begin{bmatrix} 0 & 1 & \vdots & 0 \\ 0 & 0 & \vdots & 1 \\ 0 & -5 & \vdots & -4 \end{bmatrix} \begin{bmatrix} x_1 \\ x_2 \\ x_3 \end{bmatrix} + \begin{bmatrix} 0 \\ 0 \\ 3 \end{bmatrix} r$$

The **A** matrix is partitioned, where

$$\mathbf{A}_{11} = \begin{bmatrix} 0 & 1 \\ 0 & 0 \end{bmatrix} \qquad \mathbf{A}_{12} = \begin{bmatrix} 0 \\ 1 \end{bmatrix}$$

$$\mathbf{A}_{21} = [0 \quad -5] \qquad \mathbf{A}_{22} = -4$$

Since x_1 and x_2 are measured,

$$\mathbf{y} = \begin{bmatrix} 1 & 0 & 0 \\ 0 & 1 & 0 \end{bmatrix} \mathbf{x}$$

The observer dynamics follow where

$\mathbf{D}$ is $(n - m) \times n - m = 1 \times 1$
$\mathbf{F}$ is $(n - m) \times m = 1 \times 2$
$\mathbf{G}$ is $(n - m) \times 1 = 1 \times 1$ for scalar r

$$\dot{z} = z + [f_{11} \quad f_{12}] \mathbf{y} + g_{11} r$$

$$\mathbf{T} = [-\mathbf{L} \quad \mathbf{I}] = [-l_{11} \quad -l_{12} \quad l]$$

1. The eigenvalues of **A** are at 0, $-2 + j1$, and $-2 - j1$. If **D** is -20, the error between x_3 and $\hat{x}_3$ decays with a time constant of $\frac{1}{20}$ second, one tenth that of the open-loop system. Since **D** is a scalar, then $\mathbf{D} = -20$.

2a. $\quad [-l_{11} \quad -l_{12}] \begin{bmatrix} 0 & 1 \\ 0 & 0 \end{bmatrix} + [0 \quad -5]$

$$-20[l_{11} \quad l_{12}] = [f_{11} \quad f_{12}]$$

2b. $\quad [-l_{11} \quad -l_{12}] \begin{bmatrix} 0 \\ 1 \end{bmatrix} + (-4) - (-20) = 0$

Three equations result from 2a and 2b.

Since $m = 2$, at most, $m - 1 = 1$ column of **L** can be set equal to zero:

$$-20l_{11} = f_{11} \tag{1}$$

$$-l_{11} - 5 - 20l_{12} = f_{12} \tag{2}$$

$$-l_{12} - 4 + 20 = 0 \tag{3}$$

From (3), l_{12} must be 16. From (1), both l_{11} and f_{11} can be zero (if **D, F** are controllable). From (2),

$$f_{12} = -5 - 20(16) = -325$$

Summarizing,

$$\mathbf{T} = [0 \quad -16 \quad 1]$$

$$\mathbf{F} = [0 \quad -325]$$

3. G can be obtained from

$$\mathbf{G} = \mathbf{TB} = [0 \quad -16 \quad 1] \begin{bmatrix} 0 \\ 0 \\ 3 \end{bmatrix} = 3$$

It is easy to show that **D, F** is completely controllable, so the design is finished.

The observer dynamics are

$$\dot{z} = -20z + [0 \quad -325] \begin{bmatrix} x_1 \\ x_2 \end{bmatrix} + 3r$$

$$= -20z - 325x_2 + 3r$$

Also,

$$\hat{x}_3 = z + \mathbf{L}y = z + [0 \quad 16] \begin{bmatrix} x_1 \\ x_2 \end{bmatrix} = z + 16x_2$$

The observer can be connected to the system of Fig. 7.28, resulting in Fig. 7.29. The feedback control in Fig. 7.29 will be considered next.

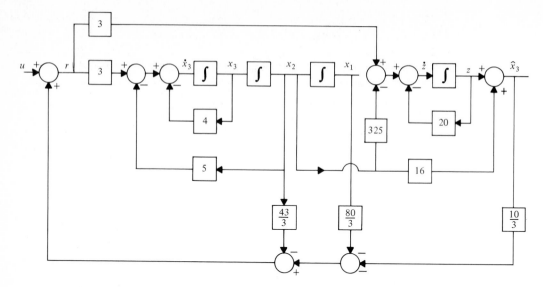

Figure 7.29   Reduced-order observer and closed-loop control for the system of Fig. 7.28.

DRILL PROBLEM

D7.19. For the systems of D7.18, design a reduced-order, first-order observer to estimate only x_2. Select the observer eigenvalue as indicated. Write the equations for the observer and create a block diagram or signal flow graph for the interconnected system and observer.

(a) The eigenvalue of **D** should be -50.

> **Ans** $\mathbf{L} = -11.5$ $\mathbf{D} = -50$
> $\mathbf{F} = 553$ $\mathbf{G} = 23$
> $\dot{z} = -50z + 23r + 553x_1$
> $\hat{x}_2 = z - 11.5x_1$

(b) The eigenvalue of **D** should be -10.

> **Ans** $\mathbf{L} = -2$ $\mathbf{F} = 13$
> $\mathbf{G} = 2$
> $\dot{z} = -10z + 2r + 13x_1$
> $\hat{x}_2 = z - 2x_1$

Control Using Observer Estimates

An observer is intended to provide measurements of those state variables that are otherwise difficult to obtain using standard sensors. All measurements are used for state variable feedback whether the measurements are from a standard sensor or from an observer.

When the system and the observer are both linear, a sort of superposition occurs. The control gains are calculated assuming that all the state variables are available. The observer is then designed for the open-loop system without considering the eventual closed-loop structure. Finally, the system, observer, and control are combined. Fortunately (due to linearity) the dynamics of the closed-loop system has eigenvalues consisting of the ideal closed-loop roots and the ideal observer roots (those of **D**) separated.

For the example system, a gain could be computed for each of the three states,

$$\mathbf{K} = [k_1 \quad k_2 \quad k_3]$$

$$\mathbf{A} + \mathbf{BK} = \begin{bmatrix} 0 & 1 & 0 \\ 0 & 0 & 1 \\ 3k_1 & 3k_2 - 5 & 3k_3 - 4 \end{bmatrix}$$

The characteristic polynomial for the closed-loop (ideal) system is

$$|s\mathbf{I} - (\mathbf{A} + \mathbf{BK})| = s^3 + s^2(4 - 3k_3) + s(5 - 3k_2) - 3k_1$$

Unlike the compensator design examples in Chaps. 5 and 6, all three closed-loop poles may be located as the designer desires. For example, the three closed-loop poles can be placed at $-2 \pm j2$ and at -10.

$$(s + 2 + j2)(s + 2 - j2)(s + 10) = (s^2 + 4s + 8)(s + 10)$$

$$= s^3 + 14s^2 + 48s + 80$$

The three gains follow easily:

$$k_1 = -\frac{80}{3}$$

$$k_2 = -\frac{43}{3}$$

$$k_3 = -\frac{10}{3}$$

However, only two of the three state variables are available using standard sensors so that k_1 and k_2 (in Fig. 7.29) operate on the actual states while k_3 must operate on the observer-generated reconstruction of the third state.

In general, a closed-loop combination including a system, reduced-order observer, and control gains takes on the structure of Fig. 7.30. The control should (ideally) be

$$\mathbf{KX} = \mathbf{K}\begin{bmatrix} \mathbf{y} \\ \mathbf{x}_u \end{bmatrix} = [\mathbf{K}_1 \quad \mathbf{K}_2]\begin{bmatrix} \mathbf{y} \\ \mathbf{x}_u \end{bmatrix}$$

$\mathbf{K}_1$ is $i \times m$

$\mathbf{K}_2$ is $i \times (n - m)$

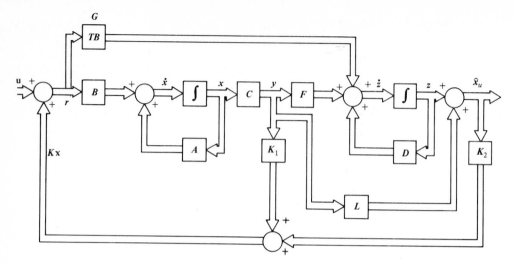

Figure 7.30   Vector block diagram of a state variable system, reduced-order observer, and closed-loop control.

Thus the ideal control gain matrix can be partitioned into one part ($\mathbf{K}_1$) operating on $\mathbf{y}$ and a second part ($\mathbf{K}_2$) operating on $\mathbf{x}_u$. Since the observer must estimate $\mathbf{x}_u$, the actual control becomes

$$[\mathbf{K}_1 \quad \mathbf{K}_2]\begin{bmatrix} \mathbf{y} \\ \hat{\mathbf{x}}_u \end{bmatrix}$$

Figure 7.30 indicates that the composite system has interconnected states consisting of $\mathbf{z}$ (from the observer) and $\mathbf{x}$ (from the plant). The closed-loop composite system has dynamics

$$\begin{bmatrix} \dot{\mathbf{z}} \\ \dot{\mathbf{x}} \end{bmatrix} = \begin{bmatrix} \mathbf{D} + \mathbf{GK}_2 & : & \mathbf{FC} + \mathbf{GK}_1\mathbf{C} + \mathbf{GK}_2\mathbf{LC} \\ \mathbf{BK}_2 & : & \mathbf{A} + \mathbf{BK}_1\mathbf{C} + \mathbf{BK}_2\mathbf{LC} \end{bmatrix}\begin{bmatrix} \mathbf{z} \\ \mathbf{x} \end{bmatrix} + \begin{bmatrix} \mathbf{G} \\ \mathbf{B} \end{bmatrix}\mathbf{u}$$

$$\begin{bmatrix} \hat{\mathbf{x}}_u \\ \mathbf{x} \end{bmatrix} = \begin{bmatrix} \mathbf{I} & : & \mathbf{LC} \\ \mathbf{0} & : & \mathbf{I} \end{bmatrix}\begin{bmatrix} \mathbf{z} \\ \mathbf{x} \end{bmatrix}$$

These equations can be written compactly as

$$\dot{\mathbf{x}}' = \mathbf{A}'\mathbf{x}' + \mathbf{B}'\mathbf{u}$$

$$\mathbf{y}' = \mathbf{C}'\mathbf{x}'$$

by defining

$$\mathbf{x}' = \begin{bmatrix} \mathbf{z} \\ \mathbf{x} \end{bmatrix}$$

$$\mathbf{y}' = \begin{bmatrix} \hat{\mathbf{x}}_u \\ \mathbf{x} \end{bmatrix}$$

It can be shown that

$$|s\mathbf{I} - \mathbf{A}'| = |s\mathbf{I} - \mathbf{D}| \, |s\mathbf{I} - (\mathbf{A} + \mathbf{BK})|$$

This means that the eigenvalues of $\mathbf{A}'$ equal the eigenvalues of $\mathbf{D}$ (from the observer) and the eigenvalues of $\mathbf{A} + \mathbf{BK}$ (from the desired closed-loop roots). The conclusion is that control and observation (estimation) are separable.

The above fact can be proven by changing variables from column $(\mathbf{z}, \mathbf{x})$ to column $(\mathbf{e}, \mathbf{x})$ where $\mathbf{e}$ is the error $\mathbf{z} - \mathbf{Tx}$.

$$\mathbf{x}'' = \begin{bmatrix} \mathbf{e} \\ \mathbf{x} \end{bmatrix} = \mathbf{T}_3 \begin{bmatrix} \mathbf{z} \\ \mathbf{x} \end{bmatrix}$$

$$\mathbf{T}_3^{-1} = \begin{bmatrix} \mathbf{I} & \mathbf{T} \\ \mathbf{0} & \mathbf{I} \end{bmatrix} \qquad \mathbf{T}_3 = \begin{bmatrix} \mathbf{I} & -\mathbf{T} \\ \mathbf{0} & \mathbf{I} \end{bmatrix}$$

The eigenvalues of $\mathbf{A}'$ are the same as the eigenvalues of $\mathbf{A}''$ where

$$\mathbf{A}'' = \mathbf{T}_3 \mathbf{A}' \mathbf{T}_3^{-1}$$

$$= \left[\begin{array}{c|c} \mathbf{D} & \mathbf{0} \\ \hline \mathbf{BK}_2 & \mathbf{A} + \mathbf{BK} \end{array} \right]$$

The zero in the upper-right quadrant of the partitioned matrix makes it possible to write the characteristic polynomial using the separate characteristic polynomials of $\mathbf{D}$ and $\mathbf{A} + \mathbf{BK}$.

For the example system of Fig. 7.29,

$$\mathbf{A}' = \begin{bmatrix} -30 & -80 & -528 & 0 \\ 0 & 0 & 1 & 0 \\ 0 & 0 & 0 & 1 \\ -10 & -80 & -208 & -4 \end{bmatrix}$$

$$\mathbf{B}' = \begin{bmatrix} 3 \\ 0 \\ 0 \\ 3 \end{bmatrix}$$

$$\mathbf{C}' = \begin{bmatrix} 1 & 0 & 16 & 0 \\ 0 & 1 & 0 & 0 \\ 0 & 0 & 1 & 0 \\ 0 & 0 & 0 & 1 \end{bmatrix}$$

$$|s\mathbf{I} - \mathbf{A}'| = s^4 + 34s^3 + 328s^2 + 1040s + 1600$$

$$= (s + 2 + j2)(s + 2 - j2)(s + 10)(s + 20)$$

This example shows how the observer eigenvalues and desired closed-loop eigenvalues are present in the final design.

DRILL PROBLEM

D7.20. (a) Assume that both states are available for the system of D7.18a. Select gains k_1 (for x_1) and k_2 (for x_2) so the closed-loop poles are at -10 and -20.

Ans $k_1 = -12; k_2 = -46$

(b) Apply the gain k_1 to x_1 but apply the gain k_2 to the observer estimate of x_2 from D7.19a. Find the $\mathbf{A}'$, $\mathbf{B}'$, and $\mathbf{C}'$ matrices. Compute and factor the characteristic polynomial

$$\mathbf{Ans} \quad \mathbf{A}' = \begin{bmatrix} -1108 & 12444 & 0 \\ -92 & 1032 & -4 \\ 0 & 1 & -4 \end{bmatrix}$$

$$\mathbf{B}' = \begin{bmatrix} 23 \\ 2 \\ 0 \end{bmatrix}$$

$$\mathbf{C}' = \begin{bmatrix} 1 & -11.5 & 0 \\ 0 & 1 & 0 \\ 0 & 0 & 1 \end{bmatrix}$$

$$|s\mathbf{I} - \mathbf{A}'| = (s + 10)(s + 20)(s + 50)$$
$$= s^3 + 80s^2 + 1700s + 10000$$

7.9 QUADRATIC OPTIMAL CONTROL

In Chap. 5, compensators and gains are selected to create a good value for steady state error and for the location of dominant roots (which, in turn, influence rise time and settling time). Each design is complete when an acceptable value results. In Chap. 6, for example, compensators and gains are intended to provide acceptable gain and phase margin. In none of these cases can the design be considered optimal (no other design is better); however, each design is, instead, acceptable.

One important class of problems (quadratic optimal control) can be solved so as to create a best (optimal) result. That class of problems generates a set of state feedback gains so that a quadratic measure of performance is minimized.

Problem Definition

The performance measure (index) takes the form

$$J = \int_0^\infty (\mathbf{x}^T \mathbf{Q} \mathbf{x} + \mathbf{r}^T \mathbf{P} \mathbf{r}) \, dt \tag{1}$$

where the upper limit of integration is often infinite. The function inside the integral sign is a quadratic form and the matrices $\mathbf{P}$ and $\mathbf{Q}$ are usually symmetric (see Appendix A for a brief review of quadratic forms and definiteness). So that the control is completely accounted for (no infinite control is allowed), $\mathbf{P}$ is usually positive definite and symmetric while $\mathbf{Q}$ is positive semidefinite and symmetric (some elements of $\mathbf{x}$ might not be weighted by $\mathbf{Q}$).

For example, if

$$\mathbf{Q} = \begin{bmatrix} 1 & 0 \\ 0 & 2 \end{bmatrix} \qquad \mathbf{P} = p \text{ (a scalar)}$$

$$\mathbf{x} = \begin{bmatrix} x_1 \\ x_2 \end{bmatrix}$$

$$\mathbf{x}^T\mathbf{Q}\mathbf{x} + \mathbf{r}^T\mathbf{P}\mathbf{r} = [x_1 \quad x_2]\begin{bmatrix} 1 & 0 \\ 0 & 2 \end{bmatrix}[x_1 \quad x_2] + pr^2$$

$$= x_1^2 + x_2^2 + pr^2$$

If

$$\mathbf{Q} = \begin{bmatrix} 1 & 0 \\ 0 & 0 \end{bmatrix} \qquad \mathbf{P} = p$$

$$\mathbf{x}^T\mathbf{Q}\mathbf{x} + \mathbf{r}^T\mathbf{P}\mathbf{r} = x_1^2 + pr^2$$

The lowest possible value for J is zero where both $\mathbf{x}$ and $\mathbf{r}$ are always zero. This type of optimal control is referred to as an optimal regulator. When the state vector must track a nonzero value, J can be redefined to create an optimal servomechanism (tracking) problem. Many of the previous control systems have been (by this terminology) servomechanisms. Regulator behavior is important for many types of control (attitude control, for example, where a zero reference should be maintained).

When $\mathbf{x}$ is not zero, a control $\mathbf{r}$ causes $\mathbf{x}$ to go toward zero. As the weight $\mathbf{Q}$ rises relative to $\mathbf{P}$, the control effort rises to reduce $\mathbf{x}$. Conversely, if a large weight is placed on $\mathbf{r}$, the control effort will diminish at the expense of larger values for $\mathbf{x}$.

For a linear system with matrices $\mathbf{A}$ and $\mathbf{B}$,

$$\dot{\mathbf{x}} = \mathbf{A}\mathbf{x} + \mathbf{B}\mathbf{r} \tag{2}$$

a number of procedures are available to solve (1). Optimization can easily become the subject for a complete textbook; therefore, only the main results are considered here. The work of mathematicians and engineers such as Hamilton, Pontryagin, Euler, Lagrange, and Kalman have resulted in a rather complete understanding of the solution to (1) where (2) defines the dynamics of $\mathbf{x}$.

Solving by Factorization

One approach uses the matrix

$$\mathbf{A}' = \begin{bmatrix} \mathbf{A} & -\mathbf{B}\mathbf{P}^{-1}\mathbf{B}^T \\ \hline -\mathbf{Q} & -\mathbf{A}^T \end{bmatrix} \tag{3}$$

The characteristic polynomial of $\mathbf{A}'$ is an even function. An even function has only even powers of s and can be decomposed into components having only RHP and (separately) LHP roots (here no **IA** roots exist).

The closed loop gains $\mathbf{K}$ for

$$\mathbf{r} = \mathbf{Kx} \tag{4}$$

cause the characteristic polynomial to equal the left half-plane eigenvalues of (3). Therefore, if

$$|s\mathbf{I} - \mathbf{A}'| = P(s)P(-s) \tag{5}$$

where $\mathbf{P}(s)$ contains only left half-plane eigenvalues

$$|s\mathbf{I} - (\mathbf{A} + \mathbf{BK})| = P(s) \tag{6}$$

The characteristic polynomial for $\mathbf{A}'$ equals

$$|s\mathbf{I} - \mathbf{A}'| = D(s)D(-s) + \left(\frac{1}{p}\right)N(s)N(-s) \tag{7}$$

when the control is a scalar and $\mathbf{P} = p$. Other useful relationships are that

$$D(s) = |s\mathbf{I} - \mathbf{A}| \tag{8}$$

and if $\mathbf{Q}$ can be decomposed into

$$\mathbf{Q} = \mathbf{g}^T\mathbf{g} \tag{9}$$

then

$$\mathbf{g}(s\mathbf{I} - \mathbf{A})^{-1}\mathbf{B} = \frac{N(s)}{D(s)} \tag{10}$$

The closed-loop gains result by computing (7) using (8) to (10), then factoring (7) into (5). Finally, the gains result from (6). Where $\mathbf{Q}$ cannot be decomposed, (7) can be derived directly from (8) and (3) (see also the method in drill problem D7.22).

As an example, the second-order system of Fig. 7.31 is to be controlled so as to minimize (1), where

$$\mathbf{Q} = \begin{bmatrix} 1 & 0 \\ 0 & 0 \end{bmatrix} \qquad \mathbf{P} = p$$

$$J = \int_0^\infty (x_1^2 + pr^2)\, dt$$

For the system of Fig. 7.31,

$$\mathbf{A} = \begin{bmatrix} 0 & 1 \\ -5 & -1 \end{bmatrix} \qquad \mathbf{B} = \begin{bmatrix} 0 \\ 5 \end{bmatrix}$$

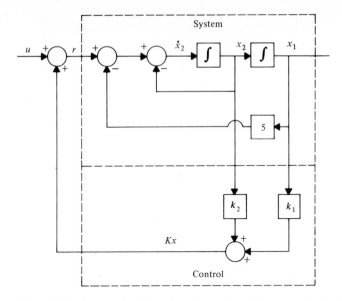

Figure 7.31 A second-order system. The gains will be computed using quadratic optimal control.

First, (7) to (10) are evaluated:

$$|s\mathbf{I} - \mathbf{A}| = s^2 + s + 5 = D(s)$$

$$\mathbf{Q} = \begin{bmatrix} 1 \\ 0 \end{bmatrix} \quad [1 \quad 0] = \begin{bmatrix} 1 & 0 \\ 0 & 0 \end{bmatrix}$$

$$\mathbf{g} = [1 \quad 0]$$

$$\mathbf{g}(s\mathbf{I} - \mathbf{A})^{-1}\mathbf{B} = \frac{5}{D(s)}$$

$$N(s) = 5$$

$$|s\mathbf{I} - \mathbf{A}'| = (s^2 + s + 5)(s^2 - s + 5) + \frac{25}{p}$$

$$= s^4 + 9s^2 + 25 + \frac{25}{p}$$

In order to factor the last equation into parts, assume that

$$P(s) = s^2 + as + b$$

$$P(s)P(-s) = (s^2 + as + b)(s^2 - as + b)$$

$$= s^4 + (2b - a^2)s + b^2$$

Finally (7), (5), and (6) are solved. For $p = 1$:

$$b = \sqrt{50} = 7.07$$

$$a = \sqrt{2b - 9} = 2.267$$

$$P(s) = s^2 + 2.267s + 7.07 = (s + 1.134 + j2.41)(s + 1.135 - j2.41)$$

The gains can be found from

$$\mathbf{K} = [k_1 \quad k_2]$$

$$\mathbf{A} + \mathbf{BK} = \begin{bmatrix} 0 & 1 \\ 5k_1 - 5 & 5k_2 - 1 \end{bmatrix}$$

$$|s\mathbf{I} - (\mathbf{A} + \mathbf{BK})| = P(s)$$

$$= s^2 + s(1 - 5k_2) + 5 - 5k_1$$

$$k_1 = -0.414$$

$$k_2 = -0.254$$

Table 7.1 shows gain values and closed-loop poles for increasing values of p.

As p increases, more importance is placed upon small control to reduce (1). The increased p reduces gain values and moves the closed-loop poles closer to the zero gain (open-loop) values of $-0.5 \pm j2.18$.

Root Square Locus

In general, a special kind of root locus (called a root square locus) can be computed using (7). An equivalent $KG(s)H(s)$ would be

$$\frac{(1/p)N(s)N(-s)}{D(s)D(-s)} = \frac{[(-1)^{n-m}(1/p)]N_L(s)N_R(s)}{D_L(s)D_R(s)} \tag{11}$$

where $D_L(s)$ and $N_L(s)$ contain only left half-plane roots; $N_R(s)$ and $D_R(s)$ contain only right half-plane roots; and the bracketed numerator term is the effective gain k.

The -1 is needed because the equivalent $G(s)H(s)$ must have a coefficient of $+1$ associated with the highest power of s in both numerator and denominator.

Table 7.1   OPTIMAL SOLUTION

p	Closed-Loop Poles	k_1	k_2
1	$-1.134 \pm j2.41$	-0.414	-0.254
10	$-0.61 \pm j2.21$	-0.049	-0.044
100	$-0.51 \pm j2.18$	-0.005	-0.005

Here n is the order of $D(s)$ and m is the order of $N(s)$. If $D(s)$ is of odd order, then (for a first-order example)

$$D(s) = s + 1$$

$$D(s)D(-s)(-1) = (s + 1)(1 - s)(-1) = (s + 1)(s - 1) = D_L(s)D_R(s)$$

The -1 is needed to clear a negative sign.

In general, if $n - m$ is even, a root square locus is plotted versus $1/p$ using the same sketching rules as for positive $K = 1/p$ from Chap. 5. When $n - m$ is odd, the root square locus acts as if $K = -(1/p)$; that is, negative K sketching rules are needed. The root square locus is always symmetrical about the vertical axis and also (but with a different shape) about the horizontal axis. The centroid is always zero. Using the above example, the effective $KG(s)H(s)$ is

$$\frac{25(1/p)}{(s^2 + s + 5)(s - s + 5)}$$

Here $n = 2$ and $m = 0$, so $n - m$ is even and positive gain sketching rules result in Fig. 7.32. Using the terminology in Chap. 4, there are four open-loop poles and zero open-loop zeros, because n and m in Chap. 4 are twice the values of n and m

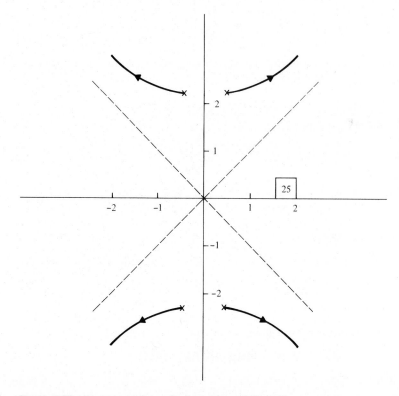

Figure 7.32 Root square locus for the system of Fig. 7.31.

used here in Chap. 7. The left half-plane roots are actually the closed-loop system roots because gain values are computed to establish the left half-plane roots. The mirror-image right half-plane roots are an artifact of the optimization process where n-order systems often result in $2n$-order equations. The matrix $\mathbf{A}'$ is $2n \times 2n$ for an $n \times n$ $\mathbf{A}$ matrix. As p becomes smaller, the effective K rises and the closed-loop poles in Fig. 7.32 move farther into the left half-plane.

A remarkable property of quadratic optimal control is that the closed-loop system is never unstable. Suppose the example system becomes open-loop unstable where

$$\mathbf{A} = \begin{bmatrix} 0 & 1 \\ -5 & +1 \end{bmatrix}$$

$$|s\mathbf{I} - \mathbf{A}| = D(s) = s^2 - s + 5$$

However, (7) and (11) remain the same because $D(s)$ and $D(-s)$ simply switch. The root square locus is the same; so, for $p = 1$, 10, or 100, the closed-loop poles of Table 7.1 remain the same (the gains change). If p rises to infinity,

$$|s\mathbf{I} - (\mathbf{A} + \mathbf{BK})| = s^2 + s(-1 - 5k_2) + 5 - 5k_1$$

$$= s^2 + s + 5$$

$$k_1 = 0$$

$$k_2 = -0.4$$

Minimum control effort for the open-loop unstable plant whose open-loop characteristic polynomial is

$$s^2 - s + 5$$

becomes (for minimum control)

$$s^2 + s + 5$$

Thus, with minimum control effort applied, any open-loop unstable roots are switched to left half-plane values reflected about the vertical axis.

Matrix Riccati Equation

Another approach to finding the gains which minimize (1) is based on finding

$$\mathbf{r} = \mathbf{Kx}$$

$$\mathbf{K} = -\mathbf{P}^{-1}\mathbf{B}^T\mathbf{M}$$

for the nonlinear matrix differential equation called the matrix Riccati equation:

$$\dot{\mathbf{M}} = -\mathbf{MA} - \mathbf{A}^T\mathbf{M} + \mathbf{MBP}^{-1}\mathbf{B}^T\mathbf{M} - \mathbf{Q} \tag{12}$$

When the upper limit of integration in (1) is infinite, (12) can be solved with the derivative equal to zero, creating the algebraic Riccati equation. A major difficulty in solving (12) is that $\mathbf{M}$ must be positive-definite and symmetric while the algebraic equation has a positive-definite (stabilizing) solution and a negative-definite (destabilizing) solution. It is best to solve (12) using a computer program so the derivative drives $\mathbf{M}$ to its correct value (actually, the differential equation is run backwards).

Although a manual solution is tedious, the second-order example can be solved. Returning to the stable $\mathbf{A}$ matrix and letting $\mathbf{M}$ be symmetric,

$$\mathbf{M} = \begin{bmatrix} m_{11} & m_{12} \\ m_{12} & m_{22} \end{bmatrix}$$

the algebraic Riccati equation requires

$$0 = 10m_{12} + \left(\frac{25}{p}\right)m_{12}^2 - 1$$

$$0 = m_{12} - m_{11} + 5m_{22} + \left(\frac{25}{p}\right)m_{12}m_{22}$$

$$0 = 2(m_{22} - m_{12}) + \left(\frac{25}{p}\right)m_{22}^2$$

Suppose $p = 1$. The quadratic in m_{12} can be solved

$$m_{12} = \frac{-10 \pm \sqrt{200}}{50} = 0.0828, -0.48$$

the quadratic in m_{22} causes

$$m_{22} = \frac{-2 \pm \sqrt{4 + 200m_{12}}}{50}$$

In order for m_{22} to have a real-valued solution,

$$m_{12} = 0.0828$$

and, in order for m_{22} to be positive,

$$m_{22} = 0.0507$$

Finally, m_{11} must be

$$m_{11} = m_{12} + 5m_{22} + 25m_{12}m_{22}$$

$$= 0.441$$

The gains are

$$\mathbf{K} = [k_1 \quad k_2] = -\mathbf{P}^{-1}\mathbf{B}^T\mathbf{M}$$

$$= -(1)[0 \quad 5]\begin{bmatrix} 0.441 & 0.0828 \\ 0.0828 & 0.0507 \end{bmatrix}$$

$$= [-0.414 \quad -0.254]$$

which agrees with the values computed previously.

If any state variables are difficult to measure, control and estimation (using an observer perhaps) can be separated, designed, and then combined.

DRILL PROBLEMS

D7.21. For each of the following systems described by **A** and **B** matrices with a quadratic integrated performance measure described by **P** and **Q**, use factorization methods to find $D(s)$, $N(s)$, and the root square locus. Compute the closed-loop poles and gains for $p = 1$ and for $p = 10$.

(a) $\mathbf{A} = -2$, $\mathbf{B} = 4$, $\mathbf{Q} = 4$, $\mathbf{P} = p$

(b) $\mathbf{A} = +2$, $\mathbf{B} = 4$, $\mathbf{Q} = 4$, $\mathbf{P} = p$

(c) $\mathbf{A} = \begin{bmatrix} 0 & 1 \\ -10 & -2 \end{bmatrix}$, $\mathbf{B} = \begin{bmatrix} 0 \\ 2 \end{bmatrix}$, $\mathbf{Q} = \begin{bmatrix} 1 & 0 \\ 0 & 0 \end{bmatrix}$, $\mathbf{P} = p$

(d) $\mathbf{A} = \begin{bmatrix} 0 & 1 \\ -10 & -2 \end{bmatrix}$, $\mathbf{B} = \begin{bmatrix} 0 \\ 2 \end{bmatrix}$, $\mathbf{Q} = \begin{bmatrix} 0 & 0 \\ 0 & 1 \end{bmatrix}$, $\mathbf{P} = p$

Ans (a) $D(s) = s + 2$, $N(s) = 8$, $-(1/p)64/(s + 2)(s - 2)$

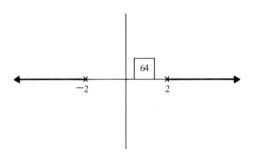

$$-8.246, \quad k = -1.562, \quad -3.225, \quad k = -0.306$$

(b) $D(s) = s - 2$, $N(s) = 8$, $-(1/p)64/(s + 2)(s - 2)$

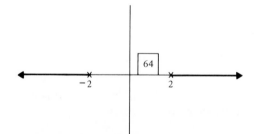

$$-8.240, \quad k = -2.562, \quad -3.225, \quad k = 1.306$$

(c) $D(s) = (s + 1 + j3)(s + 1 - j3), \quad N(s) = 2,$

$$(1/p)4/(s^2 + 2s + 10)(s^2 - 2s + 10)$$

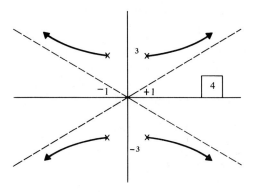

$$-1.05 \pm j3.016, \quad k_1 = -0.1, \quad k_2 = -0.048$$
$$-1.005 \pm j3.002, \quad k_1 = -0.01, \quad k_2 = -0.005$$

(d) $D(s) = (s + 1 + j3)(s + 1 - j3), \quad N(s) = 2(s),$

$$(-1/p)4s^2/(s^2 + 2s + 10)(s^2 - 2s + 10)$$

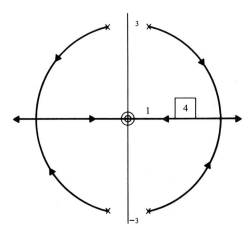

$$-1.414 \pm j2.828, \quad k_1 = 0, \quad k_2 = -0.414$$
$$-1.049 \pm j2.98, \quad k_1 = 0, \quad k_2 = -0.049$$

D7.22. When **Q** has a rank higher than 1, it is not possible to decompose **Q** by a row vector **g**, where

$$\mathbf{Q} = \mathbf{g}^T \mathbf{g}$$

(a) Show that if

$$\mathbf{x}^T\mathbf{Q}\mathbf{x} = \mathbf{x}^T\mathbf{g}^T\mathbf{g}\mathbf{x} + \frac{d}{dt}\mathbf{x}^T\mathbf{S}\mathbf{x}$$

the second term on the right can be removed from the integral so that only **g** influences the performance measure.

(b) Suppose

$$\mathbf{A} = \begin{bmatrix} 0 & 1 \\ -10 & -2 \end{bmatrix} \quad \mathbf{B} = \begin{bmatrix} 0 \\ 2 \end{bmatrix}$$

$$\mathbf{Q} = \begin{bmatrix} 1 & 0 \\ 0 & 1 \end{bmatrix} \quad \mathbf{S} = \begin{bmatrix} s_{11} & 0 \\ 0 & 0 \end{bmatrix}$$

Find **g** and **S**.

Ans $\mathbf{g} = \begin{bmatrix} 1 & 1 \end{bmatrix}, \quad \mathbf{S} = \begin{bmatrix} -1 & 0 \\ 0 & 0 \end{bmatrix}$

(c) For the problem of (b), form **A'** and factor the characteristic polynomial.

Ans $P(s) = (s^2 + 2s + 10)(s^2 - 2s + 10) - (1/p)[2(s + 1)][2(s - 1)]$

(d) Show that $N(s)$ in (c) follows from

$$\frac{N(s)}{D(s)} = \mathbf{g}[s\mathbf{I} - \mathbf{A}]^{-1}\mathbf{B}$$

Using **g** from (b).

D7.23. Repeat D7.21a to d finding the closed-loop gains from the algebraic matrix Riccati equation for $p = 1$.

Ans (a) $\mathbf{M} = 0.3904, \quad k = -1.462$

(b) $\mathbf{M} = 0.6404, \quad k = -2.562$

(c) $\mathbf{M} = \begin{bmatrix} 0.3455 & 0.0495 \\ 0.0495 & 0.0242 \end{bmatrix}$

$k = \begin{bmatrix} -0.1 & -0.048 \end{bmatrix}$

(d) $\mathbf{M} = \begin{bmatrix} 2.07 & 0 \\ 0 & 0.207 \end{bmatrix}$

$k = \begin{bmatrix} 0 & -0.414 \end{bmatrix}$

7.10 A MAGNETIC LEVITATION SYSTEM

Beginning in 1969, West Germany sought to develop a high-speed electric train system to span central Europe. Using state space analysis and aircraft technology, a train has now been designed, built, and tested for operation at speeds as high as 400 km/hr (248 mi/hr). The train is suspended in midair by magnetic fields. This type of suspension is called magnetic levitation or MAGLEV.

Figure 7.33 shows the cross section of a MAGLEV vehicle. The track is a T-shaped concrete guideway. Once underway, the train does not touch the guideway, resulting in greatly reduced friction and reduced guideway construction costs. Electromagnets are distributed along the guideway and along the length of the train in matched pairs. The magnetic attraction of the vertically paired magnets balances the force of gravity and levitates the vehicle above the guideway. The horizontally paired magnets stabilize the vehicle against sideways forces. Forward propulsion is produced by linear induction motor action between train and guideway. Only the vertical motion and control of the suspended vehicle will be considered here.

The equations characterizing the train's vertical motion are now developed. It is desired to control the gap distance d within a close tolerance in normal operation of the train. The gap distance d between the track and the train magnets is

$$d = z - h$$

Then

$$\dot{d} = \dot{z} - \dot{h}$$
$$\ddot{d} = \ddot{z} - \ddot{h}$$

where the dots denote time derivatives. The magnet produces a force that is dependent upon residual magnetism and upon the current passing through the

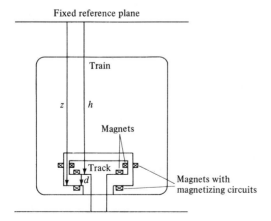

Figure 7.33   Cross section of a MAGLEV train.

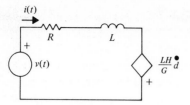

Figure 7.34   Magnetizing circuit model.

magnetizing circuit. For small changes in the magnetizing current i and the gap distance d, that force is approximately

$$f_1 = -Gi + Hd$$

where G and H are positive constants. That force acts to accelerate the mass M of the train in a vertical direction, so

$$f_1 = M\ddot{z} = -Gi + Hd$$

For increased current, the distance z diminishes, reducing d as the vehicle is attracted to the guideway.

A network model for the magnetizing circuit is given in Fig. 7.34. This circuit represents a generator driving a coil wrapped around the magnet on the vehicle. The voltage induced in the coil by the vehicle motion is represented by the term $(LH/G)\dot{d}$, for which it is assumed that the magnetic flux loss is negligible. For that circuit

$$Ri + L\dot{i} - \frac{LH}{G}\dot{d} = v$$

The three state variables

$$x_1 = d$$
$$x_2 = \dot{d}$$
$$x_3 = i$$

are convenient, and in terms of them the vertical motion state equations are

$$\begin{bmatrix} \dot{x}_1 \\ \dot{x}_2 \\ \dot{x}_3 \end{bmatrix} = \begin{bmatrix} 0 & 1 & 0 \\ \dfrac{H}{M} & 0 & -\dfrac{G}{M} \\ 0 & \dfrac{H}{G} & -\dfrac{R}{L} \end{bmatrix} \begin{bmatrix} x_1 \\ x_2 \\ x_3 \end{bmatrix} + \begin{bmatrix} 0 & 0 \\ 0 & -1 \\ \dfrac{1}{L} & 0 \end{bmatrix} \begin{bmatrix} v \\ f \end{bmatrix}$$

where

$$f = \ddot{h}$$

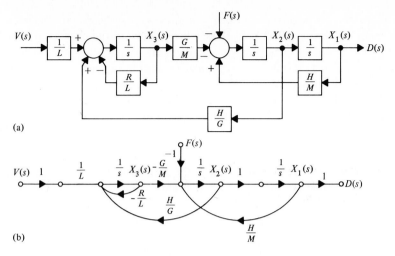

Figure 7.35 Diagrams of the state equations. **(a)** Block diagram. **(b)** Signal flow graph.

If the gap distance d is considered to be the system output, then the state variable output equation is

$$d = x_1$$

The voltage v is considered to be a control input, while guideway irregularities $f = \ddot{h}$ constitute a disturbance. Figure 7.35 shows block diagram and signal flow graph representations of the state equations.

The characteristic equation for the system, the roots of which are the transfer function poles, is given by

$$|s\mathbf{I} - \mathbf{A}| = \begin{vmatrix} s & -1 & 0 \\ -\dfrac{H}{M} & s & \dfrac{G}{M} \\ 0 & -\dfrac{H}{G} & s + \dfrac{R}{L} \end{vmatrix} = 0$$

$$= s \begin{vmatrix} s & \dfrac{G}{M} \\ -\dfrac{H}{G} & s + \dfrac{R}{L} \end{vmatrix} + \begin{vmatrix} -\dfrac{H}{M} & \dfrac{G}{M} \\ 0 & s + \dfrac{R}{L} \end{vmatrix}$$

$$= s\left(s^2 + \dfrac{R}{L}s + \dfrac{H}{M} \right) - \dfrac{H}{M}\left(s + \dfrac{R}{L} \right)$$

$$= s^3 + \dfrac{R}{L}s^2 - \dfrac{HR}{ML} = 0$$

The system is thus unstable since its characteristic polynomial has coefficients with differing algebraic signs. Also, the coefficient of s in the characteristic equation is zero. The system instability is quite understandable when one considers the action of the magnets. If the gap distance d should increase slightly, the magnetic attraction decreases, tending to further increase the gap, and so on.

To control the system, the magnetizing circuit voltage is chosen to be a linear combination of the state signals plus a tracking input $u_1(t)$:

$$v = k_1 x_1 + k_2 x_2 + k_3 x_3 + u_1(t)$$

The feedback signals are produced from sensors that monitor the state variables, namely gap distance d, gap velocity $\dot{d}$, and magnetizing current i. The resulting feedback system is described by

$$\begin{bmatrix} \dot{x}_1 \\ \dot{x}_2 \\ \dot{x}_3 \end{bmatrix} = \begin{bmatrix} 0 & 1 & 0 \\ \dfrac{H}{M} & 0 & -\dfrac{G}{M} \\ \dfrac{k_1}{L} & \dfrac{H}{G}+\dfrac{k_2}{L} & -\dfrac{R}{L}+\dfrac{k_3}{L} \end{bmatrix} \begin{bmatrix} x_1 \\ x_2 \\ x_3 \end{bmatrix} + \begin{bmatrix} 0 & 0 \\ 0 & -1 \\ \dfrac{1}{L} & 0 \end{bmatrix} \begin{bmatrix} v_1 \\ f \end{bmatrix}$$

$$d = x_1$$

Appropriate choice of the feedback gain constants k_1, k_2, and k_3, that is, the feedback gain matrix,

$$K = [k_1 \quad k_2 \quad k_3]$$

will place the system poles at any desired locations.

The parameters M, G, L, and R must be estimated in order to proceed with state variable design methods. The following values do not necessarily represent those of any specific existing system, but the methods and values are representative of the design process in general.

Suppose an engineer finds that each train car weighs 8000 kg. Each car is supported by four magnets, each of which must therefore support 2000 kg. Each subsystem can be analyzed using $M = 2000$ kg.

A static test is performed without control. The air gap is clamped shut, causing d to be zero. A -120-V source is applied to the magnetizing circuit. With a time constant of $\frac{1}{30}$ of a second, -8 amps eventually flows at steady state. A resultant force of 4000 N is measured (in addition to that of gravity).

The static test is concluded and the voltage is carefully varied until, at equilibrium, the car levitates with $d = 10$ mm under the influence of 8 amps of current.

If the magnetizing circuit is at steady state, the static test can be used to get R and L since

$$R = \frac{v}{i} = \frac{-120}{-8} = 15 \text{ ohms}$$

and, from the time constant during the static test

$$T = \frac{L}{R} \quad \text{so} \quad L = RT = \frac{15}{30} = 0.5\ H$$

The data from when the air gap was clamped shut ($d = 0$) permits G to be computed

$$f = -Gi + H \times 0$$

$$G = \frac{-f}{i} = \frac{-4000}{-8} = 500\ \text{N/Amp}$$

The data from when the car was levitated to equilibrium provides H:

$$0 = -500 \times 8 + H \times 10$$

$$H = 400\ \text{N/mm}$$

The parameter values are, therefore,

$$M = 2000 \qquad H = 400$$

$$G = 500 \qquad L = 0.5$$

$$R = 15$$

For these, the feedback system equations are

$$\begin{bmatrix} \dot{x}_1 \\ \dot{x}_2 \\ \dot{x}_3 \end{bmatrix} = \begin{bmatrix} 0 & 1 & 0 \\ 0.2 & 0 & -0.25 \\ 2k_1 & 0.8 + 2k_2 & -30 + 2k_3 \end{bmatrix} \begin{bmatrix} x_1 \\ x_2 \\ x_3 \end{bmatrix} + \begin{bmatrix} 0 & 0 \\ 0 & -1 \\ 2 & 0 \end{bmatrix} \begin{bmatrix} u_1 \\ f \end{bmatrix}$$

$$d = x_1$$

The characteristic equation for the feedback system is given by

$$\begin{vmatrix} s & -1 & 0 \\ -0.2 & s & 0.25 \\ -2k_1 & -0.8 - 2k_2 & s + 30 - 2k_3 \end{vmatrix}$$

$$= s \begin{vmatrix} s & 0.25 \\ -0.8 - 2k_2 & s + 30 - 2k_3 \end{vmatrix} + \begin{vmatrix} -0.2 & 0.25 \\ -2k_1 & s + 30 - 2k_3 \end{vmatrix}$$

$$= s^3 + (30 - 2k_3)s^2 + (0.5k_2)s + 0.4k_3 + 0.5k_1 - 6$$

The feedback gains k_1, k_2, and k_3 may be chosen to give any desired coefficients of the characteristic equation of the feedback system. For example, if it is desired to have the system poles at $s = 1 + j2, -1 - j2$, and -3, the characteristic polynomial would be

$$(s + 1 - j2)(s + 1 + j2)(s + 3) = s^3 + 5s^2 + 11s + 15 = s^3 + c_2 s^2 + c_1 s + c_0$$

which is achieved with

$$k_3 = 0.5(30 - c_2) = 12.5$$

$$k_2 = 2c_1 = 22$$

$$k_1 = 2(c_0 + 0.2c_2) = 32$$

For this choice of feedback gains, the feedback system model is

$$\begin{bmatrix} \dot{x}_1 \\ \dot{x}_2 \\ \dot{x}_3 \end{bmatrix} = \begin{bmatrix} 0 & 1 & 0 \\ 0.2 & 0 & -0.25 \\ 64 & 44.8 & -5 \end{bmatrix} \begin{bmatrix} x_1 \\ x_2 \\ x_3 \end{bmatrix} + \begin{bmatrix} 0 & 0 \\ 0 & -1 \\ 2 & 0 \end{bmatrix} \begin{bmatrix} u_1 \\ f \end{bmatrix}$$

$$d = \begin{bmatrix} 1 & 0 & 0 \end{bmatrix} \begin{bmatrix} x_1 \\ x_2 \\ x_3 \end{bmatrix}$$

The steady state output d due to a unit step disturbance input f is given by

$$0 = [\mathbf{A} + \mathbf{BK}]\mathbf{x} + \mathbf{Bu}$$

$$d = x_1 \qquad u_1 = 0 \qquad f = 1$$

So that

$$[\mathbf{A} + \mathbf{BK}]\mathbf{x} = -\mathbf{Bu} = -\begin{bmatrix} 0 \\ -1 \\ 0 \end{bmatrix} = \begin{bmatrix} 0 \\ 1 \\ 0 \end{bmatrix}$$

Cramer's rule can be used to obtain $d = x_1$. It is instructive to write the gain values in terms of the desired characteristic polynomial coefficients.

$$d = x_1 = \frac{\begin{vmatrix} 0 & 1 & 0 \\ 1 & 0 & -0.25 \\ 0 & 4c_1 + 0.8 & -c_2 \end{vmatrix}}{\begin{vmatrix} 0 & 1 & 0 \\ 0.2 & 0 & -0.25 \\ 4(c_0 + 0.2c_2) & 4c_1 + 0.8 & -c_2 \end{vmatrix}} = \frac{c_2}{-c_0} = -\frac{1}{3}$$

which depends only on the desired performance. Further study would be needed to determine if this amount of disturbance rejection from track irregularities is sufficient. The negative algebraic sign above simply means that a positive step in $f = \ddot{h}$ results in a steady state decrease in the gap distance. Other types of disturbances than constant ones should also be considered in the design.

The reference input u_1 would normally be a constant which sets the nominal gap distance. The steady state gap distance d due to a constant reference input u_1 where f is zero (level track) is given by

$$0 = [\mathbf{A} + \mathbf{BK}]\mathbf{x} + \mathbf{Bu}; \qquad \mathbf{u} = \begin{bmatrix} u_1 \\ 0 \end{bmatrix}$$

$$[\mathbf{A} + \mathbf{BK}]\mathbf{x} = -\mathbf{Bu} = \begin{bmatrix} 0 \\ 0 \\ -2u_1 \end{bmatrix}$$

$$d = x_1$$

Again, Cramer's rule can be used to obtain $d = x_1$ where it is instructive to write the gain values in terms of the desired characteristic polynomial coefficients

$$d = x_1 = \frac{\begin{vmatrix} 0 & 1 & 0 \\ 0 & 0 & -0.25 \\ -2u_1 & 4c_1 + 0.8 & -c_2 \end{vmatrix}}{-c_0}$$

$$= -\frac{0.5u_1}{c_0}$$

For a nominal gap of 10 mm, the reference input should be

$$u_1 = -(20)c_0 = -300$$

which depends only on the nominal gap and on a coefficient of the desired characteristic polynomial.

Figure 7.36 shows calculated system response where the train accelerates from a standstill and traverses an irregular guideway with a downgrade followed by an upgrade. In the West German system, the nominal airgap distance is 14 mm (about $\frac{1}{2}$ in). Improvement in disturbance rejection is obtained by modeling the track irregularities by differential equations that are included as part of an observer. Three levels of complexity are used depending on whether the track is level (actually somewhat curved between towers), following a hill, or following a curve.

Space does not permit a complete discussion of the system; however, one feature is of interest. The rate of change of the air gap is also estimated using an observer. The state vector may be reordered

$$\mathbf{x} = \begin{bmatrix} d \\ i \\ \cdots \\ \dot{d} \end{bmatrix} = \begin{bmatrix} \mathbf{y} \\ \mathbf{x}_u \end{bmatrix}$$

because d and i can be measured and the rate of change of d must be estimated by a reduced-order, first-order observer. The concepts of control and estimation can be

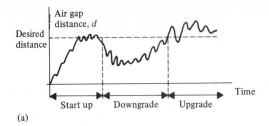

(a)

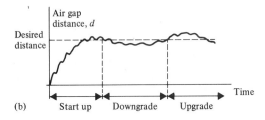

(b)

Figure 7.36 MAGLEV system response. **(a)** Response of the system with state feedback. **(b)** Improved response with disturbance modeling and feedback.

separated (as mentioned earlier in this chapter) therefore the open-loop system matrix is used to compute the observer dynamics. Due to reordering

$$\mathbf{A} = \begin{bmatrix} 0 & 0 & \vdots & 1 \\ 0 & -30 & \vdots & 0.8 \\ \cdots & \cdots & & \cdots \\ 0.2 & -0.25 & \vdots & 0 \end{bmatrix} \qquad \mathbf{B} = \begin{bmatrix} 0 \\ 2 \\ \cdots \\ 0 \end{bmatrix}$$

Following reduced-order observer design procedures,

$$\mathbf{T} = [-\mathbf{L} \quad \mathbf{I}] = [-l_{11} \quad -l_{12} \quad 1] \quad \mathbf{F} = [f_{11} \quad f_{12}]$$

The following equations allow a designer to compute the elements of $\mathbf{L}$:

$$-\mathbf{L}\mathbf{A}_{11} + \mathbf{A}_{21} + \mathbf{D}\mathbf{L} = \mathbf{F}$$

$$-\mathbf{L}\mathbf{A}_{12} + \mathbf{A}_{22} - \mathbf{D} = 0$$

Suppose the eigenvalue of $\mathbf{D}$ is to be -10. The above two matrix equations create three scalar equations in four variables:

$$0.2 - 10l_{11} = f_{11}$$

$$-0.25 + 20l_{12} = f_{12}$$

$$-l_{11} - 0.8l_{12} = -10$$

Here l_{11} can be set to zero so that

$$f_{11} = 0.2$$

$$l_{12} = 12.5$$

$$f_{12} = 249.75$$

The observer (to estimate $x_3 = \dot{d}$) is

$$\dot{z} = \mathbf{Dz} + \mathbf{Fy} + \mathbf{Gr}$$

$$\mathbf{G} = \mathbf{TB} = -25$$

$$\dot{z} = -10z + 0.2(x_1 = d)249.75(x_2 = i) - 25r$$

The estimate for x_3 is

$$\hat{x}_3 = \mathbf{z} + \mathbf{Ly} = z + 12.5(x_2 = i)$$

Figure 7.36b shows the improved performance when an observer estimates track motion. The closed-loop system also uses the observer estimate of $\dot{d}$ for feedback purposes.

7.11 SUMMARY

At least three distinct procedures exist for converting a system describable by an nth-order linear differential equation into a system having n state variables: phase/dual phase variables, canonical (diagonalized) variables, and physical (block diagram) variables. The first of these translates a system transfer function into system matrices, the second provides diagonalized system matrices, while the third preserves actual system quantities (velocity, current, temperature, for example).

The phase-variable form was shown to be especially convenient for single- and multiple-output system transfer function synthesis, and the dual phase-variable form is convenient for multiple-input system transfer functions. Simulation diagrams are not only useful in transfer function synthesis; they also give a standard, systematic, and compact description of a system.

The relationships between signals in a simulation diagram were shown to be a set of coupled first-order differential state equations and linear algebraic output equations, relating the system outputs to the state variables. These state variable equations are compactly expressed using matrix notation:

$$\dot{\mathbf{x}} = \mathbf{Ax} + \mathbf{Br}$$

$$\mathbf{y} = \mathbf{Cx}$$

System transfer functions were calculated systematically from the state variable equations using matrix algebra:

$$\mathbf{T}(s) = \mathbf{C}[s\mathbf{I} - \mathbf{A}]^{-1}\mathbf{B}$$

It was seen that all transfer functions of a system share a common characteristic polynomial,

$$|s\mathbf{I} - \mathbf{A}|$$

A nonsingular change of state variables gives a new representation for a system, but leaves the system's input-output relations, its transfer functions, unchanged. Hence a system characterized by a set of transfer functions may be represented in countless different ways, each differing in the choice of state variables.

A very special set of state variables for a system are those for which the state equations, each of first order, are decoupled from one another. A system so represented is said to be in *normal* or *diagonal* form:

$$\mathbf{A} = \begin{bmatrix} s_1 & 0 & 0 & \cdots & 0 \\ 0 & s_2 & 0 & \cdots & 0 \\ \vdots & & & & \\ 0 & 0 & 0 & & s_n \end{bmatrix}$$

Determining the change of variables which places a system in diagonal form is the *characteristic value problem* of matrix algebra. An alternative transformation method involves expansion of a system transfer function into partial fractions. Systems with repeated characteristic roots cannot be diagonalized; however, they may be placed in a related *Jordan* form, where the repeated root terms involve a distinctive nonzero "block" along the diagonal of the state coupling matrix.

Two fundamental properties of a system are that system's controllability and observability. A system is completely controllable if the system state $\mathbf{x}(t_f)$ at time t_f can be forced to take on any desired value by applying a control input $\mathbf{r}(t)$ over a period of time from t_0 until t_f. A system is completely observable if any initial state vector $\mathbf{x}(t_0)$ can be reconstructed by examining the system output $\mathbf{y}(t)$ over some period of time from t_0 until t_f.

For a system represented in diagonal form, controllability and observability are apparent from inspection of the input and output coupling matrices. For systems in other than diagonal form, simple rank tests of the controllability and observability matrices,

$$M_c = [\mathbf{B} \mid \mathbf{AB} \mid \cdots \mid \mathbf{A}^{n-1}\mathbf{B}]$$

$$M_0 = \begin{bmatrix} \mathbf{C} \\ \hline \mathbf{CA} \\ \hline \vdots \\ \hline \mathbf{CA}^{n-1} \end{bmatrix}$$

may be used.

For an nth-order system,

$$\mathbf{x}(t) = \mathbf{\Phi}(t)\mathbf{x}(0^-) + \int_{0-}^{t} \mathbf{\Phi}(t - \tau)\mathbf{Br}(\tau)\,d\tau$$

where $\mathbf{\Phi}(t)$ is the $n \times n$ state transition matrix

$$\mathbf{\Phi}(t) = \mathscr{L}^{-1}\{(s\mathbf{I} - \mathbf{A})^{-1}\}$$

The time response can be approximated by using a Taylor series to compute the state transition matrix and the convolution integral.

If all n states are available, the advantages of state variable feedback become apparent. All n closed-loop poles can be placed at desired locations in the complex plane. As an alternative, n steady state errors to power-of-time inputs can be nulled.

A combination of n closed-loop pole selections and steady state accuracy determinations can be affected by the n gains.

If some state variables are difficult to measure, a linear observer can be designed. An identity observer reconstructs all n state variables, including the m that are already measured. The identity observer is redundant but the structure is quite simply a copy of the system with an adjustive term. A reduced-order observer of minimum order reconstructs only the $n - m$ states not easily measurable. Estimation (using an observer, for example) and control are separable when a linear system is designed. The resulting closed-loop system exhibits the desired closed-loop roots and the desired observer roots as eigenvalues of the interconnected closed-loop system matrix.

The methods of quadratic optimal control include selection of a performance measure (an integrated quadratic form) followed by the optimization of that quadratic form by selecting gains. The root square locus is applied to the case where integration of the performance index extends to infinite time. The root square locus follows by factoring the characteristic polynomial of a special $2n \times 2n$ matrix. A strong feature of quadratic optimal control is that the closed-loop system is always stable, even when an open-loop unstable system operates under minimum control effort.

The design of a magnetically levitated train exemplifies state variable representation, gain selection, and observer design.

REFERENCES

Simulation Diagrams
Jackson, A. S., *Analog Computation*. New York: McGraw-Hill, 1960.
Korn, G. A., and Korn, T. M., *Electronic Analog Computers*. New York: McGraw-Hill, 1952.

State Variables
Brockett, R. W., "Poles, Zeros and Feedback: State Space Interpretation." *IEEE Trans. Auto. Contr.*, April 1965.
De Russo, P. M., Roy, R. J., and Close, C. M., *State Variables for Engineers*. New York: Wiley, 1965.
Gupta, S. C., *Transform and State Variable Methods in Linear Systems*. New York: Wiley, 1966.
Horowitz, I. C., and Shaked, U., "Superiority of Transfer Function over State Variable Methods in Linear Time-Invariant Feedback System Design." *IEEE Trans. Auto. Contr.*, February 1975.
Kalman, R. E., "Mathematical Description of Linear Dynamical Systems." *SIAM J. Contr. ser. A*, 1 (1963).
Ogata, K., *State Space Analysis of Control Systems*. Englewood Cliffs, N. J.: Prentice-Hall, 1967.
Timothy, L. K., and Bona, B. E., *State Space Analysis: An Introduction*. New York: McGraw-Hill, 1968.
Zadeh, L. A., and Desoer, C. A., *Linear System Theory: The State Space Approach*. New York: McGraw-Hill, 1963.

Matrix Algebra and the Characteristic Value Problem

Bellman, R., *Introduction to Matrix Analysis*. New York: McGraw-Hill, 1960.

Guillemin, E. A., *The Mathematics of Circuit Analysis*. New York: Wiley, 1949.

Pipes, L. A., *Matrix Methods for Engineering*. Englewood Cliffs, N. J.: Prentice-Hall, 1963.

Wylie, C. R., *Advanced Engineering Mathematics*, 4th ed. New York: McGraw-Hill, 1975.

Controllability and Observability

Gilbert, E. G., "Controllability and Observability in Multivariable Control Systems." *J. Soc. Ind. Appl. Math.*, ser. A, 1963, pp. 128–51.

Kalman, R. E., "Canonical Structure of Linear Dynamical Systems." *Proc. Nat. Acad. Sci.*, April 1962, pp. 596–600.

Stubberud, A. R., "A Controllability Criterion for a Class of Linear Systems." *IEEE Trans. Appl. Ind.* 68 (1964): 411–13.

Computational Methods

Faddeeva, D. K., and Faddeeva, V. N., *Computational Methods of Linear Algebra*. San Francisco: Freeman, 1963.

James, M. L., Smith, G. M., and Wolford, J. C., *Applied Numerical Methods for Digital Computations*, 2nd ed. New York, Harper & Row, 1977.

Melsa, J. L., and Jones, S. K., *Computer Programs for Computational Assistance in the Study of Linear Control Theory*, 2nd ed. New York: McGraw-Hill, 1973.

State Feedback

Anderson, B. D. O., and Moore, J. B., *Linear Optimal Control*. Englewood Cliffs, N. J.: Prentice-Hall, 1971.

Davison, E. J., "On Pole Assignment in Multivariable Linear Systems." *IEEE Trans. Auto. Contr.* AC-13 (December 1968): 747–48.

Wonham, W. M., "On Pole Assignment in Multi-Input Controllable Linear Systems." *IEEE Trans. Auto. Contr.* AC-12 (December 1967): 660–65.

Observers

Doyle, J. C., and Stein, G., "Robustness with Observers." *IEEE Trans. Auto. Contr.* (August 1979): 607–11.

Krogh, B., and Cruz, J. B., "Design of Sensitivity-Reducing Compensators Using Observers." *IEEE Trans. Auto. Contr.* (December 1978): 1058–62.

Luenberger, D. G., "Observers for Multivariable Systems." *IEEE Trans. Auto. Contr.* AC-11 (April 1966): 190–97.

————, "An Introduction to Observers." *IEEE Trans. Auto. Contr.* AC-16 (December 1971): 596–602.

Nuyan, S., and Carroll, R. L., "Minimum Order Arbitrarily Fast Adaptive Observers and Identifiers." *IEEE Trans. Auto Contr.* (April 1979): 289–97.

Sage, A. P., and White, C. C., *Optimum Systems Control*. Engelwood Cliffs, N. J.: Prentice-Hall, 1977.

Stefani, R. T., "Reducing the Sensitivity to Parameter Variations of a Minimum-Order Reduced-Order Observer" *Int. Journal Contr.* (1982): 983–95.

Stefani, R. T., "Observer Steady State Errors Induced by Errors in Realization." *IEEE Trans. Auto. Contr.* (April 1976): 280–82.

Quadratic Optimal Control

Kailath, T., *Linear Systems*. Englewood Cliffs, N. J.: Prentice-Hall, 1977.

Kirk, D. E., *Optimal Control Theory*. Englewood Cliffs, N. J.: Prentice-Hall, 1970.

Sage, A. P., and White, C. C., *Optimum Systems Control*. Englewood Cliffs, N. J.: Prentice-Hall, 1977.

Schultz, D. G., and Melsa, J. L., *State Functions and Linear Control Systems*. New York: McGraw-Hill, 1967.

Magnetic Levitation of Trains

Brock, K. H., Gottzein, E., Pfefferl, J., and Schneider, E., "Control Aspects of a Tracked Magnetic Levitation High Speed Test Vehicle." *Automatica*, vol. 13, no. 3, pp. 205–33 (1977).

Glatzel, K., Khurdok, G., and Rogg, D., "The Development of the Magnetically Suspended Transportation System in the Federal Republic of Germany." *IEEE Trans. Vehic. Technol.* (February 1980): 3–17.

Glatzel, K., and Schulz, H., "Transportation: The Promise of MAGLEV." *IEEE Spec.*, (March 1980) 63–66.

Gottzein, E., Meisinger, R., and Miller, L., "The Magnetic Wheel in the Suspension of High Speed Ground Transportation Vehicles." *IEEE Trans. Vehic. Technol.* (February 1980): 17–22.

Kaplan, G., "Rail transportation." *IEEE Spec.* (January 1984) 82–85.

―――――, "Transportation." *IEEE Spec.* (January 1985): 81–84.

PROBLEMS

1. Draw phase-variable form simulation diagrams for systems with the following transfer functions. Then write the state variable equations in matrix form.

(a) $T(s) = \dfrac{-2s + 8}{s^2 + 8}$

(b) $T(s) = \dfrac{10s}{s^3 + 12s^2 + 7s + 2}$

Ans
$$\begin{bmatrix} \dot{x}_1 \\ \dot{x}_2 \\ \dot{x}_2 \end{bmatrix} = \begin{bmatrix} 0 & 1 & 0 \\ 0 & 0 & 1 \\ -2 & -7 & -12 \end{bmatrix} \begin{bmatrix} x_1 \\ x_2 \\ x_3 \end{bmatrix} + \begin{bmatrix} 0 \\ 0 \\ 1 \end{bmatrix} r$$

$$y = \begin{bmatrix} 0 & 10 & 0 \end{bmatrix} \begin{bmatrix} x_1 \\ x_2 \\ x_3 \end{bmatrix}$$

(c) $T(s) = \dfrac{7s^3 - 2s^2 + s}{s^4 + 3s^3 + 9s^2 + s + 1}$

(d) Two outputs:

$$T_{11}(s) = \frac{-s^2 + 9}{s^3 + 3s^2 + s + 4}$$

$$T_{21}(s) = \frac{s^2 + s + 10}{s^3 + 3s^2 + s + 4}$$

Ans
$$\begin{bmatrix} \dot{x}_1 \\ \dot{x}_2 \\ \dot{x}_3 \end{bmatrix} = \begin{bmatrix} 0 & 1 & 0 \\ 0 & 0 & 1 \\ -4 & -1 & -3 \end{bmatrix} \begin{bmatrix} x_1 \\ x_2 \\ x_3 \end{bmatrix} + \begin{bmatrix} 0 \\ 0 \\ 1 \end{bmatrix} r$$

$$\begin{bmatrix} y_1 \\ y_2 \end{bmatrix} = \begin{bmatrix} 9 & 0 & -1 \\ 10 & 1 & 1 \end{bmatrix} \begin{bmatrix} x_1 \\ x_2 \\ x_3 \end{bmatrix}$$

2. Draw dual phase-variable form simulation diagrams for systems with the following transfer functions. Then write the state variable equations in matrix form.

(a) $T(s) = \dfrac{-2s + 8}{s^2 + 8}$

(b) $T(s) = \dfrac{2s + 8}{3s^3 + 7s^2 + 8s + 2}$

Ans
$$\begin{bmatrix} \dot{x}_1 \\ \dot{x}_2 \\ \dot{x}_3 \end{bmatrix} = \begin{bmatrix} -\frac{7}{3} & 1 & 0 \\ -\frac{8}{3} & 0 & 1 \\ -\frac{2}{3} & 0 & 0 \end{bmatrix} \begin{bmatrix} x_1 \\ x_2 \\ x_3 \end{bmatrix} + \begin{bmatrix} 0 \\ \frac{2}{3} \\ \frac{8}{3} \end{bmatrix} r$$

$$y = \begin{bmatrix} 1 & 0 & 0 \end{bmatrix} \begin{bmatrix} x_1 \\ x_2 \\ x_3 \end{bmatrix}$$

(c) $T(s) = \dfrac{-s^3 + 4s^2 - 9s + 4}{s^4 + 8s^3 + 2s^2 + s + 9}$

(d) Two inputs:

$$T_{11}(s) = \frac{3s^2 + 9}{s^3 + 3s^2 + s + 9}$$

$$T_{12}(s) = \frac{s - 4}{s^3 + 3s^2 + s + 9}$$

$$\mathbf{Ans} \quad \begin{bmatrix} \dot{x}_1 \\ \dot{x}_2 \\ \dot{x}_3 \end{bmatrix} = \begin{bmatrix} -3 & 1 & 0 \\ -1 & 0 & 1 \\ -9 & 0 & 0 \end{bmatrix} \begin{bmatrix} x_1 \\ x_2 \\ x_3 \end{bmatrix} + \begin{bmatrix} 3 & 0 \\ 9 & 1 \\ 0 & 4 \end{bmatrix} \begin{bmatrix} r_1 \\ r_2 \end{bmatrix}$$

$$y = \begin{bmatrix} 1 & 0 & 0 \end{bmatrix} \begin{bmatrix} x_1 \\ x_2 \\ x_3 \end{bmatrix}$$

3. Draw simulation diagrams to represent the following systems:

(a)
$$\begin{bmatrix} \dot{x}_1 \\ \dot{x}_2 \\ \dot{x}_3 \end{bmatrix} = \begin{bmatrix} 1 & 8 & -1 \\ 2 & 0 & 4 \\ -2 & 1 & 8 \end{bmatrix} \begin{bmatrix} x_1 \\ x_2 \\ x_3 \end{bmatrix} + \begin{bmatrix} 1 \\ 0 \\ 0 \end{bmatrix} r$$

$$y = \begin{bmatrix} 1 & -4 & 0 \end{bmatrix} \begin{bmatrix} x_1 \\ x_2 \\ x_3 \end{bmatrix}$$

(b)
$$\begin{bmatrix} \dot{x}_1 \\ \dot{x}_2 \\ \dot{x}_3 \end{bmatrix} = \begin{bmatrix} 0 & 4 & 0 \\ -1 & -1 & 4 \\ 8 & 0 & 3 \end{bmatrix} \begin{bmatrix} x_1 \\ x_2 \\ x_3 \end{bmatrix} + \begin{bmatrix} 1 \\ 8 \\ -5 \end{bmatrix} r$$

$$y = \begin{bmatrix} 0 & -3 & 6 \end{bmatrix} \begin{bmatrix} x_1 \\ x_2 \\ x_3 \end{bmatrix}$$

(c)
$$\begin{bmatrix} \dot{x}_1 \\ \dot{x}_2 \end{bmatrix} = \begin{bmatrix} 2 & 1 \\ -8 & 0 \end{bmatrix} \begin{bmatrix} x_1 \\ x_2 \end{bmatrix} + \begin{bmatrix} 1 & 0 & 8 \\ 2 & 4 & 1 \end{bmatrix} \begin{bmatrix} r_1 \\ r_2 \\ r_3 \end{bmatrix}$$

$$y = \begin{bmatrix} 1 & 0 \end{bmatrix} \begin{bmatrix} x_1 \\ x_2 \end{bmatrix}$$

(d)
$$\begin{bmatrix} \dot{x}_1 \\ \dot{x}_2 \\ \dot{x}_3 \end{bmatrix} = \begin{bmatrix} 0 & 0 & 1 \\ -1 & 2 & 4 \\ -8 & 1 & 0 \end{bmatrix} \begin{bmatrix} x_1 \\ x_2 \\ x_3 \end{bmatrix} + \begin{bmatrix} 4 & 0 \\ 0 & 0 \\ -1 & 7 \end{bmatrix} \begin{bmatrix} r_1 \\ r_2 \end{bmatrix}$$

$$\begin{bmatrix} y_1 \\ y_2 \\ y_3 \end{bmatrix} = \begin{bmatrix} 4 & 0 & 0 \\ 0 & 1 & 0 \\ 1 & 4 & 3 \end{bmatrix} \begin{bmatrix} x_1 \\ x_2 \\ x_3 \end{bmatrix}$$

4. Find state equations in matrix form for the systems described by the simulation diagrams of Fig. P7.4.

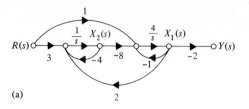

(a)

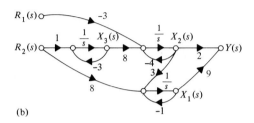

(b)

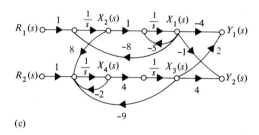

(c)

Figure P7.4

Ans (b)
$$\begin{bmatrix} \dot{x}_1 \\ \dot{x}_2 \\ \dot{x}_3 \end{bmatrix} = \begin{bmatrix} -1 & 3 & 0 \\ 0 & -4 & 8 \\ 0 & 0 & -3 \end{bmatrix} \begin{bmatrix} x_1 \\ x_2 \\ x_3 \end{bmatrix} + \begin{bmatrix} 0 & 8 \\ -3 & 0 \\ 0 & 1 \end{bmatrix} \begin{bmatrix} r_1 \\ r_2 \end{bmatrix}$$

$$y = \begin{bmatrix} 9 & 2 & 0 \end{bmatrix} \begin{bmatrix} x_1 \\ x_2 \\ x_3 \end{bmatrix}$$

5. For the following systems, find the transfer function matrices:

(a)
$$\begin{bmatrix} \dot{x}_1 \\ \dot{x}_2 \end{bmatrix} = \begin{bmatrix} -2 & 1 \\ -8 & 0 \end{bmatrix} \begin{bmatrix} x_1 \\ x_2 \end{bmatrix} + \begin{bmatrix} 2 \\ -4 \end{bmatrix} r(t)$$

$$\begin{bmatrix} y_1 \\ y_2 \end{bmatrix} = \begin{bmatrix} 1 & -1 \\ 0 & 4 \end{bmatrix} \begin{bmatrix} x_1 \\ x_2 \end{bmatrix}$$

(b) $\begin{bmatrix} \dot{x}_1 \\ \dot{x}_2 \end{bmatrix} = \begin{bmatrix} 0 & 1 \\ -3 & 8 \end{bmatrix} \begin{bmatrix} x_1 \\ x_2 \end{bmatrix} + \begin{bmatrix} 0 & 1 \\ 4 & -7 \end{bmatrix} \begin{bmatrix} r_1 \\ r_2 \end{bmatrix}$

$y = \begin{bmatrix} -1 & 4 \end{bmatrix} \begin{bmatrix} x_1 \\ x_2 \end{bmatrix}$

Ans $\begin{bmatrix} \dfrac{16s-4}{s^2-8s+3} & \dfrac{3s+24}{s^2-8s+3} \end{bmatrix}$

(c) $\begin{bmatrix} \dot{x}_1 \\ \dot{x}_2 \\ \dot{x}_3 \end{bmatrix} = \begin{bmatrix} 0 & 4 & 1 \\ 0 & 0 & 4 \\ -2 & -8 & -4 \end{bmatrix} \begin{bmatrix} x_1 \\ x_2 \\ x_3 \end{bmatrix} + \begin{bmatrix} 4 & 2 \\ 0 & 1 \\ 3 & -8 \end{bmatrix} \begin{bmatrix} r_1 \\ r_2 \end{bmatrix}$

$\begin{bmatrix} y_1 \\ y_2 \end{bmatrix} = \begin{bmatrix} 1 & 0 & 0 \\ 0 & 4 & -8 \end{bmatrix} \begin{bmatrix} x_1 \\ x_2 \\ x_3 \end{bmatrix}$

6. Find the characteristic equations of the following systems. Then for each, determine if they are stable.

(a) $\begin{bmatrix} \dot{x}_1 \\ \dot{x}_2 \end{bmatrix} = \begin{bmatrix} -1 & 3 \\ -3 & 8 \end{bmatrix} \begin{bmatrix} x_1 \\ x_2 \end{bmatrix} + \begin{bmatrix} 4 & 1 \\ -3 & 3 \end{bmatrix} \begin{bmatrix} r_1 \\ r_2 \end{bmatrix}$

$\begin{bmatrix} y_1 \\ y_2 \end{bmatrix} = \begin{bmatrix} 1 & 2 \\ 0 & 7 \end{bmatrix} \begin{bmatrix} x_1 \\ x_2 \end{bmatrix}$

(b) $\begin{bmatrix} \dot{x}_1 \\ \dot{x}_2 \\ \dot{x}_2 \end{bmatrix} = \begin{bmatrix} 0 & -2 & 3 \\ 0 & -3 & -1 \\ 0 & 1 & -8 \end{bmatrix} \begin{bmatrix} x_1 \\ x_2 \\ x_3 \end{bmatrix} + \begin{bmatrix} 2 \\ -8 \\ 4 \end{bmatrix} r$

$y = \begin{bmatrix} 4 & 1 & 6 \end{bmatrix} \begin{bmatrix} x_1 \\ x_2 \\ x_3 \end{bmatrix}$

Ans $s^3 + 11s^2 + 24s$, marginally stable

(c) $\begin{bmatrix} \dot{x}_1 \\ \dot{x}_2 \\ \dot{x}_3 \end{bmatrix} = \begin{bmatrix} -1 & -2 & 2 \\ 2 & 0 & 6 \\ -1 & 2 & -4 \end{bmatrix} \begin{bmatrix} x_1 \\ x_2 \\ x_3 \end{bmatrix} + \begin{bmatrix} 2 & -1 \\ 0 & 0 \\ 2 & 8 \end{bmatrix} \begin{bmatrix} r_1 \\ r_2 \end{bmatrix}$

$y = \begin{bmatrix} 4 & 0 & -1 \end{bmatrix} \begin{bmatrix} x_1 \\ x_2 \\ x_3 \end{bmatrix}$

7. Although the following transfer functions do not share a common denominator polynomial, they may be made to have a common denominator by multiplying their numerators and denominators by appropriate factors. Find a simulation

diagram and a matrix state variable representation for a single-input, two-output system with the following two transfer functions:

$$T_{11}(s) = \frac{4s + 1}{(s + 2)(s + 4)}$$

$$T_{21}(s) = \frac{10s}{(s + 1)(s + 4)}$$

The best solutions will involve only three integrators.

8. Use the partial fraction method to find diagonal state equations for single-input, single-output systems with the following transfer functions:

(a) $T(s) = \dfrac{-7s + 4}{s^2 + 8s + 12}$

(b) $T(s) = \dfrac{2s^2 + 3s - 7}{(s + 2)(s + 8)(s + 5)}$

Ans
$$\begin{bmatrix} \dot{x}'_1 \\ \dot{x}'_2 \\ \dot{x}'_3 \end{bmatrix} = \begin{bmatrix} -2 & 0 & 0 \\ 0 & -8 & 0 \\ 0 & 0 & -5 \end{bmatrix} \begin{bmatrix} x'_1 \\ x'_2 \\ x'_3 \end{bmatrix} + \begin{bmatrix} 1 \\ 1 \\ 1 \end{bmatrix} r$$

$$y = \begin{bmatrix} -\frac{5}{18} & \frac{97}{18} & -\frac{28}{9} \end{bmatrix} \begin{bmatrix} x'_1 \\ x'_2 \\ x'_3 \end{bmatrix}$$

(c) $T(s) = \dfrac{10}{s^3 + 8s^2 + 15s}$

9. The following transfer functions for single-input, single-output systems involve complex characteristic roots. Find diagonal state equations for these systems. Then find an alternative block diagonal representation which does not involve complex numbers.

(a) $T(s) = \dfrac{4s}{s^2 + 2s + 7}$

(b) $T(s) = \dfrac{s^2 + 3s - 8}{(s + 8)(s + 3 + j)(s + 3 - j)}$

$$\mathbf{Ans} \quad \begin{bmatrix} \dot{x}'_1 \\ \dot{x}'_2 \\ \dot{x}'_3 \end{bmatrix} = \begin{bmatrix} -8 & 0 & 0 \\ 0 & -3-j & 0 \\ 0 & 0 & -3+j \end{bmatrix} \begin{bmatrix} x'_1 \\ x'_2 \\ x'_3 \end{bmatrix} + \begin{bmatrix} 1 \\ 1 \\ 1 \end{bmatrix} r$$

$$y = \begin{bmatrix} \frac{16}{15} & 3-j9 & 3+j9 \end{bmatrix} \begin{bmatrix} x'_1 \\ x'_2 \\ x'_3 \end{bmatrix}$$

$$\begin{bmatrix} \dot{x}''_1 \\ \dot{x}''_2 \\ \dot{x}''_3 \end{bmatrix} = \begin{bmatrix} -8 & 0 & 0 \\ 0 & 0 & 1 \\ 0 & -10 & -6 \end{bmatrix} \begin{bmatrix} x''_1 \\ x''_2 \\ x''_3 \end{bmatrix} + \begin{bmatrix} 1 \\ 0 \\ 1 \end{bmatrix} r$$

$$y = \begin{bmatrix} \frac{16}{15} & 18 & 6 \end{bmatrix} \begin{bmatrix} x''_1 \\ x''_2 \\ x''_3 \end{bmatrix}$$

(c) $T(s) = \dfrac{4}{(s+2)(s^2+2s+17)}$

10. The following transfer functions for single-input, single-output systems involve repeated characteristic roots. Find block diagonal Jordan canonical form state equations for these systems.

(a) $T(s) = \dfrac{3s-1}{s^2+4s+4}$

(b) $T(s) = \dfrac{s^3-4s^2+s-2}{(s+2)(s+3)^3}$

$$\mathbf{Ans} \quad \begin{bmatrix} \dot{x}'_1 \\ \dot{x}'_2 \\ \dot{x}'_3 \\ \dot{x}'_4 \end{bmatrix} = \begin{bmatrix} -2 & 0 & 0 & 0 \\ 0 & -3 & 1 & 0 \\ 0 & 0 & -3 & 1 \\ 0 & 0 & 0 & -3 \end{bmatrix} \begin{bmatrix} x'_1 \\ x'_2 \\ x'_3 \\ x'_4 \end{bmatrix} + \begin{bmatrix} 1 \\ 0 \\ 0 \\ 1 \end{bmatrix} r$$

$$y = \begin{bmatrix} -28 & 68 & 16 & 58 \end{bmatrix} \begin{bmatrix} x'_1 \\ x'_2 \\ x'_3 \\ x'_4 \end{bmatrix}$$

(c) $T(s) = \dfrac{7s^3}{(s+2)^2(s+6)^2}$

11. The following systems have real characteristic roots. Find alternative diagonal state equations.

(a)
$$\begin{bmatrix} \dot{x}_1 \\ \dot{x}_2 \end{bmatrix} = \begin{bmatrix} -9 & 1 \\ -20 & 0 \end{bmatrix} \begin{bmatrix} x_1 \\ x_2 \end{bmatrix} + \begin{bmatrix} 1 \\ 4 \end{bmatrix} r$$

$$y = [2 \quad -3] \begin{bmatrix} x_1 \\ x_2 \end{bmatrix}$$

(b)
$$\begin{bmatrix} \dot{x}_1 \\ \dot{x}_2 \end{bmatrix} = \begin{bmatrix} 0 & 1 \\ -6 & -5 \end{bmatrix} \begin{bmatrix} x_1 \\ x_2 \end{bmatrix} + \begin{bmatrix} 1 & 0 \\ 0 & 1 \end{bmatrix} \begin{bmatrix} r_1 \\ r_2 \end{bmatrix}$$

$$y = [0 \quad 1] \begin{bmatrix} x_1 \\ x_2 \end{bmatrix}$$

Ans
$$\begin{bmatrix} \dot{x}_1' \\ \dot{x}_2' \end{bmatrix} = \begin{bmatrix} -2 & 0 \\ 0 & -3 \end{bmatrix} \begin{bmatrix} x_1' \\ x_2' \end{bmatrix} + \begin{bmatrix} 3 & 1 \\ -2 & -1 \end{bmatrix} r$$

$$y = [-2 \quad -3] \begin{bmatrix} x_1' \\ x_2' \end{bmatrix}$$

(c)
$$\begin{bmatrix} \dot{x}_1 \\ \dot{x}_2 \end{bmatrix} = \begin{bmatrix} -5 & 1 \\ -4 & 0 \end{bmatrix} \begin{bmatrix} x_1 \\ x_2 \end{bmatrix} + \begin{bmatrix} 1 & 0 \\ 1 & -1 \end{bmatrix} \begin{bmatrix} r_1 \\ r_2 \end{bmatrix}$$

$$\begin{bmatrix} y_1 \\ y_2 \end{bmatrix} = \begin{bmatrix} 1 & 1 \\ -1 & 0 \end{bmatrix} \begin{bmatrix} x_1 \\ x_2 \end{bmatrix}$$

(d)
$$\begin{bmatrix} \dot{x}_1 \\ \dot{x}_2 \\ \dot{x}_3 \end{bmatrix} = \begin{bmatrix} -3 & 1 & 0 \\ -2 & 0 & 1 \\ 0 & 0 & 0 \end{bmatrix} \begin{bmatrix} x_1 \\ x_2 \\ x_3 \end{bmatrix} + \begin{bmatrix} 1 \\ 1 \\ 1 \end{bmatrix} r$$

$$\begin{bmatrix} y_1 \\ y_2 \end{bmatrix} = \begin{bmatrix} 1 & 1 & 0 \\ 0 & 1 & 1 \end{bmatrix} \begin{bmatrix} x_1 \\ x_2 \\ x_3 \end{bmatrix}$$

Ans
$$\begin{bmatrix} \dot{x}_1' \\ \dot{x}_2' \\ \dot{x}_3' \end{bmatrix} = \begin{bmatrix} 0 & 0 & 0 \\ 0 & -1 & 0 \\ 0 & 0 & -2 \end{bmatrix} \begin{bmatrix} x_1' \\ x_2' \\ x_3' \end{bmatrix} + \begin{bmatrix} 3 \\ -3 \\ 1 \end{bmatrix} r$$

$$\begin{bmatrix} y_1 \\ y_2 \end{bmatrix} = \begin{bmatrix} 1 & 0 & -1 \\ 0 & 0 & 2 \end{bmatrix} \begin{bmatrix} x_1' \\ x_2' \\ x_3' \end{bmatrix}$$

12. The following system has a set of complex conjugate characteristic roots. Find an alternative diagonal set of state equations. Then find another alternative set of state equations where the complex root terms are placed in real number block diagonal form.

$$
\begin{bmatrix} \dot{x}_1 \\ \dot{x}_2 \\ \dot{x}_2 \end{bmatrix} = \begin{bmatrix} 0 & 1 & 0 \\ 0 & 0 & 1 \\ 0 & -17 & -2 \end{bmatrix} \begin{bmatrix} x_1 \\ x_2 \\ x_3 \end{bmatrix} + \begin{bmatrix} 2 \\ -1 \\ 0 \end{bmatrix} r
$$

$$
y = \begin{bmatrix} 1 & 1 & 1 \end{bmatrix} \begin{bmatrix} x_1 \\ x_2 \\ x_3 \end{bmatrix}
$$

13. The following system has a repeated characteristic root. Find an alternate set of state equations in Jordan form:

$$
\begin{bmatrix} \dot{x}_1 \\ \dot{x}_2 \\ \dot{x}_3 \end{bmatrix} = \begin{bmatrix} 0 & 1 & 0 \\ 0 & 0 & 1 \\ -9 & -15 & -7 \end{bmatrix} \begin{bmatrix} x_1 \\ x_2 \\ x_3 \end{bmatrix} + \begin{bmatrix} 0 \\ 2 \\ 3 \end{bmatrix} r
$$

$$
y = \begin{bmatrix} 1 & 1 & 0 \\ 0 & 1 & 1 \end{bmatrix} \begin{bmatrix} x_1 \\ x_2 \\ x_3 \end{bmatrix}
$$

14. Find diagonal state equations for systems with the following transfer function matrices:

(a) $T(s) = \begin{bmatrix} \dfrac{-6s}{s^2 + 4s + 3} & \dfrac{4}{s^2 + 4s + 3} \end{bmatrix}$

(b) $T(s) = \begin{bmatrix} \dfrac{s^2 - 4}{s^3 + 3s^2 + 2s} & \dfrac{4s - 8}{s^3 + 3s^2 + 2s} & \dfrac{s^2 + 3s - 4}{s^3 + 3s^2 + 2s} \end{bmatrix}$

Ans One possibility is the following:

$$
\begin{bmatrix} \dot{x}_1 \\ \dot{x}_2 \\ \dot{x}_3 \end{bmatrix} = \begin{bmatrix} 0 & 0 & 0 \\ 0 & -1 & 0 \\ 0 & 0 & -2 \end{bmatrix} \begin{bmatrix} x_1 \\ x_2 \\ x_3 \end{bmatrix} + \begin{bmatrix} -2 & 3 & 0 \\ -4 & 12 & -8 \\ -2 & 6 & -3 \end{bmatrix} \begin{bmatrix} r_1 \\ r_2 \\ r_3 \end{bmatrix}
$$

$$
y = \begin{bmatrix} -1 & 1 & 1 \end{bmatrix} \begin{bmatrix} x_1 \\ x_2 \\ x_3 \end{bmatrix}
$$

(c) $T(s) = \begin{bmatrix} \dfrac{3s - 1}{s^2 + 4} \\ \dfrac{-s + 8}{s^2 + 4} \end{bmatrix}$

(d)
$$T(s) = \frac{\begin{bmatrix} s \\ -3s^2 - 4 \\ 8 \end{bmatrix}}{s^3 + 3s^2 + 2s}$$

Ans   One possibility is the following:

$$\begin{bmatrix} \dot{x}_1 \\ \dot{x}_2 \\ \dot{x}_3 \end{bmatrix} = \begin{bmatrix} 0 & 0 & 0 \\ 0 & -1 & 0 \\ 0 & 0 & -2 \end{bmatrix} \begin{bmatrix} x_1 \\ x_2 \\ x_3 \end{bmatrix} + \begin{bmatrix} 1 \\ 1 \\ 1 \end{bmatrix} r$$

$$\begin{bmatrix} y_1 \\ y_2 \\ y_3 \end{bmatrix} = \begin{bmatrix} 0 & 1 & -1 \\ -2 & 7 & -8 \\ 4 & -8 & 4 \end{bmatrix} \begin{bmatrix} x_1 \\ x_2 \\ x_3 \end{bmatrix}$$

15. Find a simulation diagram and a matrix state variable representation for a two-input, two-output system with the following transfer function matrix:

$$T(s) = \begin{bmatrix} \dfrac{4s}{s^2 + 3s + 2} & \dfrac{s - 3}{s^2 + 3s + 2} \\[3mm] \dfrac{-6}{s^2 + 3s + 2} & \dfrac{s + 4}{s^2 + 3s + 2} \end{bmatrix}$$

16. A transfer function with equal numerator and denominator polynomial degrees may be expanded as a constant plus a proper remainder, as in the following example:

$$T(s) = \frac{3s^2 + 2s - 4}{s^2 + 3s + 2} = 3 + \frac{-7s - 10}{s^2 + 3s + 2}$$

It may be realized by adding a term to the system output which is proportional to the system input. The resulting state variable equations have the form

$$x = Ax + Br$$

$$y = Cx + Dr$$

Find matrices **A**, **B**, **C**, and **D** for a system with the above transfer function. For such a second-order single-input, single-output system, the state variable equations will be of the form

$$\begin{bmatrix} \dot{x}_1 \\ \dot{x}_2 \end{bmatrix} = \begin{bmatrix} a_{11} & a_{12} \\ a_{21} & a_{22} \end{bmatrix} \begin{bmatrix} x_1 \\ x_2 \end{bmatrix} + \begin{bmatrix} b_1 \\ b_2 \end{bmatrix} r$$

$$y = \begin{bmatrix} c_1 & c_2 \end{bmatrix} \begin{bmatrix} x_1 \\ x_2 \end{bmatrix} + dr$$

17. Use controllability and observability matrices to determine whether the following systems are completely controllable and whether these systems are completely observable:

(a)
$$\begin{bmatrix} \dot{x}_1 \\ \dot{x}_2 \end{bmatrix} = \begin{bmatrix} 2 & -4 \\ 0 & 1 \end{bmatrix} \begin{bmatrix} x_1 \\ x_2 \end{bmatrix} + \begin{bmatrix} 1 \\ 0 \end{bmatrix} r$$

$$y = \begin{bmatrix} 1 & 1 \end{bmatrix} \begin{bmatrix} x_1 \\ x_2 \end{bmatrix}$$

(b)
$$\begin{bmatrix} \dot{x}_1 \\ \dot{x}_2 \\ \dot{x}_3 \end{bmatrix} = \begin{bmatrix} 3 & 0 & -5 \\ -2 & 1 & 5 \\ 0 & 0 & -2 \end{bmatrix} \begin{bmatrix} x_1 \\ x_2 \\ x_3 \end{bmatrix} + \begin{bmatrix} 1 & 0 \\ 2 & 0 \\ 0 & -1 \end{bmatrix} \begin{bmatrix} r_1 \\ r_2 \end{bmatrix}$$

$$\begin{bmatrix} y_1 \\ y_2 \end{bmatrix} = \begin{bmatrix} 4 & 1 & -3 \\ 3 & 2 & -1 \end{bmatrix} \begin{bmatrix} x_1 \\ x_2 \\ x_3 \end{bmatrix}$$

Ans completely controllable but not completely observable

(c)
$$\begin{bmatrix} \dot{x}_1 \\ \dot{x}_2 \\ \dot{x}_2 \end{bmatrix} = \begin{bmatrix} 1 & 0 & -2 \\ 3 & -3 & 0 \\ 0 & 0 & 1 \end{bmatrix} \begin{bmatrix} x_1 \\ x_2 \\ x_3 \end{bmatrix} + \begin{bmatrix} 1 & -1 \\ 2 & 0 \\ 0 & 0 \end{bmatrix} \begin{bmatrix} r_1 \\ r_2 \end{bmatrix}$$

$$\begin{bmatrix} y_1 \\ y_2 \end{bmatrix} = \begin{bmatrix} 0 & 4 & 1 \\ 0 & -2 & 3 \end{bmatrix} \begin{bmatrix} x_1 \\ x_2 \\ x_3 \end{bmatrix}$$

18. Write state equations for systems, each with modes e^{2t}, e^{-3t}, $e^{(-4+j)t}$, and $e^{(-4-j)t}$ which have the following properties:

(a) The mode e^{2t} is uncontrollable.

(b) The mode e^{-3t} is unobservable.

(c) The mode e^{2t} is both uncontrollable and unobservable.

(d) The modes $e^{(-4+j)t}$ and $e^{(-4-j)t}$ are uncontrollable.

(e) The mode e^{-3t} is uncontrollable and the mode e^{2t} is unobservable.

19. The system

$$\begin{bmatrix} \dot{x}_1 \\ \dot{x}_2 \end{bmatrix} = \begin{bmatrix} -1 & 1 \\ 2 & 0 \end{bmatrix} \begin{bmatrix} x_1 \\ x_2 \end{bmatrix} + \begin{bmatrix} 0 \\ 1 \end{bmatrix} r$$

$$y = \begin{bmatrix} 1 & 1 \end{bmatrix} \begin{bmatrix} x_1 \\ x_2 \end{bmatrix}$$

is unstable. Can the instability be detected from input-output measurements? Determine whether or not the system is completely observable. Then calculate the system transfer function. A common factor in the numerator and the denominator should cancel.

Repeat if instead the output equation is

$$y = \begin{bmatrix} -2 & 1 \end{bmatrix} \begin{bmatrix} x_1 \\ x_2 \end{bmatrix}$$

20. Find a third-order system, in phase-variable form, which is not completely controllable.

21. Show that an nth-order system with n outputs is completely observable if its $n \times n$ output coupling matrix is nonsingular.

22. Solve

$$\dot{x} = -2x + r(t)$$

with

$$x(0^-) = 7$$

$$r(t) = 3e^{3t}$$

using the time-domain method involving convolution.

23. Use Laplace transform methods to find the state response of the following systems for $t \geq 0$ with the given inputs and initial conditions. Also find the system output.

(a) $\dot{x} = -3x + 2r(t)$

$y = 4x$

$x(0^-) = 7$

$r(t) = 5u(t)$, where $u(t)$ is the unit step function

(b) $\begin{bmatrix} \dot{x}_1 \\ \dot{x}_2 \end{bmatrix} = \begin{bmatrix} -3 & 1 \\ 7 & 0 \end{bmatrix} \begin{bmatrix} x_1 \\ x_2 \end{bmatrix} + \begin{bmatrix} 1 \\ 0 \end{bmatrix} r$

$y = \begin{bmatrix} 1 & 1 \end{bmatrix} \begin{bmatrix} x_1 \\ x_2 \end{bmatrix}$

$\begin{bmatrix} x_1(0^-) \\ x_2(0^-) \end{bmatrix} = \begin{bmatrix} 3 \\ 0 \end{bmatrix}$

$r(t) = \delta(t)$, the unit inpulse function

Ans $\begin{bmatrix} 1.01e^{1.54t} & +2.99e^{-4.54t} \\ 4.61e^{1.54t} & -4.61e^{-4.54t} \end{bmatrix}$; $y(t) = 5.62e^{1.54t} - 1.62e^{-4.54t}$

(c) $\begin{bmatrix} \dot{x}_1 \\ \dot{x}_2 \end{bmatrix} = \begin{bmatrix} 0 & 1 \\ -3 & -8 \end{bmatrix} \begin{bmatrix} x_1 \\ x_2 \end{bmatrix} + \begin{bmatrix} 0 & 4 \\ 1 & -1 \end{bmatrix} \begin{bmatrix} r_1 \\ r_2 \end{bmatrix}$

$y = \begin{bmatrix} 1 & 1 \end{bmatrix} \begin{bmatrix} x_1 \\ x_2 \end{bmatrix}$

$\begin{bmatrix} x_1(0^-) \\ x_2(0^-) \end{bmatrix} = \begin{bmatrix} 7 \\ -3 \end{bmatrix}$

$\begin{bmatrix} r_1(t) \\ r_2(t) \end{bmatrix} = \begin{bmatrix} e^t \\ 5 \end{bmatrix}$

(d) $\begin{bmatrix} \dot{x}_1 \\ \dot{x}_2 \\ \dot{x}_3 \end{bmatrix} = \begin{bmatrix} -3 & 1 & 0 \\ -2 & 0 & 1 \\ 0 & 0 & 0 \end{bmatrix} \begin{bmatrix} x_1 \\ x_2 \\ x_3 \end{bmatrix} + \begin{bmatrix} 0 \\ 1 \\ 0 \end{bmatrix} r$

$\begin{bmatrix} y_1 \\ y_2 \end{bmatrix} = \begin{bmatrix} 1 & 0 & 1 \\ 0 & 0 & 1 \end{bmatrix} \begin{bmatrix} x_1 \\ x_2 \\ x_3 \end{bmatrix}$

$\begin{bmatrix} x_1(0^-) \\ x_2(0^-) \\ x_3(0^-) \end{bmatrix} = \begin{bmatrix} 1 \\ 0 \\ 0 \end{bmatrix}$

$r(t) = 2u(t)$

24. The state transition matrix for a certain system is

$$\Phi(t) = \begin{bmatrix} \frac{1}{2}e^{-t} + \frac{1}{2}e^{-2t} & 2(e^{-t} - e^{-2t}) \\ 3e^{-t} - 3e^{-2t} & -e^{-t} + 2e^{-2t} \end{bmatrix}$$

Find the state $\mathbf{x}(t)$ for $t \geq 0$ if all system inputs are zero and

$$\mathbf{x}(0^-) = \begin{bmatrix} x_1(0^-) \\ x_2(0^-) \end{bmatrix} = \begin{bmatrix} -7 \\ 2 \end{bmatrix}$$

25. Calculate state transition matrices for systems with the following state coupling matrices **A**, using

$$\Phi(t) = \mathscr{L}^{-1}\{[s\mathbf{I} - \mathbf{A}]^{-1}\}$$

(a) $\begin{bmatrix} 0 & 1 \\ -6 & -4 \end{bmatrix}$

(b) $\begin{bmatrix} -1 & 4 \\ 8 & -2 \end{bmatrix}$

Ans $\begin{bmatrix} (0.4e^{-7.1t} + 0.6e^{4.1t})(0.3e^{-7.1t} - 0.3e^{4.1t}) \\ (0.6e^{-7.1t} - 0.6e^{4.1t})(0.54e^{-7.1t} + 0.46e^{4.1t}) \end{bmatrix}$

(c) $\begin{bmatrix} -5 & 1 & 0 \\ -4 & 0 & 5 \\ 0 & 0 & 0 \end{bmatrix}$

26. Show that the state transition matrix for a diagonalized system is diagonal, with the system modes along the diagonal.

27. Find state variable equations for each of the systems of Fig. P7.27. Then find the transfer function(s) from the original drawing and compare with the transfer function(s) of the state variable model.

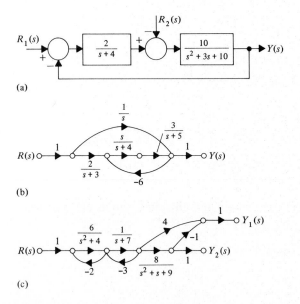

Figure P7.27

28. For the state feedback systems described by the following equations, choose the feedback gain constants k_i to place the closed-loop system poles at the indicated locations. Then, for the feedback system, find the steady state outputs due to a unit step input $u(t)$.

(a)
$$\begin{bmatrix} \dot{x}_1 \\ \dot{x}_2 \\ \dot{x}_3 \\ \dot{x}_4 \end{bmatrix} = \begin{bmatrix} 0 & 1 & 0 & 0 \\ 0 & 0 & 1 & 0 \\ 0 & 0 & 0 & 1 \\ -8 & -3 & -7 & -5 \end{bmatrix} \begin{bmatrix} x_1 \\ x_2 \\ x_3 \\ x_4 \end{bmatrix} + \begin{bmatrix} 1 \\ 0 \\ 0 \\ 1 \end{bmatrix} r$$

$$r = \begin{bmatrix} k_1 & k_2 & k_3 & k_4 \end{bmatrix} \begin{bmatrix} x_1 \\ x_2 \\ x_3 \\ x_4 \end{bmatrix} + u$$

$$y = \begin{bmatrix} 2 & -1 & 0 & 3 \\ 1 & 0 & 1 & -2 \end{bmatrix} \begin{bmatrix} x_1 \\ x_2 \\ x_3 \\ x_4 \end{bmatrix}$$

Closed-loop poles as $s = -5 \pm j3, \; -4 \pm j4$

(b)
$$\begin{bmatrix} \dot{x}_1 \\ \dot{x}_2 \\ \dot{x}_3 \end{bmatrix} = \begin{bmatrix} 0 & 1 & 0 \\ 0 & 0 & 1 \\ -10 & -5 & -2 \end{bmatrix} \begin{bmatrix} x_1 \\ x_2 \\ x_3 \end{bmatrix} + \begin{bmatrix} 0 & -1 \\ 0 & 0 \\ 1 & 7 \end{bmatrix} \begin{bmatrix} r \\ u \end{bmatrix}$$

$$r = \begin{bmatrix} k_1 & k_2 & k_3 \end{bmatrix} \begin{bmatrix} x_1 \\ x_2 \\ x_3 \end{bmatrix}$$

$$y = \begin{bmatrix} 1 & 0 & 0 \end{bmatrix} \begin{bmatrix} x_1 \\ x_2 \\ x_3 \end{bmatrix}$$

Closed-loop poles at $s = -4$ and $-4 \pm j2$

Ans $k_1 = -70, k_2 = -47, k_3 = -10, y(\infty) = -0.5625$

29. Design first-order observers of the following plants. Choose the observer eigenvalues to be at $s = -30$.

(a)
$$\begin{bmatrix} \dot{x}_1 \\ \dot{x}_2 \\ \dot{x}_3 \end{bmatrix} = \begin{bmatrix} -2 & 1 & 0 \\ -4 & 0 & 1 \\ -2 & 0 & 0 \end{bmatrix} \begin{bmatrix} x_1 \\ x_2 \\ x_3 \end{bmatrix} + \begin{bmatrix} 0 \\ 1 \\ -1 \end{bmatrix} r$$

$$y = \begin{bmatrix} 1 & 0 & 0 \\ 0 & 1 & 0 \end{bmatrix} \begin{bmatrix} x_1 \\ x_2 \\ x_3 \end{bmatrix}$$

(b) $\begin{bmatrix} \dot{x}_1 \\ \dot{x}_2 \end{bmatrix} = \begin{bmatrix} 0 & 1 \\ -4 & -2 \end{bmatrix} \begin{bmatrix} x_1 \\ x_2 \end{bmatrix} + \begin{bmatrix} 0 \\ 3 \end{bmatrix} r$

$y = x_1$

30. Design identity observers for the following plants. Choose the observer eigenvalues to be at -20.

(a) $A = \begin{bmatrix} 0 & 1 \\ -4 & -4 \end{bmatrix}, \quad B = \begin{bmatrix} 0 \\ 3 \end{bmatrix}$

$y = x_1$

(b) $A = \begin{bmatrix} 0 & 1 \\ -3 & -8 \end{bmatrix}, \quad B = \begin{bmatrix} 0 \\ 2 \end{bmatrix}$

$y = x_1$

31. For the systems of Prob. 30, design control gains to place the desired closed-loop poles at -5 and -8 assuming the measurements are available. Next, close the loop using a reduced-order observer to furnish an estimate of x_2. Show that the characteristic polynomial of the closed-loop system including the observer contains the desired closed-loop roots and the observer root.

32. For each of the following systems, use factorization to sketch the root square locus. Compute gains for $p = 0.1, 1$, and 10

(a) $A = \begin{bmatrix} 0 & 1 \\ -4 & -4 \end{bmatrix}, \quad B = \begin{bmatrix} 0 \\ 3 \end{bmatrix}$

$Q = \begin{bmatrix} 1 & 0 \\ 0 & 0 \end{bmatrix}, \quad P = p$

(b) $A = \begin{bmatrix} 0 & 1 \\ -4 & -4 \end{bmatrix}, \quad B = \begin{bmatrix} 0 \\ 3 \end{bmatrix}$

$Q = \begin{bmatrix} 0 & 0 \\ 0 & 1 \end{bmatrix}, \quad P = p$

(c) $A = \begin{bmatrix} 0 & 1 \\ -3 & -8 \end{bmatrix}, \quad B = \begin{bmatrix} 0 \\ 4 \end{bmatrix}$

$Q = \begin{bmatrix} 1 & 0 \\ 0 & 0 \end{bmatrix}, \quad P = p$

(d) $A = \begin{bmatrix} 0 & 1 \\ -3 & -8 \end{bmatrix}$, $\qquad B = \begin{bmatrix} 0 \\ 4 \end{bmatrix}$

$Q = \begin{bmatrix} 1 & 0 \\ 0 & 1 \end{bmatrix}$, $\qquad P = p$

(e) $A = \begin{bmatrix} 0 & 1 \\ -4 & 0 \end{bmatrix}$, $\qquad B = \begin{bmatrix} 1 \\ 1 \end{bmatrix}$

$Q = \begin{bmatrix} 1 & 0 \\ 0 & 0 \end{bmatrix}$, $\qquad P = p$

Ans (a) $p = 0.1, k_1 = -0.45, k_2 = -2.10$

$p = 1, k_1 = -0.08, k_2 = -0.33$

$p = 10, k_1 = -0.009, k_2 = -0.037$

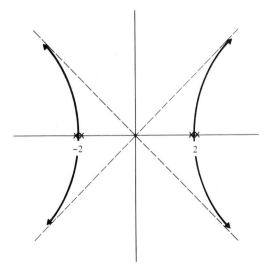

Figure P7.32

33. Find the gains for the systems of Prob. 32 using the algebraic Riccati equation.

34. For the MAGLEV system choose instead values of the feedback gain constants k_1, k_2, and k_3 to place all three of the overall system poles at $s = -5$. For this system, find the steady state response d to a unit step disturbance f and the value of constant reference input r to give a nominal gap distance $d = 15$.

35. For the open-loop MAGLEV system, suppose the vertical track elevation varies sinusoidally with time as the train is in motion, according to

$$\dot{h}(t) = 0.2 \sin \frac{\pi t}{10}$$

Find the second-order differential equation satisfied by $\dot{h}(t)$, then augment the original state equations with two more equations and two more state variables

$$x_4 = \dot{h}(t)$$
$$x_5 = \ddot{h}(t)$$

in place of the disturbance input $f = \ddot{h}$. With additional sensors for the signals x_4 and x_5 and feedback of the form

$$r = +k_1 x_1 + k_2 x_2 + k_3 x_3 + k_4 x_4 + k_5 x_5 + u_1$$

find the state equations for the feedback system in terms of the k constants.

Chapter

8

Digital Control

8.1 PREVIEW

The terms *continuous-time* and *analog* are identical in meaning when applied to signals and systems. Analog signals are functions of a continuous time variable, and analog systems are systems described in terms of such signals. Similarly, *discrete-time* and *digital* have the same meaning; they refer to signals which are defined only for specified instants of time. The use of the words *analog* and *digital* in this sense dates from an era when large-scale analog computers were common.

Past years have seen an exponential growth in the capability and application of digital computers, and there is every indication that a high rate of growth will continue far into the future. In the field of control systems, digital computers were first applied in military and space applications where the high costs were justified by new capabilities. As computer costs dropped, their use for control began in large-scale industry such as chemical processing, heavy manufacturing, and telecommunications plants which could afford the large investment. Later, mini-computers, faster, more powerful, and much less expensive than their predecessors, began to revolutionize industry everywhere. No longer was a huge central computing installation necessary; general- and special-purpose computers could be economically distributed and tailored to specific tasks. Now, with the widespread availability and low cost of microcomputers, every process is a candidate for digital control, and sophisticated control systems that were impractically expensive only a few years ago are feasible.

It is no longer farfetched to imagine a digital computer-based control system that monitors plant behavior to determine a mathematical model of the plant, then through repeated simulation or other calculation determines an optimum control strategy and proceeds to effect control according to programmed objectives.

One advantage of digital systems is that the time response can be computed by applying simple long division to a special transform, the z transform. Conversely, when continuous-time systems are analyzed, more difficult methods are needed, such as the inverse Laplace transform or the state transition matrix approach. Stability, control, observability, and controllability can all be defined and applied in the context of discrete-time control.

This chapter introduces the main concepts of discrete-time control. This material may be used as a brief terminating study or as an introduction to a more in-depth discussion of digital control systems in another course to follow.

8.2 COMPUTER PROCESSING

Computer History and Trends

A brief digital computer chronology is given in Table 8.1. Although their basic concepts are credited largely to Charles Babbage (c. 1830), today's computers became practical only after the invention and development of the transistor. Rapid technological advances in solid state physics have been the primary driving force behind the rapid evolution of digital computers since 1960. The emphasis here is upon stored program general-purpose mini- and microcomputers rather than special-purpose digital logic (which will accomplish the same purpose) because it appears that the vast majority of future control applications will be with low-cost, mass-produced hardware.

Table 8.1 DIGITAL COMPUTER CHRONOLOGY

Date	Development
B.C.	Abacus is in use. It becomes widespread in Europe and Asia.
c.1650	Pascal builds a mechanical desk calculator for addition.
c.1670	Leibniz builds a calculator also capable of subtraction, multiplication, division, and root extraction.
c.1800	Jacquard perfects automatic looms which weave designs programed by punched cards.

c.1830   Babbage develops modern computer principles, including memory, program control, and branching capabilities.

1890   Hollerith uses a punched card system for the U.S. census. Hollerith's company later becomes IBM.

1940   Aiken builds an electromechanical programed computer, Mark I, used for ballistics calculations by the U.S. Army. It is capable of several additions per second. Similar work is done by Stibitz at Bell Telephone Laboratories, who notices similarity between telephone switching and computation.

1946   Eckert and Mauchly build the first vacuum tube digital computer, ENIAC, at the University of Pennsylvania. It is capable of several thousand additions per second.

1948   Von Neumann directs construction of the IAS stored program computer at Princeton. Memory involves charge storage on cathode-ray tube targets and a rotating magnetic drum. Addition is performed in approximately 65 μsec. EDSAC at Cambridge University is completed first, becoming the first stored program computer. IBM also builds a stored program computer, the SSEC.

1950   Sperry Rand Corporation builds the first commercial data processing computer, using semiconductor diodes and vacuum tubes, the UNIVAC I.

1952   IBM begins marketing the 701 digital computer commercially.

1960   The "second generation" of computers is introduced to the market. These use solid state components in place of vacuum tubes.

1964   The "third generation" of computers begins. Integrated circuit hardware predominates as in the IBM System 360.

1965   Digital Equipment Corporation markets the PDP-8 at about $50,000. The minicomputer industry begins.

1973   Intel markets the first microcomputer system using the 8080 microprocessor chip.

1975   Ten different microprocessors are on the market, including the Fairchild F8, the Intel 8080A, the Motorola 6800, and the Signetics 2650. Others soon enter the field.

1977   Motorola, Texas Instruments, Intel, and Zilog each begin to market competing 16-bit microprocessors; 32-bit microprocessors are announced.

1978   Rockwell International markets the AIM-65 microprocessor system, priced at about $500. System includes alphanumeric display, small printer, cassette tape interface, and input-output ports. ROM-based software includes system monitor, assembler, and BASIC compiler.

1982   Very-large-scale integration (VLSI) technology is capable of integrating a million devices on a tiny semiconductor chip, approximately the number of devices in a very large computer processing unit and roughly the complexity level of primitive organisms.

1983   Personal desk top computers are comparable in storage and capability to the mainframes of the 1960s.

1984   Hand-held devices are programable in BASIC.

1986   Intense competition occurs among duplicates (clones) of IBM PCs (personal computers).

1988   The distinction between mainframes and minicomputers and between minicomputers and microcomputers blurs as capacity and architectural improvements continue.

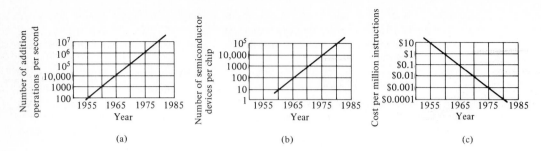

Figure 8.1 Digital computer trends. **(a)** Approximate speed increase of high-speed computers. **(b)** Approximate semiconductor device density increase. **(c)** Approximate decrease in computation cost.

The charts given in Fig. 8.1 show three dramatic trends in computer evolution. Their speed of operation has increased by about an order of magnitude every five years since 1955. The density of their electronic circuits has increased by roughly the same factor of 10 every five years since the early use of solid state components in 1960. With this compactness has come a large decrease in power requirements. The total costs of digital computation have similarly plummeted.

Whenever a digital computer becomes part of an otherwise analog system, signal conversion takes place. Each analog signal which is to be operated upon by a computer must be converted from analog form to digital form by an A/D converter. Similarly, each digital value which is to influence the analog system must be converted to analog form by a D/A converter. Since the computer output does not change until the next set of calculations and D/A conversions are completed, the analog resultant of some D/A process may be held constant during each cycle by a sample-and-hold (S/H) device. See Fig. 8.2.

Analog signals are input, analog signals are output, and in between may be placed powerful computational ability. Operations such as square-rooting, correlation, function generation, and spectral analysis which are a nightmare in analog hardware are simply and routinely done digitally. Furthermore, if a general-purpose programmable digital computer is used, changes in objectives and improvements in design may require only changes in the stored program (the software), not changes to the equipment itself (the hardware).

Digital computation is subject to numerical *roundoff* or *truncation errors* because numbers are represented by a finite number of bits. For example, the product of two 16-bit numbers involves potentially 32 bits. To represent such a product as another 16-bit number, the least significant 16 bits of the product must be eliminated. Ordinarily, finite arithmetic precision is not of great concern when the number of bits used to represent a number far exceeds the required numerical precision. However, the power of a computer to perform huge numbers of calculations in a short time means that it is possible for tiny errors to quickly accumulate to large proportions. One must also be aware that certain kinds of

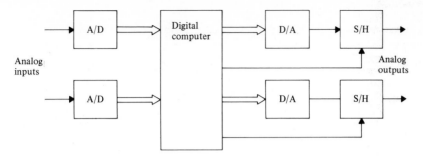

Figure 8.2 A simple digital controller.

calculations, such as forming differences between large but nearly identical numbers, are especially susceptible to error.

8.3 A/D AND D/A CONVERSION

Analog-to-Digital Conversion

An analog signal such as a voltage can be expressed as a binary number, suitable for computer processing, by assigning weights to each bit position. Table 8.2 gives a 4-bit coding of an analog signal which may range between 0 and 10 V. Each binary

Table 8.2 REPRESENTING A NON-NEGATIVE ANALOG VOLTAGE WITH A BINARY NUMBER

Analog Voltage	Binary Representation
0 to 0.625	0000
0.625 to 1.25	0001
1.25 to 1.875	0010
1.875 to 2.5	0011
2.5 to 3.125	0100
3.125 to 3.75	0101
3.75 to 4.375	0110
4.375 to 5.0	0111
5.0 to 5.625	1000
5.625 to 6.25	1001
6.25 to 6.875	1010
6.875 to 7.5	1011
7.5 to 8.125	1100
8.125 to 8.75	1101
8.75 to 9.375	1110
9.375 to 10.0	1111

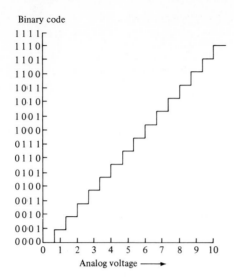

Figure 8.3   Binary coding of an analog voltage.

increment represents $2^{-4} = \frac{1}{16}$ the maximum representable voltage of 10 V. The information in the table is shown in graphical form in Fig. 8.3.

Each binary number represents a *range* of analog voltage; hence there is a *quantization error* associated with the conversion. For a 4-bit conversion, the maximum quantization error is $2^{-4} = 6.25\%$. Table 8.3 shows quantization error percentages for various numbers of bits in the digital representation. The quantization error in 16-bit conversion, for example, corresponds to a signal-to-noise ratio (SNR) of

$$\text{SNR (in dB)} = 20 \log_{10} 2^{16} = 96.3 \text{ dB}$$

Table 8.3   QUANTIZATION ERROR FOR ANALOG-DIGITAL CONVERSION

Number of Bits	Maximum Percent Error
1	50
2	25
4	6.25
6	1.56
8	0.391
10	0.0977
12	0.0244
14	0.0061
16	0.0015

Table 8.4 BIPOLAR ANALOG VOLTAGE REPRESENTATIONS

Analog Voltage	Sign and Magnitude	Offset Binary	Two's Complement
-5.0 to -4.375	1111	0000	1001
-4.375 to -3.75	1110	0001	1010
-3.75 to -3.125	1101	0010	1011
-3.125 to -2.5	1100	0011	1100
-2.5 to -1.875	1011	0100	1101
-1.875 to -1.25	1010	0101	1110
-1.25 to -0.625	1001	0110	1111
-0.625 to 0	1000	0111	1000
0 to 0.625	0000	1000	0000
0.625 to 1.25	0001	1001	0001
1.25 to 1.875	0010	1010	0010
1.875 to 2.5	0011	1011	0011
2.5 to 3.125	0100	1100	0100
3.125 to 3.75	0101	1101	0101
3.75 to 4.375	0110	1110	0110
4.375 to $+5.0$	0111	1111	0111

By comparison, typical signal-to-noise ratios in quality audio recording and reproduction are 60 to 70 dB, which may be accurately portrayed by only 12-bit coding.

For bipolar signals, the three types of binary codes shown in Table 8.4 are the most commonly used. In the sign and magnitude arrangement, the most significant bit of the binary code represents the algebraic sign of the signal, with a zero meaning a positive number. The remaining bits are the binary representation of the signal's magnitude. The offset binary code is equivalent to adding a fixed constant (or bias) to the signal to be converted so that the sum is always nonnegative. In two's complement coding, negative signals are represented as the two's complement of their magnitude, in the same manner as negative numbers are commonly manipulated in digital computers. In applications involving digital displays, binary-coded-decimal (BCD) coding may be used, where the signal is represented as *decimal* digits and then each decimal digit is individually converted to a 4-bit binary equivalent.

DRILL PROBLEM

D8.1. What is the maximum percentage error if a binary number is truncated to 10 bits? What if the number is rounded?

Ans 0.098%; 0.049%

Sample-and-Hold

A/D and D/A converters are generally used to repetitively perform conversions. For analog-to-digital conversion it is often desirable to "freeze" the analog signal while the conversion is taking place. A sample-and-hold device, with symbol given in Fig. 8.4a, may be used to hold an analog signal steady while conversion proceeds, as in Fig. 8.4b.

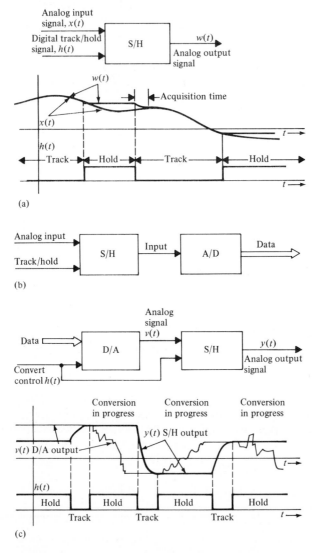

Figure 8.4 Sample-and-hold used in analog/digital conversion. **(a)** Sample-and-hold symbol and representative signals. **(b)** S/H used to "freeze" an analog signal while D/A conversion takes place.

Digital-to-Analog Conversion

In converting from digital to analog, a D/A converter output may fluctuate wildly while conversion is taking place. A sample-and-hold device is conveniently used to hold the previously converted signal while a new conversion takes place, as in Fig. 8.4c. The result is an output signal which changes in nearly a stepwise fashion each time a conversion occurs.

DRILL PROBLEMS

D8.2. A 12-bit D/A converter has minimum output voltage -10 and maximum output voltage $+10$. After the binary code 010110101001 is applied, what is the output voltage if the converter is of the following type:

(a) Sign and magnitude

(b) Offset binary

(c) Two's complement

> **Ans** (a) 7.08; (b) -2.92; (c) 7.08

D8.3. The sinusoidal signal

$$f(t) = 10 \sin t$$

is tracked for $t < 1$, held for $1 \le t < 2$, tracked for $2 \le t < 5$, then held thereafter to form the signal $g(t)$. Sketch both $f(t)$ and $g(t)$.

8.4 DISCRETE-TIME SIGNALS

Representing Sequences

Periodic samples of a continuous-time signal, as are generated by an A/D converter, form a sequence of numbers termed a *discrete-time signal*. Figure 8.5 shows a continuous-time signal $f(t)$ and the corresponding sequence of samples

$$f(t = 0), f(t = T), f(t = 2T), f(t = 3T), \ldots$$

Although it results in an ambiguity that is only resolved by context, it is common practice to denote the sequence by $f(k)$, where k is the sample number.

Some important sequences are shown in Figs. 8.6 to 8.10. All sequences consist of zero samples prior to $k = 0$. The unit pulse sequence, $\delta(k)$ in Fig. 8.6, has unit sample value for $k = 0$ and all other values zero. The unit step sequence $u(k)$, Fig. 8.7, has samples which are all unity for $k = 0$ and thereafter. The ramp, Fig. 8.8, consists of samples of the continuous-time unit ramp function.

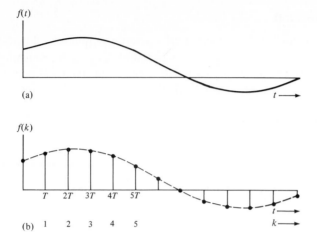

Figure 8.5 Sampling a continuous-time signal. **(a)** A continuous-time signal. **(b)** Samples of the continuous-time signal.

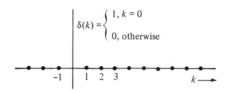

Figure 8.6 Unit sample.

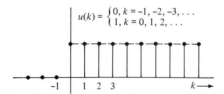

Figure 8.7 Unit step.

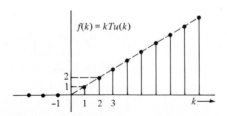

Figure 8.8 Ramp.

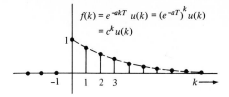

Figure 8.9 Exponential (or geometric).

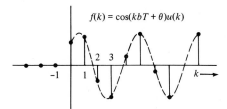

Figure 8.10 Sinusoidal.

The sampled exponential, Fig. 8.9, has samples that are progressive powers of the number

$$c = e^{-aT}$$

and so forms a geometric sequence:

$$f(0) = c^0 = 1$$

$$f(1) = c$$

$$f(2) = c^2$$

$$f(n) = c^n$$

Geometric sequences (or sampled exponential functions) have fundamental importance to discrete-time systems in the same way that exponential functions are basic to continuous-time system. A sampled sinusoidal function as in Fig. 8.10 is termed a *sinusoid sequence*.

More complicated sequences may often be represented as shifts and sums of the basic sequences. For example, $\delta(k - 2)$ is the unit pulse sequence shifted two samples to the right, as in Fig. 8.11a. The sequence of Fig. 8.11b is thus

$$f_1(k) = \delta(k) + 2\delta(k - 1) + 3\delta(k - 2) - 2\delta(k - 3) - \delta(k - 4)$$

The sequence

$$f_2(k) = \begin{cases} 10 & k = 2, 3, 4, \ldots \\ 0 & \text{otherwise} \end{cases}$$

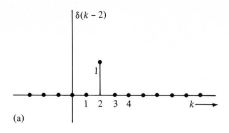

(a)

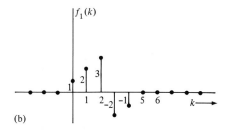

(b)

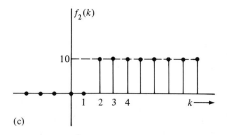

(c)

Figure 8.11 Shifting and summing sequences. **(a)** A shifted unit pulse sequence. **(b)** A finite sum of shifted pulses. **(c)** Shifted step sequence or a step sequence with samples at $k = 0$ and $k = 1$ canceled by pulses.

drawn in Fig. 8.11c, is

$$f_2(k) = 10u(k) - 10\delta(k) - 10\delta(k-1)$$

Alternatively,

$$f_2(k) = 10u(k-2)$$

DRILL PROBLEM

D8.4. Sketch the function $f(t)$ and the samples $f(k)$ for a sampling period $T = 0.5$ sec.

(a) $f(t) = e^{-0.5t}u(t)$

    Ans $f(k) = e^{-0.25k}u(k)$

(b) $f(t) = (\sin \pi t)u(t)$

Ans $f(k) = (\sin k\pi/2)u(k)$

(c) $f(t) = (\cos 2\pi t)u(t)$

Ans $f(k) = (-1)^k u(k)$

(d) $f(t) = (\sin 2\pi t)u(t)$

Ans $f(k) = 0$

z-Transformation and Properties

The z-transform of a sequence $f(k)$ is defined as the infinite series

$$Z[f(k)] = F(z) = \sum_{k=0}^{\infty} f(k)z^{-k}$$

It plays much the same role in the description of discrete-time signals as the Laplace transform does with continuous-time signals.

Table 8.5 lists basic z-transform pairs, together with the corresponding continuous-time function which gives the sequence when sampled with period T. Using the z-transform definition, the transform of the unit pulse is

$$Z[\delta(k)] = \sum_{k=0}^{\infty} \delta(k)z^{-k} = z^{-0} = 1$$

It should be noted that the unit pulse sequence $\delta(k)$, while analogous to the unit impulse $\delta(t)$, does not consist of samples of $\delta(t)$, which is infinite for $k = t = 0$.

The z-transform of the unit step sequence is as follows:

$$Z[u(k)] = \sum_{k=0}^{\infty} 1 \cdot z^{-k}$$

using the fact that for a geometric series,

$$\sum_{k=0}^{\infty} x^k = \frac{1}{1-x} \qquad |x| < 1$$

then

$$Z[u(k)] = \sum_{k=0}^{\infty} z^{-k} = \sum_{k=0}^{\infty} \left(\frac{1}{z}\right)^k$$

$$= \frac{1}{1 - 1/z} = \frac{z}{z-1} \qquad \left|\frac{1}{z}\right| < 1$$

Conditions for z-transform convergence are satisfied by all but the most pathological sequences and so will not be emphasized here.

Table 8.5 SOME LAPLACE AND z-TRANSFORM PAIRS

$f(t)$	$F(s)$	$f(k)$	$F(z)$
		$\delta(k)$, unit pulse	1
$u(t)$, unit step	$\dfrac{1}{s}$	$u(k)$, unit step	$\dfrac{z}{z-1}$
$tu(t)$	$\dfrac{1}{s^2}$	$kTu(k)$	$\dfrac{Tz}{(z-1)^2}$
$e^{-at}u(t)$	$\dfrac{1}{s+a}$	$(e^{-aT})^k u(k) = c^k u(k)$ where $c = e^{-aT}$	$\dfrac{z}{z-e^{-aT}} = \dfrac{z}{z-c}$
$te^{-at}u(t)$	$\dfrac{1}{(s+a)^2}$	$kT(e^{-aT})^k u(k) = kTc^k u(k)$	$\dfrac{Tze^{-aT}}{(z-e^{-aT})^2} = \dfrac{Tcz}{(z-c)^2}$
$(\sin bt)u(t)$	$\dfrac{b}{s^2+b^2}$	$(\sin kbT)u(k)$	$\dfrac{z\sin bT}{z^2 - 2z\cos bT + 1}$
$(\cos bt)u(t)$	$\dfrac{s}{s^2+b^2}$	$(\cos kbT)u(k)$	$\dfrac{z(z-\cos bT)}{z^2 - 2z\cos bT + 1}$
$e^{-at}(\sin bt)u(t)$	$\dfrac{b}{(s+a)^2+b^2}$	$(e^{-aT})^k(\sin kbT)u(k)$ $= c^k(\sin kbT)u(k)$	$\dfrac{z(e^{-aT}\sin bT)}{(z-e^{(-a+jb)T})(z-e^{(-a-jb)T})}$ $= \dfrac{zc\sin bT}{z^2 - (2c\cos bT)z + c^2}$
$e^{-at}(\cos bt)u(t)$	$\dfrac{s+a}{(s+a)^2+b^2}$	$(e^{-aT})^k(\cos kbT)u(k)$ $= c^k(\cos kbT)u(k)$	$\dfrac{z(z-e^{-aT}\cos bT)}{(z-e^{(-a+jb)T})(z-e^{(-a-jb)T})}$ $= \dfrac{z(z-c\cos bT)}{z^2 - (2c\cos bT)z + c^2}$

A sampled exponential function, whether decaying or expanding, is of the form

$$f(k) = e^{-kaT} = (e^{-aT})^k$$

Its z-transform is given by

$$Z[e^{-kaT}] = \sum_{k=0}^{\infty} e^{-kaT} z^{-k}$$

$$= \sum_{k=0}^{\infty} \left(\frac{1}{ze^{aT}}\right)^k = \frac{1}{1 - 1/ze^{aT}}$$

$$= \frac{z}{z - e^{-aT}}$$

Samples of an exponential function form a geometric series since by defining

$$c = e^{-aT}$$

$$f(k) = e^{-kaT} = (e^{-aT})^k = c^k$$

In terms of c, the z-transform is

$$Z[c^k] = \frac{z}{z - c}$$

The transforms of sampled sinusoids given in Table 8.5 follow easily from expanding the sinusoidal function into Euler components and applying the result for sampled exponentials. For the sampled sine,

$$Z[\sin kbT] = \sum_{k=0}^{\infty} \left(\frac{e^{jkbT} - e^{-jkbT}}{2j} \right) z^{-k}$$

$$= \frac{1}{2j} \sum_{k=0}^{\infty} e^{jkbT} z^{-k} - \frac{1}{2j} \sum_{k=0}^{\infty} e^{-jkbT} z^{-k}$$

$$= \frac{z/2j}{z - e^{jbT}} - \frac{z/2j}{z - e^{-jbT}}$$

$$= \frac{(z/2j)(z - e^{-jbT} - z + e^{jbT})}{z^2 - (e^{jbT} + e^{-jbT})z + 1}$$

$$= \frac{z\left(\dfrac{e^{jbT} + e^{-jbT}}{2j} \right)}{z^2 - 2\left(\dfrac{e^{jbT} + e^{-jbT}}{2} \right)z + 1}$$

$$= \frac{z \sin bT}{z^2 - 2z \cos bT + 1}$$

Basic z-transform properties are listed in Table 8.6. The transform of a sequence scaled by a multiplicative constant is that constant times the original z-transform. The z-transform of a sample-by-sample sum of sequences is the sum of their individual z-transforms. A sequence weighted by the step number k has z-transform

$$Z[kf(k)] = \sum_{k=0}^{\infty} kf(k)z^{-k} = \sum_{k=0}^{\infty} f(k)(kz^{-k})$$

$$= \sum_{k=0}^{\infty} f(k) \frac{zd}{dz}(-z^{-k}) = -\frac{zd}{dz} \sum_{k=0}^{\infty} f(k)z^{-k}$$

$$= -\frac{zd}{dz} F(z)$$

Table 8.6 SOME z-TRANSFORM PROPERTIES

$Z[cf(k)] = cF(z)$, c a constant

$Z[f(k) + g(k)] = F(z) + G(z)$

$$Z[kf(k)] = -\frac{z \, dF(z)}{dz}$$

$Z[c^k f(k)] = F\left(\dfrac{z}{c}\right)$, c a constant

$Z[f(k-1)] = f(-1) + z^{-1}F(z)$

$Z[f(k-2)] = f(-2) + z^{-1}f(-1) + z^{-2}F(z)$

$Z[f(k-n)] = f(-n) + z^{-1}f(1-n) + z^{-2}f(2-n)t + \cdots + z^{-n+1}f(-1) + z^{-n}F(z)$

$Z[f(k+1)] = zF(z) - zf(0)$

$Z[f(k+2)] = z^2F(z) - z^2f(0) - zf(1)$

$Z[f(k+n)] = z^nF(z) - z^nf(0) - z^{n-1}f(1) - \cdots - z^2f(n-2) - zf(n-1)$

$\quad f(0) = \lim_{z \to \infty} F(z)$

If $\lim_{k \to \infty} f(k)$ exists and is finite,

$$\lim_{k \to \infty} f(k) = \lim_{z \to 1} \left[\frac{z-1}{z} F(z)\right]$$

A sequence weighted by successive powers of a constant c has z-transform as follows:

$$Z[c^k f(k)] = \sum_{k=0}^{\infty} c^k f(k) z^{-k} = \sum_{k=0}^{\infty} f(k)\left(\frac{z}{c}\right)^k$$

$$= F\left(\frac{z}{c}\right)$$

Figure 8.12a shows an example of a sequence which is shifted one step to the right. Its z-transform is given by

$$Z[f(k-1)] = \sum_{k=0}^{\infty} f(k-1)z^{-k} = \sum_{k=-1}^{\infty} f(k)z^{-(k+1)}$$

$$= f(-1) + z^{-1}\sum_{k=0}^{\infty} f(k)z^{-k}$$

$$= f(-1) + z^{-1}F(z)$$

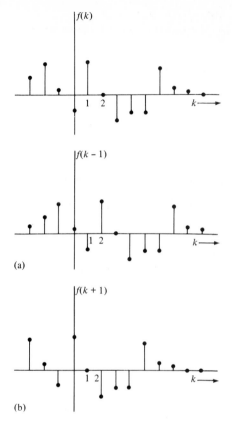

(a)

(b)

Figure 8.12 Right- and left-shifted sequences. **(a)** A sequence and the same sequence shifted right one step. **(b)** The sequence shifted left one step.

Then

$$Z[f(k - 2)] = f(-2) + z^{-1}Z[f(k - 1)]$$
$$= f(-2) + z^{-1}[f(-1) + z^{-1}F(z)]$$
$$= f(-2) + z^{-1}f(-1) + z^{-2}F(z)$$

and, similarly,

$$Z[f(k - n)] = f(-n) + z^{-1}f(1 - n) + \cdots + z^{-n+1}f(-1) + z^{-n}F(z)$$

For a left shift of the sequence, Fig. 8.12b,

$$Z[f(k + 1)] = \sum_{k=0}^{\infty} f(k + 1)z^{-k} = \sum_{k=1}^{\infty} f(k)z^{-(k-1)}$$

$$= z \sum_{k=1}^{\infty} f(k)z^{k} = z \sum_{k=0}^{\infty} f(k)z^{k} - zf(0)$$

$$= zF(z) - zf(0)$$

Similarly,

$$Z[f(k + 2)] = z[zF(z) - zf(0)] - zf(1)$$
$$= z^2 F(z) - z^2 f(0) - zf(1)$$

and

$$Z[f(k + n)] = z^n F(z) - z^n f(0) - z^{n-1} f(1) - \cdots - z^2 f(n - 2) - zf(n - 1)$$

DRILL PROBLEM

D8.5. Find the z-transforms of the following sequences:

(a) $f(k) = [(-0.5)^k - 4(0.2)^k]u(k)$

Ans $\dfrac{z}{z + 0.5} - \dfrac{4z}{z - 0.2}$

(b) $f(k) = \begin{cases} (-1)^k & k = 3, 4, 5, \ldots \\ 0 & \text{otherwise} \end{cases}$

Ans $\dfrac{-z^{-2}}{1 + z}$

(c) $f(1) = 2$, $f(4) = -3$, $f(7) = 8$, and all other samples are zero

Ans $2z^{-1} - 3z^{-4} + 8z^{-7}$

(d) $f(k) = u(k) \sin 3k - 2u(k - 4) \sin 3(k - 4)$

Ans $\dfrac{(1 - 2z^{-4})z \sin 3}{z^2 - 2z \cos 3 + 1}$

Inverse z-Transform

The sequence of samples represented by a rational z-transform may be obtained, if desired, by long division. Consider the z-transform

$$F(z) = \frac{4z}{z^2 - z + 0.5}$$

For example:

$$
\begin{array}{r}
4z^{-1} + 4z^{-2} + 2z^{-3} + \cdots \\
z^2 - z + 0.5 \overline{\smash{\big)}\ 4z \phantom{-4 + 2z^{-1}}} \\
\underline{4z - 4 + 2z^{-1}} \\
4 - 2z^{-1} \\
\underline{4 - 4z^{-1} + 2z^{-2}} \\
2z^{-1} - 2z^{-2}
\end{array}
$$

Since

$$F(z) = \frac{4z}{z^2 - z + 0.5} = 0z^0 + 4z^{-1} + 4z^{-2} + 2z^{-3} + \cdots$$

$$f(k) = 4\delta(k-1) + 4\delta(k-2) + 2\delta(k-3) + \cdots$$

and

$$f(0) = 0$$
$$f(1) = 4$$
$$f(2) = 4$$
$$f(3) = 2$$
$$\vdots$$

Repeated steps of long division do not give a closed-form expression for the sequence represented by a z-transform, although in principle as many terms in the sequence as desired may be found in this manner.

To find a formula for the sequence of samples, partial fraction expansion may be used. Rather than expanding a z-transform $F(z)$ directly in partial fractions, $F(z)/z$ is expanded so that terms of the form

$$\frac{z}{z - e^{-aT}}$$

result. For example, for the z-transform

$$F(z) = \frac{-2z^2 + 2z}{z^2 + 4z + 3}$$

$$\frac{F(z)}{z} = \frac{-2z + 2}{(z+1)(z+3)} = \frac{2}{z+1} + \frac{-4}{z+3}$$

giving

$$F(z) = \frac{2z}{z+1} + \frac{-4z}{z+3}$$

$$f(k) = 2(-1)^k - 4(-3)^k \qquad k = 0, 1, 2, 3, \ldots$$

Another example is the following:

$$F(z) = \frac{z^3 - 3}{z(z - 0.25)(z - 0.5)}$$

$$\frac{F(z)}{z} = \frac{z^3 - 3}{z^2(z - 0.25)(z - 0.5)}$$

$$= \frac{-144}{z} + \frac{-24}{z^2} + \frac{191}{z - 0.25} + \frac{-46}{z - 0.5}$$

Since

$$F(z) = -144z^0 - 24z^{-1} + \frac{-191z}{z - 0.25} + \frac{-46z}{z - 0.5}$$

$$f(k) = -144\delta(k) - 24\delta(k - 1) - 191(0.25)^k + 46(0.5)^k$$

A set of complex conjugate root terms should be manipulated into the form of the last two entries of Table 8.5:

$$F(z) = \frac{4z^2 - 3z}{z^2 + 2z + 2}$$

$$= \frac{K_1 zc \sin bT}{z^2 - 2c \cos bT + c^2} + \frac{K_2 z(z - c \cos bT)}{z^2 - 2c \cos bT + c^2}$$

Equating,

$$c^2 = 2 \qquad c = \sqrt{2}$$

$$2c \cos bT = 2\sqrt{2} \cos bT = -2 \qquad bT = \frac{3\pi}{4}$$

$$\sin bT = \sin \frac{3\pi}{4} = \frac{1}{\sqrt{2}}$$

so

$$F(z) = \frac{4z^2 - 3z}{z^2 + 2z + 2} = \frac{K_1 z + K_2 z(z + 1)}{z^2 + 2z + 2}$$

giving

$$K_2 = 4$$

$$K_1 + K_2 = -3 \qquad K_1 = -7$$

$$f(k) = 4(\sqrt{2})^k \cos \frac{3\pi k}{4} - 7(\sqrt{2})^k \sin \frac{3\pi k}{4} \qquad k = 0, 1, 2, \ldots$$

DRILL PROBLEMS

D8.6. Use long division to show that the inverse z-transform of

$$F(z) = \frac{10z}{(z - 1)^2}$$

is

$$f(k) = 10k$$

D8.7. Find the inverse z-transforms for $k \geq 0$:

(a) $F(z) = \dfrac{1}{z + 0.3}$

    Ans $\frac{10}{3}\delta(k) - \frac{10}{3}(-0.3)^k$

(b) $F(z) = \dfrac{-6z^2 + z}{z^2 + 5z + 6}$

    Ans $-19(-3)^k + 13(-2)^k$

(c) $F(z) + \dfrac{4z^2 - 3z + 2}{z^2 + 4z + 4}$

    Ans $\frac{1}{2}\delta(k) + \frac{7}{2}(-2)^k + 6k(-2)^k$

(d) $F(z) = \dfrac{3z^2 - z}{z^2 + 2z + 10}$

    Ans $3(\sqrt{10})^k \cos 1.89k - 1.33(\sqrt{10})^k \sin 1.89k$

8.5 SAMPLING

When an analog signal $f(t)$ is sampled to form the sequence $f(k)$, there is a direct relationship between the Laplace transform $F(s)$ of the analog signal and the z-transform $F(z)$ of the sequence. If a rational Laplace transform is expanded into a sum of terms of the type given in Table 8.5, the z-transform of the sample sequence is obtained by simply summing the corresponding z-transform terms from the table.

For example, for a continuous-time signal with Laplace transform

$$F(s) = \frac{4s^2 + 13s + 18}{s^3 + 5s^2 + 6s}$$

$$= \frac{3}{s} + \frac{-4}{s + 2} + \frac{5}{s + 3}$$

$$f(t) = (3 - 4e^{-2t} + 5e^{-3t})u(t)$$

For a sampling interval $T = 0.2$,

$$f(k) = (3 - 4e^{-0.4k} + 5e^{-0.6k})u(k)$$

$$[3(1)^k - 4(e^{-0.4})^k + 5(e^{-0.6})^k]u(k)$$

$$F(z) = 3\left(\frac{z}{z - 1}\right) - 4\left(\frac{z}{z - e^{-0.4}}\right) + 5\left(\frac{z}{z - e^{-0.6}}\right)$$

When the Laplace transform involves delay operations in multiples of the sampling interval T, the remainder of the transform is expanded into partial fraction terms, as in the following example, for which $T = 0.1$:

$$F(s) = \frac{e^{-0.1s} + 2}{s(s + 3)} = (e^{-0.1s} + 2)\left(\frac{\frac{1}{3}}{s} + \frac{-\frac{1}{3}}{s + 3}\right)$$

Denoting

$$G(s) = \frac{\frac{1}{3}}{s} + \frac{-\frac{1}{3}}{s + 3}$$

$$F(s) = (e^{-0.1s} + 2)G(s)$$

$$f(t) = g(t - 0.1)u(t - 0.1) + 2g(t)u(t)$$

then

$$f(k) = g(k - 1)u(k - 1) + 2g(k)u(k)$$

$$F(z) = (z^{-1} + 2)G(z)$$

$$= (z^{-1} + 2)\left(\frac{1}{3}\frac{z}{z - 1} - \frac{1}{3}\frac{z}{z - e^{-0.3}}\right)$$

$$= \frac{0.173z^2 + 0.086z}{z(z - 1)(z - 0.74)}$$

Finding the z-transform of the corresponding sequence thus involves separating the time delay operations from the rational part of the Laplace transform, then substituting z^{-1} for each unit of time delay in the delay operation and substituting from the entries of Table 8.5 for each term in the partial fraction expansion of the remainder. Another example is the following, for which the sampling period is $T = 0.05$:

$$F(s) = \frac{10}{s^2 + 4} + \frac{6e^{-0.2s}}{s^2 + 3s}$$

$$= \frac{10}{4}\left(\frac{4}{s^2 + 4}\right) + e^{-0.2s}\left(\frac{2}{s} + \frac{-2}{s + 3}\right)$$

$$F(z) = \frac{10}{4}\left(\frac{z \sin 0.1}{z^2 - 2z \cos 0.1 + 1}\right) + z^{-4}\left(\frac{2z}{z - 1} - \frac{2z}{z - e^{-0.15}}\right)$$

$$= \frac{10}{4}\left(\frac{0.0998z}{z^2 - 1.99z + 1}\right) + \frac{1}{z^4}\left(\frac{2z}{z - 1} - \frac{2z}{z - 0.86}\right)$$

DRILL PROBLEM

D8.8. For each of the analog signals with the given Laplace transform $F(s)$, find the z-transform $F(z)$ of the corresponding sample sequence with the given sampling interval T:

(a) $F(s) = \dfrac{-4s + 1}{s^2 + 7s + 12}$ $\qquad T = 0.2$

$\qquad$ **Ans** $(-4z^2 + 3.48)/(z - 0.5488)(z - 0.45)$

(b) $F(s) = \dfrac{100}{s^2 + 2s + 10}$ $\qquad T = 0.5$

$\qquad$ **Ans** $20.2z/(z^2 - 0.086z + 0.368)$

(c) $F(s) = \dfrac{e^{-0.1s} - 3e^{-0.3s} + 2}{s^2}\left(\dfrac{10}{s+1}\right)$ $\qquad T = 0.1$

$\qquad$ **Ans** $\dfrac{(2z^3 + z^2 - 3)(0.05z + 0.0045)}{z^2(z-1)^2(z - 0.905)}$

(d) $F(s) = \dfrac{se^{-s} + 4}{s^2 + 9}$ $\qquad T = 0.2$

$\qquad$ **Ans** $(z - 0.825)/z^4(z^2 - 1.65z + 1) + 0.753z/(z^2 - 1.65z + 1)$

8.6 RECONSTRUCTION OF SIGNALS FROM SAMPLES

Representing Sampled Signals with Impulses

Description of A/D conversion involves discrete-time representation of continuous-time signals, the process termed *sampling*. Describing D/A conversion requires, conversely, continuous-time representation of discrete-time signals. This process is termed *reconstruction*.

Although very often the analog signal reconstructed from digital samples is a sampled-and-held waveform, that is not always the case. A more fundamental continuous-time signal related to a sequence of samples is a train of impulses, timed periodically at intervals T, with strengths equal to the corresponding samples. Figure 8.13 shows an original analog signal $f(t)$, the corresponding sample sequence $f(k)$, and the impulse train $f^*(t)$ that is a useful representation of the samples in the analog domain. The objective of reconstruction is to recover from the samples $f(k)$ the analog signal $f(t)$ or a sufficiently close approximation to it. Of course, the signal $f(t)$ may not actually exist anywhere in the system. Typically the samples $f(k)$ are computed from combinations of samples of other signals,

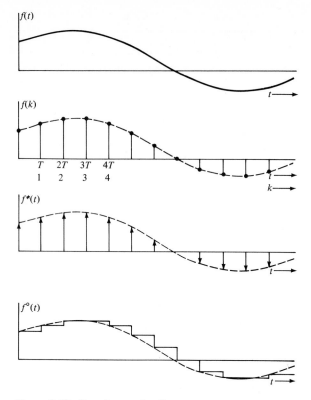

Figure 8.13 Signal reconstruction.

present and delayed. From the samples $f(k)$, it is desired to generate an analog signal which *could have been* sampled to obtain $f(k)$.

The sampled-and-held waveform $f^0(t)$ in Fig. 8.14 is a reconstructed signal with samples $f(k)$. It may be derived from the impulse train by passing the impulses through an appropriate transmittance, termed a *zero-order hold*. The impulse train is related to the samples by

$$f^*(t) = f(0)\delta(t) + f(1)\delta(t - T) + f(2)\delta(t - 2T) + \cdots$$

$$= \sum_{k=0}^{\infty} f(k)\delta(t - kT)$$

To obtain the sample-and-hold waveform from $f^*(t)$ requires a linear, time-invariant analog system with the impulse response given in Fig. 8.14a. A unit impulse input to this transmittance causes a unit rectangular pulse output of duration T as shown. A delayed impulse with different amplitude produces a delayed pulse with that amplitude, as in Fig. 8.14b, and an impulse train, Fig. 8.14c, results in the desired sampled-and-held reconstruction.

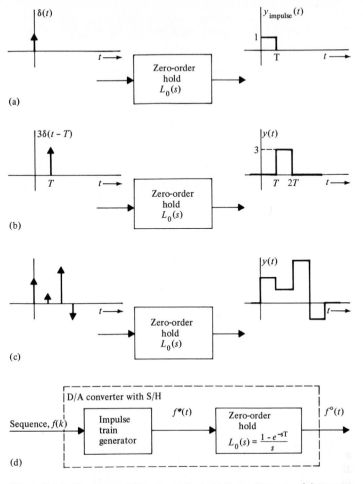

Figure 8.14 Response of the zero-order hold transmittance. **(a)** Impulse response. **(b)** Response to a scaled and delayed impulse. **(c)** Response to an impulse train. **(d)** Model of a D/A converter with S/H output.

The required impulse response of the zero order hold is

$$y_{\text{impulse}}(t) = u(t) - u(t - T)$$

and its Laplace transform is

$$Y_{\text{impulse}}(s) = \frac{1 - e^{-sT}}{s}$$

Since the Laplace transform of the unit impulse is unity, the zero-order hold transmittance, which is the ratio of the transforms, is

$$L_0(s) = Y_{\text{impulse}}(s) = \frac{1 - e^{-sT}}{s}$$

A D/A converter with sample-and-hold output may thus be modeled by the idealized impulse train generator followed by $L_0(s)$, as in Fig. 8.14d.

Relation Between the z-Transform and the Laplace Transform

The impulse train associated with a sequence of samples $f(k)$ is

$$f^*(t) = f(0)\delta(t) + f(1)\delta(t - T) + f(2)\delta(t - 2T) + f(3)\delta(t - 3T) + \cdots$$

$$= \sum_{k=0}^{\infty} f(k)\delta(t - kT)$$

The Laplace transform of the impulse train is

$$F^*(s) = \mathscr{L}[f^*(t)]$$

$$= f(0) + f(1)e^{-sT} + f(2)e^{-2sT} + f(3)e^{-3sT} + \cdots$$

$$= \sum_{k=0}^{\infty} f(k)(e^{sT})^{-k}$$

Letting

$$z = e^{sT}$$

there results

$$F^*(s)\bigg|_{e^{sT} = z} = \sum_{k=0}^{\infty} f(k)z^{-k} = Z[f(k)] = F(z)$$

One interpretation of the z-transformation is that it is the Laplace transform of the impulse train with e^{sT} replaced by z.

A sequence $f(k)$ with a rational z-transform $F(z)$ has a corresponding impulse train $f^*(t)$ with Laplace transform that may be obtained simply by substitution:

$$F^*(s) = F(z)|_{z = e^{sT}}$$

For example, the sequence with z-transform

$$F(z) = \frac{-4z^3 + 5z^2 - 6z}{z^3 + 2z^2 - z + 3}$$

has the following related impulse train transform when the sampling period $T = 0.1$:

$$F^*(s) = \frac{-4e^{0.3s} + 5e^{0.2s} - 6e^{0.1s}}{e^{0.3s} + 2e^{0.2s} - e^{0.1s} + 3}$$

The Sampling Theorem

In applications such as communications, it is especially important to establish conditions for which a signal $g(t)$ is completely specified by (and thus recoverable from) its samples. Communication signals are typically band-limited, or nearly so, meaning that they contain no frequencies higher than a certain band limit frequency f_B. The frequency content of a signal $g(t)$ is given by its Fourier transform

$$G(\omega) = \int_{-\infty}^{\infty} g(t)e^{-j\omega t}\, dt$$

a calculation similar to the Laplace transformation with $s = j\omega$ but extending over all time, not just from $t = 0$ and thereafter. A signal band-limited beyond frequency f_B is one for which

$$G(\omega) = 0 \qquad |\omega| > 2\pi f_B$$

A statement of the sampling theorem is as follows:

A signal $g(t)$ that is band-limited above (hertz) frequency f_B can be recovered from an infinite sequence of its periodic samples $g(k)$ if and only if the sampling interval T is less than $1/2f_B$.

That is, a band-limited signal must be sampled at a rate over twice that of its highest component frequency in order for the samples to be unique. The rate $2f_B$, where f_B is the highest frequency in a band-limited signal, is termed the *Nyquist rate* for that signal.

For a single sinusoidal signal of radian frequency b,

$$g(t) = A \cos(bt + \theta)$$

the sample sequence is, in terms of T,

$$g(k) = A \cos(kbT + \theta)$$

If the sampling interval is less than $1/2f_B$,

$$T < \frac{1}{2f_B} = \frac{2\pi}{2b}$$

$$bT < \pi$$

then the samples are unique, there being at least two per cycle of $g(t)$. If this condition is not met, then any higher-frequency sinusoid

$$h(t) = A \cos(b't + \theta)$$

for which

$$b'T = bT + n2\pi \qquad n = 1, 2, 3, \ldots$$

could be present, as it produces precisely the same sample sequence:

$$h(k) = A \cos[k(bT + n2\pi) + \theta]$$
$$= A \cos(kbT + kn2\pi + \theta)$$
$$= A \cos(kbT + \theta) = g(k)$$

The effects of any of these higher frequencies, being indistinguishable from those below the presumed band limit, are termed *aliasing distortion*.

A constructive statement of how to recover a band-limited signal from its samples is as follows:

To recover a suitably band-limited signal g(t) from its samples g(k), form the impulse train g*(t) and pass it through a low-pass filter that passes, unchanged, all frequencies in g(t) below its band limit frequency f_B and removes all frequencies above 1/2T.

This arrangement is shown in Fig. 8.15a. The required low-pass filter has the frequency response shown in Fig. 8.15b if the analog signal is to be reconstructed without delay. In practice, a phase shift proportional to frequency, representing a time delay in the reconstruction, is approximated. The frequency response of the zero-order hold is given in Fig. 8.15c and is seen to approximate a low-pass characteristic with time delay.

The sampling theorem does not apply directly to most control system design problems for the following reasons:

1. Many of the signals used in control system analysis, such as those involving step changes in amplitude and slope, are not band-limited.
2. For a band-limited signal, perfect reconstruction requires an infinite number of samples. Another way of stating this fact is to note that the low-pass filter required for reconstruction is a physical impossibility. It can be approximated only if a delay is introduced into the signal processing. Better approximation requires longer delays.
3. In control, good reconstruction of signals from their samples is only occasionally of primary interest compared to such concerns as stability, relative stability, and steady state errors.

Nevertheless, the sampling theorem is useful to digital control because it shows important properties of the analog-digital interface such as the following:

1. Samples of a signal uniquely determine that signal only under special circumstances, the sampling theorem stating one such situation. In particular, large transients and high-frequency oscillations in an analog signal such as a system output may not be evident from relatively widely spaced samples of that signal.
2. When A/D conversion is done on a signal, say from a sensor, containing significant frequency components above half the sampling rate, the high frequencies produce errors equivalent to the presence of lower-frequency sinusoidal components. It is thus possible for a poorly designed digital feedback system to attempt to correct presumed low-frequency errors when in fact high-frequency sensor noise is the culprit. For this reason, low-pass filters, termed *antialiasing filters*, are commonly placed before the A/D converters to greatly reduce high-frequency sensor noise in many applications.

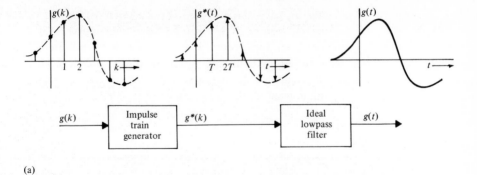

(a)

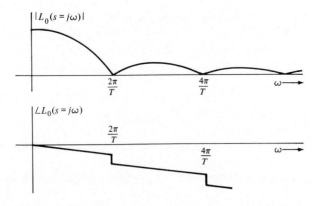

(b)

(c)

Figure 8.15 Reconstruction from impulses. **(a)** Reconstruction of a band-limited signal. **(b)** Ideal low-pass filter frequency response. **(c)** Frequency response of the zero-order hold.

3. To improve the smoothing of reconstructed signals, the equivalent impulse train-to-output transmittance should have frequency response that better approximates an appropriate low-pass filter with time delay. In practice, this is achieved with analog low-pass filters and/or higher-order hold circuits. The higher-order holds produce outputs based upon more than the single sample used by the zero-order hold.

DRILL PROBLEMS

D8.9. For the following sequences $f(k)$, find the corresponding impulse train Laplace transforms $F^*(s)$ for a sampling interval T:

(a) $f(k) = [1 - 3(-1)^k + 4(\frac{1}{2})^k]u(k)$ $T = 1$

   Ans $(2e^{3s} + 5e^{2s} - 3e^s)/[e^{2s} - 1][e^s - \frac{1}{2}]$

(b) $f(k) = \left[(-1)^k - \sin\dfrac{\pi k}{2}\right]u(k)$ $T = 0.2$

   Ans $(e^{0.6s} - e^{0.4s})/(e^{0.2s} + 1)(e^{0.4s} + 1)$

(c) $f(k) = (\frac{1}{2})^k(10 \sin 4k - 8 \cos 4k)u(k)$ $T = 0.1$

   Ans $(-6.38e^{0.1s} - 8e^{0.2s})/(e^{0.2s} + 0.65e^{0.1s} + 0.25)$

D8.10. For the continuous-time function

$$f(t) = 10 + 3 \cos \pi t - 7 \sin 6t$$

determine which of the following functions have the same sample sequence as $f(k)$ for a sampling interval of $T = 0.2$:

(a) $g_1(t) = 10 \cos 10\pi t + 3 \cos 11\pi t - 7 \sin 6t$

(b) $g_2(t) = 10 + \sin 5\pi t + 3 \cos \pi t - 7 \sin[(6 + 10\pi)t]$

(c) $g_3(t) = 10 \cos 20\pi t - 3 \cos 6\pi t + 7 \sin[(6 + 5\pi)t]$

(d) $g_4(t) = 5 + 6 \cos 10\pi t - 2 \cos 20\pi t + \cos 30\pi t + 6 \cos 11\pi t$
$\quad - 3 \cos \pi t - 7 \sin 6t$

(e) $g_5(t) = 10\sqrt{2} \sin \dfrac{170\pi t}{8} + 3 \cos 51 \pi t - 8 \sin 40\pi t - 7 \sin[(6 + 30\pi)t]$

Then find five other functions which, when sampled at this rate, have the same sample sequence.

8.7 DISCRETE-TIME SYSTEMS

Computer processing of input signal samples to produce output signal samples may be described by difference equations, analogous to the differential equations that characterize continuous-time systems. In this introductory treatment, only linear, step-invariant (or constant-coefficient) difference equations are considered.

Difference Equations and Response

Discrete-time systems are described by difference equations, of the form

$$y(k + n) + a_{n-1}y(k + n - 1) + a_{n-2}y(k + n - 2) + \cdots + a_1 y(k + 1) + a_0 y(k)$$
$$= b_m r(k + m) + b_{m-1}r(k + m - 1) + \cdots + b_1 r(k + 1) + b_0 r(k)$$

where $y(k)$ is the output sequence and $r(k)$ is the input sequence. Solving this nth-order difference equation for $y(k + n)$ gives

$$y(k + n) = -a_{n-1}y(k + n - 1) - \cdots - a_1 y(k + 1)$$
$$- a_0 y(k) + b_m r(k + m) + \cdots + b_0 r(k)$$

In other words the $(k + n)$th sample of the output is a linear combination of the previous output samples through $y(k)$ and of the input samples from step $(k + m)$ through step k.

For example,

$$y(k + 2) = 3y(k + 1) - 2y(k) + 2r(k + 1) - r(k)$$

is a discrete-time system described by a second-order difference equation. Given the input sequence $r(k)$ and two initial values of the sequence $y(k)$, the entire output sequence can be calculated recursively. If $r(k) = u(k)$, the unit step sequence, and

$$y(0) = 1$$
$$y(1) = 4$$

then

$$y(2) = 3y(1) - 2y(0) + 2u(1) - u(0)$$
$$= 12 - 2 + 2 - 1 = 11$$
$$y(3) = 3y(2) - 2y(1) + 2u(2) - u(1)$$
$$= 33 - 8 + 2 - 1 = 26$$
$$y(4) = 3y(3) - 2y(2) + 2u(3) - u(2)$$
$$= 78 - 22 + 2 - 1 = 57$$
$$\vdots$$

and so forth.

A closed-form expression for the response of a discrete-time system may be obtained by z-transform methods. For the discrete-time system

$$y(k + 1) = -0.5y(k) + 3r(k)$$

for example, with

$$y(0) = 4$$

$$r(k) = u(k)$$

the unit step sequence, z-transforming using the sequence shift relation of Table 8.6 gives

$$zY(z) - y(0) = -0.5Y(z) + 3R(z)$$

$$(z + 0.5)Y(z) = 4 + \frac{3z}{z - 1}$$

$$Y(z) = \frac{7z - 4}{(z + 0.5)(z - 1)}$$

$$\frac{Y(z)}{z} = \frac{7z - 4}{z(z + 0.5)(z - 1)} = \frac{8}{z} + \frac{-10}{z + 0.5} + \frac{2}{z - 1}$$

$$Y(z) = 8 + \frac{-10z}{z + 0.5} + \frac{2z}{z - 1}$$

$$y(k) = 8\delta(k) - 10(-0.5)^k + 2u(k) \qquad k = 0, 1, 2, \ldots$$

z-Transfer Functions

The z-transfer function of a discrete-time system is the ratio of the z-transform of the output to the z-transform of the input when all initial conditions are zero:

$$D(z) = \frac{Y(z)}{R(z)}\bigg|_{\substack{\text{initial} \\ \text{conditions}=0}}$$

For the discrete-time system

$$y(k + 3) = -0.3y(k + 2) + y(k + 1) - 0.5y(k)$$
$$+ 4r(k + 3) - r(k + 1) - 0.6r(k)$$

z-transforming with zero initial conditions gives

$$(z^3 + 0.3z^2 - z + 0.5)Y(z) = (4z^3 - z - 0.6)R(z)$$

$$D(z) = \frac{Y(z)}{R(z)}\bigg|_{\substack{\text{initial} \\ \text{conditions}=0}} = \frac{4z^3 - z - 0.6}{z^3 + 0.3z^2 - z + 0.5}$$

A discrete-time system with z-transfer function

$$D(z) = \frac{3z + 1}{z^2 - z + 2}$$

is described by the difference equation

$$y(k + 2) - y(k + 1) + 2y(k) = 3r(k + 1) + r(k)$$

If the initial conditions are zero and

$$R(z) = \frac{5}{z+4}$$

the z-transform of the output is

$$Y(z) = D(z)R(z) = \left(\frac{3z+1}{z^2-z+2}\right)\left(\frac{5}{z+4}\right)$$

Nonzero initial conditions and multiple inputs and outputs may be accommodated in a manner analogous to the transfer function development for continuous-time systems given in Chap. 1.

Block Diagrams and Signal Flow Graphs

The block diagram and signal flow graph manipulations used for continuous-time system components also apply to discrete-time systems. For example, Fig. 8.16a

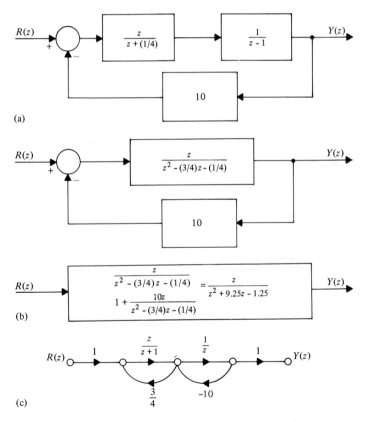

Figure 8.16 Discrete-time systems represented by block diagrams and signal flow graphs. **(a)** Block diagram of a discrete-time system. **(b)** Reduction of the block diagram. **(c)** Signal flow graph for a discrete-time system.

shows the description of a discrete-time system by a block diagram. Reduction of the block diagram to determine the overall system z-transfer function is shown in Fig. 8.16b. A discrete-time system signal flow graph is given in Fig. 8.16c. Using Mason's gain rule, the system z-transfer function is

$$D(z) = \cfrac{\cfrac{1}{z+1}}{1 - \cfrac{\frac{3}{4}z}{z+1} + \cfrac{10}{z}} = \frac{4z}{z^2 + 44z + 40}$$

Stability and the Bilinear Transformation

Table 8.7 shows the characters of sequences corresponding to various complex plane locations of the denominator roots of a z-transformed sequence $F(z)$. Denominator roots within the unit circle on the complex plane give rise to sequences which decay with k, while roots outside the unit circle represent response terms that grow in magnitude with k.

A discrete-time system is said to be stable if and only if its unit pulse response decays with k. If a system with z-transfer function $D(z)$ has a unit pulse input

$$r(k) = \delta(k) \qquad R(z) = 1$$

the system output has z-transform equal to the z-transfer function

$$Y(z) = D(z) \cdot 1 = D(z)$$

Hence the stability of a discrete-time system hinges upon whether all of the poles of its z-transfer function are within the unit circle on the complex plane.

Stability testing for a discrete-time system involves determining whether or not all of the poles of the system's z-transfer function are within the unit circle on the complex plane. One stability-testing method which avoids factoring the denominator polynomial of $D(z)$ involves a change of variables from z to W for which the region within the unit circle on the complex plane is mapped to the left half of the complex plane. Then, Routh-Hurwitz testing may be applied to determine stability. The change of variables involved is known as the bilinear transformation:

$$z = \frac{1 + W}{1 - W}$$

For example, when the bilinear change of variables is made on the z-transfer function

$$D(z) = \frac{8z^3 - 3z^2 + z}{z^3 + 0.4z^2 - 0.25z - 0.1}$$

Table 8.7 SEQUENCES CORRESPONDING TO VARIOUS z-TRANSFORM DENOMINATOR POLYNOMIAL ROOT LOCATIONS

Root location(s) on the complex plane	Corresponding sequence after $k = 0$
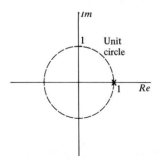	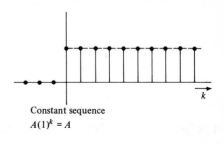 Constant sequence $A(1)^k = A$
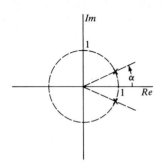	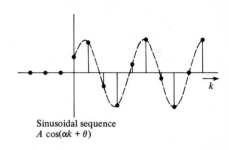 Sinusoidal sequence $A \cos(\alpha k + \theta)$
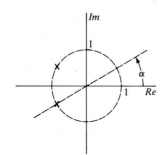	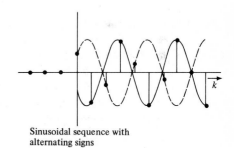 Sinusoidal sequence with alternating signs $A(-1)^k \cos(\alpha k + \theta)$
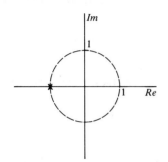	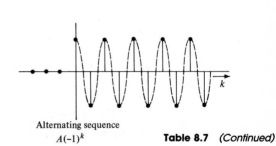 Alternating sequence $A(-1)^k$

Table 8.7 *(Continued)*

Table 8.7   *(Continued)*

Root location(s) on the complex plane	Corresponding sequence after $k = 0$

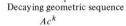

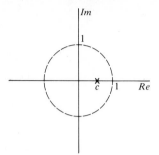
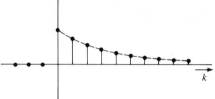

Decaying geometric sequence
$$Ac^k$$

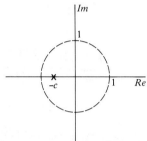
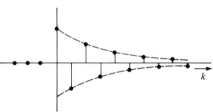

Decaying geometric sequence
with alternating signs
$$A(-c)^k$$

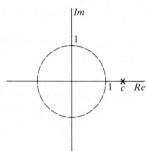
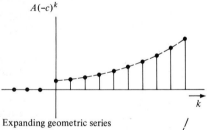

Expanding geometric series
$$Ac^k$$

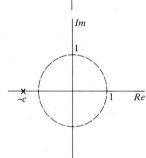
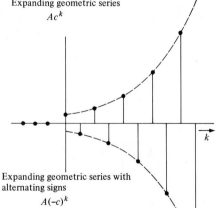

Expanding geometric series with
alternating signs
$$A(-c)^k$$

Table 8.7 *(Continued)*

Root location(s) on the complex plane	Corresponding sequence after $k = 0$

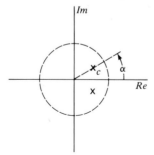

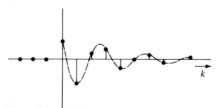

Damped sinusoidal sequence
$$Ac^k \cos(\alpha k + \theta)$$

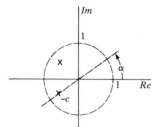

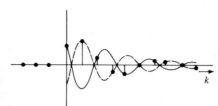

Damped sinusoidal sequence with
alternating signs
$$A(-c)^k \cos(\alpha k + \theta)$$

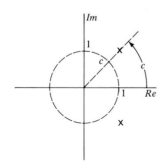

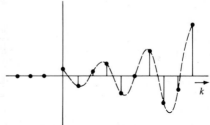

Exponentially expanding sinusoidal
sequence
$$Ac^k \cos(\alpha k + \theta)$$

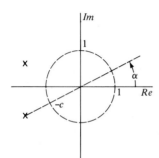

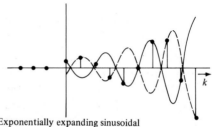

Exponentially expanding sinusoidal
sequence with alternating signs
$$A(-c)^k \cos(\alpha k + \theta)$$

there results

$$D(W) = \frac{8\left(\dfrac{1+W}{1-W}\right)^3 - 3\left(\dfrac{1+W}{1-W}\right)^2 + \left(\dfrac{1+W}{1-W}\right)}{\left(\dfrac{1+W}{1-W}\right)^3 + 0.4\left(\dfrac{1+W}{1-W}\right)^2 - 0.25\left(\dfrac{1+W}{1-W}\right) - 0.1}$$

$$= \frac{\dfrac{12W^3 + 26W^2 + 20W + 6}{(1-W)^3}}{\dfrac{0.45W^3 + 2.55W^2 + 3.95W + 1.05}{(1-W)^3}}$$

$$= \frac{12W^3 + 28W^2 + 22W + 6}{0.45W^3 + 2.55W^2 + 3.95W + 1.05}$$

Poles and zeros of $D(z)$ within the unit circle are mapped to the LHP in $D(W)$; roots of $D(z)$ outside the unit circle are mapped to the RHP in terms of $D(W)$; and roots of $D(z)$ located precisely on the unit circle are mapped to the imaginary axis in $D(W)$.

A Routh-Hurwitz test of the poles of $D(W)$ is as follows:

W^3	0.45	3.95
W^2	2.55	1.05
W^1	3.76	
W^0	1.05	

There are no left-column sign changes in the array, so all poles of $D(W)$ are in the LHP. All poles of $D(z)$ are then within the unit circle on the complex plane. The system represented by $D(z)$ is thus stable.

The bilinear transformation, because it converts a digital problem to a related analog one, is also very useful for applying root locus and frequency response methods to digital systems.

Computer Software

Programming a digital computer with A/D and D/A capability as discrete-time system is straightforward. For example, a system with z-transfer function

$$D(z) = \frac{2z^2 + 5}{z^2 + 3z + 2}$$

is described by the difference equation

$$y(k+2) = -3y(k+1) - 2y(k) + 2r(k+2) + 5r(k)$$

A Fortran program for this system is outlined in the flow diagram of Fig. 8.17 and listed in Table 8.8. The variables Y2, Y1, and Y0 are used for $y(k+2)$, $y(k+1)$, and $y(k)$ respectively, while R2, R1, and R0 represent $r(k+2)$, $r(k+1)$, and $r(k)$. The

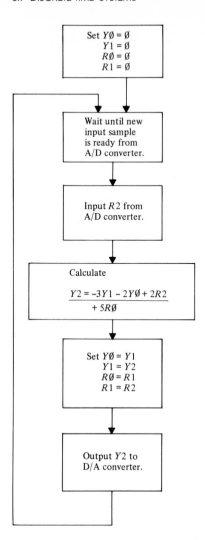

Figure 8.17 Flow diagram for a Fortran program to realize a discrete-time system.

initial conditions $Y0$, $Y1$, $R0$, and $R1$ are first set to zero. Then an input value $R2$ is read from the A/D converter. $Y2$ is calculated and the values of $Y0$, $Y1$, $R0$, and $R1$ are updated for the next calculation cycle. $Y2$ is then output to the D/A converter. Assuming that there is sufficient time between samples to perform the calculations, the program waits for a new input sample, computes the next $Y2$ sample, and so forth.

There are, of course, many other functions the computer could perform. It could limit the output signal, check that the input signal samples are "reasonable" (that is, they are within some predetermined bounds), and trigger alarms in the event that a malfunction is detected.

**Table 8.8   FORTRAN PROGRAM TO
REALIZE A DISCRETE-TIME SYSTEM**

```
100 FORMAT (F16.8)
110 Y0 = 0.
120 Y1 = 0.
130 R1 = 0.
140 R0 = 0.
ᵃ150 READ (1, 100)R2
160 Y2 = 3.*Y1 − 2.*Y0 + 2.*R2 + 5.*R0
170 Y0 = Y1
180 Y1 = Y2
190 R0 = R1
200 R1 = R2
ᵇ210 WRITE (2,100)Y2
220 GO TO 150
230 STOP
240 END
```

ᵃ The A/D converter is taken to be device number 1 with F16.8 format. It is here assumed that the processor waits at step 150 at each looping until a new sample is ready, just as it would wait for a character to be input from a keyboard device.
ᵇ The D/A converter is taken to be device number 2 with F16.8 format. It is here assumed that the D/A device contains a buffer that stores each new output sample for conversion at the sample time.

DRILL PROBLEMS

D8.11. Find the z-transfer functions of the following discrete-time systems:

(a) $y(k + 3) + 3y(k + 2) − 2y(k + 1) + y(k)$
$= r(k + 2) + r(k + 1) − 4r(k)$

Ans $D(z) = \dfrac{z^2 + z − 4}{z^3 + 3z^2 − 2z + 1}$

(b) $y(k + 4) = 0.2r(k + 4) − 0.3r(k + 3) + 0.1r(k + 2) + 0.7r(k + 1)$
$− 0.5r(k)$

Ans $D(z) = \dfrac{0.2z^4 − 0.3z^3 + 0.1z^2 + 0.7z − 0.5}{z^4}$

(c) $y(k + 3) = 0.5y(k + 2) − y(k + 1) + 0.125y(k) + 10r(k + 3)$

Ans $D(z) = \dfrac{10z^3}{z^3 − 0.5z^2 + z − 0.125}$

D8.12. For

$$y(k + 2) - 5y(k + 1) - 6y(k) = 2r(k)$$

recursively find $y(2)$, $y(3)$, $y(4)$, and $y(5)$ if

$$r(k) = (-1)^k$$

$$y(0) = -4$$

$$y(1) = \quad 7$$

Ans 13, 105, 605, 3653

D8.13. For discrete-time systems with the following z-transfer functions and input sequences, find the output sequences for $k = 0$ and thereafter if the initial conditions are zero:

(a) $D(z) = \dfrac{4}{z + 3}$

$r(k) = 5\delta(k)$

Ans $\frac{20}{3}\delta(k) - \frac{20}{3}(-3)^k$

(b) $D(z) = \dfrac{z}{z - \frac{1}{10}}$

$r(k) = u(k)$

Ans $-\frac{1}{9}(\frac{1}{10})^k + \frac{10}{9}$

(c) $D(z) = \dfrac{-8}{z + \frac{1}{3}}$

$r(k) = (\frac{1}{4})^k$

Ans $-\frac{96}{7}(\frac{1}{4})^k + \frac{96}{7}(-\frac{1}{3})^k$

(d) $D(z) = \dfrac{z}{2z - 1}$

$r(k) = \delta(k - 3)$

Ans $\frac{1}{2}(\frac{1}{2})^{k-3}u(k - 3)$

D8.14. Determine whether or not each of the following discrete-time systems is stable:

(a) $D(z) = \dfrac{-3z^2 + 1}{4z^2 + 2z - 1}$

Ans denominator roots 0.31 and -0.81; stable

(b) $D(z) = \dfrac{z^3}{z^3 + 0.3z^2 - 0.25z - 0.075}$

Ans $D(W) = \dfrac{W^3 + 3W^2 + 3W + 1}{0.525W^3 + 2.725W^2 + 3.775W + 0.975}$; stable

(c) $D(z) = \dfrac{5(z - 0.2)}{z^3 - 2.8z^2 + 1.75z - 0.3}$

Ans $D(W) = \dfrac{1.026(W + 0.667)(1 - W)^2}{W^3 + 0.538W^2 - 0.111W - 0.060}$; unstable

8.8 STATE VARIABLE DESCRIPTIONS OF DISCRETE-TIME SYSTEMS

Simulation Diagrams and Equations

Simulation diagrams for discrete-time systems involve as a basic element blocks or branches having transmittance $1/z$. A simulation diagram, in phase-variable canonical form, for a system with z-transfer function

$$D(z) = \frac{3z^2 - 2z + 8}{z^3 + 0.5z^2 - 0.25z + 0.75}$$

$$= \frac{3/z - 2/z^2 + 8/z^3}{1 + 0.5/z - 0.25/z^2 + 0.75/z^3}$$

is given in Fig. 8.18a. In terms of the indicated state variables, the z-transformed equations describing the system are as follows:

$$zX_1(z) = X_2(z)$$

$$zX_2(z) = X_3(z)$$

$$zX_3(z) = -0.75X_1(z) + 0.25X_2(z) - 0.5X_3(z) + R(z)$$

$$Y(z) = 8X_1(z) - 2X_2(z) + 3X_3(z)$$

In terms of the step, k, these equations are

$$x_1(k + 1) = x_2(k)$$

$$x_2(k + 1) = x_3(k)$$

$$x_3(k + 1) = -0.75x_1(k) + 0.25x_2(k) - 0.5x_3(k) + r(k)$$

$$y(k) = 8x_1(k) - 2x_2(k) + 3x_3(k)$$

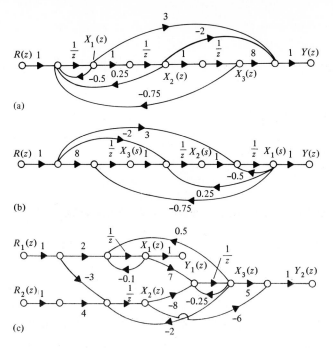

Figure 8.18 Simulation diagrams for discrete-time systems. **(a)** A system in phase-variable form. **(b)** A system in dual phase-variable form. **(c)** A multiple-input, multiple-output system.

or

$$\begin{bmatrix} x_1(k+1) \\ x_2(k+1) \\ x_3(k+1) \end{bmatrix} = \begin{bmatrix} 0 & 1 & 0 \\ 0 & 0 & 1 \\ -0.75 & 0.25 & -0.5 \end{bmatrix} \begin{bmatrix} x_1(k) \\ x_2(k) \\ x_3(k) \end{bmatrix} + \begin{bmatrix} 0 \\ 0 \\ 1 \end{bmatrix} r(k)$$

$$y(k) = \begin{bmatrix} 8 & -2 & 3 \end{bmatrix} \begin{bmatrix} x_1(k) \\ x_2(k) \\ x_3(k) \end{bmatrix}$$

A simulation diagram for a system with the same transfer function but using the dual phase-variable form is given in Fig. 8.18b. For this system,

$$X_1(z) = \frac{1}{z}[-0.5X_1(z) + X_2(z) + 3R(z)]$$

$$X_2(z) = \frac{1}{z}[0.25X_1(z) + X_3(z) - 2R(z)]$$

$$X_3(z) = \frac{1}{z}[-0.75X_1(z) + 8R(z)]$$

$$Y(z) = X_1(z)$$

or

$$\begin{bmatrix} x_1(k+1) \\ x_2(k+1) \\ x_3(k+1) \end{bmatrix} = \begin{bmatrix} -0.5 & 1 & 0 \\ 0.25 & 0 & 1 \\ -0.75 & 0 & 0 \end{bmatrix} \begin{bmatrix} x_1(k) \\ x_2(k) \\ x_3(k) \end{bmatrix} + \begin{bmatrix} 3 \\ -2 \\ 8 \end{bmatrix} r(k)$$

$$y(k) = \begin{bmatrix} 1 & 0 & 0 \end{bmatrix} \begin{bmatrix} x_1(k) \\ x_2(k) \\ x_3(k) \end{bmatrix}$$

A simulation diagram for a multiple-input, multiple-output system is given in Fig. 8.18c. It consists of constant transmittances and unit delay transmittances $1/z$. The output of each delay is a state variable, and each delay input consists of a linear combination of the inputs and the state variables. For the example system,

$$\begin{cases} zX_1(z) = -0.1X_1(z) + 0.5X_3(z) + 2R_1(z) \\ zX_2(z) = -2X_3(z) - 3R_1(z) + 4R_2(z) \\ zX_3(z) = 7X_1(z) - 8X_2(z) - 0.25X_3(z) \\ Y_1(z) = X_1(z) \\ Y_2(z) = -6X_2(z) + 5X_3(z) \end{cases}$$

$$\begin{bmatrix} x_1(k+1) \\ x_2(k+1) \\ x_3(k+1) \end{bmatrix} = \begin{bmatrix} -0.1 & 0 & 0.5 \\ 0 & 0 & -2 \\ 7 & -8 & -0.25 \end{bmatrix} \begin{bmatrix} x_1(k) \\ x_2(k) \\ x_3(k) \end{bmatrix} + \begin{bmatrix} 2 & 0 \\ -3 & 4 \\ 0 & 0 \end{bmatrix} \begin{bmatrix} r_1(k) \\ r_2(k) \end{bmatrix}$$

$$\begin{bmatrix} y_1(k) \\ y_2(k) \end{bmatrix} = \begin{bmatrix} 1 & 0 & 0 \\ 0 & -6 & 5 \end{bmatrix} \begin{bmatrix} x_1(k) \\ x_2(k) \\ x_3(k) \end{bmatrix}$$

In general, the state equations for a discrete-time system are of the form

$$\mathbf{x}(k+1) = \mathbf{F}x(k) + \mathbf{G}r(k)$$
$$\mathbf{y}(k) = \mathbf{H}\mathbf{x}(k) \tag{8.1}$$

When written out, these are

$$\begin{bmatrix} x_1(k+1) \\ x_2(k+1) \\ x_3(k+1) \\ \vdots \\ x_n(k+1) \end{bmatrix} = \begin{bmatrix} f_{11} & f_{12} & f_{13} & \cdots & f_{1n} \\ f_{21} & f_{22} & f_{23} & & f_{2n} \\ f_{31} & f_{32} & f_{33} & & f_{3n} \\ \vdots & & & & \vdots \\ f_{n1} & f_{n2} & f_{n3} & \cdots & f_{nn} \end{bmatrix} \begin{bmatrix} x_1(k) \\ x_2(k) \\ x_3(k) \\ \vdots \\ x_n(k) \end{bmatrix} + \begin{bmatrix} g_{11} & g_{12} & \cdots & g_{1i} \\ g_{21} & g_{22} & \cdots & g_{2i} \\ g_{31} & g_{32} & \cdots & g_{3i} \\ \vdots & & & \\ g_{n1} & g_{n2} & \cdots & g_{ni} \end{bmatrix} \begin{bmatrix} r_1(k) \\ r_2(k) \\ \vdots \\ r_i(k) \end{bmatrix}$$

$$\begin{bmatrix} y_1(k) \\ y_2(k) \\ \vdots \\ y_m(k) \end{bmatrix} = \begin{bmatrix} h_{11} & h_{12} & \cdots & h_{1n} \\ h_{21} & h_{22} & \cdots & h_{2n} \\ \vdots & & \vdots & \\ h_{m1} & h_{m2} & \cdots & h_{mn} \end{bmatrix} \begin{bmatrix} x_1(k) \\ x_2(k) \\ \vdots \\ x_n(k) \end{bmatrix}$$

Response and Stability

The response of a discrete-time system may be calculated recursively, starting with an initial state and repeatedly using the state equations (8.1). From $\mathbf{x}(0)$ and $\mathbf{r}(0)$, $\mathbf{x}(1)$ may be calculated:

$$\mathbf{x}(1) = \mathbf{F}\mathbf{x}(0) + \mathbf{G}\mathbf{r}(0)$$

Then, using $\mathbf{x}(1)$ and $\mathbf{r}(1)$, $\mathbf{x}(2)$ is calculated:

$$\mathbf{x}(2) = \mathbf{F}\mathbf{x}(1) + \mathbf{G}\mathbf{r}(1)$$

$$= \mathbf{F}^2\mathbf{x}(0) + \mathbf{F}\mathbf{G}\mathbf{r}(0) + \mathbf{G}\mathbf{r}(1)$$

Continuing,

$$\mathbf{x}(3) = \mathbf{F}\mathbf{x}(2) + \mathbf{G}\mathbf{r}(2)$$

$$= \mathbf{F}^3\mathbf{x}(0) + \mathbf{F}^2\mathbf{G}\mathbf{r}(0) + \mathbf{F}\mathbf{G}\mathbf{r}(1) + \mathbf{G}\mathbf{r}(2)$$

$$\vdots$$

$$\mathbf{x}(k) = \mathbf{F}^k\mathbf{x}(0) + \mathbf{F}^{k-1}\mathbf{G}\mathbf{r}(0) + \mathbf{F}^{k-2}\mathbf{G}\mathbf{r}(1)$$

$$+ \cdots + \mathbf{F}\mathbf{G}\mathbf{r}(k-2) + \mathbf{G}\mathbf{r}(k-1)$$

$$= \mathbf{F}^k\mathbf{x}(0) + \sum_{n=1}^{k} \mathbf{F}^{k-n}\mathbf{G}\mathbf{r}(n-1)$$

As a numerical example, the system

$$\begin{bmatrix} x_1(k+1) \\ x_2(k+2) \end{bmatrix} = \begin{bmatrix} 0 & -2 \\ -1 & 3 \end{bmatrix}\begin{bmatrix} x_1(k) \\ x_2(k) \end{bmatrix} + \begin{bmatrix} 2 \\ 1 \end{bmatrix}r(k)$$

$$y(k) = \begin{bmatrix} 3 & -2 \end{bmatrix}\begin{bmatrix} x_1(k) \\ x_2(k) \end{bmatrix} \tag{8.2}$$

with

$$\begin{bmatrix} x_1(0) \\ x_2(0) \end{bmatrix} = \begin{bmatrix} 5 \\ -7 \end{bmatrix}$$

and

$$r(k) = \delta(k)$$

has response as follows:

$$y(0) = [3 \quad -2]\begin{bmatrix} 5 \\ -7 \end{bmatrix} = 29$$

$$\begin{bmatrix} x_1(1) \\ x_2(1) \end{bmatrix} = \begin{bmatrix} 0 & -2 \\ -1 & 3 \end{bmatrix}\begin{bmatrix} 5 \\ -7 \end{bmatrix} + \begin{bmatrix} 2 \\ 1 \end{bmatrix} = \begin{bmatrix} 16 \\ -25 \end{bmatrix}$$

$$y(1) = [3 \quad -2]\begin{bmatrix} 16 \\ -25 \end{bmatrix} = 98$$

$$\begin{bmatrix} x_1(2) \\ x_2(2) \end{bmatrix} = \begin{bmatrix} 0 & -2 \\ -1 & 3 \end{bmatrix}\begin{bmatrix} 16 \\ -25 \end{bmatrix} + \begin{bmatrix} 0 \\ 0 \end{bmatrix} = \begin{bmatrix} 50 \\ -91 \end{bmatrix}$$

$$y(2) = [3 \quad -2]\begin{bmatrix} 50 \\ -91 \end{bmatrix} = 332$$

$$\begin{bmatrix} x_1(3) \\ x_2(3) \end{bmatrix} = \begin{bmatrix} 0 & -2 \\ -1 & 3 \end{bmatrix}\begin{bmatrix} 50 \\ -91 \end{bmatrix} + \begin{bmatrix} 0 \\ 0 \end{bmatrix} = \begin{bmatrix} 182 \\ -323 \end{bmatrix}$$

$$y(3) = [3 \quad -2]\begin{bmatrix} 182 \\ -323 \end{bmatrix} = 1192$$

$$\vdots$$

A discrete-time system's z-transfer function is found by z-transforming the state equations (8.1) with zero initial conditions and solving for the ratio of output to input transforms:

$$\begin{cases} z\mathbf{X}(z) = \mathbf{F}\mathbf{X}(z) + \mathbf{G}\mathbf{R}(z) \\ \mathbf{Y}(z) = \mathbf{H}\mathbf{X}(z) \end{cases}$$

$$(z\mathbf{I} - \mathbf{F})\mathbf{X}(z) = \mathbf{G}\mathbf{R}(z)$$

$$\mathbf{X}(z) = (z\mathbf{I} - \mathbf{F})^{-1}\mathbf{G}\mathbf{R}(z)$$

$$\mathbf{Y}(z) = \mathbf{H}\mathbf{X}(z) = \mathbf{H}(z\mathbf{I} - \mathbf{F})^{-1}\mathbf{G}\mathbf{R}(z)$$

The z-transfer function matrix of the system is thus

$$\mathbf{D}(z) = \mathbf{H}(z\mathbf{I} - \mathbf{F})^{-1}\mathbf{G} = \frac{\mathbf{H} \text{ adj}(z\mathbf{I} - \mathbf{F})\mathbf{G}}{|z\mathbf{I} - \mathbf{F}|}$$

and the system is stable if and only if all of the roots of the characteristic polynomial

$$|z\mathbf{I} - \mathbf{F}| = 0$$

are within the unit circle on the complex plane.

For the single-input, single-output example system (8.2) the transfer function is

$$D(z) = [3 \quad -2] \begin{bmatrix} z & 2 \\ 1 & z-3 \end{bmatrix}^{-1} \begin{bmatrix} 2 \\ 1 \end{bmatrix}$$

$$= \frac{[3 \quad -2] \begin{bmatrix} z-3 & -2 \\ -1 & z \end{bmatrix} \begin{bmatrix} 2 \\ 1 \end{bmatrix}}{z^2 - 3z - 2}$$

$$= \frac{[3 \quad -2] \begin{bmatrix} (2z-8) \\ (z-2) \end{bmatrix}}{z^2 - 3z - 2}$$

$$= \frac{4z - 20}{z^2 - 3z - 2}$$

This system is unstable, since the roots of the characteristic equation are

$$z^2 - 3z - 2 = (z - 1)(z - 2) = 0$$

$$z_1, z_2 = 1, 2$$

which do not lie within the unit circle on the complex plane.

Controllability and Observability

For discrete-time systems with nonrepeated characteristic roots, an appropriate change of state variables

$$\mathbf{x}' = \mathbf{P}x \qquad \mathbf{x} = \mathbf{P}^{-1}\mathbf{x}'$$

determined as for continuous-time systems with the methods of Sec. 7.4, will decouple the state equations:

$$\mathbf{x}'(k + 1) = \mathbf{PFP}^{-1}\mathbf{x}'(k) + \mathbf{PGr}(k) = \mathbf{F}'\mathbf{x}'(k) + \mathbf{G}'\mathbf{r}(k)$$

$$\mathbf{y}(k) = \mathbf{HP}^{-1}\mathbf{x}'(k) = \mathbf{H}'\mathbf{x}'(k)$$

where

$$\mathbf{F}' = \begin{bmatrix} z_1 & 0 & 0 & \cdots & 0 \\ 0 & z_2 & 0 & \cdots & 0 \\ \vdots & & & & \\ 0 & 0 & 0 & \cdots & z_n \end{bmatrix}$$

In terms of the new state variables $\mathbf{x}'$, the state coupling matrix $\mathbf{F}'$ is diagonal, with the system's characteristic roots along the diagonal.

If, when the state equations for a discrete-time system are diagonalized, any row of the new input coupling matrix

$$\mathbf{G}' = \mathbf{PG}$$

is zero, the corresponding discrete-time system mode is uncontrollable. If any column of

$$\mathbf{H'} = \mathbf{HP}^{-1}$$

is zero, the corresponding system mode is unobservable. The same rank tests for complete controllability and complete observability that apply to continuous-time systems apply here because controllability and observability are algebraic properties of the system matrices.

An nth-order discrete-time system (8.1) is completely controllable if and only if its controllability matrix

$$\mathbf{M}_c = [\mathbf{G} \mid \mathbf{FG} \mid \mathbf{F}^2\mathbf{G} \mid \cdots \mid \mathbf{F}^{n-1}\mathbf{G}]$$

is of full rank. The system is completely observable if and only if the observability matrix

$$\mathbf{M}_o = \begin{bmatrix} \mathbf{H} \\ \hline \mathbf{HF} \\ \hline \mathbf{HF}^2 \\ \hline \vdots \\ \hline \mathbf{HF}^{n-1} \end{bmatrix}$$

is of full rank.

DRILL PROBLEMS

D8.15. Draw discrete-time simulation diagrams in phase-variable canonical form for systems with the following z-transfer function:

(a) $D(z) = \dfrac{4z}{z^2 + z + 0.5}$

(b) $D(z) = \dfrac{10z^3 - 4z^2 + 5z}{z^3 + 0.5z^2 - 0.2z + 0.3}$

D8.16. Find state equations in matrix form for the systems with the following discrete-time simulation diagrams.

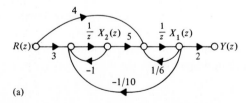

(a)

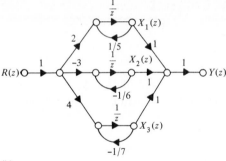

(b)

Ans (a) $\begin{bmatrix} x_1(k+1) \\ x_2(k+1) \end{bmatrix} = \begin{bmatrix} \frac{1}{6} & 5 \\ -\frac{1}{10} & -1 \end{bmatrix} \begin{bmatrix} x_1(k) \\ x_2(k) \end{bmatrix} + \begin{bmatrix} 4 \\ 3 \end{bmatrix} r(k)$

$$y(k) = \begin{bmatrix} 2 & 0 \end{bmatrix} \begin{bmatrix} x_1(k) \\ x_2(k) \end{bmatrix}$$

(b) $\begin{bmatrix} x_1(k+1) \\ x_2(k+1) \\ x_3(k+1) \end{bmatrix} = \begin{bmatrix} \frac{1}{5} & 0 & 0 \\ 0 & -\frac{1}{6} & 0 \\ 0 & 0 & -\frac{1}{7} \end{bmatrix} \begin{bmatrix} x_1(k) \\ x_2(k) \\ x_3(k) \end{bmatrix} + \begin{bmatrix} 2 \\ -3 \\ 4 \end{bmatrix} r(k)$

$$y(k) = \begin{bmatrix} 1 & 1 & 1 \end{bmatrix} \begin{bmatrix} x_1(k) \\ x_2(k) \\ x_3(k) \end{bmatrix}$$

D8.17. For the system

$$\begin{bmatrix} x_1(k+1) \\ x_2(k+1) \end{bmatrix} = \begin{bmatrix} -2 & 1 \\ 1 & -3 \end{bmatrix} \begin{bmatrix} x_1(k) \\ x_2(k) \end{bmatrix} + \begin{bmatrix} 1 \\ 2 \end{bmatrix} r(k)$$

$$y(k) = \begin{bmatrix} -1 & 1 \end{bmatrix} \begin{bmatrix} x_1(k) \\ x_2(k) \end{bmatrix}$$

with

$$\begin{bmatrix} x_1(0) \\ x_2(0) \end{bmatrix} = \begin{bmatrix} 2 \\ 0 \end{bmatrix}$$

and $r(k) = u(k)$, find $y(0)$, $y(1)$, $y(2)$, and $y(3)$.

Ans $-2, 7, -24, 86$

8.9 DIGITIZING CONTROL SYSTEMS

Step-Invariant Approximation

An important technique in the design of digital control systems is to require that the unit step sequence response of a digital transmittance be samples of the continuous-time unit step response of a model analog transmittance. Often in practice, the analog transmittance is a working component of the system and it is desired to replace the analog component with a digital one that performs similarly. In Fig. 8.19a, an analog transmittance $G(s)$ and its unit step response $f_{step}(t)$ are indicated. The step-invariant digital approximation to $G(s)$, Fig. 8.19b, has a unit step response sequence $f_{step}(k)$ that consists of samples of the analog step response $f_{step}(t)$.

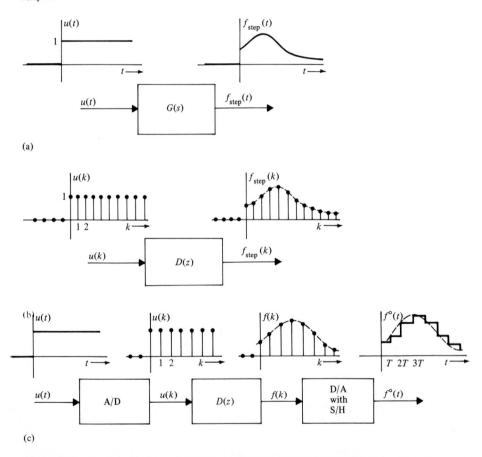

Figure 8.19 Step-invariant approximation. **(a)** Model analog transmittance and unit step response. **(b)** Digital system derived from $G(s)$ using the step-invariant approximation.

The conversion from $F_{step}(s)$ to $F_{step}(z)$, where the sequence consists of samples of the time function, is the sampling conversion of Sec. 8.5. For example, the analog transmittance

$$G(s) = \frac{-s + 2}{s^2 + 3s + 2}$$

has unit step response given by

$$F_{step}(s) = \frac{1}{s} G(s) = \frac{-s + 2}{s(s^2 + 3s + 2)}$$

$$= \frac{1}{s} - 3\left(\frac{1}{s + 1}\right) + 2\left(\frac{1}{s + 2}\right)$$

Samples of this step response at a sampling interval $T = 0.1$ are given by

$$F_{step}(z) = \frac{z}{z - 1} - 3\left(\frac{z}{z - e^{-0.1}}\right) + 2\left(\frac{z}{z - e^{-0.2}}\right)$$

$$= \frac{z}{z - 1} + \frac{-3z}{z - 0.905} + \frac{2z}{z - 0.82}$$

so the step-invariant digital system with $T = 0.1$ is to have unit step sequence response

$$F_{step}(z) = U(z)D(z) = \frac{z}{z - 1} D(z)$$

The step-invariant z-transmittance for $T = 0.1$ is thus

$$D(z) = 1 + \frac{-3(z - 1)}{z - 0.905} + \frac{2(z - 1)}{z - 0.82}$$

$$= \frac{-0.07z + 0.092}{(z - 0.905)(z - 0.82)}$$

The digital system with analog input and analog output in Fig. 8.19c then approximates the performance of $G(s)$. Generally, the smaller the sampling interval, the better the approximation. Further improvement in the approximation may be obtained if desired by further smoothing of the output waveform rather than simply holding it between samples. Step-invariance design has the properties that the digital transmittance is stable if the analog model is stable and that the resulting z-transmittance is of the same order as the original continuous-time transmittance.

Step invariance is but one of several useful approximations of an analog transmittance by a digital one. Other commonly used approximations include impulse invariance, ramp invariance, matched z-transformation, and bilinear transformation.

DRILL PROBLEM

D8.18. Find the step-invariant approximations, for the given sampling interval T, to the following continuous-time transmittances:

(a) $G(s) = \dfrac{1}{s}$ $T = 0.5$

  Ans $0.5/(z - 1)$

(b) $G(s) = \dfrac{2}{s + 4}$ $T = 0.1$

  Ans $0.165/(z - 0.67)$

(c) $G(s) = \dfrac{2}{s + 4}$ $T = 0.03$

  Ans $0.057/(z - 0.887)$

(d) $G(s) = \dfrac{e^{-s}}{s + 3}$ $T = 0.1$

  Ans $0.086/z^{10}(z - 0.74)$

z-Transfer Functions of Systems with Analog Measurements

Often in digital control system analysis and design, the situation of Fig. 8.20a occurs, where it is desired to find the z-transfer function of a system or subsystem with discrete-time input and output but intervening analog components. For this basic situation, the Laplace transform of the impulse train is

$$F^*(s) = F(z)|_{z = e^{sT}}$$

and the analog output $y(t)$ is given by

$$Y(s) = F^*(s)D(s) = [F(z)|_{z = e^{sT}}]D(s)$$

To obtain the z-transform $Y(z)$ of the output sequence, the time shift terms are separated from the rational part of the transform and the usual z-transform term substitutions made. The e^{sT} terms involved in $F^*(s)$ are part of the time shift portion of $Y(s)$, so in forming $Y(z)$, the e^{sT} terms in $F^*(s)$ are simply converted back to $e^{sT} = z$, giving

$$Y(z) = F(z)D(z)$$

The z-transmittance of the basic subsystem is thus

$$\frac{Y(z)}{F(z)} = D(z)$$

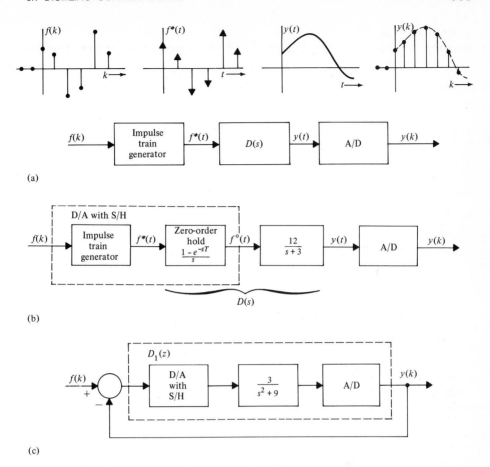

Figure 8.20 Digital subsystems with analog components. **(a)** Basic arrangement. **(b)** An example subsystem. **(c)** An example system with digital feedback.

That is, the analog transmittance $D(s)$ is converted to the corresponding z-transmittance just as in the sampling process where a signal given by $D(s)$ is sampled, yielding a z-transform $D(z)$.

The example digital-input, digital-output subsystem of Fig. 8.20b involves sample-and-hold and so has the intervening analog transmittance

$$D(s) = (1 - e^{-sT})\left[\frac{12}{s(s+3)}\right] = (1 - e^{-sT})\left(\frac{4}{s} - \frac{4}{s+3}\right)$$

for which

$$D(z) = (1 - z^{-1})\left[4\left(\frac{z}{z-1}\right) - 4\left(\frac{z}{z - e^{-3T}}\right)\right]$$

where T is the sampling interval.

For the system of Fig. 8.20c, the z-transmittance

$$D_1(z) = \frac{Y(z)}{E(z)}$$

is given by

$$D_1(s) = \frac{1 - e^{-sT}}{s}\left(\frac{3}{s^2 + 9}\right)$$

$$= (1 - e^{-sT})\left[\frac{\frac{1}{3}}{s} + \frac{-\frac{1}{3}s}{s^2 + 9}\right]$$

$$D_1(z) = (1 - z^{-1})\left\{\frac{1}{3}\left(\frac{z}{z - 1}\right) - \frac{1}{3}\left[\frac{z(z - \cos 3T)}{z^2 - 2z \cos 3T + 1}\right]\right\}$$

$$= \frac{1}{3}\left(\frac{1 - z}{z}\right)\frac{(1 - \cos 3T)(z^2 + z)}{(z - 1)(z^2 - 2z \cos 3T + 1)}$$

$$= \frac{\frac{1}{3}(1 - \cos 3T)(z + 1)}{z^2 - 2z \cos 3T + 1}$$

The overall feedback system z-transfer function is

$$D(z) = \frac{Y(z)}{F(z)} = \frac{D_1(z)}{1 + D_1(z)}$$

$$= \frac{\frac{1}{3}(1 - \cos 3T)(z + 1)}{z^2 + [\frac{1}{3} - \frac{7}{3}\cos 3T]z + [\frac{4}{3} - \frac{1}{3}\cos 3T]}$$

DRILL PROBLEM

D8.19. Find the z-transfer functions of the following systems in terms of the sampling interval T.

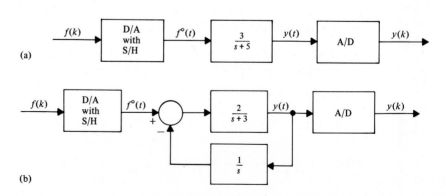

(a)

(b)

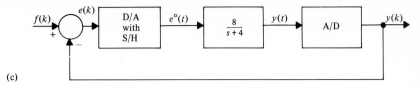

(c)

Ans (a) $0.6(e^{-5T} - 1)/(z - e^{5T})$

(b) $\dfrac{2(z - 1)(e^{-T} - e^{-2T})}{(z - e^{-T})(z - e^{-2T})}$

(c) $(2e^{-4T} - 2)/(-z + 3e^{-4T} - 2)$

A Design Example

Figure 8.21a shows a unity feedback continuous-time system consisting of a controller with compensator $G_1(s)$ and the plant $G_2(s)$. The system transfer function is

$$T(s) = \frac{G_1(s)G_2(s)}{1 + G_1(s)G_2(s)} = \frac{1}{s^2 + s + 1}$$

Its unit step response is given by

$$Y_{\text{step}}(s) = \frac{1}{s} T(s) = \frac{1}{s(s^2 + s + 1)}$$

$$= \frac{1}{s} + \frac{0.58e^{-j150°}}{s + 0.5 + j0.866} + \frac{0.58e^{j150°}}{s + 0.5 - j0.866}$$

$$Y_{\text{step}}(t) = [1 + 1.16e^{-0.5t} \cos(0.866t + 150°)]u(t)$$

To convert the analog controller to a digital one, analog-to-digital converters are used to sample the input and the output signals, a digital-to-analog converter drives the plant, and the compensator transmittance is replaced by a z-transmittance as in Fig. 8.21b. The unit step response of the original analog is given by

$$F_{\text{step}}(s) = \frac{1}{s}\left(\frac{1}{s + 1}\right) = \frac{1}{s} - \frac{1}{s + 1}$$

$$f_{\text{step}}(t) = (1 - e^{-t})u(t)$$

Requiring that the unit step response of the discrete-time compensator consist of samples of the analog unit step response gives, in terms of the sampling interval T,

$$f_{\text{step}}(k) = (1 - e^{-kT})u(k) = u(k) - (e^{-T})^k u(k)$$

$$F_{\text{step}}(z) = \frac{z}{z - 1} - \frac{z}{z - e^{-T}}$$

$$= \frac{z}{z - 1}\left(\frac{z - 1}{z - e^{-T}}\right) = U(z)D_1(z)$$

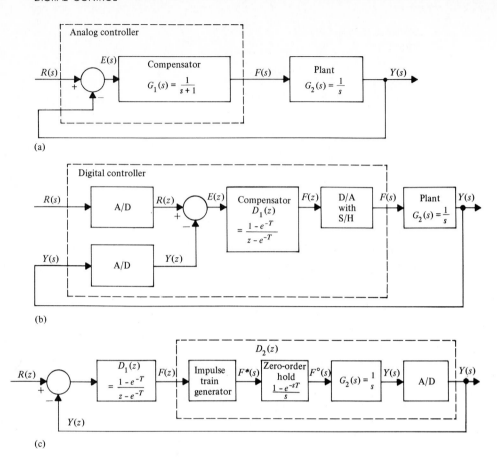

Figure 8.21 Converting analog-to-digital control. **(a)** A continuous-time feedback system. **(b)** Conversion to digital control. **(c)** Relations between the digital signals.

Hence the step-invariant design for $D_1(z)$ is

$$D_1(z) = 1 - \frac{z+1}{z - e^{-T}} = \frac{1 - e^{-T}}{z - e^{-T}}$$

The relationships between the digital signals in the system with digital control are shown in Fig. 8.21c. The z-transmittance relating the sequence $y(k)$ to $f(k)$ is given by

$$D_2(s) = \frac{1 - e^{-sT}}{s}\left(\frac{1}{s}\right) = (1 - e^{-sT})\frac{1}{s^2}$$

$$D_2(z) = (1 - z^{-1})\frac{Tz}{(z-1)^2} = \frac{T}{z-1}$$

where T is the sampling interval. The feedback system thus has overall z-transfer function

$$D(z) = \frac{Y(z)}{R(z)} = \frac{D_1(z)D_2(z)}{1 + D_1(z)D_2(z)}$$

$$= \frac{\dfrac{1 - e^{-T}}{z - e^{-T}}\left(\dfrac{T}{z - 1}\right)}{1 + \dfrac{1 - e^{-T}}{z - e^{-T}}\left(\dfrac{T}{z - 1}\right)}$$

$$= \frac{T(1 - e^{-T})}{z^2 - (1 + e^{-T})z + T(1 - e^{-T}) + e^{-T}}$$

Response samples of the step response of the digitized system for various sampling rates are plotted in Fig. 8.22, where they are compared with the

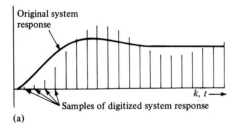

Original system response

Samples of digitized system response

$k, t \longrightarrow$

(a)

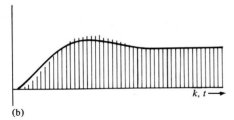

$k, t \longrightarrow$

(b)

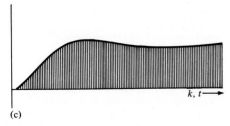

$k, t \longrightarrow$

(c)

Figure 8.22 Response of the digitized system. **(a)** $T = 0.5$. **(b)** $T = 0.2$. **(c)** $T = 0.05$.

continuous-time step response of the original analog system. For a one-half second sampling interval, $T = 0.5$,

$$Y(z) = U(z)D(z) = \frac{0.197z}{z^3 - 2.607z + 2.41z - 0.804}$$

$$= 0.197z^{-2} + 0.51z^{-3} + 0.86z^{-4} + \cdots$$

For $T = 0.2$,

$$Y(z) = U(z)D(z) = \frac{0.036z}{z^3 - 2.819z^2 + 2.674z - 1.655}$$

$$= 0.036z^{-2} + 0.102z^{-3} + 0.191z^{-4} + \cdots$$

For $T = 0.05$,

$$Y(z) = U(z)D(z) = \frac{0.0024z}{z^3 - 2.95z^2 + 1.95z + 0.954}$$

$$= 0.0024z^{-2} + 0.00708z^{-3} + 0.014z^{-4} + \cdots$$

The higher the sampling rate, the more closely the digital system approximates the behavior of the original analog system.

Since $G_1(s)$ is an integrator driven by the piecewise constant sample-and-hold (S/H) waveform $f^0(t)$ and since the continuous-time output $y(t)$ passes through the sample points $y(k)$, $y(t)$ is a piecewise linear signal through the sample points.

The final value of the unit step response is, from the formula in Table 8.6,

$$\lim_{k \to \infty} y(k) = \lim_{z \to 1} \left[\frac{z-1}{z} Y(z) \right]$$

$$= \lim_{z \to 1} \left[\frac{z-1}{z} \frac{z}{z-1} D(z) \right]$$

$$= \lim_{z \to 1} \left[\frac{T(1 - e^{-T})}{z^2 - (1 + e^{-T})z + T(1 - e^{-T}) + e^{-T}} \right]$$

$$= \frac{T(1 - e^{-T})}{T(1 - e^{-T})} = 1$$

8.10 DIRECT DIGITAL DESIGN

Steady State Response

If a discrete-time signal $f(k)$ reaches a finite, constant steady state value, it is given by

$$\lim_{k \to 0} f(k) = \lim_{z \to 1} \frac{z-1}{z} F(z)$$

which is the discrete-time version of the final value theorem. If $F(z)/z$ were expanded into partial fractions,

$$\frac{F(z)}{z} = \frac{K_1}{z-1} + \cdots$$

the residue of the $(z-1)$ pole represents the constant term in the sequence $f(k)$, which may be found by multiplying $F(z)/z$ by $(z-1)$ and evaluating at $z=1$.

For example, the sequence with z-transform

$$F(z) = \frac{3z^2 + 4}{z^2 - \frac{1}{2}z - \frac{1}{2}}$$

$$= \frac{3z^2 + 4}{(z-1)(z+\frac{1}{2})} = \frac{K_1}{z-1} + \frac{K_2}{z+\frac{1}{2}}$$

has steady state value given by

$$\lim_{k \to 0} f(k) = \lim_{z \to 1} \frac{z-1}{z} F(z) = \lim_{z \to 1} \frac{3z^2 + 4}{z^2 + \frac{1}{2}z}$$

$$= \frac{14}{3}$$

The sequence with z-transform

$$F(z) = \frac{z^3 - 4z + 1}{z^3 + z^2 - 2z} = \frac{K_1}{z-1} + \frac{K_2}{z} + \frac{K_3}{z+2}$$

does not reach a finite steady state because the $K_3/(z+2)$ term in its partial fraction expansion represents an expanding term in the sequence $f(k)$. The limit exists and is finite, however:

$$\lim_{z \to 1} \frac{z-1}{z} F(z) = \lim_{z \to 1} \frac{z^3 - 4z + 1}{z^3 + 2z^2} = -\frac{2}{3}$$

For a system with stable z-transfer function

$$D(z) = \frac{4z^2 + 8z - 1}{(3z+2)(4z-1)}$$

the error between input and output, when the input is a unit step sequence, is

$$E(z) = R(z) - Y(z) = R(z)[1 - D(z)]$$

$$= \frac{z}{z-1}\left(1 - \frac{4z^2 + 8z - 1}{12z^2 + 5z - 2}\right)$$

which has a final value

$$\lim_{k \to 0} e(k) = \lim_{z \to 1} \frac{z-1}{z} E(z)$$

$$= \lim_{z \to 1} \frac{8z^2 - 3z - 1}{12z^2 + 5z - 2}$$

$$= \frac{4}{15}$$

DRILL PROBLEM

D8.20. Each of the following systems is stable. Find the steady state error between input and output for each when the input is a unit step sequence.

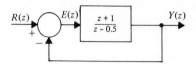

(a)

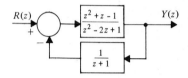

(b)

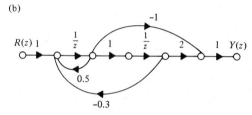

(c)

Ans (a) 0.2
(b) 0
(c) -0.25

Deadbeat Systems

A digital system with all of its poles at $z = 0$ is termed *deadbeat*. Such systems have a remarkable property: Their pulse response is zero after n steps, where n is the order of the system. For example, the system with z-transfer function

$$D(z) = \frac{z^3 + 3z^2 - 2z + 4}{z^3}$$

is deadbeat. Its unit pulse response is

$$Y_{\text{pulse}}(z) = 1 \cdot D(z)$$
$$= 1 + 3z^{-1} - 2z^{-2} + 4z^{-3}$$
$$y_{\text{pulse}}(k) = \delta(k) + 3\delta(k - 1) - 2\delta(k - 2) + 4\delta(k - 3)$$

Deadbeat systems are also commonly termed *finite-duration impulse response* (FIR) systems.

As pulse response is representative of the natural component of a system's response in general, deadbeat systems have a natural response that goes to zero after n steps. There is no counterpart in analog systems; analog natural response may only decay asymptotically to zero. In many practical situations, digital systems are designed to be deadbeat if possible.

DRILL PROBLEM

D8.21. Find the response $y(k)$ of each of the following systems if the input is a unit step sequence and the initial conditions are zero.

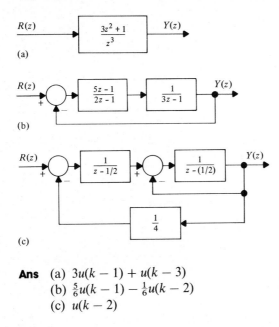

$R(z)$ → $\dfrac{3z^2 + 1}{z^3}$ → $Y(z)$

(a)

(b)

(c)

Ans (a) $3u(k-1) + u(k-3)$
(b) $\frac{5}{6}u(k-1) - \frac{1}{6}u(k-2)$
(c) $u(k-2)$

A Design Example

Figure 8.23a shows a simplified model of a satellite-tracking control system. Computer-generated digital commands, based on orbital calculations, are applied to the system consisting of a digital controller driving the analog positioning subsystem. The diagram is rearranged in Fig. 8.23b to show the relations between digital signals. The sampling interval is $T = 0.1$. It is desired to design the controller transmittance $D_1(z)$ so that, if possible,

1. The overall digital system $D(z) = Y(z)/R(z)$ is deadbeat.
2. The error between output and input, $E(z)$, to a step input $R(z)$ is zero in steady state.

In addition, it is important that the analog position output $y(t)$ be well behaved between samples, particularly as a constant steady state is approached.

Digital command
sequence

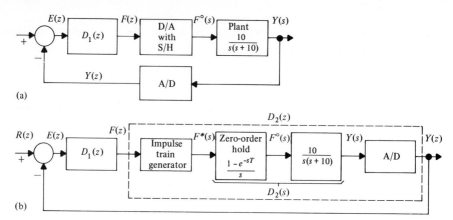

(a)

(b)

Figure 8.23   Satellite tracking system.   **(a)** System block diagram.   **(b)** Block diagram rearranged to show digital relations.

The analog transmittance $D_2(s)$ in Fig. 8.23b is

$$D_2(s) = \frac{10(1 - e^{-0.1s})}{s^2(s + 10)} = (1 - e^{-0.1s})\left(\frac{-\frac{1}{10}}{s} + \frac{1}{s^2} + \frac{\frac{1}{10}}{s + 10}\right)$$

hence the corresponding z-transmittance is

$$D_2(z) = (1 - z^{-1})\left[\frac{-\frac{1}{10}z}{z - 1} + \frac{0.1z}{(z - 1)^2} + \frac{\frac{1}{10}z}{z - 0.368}\right]$$

$$= \frac{(z - 1)(0.0368z^2 + 0.0264z)}{z(z - 1)^2(z - 0.368)}$$

$$= \frac{0.0368z + 0.0264}{(z - 1)(z - 0.368)}$$

In terms of the controller transmittance $D_1(z)$, the overall system transfer function is

$$D(z) = \frac{D_1(z)D_2(z)}{1 + D_1(z)D_2(z)}$$

$$= \frac{D_1(z)\dfrac{0.0368z + 0.0264}{(z - 1)(z - 0.368)}}{1 + D_1(z)\dfrac{0.0368z + 0.0264}{(z - 1)(z - 0.368)}}$$

Let $D(z)$ have n poles, all of which are at $z = 0$. Then

$$1 + D_1(z)\frac{0.0368z + 0.0264}{(z - 1)(z - 0.368)} = \frac{z^n}{\text{polynomial}}$$

where "polynomial" denotes an as-yet-unspecified polynomial in z. Solving for $D_1(z)$,

$$D_1(z) = \frac{(z-1)(z-0.368)}{0.0368z + 0.0264}\left(\frac{z^n}{\text{polynomial}} - 1\right)$$

Any such z-transmittance $D_1(z)$ will result in an overall system that is deadbeat:

$$D(z) = \frac{\dfrac{z^n}{\text{polynomial}} - 1}{1 + \left(\dfrac{z^n}{\text{polynomial}} - 1\right)}$$

$$= \frac{z^n - \text{polynomial}}{z^n}$$

The error between input and output is

$$E(z) = R(z) - Y(z) = R(z)[1 - D(z)]$$

$$= R(z)\left(1 - \frac{z^n - \text{polynomial}}{z^n}\right)$$

$$= R(z)\left(\frac{\text{polynomial}}{z^n}\right)$$

For a step input

$$R(z) = \frac{z}{z-1}$$

the error signal has z-transform

$$E(z) = \frac{1}{z-1}\left(\frac{\text{polynomial}}{z^n}\right)$$

and steady state value given by

$$\lim_{k \to \infty} e(k) = \lim_{z \to 1}\left(\frac{\text{polynomial}}{z^n}\right)$$

$$= \lim_{z \to 1}(\text{polynomial})$$

To summarize:

1. $D_1(z)$ must be of the form

$$D_1(z) = \frac{(z-1)(z-0.368)}{0.0368z + 0.0264}\left(\frac{z^n}{\text{polynomial}} - 1\right)$$

in order for the overall system to be deadbeat.

2. The unspecified polynomial must have the property

$$\lim_{z \to 1} (\text{polynomial}) = 0$$

for the system's steady state step error to be zero.

The choice of $n = 1$ and

$$\text{polynomial} = z - 1$$

is simple, has property 2, and results in a simplification of $D_1(z)$:

$$D_1(z) = \frac{z - 0.0368}{0.0368z + 0.0264}$$

The overall transfer function for this choice is

$$D(z) = \frac{Y(z)}{R(z)} = \frac{1}{z}$$

The digital step response of this system is given by

$$Y(z) = R(z)D(z) = \frac{z}{z - 1}\left(\frac{1}{z}\right)$$

$$\frac{Y(z)}{z} = \frac{1}{z(z - 1)} = \frac{-1}{z} + \frac{1}{z - 1}$$

$$Y(z) = -1 + \frac{1}{z - 1}$$

$$y(k) = -\delta(k) + u(k)$$

For a step command input, the output reaches its final value in one step, as shown in Fig. 8.24a.

For the step input, the controller's digital signal $F(z)$ is given by

$$F(z) = [R(z) - Y(z)]D_1(z)$$

$$= \left(\frac{z}{z - 1} - \frac{1}{z - 1}\right)\frac{z - 0.368}{0.0368z + 0.0264}$$

$$= \frac{z - 0.368}{0.0368z + 0.0264}$$

$$\frac{F(z)}{z} = \frac{27.17z - 10}{z(z + 0.717)} = \frac{-13.95}{z} + \frac{+41.1}{z + 0.717}$$

$$F(z) = -13.95 + 41.1\frac{z}{z + 0.717}$$

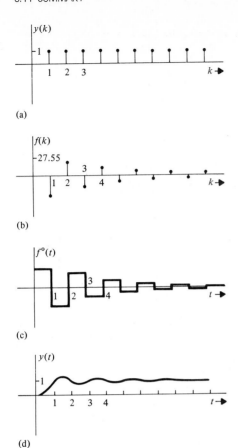

(a)

(b)

(c)

(d)

Figure 8.24 Tracking system step response. **(a)** Samples of the output position. **(b)** Samples of the controller output. **(c)** Sample-and-held controller output. **(d)** Output position as a function of time.

8.11 SUMMARY

The relation between an analog signal $f(t)$ and its sample sequence

$$f(k) = F(t = kT)$$

is found by relating the z-transform $F(z)$ of the sequence to the Laplace transform $F(s)$ of the analog signal. Each term in the partial fraction expansion of $F(s)$ is replaced by the corresponding z-transform expansion term from Table 8.5 to form $F(z)$.

Approximate reconstruction of an analog signal $f(t)$ from its samples is commonly performed by a D/A converter with sample-and-hold. The S/H waveform $f^o(t)$ is conveniently represented by the conversion of digital samples to an analog impulse train, followed by a zero-order hold transmittance. More accurate reconstruction may be done with higher-order holds, involving more than a single sample, and low-pass filtering. The sampling theorem states conditions for which a band-limited signal is uniquely represented by its samples.

Discrete-time systems are described by difference equations,

$$y(k + n) + a_{n-1} y(k + n - 1) + a_{n-2} y(k + n - 2) + \cdots$$

$$+ a_1 y(k + 1) + a_0 y(k) = b_m r(k + m) + b_{m-1} r(k + m - 1)$$

$$+ \cdots + b_1 r(k + 1) + b_0 r(k)$$

where $r(k)$ is the input sequence and $y(k)$ is the output sequence. The z-transfer function of such a system is

$$D(z) = \frac{Y(z)}{R(z)}\bigg|_{\substack{\text{initial} \\ \text{conditions}=0}}$$

$$= \frac{b_m z^m + b_{m-1} z^{m-1} + \cdots + b_1 z + b_0}{z^n + a_{n-1} z^{n-1} + a_{n-2} z^{n-2} + \cdots + a_1 z + a_0}$$

Block diagrams and signal flow graphs for discrete-time systems, described in terms of z-transforms, are manipulated in the same way as are continuous-time system descriptions in terms of Laplace transforms.

A discrete-time system is termed stable if and only if all poles of its z-transfer function are within the unit circle on the complex plane. The bilinear transformation

$$z = \frac{1 + W}{1 - W}$$

maps the unit circle to the left half-plane on the complex plane and so is very useful in relating discrete-time situations to equivalent continuous-time ones. With the above substitution for z in a rational z-transfer function, Routh-Hurwitz methods may be applied to determine stability.

State equations offer a systematic and powerful method of system representation. In terms of these,

$$\mathbf{x}(k + 1) = \mathbf{F}\mathbf{x}(k) + \mathbf{G}r(k)$$

$$y(k) = \mathbf{H}\mathbf{x}(k)$$

The z-transfer function matrix of a system is

$$\mathbf{D}(z) = \mathbf{H}(z\mathbf{I} - \mathbf{F})^{-1}\mathbf{G}$$

and the system is stable if and only if all roots of the characteristic equation

$$|z\mathbf{I} - \mathbf{F}| = 0$$

are within the unit circle on the complex plane. An nth-order system is completely controllable if and only if

$$\mathbf{M}_c = [\mathbf{G} \mid \mathbf{F}\mathbf{G} \mid \mathbf{F}^2\mathbf{G} \mid \cdots \mid \mathbf{F}^{n-1}\mathbf{G}]$$

is of full rank and is completely observable if and only if

$$\mathbf{M}_o = \begin{bmatrix} \mathbf{H} \\ \hline \mathbf{HF} \\ \hline \mathbf{HF}^2 \\ \hline \vdots \\ \hline \mathbf{HF}^{n-1} \end{bmatrix}$$

is of full rank.

Change of stable variables

$$\mathbf{x}'(k) = \mathbf{P}\mathbf{x}(k)$$

gives alternative state variable representations of a system and shows available design freedom. If the system characteristic roots are not repeated, the methods of Sec. 7.4 may be used to find a set of state variables for which the state equations are decoupled from one another.

A common analysis problem involves a system containing analog components but with discrete-time input and output. To find the z-transmittance, the system is converted to the form where an impulse train generator drives a continuous-time transmittance $D(s)$, followed by a sampler. The transmittance $D(s)$ is separated into delay terms and rational terms, with the rational terms expanded into partial fractions. Substitutions

$$e^{sT} \to z$$

into the delay terms and

$$\frac{K}{s + a} \to \frac{Kz}{z - e^{aT}}$$

for the partial fractions yields the z-transmittance $D(z)$.

One design method for digital control involves replacing a model analog subsystem with a digital subsystem consisting of A/D converter, z-transmittance, and D/A converter. For the step-invariant approximation, the z-transmittance is required to have a step response consisting of samples of the step response of the analog subsystem.

Digital control system design may also be done directly, in terms of overall system performance requirements and objectives.

REFERENCES

Computer Processing

Gothmann, W. H., *Digital Electronics, an Introduction to Theory and Practice*. Englewood Cliffs, N.J.: Prentice-Hall, 1977.

Mano, M. M., *Digital Logic and Computer Design*. Englewood Cliffs, N.J.: Prentice-Hall, 1979.

Osborne, A., *An Introduction to Microcomputers*, 2nd ed. New York: McGraw-Hill, 1980.

Peatman, J. B., *Microcomputer-Based Design*. New York: McGraw-Hill, 1977.

Sampling and Reconstruction

Cadzow, J. A., *Discrete-Time Systems, an Introduction with Interdisciplinary Applications*. Englewood Cliffs, N.J.: Prentice-Hall, 1973.

Hamming, R. W., *Digital Filters*. Englewood Cliffs, N.J.: Prentice-Hall, 1977.

Jury, E. I., *Theory and Application of the z-Transform Method*. New York: Wiley, 1964.

Oliver, B. M., Pierce, J. R., and Shannon, C. E., "The Philosophy of PCM." *Proc. IRE* 36 (November 1948): 1324–31.

Stearns, S. D., *Digital Signal Analysis*. Rochelle Park, N. J.: Hayden, 1975.

Tretter, S. A., *Introduction to Discrete-Time Signal Processing*. New York: Wiley, 1976.

Discrete-Time Systems and Filtering

Chen, C.-T., *One-Dimensional Digital Signal Processing*. New York: Marcel Dekker, 1979.

Oppenheim, A. V., and Schafer, R. W., *Digital Signal Processing*. Englewood Cliffs, N. J.: Prentice-Hall, 1975.

Peled, A., and Liu, B., *Digital Signal Processing: Theory, Design and Implementation*. New York: Wiley, 1976.

Rabiner, L. R., and Gold, B., *Theory and Application of Digital Signal Processing*. Englewood Cliffs, N.J.: Prentice-Hall, 1975.

Schwartz, M., and Shaw, L., *Signal Processing*. New York: McGraw-Hill, 1975.

Stanley, W. D., *Digital Signal Processing*. Reston, Va: Reston, 1975.

Digital Control

Cadzow, J. A., and Martens, H. R., *Discrete-Time and Computer Control Systems*. Englewood Cliffs, N.J.: Prentice-Hall, 1970.

Franklin, G. F., and Powell, J. D., *Digital Control of Dynamic Systems*. Reading, Mass: Addison-Wesley, 1980.

Freeman, H., *Discrete-Time Systems*. New York: Wiley, 1965.

Kuo, B. C., *Digital Control Systems*. New York: Holt, Rinehart and Winston, 1980.

Monroe, A. J., *Digital Processes for Sampled Data Systems*. New York: Wiley, 1962.

Ragazzini, J. R., and Franklin, G. F., *Sampled Data Control Systems*. New York: McGraw-Hill, 1958.

Schwartz, R. J., and Friedland, B., *Linear Systems*. New York: Wiley, 1965.

Tou, J. T., *Digital and Sampled-data Control Systems*. New York: McGraw-Hill, 1959.

VanLandingham, H. F., *Introduction to Digital Control Systems*. New York: Macmillan, 1985.

Digital Processing and Control Applications

Allan, R., "Busy Robots Spur Productivity." *IEEE Spec.*, September 1979, pp. 31–36.

———, "The Microcomputer Invades the Production Line." *IEEE Spec.*, January 1979, pp. 53–57.

Andrews, H. C., and Hunt, B. R., *Digital Image Restoration*, Englewood Cliffs, N.J.: Prentice-Hall, 1977.

Kahne, S., "Automatic Control by Distributed Intelligence." *Sci. Amer.*, June 1979, pp. 78-109.

Rabiner, L. R., and Schafer, R. W., *Digital Processing of Speech Signals*. Englewood Cliffs, N.J.: Prentice-Hall, 1978.

Oppenheim, A. V., ed., *Applications of Digital Signal Processing*. Englewood Cliffs, N.J.: Prentice-Hall, 1978.

PROBLEMS

1. List five home appliances for which microprocessor control is today useful and economically feasible. For each, describe the functions performed by the digital system.

2. Carefully describe functions that a microprocessor-based control system might perform in each of the following. Specify the signals to be sensed and the quantities to be controlled.

(a) An automobile

(b) A hotel

(c) Aboard a ship

(d) An electric power generating plant

(e) A hospital operating room

3. The analog signal

$$f(t) = 3 + 4 \cos 50t$$

is sampled at 0.01-sec intervals by an A/D converter preceded by a sample-and-hold (S/H) that freezes the sample while conversion takes place. Then the samples are reconverted to an analog signal with S/H. Sketch your visualization of signal $f(t)$, the first S/H waveform, the A/D samples, and the output waveform.

4. A bank account pays 9% interest per year, compounded monthly. Initially, a deposit of $1000 is made and thereafter $65 per month is deposited into the account each month. Describe the monthly bank balance as a function of the month k after the initial deposit.

5. Sketch the functions $f(t)$ and samples $f(k)$ with the given sampling interval T:

(a) $f(t) = 3e^{-10t}u(t)$ $T = 0.1$

(b) $f(t) = \left(3e^{-2t} \cos \dfrac{5\pi t}{4}\right)u(t)$ $T = 0.2$

(c) $f(t) = \left(2 \sin \dfrac{\pi t}{8}\right)u(t)$ $T = 1$

(d) $f(t) = (4 + 3e^{0.1t})u(t)$ $T = 0.5$

6. Find the z-transforms of the following sequences:

(a) $f(k) = u(k) - u(k - 4)$

(b) $f(k) = 0.1ku(k) - 0.1ku(k - 4)$

   Ans $0.1(1 - z^{-4})z/(z - 1)^2$

(c) $f(k) = \left[\cos\left(0.1k + \dfrac{\pi}{4}\right)\right]u(k)$

(d) $f(k) = [(-1)^k \sin 0.5k]u(k)$

   Ans $-0.48z/(z^2 + 1.76z + 1)$

7. Find the z-transforms of the sequences consisting of samples of the following functions $f(t)$ with the given sampling interval T:

(a) $f(t) = 5tu(t)$ $T = 0.5$

(b) $f(t) = 3u(t) - 4u(t) \sin 3t$ $T = 0.2$

   Ans $\dfrac{3z}{z - 1} - \dfrac{2.26z}{z^2 - 1.65z + 1}$

(c) $15t^2e^{-10t}u(t)$ $T = 4$

(d) $5u(t)e^{-3t} \cos(\pi t + 45°)$ $T = 0.1$

   Ans $\dfrac{3.54(z^2 - 0.93z)}{z^2 - 1.4z + 0.55}$

8. Use long division to find $f(0), f(1), f(2),$ and $f(3)$ for each of the discrete-time signals with the following z-transforms:

(a) $F(z) = \dfrac{4z^2 - 3z + 6}{2z^2 + z - 1}$

(b) $F(z) = \dfrac{2z - 3}{z^2 - 0.5z + 1}$

   Ans $0, 2, -2, -3$

(c) $F(z) = \dfrac{z}{0.3z^3 - 0.1z^2 + 0.2z + 1}$

9. Find the inverse z-transforms:

(a) $F(z) = \dfrac{4}{z + \frac{1}{3}}$

(b) $F(z) = \dfrac{3z^2 - 2z}{4z^2 + 5z + 1}$

   Ans $-(\frac{11}{3})(-\frac{1}{4})^k u(k) + \frac{5}{3}(-1)^k u(k)$

(c) $F(z) = \dfrac{z^2}{(2z + 1)^2}$

(d) $F(z) = \dfrac{3z - 2}{z^2 + 1.5z + 0.5}$

   Ans $-4\delta(k) - 10(-1)^k u(k) + 14(-\frac{1}{2})^k u(k)$

(e) $F(z) = \dfrac{1}{z^3 - 0.25z}$

10. For the following functions $f(t)$ and sampling periods T, find $f(k)$:

(a) $f(t) = 3e^{-10t}u(t)$ $\qquad T = 0.1$

(b) $f(t) = 3e^{-2t}u(t) \cos \dfrac{5\pi t}{4}$ $\qquad T = 0.2$

   Ans $3(0.67)^k u(k) \cos(0.785k)$

(c) $f(t) = 2u(t) \sin \dfrac{\pi t}{8}$ $\qquad T = 1$

(d) $f(t) = (t^2 e^{-7t} + 5e^{-6t} \sin 10t)u(t)$ $\qquad T = 0.01$

   Ans $[10^{-4}k^2(0.93)^k + 5(0.94)^k \sin 0.1k]u(k)$

11. For each of the following analog signals with given Laplace transform $F(s)$, find the z-transform $F(z)$ of the corresponding sample sequence with the given sampling interval T:

(a) $F(s) = \dfrac{16}{s^2 + 4}$ $\qquad T = 0.1$

(b) $F(s) = \dfrac{-2s + 20}{s^3 + 9s^2 + 20s}$ $T = 0.05$

Ans $\dfrac{z}{z - 1} - \dfrac{7z}{z - 0.819} + \dfrac{6z}{z - 0.789}$

(c) $F(s) = \dfrac{4e^{-0.5s}}{s^2 + 2s + 1}$ $T = 0.25$

(d) $F(s) = \dfrac{3e^{-s}}{s} - \dfrac{4}{s + 2} + \dfrac{se^{-2s}}{s^2 + 9}$ $T = 0.2$

Ans $\dfrac{3}{z^5 - z^4} - \dfrac{4z}{z - 0.67} + \dfrac{z - 0.825}{z^{11} - 1.65z^{10} + z^9}$

(e) $F(s) = \dfrac{1 - e^{-2s} - e^{-3s}}{s^2 + 2s + 10}$ $T = 0.5$

12. For the following sequences $f(k)$, find the corresponding impulse train Laplace transforms $F^*(s)$ for the given sampling interval T:

(a) $f(k) = (3 \cos 2k - 4 \sin 2k)u(k)$ $T = 0.5$

(b) $f(k) = [(-1)^k - 1]u(k)$ $T = 0.1$

   Ans $-2e^{-0.1s}/(e^{-0.2s} - 1)$

(c) $f(k) = 6(-0.5)^k u(k) \cos 3k$ $T = 1$

(d) $f(k) = 10ke^{-0.1k}u(k - 3)$ $T = 0.5$

   Ans $9.05/e^{-s}(e^{0.5s} - 0.905)^2$

(e) $f(k) = ku(k) - ku(k - 5)$ $T = 0.3$

13. The continuous-time function

$$f(t) = 20 - 15 \cos 1000t$$

is sampled at the rate of 100 samples/sec. It is then reconstructed from the impulse train $f^*(t)$, according to the sampling theorem, as if it were a signal band-limited at 50 Hz. Find the signal that results.

14. Find the z-transfer functions of the following discrete-time systems:

(a) $y(k + 2) + 0.2y(k + 1) - 0.5y(k) = r(k)$

(b) $y(k + 3) - y(k) = 4r(k + 2) - 3r(k)$

   Ans $(4z^2 - 3)/(z^3 - 1)$

(c) $y(k + 3) = 0.75r(k + 3) + 0.25r(k + 2) - 0.25r(k + 1) - 0.75r(k)$

15. For

$$y(k + 2) - 0.5y(k) = r(k + 1) - 2r(k)$$

recursively find $y(2)$, $y(3)$, $y(4)$, and $y(5)$ if

$$r(k) = (\tfrac{1}{2})^k$$

$$y(0) = 0$$

$$y(1) = 3$$

16. For discrete-time systems with the following z-transfer functions and input sequences, find the output sequences for $k = 0$ and thereafter if the initial conditions are zero:

(a) $D(z) = \dfrac{z}{z + \frac{1}{2}}$

$r(k) = 2\delta(k)$

(b) $D(z) = \dfrac{1}{z - 2}$

$r(k) = 3u(k)$

Ans $-3u(k) + 3(2)^k u(k)$

(c) $D(z) = \dfrac{z}{z^2 + 1}$

$r(k) = (-1)^k$

(d) $D(z) = \dfrac{1}{z^2 - \frac{1}{2}z}$

$r(k) = (\tfrac{1}{2})^k$

Ans $4\delta(k) - 4(\tfrac{1}{2})^k u(k) + 4k(\tfrac{1}{2})^k u(k)$

(e) $D(z) = \dfrac{z^2}{8z^2 + 6z + 1}$

$r(k) = 3(-\tfrac{1}{2})^k$

17. Reduce the block diagrams of Fig. P8.17, finding the system z-transfer functions.

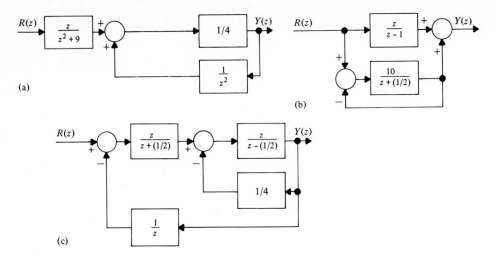

(a)

(b)

(c)

Figure P8.17

Ans   (b) $(2z^2 + 41z - 20)/(2z^2 + 19z - 21)$

18. Use Mason's gain rule to find the z-transfer functions of the systems of P8.18.

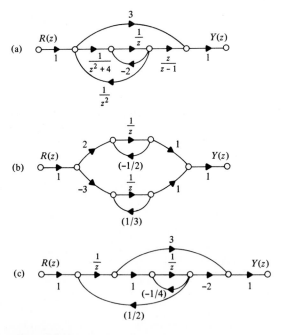

Figure P8.18

Ans   (b) $(-6z - 13)/(6z^2 + z - 1)$

19. Use the bilinear transformation and Routh-Hurwitz tests to determine whether each of the discrete-time systems with the given z-transfer functions is stable:

(a) $D(z) = \dfrac{4z^3 - 3z}{z^3 - 0.8z^2 + 0.17z - 0.01}$

(b) $D(z) = \dfrac{0.63z^2 - 8.73z + 1}{z^3 - 2.3z^2 + 0.62z - 0.04}$

Ans unstable

(c) $D(z) = \dfrac{0.02}{z^4 - 2.8z^3 + 1.77z^2 - 0.35z + 0.02}$

(d) $D(z) = \dfrac{z^4 + 0.005}{z^4 - 1.3z^3 + 0.57z^2 - 0.095z + 0.005}$

Ans stable

20. Draw discrete-time simulation diagrams in phase-variable canonical form for systems with the following z-transfer functions:

(a) $D(z) = \dfrac{z^2 - z + 2}{z^3 - 0.5z^2 + z}$

(b) $D(z) = \dfrac{10z}{z^4 + 0.1z^3 - 0.2z^2 + 0.3z + 0.4}$

(c) $D(z) = \dfrac{4z^2 - 6z + 8}{5z^4 + 3z^3 - 2z^2 + z + 1}$

21. Find state equations in phase-variable matrix form for the discrete-time systems with the following z-transfer functions:

(a) $D(z) = \dfrac{3z + 2}{z^2 + z + 4}$

(b) $D(z) = \dfrac{8}{6z^3 + z^2 - z}$

Ans
$$\begin{bmatrix} x_1(k+1) \\ x_2(k+1) \\ x_3(k+1) \end{bmatrix} = \begin{bmatrix} 0 & 1 & 0 \\ 0 & 0 & 1 \\ 0 & \frac{1}{6} & -\frac{1}{6} \end{bmatrix} \begin{bmatrix} x_1(k) \\ x_2(k) \\ x_3(k) \end{bmatrix} + \begin{bmatrix} 0 \\ 0 \\ 1 \end{bmatrix} r(k)$$

$$y(k) = \begin{bmatrix} \frac{4}{3} & 0 & 0 \end{bmatrix} \begin{bmatrix} x_1(k) \\ x_2(k) \\ x_3(k) \end{bmatrix}$$

(c) $D(z) = \dfrac{4z^2 - 3}{z^3 - 0.5z^2 + 1}$

22. For each of the following discrete-time systems, with the indicated input and initial conditions, recursively find the state vectors $\mathbf{x}(1)$, $\mathbf{x}(2)$, and $\mathbf{x}(3)$ and the outputs $y(0)$, $y(1)$, $y(2)$, $y(3)$:

(a) $\begin{bmatrix} x_1(k+1) \\ x_2(k+1) \end{bmatrix} = \begin{bmatrix} 1 & -2 \\ -1 & 1 \end{bmatrix} \begin{bmatrix} x_1(k) \\ x_2(k) \end{bmatrix}$

$y(k) = \begin{bmatrix} 3 & 0 \end{bmatrix} \begin{bmatrix} x_1(k) \\ x_2(k) \end{bmatrix}$

$\begin{bmatrix} x_1(0) \\ x_2(0) \end{bmatrix} = \begin{bmatrix} -4 \\ 0 \end{bmatrix}$

(b) $\begin{bmatrix} x_1(k+1) \\ x_2(k+1) \\ x_3(k+1) \end{bmatrix} = \begin{bmatrix} 0 & 1 & 0 \\ -1 & 0 & 1 \\ 2 & 1 & -2 \end{bmatrix} \begin{bmatrix} x_1(k) \\ x_2(k) \\ x_3(k) \end{bmatrix} + \begin{bmatrix} 0 \\ 1 \\ -1 \end{bmatrix} (-1)^k$

$y(k) = \begin{bmatrix} 3 & 0 & 4 \end{bmatrix} \begin{bmatrix} x_1(k) \\ x_2(k) \\ x_3(k) \end{bmatrix}$

$\begin{bmatrix} x_1(0) \\ x_2(0) \\ x_3(0) \end{bmatrix} = \begin{bmatrix} 0 \\ 0 \\ 0 \end{bmatrix}$

Ans $\begin{bmatrix} 0 \\ 1 \\ -1 \end{bmatrix}, \begin{bmatrix} 1 \\ -2 \\ 4 \end{bmatrix}, \begin{bmatrix} -2 \\ 4 \\ -7 \end{bmatrix}$, 0, -4, 19, -34

23. Find step-invariant discrete-time system approximations to the following analog transmittances, using the indicated sampling intervals T:

(a) $G(s) = \dfrac{1}{s}$ $T = 0.01$

(b) $G(s) = \dfrac{1}{s+1}$ $T = 0.5$

  Ans $0.393/(z - 0.607)$

(c) $G(s) = \dfrac{s}{s^2 + 1}$ $T = 0.02$

(d) $G(s) = \dfrac{1 - e^{-0.5s}}{s+4}$ $T = 0.1$

  Ans $(0.0825)(z^5 - 1)/(z^5 - 0.67z^4)$

(e) $G(s) = \dfrac{10(1 - e^{-0.1s})}{s^2 + 4s}$ $T = 0.01$

24. An *impulse-invariant* approximation of an analog transmittance by a digital system is the digital system with unit pulse response which consists of samples of the unit impulse response of the analog transmittance. Find the z-transfer function of the impulse-invariant approximation to

$$G(s) = \frac{10}{s + 4}$$

with sampling interval $T = 0.2$.

25. Find the z-transfer functions of the systems of Fig. P8.25. For each system, the sampling interval is $T = 0.3$.

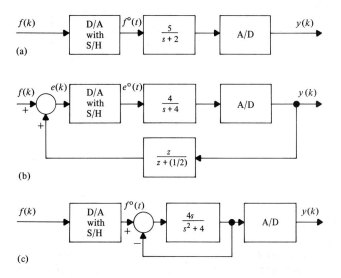

Figure P8.25

Ans (b) $(0.7z + 0.35)/(z^2 - 0.5z - 0.15)$

26. For the system of Fig. P8.26, with sampling interval $T = 0.2$, it is desired that the z-transfer function be

$$D(z) = \frac{Y(z)}{F(z)} = \frac{z}{z - 1}$$

Find, if possible, the necessary analog transmittance $G(s)$.

Figure P8.26

27. For each of the systems of Fig. P8.27, find the *z*-transfer function relating *y(k)* to *r(k)*. The sampling interval is $T = 0.2$. In Fig. P8.27b, c, note that an equivalent system is one in which $Y(s)$ and $R(s)$ are first A/D-converted then summed, as in Fig. P8.27a.

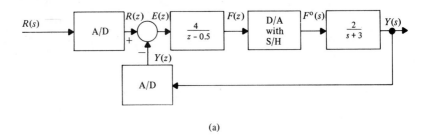

(a)

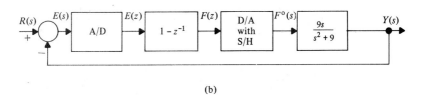

(b)

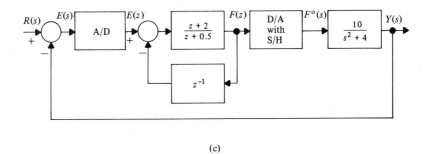

(c)

Figure P8.27

Ans (b) $\dfrac{1.69(z-1)^2}{z(z^2 - 1.65z + 1) + 1.69(z-1)^2}$

28. Determine whether the systems of Fig. P8.28 are stable. For each system, the sampling interval is $T = 0.5$.

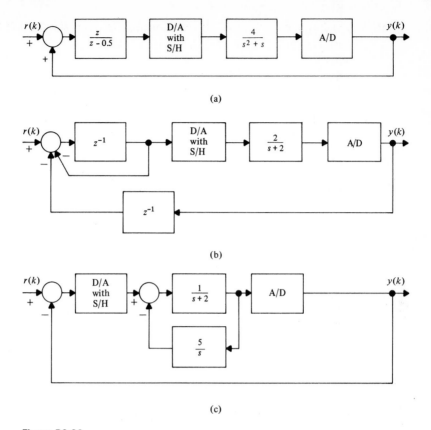

(a)

(b)

(c)

Figure P8.28

Ans (b) stable

29. For systems with the following z-transfer functions, find the normalized steady state error, between output and input, for a step input:

(a) $D(z) = \dfrac{20z^2}{12z^2 + 7z + 1}$

(b) $D(z) = \dfrac{0.25z^2 - 3z}{z^2 + 1.6z + 0.48}$

Ans infinite error; system unstable

(c) $D(z) = \dfrac{2}{2z^3 - z^2 + z}$

30. For the system of Fig. P8.30, find a transmittance $H(z)$ so that the overall transfer function $D(z) = Y(z)/R(z)$ is deadbeat.

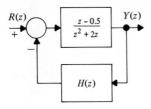

Figure P8.30

Matrix Algebra

A.1 PREVIEW

The modern approach to control system analysis depends heavily on the manipulation of matrices and functions of matrices. This appendix reviews matrix addition, subtraction, multiplication, transposition, inversion, and differentiation. This appendix is not intended to be a primary source of knowledge regarding matrix algebra. Instead, this review provides a handy reference to operations required in several chapters of the text.

A.2 NOMENCLATURE

Uppercase letters ($\mathbf{A}$, $\mathbf{B}$, $\mathbf{C}$) denote matrices of dimension n by m (n rows and m columns) where both n and m exceed 1. Lowercase letters ($\mathbf{x}$, $\mathbf{y}$) denote column vectors of dimension n by 1. A row vector would be denoted by $\mathbf{x}^T$. Where n is 1, the vector becomes a scalar. The symbol T denotes transpose while -1 signifies inversion. The general element of matrix $\mathbf{A}$ is denoted by a_{ij} (the element in row i and column j).

A.3 ADDITION AND SUBTRACTION

Two matrices or two column vectors may be added if and only if both components are of the exact same dimension (the number of rows and columns are equal). If two matrices **A** and **B** are to be added to yield **C**, then

$$\mathbf{C} = \mathbf{A} + \mathbf{B}$$

$$c_{ij} = a_{ij} + b_{ij}$$

For example, let

$$\mathbf{A} = \begin{bmatrix} 1 & 2 \\ 3 & 4 \end{bmatrix}$$

$$\mathbf{B} = \begin{bmatrix} 2 & 1 \\ 2 & 3 \end{bmatrix}$$

Then

$$\mathbf{C} = \begin{bmatrix} 3 & 3 \\ 5 & 7 \end{bmatrix}$$

Subtraction requires

$$\mathbf{C} = \mathbf{A} - \mathbf{B}$$

$$c_{ij} = a_{ij} - b_{ij}$$

For the **A** and **B** matrices above

$$\mathbf{C} = \begin{bmatrix} -1 & 1 \\ 1 & 1 \end{bmatrix}$$

A.4 TRANSPOSITION

The transpose of a matrix follows by interchanging rows and columns. The general element for the transpose of **A** follows from

$$a_{ij}^T = a_{ji}$$

For A and B above

$$\mathbf{A}^T = \begin{bmatrix} 1 & 3 \\ 2 & 4 \end{bmatrix}$$

$$\mathbf{B}^T = \begin{bmatrix} 2 & 2 \\ 1 & 3 \end{bmatrix}$$

A.5 MULTIPLICATION

The product $\mathbf{C}$ of two matrices $\mathbf{A} \times \mathbf{B}$ has a general element in row i and column j of $\mathbf{C}$ that results from multiplying each element of row i in $\mathbf{A}$ by each element in row j of $\mathbf{B}$ and summing the products. If one wants to compute

$$\mathbf{C} = \mathbf{A} \times \mathbf{B}$$

then the number of columns of A must equal the number of rows of B (so that an equal number of elements may be multiplied together). In general,

$$c_{ij} = \sum_{k=1}^{m} a_{ik} b_{kj}$$

If $\mathbf{A}$ is n by m then $\mathbf{B}$ must be m by (say p) so that $\mathbf{C}$ is n by p. For $\mathbf{A}$ above (2 by 2) and $\mathbf{B}$ (2 by 2), $\mathbf{C} = \mathbf{A} \times \mathbf{B}$ becomes 2 × 2.

$$C = \begin{bmatrix} 1 \times 2 + 2 \times 2 & 1 \times 1 + 2 \times 3 \\ 3 \times 2 + 4 \times 2 & 3 \times 1 + 4 \times 3 \end{bmatrix} = \begin{bmatrix} 6 & 7 \\ 14 & 15 \end{bmatrix}$$

DRILL PROBLEM

DA.1. Use $\mathbf{A}$, $\mathbf{B}$, and $\mathbf{C}$ to compute the indicated matrix functions.

$$\mathbf{A} = \begin{bmatrix} 1 & 2 & 3 \\ 0 & 1 & 2 \end{bmatrix} \qquad \mathbf{B} = \begin{bmatrix} 4 & 6 & 5 \\ 1 & 0 & 3 \end{bmatrix} \qquad \mathbf{C} = \begin{bmatrix} 2 & 1 \\ 0 & 3 \end{bmatrix}$$

(a) $\mathbf{A} + \mathbf{B}$

(b) $\mathbf{B} - \mathbf{A}$

(c) $\mathbf{A}^T$

(d) $\mathbf{A}^T\mathbf{B}$

(e) $\mathbf{B}^T\mathbf{C}$

(f) $\frac{1}{2}(\mathbf{C} + \mathbf{C}^T)$

Ans (a) $\begin{bmatrix} 5 & 8 & 8 \\ 1 & 1 & 5 \end{bmatrix}$; (b) $\begin{bmatrix} 3 & 4 & 2 \\ 1 & -1 & 1 \end{bmatrix}$; (c) $\begin{bmatrix} 1 & 0 \\ 2 & 1 \\ 3 & 2 \end{bmatrix}$

(d) $\begin{bmatrix} 4 & 6 & 5 \\ 9 & 12 & 13 \\ 14 & 18 & 21 \end{bmatrix}$; (e) $\begin{bmatrix} 8 & 7 \\ 12 & 6 \\ 10 & 14 \end{bmatrix}$; (f) $\begin{bmatrix} 2 & \frac{1}{2} \\ \frac{1}{2} & 3 \end{bmatrix}$

A.6 DETERMINANTS AND COFACTORS

The determinant of a matrix $\mathbf{A}$ (denoted by $|\mathbf{A}|$) is computed by adding all possible products of combinations of the matrix elements where each combination can contain only one element from each row and one element from each column. The determinant of a 2 by 2 matrix follows from the pattern

$$= a_{11}a_{22} - a_{21}a_{12}$$

For $\mathbf{A}$ and $\mathbf{B}$ above,

$$|\mathbf{A}| = 1 \times 4 - 3 \times 2 = 4 - 6 = -2$$

$$|\mathbf{B}| = 2 \times 3 - 2 \times 1 = 6 - 2 = 4$$

Where multiplication is valid,

$$|\mathbf{C}| = |\mathbf{A} \times \mathbf{B}| = |\mathbf{A}||\mathbf{B}|$$

Thus the determinant of a product (of matrices) is the product of the determinant. From Sec. A.5,

$$|\mathbf{C}| = 6 \times 15 - 14 \times 7 = 90 - 98 = -8$$

$$= -2 \times 4 = |\mathbf{A}||\mathbf{B}|$$

The determinant of a 3 by 3 matrix follows using a pattern of multiplications similar to the 2 by 2 pattern where the first two columns of the 3 by 3 are repeated. If one needs

$$|\mathbf{E}|$$

$$\mathbf{E} = \begin{bmatrix} 1 & 3 & 2 \\ 0 & 1 & 2 \\ 1 & 0 & 4 \end{bmatrix}$$

Then rewrite E and follow the indicated pattern.

$$= 4 + 6 + 0 - 2 - 0 - 0 = 8$$

The determinant of a higher-order matrix (and, as we shall see, the inverse of a matrix) requires the use of cofactors and minors. Let $\mathbf{De}(i, j, \mathbf{A})$ denote a matrix resulting from $\mathbf{A}$ where row i and column j are deleted. The determinant of each $\mathbf{De}$ is a minor of $\mathbf{A}$. Each such minor must be multiplied by $(-1)^{i+j}$ to yield a cofactor of $\mathbf{A}$ (a signed minor). Denote each cofactor by $\mathrm{Co}(i, j, \mathbf{A})$. For $\mathbf{E}$ above we have

$$\mathbf{De}(1, 1, \mathbf{E}) = \begin{bmatrix} 1 & 2 \\ 0 & 4 \end{bmatrix}$$

$$\mathrm{Co}(1, 1, \mathbf{E}) = (-1)^{1+1}|\mathbf{De}(1, 1, \mathbf{E})| = 1 \times 4 = 4$$

Similarly,

$$\mathbf{De}(1, 2, \mathbf{E}) = \begin{bmatrix} 0 & 2 \\ 1 & 4 \end{bmatrix}$$

$$\mathrm{Co}(1, 2, \mathbf{E}) = (-1)^{1+2}|\mathbf{De}(1, 2, \mathbf{E})| = -(-2) = 2$$

The remaining cofactors are

$$\mathrm{Co}(1, 3, \mathbf{E}) = -1$$
$$\mathrm{Co}(2, 1, \mathbf{E}) = -12$$
$$\mathrm{Co}(2, 2, \mathbf{E}) = +2$$
$$\mathrm{Co}(2, 3, \mathbf{E}) = +3$$
$$\mathrm{Co}(3, 1, \mathbf{E}) = +4$$
$$\mathrm{Co}(3, 2, \mathbf{E}) = -2$$
$$\mathrm{Co}(3, 3, \mathbf{E}) = +1$$

The determinant of $\mathbf{E}$ may be found by expanding $\mathbf{E}$ along any row or column. By expanding along row i is meant the computation of

$$|\mathbf{E}| = \sum_{k=1}^{m} e_{ik} \, \mathrm{Co}(i, k, \mathbf{E})$$

For row 2, we have

$$|\mathbf{E}| = e_{21} \, \mathrm{Co}(2, 1, \mathbf{E}) + e_{22} \, \mathrm{Co}(2, 2, \mathbf{E}) + e_{23} \, \mathrm{Co}(2, 3, \mathbf{E})$$

$$= 0 \times (-12) + 1 \times 2 + 2 \times 3 = 2 + 6 = 8$$

This agrees with the determinant computed above. If one expands along the rightmost column,

$$|\mathbf{E}| = 2 \times (-1) + 2 \times 3 + 4 \times 1 = 8$$

A.7 INVERSE

The inverse of a matrix A (if the inverse exists) must satisfy

$$\mathbf{AA}^{-1} = \mathbf{A}^{-1}\mathbf{A} = \mathbf{I}$$

where $\mathbf{I}$ is an identity matrix. $\mathbf{A}$ must be square. The determinant of $\mathbf{A}$ must not be zero; hence $\mathbf{A}$ must be nonsingular (a singular matrix is one with a zero-valued determinant). An identity matrix has ones along the main diagonal and zeros elsewhere. For example, a 3×3 identity matrix would be

$$\mathbf{I} = \begin{bmatrix} 1 & 0 & 0 \\ 0 & 1 & 0 \\ 0 & 0 & 1 \end{bmatrix}$$

Denote the adjoint of a matrix $\mathbf{A}$ by $\mathbf{Adj(A)}$, the general element of which is the cofactor $\mathrm{Co}(j, i, \mathbf{A})$. The inverse of $\mathbf{A}$ is given by

$$\mathbf{A}^{-1} = \frac{\mathbf{Adj(A)}}{|\mathbf{A}|}$$

For the matrix $\mathbf{E}$,

$$\mathbf{Adj(E)} = \begin{bmatrix} 4 & -12 & 4 \\ 2 & 2 & -2 \\ -1 & 3 & 1 \end{bmatrix}$$

$$\mathbf{E}^{-1} = \frac{\mathbf{Adj(E)}}{|\mathbf{E}|} = \begin{bmatrix} \frac{1}{2} & -\frac{3}{2} & \frac{1}{2} \\ \frac{1}{4} & \frac{1}{4} & -\frac{1}{4} \\ -\frac{1}{8} & \frac{3}{8} & \frac{1}{8} \end{bmatrix}$$

Note that $\mathbf{E}^{-1}\mathbf{E}$ and $\mathbf{EE}^{-1}$ both equal $\mathbf{I}$.

A.8 SIMULTANEOUS EQUATIONS

Suppose a set of simultaneous equations is given by

$$\mathbf{Ax} = \mathbf{y}$$

where $\mathbf{x}$ is an unknown vector, $\mathbf{y}$ is a known vector, and $\mathbf{A}$ is a matrix of coefficients (also known). There exists a unique solution where $\mathbf{A}$ is n by n with n independent equations (the determinant is not zero). If both sides of the matrix equation are multiplied by $\mathbf{A}^{-1}$,

$$\mathbf{x} = \mathbf{A}^{-1}\mathbf{y}$$

Suppose

$$\mathbf{A} = \begin{bmatrix} 1 & 2 \\ 3 & 4 \end{bmatrix} \qquad \mathbf{y} = \begin{bmatrix} 2 \\ 1 \end{bmatrix}$$

Then

$$\mathbf{Adj(A)} = \begin{bmatrix} 4 & -2 \\ -3 & 1 \end{bmatrix}$$

$$|\mathbf{A}| = -2$$

$$\mathbf{A}^{-1} = \begin{bmatrix} -2 & 1 \\ \frac{3}{2} & -\frac{1}{2} \end{bmatrix}$$

$$\mathbf{x} = \mathbf{A}^{-1}\mathbf{y} = \begin{bmatrix} -3 \\ \frac{5}{2} \end{bmatrix}$$

Cramer's rule provides an alternate way to solve the equation for x.

$$x_1 = \frac{\begin{vmatrix} 2 & 2 \\ 1 & 4 \end{vmatrix}}{\begin{vmatrix} 1 & 2 \\ 3 & 4 \end{vmatrix}} = \frac{8-2}{-2} = -3$$

$$x_2 = \frac{\begin{vmatrix} 1 & 2 \\ 3 & 1 \end{vmatrix}}{\begin{vmatrix} 1 & 2 \\ 3 & 4 \end{vmatrix}} = \frac{-5}{-2} = \frac{5}{2}$$

In general x_i is computed using a determinant ratio. The numerator matrix follows by replacing column i of $\mathbf{A}$ by $\mathbf{y}$. The denominator is always the determinant of $\mathbf{A}$. The $\mathbf{x}$ vector becomes

$$\mathbf{x} = \begin{bmatrix} x_1 \\ x_2 \end{bmatrix}$$

DRILL PROBLEMS

DA.2. Perform the indicated operations on

$$\mathbf{A} = \begin{bmatrix} 1 & 2 & 1 \\ 0 & 2 & 3 \\ 1 & 0 & 4 \end{bmatrix}$$

(a) Find Co(1, 1, $\mathbf{A}$).

(b) Find Co(3, 1, $\mathbf{A}$).

(c) Find the determinant of $\mathbf{A}$ by expanding down the first column.

(d) Find the determinant of $\mathbf{A}$ using the pattern for a 3 by 3 in A.6.

(e) Find $\mathbf{Adj(A)}$.

(f) Find A^{-1} using the adjoint.

Ans (a) 8; (b) 4; (c) $1 \times 8 + 1 \times 4 = 12$; (d) 12;

(e) $\begin{bmatrix} 8 & -8 & 4 \\ 3 & 3 & -3 \\ -2 & 2 & 2 \end{bmatrix}$; (f) $\begin{bmatrix} \frac{2}{3} & -\frac{2}{3} & \frac{1}{3} \\ \frac{1}{4} & \frac{1}{4} & -\frac{1}{4} \\ -\frac{1}{6} & \frac{1}{6} & \frac{1}{6} \end{bmatrix}$

DA.3. Perform the indicated operations on

$$A = \begin{bmatrix} 3 & 0 & 1 \\ 2 & 0 & 3 \\ 0 & 1 & 1 \end{bmatrix}$$

(a) Find the determinant of A by expanding down the second column.

(b) Find Co(2, 3, A).

(c) Find A^{-1} using the adjoint.

Ans (a) $1 \times (-7) = -7$; (b) -3; (c) $\begin{bmatrix} \frac{3}{7} & -\frac{1}{7} & 0 \\ -\frac{2}{7} & -\frac{3}{7} & 1 \\ -\frac{2}{7} & \frac{3}{7} & 0 \end{bmatrix}$

DA.4. For the A matrix of DA.2, solve

$$A \begin{bmatrix} x_1 \\ x_2 \\ x_3 \end{bmatrix} = \begin{bmatrix} 0 \\ 2 \\ 3 \end{bmatrix}$$

(a) By using the inverse previously computed

(b) By using Cramer's rule

Ans $\begin{bmatrix} -\frac{1}{3} \\ -\frac{1}{4} \\ \frac{5}{6} \end{bmatrix}$; $x_2 = \dfrac{\begin{vmatrix} 1 & 0 & 1 \\ 0 & 2 & 3 \\ 1 & 3 & 4 \end{vmatrix}}{\begin{vmatrix} 1 & 2 & 1 \\ 0 & 2 & 3 \\ 1 & 0 & 4 \end{vmatrix}}$

DA.5. Repeat DA.4 using A from DA.3.

Ans $\begin{bmatrix} -\frac{2}{7} \\ 2\frac{1}{7} \\ \frac{6}{7} \end{bmatrix}$; $x_1 = \dfrac{\begin{vmatrix} 0 & 0 & 1 \\ 2 & 0 & 3 \\ 3 & 1 & 1 \end{vmatrix}}{\begin{vmatrix} 3 & 0 & 1 \\ 2 & 0 & 3 \\ 0 & 1 & 1 \end{vmatrix}}$

A.9 EIGENVALUES AND EIGENVECTORS

It may be of interest to find a scalar λ and a vector $\mathbf{x}$ associated with a matrix $\mathbf{A}$ so that

$$\lambda \mathbf{x} = \mathbf{A}\mathbf{x}$$

The values of λ satisfying that equation are those λ (eigenvalues) for which

$$|\lambda \mathbf{I} - \mathbf{A}| = 0$$

Associated with each eigenvalue is a vector x (an eigenvector). If $\mathbf{A}$ is an n by n matrix, there are n eigenvalues. Associated with eigenvalue λ_i is an eigenvector $\mathbf{x}_i$ where

$$[\lambda_i \mathbf{I} - \mathbf{A}]\mathbf{x}_i = \mathbf{0}$$

No eigenvector is unique. Any eigenvector can be multiplied by a scalar and the result also satisfies the eigenvector equation. The fact that the $\mathbf{x}_i$ are not unique follows from the fact that $|\lambda_i \mathbf{I} - \mathbf{A}| = 0$.

Suppose

$$\mathbf{A} = \begin{bmatrix} -1 & 0 \\ 1 & -2 \end{bmatrix}$$

Then

$$|\lambda \mathbf{I} - \mathbf{A}| = \begin{vmatrix} \lambda + 1 & 0 \\ -1 & \lambda + 2 \end{vmatrix} = (\lambda + 1)(\lambda + 2)$$

The two eigenvalues are $\lambda_1 = -1$ and $\lambda_2 = -2$. (One could also write $\lambda_1 = -2$ and $\lambda_2 = -1$.) For the former choice eigenvectors are sought. For $\lambda_1 = -1$,

$$\begin{bmatrix} 1 & 0 \\ -1 & 1 \end{bmatrix}\mathbf{x}_1 = \begin{bmatrix} 0 & 0 \\ -1 & 1 \end{bmatrix}\begin{bmatrix} x_{11} \\ x_{21} \end{bmatrix} = \begin{bmatrix} 0 \\ 0 \end{bmatrix}$$

As long as the elements of $\mathbf{x}_1$ are such that as $x_{11} = x_{12}$, any such $\mathbf{x}_1$ vector is an eigenvector for λ_1. For example,

$$\mathbf{x}_1 = \begin{bmatrix} 1 \\ 1 \end{bmatrix}$$

is an eigenvector of λ_1 (or any multiple of x_1). For $\lambda_2 = -2$,

$$\begin{bmatrix} -1 & 0 \\ -1 & 0 \end{bmatrix}\mathbf{x}_2 = \begin{bmatrix} -1 & 0 \\ -1 & 0 \end{bmatrix}\begin{bmatrix} x_{12} \\ x_{22} \end{bmatrix} = \begin{bmatrix} 0 \\ 0 \end{bmatrix}$$

As long as x_{12} is 0, an $\mathbf{x}_2$ vector with any x_{22} will be an eigenvector for λ_2. For example,

$$\mathbf{x}_2 = \begin{bmatrix} 0 \\ 1 \end{bmatrix}$$

(or any multiple thereof) is acceptable.

If the eigenvalues are distinct, a matrix can be formed by writing the eigenvectors side by side. If one defines

$$\mathbf{P}^{-1} = [x_1 \mid x_2]$$

then (for distinct values of λ_1 and λ_2)

$$\mathbf{PAP}^{-1} = \begin{bmatrix} \lambda_1 & 0 \\ 0 & \lambda_2 \end{bmatrix}$$

In this example

$$\mathbf{P}^{-1} = \begin{bmatrix} 1 & 0 \\ 1 & 1 \end{bmatrix}$$

Therefore,

$$\mathbf{P} = \begin{bmatrix} 1 & 0 \\ -1 & 1 \end{bmatrix}$$

$$\mathbf{PAP}^{-1} = \begin{bmatrix} 1 & 0 \\ -1 & 1 \end{bmatrix}\begin{bmatrix} -1 & 0 \\ 1 & -2 \end{bmatrix}\begin{bmatrix} 1 & 0 \\ 1 & 1 \end{bmatrix} = \begin{bmatrix} -1 & 0 \\ 0 & -2 \end{bmatrix}$$

DRILL PROBLEMS

DA.6. Perform the indicated operations on

$$\mathbf{A} = \begin{bmatrix} 1 & 2 \\ 0 & 3 \end{bmatrix}$$

(a) Obtain the characteristic polynomial $|\lambda\mathbf{I} - \mathbf{A}| = 0$.

(b) Substitute $\mathbf{A}$ for λ in the characteristic polynomial and show that $\mathbf{A}$ satisfies its own characteristic polynomial. Zero here implies a 2 by 2 matrix of zeros. Multiply the λ^0 coefficient by $\mathbf{I}$.

(c) Solve the characteristic polynomial for λ_1 and λ_2.

(d) Obtain any eigenvector $\mathbf{x}_1$ for λ_1 and $\mathbf{x}_2$ for λ_2.

> **Ans** (a) $\lambda^2 - 4\lambda + 3$; (b) $\mathbf{A}^2 - 4\mathbf{A} + 3\mathbf{I} = \begin{bmatrix} 0 & 0 \\ 0 & 0 \end{bmatrix}$;
> (c) $\lambda_1 = +1$, $\lambda_2 = +3$ (or $\lambda_1 = +3$ and $\lambda_2 = +1$);
> (d) $\mathbf{x}_1 = $ any multiple of $\begin{bmatrix} 1 \\ 0 \end{bmatrix}$; $\mathbf{x}_2 = $ any multiple of $\begin{bmatrix} 1 \\ 1 \end{bmatrix}$

DA.7. Repeat DA.6 with

$$\mathbf{A} = \begin{bmatrix} 3 & 0 \\ 1 & -2 \end{bmatrix}$$

(a) $\lambda^2 - \lambda - 6 = 0$

(b) $\mathbf{A}^2 - \mathbf{A} - 6\mathbf{I} = \begin{bmatrix} 0 & 0 \\ 0 & 0 \end{bmatrix}$

(c) $\lambda_1 = 3 \quad \lambda_2 = -2$

(d) $\mathbf{x}_1 =$ any multiple of $\begin{bmatrix} 5 \\ 1 \end{bmatrix}$

$\mathbf{x}_2 =$ any multiple of $\begin{bmatrix} 0 \\ 1 \end{bmatrix}$

A.10 DERIVATIVE OF A SCALAR WITH RESPECT TO A VECTOR

The gradient vector is defined as the derivative of a scalar function of a vector taken with respect to that vector. Suppose the scalar is called $f(\mathbf{x})$, where $\mathbf{x}$ is an n by 1 vector. Then

$$\frac{df(\mathbf{x})}{d\mathbf{x}} = \begin{bmatrix} \dfrac{df(\mathbf{x})}{dx_1} \\[2mm] \dfrac{df(\mathbf{x})}{dx_2} \\[2mm] \vdots \\[2mm] \dfrac{df(\mathbf{x})}{dx_N} \end{bmatrix}$$

For example, if

$$f(\mathbf{x}) = x_1^2 + 2x_1 x_2 + 2x_2^3$$

then the gradient vector is

$$\frac{df(\mathbf{x})}{d\mathbf{x}} = \begin{bmatrix} 2x_1 + 2x_2 \\ 2x_1 + 6x_2^2 \end{bmatrix}$$

A special case of the gradient pertains to

$$f(\mathbf{x}) = \begin{bmatrix} y_1 & y_2 \end{bmatrix} \begin{bmatrix} x_1 \\ x_2 \end{bmatrix} = \mathbf{y}^T \mathbf{x} = y_1 x_1 + y_2 x_2$$

where $\mathbf{y}$ is a vector also. From the definition of the gradient vector,

$$\frac{df(\mathbf{x})}{d\mathbf{x}} = \begin{bmatrix} y_1 \\ y_2 \end{bmatrix} = \mathbf{y}$$

Notice also that

$$\frac{df(\mathbf{y})}{d\mathbf{y}} = \begin{bmatrix} x_1 \\ x_2 \end{bmatrix} = \mathbf{x}$$

In general then, for

$$f(\mathbf{x}) = \mathbf{y}^T\mathbf{x}$$

$$\frac{df(\mathbf{x})}{d\mathbf{x}} = \mathbf{y}$$

$$\frac{df(\mathbf{y})}{d\mathbf{y}} = \mathbf{x}$$

This result may be applied to quadratic forms.

DRILL PROBLEMS

DA.8. What is the gradient vector for $f(\mathbf{x}) = 2x_1x_2 + e^{3x_2}$ computed for $x_1 = 1$, $x_2 = -1$?

Ans $\begin{bmatrix} 2x_2 \\ 2x_1 + 3e^{3x_2} \end{bmatrix}\begin{bmatrix} 1 \\ -1 \end{bmatrix} = \begin{bmatrix} -2 \\ 2 + 3e^{-3} \end{bmatrix}$

DA.9. What is $\dfrac{df(\mathbf{x})}{d\mathbf{x}}$ for $f(\mathbf{x}) = \mathbf{y}^T\mathbf{x}$, where

$$y = \begin{bmatrix} 1 \\ 3 \\ 4 \end{bmatrix} \qquad x = \begin{bmatrix} x_1 \\ x_2 \\ x_3 \end{bmatrix}$$

Ans $\begin{bmatrix} 1 \\ 3 \\ 4 \end{bmatrix}$

A.11 QUADRATIC FORMS AND SYMMETRY

A quadratic form is a special scalar function of a vector x where

$$f(\mathbf{x}) = \mathbf{x}^T\mathbf{Q}\mathbf{x}$$

In order to be able to compute $f(\mathbf{x})$ where $\mathbf{x}$ is an n by 1 vector, the matrix $\mathbf{Q}$ must be n by n (that is, square). It can be shown that no loss of generality ensues by requiring $\mathbf{Q}$ to be symmetrical.

A symmetrical matrix is equal to its own transpose, that is, $\mathbf{Q}$ is symmetrical if and only if

$$\mathbf{Q}^T = \mathbf{Q}$$

The general element requires

$$q_{ij}^T = q_{ji} = q_{ij}$$

For example,

$$\mathbf{Q} = \begin{bmatrix} 1 & 2 \\ 2 & 3 \end{bmatrix}$$

is symmetrical because $q_{12} = q_{21}$. However,

$$\mathbf{Q} = \begin{bmatrix} 1 & -2 \\ 4 & 3 \end{bmatrix}$$

is not symmetrical because $q_{12} \neq q_{21}$. A matrix

$$\mathbf{Q}_s = \tfrac{1}{2}(\mathbf{Q} + \mathbf{Q}^T)$$

is always symmetrical because

$$\mathbf{Q}_s^T = \mathbf{Q}_s$$

Notice that any Q can be written as

$$\mathbf{Q} = \tfrac{1}{2}(\mathbf{Q} + \mathbf{Q}^T) + \tfrac{1}{2}(\mathbf{Q} - \mathbf{Q}^T)$$

$$= \mathbf{Q}_s + \mathbf{Q}_{sk}$$

$\mathbf{Q}_s$ is symmetrical. $\mathbf{Q}_{sk}$ is called a skew symmetric matrix because

$$q_{sk_{ij}} = \tfrac{1}{2}(q_{ij} - q_{ji}) = -\tfrac{1}{2}(q_{ji} - q_{ij})$$

$$= -q_{sk_{ji}}$$

$$q_{sk_{ii}} = 0$$

that is, the diagonal elements of $\mathbf{Q}_{sk}$ are 0 while the off-diagonal elements are equal and opposite. For example,

$$\begin{matrix} \mathbf{Q} & = & \mathbf{Q}_s & + & \mathbf{Q}_{sk} \end{matrix}$$

$$\begin{bmatrix} 1 & -2 \\ 4 & -3 \end{bmatrix} = \begin{bmatrix} 1 & 1 \\ 1 & 3 \end{bmatrix} + \begin{bmatrix} 0 & -3 \\ 3 & 0 \end{bmatrix}$$

It can be shown for any $\mathbf{Q}$ that

$$x^T\mathbf{Q}x = x^T[\mathbf{Q}_s + \mathbf{Q}_{sk}]x = x^T\mathbf{Q}_s x$$

since the skew symmetric part of $\mathbf{Q}$ does not contribute to $f(\mathbf{x})$. For example,

$$[x_1 \quad x_2]\begin{bmatrix} 0 & -3 \\ 3 & 0 \end{bmatrix}\begin{bmatrix} x_1 \\ x_2 \end{bmatrix} = 0$$

For this reason, $\mathbf{Q}$ is always a symmetric matrix when $f(\mathbf{x})$ is a quadratic form.

In order to differentiate $f(\mathbf{x})$ with respect to $\mathbf{x}$, it is useful to form the derivative with respect to each occurrence of $\mathbf{x}$ while holding the other part constant.

$$\frac{df(\mathbf{x})}{d\mathbf{x}} = \frac{d(\mathbf{y}^T\mathbf{x})}{d\mathbf{x}} + \frac{d(\mathbf{x}^T\mathbf{y}_1)}{d\mathbf{x}}$$

where

$$\mathbf{y} = \mathbf{Q}^T\mathbf{x}$$

$$\mathbf{y}_1 = \mathbf{Q}\mathbf{x}$$

Using the results from the end of Sec. A.10,

$$\frac{df(\mathbf{x})}{d\mathbf{x}} = \mathbf{y} + \mathbf{y}_1 = (\mathbf{Q}^T + \mathbf{Q})\mathbf{x}$$

Assuming that $\mathbf{Q}$ is symmetric,

$$\frac{df(\mathbf{x})}{d\mathbf{x}} = 2\mathbf{Q}\mathbf{x}$$

A.12 DEFINITENESS

The definiteness of a scalar quadratic form $f(\mathbf{x})$ depends on potential values of $f(\mathbf{x})$ for all possible nonzero values of the vector $\mathbf{x}$. A nonzero value of $\mathbf{x}$ implies that at least one element of $\mathbf{x}$ is nonzero.

$f(\pmb{x})$	Type of Definiteness
> 0	Positive definite
≥ 0	Positive semidefinite
< 0	Negative definite
≤ 0	Negative semidefinite
None of above	Indefinite

The definiteness of a symmetric matrix $\mathbf{Q}$ (associated with a quadratic form $\mathbf{x}^T\mathbf{Q}\mathbf{x}$) depends on the real part σ_i of eigenvalues $\lambda_i = \sigma_i + j\omega_i$.

σ_i	Type of Definiteness
> 0	Positive definite
≥ 0	Positive semidefinite
< 0	Negative definite
≤ 0	Negative semidefinite
None of above	Indefinite

The definiteness of an n by n symmetric matrix $\mathbf{Q}$ may be tested by examining the value of a sequence of n determinants, det $(i, \mathbf{Q})$, each defined as the determinant of the upper left-hand i by i submatrix of $\mathbf{Q}$.

The index i varies from 1 to n.

Test	Type of Definiteness
det $(i, \mathbf{Q}) > 0$	Positive definite
det $(i, \mathbf{Q}) \geq 0$	Positive semidefinite
det $(i, -\mathbf{Q}) > 0$	Negative definite
det $(i, -\mathbf{Q}) \geq 0$	Negative semidefinite
None of above	Indefinite

Suppose

$$\mathbf{Q} = \begin{bmatrix} -1 & 0 & 0 \\ 0 & -\frac{5}{2} & \frac{1}{2} \\ 0 & \frac{1}{2} & -\frac{5}{2} \end{bmatrix}$$

$$|\lambda \mathbf{I} - \mathbf{Q}| = (\lambda + 1)(\lambda + 2)(\lambda + 3)$$

The test of the real parts of the λ_i indicates the $\mathbf{Q}$ is negative definite since the real parts are all negative. Using the determinant test, one first examines

$$\det (1, \mathbf{Q}) = |-1| = -1$$

Since this determinant is negative, the matrix $\mathbf{Q}$ cannot be positive definite or positive semidefinite. The determinant test now examines $-\mathbf{Q}$.

$$\det (1, -\mathbf{Q}) = |1| = 1$$

$$\det (2, -\mathbf{Q}) = \begin{vmatrix} -1 & 0 \\ 0 & -\frac{5}{2} \end{vmatrix} = \frac{5}{2}$$

$$\det (3, -\mathbf{Q}) = |-\mathbf{Q}| = 6$$

Since all three determinants are positive, the matrix is identified as negative definite.

DRILL PROBLEMS

DA.10. Use the determinant test to establish the definiteness of

(a) $\begin{bmatrix} 6 & 3 & 0 \\ 3 & 2 & 0 \\ 0 & 0 & 4 \end{bmatrix}$

$$(b) \quad \begin{bmatrix} -2 & 0 & 2 \\ 0 & -3 & 0 \\ 2 & 0 & -4 \end{bmatrix}$$

Ans   (a) positive definite;   (b) negative definite

DA.11. What is $\dfrac{df(\mathbf{x})}{d\mathbf{x}}$, where

$$f(\mathbf{x}) = \begin{bmatrix} x_1 \\ x_2 \end{bmatrix}^T \begin{bmatrix} 6 & 2 \\ 2 & 1 \end{bmatrix} \begin{bmatrix} x_1 \\ x_2 \end{bmatrix}$$

Ans $\begin{bmatrix} 12 & 4 \\ 4 & 2 \end{bmatrix}$

A.13 RANK

The rank of a matrix $\mathbf{A}$ indicates the number of linearly independent pieces of information contained in $\mathbf{A}$. If $\mathbf{A}$ is a square n by n matrix, the maximum possible rank is n. If $\mathbf{A}$ is not square with dimension n by m where n and m are not equal, the maximum possible rank is the smaller of n or m denoted by $\min(n, m)$.

In order to formulate a test for rank, it is useful to define $\mathbf{Pe}(i, \mathbf{A}, j)$ to be permutation j where an i by i submatrix of $\mathbf{A}$ is generated. A submatrix occurs when the rows and columns of $\mathbf{A}$ are (or are not) exchanged and a contiguous i by i matrix is selected. For a 3 by 3 matrix, several 2 by 2 submatrices can be generated but only one 3 by 3 would be used. The rank of $\mathbf{A}$ is the largest value of i for which the determinant of at least one $\mathbf{Pe}$ is nonzero. The test begins by testing for the largest possible rank (n or $\min(n, m)$). Should all $\mathbf{Pe}$ yield a zero-valued determinant for that i, i is decremented (reduced by 1) and the test continues. Suppose

$$\mathbf{A} = \begin{bmatrix} 1 & 2 & 1 \\ 0 & 1 & 3 \\ 0 & 2 & 6 \end{bmatrix}$$

Since $\mathbf{A}$ is square and 3 by 3, the test first examines the determinant of $\mathbf{A}$ since, for all $j | \mathbf{Pe}(3, \mathbf{A}, j)|$ will be the same. However, $|\mathbf{A}|$ is zero, so the rank is not 3. For the first permutation of a 2 by 2 submatrix let

$$\mathbf{Pe}(2, \mathbf{A}, 1) = \begin{bmatrix} 1 & 2 \\ 0 & 1 \end{bmatrix}$$

Obviously that determinant is 1; hence that rank is 2, since at least one 2 by 2 submatrix has a nonzero determinant.

DRILL PROBLEMS

DA.12. What is the rank of

$$A = \begin{bmatrix} 1 & 2 & 1 \\ 1 & 2 & 3 \\ 1 & 2 & 2 \end{bmatrix}$$

Ans 2

DA.13. What is the rank of

$$A = \begin{bmatrix} 1 & 0 & 0 \\ 0 & 2 & 0 \\ 0 & 0 & 3 \end{bmatrix}$$

Ans 3

DA.14. What is the rank of

$$A = \begin{bmatrix} 1 & 0 & 2 \\ 0 & 0 & 0 \\ 2 & 0 & 4 \end{bmatrix}$$

Ans 1

A.14 PARTITIONED MATRICES

When a matrix is partitioned into component parts, the normal rules of addition and multiplication may be applied to the components (including compatibility of dimensions). If

$$A = \begin{bmatrix} a_{11} & a_{12} \\ a_{21} & a_{22} \end{bmatrix} \qquad B = \begin{bmatrix} b_{11} \\ b_{21} \end{bmatrix}$$

$$C = \begin{bmatrix} c_{11} & c_{12} \\ c_{21} & c_{22} \end{bmatrix}$$

then

$$A + C = \begin{bmatrix} a_{11} + c_{11} & a_{12} + c_{12} \\ a_{21} + c_{21} & a_{22} + c_{22} \end{bmatrix}$$

$$AB = \begin{bmatrix} a_{11}b_{11} + a_{12}b_{21} \\ a_{21}b_{11} + a_{22}b_{21} \end{bmatrix}$$

For example, if

$$\mathbf{A} = \begin{bmatrix} 1 & 2 & 0 \\ 4 & 5 & 6 \\ 7 & 8 & 9 \end{bmatrix} \qquad \mathbf{B} = \begin{bmatrix} 1 \\ 2 \\ 3 \end{bmatrix}$$

$$\mathbf{C} = \begin{bmatrix} 9 & 8 & 7 \\ 6 & 5 & 4 \\ 3 & 2 & 1 \end{bmatrix}$$

then the above rules result in values for $\mathbf{A} + \mathbf{C}$ and $\mathbf{AB}$

$$\mathbf{A} + \mathbf{C} = \begin{bmatrix} 10 & 10 & 7 \\ 10 & 10 & 10 \\ 10 & 10 & 10 \end{bmatrix}$$

$$\mathbf{AB} = \begin{bmatrix} 1 + 4 \\ \begin{bmatrix} 4 \\ 7 \end{bmatrix} + \begin{bmatrix} 28 \\ 43 \end{bmatrix} \end{bmatrix} = \begin{bmatrix} 5 \\ 32 \\ 50 \end{bmatrix}$$

The determinant of a $\mathbf{A}$ can be found using the components.

$$|\mathbf{A}| = |\mathbf{a}_{11}||\mathbf{a}_{22} - \mathbf{a}_{21}\mathbf{a}_{11}^{-1}\mathbf{a}_{12}|$$

The inverse of $\mathbf{A}$ is given by

$$\mathbf{D} = \begin{bmatrix} \mathbf{d}_{11} & \mathbf{d}_{12} \\ \mathbf{d}_{21} & \mathbf{d}_{22} \end{bmatrix} = \mathbf{A}^{-1}$$

$$\mathbf{d}_{22} = [\mathbf{a}_{22} - \mathbf{a}_{21}\mathbf{a}_{11}^{-1}\mathbf{a}_{12}]^{-1}$$

$$\mathbf{d}_{12} = -\mathbf{a}_{11}^{-1}\mathbf{a}_{12}\mathbf{d}_{22}$$

$$\mathbf{d}_{21} = -\mathbf{d}_{22}\mathbf{a}_{21}\mathbf{a}_{11}^{-1}$$

$$\mathbf{d}_{11} = \mathbf{a}_{11}^{-1}[\mathbf{I} - \mathbf{a}_{12}\mathbf{d}_{21}]$$

For example, the determinant and inverse of a 3 by 3 matrix can be found by partitioning

$$\mathbf{A} = \begin{bmatrix} 1 & 2 & 0 \\ 4 & 5 & 6 \\ 7 & 8 & 9 \end{bmatrix}$$

then

$$|\mathbf{A}| = |1|\left| -\begin{bmatrix} 4 \\ 7 \end{bmatrix}(1)[2 \quad 0] + \begin{bmatrix} 5 & 6 \\ 8 & 9 \end{bmatrix} \right|$$

$$= \left| -\begin{bmatrix} 8 & 0 \\ 14 & 0 \end{bmatrix} + \begin{bmatrix} 5 & 6 \\ 8 & 9 \end{bmatrix} \right| = \begin{vmatrix} -3 & 6 \\ -6 & 9 \end{vmatrix} = -27 + 36 = +9$$

The inverse follows because

$$\mathbf{d}_{22} = \left[\begin{bmatrix} 5 & 6 \\ 8 & 9 \end{bmatrix} - \begin{bmatrix} 4 \\ 7 \end{bmatrix} (1) [2 \quad 0] \right]^{-1}$$

$$= \begin{bmatrix} -3 & 6 \\ -6 & 9 \end{bmatrix}^{-1} = \begin{bmatrix} 1 & -\frac{2}{3} \\ \frac{2}{3} & -\frac{1}{3} \end{bmatrix}$$

$$\mathbf{d}_{12} = -(1)[2 \quad 0] \begin{bmatrix} 1 & -\frac{2}{3} \\ \frac{2}{3} & -\frac{1}{3} \end{bmatrix} = [-2 \quad \frac{4}{3}]$$

$$\mathbf{d}_{21} = -\begin{bmatrix} 1 & -\frac{2}{3} \\ \frac{2}{3} & -\frac{1}{3} \end{bmatrix} \begin{bmatrix} 4 \\ 7 \end{bmatrix} (1) = \begin{bmatrix} \frac{2}{3} \\ -\frac{1}{3} \end{bmatrix}$$

$$\mathbf{d}_{11} = (1) \left[1 - [2 \quad 0] \begin{bmatrix} \frac{2}{3} \\ -\frac{1}{3} \end{bmatrix} \right] = -\frac{1}{3}$$

$$\mathbf{A}^{-1} = \left[\begin{array}{c|cc} -\frac{1}{3} & -2 & \frac{4}{3} \\ \hline \frac{2}{3} & 1 & -\frac{2}{3} \\ -\frac{1}{3} & \frac{2}{3} & -\frac{1}{3} \end{array} \right]$$

As a check, notice that

$$\mathbf{AA}^{-1} = \mathbf{I}$$

The partitioned approach reduces the dimension of the inversion process. For example, the 3 by 3 inversion example was reduced to a 2 by 2 inversion. A 4 by 4 inversion can also be reduced to a 2 by 2 inversion.

DRILL PROBLEMS

DA.15. $\mathbf{A} = \left[\begin{array}{c|cc} 2 & 1 & 3 \\ \hline 0 & 1 & 0 \\ 1 & 0 & -2 \end{array} \right]$ $\mathbf{B} = \left[\begin{array}{c} 2 \\ \hline 1 \\ 6 \end{array} \right]$

Answer the following where **A** and **B** are partitioned as above.

(a) What is **AB**?

(b) What is $|\mathbf{A}|$?

(c) What is $\mathbf{A}^{-1}$?

Ans (a) $\left[\begin{array}{c} 23 \\ \hline 1 \\ -10 \end{array} \right]$; (b) $2 \begin{vmatrix} 1 & 0 \\ -\frac{1}{2} & -\frac{7}{2} \end{vmatrix} = -7$; (c) $\begin{bmatrix} \frac{2}{7} & -\frac{2}{7} & \frac{3}{7} \\ 0 & 1 & 0 \\ \frac{1}{7} & -\frac{1}{7} & -\frac{2}{7} \end{bmatrix}$

DA.16. Repeat DA.15c, where

$$A = \begin{bmatrix} 2 & 1 & 3 \\ 0 & 1 & 0 \\ 1 & 0 & -2 \end{bmatrix}$$

Ans $A^{-1} = \begin{bmatrix} \frac{2}{7} & -\frac{2}{7} & \frac{3}{7} \\ 0 & 1 & 0 \\ \frac{1}{7} & -\frac{1}{7} & -\frac{2}{7} \end{bmatrix}$

PROBLEMS

1. Use A and x to compute the indicated matrix functions

$$A = \begin{bmatrix} 3 & 2 \\ 1 & 0 \end{bmatrix} \qquad x = \begin{bmatrix} 1 \\ 2 \end{bmatrix}$$

(a) $A^T A$

(b) Ax

(c) $x^T A^T A x$

(d) $x^T x$

Ans   (a) $\begin{bmatrix} 10 & 6 \\ 6 & 4 \end{bmatrix}$;   (b) $\begin{bmatrix} 7 \\ 1 \end{bmatrix}$;   (c) 50;   (d) 5

2. Use A, B, x to compute the indicated matrix functions.

$$A = \begin{bmatrix} 4 & 3 \\ 6 & 2 \end{bmatrix} \qquad B = \begin{bmatrix} 2 & 1 \\ 3 & 6 \\ 4 & 2 \end{bmatrix} \qquad x = \begin{bmatrix} 1 \\ 3 \\ 6 \end{bmatrix}$$

(a) $B^T B$

(b) BA

(c) $x^T B$

(d) $x^T BA$

3. Perform the indicated operations on

$$A = \begin{bmatrix} 2 & 1 & 6 \\ 3 & 1 & 2 \\ 4 & 2 & 12 \end{bmatrix}$$

(a) $Co(3, 1, A)$

(b) $Co(2, 2, A)$

(c) determinant of A

(d) adjoint of **A**

(e) Does $\mathbf{A}^{-1}$ exist? If so, find $\mathbf{A}^{-1}$.

(a) **Ans** (a) -4; (b) 0; (c) 0; (d) $\begin{bmatrix} 8 & 0 & -4 \\ -28 & 0 & 14 \\ 2 & 0 & -1 \end{bmatrix}$; (e) no

4. Repeat Prob. 3, where

$$\mathbf{A} = \begin{bmatrix} 1 & 0 & 2 \\ 1 & 1 & 0 \\ 0 & 2 & 3 \end{bmatrix}$$

5. Find $\mathbf{A}^{-1}$ for

$$\mathbf{A} = \begin{bmatrix} 3 & 0 & 2 \\ 1 & 0 & 0 \\ 0 & 1 & 0 \end{bmatrix}$$

Ans $\begin{bmatrix} 0 & 1 & 0 \\ 0 & 0 & 1 \\ \frac{1}{2} & -\frac{3}{2} & 0 \end{bmatrix}$

6. Find x using Cramer's rule.

$$3x_1 + 2x_2 = 4$$
$$x_1 - 3x_2 = 2$$

7. Find x using the inverse from Prob. 5.

$$\mathbf{A} \begin{bmatrix} x_1 \\ x_2 \\ x_3 \end{bmatrix} = \begin{bmatrix} 2 \\ 1 \\ 2 \end{bmatrix}$$

Ans $\begin{bmatrix} 1 \\ 2 \\ -\frac{1}{2} \end{bmatrix}$

8. Does the following have a unique solution?

$$x_1 + 3x_2 + x_3 = 7$$
$$x_1 - x_2 = 3$$
$$2x_1 - 2x_2 = 2$$

9. What is the characteristic polynomial for

$$\mathbf{A} = \begin{bmatrix} 2 & 1 \\ 3 & 2 \end{bmatrix}$$

Ans $\lambda^2 - 4 + 1$

10. What are the characteristic polynomials for **A**, **B**, **AB**, where

$$A = \begin{bmatrix} 1 & 2 \\ 1 & 0 \end{bmatrix} \qquad B = \begin{bmatrix} 2 & 0 \\ 0 & 2 \end{bmatrix}$$

11. Show that the matrix in Prob. 9 satisfies its own characteristic polynomial.

12. What are the eigenvalues of

$$A = \begin{bmatrix} 2 & 0 & 0 \\ 0 & 1 & 2 \\ 0 & 3 & 2 \end{bmatrix}$$

13. What are the eigenvectors of

$$A = \begin{bmatrix} 1 & 0 \\ 1 & 2 \end{bmatrix}$$

Ans 1, 2; for $\lambda = 1$ any multiple of $\begin{bmatrix} 1 \\ -1 \end{bmatrix}$; for $\lambda = 2$ any multiple of $\begin{bmatrix} 0 \\ 1 \end{bmatrix}$

14. Find a matrix **P** such that PAP^{-1} is diagonal using the matrix **A** from Prob. 13.

15. What is the gradient vector of $f(\mathbf{x}) = \sin 2x_1 + x_2 e^{2x_1}$?

Ans $\begin{bmatrix} 2\cos 2x_1 + 2x_2 e^{2x_1} \\ e^{2x_1} \end{bmatrix}$

16. Show that

$$\frac{df(\mathbf{x})}{d\mathbf{x}} = \mathbf{y}$$

$$f(\mathbf{x}) = \mathbf{y}^T\mathbf{x} = y_1 x_1 + y_2 x_2 + y_3 x_3$$

17. For the quadratic form

$$f(\mathbf{x}) = 2x_1^2 + 3x_1 x_2 + 4x_2^2$$

What are

(a) **Q** for $f(\mathbf{x}) = \mathbf{x}^T\mathbf{Q}\mathbf{x}$

(b) $\dfrac{df(\mathbf{x})}{d\mathbf{x}}$

(c) Definiteness of **Q**

Ans (a) $Q = \begin{bmatrix} 2 & 1.5 \\ 1.5 & 4 \end{bmatrix}$; (b) $\begin{bmatrix} 4 & 3 \\ 3 & 8 \end{bmatrix}$; (c) positive-definite

18. What is the definiteness of

$$\mathbf{Q} = \begin{bmatrix} 1 & 2 & 0 \\ 2 & 1 & 0 \\ 0 & 0 & 8 \end{bmatrix}$$

19. For the **A** matrix of Prob. 5 show that

$$\mathbf{x}^T \mathbf{A} \mathbf{x} = \mathbf{x}^T \mathbf{A}_s \mathbf{x}$$

$$\mathbf{A}_s = \tfrac{1}{2}[\mathbf{A} + \mathbf{A}^T]$$

20. What is the rank of the **A** matrix in Prob. 3?

21. What is the rank of the **A** matrix in Prob. 5?

Ans 3

22. Answer the following questions using the indicated partitioned matrices.

$$\mathbf{A} = \begin{bmatrix} 1 & 0 & 1 \\ \hline 2 & 1 & 1 \\ 3 & 2 & 6 \end{bmatrix} \qquad \mathbf{B} = [1 \mid 3 \quad 2]$$

(a) What is **BA**?

(b) What is $\mathbf{A}^T \mathbf{B}^T$?

(c) What is $|\mathbf{A}|$?

(d) What is $\mathbf{A}^{-1}$?

23. Obtain the inverse of the following 4×4 matrix by partitioning into 2×2 matrices

$$\mathbf{A} = \begin{bmatrix} 0 & 1 & 2 & 3 \\ 2 & 1 & 0 & 0 \\ 1 & 0 & 2 & 1 \\ 0 & 1 & 1 & 2 \end{bmatrix}$$

Ans $\mathbf{A}^{-1} = \begin{bmatrix} 3 & 1 & -1 & -4 \\ -6 & -1 & 2 & 8 \\ \hline -4 & -1 & 2 & 5 \\ 5 & 1 & -2 & -6 \end{bmatrix}$

Appendix
B

Computing Phase Crossover Frequency, Gain Crossover Frequency, and Breakaway Points

B.1 PREVIEW

There are several places in this textbook where specific numbers must be computed to fit a condition that may require the solution of some complicated function. Trial and error could be employed; however, in this section a procedure is developed by which a value can be found automatically. A computer can generate the needed number.

The method is developed from a numerical analysis procedure called Newton-Raphson. The equations are written so as to be programmable in a high-level language like Fortran, Basic, and Pascal. The equations can even be implemented by keystroke programming of a pocket-sized calculator.

B.2 NEWTON-RAPHSON

The Newton-Raphson procedure tries to find a value of a variable x so as to cause some function $f(x)$ to be zero. The function may be chosen to fulfill some condition. For example, the square root of 2 results if the requirement is

$$f(x) = x^2 - 2 = 0$$

Similarly, the fourth root of 10.637 would result if

$$f(x) = x^4 - 10.637 = 0$$

The angle whose tangent is 1.56 results from

$$f(x) = \tan x - 1.56 = 0$$

The problem, of course, is that x is unknown. A procedure to search for x is needed. A Taylor series can be written if a guess at the solution x_0 is available. A Taylor series for $f(x)$ in the vicinity of x_0 is

$$f(x) = f(x_0) + \frac{df(x)}{dx}\bigg|_{x=x_0} (x - x_0) + \text{other terms}$$
$$= 0 \tag{1}$$

If the higher-order terms are ignored, an approximation for x can be computed. That approximation can be called x_1 so that

$$x_1 = x_0 - \frac{f(x_0)}{\dfrac{df(x)}{dx}\bigg|_{x=x_0}} \tag{2}$$

In some cases the derivative is easy to find. In order to find the square root of 2,

$$\frac{df(x)}{dx} = 2x\bigg|_{x=x_0} = 2x_0$$

$$x_1 = x_0 - \frac{x_0^2 - 2}{2x_0} = \frac{1}{2}x_0 + \frac{1}{x_0}$$

If the initial guess starts at 1, a sequence of numbers results from repeating the equation where x_0 is replaced by each new x_1. The sequence (to four significant digits) is 1.5, 1.417, 1.414, 1.414 (converged).

If $f(x)$ is more complicated, the derivative can be approximated as in the next section.

B.3 COMPUTING GAIN AND PHASE CROSSOVER FREQUENCIES

In Chap. 6, the gain and phase crossover frequencies provide a means to compute phase and gain margins, respectively. Sometimes the expressions for dB and phase are easy to solve, sometimes not. For example, the gain crossover frequency requires solution of

$$x = \omega$$

$$f(x) = 20 \log_{10} A(x) = 0 \tag{3}$$

Similarly, the phase crossover frequency could be solved if

$$x = \omega$$

$$f(x) = P(x) + 180° = 0 \tag{4}$$

In the above, $A(x)$ is the magnitude and $P(x)$ is the phase. In either case, the Newton-Raphson method can be applied by approximating the derivative for small d.

$$\frac{df(x)}{dx} \simeq \frac{f(x + d) - f(x)}{d} \tag{5}$$

If d is set equal to 1 percent of x_0, then (2) becomes

$$x_1 = x_0 - \frac{(0.1x_0)f(x_0)}{f(1.01x_0) - f(x_0)} \tag{6}$$

The following procedure can be used to find the gain or phase crossover frequency by applying (6) to either (3) or (4). The lines are numbered as they would be in a normal computer program to make it easy to convert the procedure to some available language (or even to use keystroke programming on a pocket-sized calculator). The indentation of certain lines is intended to make the program easier to read.

Sometimes the following scheme will run away, rather than converge to the correct answer. The initial guess x_0 should be somewhere near to the eventual answer. If the scheme does not converge, a different initial guess should be tried.

The threshold T determines the number of significant digits in the answer. For example, if $T = 0.0001$, then the first three digits after the decimal will be correct for x. If x is around 9 and T is 0.0001, the final value of x would be good to four significant digits: one before the decimal and three after.

To get the phase crossover frequency, F is (3) and G is (4), the resulting gain margin. To get the gain crossover frequency, F is (4) and G is (3), the resulting phase margin.

Procedure

```
10   XO= (starting value for x)
20   T = (threshold to set desired accuracy)
30   X = 1.01*XØ
40      GO TO the subroutine that computes F (line 200)
50      F1 = F
60   X = XØ
70      GO TO the subroutine that computes F (line 200)
80      FØ = F
90   D1 = F1-FØ
100     X1 = XØ-0.01*XØ*FØ/D1
110     D2= absolute value of (X1-XØ)
120     If D2 > T THEN GO TO line 150
130         PRINT XØ, FØ, and G
140         STOP
150     XØ=X1
160         GO TO line 30
200     F = 20log₁₀A(X) or P(X)+180°
210     G=P(X)+180° or -20log₁₀A(X)
220 RETURN to the calling statement
```

DRILL PROBLEMS

DB.1. Use the computer-aided procedure to verify the following results:

(a) $G(s)\,H(s) = \dfrac{2000}{(s+2)(s+7)(s+16)}$

Phase crossover frequency = 12.57 rad/sec
Gain margin = 5.4 dB
Gain crossover frequency = 9.18 rad/sec
Phase margin = 19.8°

(b) $G(s)\,H(s) = \dfrac{20}{s(s^2+7s+140)}$

Phase crossover frequency = 11.83 rad/sec
Gain margin = 33.8 dB
Gain crossover frequency = 0.143
Phase margin = 89.6°

B.4 COMPUTING BREAKAWAY POINTS

In Chaps. 4 and 5, a trial-and-error procedure can be used to obtain a point where the root locus leaves the real axis (maximum K in that real-axis region) or enters the real axis (minimum K in that real-axis region). In either case, the Newton-Raphson procedure can be applied where

$$x = s$$

$$f(x) = \frac{dK(x)}{dx} = 0$$

$$\frac{df(x)}{dx} = \frac{d^2 K(x)}{dx^2}$$

The function $f(x)$ can be approximated by applying a small change to x.

$$f(x) = \frac{K(x + d) - K(x)}{d}$$

The derivative of $f(x)$ can be approximated by

$$\frac{df(x)}{dx} \approxeq \frac{f(x + d) - f(x)}{d}$$

$$\approxeq \frac{K(x + 2d) + K(x) - 2K(x + d)}{d^2}$$

If d equals a 1 percent change in x_0, then (6) becomes

$$x_1 = x_0 - \frac{0.01 x_0 [K(1.01 x_0) - K(x_0)]}{K(1.02 x_0) + K(x_0) - 2K(1.01 x_0)}$$

The following procedure requires K to be evaluated three times per computational cycle. As in the last section, the initial guess of x affects convergence and the threshold T sets the accuracy.

```
Procedure

10   X0= (starting value for x)
20   T = (threshold to set desired accuracy)
30   X = 1.02*X0
40      GO TO the subroutine that computes K (line 200)
50      K2 = K
60   X = 1.01*X0
70      GO TO the subroutine that computes K (line 200)
80      K1 = K
90   X = X0
100     GO TO the subroutine that computes K (line 200)
110     K0=K
```

```
120   D2= K2+K0-2*K1
130    D1 = K1-K0
140    X1 = X0-0.01*X0*D1/D2
150    D3= absolute value of (X1-X0)
160    If D3 > T THEN GO TO line 190
170       PRINT X0, K0,
180       STOP
190    X0=X1
195       GO TO line 30
200    K=1/1G(X)H(X)1
210       RETURN to the calling statement
```

DRILL PROBLEMS

DB.2. Use the computer-aided procedure to verify the following:

(a) From Fig. 4.13 (replacing s with X),

$$K(X) = \frac{X(X + 5)}{10(X + 10)}$$

To three significant digits the breakaway point is at -2.91 where K is 0.0858. To three significant digits, the entry point is at -17.0 where K is 2.91.

(b) From Fig. 4.18 (replacing s with X),

$$K(X) = X(X + 3)(X^2 + 6X + 64)$$

To three significant digits, the breakaway point is at -1.43 where K is 129.016 (in increments of 0.05 the breakaway point is -1.45 which represent $2\frac{1}{2}$ significant digits).

REFERENCES

Newton-Raphson
Booth, T. L., and Chien, Y. T., *Computing: Fundamentals and Applications*. Santa Barbara, Calif.: Hamilton Publishing Co., 1974.

James, M. L., et al., *Applied Numerical Methods for Digital Simulation*. New York: Harper & Row, 1985.

INDEX

$$\frac{-b \pm \sqrt{b^2 - 4ac}}{2a}$$

1/10/92

Perspectives in Chemical Engineering
Research and Education

ADVANCES IN CHEMICAL ENGINEERING

Volume 16

ADVANCES IN CHEMICAL ENGINEERING

Volume 16

Editor-in-Chief
JAMES WEI

Department of Engineering
Princeton University
Princeton, New Jersey

Editors

JOHN L. ANDERSON

Department of Chemical Engineering
Carnegie-Mellon University
Pittsburgh, Pennsylvania

KENNETH B. BISCHOFF

Department of Chemical Engineering
University of Delaware
Newark, Delaware

JOHN H. SEINFELD

Department of Chemical Engineering
California Institute of Technology
Pasadena, California

Perspectives in Chemical Engineering
Research and Education

Edited by
CLARK K. COLTON

Department of Chemical Engineering
Massachusetts Institute of Technology
Cambridge, Massachusetts

ACADEMIC PRESS, INC.
Harcourt Brace Jovanovich, Publishers
Boston   San Diego   New York
London   Sydney   Tokyo   Toronto

This book is printed on acid-free paper. ∞

Copyright © 1991 by Academic Press, Inc.
All rights reserved.
No part of this publication may be reproduced or
transmitted in any form or by any means, electronic
or mechanical, including photocopy, recording, or
any information storage and retrieval system, without
permission in writing from the publisher.

ACADEMIC PRESS, INC.
1250 Sixth Avenue, San Diego, CA 92101

United Kingdom Edition published by
ACADEMIC PRESS LIMITED
24–28 Oval Road, London NW1 7DX

Library of Congress Catalog Card Number: 56–6600

ISBN 0–12–008516–X (alk. paper)

Printed in the United States of America
91 92 93 94   9 8 7 6 5 4 3 2 1

CONTENTS

v

SECTION III
Thermodynamics   123

SECTION IV
Kinetics, Catalysis, and Reactor Engineering   203

SECTION V
Environmental Protection and Energy   265

SECTION VI
Polymers   319

SECTION VII
Microelectronic and Optical Materials   371

SECTION VIII
Bioengineering   423

SECTION IX
Process Engineering   497

SECTION X
The Identity of Our Profession 563

PREFACE

A century has passed since the field of chemical engineering became formalized as an academic discipline. To mark this occasion, the Department of Chemical Engineering at the Massachusetts Institute of Technology sponsored a celebration on October 5-9, 1988 consisting of a Centennial Symposium of Chemical Engineering and an Alumni Convocation. This volume is a permanent record of all of the presentations and discussions at the symposium together with several papers presented at the convocation.

The symposium was divided into nine sections. The first eight covered topical areas of chemical engineering: fluid mechanics and transport; thermodynamics; kinetics, catalysis, and reactor engineering; environmental protection and energy; polymers; microelectronic and optical materials; bioengineering; and process engineering. Two important areas of chemical engineering, separations and surface and colloidal phenomena, were not explicitly included as separate topics but are nonethless included implicitly within several of the topical areas. The last section contained a self-examination of our identity as a profession. For each topical area, one symposium participant was asked to author a major review that included its historical development, current and future research directions, and role in education. It was to be aimed not at the expert but rather the broadly trained and well educated chemical engineer. Drafts of these manuscripts were made available to participants prior to the symposium. In addition, for each topical area one or two participants were asked to serve as discussants to critique the papers and add new material from their own perspective. The authors were given opportunity to edit and update their manuscripts, and final drafts from authors and discussants were received after the symposium. The discussants graciously accepted their role, as reflected in their shorter papers in this volume. A general discussion amongst all participants took place after the presentations in each area. These discussions, which were taped and subsequently transcribed and edited, are also included in this volume. Three papers presented at the convocation are included in Section I, most notably the masterful treatise by Skip Scriven on the intellectual roots and early evolution of chemical engineering.

The 23 chapters in sections II through IX attest to the extraordinary diversity of activities that characterizes chemical engineering today. Although these papers deal primarily with research, they also reflect the mutually reinforcing interaction between research and education—the creation of new knowledge and the infusion of that knowledge into a continuously evolving curriculum. By contrast, the general discussions at the end of each section are concerned to a greater degree with education. This is not surprising, since, as noted by Jim Wei in the discussion of the last section, the educational process is the glue that binds the profession together.

A wide variety of problems and issues are considered in this volume. One that arises repeatedly is the diversification and increased specialization we are now undergoing. This puts increased demands for incorporation of new subject matter into an already overburdened curriculum. Furthermore, some express concern that this diversification is leading to fragmentation and loss of collegiality. Others are less concerned, recognizing that these pressures are simply the irritants that are the source of evolution.

The stresses and strains we observe are certainly not unique to chemical engineering. Any vibrant profession has the capacity to leave its main stream and form new branches. Our sister professions have faced the same problems that we have, as illustrated by the following quotation:

> "How to keep education sound in an environment of continued change brought about by scientific and technological progress is one of the themes of this volume.... Periodically, as the field has grown and become too diversified, it has been necessary to hammer out new underlying unities, new core curricula, new paradigms to keep the field together and to keep pace with change. The participants in these struggles were concerned with how to keep the field open-ended and the educational process vital and relevant; they were concerned with how to educate students for the long term, since students and educators alike are always in danger of being out paced by events."

This quotation, which so aptly applys to chemical engineering today, is from a volume by K. L. Wildes and N. A. Lindgren prepared for the Centennial of MIT's Department of Electrical Engineering and Computer Science in 1982!

The tapes of the discussions were transcribed and edited, along with the manuscripts, in Cambridge. The entire book was entered electronically and made ready for publication in Cambridge. A project of this scope requires the cooperation of many individuals, and I would like to thank several people who made it possible. Ms. Barbara Driscoll helped in preparations for the symposium. Mr. Scott Anthony of Academic Press provided needed help in final formatting of the manuscript. Last, but not least, my editorial assistant, Ms. Jane Ewing, played an invaluable role in all aspects of this adventure at the frontiers of electronic publishing, including tape transcription and editing of the entire manuscript. Without her tireless efforts this book would not have been possible.

Lastly, I acknowledge my family—Ellen, Jill, Jason, Michael, and Brian—for their continued love, support, and patience through projects such as this.

Clark K. Colton
Cambridge, Massachusetts

PARTICIPANTS

Professor Andreas Acrivos, *The Levich Institute of PCH,The City College of CUNY, New York, New York 10031*

Professor James E. Bailey, *Department of Chemical Engineering, California Institute of Technology, Pasadena, California 91125*

Professor Alexis T. Bell, *Department of Chemical Engineering, University of California, Berkeley, Berkeley, California 94720*

Professor Stuart W. Churchill, *Department of Chemical Engineering, University of Pennsylvania, Philadelphia, Pennsylvania 19104*

Professor Stuart L. Cooper, *Department of Chemical Engineering, University of Wisconsin, Madison, Wisconsin 53706*

Professor H. Ted Davis, *Department of Chemical Engineering and Materials Science, University of Minnesota, Minneapolis, Minnesota 55455*

Professor Morton M. Denn, *Department of Chemical Engineering, University of California, Berkeley, Berkeley, California 94720*

Professor James M. Douglas, *Department of Chemical Engineering, University of Massachusetts, Amherst, Amherst, Massachusetts 01003*

Professor Thomas F. Edgar, *Department of Chemical Engineering, University of Texas, Austin, Texas 78712*

Professor Keith E. Gubbins, *Department of Chemical Engineering, Cornell University, Ithaca, New York 14853*

Dr. L. Louis Hegedus, *W.R. Grace & Company-Conn., Washington Research Center, Columbia, Maryland 21044*

Professor Arthur E. Humphrey, *Center for Molecular Bioscience and Biotechnology, Lehigh University, Bethlehem, Pennsylvania 18015*

Dr. Sheldon E. Isakoff, *Engineering Research and Development Division, DuPont Company, Newark, Delaware 19714-6090*

Dr. James R. Katzer, *Mobil Research and Development Corp., Paulsboro Research Laboratory, Paulsboro, New Jersey 08066*

Professor L. Gary Leal, *Department of Chemical and Nuclear Engineering, University of California, Santa Barbara, Santa Barbara, California 93106*

Professor Edwin N. Lightfoot, *Department of Chemical Engineering, University of Wisconsin, Madison, Wisconsin 53706*

Professor Dan Luss, *Department of Chemical Engineering, University of Houston, Houston, Texas 77004*

Professor Manfred Morari, *Department of Chemical Engineering, California Institute of Technology, Pasadena, California 91125*

Dr. J.R. Anthony Pearson, *Schlumberger Cambridge Research, Cambridge CB3 0EL, England*

Professor John M. Prausnitz, *Department of Chemical Engineering, University of California, Berkeley, Berkeley, California 94720*

Professor Channing R. Robertson, *Department of Chemical Engineering, Stanford University, Stanford, California 94305*

Professor Eli Ruckenstein, *Department of Chemical Engineering, State University of New York at Buffalo, Buffalo, New York 14260*

Professor William B. Russel, *Department of Chemical Engineering, Princeton University, Princeton, New Jersey 08544*

Professor T.W. Fraser Russell, *Department of Chemical Engineering, University of Illinois, Urbana, Illinois 61807*

Professor William R. Schowalter, *Dean of the School of Engineering, University of Illinois, Urbana, Illinois 61807*

Professor L.E. Scriven, *Department of Chemical Engineering and Materials Science, University of Minnesota, Minneapolis, Minnesota 55455*

Professor John H. Seinfeld, *Department of Chemical Engineering, California Institute of Technology, Pasadena, California 91125*

Professor Reuel Shinnar, *Department of Chemical Engineering, City College of New York, New York, New York 10031*

Dr. Larry F. Thompson, *AT&T Bell Laboratories, Murray Hill, New Jersey 07974*

Professor Matthew V. Tirrell, *Department of Chemical Engineering and Materials Science, University of Minnesota, Minneapolis, Minnesota 55455*

Professor Arthur W. Westerberg, *Engineering Design Research Center, Carnegie-Mellon University, Pittsburgh, Pennsylvania 15213*

From the Department of Chemical Engineering at the Massachusetts Institute of Technology:

Professor Robert C. Armstrong
Professor Janos M. Beer
Professor Daniel Blankschtein
Professor Howard Brenner
Professor Robert A. Brown
Professor Robert E. Cohen
Professor Clark K. Colton
Professor Charles L. Cooney
Professor William M. Deen
Professor Lawrence B. Evans
Professor Karen K. Gleason
Professor T. Alan Hatton
Professor Hoyt T. Hottel
Professor Jack B. Howard
Professor Marcus Karel
Professor Mark A. Kramer
Professor Robert S. Langer
Professor John P. Longwell
Professor Herman P. Meissner
Professor Edward W. Merrill
Dr. C. Michael Mohr
Professor Adel F. Sarofim
Professor Charles N. Satterfield
Professor Herbert H. Sawin
Professor George Stephanopoulos
Professor Gregory Stephanopoulos
Dr. Maria-Flytzani Stephanopoulos
Professor Jefferson Tester
Professor Preetinder S. Virk
Professor Daniel I.C. Wang
Professor James Wei
Professor Glenn C. Williams

Professor Kenneth A. Smith, *Associate Provost and Vice President for Research*

Professor Gerald L. Wilson, *Dean of Engineering*

SECTION I
Historical Perspective and Overview

1

On the Emergence and Evolution of Chemical Engineering

L. E. Scriven
Department of Chemical Engineering and Materials Science
Institute of Technology
University of Minnesota
Minneapolis, Minnesota

I. Introduction

The Industrial Revolution opened technological continents that engineering soon radiated into from its military and civil origins. In the 19th century, while military and civil engineering themselves evolved rapidly, the needs of proliferating growth industries selected for vigorous mutations and hybrids that became established as mining and mechanical engineering. Electrical engineering followed, then metallurgical and chemical engineering, and later aeronautical engineering. From these early forms have descended today's major engineering species and the many related subspecies that fill narrower niches.

This paper deals with chemical engineering as a dynamic, evolving discipline and profession and highlights what I have learned, or surmised, or wondered about chemical engineering. Most of that was in three episodes. One was nearly 30 years ago when, with several years of industrial experience (pilot plants, design, research) and certification in the engineering science movement (interfacial phenomena, fluid mechanics, transport and reaction processes), I was readying to take over as lead professor in team teaching the first course in chemical engineering that undergraduates at

ADVANCES IN CHEMICAL ENGINEERING, VOL. 16

Minnesota enrolled in. The second episode was a few years ago, when I was helping steer the studies of research needs and opportunities for the National Research Council Report on *Frontiers of Chemical Engineering*— the Amundson Report. The third was intermittent reading and reflection in preparation for the speech at the Centennial Convocation of Chemical Engineering education that is elaborated here.

My purpose here is not to compile or rework history, although I hope to challenge those who do and to stimulate interest in their syntheses and interpretations. My purpose is to suggest why chemical engineering is, how it has changed over the past hundred years, and what factors are likely to be important for the next hundred. Regrettably, there has not been time to distill into the text more than a fraction of what I think is relevant and interesting.

II. The Need for Chemical Processing Engineering

Chemical industries in the 19th century first evoked analytical chemists, whose bailiwick was product, and then industrial chemists, whose province included processing but whose skills rarely included engineering. Devising equipment and improving its operation were left to chemists of that sort, or to engineers who generally lacked adequate knowledge of chemistry, or to inventors who often comprehended little of either chemistry or engineering.

Apparently the value of combining skill in both chemistry and engineering in the same individual was first strongly felt in England around the time that country began to lose its lead of the Industrial Revolution and to sense foreign competition where there had been none. The term "chemical engineer" seems to have been first used in England, by 1880. From around that time in the ever more rapidly developing United States, there began to be clear calls from entrepreneurs and industrialists for engineers well prepared to deal with chemical and metallurgical processes on larger and larger scales, with greater and greater complexity, and of bigger and bigger capital investment. Somehow the idea of "chemical engineer" gained circulation and universities conferred degrees with that name long before the discipline took form between 1908 and 1928. The reason surely is that from the beginnings the need was growing for the essence of what has come to be chemical engineering.

That essence (so far as I can tell) is the conception or synthesis, the design, testing, scale-up, operation, control, and optimization of commercial processes that change the state, microstructure, and, most typically, the chemical composition of materials, by physicochemical separations and above all by chemical reactions. What distinguishes chemical engineering from the other major branches of engineering is the centrality of physicochemical and chemical reaction processing, and with that centrality the intimate relation with the science of chemistry (and the sometimes love–hate relationship with the profession of chemistry).

It can be argued along lines that can be glimpsed in what follows that the profession and the discipline emerged precisely because of the urgent need for this kind of engineering. The counterargument is that the same need was largely satisfied until relatively recently by a kind of partnership between industrial chemistry and process–oriented mechanical engineering in Germany and other industrialized countries. The contrast is fascinating and deserves deeper examination than it seems to have received.

III. The Role of Change

Just as industrialization brought about the emergence of chemical engineering, so also has technological and economic change shaped its development. Successive processing technologies follow a life cycle. They spring variously from scientific discovery and worldly invention. They are developed, commercialized, and perhaps licensed or sold off. That is, they rise and flourish, then diffuse to other owners—and to other nations—gaining broader applications, engendering greater competition and lessened profitability. Later they subside into the industrial background of mature technologies, or they are displaced by newly emerging processes or products. Some persist, perhaps metamorphosed; some disappear. All evolve, as is brought out below, by taking backward looks at some of the high technologies of the past, beginning with the earliest ones.

In each era, there are the established chemical technologies, and they tend to be called "traditional." At their stage in the life cycle they face rising competition, falling profitability, and then renewal by metamorphosis, subsidence into the background, or outright displacement. They exert one set of influences on profession and discipline. But there are other chemical processing industries where chemical engineering has not been as deeply involved, and, particularly where those industries are growing with new technologies, often the "high technologies" of the day, there is a different pattern. That pattern is novel products that are for a time highly profitable regardless of the way they are made; yet, as the scale of manufacture grows and competition by cost and quality intensifies, concern turns to the manufacturing process itself. Then the need for chemical engineering, or for more and better chemical engineering, is discovered.

So it is that chemical engineering has to adapt and evolve to serve the needs for competitiveness in not only the currently established technologies of the chemical and process industries but also the frontier technologies—today those in such fields as biochemicals, electronics, and advanced materials. These may benefit from chemical engineers' participation in product development as well as all aspects of processing. In any case, the challenges of the frontier stimulate research in, and development of, the discipline of chemical engineering, and often fresh hybridizations with applicable sciences—today biochemistry, electronic materials, polymer science, ceramic science—and neighboring areas of engineering. The challenges of the frontier may also lead to fresh infusions from chemistry, biology, physics, and mathematics.

Though it has missed some opportunities in the past, the profession has coped pretty well all through its evolution over nearly a century. The only danger is an episode of such fast change in its economic and technological environment, or dulling of its response mechanisms, that it is unable to keep up, an unlikely circumstance in the light of its responses a few years ago to the convulsions of the established petroleum and commodity chemical industries that accompanied evolution of international and national economies [1].

IV. When Heavy Chemicals Were the Frontier

The first chemical manufactures to be innovated in the modern sense were the chamber process for sulfuric acid in the mid-18th century in England and the Leblanc process for sodium carbonate, which originated in France in the early 19th century and soon diffused to England and throughout the Continent. Both stimulated other technological developments, some stemming from what would today be called their environmental impact. Both drew competition: the one from vapor-phase catalytic processes for sulfuric acid (invented by Phillips in 1831 but not commercialized for more than 50 years), which finally replaced it early in the 20th century; the other from the marvelously inventive Solvay process, which more rapidly replaced the earlier Leblanc technology. Both of the newer technologies, heavily metamorphosed, are in the industrial background today: chemical engineers at large pay them little heed, though sulfuric acid and sodium carbonate are indispensable [2].

Ernest Solvay's 1872 ammonia–soda process was a breakthrough. He divided the process into distinct operations of gas–liquid contacting, reaction with cooling, and separations; he invented new types of equipment for carrying them all out continuously on a large scale; and he himself dealt with the chemistry, the materials handling, the process engineering, and the equipment design. In short, the Belgian with no university education performed as what would come to be called a chemical engineer. Though this was not evident to his contemporaries, his performance did catch some attention in England, and it surely impressed the aggressive Americans. They soon licensed the process, integrating it and its principles into a fast-developing inorganic chemicals industry that would be invading European markets around the turn of the century.

In these two heavy-chemical fields are the roots of chemical engineering. They brought need for chemists and engineers in chemical manufacture. They also gave rise to George E. Davis, English consultant and entrepreneur in chemical manufacture, who in 1880 turned environmental protection act enforcer (chemical plants had become dreadful polluters) [3]. In that year, he was also calling publicly for *chemical engineering* and a society to institutionalize it. The society chose instead to be the Society of Chemical Industry, but 14 of its first 297 members in 1881 described themselves as *chemical engineers*, not industrial chemists, and in the society and its internationally circulated journal the idea of chemical engineering continued to be discussed. In an 1886 article on the international competitiveness of the British chemical industries, Davis' colleague Ivan Levinstein defined chemical engineering as the conversion of laboratory processes into industrial ones, called Ernest Solvay a chemical engineer, and proclaimed that professors who combined scientific attainments, practical knowledge, and industrial contacts were needed to train such men so that Britain could, among other things, meet the rising German competition in coal tar dyes [4].

Then, in 1887, George Davis became some sort of adjunct professor in Manchester and gave a course of published lectures on his new subject. Finally, in 1901, he turned these into the first book on the discipline of chemical engineering. His preface highlighted the mounting competition from America (the British viewed the United States then rather as the Americans view Japan today) in heavy chemicals and the "wonderful developments in Germany of commercial organic chemistry." An expanded, two-volume second edition of *A Handbook of Chemical Engineering* was published in 1904. It is pictured in Fig. 1. You see the radical departure from the earlier textbooks and handbooks on industrial chemistry, which covered each chemical industry separately. Davis had recognized that the basic problems were engineering problems and that the principles for dealing with them could be organized around *basic operations common to many*: fluid flow, solids treating, heat or cold transfer, extraction, absorption, distillation, and so on [5]. (And, as Warren Lewis noted much later, the concept was developed as quantitatively as the resources of the 1880s had allowed.)

Very few people were ready to act on George Davis' vision of a discipline of chemical engineering, either in 1887 or in 1901. It fit no university curriculum of that time. But it did not go unnoticed in the United States, where there were others who were calling themselves chemical engineers and there were university curricula called chemical engineering. Just how the name gained currency is not yet clear, but it is certain that the *Journal*

A HANDBOOK

OF

CHEMICAL ENGINEERING

ILLUSTRATED WITH WORKING EXAMPLES
AND
NUMEROUS DRAWINGS FROM
ACTUAL INSTALLATIONS.

BY

GEORGE E. DAVIS,
CHEMICAL ENGINEER,

Formerly Inspector under the Alkali, &c. Works Regulation Act for East Lancashire, Yorkshire, the Midlands and South of England.

SECOND EDITION.

VOLUME I.

MANCHESTER:
DAVIS BROS., 32, BLACKFRIARS STREET.
1904.

VOLUME I	CONTENTS
CH. 1	INTRODUCTION
CH. 2	THE TECHNICAL LABORATORY
CH. 3	MATERIALS USED IN THE CONSTRUCTION OF PLANT
CH. 4	WEIGHING AND MEASURING
CH. 5	STEAM PRODUCTION AND DISTRIBUTION
CH. 6	POWER AND ITS APPLICATION
CH. 7	MOVING SOLIDS, LIQUIDS AND GASES
CH. 8	TREATING AND PREPARING SOLIDS

VOLUME II	CONTENTS
CH. 1	THE APPLICATION OF HEAT AND COLD
CH. 2	SEPARATING SOLUBLES FROM INSOLUBLES
CH. 3	ABSORBING AND COMPRESSING GASES
CH. 4	EVAPORATION AD DISTILLATION
CH. 5	CRYSTALLISATION AND DIALYSIS
CH. 6	THE APPLICATIONS OF ELECTRICITY
CH.7	THE CONSTRUCTION OF PACKAGES
CH. 8	ORGANISATION AND BUILDING

Figure 1. *A Handbook of Chemical Engineering*, 2nd Ed., 1904, by George Edward Davis. Courtesy of L. E. Scriven and L. R. Lundsten.

of the Society of Chemical Industry had avid readers on this side of the Atlantic. It did on the other side of the Pacific, where Japanese like U.S.-educated research chemist-turned-chemical engineer Shotaro Shimomura, a Society member, were continuing to select Western technologies and university curricula for the imperative industrialization of their country [6].

V. When Organic Chemicals from Coal Tar Were the Frontier

The next chemical manufactures to be innovated were *batch* processes for small-volume production of high value-added dyes and other coal tar derivatives. These began with Perkins' mauve in England. Very soon this field was dominated by German research prowess in organic chemistry. The rise of that prowess was aided by university–industry research cooperation, an innovation triggered by British example; and it was integrated into an industrial juggernaut that before long controlled international markets. (Early pharmaceuticals followed the same route.) Here, though, the research chemist's laboratory methods were turned over to mechanical engineers to scale up directly, and many of them became skilled enough that in 1886 Levinstein thought some to be scientific chemical engineers. Still, this was no harbinger of chemical engineering as a discipline, nor did it lead the way to continuous processing and the economies that *that* could bring when markets expanded and competition demanded [7].

Meanwhile the sugar industries, distillation industries, and many others were evolving in Europe. Process engineering and industrial chemistry curricula were being installed in the technical universities (Technische Hochschule) that had appeared in Germany, Switzerland, Austria, and even Hungary [8]. A few of the professors and their counterparts, the chief engineers in certain companies, highly educated in science and mathematics, took to analyzing common, constituent operations like heat transfer (Peclét in France), vaporization, condensation and drying (Hausbrand in Germany), and distillation (by Hausbrand and by Sorel in France). Peclét's seminal monograph went through several editions earlier in the century; monographs by the rest of these authors were published between 1890 and 1899. These too were important forerunners of the discipline of chemical engineering. So also were atlases of chemical manufacturing equipment organized by basic operations rather than industries, notably the extensive one published by Wolfram in Germany in 1903 [9].

VI. When Electrochemicals Were the Frontier

Electrochemical processing rose in England, Germany, and France in the decades before 1900 but diffused to America, where cheap electricity generated from cheap coal, and mass production in the new tradition of iron, steel, copper, nickel, and tin, enabled the Americans to compete successfully, even invading international markets—for example, with electrolytic caustic soda and chlorine. Though not much American *science* came up to European standards of the era, "Yankee ingenuity," especially in improving on European inventions and technologies, was very much in evidence. Charles Hall invented in 1886 the most successful process for producing aluminum, and such innovations and commercializations were plentiful before a U.S. scientist had received a Nobel Prize [10].

In 1895 came the hydroelectric development at Niagara Falls; by 1910 that was the location of "the world's greatest center of electrochemical activity," not only of production but also of process research and product development. Outstanding was Frederick Becket and his Niagara Research Laboratories, where he invented processes for making carbon-free chromium, tungsten, molybdenum, and vanadium by direct reduction of their oxides, and other important processes as well. This was one of the very first industrial research labs in America, and it and Becket were soon bought up by Union Carbide to be theirs (a story featured by the historian of the electrochemical industry, Martha Trescott [11]).

Something else about Becket: educated as an electrical engineer, when he got into electrochemical processing he went back to school at Columbia University to get physical chemistry and industrial chemistry, earned an M.S. in 1899 and all but completed a Ph.D.—*a U.S. Ph.D.!* —in electrochemistry in 1902, several years before he went to Niagara.

Electrochemistry was the glamour science and emerging technology of the era. The Electrochemical Society was the meeting ground for leaders of the new science of physical chemistry like Ostwald, Nernst, Bancroft of Cornell, and Whitney—who moved from Arthur Noyes' circle at the Massachusetts Institute of Technology to Schenectady to head General Electric's brand-new research laboratory, the country's first corporate research establishment [12]. It was the meeting ground for leading electrical and electrochemical engineers like Steinmetz of GE, Tesla the independent, J. W. Richards of Lehigh, Burgess of the University of Wisconsin and Burgess Battery; for educated inventor–entrepreneurs like Elmer Sperry, then of National Battery, and Herbert Dow of Midland; and for prominent industrial chemists and chemical engineers like Samuel Sadtler of Philadelphia, William Walker of the partnership of Little & Walker in Boston, and Fritz Haber of Karlsruhe Technical University in Germany. As a matter of fact, all save Dow and Sadtler presented papers at the semiannual meetings of the newly formed American Electrochemical Society in 1902, 1903, or 1904 [13].

VII. The First Chemical Engineering Curriculum

With the mention of William Walker, it is fitting to go back to the 1880s and the scene then in America's colleges and universities. Industrialization of the country was accelerating, and with it the need for engineers and, to a lesser extent, chemists. From decades earlier there was popular demand for *relevant* college education, and this had been answered by the appearance of engineering schools like Rensselaer Polytechnic Institute in Troy, New York, Brooklyn Polytechnic, and the Massachusetts Institute of Technology; by scientific schools at Yale, Harvard, Dartmouth, Columbia, and so on; and by the Morrill Act of 1862, which enabled the states to establish land-grant colleges, one purpose of which was education in "the mechanic arts," which turned out to be engineering–civil, mining, and mechanical; then electrical, metallurgical, and chemical [14].

By the 1880s, strong curricula in science and engineering had sprung up in many land-grant schools, among them Pennsylvania State College. There young William Hultz Walker enrolled in 1886 and graduated in chemistry in 1890. Having been a good student, he set off for graduate study in Germany—as did some 1000 other top graduates in chemistry between 1850 and World War I and perhaps 9000 more in other fields. Germany then was the center of freedom of learning, freedom of teaching, academic research, and chemistry. Returning from Göttingen with a Ph.D. in 1892, Walker taught for a couple of years at Penn State, moved to MIT, then resigned in 1900 to join MIT chemistry graduate Arthur Dehon Little in an industrial chemistry consulting partnership. (Little had lost his original partner Griffin in a laboratory explosion [15].)

At MIT, as well as at many other schools in the United States and Europe, individual courses and laboratories in industrial chemistry (largely descriptions of the manufacture of diverse chemicals) were being taught in the 1880s. The teacher at MIT was Lewis Mills Norton, an MIT graduate and 1879 Göttingen Ph.D. in chemistry. On returning to the United States, he spent a few years as a textile mill chemist and came to appreciate engineering problems not only of "the textile industry" but also of "furnace construction and regulation" and of "the manufacture of organic products." How aware he was of the Society of Chemical Industry and its *Journal* seems not to have been recorded. Impelled, he wrote, by a demand for engineers skilled in chemistry, in 1888 he proposed a new curriculum, Course X, to be mainly the *mechanical engineering* curriculum, but with *industrial chemistry* courses and laboratory in the third year. Norton lined up Professor Peabody to teach a *special course* on "chemical machinery from an engineering point of view: pumps, refrigerating machinery, filter presses and methods for evaporation in vacuo," and he himself began developing a companion fourth-year *special course* he called applied chemistry. (Vacuum evaporation, like refrigeration, was an important new technology–a high technology of the time [16].)

The name Norton chose for the new curriculum was *chemical engineering,* though it contained no course called chemical engineering and was taught by no one called a chemical engineer. (The same name had been chosen in 1885 by Henry Edward Armstrong for a similar but short-lived course at Central Technical College in London [17].) The proposal was approved by the Faculty and the Corporation; 11 sophomores enrolled in the new curriculum and embarked on it over 100 years ago, in September 1888. Its name has not changed since.

Lewis Norton died young, less than two years later, and the curriculum reverted to mechanical engineering plus industrial chemistry, one of the combinations that began to be called *chemical engineering* elsewhere. At the Royal Technical College in Glasgow, Scotland, the applied chemistry curriculum in 1888 took the name chemical engineering for a time [17]. For another example, at Minnesota the first curriculum in chemistry, established in 1891, was named chemical engineering, and the first four graduates (in 1897) received the degree Chemical Engineer, but that designation was not repeated for a decade [18]. The University of Pennsylvania in 1892 was apparently second in establishing permanently a curriculum called chemical engineering, Tulane in 1894 third, and Michigan in 1898 fourth [19]. So begins an acorn to grow, Arthur Little may later have said. (He was big on acorn growth. The consulting firm he founded stands on grounds called Acorn Park.)

VIII. The Emergence of Chemical Engineering

To return forward to 1902: in that year William Walker (Fig. 2) not only became a charter member of the American Electrochemical Society, but also accepted an appointment in the Chemistry Department at MIT to head Course X, though he continued his partnership with Arthur Little until 1905. To his new position he brought his recollections of Norton's ideas; whatever he had picked up from such writings as Ivan Levinstein's published lecture in the *Journal of the Society of Chemical Industry* in 1886 and George Lunge's in the *Journal of the American Chemical Society* in 1893; his own annotated copy of George Davis' book on *Chemical Engineering*; and his ideas for transforming the chemical engineering curriculum. Those ideas can be glimpsed in his 1905 article, "What Constitutes a Chemical Engineer," in Richard Meade's partisan magazine, *The Chemical Engineer*, which was addressed to technical chemists and engineers in the laboratories and supervision of the operations of the "great chemical and metallurgical industries" of the United States. From what happened it's plain that Walker's program was to incorporate the new sciences of physical chemistry and thermodynamics, the new engineering sciences—as they could have been called—of heat transfer, distillation, evaporation, and fluid mechanics that had emerged in Europe, and his own speciality, corrosion; to organize a chemical engineering course including laboratories around Davis' (and the German–British–Swiss George Lunge's) conception of what would come to be known as "the unit operations"; and to develop research through student theses and industrial interactions. As it happened, in the same department his chemist colleague Arthur Amos Noyes, an 1890 Ph.D. with Ostwald in Leipzig, had just developed a course in physical chemistry that emphasized, among other things, problem solving. (Possibly this development was aided by James Walker's 35-chapter *Introduction to Physical Chemistry*, of which a new edition was published in London almost every second year beginning in 1889.) Noyes urged, and William Walker insisted, that every chemical engineering student take it [20].

Figure 2.   William Hultz Walker. Courtesy of The MIT Museum.

A complementary source of the emerging discipline's emphasis on education by problem solving was undoubtedly the three-volume "Metallurgical Computations" compiled from the 1905–1908 articles in *Electrochemical and Metallurgical Industry* by electrochemical engineer and metallurgist J. W. Richards of Lehigh University, a prominent colleague of Walker's who had been the first president of the American Electrochemical Society [21].

Figure 3. Warren Kendall Lewis. Courtesy of The MIT Museum.

The success of Walker's program was assured when he arranged for a 1905 graduate, Warren Kendall Lewis (Fig. 3), to go to Germany for a Ph.D. in chemistry at Breslau. For upon returning in 1908, Lewis joined Walker and, besides teaching, embarked on a series of analyses of distillation, filtration, fluid flow, countercurrent contacting, heat transfer, and so on. Most of them involved bachelor's and master's theses and many were published in the young *Journal of Industrial and Engineering Chemistry*, which the American Chemical Society started in 1909 (as a counter to the AIChE!). These researches developed principles of chemical engineering, which were expounded in mimeographed notes for classes and then in the famous 1923 book, *Principles of Chemical Engineering*, by Walker, Lewis, and the younger McAdams, who joined them in 1919. The *first textbook* and a fine one (apart from a severe shortage of references to the literature), it of course shaped the discipline and helped define the profession. So did another out-

Figure 4.   Arthur Dehon Little. Courtesy of The MIT Museum.

growth of mimeographed class notes, Lewis and Radasch's 1926 *Industrial Stoichiometry*, which built on J. W. Richard's earlier example [22].

In the same period, Walker and Lewis, along with Columbia University's Milton C. Whitaker, a Coloradan who alternated between industrial and academic careers, were the vanguard of those developing the laboratory course for the emerging discipline. What they wrought came to be known later as the "unit operations laboratory" [23].

Walker's program was further reinforced by his former partner Arthur D. Little (Fig. 4). In 1908 they took part in an *ad hoc* meeting provoked by Richard Meade through his magazine *The Chemical Engineer*. The outcome was a recommendation for organizing an American Institute of Chemical Engineers, against the opposition of the American Chemical Society. (Although chemical engineering benefited greatly, on balance, by developing in close connection with chemistry, the association was not a uniformly positive one.) Incidentally, it appears that no one with a degree in chemical engineering was among those trained as industrial chemists and mechanical engineers who founded the AIChE. Anyway, from 1908, Little took increasing interest in chemical engineering education and by 1915 was the chairman of a Visiting Committee for Chemical Engineering at MIT. Little was famous as a fine speaker and writer and was particularly eloquent regarding the need for industrial research in America. In that year's report to the MIT administration and Corporation, he coined a name that stuck [24].

The name was "unit operations" for those basic physical operations in chemical manufacture that Davis had written about decades earlier. George Lunge too had spotlighted them in an influential 1893 speech to the World's Congress of Chemists at America's first great international exposition, the Columbian Exposition in Chicago, the speech that appeared in the fledgling *Journal of the American Chemical Society* as already mentioned. Some professors and other members of the newly organized profession were talking and writing about those operations, and many in the chemical and metallurgical industries were of course practicing them in the course of their work [25].

What nobody got around to identifying were the basic kinds of *chemical* operations in chemical manufacture, although Walker by 1905 had epitomized the chemical engineer as one "who can devise, construct, and operate industrial plants based on *chemical reactions*," and in Germany the rudiments of chemical reaction engineering were being uncovered by Knietsch, Haber, Bosch, and others. In Germany, there was neither a discipline nor an organized profession to follow up; only in the 1930s did Eucken and Damköhler take up the line. In the United States, a young metallurgical engineer (Columbia University) turned physical chemist (1906 Ph.D. with Nernst at Göttingen) named Irving Langmuir also early worked out some fundamentals of continuous flow reactors and then went over to pure and applied science [12]. Chemical engineers might wonder how the discipline would have developed had Noyes attracted Langmuir (Fig. 5) to MIT, there to fall in with Lewis as both were starting their careers. As it was, the American story of chemical reaction engineering, apart from the chapter on combustion, got stuck in the "unit processes" idea—industrial chemistry at one remove—and didn't break out until Olaf Hougen of Wisconsin, and Kenneth Watson after leaving Universal Oil Products, readied the way with their 1947 text on *Kinetics and Catalysis*, Part III of *Chemical Process Principles* [26].

A unique development at MIT was the School of Chemical Engineering Practice, an industry–academia cooperative master's course at a high, practical level: it was proposed by Little in the same 1915 report, backed by George Eastman of Eastman Kodak, tried out under Walker's direction, interrupted by U.S. entry into the war, and restarted in 1920. Still an expensive educational scheme, it continues alone to this day. Its closest relatives are the relatively few progeny of another innovation of that era, the coop-

Figure 5. Irving Langmuir. Reprinted with permission from Biographical Memoirs, Vol. XLV, National Academy Press, Washington, D.C., 1974.

erative program in which undergraduates alternate periods of on-campus study and off-campus industrial internship [27].

Membership in the AIChE for years was highly restricted, chiefly because of the rival American Chemical Society's Industrial and Engineering Chemistry Division, which was formed as a counter. Remarkably, neither Walker nor Little was a founding member though both subsequently joined, and Little, ACS president from 1912 to 1914, was elected AIChE president for 1919. (He was also president of the British-based Society of Chemical Industry for 1928–1929.) Nevertheless, the AIChE's 12-year efforts to *define* chemical engineering, prescribe a basic curriculum, and establish a national accrediting scheme—the first in engineering—all succeeded ultimately. Arthur Little himself led the later stages of the effort to establish the curriculum, the core of the discipline [28]. The principles embodied in that curriculum have prevailed ever since.

IX. Chemical Engineering vs. Chemical Science

By the turn of the century, major American industries were being consolidated through mergers and acquisitions, partly because of the stress of
international competition. Frank A. Vanderlip observed in *Scribner's
Magazine* in 1905 that "as combinations are made in the industrial field,
the possibility of employing highly trained technical experts rapidly increases.
. . . Technical training is therefore becoming of vastly more importance than
ever before, and those nations which are offering the best technical training to their youths are making the most rapid industrial progress . . . the
relative efficiency of nations was never before so largely influenced by the
character of their educational facilities" [29].

The transformation Walker and his associates had begun working at MIT
was abundantly successful in meeting the needs of American society in the
decade of World War I, though not without notable stress and strain. The
stress and strain arose from chemistry and chemical engineering vying in
one department, and over the role of contract research in an institution with
university-like aspirations. The contenders were a leading scientist, physical
chemist Arthur Noyes, who had been the dominant figure until 1912 or 1914
or so, and William Walker, who took a leave of absence to engineer
America's response to Haber *et al.'s* war gas innovation. When Walker
returned at war's end, he prevailed and forced MIT to choose between his
views and Noyes'. Noyes resigned in 1919 and went to Pasadena, California. Around 1913, he had become interested in the project of a friend from
student days. The friend was George Hale, Director of the Mount Wilson
Observatory; the project, in which Hale had also interested Robert Millikan
of the University of Chicago, was to transform Throop Polytechnic Institute into the California Institute of Technology; and Noyes was already
involved part-time. So MIT chemical engineers inadvertently gave young
Caltech a big boost [30].

Ironically, there is evidence that Noyes foresaw at least as much of future engineering as did Walker: already in a 1915 address to the Throop
Assembly, "What Is Engineering?," he had delineated *engineering science*,
highlighted the *science of materials*, and harped on the primacy of *principles* over "specific industrial applications or technical methods." And in
an echo of what Walker had found so attractive, "Of especial importance
in developing the power [to apply fundamental knowledge] is the independent
solution of the numerous problems which constitute a large part of the work
of any good engineering school" [31].

Surprisingly, the winner of the contest, Walker, resigned the next year,
and the MIT Chemistry Department was split in two, Warren K. Lewis
heading the new and separate Department of Chemical Engineering.

X. When Petroleum Refining Was the Frontier

Notwithstanding such intradisciplinary conflicts, the transformation of chemical engineering that Walker led was remarkably successful, and he is widely regarded as the father of the discipline. Testimony to the success were the impacts of graduates, ideas, approaches, and research results on the chemical and petroleum industries of the United States. That country had come out of World War I the world's most powerful economy [32].

The relationship with petroleum refining was particularly significant. Refining, which had developed outside the mainstream of chemical processing, was a growth industry of the motor age, a large and expanding area of opportunity for chemical engineering. The Burton process of thermal cracking, a crude kind of reaction processing, had been innovated by Standard Oil in 1912; catalytic cracking was a couple of decades in the future. Distillation separations, incredible as it may seem today, were a *batch* process in refineries until continuous flow pipe stills replaced shell stills and fractionators came in after World War I. Abruptly, there was high demand for analysis and design of distillation equipment as well as cracking furnaces. Petroleum refining became the frontier. When Jersey Standard moved to set up a research and development department to apply chemistry, physics, and chemical engineering to the oil industry, it retained as advisors chemist Ira Remsen, president emeritus of Johns Hopkins, physicist Robert Millikan of Caltech, and Warren Lewis [33].

At MIT, Lewis pushed combustion, the only sort of reaction to draw concerted investigation by chemical engineers of that era, and launched Robert Haslam and Robert Russell, whose researches were capped by their 1927 volume on *Fuels and Their Combustion*. Haslam subsequently became development manager at Standard Oil Development Co. and later vice president and a director of Jersey Standard. Clark Robinson, who had graduated in 1909, spent five years in industry, and returned to do a master's and join the staff, answered demand in 1922 with the first physical chemistry-based text on *The Elements of Fractional Distillation*, a best-seller that went through several editions, later with younger Edwin Gilliland as coauthor. In the same year Robinson also published a book on *The Recovery of Volatile Solvents*. In the next year, he coauthored a seminal little textbook with Professor Frank Hitchcock of the MIT mathematics department (it was based on a course they had begun in 1919) [34]. A remarkable outburst!

The little book, *Differential Equations in Applied Chemistry* (Fig. 6), has been unsung for years, though all through an era when few engineers truly understood calculus it showed off powerful mathematical tools, among them (in the 1936 second edition) numerical solution methods. It connected with chemist J. W. Mellor's 1902 *Higher Mathematics for Students of Chemistry*

DIFFERENTIAL EQUATIONS

IN

APPLIED CHEMISTRY

BY

FRANK LAUREN HITCHCOCK, Ph.D.
Associate Professor of Mathematics in the Massachusetts Institute of Technology

AND

CLARK SHOVE ROBINSON, S.M.
Assistant Professor of Chemical Engineering in the Massachusetts Institute of Technology; author of " Elements of Fractional Distillation " and of " Recovery of Volatile Solvents "

UNIVERSITY OF
MINNESOTA
LIBRARY

NEW YORK
JOHN WILEY & SONS, Inc.
London: CHAPMAN & HALL, Limited
1923

CONTENTS

CHAPTER I

Introduction

Figure 6. *Differential Equations in Applied Chemistry*, 1923, by Frank Lauren Hitchcock and Clark Shove Robinson. Courtesy of L. E. Scriven and L. R. Lundsten.

and Physics—with Special Reference to Practical Work; it informed subsequent texts on mathematics for engineers in general; and it inspired at least a few scientifically minded young chemical engineers, among them T. K. Sherwood, C. E. Reed, probably A. B. Newman and T. B. Drew, and certainly R. L. Pigford, to develop more accurate mathematical descriptions of some of the basic operations that fell within their parts of the discipline [35]. The argument can be raised that *chemical reaction engineering* and *systematic process engineering*, including comprehensive methods of addressing *control* and *optimization*, could not develop until adequate mathematical tools had been assimilated into the discipline, which took decades.

Petroleum refining was but one of the high technologies of the time that came to need chemical engineering. Courses were strengthened or installed in universities across the United States, and war veterans swelled enrollments in the young discipline to 6000. The country's output of B.S. chemical engineers rose through the 1920s. Ph.D. programs, an inheritance from chemistry, not mechanical engineering, began to take hold. Apparently the first doctorates in chemical engineering were awarded at the University of Wisconsin starting in 1905, when Burgess and applied electrochemistry held sway. By 1920, thirty had been granted nationwide: the flow of chemistry graduates, pure and industrial, to Germany for postgraduate study had been staunched. The total of U.S. Ph.D.'s and Sc.D's in the following decade was 120, to which MIT began contributing in 1924 (John Keats; Ernest Thiele was the fifth, in 1925, the year of Wisconsin's fourth, Olaf Hougen). Many went to industry, others—the split is not clear—to chemical engineering faculties, which continued growing and which AIChE accreditation standards expected to be engaged in research. In the ensuing decades, interestingly, the largest supplier of chemical engineering doctorates to teaching was the long-established course at University of Michigan; Wisconsin, Minnesota, and Massachusetts Institute of Technology were distant seconds. Petroleum refining consumed a lot of university outputs, but so did the American chemical industry, already in aggregate the world's largest and still growing lustily [36].

XI. When High-Pressure and Catalytic Processes Became the Frontier

In the decades before the war, Haber, Nernst, and others in Germany had taken up gas-phase reactions—Haber published his instructive book on the thermodynamics of such reactions in 1906 –and were soon pursuing nitrogen fixation by ammonia synthesis, a goal of tremendous economic and geopolitical significance. Haber got the breakthrough by discovering a catalyst in 1908. A leading German corporation, Badische Anilin-und Sodafabrik (BASF) took on the development, assigning to it a self-made chemical engineer, Carl Bosch, a mechanical engineering graduate of Leipzig University with a Ph.D. in industrial chemistry, and to a methodical chemist, Alwin Mittasch, who reportedly oversaw 20,000 trials with different catalyst formulations to find the best (the irreverent assert that catalyst design hasn't changed in three-quarters of a century). Before 1900, Rudolf Knietsch of BASF had systematically analyzed and designed reactors for vapor-phase oxidation of sulfur dioxide to make sulfuric acid (Phillips' old invention). Building on this earlier pioneering, Bosch designed the first high-pressure, continuous flow tubular reactor—actually a tube bundle—and the rest of the plant. By 1913, they had the Haber–Bosch process for synthetic ammonia in operation, a signal accomplishment like Solvay's 50 years earlier [37].

German technology was a war prize. The war had also stimulated the U.S. chemical industry: explosives to be sure, but most notably dyestuffs by Dow and Du Pont and nitrogen fixation and nitric acid manufacture by government-financed plants at Muscle Shoals, Alabama. Gasoline demand for proliferating motorcars provoked improvements in thermal cracking. Easily refined petroleum deposits began to seem quite limited. There were also wartime profits to invest in upgraded and new technologies. So it was that in America in the decade after World War I, process innovation turned to high pressure and often high temperature, capitalizing on German advances in vapor-phase catalytic reaction processes and on advances in metallurgy and fabrication.

A few months after the war's end, the U.S. government organized the Fixed Nitrogen Research Laboratory, which pioneered high-pressure property and process research and ammonia catalyst development for the nation's needs. Led in succession by Dr. A. B. Lamb, Prof. R. C. Tolman, and Dr. F. G. Cottrell, outstanding and practical scientists all, it recruited talented chemists and engineers who were periodically raided by companies assembling cadres for industrial research and development—a classical mode of technology transfer. Moreover, in 1925 and 1926, industry's Chemical Foundation started financing high-pressure laboratories in the chemical engineering departments of Massachusetts Institute of Technology (Prof. W. G. Whitman), Yale University (Prof. B. F. Dodge), and University of Illinois (Prof. N. W. Krase)– each laboratory a beautifully small center of engineering research [38]!

Among recipients of the research and personnel was Du Pont, which had been growing fast and diversifying since its reorganization in 1902, the year it created its corporate engineering department and first ventured into corporate research as well. Du Pont actually purchased ammonia synthesis technology from French and Italian interests but launched a nitric acid program in its Research Department in 1924. Staffed by Guy B. Taylor, a 1913 physical chemistry Ph.D. from Princeton, Fred C. Zeisberg, a chemist turned chemical engineer expert in acid manufacture, and Thomas C. Chilton, a Columbia University chemical engineering graduate with three years of research experience, the program was a roaring success. It led to Du Pont's high-pressure nitric acid process, which was scaled up, commercialized, and licensed to other chemical companies within four years, testimony to the astuteness of Vanderlip's 1905 observation in *Scribner's Magazine*. The great impact within Du Pont was heightened by contributions the seven or eight other chemical engineers (half of them postwar graduates from MIT) in the Research Department made to process technology for producing ammonia, methanol, and other products. The research director, Charles M. A. Stine, already a respected advocate of fundamental research and spokesman for chemical engineering, responded by organizing half his chemical engineers into a group with the goal of getting their discipline onto a sound scientific footing and simultaneously serving as internal consultants. He put Tom Chilton in charge [39].

Preceding Du Pont's cheaper nitric acid, the first products from the high-pressure high technology of the mid- and late-1920s were methanol and other commercial solvents, setting the stage for what Arthur D. Little in 1928 foresaw: the coming era of chemicals –"new solvents and organic chemicals in great variety"—from petroleum feedstocks. In particular there were the refinery off-gases that were being burned as fuel, rising amounts of ethylene and other olefins among them. More and more petroleum was being thermally cracked to increase gasoline yield, and olefins were by-products. The economic environment was ready for the next high technologies to evolve [40].

And evolve they did. Union Carbide went from light hydrocarbons to liquefying them (liquified petroleum gas, LPG) to ethylene oxidation to Cellosolve solvent, a glycol ether marketed in 1926; then "permanent antifreeze" in 1928 and synthetic ethyl alcohol in 1930. The international Shell Group set up its Development Company in 1928 near Berkeley, California under the direction of Dr. E. Clifford Williams, who had just before established a chemical engineering curriculum at the University of London. Shell Development became the leading pioneer of petrochemical process development—alcohols, ketones, and glycerine, and aviation gasoline to boot [41].

On the other hand, the Shell Group as well as Jersey Standard (part of Exxon today) and Anglo-Iranian Oil declined in 1930 to back Eugene J. Houdry, a French mechanical engineer pursuing his vision of catalytic rather than thermal cracking of heavy crudes. Losing French government financing despite good progress in developing his process, Houdry managed to interest Vacuum Oil Company (part of Mobil today) and then Sun Oil Company in pilot plant work, moved his efforts to Paulsboro, New Jersey, and together with his partners (which came to include Badger Engineering Company) achieved in 1937 the first commercial cat cracker, a fixed-bed, regenerative unit. This development owed little to German chemical technology or to U.S. corporate research, though it spawned a lot of the latter. Applicable *scientific* research on catalytic cracking was led by Russian petroleum chemists, but soon the Universal Oil Products Company, stoked by émigré Vladimir N. Ipatieff, had pulled ahead, and it was owned by major U.S. oil companies [42].

Jersey Standard itself had been slowly developing German technology for hydrogenation of heavy crudes, to which it had access through agreements that Robert Haslam helped negotiate with I. G. Farbenindustrie, the cartel forged by Carl Bosch, who was by 1925 the top executive in BASF. The Houdry process and Kellogg Engineering Company's concept of a moving-bed cracker spurred formation of another development combine,

Catalytic Research Associates, a multinational one that included I. G. Farben, notwithstanding the rising war clouds. The combine had competition from Socony-Vacuum (Mobil), which was first to commercialize a moving-bed cracker. Moreover, it was upstaged by Jersey Standard's own Standard Oil Development Company, still advised by Warren Lewis and larded with former students from MIT. There the concept of gas lift was borrowed from grain transport practice, and the concept of fluidized beds was adapted from Fritz Winkler's 1921 development of high-pressure coal gasification (some reports attribute both to Lewis and co-consultant Gilliland on a train ride home from Standard's Baton Rouge laboratories). The upshot, of course, was fluidized bed catalytic cracking, in its time a major frontier of chemical engineering [43].

Like nitrogen fixation on the eve of World War I, catalytic cracking had tremendous economic and geopolitical significance. Through 1943, Houdry units and after that fluidized bed units accounted for most of the aviation gasoline available to the United States and its allies in World War II.

Between the wars, the growing numbers of continuous catalytic processes—in other manufactures as well as petroleum refining and petrochemicals—absorbed more and more chemical engineers. They brought incentives to focus as much on selective reactions of flowing fluids and suspensions as on the separations and particulate solids-processing methods that constituted the unit operations. They also became nuclei of all sorts of opportunities for the chemical engineering profession.

XII. The Consolidation of Chemical Engineering

The founding of the American Electrochemical Society in 1902 was accompanied by the birth of an independent yet related magazine, *Electrochemical Industry*, which by 1918 was *Chemical and Metallurgical Engineering*—reflecting how the frontiers of process engineering were moving (by 1946 the title had become just *Chemical Engineering*: see Fig. 7). In 1928 its editor, Howard C. Parmelee, striving for "a chemical engineering consciousness" in industry, devoted the January issue to a 25-year stocktaking. Against a backdrop that compared America's largest chemical firms, Allied Chemical & Dye Corporation, Du Pont, and Union Carbide & Chemical Corporation, with Britain's somewhat smaller Imperial Chemical Industries and Germany's already enormous I. G. Farben cartel, Parmelee published accounts by industry leaders (Arthur Little first) of "the relative penetration of chemical engineering" in their respective industries. It was 100% in heavy chemicals, electrochemicals, fine chemicals, coal tar products, and explosives; 75% in coal processing and petroleum refining; 60% in pulp and paper; and so on. Parmelee observed that "Development

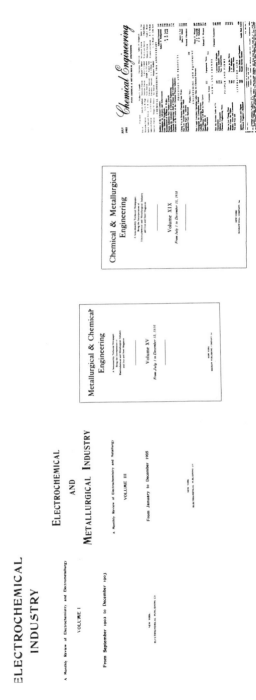

Figure 7. From *Electrochemical Industry* to *Chemical Engineering.*

CHEMICAL ENGINEERING SERIES

A SERIES OF TEXTS AND REFERENCE WORKS
OUTLINED BY THE

Following Committee

H. C. PARMELEE, *Chairman*,
    Editorial Director, McGraw-Hill Publications.

S. D. KIRKPATRICK, *Consulting Editor*,
    Editor, Chemical and Metallurgical Engineer-
    ing.

L. H. BAEKELAND,
    President, Bakelite Corporation.

HARRY A. CURTIS,
    Professor of Chemical Engineering,
    Yale University.

J. V. N. DORR,
    President, The Dorr Company.

D. D. JACKSON,
    Executive Officer, Department of Chemical
    Engineering, Columbia University.

J. H. JAMES,
    Professor of Chemical Engineering,
    Carnegie Institute of Technology.

W. K. LEWIS,
    Professor of Chemical Engineering,
    Massachusetts Institute of Technology.

ARTHUR D. LITTLE,
    President, Arthur D. Little, Inc.

R. S. McBRIDE,
    Consulting Chemical Engineer.

JAMES F. NORRIS,
    Director of the Research Laboratory of Organic
    Chemistry, in Charge, Graduate Students in
    Chemistry, at the Massachusetts Institute of
    Technology.

CHARLES L. REESE,
    E. I. du Pont de Nemours Company.

W. H. WALKER,
    Consulting Chemical Engineer.

E. R. WEIDLEIN,
    Director, Mellon Institute of Industrial
    Research.

M. C. WHITAKER,
    Consulting Chemical Engineer.

A. H. WHITE,
    Professor of Chemical Engineering,
    University of Michigan.

Figure 8.   The McGraw-Hill Editorial Advisory Committee, 1925. From Kirkpatrick
[45].

of chemical engineering to its present professional status has resulted from the harmonious coordination of education, technical literature and useful service to American industry" [44].

Parmelee and his publisher, James H. McGraw, really fostered that technical literature. In 1925, they had brought together 15 prominent chemical engineers (Fig. 8), Walker, Lewis, Whitaker, and the ubiquitous Little among them (but not one from west of Pittsburgh and Detroit), to plan a series of reference works and textbooks, following on the ones by Walker, Lewis, and McAdams and by Lewis and Radasch. Within five years, the series took off, surely a major factor in the consolidation of the discipline and no doubt a profit-maker for McGraw-Hill. Chemical engineering enrollments and employment held up quite well through the Great Depression, thanks to the strength of American chemical and petroleum industries [45].

Another series was subsequently launched by John Wiley & Sons, who also assembled a distinguished advisory board (Fig. 9), theirs consisting of Tom Chilton and his visionary subordinate, Tom Drew, from Du Pont, Donald B. Keyes of the University of Illinois, Kenneth M. Watson of Universal Oil Products near Chicago, and Olaf A. Hougen of the University of Wisconsin. The latter two authored in 1931 a salutary text on *Industrial Chemical Calculations*, precursor of their pivotal three volumes on *Chemical Process Principles*. Hougen and Watson held that process problems are primarily chemical and physicochemical in nature, whereas unit-operation problems are for the most part physical [46]. (Incidentally, Wiley & Sons did not claim for their series even the second edition of Hitchcock and Robinson's book, which they published.)

Nothing like these series of books appeared in Europe, where there were neither comparable university courses, nor numbers of students, nor professional society, nor committed industrial executives.

The early American Institute of Chemical Engineers was small, elite, and prudent in dealing with others that also catered to chemical engineers, in particular the chemical, electrochemical, and mining and metallurgy societies (all had superior journals that attracted chemical engineering research). The older American Society of Mechanical Engineers, having early lost several contingents—Heating and Ventilating in 1894, Refrigeration in 1904, Automotive in 1905—was committed to specialty divisions and growth. In 1930, 20,000 members strong, it took aim on creating a new Process Industry Division that would appeal to chemical engineers. This threat and Depression-generated deficits propelled the 900-member AIChE to liberalize its stringent membership requirements, hire a half-time executive secretary, campaign to recruit young graduates, strengthen ties to its Student Chapters in universities, and improve its meeting programming and *Transactions* publication. The latter

BOOKS IN CHEMICAL ENGINEERING

ADVISORY BOARD

T. H. Chilton, Chem. E.
Engineering Department
Experimental Station
E. I. du Pont de Nemours and Company

T. B. Drew, S.M.
Engineering Department
Experimental Station
E. I. du Pont de Nemours and Company

O. A. Hougen, Ph.D.
Professor of Chemical Engineering
University of Wisconsin

D. B. Keyes, Ph.D.
Professor of Chemical Engineering and Head of Division
University of Illinois

K. M. Watson, Ph.D.
Chemical Engineer
Universal Oil Products Company

Thermodynamics
By H. C. Weber, D.Sc., Associate Professor of Chemical Engineering, Massachusetts Institute of Technology

Industrial Chemical Calculations
By O. A. Hougen, Ph.D., Professor of Chemical Engineering, University of Wisconsin; and K. M. Watson, Ph.D., Chemical Engineer, Universal Oil Products Company

Distillation (**in preparation**)
By A. J. V. Underwood, D.Sc., Consulting Chemical Engineer

Instruments (**in preparation**)
By John J. Grebe, Physical Research Laboratory, Dow Chemical Co., Henry S. Winnicki, Clarkson College of Technology, and others.

Figure 9. The Wiley Advisory Board, ca. 1931. From Hougen and Watson [46].

had begun to draw some of the quality technical papers and research reports that had gone to the ACS's *Journal of Industrial and Engineering Chemistry* and elsewhere [47].

These developments signaled the consolidation of the profession, and on its 25th anniversary in 1933 the AIChE undertook a stocktaking, a volume edited by Sidney D. Kirkpatrick, Parmelee's successor at *Chemical and Metallurgical Engineering* and later a president of the AIChE. Fascinating reading, the book opens with—whom else?—Arthur Little on "Chemical Engineering Research: Lifeblood of American Industry"; runs through heavy chemicals, fine chemicals, electrochemicals and electrometallurgy, coal processing, petroleum refining, solvents, modern plastics (a new entry), and the rest of the gamut; and closes on "Chemical Engineering Education: Building for the Future of the Profession." There is a late chapter on "Stream Pollution and Waste Disposal" where lives the ghost of George Davis, still "disgusted with methods of manufacture which polluted the air, rivers and water courses" [48].

The most significant event of all was publication, a year later, in 1934, of the *Chemical Engineer's Handbook*, the prime reference work in the McGraw-Hill series. Seven years in the making, it had been entrusted to John H. Perry of Du Pont as editor-in-chief of 62 specialists drawn from industry and academia (over one-quarter of them from Du Pont and no more than a handful from or associated with the petroleum and gas industry). The *Handbook's* 2600 small pages were addressed to students as well as practicing engineers (the latter were no doubt the targets of the section on patent law and of the 10 pages on arithmetic). It codified the discipline.

Thus Charles Stine's goal at Du Pont of getting chemical engineering onto a sound scientific footing led first to codification. Simultaneously, the center of mass of unit-operations research shifted southwestward from New England toward Wilmington, Delaware, where the exceptional research engineer, Allan P. Colburn, had joined the others in Tom Chilton's team. The remarkable output of incisive papers that followed was, according to Chilton, a by-product of improving the company's existing technologies and engineering new ones, with a premium on fast and accurate process design and scale-up. Younger talent attracted to the team heightened its impact on the blending of unit operations and chemical processing that was already started in the high-pressure process for nitric acid; with that blending, a discipline better based in science, mathematical modeling, and rigorous computational methods emerged—to be codified, too slowly perhaps, in succeeding editions of the *Chemical Engineer's Handbook* [49].

Though it was a bellwether, Chilton's team was, of course, far from alone. Competition came not only from the United States, but also Germany, where *scientific* engineering was already especially well established. Colburn, for example, had built extensively on researches by Prandtl, Nusselt, von Kármán (in Germany), Schmidt, and Jakob. Just how well he and his American colleagues kept up with developments in Germany is not clear. A chemical engineering discipline was being compiled there too.

Chemische Ingenieur-Technik, a 2200-page treatise intended as something between a textbook and a reference work, appeared in 1935. Its 24 authors were drawn mostly from industry, a handful from technical universities, and a couple from the more prestigious universities. It is certainly more scholarly than Perry's *Handbook*, and it avoids ancillary areas like patent law,

Figure 10. *Der Chemie-Ingenieur* behind and the *Chemical Engineer's Handbook* in front. Courtesy of L. E. Scriven and L. R. Lundsten.

accounting and cost finding, safety and fire protection, and report writing. Its editor, Ernst Berl, is listed as formerly professor of technical chemistry and electrochemistry at the Darmstadt Technical University, but currently professor at Carnegie Institute of Technology, Pittsburgh, U.S.A. No champion remained. German chemical engineering research proceeded chiefly within the I. G. Farben cartel; education was split between Verfahrenstechnik in engineering and Technischen Chemie in chemistry [50].

Der Chemie-Ingenieur was far more imposing, an encyclopedia about physical operations in chemical and related process industries. It totaled 12 volumes published between 1932 and 1940 under the editorship of Professor Arnold Eucken of Göttingen and, initially, Professor Max Jakob of Berlin, who soon departed for the United States. Figure 10 shows the encyclopedia, with Perry's *Handbook* in front. Eight volumes treat unit operations on a scientific footing, with abundant citations of sources, frequent use of mathematical tools, and sophisticated accounts of the best practice. Then came the add-on volumes, which treated physicochemical and economic aspects of chemical *reaction* operations, i.e., chemical process principles;

the chapters include one by Gerhard Damköhler that is now celebrated in chemical reaction engineering. The encyclopedia's title apparently was inspired by Walker, Lewis, and McAdams' text, yet Eucken was a physical chemist who, with I. G. Farben support, championed Verfahrenstechnik as a field of engineering science. Indeed, it was he who in 1934 steered Damköhler into the study of flow reactors [51].

So it was that although chemical engineering was *practiced* very well within segments of German industry, and the *discipline* had been compiled in fine scientific form, it could not propagate without an academic base, nor could a *profession* emerge without an industry that wanted it, or journals for its voice, or a society for its organization.

XIII. New Frontiers

Even as chemical engineering consolidated half a century ago, it continued evolving, with such developments as specialty lubricants, synthetic detergents, new solvents and plastics, and synthetic polymer fibers. During World War II, some frontiers closed, others passed, and some opened up, most notably catalytic cracking, synthetic rubber, and large-scale isotope separation. These and the many subsequent ones are highlighted by *Chemical Engineering* in its 50th anniversary issue in July 1952. The AIChE followed suit on its 50th in *Chemical Engineering Progress* for May 1958, when petrochemical development was getting into full swing and the engineering science movement was gaining momentum.

Elsewhere my view of the role of past, current, and future technologies is reported, along with comments on today's frontiers and trends of change (December 1987 issue of *Chemical Engineering Progress)*. In brief, the practice of chemical engineering, like seasonal foliage, changes; like individuals, the subdisciplines grow, mature, and give birth to others; the discipline like a species evolves, but the *essence* like a tree *is invariant*. For the better part of a century, the profession in the United States has broadened its base–now rejoining materials science (née metallurgy–metallurgical and electrochemical engineering)–and built on it successfully to fulfill the needs of both the existing and the emerging chemical process technologies of each era. As past high technologies have matured, and turned senescent or moribund, the profession has again and again moved on to new frontiers, rapidly enough to avoid any danger of extinction. What factors are likely to be important for the next hundred years? Primarily those that have been important over the past hundred. I hope many of them can be glimpsed in the foregoing. My encounters with them leave me with two deep questions that remain largely unanswered. They in turn raise a flurry of subsidiary questions.

XIV. Closing

What *constitutes* an engineering discipline like chemical engineering? And what *maintains* the associated profession?

How do discipline and profession depend on the industries served and the way their technologies evolve?

How do they depend on neighboring principal disciplines, like chemistry, metallurgy, ceramics and now materials science, and mechanical engineering and mining engineering? Neighboring disciplines do compete!

How do they depend on buffer and splinter disciplines, like electrochemical engineering, combustion engineering, plastics and polymer engineering, petroleum engineering, natural gas engineering, food engineering, biochemical engineering, nuclear engineering? These too compete.

How do they depend on publication of textbooks, handbooks, monographs, professional magazines, and research journals? These surely are essential ingredients!

How do they depend on university curricula and accrediting bodies? On university department structure? On faculty qualifications and their relations to source sciences like chemistry, physics, mathematics, biology, and economics? No profession survives without a steady supply of young recruits schooled in its discipline, imbued with high standards and tuned for the future.

How do they depend on a professional society, or contending professional societies as in the case of chemical engineering? Without an association of the practitioners to define it, maintain it, and renew it, no profession can survive.

How do they depend on the *interactions* of the universities and the industries served–the established ones, the emerging ones, the expiring ones?– the intellectual interactions above all! An engineering discipline depends on its profession and the needs of society that profession serves, just as society and profession depend on the discipline.

It seems to me this last is the most important factor of all. Surely these questions of what makes chemical engineering tick are not only interesting today but also relevant to the next hundred years.

Acknowledgments

I am indebted to Elizabeth M. McGinnis, a senior in chemical engineering, for help in tracking down materials in the University of Minnesota libraries; to the late Professor Robert L. Pigford of the University of Delaware for a long afternoon of inspiring conversation in April 1988; to Professor Clark K. Colton of MIT and Michael Yeates of the MIT libraries for locating and supplying materials from there; and to unseen helpers in the California Institute of Technology library. Clark's interest and enthusiasm were marvelous tonics in the early going and his patience a needed pain-reliever in the final haul.

Notes and References

1. The role of change is a theme in the writer's "The Role of Past, Current, and Future Technologies in Chemical Engineering," *Chem. Eng. Prog.* **83** (12), 65–69 (December 1987).

2. The histories of the Leblanc, Solvay, chamber, and contact processes are told by many authors, among them Ludwig F. Haber, *The Chemical Industry During the Nineteenth Century*, Oxford University Press, Oxford, 1958. Particularly interesting sources are George Lunge's authoritative articles on "Alkali Manufacture" and "Sulfuric Acid" in *Encyclopaedia Britannica*, 11th Ed., Encyclopaedia Britannica Inc., New York, 1910. About Ernest Solvay, see the article on him in *Great Chemists*, Eduard Farber, ed., Interscience Publishers, New York, 1961.

3. On George E. Davis see Norman Swindin, "The George E. Davis Memorial Lecture," *Trans. Inst. Chem. Eng. 31*, 187–200 (1953). On Swindin, see D. C. Freshwater, "George E. Davis, Norman Swindin and the Empirical Tradition in Chemical Engineering," in *History of Chemical Engineering* (W. F. Furter, ed.), pp. 97–111, American Chemical Society, Washington, D.C., 1980.

4. Ivan Levinstein, "Observations and Suggestions on the Present Position of the British Chemical Industries, with Special Reference to Coal-Tar Derivatives," *J. Soc. Chem. Ind.* **5**, 351–359 and 414 (1886).

5. George E. Davis, *A Handbook of Chemical Engineering*, 2nd Ed., in two volumes, Davis Bros., Manchester, 1904.

6. G. Jimbo, N. Wakao, and M. Yorizane, "The History of Chemical Engineering in Japan," pp. 273–281 in W. F. Furter, ed., *History of Chemical Engineering* [3].

7. The history of coal tar-based industry is told in many places, for example L. F. Haber, *The Chemical Industry During the Nineteenth Century* [2], and *The Chemical Industry 1900–1930*, Clarendon Press, Oxford, 1971. There is a nice synopsis by Peter H. Spitz, *Petrochemicals: The Rise of an Industry*, Wiley, New York, 1988. See also George Lunge's article on "Coal Tar" in *Encyclopaedia Britannica*, 11th Ed. [2]. George Lunge, retiring Professor of Technical Chemistry at the cosmopolitan Swiss Federal Polytechnic (E. T. H.) in Zurich and preeminent international consultant, pointed out in his wide-ranging retrospective discourse before the Royal Institution in London in 1907 that university–industry cooperation in Germany was sparked by earlier British examples of combining science and practice, as transmitted through young German chemists like himself who had worked in England, and nationalistic German visitors to the first great International Exhibition in London in 1851 and its splendid sequel there in 1862. His speech was printed in *The Chemical Engineer 6*, 59–70, 115–122 (1907).

8. The development of the technical universities is described by L. F. Haber, *The Chemical Industry 1900–1930* [7], pp. 42–47, 72–75; and William E. Wickenden, *A Comparative Study of Engineering Education in the United States and in Europe*, Society for the Promotion of Engineering Education, 1929. It is interestingly touched on by Karl Schoenemann, "The Separate Development of Chemical Engineering in Germany," pp. 249–271 in W. F. Furter, ed., *History of Chemical Engineering* [3]. There is an account in English by F. Szabadváry and L. G. Nagy about "The Faculty of Chemical Engineer-

ing of the Technical University Budapest," *Periodica Polytechnica* (Budapest), ca. 1982 (kindly furnished by C. Horvath and L. Louis Hegedus).

9. The cited monographs are: Eugene Peclét, *Traité de la Chaleur considérée dans ses applications*, 4me éd., 3 tomes, G. Masson, Paris, 1878 (first edition from 1828); E. Hausbrand, *Die Wirkungsweise der Rektificir- und Destillir-Apparate*, Springer, Berlin, 1893. See also *Principles and Practice of Industrial Distillation*, 6th Ed. (translated by E. H. Tripp), Chapman & Hall, London, 1926; E. Hausbrand, *Das Trocknen mit Luft und Dampf*, Springer, Berlin, 1898. See also *Drying by Means of Air and Steam* (translated by A. C. Wright), Scott, Greenwood & Son, London, 1901; E. Hausbrand, *Verdampfen, Kondensiren und Kühlen*, Springer, Berlin, 1899. See also *Evaporating, Condensing and Cooling Apparatus* (translated by A. C. Wright), Scott, Greenwood & Son, London, 1908; E. Sorel, *La Rectificacíon de l'alcohol*, Gauthier-Villars et fils, Paris, 1893; compare his articles in *Comp. Rend.* **108**, 1128–1131, 1204–1207, and 1317–1320 (1889). See also J. P. Kuenen, *Theorie der Verdampfung und Verflüssigung von Gemischen und der Fraktionierten Destillation*, Barth, Leipzig, 1906.

10. On the electrochemical industry see L. F. Haber's volumes [7] and Williams Haynes, *American Chemical Industry*, Vol. 1, Van Nostrand, New York, 1954. There are plentiful reminders of "Yankee ingenuity." A contemporary one is James C. Abbeglen and George Stalk, Jr., *Kaisha, The Japanese Corporation*, p. 130, Basic Books, New York, 1985.

11. Martha M. Trescott, *The Rise of the American Electrochemicals Industry, 1880–1910*, Greenwood Press, Westport, Conn., 1981. Besides Trescott's discussions of Frederick Becket, see his articles, "Fifty Years' Progress in Research," *Trans. Electrochem. Soc.* **46**, 43–47 (1936) and "A Few Reflections on Forty Years of Research," *ibid.* **47**, 14–25 (1937).

12. Leonard S. Reich, *The Making of American Industrial Research*, Cambridge University Press, Cambridge, 1985, focuses on General Electric and American Telephone and Telegraph but in his first two chapters masterfully summarizes American science, technology, and industry in the 19th century. Kendall Birr's first two chapters in his *Pioneering in Industrial Research*, Public Affair Press, Washington, D.C., 1957, which is about the General Electric Research Laboratory, are valuable in the same regard and provide more of the European backdrop. Willis R. Whitney, a founder of science-based industrial research, into which he recruited such people as electrical engineer–physicist–physical chemist William D. Coolidge and metallurgical engineer–physical chemist Irving Langmuir, had turned down partnership with his lifelong friend, Arthur D. Little, before the latter was joined by William Walker.

13. The status of electrochemistry is evident in the first volumes of the *Transactions of the American Electrochemical Society*, Philadelphia, 1902 et seq. See also Frederick Becket's 1937 article and pp. 287–293 of Martha Trescott's book [11].

14. The development of American universities and their curricula in science and engineering is a well-studied subject. Useful accounts in the present context are Earle D. Ross, *Democracy's College*, Iowa State College Press, Ames, 1942; Edward H. Beardsley, *The Rise of the American Chemistry Profession*, University of Florida Press, Gainesville, 1964; and the Wickenden Report [8]. Of particular interest is Ezra S. Carr's

Inaugural Address (as professor of chemistry and natural philosophy at the University of Wisconsin), Calkins & Proudfit, Madison, 1856 (available from the State Historical Society of Wisconsin, Madison). Carr around 1868 moved further west, to the University of California. See John Muir, *Letters to a Friend*, reprinted by Norman S. Berg, Publisher, Dunwoody, 1973.

15. William H. Walker: see *The National Cyclopaedia of American Biography*; current Volume A, pp. 167–168, James T. White & Company, New York, 1926; also *The Improbable Achievement: Chemical Engineering at MIT* (60th anniversary commemorative), privately printed ca. 1980 and abridged by H. C. Weber in W. F. Hurter, ed., *History of Chemical Engineering* [3]; and Warren K. Lewis, "William Hultz Walker," *Tech Engineering News* (MIT), p. 7 (January 1921), and "Reminiscences of William H. Walker," *Chem. Eng.* **59**, 158–159, 178 (July 1952). Graduate study in Germany is described by Edward H. Beardsley in Chapter 2 of his book [14]. Little's loss of his original partner is noted in the Centennial Booklet of Arthur D. Little, Inc., Acorn Park, Cambridge, Mass., 1986. So is his firm's creation, in 1921, of a silk purse from a sow's ear (actually two silk purses from 100 pounds of sows' ears). See also E. J. Kahn, Jr., *The Problem Solvers: A History of Arthur D. Little, Inc.*, Little, Brown, Boston, 1986.

16. Massachusetts Institute of Technology Presidents' Reports, 1888–1908, and Catalogs 1887–1888 et seq.; T. M. Drown, "Lewis Mills Norton," *J. Amer. Chem. Soc.* **15**, 241–244 (1893) (an obituary note); Tenney L. Davis, "The Department of Chemistry," in *A History of the Departments of Chemistry and Physics at the M.I.T.*, Technology Press, Cambridge, Mass., 1933; Warren K. Lewis, "Evolution of the Unit Operations," *Chem. Eng. Symp. Series* **55**(26), 1–8 (1959).

17. S. R. Tailby, "Early Chemical Engineering Education in London and Scotland," in *A Century of Chemical Engineering* (W. F. Furter, ed.), pp. 65–126, Plenum Press, New York, 1982.

18. Lilian Cohen, "The Story of Chemistry at Minnesota," unpublished typescript, ca. 1936 (deposited in University of Minnesota Library).

19. James W. Westwater, "The Beginnings of Chemical Engineering Education in the USA," pp. 141–152 in W. F. Furter, ed., *History of Chemical Engineering* [3].

20. Dissolution of the partnership: "Personal," *Electrochemical and Metallurgical Industry* **3**, 482 (1905). Levinstein's article: [4]. Annotated copy of Davis's book: W. K. Lewis, "Evolution of the Unit Operations" [16]. Lunge's article: George Lunge, "The Education of Industrial Chemists," *J. Amer. Chem. Soc.* **15**, 481–501 (1893). Tenney Davis [16] records that Lunge, a Bunsen Ph.D. from Breslau and preeminent industrial chemist, visited MIT sometime between 1888 and 1893. Walker's article, "What Is a Chemical Engineer?" *The Chemical Engineer* **2**, 1–3 (1905), was one of a series. Another of interest: Alfred H. White, "The Course in Chemical Engineering at University of Michigan," *ibid.*, 262–264. Physical chemistry and problem solving: W. K. Lewis, "Evolution of the Unit Operations" [16]. Noyes: Frederick G. Keyes, "Arthur Amos Noyes," *The Nucleus*, October 1936; Linus Pauling, "Arthur Amos Noyes," *Biographical Memoirs*, National Academy of Sciences **31**, 322–346 (1958). The other Walker's textbook: James Walker, *Introduction to Physical Chemistry*, 6th Ed., Macmillan, London, 1910 (first edition from 1899). Noyes began developing his own

text in 1902 and eventually published, with Miles S. Sherrill, *An Advanced Course of Instruction in Chemical Principles*, Macmillan, New York, 1922.

21. Joseph W. Richards, *Metallurgical Calculations*, 3 vols., McGraw, New York, 1906–1908.

22. Lewis: He was elected to the National Academy of Sciences in the era before there was an Academy of Engineering, but no biographical memoir has come to the writer's hand. An informative article is Edwin R. Gilliland, "'Doc' Lewis of MIT," *Chemical Engineering Education* **4**(4), 156–160 (Fall 1970). See also Glen C. Williams and J. Edward Vivian's and H. C. Lewis's sketches in W. F. Furter, ed., *History of Chemical Engineering* [3]. American Chemical Society vis-à-vis American Institute of Chemical Engineers: Terry S. Reynolds, *75 Years of Progress*, Chapter 1, American Institute of Chemical Engineers, New York, 1983. The book, W. H. Walker, W. K. Lewis, and W. H. McAdams, *Principles of Chemical Engineering*, McGraw-Hill, New York, 1923, was later revised by W. H. McAdams and E. R. Gilliland for a third edition in 1937 (which incorporated a limited number of references, mostly to domestic papers). The other book is W. K. Lewis and A. H. Radasch, *Industrial Stoichiometry*, McGraw-Hill, New York, 1926.

23. Milton C. Whitaker, in "The Training of Chemical Engineers," *Trans. AIChE* **3**, 158–168 (1910), advocated a laboratory containing real working models and staffed by technical experts from industries. Almost the whole address was also published in the rival *J. Ind. Eng. Chem.* **3**, 36–39 (1910). What he actually did is reported in "The New Chemical Engineering Course and Laboratories at Columbia University," *Trans. AIChE* **5**, 150–169 (1912). For more on Whitaker, see C. F. Chandler's presentation, *J. Ind. Eng. Chem.* **15**, 199 (1923). The views of William H. Walker and Warren K. Lewis were recorded in "A Laboratory Course for Chemical Engineering," *J. Amer. Chem. Soc.* **33**, 618–624 (1911).

24. Richard K. Meade, "Our Introduction," *The Chemical Engineer* **1**, 44–48 (1904). (In the same issue he notes that the Society for Chemical Industry held its annual meeting in the United States for the first time, in New York City in September, with distinguished foreigners present.) On the founding see Terry S. Reynolds, *75 Years of Progress* [22] and F. J. Van Antwerpen, *Chem. Eng. Prog.* **54**, 55–62 (May 1958). It is instructive to compare the membership lists from 1908 to 1914 in *Trans. AIChE*. On Little's talents, see Chapter 9 of Maurice Holland, *Industrial Explorers*, Harper & Brothers, New York, 1928; compare his interesting presidential address to the ACS, "Industrial Research in America," *J. Ind. Eng. Chem.* **5**, 793–801 (1913).

25. Lunge's article is referenced in [20]. Another article from the Chicago conference explicitly proposes lectures on basic operations; a textbook devoted not to products but to processes, apparatus, and methods of treatment; and a laboratory to illustrate points and teach "the art of making mistakes on the small scale," Henry Pemberton (Philadelphia), "The Education of Industrial Chemists," *J. Amer. Chem. Soc.* **15**, 627–635 (1893). That "unit operations" were abundantly being practiced is obvious and has been emphasized by Martha M. Trescott, "Unit Operations in the Chemical Industry," pp. 1–18 in W. F. Furter, ed., *A Century of Chemical Engineering* [17], and in her book [11].

26. W. H. Walker, "What Is a Chemical Engineer?" [20]. Langmuir's research, which drew on German reaction studies and Mellor's *Higher Mathematics*, was published from

Stevens Institute of Technology, where he taught for three years before joining General Electric in 1909: I. Langmuir, "The Velocity of Reactions in Gases Moving Through Heated Vessels and the Effect of Convection and Diffusion," *J. Amer. Chem. Soc.* **30**, 1742–1754 (1908). C. Guy Suits speculated on how Langmuir's career might have developed had he remained in academia: see his Foreword to *The Collected Works of Irving Langmuir*, Pergamon Press, New York, 1961. The unsuccessful champion of "unit processes" stated his case: R. Norris Shreve, "Unit Operations and Unit Processes," in *Encyclopedia of Chemical Technology*, Vol. 14, pp. 422–426, Interscience Publishers, New York, 1955.

27. W. H. Walker, "A Master's Course in Chemical Engineering," *J. Ind. Eng. Chem.* **8**, 746–748 (1916); W. K. Lewis, "Effective Cooperation Between the University and the Industry," *ibid.*, 769–770; R. T. Haslam, "The School of Chemical Engineering Practice of the Massachusetts Institute of Technology," *J. Ind. Eng. Chem.* **13**, 465–468 (1921). On the School vis-à-vis cooperative courses of study and summer employment, see MIT's Henry P. Talbot, "The Relation of the Educational Institutions to the Industries," *ibid.* **12**, 943–947 (1920); the issues he raised seem eternal.

28. The AIChE efforts are well documented by Terry S. Reynolds, pp. 9–15 of *75 Years of Progress* [22]. Early discussions were published in *Trans. AIChE*; the *Report of Committee on Chemical Engineering Education of the American Institute of Chemical Engineers 1922* was privately printed by the AIChE Secretary at the Brooklyn Polytechnic Institute and distributed in advance of a May 16, 1922 special conference there (copy deposited in University of Minnesota Library). An account of the report and its impact: Alfred H. White, "Chemical Engineering Education in the United States," *Trans. AIChE* **21**, 55–85 (1928), including a lengthy commentary on the six-year-old Institution of Chemical Engineers by John W. Hinchley, author of the entry on "Chemical Engineering" in *Encyclopaedia Britannica*, 13th Ed., 1922.

29. Frank A. Vanderlip, "Political Problems of Europe as They Interest Americans. Third Paper," *Scribner's Magazine* **37**, 338–353 (1905).

30. The story at MIT is engrossingly told and massively documented by John W. Servos, "The Industrial Relations of Science: Chemical Engineering at MIT, 1900–1939," *Isis* **71**, 531–549 (1980). As Servos tells it, the story was a prelude to the changeover of MIT presidents in 1930 and to subsequent Depression pressures. But there had also been a national debate: see "A Symposium upon Co-operation in Industrial Research," *Trans. Amer. Electrochem. Soc.* **29**, 25–58 (1916), in which W. H. Walker and W. R. Whitney *inter alia* took part; and "The Universities and the Industries," *J. Ind. Eng. Chem.* **8**, 59–65 (1916), which quotes Richard C. Maclaurin, president of MIT, Henry P. Talbot, professor of chemistry at MIT, W. H. Walker, and A. D. Little. Fritz Haber's role in the unleashing of war gas is remarked by Peter H. Spitz, *Petrochemicals: The Rise of an Industry* [7], p. 28 and sources listed on p. 61.

31. Arthur A. Noyes, "What Is an Engineer?" *Throop College of Technology Bulletin* **24**(69), 3–9 (1915); see also his address on "Scientific Research in America," *ibid.*, 10–16, in which he spoke of university contributions to industrial research by actually carrying out industrial research, as in the Mellon Institute at the University of Pittsburgh; and by sending out trained men, like Professor Willis R. Whitney from MIT to the General Electric Company.

32. Page 277 in Paul Kennedy's magnificent book, *The Rise and Fall of the Great Powers: Economic Change and Military Conflict from 1500 to 2000*, Random House, New York, 1987. Chapters 4 and 5 illuminate the world scene, where industrialization gave advantage successively to Britain, Germany, and the United States; Chapter 6, the interwar situation; and Chapters 7 and 8, evolutionary change since World War II.

33. Harold F. Williamson *et al.*, *The American Petroleum Industry. The Age of Energy 1899–1959*, Northwestern University Press, Evanston, Ill., 1963, in particular Chapter 4, "Refining in Transition." On the situation in Jersey Standard and the postwar creation of the department that became Standard Oil Development Co., see Edward J. Gornowski, "The History of Chemical Engineering at Exxon," pp. 303–311 in W. F. Furter, ed., *History of Chemical Engineering* [3].

34. Robert T. Haslam and Robert P. Russell, *Fuels and Their Combustion*, McGraw-Hill, New York, 1926. Haslam was then professor-in-charge of chemical engineering practice and a member of the American Chemical Society, American Institute of Chemical Engineers, and American Society of Mechanical Engineers. Clark Shove Robinson's books are *The Elements of Fractional Distillation*, McGraw-Hill, New York, 1922; *The Recovery of Volatile Solvents*, Chemical Catalog Co., New York, 1922; with Frank Lauren Hitchcock, *Differential Equations in Applied Chemistry*, Wiley, New York, 1923; with Alfred L. Webre, Evaporation, Chemical Catalog Co., New York, 1926; *The Thermodynamics of Firearms*, McGraw-Hill, New York, 1943; and *Explosives, Their Anatomy and Destructiveness*, McGraw-Hill, New York, 1944.

35. Thomas K. Sherwood and C. E. Reed, *Applied Mathematics in Chemical Engineering*, McGraw-Hill, New York, 1939; William R. Marshall and Robert L. Pigford, *The Application of Differential Equations to Chemical Engineering Problems*, University of Delaware, Newark, 1947; A. B. Newman, "Temperature Distribution in Internally Heated Cylinders," *Trans. AIChE* **24**, 44–53 (1930); T. B. Drew, "Mathematical Attacks on Forced Convection Problems: A Review," *Trans. AIChE* **26**, 26–79 (1931); Arvind Varma, "Some Historical Notes on the Use of Mathematics in Chemical Engineering," pp. 353–387 in W. F. Furter, ed., *A Century of Chemical Engineering* [17].

36. On early doctorates, compare Olaf A. Hougen, "Seven Decades of Chemical Engineering," *Chem. Eng. Prog.* **73**, 89–104 (January 1977) and J. O. Maloney, "Doctoral Thesis Work in Chemical Engineering from the Beginning to 1960," pp. 211–223 in W. F. Furter, ed., *A Century of Chemical Engineering* [17]. On the state of the American chemical industry, see Williams Haynes, *American Chemical Industry*, Vol. 2, Van Nostrand, New York, 1945, Introduction.

37. Fritz Haber, *Thermodynamik Technischer Gasreaktionen*, Verlag R. Oldenbourg, Munich and Berlin, 1905; L. F. Haber, *The Chemical Industry 1900–1930* [7], Chapters 4 and 5; Peter H. Spitz, *Petrochemicals: The Rise of an Industry* [7], Chapter 1; Karl Schoenemann, "The Separate Development of Chemical Engineering in Germany," pp. 249–271 in W. F. Furter, ed., *History of Chemical Engineering* [3]; Max Appl, "The Haber–Bosch Process and the Development of Chemical Engineering," pp. 29–53 in W. F. Furter, ed., *A Century of Chemical Engineering* [17].

38. P. H. Spitz, *Petrochemicals: The Rise of an Industry* [7], pp. 32–34; Williams Haynes, *American Chemical Industry*, Vol. 2 [36], Chapters 8–10; Norman W. Krase *et al.*, "High-Pressure High-Temperature Technology," *Chem. Met. Eng.* **37**, 530–560 (1930).

39. John K. Smith, "Developing a Discipline: Chemical Engineering Research at Du Pont," in David A. Hounshell and J. K. Smith, *Science and Corporate Strategy: Du Pont R & D, 1902–1980,* Chapter 14, Cambridge University Press, New York, 1988 (kindly made available in advance of publication); Charles M. A. Stine, "Chemical Engineering in Modern Industry," *Trans. AIChE* **21**, 45–54 (1928).

40. Arthur D. Little, "Chemical Engineering Pervades All Petroleum Technology," *Chem. Met. Eng. 35,* 12–14 (1928); P. H. Spitz, *Petrochemicals: The Rise of an Industry* [7], pp. 63–69; H. F. Williamson *et al., The American Petroleum Industry. The Age of Energy 1899–1959* [33], pp. 423–430.

41. P. H. Spitz, *Petrochemicals: The Rise of an Industry* [7], pp. 69–115 (covering early developments by Dow Chemical and Standard Oil Company of New Jersey–Jersey Standard–besides Union Carbide and Shell).

42. H. F. Williamson *et al., The American Petroleum Industry. The Age of Energy 1899–1959* [33], pp. 612–620, 629–633; P. H. Spitz, *Petrochemicals: The Rise of an Industry* [7], pp. 123–128, 172–175.

43. H. F. Williamson *et al., The American Petroleum Industry. The Age of Energy 1899–1959* [33], pp. 620–629; P. H. Spitz, *Petrochemicals: The Rise of an Industry* [7], pp. 128–138; E. J. Gornowski, "The History of Chemical Engineering at Exxon" [33]. The train ride version was recorded by James W. Carr, Jr. in a post-1973 memo, "Exxon and MIT–How They Influenced Each Other and the Process Industries in the Chemical Engineering Arena," Exxon Research & Engineering Company. Robert T. Haslam's role came to light in Hoyt C. Hottel's verbal recounting of why he never turned in a doctoral dissertation, at lunch on March 2, 1989.

44. Howard C. Parmelee, "Inter-Dependence Through Chemical Engineering," *Chem. Met. Eng. 35,* 1–11 (1928).

45. Sidney D. Kirkpatrick, "Building the Literature of Chemical Engineering," *Chem. Eng.* **59**, 166–173 (1952); compare the endsheets of Walter L. Badger and Warren L. McCabe, *Elements of Chemical Engineering,* McGraw-Hill, New York, 1931–with an Introduction by Arthur D. Little!–and of Perry's *Handbook* [49].

46. The Advisory Board is listed in the frontispiece of Olaf A. Hougen and Kenneth M. Watson, *Industrial Chemical Calculations: The Application of Physico-Chemical Principles and Data to Problems of Industry,* Wiley, New York, 1931. Their later three volumes are *Chemical Process Principles. Part One: Material and Energy Balances,* 1943; *Part Two: Thermodynamics,* 1947; and *Part Three: Kinetics and Catalysis,* 1947, all published by Wiley.

47. Terry S. Reynolds, *75 Years of Progress* [22], Chapter 2. The change in nature of the papers is evident in successive volumes of *Transactions of the American Institute of Chemical Engineers* in the late 1920s and early 1930s.

48. Sidney D. Kirkpatrick, ed., *Twenty-five Years of Chemical Engineering Progress,* Silver Anniversary Volume, American Institute of Chemical Engineers, New York, 1933 (reprinted by Books for Libraries Press, Freeport, N.Y., 1968). George Davis's disgust was recorded by his protégé Norman Swindin in "The George E. Davis Memorial Lecture" [3].

49. John H. Perry, ed., *Chemical Engineer's Handbook,* McGraw-Hill, New York, 1934;

J. K. Smith, "Developing a Discipline: Chemical Engineering Research at Du Pont" [39]; Vance E. Senecal, "Du Pont and Chemical Engineering in the Twentieth Century," pp. 283–301 in W. F. Furter, ed., *History of Chemical Engineering* [3].

50. Ernst Berl, *Chemische Ingenieur-Technik*, 3 Bände, Verlag von Julius Springer, Berlin, 1935; Karl Schoenemann, "The Separate Development of Chemical Engineering in Germany" [37, 3].

51. A. Eucken and M. Jakob, *Der Chemie-Ingenieur: Ein Handbuch der physikalischen Arbeitsmethoden in chemischen und verwandten Industriebetrieben*, Akademische Verlagsgesellschaft, Leipzig, 1933 et seq., with a foreword by F. Haber; Klaus Buchholz, "Verfahrenstechnik (Chemical Engineering)–Its Development, Present State and Structure," *Social Studies of Science* **9**, 33–62 (1979). Ewald Wicke, "Gerhard Damköhler–Begründer der Chemischen Reaktionstechnik," *Chem.-Ing.-Tech.* **56**, A648–650 (1984), translated as "Gerhard Damköhler–Founder of Chemical Reaction Engineering," *Int. Chem. Eng.* **25**, 770–773 (1985). Irving Langmuir's reactor analysis was too early and too little, his adsorption and catalysis studies too pure and too narrow, to connect him to the engineering discipline then.

2

Academic–Industrial Interaction in the Early Development of Chemical Engineering at MIT

Ralph Landau
Department of Economics
Stanford University
Stanford, California

The early rise of chemical engineering as an academic discipline was characterized by an intimate association between the university and industry in several different ways and at various critical periods in U.S. history. This interaction was especially important in influencing the Course in Chemical Engineering at the Massachusetts Institute of Technology, the centennial of which we celebrate. In this paper, I trace the development of the program at MIT and the role played by academic–industrial interaction. I have divided this early development into two time periods, 1888–1920 and 1920–1941, because each was associated with a single dominant personality, William H. Walker and Warren K. Lewis, respectively.

I. The Early Years, 1888–1920

In 1888, the year Course X was established, MIT was a simple undergraduate engineering school, and student tuition supplied around 90% of the university payroll. What appeared in lectures by Professor Lewis M. Norton that fall

ADVANCES IN CHEMICAL ENGINEERING, VOL. 16

was essentially descriptive industrial chemistry in an era of primarily heavy inorganic chemical manufacture—sulfuric acid, soda ash, chlorine, etc.—plus coal tar-based organics. A harbinger of MIT's future seminal texts in chemical engineering, *Outlines of Industrial Chemistry* by Professor Frank H. Thorpe, was published in 1898. At that time, industrial chemists focused on the production of a single chemical from beginning to end and broke industrial processes into their chemical steps, with little attention to production methods. Few fundamental principles unifying the manufacture of different chemicals existed. Mechanical engineers, who worked with the industrial chemists, focused primarily on the process machinery, as they have since. Industrial companies alone combined the two areas and possessed the know-how to design and operate large-scale plants.

At the turn of the century, Professor Arthur A. Noyes, an MIT graduate and Ph.D. in chemistry from Leipzig, sought to convert MIT into a science-based university that included a graduate school oriented toward basic research. To train properly for careers in industry, he believed that graduates should also know the principles of the physical sciences, using the problem-solving method that he pioneered. In 1903, he established the Research Laboratory of Physical Chemistry, modeled after similar laboratories in Germany, and personally financed about half of it. It produced the first Ph.D.'s from MIT not long afterward. The Graduate School of Engineering Research was also established in 1903.

However, a different vision possessed Professor William H. Walker, a chemistry graduate from Pennsylvania State with a Ph.D. from Göttingen. Working closely with Arthur D. Little, who had already established his consulting firm, Walker maintained that MIT should remain a school of engineering and technology, with the goal of training the builders and leaders of industry through an emphasis on applied sciences. A sharp disagreement arose between Noyes and Walker, one that lasted for many years. Walker believed that learning physical science principles was not enough; only through an understanding of problems drawn from industry could a student move from theory to practice. He felt that, in this way, the problems of large-scale production could be understood by the young engineer, given the constraints imposed by materials, costs, markets, product specifications, safety, and the many other considerations that up to then were solely the province of industrial companies. Furthermore, the manner in which the large domestic market was growing followed along America's tradition of mass production, rather than the German-style craft traditions of chemical production. Although he was primarily involved in chemistry and chemical engineering, Walker's vision was applicable to other engineering disciplines as well.

Walker reorganized the languishing industrial chemistry program according to his concepts, transforming it from a heterogeneous collection of chemistry and chemical engineering courses into a unified program in chemical engineering. Its focus was increasingly based on the study of "unit operations." This terminology was first used by Arthur D. Little in 1915 and was based on a concept developed over a number of years [1]. The concept was reduction of the vast number of industrial chemical processes into a few basic steps such as distillation, absorption, heat transfer, fluid flow, filtration, and the like. Hence, Walker and Little developed their own scientific analysis of the principles of chemical engineering based on unit operations. Unlike Noyes, they made use of concepts neither employed nor needed for studying basic chemistry in the laboratory because they dealt with how to design industrial-sized plants and equipment.

As a further emphasis of his disagreement with Noyes' approach, Walker established the Research Laboratory of Applied Chemistry (RLAC) in 1908 to obtain research contracts with industrial firms, thereby providing not only income but also experience in real problems for both faculty and students. It would later also serve as the basis for graduate studies in chemical engineering and would act as a link between MIT and industry, mutually benefiting both. In the beginning of the 20th century, few American firms had their own research facilities, with General Electric (1900) and Du Pont (1902) among the pioneers in this area. Walker had observed that the rapid growth of the German chemical industry was closely tied to cooperation between industry and academia and the existence of industrial research laboratories.

To more firmly establish exposure of students to industrial problems, Walker and his younger colleague, Warren K. "Doc" Lewis, an MIT chemistry undergraduate with a Ph.D. from Breslau, founded the School of Chemical Engineering Practice in 1916, again with support from Arthur D. Little. Students gained access to the expensive industrial facilities required to relate classroom instruction in unit operations to industrial practices while still retaining faculty supervision.

The differing viewpoints of Walker and Noyes regarding MIT's direction proved irreconcilable. By the end of World War I, Noyes' influence in the chemistry department had declined, while Walker's program in chemical engineering—still situated in the chemistry department—grew in popularity. In 1909, more bachelor's degrees were awarded in chemistry than in chemical engineering; by 1919, the latter was roughly three times larger. The rapid expansion of the undergraduate chemical engineering program paralleled and was derived from the expansion of the American chemical industry, the latter a result of the war and the loss of German chemical imports. This was the first clear demonstration of the synergism between the developing discipline of chemical engineering and the growth of the chemical industry, a link that is still evident today. A foreshadowing of this relationship had occurred earlier in the decade when the du Ponts and George Eastman gave major donations to MIT; these combined gifts exceeded MIT's entire endowment and plant value by three times, supplementing tuition income as a means for funding research. By 1930, MIT had the fifth largest university endowment in the country.

MIT President Richard C. MacLaurin could not resolve the dispute between Noyes and Walker with a compromise; finally, forced to choose between the two, the Institute backed Walker's concept of stressing the engineering approach based on a systematic analysis of real industrial problems. Noyes resigned in 1919, later founding the modern California Institute of Technology with George Hale and Robert Millikan, both eminent scientists. As the only trustee of both institutions, I can still discern today the influence of these divergent views.

Walker's vision had triumphed at MIT. In 1920, he became director of the new Division of Industrial Cooperation and Research, which, in its various forms, has survived to this day. In the fall of that year, chemical engineering split off from chemistry and an independent chemical engineering department was formed with Warren K. Lewis as its first chairman. However, there was faculty opposition at MIT to what was perceived as Walker's aggressiveness and his excessive reliance on industrial ties. Ironically, his own earlier emphasis on a fundamental approach to chemical engineering tied to real problems had eroded as a result of the postwar difficulties in obtaining industrial sponsors for significant and instructive research contracts. The necessity of securing adequate funds overwhelmed selectivity in the choice of research problems. Combined with resistance directed at him from other sources, Walker felt frustrated enough to resign effective January 1, 1921. His legacy, however, has proved enduring:

1. Walker firmly linked engineering education to research, and teaching to real industrial problems, by creating strong ties with industry through the establishment of the RLAC.

2. He created the modern concept of chemical engineering as a unified discipline for a very wide range of processes and products.

3. He founded the unique School of Chemical Engineering Practice.

4. His protégé, Warren K. Lewis, carried on after Walker in an extraordinarily influential role for MIT, for chemical engineering, for the chemical process industries, for me personally, and for others like me.

II. Chemical Engineering and Petroleum Refining, 1920–1941

"Doc" Lewis inherited the RLAC, a growing source of serious problems both to MIT and to the chemical engineering department. Finances were always problematic, and short-term projects were haphazardly selected. Industrial companies were creating their own research laboratories, thereby needing the RLAC less, and salaries for RLAC staff were not competitive with industrial salaries. The postwar recession reinforced the reality that external economic conditions dominated the intellectual environment. This last issue was of particular concern to Lewis, who became increasingly aware of the need for disciplinary independence from industry. Engineering education grew in popularity after World War I, but interest soon declined somewhat as the wartime glow disappeared. Furthermore, chemical engineering was becoming more abstract, moving to quantify and extend the knowledge of fundamentals such as heat transfer, high-temperature and -pressure reactions, and absorption. While chemical engineering research arose from industrial problems, university researchers themselves often pursued more long-range, academic problems that had no immediately practical results and, consequently, were of less interest to industrial sponsors. Funding for the RLAC dried up during the Depression, and it closed in 1933.

Meanwhile, MIT was as a whole perceived by other research university scientists as no longer having as high a reputation as a center of both basic and applied research. For example, between 1919 and 1930, twice as many National Research Fellows in chemistry chose to do graduate work at Caltech rather than MIT. However, as long as industry supplied a major portion of MIT's revenues, little could be done to overcome this. Karl Compton's tenure as president of MIT, beginning in 1930, improved the balance between engineering and basic science. Unfortunately, industry was never an enthusiastic supporter of basic science at MIT, unlike Caltech's experience; it was only when federal funding became MIT's major source of research money after World War II that this balance was eventually achieved.

Thanks to the leadership of Lewis and his colleagues, the chemical engineering department became an intellectually powerful center during this period. Its preeminence was fostered by the publication of the pioneer text in 1923, *Principles of Chemical Engineering*, by Walker, Lewis, and William H. McAdams. At the same time, Lewis developed a new kind of relation with industry, one that proved to have enormous impact on the chemical process industries and on teaching and research in academia. In fact, petroleum refining was the industry that provided the key to many future developments in chemical engineering.

In 1911, an antitrust judgment split the Standard Oil Company into a group of competitive companies; none of these smaller companies had much in the way of research, development, or engineering. Before World War I, Henry Ford had created the assembly line for manufacturing automobiles aimed at a mass market. Although the war interrupted the spread of the automobile, the oil companies correctly foresaw that gasoline demand would boom after the war and realized they were not adequately equipped to develop and install improved continuous processes for large-scale gasoline production. The two largest refiners were Standard Oil Company (Indiana) and Standard Oil Company (New Jersey)— now Amoco and Exxon, respectively. The largest group of researchers was at Amoco. They developed a thermal cracking process called Burton–Humphreys, a process that was widely licensed but suffered from carbon (coke) deposits that required periodic shutdowns for burnout. Exxon initially lacked a research and development facility, so they lured E. M. Clark away from Amoco and revised the cracking process to the tube and tank process. Patent coverage appeared to be critical, so Exxon hired Frank A. Howard, a patent attorney who had coordinated Amoco's innovative efforts. A new Development Department, with Howard as its head, was created in 1919.

One of Howard's first acts was to engage the best consultant for Exxon that he could find, and that was none other than Warren K. Lewis. A partnership was formed that would have a profound influence on Exxon and on chemical engineering. Lewis' first efforts as a consultant were aimed at developing precision distillation and *making the process continuous and automatic.* Batch processes were clearly inadequate for the rising market demand for gasoline. By 1924, Lewis had helped increase oil recovery by using vacuum stills. From 1914 to 1927, the average yield of gasoline rose from 18 to 36% of crude throughput. This work and his earlier bubble tower designs became refinery standards. At the same time, Lewis brought back to campus the new insights he gained at Exxon. Course work at MIT changed to incorporate these new concepts and their *design* principles.

In 1927, Howard initiated a series of agreements with the German chemical giant, I. G. Farben Industrie. Howard wanted access to their work on hydrogenation and synthetic substitutes for oil and rubber from coal, thinking this might increase gasoline yields and increase Exxon's activity in chemicals. To exploit the application of these technologies for Exxon's purposes, Howard needed a whole new research group, particularly one knowledgeable about *chemical process technology.* Again, Lewis was consulted; he introduced Robert Haslam, head of the School of Chemical Engineering Practice and director of the RLAC, to Howard. Haslam, on leave from MIT, formed a team of 15 MIT staffers (many from the RLAC) and graduates, who set up a research organization at Baton Rouge, Louisiana.

Six of these employees eventually became high-ranking executives and board members of Exxon, so great were their talent and influence. Haslam himself left MIT and became vice president of the Exxon Development Department in 1927. Another staffer, R. E. Wilson, eventually became chairman and CEO of Amoco. He subsequently hired Walter G. Whitman from MIT; Whitman later returned to MIT in 1933 as chairman and was my thesis adviser.

By exploiting the German technology for American refining use, Exxon had covered much of the large gap that had so long existed between applied and creative or innovative research in the petroleum industry, including introducing the use of chemical catalysis in petroleum refining. Much of what took place in modern petroleum processing until World War II originated in Baton Rouge, and it was basically an MIT group that initiated and guided this effort. With the continuing advice of Lewis, and later (1935) of Edwin R. Gilliland, Baton Rouge produced such outstanding process developments as hydroforming, fluid flex coking, and fluid catalytic cracking, ultimately the most important raw material source for propylene and butane feedstocks to the chemical industry. Lewis and Gilliland remained consultants for Exxon until their deaths in 1975 and 1973, respectively. MIT Professor Hoyt Hottel also consulted for Exxon as early as 1928 on problems in combustion; this experience is reflected in Hottel's enduring influence in teaching combustion theory and practice.

Frank Howard was also instrumental in the process research for Midgley's tetraethyllead antiknock additive, which led to Exxon's 50% ownership with General Motors of the Ethyl Corporation. Charles A. Kraus, an MIT staff member just before World War I, was its developer, and it appears that "Doc" Lewis found him for Exxon. Another Lewis idea later resulted in Exxon's fluidized bed iron ore process, commercialized after World War II only in Venezuela by the Arthur F. McKee Company. Someday, such a process might be incorporated in integrated steel minimills.

By working with Frank Howard, Lewis created a very different approach from the earlier Walker programs. Instead of bringing industry to the campus, as with the RLAC, he brought the campus to industry. Unlike the Practice School, which also did this, Lewis helped solve the big problems of industry. He, too, left an enduring legacy:

1. He established that industrial consultation by professors enriched both the employers and the Institute's educational program; this practice is now quite common.

2. He created the precedent that talented MIT faculty and students should and could go into industry, simultaneously applying their learning and providing valuable experience for MIT.

3. Above all, he focused the discipline of chemical engineering on *design* of continuous automated processing of a huge variety of products, first in petroleum refining and then in chemicals. It became an essential ingredient in the growth of the world petrochemical industry [2].

Even though the RLAC eventually disappeared, a victim of the Depression, and, as it turned out, was the wrong way for industry and academic chemical engineering to interact, it had created a pool of talented people who moved industry along. "Doc" Lewis took chemical engineering well beyond the RLAC. He forcefully demonstrated that professors with hands-on experience in industry were both excellent teachers and creators of the new discipline. Hence, MIT's chemical engineering department through the 1920s and 1930s continued to enjoy great prestige, not only at MIT but throughout the world. Other departments of chemical engineering in the United States also flourished, but MIT in particular attracted the brightest students from all over the United States and abroad, who in turn moved into high positions throughout the petroleum and chemical industries. This was true even in the era when technology in general was blamed as the cause of high unemployment of the Depression. Other engineering disciplines (with the exception of electrical power engineering and possibly aeronautical engineering), lacking this history of direct involvement with a growing major industry, never developed such an overall systems approach to the design of continuously operating production plants. Rather, those disciplines tended to develop tools, while chemical engineers solved real problems. Such a difference is at the root of American difficulties in competitive manufacturing of discrete mechanical goods.

Two other aspects of the rise of chemical engineering prior to World War II deserve note. One was the flourishing of process engineering design firms that contained all the skills necessary to build complete plants. The earliest and still among the most prominent firms emerged early in this century— Universal Oil Products (UOP) and M. W. Kellogg Company. Because of the obvious need for design of continuous, automatically controlled units, these companies became dominated by chemical engineers, many trained at MIT. At UOP, there were important figures like Vladimir Haensel and Donald Broughton; at Kellogg, the chief technical officer was the brilliant P. C. Keith. "Doc" Lewis would recommend able MIT graduates to these companies as well, and I was one of them. I learned the real art of process design from Keith and his engineers.

The second aspect was the establishment in 1929 of a unique group headed by Thomas H. Chilton to carry out fundamental research in chemical engineering at Du Pont [3]. McAdams of MIT was a Du Pont consultant and stimulated research in this direction. This group attracted some of the best young minds in chemical engineering from around the country. One of the first employees was Allan P. Colburn, who completed his Ph.D. under Olaf A. Hougen at Wisconsin in 1929. Others who joined later included Thomas B. Drew, W. Robert Marshall, and Robert L. Pigford, all of whom eventually left Du Pont for distinguished careers in academia. During the 1930s, this industrial group, especially the collaboration of Chilton and Colburn, produced influential publications that figured prominently in the development of the discipline of chemical engineering.

Although a number of individuals contributed to the early development of chemical engineering, many of whom I do not have space to mention here, one of them stands above the rest. In my opinion, "Doc" Lewis virtually single-handedly created modern chemical engineering and its teaching methodology. He was the inspiration of chemical engineers throughout the world, including myself.

Acknowledgments

I owe much of the factual background in this paper to an article by John W. Servos, "The Industrial Relations of Science: Chemical Engineering at MIT, 1900–1939," *Isis* **71**, 531–549 (1980), and to James W. Carr, a retired Exxon engineer. Additional material was obtained from Terry S. Reynold's *75 Years of Progress—A History of the American Institute of Chemical Engineers, 1908–1983*, American Institute of Chemical Engineers, New York, 1983; W. F. Furter's *History of Chemical Engineering*, published by the American Chemical Society as *Advances in Chemical Engineering*, Vol. 190, 1979; and *To Advance Knowledge—The Growth of American Research Universities 1900–1940* by Roger L. Geiger, Oxford Press, New York, 1986.

References

1. Scriven, L. E., in *Perspectives in Chemical Engineering: Research and Education* (C. K. Colton, ed.), p. 3. Academic Press, San Diego, Calif., 1991 (*Adv. Chem. Eng.* **16**).

2. Landau, R., *Chem. Eng. Prog.* **85**, 25 (September 1989).

3. Hounshell, D. A., and Smith, J. K., Jr., *Science and Corporate Strategy, Du Pont R & D, 1902–1980*. Cambridge University Press, New York, 1988.

3

Future Directions of Chemical Engineering

James Wei
Department of Chemical Engineering
Massachusetts Institute of Technology
Cambridge, Massachusetts

Even as we celebrate the occasion of 100 years of chemical engineering education in the United States and our splendid past achievements in educating generations of leaders for the chemical and petroleum industries, we hear disturbing rumbles that we must change in order to survive. The effortless technological dominance that America has enjoyed since World War II is gone, manufacturing will locate anywhere in the globe that has the best technology and lowest cost, and American industry has seemingly lost the ability to manufacture what the world wants to buy. In the long run, we can live well only by producing well. In 1987, only two American manufacturing industries had a significant surplus of exports over imports: commercial aircraft and chemicals. In all other industries, we import more than we export. Moreover, plants are closing, from sunset industries of shipbuilding, textiles, and steel, to sunrise industries of electronics, robotics, semiconductors, computers, and copiers. Can we congratulate ourselves that our chemical industry is still above water, when we may be simply one step ahead of the others? Besides, to whom would the chemical industry sell if our clothing, housing, automobile, and communication industries were all shut down? Chemical engineering academics are needed now to strengthen and save our industries.

ADVANCES IN CHEMICAL ENGINEERING, VOL. 16

In the last century, the discipline of chemical engineering has gone through three periods. In the period of 1888 to 1920, the core or paradigm of chemical engineering was not yet developed, and education and research were concerned with numerous and fragmented industrial technologies without discernible common threads. Chemical engineers had not yet discovered concepts and tools that were generally applicable to many industrial processes. The second period opened with "unit operations," the first chemical engineering paradigm, and the most important landmark was the textbook *Principles of Chemical Engineering*, published in 1923 by Walker, Lewis, and McAdams. In this second period, the chemical engineers had a job to do, which they did by organizing and generating knowledge that was generally applicable and effective to improve the process efficiencies of the mass production of commodity chemicals and oil refining. In the third period after World War II, the successes of the scientists in engineering the bomb and radar led to an awareness that a more scientific and fundamental approach to engineering was necessary. America's effortless technological dominance also gave us the leisure to be more academic and inward-looking, to do more analysis, and to develop more theoretical tools and concepts. The most important landmark of that period may be the textbook *Transport Phenomena*, published in 1960 by Bird, Stewart, and Lightfoot. We are now entering the fourth period, when many of us need to leave our ivory towers and to swing back from the world of thought to the world of action. We are needed to address critical problems of manufacturing: how to generate new industries stimulated by new advances in sciences, how to rejuvenate traditional industries and to keep them world competitive, how to protect the environment and health. Of course, we should never abandon the task of generating more powerful concepts and tools, but it is time for us to look outward and make a greater impact on the quality of life.

The eight themes chosen for this symposium represent the wide spectrum of mainstream chemical engineering efforts. The first three topics—fluid mechanics and transport, thermodynamics, and reaction engineering—constitute the intellectual core of the chemical engineering curriculum. The next four topics—environment, polymers, electronic materials, and biotechnology—are relatively new application areas and involve multidisciplinary cooperations. The last topic—process engineering—brings us together again, with synthesis and integration of our knowledge to produce useful products for society.

Gary Leal pointed out that fluid mechanics and transport is an intellectually mature field, in which the outstanding problems are well defined and the governing equations are known. There are many triumphant past cases of problems solved, proofs derived elegantly, measurements made more precise, knowledge better codified in this field, but there is now a wide divergence between research frontiers and education for pragmatic engineering, with the former having little impact on the latter. Keith Gubbins noted that in the thermodynamics of small molecules encountered in petrochemicals and oil refining, precise measurements and correlations have been very effective in the past. But for the complex molecules and aggregates in the frontier industries of biotechnology and materials, such classical methods require supplementation by molecular theories and computation methods.

Another consideration is the maturity of the industry that we serve. It is well known that generals always design their battle plans to suit the last war instead of the next. Chemical engineers must avoid the same pitfall. We are used to dealing with the mature industries of commodity chemicals and petroleum refining, where the products have long life cycles and stable technologies. Because they are produced in large quantities, process efficiency is a key competitive advantage (even 1% improvement in process efficiency would be lucrative), and there is patience to invest millions of dollars and tens of years for process research. But the new technologies involve specialty chemicals with short life cycles that are rapidly obsolete and produced in small quantities, so that processing costs are not the most important considerations and cannot command such patience in terms of research dollars or time horizons. Competitive advantages are gained primarily by the speed of introduction of new products. Precise measurements and detailed models on small volume and/or quickly obsolete products may be difficult to justify and to finance. Chemical engineers need to retool and to develop a new strategy to deal with this new situation.

Alex Bell gave an accurate description of the state of current research activities in chemical reaction engineering, which are narrowly focused on modeling of existing designs. The research community has not seriously addressed the critical problems of the invention and synthesis of new reactors, characterization of reactors, and actual pathologies suggested by reactor failures that result in loss of revenue and lives. There have not been new fundamental concepts in chemical reaction engineering for many years, and its most vigorous fields are now application areas such as catalysis, biotechnology, chemical vapor deposition, and pollution. In critical problems such as multiphase flow in trickle beds, reaction engineers must learn fluid mechanics and large-scale computation before they can effectively develop better solutions.

In the four application areas, we are at a much earlier stage of technical maturity. In many cases, the industries are either quite new or not yet in existence, requiring exploration to define the most important new problems, to establish the most important quantities to measure, to propose the governing equations, and to use statistics to separate the data noise from significant trends. Learning new science and multidisciplinary cooperation are also required.

The original concept of an American chemical engineer was a person knowledgeable in chemistry and the behavior of molecules as well as mechanical engineering and equipment. The success of American chemical engineers stems from this creation of a new breed with dual competence, who can be innovative on both fronts and can rise to be leaders of the chemical and petroleum industries. Today, a new breed would require a more strenuous dual education program, in which the student must learn all of chemical engineering as well as big chunks of biochemistry, genetics, polymer synthesis and characterization, semiconductor physics, quantum mechanics, statistical mechanics, bifurcation theory, and artificial intelligence. Can all this be squeezed into a four-year B.S. program? In many universities, the actual average length for a B.S. degree is already increasing toward five years, as the curriculum is too rigid to permit any deviations for outside interests such as athletics and band, changing majors, dropping a course, junior year abroad, or other noncurricular experiences. It is impractical to offer new required courses in frontier applications. A solution is to offer them as electives only, but to enrich the core courses of transport and reaction engineering with problems drawn from these frontier applications. In this case, specialization would be delayed until graduate school, and perhaps we should accredit the M.S. rather than the B.S. as the first professional degree.

The alternative model to the dual-competent chemical engineer is the chemical engineer as a member of a multidisciplinary team, one who knows the vocabulary of the other team members but not their theories. Undoubtedly, this model may be necessary at times, but such a member would have little chance to become the dominant creative mind or project leader. For less investment, one gets less results.

Art Westerberg discussed the revolution in process engineering brought about by computers. At one time, we divided the world's problems into two piles: a small pile that could be solved by analytical solutions and slide rules, and a much larger pile of "unsolvable" problems. We are now reexamining many of these unsolvable problems and chewing them up with supercomputers. It has often been argued that analytical solutions are generally applicable for a wide variety of parameter values and asymptotic conditions and give insight into system behavior in general, but computer solutions are strictly local for one set of parameter values alone and give no insight. Computer-oriented researchers must address these complaints, perhaps by exploring ranges of parameter values to generate insight.

Mort Denn brought up a most agonizing problem, that of the fragmentation of our profession. We live in a world of increasing fragmentation in interests and services. Some of us are old enough to remember that a magazine stand used to be dominated by the *Life* and *Look* magazines, but now hundreds of specialized magazines such as *Running* and *PC* exist. The disciplines of electrical engineering, mechanical engineering, and civil engineering have all been divisionalized. The American Chemical Society has long been split into several divisions; the chemists still have a common bachelor's degree, although it is not really considered a professional degree. Chemical engineering remains the most cohesive engineering discipline, and we still claim to have a required core discipline of thermodynamics, fluid mechanics and transport, kinetics and reaction engineering, separation, design, and economics. But there is an increasingly strong pull for multidisciplinary research (such as biotechnology, microelectronics, polymers, catalysis, combustion, and environmental research) and special-purpose research (such as non-Newtonian fluid mechanics and high-vacuum studies), requiring that chemical engineers split off to go to a wide variety of meetings or read journals not shared by most other chemical engineers. A potential model is the "dual citizenship" mode in which each chemical engineer belongs to the AIChE, goes to the annual AIChE meeting in November, and reads the *AIChE Journal* and *Chemical Engineering Science;* but that engineer also, for example, belongs to the North American Catalysis Society, goes to annual meetings of the Catalysis Society, and reads the *Journal of Catalysis.* The only time that chemical engineering educators of all specializations get together is in discussing curriculum and accreditation, in recruiting graduate students and junior faculty, and in handing out AIChE awards.

There are several themes that are not addressed in this symposium. How do we educate our students to broaden their minds and embrace the bigger picture? We boast that chemical engineers excel because they can operate in all three scales: the microscale of the chemistry and physics of molecules and aggregates; the mesoscale of process equipment; and the macroscale of systems, productivity, world competition, environment and health protection, management of resources, ethics, and the impact of technology on society. Where do we teach the big picture in our curriculum so that our graduates can effectively manage the health and prosperity of our industries in harmony with society? Should we abdicate that role to management schools and perhaps lose our best graduates to MBA programs?

How can we turn our knowledge of the natural and engineering sciences, learned via analysis, into inventing and improving better processes and products? How do we teach engineering synthesis and the joy of creative engineering? Are our current design courses suitable in the long run? In our research-oriented culture, most of the faculty would fear an assignment to the senior capstone design course, as their tools of analysis are not particularly suited to the task of synthesis.

How can we teach our students to be better team players? Dean Lester Thurow of the MIT Sloan School of Management said that a professor of economics in Germany receives 30% more pay than a professor at MIT, who in turn gets 60% more than a professor at Oxford. According to Thurow, the German professor does not necessarily know more economics than the Oxford professor, but he plays on a better team. We must develop better teams in America, and we chemical engineers must learn to be effective team players. Where do we teach these nonacademic characteristics, such as personal social skills, sensitivity to others' feelings and needs, working under the leadership of others, accepting responsibility and leadership, timely delivery of results, written and oral communication skills, and knowing the vocabulary and the basic outlook of people in management and of other professionals, such as biologists, and the nonprofessional foremen and operators? Should we let industry teach this?

There is an old Chinese greeting, "May you live in uninteresting times." Indeed, if life were unchanging and less challenging, we could slip into autopilot or cruise control. But the second century of chemical engineering will not be dull. We will have splendid opportunities to conquer new frontiers, but we will have to work hard to get there. See you at the frontiers!

SECTION II
Fluid Mechanics and Transport

Centennial Symposium of Chemical Engineering

Opening Remarks

James Wei
Department of Chemical Engineering
Massachusetts Institute of Technology
Cambridge, Massachusetts

Welcome to all of you to the symposium. It is a great pleasure for me to host this gathering of luminaries of the profession. The symposium is very timely: not only is this the 100th year of chemical engineering education, it is also a time of turbulent change. We have an opportunity to sit together and think about where we've been and where we'd like to go. I have read the main chapter from each of the nine sections and the commentaries from the various discussants, and I have learned a lot. In addition to these papers, our discussions here will be published in *Advances in Chemical Engineering*, which will be widely distributed and made available to another, bigger audience.

Our challenge today is to develop insight and illumination into our profession—to eng age in intellectual debate about the changes that have occurred and what we have accomplished, the impact and utility of our science, and what else we should do and how. Let us be critical about our successes as well as our failures. The art historian Erwin Panofsky described three stages of development in the intellectual mind. In stage one, you sit at the foot of the masters; you learn the traditions, the crafts, the techniques; and you solve problems that the masters posed, using tools that the masters developed. In stage two, you're the master yourself, you develop your own style, you pose your own problems, you develop your own tools, and you become a big success in your field. This is the full flower of your success, and you have groups of admirers and followers. Then you proceed to do your science, you invent new problems for your followers or students to investigate, you develop new experiments, you test your models with more precise measurements. This is what we call normal science. In stage three, after you have become a master, you have developed your own style, and you can afford to lean back and to reflect on the whole thing. What is the impact of the work of yourself and your fellows on the world? Did you actually create new science, and does it influence the scientific work of people in other fields? Has the field developed an influence on many students and made them more effective in the world? Has it led to new technologies to better our life and environment? Have we utilized the new tools of science and concepts developed? Have we overlooked the importance of choice? In short, we should raise our sights, aim higher, and be good at our work. The question is, what impact do we have on the world? All of you are masters of stage two, or you wouldn't be here. You've developed your classical work, and you have your followers.

Let us rise to the third level here today and critically evaluate what we have done, what we haven't done, and what we should be doing. We should not hesitate to speak out on what we are not doing as well as we could out of politeness to the speaker or to our profession. I promise that I will not be a polite host myself as we wrestle for the soul of chemical engineering.

4

Challenges and Opportunities in Fluid Mechanics and Transport Phenomena

L. G. Leal
Department of Chemical and Nuclear Engineering
University of California, Santa Barbara
Santa Barbara, California

I. Introduction

This paper was written as a personal view of fluid mechanics and transport phenomena from the viewpoint of someone who has been actively involved in teaching and research in a university chemical engineering department for the past 18 years. This is not intended as a history or a blueprint for the future. Rather, I hope that it may play a role, along with the input of others with different perspectives, in shaping some future directions for research and in exploring their implications for teaching fluid mechanics and transport phenomena to chemical engineers.

I begin with the relationship between this field, as seen from a chemical engineer's perspective, and the fluid mechanics and transport phenomena characteristic of other branches of science and engineering. As research and teaching become more fundamental, the differences characteristic of an earlier focus on specific processes become smaller and the emphasis is increasingly on a generic understanding of phenomena that may affect many processes and many fields. Nevertheless, the fluid mechanics and transport phenomena of the chemical engineer do retain unique elements and a unique flavor. Perhaps the most important element is that fluid mechanics in chemical engineering applications is almost always coupled with heat or mass transfer

ADVANCES IN CHEMICAL ENGINEERING, VOL. 16

processes, which are the primary focus of interest. Indeed, convective and diffusive mass transfer remains almost the exclusive province of the chemical engineer. A second factor that distinguishes fluid mechanics and transport in chemical engineering is the common occurrence of complex materials, specifically so-called heterogeneous fluids such as suspensions and emulsions and macromolecular fluids such as polymer solutions and melts. The essential lack of understanding of the constitutive behavior of these materials strongly affects both teaching and research endeavors for a major portion of the fluids and transport problems important to the chemical engineer. Even when the fluids are Newtonian, viscous effects tend to be a much more important component of fluid mechanics for chemical engineering than for other scientific or engineering disciplines. Finally, the problems encountered span a much broader range of length scales than in the other disciplines, from the macroscopic scale of some processing equipment to the microscale of colloidal particles, individual biological cells, thin films, or even macromolecules, where nonhydrodynamic forces play a critical role. In many of these latter cases, the chemical engineer is uniquely suited to make significant contributions, requiring an intimate blend of continuum mechanics, statistical mechanics, and microscale physics.

But, I'm getting ahead of the story. First, I'll examine the historical evolution of the subject of fluid mechanics and transport phenomena in a chemical engineering context. Then I will turn to a personal view of a likely future and, in so doing, return to some of the topics that have been mentioned above.

II. A Brief Historical Perspective

Because my career spans only 20 years, I cannot hope to capture the flavor of much of the past—neither technological accomplishments nor the winds of progress that contributed to the evolution of the subject. For the most part, I can only paraphrase the historical accounts of others. Hence I shall be brief.

Much of the chemical industry existed before the 100–year span of chemical engineering as a profession that this paper commemorates. In those early days, knowledge was organized around each specialized industry; specific production techniques were regarded as largely unique to individual products and were transmitted through studies of specific industrial technologies. The pioneering text *Principles of Chemical Engineering*, published in 1923 by Walker, Lewis, and McAdams [1], was the first to emphasize the concept of unit operations (UO) by formally recognizing that certain physical or chemical processes, and corresponding fundamental principles, were common to many disparate industrial technologies. A natural outgrowth of this radical new view was the gradual appearance of fluid mechanics and transport in both teaching and research as the fundamental basis for many unit operations. Of course, many of the most important unit operations take place in equipment of complicated geometry, with strongly coupled combinations of heat and mass transfer, fluid mechanics, and chemical reaction, so that the exact equations could not be solved in a context of any direct relevance to the process of interest. As far as the large-scale industrial processes of chemical technology were concerned, even at the unit operations level, the impact of fundamental studies of fluid mechanics or transport phenomena was certainly less important than a well-developed empirical approach. This remains true today in many cases. Indeed, the great advances and discoveries of fluid mechanics during the first half of this century took place almost entirely without the participation or even knowledge of chemical engineers.

Gradually, however, chemical engineers began to accept that the generally blind empiricism of the "lumped parameter" approach to transport processes at the unit operations scale should at least be supplemented by attempts to understand the basic physical principles. This finally led, in 1960, to the appearance of the landmark textbook of Bird, Stewart, and Lightfoot (BSL) [2]. This text not only introduced the idea of detailed analysis of transport processes at the continuum level, but also emphasized the mathematical similarity of the governing field equations, along with the simplest constitutive approximations for fluid mechanics, heat, and mass transfer. As an undergraduate from 1961 to 1965, I can still remember the quarter-long course from BSL, regarded as quite a leap (forward?) by the faculty, and that it came after a full year of the "real transport" from another well-known textbook by Bennett and Myers [3] (itself published for the first time in 1962). The connection between UO-oriented transport and the more fundamental approach of BSL was tenuous at best, but at least we felt that we knew the basis of description in terms of fundamental governing equations and boundary conditions. The presentation of BSL was overly idealized in places, with derivations based on flux balances through fixed control volumes of rectangular or square shape; the text itself primarily presented simple results and solutions rather than methods of solution or analysis.[1] Whatever the shortcomings of BSL, however, its impact has endured over the years. The combination of the more fundamental approach that it pioneered and the appearance of a few chemical engineers with very strong

[1]A general, tensorially based description of the basis of the governing equations of fluid mechanics soon appeared in the Prentice-Hall series of texts for chemical engineers [4], and this material was later incorporated in several widely used undergraduate textbooks for chemical engineers [see 5, 6].

mathematics backgrounds who could understand the theoretical development of asymptotic methods by applied mathematicians produced the most recent revolution in our way of thinking about and understanding of transport processes.

This revolution is the application of the methods of asymptotic analysis to problems in transport phenomena. Perhaps more important than the detailed solutions enabled by asymptotic analysis is the emphasis that it places on dimensional analysis and the development of qualitative physical understanding (based upon the mathematical structure) of the fundamental basis for correlations between dependent and independent dimensionless groups [cf. 7–10]. One major simplification is the essential reduction in detailed geometric considerations, which determine the magnitude of numerical coefficients in these correlations, but not the form of the correlations. Unlike previous advances in theoretical fluid mechanics and transport phenomena, in these developments chemical engineers have played a leading role.

A second major initiative sprang from the detailed theoretical studies of creeping flows (i.e., viscous-dominated flows), again introduced to chemical engineers by a textbook, *Low Reynolds Number Hydrodynamics*, by Happel and Brenner [11]. In this case, the result is powerful techniques based on the tensorial structure of the problems and the superposition allowed by linearity, which again reduce the importance of detailed geometry to considerations of symmetry and the calculation of numerical coefficients and allow very general problems to be characterized by the solutions of a discrete set of fundamental problems [cf. 12, 13]. Although its foundations are now nearly 20 years old, its application to many modern areas of research involving small-particle dynamics is still under way today. Indeed, the groundwork laid by the two major initiatives described above, together with the changing nature of the important problems and technologies, has catapulted chemical engineering research to the forefront of many areas of fluid mechanics and transport phenomena.

Of course, no subject with the long-term vitality and impact of fluid mechanics and transport phenomena will remain for long without the appearance of major new directions, tools, and challenges. These developments will be the focus of the last section of this paper. However, before leaving this brief synopsis of the past, a few clarifications may be worthwhile.

First, what I have traced is basically the development of generic tools and general approaches or ways of thinking, rather than specific problems or technologies. In this regard, it is important to note that few of the major fluids and transport-oriented problem areas of concern to chemical engineers have been affected by modern, post-BSL developments. This is particularly true of the huge class of problems involving complex materials—suspensions, emulsions, polymeric liquids, etc. Although the significance of dimensional reasoning is clearly recognized, the potential of asymptotic analysis in either a detailed or generalized sense is inhibited by two factors. One is the lack of generally accepted governing equations. However, even if we were simply to accept one of the existing equations (or it actually turned out to be correct for some class of complex fluids), a second deficiency is the lack of a sufficiently developed body of detailed experimental data and, for any specific equation, of qualitative insight into the solution behavior to provide the guidance necessary for asymptotic theory. At the moment, we are left only with empirical (lumped parameter) approaches or, in principle, full-scale numerics. It should be noted, though, that the lack of qualitative understanding of any asymptotic scaling behavior greatly complicates even this endeavor.

Another general comment on the substantial fundamental progress of the past 20 to 30 years concerns the impact of fluid mechanics and transport on the technologies and unit operations processes that motivated its initial introduction into chemical engineering. Though no one would question the often spectacular progress made in developing the increasingly advanced understanding of the underlying microscale phenomena, the pertinent question is whether all this effort has substantially contributed to the technological areas of application. My personal view is that the impact to date has been remarkably small. As a profession, we have not done a good job of translating our considerable progress at the fundamental level to the important technological problems that are the foundation of the classical chemical processing industries. What we have done is set the stage for dramatic contributions in areas of new technology that are characterized by the need for understanding at a "microscopic" level. I will have more to say about this shortly.

Finally, there is the important question of how, or whether, all of these advances on the research frontiers are reflected in chemical engineering curricula. Again, it is difficult to be too positive. At the graduate level, some of the lessons and understanding generated via the powerful scaling techniques that underlie asymptotic methods are slowly finding their way into general (i.e., nonresearch-oriented) classes in transport phenomena. Courses that provide a modern view of low-Reynolds-number hydrodynamics are becoming less rare. Both of these "initiatives" may be facilitated in the near future by new textbooks that are currently in various stages of preparation/ publication. Not surprisingly, there is much less impact at the undergraduate level. The methods of asymptotic analysis seem to demand a degree of mathematical sophistication that is beyond the grasp of typical undergraduates, even today. However, the critical message that we should convey is not so much in the details of the analysis itself, but in the qualitative physical understanding and the way of thinking inherent in the advanced dimensional analysis underlying the asymptotic approach. At the least sophisticated level, we have conventional nondimensionalization of the governing equations and boundary conditions. This identifies the independent dimensionless parameters. The next stage is "scaling," which ultimately leads to the form of the relationship between dependent and independent dimensionless groups. In the absence of asymptotics, scaling requires a substantial degree of physical insight and intuition, and it is not likely to be accomplished successfully by any but the cleverest people. Asymptotics provides a systemization (albeit incomplete) of the scaling process that makes it accessible to the rest of us. Most undergraduate courses stop at the level of simply identifying dimensionless groups. The challenge is to convey the essence of this very important advance without becoming hopelessly enmeshed in a complex web of mathematical detail. To date, this has not been accomplished successfully. However, this problem of conveying complicated information in an undergraduate curriculum is not unique to modern *theoretical* developments. A similar difficulty has always been evident in transmitting the insight of experimental studies on topics that do not lend themselves to analytical investigation. The quintessential example is turbulent flows and related transport processes, where experimental studies over the past 10–15 years have revolutionized understanding at the research level of the nature of turbulent flows near boundaries [14], without any noticeable impact on graduate or undergraduate curricula.

III. The Future—Challenges and Opportunities in Research

The most difficult task in preparing this paper was the prognostication of the future; I can only offer my own personal views. However, my perception is that we are at the threshold of major new challenges and opportunities for chemical engineers in fluid mechanics and transport phenomena.

In fact, within the next 10–15 years, there are at least two distinct lines of development with revolutionary potential. One is the evolution of dramatic new classes of tools for investigation. The most obvious is the emergence of direct, large-scale computational simulations of both fluid flows and transport systems due to the combined development of new, more efficient numerical algorithms and the general accessibility of cheaper and faster computer cycles. The second, less widely recognized development is the awakening of chemical engineers (and much of Western science and technology) to the revolutionary mathematical theories of "nonlinear dynamics" [14–16]. This development may be even more important than "brute force" computing in shaping our ability to analyze new problems, as we are challenged to understand the consequences of nonlinear behavior, rather than restricting our attention to linear (or linearized) systems.

As important, and even revolutionary, as these two developments may be, they appear in the present context only as more powerful tools. They may facilitate more detailed investigation, or even change our approach to understanding many problems, but they do not provide the intellectual or technological imperative that will ultimately determine the future vitality of our subject. As for the intellectual and technological foundation, the natural evolution of the trends of the immediate past suggests that the future will bring an ever increasing focus on more fundamental, and microscopic, aspects of fluid flow and transport systems. Although no one would question the wealth of intellectual challenges to be found along this road, an obvious question is whether it fills any technological need. In my own view, at least three distinct motivations can be cited to support the prognostication of a more intensive push toward the study of fluid dynamics and transport phenomena at the microscale. First is the growing interest and critical national need for the development of novel materials. Particularly relevant to the present discussion is the general class of "microstructured fluids" such as suspensions, foams, and polymeric liquids, where a primary objective must be to understand the details of interaction between the microstructural state and the flow. Second is the increasing emphasis on new technologies which require materials to be processed and transport processes to be controlled with increasing precision, most often in systems of progressively smaller size. Third, and finally, there is a growing realization that local flow structures are critically important in determining the global behavior of a flow or transport system. An example is the presence or absence of fixed points that may predispose a flow to chaotic mixing if it is perturbed.

Lest we in the fluids and transport community within chemical engineering become too smug at being situated in the right intellectual position to respond to the persuasive technological call for a more *microscopic* view of transport systems, we should recall that the roots of our increasingly fundamental endeavors of the past 25 years were the *macroscopic* transport systems of the traditional chemical processing industries. Ironically, as time passes, we actually find work becoming increasingly relevant to present and future technological objectives. However, this is not because we have made an effective transition from our fundamentals back to the large-scale systems that presumably provided our initial motivation. In fact, we have largely failed in that endeavor. We are simply the fortunate beneficiaries of an accidental confluence of our natural tendency (especially in academia) toward increasingly detailed, small-scale, and fundamental investigations and the emergence of new applications from new technologies (for the chemical engineer) such as semiconductor and electronics material processing and biochemical engineering (with such critical needs as the development of high-efficiency, low-damage separations techniques for biological cells), which require a more microscopic view.

In the remainder of this section, I discuss the significance of these future trends in fluid mechanics and transport phenomena, beginning with the evolution of new theoretical tools and concluding with a more detailed description of future objectives, challenges, and opportunities for fluid mechanics research at the microscale. Some of the latter discussion is drawn from a more comprehensive report, "The Mechanics of Fluids with Micro-structure" written in 1986 by Professor R. A. Brown (MIT) and myself (with substantial input from the chemical engineering research community) as part of a general NSF-sponsored study of *The Future of Fluid Mechanics Research*.

A. Nonlinear Systems

It is my opinion that recent developments in the mathematical description of nonlinear dynamical systems have the potential for an enormous impact in the fields of fluid mechanics and transport phenomena. However, an attempt to assess this potential, based upon research accomplishments to date, is premature; in any case, there are others better qualified than myself to undertake the task. Instead, I will offer a few general observations concerning the nature of the changes that may occur as the mathematical concepts of nonlinear dynamics become better known, better understood, more highly developed, and, lastly, applied to transport problems of interest to chemical engineers.

First, and most important, nonlinear dynamics provides an intellectual framework to pursue the consequences of nonlinear behavior of transport systems, which is simply not possible in an intellectual environment that is based upon a "linear mentality," characterized by well-behaved, regular solutions of idealized problems. One example that illustrates the point is the phenomenon of hydrodynamic dispersion in creeping flows of nondilute suspensions. It is well known that Stokes flows are exactly reversible in the sense that the particle trajectories are precisely retraced when the direction of the mean flow is reversed. Nevertheless, the lack of reversibility that characterizes hydrodynamic dispersion in such suspensions has been recently measured experimentally [17] and simulated numerically [18]. Although this was initially attributed to the influence of nonhydrodynamic interactions among the particles [17], the numerical simulation [18] specifically excludes such effects. A more general view is that the dispersion observed is a consequence of (1) deterministic chaos that causes infinitesimal uncertainties in particle position (due to arbitrarily weak disturbances of *any kind—*

Brownian fluctuations, inertia, nonhydrodynamic interactions, etc.) to lead to exponential divergence of particle trajectories, and (2) a lack of predictability after a dimensionless time increment (called the "predictability horizon" by Lighthill) that is of the order of the natural logarithm of the ratio of the characteristic displacement of the deterministic mean flow relative to the RMS displacement associated with the disturbance to the system. This weak, logarithmic dependence of the predictability horizon on the magnitude of the disturbance effects means that *extremely* small disturbances will lead to irreversibility after a very modest period of time.

At this early date, I am aware of only two areas of *direct* application of the mathematical framework of nonlinear dynamics to transport problems of interest to chemical engineers, namely the consequences of time-periodic perturbations in mixing flows [19–21] and the dynamics of gas bubbles subjected to periodic variations in pressure or in the local velocity gradient [22–25]. A critical observation in both cases is the potential for transition from regular to chaotic dynamics—with a clear implication that regular dynamical behavior should be recognized as the exception rather than the rule. If this is true of a more general class of problems, as I believe it is, then the regular, well-behaved solutions that form the core of most current academic courses in transport phenomena must be viewed as providing the student with a false sense of the true nature of nonlinear dynamical systems [15]. It seems clear that profound changes will ultimately occur in the standard curricula in fluid mechanics, and transport phenomena that will reflect this fact, but it is too soon to see what the details of this transition might be.

A second, even more speculative point is that the mathematical framework of nonlinear dynamics may provide a basis to begin to bridge the gap between local microstructural features of a fluid flow or transport system and its overall meso- or macroscale behavior. On the one hand, a major failure of researchers and educators alike has been the inability to translate increasingly sophisticated fundamental studies to the larger-scale transport systems of traditional interest to chemical engineers. On the other hand, a basic result from theoretical studies of nonlinear dynamical systems is that there is often an intimate relationship between local solution structure and global behavior. Unfortunately, I am presently unable to improve upon the necessarily vague notion of a connection between these two apparently disparate statements.

B. *Computational Simulation*

We can anticipate more easily, and speculate less, on the potential generic impact of the recent explosive increase in computational capability, because the subject has enjoyed a longer, more widely recognized, period of development. Nevertheless, we have only just begun to scratch the surface in terms of its potential contributions in fluid mechanics and transport processes. Numerical methods and computer hardware have both only begun to reach a stage of mutual development where solutions of full 3D, nonlinear problems, on more or less arbitrary domains, are within reach of much of the research community.

The capability of rapid (or inexpensive) solution of fully nonlinear problems provides a basis for two equally important (and previously impossible) types of research contribution. On the one hand, there are many extremely important problems where the governing equations and boundary conditions are known, but solution is impossible by analytical means. In this case, computational studies can provide a unique and exceedingly valuable tool for investigation in a mode that is a close analog of an experimental study. One obvious advantage of the "numerical experiment" in unraveling the underlying physical phenomena is the possibility of varying parameters one at a time over a wide range of values, when a physical experiment would often involve simultaneous changes in these same parameters, with a range of accessible values that is relatively narrow. Even more fundamental, however, is the fact that the numerical experiment can easily resolve phenomena that are characterized physically by exceedingly small length and/or time scales that may be impossible to resolve in the laboratory. One example that I am familiar with is the collapse of a cavitation bubble, where cinematography at 10^6 frames per second was found to be inadequate to reveal some details that are easily captured via numerical simulation [26]. In the hands of a competent and fair investigator, the ready capability for direct numerical solution of complicated, nonlinear problems essentially provides a new branch of scientific investigation, which is completely complementary to the traditional analytical and experimental approaches. The ability to vary parameters independently, to eliminate nonidealities of a laboratory system, and to resolve arbitrarily small spatial and temporal scales means that this branch of investigative methodology is uniquely suited to microstructural studies of flows and to some aspects of the dynamics of fluids with microstructure as discussed below.

The use of computational simulations as a branch of research methodology is relevant to any field where the underlying physical principles or laws are known and the objective of investigation is an understanding of physical phenomena associated with a particular application. However, a current characteristic of many fluids and transport problems in chemical engineering is that the objective is actually to determine the governing equations, or boundary conditions, either in general or at least for some class of motions. A typical case is almost any of the so-called fluids with microstructure, including polymer solutions and melts, where rheologists have proposed a variety of constitutive equations relating stress and dynamical variables, but none are currently known to describe the behavior of real materials in flows of technological interest. This current status of non-Newtonian fluid mechanics is well known. However, even for Newtonian fluids, the boundary conditions at a fluid interface are, at best, an approximation, which may not describe real systems. Thus, for many of the fluid mechanics and transport areas of greatest importance to chemical engineers, the mathematical formulation of the problem is itself unknown. Until this problem is resolved, there can be no development of a truly predictive branch of theoretical fluid mechanics or transport phenomena for this class of problems. Nevertheless, numerical solutions of proposed models for constitutive equations or boundary conditions can play a crucial role. By comparing predictions from such solutions with experimental data on real fluids, we can provide meaningful tests of the many proposed models and thereby facilitate the development (where necessary) of new, more realistic ones. This application of computational capabilities is so obvious that its significance may not initially impress the reader. However, the ability to test and explore the capabilities of proposed material models by direct solution of the resulting equations is a truly revolutionary development that will dramatically change the time scale and the methodology of development of meaningful theoretical models for complex fluids. The alternative, short of a true molecular theory, is a comparison of analytical solutions with experimental data on flows that are sufficiently simple to actually yield analytical solutions from the nonlinear governing equations. This comparison with simple flows is the route attempted by rheologists for the past 30–40 years at least. The problem is that any particular simple flow, including the simple shear flow studied by rheologists, is likely to be so simple that the material response depends only on a very limited part of the behavior of a particular constitutive model. If the model is complex, with several material parameters to be chosen by comparison with experiment, the result may be extremely accurate predictions for the test flow. But the model capabilities for almost any other problem remain unknown. Although a collection of simple flows should, in principle, provide the ideal diagnostic of model assumptions, there are too few experimentally realizable flows that also allow analytic solution of the governing equations to make this a reality.

Clearly, the capability of relatively rapid and inexpensive numerical solutions of a proposed set of governing equations or boundary conditions that can be used in an interactive way with experimental observation represents a profound new opportunity that should greatly facilitate the development of a theoretical basis for fluid mechanics and transport phenomena for complex or multiphase fluids that are the particular concern of chemical engineers.

C. Fluids with Microstructure

Finally, I return to the intellectual and technical future of fluid mechanics and transport phenomena. I have already expressed the view that many of the most interesting, challenging, and technically important problem areas for the next 10–15 years will come from fundamental investigations of fluids and fluid flows at the microscale. It is impossible to do justice to all facets of this claim in the space available. Therefore, I will concentrate on only one topic—namely, the dynamics and transport characteristics of fluids with microstructure.

Fluid mechanics has traditionally been concerned with understanding complex physical phenomena for relatively simple materials whose mechanical behavior can be approximated as Newtonian. Indeed, there has been a regrettable tendency for fluid dynamicists to define their subject entirely in terms of Newtonian fluids and to regard attempts to describe the motions or transport characteristics of more complicated materials as unjustified in view of the lack of a firm theoretical basis. Of course, with this narrow point of view, the early founders of our subject might never have discovered viscosity, nor settled on the familiar no-slip condition for rigid boundaries. The fact is that much of fluid mechanics that is relevant to modern and emerging technological needs involves fluids with microstructure (i.e., polymeric solutions and melts, suspensions, colloidal dispersions, emulsions, foams, granular media, and many biological fluids), whose behavior often cannot be approximated or even intuitively understood using Newtonian fluids theories.

To understand the challenges and opportunities of future research on fluids with microstructure, an expanded list of materials is necessary to explain more generically what is meant by microstructured fluids. Obviously, all fluids have structure if viewed on a sufficiently short time or length scale. However, the molecular scale of Newtonian fluids is sufficiently small that the microstructural state is dominated by random thermal fluctuations. Thus, in the presence of flow (or deformation), a Newtonian fluid remains statistically isotropic and unchanged in structure, and its macroscopic properties are unchanged from the equilibrium state. The critical difference for the class of materials that we denote as microstructured fluids is that they are heterogeneous on a sufficiently large scale that their microstructural state can be modified substantially when the material undergoes a macroscopic flow (or deformation). Thus, the macroscopic properties of the material, which depend upon this state, are also modified in flow, and there is a very strong interaction between the flow, which modifies material properties, and the changes in properties which, in turn, modify the flow! The intimate coupling between microstructure and flow is responsible for both the challenges and the opportunities for innovative process engineering for this class of materials.

Of course, the details of the interaction between microstructure and flow depend strongly on the particular material and on the configuration of the system in which the flow occurs. Indeed, interactions between microstructural elements, or between these elements and the suspending fluid, can range from purely hydrodynamic to electroviscous or even molecular in polymeric liquids, while the motions at the microstructural level can range from purely deterministic to random and/or Brownian in nature. We can also distinguish two classes of problems depending on the relative size of the flow system and the microstructural elements. If the flow system is much larger than the length scales characteristic of the microstructure, we seek to describe the properties of the material using an averaged equation approach in which the microstructural state appears only in a statistical sense. If the length scales of the flow and the fluid structure are comparable, on the other hand, we must attempt to describe the deformation and motions of the individual components.

So far, I have described only the class of materials and problems to be considered, without providing any insight into the challenges and opportunities that should be the focus of future research.

Let me first briefly consider the challenges. We have seen that the generic characteristic of these materials is the strong coupling between flow and microstructure, and vice versa. Although the goal of research, in some instances, may be a description of material properties in a phenomenological continuum mechanical framework, it is my view that the research itself must begin at the microstructural level, to obtain an understanding and basis for prediction of the interactions between the microstructure and flow. The second (and generally easier) step is to determine the relationship between the microstructural state of the material and its bulk or macroscopic physical properties.

In all but the most basic cases of very dilute systems, with microstructural elements such as rigid particles whose properties can be described simply, the development of a theory in a continuum context to describe the dynamical interactions between structure and flow must involve some degree of modeling. For some systems, such as polymeric solutions, we require modeling to describe both polymer-solvent and polymer-polymer interactions, whereas for suspensions or emulsions we may have an exact basis for describing particle-fluid interactions but require modeling via averaging to describe particle-particle interactions. In any case, the successful development of useful theories of microstructured fluids clearly requires experimental input and a comparison between experimental data and model predictions. With some exceptions, however, each of these steps is itself a major research endeavor.

In the first place, the averaged model equations are highly nonlinear and require sophisticated numerical analysis for solution. For example, the attempt to obtain numerical solutions for motions of polymeric liquids, based upon "simple" continuum, constitutive equations, is still not entirely successful after more than 10 years of intensive effort by a number of research groups worldwide [27]. It is possible, and one may certainly hope, that model equations derived from a sound description of the underlying microscale physics will behave better mathematically and be easier to solve, but one should not underestimate the difficulty of obtaining numerical solutions in the absence of a clear qualitative understanding of the behavior of the materials.

Successful experimental studies are, in many ways, the greatest research challenge. Many of the techniques developed to obtain noninvasive flow measurements are optical in nature (e.g., laser Doppler velocimetry [LDV]), and this creates an immediate difficulty for application to materials that are opaque. Additionally, the set of variables is larger and more difficult to measure than for dynamical studies of simple, homogeneous Newtonian fluids. At the minimum, we require data on the overall velocity field and the concentration distribution for the structural elements—the particles, drops, macromolecules, etc. Beyond this, however, a successful understanding of the basis for success or failure of the proposed flow and transport models requires experimental characterization of the microstructural state as a function of position in the flow—for example, the state of alignment or deformation. Ironically, for polymeric liquids, which require the greatest number of *ad hoc* assumptions in modeling the microstructural dynamics, the experimental objectives are most easily met. Many polymeric liquids are clear, so that standard LDV techniques can be used for measurement of velocity fields [28], and dynamic light scattering can be used for direct pointwise measurement of velocity gradients [3, 29]. Other optical techniques, such as birefringence and dichroism, provide an effective experimental probe of the microstructural state [30–34]. Measurement techniques for other microstructured fluids are less advanced, except for the relatively rare but valuable instances where the same type of optical techniques can be applied.

The real intellectual challenge of research on microstructured fluids is the necessity for a fully integrated program comprised of theoretical model development, numerical or analytic solution of the model equations, and comparison with experimental data on both dynamics and structure, with the latter hopefully contributing to increased qualitative understanding and improved models. The key here is strong coupling between structure and flow. As soon as the structure is changed in any significant way from its equilibrium state by a flow, the flow itself undergoes a significant change. Developments to date have too frequently focused upon only one or two aspects of the total problem. Although the nature of the intellectual challenge may be clear, it is important to consider the numerous and exciting opportunities that exist for improved technology if the intellectual endeavor is successful. Of course, a most important consequence would be the existence of a theoretical framework that could be used in the design of new processes and/or improvements in the performance or control of existing processes. Beyond this, however, the generic feature of strong coupling between flow and structure provides a unique opportunity for process and materials engineering brought about by manipulation of the microstructure through a combination of external fields and the flow itself. In principle, this can be used for several purposes:

1. *To manipulate and control ultimate properties of processed materials* (for example, the strength, electrical, or optical properties of suspensions or polymeric liquids can be modified via flow-induced orientation, dispersion or crystallization);

2. *To control features of the flow itself* (examples include drag reduction by addition of polymer or microbubbles, magnetic stabilization of fluidized beds, foam flow in porous media for mobility control, antimisting or cavitation suppression via polymer additives); and, finally,

3. *To optimize processes that are based upon the interaction between microstructure and flow* (for example, proppant placement in hydraulic fracture of geologic formations [oil recovery], separations processes for biological materials, mixing and dispersion of additives in blenders, crystal growth and solidification processes).

It is evident that the complications of microstructure-flow interactions present not only a formidable challenge to the fluid dynamicist and engineer but also a unique *opportunity for process and materials engineering through fluid mechanics.* It is this area of *engineering technology* which affects the future of American technology. It is critical in the next 5 to 10 years to develop a fundamental base of understanding that will allow these opportunities to be exploited—leading to a critical competitive edge based upon advanced scientific knowledge rather than art.

This brings us, finally, to the statement of realistic technological goals for research on fluids with microstructure:

1. *The development of predictive capability for industrially relevant flows, including* (a) scaling laws that allow for a transition between laboratory or pilot scale studies and flow behavior in a full-scale system; (b) an understanding and control of flow instabilities, which adversely affect product quality and limit rates in many materials processing technologies; and (c) optimization of technological processes that are based upon microstructural manipulation via flow, such as separations, molding, coating, blending, and dispersion.

2. *The development of processing strategies in which the interaction between fluid motion and microstructural conformation can be used for* (a) the development of microstructured materials with desired macroscopic properties and (b) active control of the flow itself, as in drag reduction and fluidized bed stabilization.

IV. The Future—Challenges in Education

A critical question for the profession is how the many changes in research on fluid mechanics and transport processes should be reflected in undergraduate or graduate curricula in transport phenomena. Personally, though I have taught transport phenomena at one level or the other for nearly 20 years, I find this a most perplexing and difficult question.

The challenges that I see are as follows:

1. To determine the appropriate relationship between the fundamental, microscale emphasis in research and the meso- or macroscale transport problems of technology in a traditional chemical engineering environment. This is not a new problem, but it has been increasingly prevalent in the post-BSL period. In large part, the difficulty is that the transition between fundamentals and applications has not been made successfully at the research level either. Previous generations of educators have simply continued to teach parallel and largely nonintersecting courses at the undergraduate level, one a unit processes-oriented course on "real" transport processes and the other a briefer course on transport fundamentals. The underlying philosophy was more or less that real practicing engineers needed the first course, while prospective graduate school students needed the second course, which would also benefit other students in some unspecified way, despite the lack of any obvious connection with real-world problems. As chemical engineers at all degree levels become more involved in areas of new technology, such as microelectronics processing and biotechnology, this somewhat negligent attitude will come under increasing pressure. We will be forced to reconsider our options in an environment where time may not permit a more intensive undergraduate transport course without significant sacrifice of more applied counterparts. At the graduate level, many schools provide course work in fundamentals with little or no counterpart toward a more sophisticated view of real transport processes. As a consequence, the typical postbaccalaureate student views large-scale transport processes as outside the realm of areas that he or she is responsible for understanding. In this environment, it is not hard to understand why so little real progress has been made in translating fundamental progress and knowledge back to the transport processing arena.

2. To develop visual aids or other means to convey greater understanding at the research level when many developments are no longer an outgrowth of analytical theories that can be translated into an acceptable package for blackboard presentation. Again, this is an old problem made more conspicuous by new developments. There has always been a serious problem in translating a knowledge base derived from experimental observations into the traditional classroom framework, apart from empirical lumped parameter correlations that apply on a system-by-system

basis. To some extent, in fluid mechanics, this difficulty has been addressed with the preparation of an excellent film series that is widely available [35]. Now, however, an increasing fraction of our new understanding of transport problems at the research level is coming from numerical experiments based upon large-scale computer systems that are not readily transferrable to smaller-scale computers that a student could access. Perhaps the answer is simply a new version of the above-cited famous film series.

3. To address the growing gap between the theoretical tools of research and the mathematical tools that are accessible to undergraduate or graduate course work. Once again, this is an old problem that is gradually getting worse. For years, researchers have used the methods of dimensional analysis and scaling, as practiced in asymptotic theories, to great advantage in developing qualitative physical insights for transport problems. Yet, for the most part, these techniques have not been taken into the teaching environment because it is perceived that they require a level of sophistication, vis-à-vis the solution of partial and ordinary differential equations, that is beyond the standard training of chemical engineering students. Increasingly, however, researchers are relying on methods of numerical analysis and the mathematical framework of nonlinear dynamics to develop new insights and understanding. Although these latter methods of analysis complement the existing analytic and asymptotic techniques, and one would not replace one topic with the other in the teaching curriculum, a natural question is whether some realistic element of understanding of the new methodology is not a necessity in future transport phenomena curricula. This is especially true of computational and numerical techniques, which are becoming widely available in the form of software packages that are advertised as reducing the solution of fluid-flow problems to a "black-box" routine accessible to engineers at all degree levels. How will the practicing engineer, or even the nonspecialist researcher, assess the reliability and usefulness of such numerical process simulation without any background in the underlying mathematical techniques, especially when they often lack the physical intuition necessary to evaluate the results?

4. In the longer term, a greater challenge to all of mechanics and related subjects is the transformation of curricula to include a realistic view of the dynamical consequences of nonlinearity. The Western tradition in most areas of science and technology has been to try to explain nonlinear phenomena via nearly linear (or even linear) models, where the physics is dominated by well-behaved (regular or integrable) solutions. Inevitably, however, researchers (or even readers of Soviet and Eastern-bloc literature) are pushing us to accept the fact that "it is a nonlinear world after all"—and this must ultimately have a profound impact upon the teaching curricula in many areas, including fluid mechanics and transport phenomena for chemical engineers.

Unfortunately, I have relatively few answers to these challenging curricula problems. One hopes that the discussions of distinguished gatherings such as this will begin to focus on these issues.

V. Conclusions

I have tried to provide some perspective about the current and future state of teaching and research in fluid mechanics and transport phenomena in chemical engineering. If this paper accomplishes nothing else, I hope that it serves to indicate the very broad spectrum of important and exciting problems remaining to be solved, especially from the standpoint of a chemical engineer. In many of the other branches of science and engineering that share fluid mechanics and/or transport phenomena, it would seem difficult to avoid characterization of the subject as mature, with an uncertain future that must inevitably lack the excitement of discovery and progress of a new field. It is unfortunate that this characterization is sometimes carried over to chemical engineering. Many of the topics of greatest concern to us are anything but mature, and we are heavily involved in some of the most exciting new scientific developments in computational and nonlinear physics and mathematics. In my personal view, the field is ripe for major new developments from the bright and innovative young minds that I hope will continue to be attracted to the field.

References

1. Walker, W. H., Lewis, W. K., and McAdams W. H., *Principles of Chemical Engineering*. McGraw-Hill, New York, 1923.
2. Bird, R. B., Stewart, W. E., and Lightfoot, E. N., *Transport Phenomena*.Wiley, New York, 1960.
3. Bennett, C. O., and Myers, J. E., *Momentum, Heat and Mass Transfer*. McGraw-Hill, New York, 1962.
4. Aris, R., *Vectors, Tensors and the Basic Equations of Fluid Mechanics*. Prentice-Hall, Englewood Cliffs, New Jersey, 1962.
5. Whitaker, S., *Introduction to Fluid Mechanics*. Prentice-Hall, Englewood Cliffs, New Jersey, 1968.
6. Denn, M. M., *Process Fluid Mechanics*. Prentice-Hall, Englewood Cliffs, New Jersey, 1980.
7. Frankel, N. A., and Acrivos, A., *Phys. Fluids* **11**, 1913 (1968).
8. Goddard, J. D., and Acrivos, A., *J. Fluid Mech.* **23**, 273 (1965).
9. Goddard, J. D., and Acrivos, A., *J. Fluid Mech.* **24**, 339 (1966).
10. Taylor, T. D., and Acrivos, A., *Phys. Fluids* **5**, 387 (1962).
11. Happel, J., and Brenner, H., *Low Reynolds Number Hydrodynamics*. Prentice-Hall, Englewood Cliffs, New Jersey, 1965.
12. Brenner, H., *Adv. Chem. Eng.* **6**, 287 (1966).
13. Brenner, H., *J. Fluid Mech.* **111**, 197 (1981).
14. Cantwell, B. J., *Ann. Rev. Fluid Mech.* **13**, 457 (1981).
15. Lighthill, J., *Proc. Roy. Soc. London A* **407**, 35 (1986).

16. Wiggins, S., *Nature* **333**, 395 (1988).
17. Leighton, D., and Acrivos, A., *J. Fluid Mech.* **177**, 109 (1987).
18. Bossis, G., and Brady, J. F., *J. Chem. Phys.* **87**, 5437 (1987).
19. Chien, W. L. , Rising, H., and Ottino, J. M., *J. Fluid Mech.* **170**, 355 (1986).
20. Khakhar, D. V., Rising, H., and Ottino, J. M., *J. Fluid Mech.* **172**, 519 (1986).
21. Rom-Kedar, V., Leonard, A., and Wiggins, S., *J. Fluid Mech.* **214**, 347 (1990).
22. Chang, H. C., and Chen, L. H., *Phys. Fluids* **29**, 3530 (1986).
23. Kang, I. S., and Leal, L. G., *J. Fluid Mech.* **218**, 41 (1990).
24. Natarajan, R., and Brown, R. A., *Proc. Roy. Soc. London A* **410**, 209 (1987).
25. Smereka, P., Birnir, B., and Banerjee, S., *Phys. Fluids* **30**, 3342 (1987).
26. Lauterborn, W., and Timm, R., in *Cavitation and Inhomogeneities in Underwater Acoustics* (W. Lauterborn, ed.), p. 42. Springer, Berlin, 1980.
27. Leal, L. G., Denn, M. M., and Keunings, R., eds., Proceedings of the Fifth International Workshop on Numerical Methods for Viscoelastic Flows (Lake Arrowhead), *J. Non-Newtonian Fluid Mech.* **29** (1990).
28. Lawler, J. V., Muller, S. J., Armstrong, R. C., and Brown, R. A., *J. Non-Newtonian Fluid Mech.* **20**, 51 (1986).
29. Fuller, G. G., Rallison, J. M., Schmidt, R. L., and Leal, L. G., *J. Fluid Mech.* **100**, 555 (1980).
30. Dunlap, P. N., and Leal, L. G., *J. Non-Newtonian Fluid Mech.* **23**, 5 (1987).
31. Fuller, G. G., and Leal, L. G., *J. Polym. Sci.* (Polym. Phys. Ed.) **19**, 557 (1981).
32. Ottino, J. M., Leong, C. W., Rising, H., and Swanson, P. D., *Nature* **333**, 419 (1988).
33. Chow, A. W., and Fuller, G. G., *J. Non-Newtonian Fluid Mech.* **17**, 233 (1985).
34. Chow, A. W., and Fuller, G. G., *J. Rheology* **28**, 23 (1984).
35. National Committee for Fluid Mechanics Films, *Illustrated Experiments in Fluid Mechanics*. M.I.T. Press, Cambridge, 1972.

5

Fluid Mechanics and Transport Research in Chemical Engineering

William B. Russel
Department of Chemical Engineering
Princeton University
Princeton, New Jersey

I. Introduction

As noted by Leal [1], chemical engineers have moved to the forefront of research in fluid mechanics and transport phenomena over the past three decades, participating in and even leading a number of different areas relevant to chemical processes and phenomena. The numerous authoritative contributions to *Annual Reviews in Fluid Mechanics* clearly establish this fact.

During this period, theoretical advances in low-Reynolds-number fluid mechanics, asymptotic methods and order of magnitude analyses, and numerical techniques have produced a much deeper and more extensive understanding of various phenomena: (1) the dynamics of small particles, including the effects of particle shape and interactions on coagulation processes [2], electrokinetic phenomena [3], and transport properties such as viscosity [4], sedimentation velocity [5], and diffusion coefficient [6]; (2) the breakup of drops and bubbles in shear, extensional, and time-dependent flows [7]; (3) instabilities affecting processes as disparate as fluid displacement in porous media [8] and the drawing of polymer fibers [9]; (4) free surface phenomena associated with microscopic interfaces in porous media and complex high-speed coating flows [10]; and (5) complex two- and

ADVANCES IN CHEMICAL ENGINEERING, VOL. 16

three-dimensional flows of Newtonian and, to some extent, non-Newtonian fluids [11–14].

Complementary experimental developments also emerged, particularly the quantitative optical techniques [15] such as (1) laser Doppler velocimetry for detecting local velocities, (2) dynamic light scattering for following particle motions or velocity gradients, (3) birefringence for mapping molecular orientation or stress fields in polymeric liquids, and (4) dichroism for characterizing orientation of anisotropic particles or the extension of macromolecules. These have facilitated detailed probing of flows at both the macroscopic and microscopic scales. In the area of mass transfer, a growing body of work demonstrates how to capitalize on chemical reactions and thermodynamic nonidealities to enhance rates or selectivity of separations, e.g., via facilitated transport through liquid membranes, field flow fractionation and other advanced chromatographies [16], and precisely tailored polymeric membranes.

One sign of progress is the extent to which sophisticated research on transport phenomena, particularly mass transfer, has penetrated several other fields, including those described in later papers of this volume. Examples include fundamental work on the mechanics of trickle beds [17] within reactor engineering; studies of dispersion in laminar flows [18] in the context of separations important to biotechnology; coupling between fluid flows and mass transfer in chemical vapor deposition processes for fabrication of semiconductor devices [19] and optical fiber preforms [20]; and the simulation of flows in mixers, extruders, and other unit operations for processing polymers.

Nonetheless, some significant shortcomings exist in the direction of the overall effort. Leal characterizes this as a failure to translate progress in understanding the science into improvements or innovations in chemical processes. An alternative summary of the problem might be as overemphasis on *reductionism*, i.e., decomposition of problems into isolated parts for detailed and painstaking analysis, and too little interest in the *synthesis* of these components into a description of the larger picture (Fig. 1). Certainly some degree of abstraction, simplification, and decoupling is essential for engineering science to move forward. But this approach inevitably leads away from the phenomena or process originally motivating the effort and implicitly presumes either a superposition of effects or serial coupling among various parts.

This process of decomposition requires critical scrutiny after each step in the analysis, to ascertain when attention should be redirected toward synthesis of the parts and understanding of the couplings. Lack of such an assessment too often leads to unnecessary refinement of solutions to old problems, continuous generation of workable variations on the same theme,

PROBLEMS   SUB-PROBLEMS   SOLUTION

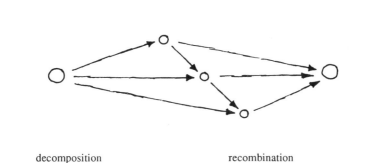

decomposition   recombination

Figure 1. Decomposition of complex problem into subproblems for analysis, followed by recombination of parts into description of original problem.

and experiments designed to demonstrate idealized theories. Appropriate concern for the real physical problem, and genuine interest in new ones arising from emerging technologies, should lead to the identification and abstraction of new classes of interesting and worthy subproblems. Of course, many such phenomena are complex, requiring carefully designed experiments to isolate the controlling mechanisms and nonlinear theory to provide realistic descriptions. But recognizing the importance of this phase of research is critical to the health of the enterprise.

This issue has an important educational component as well. Our students are educated properly only if they learn to appreciate the whole problem, as well as to dissect it, and are exposed to the process of defining meaningful physical problems from observations of complex phenomena. Those who grasp both the concept and the details succeed in either industry or academia; those who do not often struggle. This education should begin at the undergraduate level, via order of magnitude reasoning to assess the dominant features of complex problems, and continue during graduate school through both courses and examples provided by faculty in choosing and motivating thesis problems.

II. Directions for the Future

For the future, I offer several thoughts about productive directions and approaches. First, the shift of research toward complex fluids and fluid mechanics in the context of materials processing [1] necessitates fuller recognition of the coupling between microscopic and macroscopic phenomena. This means (1) addressing the process scale while still recognizing the importance of microstructure, whether colloidal, macromolecular, or molecular; (2) dealing with phase transitions, nonequilibrium and anisotropic microstructures, and other highly nonlinear phenomena; and (3) accounting for the nonideality of transport coefficients, which dictates a strong coupling between the momentum equation and either microstructural equations or species conservation equations.

The task is clearly not easy, since exact constitutive equations are elusive and the governing equations are inevitably nonlinear. However, solution of boundary value problems at the process scale, with relatively simple constitutive relations containing the relevant physics, can generate considerable insight and identify critical experiments. Although the solutions generally must be numerical, asymptotic analyses assist in confirming their validity, classifying the behavior semiquantitatively, and identifying unifying factors.

Second, chemical engineers must remain alert and receptive to significant new developments in the relevant sciences. Renormalization group theories and the theory of dynamical systems promise insight into the mixing of fluids in the absence of inertia and diffusion [21] and, perhaps, turbulence. Similarly, fractal concepts [22], kinetic theory [23], nonequilibrium statistical mechanics [24], Brownian dynamics simulations [25], and scattering techniques capable of characterizing dispersion microstructure [26] promise greater understanding of concentrated and flocculated dispersions and particulate systems. Other techniques for morphological characterization of complex fluids and solids, such as electron microscopy, x-ray scattering and scanning, and nuclear magnetic resonance (NMR) imaging, must become commonly available tools as well, as already recognized by a number of departments. The intermingling of basic science and engineering science in journals and technical meetings encourages the exchange, but penetration of the techniques and results into chemical engineering journals is essential as well. Indeed, knowledge flows both ways, with chemical engineers stimulating and often leading the advancement of the science.

Third, a serious need exists for a data base containing transport properties of complex fluids, analogous to thermodynamic data for nonideal molecular systems. Most measurements of viscosities, pressure drops, etc. have little value beyond the specific conditions of the experiment because of inadequate characterization at the microscopic level. In fact, for many polydisperse or multicomponent systems sufficient characterization is not presently possible. Hence, the effort probably should begin with model materials, akin to the measurement of viscometric functions [27] and diffusion coefficients [28] for polymers of precisely tailored molecular structure. Then correlations between the transport and thermodynamic properties and key microstructural parameters, e.g., size, shape, concentration, and characteristics of interactions, could be developed through enlightened dimensional analysis or asymptotic solutions. These data would facilitate systematic studies of important processes, for example the colloidal processing of ceramics, instabilities associated with surface effects in polymer extrusion, and novel biological separations.

III. Examples from the Present

To illustrate these ideas of productive approaches for the future, I will describe several examples from current research.

A. *Single Crystal Growth from the Melt*

This effort, exemplified by the program of Bob Brown of MIT [29], focuses on the role of fluid mechanics and heat and mass transfer in the growth of large semiconductor crystals from the melt—one of the unit operations of electronic materials processing. Several characteristics of the approach are relevant to this discussion. First is the philosophy that "it must be clear why, and on what levels, modeling can play a role in the optimization and control of systems for the growth of single crystals from the melt." This, of course, requires knowing the real processes and assessing critically where transport contributes or limits. Second, the importance of different levels of analysis "from thermal analysis of an entire crystal growth system to analysis of the dynamics of defects in the crystal lattice" is recognized (Fig. 2). Third,

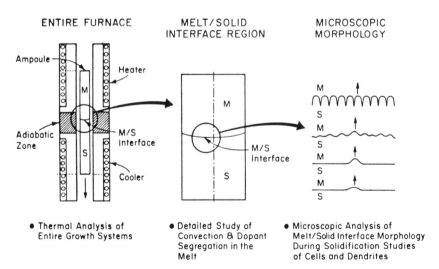

Figure 2. Three spatial scales for modeling melt crystal growth, as exemplified by the vertical Bridgman process. From "Theory of Transport Processes in Single Crystal Growth from the Melt," by R. A. Brown, *AIChE Journal*, Vol. 34, No. 6, pp. 881–911, 1988, [29]. Reproduced by permission of the American Institute of Chemical Engineers copyright 1988 AIChE.

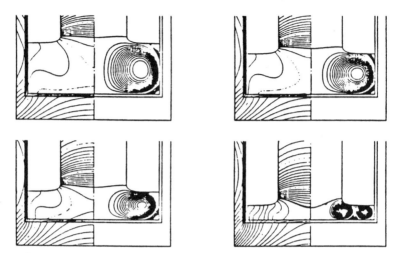

Figure 3. Streamlines (on right) and isotherms (on left) for growth of Si in a prototype Czochralski system. The volume of the melt, at the bottom in each drawing, changes among the calculations, affecting the qualitative form of the convection cell and the shape of the crystal interface. From "Theory of Transport Processes in Single Crystal Growth from the Melt," by R. A. Brown, *AIChE Journal*, Vol. 34, No. 6, pp. 881–911, 1988 [29]. Reproduced by permission of the American Institute of Chemical Engineers copyright 1988 AIChE.

dimensional analysis, order of magnitude estimates, and asymptotic solutions, taken in many cases from the broader scientific literature, are used to complement detailed numerical solutions. Fourth, sophisticated numerical techniques are adapted and developed to obtain accurate solutions in the relevant ranges of conditions, with suitable recognition of the uncertainly arising from imperfect data on physical properties and the limitations of the numerics. Comparison with observations both validates the computations and helps to interpret microscopic features of the experiments (Fig. 3). In addition, questions of process stability, control, and optimization are explored. The result is outstanding engineering science in the fullest sense of the term.

B. *Mixing in Deterministic Flows*

As Leal [1] notes, Aref [21] and others [30] interested in nonlinear dynamics have established the existence of chaotic behavior even in time-periodic laminar flows. The recent work of Julio Ottino is an innovative approach to a classical problem [31–33] that "exploits the connection between the kinematics of mixing and the theory of dynamical systems." The applica-

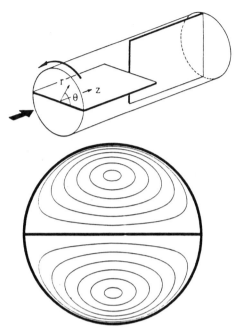

Figure 4. A schematic of one period of the partitioned pipe mixer, along with the form of the streamlines in one cross section. Reprinted with permission from *Chem. Eng. Sci.*, vol. 42, p. 2909, D. V. Khakhar, J. G. Franjione, and J. M. Ottino, "A Case Study of Chaotic Mixing in Deterministic Flows: The Partitioned Pipe Mixer," copyright 1987 [32], Pergamon Press PLC.

tion is to static or inertialess mixers which operate by stretching material lines. Indeed the "exponential separation of trajectories that are initially close together is one of the main characteristics (and also a definition) of chaotic dynamical systems." This approach incorporates a simplified geometrical model for the basic process (Fig. 4) and a quantitative description of the fluid mechanics which, although approximate, preserves the essential physics. Calculation of trajectories of fluid elements then defines fundamental parameters characterizing the local mixing—Liapunov exponents, average rates of stretching, and efficiencies—as well as residence time distributions, the conventional integral measure. Comparison among the various quantities highlights the shortcomings of the integral measure, particularly for laminar flows with well-mixed regions adjacent to unmixed ones (Fig. 5). Complementary experiments support the theory but also reveal some bifurcations that are difficult to capture in simulations. The results, though only semiquantitative, raise many interesting practical and fundamental questions for future calculations and experiments to pursue.

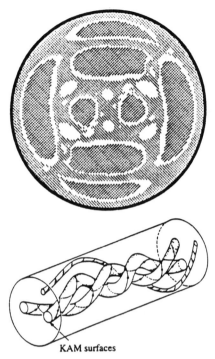

KAM surfaces

Figure 5. Plot of the initial location in the cross section of material elements with a
stretch ratio greater than 10^4 (light shading), less than 10^3 (dark shading), and
intermediate between these values (white), together with the invariant tubes
responsible for this mixing pattern. KAM refers to the Kolmogorov–Arnold–
Moser theorem. Reprinted with permission from *Chem. Eng. Sci.*, vol. 42, p.
2909, D. V. Khakhar, J. G. Franjione, and J. M. Ottino, "A Case Study of
Chaotic Mixing in Deterministic Flows: The Partitioned Pipe Mixer,"
copyright 1987 [32], Pergamon Press PLC.

C. Flow of Granular Materials

The mechanics of granular materials vitally affects chemical and materials processing as well as large-scale geophysical phenomena such as land-
slides and planetary rings. Nonetheless, the multiple modes of momentum
transfer, the many-body nature of the microscopic dynamics, and the highly
nonlinear macroscopic behavior have stalled progress. Within the past decade,
though, the work of a variety of engineers and scientists, e.g., Jackson [34–
37], Haff [38], Savage [23], and Walton [39], has unraveled many aspects
of the phenomena and opened the way toward a general understanding of
particulate systems.

The research of Roy Jackson combines theory and experiment in a distinctive fashion. First, the theory incorporates, in a simple manner, inertial collisions through relations based on kinetic theory, contact friction via the classical treatment of Coulomb, and, in some cases, momentum exchange with the gas. The critical feature is a conservation equation for the "pseudothermal temperature," the microscopic variable characterizing the state of the particle phase. Second, each of the basic flows relevant to processes or laboratory tests, such as plane shear, chutes, standpipes, hoppers, and transport lines, is addressed and the flow regimes and multiple steady states arising from the nonlinearities (Fig. 6) are explored in detail. Third, the experiments are scaled to explore appropriate ranges of parameter space and observe the multiple steady states (Fig. 7). One of the more striking results is the

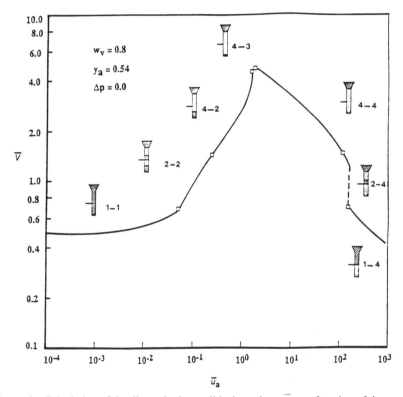

Figure 6. Calculation of the dimensionless solids throughput $\bar{V}$ as a function of the aeration rate for sand of 0.15 mm diameter in a standpipe aerated about midway down and operated with a modest backpressure. Note the existence of two steady values of $\bar{V}$ for some ranges of $\bar{U}_a$; the dashed line represents an unstable portion of the curve. From Mountziaris [36].

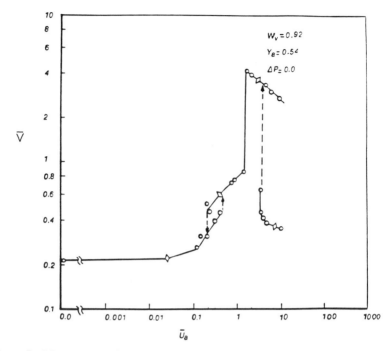

Figure 7. Measurements for conditions similar to those in Fig. 6 demonstrating the hysterisis loop for $0.2 < \bar{U}_a < 0.5$ and the jump across an unstable portion of the curve at $\bar{U}_a$ between 2 and 3. From Mountziaris [36].

"granular jump" (Fig. 8) across which an energetic, dilute phase flow decelerates into a dense sliding flow, indicating the potential importance of both frictional and collisional mechanisms of momentum transfer.

D. Optical Characterization of Complex Fluids

In some areas, progress has been impeded by the absence of suitable experimental techniques. For example, our ability to characterize quantitatively the microstructure of complex fluids during flow was limited. The basic optical techniques—light scattering, birefringence, and dichroism—were generally known but difficult to implement a decade ago. The advances have been achieved, principally by Gerry Fuller [15], through an understanding of the optics, greater facility in interfacing with computers to follow rapid, time-dependent flows, creative use of the variable-wavelength light sources, and knowledge of the fluid mechanics necessary to interpret the observations.

Fuller's effort began with polymeric systems, wherein detecting birefringence and/or dichroism within a known homogeneous flow field determines

Figure 8. Photograph of a transition in a chute flow, termed a "granular jump," from a dilute, high-speed flow upstream to a dense, low-speed flow downstream. From Johnson [37].

the magnitude and orientation of the macromolecular deformations as functions of the shear rate. Subsequent application of these measurements, or low-angle light scattering, to dilute suspensions of anisotropic particles in a shear flow yielded the aspect ratio and polydispersity. The success of these studies has convinced others to acquire or build similar instrumentation, shifting the emphasis from technique development toward studies of technologically or scientifically interesting systems, e.g., concentrated dispersions, electrorheological fluids, liquid crystals, and fiber suspensions. Quantitative data of this type, indicating the effect of flow on the structure of complex fluids, is essential to the development and testing of theories relating non-Newtonian rheology to the microscopic characteristics of the materials.

E. Finite Element Solutions of Free Surface Flows

In the 1970s, the potential of finite element and related techniques for solving the Navier-Stokes equations in complex geometries began to be realized, particularly with free surfaces. However, the application posed difficulties not present in the analyses of structures and flow in porous media then prevalent. Thus, Scriven and his group [40–42] at Minnesota adapted and extended the technique to accommodate the nonlinearities represented by boundary conditions at free surfaces, i.e., balancing surface tension and viscous stresses to determine the shape and position of the interface, and the convective terms in the governing equations. From this emerged a powerful capability whose impact extended to other research groups at Minnesota and spawned attacks on non-Newtonian fluid mechanics [11, 12], reservoir modeling, and coating flows elsewhere.

The specific studies of free surfaces began with static interfaces in classical and novel geometries, then moved on to a series of flows in various roll, curtain, and fountain coaters with Newtonian and power law fluids. The last dealt with fluid inertia, contact lines, and film splitting (Fig. 9), successfully predicting quantities such as film thickness in quantitative agreement with experimental results (Fig. 10). Comparisons with classical lubrication analyses and matched asymptotic expansions elucidate the range of validity of these approximations and suggest improved analytical treatments. Hence, the numerical schemes provide the means for detailed design of complex processes, as well as shortcut techniques.

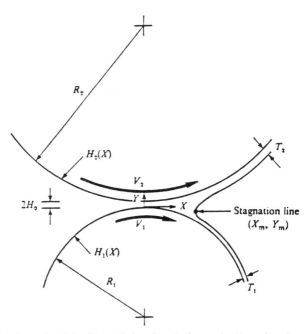

Figure 9. A schematic of the film splitting flow in forward roll coating. Reproduced from Coyle *et al.*, copyright 1986 [40]. Reprinted with the permission of Cambridge University Press.

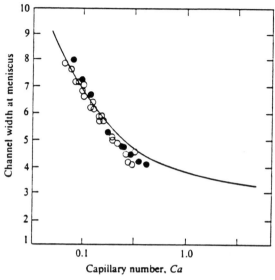

Figure 10. Comparison between the theoretical prediction (−) and measurements of the film split location expressed in terms of a dimensionless channel width for the forward roll coating process. Reproduced from Coyle et al., copyright 1986 [40]. Reprinted with the permission of Cambridge University Press.

F. Simulations of the Dynamics of Dense Dispersions

As noted by Gubbins [43], molecular dynamics simulations can successfully predict the thermodynamic and transport properties of dense fluids composed of nontrivial molecules. With dispersions of small particles in a liquid, simulations are also appealing, but the intervening viscous medium profoundly alters the dynamics and, therefore, the technique [25]. First, the particles move very little in response to individual fluctuations of the fluid molecules. Hence, the equations of motion can be integrated analytically from the molecular time scale to the diffusion time scale for the particles, producing equations containing random Brownian displacements, translations due to applied forces, and convection due to externally imposed flows. Second, the particles interact hydrodynamically at large separations. Thus, the resistance coefficients relating the various displacements to the driving forces are complex, unknown functions of the positions of all the particles.

Consequently, realistic simulations were not possible until Brady and Bossis [4] developed accurate approximations for the many-body hydrodynamics. By first preserving lubrication effects in the near field through a pairwise additive force approximation and then capturing the far field through truncated multipole expansions, they have constructed a robust and accurate, albeit time-consuming, description. Comparisons of the predictions with exact solutions for periodic arrays and experimental data for the sedimentation and diffusion of hard spheres over a range of volume fractions attest to the success.

Now the technique provides the basis for simulating concentrated suspensions at conditions extending from the diffusion-dominated equilibrium state to highly nonequilibrium states produced by shear or external forces. The results to date, e.g., for structure and viscosity, are promising but limited to a relatively small number of particles in two dimensions by the demands of the hydrodynamic calculation. Nonetheless, at least one simplified analytical approximation has emerged [44]. As supercomputers increase in power and availability, many important problems—addressing non-Newtonian rheology, consolidation via sedimentation and filtration, phase transitions, and flocculation—should yield to the approach.

Though somewhat cryptic, these examples demonstrate the necessity for recognizing important problems, understanding developments in other fields that suggest appropriate theoretical and experimental tools, pursuing both the macroscopic and microscopic aspects, and coupling theory with experiment.

References

1. Leal, L. G., in *Perspectives in Chemical Engineering: Research and Education* (C. K. Colton, ed.), p. 61. Academic Press, San Diego, Calif., 1991 (*Adv. Chem. Eng.* **16**).
2. Schowalter, W. R., *Ann. Rev. Fluid Mech.* **16**, 245 (1984).
3. Saville, D. A., *Ann. Rev. Fluid Mech.* **9**, 321 (1977).
4. Brady, J. F., and Bossis, G., *Ann. Rev. Fluid Mech.* **20**, 111 (1988).
5. Davis, R. H., and Acrivos, A., *Ann. Rev. Fluid Mech.* **17**, 91 (1985).
6. Russel, W. B., *Ann. Rev. Fluid Mech.* **13**, 425 (1981).
7. Rallison, J. M., *Ann. Rev. Fluid Mech.* **16**, 45 (1984).
8. Homsy, G. M., *Ann. Rev. Fluid Mech.* **19**, 271 (1987).
9. Denn, M. M., *Ann. Rev. Fluid Mech.* **12**, 365 (1980).
10. Ruschak, K. J., *Ann. Rev. Fluid Mech.* **17**, 65 (1985).
11. Beris, A. N., Armstrong, R. C., and Brown, R. A., *J. Non-Newtonian Fluid Mech.* **22**, 129 (1987).
12. Brown, R. A., Armstrong, R. C., Beris, A. N., and Yeh, P. W., *Comput. Meth. Appl. Mech.* **58**, 201 (1966).
13. Dupret, F., Marchal, J. K., and Crochet, M. J., *J. Non-Newtonian Fluid Mech.* **18**, 173 (1985).
14. Keunings, R., *J. Non-Newtonian Fluid Mech.* **20**, 209 (1986).

15. Fuller, G. G., *Ann. Rev. Fluid Mech.* **22**, 387 (1990).
16. Lightfoot, E. N., Chiang, A. S., and Noble, P. T., *Ann. Rev. Fluid Mech.* **13**, 351 (1981).
17. Sundaresan, S., *AIChE J.* **33**, 455 (1987).
18. Lenhoff, A. M., and Lightfoot, E. N., *Chem. Eng. Sci.* **41**, 2795 (1986).
19. Jensen, K., in *Perspectives in Chemical Engineering: Research and Education* (C. K. Colton, ed.), p. 395. Academic Press, San Diego, Calif., 1991 (*Adv. Chem. Eng.* **16**).
20. Walker, K. L., Homsy, G. M., and Geyling, F. T., *J. Colloid Interface Sci.* **69**, 138 (1979).
21. Aref, A., *Ann. Rev. Fluid Mech.* **15**, 345 (1983).
22. Weitz, D., Lin, M. Y., and Huang, J. S., in *Physics of Complex and Supramolecular Fluids* (S. A. Safran and N. A. Clark, eds.), p. 509. Wiley, New York, 1987.
23. Savage, S. B., and Jeffrey, D. J., *J. Fluid Mech.* **110**, 225 (1981).
24. Felderhof, B. U., and Jones, R. B., *Physica* **146A**, 417 (1987); **147A**, 203, 533 (1988).
25. Ermak, D. L., and McCammon, J. A., *J. Chem. Phys.* **69**, 1352 (1978).
26. Vrij, A., Jansen J. W., Dhont, J. K. G., Pathamamanoharan, C., Kops-Werkhoven, M. M., and Fijnaut, H. M., *Disc. Faraday Soc. London* **76**, 19 (1983).
27. Colby, R. H., Fetters, L. J., and Graessley, W. W., *Macromolecules* **20**, 2226 (1987).
28. Tirrell, M., *Rubber Chemistry and Technology* **57**, 523 (1984).
29. Brown, R. A., *AIChE J.* **34**, 881 (1988).
30. Wiggin, S., *Nature* **333**, 395 (1988).
31. Chien, W. L., Rising, H., and Ottino, J. M., *J. Fluid Mech.* **170**, 355 (1986).
32. Khakhar, D. V., Franjione, J. G., and Ottino, J. M., *Chem. Eng. Sci.* **42**, 2909 (1987).
33. Ottino, J. M., Leong, C. W., Rising, H., and Swanson, P. D., *Nature* **333**, 419 (1988).
34. Johnson, P. C., and Jackson, R., *J. Fluid Mech.* **176**, 67 (1987).
35. Johnson, P. C., Nott, P., and Jackson, R., *J. Fluid Mech.* **210**, 501 (1990).
36. Mountziaris, T. J., "The effects of aeration on the gravity flow of particulate materials in vertical standpipes." Ph.D. dissertation, Princeton University, 1989.
37. Johnson, P. C., "Frictional-collisional equations of motion for particulate flows with applications to chutes and shear cells," Ph.D. dissertation, Princeton University, 1987.
38. Haff, P. K., *J. Fluid Mech.* **134**, 401 (1983).
39. Walton, O. R., and Braun, R. L., *J. Rheol.* **30**, 949 (1986).
40. Coyle, D. J., Macosko, C. W., and Scriven, L. E., *J. Fluid Mech.* **171**, 183 (1986).
41. Kistler, S. F., and Scriven, L. E., *Int. J. Numer. Meth. Fluids* **4**, 207 (1984).
42. Silliman, W. J., and Scriven, L. E., *J. Comp. Phys.* **34**, 287 (1980).
43. Gubbins, K. E., in *Perspectives in Chemical Engineering: Research and Education* (C. K. Colton, ed.), p. 125. Academic Press, San Diego, Calif., 1991 (*Adv. Chem. Eng.* **16**).
44. Brady, J. F., and Durlofsky, L. J., *Phys. Fluids* **31**, 717 (1988).

6

Fluid Mechanics and Transport Phenomena

J. R. A. Pearson
Schlumberger Cambridge Research
Cambridge, England

I. Introduction

Much of what I have to say will merely reinforce what has already been well said by L. G. Leal in his paper [1]. In particular, I would emphasize that chemical engineers, indeed many other engineers also, are more concerned with transport processes than with fluid mechanics as such. It is true that in many processes the convective motions dominate, but their effect is largely a kinematic and not a dynamical one. Chemical engineers study dynamics as a means to an end; relatively infrequently are the absolute stresses caused by fluid motion their prime interest. More often, issues in heat and mass transfer are most important.

The fluids of most interest (because they cause the most problems) to chemical engineers are complex in terms of their rheological and transport properties. The study of Newtonian fluid mechanics, which has dominated aerodynamics and hydraulics, is giving way to a study of the mechanics of nonlinear, elastic, time-dependent fluids. The search for adequate, robust, but not overcomplicated rheological equations of state (constitutive equations) that will characterize polymer melts, polymer solutions, emulsions, suspensions, and multiphase flows generally now represents a substantial activity in engineering departments. There are encouraging signs that applied mathematics, applied mechanics, and physics departments recognize

ADVANCES IN CHEMICAL ENGINEERING, VOL. 16

the significance of these activities and are joining in. The feature that distinguishes this endeavor most from traditional fluid mechanics is the strong coupling that occurs between rheological behavior, flow field kinematics, temperature, and composition in many industrially important circumstances. Furthermore, there are fewer cases where inertial forces are dominant, and so many traditional results are inapplicable even qualitatively. We see a wide range of new phenomena.

This complexity of fluid structure, together with the widely different nature of various contributions to heat and mass transport, means that a very broad range of length and time scales is involved. This implies that the full processes display a range of phenomena over an equal range of length and time scales; this is at the heart of many of the difficulties experienced in practice, when we seek to modify and control these processes on an industrial scale.

II. The Role of Computers

The most important contribution now being made by the computer in engineering mechanics is its ability to deal with geometrical complexity. Using finite element or boundary element techniques, the modern analyst—and hence the designer—is freed from many of the restrictions placed upon the specification of boundaries that limited the traditional mathematical analyst. The geometry of flow fields was idealized because that was all the theorist could cope with, even though the practical engineer was well aware that small changes to boundary shapes, usually decided upon intuitively, could greatly ameliorate flow and transfer problems encountered in industrial equipment. Only in the last few years, the computer screen has replaced the laboratory flow loop for a large number of innovative engineers.

Recent developments in computer hardware promise that computer simulation of fully three-dimensional unsteady flows will be possible and that it will be available to the engineer at his or her desk on a workstation. However, this in itself will not solve a single design problem. Indeed, it may in some cases be a distracting influence on the designer because of the vast amount of data that will be created and the ease with which a sterile series of ill-thought-out changes can be made to a proposed design.

The real challenge for the future is providing the framework within which the results of extensive computer simulation can be interpreted and put to use. The solution will not be found by simply relying on global optimization techniques, because they have to be provided with well-specified "cost functions," which are not easy to determine from the start, and with only a limited number of input variables over which to optimize. The greater the complexity with which a simulation can be carried out, the greater the number of design variables that are in principle available to the designer, who is thus faced with the need to constrain optimization rather strongly. This helps the optimization process but hands back the designer's traditional role. This is a point that is not sufficiently emphasized, and it suggests that chemical engineers should concentrate more on using computer systems for which they have written the specifications than on developing the systems themselves.

III. The Role of Empiricism and Variable Definition

Empiricism—at one level little more than a synonym for reliance on experimental observation—need not be blind. In many cases, it is the crucial starting point for detailed analysis; it is the source of the critical hypotheses which underlie successful theories, whose predictions are later verified by further experimental observation. Few scientists would contest this last statement. This has relevance to the comments made above about computer simulation, in that the latter can only be as good as the mathematical models on which it is based. In any new engineering situation, there may well be areas of uncertainty, say about the nature of the fluid in question, which mean that the relations required by a fully general computational fluid mechanics simulator program are not known.

At the lowest level, empirical observation provides information about process and material length scales. In the case of an example used at the Symposium, when an eroded mud-driven-turbine stator was handed round, observation and measurement would identify length scales for the particle size distribution, the overall flow geometry, the erosion patterns, and the depth of erosion. It would also provide the time scale for flow periodicity and erosion.

At the next level, it alerts the analyst to the probable presence of "hidden" length and time scales in the problem. In the case chosen, these would be associated with visible signs of major mean flow vortices and turbulence.

At the deepest level, empirical observation forces on the theorist the need to define carefully the quantities of interest and importance in the problem or process being investigated. This may seem a trivial point, but it proves to be of considerable significance for complex flows of structured fluids exhibiting a wide range of length and time scales. Engineers are used to working with continuum (thermodynamic) variables, such as concentration, stress, velocity, and temperature, which imply means over space, time, or ensembles. In practice, it is the first two, or a combination of them, that are measured.

Formal scientific theories tend to operate at a single scale. (Physics has long been beset with the problem of unifying field theories that have grown up independently because of the different length scales over which the relevant forces act.) Consequently, there is a natural progression in engineering whereby the results of one theory are embedded in another characterized by a larger scale. Confusion arises when the concepts and definitions themselves do not change as the theories are nested, while the relations between the variables do change. This is peculiarly noticeable in rock mechanics, where it is well known that mechanical properties (certainly failure criteria) are scale dependent. The subtleties involved in the choice of scales for averaging deserve more attention in the curricula for engineering students.

It is worth noting that one of the uses to which computers can be put is to derive continuum models for highly structured systems. Thus in the case of drilling muds, simulations on the scale of clay platelets can be used to provide rheological models for use in finite element models for flow mechanics, which are then used to derive Nusselt number estimates for uniaxial mean temperature predictions in oil wells. The range of length scales goes from 1 nm to 1000 m!

IV.   Nonlinear Dynamics

Leal has already made the point that this is a key area for future development. Constitutive relations for fluids are intrinsically nonlinear, while analysis of processing flows demonstrates the importance of nonlinear dynamical effects. The role of the chemical engineer is to select the physical scales (length and time) of interest and to derive the important dimensionless groups that characterize different but related flow and transport phenomena. Empirical observation can often help in deciding when a group is large or small enough for the terms associated with it in a mathematical formulation to be neglected, or else treated by perturbation methods.

Solving sets of nonlinear equations should, however, be regarded as a specialist activity, in many cases best left to mathematicians. Making sense of the results should be regarded as the province of both the engineer and the mathematician, because the detailed questions the former might ask are often most quickly answered by the latter. An issue arises over the question of asymptotic analysis, and how far the engineer should be concerned with the mathematics involved in matching or patching expansions used to cover different portions of space and time.

Modern developments in bifurcation and stability theory, which lead on to chaos or turbulence in dynamical systems, only serve to show why it is worth embedding as much complexity as possible within the constitutive description of material properties. There is a sound intellectual case for trying to subsume fine-scale motions of polymer melt molecules within an Oldroyd model, say, or the small-scale eddies of a turbulent airflow within the K-ε model for turbulent flows, or the pore scale motion in a porous medium within Darcy's law.

V. Examples

Three examples are given below for different areas of interest which illustrate the arguments developed earlier.

A. *Polymer Melt Processing*

The objective in polymer processing is largely one of forming homogeneous polymers into required shapes. Obvious techniques are molding, extrusion, and drawing (fiber or sheet); continuous processes are used where possible. All involve careful control of temperature: in the melting (or softening) phase, convective thermal mixing is essential on grounds of cost; in the freezing (or hardening) phase, variations in thermal history determine product properties and distortion. Polymer melts are extremely viscous at processing temperatures and degrade or cross-link easily. This means that very high stresses are involved, that there is high internal heat generation, and that relevant rheological and chemical time scales overlap process time scales. In general the flows are slow, steady (or quasi-steady), and laminar. Thermal conductivity, which is very low, governs the very slow transfer of heat across streamlines; along streamlines the temperature is therefore given largely by adiabatic considerations.

The geometry of processing flows is usually such that the characteristic length along stream lines differs by several orders of magnitude from that across them—really a consequence of the need to transfer heat across stream lines to metal walls.

Early modeling of these processing flows was aimed at the fluid mechanics of the flow field and was based on a temperature-independent inelastic fluid rheology. Though illustrative, this could never capture the real essence of processing problems or behavior. Formal attempts to write down and solve a full set of coupled equations for temperature-sensitive elasticoviscous fluids led to intractable complexity, for which even the most modern computers offered no panacea.

Progress has been made [2] by carefully dividing up the flow field involved into manageable segments and selecting an appropriate rheological equation of state for each portion: in the simplest sense, this has meant distinguishing between simple shear (weak) flow regions and extensional (strong) flow regions. In the former, an inelastic, temperature-dependent power-law approximation and, in the latter, a uni- or biaxial elasticoviscous, temperature-dependent Oldroyd-type approximation are used. A clear distinction then has to be made between cases where resistance to heat transfer lies within the flowing polymer, such as confined flow in metal machinery, and where it does not, such as extensional flow in air. Finally, the importance of entry conditions, most easily understood in terms of a development length, has to be determined. Only when maximum use has been made of the most sweeping approximations can solutions be obtained. Despite the commercial importance of the processes to manufacturing industry, these relatively simple engineering analyses have taken nearly 40 years to evolve. In this the role played by asymptotic analysis and suitable dimensionless groups (including Reynolds, Froude, Weber, Weissenberg, Deborah, Brinkman, Peclet, Graetz, Nahme-Griffith, Biot, Nusselt, and Fourier numbers) has been invaluable.

This modeling is forward-predictive: if equipment geometry, material properties, and operating conditions are specified, then the output flow rates, shapes, and temperatures are given as the solution of the model equations. As yet, the inverse, or design, problem has to be solved by the user selecting progressively more appropriate values of the variable inputs and running the model iteratively. Success depends to a large extent on intuitive evolution from earlier designs.

A key role played by mathematical modeling and analysis has been in explaining unexpected instabilities and asymmetries, which often arise when certain critical flow rates are exceeded. Thus "melt fracture" in extrusion, "draw" resonance in fiber spinning, pulsatile flow in bubble blowing, film thickness variations in film blowing and casting, and surface irregularities or distortion of moldings have all been studied, and to some extent explained, in terms of bifurcations of nonlinear systems.

It is true to say that the selection of suitable mathematical models has involved a clear physical understanding of the processes involved and so can be viewed as the province of the chemical engineer; it is also true that progress would have been more rapid if the actual mathematical and computational analyses had been carried out by experts, rather than by chemical engineering graduate students or postdoctoral workers.

B. Two-Phase Flow in Porous Media

This is essentially an oil field reservoir phenomenon and has attracted a lot of attention among chemical engineers. It is dominated in practice by the inhomogeneities that occur over a wide range of length scales in the porous medium itself. Reservoir rocks, being sedimentary, often show rapid vertical (bedding) variation and slower, but equally important, variation within beds. The main structural component is typically a sand grain, of diameter 100 μm say, with other mineralogical components like clay platelets having one dimension as small as 10 nm. The effective pores vary from

<1 to 50 mm. Vertical and horizontal permeabilities can vary by an order of magnitude over 10 mm. Gross features vary on all scales up to that characterizing the extent of a reservoir (1 km for example). Fractures, natural or man-made, add to the complexity of the geometry and have a full range of length scales.

The transfer processes that arise during oil field operations occur when one fluid system residing (at equilibrium) within the pore structure is displaced by another. The flows are largely horizontal, between one vertical well and another; the typical well diameter is 200 mm, with entrance perforations of diameter about 10 mm; interwell distances are of the order of 500 m. This means that mean flow velocities vary by five orders of magnitude within the field. A further two orders of magnitude can arise between different stages of production in different wells.

The displacement flows can be miscible (brine after polymer solution, CO_2 after oil, steam after water) or immiscible (water after oil). In the former case, it is the mixing process itself which has to be understood and modeled; steam recovery requires the thermal transport problem to be accurately modeled. In the latter case, the two fluid phases coexist within the porous medium; their relative proportions are determined not only by flow and mixing processes, but equally by interfacial and surface tensions between the three phases (matrix material included). Local (capillary) variations in pressure between the two fluid phases become important. The overall flow field is determined by large-scale pressure gradients.

The goal of enhanced oil recovery is to reduce the amount of oil left behind in the reservoir after one or more displacement sweeps. The challenge is to adjust the sequence and composition of the fluids used for displacement so as to mobilize the oil that remains after the first *in situ* pore-pressure-driven production phase. Laboratory investigations on cores together with direct field observations have shown how complex are the issues involved: the movement of contact lines along solid surfaces is governed by processes on a 1 nm scale; capillary effects, which are affected by molecular mass transfer of surface-active agents, take place on both the pore scale, down to <1-mm, and the "frontal" scale (i.e., the distance over which the liquid phase ratios change), which may be as large as that of the well spacing.

There are many hazards involved in such operations: untoward chemical reactions can arise between fluid and solid phases, leading to permeability impairment; stresses may exceed the mechanical strength of the reservoir matrix; vertical flows may arise because of fluid density differences that lead to rapid breakthrough of unwanted fluid.

All of these effects and phenomena have to be modeled. It is doubtful whether even the most elaborate reservoir simulators cover more than a small fraction of the possibilities. To that extent they are all of limited applicability.

A further difficulty that faces oil field modelers is the lack of information they have about downhole conditions and reservoir characteristics. Forward predictions about production are distressingly uncertain. It is therefore common to fit observed production data retrospectively—a procedure known as history matching—and to infer reservoir parameters, particularly permeabilities and relative permeabilities.

In doing this the modeler is implicitly choosing a scale, the finite element grid scale, over which averages are taken: the mean variables involved will be saturations, capillary pressures, pressure gradients, and flow rates. All of the models for heterogeneous reservoir formations show that these values will be scale dependent: there is not one single continuum scale; this scale can range from 100 mm to 100 m. With two-phase flow, a displacement front usually exists, across which there are effective discontinuities in the mean saturations (e.g., as arise in the Buckley-Leverett approximation). However, in the case of higher-mobility fluid displacing a lower-mobility fluid, this interface becomes unstable, and so "hidden" length scales show up in a developing displacement flow.

The challenge is therefore to find a theoretical expression for these scaling laws. It will in any case depend upon scaling laws for the statistical distribution of fundamental geometrical reservoir properties. It will also depend upon these "hidden" processes that arise because of the nonlinear nature of movable boundary flows (quite apart from nonlinearities intrinsic to the continuum relations themselves). There have been some remarkable pioneering attempts to predict continuum properties of porous media from fundamental parameters, mainly by chemical engineers (of whom I wish to single out Howard Brenner and co-workers) and physicists, but they have as yet made little impact on the oil industry.

C. Heat and Mass Transfer in Pipe Flow

At first sight, this represents a much simpler situation than that of the previous example. However, in the case of (oil field) wellbore flows, the fluids are complex (or multiphase) and often inhomogeneous along the length of the pipe (as when one fluid is displacing another); the pipe (the wellbore itself—cased or open, a drill pipe or a production string) is of variable geometry at joints and may contain moving internal obstructions, as with a drillpipe moving within a cylindrical well; the ratio of length to hydraulic radius can commonly be as high as 10,000, a value rarely achieved in the laboratory for high-Reynolds-number flow!

Small perturbations to cylindrical flow can have large effects on heat and mass transfer, even though they may have negligible effect on momentum transfer. Thus the textbook, idealized, models for these transfers prove in practice to be very wide of the mark: the engineer must beware of the formal classroom calculation even when Peclet and Graetz numbers are obligingly introduced. Particularly when flow rates are near transition values, and there is coupling—through density or viscosity—between the flow field and the temperature and composition fields, it is often impossible to predict transfer coefficients; they have to be derived empirically.

Admittedly, great progress has been made in obtaining extensive computer solutions to complicated situations, but even these have difficulty in dealing with flows that, through nonlinear coupling, have exhibited multiple bifurcation about the most steady symmetric flow field consistent with initial and boundary conditions, without becoming wholly turbulent. Until a solution has been obtained, it is often difficult to decide what development lengths will be, and so the choice made in choosing grids for finite difference or finite element methods may be inappropriate for several iterations.

G. I. Taylor's concept of the effective axial diffusion coefficient, which has proved so useful in combining variable axial advection with radial transfer into one parameter, works best when there is no exchange of a passive tracer with the pipe walls. An analogue of his method, which should be applicable when development lengths are large and there is exchange at the wall, has yet to be provided. It would be of great value.

VI. Teaching

My comments on chemical engineering education will go beyond the narrow limits of fluid mechanics and transport processes. This is inevitable because an engineer has to deal with the whole problem and not just any one aspect of it. One of the themes of this Symposium has been the need for interdisciplinary work in the emerging areas of chemical engineering, and there have been many calls for wider curricula, to accommodate biological aspects of bioengineering, solid-state physics, polymer science, and computer science. Even accepting an extension of the core course to the master's level, so that the professional qualification for a chemical engineer might require two further years of study after a bachelor's degree, any attempt to provide complete coverage of all the topics mentioned as well as the traditional ones would be excessive.

At the same time, the question has been raised whether the current graduate education of chemical engineers is too scientific and not oriented enough toward problem solving and innovation at an industrial level. I believe it is, not because I have any doubt whatever about the importance of basic scientific studies, but because these studies can only have any industrial impact if there exists in industry a sufficiently large number of highly skilled, broadly based, engineers strongly motivated to applying them and able to lead and work with multidisciplinary teams. It is often more difficult to apply new ideas than to have them; the talents of a successful engineer are probably rarer than those of a successful scientist, not least because of their relative cultural standings in the Western world.

This imbalance between science and engineering must be tackled at the undergraduate level, at least for those who aspire to be leaders in their profession. It is important at the earliest stage to introduce current opportunities and unsolved problems, rather than to concentrate on well tried and tested analyses that relate to old problems and now-standard design methods. It is simply not practicable to train engineers in all the standard methods that will be expected of them during a 20- to 40-year career; they will inevitably have to learn on the job as their tasks alter.

My first point has therefore to be aimed directly at employers: they must recognize the paramount need for continuing education of practicing engineers. Universities and professional institutions are more than willing to provide courses and opportunities for mature study. Short courses on highly specific topics, chosen to give an up-to-date survey on current techniques, deserve more support than they get at present.

Undergraduate chemical engineering education could then concentrate on the following:

1. A good grounding in basic physical science and mathematics. There is every reason for the same curricula to be used as for the basic grounding of pure scientists; scientists and engineers must share a common approach to fundamental science if a culture gap is to be avoided. This should include experimental work.

2. A limited number of examples of engineering analyses that are well enough honed to be reliably predictive; there will be many occasions when prescriptive methods are called for, even if they are based on empirical data. In the cases where these refer to idealized situations, emphasis should be placed on explaining why they are adequate in practice; error estimation and the use of bounds are just as important as the learning of particular applied mathematical methods. Above all, an engineer must be taught how to make judgments on the adequacy of theoretical or experimental techniques.

3. A greater use of open-ended current problems and market opportunities to illustrate and develop an engineering approach. Too much formal training, with well-posed textbook problems and highly structured examinations, can be sterile. The excitement of engineering comes from making progress on unsolved problems, even if one cannot often implement solutions in a teaching environment. This is best done with a group of students, because it can engender a team approach. Such an approach is already part of most engineering curricula; what I am arguing for is a greater reliance on studies of this kind.

One of the advantages scientists have in recruiting good graduate students, as well as the reason for so much science being done in engineering departments, is that they can see the fruit of their endeavors directly, in their own academic environment. The staff member who is strongly tied in with industry is sometimes seen as a pluralist, who is in some way thought to be neglecting his or her proper duties. (It is as well that the medical profession does not act in this way.) Financial pressures are forcing many departments to open themselves to industry, as they were 50 to 100 years ago. We should view this as an opportunity to reorient our teaching, rather than as a distraction from our academic arrangements.

I want to end on a strong note as far as the fundamentals are concerned. First priority must go, as indeed it always has as far as the engineers I most respect are concerned, toward an intellectually sound, comprehensive foundation in the basic sciences. The postulates of Newtonian mechanics and Maxwellian electromagnetism; of quantum, atomic, and molecular theory; of continuum mechanics and thermodynamics; of statistical mechanics; of chemical rate processes; and of material science must be presented and insisted upon in their full rigor. No easy familiarity with computer packages can yet replace the understanding that is gained from a knowledge of the fundamental physical and chemical processes or laws that govern nature and our technologies.

References

1. Leal, L. G., in *Perspectives in Chemical Engineering: Research and Education* (C. K. Colton, ed.), p. 61. Academic Press, San Diego, Calif., 1991 (*Adv. Chem. Eng.* **16**).
2. Pearson, J. R. A., *Mechanics of Polymer Processing*. Elsevier Applied Science, London, 1985.

General Discussion

Fluid Mechanics and Transport

Clark Colton: In his paper, Gary Leal described the revolution in our way of thinking about transport processes that was associated with application of the methods of asymptotic analysis. He pointed out the emphasis it places on dimensional analysis and on development of qualitative physical understanding of the fundamental basis for correlations. He further indicated that these powerful scaling techniques are slowly finding their way into transport phenomena courses at the graduate level. I wish to point out, for the record, that these concepts of dimensional analysis and scaling, which go back at least as far as Prandtl and were extensively employed by Schlichting in his book *Boundary Layer Theory* (first published in English in 1955) were first introduced into chemical engineering by Andy Acrivos. Andy used these concepts in his reasearch and teaching at Berkeley in the late 1950s amd then at Stanford, and they were propagated by some of his earliest students such as Joe Goddard and Gary Leal. These same concepts were contained in the chemical engineering curriculum at MIT 25 years ago. Harold Mickley taught these methods in the early 1960s. I learned them from Ken Smith in the mid-1960s, and they were introduced at the undergraduate level shortly thereafter. At that time, Ken was recently returned from a postdoc at Cambridge, England, where matched asymptotic analysis was much in vogue. There may well have been other places where this approach was taught. It is surprising how long it has taken for their significance to be generally appreciated.

Fraser Russell: I'd like to say a bit about multiphase fluid mechanics. If one is concerned with design and, as Gary said, focusing on the macroscale (Fig. 1), one needs a lot more information about the creation and/or movement

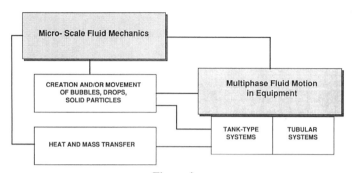

Figure 1.

of bubbles, drops, and solid particles on the microscale. For example, bubbles tend to break up and do peculiar things. The bubble breakup process is a function not only of the scale of turbulence but also of how long the bubble spends in the turbulent field. In the design of either a tubular or tank-type system, one needs to know about the breakup process in detail. The breakup process occurs on a time scale that must be considered in the design of a piece of equipment, i.e., something on the order of a tenth of a second to several seconds. If one does not understand the microscale fluid mechanics concerning the breakup of bubbles and drops, or the movement of bubbles, drops, and solid particles, then one can't properly carry out rational scale-up and design. We don't teach anything at all in multiphase fluid mechanics at the undergraduate level, yet most of the problems we face in process design involve multiphase fluid mechanics. We should pay more attention to this area.

Stuart Churchill: I have no real quarrel with anything that has been said. I would like to address a couple of problems that were alluded to and emphasize them a little more. One is that the gap between the practice and the frontier of our profession is probably growing rather than decreasing, and this is most evident in the undergraduate curriculum. Few of our students are well prepared enough when they graduate to read much of the work that is now published in our journals. I challenge our first-year grad students by giving them an issue of the AIChE journal and asking them to tell me how many of the articles they can read. Although they don't like to admit it, their preparation in mathematics is not adequate to the task. Our graduate students are well prepared technically and are of course doing work at the same level as published in our journals, but only in some narrow area. We're still training our undergraduates for the practice of the past, instead of the frontier. Tony Pearson alluded to this by saying that he didn't consider a chemical engineer educated who did not know advanced chemistry or statistical mechanics. A large fraction of our graduating seniors have not been exposed to these subjects at all. We need to think very hard about the preparation of our undergraduates and for what we are preparing them. We do very well if we get a second chance at them as graduate students, but I'm not so sure that we're doing as well at the undergraduate level. Asymptotic analysis is one subject we can teach to undergraduates; it's a very practical first step in the direction of higher mathematics and analysis. We can also emphasize nonlinear processes, even though we may not teach them much about the details of solving such problems.

In numerical work and use of computers, the critical difficulty is in determining whether or not we have the right model. That can only be assessed through a comparison with experiments. That means we have to do the experiments or look for somebody else to do them, and the latter rarely occurs. Our ability to compute has outrun our ability to run critical tests to see whether or not the model is valid. If our measurements don't agree with the computations, then we've made a breakthrough or something is wrong with the model. That aspect needs more attention.

Gary said that the field of turbulence was dead, but those who work in combustion know that's not so. It's one of the real frontiers, even though we haven't made much progress. If you work in plasma processing or detonations, the time scale of the turbulent interactions is very important. We need to continue to look at turbulence. I'm just finishing a book on turbulent fluid mechanics, and it's forced me to reexamine the literature for the past 50 years. It's disheartening that progress on so many problems, particularly at the macroscale, has been very slight during the past 20 years. I'm a little embarrassed to find that most of the topics I am writing about were done a long time ago. Even though this area has become somewhat of a backwater, most of the problems are not really solved. What's even more important is that the theoretical aspects of fluid mechanics that were known 20 years ago are not used to their full extent in industry today. We in this room have to take some responsibility for that.

Edwin Lightfoot: I've been impressed by the talks this morning. I urge fluid mechanicists to study separation processes because they are dominated by fluid mechanics. What you want is to maximize mass transfer and minimize momentum transfer. The sensitivity of momentum transfer to geometry is so much greater than the sensitivity of mass transfer that the fluid mechanics problems tend to dominate. If you look at the asymptotic results that Gary talked about, you can get useful insights into practical problems that don't require much time to develop. For example, if you examine the old problem of packed-column gas absorbers, you can show with relatively simple work at the Chilton-Colburn level that you can greatly improve packing efficiencies. You want to avoid form drag, and there are new packings that do this, but nobody's acknowledged publicly that this is the reason for their success. In many cases of chromatographic bio-separations, the efficiency of separation depends on minor departures from the one-dimensional idealization used in describing chromatographic column flows. This is a fluid mechanics problem that needs further attention, but now we have to go beyond the first-level approximations. If you look at electrophoresis, which so far has not gotten out of the analytical stage, the big problem is to produce the right kind of convective dispersion; you want it only in one direction, that is to say perpendicular to the applied electric field. This is another fluid mechanics problem that deserves attention, primarily because approximations are made that are difficult to justify in detail.

I'd like to speak last to a point Gary made—is biotransport a separate field or should it be discussed as part of transport in general? I would argue that this is a separate field. Although the same general transport principles operate everywhere, the peculiarities of biological systems are so striking that I don't think you can do useful work unless you know the biology. I'll cite one example, the movements of solutes around the interior of cells. Length scales are on the order of 10 mm, and particles range from small solutes like phosphate ions up to formed elements like mitochondria. If you look at the relative strengths of convective transport and Brownian motion, you'll find a break point around the size of protein molecules where they're about equally important. Below that size, Brownian motion overwhelms convection. Above the size of protein molecules, convective transport can dominate, and biological systems have developed peculiar ways of handling this problem—for instance, transfer along microtubules. Energy-consuming processes are used to move formed elements along small tubes inside the cell. Those have been known to exist for only a few years. The chemical engineer interested in transport who does not know about those peculiar substructures would be totally lost. Not very many chemical engineers need be involved, but those involved have got to learn the biology.

Robert Brown: Along those lines, a recurring theme is that in applying chemical engineering fundamentals—thermodynamics, transport, reaction engineering—to one of these frontier areas, such as materials processing technologies or biotransport, one must get to the interface between the fundamentals and the area of application. I think that's the summary of what Ed just said. You must sit on that interface.

L. E. Scriven: I have two observations in the wake of comments by Gary Leal, Bill Russel, and Anthony Pearson. In regard to Gary's description of the dichotomy between fundamentals research and applications teaching,

I would point to the importance of the individual's professional orientation. In the case of fundamentals in research, the orientation nowadays is most often to the world of science, quality research, and scientific disciplines. On the other hand, applications in teaching go more to the world of engineering practice and industrial technology. Scrambling, blending, or uniting these two orientations is at the crux of the matter for professors of chemical engineering. Choosing and balancing are crucial issues each of us has had to face, issues worth reflecting on.

The second observation points ahead to Alex Bell's topic. Are not chemical engineers as undergraduates exposed to nonlinearity, to hard-core nonlinearity, more than any other engineering discipline, by virtue of chemical reaction engineering and reactor analysis? The modern version of this part of our discipline would have seemed quite inaccessible to undergraduates before the pioneering teaching and textbook writing of the very late 1950s and the 1960s. Indeed, there are those who see the modern version as still too mathematically presented to be accessible. Yet much has been accomplished in bringing undergraduate chemical engineers to the nonlinearity that figured in all three of our discussors' remarks.

William Schowalter: I want to make three points, all stimulated by this morning's presentations or the papers we read prior to today. Jim in his opening remarks spoke about looking critically at our successes as well as our failures. We should to a degree, but we should not be overly critical, so my first observation is a rhetorical one. Gary said that chemical engineers came into the fluid mechanics–transport area late and were thought of as insular and not up to speed with what was going on in the field. I think in fluid mechanics, per se, that's a correct assessment. However, in the areas in which we have agreed already that chemical engineers have a special stake, namely coupled transport, we may not be doing ourselves justice. If one goes way back to the early analogies of Colburn, Chilton, etc., some of the early fathers of our field, the essence of transport was in those analogies, empirical or semiempirical as they were. Indeed, they were picked up later and made rational by persons other than chemical engineers. The works of Spalding and of Kayes, where these analogies have been developed into textbooks, were actually built on innovations from the chemical engineering community. I don't think we need to be overly critical or ashamed of our background in the coupled transport area.

Second, Gary referred, at least implicitly, to the dichotomy of what he called good science versus useful engineering—the versus might be mine. There is clearly a gap between the two. Yet, in my own experience, I know cases where good science has been put to useful results by enlightened industrial managers who are patient, together with scientists and engineers who know the needs of the application. In fiber spinning, for example, in the design of spinnerettes and of filters ahead of the spinnerettes, there's a wealth of knowledge, much of it proprietary, but much of it based on what I would call good fundamental science. Both the industrial and the academic sides need to strive more than we have in the past to bring the two communities together productively. This may sound strange coming from a person such as myself, but I think we have to ensure that faculty have some acquaintance with industrial practice, more than we have in the last generation. This doesn't mean that if a person works a summer in a plant he or she will go back to his or her university and teach in a more enlightened way. There are plenty of negative as well as positive examples of this, but both sides have to work much harder to join good science with useful engineering. Without that link, we can talk forever about bringing more examples into classroom experience, and the examples will be artificial because the faculty members themselves won't have the application or background that I think is sorely needed.

The last point pervades the fluid mechanics and transport area. It will come up in many of our discussions these two days, and it concerns me. What will be the role of research groups in a university, and how do we link together research accomplishment and graduate training in an optimal way? In chemical engineering, we've done this very well in the past, primarily by a faculty member supervising a relatively small research group, where small would mean a number such as 3, large would be a number such as 8 to 12, and astronomical would mean a number such as 15. From what's been said already this morning, particularly regarding the numerical side versus the physical side, this mode of operation will be increasingly difficult to follow if we're to have research accomplishments. It is asking too much of a graduate student to be simultaneously at the cutting edge of advanced numerics and of the physics behind the problem that the student is applying the numerics to. A solution is to form a group at an engineering research center. If one looks at the rhetoric behind the ERCs, which many of us criticize these days, I think much of that rhetoric is defensible. We need to find ways to encourage group research without losing the master/apprentice relationship that has served us so well in graduate education.

Reuel Shinnar: Let me break out of the mold from the previous discussion and raise a question that's bothered me for a long time and for which I don't have the answer. I was once a fluid dynamicist, and they've never forgiven me for leaving the field. I don't publish anymore in fluid mechanics, but I still need a lot of practical fluid mechanics in my consulting. Look at my field of chemical reactor design. The major problems we face in chemical reactor design are in fluidized bed reactors, entrained bed reactors, and trickle bed reactors. We use very little advanced fluid mechanics in their design.

We mostly use very simple empirical models. If you want to use fluid mechanics to describe these reactors, the state of our knowledge is so far off that we don't know what to do with it. We don't even know how to solve a basic problem such as the stability of flow in a trickle bed. There are instabilities in trickle beds that are quite similar to those in a fluid bed. In fluid beds, nice work has been done on those beautiful bubbles, using rigorous fluid mechanics, but we don't design reactors where these results have any relevance. Real fluidized beds don't have the type of bubbles that we see in those beautiful studies—that are stable and rise up nicely. These are industrially important problems, but if a young faculty member came to me and asked, "Should I work on these problems?" I don't know if I should be encouraging. Despite the importance of the problem, for the faculty member this might be professional suicide. The money isn't there, the support isn't there, and the difficulty in achieving results is substantial. A lot of useful work could be done if we look for better and more scientifically based correlations, even if we can't solve the problem rigorously. The question is, how do we get the support or the interest to do this work? If we want to study a trickle bed, the experiments required are not cheap. The equipment has to be fairly large, and you can't use only water with air. We need data at high pressures with liquids and vapors that more closely simulate those in a real reactor. We also need to be able to study reactions under such conditions. This high cost is a real problem facing the extension of fluid mechanics to such reactors because we now have some other areas relevant to chemical engineering and material science where things are easily definable and experiments easier and cheaper to do, and the support is there. Still, we are leaving out an important area at the heart of fluid mechanics relevant to chemical engineering that is staying in the 1950s, and, as a profession, we should be concerned about it.

Edwin Lightfoot: I think he's right on a number of points. It seems difficult for us to connect theory, which is probably the natural focus of academics, with practical problems. We really haven't solved this. We need, in close contact with our chemical engineering departments, practitioners of chemical engineering with whom we can talk. When a company like Du Pont retires professionals early, those are the people to attract to academia because they would have the background that we might need. I think this is a very pressing need in our university and I suspect in many other places as well.

Stuart Cooper: I have a small comment on something Tony brought up, namely his example of an eroded turbine material. What is important in our training of chemical engineers is that, while a solution to this problem might well exist by analyzing the flow and perhaps redesigning the metal piece itself to avoid vortex formation and the like, quite likely the solution will come from metallurgy and material science, maybe by ion implantation or some technique for hardening the material. Exposure to a multidisciplinary approach is essential in attacking a problem like that.

Louis Hegedus: I was delighted to hear several people, including Bill Russel, Fraser Russell, and Bill Schowalter, emphasize complex fluids and heterogeneous flows. In the processing of heavily formulated specialty chemicals, these are the systems we are dealing with. We have to compound, extrude, mold, grind, suspend, and slurry mixtures of ceramics, catalysts, and various polymers, heavily filled. We have to generate flow-stable suspensions, and we have to mix asphalt and latex and inorganic particles together in a complicated way to achieve the performance that is the end goal of this type of processing. For this work, we find very little to reach back to in fluid mechanic wisdom. To the extent that the chemical industry is increasingly interested in specialty-type fluids and processing, I certainly hope that more fundamental interest is generated in the fluid mechanics of those systems.

Robert Brown: There is a question evolving, and we need to find a solution. How do we teach or how do we introduce students to the idea of working in interdisciplinary areas where they take fundamental transport, thermodynamics, etc. and apply them to problems? How do we achieve that without compromising the basic level of knowledge we want to impart to them?

Eli Ruckenstein: I have read Leal's paper, and I agree with what he has written; indeed, I believe that two of the important directions of research for the future are nonlinear transport phenomena—including the deterministic theory of turbulence—and complex fluids, or fluids with microstructures. The latter direction is at the intersection of a number of fields such as surface chemistry, colloidal chemistry, liquid crystals, and fluid mechanics. Consider, for instance, systems that contain surfactants. Their rheology is very complicated because the surfactant molecules aggregate, and the process of aggregation is affected by the stresses generated by the velocity field. Similarly, in concentrated colloidal dispersions, a number of complex phenomena occur. Changing the nature and amount of dispersant changes the rheological characteristics of the system in major ways, ranging from liquidlike to solidlike. Another problem I find interesting is the coupling of the traditional macroscopic fluid mechanics with phenomena which are relevant on a much smaller scale—on the order of 10^2 to 10^3 Å. The dynamics of wetting and the stability of foams and emulsions involve such a coupling.

My main comment, however, is related to a problem of a different nature. Being from an older generation, I first learned fluid mechanics from papers published in the old chemical engineering journals. I agree that compared to the work of the fluid mechanicists, these contributions might be considered minor. However, from the point of view of our profession, the authors of those papers could be considered pioneers because they contributed to a change in the chemical engineers' mentality. It is difficult to change one's mentality and particularly difficult to change the mentality of others. The outstanding contributions to fluid mechanics by current researchers with chemical engineering backgrounds are partly a result of that change in mentality. It is also fair to say that there are some outstanding contributions in transport phenomena in the old chemical engineering literature even when judged from an absolute point of view. I will start with the 1951 paper of Danckwerts in which turbulent mass transfer was treated in a novel way. A physical model which described the turbulence on the basis of "chemical" intuition has provided insight into the problem. Let me continue with Sternling and Scriven's paper published in 1959 which explained interfacial turbulence as the result of an instability caused by the Marangoni effect. This paper had a tremendous impact not only on chemical engineers but also on physical chemists and physicists. Next is the outstanding work by Roy Jackson in 1963 concerning the stability of fluidized beds and the formation of bubbles in fluidized beds. Another important contribution was the 1966 paper of Metzner, White, and Denn concerning the viscoelastic liquids and the meaning and applications of the Deborah number. Because new physical effects are rarely discovered, let me close my list with the 1968 paper of Marshall and Metzner in which the authors observe that polymeric solutions exhibit a pressure drop in a porous medium greater than the normal one—an effect opposite to drag reduction in turbulent flow. There were many other papers, published particularly after 1960, but my comment is already too long.

Howard Brenner: I'd like to reinforce some of the comments made by Anthony Pearson regarding our current computational ability to solve complex transport problems. As the speaker observed, now that we can solve these problems numerically, the issue is that of providing a physical framework within which these results can be interpreted. Anyone who owns a copy of Bird, Stewart, and Lightfoot knows what are the pertinent conservation and constitutive equations, as well as the relevant boundary and initial conditions governing the various transport processes. While we've known some of these for a long time, we were unable until recently to deal with complex geometries involving many particles or with very tortuous passages, such as are encountered in porous media, or with nonlinearities. Given banks of supercomputers, parallel processors, and other such hardware (and software), we can now solve these problems, at least numerically. This results in exhaustively detailed solutions furnishing knowledge of the temperature, species concentration, velocity, pressure, etc., at every single point of these complex systems, at every single instant of time.

In this specific sense, we thereby know everything there is to know about this system—precisely and exactly. In another sense, however, we don't understand anything about the system. That has to do philosophically with the difference between knowing and understanding. Exquisitely detailed "microscale" (i.e., pointwise and instantaneous) data emanating from such numerical computations—namely, the "knowing" aspect of the analysis— needs to be interpreted and placed in some intuitive context that human beings, particularly engineers, can understand and hence utilize. Engineers generally think and reason macroscopically; that is, they are usually interested only in broad, generic concepts and applications thereof. The kind of detailed information issuing from computers is thus only of limited value in its raw form. This fact has, of course, always been appreciated, especially in graphical and other visual modes of data presentation.

Before the advent of high-speed computers, one could solve only the most elementary problems, generally linear, and then involving only the simplest geometries (plane, cylinder, sphere). Nevertheless, this often furnished valuable "understanding" and, concomitantly, physical insights into the phenomenon under study. Now, ironically, we're threatened with losing some of this understanding by virtue of our fantastic ability to compute! Students who focus on the acquisition of computational and related computer-graphic skills don't have the cultural advantages of having been there at the outset. In particular, in the course of learning to appreciate the obvious advantages of this new-found tool, they often fail to recognize its intellectual disadvantages! We need somehow to educate them to recognize that the elegant, multicolor computer-graphic outputs that issue from these machines, along with the obvious aesthetic appeal of seeing such abstractions as streamlines, may have virtually no engineering utility to them whatsoever, unless they also learn what to do with this information. That may be the hardest thing to teach them!

There needs to be some countergroup that debates these issues with computationally mesmerized students, that tells them about the other side of things, and that also works conceptually to develop a macroscale context in which to embed the microscale information generated by computers. Various groups are, of course, working on this. Anthony Pearson is one of the leaders of this counterforce. His contacts with complex polymer processing and multiphase porous media transport problems have endowed him with special insights into the limitations of pure computing in the transport field. His criticisms need continuously to be made— perhaps with an emphasis that increases in intensity with time at a rate at least as fast as our computational speed increases.

Morton Denn: Let me make two quick reinforcing comments. Bill emphasized that we don't want to downplay our successes in using fluid mechanics in applications. I think one of the great successes is what Anthony Pearson has done in his two books on polymer processing, in which very elegant mathematical analyses of fluid mechanics are applied in a very practical way to real processes.

The issue of fluids with microstructure keeps coming up. We should simply take as a given that that is an essential part of modern chemical engineering. There are at least 12 people in this room whose research is primarily concerned with the mechanics of fluids with microstructure. Clearly we're there, and we're going to be there for quite some time. The industrial applications are obvious.

What struck me most about Gary's paper is an educational issue that none of us have thought enough about. How do we incorporate the obvious importance—although it wasn't so obvious to me until I read his paper— of modern, nonlinear dynamics in our curriculum? There are very few institutions right now that I know of in which even graduate students are exposed to modern nonlinear dynamics unless they take specialized courses. Yet it's clear that this area is at the heart of much of the research that we'll do and that our students will need in order to read the literature. We don't teach nonlinear dynamics at Berkeley, and I'd like to know how other people are doing it.

James Douglas: Julio Ottino at the University of Massachusetts has just completed a book on the subject that should make this material easier to teach. The title of the book is *The Kinetics of Mixing: Stretching, Chaos and Transport*, and it's published by Cambridge University Press.

Robert Brown: There are a number of departments that have courses, but they're not integrated into the required curriculum of graduate course work.

Sheldon Isakoff: Gary posed some interesting questions at an abstract level regarding good science versus useful engineering. He raised the question, are we teaching a realistic view of the world to our students? One key issue to focus on as we apply our science and mathematics in our model building is embodied in the word "usefulness." I mean usefulness in the sense of making progress on problems in the real world that have economic significance. The real world is very complicated, as I'm sure you all know. The problems one faces in industry generally do require multidisciplinary approaches; they don't fit neatly into fluid mechanics or any other single discipline. The real key is that our work must be useful. If one can generalize by his research results, then so much the better. But I wouldn't be at all discouraged if one makes incremental progress through the kind of research discussed here and has to set bounds on the generality or applications of the resulting theory. There's nothing wrong with that, provided that the results one gets can be applied to real problems economically and in a useful fashion. To me that is the core of what chemical engineering is about.

Dan Luss: I want to expand on the point raised by Mort Denn and by Gary Leal concerning the recent advances in nonlinear dynamics. This point concerns research in fluid dynamics as well as in other areas such as reaction engineering. The newly developed paradigm of nonlinear science will certainly have a major role in future research and lead to many technological applications. The new tools and methodology enable the solution of many problems that were considered unsolvable in the past. The major problem is that the understanding and use of the tools require a mathematical background and sophistication that most chemical engineers do not have. The educational dilemma that we face is how to train an engineering student to be an expert researcher in this area as well as in the traditional chemical engineering subjects.

Robert Brown: Dan, your comment relates to a comment by Anthony Pearson. He said we don't want to put out computer scientists for people who are doing numerical simulation. We don't want to create mathematicians.

Dan Luss: How do you want to solve the problem?

Robert Brown: With a team of people who want to solve the problem. I think that's the key.

Gary Leal: There's a difference between creating new math and using it.

I want to summarize what I've gotten out of the discussion. Several comments referred to either inter- or multidisciplinary activities, particularly how to respond to the need for broadening of the knowledge necessary for our students, the need to understand more detailed scientific or mathematical questions, and the need to understand the practical framework in which their work is being done. I've heard four ideas that we might want to expand upon. First, we need to do a better job of making the connection with the practical world. Ed Lightfoot suggested that we should bring people with practical experience into the university environment. That sounds like a good idea in a sense, but it will require some major changes in thinking at the university level, not at the chemical engineering department level, about what it takes to be successful in the university environment.

Second, Bill Schowalter suggested that we might have to adopt a different mode of research, with larger groups of more specialized people, rather than the smaller individual investigator groups with one, or a few, students working on a problem that may require a great deal of peripheral knowledge of numerical methods and the like to make progress. In that regard, we must carefully balance the goals of education versus research productivity. Having a group of specialized people all working on separate parts of a problem may result in an unsatisfactory situation, that is, a master with an overview of where the whole project is going, maybe, and a lot of people working on the pieces without the overview. From an educational point of view, one has to ask where will the students make the transition from level one, defined by Jim Wei, to level two or, maybe eventually, to level three? Our present mode of operation with a small group of students has worked very well; the students not only are involved in the detailed research but also have some insight about how their work fits into some bigger picture. We may lose that perspective in this new model.

Third, Ed Lightfoot also suggested that we might have to split disciplines into subdisciplines because of peculiarities that exist. He was referring to biotransport, in particular. That's reasonable only if the bio area is particularly peculiar, more peculiar than other specialty areas to which one might want to try to apply transport principles, or something else. That may be true, but I'm not convinced.

Finally, we've heard people talk about how to modify curriculum or teaching. How do we transmit information that students need to understand, such as the context in which they're working or the resources that are available to apply? It's fine to say we're going to work in a group or we're going to study interdisciplinary topics. The first step toward that is knowing what resources are available to use. We can make use of science and mathematics and bring all these concepts to bear on our problems if we know they exist and what their potential is, if we know the questions to ask and the right people to approach. At least that part must somehow be incorporated into the education.

Kenneth Smith: Jimmy opened by challenging us to be impolite, and I think I intend to oblige. I've been terrified by most of this discussion. Most of what's been said is that all we've got to do is synthesize a little better, subdivide a little better, teach a little better, integrate a little better. It all says there's nothing wrong. It all says that there isn't anything we can't do, that there isn't any major challenge that would make a big difference if we did a little better. That's a pretty damning statement about this aspect of our profession if that in fact is true. What are the big challenges, what are the big things we can't do? Some of the challenges were implied in Anthony's discussion, but most of what we've said was that things are terrific, we can really do almost anything we want now. I hope that's not true. If we are uncoupled from important phenomena, if there aren't phenomena that we simply do not understand, I'm afraid most of the fun's out of it. Maybe we're in the fourth stage, Jim.

James Wei: What's that?

Kenneth Smith: Senescence.

Edwin Lightfoot: Ken, you made a good point. Most people involved in areas like transport and fluid mechanics have implicitly assumed that we have a complete model for our system. For the really interesting problems we don't have a complete model, and we haven't come to grips with this. I'd like to cite one quote from Olaf Hougen from almost three decades ago: "As we get more and more involved in the science of chemical engineering, we get more and more involved in the trivia of chemical processing." We've got to put practical impact first, rather than elegant models that we can deal with in a complete way. That's something we have to face up to more than we have in the past.

Robert Brown: Before we break, the organizers would like to say we're off to a very good start. We probably have to get a little nastier to get up to Ken's standards.

SECTION III
Thermodynamics

7

Thermodynamics

Keith E. Gubbins
Department of Chemical Engineering
Cornell University
Ithaca, New York

I. Introduction

For the past 50 years, research in chemical engineering thermodynamics has been largely concerned with problems related to the oil and petrochemical industries. The primary methods of attack have been direct experiment and macroscopic thermodynamic treatments of the data. Examples of the latter are empirical equations of state or activity coefficient expressions, group contribution methods, and macroscopic corresponding states correlations. These methods provide a convenient way to correlate large amounts of experimental data, interpolate between different state conditions or between different but structurally similar molecules, and make minor extrapolations from experimentally measured conditions. They are most successful for fluids of relatively simple molecules, such as the constituents of natural gas or of low-molecular-weight hydrocarbon mixtures. However, they offer little insight into the relation between the desired properties and the underlying intermolecular forces or molecular structure and so are of little predictive value. They will be much less useful for many of the new technologies, such as the processing or design of electronic, photonic, or ceramic materials, for biochemical processes, or for predicting the behavior of matter at and near surfaces (e.g., in micelles, porous materials, or thin films).

ADVANCES IN CHEMICAL ENGINEERING, VOL. 16

A molecular approach based on statistical mechanics, combined where possible with experiment and molecular simulation methods, will be a necessary starting point for many of the most challenging problems of the present and future. The rapidly increasing power of computer simulations based on molecular modeling will play an important role in many applications, and an interdisciplinary approach will be essential for the new technologies. In this paper, I shall focus on the present and future status of the molecular theory and simulation approaches, rather than direct experiment. This emphasis has been chosen because molecular theory and simulation seem to me the areas that will move most rapidly in the next few years; while experiment will continue to play a central role, the field of thermodynamics has not been dominated by new experimental techniques in the way that surface science, materials, and biotechnology have, for example. I first provide a brief overview of the methods that are most likely to prove useful for chemical engineering applications (Sect. II), followed by some examples of current problems (Sect. III). Finally, some areas that are likely to be important in the future are considered, together with implications for the teaching of chemical engineering thermodynamics (Sect. IV).

II. Modern Methods

Thermodynamic or transport properties can be studied by experiment, macroscopic correlations, molecular theory, or computer simulation (Fig. 1). In both the theory and simulation methods, one must specify an equation for the intermolecular potential energy and then calculate the macroscopic properties using the laws of statistical mechanics. The equations involve N-body integrals (where N is the number of molecules) over functions of the intermolecular potential energy which can be solved exactly in only a few special cases (e.g., the ideal gas or gas adsorption at low pressure). The theoretical methods therefore involve approximations to make the equations tractable. The simulations, on the other hand, solve the full N-body equations by numerical methods; if sufficiently long computer runs are carried out, the results should be exact for the model substance studied. A comparison of the simulation and theoretical results then provides a direct test of the approximations used in the theory, without any uncertainty as to the form of the potential. Thus, an important use of the simulations is to evaluate new theories; those theories that meet this test can then be compared with experimental data for real substances. If one neglects the test against simulation and proceeds to compare theory directly with experiment, it is generally necessary to fit parameters in the potential (and often the theory) to experimental data; this fitting often obscures defects in the theory, and much

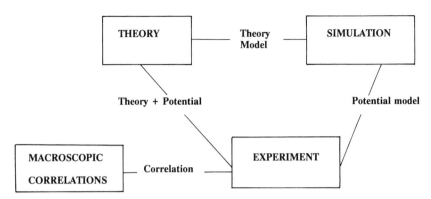

Figure 1. The four methods for studying physical properties and what may be learned
from a comparison of any two of them.

time can be wasted. The same difficulty arises with testing the macroscopic
correlations. An example is provided by the large amount of work done on
molecular theories of liquid mixtures (various 1-, 2-, and 3-fluid confor-
mal solution theories, cell theories, random mixture theory, etc.) in the period
1936–1970, before the first simulations of mixtures were reported. Although
comparisons of theory and experiment had indicated good agreement, comparisons with computer simulations showed that most of the theories were
quite inaccurate, in some cases not falling on the same piece of graph paper
as the experimental results. A second important use of simulations is to evaluate intermolecular potential models by making comparisons with experimental data; in this case, the statistical mechanics is exact in both simulation
and experiment, so the only source of error is the potential model assumed
in the simulation. In some important applications, computer simulation may
provide the best way to obtain a detailed understanding of the molecular
behavior because suitable experiments cannot be devised at the present time;
examples are the breakdown of some classical thermodynamic equations
for small drops, the study of phase transitions in narrow pores, and the
dynamics and thermodynamics of protein folding.

A. Macroscopic Correlation Methods

These methods make use of available experimental knowledge combined
with correlation techniques that may or may not be partially based on
molecular concepts. They range from methods that are completely empirical
to ones in which there is significant input from statistical mechanics (see
Fig. 2). As an example of an approach that is almost fully empirical, one
can cite the many modifications of the Benedict-Webb-Rubin equation of

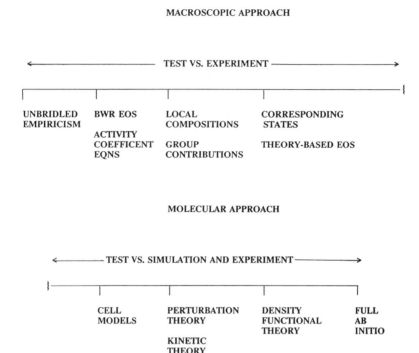

Figure 2. Examples of macroscopic and statistical mechanical methods for studying physical properties. The degree of molecular basis increases from left to right.

state, which involves expansions in density and temperature, with one or more nonlinear terms often added for good measure. For a fluid of spherical Lennard-Jones or argon molecules, it is found that at least 34 adjustable constants are needed to obtain a fit to existing data over a reasonably wide range of temperatures and pressures. To fit more complex fluids such as water, many more terms are needed, and the equation still fails in the critical region. Such methods are useful when one wants to fit a large amount of data for a pure fluid; however, they cannot be extended to mixtures in any systematic way and are unreliable even for phase regions slightly outside the range of fit.

Many of the macroscopic methods incorporate some features from molecular theory and are then usually more useful for prediction and extrapolation. Examples are equations of state of the modified van der Waals type (Redlich-Kwong, etc.), local composition methods (which attempt to account for the fact that the composition around a molecule of a particular species differs from the mean composition of the mixture), group contributions (which approximates the intermolecular forces as a sum of group-group interactions), and corresponding states. A common feature of all of these methods is that the desired macroscopic properties are not related explicitly to some expression for the intermolecular forces, in contrast to the molecular theories described in the following section. Because of this, it is not possible to test the macroscopic methods against simulation results, but only against actual experimental data. Since much fitting to this data has often been involved, such comparisons are usually a weak test.

Before leaving these methods, we should note that it is often possible to make a compromise between the molecular and macroscopic approaches by starting from a sound molecular theory and adopting approximations to obtain a semiempirical correlation that does not involve the intermolecular potential or molecular correlation functions. Many examples occur in the chemical engineering literature, and I mention only one, taken from recent work by Bryan and Prausnitz [1] on developing an equation of state for polar fluids. They write the equation for the compressibility factor $Z = PV/RT$ in the usual way as $Z = Z_{ref} + Z_{pert}$, but in place of the hard-sphere reference term they take Z_{ref} to be the equation of state for polar hard spheres, for which an accurate statistical mechanical theory (i.e., one that agrees closely with the computer simulation results for polar hard spheres) exists. Such a reference fluid is much closer to the real fluid of interest than a hard-sphere one, so that the perturbation term Z_{pert} is much smaller. Bryan and Prausnitz used the mean field term of van der Waals, $Z_{pert} = -a/RTv$, where a is the van der Waals attraction constant and v is the molar volume. This approach gives good results for polar fluids (see Fig. 3), in contrast to the many attempts to doctor the perturbation term while retaining the traditional hard-sphere reference.

B. Molecular Theory

The configurational part (the part involving the intermolecular forces) of the Helmholtz free energy for a system of N molecules in volume V at temperature T is given by

$$A_c = -kT \ln \left(\int \cdots \int dr^N d\omega^N \exp(-\frac{U}{kT}) \right) \tag{1}$$

where $dr^N = dr_1 dr_2 \cdots dr_N$, $d\omega^N = d\omega_1 d\omega_2$, U is the intermolecular potential energy (which depends on all the r_i and ω_i), and the integrations over

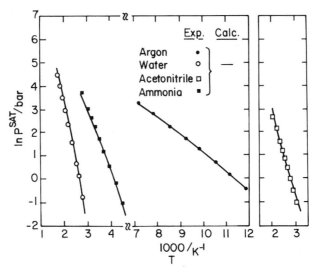

Figure 3. Calculated and observed vapor pressures for argon and several polar fluids
from an equation of state that uses a reference fluid of polar hard spheres, for
which an accurate statistical mechanical theory exists. Reprinted from Bryan
and Prausnitz [1] with the permission of Elsevier Science Publishers.

r_i are over the volume V of the system. Here r_i is the position of the center of molecule i, and ω_i ($= \vartheta$, ϕ for linear or ϕ, ϑ, χ for nonlinear molecules) is its orientation relative to some space-fixed set of axes. Equation (1) is valid for nonflexible molecules in which translational quantum effects are negligible. The integrations can be easily carried out for noninteracting molecules (e.g., ideal gases), moderately dense gases, or crystals, but for most other cases approximations are necessary. The principal theories are [2] corresponding states theory, perturbation and cluster expansions, density functional theory, lattice models, and integral equation theory.

The first three of these theories are of particular interest in chemical engineering. The molecular principle of corresponding states is based on the idea that one can identify a group of substances, all of which obey a single intermolecular potential law; they differ only in the values of the potential parameters. It provides the foundation for many existing correlations of thermodynamic and transport properties, but these can be expected to work only for groups of similar substances such as simple inorganics or low-molecular-weight hydrocarbons.

The perturbation theories [2, 3] go a step beyond corresponding states; the properties (e.g., A_c) of some substance with potential U are related to those for a simpler reference substance with potential U_0 by a perturbation expansion ($A_c = A_0 + A_1 + A_2 + \cdots$). The properties of the simple reference fluid can be obtained from experimental data (or from simulation data for model fluids such as hard spheres) or corresponding states correlations, while the perturbation corrections are calculated from the statistical mechanical expressions, which involve only reference fluid properties and the perturbing potential. Cluster expansions involve a series in molecular clusters and are closely related to the perturbation theories; they have proved particularly useful for moderately dense gases, dilute solutions, hydrogen-bonded liquids, and ionic solutions.

Density functional theories [2, 4] are similar in spirit to the perturbation theories, but are of particular value for problems involving nonuniform systems in which the density (or molecular orientation) varies with position (direction) in the system—surface phenomena, solidification and melting, thin films, liquid crystals, polydisperse systems, and so on. In this approach, one starts from the fact that the free energy density $a(\mathbf{r})$ at some point $\mathbf{r}$ in the system is a functional of the density profile $\rho(\mathbf{r}')$. Using variational methods, one finds the global minimum of the free energy density with respect to the density profile and so determines the density profile itself. The success of the method hinges on the accuracy of the expression used for the free energy density functional $a(\mathbf{r})$. Usually this is written as a sum of repulsive and attractive force contributions, the two terms being treated by different approximations. Several versions of the theory exist, differing in the approximations used for the repulsive term in $a(\mathbf{r})$, and these have been successfully applied to problems in adsorption, micelles, fluids near charged walls, and melting. This approach is likely to prove valuable for the study of many of the interfacial problems to be met in the new technologies associated with electronic and microstructured materials, thin films, etc.

Lattice models were used extensively to describe fluids from the 1930s to the 1970s, but really describe solids rather than fluids, and they have been superseded for most applications by the more sophisticated theories described above. An exception is the study of the critical point, a singular point in the phase diagram where conventional mean field and perturbation theories fail. The Ising model, a lattice theory that can be solved essentially exactly, can be used successfully in this region and, together with some related cell theories that can be mapped onto the Ising model (e.g., the decorated lattice gas model), gives valuable information on the equation of state in the critical region.

Integral equation methods provide another approach, but their use is limited to potential models that are usually too simple for engineering use and are moreover numerically difficult to solve. They are useful in providing equations of state for certain simple reference fluids (e.g., hard spheres, dipolar hard spheres, charged hard spheres) that can then be used in the perturbation theories or density functional theories.

C. Computer Simulation

In molecular simulation [5, 6], one starts from a molecular model and an equation for the intermolecular forces and calculates the macroscopic properties by a numerical solution of the equations. Two methods are in common use, the Monte Carlo and molecular dynamics techniques (Fig. 4). The Monte Carlo (MC) method makes use of a random number generator to "move" the molecules in a random fashion. Statistical mechanics tells us that, for a fixed temperature and density, the probability of a particular arrangement of the molecules is proportional to $\exp(-U/kT)$, where U is the total energy of the collection of molecules, k is the Boltzmann constant, and T is the temperature. In MC, the random moves are accepted or rejected according to a recipe that ensures that the various molecular arrangements that are generated appear with probabilities given by this law. After generating a long series of such arrangements, they can be averaged to obtain the various equilibrium properties of the system of molecules.

These two techniques have several features in common. Accurate results can be expected, provided that the simulation runs are carried on long enough and that the number of molecules is large enough. In practice, the results are limited by the speed and storage capacity of current supercomputers. Typically, the number of molecules in the sample simulated can range up to a few thousand or tens of thousands; for small molecules, the real time simulated in MD is of the order of a nanosecond.

In order to minimize boundary effects in such small samples, it is customary to use periodic boundary conditions; that is, the sample is surrounded on all sides by replicas of itself, so that when a molecule moves through a boundary and so out of the sample, it is automatically replaced by a molecule moving into the sample through the opposite face of the box. (Anyone who has played Pacman, Asteroids, or similar video games is familiar with this periodic boundaries trick.) Although molecular simulation can be successfully applied to a wide range of problems, difficulties can arise with some applications because of storage or speed limitations. Examples include ionic fluids, such as plasmas and electrolytes, in which the range of the intermolecular forces is very large, necessitating a large number of molecules. Difficulties also arise with substances in which long-range fluctuations

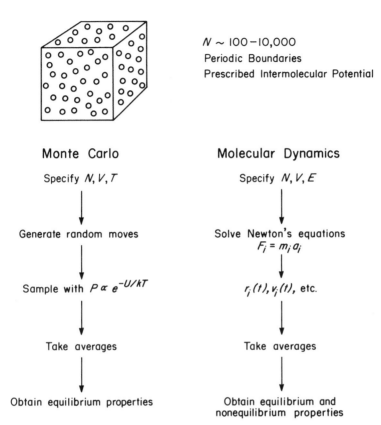

$N \sim 100 - 10,000$
Periodic Boundaries
Prescribed Intermolecular Potential

Monte Carlo	Molecular Dynamics
Specify N, V, T	Specify N, V, E
↓	↓
Generate random moves	Solve Newton's equations $F_i = m_i a_i$
↓	↓
Sample with $P \propto e^{-U/kT}$	$r_i(t), v_i(t)$, etc.
↓	↓
Take averages	Take averages
↓	↓
Obtain equilibrium properties	Obtain equilibrium and nonequilibrium properties

Figure 4. Two methods of molecular simulation. Typically, both treat a sample of N molecules in a box of volume V (here shown as being of cubic shape). In the Monte Carlo method it is common, but not necessary, to choose N, V, and temperature T as the independent variables and to keep these fixed during the simulation. Molecules are moved randomly, generating a new molecular arrangement with the intermolecular potential energy U. These new arrangements are accepted or rejected in such a way that those accepted occur with the probability distribution that is required by the laws of statistical mechanics. In molecular dynamics the energy E, rather than the temperature T, is fixed. The molecular positions r and velocities v (and also orientations and angular velocities) for each molecule i are followed in time. In the molecular dynamics (MD) method, the molecules are allowed to move naturally under the influence of their own intermolecular forces. The positions and velocities of each molecule are followed in time by solving Newton's equation of motion (force equals mass times acceleration, a second-order differential equation) using standard numerical methods. The macroscopic properties are calculated by averaging the appropriate functions of molecular positions and velocities over time.

occur; these are associated with substances very near critical points and those exhibiting certain surface phenomena. Long-time phenomena are also apt to make simulation difficult.

There are not only common features but also significant *differences* between the MC and MD methods. MC is easy to program and can be easily adapted to different conditions; it is adaptable, for example, to mixture studies at constant pressure or to adsorption at constant chemical potential. MD is more difficult to program and, in its conventional form, energy must be conserved, providing a less convenient set of state variables to work with for some applications. However, it is now possible to overcome this problem and to carry out MD calculations at constant temperature or pressure [5, 6]. MD has two important advantages over MC; it can be used to study time-dependent phenomena and transport processes, and the molecular motions are natural and therefore can be observed and photographed easily using computer graphics.

Several specialized simulation techniques exist for particular applications [5]: nonequilibrium MD for the study of transport properties and nonlinear response, Brownian dynamics for the study of large molecules (e.g., proteins) in solution, and quantum simulations for the study of non-classical fluids and solids. Simulated annealing is a Monte Carlo technique for optimization subject to a set of constraints and is finding widespread use in design of chemical processes, circuit design (especially VLSI), image processing, and protein engineering (see Sect. IV. A).

D. Computer Graphics

A single molecular simulation provides a vast amount of detailed information—typically 10^8 coordinate positions and an equal number of orientations and linear and angular velocities. Even after averaging to obtain macroscopic properties, the amount of molecular data (spatial, angular, and time correlation functions, diffusion rates, and so on) is difficult to assimilate from graphs or tables. Computer graphics provides a way to present such large arrays of data and is particularly useful in visualizing such physical processes as nucleation and phase separation, molecular motion near a surface, adsorption and hysteresis in porous materials, and the folding of proteins. In the simulation of very large molecules or materials, computer graphics can help decide which structures or motions are important and which functions would best characterize them. The ability to rotate the figure and to zoom onto regions of particular interest is a great help in viewing such processes. Although computer graphics is a rapidly advancing area, several problems remain. The graphics workstations commonly available to university research groups are still too slow to represent molecular motions or rotations of the system in real time for most problems, and one must be satisfied to view a series of still shots, or make a movie of such a series to see the motions. Moreover, the software needed for such displays must usually be prepared locally and is not easily portable to different hardware.

E. Intermolecular Potentials

A key question about the use of any molecular theory or computer simulation is whether the intermolecular potential model is sufficiently accurate for the particular application of interest. For such simple fluids as argon or methane, we have accurate pair potentials with which we can calculate a wide variety of physical properties with good accuracy. For more complex polyatomic molecules, two approaches exist. The first is a full *ab initio* molecular orbital calculation based on a solution to the Schrödinger equation, and the second is the semiempirical method, in which a combination of approximate quantum mechanical results and experimental data (second virial coefficients, scattering, transport coefficients, solid properties, etc.) is used to arrive at an approximate and simple expression.

The *ab initio* approach, while holding great promise for the future, suffers from several difficulties at present. First, the calculations are very lengthy, requiring long runs on very fast machines and massive data storage. The computing power needed increases as N^4 for Hartree-Fock calculations (zeroth-order perturbation theory), where N is the number of electrons per molecule; the rate of increase with N is even greater if electron correlations past the Hartree-Fock level (i.e., higher-order perturbation terms) are included, up to N^7 or N^8, so that the calculations become rapidly more difficult as molecular size increases. Also, if the calculation is done for just two molecules, the resulting pair potential takes no account of the many-body forces that exist in dense gases or liquids; these are often significant for dense fluids, particularly when induction forces (due to molecular distortion in the presence of an electric field, e.g., due to a neighboring polar molecule or ion) are present. These multibody forces can be included in the *ab initio* calculations by increasing the number of molecules present in the cluster, but at the cost of greatly increased computer time and storage. Among the many pair potential models proposed for water are several that come from *ab initio* calculations. None of these are able to predict a wide range of physical properties with the accuracy required for engineering calculations at the present time. For example, the MCY (Matsuoka-Clementi-Yoshimine) and CC (Caravetta-Clementi) potentials both predict pressures at specified temperatures (or energy) and density (or volume), that are considerably higher than the experimental values when used in computer simulations, although they do predict correctly many of the anomalous properties of water [7]. For molecules with more electrons than water, the situation is generally worse.

Because of the present difficulties facing the *ab initio* calculations, most workers in this field use semiempirical expressions for the intermolecular potentials. Among the older and more familiar ones are the Lennard-Jones and Stockmayer models for spherical nonpolar and polar molecules, respectively, and the Kihara model for molecules of nonspherical shape. More recent developments have included group contribution models in which the pair potential is made up of a sum of interactions between sites in the molecules; the sites may be the nuclei of the atoms, groups of atoms (CH_3, NO_2, etc.), lone pair electrons, or bonds, and it is often assumed that such site interactions are the same for different molecules having the same groups. Several versions of this idea now exist and have been applied to large organic molecules, particularly proteins and polypeptides. Examples include the TIPS (transferable intermolecular potential functions) model, which works well for water, alcohols, ethers, etc.; a modified form of TIPS is the OPLS (optimized potentials for liquid simulations) model, in which the sites are taken to be the nuclei or CH_n groups [8]. Other examples of this approach, developed mainly for conformational energy studies of proteins or polypeptides, but also applicable to a wide range of organics, include the ECEPP

Table 1. Liquids Simulated with OPLS Potential Functions[a]

Liquid	T (°C)	Liquid	T (°C)
$HCONH_2$	25	pyrrole	25
$HCON(CH_3)_2$	25,100	pyridine	25
$CH_3CONHCH_3$	100	CH_4	-161
CH_3OH	25	C_2H_6	-89
C_2H_5OH	25	C_3H_8	-42, 25
n-C_3H_7OH	25	n-C_4H_{10}	-0.5, 25
i-C_3H_7OH	25	i-C_4H_{10}	25
t-C_4H_9OH	25	n-C_5H_{12}	25
CH_3SH	6	i-C_5H_{12}	25
C_2H_5SH	25	neo-C_5H_{12}	25
$(CH_3)_2S$	25	c-C_5H_{10}	25
C_2H_5SCH	25	n-C_6H_{14}	25
$(C_2H_5)_2S$	25	$CH_3CH_2CH=CH_2$	25
CH_3SSCH3	25	t-$CH_3CH=CHCH_3$	25
$(CH_3)_2O$	-25	c-$CH_3CH=CHCH_3$	25
$C_2H_5OCH_3$	25	$(CH_3)_2C=CH_2$	25
$(C_2H_5)_2O$	25	benzene	25
THF	25	$CH_3CO_2CH_3$	25

[a] From Jorgensen and Tirado–Rives [8].

[9], CHARMM [10], AMBER [11], and DISCOVER [12] models. These models differ in how they treat the intramolecular modes (some treat the bond lengths as fixed, while others allow them to vary, for example) and the kind of data (solid or liquid) used to fit parameters.

We briefly consider only the OPLS model, since it has been used to study organic liquid properties. In OPLS, the sites are taken to be point charges and/or Lennard-Jones centers placed on the nuclei or CH_n groups, and the total potential is a sum of Coulombic and Lennard-Jones terms. Such a potential approximately reproduces the shape and the electrostatic and dispersion forces. The necessary potential parameters for neutral molecules are obtained by fitting directly to liquid property data, which requires numerous simulations. For ions the parameters are derived by fitting to structural and energetic results from *ab initio* calculations on ion-water complexes. Some results of computer simulations using this potential model for the 36 organic liquids listed in Table 1 are shown in Figs. 5 and 6. The model works quite well for a variety of aliphatic and aromatic hydrocar-

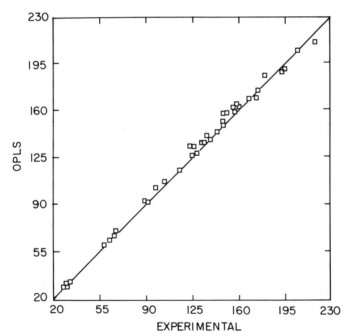

Figure 5. Comparisons of calculated and experimental volumes per molecule in $Å^3$ for the liquids in Table 1 and TIP4P water. Calculated values are from computer simulation using the OPLS potential method. Reprinted with permission from W. L. Jorgensen and J. Tirado-Rives, *J. Am. Chem. Soc.* **110**, 1657 (1988) [8]. Copyright 1988 American Chemical Society.

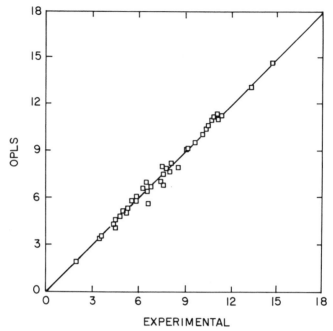

Figure 6. Comparison of calculated (computer simulation results for the OPLS model) and experimental heats of vaporization in kcal mol^{-1} for the liquids in Table 1 and TIP4P water. Reprinted with permission from W. L. Jorgensen and J. Tirado-Rives, *J. Am. Chem. Soc.* **110**, 1657 (1988) [8]. Copyright 1988 American Chemical Society.

bons, alcohols, amines, sulfur compounds, ethers, and so on, and has been successfully applied to protein crystal structures and energy minimization [8].

The models described so far use *isotropic* site-site interactions. They give a good description of the molecular shape, but in most cases neglect the rearrangement of the valence electrons that occurs on bonding. This shift of the electrons into bonds, π orbitals, and lone pairs has a significant effect on the intermolecular forces, and its description requires the use of *aniso-tropic* site-site interactions [13]. The electrostatic interactions between molecules can be qu ite accurately described by using a series of sites within the molecule (i.e., the nuclei of the atoms or CH_n groups), each of which interacts with sites on neighboring molecules with multipole forces (point charge, dipole, quadrupole, and so on); an approach called distributed multipole analysis [13, 14] can be used to determine the best location of the sites and the multipole moments needed from *ab initio* calculations. Such anisotropic site-site potentials have been successfully applied to a variety of inorganic and aromatic molecules and give distinctly better results than the isotropic site-site models. It is often possible to use fewer sites per molecule, so that computational times may not be significantly greater than with the simple isotropic site-site models. Also, the anisotropic atom-atom potential can be more confidently transferred to other molecules where it has a similar charge density.

III. Current Problems: Some Examples

Some current research areas using the methods described in the previous section are shown in Table 2. Many of these involve molecular simulations that exploit the limit of speed and storage of currently available supercomputers [6]. In this section I shall consider two examples from among these: (1) the determination of phase equilibria by computer simulation and (2) the behavior of fluids in microporous materials.

A. Phase Equilibria of Complex Systems

An understanding of phase equilibria is of fundamental importance in a wide variety of chemical processes, as well as in such diverse areas as biology,

Table 2. Some Current Research Areas

Continuous mixtures
Phase equilibria of mixtures
Polar and associating liquids
Electrolyte solutions
Solvation, folding of biological molecules
Gas hydrates (clathrates)
Polymers, advanced materials
Micelles, colloids, vesicles
Adsorption problems
Surfactants
Fluid behavior in porous materials
Nucleation
Thin films (Langmuir-Blodgett, etc.)
Chemical equilibria
Polydisperse fluids
Zeolites

geology, and the study of planetary atmospheres. Since the range of possible compositions, temperatures, and pressures that are met in practice is enormous, it is not feasible to carry out experiments for more than a small fraction of the systems of interest, and there is therefore much benefit to be gained from developing computer simulation and theoretical prediction methods. Some approximate estimates of the cost and time needed for such calculations and experiments at the present time are shown in Table 3. Theoretical and empirical correlation methods are satisfactory only for rather simple mixtures at the present time. For more complex fluids, simulation or experiment is more reliable. Estimates of the cost and time for simulations are strongly dependent on the complexity of the molecular model and the accuracy desired and thus are difficult to make. At present, the simulations are cheaper than experiments for simple mixtures but are still relatively expensive for complex fluids, e.g., hydrogen-bonded ones; these costs will decrease as faster machines become available. The simulations are in general considerably faster than the experiments.

It is not straightforward to calculate phase equilibria or chemical potentials in a simulation, and special techniques must be used. This has been an active research area over the past few years [15]. Several methods have been proposed, and these can be divided into *direct* methods, in which the coexistence properties of the phases are calculated directly, and *indirect* methods, where the chemical potential is first calculated and then used to determine the phase equilibrium conditions (Table 4).

The most straightforward direct method is to simulate a two-phase system and allow it to equilibrate. This approach is valuable for studying the properties of the interface but is less satisfactory for determining the properties of the bulk phases themselves because of slow diffusion across the interface; the results are also sensitive to the interfacial area. An important new development is the Gibbs ensemble Monte Carlo method [16], a direct method that avoids the interfacial diffusion problem and is much faster than simulating the two-phase system, especially for mixtures. The method is illustrated in Fig. 7 and involves setting up two homogeneous phases (I and II) that are in thermodynamic equilibrium but not in physical contact. Equilibration is achieved by allowing changes in the volumes and number of molecules in each phase (keeping the total volume and number of molecules for the two-phase system constant), together with the usual Monte Carlo moves of molecules in each box to obtain equilibrium. Some typical results for mixtures are shown in Fig. 8. Agreement with results from other (indirect) methods is good, and the Gibbs method offers a great improvement in speed because no interface is involved. Typically, for a binary mixture of 600 molecules, the time required is five CPU minutes per mil-

Table 3. Approximate Costs of Binary Vapor–Liquid Equilibria
Determination[a]

Method	Cost	Time
Redlich-Kwong eq.	$10	0.1 hour
Perturbation theory	$100–1000	1 hour

[a] The figures for the Redlich-Kwong equation and experiment are
taken from B. Moser and H. Kistenmacher, Fluid Phase Equilibria,
34, 195 (1987). The estimated costs and times for computations are
for a currently available supercomputer and are for a complete binary vapor– liquid phase diagram with about four or five isotherms.

lion molecular configurations on a Cray XMP supercomputer. The Gibbs
method can also be applied to equilibria in membranes or pores.

The major indirect methods [15] are (1) the test particle and grand canonical ensemble methods, which yield a chemical potential (μ) value in
a single simulation at the state point of interest, and (2) thermodynamic
integration, where one carries out a series of simulations at different state
points (e.g., different densities along an isotherm) and uses the standard
relations of classical thermodynamics to integrate along a path to get μ at
the point of interest. The test particle method involves using an imaginary
test molecule (a ghost-like observer molecule) to measure the intermolecular
potential U_t due to the other (real) molecules; an exact equation relates μ
to an integral over $\exp(-U_t/kT)$, so that attractive molecular configurations
(negative U_t) will be the dominant contributions to μ. In the grand canonical
method, one uses as independent variables μ, V, and T and calculates the
density in the simulation. This method is useful near critical points, e.g.,
in studies of supercritical extraction (an example is shown in Table 5), since

Table 4. Methods for Determining Phase Equilibria by
Computer Simulation[a]

Direct Methods	Simulate 2-phase system
	Gibbs ensemble Monte Carlo
Indirect Methods	Test particle method
	Grand canonical ensemble method
	Biased sampling methods
	Thermodynamic integration

[a] From Abraham [6].

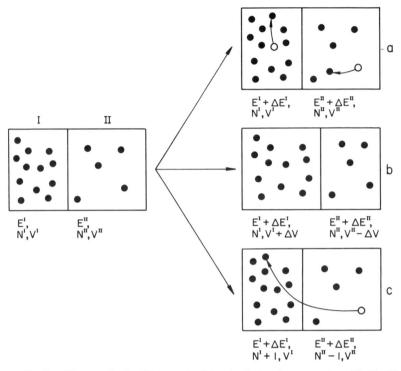

Figure 7. Possible steps in the Gibbs method for simulating the properties of fluids. The schematic illustrates the initial system configuration and three steps: (a) particle displacement, (b) volume change, and (c) particle transfer. Variables are defined as in Fig. 4. Reprinted with permission from W. L. Jorgensen and J. Tirado-Rives, *J. Am. Chem. Soc.* **110**, 1657 (1988) [8]. Copyright 1988 American Chemical Society.

it permits large density fluctuations in the fluid. The test particle, grand canonical, and Gibbs ensemble methods work well at moderate densities but become difficult to use at high densities, e.g., a dense liquid near its triple point or a solid. The reason is most easily seen for the test particle method, since the random insertion of a test molecule in a dense fluid will almost certainly result in molecular overlap and consequently a large positive value of U_t and a very small contribution to the integral that determines μ. The same problem occurs in the Gibbs and grand canonical methods with the addition of new molecules to the system. What one needs to do is develop some way of guiding the test particles or molecules toward any hole that may be present in the fluid or solid. Several "biased sampling" methods have

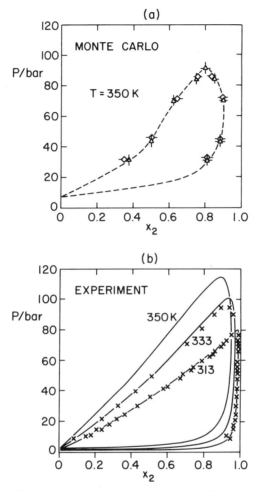

Figure 8. (a) Vapor-liquid coexistence curves for a mixture of Lennard-Jones molecules with parameters chosen to approximate acetone (1) – carbon dioxide (2): $\Diamond$ = Gibbs method at constant pressure; Δ = test particle method at constant volume. Horizontal and vertical lines are error bars. (b) Experimental ($\times$) and empirical equation of state (–) results for acetone-carbon dioxide mixtures. Reprinted with the permission of Taylor & Francis Ltd. from Panagiotopoulos *et al.* [16]; and with permission from A. Z. Panagiotopoulos, U. W. Suter, and R. C. Reid, *Ind. Eng. Chem. Fundam.* **25**, 525 (1986). Copyright 1986 American Chemical Society.

Table 5.   Effect of Water Addition on the Solubility of Naphthalene (1) in Compressed Carbon Dioxide (2) at 340 K from grand canonical MC simulations [a,b]

B_1	-5.0	-5.0	-5.0
B_2	3.1	3.1	3.1
N_{H_2O}	0	1	2
$\rho\sigma^3$	0.15 ± 0.01	0.16 ± 0.01	0.36 ± 0.25
$<N_1>$	0.17 ± 0.03	0.23 ± 0.04	10.6 ± 8.4
$<N_2>$	44.7 ± 3.2	46.4 ± 3.5	92.8 ± 30.5
$y_1 = N_1/N$	0.0038	0.0049	0.10

[a] Here $B_i = \mu_{ir}/kT + \ln N_i$, where μ_{ir} is the residual (nonideal gas) part of the chemical potential.
[b] M. Nouacer and K. S. Shing, *Molec. Simul.* 2, 55 (1989).

been developed for doing this; they extend the density range in which these methods can be used. However, they still usually fail for solids or dense fluids of highly nonspherical molecules (e.g., liquid crystals). For these more difficult situations, one must resort to thermodynamic integration.

B.   Fluid Behavior in Micropores

In microporous materials such as activated carbons, silica gel, porous glasses and ceramics, clays, and zeolites, adsorbed fluids show many unusual and interesting properties: (1) preferential adsorption of certain fluid components, (2) hysteresis effects, and (3) a variety of new or unusual phase transitions, including capillary condensation, wetting and prewetting phenomena, layering transitions, and two-dimensional solid melting. For very small pores, the behavior becomes characteristic of one-dimensional (cylindrical geometry—no phase transitions) or two-dimensional (parallel plate geometry) systems. These phenomena are poorly understood but play an important role in separation and purification processes, catalysis, and many geophysical and biological processes. They arise because of the strong intermolecular forces between the pore walls and the fluid molecules, which dominate the fluid behavior for small pores; the fluid thus behaves quite differently from the bulk. They are difficult to study experimentally because the characteristics of most porous materials are not well defined; computer simulation and molecular theory have therefore played an important role in understanding them over the last two or three years. The long-term aim of this work is to design adsorbents and membranes that will give improved performance for specific applications in chemical processing.

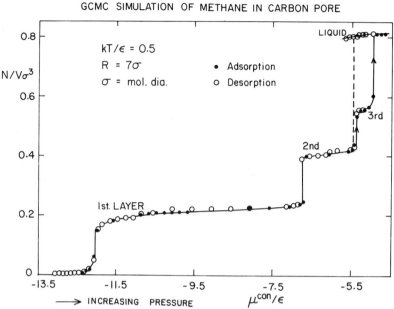

Figure 9. Adsorption isotherm for a pure Lennard-Jones fluid in a cylindrical pore of radius $R = 7\sigma$, from grand canonical Monte Carlo simulations. The system is a crude model that approximates a methane-like fluid in a carbon pore at a temperature in the liquid range; N is the number of fluid molecules, V is the pore volume, μ^{con} is the configurational part of the chemical potential, and ε is the potential well depth for the fluid molecules. Increasing μ^{con} corresponds to increasing the pressure. On raising the pressure the system passes through three layering transitions and finally condenses to liquid. The third layering transition is believed to be a metastable state, and it as well as liquefaction is accompanied by hysteresis. At low pressures the first layer is a two-dimensional liquid, and this solidifies to a two-dimensional solid at somewhat higher pressures (at $\mu^{con}/\varepsilon \sim -11$). Reprinted with permission from B. K. Peterson, G. S. Heffelfinger, F. van Swol, and K. E. Gubbins, *J. Chem. Phys.* **93**, 679 (1990).

The principal tools have been density functional theory and computer simulation, especially grand canonical Monte Carlo and molecular dynamics [17–19]. Typical phase diagrams for a simple Lennard-Jones fluid and for a binary mixture of Lennard-Jones fluids confined within cylindrical pores of various diameters are shown in Figs. 9 and 10, respectively. Also shown in Fig. 10 is the vapor-liquid phase diagram for the bulk fluid (i.e., a pore of infinite radius). In these examples, the walls are inert and exert only weak forces on the molecules, which themselves interact weakly. Nevertheless,

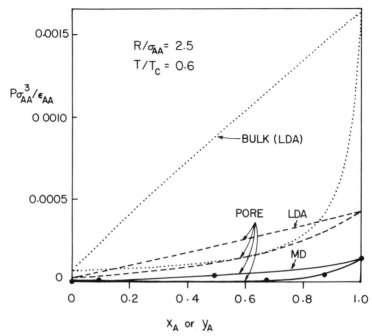

Figure 10. Vapor-liquid equilibria for an argon-krypton mixture (modeled as a Lennard-Jones mixture) for the bulk fluid ($R^* = \infty$) and for a cylindrical pore of radius $R^* = R/\sigma_{AA} = 2.5$. The dotted and dashed lines are from a crude form of density functional theory (the local density approximation, LDA). The points and solid lines are molecular dynamics results for the pore. Reprinted with permission from W. L. Jorgensen and J. Tirado-Rives, *J. Am. Chem. Soc.* **110**, 1657 (1988) [8]. Copyright 1988 American Chemical Society.

condensation occurs at pressures far below the vapor pressure of the bulk fluid (capillary condensation), and the critical temperature and pressure are reduced substantially. For pores of intermediate size, e.g., a radius of 7.0σ (σ is the molecular diameter), there are so-called layering transitions; these are first-order phase transitions between adsorbed layers of one molecular thickness and two, between two molecular layers and three, and so on. Such transitions depend strongly on the strength of the wall forces, the temperature, and the pore radius; for activated carbon pores, which interact more strongly with the fluid molecules, the layering transitions would be more pronounced than those shown in Fig. 9. For very small pores, they are inhibited by molecular packing effects, while in very large pores they have less effect on the overall phase diagram because of the large amount of bulk fluid present. When the pore radius becomes very small, a cylindrical pore approaches a one-dimensional limit in which the fluid molecules can only move along a line. Exact statistical mechanics tells us that such a system cannot show any phase transitions. This seems to occur at a radius between 1σ and 2σ. For mixtures, confinement within a pore will lead to large shifts in relative adsorption and volatility, in addition to these effects (see Fig. 10). Much remains to be done to understand these complex phase diagrams and adsorption effects in terms of the underlying intermolecular forces, particularly for the more complex fluids and porous materials of technological interest.

IV. The Future

A. Research

Chemical engineering thermodynamics has been undergoing a transition from an empirical, experimental field to one in which soundly based molecular ideas play an increasingly prominent role. This trend has been made possible by the rapid development of computers and of computer simulation and molecular theory. Computers are expected to increase in power by a factor of 10^3 to 10^6 over the next decade; such developments are bound to have a major effect on the way we approach thermodynamics research. The semiempirical method will continue to remain useful for many of the traditional applications for thermodynamics research and will be most successful when it has some molecular basis. However, the macroscopic approach will not be sufficient to meet the challenge of such new areas as biotechnology, microstructured materials, and technologies that exploit the unique properties of surfaces [20]. These areas will call for an interdisciplinary, molecular approach which combines the methods of simulation, theory, and experiment. Among the areas that are likely to provide challenges and opportunities for thermodynamicists in the near future are the following:

1. Computer-Aided Materials Design

We are moving toward an era in which simulation and molecular theory will play an important role in the design of new polymers, composites, ceramics, and electronic and photonic materials [21, 22]. Much of the theoretical and simulation work to date has asked questions of the form: what are the properties of this particular model substance? We need to invert this question and ask: here is a specific need, for example a difficult or expensive separation—how can we use our theory and simulation techniques to design a material or process to best meet this need [22]? One example would be the design of improved porous materials as adsorbents, their composition and surface characteristics being guided by a combination of simulation, theory, and experiment. Molecular simulation techniques are already at a stage where they can provide an understanding of the relation between structure and properties of these materials and can point the way to new designs. The design of polymers having specific desired properties is also foreseeable in the next decade [23]. The engineering of "designer" polymer molecules is in some ways more difficult than that of biological molecules (see below), since in the latter case the structure of a single molecule or the interaction of a pair of molecules (e.g., an enzyme and an inhibitor) is often the main concern, whereas for polymers the bulk properties (mechanical, thermal, etc.) are of major interest, so that it is necessary to simulate a sample containing a reasonable number of molecules. In many of the materials applications, it will be necessary to combine molecular simulation methods with quantum mechanical calculations of electronic structure [24].

I will give two examples drawn from recent work. The first is catalyst design. It is now possible, using *ab initio* methods (generalized valence bond with configuration interaction calculations), to estimate the energetics of simple surface reactions to good accuracy [25]. Such calculations can help to guide experiments on new catalysts and promoters, as well as in discrimination among various reaction mechanisms. These methods have recently been used to study the epoxidation of olefins on silver catalysts [26]. The next step in such work will be to use the energetics derived from the *ab initio* calculations as input to molecular dynamics stimulations to study reactions on catalytic surfaces [27]. With such methods it should eventually be possible to study the rates of elementary catalytic reactions. To do this successfully, one must learn how to deal with the different time scales involved in such reactions, a problem that will require some novel developments in the simulation techniques. The second example is the search for high-temperature superconductors. This field has been almost entirely empirical, but recently developed theories [27], though not yet fully tested, may point the way to better materials and greatly speed up the search.

2. Surface Phenomena

Many new technologies rely on the unusual properties of interfaces—Langmuir-Blodgett and other films, micelles, vesicles, small liquid drops, and so on. Classical thermodynamics is often inadequate as a basis for treating such systems because of their "smallness," and experimental probes of the interface are limited, especially for fluid systems. Computer simulation can play an important role here, both in understanding the role of intermolecular forces in obtaining desired properties and, in combination with experiment, in designing better materials and processes [6, 28].

3. Biotechnology and Biomedicine

Thermodynamic stability plays a major role in the protein folding problem, in predicting the three-dimensional structure of enzymes from their chemical composition, in predicting the limits of protein stability under a variety of conditions, and in designing enzyme inhibitors [6, 21, 29]. Quite good potential energy functions now exist (OPLS, CHARMM, ECEPP, etc.— see Sect. II,E) for the configurational and intermolecular energies of many biopolymers, including polypeptides, proteins, and nucleic acids, and these can be used to predict the energies of the vast number of possible molecular conformations of these molecules [30]. A currently active research area is the determination of the low-potential-energy conformations (i.e., local and global minima) of such large flexible molecules, using one of the potential energy functions mentioned above together with Monte Carlo or some other systematic trial method; such calculations are also important for biopolymers in solution, where it is the *free* energy minima that are important. These calculations will be particularly valuable when combined with high-resolution proton NMR measurements on biopolymer molecules in solution. Such measurements can accurately determine the distances between hydrogen nuclei [31] and, when coupled with simulation methods, can yield the molecular structure in solution. It is now possible, using free energy perturbation methods, to calculate changes in solvation energy and binding constants for molecules of different structures [21, 32, 33]. This technique is beginning to play an important role in the design of new drugs which are enzyme inhibitors. The free energy perturbation method involves a Monte Carlo or molecular dynamics simulation with the potential drug molecule bound to the active site on the enzyme (generally it is necessary only to simulate a substructure of the enzyme); if the molecule is strongly bound, it prevents the normal action of the enzyme [34]. Computer simulation of such binding activity can be used to help design the drug molecule and to guide experiments. A somewhat different method, simulated annealing [35], can also prove useful in determining the most likely conformations of biological molecules [36]. The system is first simulated at a high temperature, where thermodynamic equilibrium is easily attained, and then slowly cooled ("annealed"); this helps to avoid becoming trapped in local minima.

Such calculations should eventually be able to provide a better understanding of the thermodynamic fundamentals of biochemical separation methods, particularly for such processes as membrane and chromatographic techniques; a clearer understanding of the causes of molecular clustering will be important in this area. In the next five to ten years, these methods should be much more sophisticated and will play a major role in the design of drugs, affinity agents, and proteins that fold into desired patterns.

Somewhat later it should be possible to design molecules with desired biological properties.

4. Phase Equilibria and Fluid Properties

For the traditional oil and chemical industries, many of the most important thermodynamic problems will continue to involve calculations of fluid phase equilibria and physical properties of multicomponent fluid mixtures. Much improved theories or theoretically based correlations are needed for mixtures containing molecules of disparate types—hydrocarbons with associating molecules and ions, for example. This is an area where we already have a good understanding of the physical phenomena involved, but because of the size of these industries small improvements in our predictive ability may lead to large economic gains. The computer simulation approach is still some way from providing these small improvements because of the slowness of current computers and the sensitivity of the results to small errors in the intermolecular potentials. The development of the Gibbs method for studying multicomponent phase equilibria [16] is a major step forward. For simulation calculations to reach the stage where they can supplement or replace experiments in a major way, we will need increases of computing speed of 10^2 or more coupled with better potentials, probably of the aniostropic site-site form discussed in Sect. II,E. On the experimental side, there have been significant improvements in techniques [37], particularly for measurements at high pressures and over wide ranges of temperature. Phase equilibrium experiments are time-consuming, particularly because of the time needed for equilibration; techniques for reducing the sample size, while keeping surface effects minimal, are needed to improve this situation.

To summarize, molecular theory and computer simulation, in combination with experiment, will play a central role in many of the new areas, particularly in helping to produce "designer microstructures" and "designer molecules." This will be particularly noticeable in many of the materials areas in the next five years, for example, the design of porous materials, polymers, and catalysts for simple reactions, the design of proteins and drugs for particular purposes, and the molecular engineering of some thin films. In these areas, classical or approximate quantum mechanical methods are used, so that relatively modest improvements in computing speed and in the accuracy of intermolecular potentials will have considerable impact. However, the use of the full *ab initio* approach, involving a rigorous quantal treatment, seems to be much further away for problems of interest to chemical engineers. Because the computing requirements rise so rapidly with the size of the molecular species (typically, as the fifth or sixth power of the number of electrons), such calculations require much more computing power;

in conventional simulations, where the molecules obey classical mechanics, the computing power needed rises only as N^2, where N is the number of molecules. Nevertheless, quantum Monte Carlo methods based on Feynman path integral techniques [38] are now being used to treat electron transfer processes and in the long term should have an important impact on electrode processes, semiconductor properties, electron transfer reactions, and the use of lasers to transform matter. A spinoff of the molecular simulation work has been the development of simulated annealing [35], widely used in optimization of process and electrical circuit design but of future interest in some of the new research areas; a key advantage of this technique is the avoidance of local minima in the function to be minimized (e.g., the free energy).

Chemical engineers working in these newer areas will need to become versed in new fields and in many cases to work with scientists and engineers from other disciplines. To deal successfully with biological systems requires a group familiar with the intricacies of cell biology, biophysics, and biochemistry, for example. Among the major international conferences in applied thermodynamics are the International Conferences on Fluid Phase Equilibria and Physical Properties for Chemical Process Design [39], held every three years and alternating between Europe and North America, and the IUPAC Conference on Chemical Thermodynamics [40]. These conferences will need to give more emphasis to these new areas and approaches than they have in the past.

B. Teaching

These developments will call for a restructuring and rethinking of the teaching of chemical engineering thermodynamics. Many widely used texts concentrate exclusively on the classial approach, the word molecule never appearing in the course. Statistical mechanics and quantum mechanics are usually taught as part of the chemistry or physics sequence, but the student rarely if ever sees any applications of this material in chemical engineering courses. Implicitly, these subjects are treated like Greek mythology—part of the student's general education, but not of great importance in chemical engineering. Such an approach may prevent our graduates from involvement in important technologies of the future. We need to introduce examples into our existing chemical engineering courses that apply the fundamental knowledge of these topics that the student has gained in chemistry or physics. A problem at present is the lack of suitable textbooks at the undergraduate level. Books on statistical mechanics written by chemists usually do not progress past applications to the ideal gas and other systems of noninteracting particles, so that the student gets little feel for the relevance of the subject to modern areas of technology. The challenge is to produce a book that is usable at the undergraduate level but that has well thought out examples, including application of the principles to nonconventional processes such as thin films, materials, surface phenomena, and so on. It would be particularly valuable to have a book that covers both classical and statistical thermodynamics at the junior or senior level. There is also a need for a graduate level textbook on computer simulation methods. A possible approach to the need for interdisciplinary education would be to develop more MS course programs that cross traditional departmental lines. Obvious examples would be degrees in chemical engineering and microelectronics and in chemical engineering and biology.

Acknowledgments

I am grateful to many colleagues for helpful discussions and for putting up with many questions on areas outside my own direct experience. In particular, I thank F. H. Arnold, B. J. Berne, J. C. G. Calado, E. A. Carter, P. T. Cummings, W. A. Goddard, W. L. Jorgensen, F. Kohler, J. A. McCammon, A. Z. Panagiotopoulos, N. Quirke, H. A. Scheraga, W. C. Still, and D. N. Theodorou. Part of this work was supported by grants from the National Science Foundation and the Gas Research Institute.

References

1. Bryan, P. F., and Prausnitz, J. M., *Fluid Phase Equilibria* **38**, 201 (1987).

2. See, for example: Hansen, J. P., and McDonald, I. R., *Theory of Simple Liquids*, 2nd Ed. Academic Press, London, 1986; Gray, C. G., and Gubbins, K. E., *Theory of Molecular Fluids*. Clarendon Press, Oxford, 1984; Lee, L. L., *Molecular Thermodynamics of Nonideal Fluids*. Butterworths, Boston, 1987.

3. Chapman, W. G., Gubbins, K. E., Joslin, C. G., and Gray, C. G., *Pure Appl. Chem.* **59**, 53 (1987).

4. Evans, R., *Adv. Phys.* **28**, 43 (1979).

5. Allen, M. P., and Tildesley, D. J., *Computer Simulation of Liquids*. Clarendon Press, Oxford, 1987.

6. Abraham, F. F., *Adv. Phys.* **35**, 1 (1986).

7. Kataoka, Y., *J. Chem. Phys.* **87**, 589 (1987).

8. Jorgensen, W. L., and Tirado-Rives, J., *J. Am. Chem. Soc.* **110**, 1657 (1988), and references therein.

9. Momany, F. A., McGuire, R. F., Burgess, A. W., and Scheraga, H. A., *J. Phys. Chem.* **79**, 2361 (1975); Nemethy, G., Pottle, M. S., and Scheraga, H. A., *J. Phys. Chem.* **87**, 1883 (1983); Sippl, M. J., Nemethy, G., and Scheraga, H. A., *J. Phys. Chem.* **88**, 6231 (1984).

10. Brooks, B. R., Bruccoleri, R.E., Olafson, B. D., States, D. J., Swaminathan, S., and Karplus, M. J., *J. Comp. Chem.* **4**, 187 (1983). *See also* Brooks *et al.* [30].

11. Weiner, S. J., Kollman, P. A., Nguyen, D. T., and Chase, D. A., *J. Phys. Chem.* **7**, 230 (1986).

12. Dauber-Osguthorpe, P., Roberts, V. A., Osguthorpe, D. J., Wolff, J., Genest, M., and Hagler, A. T., *Proteins* **4**, 31 (1988).

13. Price, S. L., *Mol. Simul.* **1**, 135 (1988), and references therein.

14. Stone, A. J., and Alderton, M., *Mol. Phys.* **56**, 1047 (1988); Price, S. L., Stone, A. J., and Alderton, M., *Mol. Phys.* **52**, 987 (1984).

15. See Gubbins, K. E., *Mol. Simul.* **2**, 223 (1989), and references therein.

16. Panagiotopoulos, A. Z., *Mol. Phys.* **61**, 813 (1987); Panagiotopoulos, A. Z., Quirke, N., Stapleton, M., and Tildesley, D. J., *Mol. Phys.* **63**, 527 (1988).

17. See, for example: Peterson, B. K., Gubbins, K. E., Heffelfinger, G. S., Marini Bettolo Marconi, U., and van Swol, F., *J. Chem. Phys.* **88**, 6487 (1988); Heffelfinger, G. S., Tan, Z., Gubbins, K. E., Marini Bettolo Marconi, U., and van Swol, F., *Mol. Simul.* **2**, 393 (1989), and references therein.

18. Magda, J. J., Tirrell, M., and Davis, H. T., *J. Chem. Phys.* **83**, 1888 (1985); Bitsanis, I., Magda, J. J., Tirrell, M., and Davis, H. T., *J. Chem. Phys.* **87**, 1733 (1987).

19. Walton, J. P. R. B., and Quirke, N. P., *Mol. Simul.* **2**, 361 (1989).

20. National Research Council, Committee on Chemical Engineering Frontiers: Research Needs and Opportunities. *Frontiers in Chemical Engineering. Research Needs and Opportunities.* National Academy Press, Washington, D. C., 1988; Krantz, W. B., Wasan, D. T., and Nerad, P. V., eds., *Interfacial Phenomena in the New and Emerging Technologies*, Proceedings of workshop held at University of Colorado, May 29–31, 1986, National Science Foundation, Division of Engineering, Washington, D. C., 1987.

21. McCammon, J. A., *Science* **238**, 486 (1987).

22. See for example: Proceedings of the Conference on Industrial Applications of Molecular Simulation, *Mol. Simul.*, Vol. 2 and 3 (1989).

23. For some recent work in this area see: Theodorou, D. N., and Suter, U. W., *Macromolecules* **19**, 139, 379 (1986); Mansfield, K. F., and Theodorou, D. N., "Atomistic Simulation of Glassy Polymer Surfaces and Glassy Polymer Solid Interfaces," Annual AIChE Meeting, Washington, D. C., Nov. 1988.

24. For an application to the properties of crystalline silicon see: Carr, R., and Parrinello, M., *Phys. Rev. Lett.* **55**, 2471 (1985). For such treatments for a variety of materials, see: Proceedings of the CCP5-CCP9 Conference on Computer Modeling of New Materials, University of Bristol, UK, Jan. 4–6, 1989; *Mol. Simul.* **4**, Nos. 1–3 (1989).

25. Carter, E. A., and Goddard, W. A., *J. Chem. Phys.* **88**, 3132 (1988); Carter, E. A., and Goddard, W. A., *J. Catal.* **112**, 80 (1988); Carter, E. A., and Goddard, W. A., *Surface Sci.* **209**, 243 (1988).

26. Carter, E. A., private communication (1988).

27. Guo, Y., Langlois, J.-M., and Goddard, W. A., *Science* **239**, 896 (1988); Chen, G., and Goddard, W. A., *Science* **239**, 899 (1988). See also *Science* **242**, 31 (1988). This theory explains high-T superconductivity in terms of magnetic interactions of electrons and pairs of copper atoms, which leads to electron pairing; it predicts that the highest critical temperature for the copper oxide superconductors under development is likely to be about 225 K, about 100 K higher than it is now. The theory may also point the way to improved superconductors based on materials other than copper oxide. For a discussion of other theories, see: Emery, V. J., *Physics Today* Jan. 1989, pp. 5–26; Little, W. A., *Science* **242**, 1390 (1988).

28. For an example of the use of computer simulation to study micelles, see: Woods, M. C., Haile, J. M., and O'Connell, J. P., *J. Phys. Chem.* **90**, 1875 (1986). See also Davis, H. T., in *Perspectives in Chemical Engineering: Research and Education* (C.K. Colton, ed.), p. 169. Academic Press, San Diego, Calif., 1991 (*Adv. Chem. Eng.* **16**).

29. Bash, P. A., Singh, U. C., Langridge, R., and Kollman, P. A., *Science* **236**, 564 (1987); Kollman, P. A., *Ann. Rev. Phys. Chem.* **38**, 303 (1987).

30. For a comparison of these methods, see: Hall, D., and Pavitt, N., *J. Comp. Chem.* **5**, 441 (1984); also, Brooks, C. L., III, Karplus, M., and Pettitt, B. M., *Adv. Chem. Phys.* **71** (1988).

31. Wüthrich, K., *NMR of Proteins and Nucleic Acids*. Wiley, New York, 1986.

32. van Gunsteren, W. F., and Berendsen, H. J. C., *J. Comput.-Aided Molec. Des.* **1**, 171 (1987).

33. McCammon, J. A., and Harvey, S. C., *Dynamics of Proteins and Nucleic Acids*. Cambridge University Press, New York, 1987.

34. Wong, C. F., and McCammon, J. A., *J. Amer. Chem. Soc.* **108**, 3830 (1986).

35. Kirkpatrick, S., Gelatt, C. D., and Vecchi, M. P., *Science* **220**, 671 (1983).

36. See, for example: Brünger, A. T., Kuriyan, J., and Karplus, M., *Science* **235**, 458 (1987).

37. See Section 7 of the Proceedings of the 9th IUPAC Conference on Chemical Thermodynamics, Lisbon, July 1986. Published in: *Pure Appl. Chem.* **59** (1987).

38. Berne, B. J., and Thirumalai, D., *Ann. Rev. Phys. Chem.* **37**, 401 (1986).

39. The last of these conferences was held in Banff, Alberta, in May 1989. The conference proceedings appeared in *Fluid Phase Equilibria* **52** (1989).

40. The most recent of these conferences was held in Prague, Aug. 29–Sept. 2, 1988. The proceedings appeared in *Pure Appl. Chem.* **61** (1989).

8

Chemical Engineering Thermodynamics: Continuity and Expanding Frontiers

J. M. Prausnitz
Department of Chemical Engineering
University of California, Berkeley
Berkeley, California

Before attempting to assess the likely future of thermodynamics in chemical engineering, it may be useful briefly to recall the past. In chemical engineering, the primary use of thermodynamics was, and still is, concerned with application of the first law (conservation equations), in particular, with energy balances that constitute an essential cornerstone of our discipline. Another primary use was, and still is, directed at description of fluid behavior, as in nozzles, heat engines, and refrigerators. The fundamentals of these important applications were extensively developed in the first third of this century.

Until about 1930, most distillation calculations were based on Raoult's law; in extreme cases, where ideal-mixture behavior was clearly incorrect, distillation calculations required experimental data coupled with graphical methods, as in the McCabe-Thiele diagram.

About 50 years ago, the thermodynamics of multicomponent systems began to have some influence on chemical engineering design, but the full utility of that influence was not possible until the rise of computers after the Second World War. Computers, coupled with semitheoretical molecular-

155

ADVANCES IN CHEMICAL ENGINEERING, VOL. 16

thermodynamic models, have enabled chemical engineers to perform realistic calculations for the equilibrium properties of the many fluid mixtures typically encountered in chemical engineering design. While these equilibrium properties are also useful for calculation of chemical equilibria, they are primarily used for computer-aided design of separation operations such as distillation, absorption, and extraction. Computer-aided design calculations, combined with molecular-thermodynamic models, are now performed routinely by chemical engineers throughout the world, especially by those in the natural-gas, petroleum, and petrochemical industries.

As emphasized in Keith Gubbins' paper [1], the meteoric rise of computers during the last 20 years has produced two major developments. First, it has facilitated application of statistical mechanics to build more sophisticated models for representing properties of fluids and fluid mixtures. Second, we can now perform computer simulation for assemblies of molecules, leading to calculation of macroscopic properties of fluids under extreme conditions, e.g., fluids at very high pressures or fluids in narrow pores, which cannot easily be studied experimentally. These developments, as stressed by Gubbins, bear much promise for the future of chemical engineering science. As yet, however, their influence on chemical engineering practice is limited because the systems that can now be described by these advanced methods are primarily simple systems, often remote from those that we encounter in practice. This limitation is certain to decline as more powerful computers become available. Nevertheless, present limitations of applied statistical mechanics suggest that Gubbins may have overstated his case, at least for the short run. At the same time, the awesome potential of statistical mechanics for application to nonequilibrium problems suggests that he may have been too modest.

While I have no significant disagreement with Gubbins' paper, I would like to expand upon three topics that he only briefly mentioned. The first concerns a continuing need to develop semiempirical correlations of mixture properties to improve design calculations for contemporary or emerging processes where marginal improvements can often produce significant economic rewards. This need is especially pressing for mixtures beyond those typically encountered in the petroleum and petrochemical industries. The second concerns promising new applications of molecular thermodynamics to nonequilibrium phenomena, that is, to provide guidance toward designing optimum materials for rate processes, including catalysis. The third concerns the orientation of academic thermodynamicists, which is required if progress in thermodynamics teaching and research is to attain its maximum potential for service to chemical engineering science.

I. Semiempirical Correlations for Conventional Processes

For research-oriented chemical engineering professors, the challenge of truly new technology is more exciting than the marginal improvement of conventional technology; therefore, it is not surprising that Gubbins emphasizes the development of novel methods based on statistical mechanics and computer simulation. From an economic point of view, however, conventional processes will continue to dominate American chemical industry for many years. Since conventional processes often apply to very large-volume products, marginal improvements can translate into very large financial rewards. Because the novel methods described by Gubbins are not as yet applicable to many real systems, it is likely that semiempirical correlations will be of major importance for many years to come.

To illustrate, consider the well known process of wet-air oxidation, shown in Fig. 1. To design a wet-air oxidation process, we require volumetric, enthalpic, and vapor-liquid-equilibrium information. For close design, this information must be highly accurate. Sophisticated statistical-mechanical methods are as yet unable to supply the desired information. However, as shown by Heidemann [2], high accuracy can be achieved by using readily available ideal-gas heat-capacity data, correlated Henry's law constants and partial molar volumes for common gases in dilute aqueous solutions, the vapor pressure of water, and a suitably modified Redlich-Kwong equation to describe the properties of high-pressure vapor mixtures containing water in addition to common gases. Using these tools and a simple computer program, Heidemann generated the required thermodynamic information in a readily useful form. Heidemann's correlation is now in standard use by design engineers throughout the world.

A second example is provided by a semiempirical correlation for multicomponent activity coefficients in aqueous electrolyte solutions shown in Fig. 2. This correlation, developed by Fritz Meissner at MIT [3], presents a method for scale-up: activity-coefficient data for single-salt solutions, which are plentiful, are used to predict activity coefficients for multisalt solutions for which experimental data are rare. The scale-up is guided by an extended Debye–Hückel theory, but essentially it is based on enlightened empiricism. Meissner's method provides useful estimates of thermodynamic properties needed for process design of multieffect evaporators to produce salts from multicomponent brines. It will be many years before sophisticated statistical mechanical techniques can perform a similar scale-up calculation. Until then, correlations such as Meissner's will be required in a conventional industry that produces vast amounts of inexpensive commodity chemicals.

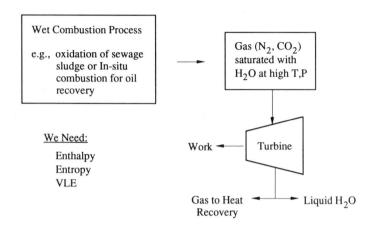

* Modified Redlich-Kwong EOS for Vapor Phase

    EOS Constant

    $$a_{H_2O} = a_{H_2O}(\text{non-polar}) + a_{H_2O}(\text{polar})\,\mathscr{Q}\,(T)$$

    $$i = N_2 \text{ or } CO_2$$

    $$a_{i\text{-}H_2O} = [a_i a_{H_2O}(\text{non-polar})]^{1/2}$$

* Krichevsky-Kasarnovsky Equation for Solubilities of Gases
  in Liquid Water

Figure 1.   Thermodynamic properties of moist gases from a wet-air oxidation process.

Semiempirical thermodynamic methods are also important for emerging technology slightly beyond contemporary chemical engineering. To illustrate, Fig. 3 gives a schematic representation of an extraction process for separating a dilute aqueous mixture of biomolecules downstream from a biochemical reactor. This extraction process is well known in biotechnology, but, at present, process design is essentially by trial and error because little is known about the fundamental thermodynamics of two-phase aqueous systems containing water-soluble polymers, salts, and biomolecules which may be electrically charged, depending on pH. To provide a basis for rational design, a semiempirical correlation can be helpful.

First, it is necessary to establish the phase diagram of the aqueous two-phase system formed by water and two water-soluble polymers. Second, a method must be established for calculating the distribution coefficient of a biomolecule that partitions between the two aqueous phases. A simple molecular-thermodynamic description is provided by the osmotic virial

Meissner's Correlation for Activity Coefficients
of Electrolytes in Water

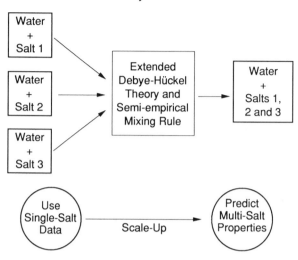

e.g., Precipitating soda ash from Trona or
from evaporation step of Solvay process.

U.S. production of soda ash is $\cong 10^7$ ton/yr.

Figure 2. Multicomponent aqueous salt solutions.

equation coupled with some elementary concepts from the thermodynamics of aqueous electrolytes. The essentials of this model are shown in Fig. 4.

To reduce this model to practice, osmotic second virial coefficients for three polymer-polymer interactions (2–2, 3–3, and 2–3) must first be measured; these are needed to determine the phase diagram of the three-component (biomolecule-free), two-phase system. Second, to determine the distribution coefficient, it is necessary to measure osmotic second virial coefficients for the two biomolecule-polymer interactions (4–2 and 4–3). In addition, the difference in electric potential between the two phases must be determined. The required osmotic virial coefficients are obtained from low-angle-light-scattering measurements and the very small (but extremely important) difference in electric potential is measured with microelectrodes, as shown recently by Robert King at Berkeley [4]. The bottom of Fig. 3 shows a schematic phase diagram for an aqueous mixture of polyethylene glycol and dextran. The length of the tie line serves as a convenient measure to characterize the chemical difference between the coexisting phases;

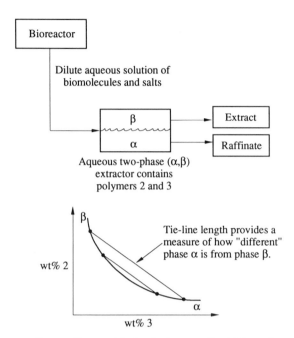

Ternary diagram calculated using osmotic virial equation

Figure 3. Two-phase aqueous extraction system for separating a mixture of biomolecules.

$$K_4 = m_4^\beta / m_4^\alpha$$

m = molality

At high dilution,

$$\ln K_4 = B_{24}(m_2^\beta - m_2^\alpha) + B_{34}(m_3^\beta - m_3^\alpha)$$

$$+ z_4 F(\phi^\beta - \phi^\alpha)/RT$$

$\phi^\beta - \phi^\alpha$ = Difference in electric potential
z_4 = Charge on biomolecule
F = Faraday's constant
B_{ij} = Osmotic second virial coefficient for i-j interaction in solution is obtained from low-angle light-scattering measurements

Figure 4. Distribution coefficient *K* for biomolecule.

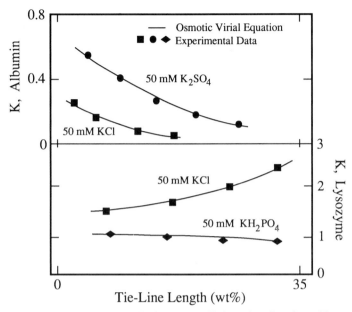

Figure 5. Predicted and observed distribution coefficients for albumin and lysozyme.

that measure can then be used as the independent variable for correlating the distribution coefficient (at high dilution) as shown in Fig. 5. Such correlations provide a useful tool for designing a two-phase aqueous extraction process. Someday it is likely that this semiempirical molecular-thermodynamic method will be replaced by computer-generated calculations, but for practical bioengineering, semiempirical methods similar to the one shown here are likely to be useful for many more years.

II. Thermodynamics for Rate Processes, Including Catalysis

During the past half century, chemical engineering thermodynamics has been concerned primarily with applications to design of separation operations. These applications will continue to be a prime focus, but the next stage that I see on the horizon is a radically different aspect, viz., determination of material structure aimed at improved design of rate processes.

Thermodynamics provides us with tools to determine the equilibrium state, i.e., the most probable state that exists under specified conditions such as temperature, pressure, and composition. The happy message provided by classical thermodynamics is that the most probable state is that which has a minimum free energy, and the happy message provided by molecular theory is that we can calculate the free energy of a material as a function of its microscopic structure.

Microscopic structure, in turn, determines rate processes. For example, transport of gases through polymers, or transport of drugs through membranes, or catalytic activity of an enzyme depends on how molecules are arranged in space. Therefore, a highly promising application of thermodynamics is to provide clues for determining the optimum structure of a material used in a desired rate process such as transport or catalytic reaction.

In traditional applications of thermodynamics, the goal is to find the free energy and its derivatives for direct use in chemical engineering design; a classic example is the equilibrium curve in a McCabe-Thiele diagram. In emerging applications of thermodynamics, the free energy and its derivatives are not of interest in themselves but serve only as intermediates to find the material structure which provides optimum conditions for some particular process.

To illustrate, Fig. 6 shows results of sophisticated computer-simulation calculations obtained by Larry June, who is working with Doros Theodorou and Alexis Bell at Berkeley to provide rational criteria for developing more effective zeolites. The objective of these simulations is to find the most probable spatial distribution of adsorbed polyatomic molecules within a heterogeneous zeolite matrix; under given conditions, the most probable state is determined by finding the state corresponding to minimum free energy. Theodorou and June studied adsorption of paraffins on orthorhombic silicalite. Because their studies are highly detailed, they are able to calculate much more than the usual macroscopic thermodynamic quantities; in addition, they can calculate not only how much adsorbs under given conditions but also the distribution of adsorbed molecules within the heterogeneous pores.

Figure 6 shows adsorption results for n-hexane and for 2-methylpentane at 400 K. The sinusoidal pores are shown going from left to right with a positive slope; the straight pore goes from right to left with a negative slope. Results are normalized such that the amount adsorbed is the same for both paraffins. The plots show regions where 50% of the adsorbed centers of mass are located. The dark areas shown are *not* an indication of how many molecules are adsorbed; rather, they indicate those regions within the pores where the adsorbed molecules are most likely to be found. The overall extent of adsorption is indicated by ρ_{cm} which stands for the density of the centers of mass per unit of pressure. (In effect, ρ_{cm} is the inverse of Henry's constant.) The results show that normal hexane prefers to sit in the sinusoidal pores while branched 2-methylpentane prefers to sit at the intersections between straight and sinusoidal pores. These results show the large

Contours include 50% of the sampled centers of mass

n-Hexane

$\rho_{cm} > 7.7 \times 10^{-4}$ (molecule/$\mathring{A}^3$ Pa)

2-Methylpentane

$\rho_{cm} > 8.8 \times 10^{-4}$ (molecule/$\mathring{A}^3$ Pa)

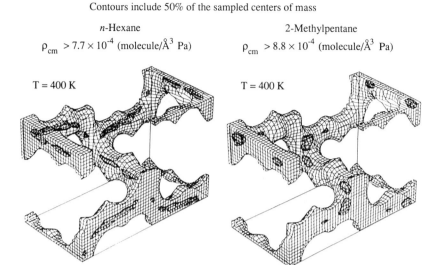

Figure 6. Effect of molecular shape on local concentration at 400 K. Sorption of n-hexane and 2-methylpentane in a zeolite. Reprinted with permission from R. L. June, A. T. Bell, and D. N. Theodorou, *J. Phys. Chem.* **94**, 1508 (1990). Copyright 1990 American Chemical Society.

importance of steric effects where the branched paraffin can only sit in the larger voids, whereas the normal paraffin likes the snug fit in a narrow channel.

These results suggest that if we want to design a molecular sieve to separate a mixture of normal hexane and 2-methylpentane, we should use a zeolite with normal sinusoidal pores but small-diameter straight pores; such a zeolite will preferentially accept normal n-hexane and preferentially reject 2-methylpentane. More important, perhaps, these results have implications for selective catalysis. If we want n-hexane to react but not 2-methylpentane, then we should make a zeolite where the catalytically active centers (aluminum oxides) are situated in the sinusoidal pores but not in the straight pores. Conversely, if we desire preferential catalysis for the branched isomer, we want a zeolite where the active centers are at the intersections between straight and sinusoidal pores.

Figure 7 shows the effect of temperature on adsorption of normal hexane. The results are normalized such that the amount adsorbed is the same for 400 and 200 K. Again, the dark areas do *not* show the amount adsorbed but the regions where the adsorbed molecules are situated. We see that at 200 K, adsorption is highly localized; all the adsorbed molecules sit at

Contours include 50% of the sampled centers of mass

$\rho_{cm} \geq 7.7 \times 10^{-4}$ (molecule/Å³ Pa) $\rho_{cm} \geq 3.0 \times 10^{7}$ (molecule/Å³ Pa)

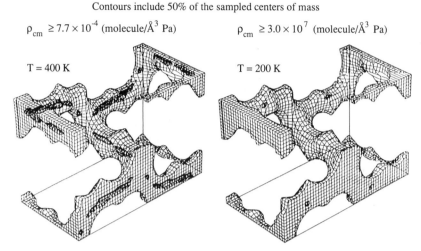

T = 400 K   T = 200 K

Figure 7. Effect of temperature on localization of n-hexane sorption in a zeolite. Reprinted with permission from R. L. June, A. T. Bell, and D. N. Theodorou, *J. Phys. Chem.* **94**, 1508 (1990). Copyright 1990 American Chemical Society.

energetically favorable sites. At 400 K, adsorption is less localized; because entropic effects are now more important, we see thermal randomization, that is, the molecules are now more "tolerant" and not as "fussy" concerning where they are willing to sit. These results suggest that for low-temperature catalysis, it will be important to place the catalytically active sites very near those sites which are energetically favorable for adsorption. At higher temperatures, the precise location of the active sites is less important.

Another promising new application of thermodynamics is directed at enzyme catalysis. While numerous simplifying assumptions are required for the present state of the art, computer simulations can be used to indicate how a directed mutation in an enzyme is likely to influence the kinetics of a particular enzyme-catalyzed reaction. To illustrate, Fig. 8 shows the structure of subtilisin, a readily available enzyme with a molecular weight of about 28,000. Subtilisin is used in detergents for washing dishes or clothing because it catalyzes hydrolysis of peptides typically found in common foods that soil our dishes and stain our clothes. Unfortunately, subtilisin's stability decreases with rising temperatures; it is therefore of practical interest to consider a mutant of subtilisin which is likely to be stable at temperatures used in a washing machine. At the Pharmaceutical Sciences Department of the University of California (San Francisco), Peter Kollman and co-workers investigated the effect of substituting alanine for asparag-

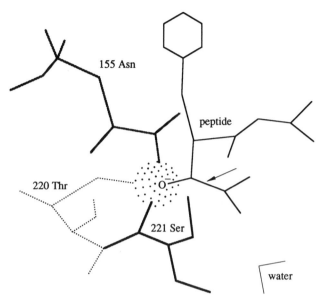

Figure 8. Active site on the transition state for hydrolysis of a peptide bond by subtilisin. The dots represent the van der Waals radius of the oxygen of the scissile peptide bond (indicated by an arrow). Shown only are those residues containing a hydrogen atom which interacts with that oxygen.

ine at the 155 position in subtilisin [5, 6]; such substitution increases stability at higher temperatures. But what does this substitution do to the kinetics of peptide hydrolysis?

Kollman's computer-simulation calculations are schematically summarized in Figs. 9 and 10. Native enzyme E reacts with substrate S in several steps. First, a nonbonded complex ES is formed; this is the binding step. Next, a transition state ETS is formed; this is the activation step. Finally, the activated state decays to products.

The top and middle of Fig. 9 indicate results of a perturbation calculation; in this case, the perturbation is provided by substituting mutant enzyme E' for native enzyme E. (For illustrative purposes, the diagram optimistically assumes that the line for the mutant enzyme lies below that for the native enzyme.) Computer simulations are used to calculate the free energy changes ΔG_1, ΔG_2, and ΔG_3. Since free energy is a state function, these calculated quantities then give the change in the free energy of binding ($\Delta\Delta G_b$) and the change in the free energy of activation ($\Delta\Delta G^*$). The first of these changes is negligible, but the second is not. Kollman's calculations

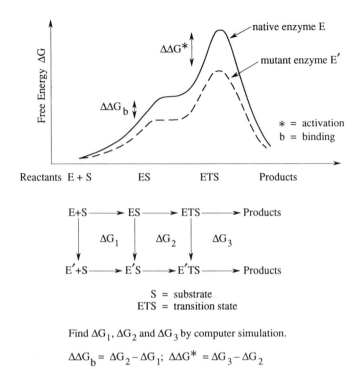

Find ΔG_1, ΔG_2 and ΔG_3 by computer simulation.

$$\Delta\Delta G_b = \Delta G_2 - \Delta G_1; \quad \Delta\Delta G^* = \Delta G_3 - \Delta G_2$$

Figure 9. Free energy calculation for a perturbed (mutant) enzyme-catalyzed reaction.

show that the dashed line in Fig. 9 should lie above the continuous line; while the substitution of alanine for aspargine at position 155 may increase thermal stability, the rate of hydrolysis decreases by a factor of about 600. This prediction was later confirmed by kinetic experiments.

III. Integration of Thermodynamics into the Broad Concerns of Chemical Engineering

The preceding examples illustrate the power of thermodynamics to contribute to chemical process design. These examples also suggest that, to maximize its engineering utility, thermodynamics must be related to product and process development, lest the new developments in thermodynamics go off on a tangent and become isolated from chemical engineering science and practice.

Everywhere in the scientific community, we see increasing fragmentation, increasing specialization, and, as a result, alienation. At technical

E = Native Subtilisin   S = Ala-Ala-Phe
E' = Subtilisin with one mutation in the
       155 position:

Asparagine (Asn155) replaced by Alanine (Ala155)

	$\Delta\Delta G_b$ (kcal/mole)	$\Delta\Delta G^*$ (kcal/mole)
Simulation[a]	0.11±0.80	3.40±1.13
Kinetic data[b]	0.41	3.67

[a]In this case, the mutant enzyme is <u>less</u> effective (ratio of rate constants is about 600).
[b]Kinetic data, obtained <u>after</u> simulation, agree with the prediction.

Figure 10. Hydrolysis of a tripeptide catalyzed by subtilisin.

meetings, thermodynamicists talk primarily to other thermodynamicists, just as fluid dynamicists talk primarily to other fluid dynamicists. Increasingly, publications in chemical engineering thermodynamics appear in specialized journals where relevance to chemical engineering is often neither stressed by the author nor appreciated by the editor.

For chemical engineering, thermodynamics is an extraordinarily useful tool, but there is a growing tendency by thermodynamicists to regard their subject as an independent entity. Consequently, much of what is published today presents "far out" results, seemingly unrelated to developments elsewhere in the chemical engineering world. I say "seemingly" because in many cases there is a potential relation that, however, the author all too often does not bother to develop. In many potentially important publications, the regrettable attitude seems to be, "Here is the general idea, and in the appendix I give essential mathematical details. Let others figure out how my work may be put to good use."

In research and teaching, the problem of alienation and fragmentation is not new, but it is becoming increasingly severe. It has not received the attention that it deserves, indeed that it must have, to reduce the gulf between science and engineering.

If new research in thermodynamics is to serve new chemical engineering science, it is not enough for thermodynamicists to explore new phenomena and to develop new computational techniques. Thermodynamic researchers must also make an earnest effort to listen to what goes on in other chemical engineering areas and relate their work to these areas. I do not agree with those researchers who are content to impress only their closest, like-minded colleagues. Chemical engineering thermodynamics must be good science but, to be truly effective, it must be good science in a chemical engineering context.

Let me close with a crude analogy. A few years ago there was a popular poster which shows a little child sitting on a potty. The caption, evidently coming from the child's mother, said, "You're not finished till the paper work is done." I certainly do not want to take this analogy too far. But I urge my thermodynamics colleagues to tidy up their work by relating it to the wide world of chemical engineering.

Acknowledgments

For sustained financial support of his research, the author is grateful to the National Science Foundation, the Lawrence Berkeley Laboratory (U.S. Dept. of Energy), and the Petroleum Research Fund (American Chemical Society). Special thanks go to Mr. Larry June and Professor Doros Theodorou (Berkeley) for permission to quote their computer-simulation results prior to publication, to Eric Anderson and Juan de Pablo for assistance in preparing the figures, and to Dr. E. de Azevedo (Lisbon) and Prof. P. Virk (MIT) for helpful comments.

References

1. Gubbins, K., in *Perspectives in Chemical Engineering: Research and Education* (C. K. Colton, ed.), p. 125. Academic Press, San Diego, Calif., 1991 (*Adv. Chem. Eng.* **16**).

2. Heidemann, R. A., and Prausnitz, J. M., *Ind. Eng. Chem. Proc. Des. Dev.* **16**, 375 (1977).

3. Meissner, H. P., and Manning, M. P., in *Chemical Engineering Thermodynamics* (S. A. Newman, ed.), p. 339. Ann Arbor Science, Ann Arbor, Mich., 1983.

4. King, R. S., Blanch, H. W., and Prausnitz, J. M., *AIChE J.* **34**, 1585 (1988). See also Haynes, C. A., Blanch, H. W., and Prausnitz, J. M., *Fluid Phase Equilibria* **53**, 463 (1989).

5. Rao, S. M., Singh, U. C., Vash, P. A., and Kollman, P. A., *Nature* **320**, 551 (1987).

6. Washel, A., and Sussman, F., *Proc. Nat. Acad. Sci.* **83**, 3806 (1986).

9

Future Opportunities in Thermodynamics

H. Ted Davis
Department of Chemical Engineering & Materials Science
University of Minnesota
Minneapolis, Minnesota

I. Introduction

My job of discussing Keith Gubbins' paper [1] would be easier if I disagreed with his summary of the current status and future opportunities in the use of thermodynamics in chemical engineering. Unfortunately for me, his is a fine paper, and it touches on most of the important areas that are likely to prosper during the next couple of decades. However, there are some areas that I believe he underemphasizes or whose importance he underestimates.

I agree with Gubbins that the future advancement in our understanding will come through the interplay of molecular theory, computer simulation, and experimentation (Fig. 1). The rapidly increasing speed and memory capacity of computers enable the simulation of more realistic molecular systems and the solution of more rigorous molecular theories.

The next two decades will witness a substantial increase in the number of chemical engineers using computer-aided molecular theory and molecular dynamics as routine tools for elucidation of thermodynamic mechanisms and for prediction of properties. Chemical engineers already have a considerable presence in the area. Two of the more significant developments in computer simulations of the last decade have been accomplished by young

ADVANCES IN CHEMICAL ENGINEERING, VOL. 16

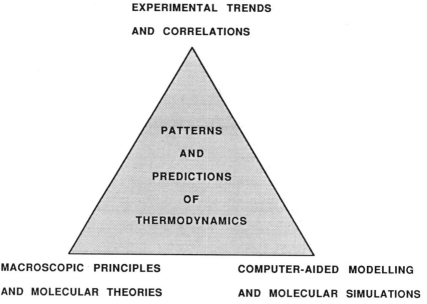

Figure 1. Developments in thermodynamics.

chemical engineers. One is the Gibbs ensemble Monte Carlo method, created by A. Panagiotopoulos [2], which is especially potent in simulating
multiphase equilibria of mixtures. The other is the excluded volume map,
a technique invented by G. Deitrick [3] to reduce by orders of magnitude
the time required to compute chemical potential in molecular dynamics
simulations.

An example drawn from Deitrick's work (Fig. 2) shows the chemical
potential and the pressure of a Lennard-Jones fluid computed from molecular
dynamics. The variance about the computed mean values is indicated in the
figure by the small dots in the circles, which serve only to locate the dots.
A test of the "thermodynamic goodness" of the molecular dynamics result
is to compute the chemical potential from the simulated pressure by integrating the Gibbs-Duhem equation. The results of the test are also shown
in Fig. 2. The point of the example is that accurate and affordable molecular
simulations of thermodynamic, dynamic, and transport behavior of dense
fluids can now be done. Currently, one can simulate realistic water, electrolytic solutions, and small polyatomic molecular fluids. Even some of the
properties of micellar solutions and liquid crystals can be captured by idealized models [4, 5].

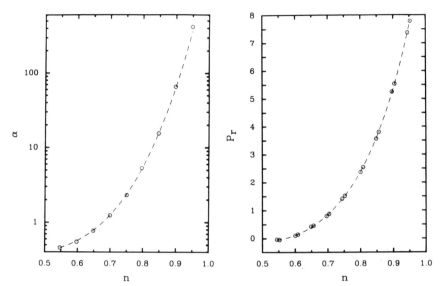

Figure 2. Fugacity a and pressure P_r as a function of density n for a 6–12 Lennard-Jones liquid. Circles denote molecular dynamics results; dashed curves were derived from integration of the Gibbs-Duhem equation using molecular dynamics data. Convenient molecular dimensions were used. Reprinted with permission from Deitrick *et al.* [3].

In the future, experimentation will play an increasingly important role in understanding thermodynamic behavior. Current and emerging instrumentation will enable study of structure and dynamics on the molecular or colloidal scale (angstroms to hundreds of nanometers). With the aid of the rotating anode and synchrotron radiation sources and sophisticated computer-aided data analysis, x-ray scattering now provides an accurate noninvasive probe of molecular-level organization. At national and international facilities, neutron scattering competes with x-ray scattering in structural analysis and provides dynamical data on molecular rearrangement not accessible by x-rays. Another modern tool, the video-enhanced light microscope, can be used to study structure and transformation kinetics on the scale of tenths to tens of microns. The transmission electron microscope, extended by fast-freeze cold-stage technology, enables photography and examination by eye or computer-aided image analysis of molecular organization on a scale ranging from a few nanometers to tens or even hundreds of nanometers. Scanning electron microscopy spans the gap between transmission electron microscopy and ordinary light microscopy. The recently invented scanning tunneling microscopy has begun to yield detailed images of the atomic topography of biological membranes and proteins. The turns in the helical structure of DNA are clearly visible in several recent publications of scanning tunneling micrographs [6]. This remarkable microscope can be operated on a laboratory bench without the aid of clean room heroics and can be open to the atmosphere—in fact, it can be operated under liquid. The emerging atomic force microscope, which does not require proximity of an imaged object to a conducting substrate, could be even more powerful than the scanning tunneling microscope.

Modern spectroscopies, such as fast Fourier transform infrared spectroscopy, picosecond laser spectroscopy, x-ray photoemission spectroscopy, electron spin resonance, and nuclear magnetic resonance, based on improved instrumentation and high-speed computer hardware and sophisticated software, provide much information concerning molecular structure and dynamics. For example, with the aid of computer-controlled magnetic field pulsing sequences, nuclear magnetic resonance will give a fingerprint spectrum, which identifies the phase of the subject system, and magnetic relaxation rates, from which local molecular rotational behavior can be deduced and tracer diffusivities of every component of the material can be determined simultaneously. Furthermore, NMR imaging can be used as a noninvasive tool to map out the microstructure of fine composites, biological organisms, and polymeric materials.

The surface force apparatus is now being used routinely to study the equation of state of solutions confined between opposed, molecularly thin solid films. The apparatus is also used in one laboratory to study electrochemistry of thin films at electrodes a few nanometers thick and in a few other laboratories to study the behavior of molecularly thin films subjected to shear and flow [7].

An important point is often overlooked with regard to the role of experimentation. It is not unusual in thermodynamics to think of experiment as the servant of theory and to consider theory as that which is derived from the equations of thermodynamics or a statistical mechanics-based hierarchy of mechanics (e.g., classical, relativistic, quantum). In my opinion, experimental thermodynamics has a more proactive role to play in establishing the conceptual basis on which theoretical models will be built. For example, it has been observed that certain patterns of phase behavior are generic, functions of field variables (intensive thermodynamic variables which are the same in all coexisting phases at equilibrium). Examples include the law of corresponding states, behavior near a critical point, and the sequence of phases that appear in scans of temperature, activity of a solute, carbon number in a homologous series, and the like. The canonization of generic behavior or patterns shared by a large class of thermodynamic systems is indeed theory. So is the concept that the atomics of a solid sit at three-dimensionally periodic sites. If the behavior can also be predicted from model equations of state or statistical mechanics, so much the better.

In his paper, Gubbins identifies the corresponding states theory, perturbation and cluster expansion theory, and density functional theory as those of particular interest to chemical engineers. I agree that these are important theories for future growth in chemical engineering. However, during the next decade or so, the best one can hope to accomplish with these theories is handling molecular solutions of not too complex polar, polyatomic molecules. Microstructural materials such as micellar solutions, liquid crystals, microemulsions, protein solutions, and block copolymer glasses will require approaches not emphasized in Gubbins' paper. One of these approaches is lattice theory, which, contrary to his assertion, is likely to be very useful in understanding and predicting properties of microstructured materials. Another approach neglected by Gubbins is what I have chosen to call the *elemental structures model*. In what follows, I will present the phase behavior of oil, water, and surfactant solutions as a case study to illustrate the special problems posed by microstructured fluids and to indicate how lattice and elemental structures models useful to chemical engineers might be developed.

II. Microstructured Fluids: A Primer

A surfactant molecule is an amphiphile, which means it has a hydrophilic (water-soluble) moiety and a hydrophobic (water-insoluble) moiety separable by a mathematical surface. The hydrophobic "tails" of the most common surfactants are hydrocarbons. Fluorocarbon and perfluorocarbon tails are, however, not unusual. Because of the *hydrophobic tail*, a surfactant resists forming a molecular solution in water. The molecules will tend to migrate to any water-vapor interface available or, at sufficiently high concentration, the surfactant molecules will spontaneously aggregate into association colloids, i.e., into micelles or liquid crystals. Because of the *hydrophilic head*, a surfactant (with a hydrocarbon tail) will behave similarly when placed in oil or when put in solution with oil and water mixtures. Some common surfactants are sodium or potassium salts of long-chained fatty acids (soaps), sodium ethyl sulfates and sulfonates (detergents), alkyl polyethoxy alcohols, alkyl ammonium halides, and lecithins or phospholipids.

The special property of surfactants in solution is that they associate into a monolayer or sheetlike structure with the water-soluble moieties (hydrophilic heads) on one side of the sheet and water-insoluble moieties on the other side [8]. These sheetlike structures provide the building blocks for a rich variety of fluid microstructures, which, depending on thermodynamic conditions, assemble spontaneously. Several known examples of surfactant fluid microstructures are illustrated in Fig. 3.

At an air-water interface, a monolayer forms with heads lying down and tails up (toward air), whereas at an air-hydrocarbon interface the monolayer lies with tails down. By closing on the tail side, the sheetlike structure can be dispersed in aqueous solutions as spherical, rodlike, or disklike micelles (Fig. 3). Closure on the head side forms the corresponding inverted micelles in oil. Oil added to a micellar solution is incorporated into the interior of the micelle to form a swollen micellar solution. Thus, surfactant acts to solubilize substantial amounts of oil into aqueous solution. Similarly, a swollen inverted micellar solution enables significant solubilization of water in oil.

Micellar solutions are isotropic microstructured fluids which form under certain conditions. At other conditions, liquid crystals periodic in at least one dimension can form. The lamellar liquid crystal phase consists of periodically stacked bilayers (a pair of opposed monolayers). The sheetlike surfactant structures can curl into long rods (closing on either the head or tail side) with parallel axes arrayed in a periodic hexagonal or rectangular spacing to form a hexagonal or a rectangular liquid crystal. Spherical micelles or inverted micelles whose centers are periodically distributed on a lattice of cubic symmetry form a cubic liquid crystal.

In addition to the geometrical order associated with the relative arrangements of the surfactant fluid microstructures, the sheetlike assembly of the surfactant imposes a topological order. The two sides of the sheetlike monolayer are different, water-rich fluids always lying on the head side and oil-rich fluids always lying on the tail side. Swollen micellar solutions are water-continuous and oil-discontinuous; i.e., the water-rich region continuously spans a sample of the solution whereas the oil-rich regions are disjoint and lie inside surfaces formed by surfactant monolayers closed on their tail side. Swollen inverted micellar solution, on the other hand, is oil-continuous and water-discontinuous. Similarly, the cubic liquid crystals described above are water-continuous if it is micelles that are periodically arrayed, and oil-continuous if it is inverted micelles that are periodically arrayed. As shown in Fig. 3, topologically different microstructures are possible, namely, bicontinuous microstructures. In these microstructures, the sheetlike surfactant structure separates a sample-spanning water-rich region from a sample-spanning oil-rich region. As indicated in Fig. 3, bicontinuous microstructures occur in liquid crystalline phases (bicontinuous cubic) and in isotropic phases (some microemulsions having substantial amounts of oil and water). The existence of bicontinuous microstructures was postulated by Scriven in 1976 [9] and after several years of controversy has now been experimentally verified.

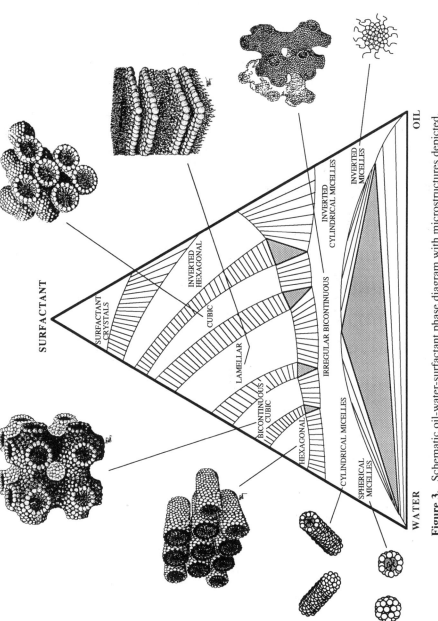

Figure 3. Schematic oil-water-surfactant phase diagram with microstructures depicted.

Proof of the geometric symmetries of the liquid crystals shown in Fig. 3 has been ambiguously established by x-ray scattering. However, unequivocal demonstration of the bicontinuity of some cubic liquid crystals and certain microemulsions requires more than x-ray scattering. The microemulsions story is especially interesting since it represents a case of potential engineering application, namely enhanced oil recovery, driving basic scientific discovery. The triangular phase diagram shown in Fig. 3 is but a single projection out of multivariable parameter space. By varying a field variable—temperature, pressure, activity of a fourth component, carbon number of a homologous series of the oil or surfactant, etc.—a triangular prismatic phase diagram results. The generic pattern of microemulsion phase splits as a function field variable, which has been identified by the Minnesota group [10] and by Kahlweit, Strey, and co-workers [11], is illustrated in Fig. 4. Shown there is temperature-composition diagram indicating the isotropic phase equilibria that occur in mixtures of pentaethylene glycol dodecyl ether ($C_{12}E_5$), water, and octane. At low temperatures two phases coexist, the lower

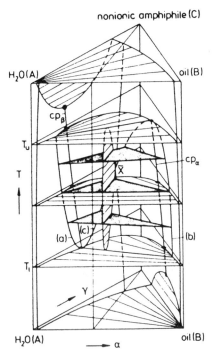

Figure 4.   Generic phase prism of a nonionic surfactant, water, and oil system. Reprinted with permission from Kahlweit *et al.* [11].

phase (in a test tube) being a water-rich microemulsion. As temperature is decreased, a third phase appears at a critical end point (CEP) at the water-rich end of the two-phase tie line. With continued increase in temperature, the middle phase microemulsion increases in oil concentration until the middle phase disappears at a CEP at the oil-rich end. This pattern of phase behavior has been designated by Knickerbocker *et al.* [10] as a $2,3,\bar{2}$ sequence: 2 denotes a two-phase system in which the microemulsion is the lower, water-rich phase; 3 denotes a three-phase system in which the microemulsion is the middle phase and contains appreciable amounts of oil and water; and $\bar{2}$ denotes a two-phase system in which the microemulsion is the upper, oil-rich phase.

The special significance of the $2,3,\bar{2}$ microemulsion sequence to petroleum recovery is that, under certain conditions (of temperature, pH, choice of surfactant, addition of cosurfactant, salt activity, etc.), a middle phase microemulsion has ultralow tension against both oil-rich and water-rich phases [12]. In the example illustrated in Fig. 5, a $2,3,\bar{2}$ sequence of phase splits can be generated as a function of salt activity in mixtures of oil, brine, and ionic surfactants. The ultralow-tension microemulsion is obtained by selecting a particular surfactant and adding a small amount of a low-molecular-weight alcohol—all of which amount to locating the right region of field variable parameter space to find the desired pattern of behavior.

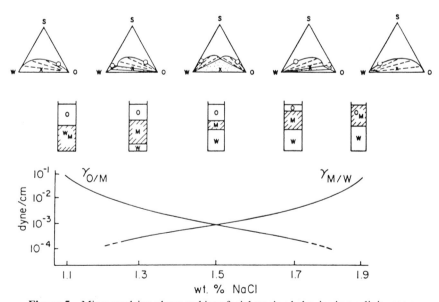

Figure 5. Microemulsion phase and interfacial tension behavior in a salinity scan.

The sequence shown in Fig. 5 has played a prominent role in resolving the question of the microstructure of the mid-range microemulsion. At low salinity, the microemulsion has a low oil concentration and thus is a water-continuous swollen micellar solution. At high salinity, the microemulsion has a low water concentration and is an oil-continuous swollen inverted micellar solution. Somewhere between low and high salinities, the microemulsion undergoes a transition from a water-continuous to an oil-continuous phase. This could happen by sudden inversion of micellar droplets or through the occurrence of bicontinuous microstructure [13] (Fig. 6). The sequence of structures hypothesized in Fig. 6 would be accompanied by very different diffusion behavior (Fig. 7). In the droplet transition case, the self-diffusion coefficient of water would be high until inversion to oil-continuous microemulsion, after which water and surfactant would diffuse together as a relatively slow swollen inverted micelle. Oil and surfactant diffusivities would be low and equal in the water-continuous region and oil diffusivity would be high in the oil-continuous region. Alternatively, if the transition is through bicontinuous microemulsions, there will be an intermediate salinity region in which oil and water diffusivities are comparable while the surfactant diffusivity is lower than either, since surfactant is constrained to move along the sheetlike surface layers separating oil-rich and water-rich regions.

Using pulsed field gradient spin echo NMR, Guering and Lindman [14] and, independently, Clarkson et al. [15] measured the self-diffusion coefficients of the components of microemulsions of sodium dodecyl sulfate (SDS), toluene, butanol, and NaCl brine. The results (Fig. 8) establish unequivocally the existence of bicontinous microemulsion.

It appears that the role of increasing salinity is to change the mean curvature of the surfactant sheetlike structure from a value favoring closure on the oil-rich regions (swollen inverted micelles). In between, in bicontinous microemulsion having comparable amounts of oil and water, the preferred mean curvature must be near zero.

The globular-to-bicontinuous transition of a microemulsion along the $2,3,\bar{2}$ sequence of phase splits is of special interest because of its relationship to enhanced oil recovery technology. However, the microstructural transition is not controlled by phase transitions. If one considers a constant oil/water plane in the prismatic phase diagram given in Fig. 4, the resulting temperature-surfactant composition phase diagram [15] (Fig. 9) has a one-phase (10) corridor lying between a liquid crystalline phase (L_α) to the right and two- and three-phase regions to the left. The water/octane weight percentages are fixed at 60/40 in Fig. 9. This corresponds to about equal volumes of water and octane. Thus, along the corridor, the states A, B, ..., A' represent a progression of one-phase microemulsion of about equal water and

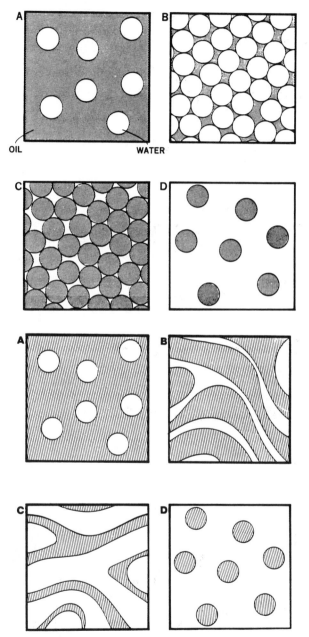

Figure 6. Schematic diagram comparing the droplet inversion transition with the bicontinuous transition of microemulsions in the $\underline{2},3,\overline{2}$ sequence of phase splits. Reprinted with permission from Kaler and Prager [13].

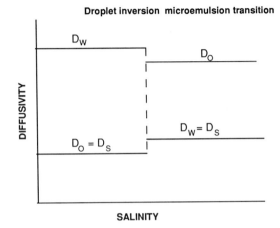

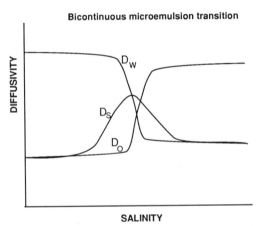

Figure 7. Schematic diagram comparing the behavior of self-diffusion coefficients of oil (D_o), water (D_w), and surfactant (D_s) expected for the droplet inversion transition and the bicontinuous transition of microemulsion depicted in Fig. 6.

oil volumes. Diffusivities (Fig. 10) measured along this corridor by NMR show plainly that, between G and H, a transition from globular to bicontinous microstructure occurs [16].

Although NMR results provide perhaps the most convincing evidence of the bicontinuous structure of some microemulsions, many other techniques support their existence. These techniques include electrical conductimetry, x-ray and neutron scattering, quasielastic light scattering, and electron

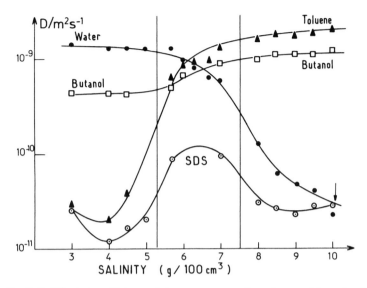

Figure 8. Self-diffusion coefficients of the components of a microemulsion of sodium dodecyl sulfate (SDS), butanol, toluene, and NaCl brine. Vertical lines denote 2,3 and 3,2̄ phase transitions. Reprinted with permission from P. Guering and B. Lindman, *Langmuir* **1**, 464 (1985) [14]. Copyright 1985 American Chemical Society.

microscopy. For example, a tantalum-tungsten replica of a fracture surface of frozen microemulsion viewed in the transmission electron microscope [17] shows a cross section consistent with a structure of intertwined water and oil-continuous regions (Fig. 11 is a TEM micrograph of $C_{12}E_5$, water, octane microemulsion at state J in Fig. 9). In the globular regime, TEM micrographs reveal disjoint spherical structures quite different from the structures seen in Fig. 11 [16].

III. Microstructured Fluids: Theory and Simulation

At its most satisfying level, a statistical thermodynamic theory would begin by specifying realistic interaction potentials for the molecular components of a complex mixture and from these potentials the thermodynamic functions and phase behavior would be predicted without further approximation. For the next decade or so, there is little hope to accomplish such a theory for microstructured fluids. However, predictive theories can be obtained with the aid of elemental structures models. Also, lattice models

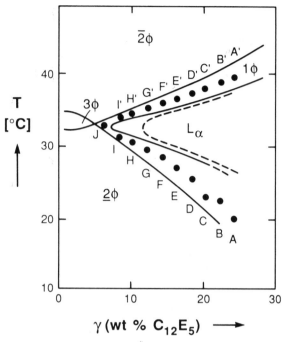

Figure 9. Temperature-composition phase diagram of the system pentaethylene glycol ether ($C_{12}E_5$), water, and *n*-octane. The weight ratio of water to octane is fixed at 60 to 40. Diffusivities were measured in samples denoted by filled circles. Phase diagram determined by Kahlweit *et al.* Reprinted with permission from Kahlweit *et al.* [11].

provide an affordable means to probe, primarily by computer simulation, the connection between thermodynamic behavior and molecular interactions.

The ideas underlying elemental structures models are to establish micro-structures experimentally, to compute free energies and chemical potentials from models based on these structures, and to use the chemical potentials to construct phase diagrams. Jönsson and Wennerström have used this approach to predict the phase diagrams of water, hydrocarbon, and ionic surfactant mixtures [18]. In their model, they assume the surfactant resides in sheetlike structures with heads on one side and tails on the other side of the sheet. They consider five structures: spheres, inverted (reversed) spheres, cylinders, inverted cylinders, and layers (lamellar). These struc-tures are indicated in Fig. 12. Nonpolar regions (tails and oil) are cross-hatched. For these elemental structures, Jönsson and Wennerström include in the free energy contributions from the electrical double layer on the water

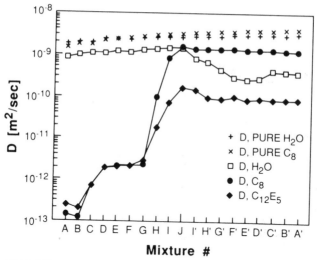

Figure 10. NMR diffusivities for the components of the samples denoted in Fig. 9 by filled circles. The samples are microemulsions lying along the one-phase corridor in Fig. 9. Reprinted with permission from Bodet *et al.* [16].

side of the surfactant sheets, the entropy of mixing of micellar aggregates and of the molecular components in aqueous and oleic regions, and the surface energy of the surfactant sheets.

Figure 13 compares the observed ternary phase diagram of water, octanol, and potassium caprate with that calculated by Jönsson and Wennerström. The agreement, though not quantitative in every detail, is remarkable given the complexity of the phase behavior. The agreement is certainly good enough to encourage future development of theory along these lines for other mixtures. Wennerström and Jönsson have in fact extended their analysis to a few other systems.

In solutions of water and surfactant, the surfactant monolayers can join, tail side against tail side, to form bilayers, which form lamellar liquid crystals whose bilayers are planar and are arrayed periodically in the direction normal to the bilayer surface. The bilayer thickens upon addition of oil, and the distance between bilayers can be changed by adding salts or other solutes. In the oil-free case, the hydrocarbon tails can be fluidlike (L_α) lamellar liquid crystal or can be solidlike (L_β) lamellar liquid crystal. There also occurs another phase, P_β, called the modulated or rippled phase, in which the bilayer thickness varies chaotically in place of the lamellae. Assuming lamellar liquid crystalline symmetry, Goldstein and Leibler [19] have constructed a Hamiltonian in which (1) the intrabilayer energy is calculated

Figure 11. Freeze-fracture transmission electron micrograph of the microemulsion denoted as sample J in Fig. 9. Reprinted with permission from W. Jahn and R. Strey, *J. Phys. Chem.* **92**, 2294 (1988) [17]. Copyright 1988 American Chemical Society.

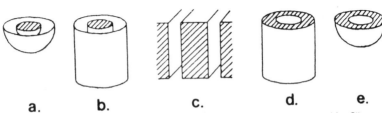

    a.   **b.**   **c.**   **d.**   **e.**

Figure 12. Schematic representation of the five elemental structures used by Jönsson and Wennerström [18]: (a) spherical, (b) cylindrical, (c) lamellar, (d) inverted cylindrical, and (e) inverted spherical. Nonpolar regions are crosshatched.

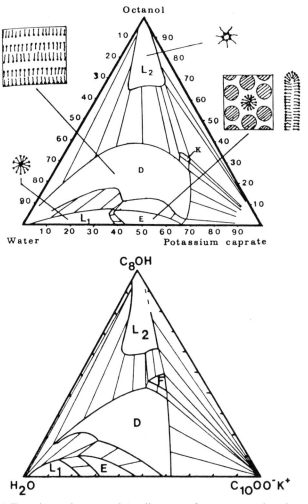

Figure 13. (a) Experimental ternary phase diagrams of water, octanol, and potassium caprate at 20°C. (b) Phase diagram predicted by Jönsson and Wennerström [18] using an elemental structures model. Reprinted with permission from B. Jönsson and H. Wennerström, *J. Phys. Chem.* **91**, 338 (1987) [18]. Copyright 1987 American Chemical Society.

from a Landau-Ginzburg Hamiltonion accounting for the energy of thickening, stretching, and bending of the bilayer and (2) the interbilayer interactions are approximated by the van der Waals attractive potential plus the hydration repulsion potential. The hydration potential has been found empirically to be an exponential function of the distance between adjacent

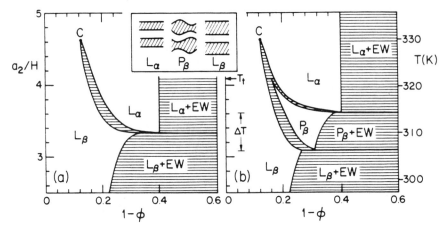

Figure 14. Predicted (Goldstein and Leibler [19]) temperature-composition phase diagrams of lamellar phases in water and phospholipid mixtures. L_α and L_β are periodic liquid crystals and P_β is a modulated or rippled lamellar phase. a_2 is a model parameter that scales as temperature. f is the volume fraction of phospholipid. Reprinted with permission from Goldstein and Leibler [19].

bilayers. Goldstein and Leibler's elemental structures model yields a phase diagram (Fig. 14) similar to those observed for mixtures of water and phospholipids, which are the well-known surfactants in biological membranes.

Many of the special properties of microemulsions have been predicted with the aid of the following elemental structures model advanced by Talmon and Prager [20]: (1) oil- and water-rich domains are randomly interspersed in space, (2) the surfactant lies entirely in monolayers separating the oil- and water-rich domains, and (3) the free energy arises from the entropy of interspersion of the domains and from the energy of curvature of the surfactant layers. Talmon and Prager used a random subdivision of space, namely the Voronoi polyhedral tessellation, to provide a means of randomly interspersing oil and water domains. In later variations of the Talmon-Prager model, de Gennes and Taupin [21], Widom [22], and Safran *et al.* [23] used the simpler cubic tessellation of space to model the random interspersion. Qualitatively, the choice of tessellation does not appear to make much difference; from the micrograph shown in Fig. 11, it is clear that no polyhedral tessellation will capture the detailed geometry of a microemulsion, although it might adequately account for the bicontinuous topology. The transition from globular to bicontinuous microemulsion is predicted by the Talmon model, and the electrical conductivity observed in a $2,3,\bar{2}$ salinity scan is quantitatively predicted by that model. The $2,3,\bar{2}$ sequence of

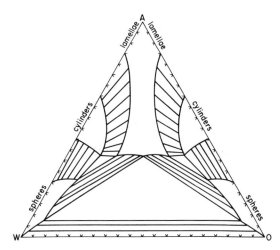

Figure 15. Monte Carlo simulation of the phase diagram of the amphiphile H_4T_4, water, and oil modeled on a cubic lattice. Reprinted with permission from Larson [24].

phase splits and the low-tension behavior are qualitatively accounted for by various versions of the model. Recent versions of the model [22, 23] also focus attention on the importance of the preferred curvature and bending energy of the surfactant sheets for the patterns of phase behavior of surfactant microstructured fluids.

As I stated earlier, lattice models, aided by computer simulations, are likely to be very useful in predicting patterns of thermodynamic behavior and in understanding molecular mechanisms in microstructured fluids. Perhaps the simplest example of such a lattice model is the one employed recently by Larson [24] to investigate patterns of liquid crystal formation by Monte Carlo simulation. He assumes that the molecules occupy sites of a cubic lattice. Water and oil molecules occupy single sites in the lattice. The surfactant molecules (H_mT_n) consist of a string of m head segments and n tail segments. A surfactant molecule occupies a sequence of adjacent sites, either nearest neighbor or diagonal nearest neighbor. In terms of pair interaction energies, water and a head segment are equivalent, and oil and a tail segment are equivalent. A simulated phase diagram for H_4T_4 is shown in Fig. 15. For this symmetric surfactant (equal lengths of head and tail moieties), no cubic-phase liquid crystals are observed. Nevertheless, a surprisingly complex phase diagram is found, with many of the features depicted in Fig. 3. Larson found the bicontinuous liquid crystal phase with the asymmetric surfactant H_2T_6.

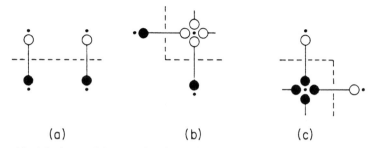

Figure 16. A lattice model accounting for the bending energy of surfactant layers. Filled circles denote the head group and unfilled circles the tail group of the surfactant. Reprinted with permission from Widom [25].

Widom has recently formulated a lattice model that takes into account the amphiphilic nature of surfactant and introduces molecular interactions that explicitly affect curvature of surfactant sheetlike structures [25].

The three configurations of adjacent surfactant molecules shown in Fig. 16 have different energies of interaction. In addition to these interactions, the model allows pair interactions among water, oil, surfactant head groups, and surfactant tail groups. Taking advantage of the relationship between Widom's model and an equivalent Ising model, Dawson et al. [26] have computed the phase diagram for oil, water, surfactant solutions from known results for the Ising model and from mean field theory. They find lamellar liquid crystalline phases, micellar solutions, and bicontinuous microemulsion phases. For the model microemulsion, Dawson [27] has predicted phases exhibiting ultralow tensions against oil-rich and water-rich phases having compositions very different from that of the microemulsion. This is in agreement with experiment. Further mean field theory and computer simulations based on this model should be fruitful.

There are presently several groups around the world conducting molecular dynamics simulations of micellization and liquid crystallization of more or less realistic models of water, hydrocarbon, and surfactants. The memory and speed of a supercomputer required to produce reliably equilibrated microstructures constitute a challenge not yet met, in my opinion. By taking advantage of identified or hypothesized elemental structures one can, however, hope to learn a great deal about the dynamics and stability of the various identified microstructures.

References

1. Gubbins, K. E., in *Perspectives in Chemical Engineering: Research and Education* (C. K. Colton, ed.), p. 125. Academic Press, San Diego, Calif., 1991 (*Adv. Chem. Eng.* 16).

2. Panagiotopoulos, A. Z., *Mol. Phys.* 61, 813 (1987).

3. Deitrick, G. L., Scriven, L. E., and Davis, H. T., *J. Chem. Phys.* 90, 2370 (1989).

4. Jönsson, B., Edholm, O., and Teleman, O., *J. Chem. Phys.* 85, 2259 (1986).

5. Egberts, E., and Berendsen, H. J. C., *J. Chem. Phys.* 89, 3718 (1988).

6. Lee, G., Arscott, P. G., Bloomfield, V. A., and Evans, D. F., *Science* 244, 475 (1989).

7. Israelachvili, J. N., McGuiggan, P. M., and Homola, A. M., *Science* 241, 795 (1988); Van Alsten, J., and Granick, S., *Phys. Rev. Lett.* 61, 2570 (1988).

8. Davis, H. T., Bodet, J. F., Scriven, L. E., and Miller, W. G., in *Physics of Amphiphilic Layers* (J. Meunier, D. Langevin, and N. Boccara, eds.), p. 310. Springer-Verlag, Berlin, 1987.

9. Scriven, L. E., *Nature* 283, 123 (1976).

10. Knickerbocker, B. M., Pesheck, C. V., Scriven, L. E., and Davis, H. T., *J. Phys. Chem.* 83, 1984 (1979); *J. Phys. Chem.* 86, 393 (1982).

11. Kahlweit, M., Strey, R., Haase, D., Kunieda, H., Schemling, T., Faulhaber, B., Borkovec, M., Eicke, H.-F., Busse, G., Eggers, F., Funck, Th., Richmann, H., Magid, L., Söderman, O., Stilbs, P., Winkler, J., Dittrrich, A., and Jahn, W., *J. Colloid Interface Sci.* 118, 436 (1987).

12. Healy, R. N., Reed, R. L., and Stenmark, D. G., *Soc. Pet. Eng. J.* 16, 147 (1976).

13. Kaler, E. W., and Prager, S., *J. Colloid Interface Sci.* 86, 359 (1982).

14. Guering, P., and Lindman, B., *Langmuir* 1, 464 (1985).

15. Clarkson, M. T., Beaglehole, D., and Callaghan, P. T., *Phys. Rev. Lett.* 54, 1722 (1985).

16. Bodet, J. F., Bellare, J. R., Davis, H. T., Scriven, L. E., and Miller, W. G., *J. Phys. Chem.* 92, 1898 (1988).

17. Jahn, W., and Strey, R., *J. Phys. Chem.* 92, 2294 (1988).

18. Jönsson, B., and Wennerström, H., *J. Phys. Chem.* 91, 338 (1987); Wennerström, H., and Jönsson, B., *J. Phys. France* 49, 1033 (1988).

19. Goldstein, R. E., and Leibler, S., *Phys. Rev. Lett.* 61, 2213 (1988).

20. Talmon, Y., and Prager, S., *J. Chem. Phys.* 69, 2984 (1978).

21. de Gennes, P. G., and Taupin, C., *J. Phys. Chem.* 86, 2294 (1982).

22. Widom, B., *J. Chem. Phys.* 84, 6943 (1986).

23. Andelman, D., Cates, M. E., Roux, D., and Safran, S. A., *J. Chem. Phys.* 87, 7229 (1987).

24. Larson, R. G., *J. Chem. Phys.* 89, 1642 (1988).

25. Widom, B., *J. Chem. Phys.* 84, 6943 (1986).

26. Dawson, K. A., Lipkin, M. D., and Widom, B., *J. Chem. Phys.* 88, 5149 (1988).

27. Dawson, K. A., *Phys. Rev. A* 35, 1766 (1987).

General Discussion

Thermodynamics

Morton Denn: Keith Gubbins made a throw-away comment that has a good deal of educational significance. He implied that we don't teach statistical mechanics in our thermodynamics courses, therefore students don't relate it to chemical engineering. One conclusion to draw from that comment, if I understood it correctly, is that we're responsible for making our students understand that this topic is important. It's obvious from the direction the curriculum is moving these days that we have to be as efficient as possible in using what's taught elsewhere. Clearly, statistical mechanics should be taught in other courses as well. I don't believe that we want to add it to our undergraduate thermodynamics curriculum, but we want to make sure that wherever else it's relevant, the students are reminded that it's important and useful in chemical engineering.

James Wei: Thermodynamics in service to oil and gas and commodity chemicals is a very successful model. It's a mature technology on the S curve. The problem is to know the material properties in very great detail so we can optimize each percent improvement in efficiency, which represents a lot of money. Early in his lecture, Keith mentioned new materials and speciality chemicals. In these applications, we're going into a different mode, and a different type of a commitment is needed. For instance, many biotechnology products and some of the microstructured fluids and solids are produced in very small volume, they have a short product cycle, and we cannot be assured that the same product will still be produced ten years from now. It's important to get these specialty chemical products on the market speedily. Because they are produced in very small volume, we cannot afford large amounts of time or research money. Let's say that one percent of the sales volume is a reasonable research dollar. Instead of methane, propane, and butane, which you can study for 30 years in order to understand everything about them, think of one of the new growth hormones that may be here now and replaced by another growth hormone in five years. You cannot study them in the same way. With respect to thermodynamics in service to these new speciality chemicals, a different type of information and a different speed of delivery are needed. It cannot be business as usual. Ideas of molecular theory and computer simulation may be necessary, but I'm not sure that they are sufficient to face the challenge of how to put thermodynamics in service to these new industries.

Edwin Lightfoot: That may be a key problem, Jim. We've already heard in the first session about several subjects that should be in the curriculum that would be advantageous to the master's degree, perhaps even the Ph.D. In biotechnology, there is an enormous amount of factual information, and the biologists and biochemists have got to teach it themselves. Our students taking courses in those departments suddenly realize that it's a much bigger world than they thought. We're seeing chemical engineering expand, and the challenge is to give our students a working knowledge of where to look for the information they might be applying. We can't give them detailed descriptions of the kind that Ted just ran through, except in specialized courses. This means writing much better textbooks than ever before. Yet there's very little time available to do that, and that may be our single greatest challenge.

James Wei: I think it's a research problem first, before it becomes an educational problem.

Edwin Lightfoot: Of course, but it ultimately has to be packaged in such a way that the students have access to the information available without a great deal of detailed study. Understanding orders of magnitude is clearly important. We talk about all the topics that must be in the curriculum, but there's no room for the kind of detail that experts feel is necessary.

Jefferson Tester: I completely agree on the textbook issue. Keith points it out nicely in his paper. I wonder what they're doing at Cornell to solve this problem. Keith, are you historically separated in physical chemistry from the undergraduate curriculum?

Keith Gubbins: At the undergraduate curriculum level what's happened is a lot of sound and not much action. We're discussing it. The types of textbooks that you'd like are not there. The most widely used thermodynamics textbook in this country doesn't mention the word molecule from beginning to end, as far as I know. That worries me, because I don't see any other textbooks that bring out the uses of molecular theory and statistical mechanics with the kind of examples that you would like at a level that an undergraduate can understand. We don't get very far when we talk to our chemistry and physics colleagues about putting these topics into their curriculum for our students. The fact is, it's very difficult to think up nice examples of the uses of statistical mechanics for applications other than ideal gases and teach it to undergraduates in the time available. I think that's a problem that has to be solved, a very important problem.

Jefferson Tester: We find ourselves at MIT with the exact same dilemma in dealing with a physical chemistry textbook like Castellan, and an undergraduate favorite, Smith and Van Ness. The fourth edition of Smith and Van Ness does include a very brief treatment of statistical mechanics. Although it's a start, a more effective route we have found is to integrate statistical and molecular concepts from a physical chemistry text into our introductory chemical engineering thermodynamics course. In this situation, we have a two-semester sequence that uses both Castellan and Smith and Van Ness as textbooks.

Alexis Bell: We've recently revised the undergraduate curriculum at Berkeley, and very heavy consideration was given to what is taught in both the courses that we teach and in the service courses. We've implemented a new physical chemistry sequence that was developed by the chemists. One of the two courses is largely devoted to statistical thermodynamics and the introduction of thermodynamics at the molecular level, then going up to the continuum level. In the future, our students will see the molecular picture as taught by chemists and the continuum picture in a separate course taught by our own faculty.

L.E. Scriven: There is a story that has to be apocryphal. The story runs that in the 16th century, if one wanted to learn the newly invented long-division algorithm, which we now pick up in—where? third, fourth, fifth grade— one had to go to Germany to study with the specialist professor in the University of—I'm not sure just where—to learn the new-fangled technique. It was beyond ordinary mortals. This apocryphal story of course points up the eternal problem of curriculum evolution and of textbook preparation. What we need, as every generation needs, are inspired teachers who can reduce seemingly complex and multifarious notions to easily assimilable form. Obviously, we need to promote the process in molecular thermodynamics and every other part of our discipline.

Eli Ruckenstein: We have had three speakers who have discussed the problems of thermodynamics. They have different backgrounds and work in different directions. One is interested in molecular dynamics, another in enlightening correlations, and the third in fluid microstructures. All three are doing excellent research, which demonstrates that there are many good ways to do research in a given area.

Let me comment on molecular dynamics. There is no doubt in my mind that our students should learn this discipline. The progress in this direction in the next two or so decades will be so great that its use will become routine. The perception about doing research will be affected enormously by the capabilities provided by computers. Still, I feel that we should not entirely renounce simple models because they can be illuminating. Let me use as an example capillary condensation, the same phenomenon Professor Gubbins has used to illustrate the molecular dynamics technique. Of course, the results obtained by molecular dynamics are exact. If the interaction potential used in the calculations is accurate, one can obtain good results. This problem was also treated, with the same expression for the interaction potential, by de Boer and his co-workers, in a group of papers published in *J. Catalysis* and *J. Colloid and Interface Science* more than two decades ago. These authors employed simple models and obtained simple expressions which illuminate the phenomena that occur in capillary condensation. I would like to emphasize that not only the sophistication of the theory and the accuracy of the numbers provided should matter, but so should the insight gained into the problem and the general insight which affects our overall outlook. The main achievements of engineers have rarely been a result of exact calculations, but a result of their engineering instincts, instincts generated by the insight gained earlier in various ways.

Let me close with two more familiar examples. We all know that Flory's theory of polymeric solutions is not in agreement with experiment. However, the insight on polymer compatibility gained from the theory was enormous. The same thing may be said about the van der Waals equation of state. We all know that this equation is very approximate, but something remains in our intuition that is extremely helpful.

John Prausnitz: I'd like to make one quick addendum. I want to defend the use of molecular simulations because we have gained insight from them as well. Let me mention one outstanding example. Until about 20 years ago, we believed that you could only condense a phase with attractive forces, and no one ever questioned that myth. Then computer simulations were done in the 1960s by Berni Alder and his associates. The results showed that even for hard spheres, without any attractive forces, you can get a phase transition. This was never present in the van der Waals theory. I want to emphasize that simulations also add to our conceptual knowledge.

Eli Ruckenstein: I don't disagree with you. People can gain insight from everything if they have sufficient experience. I am sure that people with experience in molecular dynamics have gained a lot of insight from their work. What I mean is that simple models can also provide insight and that they remain useful.

Matthew Tirrell: The emphasis in the talks on thermodynamics has been exclusively on equilibrium thermodynamics, on the energetics, and not on the dynamics. There's a great opportunity to get involved in the study of the dynamics, especially of phase transitions, if we're interested in material processing. In the polymer area, the time scale and the length scale of the dynamics of these phase separation processes are ripe for study. In processing operations, we impose the time scale on the system; if the phase separation dynamics can't keep up with that time scale, we're quite likely to be trapped in kinetically nonequilibrium states, and that will affect the structure and the properties of the materials that we create. It's true for polymers and for other kinds of materials. Polymers themselves have other advantages for study, in that they are good mean field objects because they're big, they average locally quite a bit, they interact with many of their neighbors, and we have a good understanding of the equilibrium as a starting point to study the dynamic processes. Interactions among molecules that produce interesting equilibrium phase behavior also influence the rates of phase transition.

Clark Colton: All the comments about molecular simulations have emphasized their power and the computational aspects. Nothing's been said about the intermolecular potential energy functions. What we have is fine for small molecules, but once we get to large, complicated molecules, such as charged proteins in an aqueous electrolyte solution, we have difficulty writing the intermolecular potential energy function. Why hasn't that issue been identified?

John Prausnitz: May I respond to that? That's handled by breaking up your molecule into segments of parts. You then write segment-segment potentials that you think are reasonable. You write a form without necessarily knowing what the numbers are in the equation. You then treat those as adjustable parameters. In other words, you have some macroscopic experimental information, and you use that as a boundary condition to fit your calculations. From the macroscopic information, you extract the adjustable parameters of the potential. This has been very successful in modeling relatively complicated molecules. We are far beyond the argon-xenon stage. Of course, for a purist, that wouldn't be good enough. Physicists wouldn't like it. But for engineering, this is a perfectly reasonable way to do it.

Jefferson Tester: I think a comment on *ab initio* calculations and the fundamental nature of the intermolecular potential function would be appropriate. Perhaps Ted Davis would respond.

Ted Davis: Potentials now exist that are good for water, in the sense that you name those properties of water that you want your simulation to contain, and there are potentials that people have labored over that will reproduce those properties of water in a simulation. The molecular segment approach that John's talking about has been applied to surfactants as large as at least eight carbon atoms. Sodium octanoate in water has been simulated. Micelles of sodium octanoate and water have been simulated. One can literally watch the molecular movements and see the relative movement of hydrated water versus water in bulk and the changes of configurations in the hydrocarbon tail. In fact, with good computer graphics, I think one can very soon sit and look at a real-time or some slowed-scale-time scan of the molecular processes involved in relatively complicated molecules in water and in other solvents. Adding another salt, say sodium chloride, is easy, but you pay for everything you put in there. The molecular dynamics don't get more complicated, but the price goes up.

Keith Gubbins: I want to add a comment about *ab initio* calculations; I'm certainly not an expert in that, but there is a lot of industrial interest now in combining electronic structure calculations with large-scale simulations. There was a conference on this in Britain in January that I went to, called "Industrial Applications of Computer Simulation," and I was surprised to see many more people from industry than academia. As a result of the industrial interest, their next conference in January will get together the electronic structure people with the simulators, but I wouldn't hazard a guess as to when this would be sufficiently practical, say, to design a new catalyst. Whether we're close to that in the next five or ten years, I'm not really certain.

L. E. Scriven: I'd like to share a piece of information. Not so long ago, I ran across a physical organic chemist whose motivation was synthesis of new molecules, but in his theoretical studies he was in fact using *intra*molecular potentials to deal with complicated molecules, potentials that looked for all the world like proper *inter*molecular potentials. I thought I glimpsed there a view of what's coming.

James Katzer: In John Prausnitz's talk, I was happy to hear that we'd reached an important milestone. We've now designed something in an enzyme structure. That's a most important direction. I was also happy that Keith Gubbins focused on the molecular scale. In the catalysis area, chemical engineers had a major impact initially; then they reached an asymptotic level when they did not bring in the chemistry. When we start integrating the chemistry with the chemical engineering, and that's frequently occurring now, we will see major advances.

We should also look at frontiers and worry about where those frontiers are headed. I'd like to step back to the discussion of this morning. In my estimation, one thing missing from the fluid dynamics session is that nobody worried about approaching it from the molecular point of view. We should be able to start with the molecules and with molecular dynamics and predict some of those parameters in the constitutive relations that we now have to measure. There are certainly high-tech areas today where we can use that approach in design. One area that comes to mind is lubrication, where you want to design molecules to have specific properties. Right now, lubrication is largely empirical. If we can bring in molecular dynamics, we can close the loop and do *ab initio* design. In multilayer surface coating and deposition, as in fluid mechanics, we have the opportunity to carry out design and investigate new applications from a very fundamental level, rather than just generating a lot of hard-to-interpret empirical data.

I have one other comment: We need to advance the quality and the structure of our undergraduate and graduate education. At the graduate level, what's important is not larger groups necessarily but more integrated groups. Say that you have a biochemist, a chemical engineer, and maybe a theoretical chemist or physicist working in the same group. In that environment, the chemical engineering student doesn't have to learn everything, which he can really never do. A lot of cross-fertilization can occur with multidisciplinary interactions, but it is still a struggle in the academic world to see this approach take off. We still compartmentalize to a high degree. Although I think most of you work at trying to interact, there's still more lip service than real concrete activity.

Anthony Pearson: I wish to talk about thermodynamics, I hope, because I don't really know any thermodynamics. First, if I can address the question about fluid mechanics and molecules: I had cause to give a lecture not a month ago to rheologists—and I hope you'll appreciate that a rheologist does look at molecular effects—in which I talked about length scales in rheology. First, there is the rheologist's microscale at the molecular level, and there's a lot being done with it. Next, there is an area of particulate rheology, which I think is where Gary Leal has worked, that I call the mesoscale, from about 100 nanometers to about 100 micrometers. Then there is the rheologist's macroscale, which is the one in which he produces his constitutive laws. If I had an applied rheologist, he'd be displaced down there in the street.

Now to the thermodynamics. Like Matt Tirrell, I was a bit upset that we hear only from the thermodynamicists about equilibrium thermodynamics. I did struggle once to try to understand irreversible thermodynamics, and it seemed to me rather trivial. Nevertheless, there is a thing that I think could be called nonequilibrium thermodynamics, which nobody has talked about. So let me explain why it seems to me it might have been discussed so far, and then you can shoot me down when I've made an ass of myself.

I tried to explain about the microscale and the mesoscale. Brownian motion occurs at the microscale, and it is crucial in understanding what happens. One is in fact modeling Brownian motion with molecular theories. If you go to the mesoscale, then in many of the processes—in terms of the rates and the scales that we're looking at—you do not get Brownian motion. The system does not have a chance to reach what I believe is regarded as thermodynamic equilibrium, and since I'm a rheologist, I'm talking about continuously deforming materials. Now, maybe somebody will be able to tell me whether thermodynamics in any form can apply to continuously deforming materials. What I do know is that there is a very successful equilibrium thermodynamic theory of rubber elasticity, but that has in it something nobody has mentioned today—perhaps chemical engineers regard it as irrelevant—which is the tensor quantity known as stress. You teach your students only the effect of one mechanical parameter, that is, the isotropic component of stress.

James Douglas: Can you be more specific?

Anthony Pearson: The deviatoric stress is an important feature. I used the term stress there, but when one does these calculations on multiphase mixtures, suspension, emulsions—one is really looking not at the stresses initially, but one tends to be looking at rates of deformation. Although you get no Brownian motion, you do get very considerable structure development. My question is, is there any way in which thermodynamics can deal with structure development in a nonequilibrium state? If you stop shearing the material, the structure disappears.

William Russel: May I follow up on that and sharpen the issue a bit? In the complex fluids that we have talked about, three types of nonequilibrium phenomena are important. First, phase transitions may have dynamics on the time scale of the process, as mentioned by Matt Tirrell. Second, a fluid may be at equilibrium at rest but is displaced from equilibrium by flow, which is the origin of non-Newtonian behavior in polymeric and colloidal fluids. And third, the resting state itself may be far from equilibrium, as for a glass or a gel. At present, computer simulations can address all three, but only partially. Statistical mechanical or kinetic theories have something to say about the first two, but the dynamics and the structure and transport properties of the nonequilibrium states remain poorly understood, except for the polymeric fluids.

Eli Ruckenstein: Various time scales have to be compared to obtain an answer to this question. If the relaxation time to attain equilibrium is short compared to the time scale determined by the fluid mechanics, "local" thermodynamic equilibrium can be assumed. Such an assumption can be made for micellar solutions because the surfactant aggregation and the dissociation of aggregates are relatively rapid processes. It is, however, expected that the size of the aggregate will depend on the stress. For concentrated dispersions, the problem may be more complicated because the processes of aggregation and dissociation of aggregates probably are much slower.

William Russel: But the question is, how do you know what the time scale is for the relaxation?

Eli Ruckenstein: Only kinetic theories can provide such information. Dr. Mewis and I have applied such a treatment to thixotropy, and the theory developed accounted for the formation and destruction of various microstructures. Something similar could be extended to concentrated dispersions.

William Schowalter: We've heard that it's a matter of time scales, that there are many important cases where these time scales are of comparable magnitude and are interfering with each other. I think the question Anthony is asking, and certainly one I would ask, is given that you have this confounding of time scales in important applications, are there people in the arena of thermodynamics doing something that might help those of us in fluid mechanics who don't know thermodynamics?

Eli Ruckenstein: For surfactant aggregation there are both theories and experiments that address the problem of relaxation to equilibrium. The experimental methods have been borrowed from rapid-reaction kinetics. The first theory was developed by Aniansson and Wall (*J. Phys. Chem.* 1974). I am not familiar with experiments regarding relaxation in concentrated dispersions. For a sufficiently dilute colloidal solution, such experiments and theories probably could be carried out.

Jefferson Tester: I would like to shift gears and direct a question to John Prausnitz regarding his comments. You didn't talk very much about some of your own contributions and those of your students, for instance, the NRTL equation and UNIQUAC-UNIFAC models for nonideal solutions now in widespread use throughout the chemical industry and certainly employed by many people making practical calculations. There have been extensions of that local composition approach, in particular to electrolyte systems, by C. C. Chen and others. I wonder how you personally feel about that and whether that is a direction you would suggest that research might go as time proceeds.

John Prausnitz: My feeling is that although it's now badly out of date, it's still a good engineering tool. We also know the assumptions used to derive those equations were in serious error, and molecular simulations have shown that many of our assumptions were not right. It works as well as it does in practice as a consequence of a well-known rule: if you must make assumptions, make many assumptions because the associated errors might cancel out. That's precisely what was done with the NRTL and the UNIQUAC equation. One reason why they're used so much is because they have nice names; one must always package things correctly. They also allow you to use the group contribution method, which is probably the most popular thing that's ever come out of my group at Berkeley. People in industry just love it because they don't have to think at all; they just push the button. My own feeling is that we ought to stop that and go on to something new.

James Wei: Earlier I made the statement that the method of doing thermodynamics by detailed study of properties will not work for the low volume and short product cycle chemicals, and a new method is needed. I was waiting either to be shot down in flames or to find this is common knowledge, and everybody agrees with it.

Clark Colton: I'd like to respond, because I disagree with you.

James Wei: Okay, good. Tell me why.

Clark Colton: There are areas where our knowledge is primitive, where we need general understanding and also a better sense of what causes exceptions. An example would be thermodynamic properties of some of the high-value proteins you mentioned. Although produced in low volume, these materials may have extremely high value, and their life cycle is not necessarily short. Our knowledge of thermodynamic properties of protein solutions is incredibly poor. Jay Bailey will talk about this tomorrow. We have little data and very few good rules, even *ad hoc* rules. It's not a matter of learning about a specific protein, it's learning general properties and how they relate to protein chemistry and structure. Later we'll worry about specific exceptions. There are too few studies of any kind going on. It's an area that needs some work, and in-depth studies are certainly justified.

James Wei: Clark, in that case, what the community should do is to select one or two to study in depth. It's just like what the people did in genetics. They said we'll only study *E. coli* and nothing else. Then later maybe they will investigate different bacteria. If I study one thing and he studies another, that is no common ground. We don't know anything. If there is anything to be studied, it should be something which is a model that represents a lot, and we should concentrate on that.

Morton Denn: I have a mild problem with the "pick one thing and study it to death" notion. There's at least one good counterexample to *E. coli*, and that's polymer rheologists who invested about one and a half decades studying polyethylene to death, until it was finally discovered that polyethylene is sufficiently atypical of all other polymers that most of what went on just couldn't be extrapolated.

Edwin Lightfoot: Mort, I disagree, because there are substructures in proteins that are pretty much ubiquitous, and I think it's better to study those.

Jefferson Tester: You're suggesting a set of model compounds?

James Wei: You study model compounds and model bacteria. We have to have models because there are too many substances to study.

James Douglas: Let me add to Jimmy Wei's comment. When we design new processes, we'll need a hierarchy of models. During initial design when you're trying to see if a process will make money, you want to use the simplest models you can. On the other hand, you don't understand the nature of approximation unless you understand the nature of the rigorous solution. We will have to do a lot of fairly detailed studies to develop shortcut models that are useful in these new areas. Designers won't be able to wait for that, and therefore we're going to design a lot of new processes without much understanding at all, which we've always done in the past. Some of these systems will work, and some of them won't. Like John Prausnitz says, if you make enough assumptions, sometimes they cancel out and things work out fine.

James Wei: You don't even need thermodynamic information?

James Douglas: No, you rely on a minimum number of experiments instead.

SECTION IV
Kinetics, Catalysis, and Reactor Engineering

10

Reflections on the Current Status and Future Directions of Chemical Reaction Engineering

Alexis T. Bell
Department of Chemical Engineering
University of California, Berkeley
Berkeley, California

I. Introduction

Virtually all of the fuels, chemicals, and materials used by modern societies are obtained through chemical transformations. The subfield of chemical engineering that deals with how to carry out these transformations in a practical, economical, and safe fashion is known as chemical reaction engineering. The range of subjects encompassed by chemical reaction engineering is broad and includes reaction kinetics, catalysis, and reactor modeling. While research on kinetics and catalysis considerably predates that on reactor modeling, today all three subjects are closely related. In fact, it is now inconceivable to consider discussing the performance of a reactor without giving thorough consideration to the relevant reaction kinetics and the influence of catalyst composition and structure. Moreover, it is increasingly recognized that, to achieve a given process objective, it is possible to apply the concepts of reactor design to the design of the catalyst.

Since its appearance as a distinct part of chemical engineering about 45 years ago, chemical reaction engineering has made extensive progress. This chapter will review briefly the history of chemical reaction engineering for

ADVANCES IN CHEMICAL ENGINEERING, VOL. 16

the purpose of highlighting its evolution and sphere of application. The current status of the field will then be summarized, and the contemporary challenges for reaction engineers will be identified. Several generic problems requiring research in the coming five to ten years will be discussed, and illustrations drawn from the recent literature will be presented. The chapter will conclude with the author's views regarding the direction that the teaching of chemical reaction engineering should take in view of current and future developments in the field.

II. Historical Perspectives

Chemical reaction engineering evolved as a distinct branch of chemical engineering in the mid-1940s and developed rapidly during the 1950s and 1960s, stimulated by the expansion of the petroleum and petrochemical industries [1]. The concept of ideal continuous stirred tank reactors (CSTRs) and plug flow reactors (PFRs) was first introduced by Denbigh [2]. This pioneering contribution demonstrated that a rudimentary description of reactor performance could be achieved by combining a statement of reaction kinetics with a species balance written on a suitable control volume. Subsequent studies by Danckwerts and others [3] demonstrated through the use of tracers that the flow through practical reactors is not ideal and that observed residence time distributions are distinctly different from those expected for CSTRs and PFRs. These efforts led to the concept of axial dispersion to account for axial mixing in tubular reactors.

Studies with porous catalyst particles conducted during the late 1930s established that, for very rapid reactions, the activity of a catalyst per unit volume declined with increasing particle size. Mathematical analysis of this problem revealed the cause to be insufficient intraparticle mass transfer. The engineering implications of the interaction between diffusional mass transport and reaction rate were pointed out concurrently by Damköhler [4], Zeldovich [5], and Thiele [6]. Thiele, in particular, demonstrated that the fractional reduction in catalyst particle activity due to intraparticle mass transfer, η, is a function of a dimensionless parameter, ϕ, now known as the Thiele parameter.

More rigorous mathematical models of reactors were developed during the 1950s and 1960s. In industry these efforts were spearheaded by Prater at Mobil and by Van Deemter at Shell, and in academia by Amundson and Aris at the University of Minnesota. This work showed that external mass transfer effects could become important, particularly in reactors involving gas–liquid and liquid–solid contacting. The importance of thermal gradients in catalyst particles and reactors was also recognized during this period and the consequences of such gradients were investigated both experimentally and theoretically. The widespread availability of computers in the mid-1960s and the development of software for solving sets of differential equations opened the way for evaluating reactor models which would have been too complex to handle by analytical methods. Aided by increasingly more powerful methods of numerical analysis and faster computers, the application of chemical reaction engineering principles expanded rapidly and soon virtually every type of gas–solid, gas–liquid, and gas–liquid–solid reactor was examined [7]. These efforts demonstrated that when all aspects of transport and kinetics are treated properly, it is possible to achieve an accurate representation of actual reactor behavior. An excellent example of such a model is KINPtR, which was developed by Mobil for the design and performance analysis of catalytic reformers [8]. Since 1974, this model has been used extensively throughout Mobil to optimize commercial reformer performance, to monitor Mobil's worldwide reformers, to evaluate commercial catalysts, to design new reformers, and to predict aromatics yield for petrochemical feedstocks.

The decade of the 1970s saw the application of chemical reaction engineering principles in a number of new directions. Perhaps the most significant of these was the analysis of catalytic converters for the abatement of auto exhaust emissions. The need to control simultaneously the emissions of CO, hydrocarbons, and NO under inherently unsteady-state conditions and the limited space available for the reactor posed a formidable set of design challenges. Studies carried out primarily at General Motors [9, 10] demonstrated that chemical reaction methods could be used to determine the optimal placement of the active components of the catalyst (i.e., Pt, Pd, and Rh) on the support, the optimal configuration of the support (e.g., beads versus monoliths), and the light-off and transient response behavior of the overall reactor.

The application of reaction engineering concepts to electrochemical and plasma chemical systems was also initiated in the 1970s and developed rapidly thereafter [11–14]. The participation of charged species in these systems requires that the transport and conservation equations for individual species be modified to include the effects of electric (and in some instances magnetic) fields and that Poisson's equation be solved to determine the distribution of field strength. For low-pressure plasmas, the electron gas does not equilibrate with the neutral gas species, and as a consequence, additional equations are required to fully describe the system. Principal among these are number and energy balances written on the electron gas. The latter of these equations is particularly difficult to solve because the electron energy distribution in nonequilibrium plasmas is not Maxwellian. Some authors have dealt with this problem by assuming a Maxwellian distribution as an approximation, while others have attempted to solve the Boltzmann equation in order to determine the energy distribution function.

Another subject that captured the attention of researchers in the 1970s was the identification of reaction conditions under which catalyzed and uncatalyzed reactions exhibit multiple steady states and/or oscillatory behavior. Theoretical investigations demonstrated that such behavior could arise from the nonlinear character of the reaction kinetics or from an interplay between the kinetics of a reaction and mass transport processes. A rich body of literature has now emerged detailing the space of reaction conditions and parameters within which multiple steady states and oscillations can be expected [15].

The rapid development of biotechnology during the 1980s provided new opportunities for the application of reaction engineering principles. In biochemical systems, reactions are catalyzed by enzymes. These biocatalysts may be dispersed in an aqueous phase or in a reverse micelle, supported on a polymeric carrier, or contained within whole cells. The reactors used are most often stirred tanks, bubble columns, or hollow fibers. If the kinetics for the enzymatic process is known, then the effects of reaction conditions and mass transfer phenomena can be analyzed quite successfully using classical reactor models. Where living cells are present, the growth of the cell mass as well as the kinetics of the desired reaction must be modeled [16, 17].

Catalysis and reaction engineering became entwined in the late 1930s with the realization that the cracking of petroleum could be achieved most effectively using silica–alumina catalysts. With time, the connection between these two areas grew stronger as more and more catalytic processes were developed for the refining of petroleum, the production of petrochemicals, and the synthesis of polymers.

Research on catalysts has traditionally followed two parallel lines, distinguishable by the goals of the research. One line of work has been aimed at identifying and developing highly active, selective, and stable catalysts. This type of research involves the preparation, characterization, and evaluation of catalysts and can be highly time and labor intensive. The introduction of automated catalyst testing systems, in which dozens of reactors can be operated simultaneously under the control of a computer, has greatly accelerated the evaluation of new catalysts.

The second line of research in catalysis has been aimed at determining the structure and composition of catalysts and interpreting catalysis in terms of the elementary processes occurring on the catalyst surface. Pursuit of these objectives has depended heavily on the availability of instrumentation and has advanced at a rate closely correlated with the rate of introduction of

Table 1. Experimental Techniques for Characterizing Catalyst Surfaces and Adsorbed Species

Technique	Acronym	Type of Information
Low energy electron diffraction	LEED	2D Structure and registry with metal surface
Auger electron spectroscopy	AES	Elemental analysis
X-ray photoelectron spectroscopy	XPS	Elemental analysis and valence state
Ion scattering spectroscopy	ISS	Elemental analysis
UV photoelectron spectroscopy	UPS	Electronic structure
Electron energy loss spectroscopy	EELS	Molecular structure
Raman spectroscopy	—	Molecular structure
X-ray diffraction	XRD	Bulk crystal structure
Transmission electron microscopy	TEM	Crystal size, shape, morphology, and structure
Scanning transmission microscopy	STEM	Microstructure and composition
UV spectroscopy	—	Electronic state
Mössbauer spectroscopy	—	Ionic state
Nuclear magnetic resonance spectroscopy	NMR	Molecular structure and motion

new analytical techniques [18]. Table 1 lists many of the techniques currently in use for the study of catalysts and adsorbed species and the type of information obtainable from each technique. By using a combination of these techniques, tremendous strides have been made in understanding the structure and composition of catalysts, as well as the manner in which reactants and products interact with the surface of a catalyst. It is now possible to obtain an atomic scale view of a catalyst surface and to determine the nature and concentration of surface defects, the composition and structure of a catalyst, and the dependence of the kinetics of elementary rate processes on adsorbate coverage as well as temperature. This information has proved highly useful for interpreting catalytic processes and developing a deeper understanding of how surface chemistry affects the performance of a catalyst.

III. Contemporary Challenges

The principles underlying chemical reaction engineering are now well established and their successful application to the description of a large number of systems has been demonstrated. There remain, nevertheless, a large number of unanswered questions and an increasing number of new challenges to be met. The subjects deserving future attention can be divided into three areas: representation of rate and equilibrium processes, reactor modeling, and catalysis. Each of these areas will be discussed with attention given to identifying the key issues requiring research. Examples taken from the recent literature will be used to illustrate what might be accomplished.

A. Representation of Rate and Equilibrium Processes

The kinetics of elementary reactions occurring in the gas phase are well represented by expressions of the form

$$r = k_0 \exp(-\frac{E}{kT})\theta_i\theta_j \tag{1}$$

where k_0 is the preexponential factor, E is the activation energy, T is the temperature, and θ_i and θ_j are the concentrations of reactants i and j. Measurements of k_0 and E have been made for a very large number of elementary processes, particularly those involved in the combustion of hydrocarbons and those contributing to the pollution of the atmosphere, and are tabulated in several compendia [19]. Methods for estimating k_0 and E have been developed, but these techniques have not yet advanced to the level where they can be used routinely by the reaction engineer [20]. Thus, while it is possible to make reasonable estimates of k_0 from absolute rate theory, the quality of these estimates is highly dependent on the knowledge of the transition state configuration and vibrational spectrum. Impressive strides have been made in the use of quantum mechanics to describe the potential surfaces involved in chemical reactions, and molecular dynamics calculations made using these potentials have been shown to give accurate representations of the rate coefficient for an elementary reaction. Unfortunately, though, such calculations tend to be system-specific and hence do not have broad applicability. It is the opinion of the author that it may be profitable to investigate what could be done to systematize existing methods and bring them to a point where they can be used for engineering calculations.

The kinetics of elementary reactions occurring at a solid surface cannot be described with anything near the accuracy available for gas and liquid phase reactions, and, as a consequence, it is often necessary to describe the kinetics of an overall reaction using power-law or Langmuir–Hinshelwoodtype expressions. While the global kinetics represented by such rate expressions can be rationalized on the basis of a postulated reaction mechanism, such efforts must be viewed with caution, since in most instances it is assumed that the rates of elementary surface processes can be represented by expressions of the form

$$r = k_0 \exp(-\frac{E}{kT}) n_i \theta_j \qquad (2)$$

and

$$r = k_0 \exp(-\frac{E}{kT}) \theta_i \theta_j \qquad (3)$$

where θ_i and θ_j are the fractions of the surface covered by species i and j. The problem with expressions such as equations (2) and (3) is that the rate parameters k_0 and E are assumed to be coverage independent, whereas recent experimental investigations [21] show both parameters to be strongly coverage dependent. If physically accurate descriptions of surface processes are to be achieved, then attention must be given to developing models for describing the coverage dependence of both rate parameters.

An example of how one might model the effects of surface coverage on the activation energy for the desorption of an adsorbate has been given recently by Lombardo and Bell [22]. The local heats of adsorption and the local activation energy for desorption were calculated using the bond order conservation–Morse potential approach developed by Shustorovich [23]. This method takes into account the effects of nearest- and next-nearest-neighbor adsorbates. By using a Monte Carlo method to describe the kinetics of desorption, simulations were obtained of temperature-programmed desorption (TPD) spectra. Figures 1 and 2 illustrate the predicted TPD spectrum and the dependence of the activation energy, respectively, on adsorbate coverage for the desorption of hydrogen from a Mo (100) surface. The occurrence of three well-defined desorption peaks is in excellent agreement with experiments, as is the manner in which the activation energy decreases with increasing hydrogen atom coverage. Techniques similar to those described by Lombardo and Bell [22] can be used to describe the kinetics of simple surface reactions and to account for the effects of poisons and promoters on the dynamics of surface processes.

An analysis of the kinetics of CO oxidation by O_2 over Rh catalysts based on a model which accounts for the individual elementary reaction steps has been presented by Oh et al. [24]. A significant feature of the model is that the rate expressions used to describe the elementary steps and the rate parameters associated with these steps are all drawn from surface science

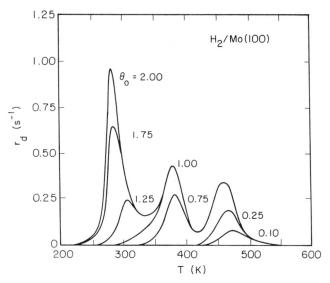

Figure 1. Simulated TPD spectrum for H_2 desorption from a Mo (100) surface. Reprinted with permission from Lombardo and Bell [22].

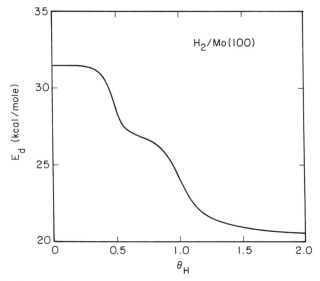

Figure 2. Predicted variation in the activation energy for H_2 desorption from a Mo (100) surface as a function of H atom coverage. Reprinted with permission from Lombardo and Bell [22].

studies reported in the literature. The mechanism of CO oxidation is represented by the reaction sequence

$$CO \rightleftharpoons CO_a$$
$$O_2 \rightarrow 2O_a$$
$$CO + O_a \rightarrow CO_2$$

The subscript "a" refers to species adsorbed on the Rh surface, while the variables without a subscript denote gas-phase species. For the majority of the reaction conditions considered, the CO surface coverage is near unity, and the overall rate of CO_2 formation can be expressed as

$$R_{CO_2} = \frac{k_{O_2,ads} \, k_{CO,des}}{k_{CO,des}} \left(\frac{P_{O_2}}{P_{CO}} \right) \tag{4}$$

where $k_{O_2,ads}$ is the rate coefficient for O_2 adsorption, $k_{CO,ads}$ is the rate coefficient for CO adsorption, $k_{CO,des}$ is the rate coefficient for CO desorption, P_{O_2} is the partial pressure of O_2, and P_{CO} is the partial pressure of CO. Comparisons of the rate of CO_2 formation predicted by equation (4) with those observed over Rh (111) and Rh/Al_2O_3 are presented in Figs. 3–5. For a wide range of conditions, the model developed by Oh et al. [24] fits the experimental data quite closely.

The description of reaction kinetics for circumstances where an exceptionally large number of species are involved or where well-defined molecular entities cannot be specified poses a particularly difficult challenge. Such situations can arise, for example, in the cracking of petroleum, the pyrolysis of biomass, and the liquefaction of coal. The best way to treat such systems is to either lump classes of reactions or follow the dynamics of functional groups rather than specific molecular species.

McDermott and Klein [25] have proposed yet another approach to dealing with complex systems, which they refer to as stochastic kinetics. The basic principle behind their idea is that a reactive species has an associated probability of reacting within a given time interval. This interpretation of kinetics can then be used in a Monte Carlo algorithm to describe the overall reaction of a large collection of species. McDermott and Klein [25] have used this approach successfully to describe the kinetics of lignin depolymerization. The reaction pathways for lignin depolymerization and the kinetics for elementary steps were deduced from independent studies of the model compounds veratrylgylcol-β-guaiacyl ether (VGE) and guaiacylglycol-β-guaiacyl ether (GGE). Figure 6 illustrates the temporal variation of key products of lignin (polyVGE) pyrolysis at 380°C. The results shown are in

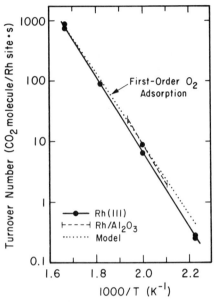

Figure 3. Comparison of the specific rates of the CO–O_2 reaction measured over Rh (111) and Rh/Al_2O_3 at $P_{CO} = P_{O_2} = 0.01$ atm with that predicted theoretically. Reprinted with permission from Oh *et al.* [24].

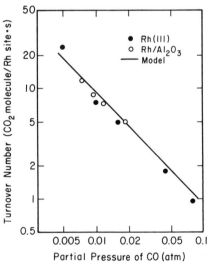

Figure 4. Measured and predicted rates of the CO–O_2 reaction over Rh (111) and Rh/Al_2O_3 as a function of the partial pressure of CO. The partial pressure of O_2 (0.01 atm) and the temperature (500 K) were held constant. Reprinted with permission from Oh *et al.* [24].

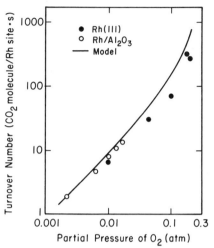

Figure 5. Measured and predicted rates of the CO–O$_2$ reaction over Rh (111) and Rh/ Al$_2$O$_3$ as a function of the partial pressure of O$_2$. The partial pressure of CO (0.01 atm) and the temperature (500 K) were held constant. Reprinted with permission from Oh *et al.* [24].

very general agreement with available lignin pyrolysis data, if acetovanillone can be considered as representative of single-ring products.

Many reaction processes of industrial importance occur in microporous solids (catalysis, electrolysis, shale retorting, etc.). Access to the interior of the solid is by diffusional transport and the transport of mass is usually described by Fick's law. The parameter describing the ease of mass transport is the effective diffusivity, D_e, where

$$D_e = \frac{D_M}{\varepsilon \tau} \tag{5}$$

or

$$D_e = \frac{D_K}{\varepsilon \tau} \tag{6}$$

In equations (5) and (6), D_M and D_K are the molecular and Knudsen diffusivities, respectively, and ε and τ are the void fraction and the tortuosity of the porous solid, respectively. For pore dimensions significantly larger than the mean free path of the diffusant in the gas phase, the diffusivity is governed by molecular diffusion, but when the pore diameter becomes smaller than the mean free path, diffusion is properly described by Knudsen diffusion. When the pore diameter approaches that of the diffusing species, around 10, one enters the configurational regime.

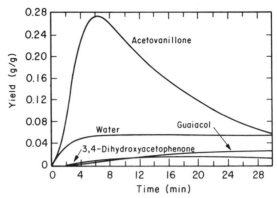

Figure 6.   Simulated pyrolysis of polyVGE at 380°C. Reprinted with permission from
Chem. Eng. Sci., vol. 41, p. 1053, J. B. McDermott and M. T. Klein,
"Chemical and Probabilistic Modelling of Complex Reactions: A Lignin
Depolymerization Example," copyright 1986 [25], Pergamon Press PLC.

The calculation of D_M and D_K can be estimated using the kinetic theory
of gases and requires knowledge only of the average pore diameter and the
thermal velocity of the diffusing species. The determination of configurational diffusivities cannot be done as easily. Recent studies have shown,
though, that good estimates of the configurational diffusivity can be obtained
from the solutions of Newton's laws of motion for the diffusing species
moving in the force field exerted by the atoms of the porous solid. The
configurational diffusivity is calculated from the mean square displacement
using the Einstein relation. This approach has been used to determine the
configurational diffusivity of small molecules moving through slits [26] and
is currently being used to determine the diffusivity of hydrocarbons in zeolites
[27].

Another illustration of the power of molecular dynamics simulation can
be drawn from the sphere of enzyme catalysis. Many enzyme-catalyzed
reactions proceed at a rate that depends on the diffusion-limited association of the substrate with the active site. Sharp *et al.* [28] have carried out
Brownian dynamics simulations of the association of superoxide anions with
superoxide dismutase (SOD). The active center in SOD is a positively charged
copper atom. The distribution of charge over the enzyme is not uniform,
and so an electric field is produced. Using their model, Sharp *et al.* [28] have
shown that the electric field enhances the association of the substrate with
the enzyme by a factor of 30 or more. Their calculations also predict correctly
the response of the association rate to changes in ionic strength and amino

acid modification and provide suggestions for enhancing the enzyme turnover rate through site-directed mutagenesis.

When dimensional constraints (surface reactions) or topological constraints (solid-state reactions) are imposed, a reaction can exhibit fractal kinetics [29]. The hallmarks of "fractal-like" reactions are anomalous reaction orders and time-dependent reaction rate "constants." These anomalies stem from the nonrandomness of the reactant distribution in low dimensions. For homobimolecular reactions, the distribution is partially ordered, for example, quasi-periodic. However, for heterobimolecular reactions, the reactants segregate. The application of fractal kinetics as a tool for interpreting reaction kinetics is still in its early stages, but the initial results hold promise for explaining phenomena occurring on surfaces, in the solid state, and in biological systems.

B. Reactor Modeling

The fundamental concepts involved in reactor modeling have been clearly defined, and, as a consequence, emphasis has shifted to the development of more accurate reactor models and the description of novel types of reactors. The availability of increasingly fast computers has prompted the evaluation of models incorporating highly detailed descriptions of fluid flow and reaction kinetics. However, the development of detailed reactor models has in some instances led to diminishing returns, particularly in situations where the observables do not permit discrimination between more and less complete descriptions. Increasing the level of detail in a model often comes at the price of an increase in the number of parameters to be evaluated. If all of the parameters can be estimated or measured independently, then it is possible to test the predictive properties of the model, since it is not necessary to use the model to evaluate the parameters. If, on the other hand, a significant fraction of the parameters must be adjusted by causing a fit between the output of the model and experimental observation, it is less certain how accurately the model will predict reactor performance for operating conditions significantly different from those used to determine the unknown parameters. Consequently, an important challenge for reaction engineers in the future will be to determine the appropriate level at which a given reactor should be modeled.

It is the opinion of the author that future research on reactor modeling should focus on developing models for reactors and processes which are as yet poorly understood. Examples of such systems include fluidized-bed reactors, membrane reactors, reactors used for the synthesis of ceramics, and low-pressure reactors used for the deposition and etching of thin films.

Several illustrations of the physical insights that can be obtained from modeling such systems are presented below.

An excellent illustration of the impact of fluid dynamics on reactor performance has been presented by Gidaspow *et al.* [30], who have modeled coal gasification in a fluidized bed. The model accounts for heat, mass, and momentum transfer in both the gas and solid phases and for both heterogeneous and homogeneous reactions. The results of the calculations provide a detailed view of bubble formation and collapse, as well as the spatial and temporal distributions of gas composition, gas temperature, and bed porosity. These calculations are of considerable value in predicting the location and magnitude of the maximum solid temperature and of oxygen penetration into the bed, characteristics of importance to the design of fluidized-bed coal gasifiers.

Kee *et al.* [31] have described a very detailed model of the chemical vapor deposition of silicon from silane. The physical system considered is shown in Fig. 7. The reactor consists of a rotating-disk susceptor maintained at high temperature and exposed to a downflow of silane. Decomposition of silane on the susceptor leads to the deposition of silicon and the release of hydrogen. The authors note that a general treatment of the problem involving simultaneous simulation of multidimensional fluid transport, heat transfer, and chemical kinetics is impractical, even with currently available supercomputers. Instead, they break down the problem into pieces. From an analysis of the fluid mechanics in the absence of chemical reactions, they established conditions under which a similarity transformation could be used to reduce the three-dimensional fluid flow problem to a one-dimensional problem. The similarity solution for the fluid velocity was incorporated into the reactor model. The gas-phase chemistry was described in terms of 27 elementary reactions involving 17 chemical species. Solid film formation is assumed to occur by the diffusion of SiH_2, Si_2H_4, and SiH_4 to the susceptor.

Examples of the spatial distribution of gas-phase species and the net Si deposition rate as a function of temperature are given in Figs. 8 and 9, respectively. From an analysis of the simulation shown in Fig. 8, the authors determined that 57% of the silicon deposition comes from the flux of SiH_2 to the surface, 24% from the flux of Si_2H_4, and 13% from SiH_4. By doing computations with and without thermal diffusion, it was possible to establish that thermal diffusion decreases the silicon-containing fluxes at the surface by about 20%. Figure 9 shows that at low temperatures the deposition rate is reaction limited. However, as the temperature increases, the reactions become faster, allowing transport processes to become rate limiting. The fact that the deposition rate is faster in helium than in hydrogen is a chemical effect. Hydrogen is one of the reaction products as the silane

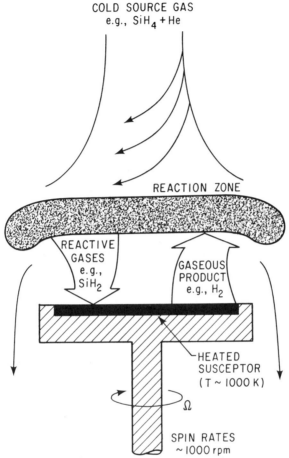

COLD SOURCE GAS
e.g., $SiH_4 + He$

REACTION ZONE

REACTIVE
GASES
e.g.,
SiH_2

GASEOUS
PRODUCT
e.g., H_2

HEATED
SUSCEPTOR
$(T \sim 1000 \text{ K})$

Ω

SPIN RATES
~ 1000 rpm

Figure 7. Schematic of the rotating-disk chemical vapor deposition reactor. Reprinted with permission from R. J. Kee, G. H. Evans, and M. E. Coltrin in *Supercomputer Research in Chemistry and Chemical Engineering* (K. F. Jensen and D. G. Truhlar, eds.), p. 334. ACS Symposium Series 353, American Chemical Society, Washington, D.C., 1987 [31]. Copyright 1987 American Chemical Society.

decomposes to form species that react with the surface. Thus, when hydrogen is the carrier, it works to inhibit the progress of these reactions and consequently impedes deposition.

Hlavacek and co-workers [32] have simulated the synthesis of refractory metal nitrides by adiabatic combustion. The reactions involved occur at 2500

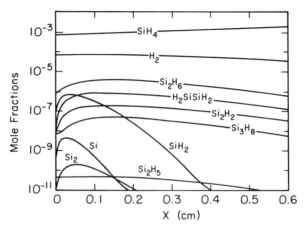

Figure 8. Species profiles in a rotating-disk CVD reactor. Inlet gas is 0.1% silane in carrier of 99.9% helium. The disk temperature is 1000 K and the spin rate is 1000 rpm. Reprinted with permission from R. J. Kee, G. H. Evans, and M. E. Coltrin in *Supercomputer Research in Chemistry and Chemical Engineering* (K. F. Jensen and D. G. Truhlar, eds.), p. 334. ACS Symposium Series 353, American Chemical Society, Washington, D.C., 1987 [31]. Copyright 1987 American Chemical Society.

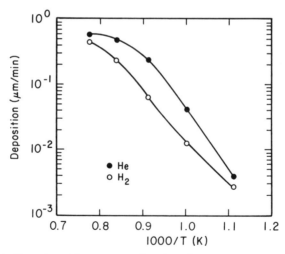

Figure 9. Net silicon deposition rates as a function of susceptor temperature for both hydrogen and helium as the carrier gas. Rotation is 1000 rpm. Reprinted with permission from R. J. Kee, G. H. Evans, and M. E. Coltrin in *Supercomputer Research in Chemistry and Chemical Engineering* (K. F. Jensen and D. G. Truhlar, eds.), p. 334. ACS Symposium Series 353, American Chemical Society, Washington, D.C., 1987 [31]. Copyright 1987 American Chemical Society.

to 4500°C and are characterized by very high activation energies. As a result, the nitridation front represents a strongly nonlinear combustion wave which may be extremely thin and corrugated. A one-dimensional simulation was carried out using adaptive regridding techniques to reduce the computational time. Figure 10 shows the calculated temperature profiles. The temperature in the reaction zone fluctuates wildly. As a result, a spatially inhomogeneous product is obtained, since some of the material is formed at high temperatures and some of it at lower temperatures. Further insights were obtained from a two-dimensional model. These calculations revealed not only temporal oscillations in temperature but also a breakup of the planar combustion front. Moreover, hot spots were found to appear along the front, disappear, reappear, etc., but in positions different from before.

Opportunities also exist for the use of chemical reaction engineering principles to explore options and opportunities for the design of novel types of reactors. Of particular interest are membrane reactors, distillation reactors, and crystallization reactors, all of which combine product formation and separation into one unit. Properly designed, such reactors could lead to the achievement of higher yields and/or selectivities than can be attained with more conventional designs. The development of novel reactor configurations for highly exothermic reactions (methanol synthesis, ethylene oxidation, etc.) is also needed. One possible concept for investigation is that of concurrent reaction and boiling heat transfer. By proper selection of a

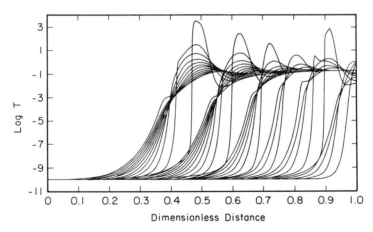

Figure 10. Temperature profiles produced during the formation of nitrides by adiabatic combustion. Reprinted with permission from J. Degreve, P. Dimitriou, J. Puszynzki, V. Hlavacek, and R. Behrens in *Supercomputer Research in Chemistry and Chemical Engineering* (K. F. Jensen and D. G. Truhlar, eds.), p. 376. ACS Symposium Series 353, American Chemical Society, Washington, D.C., 1987 [32]. Copyright 1987 American Chemical Society.

solvent for the reactants and products, it might be possible to maintain the reaction at a desired temperature by using the heat of reaction to boil the solvent. The solvent vapor would be condensed and the condensate returned to the reactor.

An example of the advantages of combining reaction and concurrent separation is represented by the work of Barbosa and Doherty [33]. These authors have carried out a theoretical analysis of the influence of equilibrium chemical reactions on vapor–liquid phase diagrams and have developed design procedures for describing the performance of single- and double-feed multicomponent reactive distillation columns. They show that two main advantages result from the use of reactive distillation: one is the possibility of carrying out equilibrium-limited chemical reactions to completion; the other is the simultaneous separation of the reaction products in one unit. This reduces or eliminates reactor and recycle costs.

C. Catalysis

The complexity of industrial catalysts makes it very difficult to describe such materials in terms of atomistic models. As a consequence, it is not possible to predict the performance of catalysts from first principles. Nevertheless, fundamental understanding of catalysis has contributed to the development of practical catalysts. This has occurred to a large extent through the application of analytical techniques such as those listed in Table 1. By characterizing industrial catalysts with these techniques, it has been possible in many instances to establish how a catalyst has been altered by a change in preparative procedure. Such information can help in understanding how changes in catalyst composition and structure translate into changes in catalyst performance. The net result then is that catalyst development becomes less empirical.

One of the challenges for the future is to refine existing analytical techniques and to develop new ones for characterizing catalysts and species adsorbed on catalyst surfaces. Of particular need are methods that allow the observation of a catalyst under actual working conditions, because the structure and composition of a catalyst surface in the working environment are often different from those existing prior to reaction. Examples of such effects include the reconstruction of metal surfaces, the appearance of defects in oxides, and the deposition of poisons.

Among the techniques ideally suited for *in situ* studies are infrared, Raman, and nuclear magnetic resonance (NMR) spectroscopies and extended x-ray absorption fine structure (EXAFS). While still relatively new, the scanning tunneling and atomic force microscopes are expected to play an increasingly important role in catalyst characterization. Both instruments permit visualization of a catalyst surface at the atomic level and hold the potential of showing how atoms and molecules interact with a surface.

While the intrinsic activity and selectivity of a catalyst establish its performance in the absence of mass transfer effects, it is well known that the placement of the active components and access to these components by reactants can play a major role in the performance of practical catalysts. One of the challenges for reaction engineers is to develop models for predicting the distribution of active components in a catalyst and the effects of this distribution, together with the pore size distribution and particle size and shape, on the performance of a catalyst.

The application of reaction engineering principles to the prediction of metal profiles in a catalyst is nicely illustrated by the work of Chu et al. [34]. These authors present a transport model for the coimpregnation of γ-Al_2O_3 with $NiCl_2$ and HCl. The equilibrium adsorption is controlled by hydrolysis of surface hydroxyl groups and ion binding on these groups. An important feature of the equilibrium adsorption model is the generation of protons upon nickel adsorption. As nickel diffuses in and adsorbs, a local excess of H^+ ions is generated, producing a pH profile inside the pellet and reducing the nickel uptake. This blocking phenomenon, which cannot be described by Langmuir isotherms, produces unique concave–convex metal profiles in the pellet. The success of this model in predicting experimentally observed profiles is demonstrated in Fig. 11.

The use of reaction engineering concepts to design an optimal catalyst structure for the removal of vanadium from petroleum feedstocks is illustrated by the work of Pereira et al. [35]. Catalyst pellets were modeled as

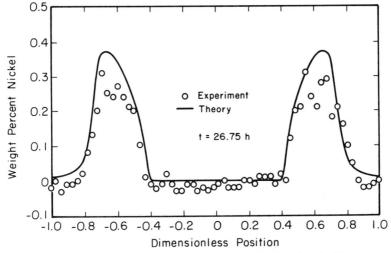

Figure 11.   Nickel profile in a γ-alumina pellet after 26.75 h. Reprinted with permission from Chu et al. [34].

an aggregate of microporous grains. Vanadium deposition was represented by a pseudo-first-order rate law, and the deposition rate was taken to be proportional to the available surface area. The reaction rate constant was dependent on the fractional coverage of the catalytic surface by deposits. Metal deposits caused a spatially varying decrease in pore dimensions, and metal-bearing molecules were subject to configurational diffusion. A fixed-bed hyrodemetallation reactor was modeled as a series of mixing cells. This model enabled catalyst pore structure to be optimized for some target performance feature (e.g., maximum life with the constraint of a minimum level of conversion of metals in the case of constant reactor temperature operation). Using this model, the authors were able to design a Minilith catalyst particle that exhibited about 50% higher vanadium removal activity than a commercial catalyst after 30 days on stream.

IV. Implications for Education

Graduate-level courses on chemical reaction engineering have focused traditionally on issues of reaction kinetics and reactor design, with the latter subject usually taking up the bulk of the time. Considerable attention is given to dealing with intra- and interparticle mass transfer in various types of reactor configuration. Less time is spent, though, on parameter estimation and evaluation and the interplay between parameter and model accuracy. It has also been fashionable in recent years to include several lectures on the pathological behavior of reactors (multiple steady states, oscillations, chaos, etc.).

It is safe to say that most graduate courses in chemical reaction engineering today suffer from an excess of mathematical sophistication and insufficient contact with reality. Because of the complexity of many reaction engineering models, it is essential that students be given a balanced and realistic view of what can and cannot be achieved. For example, they must learn that if the intrinsic kinetics of a reaction are not known accurately, this deficiency cannot be made up by a more detailed understanding of the fluid mechanics. In this connection, it would be useful pedagogically to take a complex model and illustrate its sensitivity to various aspects, such as the assumptions inherent in the model, the reaction kinetics, and the parameter estimates.

There is also a need for chemical reaction engineering courses to deal more thoroughly with the chemistry of the process under consideration. This is particularly important when both product quality and yield are the performance targets. The use of modern concepts of physical chemistry to make predictions of transport and rate parameters should also be emphasized, since such concepts show how the properties of a system affect these parameters.

Finally, where catalytic reactions are concerned, students should be made aware that catalysis can be understood through the use of fundamental principles and that pure empiricism is insufficient.

Acknowledgment

This work was supported by the Director, Office of Basic Energy Sciences, Chemical Science Division of the U.S. Department of Energy under contract DE-AC03-76SF00098.

References

1. Dudukovic, M. P., *Chem. Biochem. Eng. Q.* **1**, 127 (1987).
2. Denbigh, K. G., *Trans. Faraday Soc.* **40**, 352 (1944).
3. Danckwerts, P. V., *Chem. Eng. Sci.* **2**, 1 (1953).
4. Damköhler, G., *Chemieingenieur* **3**, 359 (1937).
5. Zeldovich, Y. B., *Acta Phys. Chim. USSR* **10**, 58 (1939).
6. Thiele, W. E., *Ind. Eng. Chem.* **31**, 916 (1939).
7. Carberry, J. J., and Varma, A., *Chemical Reaction and Reactor Engineering*. Marcel Dekker, New York, 1987.
8. Ramage, M. P., Graziani, K. R., Schipper, P. H., Krambeck, F. J., and Choi, B. C., *Adv. Chem. Eng.* **13**, 193 (1987).
9. Taylor, K. C., in *Catalysis Science and Technology*, Vol. 5 (J. R. Anderson and M. Boudart, eds.), p. 119. Springer Verlag, New York, 1984.
10. Hegedus, L. L., and McCabe, R. W., *Catal. Rev. Sci. Eng.* **23**, 377 (1981).
11. Newman, J. S., *Electrochemical Systems*. Prentice-Hall, Englewood Cliffs, N.J., 1973.
12. Trost, G. G., Edwards, V., and Newman, J. S., in *Chemical Reaction and Reactor Engineering* (J. J. Carberry and A. Varma, eds.), p. 923. Marcel Dekker, New York, 1987.
13. Bell, A. T., in *Techniques and Applications of Plasma Chemistry* (A. T. Bell and J. R. Hollahan, eds.), p. 1379. Wiley, New York, 1974.
14. Graves, D. B., *AIChE J.* **35**, 1 (1989).
15. Morbidelli, M., Varma, A., and Aris, R., in *Chemical Reaction and Reactor Engineering* (J. J. Carberry and A. Varma, eds.), p. 973. Marcel Dekker, New York, 1987.
16. Bailey, J. E., and Ollis, D. F., *Biochemical Engineering Fundamentals*. McGraw-Hill, New York, 1977.
17. Erickson, L. E., and Stephanopoulos, Geo., in *Chemical Reaction and Reactor Engineering* (J. J. Carberry and A. Varma, eds.), p. 779. Marcel Dekker, New York, 1987.
18. Delgass, W. N., and Wolf, E. E., in *Chemical Reaction and Reactor Engineering* (J. J. Carberry and A. Varma, eds.), p. 151. Marcel Dekker, New York, 1987.
19. JANAF Thermochemical Tables, Natl. Stand. Ref. Data Ser. NSRDS-NBS 37 (1971).
20. Laidler, K. J., *Chemical Kinetics*, 3rd Ed. Harper & Row, New York, 1987.
21. Seebauer, E. G., Kong, A. C. F., and Schmidt, L. D., *Surface Sci.* **193**, 417 (1988).
22. Lombardo, S. J., and Bell, A. T., *Surface Sci.* **206**, 101 (1988).
23. Shustorovich, E., *Surface Sci. Repts.* **6**, 1 (1986).
24. Oh, S. H., Fisher, G. B., Carpenter, J. E., and Goodman, D. W., *J. Catal.* **100**, 360

(1986).

25. McDermott, J. B., and Klein, M. T., *Chem. Eng. Sci.* **41**, 1053 (1986).

26. Schoen, M., Cushman, J. H., Diestler, D. J., and Rhykerd, C. L., Jr., *J. Chem. Phys.* **88**, 1394 (1987).

27. June, R. L., Theodorou, D., and Bell, A. T., *J. Phys. Chem.* **94**, 1508 (1990).

28. Sharp, K., Fine, R., and Honig, B., *Science* **236**, 1460 (1987).

29. Kopelman, R., *Science* **241**, 1620 (1988).

30. Gidaspow, D., Ettehadieh, B., and Bouillard, J., *AIChE Symp. Ser.* **241**, 57 (1985).

31. Kee, R. J., Evans, G. H., and Coltrin, M. E., in *Supercomputer Research in Chemistry and Chemical Engineering* (K. F. Jensen and D. G. Truhlar, eds.), p. 334. ACS Symposium Series 353, American Chemical Society, Washington, D.C., 1987.

32. Degreve, J., Dimitriou, P., Puszynzki, J., Hlavacek, V., and Behrens, R., in *Supercomputer Research in Chemistry and Chemical Engineering* (K. F. Jensen and D. G. Truhlar, eds.), p. 376. ACS Symposium Series 353, American Chemical Society, Washington, D.C., 1987.

33. Barbosa, D., and Doherty, M. F., *Chem. Eng. Sci.* **43**, 529, 541, 1523, 2377 (1988).

34. Chu, P., Petersen, E. E., and Radke, C. J., *J. Catal.* **117**, 52 (1989).

35. Pereira, C. J., Donnelly, R. G., and Hegedus, L. L., in *Catalyst Deactivation*, p. 315. Marcel Dekker, New York, 1987.

11

Frontiers in Chemical Reaction Engineering

James R. Katzer and S. S. Wong
Mobil Research and Development Corp.
Princeton, New Jersey

I. Introduction

The basis of chemical engineering is application of scientific and engineering principles to solve problems of industrial and societal importance. Chemical reaction engineering is unique to chemical engineering and is at the core of its identity as a separate discipline because it combines chemistry, chemical kinetics, catalysis, fluid mechanics, and heat and mass transfer. Other disciplines involve some of these aspects, but none brings the reaction chemistry and the transport together in the same way.

We distinguish between two types of "frontiers" in chemical reaction engineering. One focuses on problems in newly emerging areas, such as microelectronics, semiconductors, materials, and biotechnology. These frontiers represent a stimulus for the expansion and renewal of the intellectual foundation of our profession. The other frontier is the boundary between what we know and understand well and what we do not know and cannot yet quantify in areas where chemical engineers have traditionally played an important role. These areas, from which chemical reaction engineering grew, include chemical reactor design and analysis, chemical kinetics, and catalysis and will be referred to as the "traditional" areas of chemical reaction engineering. As the problems that were yesterday's frontiers are solved, the

ADVANCES IN CHEMICAL ENGINEERING, VOL. 16

boundary advances to areas that are still unresolved. This frontier is no less challenging or less important than that associated with new areas. In general, problems at this latter frontier are progressively moving from the mesoscale toward both the microscopic and more molecularly oriented scales and to the broader macro- or systems scales. As such, the frontiers are at once becoming progressively more fundamental in nature and more global. This paper focuses on the frontier of the more traditional areas in chemical reaction engineering.

Chemical reactor modeling will be considered first, followed by kinetic modeling, and then the integration of various elements of chemical reaction engineering into a more useful whole. The status of each area will be briefly reviewed. Then current and future frontiers will be discussed, emphasizing those that provide the most challenge and the greatest potential impact.

II. Chemical Reactor Modeling

The foundation of chemical reactor modeling was laid in the early 1940s and 1950s by Denbigh [1], Hougen and Watson [2], Danckwerts [3], Levenspiel [4, 5], and others. The early models involved macroscopic considerations of reactor performance and simplified homogeneous rate expressions, independent of the presence of a heterogeneous (solid) catalyst. Tubular and stirred-tank reactors received the most emphasis. As batch processing was replaced by continuous-flow tubular reactor systems, chemical reaction engineering focused on packed-bed tubular reactors, then gaining increased industrial attention and large-scale application. The relevance of early simple models was readily demonstrated, and the level of sophistication for single-fluid phase reactor modeling advanced as rapidly as did the development of mathematical tools and the speed and power of computers, incorporating more sophisticated reaction kinetic expressions and simultaneous heat and mass transfer and energy balances. This led in about 20 years to the capability to predict reactor performance for relatively complex single-phase reaction systems. Now, almost any reaction system can be handled in a quantitative manner, including both steady-state and transient reactor behavior. This is illustrated by the ability to adequately predict hot-spot formation in a hydrocracker using a single-phase flow model with simple kinetics, as shown in Fig. 1 [6]. The current frontiers in this area have moved to definition of kinetic expressions and the understanding and design of catalysts.

However, as soon as we move from these relatively well defined, single-fluid-phase tubular systems to more complex geometry and more complex multiphase systems, the frontiers still involve developing an adequate de-

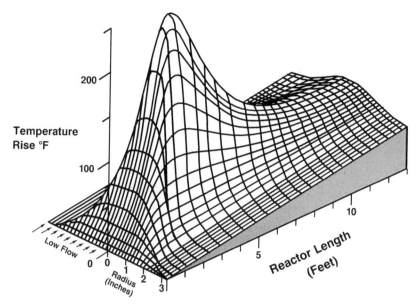

Figure 1. Calculated hot-spot formation in modeling hydrocracker operation. Reprinted with permission from S. B. Jaffe, *Ind. Eng. Chem. Process Dev.* **15**, 410 (1976). Copyright 1976 American Chemical Society.

scription of the reactor performance and require a substantial understanding of the hydrodynamics of the system. An example is the modeling of turbulent flow combustion, where flow description and flame chemistry/ kinetics are the current frontier areas. Two areas will be used to illustrate the intellectual status of these frontiers.

The first is fixed-bed, three-phase catalytic reactors, frequently called trickle-bed reactors. Large-scale commercial units were in operation for many years before any understanding evolved. Academic interest developed in the late 1960s. Slowly thereafter, empirical physical and chemical correlations were developed and adapted to allow adequate prediction of the relevant rate processes and provide understanding of the important and limiting phenomena that determine trickle-bed reactor performance [7, 8]. Most of these correlations were developed in small-scale equipment or were adaptations of heat and mass transfer correlations developed from single-particle studies or from large-particle, packed-bed studies. Thus, they had to be extended beyond their range of validity or modified to account for geometric and operating differences. Examples include heat [9] and mass [10] transfer correlations and liquid holdup relations [11]. Considerable kinetic information has also been generated to describe the catalytic rate phenomena that occur in these types of reaction systems, particularly for hydroprocessing reaction chemistry. As a result, a localized description of trickle-bed reactor processes is now satisfactory. However, scale-up and global performance of trickle-bed reactors cannot consistently be predicted because an adequate hydrodynamic model of trickle-bed operation is not available, and several flow regimes may be operative, depending on the fluids involved and their relative and absolute mass flow rates, as shown in Fig. 2 [12]. Developing an adequate hydrodynamic model will require reintegration of fluid mechanics with other aspects of chemical reaction engineering. With an adequate hydrodynamic model, heat and mass transfer, energy balance, and chemical kinetics can be incorporated to produce fully predictive models which may be validated by comparison with commercial-scale data. Ultimately, full-scale design and optimization can be carried out, converting trickle-bed chemical reaction engineering from an analytical to a synthesis/design mode and moving the frontiers to the understanding, quantification, and design of catalysts similar to the situation today for single-fluid-phase tubular reactors.

The second major reactor type that requires much further quantification is the fluid-bed chemical reactor, which is of tremendous industrial importance, as indicated by Table 1. A related reactor is the fluid-bed combustor that is employed for combustion of relatively coarse solids with reduced emissions.

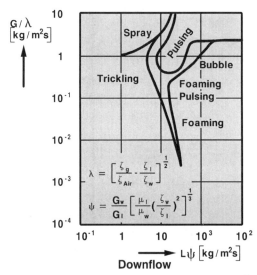

Figure 2.   Flow map for trickle-bed reactor operation.

Table 1. Fluid Bed Reactors

Process	Status
Fluid catalytic cracker (1942)	10×10^6 barrels per day, >350 units
Phthalic anhydride (1945)	0.3×10^9 pounds per year (U.S.)
Fischer-Tropsch synthesis (1955)	3 units
Chlorinated hydrocarbons and chlorine (early 1950s)	Large number of units
Acrylonitrile (1960)	$>6 \times 10^9$ pounds per year, >50 units
Polyethylene high density (1968) low density (1977)	>15 units

The scale-up and design configurations of fluid-bed chemical reactors have evolved rapidly and empirically. An example is fluid catalytic cracking (FCC) [13]. The general fluid-bed concepts developed early. However, the correlations describing the various rate processes and other operational phenomena developed slowly because they could not easily be related back to already established data bases developed for other systems; in the case of trickle-bed reactors, data developed for packed-bed absorption towers were utilized.

Fluid-bed reactors are currently designed by scaling from prior experience because they cannot be described quantitatively and thus cannot be adequately modeled. As with trickle-bed reactors, hydrodynamic models are needed as a basis for reactor model development for use in reactor design and optimization; such hydrodynamic models are almost nonexistent at this point. This is clearly a frontier area in chemical reaction engineering. Figure 3 shows the evolution of hardware and reactor operation in the development of fluid-bed catalytic cracking. Actual operation went from a fast fluidized bed (left side) to a true fluid bed (middle) with freeboard. Driven by the development and application of zeolite catalysts, the technology evolved to full riser cracking, where essentially all the reaction occurs in the riser operating with very short contact times (right side).

As can be seen from Fig. 4 [14], today's high-efficiency riser–fluidized bed FCC units contain all major fluidization regimes, creating a significant challenge to the integration of fluid mechanics with chemical reaction engineering for development of a hydrodynamic model of the whole system or even of limited portions thereof. With an adequate hydrodynamic model, the heat and mass transfer, energy balance, and chemical kinetics could be successfully integrated to give a model with full predictive ca-

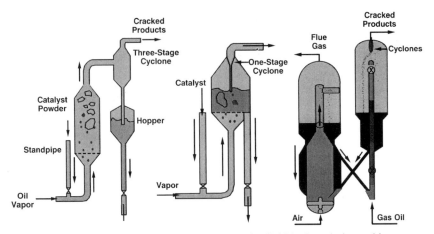

Figure 3. Evolution of fluid-bed reactors for fluid-bed catalytic cracking.

pability. Commercial data could then be used to validate the model's predictive capability for scale-up, design, and optimization purposes.

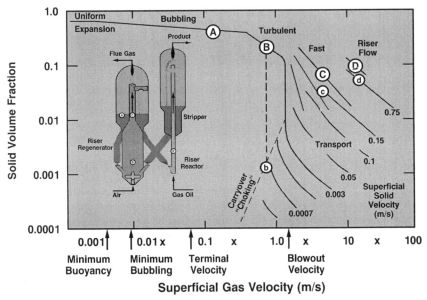

Figure 4. Fluid-bed flow regimes in a modern FCC unit. Reprinted with permission from Avidan *et al.* [13].

III. Kinetic Modeling

Only a few complex reaction systems have been modeled well. More complete and accurate information concerning reaction kinetics is needed to develop better models of catalytic processes. An example of the benefits of good reaction kinetic modeling is shown by Mobil's KINPtR–Kinetic Reforming Model [15]. To develop good process kinetics, for example in reforming, the reaction network and associated rate constants had to be determined for a nonlinear system containing a large number of chemical components. The current solution to this problem requires lumping the large number of components into smaller sets of kinetic lumps. This frontier area is becoming progressively more basic. The fundamental kinetics of many reactions (including elementary reactions) occurring at the solid surface are still unclear and cannot be accurately described. Improved techniques for estimating kinetic parameters are required. A better determination of the structure and composition of catalysts and quantification of their impact on fundamental reaction kinetics is also important. With the development of improved kinetic parameter estimation techniques, better correlations with catalyst characteristics can then be developed, providing a better understanding of the critical catalyst parameters that affect process kinetics. Eventually, we could design better catalysts for specific reaction schemes. The potential of prediction based on measurements of the correct quantities is shown in Fig. 5 [16].

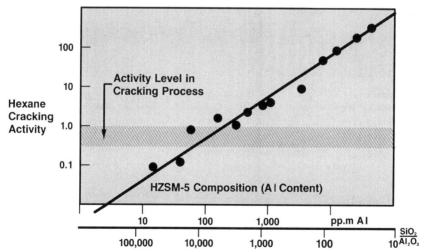

Figure 5. Direct relationship between zeolite aluminum content and hexane cracking activity for HZSM-5. Reprinted with permission from *Nature* vol. 309 p. 589 [16]. Copyright (c) 1984 Macmillan Magazines Ltd.

There have been significant advances in analytical capabilities (including high-vacuum surface spectroscopies and *in situ* spectroscopies) that can elucidate the structure and composition of catalysts, as well as the manner in which the reactants and products interact with the catalyst surface. Advanced supercomputers can facilitate quantum chemical calculations which should have predictive capabilities. Integration of spectroscopic characterization, quantum chemistry, and supercomputing is an important frontier area.

IV. Integration

Operating scale (laboratory vs. industrial) affects the behavior of chemical reaction systems. It is critical that we develop hydrodynamic models for those systems that are scale sensitive. This will require a collaboration between academic and industrial groups to collect data necessary for commercial-scale equipment. Once the hydrodynamic models have been developed and validated, kinetic models can be integrated with them.

Another frontier is the integration of process models into control schemes so that process optimization and simulation (driven by rapid optimizers) can be employed in commercial operations. Artificial intelligence approaches, such as expert systems, can then be applied to model building and process control, with integration extended to encompass an entire plant composed of several interacting processes. In the petroleum industry, its ultimate extension could encompass the purchase of crude oils and their allocation to a network of refineries, each with different processing technology, with the objective of maximum utilization of the different process technologies and different crude compositional properties to make the desired products.

Development of fundamental kinetics for improved understanding of complex reaction systems is another frontier. More advanced catalyst characterization tools, including on-line and in-line measurements, need to be developed to provide better understanding of critical catalyst parameters. This should involve application of predictive chemistry capability to design better catalysts which carry out desired conversions in complex reaction systems.

V. Conclusions

There are still important frontiers in traditional chemical reaction engineering to be conquered. They are both scientifically and intellectually challenging and are critical to the health of the profession, industry, and society. These traditional areas must not be abandoned in favor of research in newly evolving technological areas. Rather, the new areas must be folded into the ever-expanding sphere that chemical engineering encompasses.

References

1. Denbigh, K. G., *Trans. Faraday Soc.* **40**, 352 (1944).
2. Hougen, O. A., and Watson, K. M., *Chemical Process Principles.* Part 3: *Kinetics and Catalysis.* Wiley, New York, 1947.
3. Danckwerts, P. V., *Chem. Eng. Sci.* **2**, 1 (1953).
4. Levenspiel, O., *Chemical Reaction Engineering.* Wiley, New York, 1962. *Chem. Eng. Sci.* **35**, 1821 (1980).
5. Levenspiel, O., *Chem. Eng. Sci.* **43**, 1427 (1988).
6. Jaffe, S. B., *Ind. Eng. Chem. Process Des. Dev.* **15**, 410 (1976).
7. Herskowitz, M., and Smith, J. M., *AIChE J.* **29**, 1 (1983).
8. Ng, K. M., and Chu, C. F., *Chem. Eng. Prog.* **83**, 55 (November 1987).
9. Specchia, V., *Chem. Eng. Commun.* **3**, 483 (1979).
10. Speechia, V., and Baldi, G., *Ind. Eng. Chem. Proc. Des. Dev.* **17**, 362 (1978).
11. Speechia, V., *Chem. Eng. Sci.* **32**, 515 (1977).
12. Hofmann, H. P., *Catal. Rev. Sci. Eng.* **17**, 71 (1978).
13. Avidan, A. A., Edwards, M. S., and Owen, H., *Oil & Gas J.* p. 33 (Jan. 8, 1990).
14. Squires, A. M., Kwauk, M., and Avidan, A. A., *Science* **230**, 1329 (1985).
15. Ramage, M. P., Graziani, K. R., Schipper, P. H., Krambeck, F. J., and Choi, B. C., *Adv. Chem. Eng.* **13**, 193 (1987).
16. Haag, W. O., Lago, R. M., and Weisz, P. B., *Nature* **309**, 589 (1984).

12

Catalyst Design

L. Louis Hegedus
W. R. Grace & Co.- Conn.
Columbia, Maryland

I. Introduction

Professor Bell's preceding paper views the world of kinetics, catalysis, and reactor engineering from a primarily microscopic, chemical perspective. This permits him to take a fresh approach in covering past developments and to fold a broad range of catalytic phenomena into the realm of chemical reaction engineering. He and I agree that chemical reaction engineering is a way of integrating chemical and physical phenomena into systematic models (mathematical or otherwise) of catalyst performance. These models have to contain parameters with physical and chemical meaning; their optimum combination would then yield the optimum design for a catalyst.

There are strong economic forces that demand rational methods for the development of new catalysts. These were discussed in the introductory chapter of a recent book on catalyst design [1]. Although catalysts represent only about $1.5 billion in sales volume, they contribute to the manufacture of about $300 billion in product value (U.S., 1985), an economic leverage of about 200 : 1. Table 1 shows some of the numbers in perspective; Table 2 [1] shows a breakdown of catalyst consumption (U.S., 1984).

Driving forces that demand more efficient methodologies for catalyst development include new business opportunities (in all three major application areas of emission control, petroleum conversion, and chemicals), the high cost of empirical developments, and the increasing competitiveness

ADVANCES IN CHEMICAL ENGINEERING, VOL. 16

Table 1. Economic Impact of Catalysts (U.S., 1985, $billions)

Gross national product	3,853
Manufacturing	773
Chemicals	168
Specialty chemicals	34
Catalysts	1.53
Catalytically manufactured products	298

Table 2. U.S. Consumption of Catalysts in 1984, Including Captive Consumption

Chemical processing[a]	
Polymerization	235
Organic syntheses	85
Oxidation, ammoxidation, and oxychlorination	80
Hydrogen, ammonia, and methanol syntheses	50
Hydrogenation	30
Dehydrogenation	10
TOTAL	490
Petroleum refining[a, b]	
Catalytic cracking	255
Hydrotreating	75
Hydrocracking	40
Catalytic reforming	20
TOTAL	390
Emission Control[a]	
Automotive[c, d]	445
Other	5
TOTAL	450
TOTAL CATALYST CONSUMPTION	1,330

[a]Strategic Analysis, Inc. (1985).
[b]Excluding $155 million for HF and H_2SO_4 alkylation catalysts.
[c]Including an estimated $160 million for noble metals.
[d]Estimated.

in the catalyst industry, which demands speedier routes toward commercialization. Concomitant with these needs are the opportunities arising from major advances in computation such as supercomputers; the emergence of user-friendly software for scientific computing; the evolution of new, powerful numerical techniques; and the evolution of new architectures such as parallel processors, vector processors, neural networks, hypercubes, and the like. These advances promise to replace certain types of catalytic experimentation by mathematical modeling, similar to what happened in areas such as aircraft or automobile design.

II. The Catalyst Design Problem

Ideally, optimum design requires the identification of the pertinent design variables, the identification of the objective function or functions to be maximized or minimized, and the identification of bounds (minimum and maximum values) on the design variables to satisfy a set of constraints, implicit and explicit. Again, ideally this can be done in the form of a sufficiently powerful mathematical model which would then yield the optimum values of the design variables and the best attainable value of the objective function, within the constraints imposed [2].

Although such ideal systems rarely exist, it turns out that we often know enough about a catalytic system to be able to make substantial improvements via the above methodology. Several examples have been discussed in recent literature [3].

The factors that influence catalyst performance are numerous and only partially understood. What follows is a discussion of key catalyst design issues, starting with the catalytic surface and progressing through supported catalysts to shaped catalyst particles, successively incorporating new phenomena and variables as the complexity of the system increases.

III. Catalytic Surfaces

Developments in surface instrumentation over the past 20 years or so have revealed rich detail about the structure and composition of catalytic surfaces and the relationships of various surface properties to catalytic behavior. The status of our surface catalytic knowledge and the resulting opportunities for catalyst design have been reviewed by Somorjai [4].

The structure sensitivity of catalytic reactions is often striking. As an example, Spencer et al. [5] demonstrated a factor of over 400 difference between the activity of the Fe (111) surface and that of the Fe (110) surface (the former being the more active one) for ammonia synthesis. This type of investigation, while of course not predictive in nature, has the promise

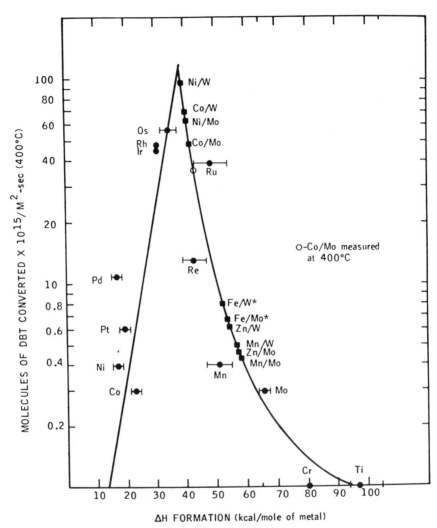

Figure 1. Volcano plot showing the effect of the heat of formation of metal sulfides on the dibenzothiophene hydrodesulfurization activity of various mono- and bimetallic catalysts. Adapted from Chianelli et al. [6] and reprinted with permission of John Wiley & Sons from L. L. Hegedus, ed., *Catalyst Design— Progess and Perspectives*, p. 1. Wiley, New York (1987) [1]. Copyright 1987, John Wiley & Sons.

of practical utility in the design of high-activity catalytic surfaces. The practical corollary is that the preferred surfaces have to be chemically and thermodynamically stable at reaction conditions. In the case of commercial polycrystalline Fe catalysts for ammonia synthesis, the dominant phase under reaction conditions indeed appears to be Fe (111), a finding achieved by over 70 years of empirical development.

In industrial practice, catalytic surfaces are often very complex, not only structurally but also chemically. An example is shown in Fig. 1 from Chianelli *et al.* [6] for hydrodesulfurization catalysts. The data indicate that maximum dibenzothiophene hydrodesulfurization activity is achieved at intermediate heats of formation of metal sulfides, i.e., at intermediate metal–sulfur bond strengths. Again, while such surface energetic considerations do not have *ab initio* predictive ability, they are valuable tools for catalyst synthesis and prescreening.

Another level of surface chemical complexity results from catalytic metal–catalyst support surface interactions. Table 3, taken from Bell [7], shows the surface-specific activity of Rh for CO hydrogenation as a function of

Table 3. Surface-Specific Activity of Rh for CO Hydrogenation as a Function of the Catalyst Support

Support	Relative Activity[a]	Reference
TiO_2	100	Katzer *et al.* [34]
MgO	10	
Al_2O_3	5	
CeO_2	3	
SiO_2	1	
TiO_2	100	Orita *et al.* [35]
MgO	9	
Al_2O_3	5	
TiO_2	100	Haller *et al.* [36]
SiO_2	6	
Nb_2O_5	100	Iizuka *et al.* [37]
ZrO_2	2	
SiO_2	0.3	
Al_2O_3	0.2	
MgO	<0.1	

[a]Based on surface Rh sites as determined by H_2 chemisorption.

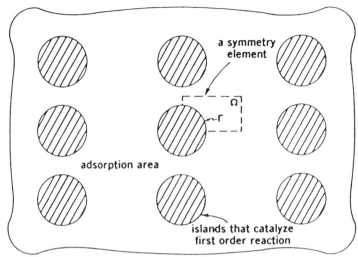

Figure 2. Regularly shaped catalyst islands on an adsorptive support. Reprinted with permission from *Chem. Eng. Sci.*, vol. 38, p. 719, D.-Y. Kuan, H. T. Davis, and R. Aris, "Effectiveness of Catalytic Archipelagos. I. Regular Arrays of Regular Islands," copyright 1983 [8], Pergamon Press PLC.

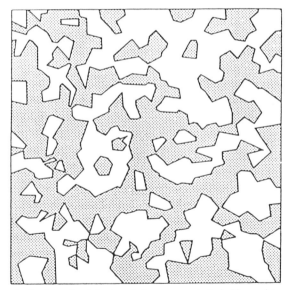

Figure 3. Irregularly shaped catalyst particles represented via Voronoi tessellation. Reprinted with permission from *Chem. Eng. Sci.*, vol. 38, p. 1569, D.-Y. Kuan, R. Aris, and H. T. Davis, "Effectiveness of Catalytic Archipelagos. II. Random Arrays of Random Islands," copyright 1983 [9], Pergamon Press PLC.

the catalyst support. Depending on the nature of the support, changes of over two orders of magnitude in surface-specific reaction rates were demonstrated. Again, these observations provide valuable information for catalyst developers, but the phenomena cannot yet be predictively modeled.

When coupled with surface adsorption and diffusion, the geometry of catalytic arrays has interesting effects on catalytic behavior; examples of this type of reaction engineering models are shown in Figs. 2 and 3, by Kuan *et al.* [8, 9], respectively.

As the above examples illustrate, the kinetics of surface-catalytic events depend on complex structural and electronic considerations that, thus far, have not been understood at the level of detail that would permit predictive mathematical modeling and therefore rational design. For this reason, the molecular-level engineering of catalytic surfaces harbors perhaps the greatest future potential for chemical reaction engineering, at least from the standpoint of the design of catalysts.

IV. Transport Within Pores

The coupling of kinetics with intra- and extraparticle transport has been the traditional focus of chemical reaction engineering; major accomplishments have been admirably summarized by Aris [10]. The effectiveness factor and Thiele parameter of diffusion-influenced catalyst particles represent a balance between their reactive and diffusive properties. In this section, we shall concentrate on the latter.

A key design consideration in catalysis is the optimum pore structure of catalyst particles. The random pore model of Wakao and Smith [11] works well—at least for aluminas, for which it has been extensively tested—with no adjustable parameters and employing integral-averaged macro- and micropore dimensions and porosities. This author's experiences with pore structure optimization employing the random pore model are discussed elsewhere [2]. More recent considerations, such as percolation theory using Bethe networks [12], allow for the influences of narrow necks, tortuous paths, and dead-ended pores, opening up many new and as yet untested opportunities for the structure optimization of catalyst pores via mathematical modeling.

Restricted transport or, by a different name, configurational diffusion [13] occurs when the diffusing molecules are comparable in size to the pores within which they diffuse. This happens, for example, in hydrodemetallation over alumina-supported Co–Ni catalysts [14]. The observation that the effective diffusivity depends on the fourth power of the molecule-to-pore size ratio is important, but it is not yet evident how to correlate complex pore size distributions with effective diffusivities in the configurational regime. An impressive effort in this direction is represented by the paper of Rajagopalan and Luss [15], who computed optimal pore size distributions for hydrodemetallation catalysts.

Extensive manifestations of configurational diffusion can be seen in catalytic zeolites. The landmark measurements by Gorring [16] of the diffusion coefficients of alkanes (in potassium T zeolites) as a function of their carbon number are shown in Fig. 4, indicating over two orders of magnitude of change in diffusivity, with a minimum at C-8 and a maximum at C-12 for unexplained reasons. Similarly, spectacular effects for more complex molecules have been observed by Haag and Chen [17] and are shown in Fig. 5. Although we do not yet have a workable correlation between zeo-

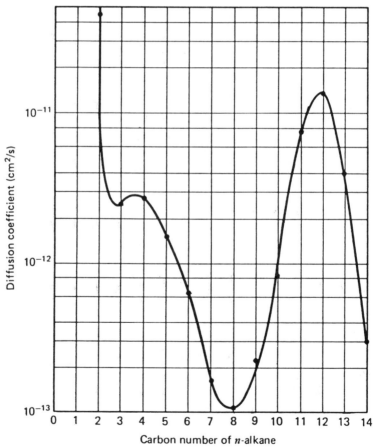

Figure 4. Diffusion coefficients of n-alkanes in potassium T zeolite at 300°C. Reprinted from Gorring [16] with the permission of Academic Press.

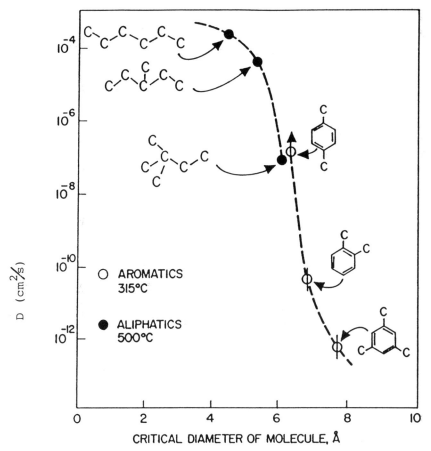

Figure 5. Diffusion coefficients in HZSM-5 for hexane isomers (at 500°C) and aromatics (at 315°C). Adapted from Haag and Chen [17] and reprinted with permission of John Wiley & Sons from L. L. Hegedus, ed., *Catalyst Design— Progress and Perspectives*, p. 1. Wiley, New York (1987) [1]. Copyright 1987, John Wiley & Sons.

lite structure, molecular structure, and diffusion properties, work in several research groups is progressing toward that point. Examples include the stochastic analysis of Mo and Wei [18], Fig. 6, and the more recent work on the atomistic modeling of sorption and transport in pentasil-type zeolites by Theodorou *et al.* [19].

Of the over 200 synthetic zeolites known today, only a few are being employed in commercial practice. The synthesis and testing of new zeolithic materials require considerable efforts. Clearly, a quantitative and predictive

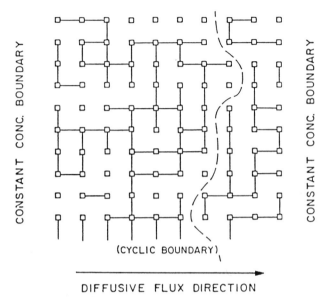

Figure 6. Example of two-dimensional 10×10 lattice model used to examine the effects of pore blockage on the effective diffusion coefficient in zeolites. Reprinted with permission from *Chem. Eng. Sci.*, vol. 41, p. 703, W. T. Mo and J. Wei, "Effective Diffusivity in Partially Blocked Zeolite Catalyst," copyright 1986 [18], Pergamon Press PLC.

understanding of diffusive transport in zeolites would be of major practical significance. Therefore, it represents an important frontier of chemical reaction engineering.

V. Catalyst Impregnation Profiles

One important catalyst design variable is the macroscopic, spatial profile of activity along the characteristic dimension of the catalyst particle. As with many new phenomena, this was first recognized in the patent literature [20, 21]. The first theoretical analysis was developed by Shadman-Yazdi and Petersen [22]. Specific applications for automobile exhaust catalysts were proposed, e.g., by the influential papers of Becker and Wei [23, 24]; these concepts were subsequently proved by experiment and used for the optimum design of automobile exhaust catalysts [25]. Figure 7 is one example of the effects that can be achieved. As Vayenas and Pavlou [26] (1988) pointed out, the theoretical analyses of optimum catalyst distributions became so popular that they are now "way ahead of experimental verifica-

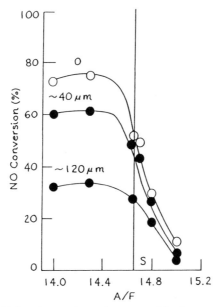

Figure 7. Effects of Rh location on the performance of fresh automobile exhaust catalyst beads as a function of engine air–fuel ratio (A/F). Adapted from Hegedus *et al.* [33] and reprinted from Hegedus and McCabe [25], p. 94, by courtesy of Marcel Dekker, Inc.

tion." Methods to achieve various intraparticle catalyst distribution profiles have been concisely summarized in the chapter by Aris [3].

VI. Catalyst Particle Shapes

Transport-influenced catalyst particles require high geometric surface areas for maximum activity; this is synonymous with short diffusive pathways resulting in small Thiele parameters or high effectiveness factors. One obvious way to achieve this is with very small catalyst particles; indeed, this is being done in slurry-phase or fluidized-bed systems. For packed-bed reactors, however, small particles may cause an intolerably large reactor back pressure. Therefore, a wide variety of catalyst shapes have been invented and are being manufactured to provide an optimum trade-off between geometric surface area (or diffusion path length) and back pressure in fixed-bed reactors. Such shaped catalysts have also gained some popularity in two-phase-flow systems (such as trickle-bed reactors) due to their favorable holdup characteristics as compared to spheres or cylinders. Figure 8 shows

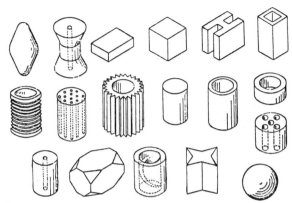

Figure 8. Illustration from an early patent on shaped catalysts. From Foster [27].

some unusual and somewhat amazing examples from an early patent on hydroprocessing [27].

The reactivity of shaped catalysts can be nicely normalized by their reaction path length [28, 29]. Back pressures correlate well with the Ergun equation [30] when effective hydraulic radii are employed. However, bed compressibilities, particle interlocking characteristics, manufacturability and manufacturing costs due to extrusion die design, die wear, and the like are, of course, beyond the realm of chemical reaction engineering methodologies while contributing in decisive ways to catalyst shape selection. The role of catalyst shape in hydrodemetallation service has been discussed by Pereira *et al.* [31], who advanced a novel catalyst design.

VII. An Industrial Example

The catalytic control of NO_x emissions from fossil fuel–fired electric power generating facilities is a commercial reality in Japan and Germany and may also gain large-scale application in the United States, pending future legislation. The process is called selective catalytic reduction (SCR), and it employs the injection of ammonia (Fig. 9). The catalyst is in the form of huge monolithic assemblies made of, e.g., vanadia-impregnated porous titania. In current practice, about 1 to 1.5 m^3 of catalyst is employed per electric megawatt of power generating capacity. That is, a typical large power plant of, say, 700 electric megawatts capacity would have an SCR de-NO_x reactor of about 1000 m^3 in volume (Fig. 10). This is assembled from extruded monoliths measuring 15 cm by 15 cm in cross section and about 1 m in length; the openings are 3 to 7 mm wide.

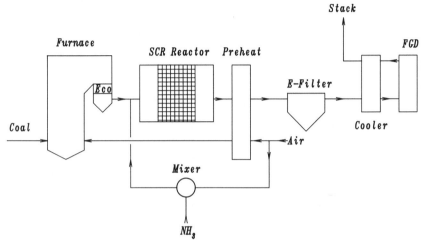

Figure 9. Selective catalytic reduction (SCR) process for NO_x control in fossil fuel–fired power plants. E-filter stands for the electrostatic dust filter, ECO for the economizer (a heat exchanger), and FGD for the flue gas desulfurizer (sulfur oxide scrubber).

The economics of the SCR process are strongly dependent on the cost and performance of the catalyst. Beeckman [32] developed a mathematical model that combines the measured kinetics of the reactions

$$4NO + 4NH_3 + O_2 \longrightarrow 4N_2 + 6H_2O \qquad (1)$$

$$2SO_2 + O_2 \longrightarrow 2SO_3 \qquad (2)$$

with diffusive transport in the porous walls, and with convective and diffusive transport in the monolith channels, to describe catalyst performance (NO and NH_3 conversion, SO_2 oxidation) as a function of catalyst pore properties (macro- and microporous radii and pore volumes) under typical stack gas operating conditions. The objective of the work was to determine what potential has been left in this technology by its original developers and how to harness this potential via pore structure optimization.

Some of the results are shown in Fig. 11 and suggest that by restructuring the catalyst (its normalized NO removal activity is 1), over 50% improvement in activity ought to be achievable. By making substantial changes in catalyst support composition (which permitted the required changes in pore structure while maintaining the required physical strength), laboratory results on extruded catalyst bodies have verified the computed improvements,

Figure 10. Selective catalytic reduction (SCR) de-NO$_x$ reactor installation in a large coal-fired power plant.

qualifying the new catalyst for pilot manufacturing. Thus far, it appears that reaction engineering support of this research may have saved several man-years of experimental effort and may have saved about two years in development time.

Acknowledgment

The author thanks Messrs. C. J. Pereira and J. W. Beeckman for stimulating discussions.

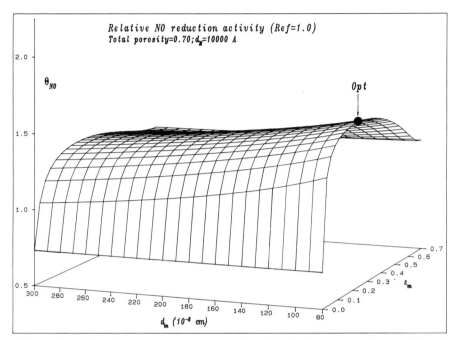

Figure 11.   Effect of catalyst pore structure on the NO reduction activity of a monolith catalyst for the SCR process. From Beeckman and Hegedus [32].

References

1. Hegedus, L. L., ed., *Catalyst Design—Progress and Perspectives*, p. 1. Wiley, New York, 1987.
2. Hegedus, L. L., *Ind. Eng. Chem. Prod. Res. Dev.* **19**, 553 (1980).
3. Aris, R., in *Catalyst Design—Progress and Perspectives* (L. L. Hegedus, ed.), p. 213. Wiley, New York, 1987.
4. Somorjai, G. A., in *Catalyst Design—Progress and Perspectives* (L. L. Hegedus, ed.), p. 11. Wiley, New York, 1987.
5. Spencer, N. D., Schoonmaker, R. C., and Somorjai, G. A., *J. Catal.* **74**, 129 (1982).
6. Chianelli, R. R., Pecoraro, T. A., Halpert, T. R., Pan, W. H., and Stiefel, E. I., *J. Catal.* **86**, 226 (1984).
7. Bell, A. T., in *Catalyst Design—Progress and Perspective* (L. L. Hegedus, ed.), p. 103. Wiley, New York, 1987.
8. Kuan, D.-Y., Davis. H. T., and Aris, R., *Chem. Eng. Sci.* **38**, 719 (1983).
9. Kuan, D.-Y., Aris, R., and Davis. H. T., *Chem. Eng. Sci.* **38**, 1569 (1983).

10. Aris, R., *The Mathematical Theory of Diffusion and Reaction in Permeable Catalysts*, Vol. I. Clarendon Press, Oxford, 1975.

11. Wakao, N., and Smith, J. M., *Chem. Eng. Sci.* **17**, 825 (1962).

12. Reyes, S., and Jensen, K. F., *Chem. Eng. Sci.* **40**, 1723 (1985).

13. Spry, J. C., and Sawyer, W. H., "Configurational Diffusion Effects in Catalytic Demetallization of Petroleum Feedstocks." AIChE 68th Annual Meeting, Los Angeles, Nov. 16–20, 1975.

14. Wei, J., in *Catalyst Design—Progress and Perspectives* (L. L. Hegedus, ed.), p. 245. Wiley, New York, 1987.

15. Rajagopalan, K., and Luss, D., *Ind. Eng. Chem. Proc. Des. Dev.* **18**, 459 (1979).

16. Gorring, R. L., *J. Catal.* **31**, 13 (1973).

17. Haag, W. O., and Chen, N. Y., in *Catalyst Design—Progress and Perspective* (L. L. Hegedus, ed.), p. 163. Wiley, New York, 1987.

18. Mo, W. T., and Wei, J., *Chem. Eng. Sci.* **41**, 703 (1986).

19. Theodorou, D. N., Bell, A. T., and June, R. L., "Counter-Diffusion of Benzene and Toluene in Zeolite ZSM5." Paper No. 35b, AIChE Annual Meeting, Washington, D.C., Nov. 27–Dec. 2, 1988.

20. Michalko, E., U.S. Patents 3,259,454 and 3,259,589 (both 1966).

21. Hoekstra, J., U.S. Patent 3,388,077 (1968).

22. Shadman-Yazdi, F., and Petersen, E. E., *Chem. Eng. Sci.* **27**, 227 (1972).

23. Becker, E. R., and Wei, J., *J. Catal.* **46**, 365 (1977).

24. Becker, E. R., and Wei, J., *J. Catal.* **46**, 372 (1977).

25. Hegedus, L. L., and McCabe, R. W., *Catalyst Poisoning*. Marcel Dekker, New York, 1984.

26. Vayenas, C. G., and Pavlou, S., "Optimal Catalyst Distribution for Selectivity Maximization in Pellets." Paper No. 72d, AIChE Annual Meeting, Washington, D.C., Nov. 27–Dec. 2, 1988.

27. Foster, A. L., U.S. Patent 2,408,164 (1946).

28. Aris, R., *Chem. Eng. Sci.* **6**, 262 (1957).

29. Miller, D. J., and Lee, H. H., *Chem. Eng. Sci.* **38**, 363 (1983).

30. Ergun, S., *Chem. Eng. Progr.* **48**, 89 (1952).

31. Pereira, C. J., Donnelly, R. G., and Hegedus, L. L., in *Catalyst Deactivation* (E. E. Petersen and A. T. Bell, eds.). Marcel Dekker, New York, 1987.

32. Beeckman, J. W., and Hegedus, L. L., "Design of Monolith Catalysts for Power Plant NO_x Emission Control." Paper No. 72e, AIChE Annual Meeting, Washington, D.C., Nov. 27–Dec. 2, 1988.

33. Hegedus, L. L., Summers, J. C., Schlatter, J. C., and Baron, K., *J. Catal.* **56**, 321 (1979).

34. Katzer, J. R., Sleight, A. W., Gajardo, P., Michel, J.B., Gleason, E. F., and McMillan, S., *Disc. Faraday Soc. London* **72**, 121 (1981).

35. Orita, H., Naito, S., and Tamaru, K., *J. Chem. Soc. Chem. Commun.* **18**, 993 (1983).

36. Haller, G. L., Henrich, V. E., McMilland, M., Resasco, D. E., Sadeghi, H. R., and Sakellson, S., Proc. 8th Int. Congr. Catal. Berlin V, 135 (1984).

37. Iizuka, T., Tanaka, Y., and Tanabe, K., *J. Molec. Catal.* **17**, 381 (1982).

General Discussion

Kinetics, Catalysis, and Reactor Engineering

James Wei: I've gathered notes from this evening's talks on what's working and not working in chemical reaction engineering. A number of you may have heard that chemical reaction engineering is dead and should be forgotten. Many of us have tried to address that issue for some time. My list contains areas of reaction engineering that I think are alive and others that I think are dead. I want to throw my ideas out for discussion, to see whether I will be shot down in flames or sustained. Reaction engineering looks like a doughnut with a dead center; what's dead is the basic theory, and it's dead because we haven't had any new ideas in a long time. I don't know that I've heard of a new idea since residence time distribution. We have invested too much in the subject of pathology; the results are repetitious by now, unless they relate to the operation and design of useful reactors. The ones that seem to be alive are some aspects of catalysis, catalyst development, and their characterization. Multiphase flow in reactors is very much alive, but here the chemical reaction engineering community seems to be stuck. The community can do very simple fluid mechanics problems, like the Ergun equation, but they don't know how to solve really tough problems. Two days ago, Skip Scriven gave a talk at MIT, and I believe it is the first time I have heard trickle-bed flow taken seriously by a chemical engineer skilled in fluid mechanics. The results were dramatic. This is what the chemical reaction engineering community needs, interdisciplinary nourishment to get them off dead center. The disciplines needed come from other parts of the chemical engineering community, from the fluid mechanics people, from the catalysis people. Another part that is alive is the integration of very large-scale systems and design integration problems, perhaps with the help of large computers and artificial intelligence. Finally, very active and alive portions are the applications to biotechnology, pollution control, chemical vapor deposition, and so forth. These also involve people from different disciplines. What I see is that chemical reaction engineering can't make progress without expanding its intellectual foundations with help from other communities.

Stuart Churchill: I'd like to say something critical. All of the speakers, including yourself, neglected one of the most active and maybe the prototype problem of chemical reaction engineering, namely combustion. Here we deal with fluid mechanics that cannot be assumed to be plug-flow or perfectly mixed. We have transport of many species, with great difficulty in estimating the diffusivities, and we deal with hundreds of reactions with increasing reality and accuracy. This poses difficult mathematical problems of stiffness. There is a real burst of work going on because, in the last decade, the chemists have finally measured enough fundamental reaction rates of free-radical reactions so that we can model without empiricism. Granted, this is still a frontier area, and the rate constants are pretty bad, but they're good enough to get meaningful results. I'm shocked that four of you would speak on reaction engineering and not refer to this problem at all.

James Wei: I completely agree with you, Stu, provided you believe that combustion is a branch of chemical reaction engineering. Traditionally, combustion has not been regarded as such.

Clark Colton: Combustion will be brought up in the next section on environmental protection and energy.

Stuart Churchill: I think it occurs in both places.

James Douglas: I think it is necessary to broaden the traditional perspective of reaction engineering. From a design perspective, some of the most exciting developments are coming in the areas of reactor separations, where you can carry out separations in the reactors. Kodak recently replaced a process with two reactors and six columns by a single reactive distillation column, with a dramatic cost savings. By focusing on reactor problems in isolation, you tend not to think of solutions of this type.

James Wei: I would accept that as a living part of the discipline.

James Douglas: What I am trying to say is that in reactor engineering courses we should spend some time on the interaction of the reactor with the process and consider the possibility of eliminating a conventional reactor.

James Wei: No reason not to, but when we compartmentalize subjects into reaction engineering and separations, it is done with a purpose in mind.

Dan Luss: I disagree with what you said about pathologies.

James Wei: I was hoping you would.

Dan Luss: I believe that there is still a need to study pathological behavior, especially in systems where it is caused by the interaction of the fluid flow and chemical reaction. For example, a major problem in the design and operation of a trickle-bed reactor is the presence of local hot spots, which have caused several major explosions. I've just been involved in litigation that resulted from a reactor explosion; it caused $15 million worth of damage at a plant in Corpus Christi. Most of the existing models in the literature are oversimplified and cannot predict this important feature. A model that will predict this behavior would be an important contribution. Another example is the self-ignition of coal piles, which is the major source of emission of sulfur dioxide into the atmosphere in South Africa. It's definitely desirable to get a better quantitative understanding of this behavior and how to prevent it.

I also believe that many new applications have appeared in recent years that require use of reaction engineering tools and methodology. Jim Douglas mentioned one process, catalytic distillation. Many of the reactors currently used in the electronic industry are poorly designed. I presume that Larry Thompson will talk about this tomorrow. The design of these reactors is an exciting reaction engineering problem. Similarly, future production of superconducting materials and high-tech ceramics will require significant efforts by reaction engineers. At present, very little information is available about solid-solid reactions, and there is an urgent need to develop design and scale-up procedures.

James Wei: You're giving me more and more examples of the living part.

Dan Luss: You just said it was mostly dead. I'm saying that there are many living parts and important problems. Consider the design of reactors in which fast mixing of reactants and a fast reaction occur simultaneously. The rapid manufacturing of automobile parts from plastic composites requires the development of reliable design and scale-up procedures for this type of reactor. I believe that reaction engineering is alive and that there exist many important and interesting problems. The challenge is to develop new tools and solve these problems, instead of working over and over again on the same problem.

James Wei: Has everyone finished shooting at me yet?

James Douglas: Let me add two more things. I think reactive crystallization and reactive extraction will become more important in the future, as has reactive distillation. Reactive distillation is receiving a lot of attention at present, but reactive separations should find more applications.

Morton Denn: Stu was absolutely correct. To suggest that combustion is not part of reaction engineering is equivalent to saying that reaction engineering automatically excludes any of the problems that we once found too difficult. I'm encouraged that some fine fluid mechanicians have been focusing on combustion problems. Combustion is an area where the integration of fluid mechanics and chemical reaction has been going ahead at a tremendous rate, apparently unknown to most of the chemical reaction engineering community, because it is all done by the AIAA people. In fact, AIAA just published a series of three books in which the content is mainstream chemical engineering and not a single chemical engineer is involved in the whole thing.

Reuel Shinnar: I didn't see or hear anything in the presentations about the need for new reactors. If you look at what's happening in the chemical and energy industries, a definite need exists for new reactors in conventional areas. Let me mention two that are extremely important. One is that we don't have any good method for removing heat from large reactors. Large tubular reactors need a lot of tubes. Fluid beds are our main tool for getting heat out, that's why we mostly design fluid beds and not many other reasons. Nobody dares to design them at high pressure. For highly exothermic or highly endothermic reactors, the coolers that we have are insufficient. We need something new.

The second area, and the most important one that I see forthcoming, is that we're doing more and more at high temperatures, which means much shorter residence times in reactors. We need very fast mixing and faster catalytic response. We don't know how to handle this. The concept of designing reactors for very short residence times and at high temperatures is a challenging problem. The fact is that we're neither inclined nor equipped as a profession to deal with problems of this type. We're doing so much engineering science that anything to do with real engineering is not respectable unless we can tie it to computers. In some areas, this is bad. Real engineering research has to become respectable. Another problem is that presently we don't have that many practitioners who know what design is. How do we solve that problem? On the occasion of this Centennial Symposium, let's look at the early days at MIT. They were well ahead of industry. If you consider one of our major achievements in the oil industry, the fluidized-bed catalytic cracker, MIT had a substantial part in that. Now such interaction between industry and academia isn't as prevalent anymore. The question is, should we get it back, or should we seek those new ideas some other way? It is not easy to do, and there's also a question of respectability. I can afford to write papers about this issue mainly because I have paid my tribute and written enough unreadable academic papers in the field for people who count. I paid my dues, so I can count my fees, I can write papers on anything I want, and editors are writing me letters, send me something, we'll take anything. If you talk about reacting to the climate, then this becomes something of a game. The issue is a major challenge to the profession. It doesn't just relate to reaction engineering, it relates to almost any of the fields that we're discussing today.

Dan Luss: Something we need to emphasize about reaction engineering, especially in teaching, is that design is usually carried out without complete information. Once I have a model, I don't care how complex it is, I can run it on a computer. The problem is estimating the parameters that go into the model and knowing what the chemical reaction is. Before I can lump reactions, I need to know the reaction scheme. That's the essence of design, and we ought to emphasize this point. Also, we don't use enough thermodynamics, and part of the problem is that most people in thermodynamics are really working in transport phenomena, i.e., estimation of transport coefficients. This morning, nobody spoke about using thermodynamics to design reactors. You can get a lot of insight by using thermodynamics: what is feasible, what kind of arrangement of reactors should be used, what kind of temperatures profiles should be used, and so on. That's a beautiful problem and a very useful one for chemical engineers, but if you look in the thermodynamic community, almost no one has made a career of it. We have to influence some of the younger people into working in that direction because it's a frontier area, and the pay off can be great. Reuel Shinnar started to show what can be done just by using basic thermodynamic principles.

James Douglas: I agree with Dan. Another problem, however, is the fracturing of the field of reaction engineering. What we have now is a field with petrochemical reaction engineers, bioreaction engineers, polymer reaction engineers, chemical vapor deposition (CVD) engineers, reaction injection molding engineers, and few people talk to each other. Is there any commonality? Why do we stress the differences, rather than the similarities?

James Wei: Jim, of course there has to be something in common among the different reactors for bio, CVD, polymers, and so forth. One of the most fascinating reactors I've ever heard about was in a lecture at Cal Tech on the evolution of the sun. It was a beautiful study of a reactor with integration of all the reaction engineering elements involved. Right now we are doing highly specialized work without enough integration. We have combustion courses at MIT, and we have reaction engineering courses, but they are separate. The reaction engineering courses are required for undergraduates, at least, and for the Ph.D. exam, but the combustion courses are not. They are not part of the core education. No reason why; it's just the way we are doing it right now.

Arthur Humphrey: Integration is key here. I say that because I'm stunned to hear Jim Douglas talk about these new separatory reactors. We heard about the first extractive fermentation in 1970, assuming that you will consider a fermentation a reaction. Even the first volume of Richardson and Coulson has an example of choosing the optimum ratio of extractant to reactant. I don't think that's new.

James Douglas: I agree that the basic ideas are not new at all, but they've often been ignored in both reaction engineering and design until recently.

Arthur Humphrey: Even within our profession, there's not enough crosstalk; we have a tendency to put each area in a little chamber by itself. There should be more integration between these compartmentalized subdisciplines of chemical engineering.

L. E. Scriven: I'd like to veer a bit in response to the talks earlier and in particular the discussions of catalysis. Our presenter and discussors focused on quite important core areas, but there are others. This morning, we heard a great deal about microstructure, and that's my point of departure. What about microstructure catalysis? That is, the catalysis of not only chemically bound microstructures, supramolecular structures, but physically bound ones as well. I'd say that the stereospecific polymerization catalysts were a harbinger. Maybe the possibilities will come up in connection with bioengineering. It strikes me that biomimetic catalysts, which are now in hot play in some circles, are examples. Antibodies can act as catalysts, and antibodies, it has now been demonstrated, can be manipulated in composition and structure and so in specificity and activity as to perform outside the original biological setting. That brings me back again to this notion of microstructure catalysis, a whole concept, it seems to me. Those like Ted Davis who are wrestling with self-assembly systems that, after self-assembling, don't disassemble are having to think about this. What is the status of the concept in reaction engineering?

James Wei: Perhaps I don't quite understand your terms. Would you consider a zeolite a microstructure catalyst?

L. E. Scriven: Yes, I had zeolites in mind, not as catalysts themselves but because of the role of templates in the synthesis of zeolites, templates that catalyze the desired supramolecular structure. That would serve as a point of discussion on this notion.

James Wei: The entire subject of enzymes, aren't they all templates?

L. E. Scriven: Yes, and not only for covalent bonding but also for molecular conformation and, I believe, for structured aggregates of molecules. I would say not just enzyme catalysis in the biochemical setting, but also in its biomimetic aspect, that is, agents that behave in the same way in inducing supramolecular microstructure.

Robert Brown: It's awkward for me to comment because I'm an outsider to this field, but after listening to the discussion, I have some questions. A simple way of looking at reaction engineering is boiling it down into kinetics, surface catalysis, and what I would call chemical reaction modeling. When I hear chemical reaction modeling defined as it has been this evening, I don't see it as distinct from analysis of transport systems. Sure, you have some reaction in the system. As long as it's not the chemistry but the overall integration of that chemistry and the transport phenomena that's the central issue, then it just becomes a case study in a specific, technologically applied area of transport phenomena. The question is, aren't we taking it too far out of context, in the sense of splitting it too far afield from its fundamental intellectual base? In a lot of our teaching, at both the undergraduate and graduate levels, we lose the coherence that we might have and then lose a lot of the connectivity to fluid mechanics, mathematical methods, and so on. That's where we need to go in chemical reaction engineering. We shouldn't keep lumping the system and doing tubular reactors, that's the message I got. That's question one. We heard tonight from Louis Hegedus of the tremendous importance of catalysis industrially. My second question is, why didn't we hear anything about clean surface science, which I understand from academic colleagues is connected with catalysis? One, is it catalysis? And second, whether it is or it isn't, should academics study catalysis? Do academics study catalysis in the same way that industrial manufacturers of catalysts do? The basic question is, are the academics relevant?

James Wei: First of all, your definition of chemical reaction engineering is wrong.

Robert Brown: Okay.

James Wei: Most work in chemical reaction engineering is like separations. There's a total job to be done, we want to get some reactions carried out, products made, that's what it's about. How is it to be done? You have all the ingredients of fluid mechanics, kinetics, transport, catalysis, etc. The purpose is to get a job done.

Robert Brown: I will accept that, but then you're telling me, it's totally product driven.

James Wei: Not product driven, it's process driven. You've got something you want to get done, you want to put it in the kitchen and cook it until it's done, that is chemical reaction engineering, my definition of it. The second issue you raised concerned the relationship of studies at high vacuum with clean surfaces to research in catalysis. The catalysis people use studies on clean surfaces to help them solve their problems; sometimes they do, sometimes they don't.

Alexis Bell: Let me add to Jimmy's last comment about surface science, since it strikes a resonant note in me. To draw an analogy, surface science plays the same relationship to catalysis as studying the fundamental thermodynamics of argon or methane does to practical thermodynamics. It provides a basis for developing physical models, testing out techniques, establishing an intellectual framework within which to work. Clearly, it's not the end of the game, but at the very beginning.

Morton Denn: Bob Brown's first question is significant from the educational point of view. I think Jimmy's response was maybe right for the wrong reasons. Whenever we talk about curricula, we generally argue that one of the reasons that we have a reaction engineering course is for the opportunity to integrate all of the basic concepts that we've developed and then to focus on the way in which they all come together. Indeed, that's also the argument that some of us used recently for the separations course. The separations course should also be a way of integrating concepts and the process point of view developed in the other courses. Jimmy's answer is the right one for the wrong reasons because that was the implication of the answer, but in fact I don't know any place where the reaction engineering courses are really taught that way. We need to rethink what's in our reaction engineering courses to carry out that integrating process-and-product–oriented focus, which indeed we all agree should be there.

Robert Brown: My recollection of how reaction engineering was taught to me and how I've seen it taught in other places, almost as a separate gospel, an independent, fundamental entity, drives me to ask a question. Where will the issue ever reach the student that the hydrodynamic complexity of the reactor is a central focus?

James Wei: But that isn't always the central focus. You see, there are many reactors where it is not the focus.

Robert Brown: I understand that. But we're talking about subjects we don't understand, not subjects we do.

Klavs Jensen: Mort Denn has said most of what I wanted to say, but in its broader sense, chemical reaction engineering is transport phenomena combined with chemical reactions, with the aim of designing a reactor or optimizing a reactor. It should be taught that way so that the students realize the importance of the fluid mechanics. Over the years, we were very successful with simple concepts like stirred-tank reactors and plug-flow reactors.

Those concepts should still be taught because they give an intuitive feel for what happens in a system. However, we have forgotten that there are other systems where the fluid mechanics are essential. We've been guilty of not looking at what happens in other fields. Combustion is a prime example of this. Combustion researchers are doing highly sophisticated monitoring of very complex reactions. They are on the forefront of gas-phase kinetics, and they put all the kinetics into realistic fluid mechanical models. The models are complicated, but they can be dealt with, and I think it's a pity that we don't teach the students that.

Robert Brown: Is that why the combustion course and reaction engineering course are not combined?

Klavs Jensen: Combustion is typically taken as a graduate topic that requires far more physical chemistry than you're willing to subject an undergraduate to. My personal view is that if we have the students taking a physical chemistry sequence, why not also use it?

Preetinder Virk: I've been a little dismayed by the three speakers today, the least by Alex Bell. Although I understand that it's their task to present what industry would like to get out of reaction engineering, from my point of view, it would seem more useful to define the fundamentals that students could use creatively within whatever task industry sets them. Within this context, the chemistry and the reaction pathways seem the most fundamental. We need an appreciation of what makes species react, what the fundamental aspects are that go into designing reaction logic and trains, and then, with that, what the consequences are for reaction engineering. This is the chemical aspect that's most worth stressing. Second, along the same lines, is the fluid mechanics, if I can hark back to this morning. Professor Brenner, for example, criticized computer-generated simulations. However, these can provide just the kind of information that one needs to incorporate into reactor designs. Whatever the fundamentals are in fluid mechanics, those too need to be defined for the students. Our students need to go away with the operational fundamentals of reaction engineering, rather than the particular important problem that industry wants solved today. That's the goal we should try to focus on.

Reuel Shinnar: I wanted to answer Bob Brown's question, because I think it is a very important one. It may show up at every step as we go along. On one hand, he is correct. The key to reaction engineering is integration of fluid mechanics and chemical reactions kinetics. Understanding the nature of the chemical reaction themselves should be an important part of teaching reaction engineering, and we don't do enough about it. On the other hand, there is much more to reaction engineering than integrating those two areas. When we design chemical reactors, we often have to design them without understanding the chemistry, without understanding catalysis, which nobody yet understands, and without understanding the fluid mechanics. In very few reactors do we have sufficient knowledge of the fluid mechanics. And if we know how, we can still design them reliably and effectively. We learn to design with minimum information by asking, what information do I really need, what can I get away with, what information would I like to have to improve my design? Reaction engineering in chemical engineering has been developing and focusing, both in research and in industry, on exactly those questions. Consider for example the complex kinetic models that are used in refinery processes. They are still very far from being accurate descriptions of the actual chemistry. But they are a great improvement on the old empirical statistical relations. They are still basically statistical correlations, but the structure for the estimation process has a sound thermodynamic and kinetic foundation. Research on the impact of lumping has greatly helped us to understand the value of such models.

There are other areas of reactor design for which proper modeling has allowed us to recognize the critical information required for safe scale-up. A thorough understanding of the interaction between transport processes and chemical reactions is essential in order to appreciate these designs and modeling concepts, but it is not enough. We need to understand what is involved in design and how to deal with processes for which only approximate models are available, because the process is too complex for an accurate mathematical description or the expense of obtaining data for a more rigorous description cannot be justified. The same approach is useful for many large complex problems that cannot be handled by rigorous analysis. The ability to handle and structure such complex problems has given the chemical engineer an edge and an advantage over the conventionally trained scientist. It is essential, in both teaching and choice of textbook, that students get a taste of what is really involved in a design problem, which is more than just fluid mechanics and kinetics. It can be very useful to teach the students what design is about, what minimum information is about, what are the problems we really want to solve. Maybe it is beyond the textbooks to do that, but it is worth a try.

Louis Hegedus: Yes, we need to understand the chemistry, and we need to understand the fluid mechanics. It's very important to realize that reaction engineering as a discipline rose out of the discovery that the sum of chemistry and fluid mechanics is not adequate. New considerations, new problems needed specific tools, and that became chemical reaction engineering. I don't want you to feel that reaction engineering is simply fluid mechanics plus chemistry. It is the interaction of the two that makes the field.

James Wei: It is integration of all these things with a particular purpose in mind; the purpose is to get a job done.

Howard Brenner: I'd like to support the use of transport phenomena combined with chemical reactions of the kind Bob Brown and Mort Denn put forward. In principle, for any reactor one can enumerate the basic transport equations, including the reaction terms, as well as the appropriate boundary conditions. If all of the phenomenological coefficients appearing therein, together with the flow patterns involved, are known, then—in principle—one can solve these equations to uncover what's going on at every single point in the reactor at every single instant of time. This is especially true now, given the current existence of high-speed computational abilities. But such exhaustively detailed information does not necessarily provide any useful understanding of the basic phenomena. That is the basic issue. We don't really know what to do with all of that microscale information, in particular how to extract macroscale understanding from it. And such understanding is, I believe, the very essence of chemical reaction engineering. Take the idea of an effective reaction rate, which may be only remotely related to the true, microscale chemical kinetics. The term "effective" constitutes a macroscopic concept, which could presumably be extracted from all the fine-scale information if only one knew how. In this macroscopic context, important progress has been made in apply generalized Taylor dispersion theory to convective–diffusive–reactive systems [see Shapiro, M., and Brenner, H., *Chem. Eng. Sci.* **41**, 1417–1433 (1986); *AIChE J.* **33**, 1155–1167 (1987); *Chem. Eng. Sci.* **43**, 551–571 (1988)].

Gary Leal: I'm responding to Howard. If you think that you understand all the details of the underlying microphysics and you simulate it, I would call that a computational experiment. I don't understand why you think that there's any less information and understanding in that than there would be in a laboratory experiment, assuming that the underlying principles that you said exist do exist. Why is there less understanding, or is there? Is that what you're saying?

Howard Brenner: Let me give a simple example of this, that derives from the generalized Taylor dispersion theory references cited in my previous comments. Think of a tubular reactor in which one has a Poiseuille flow, together with a chemical reaction occurring on the walls. One can certainly write down all the relevant differential equations and boundary conditions and solve them numerically. However, the real essence of the macrophysics is that if one examines the average velocity with which the reactive species moves down the tube, this speed is greater than that of the carrier fluid because the solute is destroyed in the slower-moving fluid streamlines near the wall. Consequently, the only reactive solute molecules that make it completely through the reactor intact are those that were smart enough to stay away from the wall. These survivors thus necessarily tend to move near to the center of the tube, where higher Poiseuille velocities prevail. This reasoning furnishes the general conception that, on average, the reactive material moves faster than the inert carrier fluid. This is a general concept, one that derives from thinking about the overall aspects of the transport process, especially the fact that the microscale velocity field, that is, the Poiseuille flow, is nonuniform over the tube cross section. As a result, this global picture provides some general understanding of the phenomena, albeit qualitative, independently of the numerical details.

George Stephanopolous: I wish that some other people in the general area of process systems engineering would speak out on some of the issues. In process systems engineering, one learns very explicitly to think contextually. For example, looking at the reactor, if there is a large economic incentive in improving 1% of the yield of the reactor, and that was the case in the fluid catalytic cracking (FCC) example given by Katzer, then one needs to carry out a detailed analysis of what is going on. In other instances, the reactor itself is the end product, but you don't know how to design it. There you have the problem of native design. We shouldn't mix the two. They are extreme examples that imply that you do need the analysis within the context of the first example, and you need creative design, which we don't teach today, within the context of the second one. How can we show someone to be creative in cases where analysis is not the driving force? For example, we have a number of reactors that include separation. Should we have the people who invented them say, take them and analyze them? That's not what we need. What we need is to teach creative design so that we can invent them. The reactor itself should be seen as an end product that will be designed and not just be analyzed. I don't know how that can be done; I don't see it being done.

James Katzer: That is in reality the situation in the FCC case. I want to understand what's going on, on a short enough time scale, so that I can design the next FCC, to improve my yield. It may operate with a 1-second contact time, and I need to know what the mixing is in the separations.

SECTION V
Environmental Protection and Energy

13

Environmental Chemical Engineering

John H. Seinfeld
Department of Chemical Engineering
California Institute of Technology
Pasadena, California

I. Introduction

Environmental protection has always had a central role in chemical engineering. Indeed, problems in gas cleaning are among the earliest addressed in the development of unit operations. As environmental regulations have expanded and the technology being controlled has become more advanced, the associated chemical engineering problems have grown more complex. Ever more stringent motor vehicle exhaust emission limits, for example, led to the development of the catalytic muffler [1]; stricter control of nitrogen oxides emissions from stationary sources such as power plants and boilers instigated the invention of the thermal de-NO_x process for homogeneous NO reduction by ammonia injection [2]. Chemical engineers have had prominent roles in the development of stack gas scrubbing technologies, in understanding the mechanisms of particle and NO_x formation in coal combustion, and in developing mathematical models of air pollution.

Chemical engineering problems in environmental protection usually require a synthesis of thermodynamics, chemical reaction engineering, and transport phenomena, often applied to systems that are not well understood at a fundamental level. An attribute of the revolution in chemical engineering is the opportunity for a deepened understanding of processes at the molecular level, which will have a significant impact on environmental problems.

ADVANCES IN CHEMICAL ENGINEERING, VOL. 16

In spite of the centrality of environmental considerations to the chemical process industries, environmentally oriented research has never occupied a premier position in chemical engineering. Perhaps this is because, from an industrial point of view, environmental protection is essentially a defensive need; pollution control equipment does not make money—it costs money. Developing a new stack gas scrubbing system is not as economically productive as developing a new catalyst. Environmental considerations in process design and operations are, nonetheless, here to stay. Attention to issues such as the global greenhouse effect and stratospheric ozone depletion by chlorofluorocarbons can only be expected to intensify in the future. Since low-pollution plants and transportation systems will be a permanent fixture of our society, the industrial challenge is to turn environmental protection from a defensive endeavor to an offensive strategy.

What constitutes the field of environmental engineering? Research that can be classified as environmental engineering occurs in academic departments of civil engineering, mechanical engineering, and chemical engineering, in addition to those actually named *environmental engineering*. In terms of numbers of faculty members, the majority is carried out in departments of civil and environmental engineering. Table 1 presents a breakdown of the principal environmental research areas in departments of civil and environmental engineering. An informal count of the number of U.S. faculty members in departments of civil and environmental engineering engaged in environmental research, using the 1988 *Peterson Guide to Graduate Study in Engineering*, places the number at about 500. Entries from the 1987 *ACS Directory of Graduate Research* were used to determine the number of chemical engineering faculty members in environmental research areas (Table 2). Comparison between the research fields in Tables 1 and 2 shows the division that has developed over the years between civil/environmental and chemical engineering environmental research programs, namely the overwhelming emphasis of the former on water pollution issues. Chemical engineers, while assuming most of the air pollution research, have also contributed in the waste and wastewater treatment areas. As for mechanical engineering departments, the major environmental focus has been combustion and control of combustion-generated pollutants, together with several important programs in chemical engineering departments. In summary, aquatic, waste, and wastewater treatments constitute the research interests of most civil and environmental engineering faculty working in the environmental field, while many chemical engineering faculty are engaged in research on issues related to air pollution: combustion emissions, stack gas cleaning, or the atmospheric behavior of pollutants. Finally, mechanical engineers have also contributed importantly in the area of combustion-generated pollutants.

Table 1. Research Areas in Environmental Engineering Departments (Non-Chemical Engineering)

Environmental engineering	Wastewater and industrial waste treatment
	Biological waste treatment
	Solid waste processing
	Soil and hazardous waste management
Hydraulics, water resources, and ocean engineering	Fluid mechanics and hydraulics
	Transport and mixing in channels, lakes, and estuaries
	Jets and plumes; density-stratified flows
	Coastal engineering; ocean engineering
	Ground water flow through porous media
	Fate of contaminants in soils and ground water
	Optimal design of water resource systems
Water chemistry	Fate of inorganic and organic pollutants in natural waters
	Analytical chemistry of natural waters and trace contaminants
	Trace metal–particulate matter interactions
	Structure–activity relationships for organic compounds
	Aquatic colloid chemistry
	Precipitation chemistry/acid rain

Table 2. Environmental Research Fields in Chemical Engineering Departments

Field	No. of Faculty Members (1987)
Air pollution control and combustion processes	
Combustion	10
Gas treatment	8
Air pollution control	6
Waste and wastewater treatment	15
Air pollution	
Atmospheric processes	9
Aerosols	14
Environmental transport	4
Miscellaneous	13
TOTAL FACULTY	79

What are the problems and opportunities for chemical engineers in environmental research? First, the problems. While research funding for waste- and water-related areas is concentrated in the NSF's Environmental Engineering Program, NSF lacks a similar focus for air pollution engineering research. (Atmospheric chemistry research is well served by the Program in Atmospheric Chemistry.) The demise of EPA's research program has virtually eliminated what should be the most significant supporter of air pollution research from the federal scene. Modest support is sometimes available from state and local agencies, although the support here tends to focus on a shorter horizon than might be most appropriate for university research.

In spite of the funding uncertainties, however, research problems in environment chemical engineering offer exciting intellectual challenges. The foundations of chemical engineering, as it pertains to environmental research, are the traditional foundations of chemical engineering itself. Unlike biochemical engineering, which has infused a biology base into our discipline, environmental chemical engineering does not require the assimilation of new areas of science and engineering into chemical engineering. Rather, its key characteristic is dealing with systems of enormous complexity and breadth.

In this paper, we attempt a narrow survey of some important problems in environmental chemical engineering to convey their flavor and challenges. The open-ended character of environmental chemical engineering problems makes them ripe for innovation. We hope to stimulate interest in attacking these problems, in terms of both research and inclusion in the basic chemical engineering curriculum.

II. Transport and Transformation in the Environment

Environmental engineering can be divided into two major regimes—all processes occurring before material is released to the environment, and everything that occurs once that release takes place.

A related way to view environmental research is synthetic versus analytic. Synthetic research is aimed at minimizing emissions—for example, by designing pollution control equipment and low-emission processes. Analytic research focuses on analysis of the behavior of pollutants in the environment. These two areas merge when one designs pollution control strategies to achieve targets for the environmental impact of emitted pollutants. In doing so, the required technology is driven by the emission–environmental quality relationship derived from the analysis of the physicochemical behavior of pollutants in the environment. Transport and transformation in and removal from the environment—air, water, and land—is the area that we address first.

More than 65,000 chemicals are used commercially in the industrialized nations of the world [3]. Many of these substances, such as pesticides, polychlorinated biphenyls (PCBs), solvents, combustion-related compounds, trace elements, and organic compounds such as polycyclic aromatic hydrocarbons (PAHs), are emitted directly or indirectly into the environment. Once released into the environment (air, land, or water), they are subjected to various physical and chemical processes that determine their ultimate environmental fate. In many respects, the environment functions as a giant chemical reactor in which species are transported, mixed, and transformed. All but the most stable of substances are chemically modified once emitted. Environmental systems encompass tremendous ranges of temporal and spatial scales. Spatial scales span some 15 orders of magnitude, from the thousands of kilometers characteristic of hemispheric and oceanic compartments to the nanometer scale of molecular dimensions. Temporal scales range over 12 orders of magnitude, from the millisecond range of fast reactions and turbulent eddy transport to the thousand-year scale for long-term waste disposal.

Pollutants released to the environment invariably end up being distributed across the atmosphere, soil, and water and among biota. Because of the difficulties in adequately monitoring pollutant distributions in all three media, mathematical models often provide the only means of predicting multimedia transport and transformation [4]. Table 3 lists the elements of the transport and transformation of pollutants in the environment.

Most existing environmental models emphasize single-medium transport; models for air pollutants include removal processes at water and soil surfaces but neglect two-way interactions with the water or soil compartments or phases. Similarly, transport models for chemicals in soil focus more or less exclusively on transport in the soil. The principles of multimedia pollutant models are straightforward; transport equations for individual environmental media are formulated and coupled through the appropriate boundary conditions. Atmospheric, aqueous, and terrestrial compartments are interconnected through processes such as wet and dry deposition (to be discussed shortly), volatilization, and resuspension. In spite of the apparent simplicity of the model structure, a great deal of information is required to describe pollutant transport across medium boundaries and to represent chemical and physical transformations within the media themselves. Simulating environmental flows such as occur in lakes, reservoirs, estuaries, streams, and the atmosphere alone occupies major research endeavors in computational fluid dynamics. A particular characteristic of multimedia models is the difference in time scales associated with the different intermedia transport processes.

Table 3. Transport and Transformation of Pollutants in the Environment[a]

Atmosphere
Transport from the atmosphere to land and water
  Dry deposition of particulate and gaseous pollutants
  Precipitation scavenging of particulate and gaseous pollutants
  Adsorption of gases onto particles and subsequent dry and wet deposition
Transport within the atmosphere
  Turbulent dispersion and convection
Atmospheric transformation
  Diffusion to the stratosphere
  Photochemical degradation
  Oxidation by free radicals and ozone
  Gas-to-particle conversion

Water
Transport from water to atmosphere, sediment, and organisms
  Volatilization
  Sorption by sediment and suspended solids
  Sedimentation and resuspension of solids
  Aerosol formation at the air-water interface
  Uptake and release by biota
Transport within water bodies
  Turbulent dispersion and convection
  Diffusion between upper mixed layer and bottom layer
Transformation
  Biodegradation
  Photochemical degradation
  Degradation by chemical processes such as hydrolysis and free radical oxidation

Soil
Transport from soil to water, sediment, atmosphere, or biota
  Solution in rainwater
  Adsorption onto soil particles and transport by runoff or wind erosion
  Volatilization from soil or vegetation
  Leaching into ground water
  Resuspension of contaminated soil particles by wind
  Uptake by microorganisms, plants, and animals
Transformations
  Biodegradation
  Photodegradation at plant and soil surfaces

[a] From Cohen [4].

Because of the large range of spatial scales inherent in environmental systems, smaller-scale phenomena usually remain unresolved in models of transport and transformation. Yet, unresolved turbulent motions, for example, can result in significant transport of momentum, energy, and species; they interact dynamically with the larger-scale eddies explicitly resolved to degrade the accuracy with which transport processes on the scale of resolution of the model are simulated. A major shortcoming of current environmental models is their lack of representation of the unresolved scales of phenomena that lie below the spatial scale of the numerical calculation. In many cases, the real turbulence occurs at scales below those resolved numerically. Developing consistent and fundamental parameterizations of unresolved motions is critically needed.

Current multimedia models are inadequate in many respects. Description of intermedia transport across the soil–air and unsaturated soil–saturated soil zones suffers from the absence of a suitable theory for multiphase transport through the multiphase soil matrix. These phenomena are crucial in describing pollutant migration associated with hazardous chemical waste sites. Existing unsaturated-zone soil transport models fail to include mass transfer limitations associated with adsorption and desorption and with absorption and volatilization processes. Rather, most models assume equilibrium among the soil–air, soil–solid, solid–water, and soil–contaminant phases.

Considerably less is known about the behavior in the atmosphere of organic species than about the major inorganic pollutants such as SO and NO_x (NO and NO_2). It has been established that numerous potentially toxic organic compounds from anthropogenic sources are dispersed worldwide [3]. Most species emitted to the atmosphere are eventually removed by natural atmospheric cleansing mechanisms. These removal processes are generally grouped into two categories: dry deposition and wet deposition. Dry deposition is the direct transfer of gaseous and particulate materials to the earth's surface in the absence of precipitation. Wet deposition encompasses all processes by which airborne species are transferred to the earth's surface in aqueous form, i.e., rain, snow, or fog. Even though wet deposition processes involve several complex physicochemical steps, current understanding of them far exceeds that of dry deposition. Dry deposition rates are extremely difficult to measure accurately. The rate of transfer of a compound between the atmosphere and the surface of the earth depends on many chemical, physical, and biological factors, the relative significance of which varies depending on the nature and condition of the exposed surface, the physicochemical characteristics of the compound, and the micrometeorological condition of the atmosphere. Ability to predict the effectiveness of wet and dry deposition processes is crucial in determining the overall atmospheric residence time of pollutants; considerable research remains to be done on the detailed chemistry and physics of these two removal mechanisms.

Pesticides, PCBs, and PAHs are examples of semivolatile organic compounds (SOCs) that exist in the atmosphere as gases or particles or are distributed between these two phases [5]. These substances have vapor pressures ranging from 10^{-4} to 10^{-11} atm at ambient temperatures, with those having vapor pressures at the high and low ends of this range existing entirely as gases and particles, respectively. Since gases and particles can have vastly different residence times in the atmosphere, predicting the atmospheric fate of SOCs requires that one be able to estimate the vapor–particle partitioning. Mercury, although not an SOC, presents a case in point. Virtually all of the mercury in coal is released to the atmosphere as mercury vapor when coal is burned. Elemental mercury vapor has an estimated atmospheric residence time of several months to 2 years. If absorbed on particulate matter, however, mercury is expected to be removed in a few weeks, corresponding to the residence time of the host particulate matter. Additional research is needed on methods to distinguish gaseous and particulate SOCs in the atmosphere and on the detailed interactions between SOC vapors and atmospheric particles.

III. Aerosols

The behavior of small particles in air remained largely a laboratory curiosity until the recognition of the role of such particles in atmospheric pollution. While environmental research has focused on the sources, behavior, and effects of unwanted airborne particles, aerosol researchers are now formulating general theories of aerosol behavior that will apply to processes as diverse as atmospheric pollution and chemical vapor deposition. Table 4 summarizes physical and chemical phenomena associated with a number of areas in aerosol research. We note that many of these areas might also be found in a chapter devoted to materials processing. It is now recognized that the body of knowledge characterized as aerosol science and technology is applicable to a remarkably diverse range of technological problems.

In this section we focus on four specific environmental problems: coal combustion aerosol formation, dynamics of atmospheric aerosols, the chemical characterization of particles, and the role of aerosols in clean room technology or so-called microcontamination control. For the reader interested in an introduction to aerosol science, we recommend three texts [6–8].

Table 4. Aerosol Science and Technology[a]

Aerosol Phenomena	Science and Technology
Gas-phase nucleation	Flame synthesis of particles (e.g., carbon black, silica); cluster formation in chemical vapor deposition; manufacture of high-purity silicon; cluster structure and energetics; plasma synthesis of refractory materials and coatings.
Gas–particle interactions	Accommodation processes on surfaces; growth of particles by condensation; atmospheric chemistry; chemical processing of aerosolized materials; aerosol catalysis; chemical vapor deposition; surface and cluster chemistry.
Coagulation	Catalyst preparation; cluster and polymer formation in chemical vapor deposition; intermolecular interactions between particles; diffusion-limited aggregation; soot formation.
Charged aerosols and aerosol charging	Microcontamination; electrostatic precipitation; atmospheric electricity; aerosol sampling and measurement; ion–particle interactions.
Interactions of electromagnetic radiation with particles	Atmospheric visibility; radiative transfer in combustors; analytical chemistry of particles; military applications
Transport and deposition	Process equipment fouling; thin–film deposition; microcontamination; filtration; kinetic theory of rarefied gases; hydrodynamics; atmospheric dry and wet deposition.
Measurement	Process stream monitoring; optical properties; electrical properties; transport properties.

[a]Adapted from a similar table suggested by W. H. Marlow.

A. Coal Combustion Aerosol Formation

Coal combustion is a major source of particulate emissions into the atmosphere, especially of particles 0.1 to 5 μm in diameter [9]. Certain toxic elements, e.g., As, Se, Sb, and Zn, are preferentially concentrated in the smaller sizes. The characteristics of coal combustion aerosols depend on a number of factors, such as the type, properties, and size of the parent coal and the furnace design and operating condition.

Coal, pulverized into a fine powder of about 50 μm diameter, is blown into the furnace with carrier air. While combustion temperatures up to 2000°C are usual, different coal particles are subjected to varying temperatures due to particle size differences and nonuniformity in mixing. As a coal particle is heated, it may mechanically break up into fragments due to thermal stresses. Volatile fractions originally present in the coal or formed by pyrolysis are vaporized. The heated particle may swell and become porous. At lower temperatures, mineral inclusions may decompose or sinter; at higher temperatures they may fuse, producing highly viscous molten ash. Whether solid or liquid, the mineral inclusions form fly ash particles. An ash species may vaporize, and reducing conditions close to the burning particle surface may produce even more volatile species that oxidize in the bulk gas phase away from the particle surface. The vaporized species may condense, either nucleating homogeneously to form very fine particles or condensing on existing particles, enriching fine particles and particle surfaces with volatile species.

Although the fly ash particle size distribution in the submicron regime is explained qualitatively by a vaporization/homogeneous nucleation mechanism, almost all of the available data indicate particles fewer in number and larger in size than predicted theoretically. Also, data on elemental size distributions in the submicron size mode are not consistent with the vaporization/condensation model. More nonvolatile refractory matrix elements such as Al and Si are found in the submicron ash mode than predicted from a homogeneous nucleation mechanism. Additional research is needed to elucidate coal combustion aerosol formation mechanisms.

B. Aerosol and Cloud Chemistry

Atmospheric aerosols are multicomponent particles ranging from 0.001 to 10 μm in diameter. Particles are introduced into the atmosphere by combustion processes and a variety of other anthropogenic and natural sources. They evolve by gas-to-particle conversion and coagulation, are augmented due to the formation of fresh particles by nucleation, and are removed by deposition at the earth's surface and scavenging by airborne droplets. Atmospheric aerosols are the main cause of the visibility degradation accom-

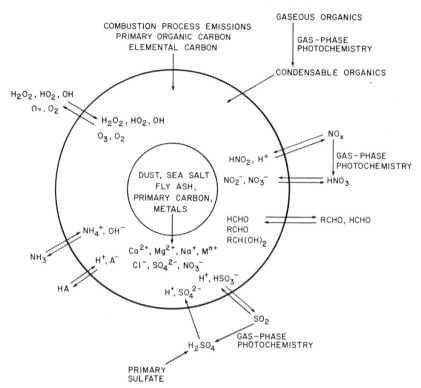

Figure 1. Chemical processes associated with an atmospheric aerosol particle.

panying elevated air pollution levels and may also carry toxic materials into the lung.

Figure 1 shows a schematic of a typical atmospheric aerosol particle (if such an entity can be assumed to exist). The particle consists of sulfates, nitrates, water, ammonium, elemental and organic carbon, metals, and dust. After a primary particle is emitted, gas-phase reactions occur, converting oxides of nitrogen to nitric acid, sulfur dioxide to sulfuric acid, and hydro-carbons to oxidized, low-vapor-pressure condensable organics.

An important current problem is attaining sufficient understanding of at-mospheric aerosol dynamics to develop mathematical models capable of relating emission reductions of primary gaseous and particulate pollutants to changes in ambient aerosol loadings and thereby to improvements in visibility and health effects. These models involve thermodynamics, transport phenomena, and chemical kinetics in an intricate equilibrium and nonequilibrium system [10]. Volatile compounds, such as ammonium, chloride, nitric acid, and water, are distributed between the gas and aerosol phases as determined by thermodynamic equilibrium. The distribution of nonvolatile species, such as sulfate, sodium, and elemental carbon, is determined by nonequilibrium transport processes.

The acidity in rain consists primarily of nitric and sulfuric acids, the gas-phase precursors of which are NO_x and SO_2. Considerable effort has gone into elucidating the gas-phase oxidation mechanisms of these species. It has become clear that homogeneous gas-phase reactions alone are insufficient to explain the levels of nitric and sulfuric acids in precipitation [11]. Heterogeneous reactions, both in clouds and in aerosol particles, appear to be involved. Heterogeneous chemistry may affect atmospheric acid formation even if the oxidation reactions themselves take place in the gas phase. The primary species oxidizing NO_x and SO_2 in the gas phase is the hydroxyl radical; yet the ambient OH concentration still has not been accurately measured. There is evidence that the hydroxal radical may be produced or destroyed by heterogeneous reactions, including the photolysis of Fe (III) complexes to produce OH, the Cu ion-catalyzed destruction of HO_2 radicals, and the transition metal ion-catalyzed conversion of NO_2 to HONO, which readily photolyzes to form OH.

Although the rate of homogeneous, gas-phase oxidation of SO_2 can be estimated reasonably accurately, one of the most persistent fundamental questions is the extent to which heterogeneous SO_2 reactions are important [12–14]. The heterogeneous chemistry is thought to involve the droplet and aerosol-phase oxidation of SO_2 in the presence of species such as dissolved oxygen, hydrogen peroxide, ozone, and certain metal catalysts. Progress has been made in elucidating the rates and mechanisms of SO_2 oxidation under conditions occurring in cloud and fog droplets [15], but there is little agreement as to the chemical routes occurring in submicron aerosol particles.

Much of the detailed work on aqueous-phase atmospheric chemistry has focused on the reactions taking place in clouds or fog droplets, where the relative humidity approaches 100%. Under these conditions, droplets are relatively dilute. For example, fog and cloud droplets have diameters of 5 to 10 μm and ionic strengths from 0.001 to 0.01 M. Although high-humidity conditions are quite readily studied using classical chemical techniques, low-humidity conditions lead to solute concentrations that are beyond the range normally amenable to wet chemical experiments. A major experimental obstacle in studying heterogeneous atmospheric chemistry, particularly under subsaturated conditions, is to reproduce ambient conditions of solution concentration and particle size in the laboratory.

To study the chemistry of highly concentrated particles in bulk solution one must avoid mass transfer limitations and the effects of container surfaces. Both of these problems are eliminated by directly using aerosol particles. Two approaches have been used to study aerosol chemistry: (1) aerosol reactors in which the evolution of a suspension of particles is followed, and (2) experiments in which the changes occurring in a single particle can be followed.

C. Nucleation

Cloud and aerosol formation in the atmosphere is initiated by homogeneous or heterogeneous nucleation. As noted in Table 4, nucleation processes are important in areas ranging from atmospheric chemistry to flame synthesis of particles and chemical vapor deposition. Considerable theoretical attention has been given to homogeneous nucleation [16]. Heterogeneous nucleation poses all the theoretical problems of homogeneous nucleation with the added complication of representing the interactions between vapor-phase molecules and foreign surfaces [17, 18]. In the classical theories of homogeneous and heterogeneous nucleation, the nucleation rate as a function of temperature and saturation ratio depends on the surface tension, collision cross section, sticking coefficient of monomers on clusters, and free energy of formation of the critical size cluster. Bulk values of surface tension and density are used to evaluate the nucleation rate. In addition, the simple geometric cross-sectional area and unit sticking coefficient are used. It is well known, however, that the sticking coefficient is dependent on both size and temperature: it decreases with decreasing size and increasing temperature. There exists no real experimental verification of the validity of using bulk properties for small clusters. Even the validity of using a size-dependent surface tension has not been established.

In spite of the widespread recognition of the theoretical inadequacies of classical nucleation theories, attempts to formulate more realistic theories have met with limited success, in part because nucleation rate measurements are notoriously difficult to make. Consequently, the available data base with which to evaluate various theories is inadequate. Molecular level approaches would seem to hold promise of providing more rigorously acceptable theories without resorting to the use of uncertain bulk properties in treating clusters that are intrinsically molecular. Furthermore, new experimental techniques, such as molecular beams and cluster spectroscopy, make the properties of small clusters amenable to investigation at the molecular level.

Nucleation processes, especially in nature, are rarely homomolecular and homogeneous; usually two species are involved (binary heteromolecular nucleation), and foreign surfaces may also be present (heterogeneous nucleation). Heterogeneous nucleation in the atmosphere encompasses numerous particulate systems, including ions, organic and inorganic species, cloud condensation nuclei, and other preexisting aerosols. Although atmospheric ions probably play a less important role in heterogeneous nucleation than other cloud condensation nuclei, they are believed to be important for aerosol formation in the stratosphere. The bonding of atmospheric molecules to ions is an important problem, especially for systems involving water and NO_3^-, H_3O^+, $H_2SO_4^-$, and NH_4^+. Although the role of organic molecules in cloud and aerosol formation has been of increasing interest, virtually no experimental or theoretical studies of the interaction between water molecules and atmospheric organics exist. Ions and uncharged (organic and inorganic) molecules enhance the formation of hydrated clusters by providing centers for creating added intermolecular attractive electric fields, that is, monopole (for ions), dipole, and induced dipole fields. In the case of heterogeneous nucleation involving other foreign nuclei, their surface further lowers the barrier height in the energy of formation due to collective attractive forces, so that lower supersaturations are required for nucleation.

The study of nucleation phenomena involves demanding experimental techniques capable of revealing the effects of saturation ratio, temperature, and molecular properties on the rate of particle formation, coupled with a chemical kinetic theory of molecular cluster dynamics. A molecular-based kinetic theory of nucleation requires a statistical mechanical description of the free energy of formation of clusters and a molecular collision dynamics treatment of sticking coefficients. A statistical mechanical approach to nucleation should circumvent uncertainties in classical theories that rely on bulk surface tension. Only recently have experimental approaches been devised to measure molecule–particle sticking coefficients [19]. The prediction of molecule–molecular cluster sticking coefficients is an open area for research, requiring, in principle, solution of the Boltzmann equation, or approximations thereto, including intermolecular potentials.

D. Chemical Characterization of Particles

The theoretical analysis of atmospheric aerosols just described focuses on a population of "typical" yet hypothetical particles that have the average chemical characteristics of the ambient mix. Yet, the nature of actual, individual particles remains largely a mystery due to the difficulties in collecting and probing single submicron particles.

Conventionally, particles are collected on a filter substrate, where they can be probed by a variety of methods. Table 5 lists the analytical measurement capabilities of the four most commonly used techniques for the analysis of particle composition, and Fig. 2 shows an idealized approach

Table 5. Characteristics of the Most Commonly Used Microsurface Techniques[a]

	XPS[b]	AES[c]	ISS[d]	SIMS[e]
Excitation	X-ray	Electron	Ion	Ion
Emission	Photoelectron	Secondary Auger electron	Primary ion	Secondary ion
Measured information	Photoelectron binding energy (single ionization)	Auger electron energy (double ionization)	Ion mass	Ion mass/charge
Analyzed volume				
Cross sectional diameter (μm)	200	0.1	20	1
Depth (μm)	0.0025	0.0015	0.0002	0.0005
Detection limit (mg/kg)	10^3	10^3	1 to 10^4	0.1 to 10^3
Detectable elements	All except H	All except H, He	All except H	All, isotopes
Chemical molecular data	Some	Some	None	Some

[a] From EPRI [53].

[b] XPS—x-ray photoelectron spectroscopy.

[c] AES—Auger electron spectroscopy.

[d] ISS—ion scattering spectroscopy.

[e] SIMS—secondary ion mass spectrometry.

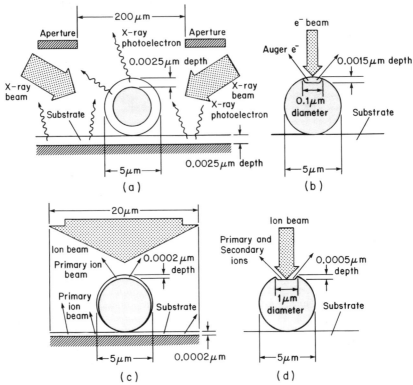

Figure 2. Analysis sequence for chemical characterization of a particle on a substrate medium: (a) XPS; b) AES; (c) ISS; (d) SIMS. Copyright © 1986. Electric Power Research Institute. EPRI TB.EAE.4.4.86 "Atmospheric Particles" [53]. Reprinted with permission.

to the analysis using these four methods. Construction of a three-dimensional chemical profile of a particle begins by bombarding the particle surface with an x-ray beam. The emitted photoelectron yields chemical information from the top few atomic or molecular layers (0 to 0.0025 μm). This technique is x-ray photoelectron spectroscopy (XPS). Next, Auger electron spectroscopy (AES) also yields chemical information from the particle surface but from a much smaller circular area. Damage to the particle is very slight. Then an ion beam, nonelastically reflected off the particle (ion scattering spectroscopy, ISS), gives the elemental composition of the top atomic monolayer (0 to 0.0002 μm). Particle damage remains slight. Finally, bombarding the particle with a more energetic ion beam results in removal of perhaps

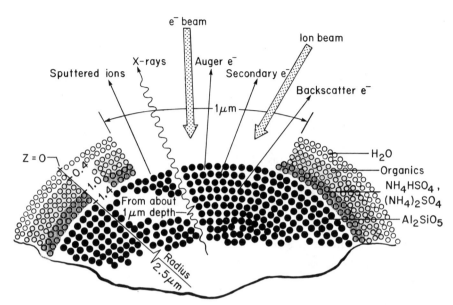

Figure 3. Depth profiling of a hypothetical particle by the techniques of Table 5. Copyright © 1986. Electric Power Research Institute. EPRI TB.EAE.4.4.86 "Atmospheric Particles" [53]. Reprinted with permission.

two atomic or molecular layers from the surface. A mass spectrometer measures these "sputtered" species by the process known as secondary ion mass spectrometry (SIMS). Figure 3 shows the depth profiling of a hypothetical particle by these techniques.

These techniques have been used to examine in-stack fly ash particles from coal-burning power plants. Although characterizing individual particles is still quite difficult, it has been established that the composition of the top 0.005 to 0.01 mm of a coal ash particle is different from that of the bulk. In particular, there is a surface enrichment of carbon and sulfur species.

An ultimate goal in particle characterization is to extract individual particles from a system and measure their size, composition, optical properties, etc., without the need to collect the particles on a surface.

The electrodynamic balance (EDB) is a modern version of the Millikan oil drop apparatus in which a charged particle is levitated in an electric field [20]. By using quadrupole focusing, it is possible to suspend a single particle in a controlled environment virtually indefinitely. The size of a levitated particle can be measured by a variety of methods, the most precise of which uses the information contained in the resonant structure of light scattered by the particle [21–23]. For a homogeneous spherical particle, the intensity of scattered light as a function of scattering angle can be inverted using Mie scattering theory to provide highly accurate values of the particle's refractive index and size. In addition to providing absolute size determination, the structure resonances of the scattered light can be used for accurate determination of size changes, with an ultimate sensitivity of order 1 part in 10^5 [21]. The electrodynamic balance is also a useful tool for the study of gas–solid and gas–liquid reactions. Surface areas can be measured gravimetrically by gas adsorption. Mass loss or mass accumulation can also be measured in order to study chemical reactions.

In the past several years, there has been considerable interest in the development of spectroscopic methods for the analysis of the composition of a single microparticle. A number of techniques have been explored, including the following:

- Raman spectroscopy [24–26]
- Stimulated Raman spectroscopy [27]
- Coherent anti-Stokes Raman scattering (CARS) [28]
- Fluorescence spectroscopy [29–33]
- Photophoretic spectroscopy [34, 35]
- Structure resonance modulation spectroscopy [34, 36–38]
- Fourier transform infrared structure resonancemodulation spectroscopy [39, 40]

An important parameter characterizing the various spectroscopies is the normalized particle size, $x = \pi D_p / \lambda$, the ratio of the particle circumference to the wavelength of light. For $x > 1$, the electromagnetic radiation can establish resonances in the particulate medium, leading to a complex interaction between the particle and the radiation as first described by Mie [41]. These resonances are exhibited most strongly in the scattering of light but also play an important role in absorption. For wavelengths in the visible regime, the resonant interactions are most pronounced for x of order 0.2 to 10. If one considers infrared radiation in the range of 5 to 10 μm, the resonant structures are absent from particles smaller than about 1 μm.

The levitation of a micron-sized particle provides the opportunity to examine the thermochemistry of the particle and phase transformations it may undergo with no interferences from foreign surfaces [42–44]. Moreover, since the diffusion times are small, the composition becomes uniform throughout the particle much more rapidly than it would in a bulk quantity of the particulate material. Applications of the EDB to systems other than those of environmental interest seem promising. Polymerization reactions in small drops can be followed by levitating a droplet of polymer precursor solution under conditions where polymerization will proceed. Reactions involving metal–organic compounds used for the synthesis of oxide powders in ceramic applications can be studied with individual microparticles. Rates of diffusion of encapsulated compounds from small particles could be measured from the mass loss of the suspended particle.

E. *Microcontamination Control*

Particle contamination is one of the most important factors limiting semiconductor production in the United States and a key issue in the manufacturing of the next generation of microelectronic devices with smaller feature sizes. This is particularly true in the manufacture of integrated circuits for computers and various microprocessors, where higher speeds of operation and lower costs require higher circuit densities and hence further miniaturization. It is anticipated that, for the U.S. semiconductor industry to remain competitive, advanced chip designs will utilize 0.5-μm features by 1990 and 0.3 μm by 1993. Thus, to manufacture these devices with high yields and high reliability, particle contamination will have to be substantially reduced from present levels and emphasis will be on controlling particles as small as 0.01 μm.

Every step in the production process potentially contributes particles that can contaminate the lot. Typically, particle sources are about equally divided between the clean room environment (e.g., people), the processing equipment (due to friction and wear), and the processing fluids (including chemicals, gases, and water). Particles from these sources are then transported to the vicinity of the product, and some are deposited on the wafer surface. The deposition rate depends on the particle size distribution, the particle concentration, and the deposition velocities due to transport mechanisms such as convection, gravitational settling, Brownian diffusion, and phoretic migration (e.g., thermal and electrical). Characterizing these processes and identifying the procedures required to control particle contamination during semiconductor processing represents a considerable technological challenge.

IV. Treatment of Hazardous Wastes

The treatment of hazardous wastes has emerged in the last 10 years or so as one of this country's most acute environmental problems. It is now recognized that the cost of dealing with the accumulated and yet-to-be-generated waste will be staggering. Unfortunately, the hard science underlying waste treatment strategies is woefully inadequate. For example, with respect to hazardous waste incineration, Senkan [45] has noted that although large-scale field test burns . . . have resulted in the accumulation of a significant amount of very expensive empirical data, these tests have contributed very little to our understanding of the underlying chemical processes that take place in incinerators.

A. *Incineration of Hazardous Wastes*

Pollutant emissions from combustion systems have been substantially reduced in recent years through the development and application of new combustion technologies. High-efficiency combustion systems offer the possibility of complete detoxification of hazardous materials by conversion to carbon dioxide, water, and contaminants such as hydrochloric acid vapor that can be removed from the exhaust gases. Incineration can, in principle, be applied to virtually any organic waste material, making the development of incineration technologies highly desirable for the vast numbers of specialty chemical products and industries [46].

In practice, complete conversion of hazardous materials to innocuous compounds is difficult and expensive to achieve, particularly in systems that rely on the waste material as the primary source of fuel to drive the degradation reactions. The highest destruction efficiencies are generally reported for incineration systems in which the wastes are exposed for long times to hot gases provided by combustion of a clean fuel, for example, in a rotary kiln, or in which plasma or electrical resistance heat sources are used. While these systems clearly demonstrate the feasibility of attaining very high destruction efficiencies, they are costly to construct and operate. For the detoxification of solid wastes such as contaminated soils, rotary kilns and circulating-bed combustion systems are promising technologies. The solids can be retained in the hot environment long enough (from seconds to hours) to be fully detoxified. The combustion gases pass through the system much more rapidly. Mixing is often incomplete, particularly in the rotary kiln, so an afterburner is frequently required to assure complete destruction of the toxic materials once they are released to the gas phase.

The requirements for combustion control of incinerators used for disposal of toxic materials are far more severe than for combustors burning more conventional fuels. Whereas a combustion efficiency of 99% may be quite acceptable for a fossil-fuel-fired energy conversion system, the carryover of 1% of the feed from an incinerator results in an unacceptable release to the environment.

Factors that influence the destruction efficiency include local temperatures and gas composition, residence time, extent of atomization of liquid wastes and dispersion of solid wastes, fluctuations in the waste stream composition and heating value, combustion aerodynamics, and turbulent mixing rates. Moreover, the waste materials to be incinerated are considerably more variable than are typical fossil fuels. The composition and even the heating value of the fuel entering an incinerator can vary dramatically and rapidly. Liquid wastes may contain both water and combustible organics. Imperfect mixing of such feedstocks can seriously impair combustion and, thereby, destruction efficiencies. The physical form of the waste material may also influence its conversion. It has been reported that dioxins are not destroyed at temperatures below about 700°C, but pure compounds are largely destroyed at 800°C. The dioxin nucleus, however, may survive intact through incineration up to 1150°C if bound to particulate matter [47].

Mere destruction of the original hazardous material does not adequately measure the performance of an incinerator. Products of incomplete combustion can be as toxic as, or even more toxic than, the materials from which they evolve. Indeed, highly mutagenic polycyclic aromatic hydrocarbons are generated in the fuel rich regions of most hydrocarbon flames [48]. Dioxin formation in the combustion of chlorinated hydrocarbons has also been reported [49].

The routine monitoring of each hazardous constituent of a waste stream and all products of incomplete combustion in the effluent gases of operating incinerators is not presently possible, particularly since the quantities of the suspect materials are relatively small in the feedstock and very small in the effluent gases. EPA has established procedures for characterizing incinerator performance in terms of selected components of the anticipated waste stream. These compounds, labeled principal organic hazardous components (POHCs), are presently designated on the basis of the estimated difficulty of incineration and the concentration in the anticipated waste stream.

The relative ease or difficulty of incineration has been estimated on the basis of the heat of combustion, thermal decomposition kinetics, susceptibility to radical attack, autoignition temperature, correlations of other properties, and destruction efficiency measurements made in laboratory combustion tests. Laboratory studies have indicated that no single ranking procedure is appropriate for all incinerator conditions. In fact, a compound that can be incinerated easily in one system may be the most difficult to remove from another incinerator due to differences in the complex coupling of chemistry and fluid mechanics between the two systems.

Materials that enter the incinerator in large sludge droplets may be cooled by evaporation for long times, thereby limiting the exposure to the hot oxidizing environment. The apparent stabilization of dioxins bound to particulate matter at temperatures significantly higher than that which the pure compound would not survive illustrates the need to understand the role of the physical form of the material in its destruction by incineration.

EPA currently uses the heat of combustion to rank the incinerability of hazardous compounds, although kinetics, not thermodynamics, govern incinerator performance. More important, because of interactions between species, rates of reaction of hazardous compounds in a mixture can be considerably different from those observed under single-compound conditions. Moreover, the hydrocarbon fuels used to cofire the incinerator can interact with the hazardous substances themselves.

Senkan [45] has illustrated this latter point by means of the destruction of trichloroethylene (C_2HCl_3) in the presence of oxygen. The reaction begins at 600-700°C. When methane is also present, however, the thermal oxidation of trichloroethylene is inhibited to such an extent that its decomposition does not begin until temperatures of about 950°C are reached. C_2HCl_3 destruction in the presence of O_2 occurs by the following reaction:

$$C_2HCl_3 \xrightarrow{\ 1\ } C_2HCl_2\bullet + Cl\bullet$$

$$C_2HCl_3 + Cl\bullet \xrightarrow{\ 2\ } C_2Cl_3\bullet + HCl$$

$$C_2Cl_3\bullet + O_2 \xrightarrow{\ 3\ } COCl_2 + COCl\bullet$$

$$COCl\bullet \xrightarrow{\ 4\ } CO + Cl\bullet$$

Reaction 1 is the initiation step, and Cl, C_2Cl_3, and COCl are the chain carriers. When methane is present, it intercepts and removes the Cl radical by

$$CH_4 + Cl\bullet \longrightarrow CH_3\bullet + HCl$$

and generates the much less reactive CH_3 radical, thus decreasing the rate of destruction of C_2HCl_3.

The development of environmentally acceptable incineration technologies for the disposal of hazardous wastes is dependent on an understanding of the roles of (1) atomization or method of introduction of the waste materials, (2) evaporation and condensed-phase reactions of the waste droplets in the incinerator environment, (3) turbulent mixing in the incinerator, (4) kinetics of the thermal degradation and oxidation of the chemical species in question, and (5) heat transfer in the incinerator.

The release of incompletely reacted toxins appears to be determined by localized deviations from the mean conditions. The development of highly efficient and reliable incineration systems is therefore contingent on an understanding of the factors that control the microscale environment to which the wastes are exposed.

B. Photocatalytic Degradation of Hazardous Wastes on Semiconductor Surfaces [50–52]

Semiconductors such as TiO_2, Fe_2O_3, CdS, and ZnS are known to be photochemical catalysts for a wide variety of reactions. When a photon with an energy of $h\nu$ matches or exceeds the band gap energy, E_g, of the semiconductor, an electron, e^-, is promoted from the valence band, VB, into the conduction band, CB, leaving a hole, h^+, behind. Electrons and holes either recombine and dissipate the input energy as heat, get trapped in metastable surface states, or react with electron donors and acceptors adsorbed on the surface or bound within the electrical double layer.

Organic alcohols and aldehydes, for example, can be photooxidized by holes on the semiconductor surfaces to yield carboxylic acids

$$RCH_2OH + H_2O + 4h^+ \xrightarrow[\text{TiO}_2]{h\nu} RCHOOH + 4h^+$$

while the acids can be further oxidized to CO_2

$$2CH_3CO_2^- + 2h^+ \xrightarrow[\text{TiO}_2]{h\nu} C_2H_6 + 2CO_2$$

The reductive pathway is important for degradation of halogenated hydrocarbons via a rapid one-electron reduction as follows:

$$CCl_4 + e^- \xrightarrow[\text{TiO}_2]{h\nu} CCl_3{\cdot} + Cl^-$$

$$CCl_3{\cdot} + O_2 \longrightarrow CCl_3O_2{\cdot}$$

$$2CCl_3\,O_2{\cdot} + 4H_2O \longrightarrow \underset{-H_2O_2}{\longrightarrow} 2CO_2 + O_2 + 6HCl$$

although degradation of trichloroethylene appears to be initiated by electron transfer to h^+.

The redox potentials of both e^- and h^+ are determined by the relative positions of the conduction and valence bands, respectively. Band gap positions are material constants; they are known for a variety of semiconductors. Electrons are better reductants in alkaline solutions, while holes have a higher oxidation potential in the acid pH range. With the right choice of semiconductor and pH, the redox potential of the e_{cb}^- can be varied from +0.5 to −1.5 V and that of the h_{vb}^+ from +1.0 to more than +3.5 V. Thus valence band electrons and conduction band holes in semiconducting metal oxides are capable in principle of reducing and oxidizing almost any organic or inorganic compound that may be present in hazardous waste effluents. This is an example of a new technology that should be explored.

V. Role of Environmental Chemical Engineering in Education and Academic Research

Unlike areas such as biochemical engineering and advanced materials, environmental chemical engineering does not require, at least at the undergraduate level, courses outside the traditional chemical engineering curriculum. In fact, it can be argued that training in chemical engineering is the ideal foundation for environmental research, which invariably involves significant elements of chemistry coupled with transport phenomena. In spite of the fact that chemical engineering provides a solid foundation for addressing environmental problems, the usual chemical engineering curriculum does not contain many interesting environmental applications, nor is it configured so that environmentally related material such as combustion, turbulence, colloids and particle dynamics, and processes in porous media is prominent. At the graduate level, one can pursue specialized courses in aquatic chemistry, atmospheric chemistry, environmental fluid mechanics, meteorology, and waste treatment, in addition to the usual graduate chemical engineering courses.

There exist significant environmental research opportunities for chemical engineers, for example, in incineration of hazardous wastes, environmental behavior of toxic substances, synthesis and design of nonpolluting plants, regional and global air pollution, and water chemistry.

As is evident from the examples we have cited, most environmental problems are highly complex and often ill-defined. At a minimum, they usually require a synthesis of virtually all elements of the chemical engineer's arsenal—thermodynamics, chemical kinetics, and transport phenomena; at the most, an interdisciplinary team is required. An insufficient understanding of molecular-scale processes is frequently the key obstacle in developing innovative approaches to environmental protection.

References

1. Wei, J., *Adv. Catal.* **24**, 57 (1975).
2. Lyon, R. K., *Environ. Sci. Technol.* **21**, 231 (1987).
3. Schroeder, W. H., and Lane, D. A., *Environ. Sci. Technol.* **22**, 240 (1988).
4. Cohen, Y., *Environ. Sci. Technol.* **20**, 538 (1986).
5. Bidleman, T. F., *Environ. Sci. Technol.* **22**, 361 (1988).

6. Friedlander, S. K., *Smoke, Dust and Haze*. Wiley, New York, 1977.

7. Hinds, W. C., *Aerosol Technology*. Wiley, New York, 1982.

8. Seinfeld, J. H., *Atmospheric Chemistry and Physics of Air Pollution*. Wiley, New York, 1986.

9. Flagan, R. C., and Seinfeld, J. H., *Fundamentals of Air Pollution Engineering*. Prentice-Hall, Englewood Cliffs, N.J., 1988.

10. Pilinis, C., and Seinfeld, J. H., *Atmos. Environ.* **20**, 2453 (1987).

11. Calvert, J. G., and Stockwell, W. R., in *SO_2, NO, and NO_2 Oxidation Mechanisms: Atmospheric Considerations* (J. G. Calvert, ed.), p. 1. Butterworth, Boston, 1984.

12. Schryer, D. R., ed., *Heterogeneous Atmospheric Chemistry*. American Geophysical Union, Washington, D.C., 1982.

13. Jaeschke, W., ed., *Chemistry of Multiphase Atmospheric Systems*. Springer-Verlag, Berlin, 1986.

14. Angeletti, G., and Restelli, G., eds., *Physico-Chemical Behaviour of Atmospheric Pollutants*. D. Reidel, Dordrecht, 1987.

15. Hoffmann, M. R., and Calvert, J. G., *Chemical Transformation Modules for Eulerian Acid Deposition Models*. Volume II: *The Aqueous-Phase Chemistry*. National Center for Atmospheric Research, Boulder, Colo., 1985.

16. Springer, G. S., *Adv. Heat Transfer* **14**, 281 (1978).

17. Lewis, B., and Anderson, J. C., *Nucleation and Growth on Thin Films*. Academic Press, New York, 1978.

18. Adamson, A., *Physical Chemistry of Surfaces*. Wiley, New York, 1982.

19. Mozurkewich, M., McMurry, P. H., Gupta, A., and Calvert, J. G., *J. Geophys. Res.* **92**, 4163 (1987).

20. Davis, E. J., *Surface Colloid. Sci.* **14**, 1 (1987).

21. Ashkin, A., and Dziedzic, J. M., *Phys. Rev. Lett.* **38**, 1351 (1977).

22. Davis, E. J., and Periasamy, R., *Langmuir* **1**, 373 (1985).

23. Davis, E. J., and Ravindran, P., *Aero. Sci. Tech.* **1**, 337 (1982).

24. Preston, R. E., Lettieri, T. R., and Semerjian, H. G., *Langmuir* **1**, 365 (1985).

25. Thurn, R., and Kiefer, W., *Appl. Optics* **24**, 1515 (1985).

26. Fung, K. H., and Tang, I. N., *Appl. Optics* **27**, 206 (1988).

27. Eickmans, J. H., Qian, S. X., and Chang, R. K., *Part. Charact.* **4**, 85 (1987).

28. Qian, S., Snow, J. B., and Chang, R. K., *Optics Lett.* **10**, 499 (1985).

29. Folan, L. M., and Arnold, S., *Optics Lett.* **13**, 1 (1988).

30. Folan, L. M., Arnold, S., and Druger, S. D., *Chem. Phys. Lett.* **118**, 322 (1985).

31. Arnold, S., and Folan, L. M., *Rev. Sci. Inst.* **57**, 2250 (1986).

32. Arnold, S., and Folan, L. M., *Rev. Sci. Inst.* **58**, 1732 (1987).

33. Ward, T. L., Zhang, S. H., Allen, T. and Davis, E. J., *J. Colloid. Interface Sci.* **118**, 343 (1987).

34. Lin, H. B., and Campillo, A. J. *Appl. Optics* **24**, 422 (1985).

35. Arnold, S., Amani, Y., and Orenstein, A., *Rev. Sci. Inst.* **51**, 1202 (1980).

36. Arnold, S., Murphy, E. K., and Sageev, G., *Appl. Optics* **24**, 1048 (1985).

37. Arnold, S., Newman, M., and Pluchino, A. B., *Optics Lett.* **9**, 4 (1984).

38. Arnold, S., and Pluchino, A. B., *Appl. Optics* **21**, 4194 (1982).
39. Grader, G. S., Flagan, R. C., Arnold, S., and Seinfeld, J. H., *Rev. Sci. Inst.* **58**, 584 (1987).
40. Grader, G. S., Flagan, R. C., Arnold, S., and Seinfeld, J. H., *J. Chem. Phys.* **86**, 5897 (1987).
41. Mie, G., *Ann. Physik* **25**, 377 (1908).
42. Cohen, M. D., Flagan, R. C., and Seinfeld, J. H., *J. Phys. Chem.* **91**, 4563 (1987).
43. Cohen, M. D., Flagan, R. C., and Seinfeld, J. H., *J. Phys. Chem.* **91**, 4575 (1987).
44. Cohen, M. D., Flagan, R. C., and Seinfeld, J. H., *J. Phys. Chem.* **91**, 4583 (1987).
45. Senkan, S. M., *Environ. Sci. Technol.* **22**, 368 (1988).
46. Oppelt, E. T., *Environ. Sci. Technol.* **20**, 312 (1986).
47. Esposito, M. P., Tierman, T. O., and Dryden, F., "Dioxins." U.S. Environmental Protection Agency report EPA 600/1-80-19 (1980).
48. Haynes, B. S., and Wagner, H., *Prog. Eng. Combust. Sci.* **7**, 229 (1981).
49. Lahanictis, E. S., Parlow, H., and Kute, F., *Chemosphere* **6**, 11 (1977).
50. Bahnemann, D. W., Kormann, C., and Hoffmann, M. R., *J. Phys. Chem.* **91**, 3789 (1987).
51. Hong, A. P., Bahnemann, D. W., and Hoffmann, M. R., *J. Phys. Chem.* **91**, 2109 (1987).
52. Hong, A. P., Bahnemann, D. W., and Hoffmann, M. R., *J. Phys. Chem.* **91**, 6245 (1987).
53. Electric Power Research Institute, Technical Brief RP1625, Palo Alto, Calif., 1988.

14

Energy and Environmental Concerns

T. W. F. Russell
Department of Chemical Engineering
University of Delaware
Newark, Delaware

I. Introduction

Figure 1 shows the predicted energy use in the United States as published in the 1974 Ford Foundation study [1]. The solid curve shows actual consumption; the dashed curves show the 1974 predictions for three scenarios:

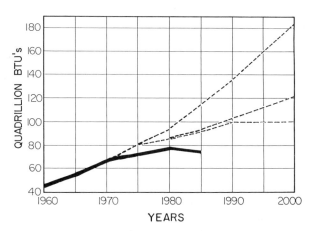

Figure 1. Predicted and actual energy consumption (United States). Adapted from [1].

ADVANCES IN CHEMICAL ENGINEERING, VOL. 16

Historical Growth (top curve), Technical Fix (middle curve), and Zero Growth (bottom curve).

It is obvious that actual energy consumption has been well below prediction. The Ford Foundation study was roundly criticized by many, including some on the Foundation's own advisory board, for suggesting that society could reduce energy use below that of Historical Growth without causing major societal dislocations. According to D. C. Burnham, chairman of Westinghouse Electric Corporation [1],

> The future demand for energy has been severely underestimated: The Technical Fix scenario appears to me to underestimate total energy requirements for the year 2000 by 25 percent, or 44 quadrillion BTUs. Our studies indicate that taking into account both conservation and technological changes in the way we use energy will require 168 quadrillion BTUs in 2000, not the 124 the project indicates.

Mr. Burnham also notes that his firm predicts 210 quadrillion BTUs in the year 2000 for the Historical Growth scenario and that energy use at the Zero Growth level would lead to "substantial social upheaval as well as economic stagnation." In addition, since the report was published, the U.S. economy has expanded by over 35%.

Table 1 shows an impressive reduction in energy intensities (megajoules per 1980 GNP dollar) in all industrial market countries between 1973 and 1985 [2]. Japan's record is particularly noteworthy, since it started with one of the world's most energy-efficient economies. Efficiency improvements have been most impressive in the industrial sector, mainly because of improvements in the chemical processing industries. Energy intensity in industry for most industrial market countries has dropped 30% since 1974. U.S. industrial energy use dropped 17% while production increased 17%. Building and transportation improvements have also been substantial. Chemical engineers have played a key role in reducing energy consump-

Table 1. Energy Intensities (Megajoules/1980 GNP $)[a]

Country	1973	1985	Change 1973–1985 (%)
Canada	38.3	36.0	−6
Japan	18.9	13.1	−31
United Kingdom	19.8	15.8	−20
United States	35.6	27.5	−23
West Germany	17.1	14.0	−18

[a] Adapted from [2].

tion and showing that we can have a healthy expanding economy with reasonable energy use. In the years ahead, the profession will again be called upon to solve new and serious problems at the energy–environment interface.

Today, with oil prices below $15/bbl, there is a tendency to believe that the "energy problem" is under control. Rose discusses the global problem with unique clarity [2], and his conclusions are worth reviewing:

1. There is an "upper" limit on energy production by fossil fuels or nuclear reaction due to the amount of heat which must be radiated into space. This upper limit is assumed to be that at which we would have a global temperature increase of 1°C.

2. Our society appears to be committed by both population growth and increases in energy use to a course of action which will bring us within a factor of 2 of the upper limit in the next couple of decades.

3. Our ultimate energy source must be solar if we are to avoid the world temperature increase problem. It is clear that we will have enough solar fuel [3], and society has been struggling for centuries with devices to capture it effectively.

4. Before society reaches the "upper thermal" limit on energy production using hydrocarbon fuels "other environmental constraints are likely to intervene."

II. Energy–Environment Interface

Rose accurately predicted that the world would face a critical problem in global warming because significant amounts of solar energy which used to be transmitted into space are now reflected back to earth. During the past year, a good deal of publication, both scientific and popular, has been devoted to this "greenhouse" effect.

Certain issues are clear-cut. Figure 2 shows the increase in CO_2 content in the earth's atmosphere [5]. The increase since 1950 is particularly noticeable, and we need to be concerned about the effect this will have on the earth's climate. Although it is important, carbon dioxide is not the only "greenhouse gas" of concern. Table 2 shows the other pertinent pollutants in the atmosphere, their 1985 concentrations, and their projected annual growth rate [5].

To predict the effect on global warming, growth rates for the production of greenhouse gases and their effect on global temperature must be estimated. Various models have been proposed, but Mintzer [5], in a careful and extensive study, uses the temperature prediction of Ramanathan *et al.*

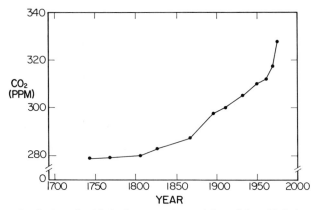

Figure 2. Carbon dioxide in the atmosphere. Adapted from Neftel *et al.* [4]

[6] and various scenarios for growth to draw the conclusions shown in Table 3. The warming commitment columns refer to the warming to which the earth is committed due to the cumulative emissions to the year 2000 or 2050. Complete definitions for Mintzer's scenarios are not repeated here. However, some idea of the range of his definitions can be obtained from the following: (1) the High Emissions scenario assumes global energy use to be about triple that in the Slow Buildup case, and (2) the energy supply in the High Emissions case is assumed to be made up with about 58% solid fuels, compared to 22% in the Slow Buildup case.

Past global temperatures and a set of predicted temperatures from another study are shown in Fig. 3. This study was prepared by the Beijer Institute in Stockholm, based on two extensive workshops sponsored by the World Meteorological Organization and the United Nations Environment Programme

Table 2. Important Greenhouse Gases[a]

Name	Residence Time (years)	1985 Conc.[b]	Annual Growth Rate (%)
Carbon dioxide	2–3	345 ppmv	0.5
Nitrous oxide	150	301 ppbv	0.25
Methane	11	1650 ppbv	1.0
CFC-11	75	0.20 ppbv	7.0
CFC-12	111	0.32 ppbv	0.07

[a]Adapted from Mintzer [5].
[b]Concentrations are on a volume basis.

Table 3. World Temperature Predictions

Scenario	Warming Commitment Relative to 1980	
	Year 2000	Year 2050
Base case	0.4–1.1°C	1.6–4.8°C
High emissions	0.5–1.4°C	3.0–8.9°C
Modest policies	0.3–1.0°C	1.3–3.9°C
Slow buildup	0.3–0.8°C	0.8–2.3°C

[7]. The upper and lower scenarios represent extremes over the middle "no change in present pattern" scenario.

A global warming of something close to 1°C by the year 2000 appears to be quite probable. The 1980s have already seen the four hottest years on record. Society faces a serious problem, the detailed consequences of which are a matter of some debate. However, it does appear that weather patterns will shift, causing drought in some areas that are now fertile and increased rain in areas that may be too fragile to cope with the increased water. A rise in sea level of 5 to 20 cm will occur by the year 2000 [7], and changes in climate will occur over decades rather than over tens of centuries as in the past. Plant and animal life will have difficulty adapting this quickly.

The world should not carry out this experiment in global warming to see what will happen. We must find a way to reduce the greenhouse gas

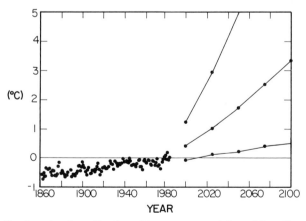

Figure 3. Actual and predicted world temperature. Adapted from Jaeger [7].

concentration in the atmosphere now. It will be necessary to restrict hydrocarbon fuel use (which supplies 90% of the CO_2) and to design for much more efficient energy production per unit mass of fuel burned. We also must find ways to control deforestation and to terminate the manufacture of chlorofluorocarbons. These are problems that chemical engineers can solve. We need to reduce world temperature increases (Fig. 3) below prediction, as we were able to do for energy use (Fig. 1).

Hydrocarbon fuel use can be reduced by employing alternative energy sources; we need to examine both solar and nuclear energy options. The nuclear option will eventually lead to global warming due to the thermal effect [8], but in addition there are major social–political problems and serious nuclear waste disposal problems. Power generation using photovoltaic panels as modules is a technically feasible technology today and is rapidly becoming economically viable.

III. Photovoltaic Power Generation

A solar or photovoltaic panel is a solid-state power generation device which uses no fuel or water and produces no effluent. Electricity is produced when sunlight falls upon the devices or cells making up the panel. The sun's energy is absorbed and creates electrical charges, which are separated by the electric field in the device and flow externally as current. Commercially available photovoltaic panels convert between 5 and 15% of the sunlight to electricity. A square meter of photovoltaic panel will produce between 50 and 150 watts of power. A 1000-MW power generation facility would require an area of about 2 miles by 2 miles, which is comparable to the safety zone around a 1000-MW nuclear plant.

A small but growing photovoltaic industry now exists with worldwide sales between $300 million and $400 million. The United States and Japan account for 75% of worldwide sales, but European firms are increasingly active, with over 17% of the market [9]. Panel prices are about $5 per peak watt. With this capital cost and today's attendant equipment costs, electric power can be generated for between 30 and 40 cents per kilowatt-hour.

Photovoltaic research and development has been an integrated, well-planned effort involving industry, universities, and the federal government. For the past decade, the photovoltaic community has supported an evolving 5-year plan which has allowed the research effort to be carried out efficiently. Federal research support has been modest (between $30 million and $40 million per year for the past 5 years) but progress has been remarkable:

1. Solar cell efficiencies have been brought up from about 5% in the early 1970s to just over 30% today. Single-crystal silicon cells with efficiencies between 18 and 20% have been made. Cells using gallium arsenide have achieved efficiencies of 22–24%, and multijunction concentrator cells have been made with over 30% efficiency. Inexpensive thin-film cells of amorphous silicon have reached efficiencies between 10 and 12% in single-junction and between 14 and 16% in multijunction devices. Inexpensive stable polycrystalline thin-film cells of $CuInSe_2$ and CdTe have reached efficiencies between 10 and 12%.

2. Thin-film modules of both amorphous silicon and $CuInSe_2$ with areas on the order of a square foot and efficiencies approaching 10% have been made commercially. With properly planned research and design efforts, panels using these materials can be manufactured at a cost approaching $1 per peak watt. This will bring electric power cost to the consumer down to 10–20 cents per kilowatt-hour.

3. There are thousands of photovoltaic panel installations worldwide producing small amounts of electricity for such applications as remote telecommunications, navigational aids (battery charging for buoys), water pumping, vaccine refrigeration, home electric supply for remote locations, battery charging or battery replacement for boats, recreational vehicles, yard lighting, emergency call boxes, electric fence, golf carts, and various consumer goods.

4. About 10 MW of utility grid-connected photovoltaic power generation is installed and effectively operating today.

Photovoltaic electric power generation must and will play a key role in the decades ahead. It is essential that we bring this technology into widespread commercial use as quickly as possible. Chemical engineers are needed to design the manufacturing facilities that will make millions of square meters per year. The scale-up problem is not a trivial one, since all laboratory-scale experiments produce small cells with areas on the order of square centimeters.

To bring manufacturing costs down and to design safe processing plants with no harmful environmental impact, the following critical issues need to be addressed:

• Deposition of the semiconductor layers
• Design of stable optimal contacts and interconnects
• Encapsulation and packaging
• Raw material and reactor effluent processing
• Process integration and optimal control

Chemical engineers have an obvious role to play in raw material and reactor effluent processing and in process integration and optimal control. They need to apply semiconductor reactor analysis to solve the critical problem

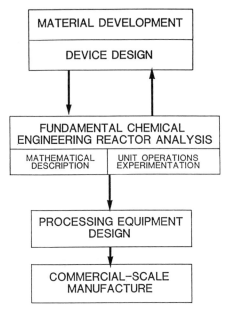

Figure 4. The role of chemical engineers in electronic material processing.

of depositing semiconductors over large areas. New and better semiconductor reactors will have to be designed and operated to accomplish this. Research emphasis should be shifted more toward reactor design and away from an overemphasis on materials and devices. The role of the chemical engineer is illustrated in Fig. 4.

References

1. Energy Policy Project of the Ford Foundation, Final Report, "A Time to Choose: America's Energy Future." Ballinger Publishing Co., Cambridge, Mass., 1974.
2. International Energy Agency, "Energy Conservation in IEA Countries." Organization for Economic Co-operation and Development, Paris, 1987.
3. Hubbert, M. King, Energy Resources: A Report to the Committee on Natural Resources, Publication 1000-D, Natl. Acad. of Sci.–Natl. Res. Council, Washington, D. C., 1962.
4. Neftel, A., Moor, E., and Oeschger, H., *Nature* **315**, 45 (May 2, 1985).
5. Mintzer, I. M., "A Matter of Degrees: The Potential for Controlling the Greenhouse Effect," Research Report No. 5, World Resources Institute, April 1987.

6. Ramanathan, V., Cicerone, R. J., Singh, H. Bir., and Kiehl, J. T., *J. Geophys. Res.* **90**, 5547 (1985).

7. Jaeger, J., "Developing Policies for Responding to Climatic Change," Proceedings of discussions and recommendations of workshops held in Villach (Sept. 28–Oct. 2, 1987) and Bellagio (Nov. 9–13, 1987) under the auspices of the Beijer Institute, Stockholm, April 1988.

8. Rose, A., CHEMTECH 11, 566–571 (1981).

9. International Solar Energy Intelligence Report, "PV Shipments Continued to Grow Modestly in 1987 to 25.4 MWp:PNL," p. 228, July 12, 1988.

15

The Role of Chemical Engineering in Fuel Manufacture and Use of Fuels

Janos M. Beer, Jack B. Howard, John P. Longwell, and Adel F. Sarofim
Department of Chemical Engineering
Massachusetts Institute of Technology
Cambridge, Massachusetts

I. Introduction

Manufacture and use of fuels is expected to be the major source of energy well into the next century. Petroleum and natural gas will continue to be important; however, the use of solid raw materials such as coal, shale, and biomass will grow with increasing constraints related to environmental impact and safety.

The manufacture and use of petroleum-based fuels has been a major factor in the growth and content of chemical engineering. Chemical engineers have pioneered advances in separations, catalytic processes such as fluid catalytic cracking, hydroprocessing, and many others. Many of the advances in chemical engineering science have been stimulated by the need for quantitative understanding of these processes.

In manufacturing, extension of our knowledge in the traditional areas of catalysis, kinetics, and transport processes will be essential; however, handling and conversion of solids offer special challenges and opportunities that will become increasingly important as use of solid raw materials grows. This area requires advances in multiphase flows, especially at high solids concentration, including advances in understanding fluidized solids phenom-

ADVANCES IN CHEMICAL ENGINEERING, VOL. 16

ena, moving fixed beds, and the like. Chemical conversion of solids as in pyrolysis, liquefaction, gasification, and combustion, while not unstudied, has many remaining challenging opportunities for research.

By definition, fuel use involves oxidation to generate heat or to destroy unwanted materials, as in incineration. Here, the major challenge is the economical control of emissions. The chemical and transport process involved in generation and destruction of mutagenic aromatics and soot, the formation and control of nitrogen oxides, and the capture and removal of sulfur compounds call for increased understanding and participation by the chemical engineering community. In these systems, the interaction between turbulent transport processes and chemistry dominates. Important advances are being made in the fluid mechanics of turbulent reacting flows. Focus has been on high-Dahmköhler-number combustion, where chemistry is fast relative to mixing, and the assumption of highly simplified chemistry is useful. For emission control, the chemical processes subsequent to the initial combustion dominate, and the assumption of fast simple chemistry is no longer valid. Turbulent transport is still of major importance, and the problems of obtaining very high conversions are a fertile area for chemical engineering research.

Fuel use and manufacture draw on most of the principles of chemical engineering and often add the challenge of extreme conditions of high temperature, sharp gradients, and short residence times. For these reasons, the subjects are excellent vehicles for teaching, recognized since the early days of the profession. Indeed, in the MIT President's Report for 1888, the description (pp. 41, 42) of the founding of Course X, which this Centennial gathering commemorates, states: "The instruction to be given ... includes an extended study of Industrial Chemistry, with laboratory practice. Special investigations into fuels and draught, with reference to combustion, will be a feature of the course." Combustion has been an integral part of the curriculum since, and currently is used to illustrate application of homogeneous and heterogeneous kinetics in our graduate course on reactor engineering, in addition to being the subject of specialized courses and seminars.

The following discussion presents our views on a few subjects of special interest: fuel process and combustion chemistry, reduction of pollutant emission, and char reactivity.

II. Combustion Chemistry

The chemistry of combustion and fuels processing may be characterized as shown in Fig. 1. The processes typically occur at high temperatures and involve gas-phase as well as gas–solid and liquid-phase reactions. The chemistry is often influenced by transport phenomena. While physics, applied

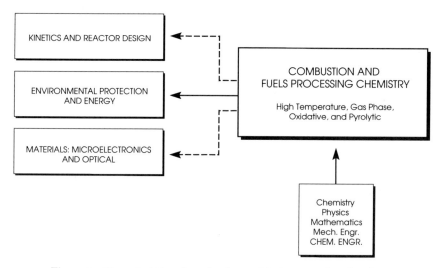

Figure 1. Characteristics of combustion and fuels processing chemistry.

mathematics, mechanical engineering, and chemistry are important to this area, no discipline is better suited than chemical engineering for dealing with both the chemistry and the physical phenomena with which the chemistry is coupled. Thus the area of combustion and fuels processing chemistry offers much opportunity and challenge for chemical engineering. At the same time, chemical engineering can continue to benefit from progress made in this area, which now includes detection and measurement of short-lived radical intermediates in reaction networks, microscopic kinetics experiments and theoretical prediction of rate coefficients, quantum chemical calculation of potential energy surfaces and thermodynamic properties, and mathematical modeling of complex reaction systems involving hundreds of reactions. The truly fundamental quantitative descriptions of gas-phase combustion chemistry that have been achieved in some systems have contributed significantly in environmental energy and materials areas and can provide guidance to improved methods for chemical engineering kinetics and reactor design.

This area of study interfaces with two of the intellectual frontiers in chemistry, namely chemical kinetics and chemical theory. Chemical reactions are studied on ever-decreasing time scales, and further advances in the understanding of chemical kinetics are likely to be made. Advanced chemical research helps to discern the most likely pathways for energy movement within molecules and the energy distribution among reaction products, thereby clarifying factors that govern temporal aspects of chemical change. Also, large computers are now used with experimental data and theoretical insights to analyze intermediate steps in combustion processes.

The advanced chemical information and techniques already in use in the area of combustion chemistry are also pertinent to several other areas of chemical reactor engineering and therefore would be of generic value in chemical engineering education. For example, the determination of kinetics and equilibrium parameters, the description of reaction mechanisms, and the prediction of reaction rates and product yields are all now much more quantitative and more fundamentally understood than before. Progress in understanding of complex chemical reactions in combustion systems has preceded that of reactions in conventional chemical reactors as a result of the significant efforts expended to understand high-output propulsion systems, propagation, and, to a lesser extent, combustion-generated pollutants. This information can be taught to graduate students as part of a course in chemical reactor engineering or covered in more detail as a course in applied chemistry for engineers.

III. Reduction of Combustion-Generated Pollution by Combustion Process Modification

Options for reducing combustion-generated pollutant emission fall into three broad categories: cleaning of the fuel, postcombustion cleaning of stack gases, and lowering the formation, and/or promoting the destruction, of pollutants in the combustion process. Chemical engineering can make important contributions to all three of these areas, but perhaps the most challenging is that of the combustion process modification. This area of study grows out of the recognition that reaction pathways and rates of formation of pollutants in flames are closely tied to the details of the combustion process.

It is especially rewarding that the solutions of practical engineering problems, such as the reduction of emissions of nitrogen and sulfur oxides and polycyclic aromatic compounds from boilers, furnaces, and combustors, are amenable to the application of chemical engineering fundamentals. Guidance for preferred temperature–concentration history of the fuel may be given by reaction pathways and chemical kinetics, and elements of combustion physics, i.e., mixing and heat transfer may be used as tools to achieve the preferred temperature–concentration history in practical combustion systems.

The following examples [1] are given to illustrate the potentials of combustion process modifications for the reduction of combustion-generated pollutants. Mechanistic pathways of nitrogen oxides formation and destruction in flames are represented in Fig. 2. Nitrogen oxides are formed in two principally different ways: the fixation of atmospheric nitrogen, N_2, and the oxidation of heterocyclic nitrogen compounds organically bound in fossil

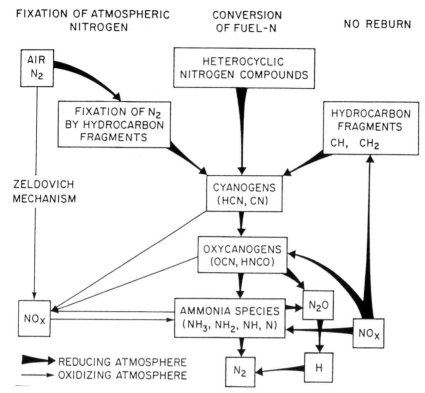

Figure 2. Formation and reduction of nitrogen oxides in combustion; mechanistic pathways.

fuels. While the rate of formation of the former is primarily temperature dependent, the oxidation of organically bound nitrogen depends more on the local stoichiometry in the flames: it will readily oxidize to NO in a fuel-lean flame zone but will convert to molecular nitrogen in fuel-rich high-temperature combustion (Fig. 2). The various practical systems of staged combustion, i.e., staged air or staged fuel injection, are all based on information on fuel-rich, high-temperature nitrogen chemistry.

In the case of air staging, the conversion of fuel nitrogen in the fuel-rich zone follows a reaction path through the rapid initial formation of cyanides, their subsequent conversion to amines, and eventual conversion to molecular nitrogen (middle column in Fig. 2). The combustion is then completed in a fuel-lean zone by injecting additional amounts of combustion air.

The reactions between nitrogen oxides and hydrocarbon radicals (right-hand column of Fig. 2) form the basis of the reduction of NO by the injection of a hydrocarbon fuel into fuel-lean combustion products of a primary fuel.

Pyrolysis reactions in fuel-rich flame zones may lead, however, to emissions of polycyclic aromatic compounds (PACs) and soot. The close correlation between the concentrations of PACs and the bioactivity of flame samples is indicative of some of the health hazards involved in the emissions of PACs (Fig. 3). Fluxes of PAC species determined in fuel-rich, natural gas turbulent diffusion flames show the build up of hydrocarbons of increasing molecular weight along flames (Fig. 4). It is postulated that PACs are formed by the successive addition of C_2 through C_5 hydrocarbons to aromatic compounds (Fig. 5).

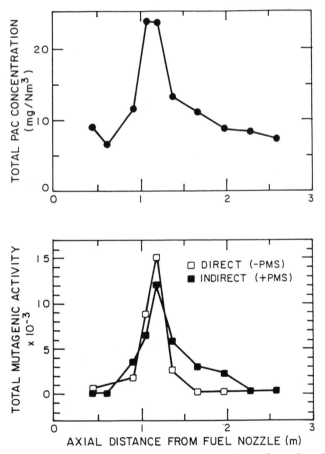

Figure 3. Total PAC concentration and total mutagenic activity of samples taken from a 1 MW fuel-rich natural gas–air flame. Reprinted from Beer [1] with permission of The Combustion Institute.

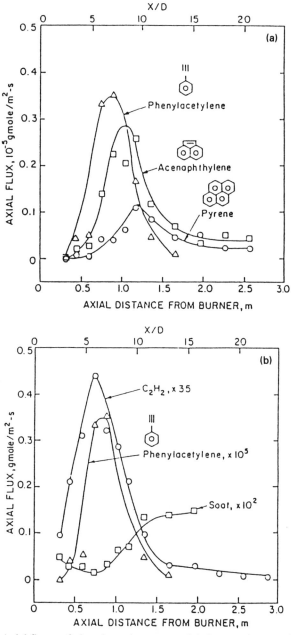

Figure 4. (a) Axial fluxes of phenylacetylene, acenaphthylene, and pyrene in a fuel-rich natural gas turbulent diffusion flame. (b) Axial fluxes of C_2H_2, phenylacetylene, and soot in a fuel-rich natural gas turbulent diffusion flame. Reprinted from Beer [1] with permission of The Combustion Institute.

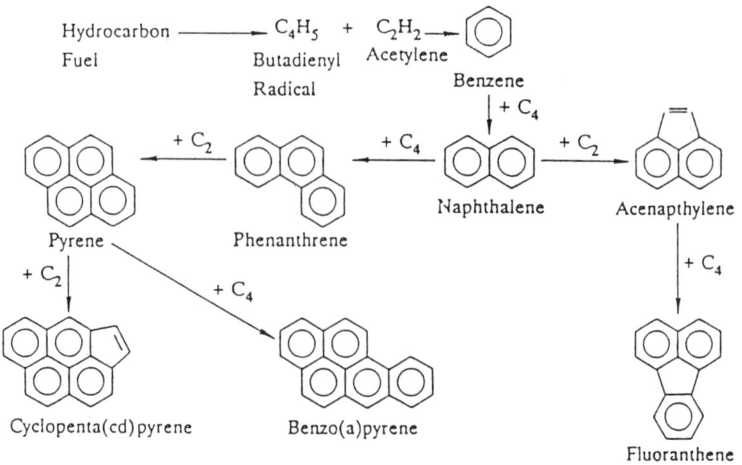

Figure 5. Mechanism of formation of polycyclic aromatic compounds during combustion: cyclization via butadienyl and acetylene, followed by addition of C_2 to C_5 hydrocarbons. Reprinted from Beer [1] with permission of The Combustion Institute.

Because of the required high destruction efficiencies of PACs in practical flames, the mixing process in the final oxidative zone of the staged combustion process must perform to very high standards. Interest is focused on the high-wave-number region of the turbulence energy spectrum and thus the smallest eddies in which turbulence energy is dissipated by molecular processes leading to chemical reaction.

The solutions to this problem require better understanding of the fine structure of chemically reacting turbulent flows. Recent advances in numerical simulation of reacting shear layers make it possible to represent the effects of rollup of the shear layer on the change in the burning velocity of the mixture. Such a model can explain the deceleration of local burning velocities or possible extinction within a shear layer due to the propagation of the flame in a region of flow with high velocity gradients. While these novel developments in modeling represent important progress, the theoretical treatment of mixing and burnout of trace fuel concentrations in turbulent flows remains a challenge to chemical engineers.

IV. Char Gasification Reactions

Coal, as our most abundant fossil fuel resource, will supply an increasing fraction of our energy needs until nonfossil energy supplies are developed—

a period that may need to be shortened in response to the problem arising from increases in atmospheric CO_2 levels. Many of the challenges to the clean and efficient utilization of coal result from the difficulties associated with the conversion of the char produced during coal conversion or combustion. Our incomplete understanding of the processes governing the physical and chemical transformations of char provide an area of research well suited for chemical engineering contributions.

One may well ask if the treatments of Thiele and Aris do not already represent the alpha and omega for chemical reactions in porous media. However, char reactions have the interesting added dimension of pores that grow in diameter with increasing conversion, resulting in an increase in surface area available for reaction until the pores overlap. Chemical engineers have again been at the forefront of developing models that take into account the recession of the reacting surface, but the experiments with coal char continue to provide data that cannot be adequately addressed by existing models. For example, shrinkage of char particles is observed under chemically controlled conditions; mesopores are opened up by mineral matter migrating into the carbon matrix, thereby providing access for the reactant gas to the microscopic grain; and particle fragmentation is observed under diffusion-controlled conditions, yielding char fragment sizes related in an as yet unquantified manner to the macropores. To complement the challenge of the physical transformations of chars, there are major unresolved issues concerning the mechanisms of char reactions—the processes that govern the reaction order, the CO/CO_2 ratio for reactions with oxygen, and the variability of the reactivity of different surface sites and how the reactivity is affected by mineral matter.

The practical motivation for understanding the microscopic details of char reaction stem from questions such as: How does the variability in reactivity from particle to particle and with extent of reaction affect overall carbon conversion? What is the interdependence of mineral matter evolution and char reactivity, which arises from the catalytic effect of mineral matter on carbon gasification and the effects of carbon surface recession, pitting, and fragmentation on ash distribution? How are sulfur capture by alkaline earth additives, nitric oxide formation from organically bound nitrogen, vaporization of mineral constituents, and carbon monoxide oxidation influenced by the localized surface and gas chemistry within pores?

Such questions are answered empirically all too often. A more fundamental approach is needed. In the area of gas-phase kinetics, the developments in the chemistry of large sets of elementary reactions and diffusion in multicomponent mixtures in a combustion context are now finding applications in chemical engineering, as mentioned above. In the area of gas–solid reactions, the information flow will be in the opposite direction. A need exists to apply new tools derived from catalysis and surface science to char oxidation, such as the application of percolation theory to model char fragmentation. Chemical mechanisms should be guided by the measurement of intermediates on the reacting surfaces using modern surface spectroscopy tools and scanning tunneling microscopy. The kinetics of key steps need to be evaluated using quantum mechanics. The diffusion in the porous structure needs to be estimated from better experimental measurements of the complex micropore structure and from models that properly account for the interaction of the diffusing molecules with the pore surfaces. In the past four decades, a transition from empirical treatments of gas-phase kinetics to detailed models has occurred. The next decades should see a similar evolution in our treatment of gas–solid reactions.

References

1. Beer, J. M. *Twenty-second Symposium (International) on Combustion*, p. 1. The Combustion Institute, 1988.

General Discussion

Environmental Protection and Energy

Klavs Jensen: Sebastian Reyes' research on percolative fragmentation, which is the work that Adel Sarofim cited, is an example of what can be achieved by interdisciplinary problem solving, in this case a combination of fluid mechanics and reaction engineering. The idea of using percolation theory came from attending seminars on enhanced oil recovery in Skip Scriven and Ted Davis' research group.

Adel Sarofim: That point was made very well in John Seinfeld's paper, and everything he says about environmental problems is applicable to the combustion field.

Sheldon Isakoff: Combustion is one of the techniques used for disposing of wastes and also for generating power. Is there research going on concerning ash disposal, particularly from combustion of material containing metals? Coal, of course, also contains metals. Ultimately, the ash also presents a disposal problem, and I wonder if this is an area for chemical engineering research?

Adel Sarofim: Some research has been carried out on metals that accompany fly ash and residue, and there is work now going on in leachability. It's an area of increasing concern. The biggest concern today, in terms of leaching, is municipal solid wastes, where you have a high content of cadmium, zinc, and the more toxic trace elements. Research can contribute to the minimization of the leaching potential by better characterizing the fate of metals during combustion. Many of the toxic metals are concentrated in the surface layers and in the smaller sizes of the fly ash as a consequence of vaporization at combustion temperatures followed by condensation, either heterogeneous or homogeneous, in the cooler sections of a furnace. Knowledge of the distribution between sizes and the chemical state of the elements can be utilized to advantage in reducing leaching by size classification or pretreatment.

Reuel Shinnar: Environmental concern has become a major driving force for industry. Once we know what we want or have to do, then it becomes a conventional chemical engineering problem. We can get rid of some of the waste products formed by combusting them with something else, or we can modify the technology so that pollutants are eliminated, using different catalysts, and so on. Such considerations are becoming a major driving force in refinery process development and should become one in chemical process development. For example, we now can build coal power plants using gasifiers that are no more polluting than gas-powered plants.

Another area of environmental concern is the products themselves. Petroleum refining in the last 10 years has been driven by requirements for the composition of gasoline and diesel fuel that are continuously changed by environmental demands. Such considerations will have even stronger impact in the future. There are similar problems in the chemical process industries. The search for an environmentally acceptable substitute for Freon is a prominent present example. Another problem is polyvinyl chloride (PVC), which causes difficulties in incinerators. The question is, should we continue to produce PVC or should we find a substitute that is easier to dispose of? Can we modify polymers in such a way that they cause fewer problems of pollution than they cause now? These are major challenges for the profession, in both academia and industry.

Another area where we could make a significant contribution is in the formulation of sensible environmental policies. Environmental policy in the United States is driven and formulated by adversarial forces. You have the environmental groups, some nuts, some reasonable, screaming about what should be done. You have Congress answering by passing laws and industry saying that nothing needs to be done, nothing should be done. Much could be improved if we could change the adversarial situation to one in which the engineering community and industry take an active part in the system, providing another point of view and explaining to the public how to formulate a reasonable science policy.

Arthur Humphrey: I just happened to have a page from the Congressional Record of 1875 on this subject. I thought it might be worthwhile. "A new source of power which burns a distillate of kerosene called gasoline has been produced by a Boston engineer." Since we're here in Boston, that's appropriate. "Instead of burning fuel under a boiler, it is exploded inside the cylinder of an engine. This so-called internal combustion engine may be used under certain conditions to supplement steam engines. Experiments are under way to use this engine to propel a vehicle. This discovery brings in a new era in the history of civilization." I guess they should have said engineering. "It may someday prove to be more revolutionary in the development of human society than the invention of the wheel, the use of metals, or the steam engine. Never in history has society been confronted with a power so full of potential danger, at the same time so full of promise for the future of man and the peace of the world." Now, this is what I think is very interesting. "The dangers are obvious. Stores of gasoline in the hands of people interested primarily in profit would constitute a fire hazard, an explosive hazard of the first rank." That's been true. "Horseless carriages propelled by gasoline engines might attain speeds of 14, even 20 miles an hour. People with vehicles of this type hurtling through our streets and the long roads, frightening our children and animals and poisoning our atmosphere, would call for prompt legislative action." It's too bad they didn't do something then. "The cost of producing gasoline for these vehicles is far beyond the financial capacity of private industry. Yet, the defense of the nation demands that an adequate supply should be produced. In addition, the development of this new power may displace the use of horses, which could wreck our agriculture. The discovery with which we are dealing involves forces of a nature too dangerous to fit into any of our usual concepts." I thought this might be appropriate here for the Centennial Symposium.

Larry Thompson: I thoroughly enjoyed John's paper. I agree that chemical engineering is not the focus of environmental engineering. Within the efforts that I have at Bell Laboratories, I've looked at how chemical engineers can be an important part of environmental engineering. To me, the most important way that we can contribute in plant design, which John Seinfeld alluded to, is in waste minimization. You do that with better processes. Probably the most hazardous part of the semiconductor industry is the materials used, like arsene, phosphene, and silane. It's like handling nitroglycerin in cold conditions. A tremendous challenge for chemical engineers is how to design reactors and reaction systems that can produce precursor material needed at the point of use so you don't have to store large quantities of the materials. The whole concept of waste minimization involves designing plants that (1) don't require storage of large quantities of hazardous materials and (2) don't produce much waste in the first place. There are all kinds of process challenges that chemical engineers are intrinsically trained to carry out. This is a frontier in chemical engineering and a challenge to chemical engineers.

Clark Colton: As someone not in the field, I marvel at the way we are potentially hurtling toward disaster in atmospheric phenomena. I'm surprised that science is doing so little about it. I don't understand why there isn't more research funding available for what are potentially disastrous global problems. If money was available, do we have adequate technology today to solve the problems? Is it strictly a political problem, or are there technical problems that have to be solved before we begin to get somewhere?

John Seinfeld: The steady increase of the atmospheric concentrations of the so-called greenhouse gases and the resulting effect on our climate is a problem of almost unimaginable proportions. Carbon dioxide is estimated to contribute about one-half the greenhouse effect at this time, and CO_2 emissions are connected with everything we do. There's practically no way to substantially cut CO_2 emissions worldwide. For stratospheric ozone, at least, we know it's the chlorofluorocarbons, and alternative chemicals can be sought. There are serious technical problems associated with predicting the climatic effects of increasing greenhouse gases, such as determining the role of the oceans and clouds. There are also problems associated with developing fuels that produce less CO_2 upon burning than those now in use.

Edwin Lightfoot: I would take heart from that, John, because only a few years ago, a meteorologist predicted that we were going to have an ice age because of the dust problem. So, if we make enough dust, we can solve the greenhouse problem.

Stuart Churchill: John, you said NSF hadn't funded the environmental field. When you listed all those people working in environmental problems, you didn't mention the UCLA Engineering Research Center, which would fall in that category.

John Seinfeld: Relative to the magnitude and importance of environmental problems, the federal funding is clearly inadequate. But yes, UCLA has an Engineering Research Center, and it's centered in chemical engineering.

Stuart Churchill: An ERC may not be the best way to fund this area, but it did come from NSF.

Morton Denn: I'd like to pick up on Klavs's comment, which I think is one of the most important we've heard today. Let me turn back to the discussion of the last session, in which we seemed to be hung up on what to call this beast—does it matter if it's reaction engineering or something else? If I can be forgiven the good-old-days syndrome, when I was a graduate student 25 years ago, two or three seminars on different subjects a week were the norm. Everybody damn well went and listened to whatever was going on and got some real sense of what was involved in a variety of areas. I think there's been a significant cultural change in this profession over the last 25 years. If there are two or three seminars a week, you now look at what each is about. If it doesn't fit into your particular niche, you don't go unless your department chairman, if he's going, comes by and says, come on you've got to go to this thing. That has caused many of our problems. Klavs's comment, "I never would have done this if I hadn't been over to Scriven's seminar," is particularly relevant, because it doesn't matter if it's reaction engineering, or environmental engineering, or anything else. Interesting problems are always going to come up, and we'll hear about them if we happen to be in the right place. If we make this conscious effort to avoid being in the right place, which our current pressures seem to impose on most of us, then we're just going to miss the interesting problems.

Andreas Acrivos: I've been chided by our hosts for remaining silent all day. In fact, I was told that in order to earn my keep, I had to make some comments. In fact, I was told that I had to be nasty. Now, I can't be nasty because I'm not nasty. So, let me philosophize about the program that we heard earlier today. I was very much impressed this morning. I thought, reflecting back on my own career, that I cannot but be very impressed by how fluid mechanics and thermodynamics have developed in chemical engineering. If one was asked, 40 years ago, to name the leaders of fluid mechanics, he would be very hard pressed to find any chemical engineers among them. Yet today, among the most active researchers of fluid mechanics, some of the more innovative research is done by chemical engineers. To a certain extent, the same is true for thermodynamics. One cannot but be tremendously impressed with the advances that have been made using molecular simulations, the kind of work that Keith Gubbins spoke about. Again, chemical engineers are very much at the forefront of the research. By contrast, I was a bit disappointed later on today. Maybe it was the hour. I have had a very close association with reaction engineering because I was in Minnesota during the early 1950s when Amundson and Bilous started their classical work on reactor stability. I've seen the field develop, but lately I get the impression that the fun has gone out of it. There is one area that was just very briefly mentioned that I think is extremely exciting, and this is the area of theoretical surface science. There's a lot of work done in surface science by chemical engineers, but to my knowledge all of it is experimental. Absolutely no theory. It's high time that some people got involved in this and tried to make some sense of what's going on. There are some chemists that understand it, and perhaps some physicists, but as far as I know, no chemical engineers are involved. As for environmental engineering, of course it's an extremely important topic, but it is so specialized. I can't agree with the conclusion in John's presentation. He said that two programs have a number of senior and graduate level courses in topics such as these, but more are clearly desired. I don't believe that.

John Seinfeld: More programs, not more courses.

Andreas Acrivos: I think somebody who has an excellent training in chemical engineering, like you, can handle the environmental problems without having to go through formal training.

SECTION VI
Polymers

16

Polymer Science in Chemical Engineering

Matthew Tirrell
Department of Chemical Engineering and Materials Science
University of Minnesota
Minneapolis, Minnesota

I. Introduction

A young colleague, newly embarked on an academic career in chemical engineering, was at a recent dinner that was part of a polymer research conference. He found himself seated across from an industrial polymer scientist with many years experience in a major international chemical company. Their getting-acquainted chatter abruptly ended, however, when the industrialist pronounced the academic a "fraud," having ascertained that the younger man was an assistant professor of chemical engineering at a world-renowned institution and (1) had never worked in the chemical industry and (2) was absorbed in a rather specific fundamental scientific pursuit. The industrialist had engaged in predinner activities that had lowered his inhibitions, allowing him to speak his mind so bluntly. Still, when we heard the story recounted, the words touched a nerve in several of us, similarly in academic chemical engineering and engaged in polymer research with goals like those of our young colleague.

The current status of polymer science, and other materials-related research, within chemical engineering demands a wider definition of what is fruitful activity within the profession than that of 20 years ago. Polymer sci-

ADVANCES IN CHEMICAL ENGINEERING, VOL. 16

ence in chemical engineering has reached far from the roots of chemical engineering, into chemistry, physics, and materials science. Much polymer research in chemical engineering derives its engineering relevance from answering fundamental questions underlying polymer technology, rather than from directly pragmatic polymer engineering activities. Polymer courses that we teach our students contain large components of synthetic organic chemistry, physical chemistry, scattering and diffraction, crystallography, and other tools and techniques not among the traditional repertoire of an engineer. Much of our research would be accepted as good fundamental work in physics or chemistry departments. We distinguish ourselves from our pure science colleagues by feeling no need to apologize for the eventual utility of what we do and by pursuing contact and collaboration with our industrial colleagues, both for intellectual stimulation and for implementation of our ideas in the technological arena. This is our response to the accusation of fraud visited upon our young colleague. Polymers are scientifically interesting and technologically relevant. As with other high-performance materials, the time span from conception of a new development in polymer materials science to implementation can be short. Engineers must therefore be close to the science in order to participate.

The worldwide polymer industry developed in the 1930s as the concept of unit operations was taking hold as the first paradigm of the chemical engineering discipline. Bitter arguments about the very existence of macromolecules in the late 1920s gave way within five years to the construction of commercial polystyrene plants by I. G. Farben, Dow, and others. High-density polyethylene and polypropylene became commercial products in the 1950s, three years after the discoveries of the mode of action of the catalysts that enabled their synthesis by the teams of K. Ziegler and G. Natta (the latter being the only chemical engineer ever to win the Nobel Prize). These plants were built by engineers who viewed polymers as new commodity chemicals.

Evidently, close contact with its scientific base has long been a characteristic of the polymer industry; however, the industry itself has changed substantially in the last 20 years. In the 1970s, vast improvements in the rational design of polymerization reactors and processes were achieved, spurred in part by advances in the general area of chemical reactor analysis and design [1]. Recognition developed that polymerization reactors should be optimized with respect to the materials properties of the polymers that they produced, rather than exclusively with respect to throughput. Some chemical engineers began to deal with the fact that, while toluene is toluene, not all polystyrene is alike. More elaborate measures of product quality had to be developed, including effects of molecular weight distribution, copolymer composition and sequence distributions, and compositional heterogeneity. This demanded more knowledge on the part of engineers of both the polymer materials science and the measurement techniques available to determine whether the desired product had been achieved.

A major accomplishment of this period was the development of the Unipol process for the production of linear, low-density polyethylene, a new process for an established commodity polymer that enabled exquisite manipulation and control over the materials properties of the product [2]. This achievement was made through the efforts of scientists to develop new catalysts and of chemical engineers armed with new means of rational analysis and design of polymerization processes.

Today, opportunities for design of new large-scale polymer production processes are rare and promise to remain so, at least until the end of this century. Several factors are responsible. One is that the economic factors weighing against a new monomer attempting to break into the arena of commodity polymers are enormous. Polyethylene, polypropylene, polystyrene, and polyvinyl chloride are the commodity polymers; certain nylons and polyesters are produced in sufficient volume to be considered in the same vein. Marginal profitability of commodity polymers is low, so that tremendous sales volumes are required to return the investment in new processes; prudence and inertia then make it difficult to justify even the initial stages of such development projects, let alone construction of large production volume. The second factor reducing activity in the development of large-scale chemical processes for polymer production is related to the first. New products developed for new applications are frequently copolymers containing commodity monomers. Development of new polymeric materials is similar to architecture at the molecular level with monomeric building blocks; like architects, synthetic chemists work with readily available materials as much as possible. Copolymers, like metal alloys, can be individually tailored to new applications. It is more straightforward to tinker with available products in this way than to design a new material from scratch. Engineering design of plants to produce commodity polymers is no longer a principal focus of polymer engineering activities. Furthermore, much polymer production is now accomplished outside the chemical process industries.

New processes and products are closely tied to applications. Areas of application that currently drive much new polymer development include adhesion, microelectronics, transportation, medicine, and communications.

Polymer technology has been the most natural (or perhaps the most familiar) entrée for chemical engineers into advanced materials processing. Polymers are mainly organic materials, made of raw materials derived from petrochemicals. The chemical engineer's background in organic chemistry is an advantage here. Polymers are mainly processed in the fluid state, bringing fluid mechanics expertise into play. Effective participation in other areas of materials technology, such as microelectronics, for example, requires knowledge of quantum mechanics and solid-state physics beyond the ken of the typically educated chemical engineer. Future advances in polymer technology will, however, require a further expansion of the science base of the chemical engineer.

On the academic side of the profession, chemical engineering departments have provided a very hospitable environment for polymer research and education. This is partly for the reasons given above and partly because our chemistry colleagues historically have been strangely hostile to polymer work, thus shaping the nature of polymer science within chemical engineering. A rough estimate is that 20% of the chemical engineering faculty members currently pursuing polymer research took their doctoral degrees outside chemical engineering, with the vast majority of those degrees being in chemistry. These people have found chemical engineering to be a more fertile environment for their intellectual endeavors than a pure science department, both in terms of the value of their work in the eyes of their colleagues and in terms of the numbers and talent of the students that they could attract to polymer research. As with all scholarly activity, this is self-perpetuating; a broad view of polymer science has taken hold in chemical engineering.

There are signs that chemistry and physics departments are awakening to the opportunities they've missed. Chemical engineering researchers in polymers are frequently invited to give colloquia in chemistry and physics departments or to participate in, for example, ACS or APS symposia, where the issues are the fundamental chemistry and physics of polymers. In fact, many polymer researchers in chemical engineering find the most critical and stimulating audience for their work outside chemical engineering. (Similar observations can be made in other new areas of chemical engineering [3].) Chemistry students in some universities are encouraged to take the course offerings of chemical engineering departments in polymers. NSF has launched a Materials Chemistry Initiative, funded in part by the Chemistry Division, which recognizes that polymers may involve some potentially interesting chemistry.

There are still questions concerning the role of polymers in chemical engineering beyond the hospitality of the research environment. How fully integrated into our discipline will polymer research and education become? Will the journals of our profession and the meetings of our professional society become more important, active forums for discussion of substantive technical issues in polymers, instead of the passive bulletin boards that they are now, serving mainly to give research exposure in the chemical engineering community? Will course work in polymers move from electives to the core of our curricula? The purpose of this article is to point out some opportunities for movement in these directions.

II. Research Opportunities in Polymers

The examples to be discussed in this section are not restricted to those that chemical engineers are pursuing, but attention will be given to how chemical engineering can participate and be influenced.

A. Synthesis of New Polymeric Materials

Polymers are no longer inexpensive, substitute materials; they are, with increasing frequency, the materials of choice based on performance criteria. Polymers offer the opportunity to design in material properties at the molecular level in the construction of the macromolecule—an avenue seized by nature in ways that we can only weakly mimic.

1. Polymers Developing Novel Microstructures

Building fine detail into the chemical structure of a macromolecule can alter the properties of the materials via more than just the introduction of a different set of covalent bonds. Many of the properties of polymer materials are determined largely by the topology (chainlike character) of the molecules and by their mutual interactions on the length scale of the entire macromolecule, rather than by local covalent bonds. Wegner [4] has termed this the "supramolecular" structure of polymer materials. This does not eliminate the synthetic chemistry as an important determinant of the properties; rather, it gives the synthetic chemist the opportunity to build into the molecule elements that will direct the supramolecular organization. An emerging philosophy is to synthesize chemically the desired supramolecular structure, instead of the more conventional route of using processing conditions on preformed polymers to manipulate the final structure. At least three different mechanisms, encompassing several classes of polymers, afford these opportunities. One can use the large-scale self-assembly properties of the macromolecules themselves (block copolymers), the more local packing properties of the chain segments or monomers (liquid crystalline polymers or Langmuir–Blodgett films), or the associative properties of parts of the chains (ion-containing polymers).

a. BLOCK COPOLYMERS.   Block copolymers share with small molecule amphiphiles surfactant properties arising from their tendencies to straddle phase boundaries [5]. Applications as adhesives, compatibilizing agents in polymer alloys, coupling agents in polymer composites, stabilizing agents in colloidal dispersions, and in various microelectronic devices are on the increase because of these desirable properties and the chemical synthesis techniques enabling these materials to be made in very well-defined block lengths. They are a tool with which the engineer can manipulate interfacial behavior in polymers. Block copolymers separate into microphases that form molecular composites, creating their own internal interfaces.

From the viewpoint of fundamental understanding of amphiphilic surfactant behavior, block copolymers possess several special characteristics that derive from their polymeric nature. The structural units of block copolymers are large enough so that many aspects of the interactions with solvents and among the chains can be understood in universal terms that transcend the specific chemical nature of the species involved [6]. These features have their origins in the chainlike character of the molecules (for example, excluded volume, entropic elasticity of the chains, nonideal free energy of mixing). Furthermore, studies of homologous block molecular-weight series provide the means of varying size systematically while holding chemical interactions constant, which is generally not possible with small-molecule surfactants. Finally, the large size of the polymer blocks dictates that, in general, the interfacial structures that are formed are of a significantly larger size than with low-molecular-weight amphiphiles.

As bulk materials, block copolymers can organize into a rich variety of microstructures, as shown in Fig. 1, driven by the tendency of the two blocks to phase-separate and the prevention of that on a macroscopic scale by covalent joining of the blocks. This, in turn, provides a means of manipulating mechanical and transport properties via chemical synthesis. Segmented polyurethanes and polyesters derive their useful, and highly controllable, mechanical behavior from these chemical characteristics [7]. Variation of block molecular weight changes the equilibrium morphology, as does changing the geometry of joining the blocks.

Thomas and co-workers [8] discovered an entirely new morphology on examination of multiple-arm star block copolymers. For the right combinations of arm lengths and compositions, these materials organize themselves into a bicontinuous structure, with each phase fully interpenetrating the other in a diamond lattice arrangement. In these star molecules, the arms were diblock polymers; a considerably wider range of structural variation is conceivable and even possible.

Anionic polymerization has been the usual route for this type of special synthesis. Cationic [9], catalytic [10], and group transfer [11] polymerizations have been developed to produce well-defined blocks from different classes of monomers. Perhaps the richest and most technologically useful future route to the production of these materials is via so-called telechelic [12] or functionalized polymers. Generically, this refers to block polymers in

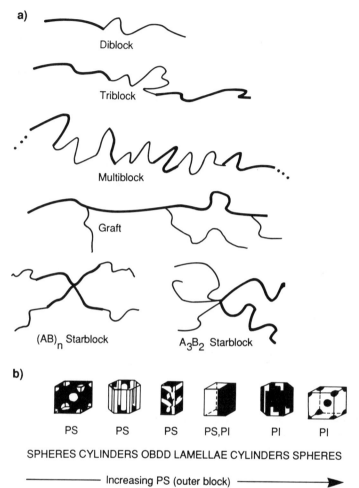

Figure 1. (a) Schematic diagrams of some structures of block copolymer molecules. (b) Schematic of compositional dependence of microdomain morphologies of poly(styrene–isoprene) star block copolymers. OBDD stands for ordered bicontinuous double diamond structure. Based on [22].

which the blocks are synthesized separately (not sequentially along the same growing chain, as in anionic polymerization) and endowed with appropriate chemical functionality, which is then used to couple them into the desired block copolymer structure.

Production of these materials requires meeting several engineering challenges. More precise control is needed for these materials than for usual polymer products, controlling not only molecular weight but also chemical composition and architecture. One needs products with uniform numbers of blocks, or arms in the case of nonlinear polymers. The chemistries for achieving this are more delicate than the workhorse free-radical polymerization. Within these types of polymers, sections may individually be rather small, that is, oligomeric. The technology of the effective production of small polymers is not fully developed [13]. Processing these materials once they are synthesized demands knowledge of the kind of microstructure desired. Of interest in these materials is that they organize their own structures, but this is an equilibration process that takes time and can easily be disrupted by the processing environment. Virtually unexplored is the nature of kinetically trapped nonequilibrium states that can be frozen into the material. Chemical engineers can play a major role in this kind of investigation, which relies on the laws of transport phenomena and thermodynamics.

b. LIQUID CRYSTAL (LC) POLYMERS.   Polymers that melt into orientationally ordered fluids can exhibit more complex behavior than classical liquid crystals formed by small molecules [14]. The largest class of polymeric liquid crystals form nematic phases, wherein rigid units of aspect ratio significantly different from unity find it most energetically favorable to align along a single direction, with no registry or orientational order in the other directions. There can be specific energetic interactions that assist this alignment but it occurs mainly as a result of the entropic driving force: the largest number of available configurations in a certain volume exists when one packs rods in a parallel arrangement.

This ordered packing tendency, combined with fluidlike flow properties, permits the development of materials with structures resembling solids and the processability of liquids. The orientation can produce interesting, anisotropic mechanical properties, such as flexural modulus. Liquid crystalline polymers have been synthesized in two structural classes: those with the liquid crystal-forming group in the main chain backbone, and those with the mesogen in the side chains. These latter materials can form smectic liquid crystal phases (in addition to nematics), where there is both uniaxial order and registry among the mesogens. Incorporation into these materials of a quantity of optically (dye), electrically, or magnetically active agent, either as a solute or covalently bound to the backbone, endows these materials with nonlinear optical, electrooptical, or magnetooptical properties that are potentially very useful. As shown schematically in Fig. 2, the liquid crystalline polymer serves as a template for organizing the agent to give the desired properties.

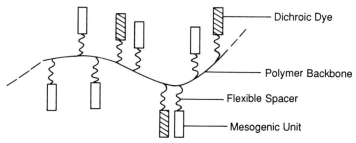

Figure 2. Schematic structure of side chain liquid crystal polymers with dichroic dyes (imparting nonlinear optical properties) and mesogenic side groups.

In addition to the opportunities for new materials synthesis and characterization along these lines, transport properties, rheology, and processing techniques for liquid crystal polymers are essentially unexplored. Experiences with synthesis of polymer structure based on these liquid crystal templates may open up other creative avenues for template synthesis, for example, inside other crystalline structures, chlathrates, or zeolites, or on surfaces [4]. Composites, alloys, or mixtures of liquid crystalline and flexible polymers may produce new materials.

c. POLYMERIZATION IN LANGMUIR–BLODGETT (LB) FILMS.   LB films, like liquid crystals, develop microstructure due to the local self-organizing properties of monomers or segments [15]. In the case of LB films, the monomers are small surfactant molecules, typically with ionic or otherwise hydrophilic heads and hydrophobic tails. To polymerize them, functionality must be built into the monomer, frequently by vinyl that can be initiated by radiation. The potential for production of ultrathin films has already been extolled [16].

Polymerization of LB films, or deposition of LB films on polymers, offers the opportunity to impart to LB films a higher degree of mechanical integrity. However, preliminary work in this direction shows a conflict between the chainlike primary structure of the polymer and the well-organized supramolecular structure [15]. One possible solution may be the insertion of flexible spacers between the main chain polymer and the side chain amphiphile, a route also employed in liquid crystal polymers. These materials belong to an interesting class of two-dimensional polymers, of which there are few examples. These toughening techniques may eventually be applied to stabilize other self-assembled microstructures, such as vesicles, membranes, and microemulsions.

With polymerically toughened LB films, further issues for engineers present themselves in the deposition techniques. These are questions that span the range between fluid mechanics and the physics of wetting [17], such as understanding the role of the spreading solvent for the LB film. Why do some films squeeze or "zip" down smoothly onto a solid substrate while relatively similar monolayers carry along solvents that drain poorly and form nonuniform, defective LB films? What are the possibilities of trapping states far from equilibrium? How does one scale up these processes when necessary?

d. SILANE AND OTHER NON-CARBON-BACKBONE POLYMERS.   Although many polymers with some atoms other than carbon in the backbone are well known, there are few significant examples to date of completely non-carbon-backbone polymers. Figure 3 shows some examples. Polysiloxanes are the outstanding exception in achieving commercial success. Interesting new opportunities are on the horizon. Although polysilanes have existed for more than 60 years, scientific and technological interest has been nil until recently,

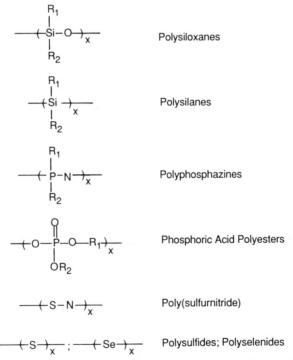

Figure 3.   Some examples of inorganic backbone polymers (R_1 and R_2 may be organic).

since all early materials were intractable and insoluble in all common solvents. There have been a number of reports [18] of syntheses of soluble polysilane derivatives that are processable and can be studied by conventional analytical means. The class receiving the most attention is the poly(di-*n*-alkylsilanes), which have been shown to have interesting optical, electronic, and adhesive properties. The side chains of hexyl and longer silanes crystallize among themselves, independently of the backbone, and in so doing enforce a conformational rigidity on the chain that is responsible for the special electronic properties of these materials.

Manipulation of the effective solubility, flexibility, and melting behavior of the main chain by attachment of the appropriate side chains is a theme from current work on liquid crystal and amphiphile-containing polymers that is repeated. Such methods are necessary for adapting non-carbon-chain polymers to the array of processing techniques available for carbon polymers. New chemistry and processing techniques are forthcoming. These non-carbon-based polymers may also serve as processable precursors in the synthesis of inorganic or ceramic materials, using high-temperature treatment to remove residual organics [19].

e. ION-CONTAINING POLYMERS.   Polymers in which every monomer segment is charged are known as polyelectrolytes and have long been studied in aqueous media, mainly to understand how the Coulombic repulsions affect the configurations of the molecules. These molecules are quite useful for manipulating the rheological properties of water-based materials.

More interesting as materials in the bulk state are molecules that have a low density of ionizable groups [20]. In nonionic media, such as organic solvents or the bulk polymer media composed mainly of their nonionic monomer neighbors, these ionic groups interact with one another, often exhibiting a strong tendency to associate. This may manifest itself in clustering or (reversible) gelation of the materials, with dramatic and useful changes in the rheological and mechanical properties of the materials. These "ionomers" have already been commercialized in several forms, as tough, impact-resistant materials (Surlyn®), as membranes (Nafion®), and in organic syntheses, ion exchange, and catalytic systems. Nonetheless, the detailed structure, flow properties, and processability of these materials are still largely unknown, mostly because of the extraordinary range over which variables such as charge density and the nature of the counterion can be varied.

2. Electrical, Optical, and Magnetic Properties of Polymers

Polymers are now being developed as active electrical, optical, and magnetic materials. Synthetic metals, produced by polymerization, have been made from a variety of conjugated, aromatic, and inorganic $[(-S-N-)_x]$ backbones. The excitement generated by the prospects of tremendous weight and cost savings relative to copper and the extraordinary circuit design possibilities with molecular wires has faded somewhat recently, largely because of the difficulty of achieving truly metallic conductivity and the intractability of these materials to processing. Routes such as side chain attachment, discussed above for other classes of polymers, are employed with progress on this front. These materials occupy a status similar to that of high-T_c ceramics, in that the need is for an engineering focus on the materials processing.

Current interest in organic, nonlinear optical (NLO) materials derives from the attractiveness of large susceptibilities, fast response times, low dielectric constants, and the intrinsic tailorability of organic structures [21]. NLO activity is typically assessed via the measurement of second-harmonic generation (SHG), in which the material is exposed to incident laser light of frequency ω and emerging light at 2ω is detected. The typical route to production of such a material is to imbed a chromophore in a matrix where it can be fixed at a certain desired alignment. The matrices being explored include both glassy polymers and liquid crystal polymers, with and without covalent bonding of the chromophore to the backbone, though the latter is increasingly the method of choice. Orientation of these materials can be accomplished in several ways, the most convenient being the application of an electric or, for LC polymers, magnetic field. Poling at high temperature, followed by quenching, is the usual procedure. As processing hurdles are surmounted, polymers will play more than passive roles in electronic and photonic devices. Langmuir–Blodgett methods or other techniques building on the self-organizing properties of the macromolecules will be important.

3. Biological Synthesis of Polymeric Materials [22]

All human-made polymeric materials to date have distributions of molecular weight, composition, and architecture. The techniques discussed for synthesis of block copolymers are the most precise available. It is interesting to speculate on the possibilities for new materials synthesis if one were to have perfect control over these factors. Most polymers synthesized under the direction of the genetic code achieve this perfection, suggesting genetic engineering routes to these materials. An intriguing possibility is to synthesize and splice into organisms genes encoded for predetermined polymeric products. Although for the present this route seems limited to polypeptide materials, this is not a very restricted category. One can write down the genetic instructions for polymers that will fold or crystallize in some specified way, expose some derivatizable chemical functionality, or aggregate or coalesce into supramolecular structure. The degree of structural regularity possible is beyond anything else currently imaginable. An example in which a polypeptide sequence designed to fold in a certain way was synthesized by these means has been published [23]. Vast possibilities for polymer materials science exist in this area.

4. Summary

The focus here on self-organizing polymers, electrical and optical properties, and biosynthesis is not even close to comprehensive in the enumeration of avenues and opportunities in new polymeric materials. Degradable plastics that do not remain in our environment forever [24] and polymer precursors to ceramics [19] and inorganic fibers are two of many more areas in which new polymeric materials will provide new challenges to engineers in production and processing.

B. Developments in Polymer Processing

1. Polymerization

As discussed in the Introduction and in connection with block copolymers, the trend in polymerization engineering is toward smaller volume and therefore batch processes, and also toward production of lower-molecular-weight functionalized polymers that are used as building blocks for microstructured polymers. The trend toward batch polymerization reverses that of the 1960s and places new demands on the design and optimal control of these reactors. Steady-state control is irrelevant; optimal trajectories in time are now the desired objective. The arsenal now available to the polymerization engineer to accomplish these objectives is much more powerful than it was 20 years ago. Recent advances in understanding the polymerization kinetics of some systems, combined with continuing increases in the power of low-cost digital computers and improved on-line polymer property-measuring devices, are making on-line control of the entire distribution of molecular weight of polymer products feasible. Figure 4, taken from Ellis *et al.* [25], shows how a combination of sporadic, time-delayed measurements, an extended Kalman filter estimator, and a physically based, detailed kinetic model can give excellent control of the entire molecular weight distribution over the complete trajectory of a free radical polymerization of methyl methacrylate.

These kinds of advances in control implementation in turn bring new life to efforts to develop detailed models of other polymerization reactions, since it is no longer such a major leap to on-line implementation. Efforts to model emulsion polymerization and heterogeneous Ziegler–Natta polymerization are among the most ambitious [26–28]. Fundamental modeling efforts on

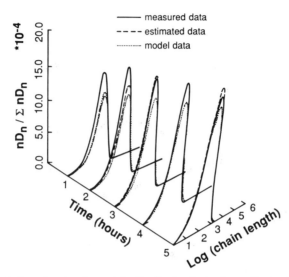

Figure 4. Evolution of molecular weight distribution in the free radical polymerization
of methyl methacrylate. These curves were obtained [25] during a polymerization controlled by an on-line computer program using a molecularly based
kinetic model [29], implemented by an extended Kalman filter, updated by
sporadic on-line gel chromatography measurements of molecular weight
distribution. D is the concentration of polymers of chain length n. From "Estimation of the Molecular Weight Distribution in Batch Polymerization" by M.
F. Ellis, T. W. Taylor, V. Gonzalez, and K. F. Jensen, *AIChE Journal*, Vol. 34,
No. 8, pp. 1341–1353, 1988 [25]. Reproduced by permission of the American
Institute of Chemical Engineers copyright 1988 AIChE.

free radical polymerization, which exhibits diffusion-limited kinetics during most of the course of the reaction, have been advanced significantly by
new measurements and theoretical understanding of diffusion in dense
polymer fluids [29].

2. Reactive Processing

The reaction injection molding (RIM) process, illustrated schematically in
Fig. 5, was the first major development moving toward reactive processing of polymer materials [30]. Reactive processing refers to processes in
which the chemical synthesis and the forming operations are accomplished
in a single step or in a single piece of process equipment. In RIM processing,
two or more liquid reactive materials are simultaneously injected into a mold,
where they polymerize and take their final shape within seconds.

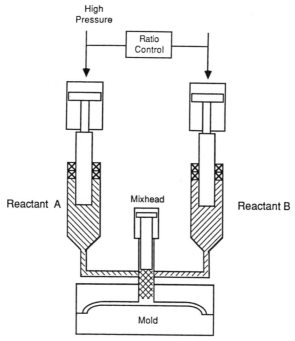

Figure 5. Schematic of a reaction injection molding machine. Reproduced with permission from Macosko, C. W., *Fundamentals of Reaction Injection Molding.* Hanser Publishers (dist. by Oxford University Press), New York, 1989 [30].

Major growth of RIM was pushed by needs in the U.S. automotive industry. Flexible polyurethane fascias on the front and rear of cars were found to meet the 1972 federal requirement that all bumpers withstand a 5-mph impact. RIM quickly proved the most economical way to make these large parts, and production began in 1974. Weight reduction of the vehicles was a secondary benefit of the introduction of RIM materials. Materials and markets have expanded to current production levels of close to 100,000 tons annually and greater than 10% growth. Table 1 lists the polymerization chemistries now applied in RIM. The low viscosity of the fluids injected is the principal advantage of RIM, eliminating the need for the enormous hydraulic clamping equipment necessary at the injection pressures of standard thermoplastic injection molding.

The success of RIM has spurred interest in other means of reactive processing. Reactive extrusion and reactive blending are two related processes with interesting potential. Reactive extrusion refers to processes in which

Table 1. RIM Chemical Systems: Polymerization
             Reactions

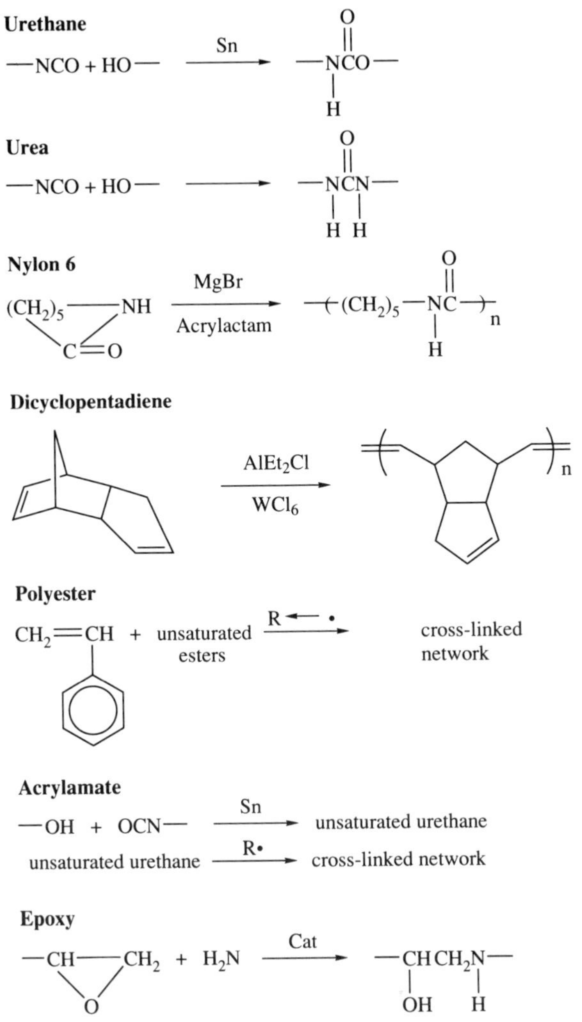

Urethane

$$-NCO + HO- \xrightarrow{Sn} \overset{\overset{O}{\|}}{-N}CO-$$
$$\underset{H}{|}$$

Urea

$$-NCO + HO- \longrightarrow \overset{\overset{O}{\|}}{-N}\underset{\underset{H\ H}{|\ |}}{C}N-$$

Nylon 6

Dicyclopentadiene

Polyester

$$CH_2{=}CH \ + \ \underset{esters}{unsaturated} \ \xrightarrow{R \longleftarrow \bullet} \ \underset{network}{cross\text{-}linked}$$

(phenyl ring)

Acrylamate

$$-OH \ + \ OCN- \ \xrightarrow{Sn} \ unsaturated\ urethane$$
$$unsaturated\ urethane \ \xrightarrow{R\bullet} \ cross\text{-}linked\ network$$

Epoxy

$$-CH{-\!-}CH_2 \ + \ H_2N \ \xrightarrow{Cat} \ -\underset{\underset{OH}{|}}{C}H\,C H_2\underset{\underset{H}{|}}{N}-$$
$$\underset{O}{\diagdown\diagup}$$

the extruder pump is also employed as a chemical reactor. It may be a
polymerization reaction or a reaction on or among preformed polymers.
Though the former case has, in principle, the same low pumping cost advantages as RIM, it has not yet advanced very far. This is due to the un-

availability of extrusion equipment designed for pumping fluids whose viscosity varies by orders of magnitude during one residence time.

Reactive blending refers specifically to processes in which alloying is enhanced by inducing a chemical reaction between the components. This produces block or graft copolymer surfactants at the interface and promotes miscibility. This approach is exciting because one can produce stable mixtures of immiscible polymers that would be impossible to produce by other means. Polyolefin–polyamide and polyester–polyamide blends [31] are two of the many possibilities that have been realized. The real potential of these processes must be assessed from an engineering standpoint. For example, is it better to make copolymer surfactants outside the extruder and then mix them in during later processing? To answer this and related questions, data on phase behavior and transport coefficients, as well as thorough studies of the end-use properties of the resulting materials, are required.

3. Processing of Polymer Composites

Traditionally, processing of polymer composites has been time-consuming and labor-intensive. Reactive processing methods seek to change that. The pultrusion process (Fig. 6) is a continuous manufacturing process for fiber-reinforced polymer composites. Liquid polymer resin impregnates fiber rovings or mats before curing in a heated die [32]. As the resin (typically unsaturated polyester) gels and hardens around the fiber reinforcements, a rigid polymer composite is formed that requires few finishing operations after leaving the die. Capital and operating costs are low and the speed and versatility of the process are attractive.

Clearly, Fig. 6 suggests a wealth of fluid mechanics and heat and mass transfer problems ripe for chemical engineering attack. But the opportunities go beyond these. There is need for studies of how processing conditions affect properties. Furthermore, like block copolymers and polymer

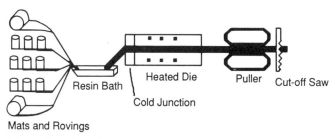

Figure 6.   Schematic of the pultrusion process for continuous production of polymer composites.

blends, composites are materials in which the properties are controlled to a large extent by interfaces. We must understand more about how interfacial agents, added to couple matrix and fibers mechanically, get to and act at the interface. Fundamental studies of adhesion at polymer–solid interfaces are very important in this connection [33].

C. Emerging Areas of Polymer Applications

Thus far, we have discussed synthesis and processing of polymeric materials. These are familiar territory to the chemical engineer, though it is evident that synthesis and processing are merging to some extent. It is clear, however, that further opportunities are available to the engineer who looks beyond familiar territory. Areas outside the chemical industry where chemical engineers with the proper training and mind set can make important contributions include the following.

1. Adhesion

In the broadest sense, understanding of adhesion means knowledge of the fundamental science and practice so that the mechanical interaction between two surfaces can be manipulated at will. This may mean developing the strongest possible epoxy joints for the aerospace industry or controlling weak adhesion so that the adhesive does not stick where it is not wanted (for example, the "wrong" side of Scotch® tape). It may mean using polymers to promote the flocculation of fine particles in waste effluent or preventing flocculation of pigment particles in coatings, using polymer agents to control the surface interaction. It could mean developing agents to improve the mechanical coupling between matrix and filler in composites or between polyimide and copper when the polymer is used as an interlevel dielectric in integrated circuits. The knowledge of the polymer behavior in each of these instances, and for the next several examples as well, must be augmented by a full appreciation of the application and of the nonchemical processes involved in each.

2. Medicine

The long quest for blood-compatible materials to some extent overshadows the vast number of other applications of polymers in medicine. Development and testing of biocompatible materials have in fact been pursued by a significant number of chemical engineers in collaboration with physicians, with incremental but no revolutionary results to date. Progress is certainly evident, however; the Jarvik-7 artificial heart is largely built from polymers [34]. Much attention has been focused on new classes of materials, such as polyetherurethanes and hydrogels, and on surface interactions (especially protein adsorption on polymer and other surfaces) by this work [35]. A generic understanding of principles is still elusive. Two other areas of current and future excitement are in controlled drug delivery using diffusion from polymeric reservoirs, including polymer-stabilized vesicles [36], and work on using cell-seeded polymeric scaffolds as templates for skin and nerve cell regeneration after traumatic injury [37].

3. Electronics and Photonics

The opportunities for chemical engineers in the microelectronics and related industries, including those involving polymers, have been thoroughly covered in the accompanying chapter by Larry Thompson [38]. However, beyond the examples discussed there, related to the use of photoresists in device manufacture are equally crucial elements of packaging, interconnections, and printed wire boards incorporating polymers that are being pushed to their limits by the drive toward miniaturization. Conducting polymers, should the processability problems be surmounted, could cause a revolution in materials for microelectronics, as could polymer-based nonlinear optical materials [39].

Had we more space, we could delve into still other applications, such as the food industry, where the combination of chemical engineering and polymer science promises to have powerful results. A brief closing example from our own work may serve more usefully to describe our vision of how work in these areas outside the chemical industry can lead to progress and enrich chemical engineering.

Over the last several years, we have been studying forces exerted between two adsorbed polymer layers brought into close proximity, using the direct force measurement technique developed by Israelachvili [40], which permits force as a function of separation to be measured with subnanometer precision. One of our motivations was to understand better the configurations of adsorbed polymers [5]. As two adsorbed layers were brought together, we found that a significant portion of the forces that we measured arose not from interactions between polymers but rather from single polymer molecules that bridged that gap between the substrate surfaces. We decided to make a direct measurement of this bridging force [41]. Figure 7 compares forces between two polymer layers with those between the polymer on one surface interacting with bare substrate. As shown, the force in the latter case is more than an order of magnitude higher, revealing the strong contribution of bridging. More important, we were able to determine for the first time, by direct mechanical measurement, the segment binding strength of the polymer on the substrate. Description of this work in appropriate industrial laboratories has led to three active industrial collabo-

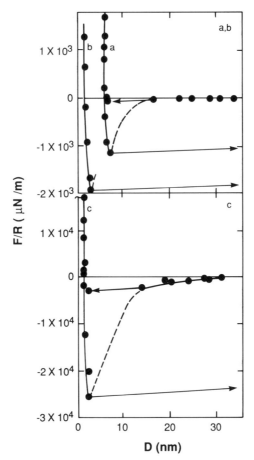

Figure 7. Force–separation profiles between two curved mica surfaces immersed in cyclohexane at 25°C [41]. The force scale is normalized by the radius of curvature to give the energy per unit area between the surfaces as a function of separation. (a) Each mica sheet bears about 3.0 mg/m^2 of adsorbed poly(α-methylstyrene). (b) Each mica sheet bears about 0.5 mg/m2 of adsorbed poly(α-methylstyrene). The forces are shorter range and more attractive due to lower adsorbed amount and consequent enhanced bridging. (c) One mica sheet bearing 3.0 mg/m2 of poly(α-methylstyrene) brought into contact with a bare mica sheet. Deep attraction results from strong bridging. Adhesive minimum corresponds to an effective monomer unit binding energy to mica of about $kT/3$ per monomer. Reprinted with permission from Granick *et al.* [41].

rations: one to examine directly interactions of polymers with silica for adhesion applications, one to examine forces between miscible polymers, and one to study new classes of colloidal stabilizers. Each of these efforts is related to a product development program within the company.

D. Techniques Enabling Recent Progress in Polymer Science

To become fully efficient in the future polymer research arena, the chemical engineer should master a set of instrumental techniques that have not traditionally been within the purview of our discipline. We simply list those that we feel have broad utility here and invite the interested but unfamiliar reader to seek out any of a number of more specialized reviews. The ACS Polymer Division has sponsored a set of instrumentation tutorials for the last eight years, so the Polymer Preprints of that Division are a good starting point.

1. Spectroscopy: infrared, UV/visible/fluorescence, NMR, and photoelectron.

2. Microscopy: electron, light (with video enhancement), scanning tunneling, atomic forces, IR microscopy, NMR imaging, thermal and magnetic domain microscopy.

3. Scattering: light, x-ray (including synchrotron sources), neutrons.

4. Birefringence, dichroism, ellipsometry, neutron and x-ray reflection.

5. Surface forces, LB trough.

Facility in a subset of these techniques is essential for the fundamental examination of microstructured materials such as polymers. Note that all of these techniques demand fundamental knowledge of optics and the interaction of radiation with matter that is not routinely dispensed in a chemical engineering education.

Though not instrumental in nature, another important technique in the polymer arsenal is large-scale computer simulation experiments. These have proved especially useful over the last several years in, for example, molecular-level simulations of polymer mechanical properties [42] and of the transport properties of concentrated polymer solutions [43]. Polymers are in many ways ideal objects for this level of simulation study; although it is difficult to have accurate detailed knowledge of local interactions, as mentioned earlier, much polymer behavior is dominated by nonlocal interactions that can be much more adequately represented.

III. Some Outstanding Problem Areas in Polymer Science and Engineering

No order of priority is implied in this brief list of outstanding polymer problems, as each has implications for chemical engineering. It should be borne well in mind that not all fruitful research endeavors can be readily formulated as outstanding problems.

1. *Transport phenomena of polymeric fluids.* Above a certain threshold of concentration and molecular weight, polymers begin to exhibit a characteristic set of dynamic behaviors known as "entanglement." This type of behavior has been well studied experimentally, and its effects are exceedingly important since it makes all transport properties very sensitive to concentration and molecular weight. Why it occurs at a certain molecular weight or concentration is unknown. While a brilliant theory, dubbed "reptation" [6], goes far to rationalize entangled behavior and to predict new phenomena, entanglement itself cannot be predicted. Much of polymer fluid mechanics and polymer processing is influenced by this behavior. At this stage, it is not even clear from what route the answer will arrive; the surest but likely the slowest route is via classical statistical mechanics and molecular transport theory.

2. *Phase separation processes in polymers.* Polymers have become very useful molecules with which to study phase separation processes for at least two fundamental reasons. Their relatively large spatial extent, and consequent simultaneous interactions with many neighbors, means that mean field theories are expected to apply and can be tested. Furthermore, their phase separation processes are slow, making experimental study of the dynamics easier. Very small interactions are multiplied by the number of segments in the chains into macroscopic effects, as Bates and Wignall have shown in their surprising demonstration that even hydrogenated and deuterated versions of the same polymer phase separate from one another [44]. With these advantages, one can hope to understand the time course of these processes in detail and begin to explore the infinite number of nonequilibrium states accessible along the way. Many polymer processing operations involve phase separation as a means of manipulating polymer microstructure.

3. *Fundamental mechanisms of adhesion.* All classical adhesion tests involve a rheological component, in the deformation of the near-interface material, and a surface chemical component. With the recent availability of microscopic techniques to study surface forces, one can possibly go after the surface chemical component, separately from the rheological component. More generally, the configurational and dynamic behavior of macromolecular interfacial regions remains a very rich area.

4. *Controlled synthesis of polymeric materials.* The hallmark of much polymer education traditionally has been concern about materials containing distributions of molecular species. Many endeavors in polymer physical science, and untold numbers of applications, would benefit enormously from the finest possible control over the molecular purity of polymeric products.

5. *Interaction of polymers with biological systems.* This problem is not new but has an enduring appeal, as do many biomedical problems. The payoff is potentially enormous in terms of the lifesaving devices that could be developed. The arsenal of techniques available to investigate the controlling surface interactions has never been stronger.

IV. Conclusions

Aside from the specifics of various interesting, provocative, or at least debatable research areas discussed here, what conclusions should one draw from the view of polymers in chemical engineering presented here?

One conclusion concerns education. At the graduate level, and to a lesser extent at the undergraduate level, students should be encouraged to take the broader view demanded by the interdisciplinary polymer field, rather than dealing with only those aspects of polymers that fit the more traditional chemical engineering mold. Studies of polymers, and other technologically important materials, can be the vehicle with which our students begin to break the mold. Polymers are molecular materials, not continua; young engineers will not get far in modern polymer rheology and transport if they rely exclusively on classical fluid mechanics and diffusion equations. Other concepts and tools appropriate to molecular materials should be introduced: statistical mechanics, quantum mechanics, spectroscopy, scattering, optics, and computer simulation should become familiar. A revolutionary revision of the curriculum is not the point here; rather we should seek to revolutionize the culture surrounding polymers in chemical engineering. Students should be encouraged to teach themselves these subjects. We must let them know that such subjects are crucial to their development as effective engineers, and we need to provide better materials so that they may accomplish this goal. Textbooks and elective or seminar courses are necessary.

The need for new books cannot be overemphasized. Books on instrumental methods appropriate for the engineer are one outstanding example. Books on polymerization engineering and polymer processing can no longer let chemical engineers go away thinking that they understand a subject well via sets of population balance or transport equations. They must know the organic chemistry of synthesis and the physical chemistry of characterization; they must know how to measure the controlling parameters of the materials they are producing. They should learn to think "small," both with respect to materials microstructure and with respect to processing scales.

The array of subjects we have laid out is daunting and could be taken by some as a recipe for superficiality. It is not the intent here to suggest that everyone can be an expert in a variety of areas twice as broad as our curriculum was 10 years ago. The point is twofold. First, these are the kinds of areas now needed to make progress; second, one is well served by a broad awareness and an interdisciplinary outlook. This kind of situation is well suited to the formation of interdisciplinary research centers where collaborative efforts pool expertise.

The final point goes back to the opening paragraph. Should we worry about what defines a chemical engineer? Not at all! Chemical engineering is the most versatile of the engineering disciplines. There are definite trends, for example, in microelectronics, from mechanical toward chemical means of manufacturing. We should do what we have always done well—that is, understand the fundamental science and use it to advance technology. This chapter points out some of these directions in polymers.

References

1. Tirrell, M., Galvan, R., and Laurence, R. L., Polymerization Reactors, in *Chemical Reaction Engineering Handbook* (J. J. Carberry and A. Varma, eds.), p. 735. Marcel Dekker, New York, 1986.
2. National Research Council, Committee on Chemical Engineering Frontiers: Research Needs and Opportunities. *Frontiers in Chemical Engineering. Research Needs and Opportunities.* National Academy Press, Washington, D.C., 1988.
3. Denn, M. M., *AIChE J.* **33**, 177 (1987).
4. Wegner, G., and Orthmann, E., *Angew. Chem. (Int. Ed.)* **25**, *1105 (1986).*
5. Patel, S., Hadziioannou, G., and Tirrell, M., *Proc. Natl. Acad. Sci. U.S.A.* **84**, 4725 (1987).
6. de Gennes, P. G., *Scaling Concepts in Polymer Physics.* Cornell University Press, Ithaca, N.Y., 1979.
7. Stevenson, J. C., and Cooper, S. L., *Macromolecules* **21**, 1309 (1988).
8. Hermann, D. S., Kinning, D. J., Thomas, E. L., and Fetters, L. J., *Macromolecules* **20**, 2940 (1987).
9. Sawamoto, M., Okamoto, C., and Higashimura, T., *Macromolecules* **20**, 2693 (1987).
10. Cannizzo, L. F., and Grubbs, R. H., *Macromolecules* **21**, 196 (1988).
11. Sogah, D. Y., and Webster, O. W., *Macromolecules* **19**, 1775 (1986).
12. Wagener, K. B., and Wanigatunga, S., *Macromolecules* **20**, 1717 (1987).
13. O'Driscoll, K. F., *Proceedings of 2nd Berlin Workshop on Polymer Reactor Engineering* (K.-H. Reichert and W. Geiseler, eds.), p. 229. Hanser, München, 1986.
14. Martin, P. and Stupp, S. I., *Macromolecules* **21**, 1222 (1988).
15. Laschewsky, A., and Ringsdorf, H., *Macromolecules* **21**, 1936 (1988).
16. *Thin Solid Films*, Vols. 132–135 (1985).
17. de Gennes, P. G., *Rev. Mod. Phys.* **57**, 827 (1985).
18. Rabolt, J. F., Hofer, D., Miller, R. D., and Fickes, G. N., *Macromolecules* **19**, 611 (1986).
19. Chien, J. C. W., Gong, B. M., Madsen, J. M., and Hallock, R. B., *Polym. Prepr. (ACS Div. Polym. Chem.)* **29**(1), 185 (1988).

20. Eisenberg, A., and King, M., *Ion Containing Polymers: Physical Properties and Structure*. Academic Press, New York, 1977.

21. Ye, C., Marks, T. J., Yang, J., and Wong, G. K., *Macromolecules* **20**, 2322 (1987).

22. This section is based largely on discussions with D. A. Tirrell, University of Massachusetts.

23. Regen, L., and DeGrado, W. F., *Science* **241**, 976 (1988).

24. Heppenheimer, T. A., *High Tech. Business* **8**(8), 30 (1988).

25. Ellis, M. F., Taylor, T. W., Gonzalez, V., and Jensen, K. F., *AIChE J.* **34**, 1341 (1988).

26. Rawlings, J. B., Prindle, J. C., and Ray, W. H., *Proceedings of 2nd Berlin Workshop on Polymer Reactor Engineering* (K.-H. Reichert and W. Geiseler, eds.), p. 1. Hüthig and Wepf Verlag, Heidelberg, 1986.

27. Ray, W. H., *Ber. Bunsenges. Phys. Chem. 90, 947 (1986).*

28. Ray, W. H., *Proceedings Internatonal Symposium on Transition Metal Catalyzed Polymerizations* (R. Quirk, ed.), p. 536. Cambridge University Press, 1988.

29. Tulig, T. J., and Tirrell, M., *Macromolecules* **14**, 1501 (1981).

30. Macosko, C. W., *RIM: Fundamentals of Reaction Injection Molding.* Hanser, New York, 1988.

31. Akkapeddi, M. K., and Gervasi, J., *Polym. Prepr. (ACS Div. Polym. Chem.)* **29**(1), 567 (1988).

32. Batch, G. L., and Macosko, C. W., "Computer Aided Analysis of Pultrusion Processing." AIChE Conference on Materials in Emerging Technologies, Minneapolis, August 1987.

33. Noolandi, J., Allara, D., Rubloff, G., Fowkes, F., and Tirrell, M., *Mater. Sci. Eng.* **83**, 213 (1986).

34. DeVries, W. C., *MRS Bull.* **8**(5), 25 (1988).

35. Evans, E., and Needham, D., *Macromolecules* **21**, 1822 (1988).

36. Borden, K. A., Eum, K. M., Langley, K. H., and Tirrell, D. A., *Macromolecules* **20**, 454 (1987).

37. Yannas, I., Presentation at symposium in honor of D. Pino, Zurich, June 1988.

38. Thompson, L. F., in *Perspectives in Chemical Engineering: Research and Education* (C. K. Colton, ed.), p. 373. Academic Press, San Diego, Calif., 1991 (*Adv. Chem. Eng.* **16**).

39. Chemla, D. S., and Zyss, J. eds., *Nonlinear Optical Properties of Organic Molecules and Crystals*. Academic Press, New York, 1987.

40. Israelachvili, J. N., *Intermolecular Forces*. Academic Press, Orlando, Fla., 1985.

41. Granick, S., Patel, S., and Tirrell, M., *J. Chem. Phys.* **85**, 5370 (1986).

42. Theodorou, D. N., and Suter, U. W., *Macromolecules* **19**, 139 (1986).

43. Bitsanis, I., Davis, H. T., and Tirrell, M., *Macromolecules* **21**, 2824 (1988).

44. Bates, F. S., and Wignall, G. D., *Phys. Rev. Lett.* **57**, 1429 (1986).

17

Chemical Engineers in Polymer Science: The Need for an Interdisciplinary Approach

Richard A. Register
Department of Chemical Engineering
Princeton University
Princeton, New Jersey

Stuart L. Cooper
Department of Chemical Engineering
University of Wisconsin–Madison
Madison, Wisconsin

I. Interdisciplinary Character of Polymer Science

Chemical engineers have worked with polymers since the first Bakelite articles were produced early in this century. Since then, the class of polymeric materials has grown to encompass a whole range of thermoset and thermoplastic resins, as well as copolymers and polymer blends, and chemical engineers have played major roles in the rise of these materials to commercial success. From production of the resins (which involves heat and mass transfer, kinetics, fluid dynamics, process design, and control) to the fabrication of final articles (involving many of the same processes, as well as some unit operations not part of "traditional" chemical engineering, such as extrusion

ADVANCES IN CHEMICAL ENGINEERING, VOL. 16

and injection molding), the broad background of chemical engineers has facilitated their involvement at all levels of polymer manufacture.

Today, the new polymers being introduced are generally small-volume specialty polymers, such as block copolymers and liquid crystalline polymers, produced by smaller-scale production techniques and novel fabrication methods. Often the outstanding problem with such materials is not how to produce huge volumes at commodity-polymer prices, but rather how to produce an article that will perform in a particularly demanding application. Fiber-reinforced polymer composites used for aircraft bodies are a well-known example: fabrication is extremely labor-intensive, and the emphasis is on producing a material with superior and enduring structural integrity [1]. To produce these special-purpose polymers, a good understanding of the structure–property–processing relationships is essential; that is, what type of chemical and/or microphase structure, combined with what type of processing, yields the desired physical properties? Elucidation of these relationships is one of the important areas where chemical engineers can participate, and one with often immediate commercial payoffs.

Considering the breadth that the field of polymer science and engineering encompasses, it is not surprising that investigators from a variety of disciplines have contributed to recent advances. For example, theories of polymer chain dynamics and entanglement effects in bulk resins, now an important area of research for many chemical engineers, were put forth initially by physicists. The synthesis of polymers with novel properties is almost exclusively the province of chemists. Most important, the most modern analytical tools used to study polymers have generally been developed by, and are still used primarily by, physicists and chemists. Therefore, it is essential that chemical engineers work more closely with colleagues in these disciplines, so that the improved understanding of polymers brought about by developments in all fields can be translated into useful products. Illustrative examples of important recent advances in each of these three areas will be discussed below; not coincidentally, chemical engineers have been involved to some degree with all of these.

II. Theory: Dynamics of Polymers in Bulk

The motion of polymers in concentrated solution and bulk is of major theoretical and practical concern. For example, the strong dependence of zero-shear viscosity on molecular weight (approximately the 3.4 power) and the marked decrease of viscosity η with shear rate γ not only bespeak some of the unusual properties of long-chain molecules but also are of essential importance in virtually every processing operation. Yet the reasons for these unusual behaviors have become clear only recently. The "reptation" concept of de Gennes [2], where polymer motion is primarily confined to slithering along the chain contour in an entangled melt, has been developed into complete kinetic theories by Doi and Edwards [3] and Curtiss and Bird [4]. While Tirrell comments [5] that the use of statistical mechanics and molecular transport theory (from which the Curtiss–Bird theory is derived) is both the surest and slowest route to correctly modeling the dynamics of polymeric liquids, the Curtiss–Bird theory has achieved a state of development at least equal to that of Doi and Edwards.

While chemical engineers are well-grounded in the mechanics of Newtonian fluids, it is the *non*-Newtonian character of polymers that controls their processing. Three striking examples [6] of the differences between Newtonian and typical polymeric liquids (either melts or concentrated solutions) are shown schematically in Fig. 1. The upper portion of the figure refers to the Weissenberg effect [7], or rod-climbing, exhibited by polymers; excellent photos may be found in Bird *et al.* [4] as well. When a rod is rotated in a Newtonian fluid, a vortex develops near the rod due to centripetal acceleration of the fluid. When the same experiment is repeated with a polymeric fluid, however, the fluid climbs the rod. In the center

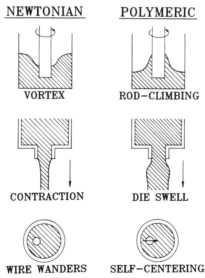

Figure 1. Schematic cross-sectional representations of the flow of Newtonian (left, N) and polymeric (right, P) fluids in select geometries. The fluid is shaded in all cases. Top: N exhibits a vortex near the rod, while P climbs the rod. Center: N contracts upon exiting the die, while P swells. Bottom: N coating does not provide restoring force to center wire being pulled through it, while P does.

of Fig. 1, the phenomenon of die swell is noted. Whereas the diameter of a jet of Newtonian fluid diminishes upon exiting a nozzle, due to redistribution of the velocity profile, a polymeric extrudate swells upon passing through the die. The final example refers to the industrially important process of wire coating. If a Newtonian coating is used, the alignment of the wire must be strictly controlled to prevent the wire's centerline from wandering relative to the tube's centerline. For polymeric fluids, on the other hand, a restoring force exists that serves to center the wire and provide a uniform coating. All three of these effects are manifestations of the elasticity of polymers, a property not shared by Newtonian fluids. Although the Doi–Edwards theory does not predict the Weissenberg effect, the Curtiss–Bird theory can. The latter can also accurately represent the decrease of η with γ, as well as several less well known but often equally important properties of polymeric fluids [4]. While theories of chain dynamics in bulk are still in a state of development, there may come a time when it will be possible to predict accurately the flow patterns and normal stresses in a complicated processing operation such as injection molding. This would minimize the need for trial-and-error determination of mold design and mold cycle needed to produce a part free of fracture-prone regions.

III. New Material Development

A. Polymer Blends

The multifaceted advances in synthetic methods in such exciting areas as liquid crystalline polymers and polysilanes have been discussed by Tirrell [5]. However, it is often possible to produce a new material with a minimum of synthetic effort by blending together two or more existing resins [8]. In general, polymers are immiscible, due to the low entropy of mixing long-chain molecules and the unfavorable enthalpy of mixing two unlike substances. However, when a specific interaction is present, such as hydrogen bonding, it may be possible to mix two homopolymers. In some cases, it may be desirable to blend in a cheaper resin to produce a lower-cost blend with satisfactory properties. In other cases, a favorable synergistic interaction is obtained so that the blend has properties that surpass those of either of the component resins. Perhaps the best-known example of this is General Electric's Noryl®, a blend of poly(2,6-dimethyl 1,4-phenylene oxide) with polystyrene [9]. The resulting blend, which is used in business machine housings and other applications that need good heat resistance while maintaining an attractive appearance, represents the quintessential value-added end product that many polymer manufacturers are striving for. While the number of homopolymer pairs that are miscible is rather limited, miscibility can be achieved when one or both of the polymers in the pair is a copolymer of two highly immiscible monomer units, even though the corresponding homopolymers may be immiscible [10–12]. Thus, the potential exists for the development of a whole new class of polymer materials without the need for the development of a new route of chemical synthesis. Often, "miscible" polymer blends are miscible only over a restricted temperature range. When the temperature is dropped or jumped into the two-phase region, these materials separate by a mechanism resembling spinodal decomposition [13, 14]. The resulting structure, if quenched into the glassy state, retains a high degree of interconnectivity and regularity, with a size scale that can be controlled through the processing history.

B. Biomaterials

A biomaterial is used in contact with living tissue, resulting in an interface which undergoes protein adsorption, cell adhesion, and often extremely specific and complex biological reactions. Widely used polymers include plasticized polyvinyl chloride (for disposable devices such as blood bags), silicone rubber, polyethylene, Teflon®, polyesters, and nylons. Some other materials attracting current interest are hydrogels (highly hydrophilic, cross-linked polymers) and polyurethanes. Polyurethanes are particularly promising as biomaterials [15]; not only are they relatively biocompatible as a class, but the versatility of polyurethane chemistry allows a broad range of chemical structures to be explored. While some useful biomaterials are systemically or pharmacologically inert, some degree of interaction between the material and its environment will often be required for a material to be effective.

An example of how a change in the chemical structure of a polyurethane can change its biocompatibility is shown in the scanning electron micrographs in Fig. 2 [16]. The polymer on the top is a typical polyurethane, composed of poly(tetramethylene oxide), methylene bis(p-phenylisocyanate), and butanediol. The polymer on the bottom is the same material, but with 20% of the urethane linkages derivatized with propyl sulfonate groups. Both materials have been exposed to canine blood for 60 minutes in an *ex vivo* shunt. The white spots on each polymer are blood platelets; the differences between the two polymers, in both number of adherent platelets and the degree of spreading of the platelets, is enormous. Several major challenges must still be met in the biomaterials area, however. In particular, recent work [17] indicates that improved biostability of polyurethanes is essential if they are to serve in long-term implants, such as the Jarvik-7 artificial heart noted by Tirrell [5]. A substantial fraction of the research on polyurethane biomaterials in the United States is being conducted by the academic chemical engineering community.

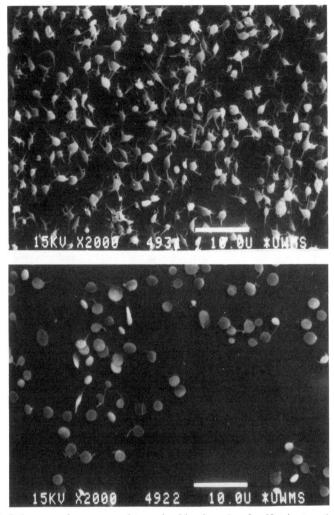

Figure 2. Polymer surfaces exposed to canine blood *ex vivo* for 60 minutes. Adherent
objects are blood platelets. Top: underivatized polyurethane; bottom:
sulfonated polyurethane. Reprinted with permission of John Wiley & Sons
from Grasel, T. G., and Cooper, S. L., "Properties and Biological Interactions
of Polyurethane Anionomers: Effect of Sulfonate Incorporation," *J. Biomed.
Mater. Res.* **23**, 311 (1989) [16]. Copyright 1989, John Wiley & Sons.

IV. Analytical Techniques

A. Synchrotron SAXS

Nowhere is the interdisciplinary nature of modern polymer science as evident as in its analytical techniques. To develop structure–property–processing relationships, knowledge of the structure of the material under investigation is essential. Numerous techniques have evolved recently that have greatly affected the study of polymers. Although Tirrell [5] has listed some of these, we feel a few illustrative examples can effectively demonstrate the diversity of current polymer research. One powerful tool is high-intensity, tunable (in energy), naturally collimated x-radiation from synchrotron sources [18]. Small-angle x-ray scattering (SAXS) has been used for decades to study the structure of materials with inhomogeneities in the range 1–100 nm, such as semicrystalline polymers (polyethylene, polypropylene, nylons, Teflon®), block and segmented copolymers (Kraton®, polyurethanes), and immiscible polymer blends. With the high brightness and increasing availability of synchrotron radiation, it is possible to perform real-time studies on the dynamics of crystallization or microphase separation. Figure 3 shows a series of SAXS patterns for a low-density polyethylene as the temperature is lowered [19]. The development of a SAXS peak reflects the formation of lamellar polyethylene crystallites. Other information obtainable from the SAXS patterns includes the interlamellar spacing and degree of crystallinity, both important factors in determining the mechanical properties of the resulting product. Since polyethylene is the highest-volume commodity polymer, with a 1987 production of over 17.6 million lb [20], and since cooling from the melt (after extrusion, molding, or sheet blowing) is always part of its processing history, it is essential to know how the crystallinity develops.

Synchrotron SAXS has also been used to find the order–disorder transition in block copolymers. While block copolymers derive their desirable physical properties from their two-phase nature, the same two-phase nature makes them very difficult to process. Tirrell [5] has already discussed the reactive processing option, and although quite useful in some cases, this approach is applicable only to certain types of polymers in certain types of processes. Another route is to heat the polymer into a region of its phase diagram where it is homogeneous (if such a region exists), process the article into the desired form, and then cool it to induce the development of the two-phase structure. The existence of such a microphase separation transition has been demonstrated for a polyurethane using synchrotron SAXS during heating [21]; at low temperatures the material is microphase-separated, while

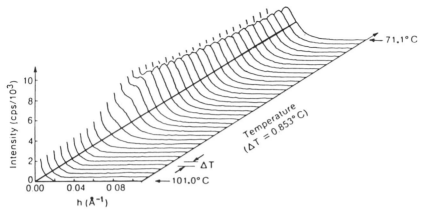

Figure 3. Series of SAXS patterns of a low-density polyethylene taken during cooling at
10°C/min. The peak develops near 93°C, indicating the onset of crystallization. Reprinted with permission of John Wiley & Sons from Russell, T. P., and
Koberstein, J. T., "Simultaneous Differential Scanning Calorimetry and Small
Angle X-Ray Scattering," *J. Polym. Sci. Polym. Phys. Ed.* **23**, 1109 (1985)
[19]. Copyright 1985, John Wiley & Sons.

above 170°C (in that example) the melt becomes homogeneous. Because
of the versatility of polyurethane elastomers, which find applications ranging
from tank tread pads to biomaterials, the possibility of improved
processability is exciting.

B. Deuterium NMR

For measuring structure or dynamics on a smaller scale than SAXS, deu-
terium nuclear magnetic resonance (^{2}H NMR) has proved valuable. Selective
deuteration of part of a polymer chain allows the observation of anisotropy
(reflecting orientation) or motion of that specific portion of the chain. For
example, Stadler [22] used a cross-linked network swollen with deuterated
oligomers to show that even when the deuterated chains are allowed to relax
completely, they still retain some orientation in the direction of strain. This
has profound implications for the dynamics of polymer chains in bulk, since
it suggests a cooperativity of chain motions. Gronski and Stoppelmann [23]
used ^{2}H NMR to probe the dynamics of polymer segments in the interfacial regions of block copolymers. This research suggests that interfacial
regions play an essential role in the mechanical behavior of block copolymers and that adhesion and wetting in composite materials may be illumi-
nated by this technique.

C. SANS

Another technique which has gained wide acceptance in polymer science is small-angle neutron scattering (SANS), which can probe the dimensions of single chains or parts of chains, as well as the microstructure in heterogeneous polymers [24]. A study of polystyrene coextruded in polyethylene billets [25] showed that the chains oriented affinely with the macroscopic strain created by the extrusion. However, for polystyrene of lower molecular weight, relaxation was fast enough for the material to lose much of its orientation before being quenched into the glassy state [26]. Drawing is an essential component of the processing of many polymers, from traditional textile fibers to the new ultra-high-modulus polyethylene fibers which are beginning to compete with graphite. Although older techniques such as birefringence have allowed some measurement of the ordering of chain segments, only SANS can probe the dimensions of the entire polymer chain, at any angle to the direction of deformation. SANS can also be used to determine chain dimensions in polymer solutions or melts undergoing deformation, which again relates to the question of polymer chain dynamics. SANS has also been used to elucidate the network structure of epoxies [27]. Because thermoset polymers are intractable and insoluble, few techniques are suitable for probing their structure on any scale. A much different area of polymer science in which SANS has made an important contribution is in the low-resolution mapping of chromosomes [28]; the area of biological macromolecules is beyond what we can cover here, however.

V. Outlook

A. Applications

Although some may consider the plastics and rubber industries mature, in reality the prospects for development of new polymers and products are bright. In the near future, growth is most likely to occur in the small-volume, high-margin area of specialty polymers—materials designed for specific, demanding applications. However, not all the demand will be in the emerging high-technology areas noted by Tirrell [5]. For example, in the structural arena, polymers are making increasing penetration into under-the-hood automotive applications [29]. Liquid crystalline polymers, with their high modulus, good chemical resistance, and ability to be used at high service temperatures, have outstanding prospects for use as structural materials [30]. The use of Kevlar® in bulletproof vests and as the reinforcing fiber in high-performance composites demonstrates the promise that liquid crystalline polymers have as structural materials; however, their extended-chain structure produces unusual rheological behavior as well [31], so that improved processing methods appropriate for these materials need to be developed.

Another promising area for polymer development, as alluded to by Tirrell [5], is microelectronics. Plasma polymerization can be used to produce a polymeric coating directly on a substrate; changing the composition of the gas feed allows a wide variation in the chemical composition of the surface produced [32]. The same technique can also be used to modify surfaces for other applications, such as to improve the blood compatibility of biomaterials. The essential processes occurring in a plasma—mass transfer and reaction kinetics—have long been the domain of chemical engineers.

In addition to the processing of polymers, their raw materials and methods of ultimate disposal will inevitably change in the future, providing more opportunities for chemical engineers. Nearly all synthetic polymers are derived from petroleum; as petroleum reserves diminish, new feedstocks must be sought, such as coal-derived or renewable biomass sources. A more immediate problem involves the disposal of today's polymer products. Though many plastics could be recycled, because of the low value of the resin compared to the reclamation cost, few are. The salvage of poly(ethylene terephthalate) used in 2-liter soda bottles in the United Kingdom is currently one of very few successful examples of recycling. To demonstrate the immediacy of this problem, a ban on all nonrecycled polystyrene began in the State of Massachusetts in June 1989 [33]. In order to minimize the disposal problem, another possibility is to produce resins that are biodegradable, such as by blending with corn starch [34].

B. Research

Many new areas of polymer research are expanding rapidly. Just to mention one related to the foregoing text, rheo-optical methods [26] such as birefringence, infrared dichroism, and particularly light, x-ray, and neutron scattering will be increasingly used to test theories of polymer chain dynamics. Another feature of research likely to become more prominent in the future is the increased use of central user facilities for experiments such as x-ray and neutron scattering and even NMR. The National Center for Small-Angle Scattering Research at Oak Ridge National Laboratory and the Stanford Synchrotron Radiation Laboratory are two well-known and heavily used facilities, and new synchrotrons and neutron sources are continuing to come on line in the United States and around the world. Although such facilities offer enormous potential for polymer research, it will be essential for chemical engineers to work closely with chemists and physicists in conducting their experiments.

C. Education

As all the essays in this book show, the term "chemical engineering" covers an enormous scope. Polymer science, like many of the emerging subfields in chemical engineering, is not routinely taught at the undergraduate level to chemical engineering students. Moreover, those working in research are finding it increasingly desirable to collaborate with physicists and chemists, while those in an industrial environment find themselves interacting with mechanical engineers and materials scientists. For chemical engineers to continue to play a major role in polymer science, increased interaction with personnel from other disciplines will be necessary, and successful interaction demands that the two parties have enough common ground to communicate. Therefore, those desiring to work in polymers would be well advised to take electives in other disciplines, as well as in interdisciplinary areas such as materials science, where available. Polymer science courses appropriate for both graduate students and undergraduates should be offered in chemical engineering departments as well. In this respect, the complements of polymer courses at MIT, Minnesota, Princeton, and Wisconsin, both in chemical engineering and in other departments, serve as admirable models.

In conclusion, chemical engineers have made essential contributions to polymer science and will certainly continue to do so in the future. As polymer science broadens in scope and permeates other fields as well, chemical engineers must recognize its interdisciplinary nature and be prepared to participate in this important technology in ever-evolving ways.

Acknowledgment

R.A.R. wishes to acknowledge the support of the Fannie and John Hertz Foundation while this discussion paper was written.

References

1. Margolis, J. M., *Chem. Eng. Prog.* **83**(12), 30 (1987).
2. de Gennes, P. G., *Scaling Concepts in Polymer Physics.* Cornell University Press, Ithaca, N. Y., 1979.
3. Doi, M., and Edwards, S. F., *Theory of Polymer Dynamics.* Oxford University Press, Oxford, 1986.
4. Bird, R. B., Curtiss, C. F., Armstrong, R. C., and Hassager, O., *Dynamics of Polymeric Liquids,* Vol. 2: *Kinetic Theory,* 2nd Ed. Wiley, New York, 1987.
5. Tirrell, M., in *Perspectives in Chemical Engineering: Research and Education* (C. K. Colton, ed.), p. 321. Academic Press, San Diego, Calif., 1991 (*Adv. Chem. Eng.* **16**).

6. Bird, R. B., and Curtiss, C. F., *Physics Today* **37**(1), 36 (1984).
7. Weissenberg, K., *Nature* **159**, 310 (1947).
8. Manson, J. A., and Sperling, L. H., *Polymer Blends and Composites*. Plenum, New York, 1976.
9. Cizek, E. P., U.S. Patent 3,383,435 (1968).
10. Kambour, R. P., Bendler, J. T., and Bopp, R. C., *Macromolecules* **16**, 753 (1983).
11. Paul, D. R., and Barlow, J. W., *Polymer* **25**, 487 (1984).
12. ten Brinke, G., Karasz, F. E., and MacKnight, W. J., *Macromolecules* **16**, 1827 (1983).
13. Han, C. C., Okada, M., Muroga, Y., McCrackin, F. L., Bauer, B. J., and Tran-Cong, Q., *Polym. Eng. Sci.* **26**, 3 (1986).
14. Inaba, N., Yamada, T., Suzuki, S., and Hashimoto, T., *Macromolecules* **21**, 407 (1988).
15. Lelah, M. D., and Cooper, S. L., *Polyurethanes in Medicine*. CRC Press, Boca Raton, Fla., 1986.
16. Grasel, T. G., and Cooper, S. L., *J. Biomed. Mater. Res.* **23**, 311 (1989).
17. Planck, H., Syré, I., Dauners, M., and Egbers, G., eds., *Polyurethanes in Biomedical Engineering II*. Elsevier, Amsterdam, 1987.
18. Elsner, G., *Adv. Polym. Sci.* **67**, 1 (1985).
19. Russell, T. P., and Koberstein, J. T., *J. Polym. Sci. Polym. Phys. Ed.* **23**, 1109 (1985).
20. *Chem. Eng. News* **66**(25), 42 (1988).
21. Koberstein, J. T., and Russell, T. P., *Macromolecules* **19**, 714 (1986).
22. Stadler, R., *Macromolecules* **19**, 2884 (1986).
23. Gronski, W., and Stoppelmann, G., *Polym. Prepr. (ACS Div. Polym. Chem.)* **29**(1), 46 (1988).
24. Stein, R. S., and Han, C. C., *Physics Today* **38**(1), 74 (1985).
25. Hadziioannou, G., Wang, L., Stein, R. S., and Porter, R. S., *Macromolecules* **15**, 880 (1982).
26. Stein, R. S., *Polym. J.* **17**, 289 (1985).
27. Wu, W.-L., and Bauer, B. J., *Polymer* **27**, 169 (1986).
28. Moore, P. B., *Physics Today* **38**(1), 63 (1985).
29. Wigotsky, V., *Plastics Eng.* **44**(7), 34 (1988).
30. Stupp, S. I., *Chem. Eng. Prog.* **83**(12), 17 (1987).
31. Fuller, G. G., Moldenaers, P., and Mewis, J., *Bull. Am. Phys. Soc.* **33**, 639 (1988).
32. Yasuda, H., *Plasma Polymerization*. Academic Press, Orlando, Fla., 1985.
33. *Plastics Eng.* **44**, 15 (1988).
34. Otey, F. H., Westhoff, R. P., and Doane, W. M., *Ind. Eng. Chem. Res.* **26**, 1659 (1987).

General Discussion

Polymers

Robert Armstrong: I want to emphasize the importance of rheology and polymer processing to chemical engineers and its implications for intellectual areas we should focus on in the future. Let me sketch a figure that provides a context for this discussion (Fig. 1). To understand polymer processing, we draw from our background in fluid mechanics, transport phenomena, and polymer chemistry, and also from polymer rheology. Polymer rheology requires input from various sources: rheometry, fundamental experiments to characterize the flow behavior of polymeric liquids; kinetic theory, to get from first principles the behavior of these non-Newtonian fluids; and finally continuum mechanics. I show this for two reasons. One is that the need to analyze polymer processing systematically is more important now than in the past. That's because a lot of polymer processing design, say, mold design, has been done by machinists who are artisans in that trade. They're experienced in dealing with certain polymers; they become adept

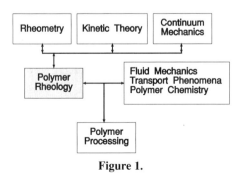

Figure 1.

at tailoring molds and adaptively designing molds. They're a dying breed now, and we have to replace them by providing engineers with a more systematic approach. That's one driving force. The second has been mentioned already, that is, the trend toward specialty chemicals in the future. As we move toward specialty chemicals, we need to shorten the time from design of the new chemical to the time when we can process it effectively. That requires much better systematic understanding of this area.

The two-way arrow between polymer rheology and fluid mechanics has not always been appreciated. Traditionally we look at polymer rheology as input to fluid mechanics, as a way to supply constitutive equations. Gary Leal pointed out the use of fluid mechanics to provide feedback to tell us whether the constitutive equation is satisfactory. In the past, we tested constitutive models by examining polymeric liquids with very simple kinematics, homogeneous flows as a rule, either simple shear or simple shear-free types of flows. We don't actually use polymers in such simple flows, and it's essential to understand whether or not these constitutive equations actually interpolate properly between those simple types of kinematics. So there's a two-way arrow that we have to pay more attention to in the future.

I would also like to list some of the challenges that will provide the foundation for where the profession has to go (Fig. 2). This is not meant to be comprehensive, but to suggest some of what we should be doing. This wish list derives from work Bob Brown and I have done on modeling flows of polymer fluids. The first item has to do with the need to understand the effects of polymer structure and rheology on flow transitions in polymeric liquids and on polymer processing operations. In the past, we've studied extensively the behavior of Newtonian fluids and how Newtonian flows evolve as, say, the Reynolds number is varied. We have tools available to

Outstanding Challenges

- Effects of polymer structure on rheology
  ➡ flow transitions ➡ processing

- Boundary conditions and behavior
  of polymers near boundaries

- Kinetic theory
  - Nonhomogeneous flows
  - Wall effects
  :

Figure 2.

teach that information to students. We need similar information for polymeric liquids. Unfortunately, we don't even know how to categorize these transitions yet. Next, the Deborah number has been a useful idea, but it's not sufficient to characterize transitions. More needs to be understood about boundary conditions, the behavior of polymers near boundaries, and the subsequent effects on the fluid mechanics and on polymer processing. Lastly, kinetic theory, which Stu Cooper brought up, is a critical area. We need to exploit the fact that we can now do a lot more systematic work in getting from polymer structure to rheology and then on to processing. We need to look at how kinetic theory is altered by nonhomogeneous flows. Matt Tirrell's made some progress on that. A lot of the flows we put polymers in are nonhomogeneous, even on a molecular scale, particularly when polymers become extended. We need to understand wall effects from a molecular viewpoint. There are a lot more items that could be put on this list that we need to identify as a profession.

Robert Cohen: There are three things I want to bring up. First, the intellectual underpinnings of polymers are broadly based. Matt correctly pointed out that it has a close connection with science-based material and clientele. In addition, the traditional tools that polymer engineers use are the ones that they learn from chemical engineers: thermodynamics, kinetics, and transport phenomena. Second, those in polymers within chemical engineering look outside the profession for their professional appreciation. The audience is different for polymers than for most other parts of chemical engineering. The differences in intellectual focus and the clientele should be kept in mind during our discussion. Lastly, how do we incorporate all that we've talked about into our chemical engineering education? Polymer specialists need the traditional chemical engineering education and they need the science, and revolutionizing the curriculum, as Matt said, is really not what's needed. How to bring the polymer culture and the chemical engineering culture together educationally is the challenge. I hope that comes out in the discussion.

Andreas Acrivos: The central problem here is a connection between the chemistry and the morphology of the polymer and its rheological–mechanical properties. My question is, how far along are we in trying to correlate these two properties? Given the morphology of a polymer, can we tell anything quantitative about its structure, mechanical properties, and polymer rheology?

Matthew Tirrell: For homogeneous materials, we know a lot about how molecular weight, branching, and that sort of thing affect the rheological properties and the flow properties. Once the materials begin to organize and phase-separate into some larger microstructure, then we don't know very much.

Andreas Acrivos: Is this an active area of research?

Matthew Tirrell: It's not very advanced, but it's active.

Robert Cohen: Structure–property relations are what most polymer specialists in chemical engineering study. That's what they're striving for. Atomistic, bond-to-bond calculations are being done now with large computers. People are looking at every kind of interaction, along chains, between chains. Putting that all together could mean predicting properties like solubility parameters. Ulie Suter and Doros Theodorou were doing that kind of work with very nice results. That's the new wave of research going on in chemical engineering.

Morton Denn: From the point of view of both research and education in chemical engineering, the polymer area provides a unique sociological framework. It is a historical accident that, regardless of background, polymer specialists tend to be in chemical engineering departments. This situation provides an opportunity to see how the problem-driven nature of the engineering outlook has brought people together with different backgrounds to deal with real engineering problems. There are a couple of examples in the room that I could cite. One is the Armstrong–Brown collaboration at MIT. It's not the only collaboration at MIT, but it shows nicely how people who came in with very different backgrounds were driven by a problem in polymer engineering to collaborate both experimentally and computationally, bringing a variety of tools together. That presumably can be integrated into curriculum matters as well. At Berkeley, one collaboration that I'm sure will surprise many people is an Alex Bell–Mort Denn collaboration, which again is driven by the need to understand polymer processing problems with a variety of tools. It's not as easy to see how that can be done in the broader context of other fields, where the talents sometimes lie inside the department and sometimes outside. Crossing departmental lines is difficult, but I think the model is there in polymers.

Robert Cohen: You say that the problem-driven nature necessarily leads to interdisciplinary or collaborative activities.

Morton Denn: Yes, and to me this is the natural course of engineering education and engineering research.

Robert Cohen: I think that's largely correct. Breaking down the departmental barriers for polymers would lead to greater collaboration, and that's good.

Morton Denn: Even within our narrower context, just what's going on in chemical engineering is a very good model.

Anthony Pearson: I've two things to say. First, from the half outside, it seems from what's been said that the expectations for engineered polymers are so high that it can be regarded as the star of this meeting. Ken Smith asked for some excitement to be generated for the future. The possibilities inherent in what Matt Tirrell spoke about overwhelm anything else I've heard from any other field. That's my first comment. The second is that, when I started my contribution, I put up words in boldface like complexity and nonlinearity. I want to come back in a minute to simplicity, because I think we're in danger of forgetting that our function, particularly when we refer to the intellectual foundations of our discipline, is to see the simple facts.

Before I go back to simplicity, I will take up the issue of what I call instability of the symmetric solutions, whether there is steadiness in time or simple symmetry in space. There was a discussion between Jimmy Wei and Dan Luss, where Jimmy said he was fed up with pathologies. If I understand pathologies as meaning the multiplicity of solutions to nonlinear problems, then I think, sir, you have it wrong. When I asked about the difference between equilibrium thermodynamics and nonequilibrium thermodynamics, I think I saw it, partly because of my mathematical background, in the fact that when you have a complex system, which is the case with polymerization reactions, you find that the initial conditions of the nonreacting system are extremely important in deciding what the frozen state will be. It is not an overwhelming equilibrium state. It is some particular frozen state that is critically dependent on the initial spatial distribution of the components and on the processing to which they have been subjected. I'm not saying anything that hasn't been said before, but I think one should not go away from here without realizing that that is something important and crucial to what we should be looking at in the future. The same might be true in other forms of reaction engineering.

Let me come back to the issue I thought we were really here to discuss, which is what are the intellectual foundations, and how do we arrange for people going into chemical engineering to be prepared in the most economical and satisfying way to do this? I won't take issue with those people who have divided our discipline into traditional areas and new areas we want to inject into it. It always looks to me as though the injection means another series of almost incomprehensible titles for the schoolchild to look at when he considers what subjects he will read at the university. I think we've got it wrong. What we should be insisting on is that what was taught by physicists when I was a schoolboy, called properties of matter—solids, liquids, and gases— is an inadequate breakdown of nature. The first exception clearly identified—it struck me when I was a student at the university—was that the polymeric state is a very special state. In those days, people used to call it the fourth state of matter. I've changed my mind, because it's not the last state of matter, it's like these new particles that you keep discovering in fundamental physics. Many of us here are concerned with structure—structured fluids, porous materials; there is something about multiphase continua which represents intellectually a different state of matter. We must know already that we are not going to be able to understand it in terms of the fine interfacial detail of what constitutes a clay, for example, or what constitutes a catalyst particle during its operation of catalyzing. We don't teach it as chemical engineers. We should persuade those who teach the students earlier, under grander and simpler titles like physics and chemistry and biology, that they must get into the minds of the young, much earlier, the idea that it is worth thinking about matter in many different states.

Eli Ruckenstein: How useful is thermodynamics in the treatment of the kind of problem described by Anthony Pearson? Is the solid polymeric material in an equilibrium state or in a metastable state? My personal experience with polymeric surfaces is that their characteristics depend upon the environmental fluid. In a hydrophilic environment, the polymers tend to expose their hydrophilic moieties to the environment, while in a hydrophobic one they will expose their hydrophobic moieties. A polymeric material equilibrated first in a hydrophobic medium and introduced later in a hydrophilic medium will undergo changes in its surface. These rearrangements are in some case rapid, but in others, very slow.

Robert Cohen: In polymers, we are used to considering time scales and understanding that equilibrium may be a long way off, but I think it's crucially important to understand what the equilibrium state can, in principle, be. Thermodynamics is always important. For polymers, there's a tension between thermodynamics and kinetics, and depending on temperature, one or the other may control.

William Schowalter: I have a question related to polymer synthesis. First, an observation. In biochemistry departments, there's been tremendous interest in the structure of large natural molecules. In chemistry departments, people have run away about as fast as they can from the synthesis of large molecules, the kinds of polymers of commerce that most of us are familiar with. I would like to address the question first to Matt. He showed us results of the physical chemistry that came out of those syntheses. The work of Ned Thomas was made possible because some clever synthetic chemists made some special kinds of molecules. I am trying to figure out how that activity can be fit in. If this research is so important, how do we make a home available for the people who do the synthesis that is necessary for it to go on? I don't see that happening in our educational structure, either in chemistry departments or in chemical engineering departments.

Matthew Tirrell: I started the criticism of chemistry for letting us down in doing the syntheses of these kinds of materials. They are responding now. What the synthetic chemists want is an interesting target to be accomplished with interesting chemistry. The rationale for a long time has been that esterification and amidation and free radical chemistry are well understood, so why should a first-rate synthetic chemist deal with them? What they instead went after were complicated natural products, which are challenging molecules to construct. I think there's now a reorientation toward the material properties as a target and how to construct them. An example I alluded to is genetically engineered polymers. Assuming you can do it, what molecule do you make first? How do you decide what material properties to go after? I believe that there are chemists around the country that are recognizing this now, and there will be a better home for this in chemistry. To the extent that we also have to worry about the production that involves processing, chemical engineering would also be a reasonable home for that kind of work.

Robert Cohen: I have a strong disagreement with Bill Schowalter about this point. As Eli said yesterday, not all people in chemical engineering have to study all the material available. Many places, certainly our department, offer synthesis as an available course, having been a subject for 20 years. Ed Merrill has written a book on the topic. You have colleagues in chemical engineering departments teaching polymer synthesis. The population in those subjects is often not only chemical engineering students. Departmental boundaries become artificial when you deal with polymers. Chemical engineers are teaching polymer synthesis in chemical engineering departments, and students from different departments are taking that subject. It is already available in our field.

William Schowalter: You're talking about a very special case, Bob. I don't think you can claim that what you just said is the norm in chemical engineering departments in this country, or in chemistry departments. I just don't think that's what we would find if we did a survey across the country.

Stuart Cooper: It would be a mistake, just as we avoid trying to teach too much math to our students, to try to teach them too much polymer chemistry. It's healthy and good to see a rekindling of interest in polymer chemistry from chemistry departments and polymer science departments at various universities around the country. Some of the most beautiful collaborations are indeed those where you see someone—maybe in an industrial laboratory or a university laboratory—synthesizing tailor-made materials, whether it be a starblock polymer or deuterated polymers, for collaboration with someone in the solid-state area who may be in a chemical engineering department interested in structural questions. In terms of intellectual foundations, I don't think chemical engineering should strive to hold on to polymer synthesis. In fact, it probably should encourage the nurturing of that discipline in more traditional departments like chemistry.

George Stephanopoulos: From what has been said in the last few minutes, the main task is product design and synthesis. The product design, unlike process design, is the hardest subject nowadays in the area of materials. What I don't hear is how do people formalize the synthesis and the design? Is it just the art that we have always come to accept as the synthetic task? Or is it more scientific, more organized, more systematized?

Robert Cohen: There's a difference between the design of the molecule and the synthesis of it in terms of organic chemistry. Bill was talking about polymer chemistry synthesis. You're talking about design of a molecule, which would mean the structure–property relationships.

George Stephanopoulos: What do people do to identify the structure of the molecule they want to produce?

Matthew Tirrell: So far, there has been no systematic approach to the development of targets for synthetic polymer materials. Structures have been chosen that, based on experience with related materials, might give so-called interesting properties. Interesting in this context has often meant technologically so, which is why some pure chemists lost interest.

Larry Thompson: I thought Bill Schowalter had a good point. There are two aspects of polymer synthesis, at least as we practice it. First is the assembly of polymers from known monomers. Chemical engineers do that in more than one department. I think Bill has a more far-reaching comment. When you don't have known monomers to assemble, you then need a much higher degree of polymer or organic synthetic expertise. I have found the way to engage synthetic polymer chemists in this is to make it intellectually challenging for them. The reason we don't have polymer synthesis and monomer synthesis carried on in chemistry departments is partially our fault, because we haven't presented the problem in an exciting way to the organic chemists.

Morton Denn: A number of these comments totally miss the essence of Bill's concern. It's not that difficult to get micrograms synthesized of very interesting new polymers, but that quantity only allows you to run one or two quick characterizations. Engineering research usually requires much larger amounts of material. There is no way that the academic system will be able to adjust to having people in any department as faculty members who provide a service role for producing sufficient amounts of interesting materials in order to do the research.

L. E. Scriven: There's vast excitement over the molecular microstructural material–property–performance relationship, which Anthony Pearson pointed to in the case of polymers. It's not restricted to covalently bonded polymers alone. I wanted to emphasize something that Matt Tirrell and Stu Cooper brought up; Anthony's comment galvanized me to raise my hand. In the earth, in rock and clay and other mineral matter, and in life, in shell and bone and the like, we see inorganic polymers that do not depend upon the carbon–carbon backbone, nor do they invariably rely on covalent bonds. That sets the backdrop for scenes just now emerging. On the way to crystalline materials, but also for performance in their own right, variously as microstructural to amorphous materials, are inorganic polymers: heat resistant, highly robust material, many with exceptional properties, such as superconductivity of late. These are materials that come to us from the domain that has been known as ceramic science and engineering. Now they are taking leading roles in inorganic polymer chemistry, inorganic polymer science, and inorganic polymer technology. One can see coming hybrid inorganic–organic polymer technology, for example, block copolymers of inorganics and organics. One sees already in this arena, from the ceramicists and the industrial innovators who are pushing way ahead in conceptualization, materials that are covalently bonded mixtures of colloidal grit and polymer molecules. That is, they contain building blocks that are already clumps that bond right into the chemically bonded backbones. There are solution polymerization processes, there are suspension gelation processes, and more, all full of opportunities.

There's a major challenge here. That's to get this burgeoning area, which I believe has exciting future potential for chemical engineers, into chemical engineering, and not to leave it to ceramic engineering as we left part of early chemical engineering to the seceding metallurgical engineering of that era.

Robert Cohen: Do you think we need to teach different things, a different kind of polymer science for a different kind of chemical engineer, to our students?

L. E. Scriven: There are many implications of what is already happening and what is likely to happen, implications for our research and for our teaching. I have not thought them through in detail. But if there's anything that we're supposed to be doing here, it's to identify these areas and start homing in on them. I think this is an area we're in danger of missing out on.

Robert Cohen: Is it because we're not equipped, we're not doing the right things?

L. E. Scriven: I think chemical engineering, for reasons that have become evident here, is eminently qualified, eminently prepared to move on this area. At the AIChE Annual Meeting next month, there will be sessions on inorganic polymers organized by our colleague Martha McCartney, who's coming to the area from the ceramics side. What is important, I judge, is that this field be embraced from the polymer side, that those of us who profess polymer science and engineering begin broadening the scope of our teaching and research if we've not started already.

Alexis Bell: Skip has provoked a comment. We mustn't divorce chemistry from the processing of materials, and there are at least two ways to approach the problem. One is to do collaborative research with your colleagues in chemistry. That can be very successful and draw upon a vast resource that's already there. The other is not to be afraid to do it yourself. Chemistry is an approachable subject and ought to be integrated into the synthesis and processing of materials. That's why chemical engineers are uniquely suited, among the engineers, to this kind of problem.

Sheldon Isakoff: My comments reinforce what has been said about the needs, challenges, and excitement of research in this area. It's an eminently practical field. There are products out there, and new ones to be invented, that depend upon this kind of technology to make progress. For example, making polymer blends is a very important field in commerce. Toughness in a material, its ability to absorb energy without failure, is a key property. Over the years, homogeneous polymers have been made tougher, but greater advances have been made by blending dissimilar polymers together, for example, polyamides, nylons, with elastomers to make supertough material. The technology to do this more effectively still remains to be developed. We are working on that in industry, and we do have products out in the marketplace. The combination of two immiscible materials, such as an elastomer and a polymer like nylon, goes through interesting phase inversions during mixing. The chemistry that we apply at the surfaces to compatibilize these materials, as we call it, to make them useful, is not well understood. Some of the research talked about here can help in that regard. Another area in which polymer blends have practical commercial importance is where diffusional properties, resistance to permeation of gases through a polymeric structure, are important. Some interesting products are coming which are going to have enormous applicability in the food packaging market and in the containment of chemicals. These polymers blended for permeation properties require a completely different kind of microstructure from those blended for tough structural materials. This is an exciting field from the scientific point of view. It also has practical importance which we, as chemical engineers, should take as encouragement. I think we have a winner in this research area.

Keith Gubbins: I'd like to ask a question concerning the intellectual frontiers of this area. If I want to design a material that has certain thermal, mechanical, and chemical properties, how do I do that? That's the question that George was asking. As I understand it, that's done now by clever and innovative experiments. I'd like to know what the community that works in this area thinks is the future. Are we going to continue to do this experimentally? What is the role of theory? If there are things that theory should be doing to make real breakthroughs in this area, what are they? Is it just a matter of getting more powerful computers? I hear numbers from the computer industry that vary between 10^3 and 10^6, depending on how conservative people are on increasing the power of computers in the next 10 or 20 years. Is that sufficient to change the way we do this research? What does the future hold in that area?

Matthew Tirrell: It seems to me very likely that, one day, we will be able to compute, *a priori*, the properties of a conceptual polymer structure before its synthesis is carried out. There exist already large data bases from which one can make group contribution-type estimates. Real theory is a long way off but, I believe, will eventually exist.

Stuart Cooper: This concerns the questions that George Stephanopoulos and Keith Gubbins raised on the rationale for the design of new polymeric materials. I think it depends on the framework which gives rise to the need. In the case of testing various new theories coming out of the polymer physics community, special diblock, star molecules, cyclic compounds, or polymers with specific interaction sites might be specified. The design issues relating to the production of new polymers for industry are concerned, for example, more with questions of increased—or decreased—stability, improved strength and toughness, and the need for polymeric materials that can open up new markets in the electronic and ceramic materials areas.

Louis Hegedus: Coming from the specialty chemicals applications, I would like to further emphasize what Sheldon Isakoff said. Take polyethylene, which our company uses extensively for a variety of applications. What do we do with it? We fill it, modify it, functionalize it, formulate it, compound it, extrude it, laminate it, and end up with such vastly different products as battery separators, membranes, construction materials, insulation materials, battery electrodes, and microfilters. It is amazing what you can do with any given polymer. Much research emphasis today is on the properties of bulk polymers themselves. There is a whole science associated with combining bulk polymers with other substances—fillers, additives, surface functionalizers—to end up with a wide variety of specialty chemical products. It is a huge area which chemical engineers are not much involved with yet.

Robert Cohen: Obviously, there's great interest in this and no shortage of opinions. I think that's healthy. It's clear that the concept of the chemical engineer as only a process engineer is gone. Now we are also, for lack of a better word, product engineers as well. We're interested in materials. Polymers are the natural ones that we feel comfortable with because we have the scientific molecular base for it. As has come out in the discussion, it's an applications-oriented field, arising out of the marketplace. There's a lot of industrial contact. That's also healthy for the chemical engineer. The molecular chemical insight that we need for the field is also crucial.

SECTION VII
Microelectronic and Optical Materials

18

Chemical Engineering Research Opportunities in Electronic and Optical Materials

Larry F. Thompson
AT&T Bell Laboratories
Murray Hill, New Jersey

I. Introduction

The character of American industry has changed dramatically over the past three decades. Much of the high-technology manufacturing carried out today is based on chemical rather than mechanical modifications of materials. These modifications are based on complex and sophisticated reactions that are employed on a much smaller scale than is customary in the traditional chemical industry. In addition to the differences in reaction scale, purity requirements, the use of highly toxic and pyrophoric reagents, and the generation of unusual toxic waste by-products all differentiate these industries from most traditional chemical process industries and represent interesting new intellectual challenges. There are, however, similarities in the basic unit operations required, the need for process integration, and the necessity for waste management.

Over the last three decades, many of these "high-technology" industries have evolved from mechanical-based manufacturing to chemical-based manufacturing. Examples include home entertainment, data storage and manipulation (computation), high-performance structural materials, and

ADVANCES IN CHEMICAL ENGINEERING, VOL. 16

telecommunications. Telecommunications has evolved from an industry that was primarily concerned with the transmission and manipulation of voice signals to one that deals with all forms of information, i.e., voice, digital data for computation, and images. These changes have resulted in the evolution of telecommunications to a much broader industry known as the *information movement and management industry*. Along with this change there is also a convergence of the telecommunications and computer industries [1]. This convergence is best illustrated not by the services offered but rather by the changes in the components, equipment, and systems used to support these industries (Table 1). This latter comparison will form the basis of this paper.

Prior to 1950, these industries were based on vacuum tube technology, and most electronic gear was assembled on metal chassis with mechanical attachment, soldering, and hand wiring. All the components of pretransistor electronic products—vacuum tubes, capacitors, inductors, and resistors—were manufactured by mechanical processes. A rapid evolution occurred after the invention of the transistor and the monolithic integrated circuit. Today's electronic equipment is filled with integrated circuits, interconnection boards, and other devices that are all manufactured by chemical processes. The medium used for the transmission of information and data over dis-

Table 1. A Comparison of Systems and Devices Manufactured for Use in the Telecommunications Industry in 1958 and 1988

System or Device	1958	1988
Electronic components	Vacuum tubes	Integrated circuits
	Rectifiers	Diodes
	Wound foil capacitors	Thin film and ceramic capacitors
	Relays	Solid state switchers
Data storage media	Relays	Magnetic thin films
	Ferrite core	Magnetic bubbles
	Paper tape	Optically active films
	Punch cards	
Transmission media	Copper wire	Light wave media (optical fiber)
	Coaxial cable	
Interconnection	Hard wiring and manual soldering	Multilayer interconnector boards
Equipment closures	Steel cabinets	Molded plastic enclosures

tances has evolved from copper wire to optical fiber, and we have rapidly entered the photonics era. It is quite likely that no wire-based transmission systems will be installed in the future. The manufacture of optical fibers, like that of microcircuits, is almost entirely a chemical process.

To achieve the required performance of today's devices and transmission media, it is necessary to modify and manipulate materials at the atomic level with an unprecedented degree of control and purity [2]. To illustrate this point, materials and material processing used to manufacture electronic devices and light wave media (optical fibers) will be discussed in some detail. Table 2 lists specific topics in these two areas where chemical engineering currently plays, and will continue to play, an important role [3].

II. Electronic Devices

Over the past three decades, enormous progress has been made in materials and materials processing used to fabricate electronic devices. This progress has made possible an astonishing increase in device complexity and decrease in the dimensions of features used to fabricate the circuit, which, in turn, has led to improved performance and reduced cost per function. Figure 1 illustrates these trends for dynamic random access memory (DRAM), historically the most complex chip produced in terms of feature size and components per chip.

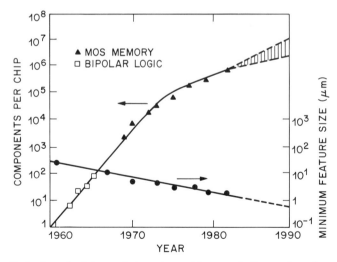

Figure 1. Reduction of minimum feature size and increase in the number of components per chip as a function of date of introduction into manufacture for DRAM devices.

Table 2. Chemical Engineering Aspects of Electronic Devices and Light Wave Media [a]

Chemical Engineering Contributions	Technologies	
	Microcircuits	Light Wave Media and Devices
Large-scale synthesis of materials	Ultrapure single-crystal silicon	Ultrapure glass
	III–V compounds	Fiber coatings
	II–VI compounds	Connectors
	Electrically active polymers	
	Chemical processing	
Engineering of reaction and deposition processes	Chemical vapor deposition	Modified chemical vapor deposition
	Physical vapor deposition	Sol-gel processing
	Plasma-enhanced chemical vapor deposition	
	Wet chemical etching	
	Plasma etching	
	Polymer resists	
Other challenges in engineering chemical processes	Effluent treatment	Fiber drawing and coating processes
		Waste recovery
Packaging and/or assembly	Process integration and automation	Cabling
	New materials for packaging	Laser packaging
	Modeling of heat transfer in design of packaging	

[a] From Ref. [3]

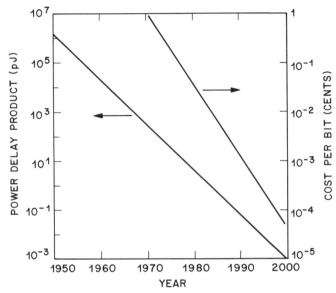

Figure 2. The power delay product and cost per bit as a function of the date of introduction into manufacture of the most complex MOS devices.

Figure 2 shows that the power delay product, a measure of device speed and thermal performance, has decreased dramatically (increasing performance) from 1950 and will continue to do so for at least two more decades. This performance increase has been accompanied by a simultaneous decrease in cost per bit from about 1 cent in 1970 to about 0.003 cent in 1988. This remarkable progress is the result of improved manufacturing efficiency and materials processing [4, 2]. The remainder of this section will deal with the manufacturing sequence of integrated circuits and describe in detail several important "unit operations" and new processes in which chemical engineering has played a significant role.

There are many elements and alloys that possess the electrical property known as band gap [5] that are useful for semiconductor devices. Silicon became the lynchpin material for device fabrication. It is currently, and will continue to be, the most important material for integrated circuit fabrication. Alloys of group III and V elements are important for optical devices such as lasers, optical detectors, and specialized high-speed circuits. In this paper, the fabrication of silicon devices will be used to illustrate the role of chemical processing in circuit manufacture.

III. Device Fabrication

An integrated circuit is a multilayer, three-dimensional structure of electrically interconnected solid-state circuit elements isolated with patterned dielectric films. The dielectric, conductor, and semiconductor films are deposited or formed by sophisticated chemical reactions. The successful growth and manipulation of these films depend heavily on the proper design of chemical reactors used in deposition and etching, the choice of appropriate chemical reagents, separation and ultrapurification, and operation of sophisticated control systems.

Microcircuitry has been made possible by our ability to use chemical reactions and processes to fabricate millions of electronic components or elements simultaneously on a single substrate, in this case, silicon. For example, a 1-million-bit DRAM device contains 1.4 million transistors and 1 million capacitors. Many of the chemically etched features on the chip are as small as 0.9 μm, and this dimension will decrease to below 0.5 μm in the next five years.

A. Silicon Production

The manufacture of silicon integrated circuits involves a sequence of inter-related steps, each designed to purify, modify, deposit, or pattern materials. Figure 3 illustrates the principal steps from silicon purification to packaging. A good review of the entire process can be found in Wolf and Tauber [6]. The requirement for ultrapure starting reagents (contaminants

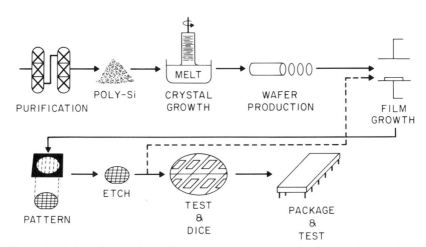

Figure 3. A flow diagram from silicon production to testing for the manufacture of an integrated circuit.

at parts per billion or less) is common to most high-technology manufacturing, and silicon circuits are no exception. The required purity of the silicon wafers shown in Table 3 is achieved by distillation of the trichlorosilane starting material. High-purity polycrystalline silicon is produced from the high-purity $SiHCl_3$ by the Siemens process using the following reaction:

$$2H_2 + 2SiHCl_3 \rightleftharpoons 2Si + 6HCl \tag{1}$$

A complete description of this process can be found in *The Chemistry of the Semiconductor Industry* by S. J. Moss and A. Ledwith [7].

The next step is to produce nearly perfect single-crystal boules of silicon from the ultrapure polycrystalline silicon. Many techniques have been developed to accomplish this, and they all rely on a similar set of concepts that describe the transport process, thermodynamically controlled solubility, and kinetics [8]. Three important methods are the vertical Bridgman–Stockbarger, Czochralski, and floating zone processes, fully described in *Fundamentals of Crystal Growth* by Rosenberger [9].

To prepare single-crystal silicon ingots suitable for use as materials in semiconductors, the Czochralski method is used [8]. Polycrystalline silicon is melted in a crucible at 1400–1500°C under an argon atmosphere. Tiny

Table 3. Impurity Levels in High-Purity Polysilicon Useful for Circuit Applications

Element	Concentration (ppba)^a
Carbon	100–1000
Oxygen	100–400
Phosporus	≤0.3
Arsenic	<0.001
Antimony	<0.001
Boron	≤0.01
Iron	0.1–1.0
Nickel	0.1–0.5
Chromium	<0.01
Cobalt	0.001
Copper	0.1
Zinc	<0.1
Silver	0.001
Gold	<0.00001

^{a}ppba is number of atoms per 10^9 Si atoms

quantities of dopants, compounds of phosphorus, arsenic, or boron, are added to the melt to achieve the desired electrical properties of the finished single-crystal wafers. A tiny seed crystal of silicon with the proper crystalline orientation is inserted into the melt and slowly rotated and simultaneously withdrawn at a precisely controlled rate, forming a large cylindrical, single-crystal boule that is typically 15 cm in diameter and about 2 m long with the desired crystalline orientation and composition. The resulting single-crystal ingots are sawed into wafers that are subsequently polished to a flatness in the range of 1 to 10 µm. The Czochralski method is shown schematically in Fig. 4. There is perhaps no crystal growth technique that has been more thoroughly studied both theoretically and experimentally than the Czochralski method for producing silicon wafers. Important processes that control the growth and perfection of the crystal include solidification kinetics, heat transfer, mass transfer, impurity segregation, melt flow, and convection. Central to our understanding has been the application of fun-

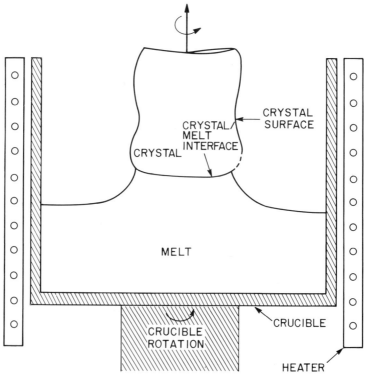

Figure 4. Schematic of the Czochralski process.

damental chemical engineering principles. An excellent review of the important phenomena that control Czochralski growth was recently published by Brown [8].

B. Lithography

The next steps in device fabrication are the sequential deposition and patterning of thin dielectric and conducting films. The wafer, with the desired thin film, is coated with a radiation-sensitive polymeric material, termed a resist, and exposed to light or other radiation through the appropriate photomask that contains the desired pattern. The purpose of the photolithographic process is to transfer the mask pattern to the thin resist film on the wafer surface. The exposed photoresist is developed with a solvent that removes unwanted portions, and the resulting image serves as a mask for chemically etching the pattern into the thin film. The resist is removed with an oxidizing agent, such as a sulfuric acid–hydrogen peroxide mixture, and the wafer is chemically cleaned and made ready for other steps in the fabrication process [10, 11]. The lithographic process is shown in Fig. 5.

Surprisingly, photolithography is still the technology used to fabricate microelectronic chips. Reduction step-and-repeat cameras and highly sophisticated one-to-one projection printers are the dominant printing technologies. There is perhaps no better example than lithography to illustrate the uncertainty associated with predicting technological direction and change. In 1976 it was believed, though not by everyone, that photolithography could not produce features smaller than about 1.5 μm with high chip yields in a production environment. The current belief is that conventional photolithography will be able to print 0.6-μm features and will be the dominant technology well into the first half of the 1990s. The same basic positive photoresist, a diazonaphthoquinone–novolac composite, will likely still be the resist of choice. The cost of introducing a new resist material and the cost associated with new hardware are strong driving forces pushing photolithography to its absolute, ultimate limits.

The technological alternatives to conventional photolithography are short-wavelength photolithography (248 nm) and x-ray or scanning electron beam lithography. This paper will not discuss the various lithographic technologies; however, no matter which alternative technology becomes dominant after photolithography has reached its limit, new resists and processes will be required, and this will necessitate enormous investments in research and process development.

The development and implementation of new resists for microlithography require considerable engineering development, and chemical engineering is central to that development. Many polymer systems have lithographic properties that appear to be useful for each of the lithographic options [12].

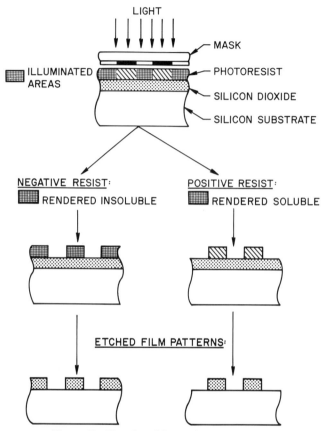

Figure 5.   The photolithography process.

To achieve a manufacturable system for sub-0.5-μm patterning, extremely precise control of the molecular properties, structure, composition, and purity of the polymer is required (Table 4). Meeting these requirements provides intellectual challenges in ultrapurification reaction engineering and chemical synthesis. An illustration of the control required in this synthesis process can be found in a negative electron beam resist, GMC.

Poly(glycidyl methacrylate-co-3-chlorostyrene) (GMC) is a random co-polymer synthesized using free radical, solution-polymerization techniques. It is highly useful as a negative-acting electron resist for microlithography [13, 14]. Cross-linking occurs through the epoxide group of the glycidyl methacrylate (GMA) monomer, whereas the monomer 3-chlorostyrene (CLS) imparts thermal stability and resistance to plasma etching environments. The electron beam lithographic response of GMC has been studied extensively and found to depend on the polymer composition (CLS/GMA ratio), weight-average molecular weight, and polydispersity. These studies resulted in the molecular design of GMC that is capable of submicron patterning. The molecular properties and specifications required for GMC are given in Table 4.

Initial synthesis of GMC for process development and optimization studies was accomplished on a small laboratory scale with synthetic runs typically yielding 5–15 g of polymer. However, in order to test GMC on a production basis and introduce it into manufacture, scale-up of the synthesis was necessary. The control of molecular properties and composition had to be considerably better than for most commercial polymers. To this end, a pilot plant for the manufacture of GMC was designed, constructed, and used to produce kilogram quantities of polymer. The scale-up of GMC provides an excellent example of how basic chemical engineering principles are employed in microcircuit fabrication, as well as some of the challenges in synthesis, process control, and purification. The major components of the pilot plant are shown in Fig. 6.

Process monitoring and control are carried out remotely with a computer located near a walk-in hood. The control system consists of a process computer, interfacing data acquisition/control unit, and manual switching console. It provides for data logging and automatic control of heat exchangers, pumps, agitators, and valves associated with the distillation, reaction, quench, and precipitation steps and the solvent metering systems. The computer is programmed to provide an on-line operating manual and operator prompts and alarms to facilitate totally automated operation of the process. Resistance temperature detectors (RTDs) are used to measure temperatures during monomer distillation and reaction.

Table 4. Polymer Property Control Required for a High Resolution Resist

Property	Typical Commercial Polymer	GMC	Lithography Property Affected
Molecular weight	$10^4–10^6 \pm 20\%$	$1.15 \times 10^5 \pm 3\%$	Sensitivity
Molecular weight distribution	$3–20 \pm 1–6$	2.0 ± 0.05	Resolution
Composition	$\pm 10–30\%$	$37\% \pm 2\%$	Sensitivity and etch resistance
Purity	$\sim 90\%$	< a few ppb	Defects and yields

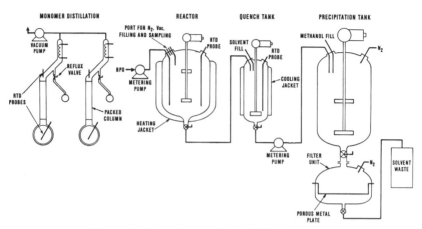

Figure 6. Flow diagram of the GMC pilot plant.

GMC is synthesized from a dilute solution of GMA and CLS monomers in toluene. Radical initiation proceeds via thermal homolysis of benzoyl peroxide (BPO) at 85°C. The initial concentration of each monomer, the initial BPO concentration, and the reaction temperature directly affect the copolymer product composition, molecular weight, and polydispersity. The reaction takes place in a vessel made of glass and Teflon®. The vessel is pressure-rated from vacuum to 15 psig. The nominal volume of the pilot reactor is 10 liters for a scale-up factor of about 25×. The reactor is evacuated and purged with N_2 several times to remove O_2. Monomers and toluene are then fed in, and the system is rapidly brought to 85°C. Polymerization is initiated with a single batch-wise feed of BPO in toluene small enough in volume to have a minimal effect on the temperature of reactor contents. A solution of BPO–toluene is continuously fed to the reactor at a rate calculated to maintain a constant concentration of BPO. The continuous BPO feed alleviates problems of rate slowdown and molecular weight distribution broadening at higher monomer conversions. The reactor is closed during operation and maintained at about 10 psig with N_2 pressure.

The kinetics of this copolymerization have been studied and found to behave according to the following expression for a constant-density system

$$\frac{dC_m}{dt} = kC_i^{1/2}C_m \tag{2}$$

where C_m is the total monomer concentration and C_i is the initiator (BPO)

Table 5. Comparison of GMC-II Properties from Lab- and Pilot-Scale Systems

Property	Lab Scale	Pilot Scale
Amount synthesized (kg)	0.055	1.16
Monomer conversion (%)	62.8	54.1
$\overline{M}_w$	114,200	114,100
Polydispersity	2.17	2.02
Mole % CLS	37.1	37.4
$D_g^{0.6}$	4.1	3.9
Contrast	1.4	1.4

concentration. The measured reactivity ratios are $r_{GMA} = 0.58$ and $r_{CLS} = 0.83$ for GMA and CLS, respectively, indicating very little compositional drift with conversion.

The critical properties of GMC-II produced from both laboratory and pilot systems are listed in Table 5. In comparing a 0.055-kg batch (62.8% monomer conversion) from laboratory synthesis and a 1.16-kg batch (54.1% monomer conversion) from the pilot synthesis, the material properties responsible for lithographic response are nearly identical.

IV. Light Wave Media (Optical Fibers)

Photonics involves the transmission of optical signals through a guiding medium, generally a glass fiber, for purposes that include telecommunications, data and image transmission, energy transmission, sensing, display, and signal processing. Optical fiber technology evolved from the laboratory to manufacturing in less than a decade, becoming a commercial reality in the early 1980s. The data-transmitting capacity of optical fiber systems has doubled every year since 1976. In fact, optical fiber systems planned on the basis of the prevailing technology at a given time are often obsolete by the time they are implemented.

Light wave media and devices include the guiding medium (optical fibers), sending and receiving devices, and associated electronics and circuitry. The transmission of light signals through optical fibers must occur at wavelengths at which the absorption of light by the fiber is minimal. Typically for SiO_2/GeO_2 glass, the best transmission windows are at 1.3 and 1.5 μm (Fig. 7). A comprehensive review of this technology has been compiled in the book titled *Optical Fiber Telecommunications* by Miller and Chynoweth [15].

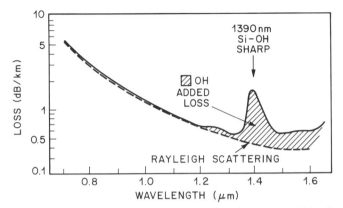

Figure 7. Loss as a function of wavelength for a GeO$_2$-doped SiO$_2$ fiber.

A. *Optical Fiber Manufacturing*

Optical fibers are made by chemical processes. The critical feature of an optical fiber that allows it to propagate light down its length is a core of high refractive index, surrounded by a cladding of lower index. The higher-index core is produced by doping silica with oxides of phosphorus, germanium, and/or aluminum. The cladding is either pure silica or silica doped with fluorides or boron oxide.

There are four principal processes that may be used to manufacture the glass body that is drawn into today's optical fiber. "Outside" processes, including outside vapor-phase oxidation and vertical axial deposition, produce layered deposits of doped silica by varying the concentration of SiCl$_4$ and dopant halides passing through a torch. The resulting "soot" of doped silica is deposited and partially sintered to form a porous silica boule. Next, the boule is sintered to a pore-free glass rod of exquisite purity and transparency. "Inside" processes, such as modified chemical vapor deposition (MCVD) and plasma chemical vapor deposition (PCVD), deposit doped silica on the interior surface of a fused silica tube. In MCVD the oxidation of the halide reactants is initiated by a flame that heats the outside of the tube (Fig. 8). In PCVD, the reaction is initiated by a microwave plasma. More than a hundred different layers with different refractive indexes (a function of glass composition) may be deposited by either process before the tube is collapsed to form a glass rod. There are numerous unit operations and chemical processes associated with the MCVD method. Two will be discussed to illustrate the role of chemical engineering: thermophoresis and reagent purification. An excellent review of the MCVD process has been written by Nagel *et al.* [16].

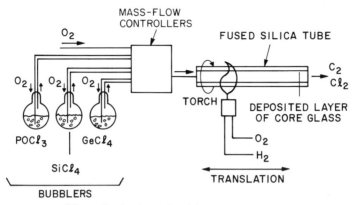

Figure 8. A schematic of the MCVD process.

B. Thermophoresis

Thermophoresis is the dominant factor in the particulate deposition mechanism in MCVD [17]. Thermophoresis describes the force on a particle suspended in a gas with a temperature gradient. The thermal gradient imparts a velocity to the particle in the direction of decreasing temperature as a result of the gas molecules impacting the particle on opposite sides with different average velocities. In the MCVD configuration, the tube that serves as the reaction vessel is heated externally by a traversing oxygen-hydrogen torch. At any point in time, this external heating results in a temperature field inside the tube. The magnitude of the temperature field is a function of the thickness of the tube wall, the outside wall temperature profile, and the reactant flow rate and properties of gases within the tube. Figure 9 shows the temperature field within an MCVD substrate tube relative to the torch position at a given point in time. Under typical operating conditions, the gas flow in the tube is laminar and the cool gas is heated as it enters the hot zone. At some critical temperature, T_{rxn}, chemical reactions take place, resulting in rapid oxidation of the reagents to form submicron oxide particles. The chemical composition of the particles is strongly affected by chemical reaction equilibria, particle growth dynamics, and transport phenomena. Once a particle has formed at a given radial position within the tube, its trajectory is determined by the thermophoretic forces generated by the temperature field as illustrated in Fig. 10. Initially, the particles move radially inward because the wall temperature is hotter than the gas. Farther downstream from the torch, the wall is cooler than the gas, and particles move toward the tube wall. Certain trajectories near the wall result in deposition, while particles near the center are swept out of the tube. A

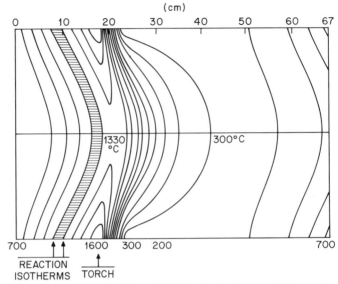

Figure 9.   Temperature field in an MCVD tube during a typical deposition run.

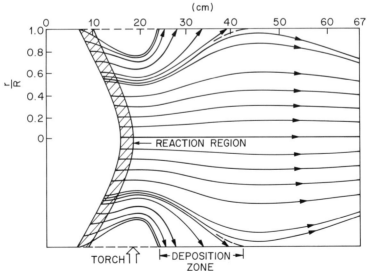

Figure 10. Particle trajectories resulting from the field in Fig. 9.

detailed mathematical model to describe thermophoretic deposition of particles in MCVD was derived by Walker *et al.* [17] and has shown that the efficiency, ε, of particle incorporation can be described by

$$\varepsilon \approx 0.8(1 - \frac{T_e}{T_{rxn}}) \tag{3}$$

where T_e is the temperature downstream to which the inside wall and the gas thermally equilibrate. Thus, a finite particle incorporation efficiency exists in the MCVD process. The value of T_e depends strongly on torch traverse length and velocity, the ambient temperature, and the tube wall thickness, and weakly on the flow rate and tube radius. The length over which deposition takes place, as depicted by the deposition zone in Fig. 10, is proportional to Q/α, where Q is the total volumetric flow and α is the thermal diffusivity of the gas mixture. Understanding this mechanism has been the key to process scale-up and optimization.

C. Ultrapurification

Kaiser has shown that minimizing the water content in fibers is extremely important in limiting the attained fiber loss, particularly in the 1.3–1.6-μm region where silica-based fiber intrinsic losses are minimal [18]. This is shown schematically in Fig. 7 for a typical multimode fiber with a GeO_2-P_2O_5-SiO_2 core. The fundamental Si–OH stretching vibration gives rise to a relatively sharp absorption band at 2.72 μm, which has overtones and combination bands in the near infrared.

There are two primary sources of OH in fibers fabricated by MCVD: (1) OH incorporated from hydrogenic impurity species that enter the gas stream during the various phases of processing, and (2) OH that thermally diffuses from the starting substrate tube into the active region of the fiber during the various phases of processing. It is very important to use ultrapure starting materials, and it is necessary to purify the $SiCl_4$ just prior to use in order to minimize contamination during prolonged storage.

The amount of OH incorporated during processing is controlled by the equilibria [19]

$$H_2O + Cl \rightleftharpoons HCl + \frac{1}{2}O_2 \tag{4}$$

$$H_2O + [Si-O-Si]_{solid} \rightleftharpoons 2[SiOH]_{solid} \tag{5}$$

H_2O is incorporated into the glass as Si–OH, while HCl is not. Solving the equilibrium equations shows that the resultant equilibrium concentration of

OH in the glass can be described by

$$C_{SiOH} \propto \frac{[P^i_{H_2O}][P_{O_2}]^{1/4}}{[P_{Cl_2}]^{1/2}} \tag{6}$$

Here $P^i_{H_2O}$ is the initial partial pressure of equivalent H_2O in the gas stream from all sources (such as HCl, SiHCl$_3$, etc.), and P_{O_2} and P_{Cl_2} are the partial pressures of oxygen and chlorine at equilibrium. Depending on the processing step, the equilibrium conditions are established at consolidation or collapse temperatures. These analyses clearly show the importance of removing *all* hydrogen-containing impurities.

A two-component process for the continuous purification of silicon tetrachloride (SiCl$_4$) has been developed for MCVD lightguide manufacturing. This system has been designed to produce a continuous supply of ultrapure SiCl$_4$. This process provides an excellent example of intellectual challenges that were encountered, including ultrapurification, process control, and the handling of hazardous materials. The continuous purification system consists of a vertically sparged, concurrent photochlorination reactor and a nitrogen-fed stripping column arranged in series. The process flow schematic is given in Fig. 11. The bubble column (BC) reactor provides for simultaneous chlorine absorption and ultraviolet light-induced chlorination of hydrogenous contaminants in epitaxial grade SiCl$_4$, primarily trichlorosilane (SiHCl$_3$), according to the following stoichiometry:

$$Cl_2 + SiHCl_3 \xrightarrow{h\nu} HCl + SiCl_4 \tag{7}$$

The stripping column is used to desorb HCl and excess Cl$_2$ from the reactor product by countercurrent contacting with nitrogen gas in a packed bed. A heat exchanger downstream of the bubble column maintains the temperature of stripper feed at about 25°C. A condenser located at the gas outlet from the stripping column is fed with coolant at −13°C for efficient recovery of SiCl$_4$ vapor. The outlet nitrogen gas stream, containing traces of HCl, Cl$_2$, and SiCl$_4$, passes into a caustic scrubber for removal of these environmentally hazardous chemicals. Downstream from stripping, the purity of the treated SiCl$_4$ is monitored by in-line infrared (IR) spectrophotometry. If the product exceeds the impurity limits for hydrogenous contaminants, a three-way valve automatically diverts flow from the product tank back to the feed tank. All materials of construction contacting SiCl$_4$ in the feed tank and downstream are polyvinylidene fluoride, Teflon®, or borosilicate glass. This system has been integrated into a large optical fiber plant and run successfully for over three years.

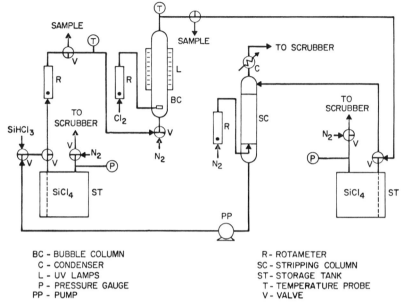

BC - BUBBLE COLUMN
C - CONDENSER
L - UV LAMPS
P - PRESSURE GAUGE
PP - PUMP

R - ROTAMETER
SC - STRIPPING COLUMN
ST - STORAGE TANK
T - TEMPERATURE PROBE
V - VALVE

Figure 11. Flow diagram for the $SiCl_4$ purification system.

V. Summary

The industries that manufacture materials and components for electronic and optical-based systems are characterized by products that are rapidly superseded in the market by improved ones. This quick turnover stems from the intense competition among these industries and results in dramatic price erosion for products, once introduced. Consequently, these industries also require rapid technology transfer from the research laboratory onto the production line. Many of their *products* cannot be protected by patents, except for minor features. Therefore, the key to their competitive success is thoroughly characterized and integrated manufacturing processes, supported by *process* innovations. Since nearly all of the processes are chemically based, chemical engineers can play an important role in ensuring the success of these companies.

Process integration is the key challenge in the design of efficient and cost-effective manufacturing processes for electronic and photonic devices. These products are currently manufactured through a series of individual, isolated steps. It is crucial that the overall manufacturing methodology be examined and that integrated manufacturing approaches be implemented. Historically, all industries have benefited economically, as well as in the quality and yield of products, by the use of integrated manufacturing methods. As individual process steps become more complex and precise, the final results of manufacturing (e.g., yield, throughput, and reliability) often depend critically on the interactions among the various steps.

The concepts of chemical engineering are easily applied to the challenge of process integration in these industries, particularly because many of the key process steps involve chemical reactions. For example, in the manufacture of microcircuits, chemical engineers provide mathematical models and control algorithms for the transient and steady-state operation of individual chemical process steps (e.g., lithography, etching, film deposition, diffusion, and oxidation), as well as interactions between process steps and ultimately between processing and the characteristics of the final device. As another example, in microcircuit manufacturing, chemical engineers can provide needed simulations of the dynamics of material movement through the plant and thus optimize the flow of devices (or wafers) through a fabrication line. Over the past decade, the number of chemical engineers in the electronics industry has tripled from 700 in 1977 to well over 2000 in 1987. The demand for chemical engineers by these industries should increase for several more decades. It is important that they remain identified with the profession and integrated into its every aspect. Although the current undergraduate curriculum in chemical engineering provides an excellent conceptual base for graduates who move into the electronics industries, it could be improved by the introduction of instructional material and example problems relevant to the challenges outlined in this chapter. This would not require the creation of new courses, but rather the addition of new material to enrich existing ones.

References

1. Mayo, J. S., in *Information Technologies and Social Transformation* (B. R. Guile, ed.), p. 7. National Academy Press, Washington, D.C., 1985.
2. Stinson, S. C., *Chem. Eng. News* **66**, 7 (1988).
3. National Research Council, Committee on Chemical Engineering Frontiers: Research Needs and Opportunities. *Frontiers in Chemical Engineering. Research Needs and Opportunities.* National Academy Press, Washington, D.C., 1988.
4. Williams, D. S., Sze, S. M., and Wagner, R. S., eds., *Proceedings of the First Electronics Materials & Processing Conference*, American Society for Metals, September 1988.
5. Sze, S. M., *Semiconductor Devices: Physics and Technology.* Wiley, New York, 1985.
6. Wolf, S., and Tauber, R. N., *Silicon Processing for the VLSI Era*, Vol. 1—*Process Technology.* Lattice Press, Sunset Beach, Calif., 1986.
7. Moss, S. J., and Ledwith, A., *The Chemistry of the Semiconductor Industry.* Blackie, Glasgow, Scotland, 1987.

8. Brown, R. A., *AIChE J.* **34**, 881 (1988).
9. Rosenberger, F., *Fundamentals of Crystal Growth.* Springer, New York, 1979.
10. Thompson, L. F., Willson, C. G., and Bowden, M. J., eds., *Introduction to Microlithography*, Vol. 219. American Chemical Society, Washington, D.C., 1983.
11. Sze, S. M., *VLSI Technology.* McGraw-Hill, New York, 1983.
12. Reichmanis, E. and Thompson, L. F., *Annu. Rev. Mater. Sci.* **17**, 235 (1987).
13. Thompson, L. F., and Doerries, L. M., *J. Electrochem. Soc.* **126**, 1699 (1979).
14. Novembre, A. E., Frackoviak, J., Kowalski, L. M., Mixon, D. A., and Thompson, L. F., *Solid State Tech.* **31**(4), 135 (1988).
15. Miller, S. E., and Chynoweth, A. G., *Optical Fiber Telecommunications.* Academic Press, New York, 1979.
16. Nagel, S. R., MacChesney, J. B., and Walker, K. L., *Advances in Optical Fiber Communication.* Academic Press, Orlando, Fla., 1985; *J. Quantum Elec. 18, 459 (1982).*
17. Walker, K. L., Homsy, G. M., and Geyling, F. T., *J. Colloid Interface Sci.* **69**, 138 (1979).
18. Kaiser, P., *Appl. Phys. Lett.* **23**, 45 (1973).
19. Wood, D. L., Kometani, T. Y., and Saifi, M. A., *J. Am. Ceram. Soc.* **62**, 638 (1979).

19

Chemical Engineering in the Processing of Electronic and Optical Materials: A Discussion

Klavs F. Jensen
Department of Chemical Engineering
Massachusetts Institute of Technology
Cambridge, Massachusetts

I. Introduction

The processing of electronic and optical materials involves scientific and engineering concepts from a multitude of disciplines, including chemistry, solid-state physics, materials science, electronics, thermodynamics, chemical kinetics, and transport phenomena. Chemical engineers have a long history of solving multidisciplinary problems in other specialized fields, such as food processing and polymer processing, and there is a growing recognition of the useful contributions that chemical engineers can make to electronic materials processing. Within the last decade, between 15 and 30% of chemical engineering graduates have taken positions in electronic materials processing companies, and chemical engineering research on the topic has grown to the point where it has become recognized in the electronic materials community.

The emergence of electronic materials processing, along with other specialized topics within the chemical engineering discipline, raises questions about research, teaching, and the profession in general, questions similar to those asked in other areas, such as biotechnology. The aim of this dis-

ADVANCES IN CHEMICAL ENGINEERING, VOL. 16

cussion is to address some of these questions and not to present an exhaustive review of the field. A more complete overview is given in the contribution to this volume by Thompson [1] and in other recent reviews [2, 3]. In particular, the issues to be addressed are research opportunities as well as undergraduate and graduate teaching. The views expressed in the following are those of the author and cannot do justice to the many different viewpoints possible in this highly interdisciplinary research area.

II. Characteristics of Electronic Materials Processing

Electronic materials processing is a chemical manufacturing process aimed at modifying materials to form microstructures with specific electronic and optical properties. For example, in the manufacturing of silicon-based integrated circuits, silicon is refined into high-purity crystals. These are sliced into wafers that serve as the foundation for the electronic devices. The subsequent process sequences involve oxidation of the wafer surface and deposition of semiconductors, dielectrics, and conductors interspersed by patterning through lithography and etching. The final microstructure is then cut from the wafer and enclosed in a ceramic or polymer-based package that provides connections to other electronic components and protects the microstructure from contamination and corrosion. Similar process steps are used in other applications of electronic materials processing, including production of optical coatings, solar cells, sensors, optical devices, magnetic disks, and optical storage media. While microelectronic applications have typically received the most attention, many other related processes present equally challenging chemical engineering problems.

Regardless of the final device, electronic materials processing involves a large variety of chemical procedures applied to a multitude of material systems. In addition to conventional chemistry, electron- and photon-driven reactions play a major role through plasma- and laser-assisted processes. Multiple length scales are involved in the fabrication. Typical sizes of reactors used for depositing and removing layers are of the order of $1 \mu m$ and the substrate wafers are 15–20 cm across. On the other hand, in electronic devices the typical feature size is of the order of $1 \mu m$ and shrinking with each new generation of devices. The active region in quantum well lasers is less than 5 nm. By way of comparison, the ability to resolve a 1-μm feature on a 20-cm wafer corresponds to being able to observe individual houses on a map of the United States. The microstructures have to be reproduced uniformly across each wafer as well as from wafer to wafer.

Electronic materials processing demands high-purity starting materials. Even minute quantities of impurities have the potential to alter or destroy the electronic and optical properties of a device. Unlike many other chemical processes, the costs of starting materials are usually insignificant in comparison to the value added during the process. Furthermore, the field changes rapidly, is highly competitive, and is based on innovation, unlike commodity chemicals, where small improvements in large-volume processes are typical.

The manufacture of even simple devices may entail more than 100 individual steps. However, many of the same concepts and procedures are invoked several times. Therefore, it is advantageous to group the process steps broadly in terms of unit operations analogous to those used successfully to conceptualize, analyze, design, and operate complex chemical plants involving a similarly large number of chemical processes and materials. Examples of these unit operations are listed in Table 1.

The use of chemical engineering concepts has already contributed significantly to crystal growth [4], thin-film formation [5–7], and plasma

Table 1. Examples of Unit Operations in Electronic Materials Processing

Unit Operation	Examples of Processes
Bulk Crystal Growth	Czochralski
	Bridgman
	Float zone
Chemical Modifications of Surfaces	Oxidation
	Cleaning
	Etching
Thin Film Formation	Liquid Phase Coating
	Physical Vapor Deposition
	Chemical Vapor Deposition
Plasma Processing	Etching
	Deposition
Lithography	Spin Coating
	Exposure
	Development
Semiconductor Doping	Solid-State Diffusion
	Ion Implantation
Packaging	Polymer Processing
	Ceramics Processing
	Metallization

processing [8, 9]. These unit operations involve the complex blend of transport phenomena and chemical reactions that chemical engineers have a unique
background for understanding and controlling. The polymer processing
aspects of lithography [10, 11] and packaging of devices [12] as well as
the ceramics processing problems related to packaging provide additional
opportunities for chemical engineering research.

III. Research Opportunities

Research in electronic materials processing must necessarily revolve around
one general question: *How do structural, electronic, and optical properties of a material or a device depend on the processing and how can they
be controlled?*

The many ways in which chemical engineers can contribute to addressing this question have been described in the Amundson report [13] in broad
terms. Therefore, the present discussion will focus on three examples: (1)
organometallic vapor-phase epitaxy of compound semiconductors, (2) plasma
processing, and (3) process control. These examples are chosen on the basis
of the author's experience to illustrate particular research issues rather than
to promote specific research topics. The first example explores research
questions with clear analogies to similar problems already solved by chemical
engineers in the areas of heterogeneous catalysis and combustion. The second
example introduces research issues related to the presence of charged species
and the control of microscopic features. The third case is intended to show
how chemical engineers could use their analysis and modeling skills in
process control of electronic materials manufacture. In addition, a discussion of general research trends appears at the end of this section.

A. *Example 1: Organometallic Vapor-Phase Epitaxy*

Organometallic (also called metalorganic) vapor-phase epitaxy (MOVPE)
is an organometallic chemical vapor deposition (MOCVD) technique used
to grow thin, high-purity, single-crystalline films of compound semiconductors such as GaAs, InGaAsP, ZnSe, and HgCdTe [14]. These films form
the basis for a wide range of optoelectronic devices, including solid-state
lasers and detectors. The technique, which is illustrated schematically in
Fig. 1, derives its name from the fact that the film constituents are transported as organometallic species in the gas phase to the heated growth surface,
where the individual metal atoms are cleaved from their organic ligands and
incorporated into the compound semiconductor lattice. For example, GaAs
can be grown by combining trimethylgallium and arsine according to the
overall reaction

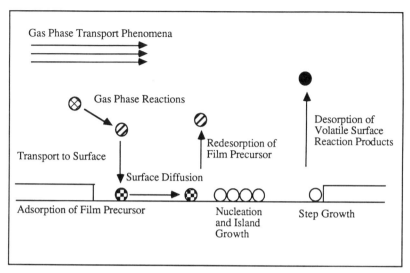

Figure 1. Schematic diagram of transport phenomena and chemical reactions underlying organometallic vapor-phase epitaxy.

$$Ga(CH_3)_3(gas) + AsH_3(gas) = GaAs(solid) + 3CH_4(gas) \qquad (1)$$

The technique has the flexibility to grow a multitude of compound semiconductor alloys by varying the composition of the source gas.

Transport processes govern the extent of gas-phase reactions and the access of the resulting film precursors to the growth interface. As the organometallic compounds approach the hot substrate, they react to form growth precursors as well as undesirable species causing unintentional doping of the growing film. Similarly, surface reactions participate in the film growth. However, parasitic reaction paths may incorporate impurities into the solid film, in particular carbon which is derived from the organometallic precursors. In the worst case, the incorporation of unintentionally added optically and electronically active impurities will render the grown semiconductor useless for device applications. This complex mixture of chemical reactions and transport phenomena is well known to chemical engineers in the context of heterogeneous catalysis, combustion, and in particular catalytic combustion. The same modeling and experimental approaches can in principle be utilized to investigate MOVPE processes. In fact, the presence of well-defined, high-purity source compounds and single-crystalline substrates used in MOVPE means that the actual process can be studied without the need for model systems, which have had to be invoked to gain insight into the more complex heterogeneous catalytic processes.

1. Experimental Investigations

Gas-phase chemical kinetics give rise to challenging experimental problems. The reaction intermediates may be short lived and difficult to observe. Efforts have been made to characterize the gas-phase chemistry of a few systems, in particular the growth of GaAs from trimethylgallium and arsine [15]. However, most of the observations have been performed *ex situ* and are affected by the sampling method used. The relatively large size of the molecules, the very low concentration of the species, and missing spectral information complicate the application of standard optical spectroscopies such as laser-induced fluorescence and spontaneous Raman scattering. Thus, the monitoring of reactive species and their chemical kinetics present opportunities for innovative experimental approaches.

Comparatively few studies attempt to understand the surface chemistry. Recent photoelastically modulated reflectance spectroscopy experiments indicate that surface chemistry plays a major role in MOVPE [16]. This technique has the advantage of observing surface changes *in situ* under standard processing conditions, but it provides no information on the actual surface species. There is need for a concerted surface spectroscopy effort similar to that used in heterogeneous catalysis to unravel surface mechanisms. The recent work by Bendt *et al.* [17] on the surface reaction mechanisms of triisobutylaluminum on aluminum is a good example of the information obtainable through such an effort. Studies of industrially relevant compound semiconductor surfaces (e.g., (100) GaAs) will be more difficult because these surfaces tend to decompose at elevated temperatures and have several possible surface reconstructions. In addition to the common high-vacuum electron spectroscopy techniques (e.g., XPS, AES, LEED), it will be necessary to develop spectroscopies for monitoring surface chemistry during actual film growth.

An understanding of gas-phase and surface chemistry is particularly important to the next generation of MOVPE processes involving selective epitaxy [18] and atomic layer epitaxy (ALE) [19]. In the first process, the compound semiconductor is deposited selectively on substrate areas opened in a suitable masking material (e.g., SiO_2). This is achieved by operating under conditions where nucleation occurs only on the substrates. Slight variations in processing environment and the presence of impurities can cause nucleation on the mask and result in loss of selectivity.

In ALE, growth proceeds by the deposition of a monolayer per cycle. As an example, in the ALE growth of GaAs, the surface is exposed to trimethylgallium until the surface is saturated with a monolayer of gallium species (the actual species is unknown). The gallium precursor is flushed out and arsine is introduced, which results in the formation of a layer of

GaAs. The gallium precursor is reintroduced and the cycle repeated until the desired film thickness is reached. This is a powerful technique allowing, in principle, molecular engineering of the active device region on a layer-by-layer basis. However, with current sources the technique works only in a limited range of operating conditions. Clearly, an understanding of the surface chemistry is critical to both selective growth and ALE.

Similar issues arise in the nucleation and growth of one material on a different substrate, e.g., GaAs on Si, known as heteroepitaxy. The early stages of film growth must be understood to minimize defect generation and to realize strained microstructures with unique optical and electronic properties. Moreover, there may be an opportunity to use surface step growth mechanisms to make specialized microstructures. The effect of strain on the chemistry is an open area. Some of the problems could be addressed through the use of scanning tunneling and atomic force microscopies.

The above discussion has stressed understanding, but innovative approaches to MOVPE play an equally important role. The development of new source chemistries can bring about improvements in selective growth which exceed that possible through understanding conventional source chemistry. Similarly, novel processing ideas may circumvent existing problems. For example, by providing layer-by-layer growth, ALE eliminates uniformity problems.

2. Models

Detailed models are necessary to understand the controlling rate processes underlying MOVPE, to identify critical experiments, and to provide a relationship between process performance and growth conditions. Simple plug flow and continuous stirred tank reactor models with simplified chemical kinetics ($A \longrightarrow B$) used traditionally in chemical reaction engineering are useful for quick estimates, but they cannot be justified as research topics. Given the rapid growth in supercomputer and engineering workstation technology, it is possible to formulate and solve models that give an accurate picture of the physical process. For the purpose of this discussion, the modeling issues can be separated into (1) gas-phase transport phenomena, (2) gas phase chemical kinetics, (3) surface reactions, and (4) surface growth modes.

Transport phenomena in MOVPE reactors operating at atmospheric and reduced pressures are affected by buoyancy-driven flows caused by large thermal and concentration gradients [15]. The buoyancy-driven flows superimpose on the main flow to yield complex mixed-convection flows, the study of which provides ample opportunities for research in computational fluid dynamics. An understanding of the origin and nature of fully three-dimensional mixed convection would be useful in designing MOVPE re-actors giving uniform deposition rates over large substrate areas as well as sharp compositional transitions between adjacent layers. Besides the technological applications, the fluid mechanical studies would add to the general knowledge base on thermal and solutal convection as well as spur the development of numerical techniques for solving large, nonlinear gas flow problems.

Although the experimental data base for gas-phase kinetics is far from complete, it is still worthwhile to formulate detailed kinetic models. These models form a conceptual framework for understanding and evaluating experimental observations. In addition, sensitivity analysis of the models is useful in identifying the essential portions of the chemical mechanism for subsequent experimental studies. Examples of this approach have been reported for Si [20, 21] and GaAs [22, 23] deposition, but challenging problems remain for key semiconductor alloys such as AlGaAs and GaInAsP.

Models of surface chemistry and growth modes will provide needed additional insight into MOVPE processes. Surface kinetic models can be used to understand the incorporation of intentionally and unintentionally added impurities into the growing semiconductor. This will be essential to predict the electrical and optical characteristics of the deposited film. Furthermore, it may help explain the long-range ordering observed in the growth of compound alloy semiconductors such as $Al_xGa_{1-x}As$ [24]. Modeling of the surface growth modes (e.g., nucleation and step growth) is necessary to understand the development of defects and develop new growth strategies for nanoscale structures such as quantum wire devices.

Molecular dynamics, Monte Carlo simulations, and step growth models aided investigation of some of these issues in the context of molecular beam epitaxy (MBE) [25, 26]. Molecular dynamics provides detailed microscopic information, but current computer technology limits it to simulations of time scales that are six orders of magnitude smaller than those characteristic of actual growth rates. Monte Carlo methods allow simulations over realistic time scales but at the expense of requiring macroscopic parameters. Step dynamic simulations may be a vehicle for incorporating microscopic growth phenomena into macroscopic reaction–transport models of MOVPE processes. The extension of the above techniques from physical vapor deposition processes such as MBE to chemical techniques raises interesting research problems with important technological applications.

Since the goal of research on MOVPE is to relate processing parameters to the optical and electronic properties of the grown films, chemical engineers must collaborate with researchers from other disciplines who have the necessary chemistry, materials science, and device background. In addition to providing needed fundamental understanding of current practice, models will have to be used to develop new processing equipment.

B. *Example 2: Plasma Processing*

Plasma processing is used extensively to deposit and, in particular, etch thin films. Plasma-enhanced chemical vapor deposition allows films to be formed under nonequilibrium conditions and relatively low process temperatures. Furthermore, the films have special material properties that cannot be realized by conventional thermally driven chemical vapor deposition processes [8, 9]. Plasma etching (dry processing) has almost totally replaced wet etching since it provides control of the shape of the microscopic etch profile [27].

Plasmas used in microelectronics processing are weakly ionized gases composed of electrons, ions, and neutral species, and they are also referred to as glow discharges. They are generated by applying an external electric field to the process gas at low pressures (1–500 Pa). Direct-current, radio-frequency, and microwave sources are used. Radio frequencies between 40 kHz and 40 MHz dominate microelectronics applications, but microwave sources are of increasing interest for a range of downstream processes including polymer etching and diamond growth [28, 9].

Plasmas contain a mixture of high-energy, "hot" electrons (1–10 eV) and "cold" ions and neutral species (400 K). The high electron energy relative to the low neutral species temperature makes discharges useful in chemical processing. Inelastic collisions between the high-energy electrons and neutral molecules result in, among other processes, electron impact ionization and molecular dissociation. The created ions, electrons, and neutral species participate in complex gas-phase and surface reactions leading to film etching or deposition. Positive-ion bombardment of surfaces in contact with the plasma plays a key role by modifying material properties during deposition and giving control of microscopic etch rate profiles. A bias potential may be applied to the excitation electrode to increase the ion energy and enhance the desired effects of ion bombardment.

Plasma deposition and etch rates are affected by a large number of process parameters and physicochemical processes, illustrated schematically in Fig. 2, making the development and operation of plasma processes difficult. Moreover, given a particular process chemistry, it is not obvious how readily accessible parameters (e.g., feed rate, pressure, power, and frequency) should be manipulated to obtain the desired film uniformity and material properties. Glow discharge physics is complex, and the chemical mechanisms are not well known, in particular those underlying the plasma–surface interactions. Consequently, there is considerable incentive for gaining

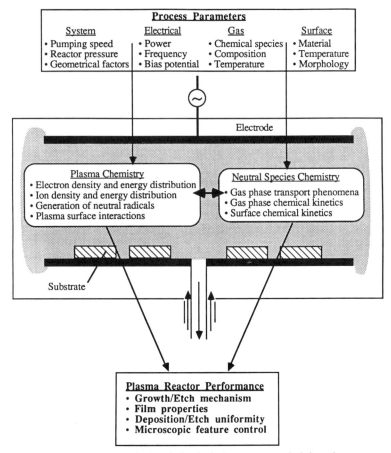

Figure 2. Interactions of chemical and physical phenomena underlying plasma processing.

insight into the underlying mechanisms through concerted experimental and modeling efforts.

1. Experimental Investigations

The experimental aspects of neutral plasma gas-phase chemistry are very similar to those discussed above in conjunction with MOVPE. The development of *in situ* diagnostics and kinetic studies are needed to unravel the complex free radical-dominated gas-phase chemistry. The unique aspects of plasma processing stem from the transport and reactions of charged species. Besides providing insight into the underlying fundamentals, techniques for measuring electric fields and electron energy distributions would have a considerable impact on plasma reactor modeling and design. The results would allow control of the electron impact reactions that form the basis of the subsequent neutral chemistry. Furthermore, the electric fields could be manipulated to achieve special material properties in addition to uniformity.

Since ion–surface and neutral molecule–surface interactions are essential to the directional etching and the unique material properties obtained in plasma reactors, they are prime research topics. These interactions have been explored in terms of understanding both etching and deposition systems [9], but many questions remain to be addressed. Examples of these include ion-induced surface chemistry, the role of surface damage in promoting reactions, and the relationship between ion energy and microstructures. As in the case of MOVPE, it will be useful to develop *in situ* surface probes besides applying the common high-vacuum surface spectroscopy techniques. Since the advantages of plasma processing derive from the formation of specific micro- or nanostructures through ion–surface interactions, it will be important to combine ion energy measurements [29] with structural probes. Application of the scanning tunneling and atomic force microscopies is difficult but promises exciting new insight into structural modifications in plasmas.

2. Modeling Approaches

Modeling of gas-phase plasma chemistry may be viewed as two interwoven problems, which are (1) to determine the electron density and energy distribution and (2), given the electron impact reactions, to predict the relative amounts and spatial distributions of neutral species. The latter problem has direct implications for the uniformity of the etching or deposition process. Since only neutral species are involved, the modeling issues are equivalent to those already discussed in connection with MOVPE. Because of the nonequilibrium nature of the discharge, the first problem presents challenging research questions. There are several possible avenues of attack: solution of the Boltzmann equation, Monte Carlo simulations of charged species transport, formulation of approximate fluid models, and simple equivalent circuit models [8, 9].

The last two approaches show promise in engineering applications requiring relatively simple models. The first method is the most fundamental, producing the electron distribution function as the result. However, it has been applied only to very simplified systems. The numerical solution of the Boltzmann equation for realistic plasma reactor configurations raises challenging computational problems even for the present generation of supercomputers. Given a statistically significant sample and adequate cross-section data, Monte Carlo simulation of the actual electron and ion trajectories may be used to obtain a realistic representation of the plasma. However, the large disparities in magnitude between electron and ion transport coefficients has so far prevented self-consistent computations. Moreover, extensions to three dimensions are computationally intensive. On the other hand, Monte Carlo techniques allow inclusion of the physical phenomena in a straightforward manner and the new generation of high-speed parallel computers may have advantages in solving this type of problem.

Models of the surface chemistry and in particular the evolution of the film microstructure during film growth will be useful in understanding the relationship between processing parameters and film properties. Because of the microscopic scale, the modeling approaches will range from molecular dynamics to Monte Carlo simulations. This relatively unexplored area could have a significant impact on the understanding and use of plasma processing.

C. Example 3: Process Control of Microelectronics Manufacturing

The fabrication of microelectronic and photonic components involves long sequences of batch chemical processes. The manufacture of advanced microstructures can involve more than 200 process steps and take from 2 to 6 weeks for completion. The ultimate measure of success is the performance of the final circuits. The devices are highly sensitive to process variations and are difficult, if not impossible, to repair if a particular chemical process step should fail. Furthermore, because of intense competition and rapidly evolving technology, the development time from layout to final product must be short. Therefore, process control of electronic materials processing holds considerable interest [30, 31]. The process control issues involve three levels:

- Plantwide management
- Materials handling
- Unit operation control

Because of the batchwise nature of electronic materials processing and the long process sequences, a supervisory system is essential. This system collects information on the state of the system; gives status of work in progress; schedules work based on process priorities, product requirement, equipment readiness, and materials availability; and controls product and raw material inventories. Materials handling concerns the physical movement of wafers through the fabrication line via various mechanical means [32] and is therefore primarily a mechanical engineering problem. The control of the individual unit operations and the incorporation of local control functions in plantwide control are topics that chemical engineers have addressed in the context of chemical plants. This experience could be applied effectively to electronic materials processing with appropriate consideration given to the rapidly changing nature of the technology.

The equipment used in the unit operations is complex and microprocessor controlled to allow the execution of process recipes. However, advanced control schemes are rarely invoked. The microprocessor adjusts set points according to some sequence of steps defined by the equipment manufacturer or the process operator. Flows, pressures, and temperatures are regulated independently by "off-the-shelf" proportional-integral-derivative controllers, even though the control loops interact strongly. For example, fluorine concentration, substrate temperature, reactor pressure, and plasma power all influence silicon etch rates and uniformity, but they are typically controlled independently.

The process control situation could be improved by development of process models and monitoring techniques. The models should provide an accurate picture of the underlying physical and chemical rate processes while being sufficiently simple for on-line control strategies. There is currently emphasis on statistically based models. However, the resulting simple polynomial relationship between performance variables and process parameters is only valid over a narrow operating range. Models rooted in the underlying physicochemical processes will allow greater flexibility and extrapolation to new operating conditions. The detailed models discussed above in connection with MOVPE and plasma processing will require more computational resources than are available in on-line control systems. However, these models could be used effectively to construct reduced-order models superior to simple statistical relations.

Whether or not a chemical process step has been successful is difficult to measure, since there are few on-line measurable electrical properties. For example, film thickness and grain structure of polycrystalline silicon can be measured after a deposition step. However, their effect on device performance might not show up until subsequent doping or patterning steps fail. Similarly, it is possible to measure etch rates on-line by laser interferometry, but the etch profiles must be checked by electron microscopy. Unexpected mask undercutting or undiscovered etch residues can result in subsequent contact and device lifetime problems.

Manufacturing processes are run in open loop, where corrective action is taken only after a problem has been discovered at a later stage, possibly even in the final circuit. These long measuring lags and feedback loops lead to wide swings in process yield and quality [33]. Therefore, there is a considerable incentive to incorporate control strategies for the individual process step into closed-loop control of the entire manufacturing process.

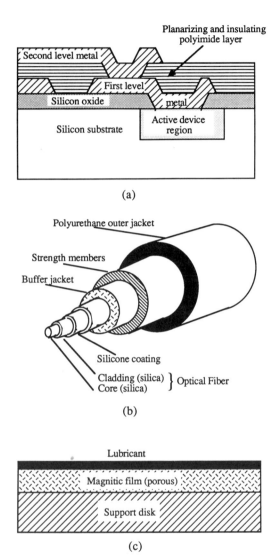

Figure 3. Examples of microstructures of dissimilar materials. (a) Microelectronic device. (b) Optical fiber. (c) Magnetic disk.

This will provide opportunities for pursuing control and artificial intelligence research analogous to work on plantwide control in the chemical industry.

D. Common Aspects of Electronic Materials Processing Research

Chemical engineering contributions to electronic materials processing clearly go beyond the three preceding examples. Other microelectronic areas with similar research issues include crystal growth, chemical vapor deposition of metals and insulators, physical vapor deposition (evaporation and sputtering), and laser-stimulated processes. Furthermore, although microelectronics processing has received considerable attention because of its rapid growth, other aspects of the broad field of electronic materials processing offer equally exciting research opportunities. Examples are optical fiber manufacture, optical and magnetic disk technology, and solar cell production. These technologies involve unit operations similar to those listed in Table 1 and are likewise characterized by the need to process complex material systems. For chemical engineering research to be successful in these areas, researchers must be willing to address large, "messy" problems that seldom have neat analytical solutions. A wide range of materials are typically involved, and the chemical engineer must know enough materials science to address the research issues effectively. For example, in the case of MOVPE growth, perfect uniformity is useless if the solution to the uniformity problem destroys the electronic properties of the grown film, for example, due to particulate contamination caused by a mixing system installed to create continuous stirred tank reactor conditions.

Electronic materials processing research essentially revolves around the reproducible formation of microstructures of dissimilar materials. For example, a typical microelectronic chip involves semiconductors, inorganic dielectrics, metals, and polymers (polyimide) as illustrated in Fig. 3a. Unless the interfaces are well controlled, the device will not function; in the worst case, the layers will lose adhesion to each other. The packaging of the final device entails the processing of metals, ceramics, and polymers into a complex mixture by using techniques similar to those used in fabrication of the device. Optical fiber technology involves a number of polymer coatings on the silica fiber (Fig. 3b), and magnetic disks are carefully engineered magnetic composites covered by a thin organic lubricant (Fig. 3c). Thus, research will have to focus on fabrication and characterization of microstructures and interfaces with highly specific mechanical, electrical, and optical properties. The advances in spectroscopic techniques over the past two decades can be utilized effectively to gain insight into the process chemistry.

Chemical engineers have the opportunity to use their analytical and modeling background to develop detailed process models. However, for these models to provide understanding of the process and be useful in optimization and control, they must include detailed descriptions of the underlying physical and chemical processes. With current computer technology, there is no justification for using oversimplified fluid mechanical and chemical kinetics models instead of realistic reaction–transport models. The necessity to predict final material properties will lead to increased emphasis on models describing microscopic phenomena.

IV. Teaching Electronic Materials Processing Concepts

Although an increasing number of chemical engineering graduates join companies to work on electronic materials processing, the topic is too large and specialized to warrant yet another required undergraduate chemical engineering course in an already crowded curriculum. Most schools already require a materials science course, which is essential for working with materials. Undergraduate students should be shown that the basic chemical engineering principles taught in transport phenomena, thermodynamics, and reaction engineering courses can be utilized effectively in electronic materials processing. This is probably best done through examples and problem assignments. Given sufficient flexibility in the undergraduate curriculum, an elective senior/first-year graduate student survey course in electronic materials processing is a good preparation for an entry-level industrial position or subsequent graduate studies in the area.

A graduate program in electronic materials processing needs to include a solid foundation in materials science, solid-state physics, and devices, besides chemical engineering core courses in transport phenomena and chemical kinetics. Analogous to the situation of undergraduate education, the inclusion of selected examples and problems related to electronic materials processing in the core curriculum will be useful to demonstrate the general utility of the course concepts as well as to illustrate the breadth of the chemical engineering profession. At least one processing course should be part of an electronic materials processing program. By requiring materials science and solid-state physics as prerequisites, it is possible to discuss the relationship between processing and materials properties without spending time on introductory materials science issues. A large number of topics could be included in such a course. A treatment of melt crystal growth, physical vapor deposition including molecular beam epitaxy, chemical vapor deposition, plasma processing, oxidation and diffusion, and polymer processing related to lithography and packaging would amply constitute a one-semester course and touch upon most of the major unit operations in electronic materials processing.

V. Conclusion

Electronic materials processing has emerged alongside other specialized topics in chemical engineering. Increasing numbers of chemical engineers are employed in electronic materials companies and chemical engineering research is making recognized contributions to the field. The traditional chemical engineering strengths in handling problems involving transport phenomena and chemical reactions provide an excellent background for addressing processing issues. To improve the undergraduate training in this area, examples and problems related to electronic materials processing need to be included in the core curriculum. At the graduate level, students should take materials science, solid-state physics, and electrical engineering courses along with at least one processing course.

A wealth of research problems exist in electronic materials processing. These revolve around the relationship between processing and materials properties. Sophisticated experimental procedures using recent developments in gas-phase and surface spectroscopies are necessary to address the problems. Detailed models taking advantage of modern computer resources can provide much-needed insight into the underlying fundamental processes as well as serve as design tools for new applications. It will be important to include accurate physical descriptions and actual chemical kinetics to make an impact on real problems. Moreover, the results must be related to the final material properties. Microscopic issues and the prediction of material characteristics will become increasingly important. Chemical engineers will need to collaborate with researchers from other disciplines to address significant materials issues and find a place in this highly interdisciplinary field.

References

1. Thompson, L. F., in *Perspectives in Chemical Engineering: Research and Education* (C. K. Colton, ed.), p. 373. Academic Press, San Diego, Calif., 1991 (*Adv. Chem. Eng.* **16**).
2. Hess, D. W., and Jensen, K. F., eds., *Microelectronics Processing—Chemical Engineering Aspects*. American Chemical Society, Washington, D.C., 1989.
3. Sze, S. M., *VLSI Technology*, 2nd Ed. McGraw-Hill, New York, 1988.
4. Brown, R. A., *AIChE J.* **34**, 881 (1988).
5. Anderson, T. J., *Adv. Chem. Ser.* **221**, 105 (1989).
6. Jensen, K. F., *Adv. Chem. Ser.* **221**, 199 (1989).
7. Russell, T. W. F., Baron, B. N., Jackson, S. C., and Rocheleau, R. E., *Adv. Chem. Ser.* **221**, 171 (1989).
8. Graves, D. B., *AIChE J.* **35**, 1 (1989).
9. Hess, D. W., and Graves, D. B., *Adv. Chem. Ser.* **221**, 377 (1989).

10. Thompson, L. F., Wilson, C. G., and Bowden, M. J., eds., *Introduction to Microlithography, Theory, Materials, and Processing.* American Chemical Society, Washington, D.C., 1983.

11. O'Brien, M. J., and Soane, D. B., *Adv. Chem. Ser.* **221**, 325 (1989).

12. Jensen, R. J., *Adv. Chem. Ser.* **221**, 441 (1989).

13. National Research Council, Committee on Chemical Engineering Frontiers: Research Needs and Opportunities. *Frontiers in Chemical Engineering. Research Needs and Opportunities.* National Academy Press, Washington, D.C., 1988.

14. Kuech, T. F., *Mater. Sci. Rept.* **2**, 1 (1987).

15. Jensen, K. F., *J. Crystal Growth* **98**, 148 (1989).

16. Aspnes, D. E., Colas, E., Studna, A. A., Bhat, R., Koza, M. A., and Keramidas, V. G., *Phys. Rev. Lett.* **61**, 2782 (1988).

17. Bendt, B. E., Nuzzo, R. G., and Dubois, L. H., *J. Am. Chem. Soc.* **111**, 1627 (1989).

18. Heinecke, H., Brauers, A., Grafehrend, F., Plass, C., Putz, N., Werner, K., Weyers, M., Lutz, H., and Balk, P., *J. Crystal Growth* **77**, 303 (1986).

19. Nishizawa, J., Kurabayashi, T., Abe, H., and Nozoe, A., *Surface Sci.* **185**, 249 (1987).

20. Coltrin, M. E., Kee, R. J., and Evans, G. H., *J. Electrochem. Soc.* **136**, 819 (1989).

21. Coltrin, M. E., Kee, R. J., and Miller, J. A., *J. Electrochem. Soc.* **133**, 1206 (1986).

22. Tirtowidjojo, M., and Pollard, R., *J. Crystal Growth* **93**, 108 (1988).

23. Jensen, K. F., Mountziaris, T. J., and Fotiadis, D. I., in *III–V Heterostructures for Electronic/Photonic Devices* (C. W. Tu, C. V. D. Mattera, and A. C. Gossard, eds.). Proc. Materials Research Society, **145**, 107, Pittsburgh, 1989.

24. Kuan, T. S., Kuech, T. F., Wang, W. I., and Wilkie, E. L., *Phys. Rev. Lett.*, **54** 201 (1985).

25. Gilmer, G. H, and Grabow, M. H., *Am. Chem. Soc. Symp. Ser.* **353**, 218 (1987).

26. Maduhkar, A., Chaisas, S. V., *CRC Crit. Rev. Solid State Mater. Sci.* **14**, 1 (1988).

27. Winters, H. F., and Colburn, J. W., *J. Vac. Sci. Technol.* **B3**, 1376 (1985).

28. Angus, J. C., and Hayman, C. C., *Science* **241**, 912 (1988).

29. Thompson, B. E., Allen, K. D., Richards, A. D., and Sawin, H. H., *J. Appl. Phys.* **59**, 1890 (1986).

30. Atherton, R. W., "The Application of Control Theory to the Automation of IC Manufacturing—Progress and Problems," Proc. Am. Control Conf., San Diego, Calif., p.753 (1984).

31. Kaempf, U., "Computer Aided Control of the Integrated Circuit Manufacturing Process," Proc. Am. Control Conf., San Diego, Calif., p.763 (1984).

32. Heim, R. C., *Solid State Technol.* **29**, 65 (1986).

33. Levy, K., *Solid State Technol.* **27**, 103 (1984).

General Discussion

Microelectronic and Optical Materials

Herbert Sawin: Electronics processing is a subset of chemical processing, and all the chemical engineering tools we could use are obvious. We have a tendency to jump in and look for the kinds of problems that we're used to in the chemical industry. The question is, are we solving the critical problems of electronics processing? Are we going to make contributions or are we just following intellectual pursuits that have no technological relevance? The latter will not educate our students to have an impact in this field. Because electronics processing is an electrical engineering-based industry, chemical engineers are largely viewed in a service role, and we'll never have an impact on that industry unless we can get out of that service role. We need to communicate better with the electrical engineers and materials scientists and to understand the problems ourselves. We can't depend on others to tell us what the critical problems are. We need some background in solid-state physics and electronics. We don't need great depth, just a depth sufficient to broaden our chemical engineering background. The critical problems are all associated with the relationship between chemical processing and the resulting electronic properties. If we're not working on processing that's related to electronic properties, we're fooling ourselves. Approximately 3% of the cost of producing an integrated circuit is associated with the raw materials, so issues about recycle, conservation, and utility of the chemicals are not the critical problems, they're superficial. How can we respond? As Larry Thompson pointed out, we can use example problems within our existing chemical engineering subjects that will acquaint the students with the type of problems they'll have to address. We can broaden our curriculum to include material science or some basic

413

electronics. Moreover, we might consider whether service courses for electrical engineers are appropriate, because they need to understand some chemical processing, and it's hard for them to fit within our regular curriculum. They don't have an appreciation of our appropriate role.

Thomas Edgar: At our school, electrical engineers take one chemistry course and that's it. They're intimidated by chemical engineering courses.

Herbert Sawin: You're right. Other than a major shake-up of the electrical engineering curriculum, the only way I know to expose them to chemical processing is to teach joint courses that involve both chemical and electrical engineers. We can sneak them through in that context so that they can start to understand some of the principles.

Sheldon Isakoff: Klavs Jensen mentioned areas in which chemical engineers have unique skills that are not in direct competition with material scientists or physicists. We, as chemical engineers, can see that very readily, but people in these other disciplines may not. My experience indicates that they need to have some first-hand collaborative work with chemical engineers to appreciate what we bring to the table. I have been involved in two joint ventures that Du Pont has undertaken: one in optical communications with British Telecom and the other in optical information storage with Philips of the Netherlands. Both these firms had little experience with chemical engineers and needed to be shown what we are able to contribute. Working together, we have been able to convince our venture partners that chemical engineers have a role to play and that we do bring skills which complement those of their physicists and materials scientists.

Keith Gubbins: Where will we find the potential chemical engineering faculty members in micro- and optical electronics? There are very few academic groups in this area now. Klavs Jensen has a nice group of students, but they don't want to go into academia. At least, that is what I'm hearing. There's a problem in implementing this idea.

Herbert Sawin: There are two solutions. One, we can produce retrofits like me. I had a surface science background. Two, we can look within our industries for people who have received degrees in chemical engineering and worked in industry and bring them back. The last possibility one might try is similar to what the University of Texas, Austin, has done, and that is to bring in a senior fellow from industry to entice, to cajole, to twist arms and get the faculty interested in the significant problems.

Robert Brown: I want to emphasize that last point, because I think we should hire people well versed in fundamentals. Looking for someone who knows where to get money in an area or how to teach the area is the wrong approach. What Herb said is exactly the way to go. To find someone who has expertise or experience in this area, you're forced to go to industry because the number of academic groups that have those people is very small. The other option is a postdoc, where you would hire someone with the appropriate intellectual background and send the person out for a number of years to gain the experience you're looking for.

Stuart Cooper: Another issue might be the staying power of the microelectronics processing industry. What would be the consequences of altering our curricula, possibly substantially, for an industry that may be moving offshore? It does seem that chemical engineering has served a robust chemical industry well in the past. The challenges in the future may be to adjust to an increasingly non-fossil-based raw materials base as well as to apply chemical engineering approaches to problem solving in rather diverse emerging technologies.

Herbert Sawin: I'd like to respond. We want to hire people of sufficient caliber so that, even if the industry declines, they can go into other areas. That is not of concern to me.

Klavs Jensen: Parts of silicon-based microelectronics manufacturing are moving overseas, but most of the research and development is staying in this country. What goes overseas are the labor-intensive tasks, like packaging and parts of the processing where it's a question of labor costs. All the tool development still remains in this country. To stay competitive with the Japanese and Europeans, we have to educate the people that run these processes. The manufacture of a particular device or chip may be moved overseas, but the whole idea of making microstructures with specific properties for mechanical, electronic, and optical use is not going to disappear. That's what one should focus on. We should not focus on a particular technology.

Larry Thompson: I couldn't disagree with Stu Cooper more. The semiconductor industry in this country is healthy. Leading-edge devices are still manufactured and produced in this country. If we lose the entire microelectronics industry, the United States will not be as technologically exciting. We have got to make that industry work. It's important to the country.

Karen Gleason: There's apparently some concern that chemical engineers will always be service people to electronics engineers. I believe that view is a consequence of our limited cultural bias on what we as chemical engineers can do. Let's go back and look at the wisdom of undergraduates who pick chemical engineering as a profession. They want to use the tools of chemistry and mathematics to produce useful products. I assert that we, as chemical engineers, put a lot of emphasis on the mathematics tools and not as much emphasis on the chemistry tools. We believe we'll leave that to the chemists. I was told by a famous chemist that chemical engineers are interested in stuff, and historically that's been true. Complex, messy problems have been left to the chemical engineers, such as heterogeneous catalysis and polymers, and I suggest that microelectronics can fit also into that realm. We have not been afraid to use advanced computations to deal with bond length distances, but we have not been as aggressive in using spectroscopy. By their nature, both spectroscopy, which is our chemistry component, and computations, our mathematical component, are expensive and long-range studies. We have to make sure that the problems we are dealing with are worth spending money on. We need to be fundamental. These problems will take a long time to solve, so that by the time we do solve them, we want to ensure that our solution will still be meaningful to someone else. Microelectronics is a fast-changing industry, and this means that chemical engineers need to study long-range projects. In this respect, chemical engineers can fill the gap in the current research that's done in microelectronics. Electrical engineers are so driven to produce useful devices that they look at very short-range projects; chemists develop long-range tools, but they want to look at very simple problems that show the beauty of their tools. I believe that a chemical engineer is positioned to understand what the processing issues are in relation to producing devices and also to understand the chemistry of the spectroscopist and actually drive the use of spectroscopy. What are the spectroscopies that we need to look at, what are the issues that we want to look at in microelectronics? If chemical engineers put some innovation into a microelectronics process, they will gain the respect of the microelectronics industry. It will not come from the bottom up by having our undergraduates take more courses. We as a profession need to make sure that we're really contributing to the industry.

James Katzer: I want to refer back to my comment in last night's discussion regarding the intellectual frontiers and expand it beyond chemical reaction engineering, because I think the issues are applicable to all chemical engineering areas. If researchers in a field are not working at the frontier, then that area declines. That happened in process control in the late 1960s and early 1970s. Process control was at the intellectual frontier, and then it went into infinite internal recycle. Academic researchers solved the same problems in greater and greater detail; the frontiers moved out and in different directions, but the researchers didn't follow them. They lacked leadership and did not follow the intellectual frontiers, and process control almost died. Research in process control is now reviving.

Reaction engineering may be experiencing the same thing. If we continue to work on fixed-bed single-phase reactors with multiple steady states in a given problem, then we are not following the frontiers. Frontiers are always moving, and we have to follow them to remain relevant. That's the reason I split frontiers between traditional and evolving technology. Wherever evolving technologies are developing, there are challenging frontiers that our discipline can follow. In the traditional areas of chemical engineering the frontiers are no less important and no less challenging, but as with all frontiers they continue to move and we must follow them. Otherwise, instead of working at the frontier, we are doing irrelevant problems, and the area stagnates. That's the key, critical issue here.

Sheldon Isakoff: I'm encouraged by the staying power of U.S. companies in the field of microelectronics. It's at the core of one of the largest and fastest-growing segments in the world economy. To a large extent, the field has been driven by new materials and materials processing technology. Hence, many of our major chemical firms, Du Pont included, have targeted it for business growth. The contributions that chemical engineering can make have been increasingly recognized. More extensive participation by traditional chemical firms in businesses based on microelectronic and optoelectronic materials will certainly expand the opportunities for chemical engineers.

Arthur Westerberg: I have a question for Larry Thompson that I hope we can discuss. I was fascinated by his comment that in 20 years the Japanese will automate integrated systems and that the United States has to respond by making use of the unique characteristics of our chemical engineering. Does that mean that chemical engineering in Japan is already going in that direction? If so, how can the Japanese do it without making the commitment to fundamentals that is occurring within the chemical engineering profession in the United States?

Larry Thompson: Chemical engineering in Japan is well integrated into the large companies that use chemicals to manufacture, for example, electronic devices. Chemistry is important in the manufacture of many high-technology structures. I think the statement that chemical engineers will be subservient to other disciplines in microelectronics is totally unrealistic. Chemical engineers have risen to top management in every industry they've been involved in. Our unique problem-solving ability gets us into the management structure of companies.

Klavs Jensen: I want to address some of the control issues that are involved because they're very different from what we traditionally have done with continuous processes. In most of these processes, materials are deposited and etched in batch processes. In the next generation of production equipment, the wafer will travel in vacuum channels from one piece of equipment to the next. Furthermore, the individual process steps are short and involve rapid changes. Ideas of adaptive schemes based on slowly varying parameters are totally worthless, and one should focus on new problems that involve combining robotics, which is how these wafers get around, with transient batch processes.

James Douglas: In response to Art Westerberg, the Japanese may beat us in 20 years on this issue because they're willing to automate traditional technology, and they take the automation issue seriously. They concentrate on vision, on robotics, on whatever will allow them to improve a process that they already understand quite well. We have better technology in many respects, but they will take that technology and just hammer it with these other approaches. The automation issue is a big one. My second point— from my interaction with some electrical engineers, I find it isn't just that they don't know chemistry, they don't understand processes either. They think if you put ethyl alcohol and water together in a beaker, the ethyl alcohol comes out first, not that it comes off as a mixture. It's a serious handicap when you're trying interact with them.

Robert Brown: Larry said that the Japanese are giving us a beating because their chemical engineers are doing better than ours. I don't think that's true. The people we run into are of widely varying backgrounds; they're electrical engineers, applied physicists, some chemical engineers, mechanical engineers, etc. One of the big differences in the Japanese industrial society is how they treat continuing education for industrial purposes. Training at the university level is not so critical to their productivity. A Japanese engineer will stay with the same company for a very long time, and when his management identifies an area in which the person needs to be trained, they send him off to the best place they can find. That's not true in our profession, nor of American engineering in general. We're having difficulty deciding what context to put these materials processing educational issues in. The Japanese do not worry about this problem. They just retrain their good people as needed. Very few of us in this room can deal with that. Only industrial people understand it.

Gerald Wilson: I'm probably considered an outsider here, because I have an electrical engineering background. I came in last night, listened to the reaction theory discussion, the environmental discussion, the discussion on polymers. We have great difficulty in engineering in general with our labels. I can't figure out what the hell a chemical engineer is anymore after all this. At one time, I thought it was the interface between the discipline of chemistry and the other disciplines of engineering. Yet, most chemical engineers do not seem to know much chemistry. I don't understand this, and I think we should be spending less time on labels. In the polymer talk this morning, it was said that chemical engineers should take more subjects that relate to polymers. I've heard chemical engineers should learn some electronics. Last night I heard that reaction engineering–catalysis is going to a smaller scale, so they need to understand phenomena on a microscale. I heard this morning that electrical engineers should take more chemical engineering. We must stop educating young people, especially early in their career, for some narrow box we want to fit them into. The most effective engineers are those that learn the broad fundamentals and learn the basic technology in enough depth to demonstrate they can deal with it; secondly, they've learned how to learn. They're not afraid to move to whatever discipline they need. In my opinion, we should change our whole undergraduate program everywhere in this country, to remove these bloody labels. Engineers have to know how to deal with technology and interface with other disciplines, not narrow technologies and narrow little boxy compartments that we defined 100 years ago. Those definitions are wrong for today. We've got to be much different and much more flexible. I don't want to go into the issue of Japanese competitiveness, but the Japanese know how to organize large groups of people to work on one project as a team much more effectively than we do.

Daniel Luss: We are now facing a situation that is rather different from what we faced several years ago. In the past, a major concern of the chemical industry was developing new and improved processes for producing bulk chemicals. At present, the trend is to develop new products, and this requires changes in training and education.

There has recently been a change in the manner in which universities and industry interact, leading to the formation of industrial consortia and centers in many academic institutions. This has led to the development of research teams and to a change in the research experience gained by graduate students who participate in these groups. Every member of a large team, which works on the development of a new product or device, interacts continuously with his colleagues. The pace and depth of his research are affected by a continuous pressure not to become the bottleneck in the development. This environment makes it more difficult to evaluate the innovative nature and contribution of any individual and requires a new attitude toward the expected research accomplishments of a Ph.D. candidate.

George Stephanopoulos: I'd like to make one observation and one comment. Starting with the remarks made by Wilson and Luss, I ask myself where we find someone to hire in an area like microelectronics processing with strong education preparation in physics, chemistry, materials, electronic devices, and processing systems, when in fact most of the academic Ph.D. programs, pressed by funding considerations, prepare students with focused education and perpetuate the division of labor that we have today. Nearly 30 years ago, and for a long time afterward, we had the best attitude in requiring that we hire the best overall athletes and then provided the environment and support that allowed them to champion new and exciting areas. We have some excellent examples of that in this room today. They became champions because they realized the opportunity and gave their personal best. What we see today and for the last few years is rather alarming; chemical engineering departments opt for the best left tackle, or running back, rather than the best athlete. Microelectronics is an area for our brightest with an excellent overall education and not for specialized but weaker ones.

Art Westerberg brought up a very important aspect, automation, all the way from design to manufacture of microelectronic devices. But automation requires a better understanding of the processing and the manufacturing steps. Then the question is, where do you put your money? Do you put your money on understanding just enough so that you can proceed with the automation, or do you spend all your money in trying to understand all the physicochemical phenomena occurring in the process itself? The point here is the context in which you try to solve the problem. In the area of process design, operations, and control, this is probably the most sensitive point. Engineering problems are always solved within a context. What is the context in which to solve the problems in materials engineering? We have been gaining some experience in the control of gallium arsenide manufacture for the last few years. I will use that as an example. Gallium arsenide is used as a material for both digital- and analog-oriented applications. The digital-oriented applications require the characteristics Klavs Jensen spoke of, and they are very difficult to achieve. We don't know the relationship between device properties and internal structure. But digital devices of gallium arsenide represent a very small part of the market. A bigger part is in the analog-oriented applications and devices, where we can move much faster and have complete examples of manufacturing automation. That alone defines a context characterizing the problem that needs to be solved in the process control area. That contextual scope is what I don't clearly see in all of the previous discussion and in many of the research projects in this area. When you're trying to do design or control, you need the help of people who are involved in the scientific analysis, so that the scope you define as designer is consistent with the underlying physics and chemistry.

John Prausnitz: I want to reply to what Dean Wilson said. We have some experience in the suggestion that you're making. You may remember that the engineering school at UCLA was founded in the 1940s, and the idea there was to have no departments. Dean Boelter said, we don't educate chemical or electrical engineers, we educate engineers. There was just one engineering department. Another example was Yale. In the early 1960s, President Brewster at Yale made the famous remark that while professors eventually retire, and even die, departments go on forever unless you kill them off. He tried to do that. He then founded a department of engineering and applied science. Both those experiments failed. We have standard engineering departments now at UCLA and at Yale. One can argue as to why this is. My own explanation is that we have them because of a deeply felt human need to be in families. The reason this experiment failed is the same reason why the communes at Berkeley failed. During the 1960s, young people tried to live together in large groups, or communes. It didn't work. All these communes are gone. There's something about human beings, I think especially about academics, that wants to congregate in small numbers. If you have a very large amorphous group, they're not happy. So the departmental structure is not something you can do away with. Perhaps you can change the name, but the small unit is required to meet a human need.

Gerald Wilson: Sir, I wasn't proposing getting rid of departments. What I'm proposing is that the undergraduate education in those departments not be confined so much to the boundaries that we defined decades ago. Engineering is becoming more and more an interdisciplinary field. If we educate civil engineers to understand only civil engineering, and they can't cross that boundary, engineering's going to fail. The discussions we've had in the last day and a half demonstrate more and more the interdisciplinary nature of engineering. I totally agree with you not to eliminate departments. But if we have departments where undergraduate education focuses only on chemical engineering, say, or on electrical engineering, I think we'll put ourselves right out of the competition in industry.

SECTION VIII
Bioengineering

20

Bioprocess Engineering

James E. Bailey
Department of Chemical Engineering
California Institute of Technology
Pasadena, California

I. Introduction

Bioengineering encompasses a myriad of biological phenomena, objectives, and engineering problems and opportunities. I will focus on bioprocessing, defined here as processing that involves cells or biological molecules as products, as raw materials, or as integral components of the process itself.

Besides suggesting some intellectual frontiers of bioprocess engineering research, I will try to identify areas in which engineers can make special contributions to bioprocessing development and practice. The main theses of this paper can be summarized as follows: (1) the central problems of bioprocess engineering derive from lack of knowledge concerning the intrinsic behavior of cells and proteins, and (2) high-leverage opportunities for chemical engineering contributions to bioprocessing involve systematic manipulation of these intrinsic characteristics, often employing genetic methods. Examples and arguments supporting these theses will be presented in the context of highlighting the characteristics of bioprocesses and their biological components.

II. Why Bioprocesses?

Cells are used because they can perform certain types of chemistry better than anything else [1, 2]. They can conduct certain multistep chemical

ADVANCES IN CHEMICAL ENGINEERING, VOL. 16

conversions with very high selectivity (example: conversion of glucose to ethanol) and can synthesize many complex organic chemicals with high efficiency (examples: L-amino acids, antibiotics). Also, cells are the premier synthesis machines for proteins, including molecules with important pharmaceutical and clinical applications. Although the emphasis of these comments is the use of bioprocesses in manufacturing, cellular activities in waste treatment are also important in current practice. The value of cells in this second context derives from their ability to degrade a variety of complex molecules and at extremely low concentrations of waste materials.

Proteins are macromolecules (molecular weights from about 5000 to 10^6 [3]). There are thousands of different types of proteins, each with a particular biological function, often extremely specific. Thus, a particular enzyme will often recognize only one or a very narrow class of compounds as reactants and catalyze reaction of those to particular products. Other enzymes, typically those that catalyze hydrolysis or other degradative reactions, recognize a particular bond type but will act upon a broad class of reactants (substrates). These protein catalysts typically operate effectively at ambient temperature and pressure. Unique catalytic capabilities give enzymes their niche in bioprocessing.

Antibodies are relatively large proteins normally produced by cells in the animal immune system. Structures at the tips of the antibody molecule endow it with highly specific recognition and binding to a particular small domain on another molecule or larger entity such as a macromolecule or a virus. This highly specific molecular-level recognition is the basis for use of antibodies in chemical analysis, in targeting of drugs and markers to diseased cells, and for specific adsorption and recovery of products in bioprocessing. Like enzymes, antibodies exhibit these extremely selective, tight binding characteristics under ambient conditions.

Since cells and proteins possess this remarkable array of catalytic and binding capabilities, why aren't they more widely used in chemical processing? Present process use of biological activities is limited by their inability to conduct the desired step; by rates that are too small; by requirements for low substrate, product, or antigen concentrations; by rapid loss of activity (termed "instability"); by excessive costs; or by some combination of these factors. Future biochemical engineering research should seek solutions to these fundamental obstacles using both process strategies and genetic manipulation. We must recognize that cells and proteins in bioprocesses are exposed to conditions much different from those in nature, so that properties described by biologists must be reexamined in process contexts. Chemical engineers have been slow to realize that the modern arsenal of genetic methods for accomplishing fundamental changes in cells and proteins is also at our disposal for accomplishing process objectives. In my opinion, one of the greatest opportunities for major chemical engineering impact on process feasibility and productivity is use of modern genetic methods by chemical engineers to create new processes and improve existing ones.

The economic feasibility of a bioreaction process clearly depends on the characteristics of the associated bioseparation process, especially in the usual case when the product is present at low concentration in a complex mixture. For example, the existence of an extremely efficient and low-cost separation process for a particular compound could significantly lower the final concentration of that compound required in the bioreactor to achieve a satisfactory overall process. After noting that special approaches and processes are needed for efficient recovery of small molecules (ethanol, amino acids, antibiotics, etc.) from the dilute aqueous product streams of current bioreactors, I shall discuss further only separations of proteins. These are the primary products of the new biotechnology industry, and their purification hinges on the special properties of these biological macromolecules.

In the next section I outline the new genetic technologies and some of their process implications. Following that, some of the important research frontiers pertaining to process applications of proteins and cells are highlighted. Each of these sections begins with a brief overview of the properties of biological materials that challenge our ability to apply them more broadly and to design and operate bioprocesses in a more systematic and rigorous fashion. After indicating key questions of protein and cell responses to bioprocess environments, I will summarize how modern genetic methods can be used to improve proteins and cells for bioprocess applications.

III. The New Genetic Technologies

Manufacture of numerous products, including antibiotics, amino acids, citric acid, steroids, and other chemicals and food components, was a multibillion-dollar industry in the 1970s. However, at that time biological processes were not widely considered to be the key to an entire new technology and many new types of products and businesses. Yet, the past 10 years have seen an explosive growth in research, development, and recently in sales of new products based on biological processes. This new industrial activity is a direct consequence of the emergence in the late 1970s of two new genetic technologies—recombinant DNA and cell fusion—which have made possible the construction of new types of cells with new characteristics. These are basic enabling technologies in the sense that they make possible cell activities that did not exist previously. In order to appreciate the engineering problems and opportunities associated with commercialization of these genetic technologies and their products, it is necessary to have a rudimentary understanding of the basic concepts and mechanisms involved. These are outlined in the next section; the reader interested in more than beginning awareness of these concepts should consult more complete sources [4–6] since the following presentation simplifies the biology somewhat in the interest of required compactness.

A. Definitions and Introduction

A protein consists of a prescribed linear sequence of amino acids (there are 20 different common, naturally occurring amino acids). The particular sequence of amino acids in a protein is the primary determinant of that protein's functional properties. A cell carries the information for the amino acid sequences of its proteins in the cell's DNA. A small segment of DNA that carries information for the amino acid sequence of one protein is called a gene. Even simple cells typically carry several thousand genes and therefore have the capability to synthesize several thousand different proteins. These proteins together conduct the cell's activities. Consequently, the bioprocess properties of any cell depend ultimately on the set of genes that cell carries and faithfully propagates.

The information for protein assembly is carried in the gene in the form of a sequence of four different building blocks called nucleotides. The process of synthesizing a particular amino acid sequence based on the nucleotide sequence of the corresponding gene is called *expression* and consists of two sequential steps (see Fig. 1). Cells do not always express all of their genes, implying the existence of expression regulation mechanisms associated with particular genes or sets of genes. Some proteins are chemically modified after expression by a variety of processes, which is called *posttranslational processing*. The nature of posttranslational processing of a particular protein depends both on that protein and on the type of cell in which the protein is made.

Biochemical engineers studying kinetics of gene expression have measured product protein activities, but the cellular content of the corresponding gene and messenger RNA (mRNA) and the rates of mRNA and protein synthesis have rarely been determined, although there are standard assays available. These must soon become part of the biochemical engineer's working tools. Kinetic models for synthesis of a particular protein have included transcription and translation, but posttranslational processing is often ignored. Moreover, except for a singular fixation on asymmetric DNA segregation (one model for which has been published at least five different times), biochemical engineers have consistently ignored, both in their experimental studies and in their kinetic models, most of the other impor-

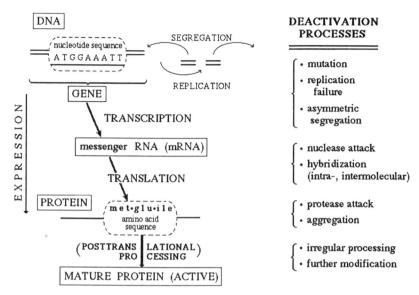

Figure 1. Schematic diagram of mature protein synthesis from the corresponding gene. The nucleotide (A, T, C, or G) sequence in the gene implies the corresponding amino acid sequence (*met*hionine, *glu*tamate, and *isoleu*cine in the inset) through transcription into mRNA and translation of that mRNA into the protein. Removal of some amino acids, addition of sugar groups (glycosylation), and phosphorylation are some of the posttranslational modifications made on certain proteins by certain cells. No posttranslational processing is done on many proteins. To retain their properties, cells must faithfully replicate the DNA and segregate a complete set of DNA to daughter cells upon cell division. The rate of production of active protein depends on the rates at which this sequence of processes occur and also on the rates of the listed deactivation processes that detract from different steps in active protein production.

tant deactivation processes listed in Fig. l. These require much more attention to identify bottlenecks and ways to relieve them.

Until the advent of genetic engineering and cell fusion technologies, the types of changes that could be made in DNA and subsequently passed on to offspring cells were usually very limited in scope and were largely random in location and effect. Thus, it was possible to find a modified bacterial strain that no longer made a certain enzyme or that made that enzyme without the normal regulation, but it was not possible to produce a cell that could synthesize an entirely new protein, much different from any protein normally made by that cell. Genetic engineering technology, also called recombinant DNA technology, provides a well-controlled way of doing this. It is now possible to introduce into a cell a new gene that (in some cases) is expressed to obtain a new protein in the cell. This is important from a manufacturing viewpoint, because genetic engineering enables us to combine the ability possessed by some cell types for economical large-scale growth in bioreactors with the ability to synthesize proteins found only in a few cells in nature that cannot be grown in culture. Therefore, for example, we can now make proteins found in the human body at exceedingly low levels in sufficient quantity for use as therapeutic agents. Besides these clearly dramatic capabilities, genetic engineering also enables other types of systematic improvements in cells for bioprocessing.

Using recombinant DNA methods, it is possible to place several new genes into a cell. However, transfer of entire functional characteristics that depend on a large number of genes into a cell via genetic engineering is not straightforward at present, and in some cases identification of all of the genes involved in some function has not been accomplished. An alternative technology, cell fusion, enables combinations of large amounts of genetic material from different cells with the occasional production of stable, novel hybrids that carry some characteristics from the parental cell types. While this type of large-scale shuffling of many genes has been practiced before (e.g., hybrid plant strains in agriculture), it entered the current realm of process systems when an antibody-producing cell was fused with a long-lived cultured cell to make a *hybridoma* cell. The antibody-producing cell cannot be grown in culture. The hybridoma cell combines the active culture growth of the cancer cell parent with the synthetic capabilities of the immune cell for making one single type of antibody molecule, called a *monoclonal antibody*. While previously it was possible to obtain only small quantities of antibody mixtures from inoculated animals, any quantity of a particular antibody can now be made in bioreactors using hybridoma cells.

Genetic engineering involves laboratory protocols and reagents that, for ease of communication among the cognoscenti, have been given shorthand names. This practice, common to most fields of science and engineering including our own, cloaks the basic ideas and methods of genetic engineering in an aura of sophistication and inaccessibility that is highly misleading. Chemical engineering undergraduates can easily learn and accomplish genetic engineering. Laboratory methodology is well established and reduced to efficient standard protocols.

Thus, it is appropriate to consider genetic engineering as a design tool for the chemical engineer. Once this principle is accepted, the chemical engineer enters a fascinating domain of regulation and manipulation of cells, the real chemical factory of our bioprocesses, and of proteins, the target products of many processes and the catalysts and specific adsorbers in others.

Process applications of genetic engineering will enable chemical engineers to make a high-leverage impact on process performance. Although use of genetic engineering methods by chemical engineers is a relatively new phenomenon, there are already significant examples of important advances.

B. *Process Implications of Current Biological Capabilities*

With presently available cells and methods, there are important limitations on the ability to express a large quantity of *active* protein in different kinds of host cells. Consequently, it is necessary to employ particular types of cells in order to maximize active protein production. In turn, the host cell used dictates many aspects of reactor design and process costs.

Some types of cells presently lack the capability of conducting certain necessary steps in protein production. Most of the important differences concern posttranslational processing. Some bacteria produce inactive, aggregated forms of the product protein. In some cases export of the protein into the external medium is desired, and some cells presently do not do this effectively for proteins from other organisms. Most bacteria do not glycosylate their proteins (add particular sugar groups at specific sites on the protein), yet many proteins that are desired for therapeutic applications are normally made in mammals, where they are glycosylated. Although there has been limited success in making suitably glycosylated proteins in yeast, the surest way to do it now is to employ genetically engineered mammalian cells in culture. The ability to obtain high levels of active larger proteins, which tend to aggregate in bacteria, using genetically engineered insect cells has recently stimulated interest in large-scale cultivation of those cells.

The ability of mammalian cells to synthesize large, relatively complex proteins in active form, with posttranslational processing resembling that in the original organism, motivates much of the current interest in engineering of mammalian cell culture. Another major factor driving current emphases on these systems is interest in production of large quantities of monoclonal antibodies using hybridoma cells, which are formulated from mammalian cell parents. Mammalian cell characteristics in culture remain poorly defined, and equipment and operating strategies for their large-scale cultivation are still limited. For example, we do not know in many cases whether certain proteins are made by growing mammalian cells or by cells after they stop growing, one of the most basic elements in a starting kinetic description. There is much to be done in defining in detail the chemical requirements for growth of mammalian cells and, moreover, the medium composition that will also give rise to maximum product accumulation. Similarly, the response of mammalian cells to dissolved oxygen and pH of the culture is also poorly understood. This is unfortunate, as many strategies including hollow-fiber bioreactors, microencapsulated cells, static maintenance reactors perfused through membranes, and other high-cell-density configurations likely in-volve gradients in oxygen concentration with effects on cell multiplication and productivity that are difficult to anticipate.

As just illustrated, the inherent biological features of different genetically engineered systems define many of the engineering questions. However, one important research frontier is alteration of the cells or culture conditions to overcome some of the current obstacles to use of rapidly growing cells in inexpensive processes. For example, a common barrier to use of genetically engineered *E. coli* for making a valuable protein is the tendency of proteins to form large aggregates in the cells that, after isolation, must be manipu-lated extensively in order to dissolve and reactivate product protein. A re-cent paper by Schein and Noteborn has shown that, by altering the operating conditions during product synthesis, aggregate formation can be avoided and much more active protein can be produced [7]. Further research is needed to identify how cultivation conditions influence the fate of the protein af-ter synthesis. This will require careful attention to basics of protein bio-physics as well as cell physiology. On a longer-term basis, it is not unreasonable to suggest that genetic engineering methods could install in bacteria or in yeast the types of posttranslational processing capabilities that currently require mammalian cells. Such an accomplishment would have dramatic effects on bioprocess economics, on the process equipment involved, and on raw material requirements.

C. Special Considerations for Protein Drugs

Engineers, especially those engaged in bioprocess research in universities, should be more aware of the impact of government food and drug regula-tory processes on engineering objectives and constraints in biotechnology. Moving from an enzyme used in a chemical process to an enzyme used in a food processing application, a protein intended for injection into an ani-mal, or a protein intended for injection into a human, one encounters an increasing maze of required regulatory procedures. These regulations im-pinge upon the point during product development when major engineering contributions to the process can be made in a cost-effective manner. To illustrate this point, I will mention some examples in the area of human drugs.

Before a new drug can be sold, it must go through a series of chemical and biological tests. The time and expense involved in such testing are extraordinary: ballpark figures are a total of 10 years and a cost of $120 million for each new drug. Because of this long lead time and because of the importance of getting to the marketplace first in order to establish product visibility and market share, any company with a new drug aims to enter and push through this process as rapidly as possible. This objective intersects with the testing process in a way that has major implications for bioprocess engineering. After certain points in the testing process, modifications in the organism or in the process require some degree of reiteration of earlier tests or reexamination of the system by regulatory authorities. Therefore, a premium exists for engineering strategies that provide effective a *priori* guidance on the most crucial components of organism, product, and process development. Because of the associated regulatory costs, fine-tuning of an existing process to achieve modest improvements in productivity is often not commercially significant. Such marginal improvements in the system do become more important, however, as competition for supply of that drug develops.

Once the drug and its manufacturing process have been approved by the appropriate regulatory authorities, it is necessary to manufacture that protein using reproducible procedures to ensure the effectiveness and safety of the product. Each product batch is carefully logged with respect to starting materials, process conditions, outputs, and so forth, through each step of the process. The documentation procedure for batch processes is so well established that introduction of continuous-flow processes for manufacture of protein drugs or other pharmaceuticals may be constrained. Intrinsic technical problems in use of some continuous-flow bioreactors for product manufacture are indicated later.

IV. Bioprocess Applications of Proteins

Numerous proteins such as interferons, interleukins, growth hormones, growth factors, and colony-stimulating factors are the primary products of new biotechnology enterprises. Other proteins are produced to be employed in bioprocessing, either as enzymes for catalysis or as antibodies for identification, analysis, and separation of particular molecules. In this section, I examine current technology and future prospects for use of proteins as an integral part of bioprocessing. In these situations, there are often no cells involved, and the protein serves as the primary biological agent in the process. Reaching the heart of this subject requires a modest background in biochemistry, which follows.

A. Background: Environment–Structure–Function Concepts

A basic principle of protein chemistry is the central relationship between three-dimensional structure and activity. Unless the linear polypeptide chain folds into a particular three-dimensional configuration, the protein is inactive. As Fig. 2 illustrates, the active form of a protein is typically a highly convoluted, globular structure in which a particular small domain is the precise locus of interaction with reactant or binding ligand.

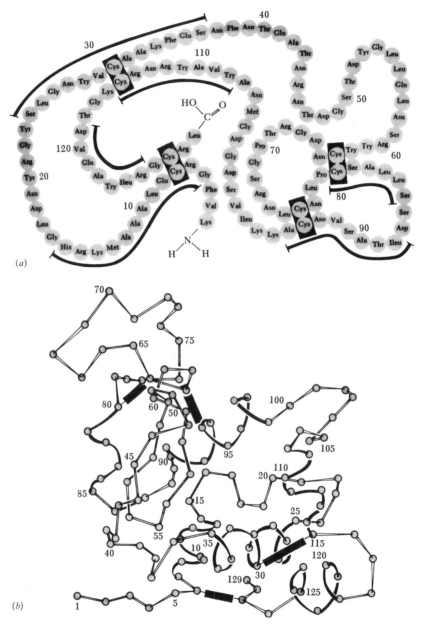

Figure 2. (a) The amino acid sequence of the enzyme lysozyme (from egg white). Blocks enclosing two cysteines (Cys) denote intramolecular covalent cross-links (disulfide bonds). This molecule in crystalline form has the three-dimensional structure sketched in part (b). Note the helical subdomains and "sheet" substructures formed by nearby extended segments. Reprinted by permission from C. C. F. Blake, "Structure of Hen Egg-White Lysozyme," *Nature* vol. 206 p. 757. Copyright (c) 1967 Macmillan Magazines Ltd.

The factors that determine the conformation of a protein are the amino acid sequence, the present environment, and previous history of the protein's environment. Many proteins, when placed in aqueous solution at pH, temperature, and ionic strength similar to those of their native environment, will fold spontaneously into an active form. This occurs because many aspects of protein folding are dictated by noncovalent interactions such that the protein chain can maneuver spontaneously to an active, presumably local free energy-minimizing, configuration. Some proteins, however, include covalent cross-links, and with these a particular sequence of folding and cross-link formation and rearrangement may be needed for proper folding [8]. Although folding pathways have not been elucidated for many proteins, available data indicate that the folding process can involve many intermediate stages, and, consequently, the success of proper folding can be strongly influenced by the environmental history experienced during the folding process. These comments also apply, in reverse, to unfolding, which usually inactivates the protein. Typically, any major change away from the conditions in which a protein functions in nature (higher temperature, much lower pH, addition of organic solvents) will disrupt its structure and cause inactivation. This is a primary consideration in applications of proteins as catalysts, adsorbents, and analytical reagents, since conditions that irreversibly inactivate the protein must be avoided. However, as indicated in examples below, recent research has shown that protein activity can be retained in some environments entirely different from native ones, suggesting applicability of protein activities in a much broader class of process situations than has been widely considered to date.

B. Molecular Recognition Using Antibodies

Antibodies with highly specific binding to particular small domains of other molecules and particles have at least two important potential process applications. One is providing a basis for a variety of highly specific sensors that can be used to analyze the composition of solutions in the bioreactor or in the downstream processing system. Interesting challenges in this area include immobilization methods for maximum activity retention and maximum operational stability, use of antibody fragments instead of entire antibodies to obtain higher densities of binding sites, and discovery of a variety of physical, chemical, and optical methods for transducing the antibody–antigen binding event to an electrical signal. The second area, antibody applications for protein separations, is discussed below.

C. Biocatalysis

Many processes, including some that date back thousands of years, employ enzymes as disposable soluble reagents. Use of costly enzymes was economically prohibitive until the invention and extensive subsequent development of numerous methods for retention of active enzymes in bioreactors. Such retention can be achieved by membrane ultrafiltration, by entrapment in polymeric gels, or by adsorption or covalent attachment of enzymes to membranes or surfaces of insoluble particles. The current knowledge base with respect to retention and reuse of simple enzymes is more broad than deep; anecdotal reports of new combinations of enzyme–immobilization chemistry–support matrix or enzyme–membrane system continue to proliferate in the literature, while in-depth investigations of the mechanisms whereby attachment to a surface can render a protein less susceptible to deactivation, or whereby enzyme activity and membrane flux decline in ultrafiltration systems, are relatively rare.

Technology for cofactor-requiring reactions is not yet generally available, nor are methods for using several enzymes together to effect a long sequence of catalyzed reactions. Finally, industrial use of enzymes outside the conditions typical of natural environments—that is, aqueous solutions, atmospheric pressure, moderate temperatures—also is very limited, but there are tantalizing indications that recent discoveries and currently available methods for manipulating proteins will provide breakthroughs into use of enzymes under very much different reaction conditions than those that are now generally practiced. The rest of this section will concentrate on these unsolved yet promising new venues for applications of enzyme catalysis.

l. Cofactor Requirements and Multiple Enzyme Catalytic Sequences

Many enzyme-catalyzed reactions of potential interest for synthetic chemistry require cofactor substrates which serve as electron donors or acceptors or phosphate donors. These chemicals are much too costly to be used as process feedstocks, so economical practice of these cofactor-requiring enzyme-catalyzed reactions necessitates recovery and chemical regeneration of the required cofactor substrate. In spite of significant research aimed at solving this problem, no general solution exists at present.

To address this problem, Kula, Wandrey, and co-workers in West Germany attached nicotinamide adenine dinucleotide (NAD) to polyethylene glycol [9]. The resulting enlarged-molecular-weight NAD can be efficiently retained, along with enzymes, in the process using ultrafiltration. It should be interesting to look for other types of modifications in these cofactors and associated cofactor recovery, recycle, and regeneration systems.

Cells possess membranes and regeneration systems that accomplish the cofactor retention and regeneration objectives now desired in bioproccessing. This suggests the use of whole cells for cofactor retention and regeneration [10]. Long-term retention of enzyme and cofactor activity in intact cells has proved extremely difficult. These chemical activities are lost by non-growing cells on a time scale too rapid for general use of these multienzyme biocatalysts in industrial processes. Many cells possess special stress response mechanisms which are activated when the cells are challenged with slow growth. Much more needs to be learned about the response of cells to starvation and to conditions that restrict growth, since those are the conditions that would ideally be used in whole-cell biocatalysis. Identifying the cell environments and the types of genetic changes in the cells which could be employed to enhance catalytic stability of nongrowing cells is an important and difficult future biochemical engineering challenge.

2. Enzymatic Catalysis Under Abiotic Conditions

Most enzymes operate in nature in an aqueous environment; this fact, combined with frequent observations of protein precipitation in the presence of organic solvents, created a general belief that an aqueous bulk medium is necessary to obtain enzymatic catalysis. In recent experiments with numerous enzymes and reactions reported by Klibanov and co-workers, the ability of small aggregates of enzymes to conduct catalysis in predominantly organic liquid has been described [11]. These experiments suggest that enzymes may require only a small amount of water. There are numerous challenges for engineering and chemical ingenuity to apply this concept, to expand it, and to discover the general principles underlying these observations. Furthermore, there are many scenarios at the microscale for creating multiphase systems in which enzymes reside in a local aqueous environment that occupies only a small volume fraction of the working volume of the reactor.

Other major expansions of the types of reaction environments where enzymes might be used are suggested by the observation by Klibanov *et al.* of enzyme catalysis in a gas-phase mixture [12] and the data of Blanch and co-workers on enzymatic cholesterol oxidation in a supercritical mixture [13]. Further evidence that enzymes can retain activity under extreme environments is provided by the discovery of microorganisms growing at high temperatures and pressures relative to ambient (e.g., [14]). Such growth requires enzyme catalysis and suggests either novel properties of the enzymes involved or retention of enzyme activity under certain combinations of high temperature and high pressure. Further knowledge of the response of enzymes and cells to extreme conditions is needed to enable systematic investigation of enzymatic and cellular catalysis in organic, high-temperature, and high-pressure environments. Successful adaptation of biocatalysts to these situations may enable synthesis of new chemicals and materials. Chemical engineers should strive to contribute to these opportunities to create new products, since this activity often has much greater technological and financial return than improvement of processes for existing products.

Another potential source of new breakthroughs in biocatalysis is the advent of catalytic antibodies and other synthetic enzymes [15]. New enzyme activities and affinities can be constructed by fusing different peptides. Further, enzymes may be altered chemically to modify their response to adverse environments.

D. Improved Biocatalysts from Protein Engineering

By genetic engineering methods, the amino acid sequence of a protein can be changed in a precise fashion, substituting a particular amino acid for another at a particular point in the protein chain. As many such changes as desired can be introduced in principle. Because both the folding characteristics of a protein and the binding or catalytic activity of its functional domain depend on interactions among amino acids in the protein with each other, with solvent, and with substrates and ligands, alteration of amino acid sequence provides a means for changing the functional properties of proteins significantly [16, 17]. That a single amino acid change could render a protein unstable at high temperature (defined as ca. 42°C) or completely destroy enzymatic activity was well known from earlier studies of spontaneous mutants. Now, using genetic engineering, we can address process requirements for protein applications at the level of protein molecular design.

The processing objectives of such design modification fall into two broad classes. By altering amino acids in the binding pocket of the protein or the catalytic site, the specificity and the activity of the enzyme or binding protein can be changed. Further, changes in this region can significantly shift the pH at which the protein exhibits maximal activity. The other type of functional alteration that is desired in an enzyme is reduction of the deactivation rate in a particular process environment or stabilization of enzyme structure, and thereby enzyme activity, in abiotic environments such as nonaqueous solvents. Several studies have shown that the thermal stability of an enzyme can be increased by particular amino acid substitutions. Often, these involve or appear to involve changes in amino acids far from the binding site which mediate interactions with other parts of the polypeptide or with solvent. Design rules for proteins stable in organic solvents suggested by Arnold indicate directions that will likely produce enzymes active and, perhaps, soluble in nonaqueous media [18], and initial examples of protein stabilization in organic solvents by protein engineering techniques have recently appeared [19]. More needs to be learned about how enzymes interact with process environments and how amino acid changes alter these interactions and the resulting functional properties of the proteins to arrive at a working strategy for protein design for bioprocessing.

V. Protein Recovery

Until genetic engineering made large-scale production of almost any protein a practical prospect, protein recovery was an important but not central interest in biochemical engineering. However, because of genetic engineering and the large markets for therapeutic proteins, protein recovery has become very important, and new extremes of product purity must be achieved on a large scale. Since separation systems often expose proteins to high salt concentrations, organic solvents, interfaces, and other unnatural conditions, we must carefully consider to what extent a particular process for isolating the protein inactivates it.

Protein purification typically involves a sequence of process steps for breaking cells (unnecessary for a secreted protein), separating proteins from other contaminants, concentrating the protein mixture, and then isolating the particular desired protein. Rather than discuss the different types of unit process involved and pertinent technological challenges with each, a more systematic approach to protein purification technology is one based on the fundamental properties of protein interactions with the different types of environments found in these processes. In my opinion, our lack of understanding about how proteins interact with each other, with different bulk phases, and with interfaces limits our ability to invent new separation concepts, to scale up laboratory observations, and to formulate reasonable first approximations to an optimized operating strategy. By becoming more cognizant of key elements of protein chemistry and by studying protein characteristics in some of the novel environments encountered in processing, biochemical engineers can move to center stage in inventing and designing future protein purification processes. The next subsections underscore some of the important generic problems in protein purification.

A. Equilibrium Partitioning of Protein Mixtures

Proteins are molecules much larger than the solvent molecules with which they interact, and proteins possess subdomains that are positively charged, negatively charged, uncharged, hydrophobic, and/or hydrophilic. The chemical potential of such a molecule in a pure solvent is complicated, and the situation is even more complex in a solvent containing ions or other soluble molecules. Further difficulties in describing the thermodynamics of these situations derive from the dynamic and changeable character of the protein molecule itself. First, the protein molecule is in motion, exposing different functionalities to solvents in a transient fashion. Further, the interaction of the protein with its environment may alter the three-dimensional structure of the protein molecule in a time-average sense, such that the groups in contact with solvent differ depending on solvent and temperature.

Shortcomings in current understanding of the thermodynamics of protein solutions can be illustrated by considering one of the oldest protein concentration methods, precipitation, and one of the newest and most technologically interesting methods for protein separation and purification, aqueous two-phase extraction. In both cases, components are added to the solution to introduce a phase change. Salts or organic solvents cause the formation of protein aggregates, and addition of two polymers or of polymer–salt combinations causes a phase separation into two aqueous phases into which different proteins partition in different ways. Present inability to predict the equilibrium solubility of a protein in a precipitation process or the equilibrium partitioning of a single protein in an aqueous two-phase system is clear testimony to the challenges, and therefore the great rewards, of quantitative understanding of the thermodynamics of these complex mixtures. In practice, one is interested not in single-component systems but rather in how multiple proteins respond to precipitation or to contact with an aqueous two-phase system. Learning how to describe equilibrium behavior in these systems, and how this behavior depends on the protein and the solutions employed, will greatly advance the opportunity to employ precipitation and aqueous two-phase extraction to concentrate and purify proteins more efficiently. A complete predictive theory may be beyond our reach; therefore, an appropriate goal is a combination of more limited theory complemented by key experimental information.

Equilibrium behavior of proteins is also crucial, for example, in understanding and improving product recovery from protein aggregates formed in genetically engineered *E. coli* bacteria. What types of solutions will most favorably partition product into solution from the aggregate? To what extent is the aggregate a nonequilibrium structure, and how would this influence processing strategies to recover active product?

B. Interactions with Interfaces

Many current protein separation operations involve exposure of a protein to interfaces, sometimes as the primary purpose of the process step and sometimes as a secondary consequence of that step. In either case, the extent to which a protein partitions between bulk solution and the interface greatly affects the process, and how multicomponent mixtures partition is even more important and even less understood and less predictable. Protein transport processes are also significant and not well understood, especially in confined or highly concentrated domains such as interstices in porous media and faces of membranes.

In many process situations, it is desirable that all or most proteins have minimal interaction with surfaces. In other cases, the surface interaction may serve as the primary mechanism of separation. What is the relationship between the physical and chemical structure of a surface and the partitioning of particular proteins to that surface? How will a protein mixture partition in the presence of a surface with a certain density and organization of charges or hydrophobic groups? What minimum of experimental data is required to define such multicomponent partitioning quantitatively? Considering both association with the surface and the subsequent conditions which must be applied to release the protein(s), to what extent does this bind–release cycle inactivate the protein? How would changing the surface or the conditions used during binding or during release influence the quantity of recovered activity? None of these questions have clear answers at present, and all are significant in terms of improved materials and engineering methods for protein purification.

Although cells possess extremely efficient systems for faithful synthesis of identical molecules of a certain protein, these mechanisms are imperfect and variant molecules can be produced at low frequency. The possibility for heterogeneity at the molecular level may be greater for some foreign proteins made in a genetically engineered cell. As noted in Fig. 1, some heterogeneity in posttranslational processing is relatively common. Because many protein products are destined for injection into humans and animals, it is important both to detect molecular-level inhomogeneity in the product protein and to isolate, to the fullest extent practically possible, a product homogeneous at the molecular level. Regnier and co-workers have shown that proteins differing only slightly in their amino acid sequence interact sufficiently differently with some surfaces to enable separation of the variants in a chromatography column [20]. Discovery of some way to generalize this observation to different proteins possessing different types of structural variants would have great impact. It is very difficult even to detect an extremely small quantity of inhomogeneous product that might cause an adverse immunological reaction in the recipient. Ideally, the ability to better detect such variants would be paralleled by a capability to isolate the desired forms economically at process scale.

Similar questions apply concerning the interaction of proteins with fluid–fluid interfaces. Lipases, enzymes that degrade fats, act at aqueous–fat interfaces. The role of the interface in activating the enzyme and how more efficient enzyme–interfacial contact might be achieved (in a membrane, for example) are intriguing challenges in multiphase contacting, transport, and protein chemistry.

C. Affinity Purification

The extremely high affinity of certain proteins for a particular ligand provides a special opportunity for protein isolation technology. Purification procedures that employ such high-affinity interactions with one or a very small class of proteins in a protein mixture are called affinity purifications. Many known ligand–protein pairs provide extremely high affinity association [21, 22]. These include antibody–antigen, inhibitor–enzyme, substrate–enzyme, protein A–immunoglobulin G, triazine dye–NAD-utilizing enzymes, and particular metals–proteins. Frequently, process exploitation of these affinity interactions involves immobilization of the ligand on a solid support for use in an affinity chromatography separation.

Many current research and development activities seek affinity interactions in combination with other separation principles. For example, affinity ligands bound to membranes allow very rapid contact between protein and the affinity ligand. Another combination system involves binding of an affinity ligand to the polymer in one phase of an aqueous two-phase extraction to increase greatly the partition coefficient for a particular protein. It is likely that imagination, ingenuity, and good applied chemistry will offer breakthroughs in the application of inexpensive affinity ligands in contexts which are amenable to large-scale operations. This is an important area for future chemical engineering research.

D. Genetic Engineering for Enhanced Separations

As noted earlier, genetic engineering methods can be used not only to make a particular protein but also to introduce amino acid substitutions into the protein in a controlled fashion. These protein engineering methods can be used to modify proteins to facilitate their purification.

Examples of this type of modification are already documented. Many involve extension of the protein of interest with some additional amino acids, conferring a distinctive charge or binding property to that protein to enhance its separation. (e.g., [23]). After isolation of this *fusion protein* involving the original molecule and the extension, particular chemistries or enzymes are employed to cut the fusion protein and obtain the desired product.

Other, more subtle applications of protein engineering to enhance the isolation of a particular protein involve amino acid substitutions, carefully chosen to avoid interference with the protein's activity, which confer special binding or partitioning properties on that protein [22]. Chemical engineers, armed with awareness of the types of interactions possible and the options and overall goals of separation processes, are well equipped to attack these protein design problems and should be increasingly involved in this area in the future.

VI. The Cell as a Catalyst and a Chemical Plant

In this section I examine some of the important properties of cells pertaining to applications in process engineering. Henceforth, the emphasis is on production of some desired compound; considerations analogous to those outlined here apply to use of cells as adsorptive agents and decomposition catalysts in waste treatment.

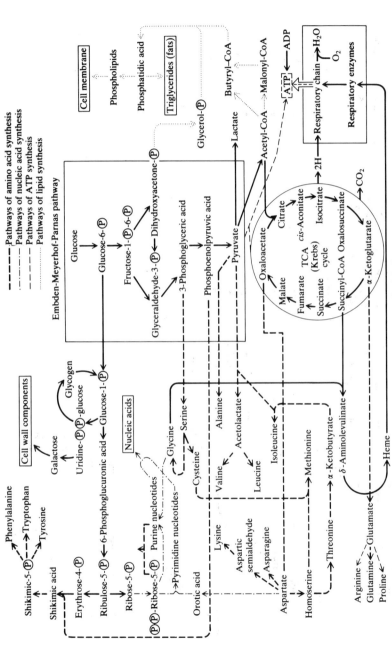

Figure 3. An abbreviated schematic diagram of the major metabolic pathways in an *E. coli* bacterial cell. Reprinted with permission from J. D. Watson, *Molecular Biology of the Gene*, 2d ed., p. 96. W. A. Benjamin, New York, 1980. Copyright (c) 1980 W. A. Benjamin.

A living cell grows and reproduces using a chemical reaction network with about one thousand different catalysts housed within selectively permeable membranes. Figure 3 presents a simplified summary of the metabolic reaction system in a bacterium (*metabolism* means the sum of all chemical activities of the cell). This system must synthesize hundreds of different molecules in the exact proportions (and locations) required to propagate more cells, meanwhile using available nutrients and energy sources with high efficiency. This requires a sophisticated, hierarchical control system implemented biochemically within the cell. Usually (unless the cells themselves are the final product), we desire cells to produce the maximal amount of some particular compound. This overproduction (relative to the cells' requirement for this compound) is achieved by manipulating the cell environment and the cells' genetic constitution.

Application of continuous stirred tank reactor (CSTR) kinetic data on cell activities to predict batch bioreactor performance (or vice versa), translation of laboratory-scale experiments to large-scale operations, and definition of operating conditions to maximize bioreactor productivity are all goals that have been pursued by biochemical engineers for the past 40 years. Success based on anything other than a trial-and-error, empirical bludgeon has been very limited; why haven't we yet found the key to these problems? The reasons are the inherent complexities of cells, which we have too long ignored in the hope that simple macroscopic descriptions of cell stoichiometry and kinetics would suffice. Present biochemical engineering research is turning increasingly toward detailed measurements and models of intracellular processes. This surely is movement in the right direction, yet it is clear that cells are too complicated and too unknown for us to define a complete, entirely mechanistic quantitative description of transport and reactions in cells. Accomplishing quantitative engineering design and analysis of cells (and thus of the bioreactors which propagate and house them) must then depend on a combination of experiment, modeling, and approximation which remains undiscovered. Progress toward these "classical" objectives of biochemical engineering will also contribute to solution of more recent questions, such as how to use slowly growing (or nongrowing) cells as efficient long-term chemical factories and how to use genetic engineering to redirect cell transport and metabolism to maximize chemicals production.

The following sections highlight some of the basic functional properties of cells that profoundly influence bioreactor performance, selected recent advances, and future challenges connected with each.

A. Transport Pathways, Barriers, and Facilitators

Each cell is surrounded by an outer envelope that exhibits different permeabilities to different chemical species. The outer envelope is a complicated, multicomponent, layered structure which includes proteins called permeases, ports, or channels that provide specific transport pathways for particular components. Some of these apply chemical energy from the cell to provide active transport that moves a component to a region of higher chemical potential. The cell includes not only transport systems for ions and small molecules but also transport mechanisms for moving proteins across membranes. In all but one class of bacteria, we also find membranes within the cell that differentiate intracellular compartments having different compositions and in which different activities occur.

Because product synthesis occurs within the cell, and the intracellular conditions are in large part determined by the transport processes and barriers that divide the cell from its environment and the intracellular compartments from each other, these transport barriers and systems are crucial components of cell function. Although a few examples of manipulation of these transport processes and their characterization in engineering terms can be cited, few biochemical engineers are sufficiently familiar with transport mechanisms in cells and their connection with metabolism and cell productivity to appreciate the importance of considering these mechanisms when choosing the cell environment, formulating a kinetic model for design and analysis, or seeking a genetic improvement in cell function.

Paul Weisz suggested in a lucid note published in 1973 that cells, and indeed even entire organisms, have evolved in a way that maintains unity effectiveness factor [24]. That is, the size of the catalytic assembly is increased in nature as the overall rate at which that assembly operates decreases, and the relationship between characteristic dimension and activity can be well approximated by the observable modulus criterion for reaction limitation. It is possible that Weisz's arguments may fail under process conditions, and internal gradients within a compartment or cell may be important. However, at present it appears that the most important transport limitations and activities in cells are those that operate across cellular membranes. Therefore, to understand and to manipulate key transport activities in cells, it is essential that biochemical engineers understand these membrane transport processes and the factors influencing their operation. A brief outline of some of the important systems and their implications in cell function and biotechnology follows.

Protons and other ionic species usually are transported across cell membranes by membrane-associated proteins. Cell energy is invested in many cases to drive ion pumps that cause a concentration difference of the ion across the membrane. Maintenance of these ionic gradients is a major component of the total energy requirement in slowly growing bacteria. That is, when growing slowly, cells must take in energy-yielding substrates such as glucose and use most of this substrate as an energy source to maintain the cell's physical and chemical integrity. This obviously detracts from efficient use of glucose as a carbon source to produce chemicals or more cells, and this maintenance tax on product yield can greatly influence the competitiveness of biological routes to chemicals for which raw material costs are significant. Little is known about maintenance metabolism from a fundamental viewpoint, hindering discovery of new genetic or operating strategies which might reduce maintenance requirements.

Ion transport is also often coupled with cellular energy production and with nutrient and product membrane transport. Aside from Papoutsakis' work on the influence of methanol transport on growth of methanol-consuming bacteria, the importance of membrane control of nutrient and product fluxes into the cell has been largely ignored by biochemical engineers [25]. Better methods for measuring the pH and electrical potential differences across cell membranes are needed, as is more careful consideration of membrane-mediated processes in cell kinetics models.

Among the current mysteries concerning mammalian cell culture is the influence of different proteins in the medium on cell growth and product formation. Providing these proteins in the form of animal serum is expensive and, perhaps more important, introduces variability into cell performance (different lots of serum can give qualitatively different cell responses) and also complicates product recovery. In moving from black art to a systematic methodology based on defined culture media, it is important to know how different proteins in the medium interact with and enter cells to influence their behavior. This is another uncharted frontier in biochemical engineering that should be explored further, building on a wealth of information in the biological science literature about the mechanisms of uptake of particular proteins and their subsequent actions inside the cell. Lauffenburger's models of growth factor receptor trafficking are a singular beacon at present on this intriguing subject [26].

In biotechnology and in enzyme manufacture, export of protein products from the cell is often desired and may be essential for practical feasibility of the process. In spite of this, there have been few biochemical engineering studies to date of the rates of protein transport out of cells and the factors influencing these rates. As cells of greater complexity are considered in terms of intracellular compartmentalization, it is important to realize that protein export is a multistep process with several intermediate stages, any one of which could significantly influence overall protein production rate. Recent research by Gregory Stephanopoulos and co-workers indicates some of the possibilities in this regard [27]. Working with a mammalian cell system in which protein processing and secretion is sensitive to a particular chemical in the medium (a "secretagogue"), they have shown how secretion activities of these complex cells can be manipulated by changing the medium composition, possibly providing engineering benefits in terms of controlling the point in time or space at which the product is released.

B. Activities in Metabolic Reaction Networks

To apply contemporary chemical reaction engineering methods to cells, we would like to know how to calculate the rate at which products are formed and raw materials are utilized as functions of the composition of the reaction mixture—i.e., the cell environment—based on knowledge of the kinetics of the actual reactions taking place. Realization of this objective has been attempted in only a few cases, and no general strategy for achieving this goal has emerged or even been suggested. By examining a single enzyme-catalyzed reaction in the cell and a simple regulated reaction network, I shall identify some of the complications involved in such a basic, systematic approach to cell kinetics.

The enzyme phosphofructokinase (PFK) catalyzes a reaction in the breakdown of glucose to pyruvate found in many types of cells:

$$\text{Fructose 6-phosphate} + \text{ATP} \xrightarrow{\text{PFK}} \text{Fructose 1,6-bisphosphate} + \text{ADP} \qquad (1)$$

The rate of this enzyme-catalyzed reaction depends not only on the concentration of the substrates but also on the concentrations of other components called effectors. For this reaction in yeast, AMP is an effector, and intracellular pH is also important.

While some activating and inhibiting compounds act only on a few enzymes, certain cellular components shuttle chemical groups between different reactions in the cell and, by acting as substrates, products, inhibitors, and activators, transmit chemical information from one part of the metabolic reaction network to another. In this reaction, ATP, ADP, and AMP are components of this type. Because we often have at least a qualitative idea of how these compounds might affect a particular metabolic step, measurement of the intracellular concentrations of these compounds on-line in a bioprocess, or at least conveniently in a research context, would be extremely useful but is not now possible. Significant improvements in our ability to observe ongoing cellular processes in detail and therefore to better control them, or to formulate an overall strategy for experimental characterization and mathematical description of cell kinetics, depend on new instrumentation developments and on more careful and systematic identification of exactly where and how different quantities we measure and manipulate influence operation of the cell machinery.

Moving from such qualitative consideration of inhibition and activation to detailed quantitative description is difficult. The yeast PFK enzyme is one of the most thoroughly studied, and the following rate expression has been found to describe the dependence of the rate of reaction 1 on substrates and effectors:

$$V_{PFK} = V_{PFK}^{max} \frac{g_R \lambda_1 \lambda_2 R^{n-1} + qL g_T c_1 \lambda_1 c_2 \lambda_2 T^{n-1}}{R^n + LT^n} \tag{2}$$

where

$$\lambda_1 = [F6P]/K_{R,F6P}$$

$$L = L_0 \left[\frac{1 + C_\gamma \gamma}{1 + \gamma} \right]^n$$

$$\lambda_2 = [ATP]/K_{R,ATP}$$

$$R = 1 + \lambda_1 + \lambda_2 + g_R \lambda_1 \lambda_2$$

$$\gamma = [AMP]/K_{R,AMP}$$

L_0 is pH dependent

$$c_j = K_{R,j}/K_{T,j}$$

$$T = 1 + c_1 \lambda_1 + c_2 \lambda_2 + g_T c_1 \lambda_1 c_2 \lambda_2$$

$$q = V_{T,max}/V_{R,max}$$

To evaluate such kinetic expressions *in the cell* obviously requires a method for determining the reaction rates and the concentrations of all the important substrates and effectors. This is now possible for some very well studied carbon breakdown pathways using whole-cell NMR spectroscopy, but extension of these methods to synthesis in cells is so far untouched. From a qualitative viewpoint, biosynthetic pathways are known to employ feedback product inhibition in order to regulate the flow of a starting precursor to a variety of different products required by the cell (see Fig. 4).

Because modulation of enzyme activities depends on metabolite concentrations, which in turn are determined by the entire metabolic network, the overall response time for these controls can be on the order of seconds. This is the same as the time scale for changes in environmental conditions (e.g., pH, dissolved oxygen concentration) encountered by cells as they circulate through the nonuniform contents of a large-scale bioreactor. Therefore, beyond the complexities of enzyme activity control in the steady state, dynamic properties of this control system are important. The circulation pattern in a bioreactor has major effects on product formation [28]. Lack of understanding of transient responses of cell metabolism is one central obstacle to systematic scale-up of laboratory results (obtained in idealized,

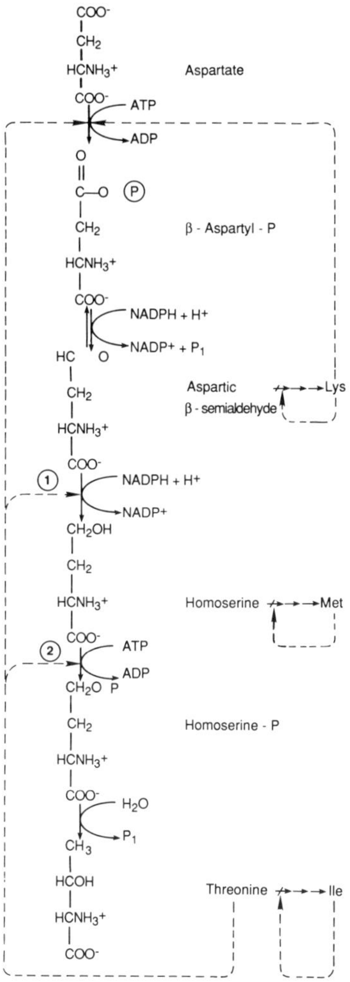

Figure 4. Summary of the metabolic reaction pathways which synthesize the aspartic family of amino acids [homoserine, lysine (Lys), methionine (Met), threonine, and isoleucine (Ile)]. Solid arrows denote reactions catalyzed by different enzymes, and dashed arrows denote enzyme activity control (the compound at the tail of the arrow inhibits the enzyme at the head of the arrow). Enzymes (1) (homoserine dehydrogenase) and (2) (hemoserine kinase) were amplified by genetic engineering in order to improve threonine production. A mutant host was used in which most of the feedback inhibition interactions shown here were disabled. Adapted with permission from W. B. Wood, J. H. Wilson, R. M. Benbow, and L. E. Hood, *Biochemistry, a Problems Approach*, 2d ed., p. 294. Benjamin/Cummings, Menlo Park, Calif., 1981. Copyright (c) 1981 Benjamin/Cummings Publishing Co.

relatively uniform mixtures) to large equipment design. Another is the fixation of most biochemical engineering practitioners on the same sparged, agitated stirred tanks used since the 1940s. These have impossibly complex patterns of internal circulation which are extremely scale sensitive. Several European groups seem to be on a more promising long-term track by developing loop and column bioreactors with relatively well-defined flow, and hence environmental trajectory, distributions and thereby suitable properties to allow predictable scale-up.

C. Regulation of Metabolic Network Structure

A fascinating property of cells, and one with important practical implications, is their regulation of the *amounts* of different enzyme catalysts and permeases present in the cell. Placed in a certain environment, a particular micoorganism will not synthesize all of the enzymes it is capable of making. The cell takes signals from its environment and determines which enzymes to make and how much of those enzymes to produce. Thus, the structure of the metabolic reaction network is variable, presenting a major challenge to mathematical description of cell kinetics. So far, a few kinetic models have considered these controls of protein amounts, but how kinetics change when levels of numerous proteins are controlled has not been investigated. In fact, very little data exists on quantities of different key enzymes and transport proteins in the cell as functions of extracellular environment and genetic makeup. Biochemical engineers should be more active in applying and improving contemporary methods of multicomponent protein analysis and in incorporating results of such measurements into cell kinetics models.

Identifying an environment that avoids induction of undesired enzymes and repression of desired ones and implementing bioreactor control systems that maintain these desired conditions in a bioprocess are subjects of future importance. For example, accumulation of a product in the cell environment can often repress synthesis of some of the enzymes required for production of that compound. Product repression and inhibition phenomena have motivated special interest recently in combined bioprocessing operations which accomplish separation simultaneously with bioreaction. By continuously removing a product that inhibits its own synthesis, production of that material is improved. Development of new selective membranes and other process strategies for accomplishing these separations is an important area for future research.

To date, biochemical engineers have concentrated heavily on modeling and analysis of the enzyme *synthesis* process and have almost completely ignored the enzyme *decomposition* processes which are important in many types of cells. For example, bacteria actively degrade many or perhaps all of their proteins with rates that depend on the particular protein being degraded and the cell environment. Further experimental studies to examine these effects are future frontiers for biochemical engineering, as is description of these protein degradative processes within cell kinetic models.

If a cell encounters transient conditions, the time scale for changing levels of different proteins via modulation of synthesis and degradation is on the order of a few minutes or more. Since cell doubling times range from 20 minutes for bacteria in rich medium to 10 to 20 hours for animal cells and 25 to 100 hours for plant cells, we expect that protein levels, and with them some major aspects of cell operation, will change during the course of a batch process (which might involve 3 to 50 doublings of cell numbers). This is a major complication and challenge in design and optimization of batch bioreactors and in relation of data and models for transient batch cultures with those for steady-state CSTR cultivation.

D. Long-Term Genetic Drift

Darwin's principle of natural selection has implications for the technical feasibility of continuous bioreactor operation and also plays a more obvious role in determining the fate of growing organisms in natural systems. When a process involves growth of cells over many generations, the possibility exists of genetic drift resulting from spontaneous mutation (or loss of whole molecules of DNA). This is important in practice because cells used in manufacturing have usually been genetically altered for maximum product formation, usually resulting in decreased cell growth rate. Thus, subsequent mutations that enhance growth rate and decrease productivity are, from the viewpoint of cell survival in a CSTR, advantageous and will eventually lead to a transition from a productive to a less productive, but more rapidly growing, type of cell. Genetic drift during long-term propagation poses challenges for continuous bioreaction processes and also for application of genetically modified cells in waste treatment, mineral processing, and agriculture. In some genetically engineered cells, these genetic changes can occur so fast as to have a significant impact even in batch processing.

Long-term genetic drift is also important in recently conceived processes for manufacturing valuable proteins using mammalian cells. Although only anecdotal observations are available at present, it is clear that the important engineering properties of these cells, such as their oxygen utilization rate, rate of product synthesis and excretion, and susceptibility to mechanical damage, can change significantly over numerous generations. Cells placed in a new environment first show some adverse reactions, then frequently adapt over many generations. We know almost nothing about the nature of these changes, the factors influencing them, and how to monitor and characterize the properties of cells used in manufacturing to ensure reproducible results from one batch to another.

E. Cell Interactions with Interfaces

Several hardware configurations have been demonstrated to produce cell masses sufficiently dense that cell–cell contact begins to occur to a significant degree or in which cell–heterogeneous surface interactions occur. These systems often employ membranes, either in a cross-flow configuration or using rotating or vibrating membranes to seek minimal cell fouling. Alternatively, a relatively stationary mass of cells can be grown on one side of a membrane, or within the pockets found in porous particles, or in the outer domains of an asymmetric membrane. Interaction with membrane surfaces and with other cells and intense hydrodynamics near the membrane are known to influence cell activities. However, there is little fundamental data on these effects.

Some configurations with stationary cells are potentially influenced by spatial gradients in nutrients and product levels arising from the kinds of diffusion–reaction phenomena familiar in chemical engineering analysis of catalyst pellet performance. While available chemical engineering methods help estimate the relative importance of mass transfer limitations, these methods are difficult to extend to quantitative design of systems with significant gradients. One reason should already be clear from the previous discussion—cells in different environments will exhibit a different mix of protein activities, so that one is dealing with a continuous distribution of different types of catalysts. Furthermore, several experiments have clearly shown that intrinsic kinetics of cells may be altered significantly when the cells reside in a dense mass or in contact with a surface. Much more needs to be done to characterize cell responses to surfaces and to contact with other cells.

Many types of mammalian cells, including some that are important in synthesis of commercial products, will not grow unless anchored to a surface. Furthermore, the cells will not grow unless there are other cells sufficiently nearby.

F. Genetic Engineering for Chemicals Production and Degradation

The first wave of genetic engineering applications is directed at valuable proteins. However, genetic engineering also provides a new technology for highly controlled genetic manipulation of the cellular chemical factory. That is, using recombinant DNA methods, it is possible in principle to introduce new enzymes and/or new permeases into a cell which expand the capabilities of the original chemical machinery in that cell. Furthermore, we can selectively eliminate certain steps in the original metabolic network or amplify the activity of a particular step already present. This capability creates a new era for cellular design. We now should regard the cellular reaction network as under our direct control through genetic engineering. We can begin to restructure this network in ways that will enhance the production of chemicals already made by the cell, enable the production of chemicals previously not synthesized by that cell, and even cause synthesis of entirely new molecules. The products made by these cells could include pharmaceuticals, specialty chemicals, bulk chemicals, and materials for use based on their surface, interfacial, or bulk properties.

It is unlikely that cells that evolved in nature for growth in a natural habitat are optimized for making particular compounds. Thus, it is expected that genetic modifications can improve, perhaps dramatically, the productivity of cells for particular chemicals. The challenge is to identify these changes and also to discover changes that create entirely new cellular reaction networks that produce new compounds of value. After surveying a few examples already reported of such metabolic engineering (a term used to describe manipulation of cellular metabolism using genetic engineering), I shall summarize some of the major challenges to broader, more systematic, and more sophisticated application of this general concept.

After transferring to *E. coli* DNA from another bacterium, Burt Ensley and co-workers at Amgen noticed cultures that were a deep blue color [29]. It was discovered that adding new enzyme activity to *E. coli* metabolism resulted in synthesis of the dye indigo. This new biological route to indigo contrasts markedly with current chemical synthesis methods, which involve toxic raw materials and wastes. Minimization of toxic wastes production by use of biological processes is one of many factors that motivates greater use of biological systems for chemicals manufacture in the future.

As part of a major corporate effort to develop a new process for production of microbial cells for animal feed, ICI realized that the organism to be used in this process employed an energy-requiring pathway to assimilate nitrogen [30]. By inserting one new enzyme into those cells via genetic engineering, an alternative pathway for nitrogen uptake without energy requirements was installed. This increased the yield of biomass per unit amount of methanol raw material consumed by 4 to 7%, a figure significant in such a commodity process and one that represents only a hint of the yield improvements that subsequent genetic enhancements might provide.

Researchers at Ajinomoto have used genetic engineering in a directed fashion to improve yields of amino acids produced by microbial fermentations [31]. For example, by increasing the amount of the enzyme for the step labeled 1 in Fig. 4, they were able to increase the final concentration of threonine in a batch cultivation from 17.5 to 25 g/liter. Subsequent amplification of the enzymatic activity for step 2 in Fig. 4 in concert with the amplification of step 1 gave further yield enhancement to 33 g/liter of culture.

We can envision in the future much more extensive genetic manipulation of cells for enhancement of chemicals production. Numerous enzyme activities will be increased, others decreased, some new enzyme activities will be added, and similar manipulations will be undertaken which modify transport across biological membranes. The obstacles to this are twofold: (1) techniques for moving large numbers of new genes into cells and maintaining them at desired relative levels must be improved, and (2) a strategy for systems analysis of cell function is needed so that, through a facile combination of experiment and theory, those steps that should be modified to most improve productivity can be identified. The chemical reaction network of a cell is too complicated and subject to too much internal regulation to anticipate, without some systematic quantitative framework, how one or several alterations in the network will change its operation. The problem is exacerbated by the extremely large number of changes that could be made in terms of levels of present enzymes, regulation of synthesis and/or activity of those enzymes, and introduction of new enzymes and permeases. The existence of such a large number of decision variables, which interact in a very complicated way in the system to be optimized, makes a rational design and optimization structure crucial.

To move toward such a framework, it is important to develop and expand presently available cell modeling concepts and methods for obtaining information on sensitivity of performance to particular steps based on a systematic model–experiment methodology. Further, instrumentation for off-line and on-line measurements of metabolite levels and rates of reactions inside cells is important to provide key data for such analysis and for development of a procedure for cell design. In addition, it will be important to observe carefully how introduction of changes in the levels of particular proteins may interact, through the cell's control system, to change the amounts and activities of other proteins. Such interaction and feedback that occur within the cell make this design problem challenging.

Even more complicated are cell manipulations that produce new molecules (other than new proteins as discussed above). When new combinations of catalytic activities are expressed in cells, they may be able to accomplish functions not previously possible. For example, bacteria have been constructed using genetic engineering that can degrade toxic hydrocarbons, although the starting organism did not possess this capability [32]. Another example is manipulation of antibiotic synthesis in the laboratory of D. Hopwood. There, by placing several genes from one antibiotic-producing organism in another antibiotic-producing organism, a new genetic construct was created that synthesized an entirely new antibiotic molecule [33]. The opportunity of producing new substances is one that has potential for breakthroughs in different venues of the chemicals and pharmaceuticals industry. Again, there are many possibilities for trial-and-error attempts, but only anecdotal and incidental success in this regard can be expected until a systematic strategy is developed for estimating and predicting what types of new structures might result when new combinations of enzyme catalysts are created inside cells.

G. Genetic Engineering for Broadly Enhanced Cell Activity

In addition to manipulation of certain particular protein activities in the cell to enhance production of a desired chemical, another class of genetic manipulation of cells can be envisioned that improves them in a general sense for process applications. For example, growth of all cells depends on effective uptake of nutrients, effective production of chemical energy for driving essential cell activities, and minimization of formation of toxic by-products. Therefore, genetic manipulations which make the cells more robust with respect to nutrient uptake, more active in energy production, and less disposed to make toxic by-products would have applications in production of numerous different compounds and cloned proteins. The goal in this arena is to make a cell that is generally improved for process environments, which would then serve as the host or the starting material for subsequent genetic manipulation directed at making a particular process.

Examples of this type of genetic enhancement are rare but intriguing. Limitation of cell activity resulting from low supplies of oxygen is a common endpoint in many bioreaction processes. A general genetic strategy for improving cell activity in low-oxygen environments (and perhaps in higher-oxygen environments also) has recently emerged from Chaitan Khosla's research in my laboratory. Khosla has added to *E. coli* the ability to synthesize a functioning hemoglobin molecule [34]. Cells producing this hemoglobin, besides being a deep red color, also respire more rapidly, grow more rapidly in low-oxygen environments, reach higher final cell densities, and synthesize protein more efficiently than do control strains that lack hemoglobin [35].

The discovery of this new genetic approach for enhancing activities of oxygen-requiring cells was triggered by observations in other laboratories that one bacterium that lived in an oxygen-poor environment synthesized hemoglobin. The success in this exercise suggests a possible paradigm for identifying genetic modifications that might render cells more active and more compatible with process environments. That is, by studying the properties of cells that grow naturally or have grown for a long time in the laboratory under conditions similar to those of process interest, it may be possible to identify particular genes that provide that cell with enhanced characteristics for the process environment.

Another example of a possible strategy for genetic enhancement of cell function in a broad sense is the research presently in progress by Valkanos, Stephanopolous, and Sinskey at MIT to reduce lactic acid production in mammalian cells by eliminating specifically the enzyme needed to produce lactic acid. This strategy may lead to cells that use carbon more efficiently and produce fewer inhibitory products in the medium. Such a cell line would be of interest in many different contexts.

This area and that discussed previously provide fascinating opportunities for chemical engineers, with their knowledge of process objectives and problems, their ability to analyze complex systems, and their quantitative skills, to interact synergistically with biologists to identify how we should change cells to make them better process organisms and then to implement those changes. The practical leverage of this genetic approach to engineering improvement of organisms for process application can be very high.

VII. Neglected Areas: Agriculture, Food Processing, and Waste Treatment

There is some danger that a narrow focus on the most familiar technology—namely bioprocessing—is excessively limiting the scope and imagination of biochemical engineering research and educational activities in the United States (and elsewhere) at present. Agriculture, food processing, and waste treatment are already vast industries and ones in which chemical engineering can make tremendous contributions. Since these are, in one sense or another, bioprocesses, a few comments on possible opportunities in these areas are provided here. Diversification of more chemical engineering research and educational activity into these industries would strengthen and stabilize the profession.

A. Chemical Engineering in Agriculture and Waste Treatment

Many important agricultural processes involve providing nutrients and chemicals to organisms to maximize their rate and yields of growth or of formation of some product such as grain. A more systematic and rational analysis of transport of the active material from the preparation applied into the soil, subsequent interaction with soil surfaces and soil organisms, and eventual uptake by the plant—all reminiscent of familiar chemical engineering fixed-bed adsorption and mass transfer systems—could aid tremendously in converting greenhouse and field testing from data gathering to an efficient systematic science. How would rainfall influence these transport processes and the biological uptake processes? How would rainfall and sunlight influence the plant metabolism relative to the action of the compound? How would an insecticide or herbicide be transported and contacted with the target pest? Clearly the interactions among multiple components, soil, microbes, insects, and plants are complex, but they can be analyzed by chemical engineers in terms of the fundamental transport, adsorption, and reaction processes involved. Also, novel techniques are needed to obtain reliable estimates of the important parameters and variables in these processes.

There is another, somewhat different arena in agricultural biotechnology where chemical engineers can contribute significantly. Because of concerns about how genetically engineered cells or their genetic elements will propagate in natural environments, opportunities to apply genetic engineering technology to produce microorganisms that can provide plants with insecticides, nutrients, or frost protection, for example, have been highly constrained. Similar barriers presently limit use of cells genetically engineered to degrade toxic waste in open environments or in waste treatment processes. Aside from the emotional positions that have influenced legal and regulatory actions on environmental release of genetically engineered cells to date, there are many important scientific questions. One major class of questions is connected with microbial ecology and population dynamics. What will happen to genetically engineered cells after they are placed in the soil or in receiving waters? Will they proliferate or will they fade in concentration as a result of direct interactions with other organisms and indirect interactions, through competition for nutrients and production of toxins or antagonists? Chemical engineers are well prepared to consider all of the attendant complications in these systems, such as nutrient exhaustion with time, inhibitory by-product accumulation, spatial gradients of nutrients and inhibitors, multicomponent adsorption, and growth of different organisms. It should be emphasized in connection with this important area of contemporary science policy that genetically engineered cells will be able to attack some pests in ways that produce much less environmental disruption than application of chemical pesticides for the same purpose. Similarly, genetically engineered cells degrading dilute toxic compounds can aid the environment substantially. The risks of environmental release of genetically engineered organisms are at present entirely hypothetical, yet concerns about these are presently blocking movement into the realm of biological agents that can provide immediate environmental benefit.

There are probably similar opportunities for chemical engineering contributions in animal science, but this author is too naive in that subject to make even starting suggestions. There is, however, one aspect of animal biotechnology that deserves special consideration here, as it may render obsolete many of the processes that now occupy much of our teaching, research, and industrial development activities. Several different research groups have demonstrated the ability to synthesize foreign proteins in plants and animals. Rats with human proteins in their milk have been produced by genetic engineering of the whole animal, and foreign proteins have also been synthesized in whole plants. The development track here is long in terms of obtaining high foreign protein yields and moving through the whole regulatory process for a therapeutic agent made with a completely new process. Still, it is possible that genetically engineered plants and animals may eventually produce, relatively cheaply and simply, the valued proteins which are now under development in microbial or mammalian cell systems. Chemical engineers may want to consider the kinetic and yield aspects of these macroorganisms for production of valued proteins and to identify the special separations requirements which may arise in isolating proteins from plants, milk, or animal fluids. Furthermore, the ability to transfer functional genes into plants and animals will certainly lead ultimately to new products. Chemical engineers could contribute significantly to the identification and implementation of this possibility.

B. Chemical Engineering in Food Processing

Many foodstuffs are produced using enzymes or growing cells at some points in the process. Yet, quantitative experimental characterization of these processes in the terms that a chemical engineer would consider necessary to define the fundamental transport, chemical reaction, and thermodynamic aspects are rare. Although some areas of food science and engineering, such as milk pasteurization, have been analyzed experimentally and mathematically using essentially chemically engineering methods and perspectives, this is not true of many aspects of food processing. It is ironic that at present many more chemical engineers in university laboratories study plasmid stability in recombinant *E. coli*, one of many important problems in this system that has only moderate economic impact at present, while practically no academic chemical engineering groups consider kinetics of growth of cheese starter cultures or the complex multicomponent flavor chemistry of brewing and how these can be manipulated by process design and operation. These and other food processing industries already have annual *multibillion-dollar* markets and are increasingly moving in the direction of more processing with more synthetic combinations of flavors and textures.

VIII. Biotechnology and Chemical Engineering Education

The potential value of the outlook, knowledge, skills, and experience of chemical engineers in inventing, developing, and improving processes involving biological components has been repeatedly emphasized in this paper. What is needed to realize the potential for chemical engineering impact in biotechnology and all the other fields in which cells and proteins are key players? How should the current situation be changed or evolved to aid bioprocessing, agriculture, biological waste treatment, and food processing in finding and utilizing the most efficient and productive processes? Clearly, we cannot fulfill the potential in these areas unless we understand the key sensitivities and opportunities involved in these processes, and, as evident from the entire structure and tone of the above presentation, I think that knowledge of the properties of the biological components in processing contexts is absolutely crucial.

While the comments above have emphasized limitations and opportunities based on biological properties, it is important to recognize that there are needs and opportunities to improve heat transfer, mass transfer, hydrodynamic features, process configurations, and other aspects of process design and operation which are highly developed fields of knowledge, research, and practice within contemporary chemical engineering. Thus, many currently practicing engineers could provide major contributions in biotechnology if only they understood the problems and the systems.

Greatly improved biological literacy in chemical engineering across the board is very important for two additional reasons. One is the identity of the chemical engineering profession. Biochemical engineering, which increasingly involves major emphasis on biological properties and components, is one of the most rapidly growing areas of chemical engineering research. Yet, communication of the aspirations and the results of this research to colleagues often proves impossible because many cannot tell the difference between an enzyme and a cell, much less the properties of either. Perhaps because of a smaller barrier of foreign terminology, chemical engineering colleagues seem much more comfortable with other emerging chemical engineering fields such as polymer science and microelectronics processing than with biotechnology. Efforts must be made to improve this situation and to expand the general biological literacy of chemical educators and practitioners. Another motivation for this is the emergence of biological science as the most dynamic and exciting realm of contemporary science. One need only look at the contents of *Science, Nature*, or *Proceedings of the National Academy of Sciences* to appreciate the extent to which advances in biology dominate the science horizon.

Turning now to education of the biochemical engineering specialist, it is essential that students intending to practice biochemical engineering in research, development, or operations have knowledge, ranging from basic to advanced as one moves backward from production to research, in molecular genetics, cell biology, and protein chemistry. The movement of bioprocessing from an interesting but fringe element of chemical engineering and industrial technology to a center position is based entirely on advances in protein chemistry and molecular genetics. If chemical engineers do not understand these subjects, cannot interact effectively with biological scientists, and do not begin to participate at the level of cell and protein design, we will remain on the fringes of this industry. Even now, many major U.S. corporations view the biochemical engineer as one who sets up and operates the manufacturing process. The valuable insights and contributions that biochemical engineers well trained in modern biology could make at the research and development levels are for the most part ignored. Perhaps even more telling is the observation that fermentation research and process development activities in many biotechnology companies are managed and staffed by microbiologists, and most protein recovery groups in those companies are led and staffed by protein chemists. These practices indicate that understanding cells and understanding proteins may be more central to maximizing the efficiency and output of a process than understanding mixing and mass transfer or fixed-bed adsorption, if one has to make the choice. On the other hand, a chemical engineer with a solid grounding in biological basics pertinent to bioprocessing has extraordinary qualifications and capabilities for developing and conducting superior bioprocesses. More than any other type of professional, such chemical engineers will make exceptional contributions in the integration of product, organism, and process design.

The forces just outlined suggest that emergence of biochemical engineering as a separate entity is a strong possibility. If in fact the greatest scientific and commercial leverage comes from an amalgamation of chemical engineering approaches and biological science, and if those on this track become increasingly isolated professionally from the rest of chemical engineering, emergence of separate curricula designed to maximize the effectiveness of education for biotechnology and related fields is a very clear option if not a likelihood. This should not be necessary. Personally, I believe that the approach to approximate quantitative description and effective utilization of complex systems embodied in contemporary chemical engineering science is an extraordinarily valuable resource in biotechnology. Thus, a major challenge in chemical engineering education is providing a core education in this chemical engineering outlook, which confers the requisite engineering and mathematical knowledge, methods, and intuition, and can then serve as a basis for further study and practice in biotechnology or other emerging new fields of chemical processing.

Acknowledgments

This paper is a statement of the author's opinions. These have been refined and occasionally corrected by helpful comments and criticisms from Frances Arnold, Georges Belfort, Harvey Blanch, Doug Clark, Clark Colton, Larry Erickson, Mike Shuler, Tony Sinskey, and all of my research group. The author is also indebted to the National Science Foundation and the Chemical and Biocatalysis Program of the U.S. Department of Energy for financial support of research in several areas considered in this commentary.

References

1. Darnell, J., Lodish, H., and Baltimore, D., *Molecular Cell Biology*. Freeman, New York, 1986.
2. Gottschalk, G., *Bacterial Metabolism*, 2nd Ed. Springer-Verlag, New York, 1986.
3. Stryer, L., *Biochemistry*, 3rd Ed. Freeman, New York, 1988.
4. Drlica, K., *Understanding DNA and Gene Cloning*. Wiley, New York, 1984.
5. Rosenfield, I., Ziff, E., and Van Loon, B., *DNA for Beginners*. Writers and Readers, Exeter, 1982.
6. Watson, J. D., Tooze, J., and Kurtz, D. T., *Recombinant DNA: A Short Course*. Freeman, New York, 1983.
7. Schein, C. H., and Noteborn, M. H. M., *BIO/TECHNOLOGY* **6**, 291 (1988).
8. King, J., *BIO/TECHNOLOGY* **4**, 297 (1988).
9. Bueckmann, A. F., Kula, M.-R., Wichmann, R., and Wandrey, C., *J. Appl. Biochem.* **3**, 301 (1981).
10. Mattiasson, B., ed., *Immobilized Cells and Organelles*, Vols. I and II. CRC Press, Boca Raton, Fla., 1983.
11. Klibanov, A. M., and Zaks, A., *J. Biol. Chem.* **263**, 3194 (1988).
12. Klibanov, A. M., Barzana, E., and Karel, M., *Appl. Biochem. Biotechnol.* **15**, 25 (1987).
13. Randolph, T. W., Clark, D. S., Blanch, H. W., and Prausnitz, J. M., *Science* **238**, 387 (1987).
14. Deming, J. W., and Baross, J. A., *Appl. Environ. Microbiol.* **51**, 238 (1986).
15. Schultz, P. G., *Science* **240**, 426 (1988).
16. Oxender, D. L., and Fox, C. F., *Protein Engineering*. Liss, New York, 1987.
17. Cantor, C. R., and Schimmel, P. R., *Biophysical Chemistry*, Part I, *The Conformation of Biological Macromolecules*. W. H. Freeman, San Francisco, 1980.
18. Arnold, F. H., *Protein Engineering* **2**, 21 (1988).
19. Arnold, F. H., *Trends in Biotechnology* **8**, 244 (1990).
20. Regnier, F. E., *Science* **238**, 319 (1987).
21. Gribnau, T. C. J., Visser, J., and Nivard, R. J. F., eds., *Affinity Chromatography and Related Techniques*. Elsevier, Amsterdam, 1982.

22. Arnold, F. H., *BIO/TECHNOLOGY* **9** (in press, 1991).

23. Sassenfeld, H. M., and Brewer, S. J., *BIO/TECHNOLOGY* **2**, 76 (1984).

24. Weisz, P. W., *Science* **179**, 433 (1973).

25. Papoutsakis, E. T., Bussineau, C. M., Chu, I.-M., Diwan, A. R., and Huesemann, M., *Ann. N.Y. Acad. Sci.* **506**, 24 (1987).

26. Lauffenburger, D. A., Linderman, J., and Berkowitz, L., *Ann. N.Y. Acad. Sci.* **506**, 147 (1987).

27. Sambanis, A., Stephanopoulos, G. N., Lodish, H. F., and Sinskey, A. J., *Biotechnol. Bioeng.* **35**, 771 (1990).

28. Osterhuis, N. M. G., *Scale-Up of Bioreactors: A Scaled-Down Approach*, Ph.D. Thesis. Delft University of Technology, Delft, The Netherlands, 1984.

29. Ensley, B. D., Ratzkin, B. J., Osslund, T. D., Simon, M. J., Wackett, L. P., and Gibson, D. T., *Science* **222**, 167 (1983).

30. Windass, J. D., Worsey, M. J., Pioli, E. M., Pioli, D., Barth, P. T., Atherton, K. T., Dart, E. C., Byrom, D., Powell, K., and Senior, P. J., *Nature* **287**, 396 (1980).

31. Morinaga, Y., Takagi, H., Ishida, M., Miwa, H., Sato, T., Nakamori, S., and Sano, K., *Agric. Biol. Chem.* **51**, 93 (1987).

32. Ramos, J. L., Wasserfallen, A., Rose, K., and Timmis, K. N., *Science* **235** , 593 (1987).

33. Hopwood, D. A., Malpartida, F., Kieser, H. M., Ikeda, H., Duncan, J., Fujii, I., Rudd, B. A. M., Floss, H. G., and Omura, S., *Nature* **314**, 642 (1985).

34. Khosla, C., and Bailey, J. E., *Nature* **331**, 633 (1988).

35. Khosla, C., Curtis, J., De Modena, J., Rinas, U., and Bailey, J. E., *BIO/TECHNOLOGY* **8**, 849 (1990).

21

Some Unsolved Problems of Biotechnology

Arthur E. Humphrey
Center for Molecular Bioscience and Biotechnology
Lehigh University
Bethlehem, Pennsylvania

I will begin by complimenting Professor Bailey on his paper [1]. Overall, I agree with most of his conclusions, but I do disagree with some of his points. In this paper, I will introduce several topics by discussing the role of chemical engineers in biotechnology research, issues on which our views differ widely.

In 1948, the McGraw-Hill Award for Achievements in Chemical Processing went to the Merck Company for "Biochemical Engineering." Although biotechnology was recognized three decades ago, it is still an emerging, entrepreneurial industry, with a relatively small process engineering knowledge base. Nearly two-thirds of the over 400 companies involved in biotechnology have been formed within the last 10 years. They lack the experience in bioprocessing of large companies, such as Du Pont or Monsanto. Moreover, federal support of biotechnology research has been directed mostly toward the sexy, basic research in molecular and cell biology, with little toward the support of research in downstream bioprocessing. In 1986, $600 million was spent on basic biotechnology research, with only $6 million spent on applied biotechnology research. That is a hundred-to-one ratio. Very little has been done to correct this imbalance despite the reports issued by the Office of Science and Technology Policy (OSTP), the

ADVANCES IN CHEMICAL ENGINEERING, VOL. 16

National Research Council (NRC), and the Congressional Office of Technology Assessment (OTA) lamenting this imbalance and its threat to rapid commercialization of biotechnology. Funding bioprocess engineering research is not taken seriously at the federal funding level, but the opposite is generally true at the state level. There are now more than 80 biotechnology centers receiving state funding. Many have pilot plants with a strong bioprocessing research focus. Some even operate under good manufacturing practices (GMP). I worry about the effect of the vastly greater availability of funding for basic research at the federal level, relative to the need for chemical engineering research in the biotechnology area. Because of this basic research focus, there has been a tendency among some chemical engineers to become involved in the fringes of molecular biology, rather than focusing on bioprocessing needs. It is on this point that I have a basic disagreement with Professor Bailey. I believe that the so-called biochemical engineers should do what chemical engineers do best, not do poorly what biochemists and molecular biologists can do better.

Dr. Bailey also notes that the chemical engineering challenges in biotechnology should include waste utilization, foods, and agriculture, and that these areas are mostly overlooked. I agree. But a key point he makes is that "every chemical engineer should have a basic understanding of protein structure, protein chemistry, cellular behavior, and the rudiments of genetics." In my opinion, this is necessary only to the extent that the chemical engineers gain a sufficient knowledge base to communicate in these areas with biotechnology researchers and be involved in bioprocess development from the very beginning. I agree with his statement that "because of the large amount of apparent foreign terminology in biotechnology, our chemical engineering colleagues seem more comfortable with other emerging fields like polymer science rather than biotechnology." My own experience has been that foreign terminology need be only a small barrier in communicating results of research in biotechnology. I disagree, however, with Professor Bailey's statement that "understanding cell and protein behavior may be more central to maximizing bioprocess efficiency than an understanding of mixing and mass transfer." My experience has shown that they are equally important in large-scale commercial biotechnology systems. When we talk about a large-scale bioreactor, we are talking about 100 to 1500 cubic meter systems. ICI built a 15,000 cubic meter bioreactor to biooxidize methanol to make biopolymers. Mixing was the key problem in optimizing the operation.

In passing, I note that biotechnology is becoming a "two-scale culture." One is based on relatively low-value, high volume products such as biochemicals, antibiotics, enzymes, etc., having a value around $10/kg, an annual production around 1000 kilotons, and produced in 10–500 cubic meter bioreactors. The other involves the very high-value products such as the interferons, interleukins, erythropoietin, etc., that have a value on the order of $1,000,000/g, an annual demand of 1 kg, and are produced in 0.05 to 1 cubic meter bioreactors. In the former, the contributions of the chemical engineer to bioreactor design and scale-up are well recognized. In the latter, molecular biologists and biochemists are not yet convinced they really need the services of the chemical engineer. However, problems of shear sensitivity, mass transfer, flow distribution, aseptic design, and control may be even more critical than in the large bioreactors. Hence, the chemical engineer can make significant contributions to process design of systems for producing these high-value biochemicals.

I take particular exception to Professor Bailey's statement that "the greatest opportunity for major chemical engineering impact on bioprocess feasibility and productivity is the use of modern genetics by *the chemical engineer* to create new processes and improve new ones." He should have said by teams of biochemists and molecular biologists that include chemical engineers. Few if any chemical engineers are sufficiently trained or experienced to create new bioprocesses on their own.

In this connection, I would like to cite three examples of chemical engineering challenges in biotechnology where, while an understanding of cell behavior is needed, it is the basic chemical engineering research that will likely make the difference. One example is from plant cell culture, the second from antibiotic production, and the third from biotechnology monitoring and control.

I. Plant Cell Cultures

One reason for interest in plant cell culture is that over 20,000 different chemicals are produced from plants, with about 1600 new plant chemicals added each year. Also, 25% of all prescribed drugs come from plants. These chemicals can be produced in a bioreactor through suspension culture. Advantages of plant cell suspension culture, as compared to agriculture, are that plant cell suspension culture can be carried out independently of weather conditions and political problems, it does not compete with other agricultural products for land use, and it is done in a controlled environment which minimizes contamination and provides easier product validation and assurance.

To estimate what conditions need to be met to make plant cell culture economical, I have utilized Professor Cooney's economic analysis of fermentation systems (see Table 1). Professor Cooney concluded after examining a number of fermentations that 12 cents per liter per day of product value must be achieved in the bioreactor to have an economical process. The

Table 1. Cash Flow Analysis of Various Fermentations (unpublished estimates from C. Cooney)

Product	Concentration (g/liter)	Productivity (g/liter-day)	Price (¢/g)	Revenue (¢/liter-day)
Citric Acid	150	50	0.17	9
Ethanol	80	80	0.07	5
Glutamic Acid	100	33	0.04	13
Gluconic Acid	300	100	0.10	11
Lactic Acid	130	65	0.19	13
Lysine	73	33	0.50	12
Penicillin	30	4	4.0	15
Protease	20	7	1.0	7
Riboflavin	10	2	6.0	10
Streptomycin	25	3	1.0	15
Xanthan	25	3	1.0	15
Vitamin B_{12}	0.06	0.02	750	15
				average: 12¢/liter-day

reported activities of two plant cell culture systems (Table 2), one producing diosgenin and the other ajmalicine, indicate that, with the best technology available today, these suspension cultures achieve only 2 to 3 cents per liter per day value. We need roughly an order of magnitude improvement in plant cell culture technology for economic feasibility.

What can be done to achieve this necessary improvement? Most plant cell products are secondary metabolites, which means that plant suspension culture systems involve a growth phase and a production phase. In any commercial system, one will probably want to separate physically these two phases. Unfortunately, one has to add a growth hormone such as 2-4-D for a successful growth phase, and these hormones must be completely removed before elicitors are added to the production phase to induce the secondary product. Optimizing this two-stage system is an engineering challenge.

Table 2. Plant Cell Systems

Product	Concentration (g/liter)	Productivity (g/liter-day)	Price (¢/g)	Revenue (¢/liter-day)
Diosgenin	0.45	0.03	67	2
Ajmalicine	0.4	0.02	150	3

Other engineering problems need to be solved to achieve successful plant cell culture. For example, suspended plant cells tend to aggregate. Aggregation can lead to intraparticle mass transfer limitations. Plant cells, like mammalian cells, are shear sensitive. Because of this shear-sensitive characteristic, most plant cells systems cannot be vigorously agitated; hence, their processes are frequently mass transfer limited. Ways must be found to maximize mass transfer while minimizing shear in these plant cultures. Furthermore, when working in small scale, one of the great difficulties is that plant cells readily adhere to glass. It is difficult to keep plant cells suspended in glass bioreactors. Alternatives to glass bioreactors must be found.

In spite of these problems, one product of plant cell suspension culture has achieved commercialization. This product, shikonin, is a red dye, produced in Japan at the level of several tons per year. Published reports suggest the optimum yield occurs at a relatively low volumetric mass transfer for oxygen, roughly 10 hr^{-1} (see Fig. 1). When higher oxygen transfer rates are used, the production rate drops.

This suggested that the plant cells producing shikonin were shear sensitive, which we have confirmed in our laboratory. We investigated plant cell suspension culture under fixed shear conditions. When insoluble oxygen carriers were added, in our case a perfluorocarbon, significant improvements occurred in the cell and product yield.

What levels of mass transfer rates are needed to achieve economical levels of plant cell culture? I have calculated some requirements in terms of $k_L a$,

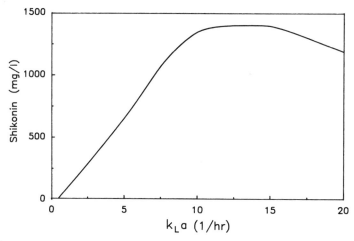

Figure 1. Relationship between the yield of shikonin and initial $k_L a$. Reprinted with permission from Tanaka [2].

the product of mass transfer coefficient and surface area permit volume, to achieve economical production of some of the most interesting plant cell products (Table 3). For example, using *Catharanthus roseus*, the *E. coli* of the plant cell world, to produce ajmalicine, the table indicates that a 40- to 50-fold improvement will be needed to achieve an economical process. To do this we will need to achieve around 25 to 50 g/liter of cells with a growth rate of 0.4 reciprocal days. That means the cells need to double every 1.6 days. This growth rate involves a $k_L a$ for oxygen in the range of 300 hr^{-1}. The chemical engineering challenge is to achieve this oxygen transfer while minimizing cell shear. Alternatively, the chemical engineer could work with the molecular biologist to find ways of developing plant cells that are not shear sensitive. Some engineering solutions to this problem would include oxygen enrichment of the aeration gas, use of oxygen carriers, or immobilization of cells in gels or other beads.

II. Antibiotic Production

I now turn to an example concerning antibiotic production, where the problem is increasing productivity in oxygen-limited, very viscous, mycelial systems. In many of these antibiotic-producing biosystems, a 5% improvement in yield represents several million dollars of increased profit per year—obviously, there is great motivation to try to get even that 5% improvement.

Most antibiotic fermentations are performed with batch processes because the highly mutated cells employed tend to revert back to a more healthy (wild) state. Such mutated cells cannot be sustained for long continuous periods of time. For the most part, the problems of antibiotic fermentation involve broth viscosity and the related gas holdup and oxygen transfer limitations. All antibiotics are secondary metabolites and the fermentations typically run 200 hours. The first 40 hours involve mostly cell growth, while the remainder involve mostly antibiotic production. The fermentation is driven essentially to oxygen choke-out, i.e., close to zero dissolved oxygen by the end of the growth period (see Fig. 2). Cells stop growing and maximum viscosity is reached. The cell concentration and activity at this point set the antibiotic production rate. In the antibiotic production phase, the substrate addition is controlled at or slightly above the maintenance requirement.

One would think a solution to increase the productivity would be to use oxygen enrichment, not just during the choke-out period but also during the remainder of the fermentation in order to sustain the productivity. This does not work because of broth viscosity and gas holdup problems. In highly mycelial systems, a 15% increase in cell mass doubles the viscosity. The volumetric oxygen mass transfer coefficient and the bubble rise velocity—

Table 3. $k_L a$ Requirements for Plant Cell Suspension Culture [a]

Plant/drug	Max. cell conc. (g/liter)	Max. growth rate (days^{-1})	yield (g/g)	oxygen demand (g/liter-hour)	$C*$ [b] (g/liter)	C_L [c] (g/liter)	Needed $k_L a$ (hr^{-1})
Serpentine	25	0.1	1	0.1	0.005	0.0015	30
Ajmalicine	15	0.35	1	0.22	—	—	63
Diosgenin	11	0.07	1	0.03	—	—	10
C. roseus	30	0.2	1	0.25	—	—	72
Maximum likely	25–50	0.4	1	1.0	—	—	300

[a] With shikonin, shear effects bein to limit productivity at a $k_L a$ of only 10 hr^{-1} in a shake flask culture. Therefore, if shear limitations can be overcome, at least a 30-fold improvement in productivity should be possible. This is the challenge!

[b] This represents 25% utilization of O_2 in sparged air.

[c] This represents 20% of saturation dissolved oxygen level.

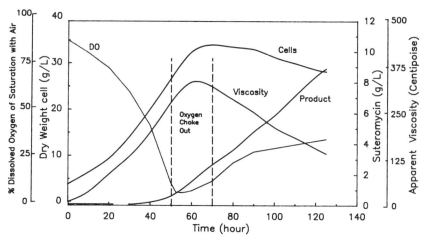

Figure 2. Time course of an antibiotic fermentation.

hence holdup—are inversely proportional to the viscosity. The problem with fungal antibiotic fermentations is that the system starts out as water, but ends up with a 200- to 1000-fold increase in viscosity. Another challenge is to contain such fermentations within the bioreactor system and not allow the volume to overexpand due to increased gas holdup. Gas holdup is a function not only of the air flow rate but also of liquid depth, liquid volume, bubble velocity, and surface tension. For low-viscosity aqueous systems, gas holdups are only on the order of 2 to 5%. For most chemical reactions, holdup problems are not a concern. In a fermentation, where viscosity can increase from 1 to 1000 centipose, holdup can be a major problem during that period of greatest viscosity. Two solutions are used to minimize the holdup and to contain the broth in the bioreactor. One is to control holdup with surface-active agents. The other is to decrease the aeration demand through oxygen enrichment.

There are differences of opinion, which should be resolved by application of simple chemical engineering considerations, on how to achieve oxygen enrichment. Two recommendations have been advanced. One suggests adding the oxygen to the aeration gas supply upstream of the air sterilization system. The other recommends that the oxygen should be sparged directly into the fermentation tank because liquid oxygen is available in pressurized tanks, and the stored energy can be used to produce finer bubbles and higher mass transfer area. Advocates of this solution point out that these bubbles will have four to five times the oxygen concentration driving force compared to that for air bubbles. However, when injected into a high-viscosity broth, fine bubbles have trouble separating from the liquid, and one can make literally a shaving cream out of the broth. With this approach, a separate gas sterilization system must also be provided. Which solution is better? This is a good chemical engineering problem, a problem in which there is money to be made on the correct solution.

III. Monitoring and Control

My third example of chemical engineering challenges in biotechnology is a problem in monitoring and control. In virtually all practical fermentations where the medium contains solids and/or where the cell concentrations are greater than 10 g/liter, it is impossible to monitor directly the cell concentration and hence cell activity. It would be most helpful to know the instantaneous cell concentration and activity in order to control the substrate feed rate, particularly in a fed-batch operation.

With the advent of reliable on-line gas spectrophotometers, it has become possible to estimate the cell concentration and activity indirectly by use of oxygen material balances around the fermentor and using crude models for cell behavior. The engineering challenge is to improve the existing models, by learning how to express the model constants in terms of measurable environmental parameters and/or by using models containing meaningful cell structure that can be measured on line. Alternatively, one can develop trend analysis for the biosystem behavior and then incorporate such analysis in an expert system for on-line control and optimization of the bioreactor.

In the existing approach, on-line oxygen balancing is used for two reasons. First, oxygen concentration is easy to measure in the entering and exit gas streams; second, oxygen is sufficiently insoluble in the biomedium (essentially water) that the accumulation term in the material balance can be neglected. Hence, the oxygen uptake rate (OUR) is simply the difference in mass flow rates between the oxygen in the gas flowing into the bioreactor less the oxygen in the gas leaving the bioreactor.

The most simplistic oxygen uptake model is one that treats the cells collectively as uniform black boxes requiring oxygen for cell growth (dX/dt), maintenance (where m is the specific maintenance uptake rate), and product formation (dP/dt), i.e.,

$$\text{OUR} = \frac{1}{Y_{x/o}} \frac{dX}{dt} + mX + \frac{1}{Y_{x/o}} \frac{dX}{dt} \tag{1}$$

Here, the Y's are stoichiometric or yield constants. To obtain an estimate of cell concentration (X) by this relationship, one must obtain an expression for the product formation rate (dP/dt) in terms of cell concentration and measurable parameters. The simplest model of product formation is that the rate is either growth associated, non-growth associated, or some combination of both, i.e.,

$$\frac{dP}{dt} = \alpha \frac{dX}{dt} + \beta X \tag{2}$$

where α and β are rate constants. Substituting equation (2) into (1), one obtains

$$\text{OUR} = A \frac{dX}{dt} + BX \tag{3}$$

where $A = (\alpha + 1 /Y_{x/o})$ and $B = (m + \beta)$. It is now possible to solve for dX/dt and X from the following relationships:

$$\frac{dX}{dt} = \frac{1}{A} (\text{OUR} - BX) \tag{4}$$

and

$$X = \int_{X(0)}^{X(t)} dX = \int_{0}^{t} \frac{(\text{OUR} - BX)}{A} \, dt \tag{5}$$

To solve equation (5) one must initialize X, i.e., determine $X(0)$. This can be estimated by knowing the cell inoculum size. It can also be estimated in the early part of a fermentation by assuming that the growth rate is close to or at the maximum growth rate.

Thus, through application of simple chemical engineering stoichiometric and kinetic principles, the biochemical engineer has the ability to estimate cell concentration and growth rate online, a task that has eluded the biologist.

IV. Conclusions

These three examples represent only a "drop in the bucket" of the myriad of challenges to the chemical engineer in biotechnology. Their solution involves chemical engineers doing what they do best. While we need to interact with molecular bioscientists, I do not believe we need to become second-class geneticists or protein chemists to contribute effectively to biotechnology problem solutions. We do need to convince the geneticists and biochemists that chemical engineers can be useful in solving many of their problems.

There are some fundamental chemical engineering challenges in biotechnology. Only by forming integrated teams of chemical engineers and molecular biologists will the best solutions be found.

References

1. Bailey, J. E., in *Perspectives in Chemical Engineering: Research and Education* (C. K. Colton, ed.), p. 425. Academic Press, San Diego, Calif., 1991 (*Adv. Chem. Eng.* **16**).
2. Tanaka, H., *Process. Biochem.* **22**, 106 (1987).

22

Chemical Engineering: Its Role in the Medical and Health Sciences

Channing Robertson
Department of Chemical Engineering
Stanford University
Stanford, California

I. Introduction

The chemical engineering profession has made significant contributions toward our understanding of the life sciences. Within the subset of the life sciences that emphasizes medical and health concerns, chemical engineers played a widely recognized role in developing the basic principles on which modern renal dialysis, or artificial kidney systems, has been designed and constructed [1]. The cumbersome plate and frame devices of the early 1970s are now relics of the past, having been replaced with compact hollow-fiber modules. What made these advances possible was a sound understanding of hemodynamics, blood–materials interactions, and water and solute transport through synthetic polymeric membranes. The purpose of this brief discourse is to review the foundation that chemical engineers have laid in the medical and health sciences and to provide some guidance for future activity in this area.

II. Historical Perspectives

The principle realms of the chemical engineer in the medical and health sciences are discussed below. While this is not intended to be all-inclusive,

ADVANCES IN CHEMICAL ENGINEERING, VOL. 16

it does represent the primary areas where chemical engineers have contributed most.

A. Hemodynamics

The flow of blood in both large and small vessels, as well as within conduits that comprise a portion of an *ex vivo* or artificial system, has interested chemical engineers. Since blood is a suspension consisting of an essentially Newtonian continuous phase and deformable particles or cells with sizes ranging from approximately 2 to 10 μm, it has been convenient to divide flow problems into two categories: those in which blood can be treated as a continuum and those in which it cannot. The former generally is possible for vessels or conduits larger than approximately 60–90 μm in diameter. While laminar flow typically is characteristic of the conditions encountered *in vivo* and *ex vivo*, there are exceptions, such as at the root of the aorta. The issues that have been studied in large-vessel hemodynamics take into account pressure and flow wave propagation arising from the elastic properties of the vessel walls. Several interesting nonlinear phenomena occur in such cases, some with clinical significance (e.g., the so-called shot sounds that can be heard through a stethoscope). Flow separation at deposits on the inner walls of vessels and at vessel bifurcations has also been a fertile area of investigation.

In small vessels or conduits, where blood can no longer be treated as a continuum, the flow problems become exceedingly complex. This complexity is exacerbated by the deformability of the cells and their tendency to distribute nonuniformly across the conduit cross section.

Complementing these hydrodynamic issues has been an ongoing effort to examine the rheological properties of blood as well as other body fluids. Blood exhibits shear thinning behavior and appears to have a yield stress owing to the formation of macroscopic structures that incorporate protein-aided bridging of red cells.

The literature is replete with both experimental and theoretical studies attempting to address the aforementioned issues. While efforts in this general area still persist, the ground is certainly well trodden.

B. Water and Solute Transport

As mentioned already, the artificial kidney is a classic example of chemical engineering prowess. The proper design of such devices requires a description of both water and solute transport to and from blood, across membranes, and to and from an adjacent fluid known as the dialysate. Variations on this theme include hemodilution, hemoconcentration, and hemofiltration. Applications of these same principles have been used to examine continuous ambulatory peritoneal dialysis. Oxygenation of blood, primarily for purposes of surgical procedures requiring cardiac cessation, is another example of interphase mass transport, in this case between a liquid and a gas phase. Denaturation of blood proteins at the gas–liquid interface is a problem; reducing or eliminating denaturation by use of an intervening gas-permeable membrane is now becoming a common practice.

Mass transfer processes responsible for the accumulation and deposition of cholesterol-carrying low-density lipoproteins in blood vessel walls have received considerable attention [2]. Such studies have enhanced our understanding of phenomena that contribute to the formation of atherosclerotic plaques.

Yet another application of solute transport to an area of clinical significance is the development of advanced drug delivery systems. These systems seek to control the rate and amount of a substance delivered to the human or animal environment. Most are based on either controlled dissolution processes or membrane transport behavior. Such an approach to drug delivery has become a major industry and has been discussed and described in the chemical engineering literature [3, 4].

C. Pharmacokinetics

Prompted by the early work with the artificial kidney by Dedrick and Bischoff [1], together with later applications to cancer chemotherapy, considerable effort has been made to develop analytical models to describe the transient fate of solutes within physiological systems. Experimental measurements obtained in laboratory animals have been used to develop a set of scaling rules useful for extending these results to humans. These models describe the time course of solutes that are either being removed or injected according to a prescribed temporal pattern.

D. Biomaterials

Many applications of engineering in the health and medical sciences involve contact between a physiological fluid or tissue and a nonphysiological artificial surface. Resulting undesirable responses include thrombogenesis, encapsulation, and activation of the immune system. It is now generally recognized that the initial consequence of exposing a virgin artificial material to blood is the deposition of plasma proteins on its surface. The type of proteins and their geometric and spatial arrangement on the surface are all thought to influence the subsequent physiological reactions. Therefore, efforts aimed at elucidating the behavior of proteins at interfaces have recently been undertaken [5]. Early results indicate that the hydrodynamic environment present at the time of adsorption or desorption of these macromolecules is an important factor in determining the ultimate configuration of the adherent protein layer.

The primary focus of previous studies has been on cell interactions with surfaces. While this remains an important and poorly understood phenomenon, it must be recognized that the appropriate surfaces to study are those to which a protein "coat" has already adhered. Furthermore, the mechanisms of cell adhesion and subsequent morphological changes may well be a result of the precise character of the protein layer to which cell adherence occurs.

E. Organ Analogs

Partial or complete replacement of natural organs with prosthetic components will someday be commonplace. For instance, the design of the total artificial heart, which has had limited clinical success, involved an application of many fundamental principles already discussed as they relate to hemodynamics, biomaterials, and control. Most would agree, however, that the materials–blood–tissue interface is the nidus for some of the most serious problems preventing the development of a safe and reliable artificial heart. This reinforces the importance of investigating at the molecular level the complex interactions that occur between artificial surfaces and the physiological environment.

Another example is the artificial pancreas. Using a combination of synthetic polymers and living tissue and guided by the principles of transport phenomena, prototype implantable devices that incorporate insulin-producing cells are now being tested. This approach to organ function replacement is particularly novel since it incorporates both physiological and nonphysiological components.

F. Blood Processing

In recent years, the fractionation and purification of blood and blood products has emerged as a significant enterprise. Separation of blood and plasma into various cellular and protein fractions has become more of a necessity, given the specific requirements of newer therapies. The first step, the separation of plasma from whole blood (a procedure known as plasmapheresis), is now carried out with a filtration process using synthetic microporous membranes. Chemical engineers pioneered the development of this process and have provided the understanding of what determines its performance in terms of fundamental transport principles.

III. Future Directions

To a considerable extent, chemical engineers have made contributions in the medical and health sciences despite the absence of a larger framework of coherent activity. Few curricula address the issues specific to solving problems in the medical or clinical arena; therefore, the chemical engineer must adapt analytical and experimental techniques learned in a very different setting to problems that initially appear to be unrelated to the mainstream of the profession. In an attempt to remedy this situation, E. N. Lightfoot and S. Middleman each wrote textbooks that addressed issues relevant to living systems, wherein the tools of chemical engineering were applied to solve problems or attain understanding of a particular phenomenon [6, 7]. However, this was not sufficient to foster the expansion of curricula embracing the peculiarities of issues associated with life processes. In addition to these efforts, other authors have occasionally published accounts of the potential role of chemical engineers in the medical and health-related sciences. Some were philosophical in nature [8], whereas others focused on specific examples [9, 10]. Although these articles and textbooks span more than a 20-year period, their impact on the profession has been minimal.

The opportunities for chemical engineers to participate in advancing the medical and health sciences fields are abundant. That the human body is a complex reactor that takes in a variety of feedstocks and catalytically converts them to new forms or to energy is so self-apparent as to sometimes go unnoticed. There is virtually no aspect of a chemical engineer's training that cannot be put to use in addressing problems associated with a living system. Consequently, the areas where chemical engineers may have the greatest future impact are organized below in terms of phenomena of the type chemical engineers are familiar with, rather than specific applications. Three broad areas in which chemical engineers could have a substantial influence in the years to come include interfacial phenomena, transport phenomena, and macromolecules; though certainly not new to the chemical engineer, they offer exceptional challenges when viewed within the context of biological systems.

A. Interfacial Phenomena in Biological Systems

1. Liquid–Solid Interfacial Phenomena

Any time a biological material comes in contact with a solid substrate, a plethora of adsorption/desorption and chemical reaction phenomena occur. As discussed earlier, this happens with blood processing, artificial organs, implantable devices and prosthetics, and certain kinds of biologically mediated measurement or diagnostic systems. A new kind of biologically oriented surface science will emerge in the years to come, with many parallels to the surface science that is currently an accepted mainstay of our profession. Rather than examining problems associated with low-molecular-weight species interacting with exceptionally well-defined solid substrates (e.g., a particular crystal face on a clean metal) under ultra-high-vacuum conditions, the biological analog is that of a large macromolecule, such as a protein, interacting with a metal, ceramic, or polymeric surface in a fluid phase under ambient conditions. Although the systems are different, the issues are fundamentally the same. How do biological molecules associate with a surface? What kind of bonds are formed or broken? Do the macromolecules undergo conformational change? If so, what are the nature and extent of such changes, and are they reversible? Do adsorbed macromolecules migrate on surfaces? Do they self-assemble or form ordered aggregates? How are the adsorption/desorption characteristics altered or influenced by the presence of local hydrodynamic forces? Do adsorbed macromolecules respond or react to the local hydrodynamic environment? To what extent are the functional properties of macromolecules altered by adsorption processes? To answer these and other questions requires the development of novel experimental methods, as well as models within whose framework experimental results can be interpreted and analyzed.

2. Liquid–Gas Interfacial Phenomena

Liquid–gas interfacial phenomena are important in blood oxygenation systems, gas exchange in the lung, and gas dissolution and evolution in hypo- and hyperbaric environments. Selective fractionation occurs at the biological liquid–gas interface. Many proteins and other biological surface-active molecules tend to concentrate at gas–liquid interfaces followed by subsequent denaturation or other undesirable results. Chemical engineers have been involved with interfacial phenomena at the gas–liquid interface where lipid molecules are present, but have given little attention to proteins, lipopolysaccharides, glycoproteins, or lipoproteins.

B. Transport Issues in Biological Systems

1. Organ Systems

The mammalian kidney has yielded most readily to the application of chemical engineering principles in elucidating the mechanisms controlling its operation and behavior. The work of Robertson and Deen, which has evolved over a period of about 15 years, is now commonplace in medical physiology textbooks [11]. This is an example of how relatively straightforward analytical techniques can be used to acquire insights into the operation and functioning of a complex organ system. Similar attempts have been made with the gallbladder, lung, heart, and liver, though these have been fairly primitive. We still await more comprehensive analyses of these and other organ systems in the human body.

2. Intracellular Systems

Molecular processing within cells is a fascinating subject where little quantification has occurred. Cells themselves are microcosms of the larger aggregate organism. The time has come for chemical engineers to address transport and chemically reactive issues systematically at the level of individual cells, for we are the best equipped to handle them. The biological sciences community has provided a vast data base with which to work. Within each cell, metabolic processes responsible for life are conducted. Because these are a result of catalyzed chemical reactions and the transport of solutes to, from, and within cells, the phenomena are well suited to study by the chemical engineer. In addition to metabolic considerations are matters of the regulation and control of transcription and translation of the genetic code within cells. The rate at which proteins are synthesized and degraded, as well as which specific proteins are needed or used, is a consequence of a precisely regulated control network having both macroscopic and microscopic features. Application of kinetic principles to such systems is just under way. Certainly, the body of knowledge gained by chemical engineers from studies of nonliving systems will have some bearing on the quantitative examination of cellular systems. The hierarchy of structure within a cell, from DNA to posttranslational processing of proteins in specific organelles, offers a multitude of well-formulated and unsolved problems.

3. Protein Transport and Dynamics

The means whereby proteins are transported from one point to another, whether across the wall of an organelle or cell or within a porous matrix, needs further investigation. Surprisingly, the chemical engineering literature is only sparsely populated with articles dealing with protein transport. However, a substantial literature exists that addresses problems of protein immobilization and reactivity as related to sequestered enzyme systems, though lately interest in this area has waned. In addition, the behavior of proteins in solution, studied by using simulation techniques, has been approached primarily by life scientists and not by chemical engineers. This is somewhat perplexing given the significant contributions that chemical engineers have made with similar simulations, applied to macromolecular polymeric systems. Perhaps a key to breaking down the barrier around these fertile areas is to recognize that proteins and other biological macromolecules have many of the same properties as some polymeric systems that chemical engineers are more familiar with. Proteins are linear polymers of nonrepeating units that take on exquisite three-dimensional structures and possess recognition and catalytic behavior. Rather than discourage participation of our profession, such complexity could actually benefit the design of future polymer systems. In fact, biomimetics is a recent area of considerable interest for study. Here, catalyst formulators are attempting to capture some of the exceptional selectivity of biological catalysts in nonbiologic structures, such as in asymmetric catalysis. Chemical engineers have much to gain from investigating natural phenomena and living systems. The result surely will be the development of analog systems that may be totally nonbiological or only partially biological with performance characteristics similar to nature's counterparts.

C. Macromolecular and Cellular Phenomena with an Emphasis on Recognition Properties

Perhaps the feature that distinguishes biological macromolecules from all others is their recognition capability. Enzymes interact with great specificity for either a single or a highly selective class of substrates. Antibodies recognize target antigens with a degree of specificity unmatched by any nonbiological system. How these characteristics are developed within a macromolecular structure is largely unknown. Yet, the successful operation and conduct of all biological systems rest on the recognition capabilities of these high-molecular-weight substances. Chemical engineers have much to learn and perhaps even more to offer in elucidating the elements of macromolecular composition essential to the construction of these entities.

1. Protein Science

With respect to catalytic phenomena, some enzymes have exquisite specificity for reactants, and many have substantial turnover numbers. Although they are labile, their ability to function under ambient conditions makes severe chemical processing environments unnecessary. A recent, unexpected finding is that enzymes can operate in nonaqueous media, though application of this observation is still not fully explored. The advent of recombinant DNA technology has permitted biological catalyst design to go from random and largely uncontrolled manipulation to deliberate modification via site-directed mutagenesis. New catalytic activities and properties can be conferred on naturally occurring enzymes, and, as a result, design and construction of unnatural enzymes will soon become a reality. Such "engineered" enzymes will find use in biomedical diagnostic devices. In addition, devices that transduce a biological signal as a measure of, say, concentration will become prevalent among diagnostic tools available to the medical profession in the future. Such devices will incorporate reactive or recognition centers to which molecules will have to be transported. The elements of these systems are easily analyzed and understood by the chemical engineer. However, to accomplish this, the relationship between structure and function of biological macromolecules must be better understood. How is it that the linear sequence of amino acids constituting the primary structure of a protein dictates the final three-dimensional conformation that is so critical to the molecule's ability to recognize or to catalyze? The so-called folding problem in biochemistry is closely related to this issue. Once understood, the design of synthetic recognition molecules will be forthcoming.

One point raised before deserves repeating. One cannot overlook the importance of understanding and delineating the mechanisms associated with interactions of proteins with surfaces. The extent to which adsorption, desorption, and surface-associated phenomena render a biological macromolecule more or less capable of its primary mission is of considerable interest. Therefore, one cannot study only recognition phenomena in solution; these interactions must be examined at surfaces as well. Indeed, it is likely that most of the engineered uses of the recognition qualities of biological macromolecules will occur in conjunction with some kind of surface.

2. Cellular Science

In addition to interactions of biological macromolecules with themselves, with surfaces, and with molecules they have been intended to recognize, similar kinds of phenomena occur at the cellular level that are no less important. Cellular recognition phenomena are profoundly important in the human immune response, and an enlightened view of how such specific interactions occur is needed. The basis of most cell–cell and cell–surface interactions occurs at the molecular level. To that extent, much of what is gained from studying issues related to protein science will have an impact on problems of cellular interactions.

IV. Summary

For the past quarter century, chemical engineers have contributed to a number of medical and health-related science and engineering issues. Generally speaking, these contributions have been at the systems or macroscopic level. Given the recent breakthroughs in both the tools and principles of molecular biology and recombinant DNA technology, chemical engineers are well positioned to address many of the problems of living systems, especially at the molecular level. There are clear parallels between the more classical activities of chemical engineers, i.e., chemical reactivity, catalysis, transport, and polymers, and those of biological systems where a one-for-one set of counterparts can be identified. It is time for the profession to accept the challenge and responsibility of translating our knowledge base to fundamental concerns that stem from the medical and health-related sciences. Not only will it become possible to mimic many of nature's exquisitely designed features, but also applications of this understanding will find their way into the emerging areas of drug delivery and therapy, biomaterials formulation, prosthetics, organ analogs, diagnostic systems, and biosensors.

References

1. Dedrick, R. L., and Bischoff, K. B., *Chem. Eng. Prog. Symp. Ser.* **64**, 32 (1968).
2. Bratzler, R. L., Chisolm, G. M., Colton, C. K., Smith, K. A., and Lees, R. S., *Atherosclerosis* **23**, 289 (1977).
3. Chandrasekaran, S. K., *AIChE Symp. Ser.* **77**, 206 (1981).
4. Sanders, H. J., *Chem. Eng. News* **63**, 30 (1985).
5. Darst, S. A., Robertson, C. R., and Berzofsky, J. A., *Biophys. J.* **53**, 533 (1988).
6. Lightfoot, E. N., *Transport Phenomena in Living Systems.* Wiley, New York, 1974.
7. Middleman, S., *Transport Phenomena in the Cardiovascular System.* Wiley, New York, 1972.
8. Leonard, E. F., *Chem. Eng. Prog.* **83**, 65 (October 1987).
9. Leonard, E. F., *Chem. Eng. Prog. Symp. Ser.* **62** (1966).
10. Caruana, C. M., *Chem. Eng. Prog.* **84**, 76 (August 1988).
11. Robertson, C. R., and Deen, W. M., in *Advances in Biomedical Engineering* (D. O. Cooney, ed.), p. 143. Marcel Dekker, New York, 1980.

General Discussion

Bioengineering

Charles Cooney: I'd like to share some observations relevant to the challenges we face in moving the frontier of biochemical engineering forward. Specifically, there are four points in the context of what we're not doing now as well as we could and what is important to do in the future to support both research and education.

First, the biochemical process industry is large and parallels the chemical process industry that we're all familiar with. Today, the excitement in biochemical processing is associated with recombinant DNA products. There was over $1 billion in sales from five recombinant DNA-derived products in 1989, and the market is growing. This industry is driven by the new scientific discoveries, which Jay Bailey pointed out very well. Applications of this new science demand new technology to translate the science of recombinant DNA, protein engineering, etc. into commercial practice. There is not only opportunity and challenge but also responsibility for chemical engineering. What should we do to meet this challenge? How do we train students in a dynamic, technological environment to understand and respond to this new science and translate it into technology? These questions have been addressed in a number of different areas, but an important point has been missed. We need to train students to solve problems, not just to understand solutions and fundamentals. Solving problems and knowing how to make use of fundamentals is a mission that we must address.

The second observation is that we have an interesting competition between genetic engineering and chemical process engineering in solving biochemical process problems. This competition has several implications. One is to identify a process problem, then have the biologist create molecular solutions via genetics to solve it. Alternatively, we can seek solutions via traditional or innovative chemical processing. It's imperative that we train our students to appreciate the biology, on the one hand, and to understand the chemistry, on the other. Most important, we need to train our students and the people in our laboratories to understand and contribute to a multidisciplinary effort. This is one of our biggest challenges, one that I don't think we've addressed adequately thus far.

My third observation is that, when looking at the range of processes for handling proteins that have evolved recently, two technologies dominate—filtration and adsorption—though occasionally other methods are used. These are the domain of the chemical engineer. In two decades, we've done a reasonable job of understanding the fluid dynamics of processes such as ultrafiltration and microfiltration. Yet, the chemistry of filtration processes is poorly understood, i.e., the interaction between proteins and the surface of the filter and between the protein molecules and other particles that bind to the surface. Consequently, there are no predictive models for filtration processes, despite the significant amount of modeling that's been done. Second, adsorption processes such as chromatography are critical to this industry. Again, our understanding of protein interactions with the surface and with each other is weak. Our understanding of the alternatives to these technologies, such as extractive processes, is also poor. We need to better understand both the alternatives and the fundamental chemistry of the process.

Lastly, process synthesis is important in this field. The biologists dominate process synthesis in the recombinant DNA and protein manufacturing business. What limitations prevent chemical engineers from making contributions? There are two: an almost complete absence of predictive models based on fundamentals and a poor understanding of the physical and chemical properties of biological materials. These issues need to be addressed in both teaching and research.

Gregory Stephanopoulos: Jay Bailey underlined the enormous opportunities for chemical engineers in biochemical engineering. We must answer the question, what can chemical engineers do here better than biologists? One example is introduction of rate processes in describing biological systems, since biologists probably are not going to do it. They are also not interested in applied biology, represented, for example, by solving the problem of hemoglobin synthesis that Jay mentioned. Does a chemical engineer require specialized training to work in molecular biology? Yes, but we do have examples of chemical engineers trained to do gene splicing. Jay mentioned one, and we have others at MIT. The important point is that these students on their own initiative learned gene splicing by talking the professors into letting them take these laboratory courses. About a year later, these students were producing results by putting a gene to do something useful into a microorganism or taking out a gene that the organism does not need and that we don't want. This technology has been around for 10 years, but biologists have not applied it to the kind of problems that we are interested in. We need courses in this area. The limitation is not the eagerness of our students nor their capacity to absorb the material. But can we supply them with the tools and the courses that they need to assimilate these new concepts and techniques? It won't be easy, because there are not many chemical engineering faculty trained in this area, and these courses are very expensive.

My final point concerns the single most important development that brought about the explosion in biological sciences—the development of techniques to separate and accurately characterize proteins and other molecules that exist in the cell. This key capability enabled the discovery of restriction enzymes and other phenomena. These separation methods have been mainly physical and physicochemical techniques, not involving organic chemistry. Biologists failed as long as they tried to accomplish these tasks with organic chemistry. Gel chromatography and blotting techniques led to these results. We have an opportunity to improve these techniques, and they are the prototypes of processes by which proteins or other macromolecules eventually will be separated on a large scale.

Edwin Lightfoot: I have a few remarks on topics that may have been missed. First, we can divide biotechnology into the areas of bioprocessing, which is truly a chemical engineering activity, and metabolic engineering, where we're concerned with the details of the living organism. Both of these have big opportunities for us.

I'll begin with an issue related to Art Humphrey's plant cell culture example. In the early days of penicillin production, it was found that you could take molds that normally grow on the surface of water, or on wet surfaces, and submerge them into a deep tank where they would still grow. The big problem was to supply oxygen to them, because oxygen has a low solubility in water. Small submerged fermentation cultures have dominated world wide biotechnology ever since, but they aren't suitable for a lot of purposes. A classic example is mold spores, which are not normally formed in submerged fermentation cultures. This suggests going a different route. One can find some good leads in looking at the diffusion capacity of water and air. Air will transport oxygen 500,000 times better than water. Thus, molds and other plants grew in air originally for a good reason. Consequently, there's increasing interest in solid substrate fermentations, of which the prototype is the malting industry, where air is blown through a damp, not wet, solid, granular mass. This method has several advantages. It is the basis for a great many foodstuffs, such as soy sauce. It can also be used for processes on the horizon, like biopulping of wood, where wood chips are inoculated with molds, and so forth. Radical changes in fermentation technology can solve some basic problems and produce substantially increased effectiveness.

The essence of living systems is the complex reaction networks that produce life processes, and those are a natural place for chemical engineers to get involved. We can't beat the biochemists on their own ground, because they're trained to handle these multitudinous biochemical reactions and to learn something about their interrelationships at a qualitative level. Until recently, few of them addressed systems aspects. There are three that I think are particularly important. One is getting material balances when you have thousands of species present. It's sometimes important to find out how many reactions are being used and at what rates. Charlie Cooney has pioneered that area. The most recent work I'm aware of is from Eastman-Kodak; John Hamer and Jim Liao have used singular value decomposition techniques to get relations between metabolites. By doing that they can find out when the fermentations are off course, and perhaps even when they become contaminated, quite early. The second is sensitivity analysis, or how to decide what enzyme out of hundreds you want to increase the activity of. The biochemists call that control analysis. It's an important area that I think Jay Bailey touched on. The final one is dynamics, and it is important to realize how many diseases are essentially dynamic and have much in common with poorly controlled manufacturing processes. Diabetes is one of the most interesting to me at the moment. The whole purpose of insulin management is to minimize the time-average sugar levels in the blood without going below a lower level too often, resulting in insulin shock. That's similar to problems in a manufacturing process, because you don't have access to all of the information you need. All you can do is measure sugar levels at intervals that are too far apart to be really controlled. That gets pretty close to the heart of modern chemical engineering.

Clark Colton: I have comments in three areas. The first comment concerns the motivation for research in biomedical areas. I believe the primary factor is that this field is rich in intellectually challenging issues that provide good problems for training chemical engineers. As Channing Robertson pointed out, living organisms are complex bioreactors. In animals, one finds that life processes are determined largely by transport phenomena, thermodynamics, and reaction kinetics, as well as an extraordinary diversity of molecular and supramolecular interactions. In his review of the intellectual origins of chemical engineering to be presented at the Convocation, Skip Scriven focused largely on our growth out of industrial chemistry. Although the lineage is far less direct, one could make a case that the intellectual forebears of modern chemical engineering include some of the physician–physiologists of the 19th century. These scientists set out to show that the functions of organisms could be explained on the basis of the physicochemical laws of nature. One example is Jean L. M. Poiseuille, who was interested in the circulation of blood through the cardiovascular system. His objective, to test the then-popular hypothesis that blood flow derived from the motive force of red cells, led him to experiments on pressure–flow relationships with pure fluids in fine glass capillary tubing, resulting in the classic law that bears his name, published in 1840. Poiseuille's experiments were conducted with great care; his measurements of the temperature dependence of what is now called viscosity agree to within 0.5% of modern values for water. Perhaps the quintessential example of these researchers is Adolph E. Fick, who began his studies in physics and mathematics, switched to medicine, and eventually became a professor of anatomy and physiology. In the course of his career during the second half of the 19th century, Fick's diverse contributions included molecular diffusion in liquids and porous membranes, solid mechanics applied to bone joints and muscles, hydrodynamics in rigid and elastic tubes, thermodynamics and conservation of energy in the body, optics of vision, sound, and bioelectric phenomena. We know him for the differential equations he developed (1855–1857), known as Fick's laws of diffusion, that are taught to all undergraduate chemical engineers. He is at least as renowned in the medical profession for another law, Fick's law of the heart. In 1870, he developed a method for calculating cardiac output from measurements of oxygen consumption and of oxygen concentration in the venous and arterial blood. His principle is nothing more than a material balance for oxygen around the pulmonary circulation! My last example is Adrian V. Hill, who received the Nobel Prize for Physiology and Medicine in 1922 for his work on the production of heat and lactic acid by muscle. In 1928, he published an extensive paper on mathematical aspects of reaction–diffusion problems associated with diffusion of oxygen and lactic acid through tissue. His paper presaged the classic works of Thiele, Damköhler, and Zeldovich a decade later. I could cite other examples, notably Hermann von Helmholtz, but I think my point is clear.

A secondary motivation for research in biomedical areas is the significant growth in hiring of chemical engineers by the health care industry in the past decade. For example, artificial kidneys are now used to treat more than 300,000 patients worldwide, and commercial sales of associated products are more than $1.5 billion per year. In the next decade, we are likely to see other new therapies open up similar markets, possibly with implanted devices. Development of these new technologies will require a much better understanding of the interaction of materials with biological macromolecules, cells, and tissues, some of which Channing referred to.

My second comment concerns labels in bioengineering. There was a time when those in biochemical engineering were concerned exclusively with production of materials by fermentations of microorganisms and those in biomedical engineering with physiological processes and medical devices. With the advent of recombinant DNA technology, especially its application to animal cells, and the growth of "biotechnology," a term which has been applied to a multiplicity of disparate areas, the distinctions have blurred. For example, some areas of research, such as animal cell bioreactors, interactions between synthetic and biological materials, and separation processes for proteins and cells, fall within both provinces. Likewise, the common knowledge base required in these areas of bioengineering has broadened to the point of substantial overlap, encompassing such fields as biochemistry, biophysics, cell biology, and immunology.

My third comment echoes the point made by Jay Bailey, that our undergraduates should be exposed to biology. My position is not based on its utility for later research or its role in professional training. Rather, I believe the time has come that individuals cannot consider themselves scientifically literate without some understanding of developments in biology. The revolution in the biological sciences that began in the 1960s has proceeded at a rapid pace, and the rate of development of new knowledge and understanding shows no signs of abating. We pride ourselves on being the most broadly trained of engineers, with backgrounds in mathematics, physics, and chemistry. I suggest that biology be added to that list, as it should be for any educated scientist.

Sheldon Isakoff: There's an issue in the production of pharmaceuticals that hasn't been brought up by the speakers, yet it's pertinent to the changing character of chemical engineering, and that is regulation. Anytime you introduce a new material that gets into the human body, you need the approval of the Food and Drug Administration. If you do experiments on a small scale and get your approvals from the agency, those approvals are not only for the material that you've produced but also for the process by which you produced them. If you change your process when you go commercial or produce larger lots of the material, you have to go through the approval process again. Timeliness in getting the product out to the marketplace is crucial in this business, as in many others. It's important for chemical engineers to be involved early so that when FDA approval is obtained, you know that you have a process that is scalable to commercial size and will turn out a product having the same properties and efficacy as the material produced at small scale.

Daniel Wang: Bioengineering is really about practicing the fundamentals of chemical engineering. We do things well quantitatively, whereas biologists don't. Take Sheldon Isakoff's point about the FDA and clinical trials. Do you realize that it costs more to put a product through a clinical trial than to make it? It's about $100 million to $150 million per product, and one out of three is successful. Any big company will have to spend at least $100 million to get out one product.

Let's look at pharmacokinetics and the work done with compartmental analysis. We should become involved with the biology of clinical trials in a quantitative way. Can we do more than just say the profile in compartments of the body follows a certain pattern? Engineers have to start looking into this other side of regulatory activities.

We also totally lack a leadership role in the proper places, such as Washington. Why does the FDA ask us questions about the process? They don't know what the three-dimensional structure is. The covalent structure is easier to define, the secondary and tertiary structures are more difficult. The only way they can be sure of what's happening is to make certain our processes are identical, because all else is unknown. Why are the biologists doing everything? Because they have a presence there. We have to make our own destiny. One example is involvement with clinical trials. We need to do more, because we can do things in a quantitative way.

Edward Merrill: I feel like Don Quixote against the windmills with respect to blood viscosity, the study of which, as Clark alluded, began with Poiseuille, a French physician in the 1700s. Would you believe that today, if you go through the index of Harrison's *Principles of Internal Medicine*, you won't find a single reference to the viscosity of anything, including blood, so the internist has no idea of the concept? For 25 years, I've been talking to physicians about the importance of viscosity, and they look at me with glazed eyes. We sometimes have to work hard on our colleagues in other disciplines to bring their attention to what we would have thought was an important concept.

My second comment is apropos of Channing Robertson's remarks, which I admired. In regard to polymers, it seems to me that the time has come to adopt again a unifying approach like that of Charles Tanford, professor of biochemistry at Duke University, who wrote the book *Physical Chemistry of Macromolecules* comparing biopolymers with synthetic polymers. In polymer synthesis, let's look at the synthesis of peptides, amino acid by amino acid, and the desynthesis of peptides by sequencing, and bring this into our curriculum.

William Deen: I'd like to emphasize a point that came up in Clark's remarks and Channing's remarks. That is, the exchange of information between chemical engineering and medicine or the life sciences is in both directions. Several examples were cited of how we can affect medical practice through development of devices or novel therapies. We can also affect medical education by contributing to its scientific base. Further, we can learn from developments in the medical sciences. Clark pointed out two very nice historical examples, Poiseuille's law and Fick's law. Another more recent example is the concept of carrier-facilitated transport, which originated to explain special properties of biological membranes and has since found application in chemical separations processes. At least until recently, membrane biophysicists have been way ahead of engineers in understanding the physics of transport through membranes. That understanding provides a basis for modern membrane separations, including ones becoming important in biotechnology. It's important for us as a profession to maintain this window, this area of focus, even though the health care industry will probably never provide a large number of career opportunities for chemical engineers. As Clark mentioned, the health sciences provide interesting problems for research, and the exchange of information occasionally inspires new ideas on how to accomplish a chemical transformation or separation.

Thomas Edgar: What percentage of B.S. chemical engineers would you estimate go into the health care, pharmaceutical, or biorelated industries today, and what percentage do you forecast 10 years from now? Bill Deen made it sound like there are not that many opportunities. I am not talking about Ph.D.'s, just B.S. engineers.

Clark Colton: In the past few years, there has been a marked increase in hiring of graduate chemical engineers by the pharmaceutical companies in the bioprocessing area and, to a smaller extent, by the health care companies. I believe there has also been an increase in hiring of B.S. chemical engineers, although I don't think the demand will ever be gigantic at that level.

Thomas Edgar: I'm just trying to get perspective. Larry Thompson showed us a curve of engineers hired by the microelectronics industry, and I'm trying to understand the relative magnitude of demand here.

Edwin Lightfoot: Tom, you're being unfair because you're not counting the Ph.D.'s, and that makes considerably less people. It takes a fair amount of specialization in an area like this.

Thomas Edgar: I'm not being unfair, I'm just asking a question.

L. E. Scriven: It's quite evident in Minneapolis, which is the home base of General Mills, Pillsbury, International Multifoods, and so on, that the process food industry has absorbed thousands of chemical engineers. They may not be doing the kind of biotechnology that you're speaking of, but their industry certainly is an important consumer of chemical engineers. The conversion from food technologists to chemical engineers has been made. There appears to be a similar trend in other agriculture-based process industries.

Arthur Humphrey: If you lump all of the B.S. students who go into medicine, veterinary science, and dental science, and add all of those going into the health care, pharmaceutical, food, and agricultural-based industries, you get up to almost 25% of the chemical engineering profession. You have to define food and agriculture on a very broad base. My own concept is that this field is probably going to utilize 25 to 30% of the chemical engineers when it reaches an equilibrium.

James Wei: The AIChE had a survey on that subject. I think that the food industry is hiring 1.7% so far.

Daniel Luss: I want to raise an issue that concerns the training of chemical engineering graduates who conduct research in biochemical engineering. Every major department now has a research program in biochemical engineering, and they attract many of the most talented students who enter the graduate programs. The proper training of chemical engineering students carrying out research in this area requires that they take a large number of courses in the biological sciences. Thus, if they work in a traditional chemical engineering area, a large fraction of their training will not be utilized.

Clark Colton: I disagree. It's true that students doing graduate research in bioengineering may have to take some specialized subjects in the biological or medical sciences, but this is no more limiting than for any of the other relatively new areas we've been discussing. It's important that students are exposed to sufficient graduate chemical engineering course work and that there be substantial chemical engineering content in their research. With this proviso, these students don't lose the breadth of their undergraduate training for applications in industry. The limitations are greater if one goes into academia, because one's thesis often fixes one's research area, and it's difficult to change fields, especially as an assistant professor. When I started out, there were few if any industrial jobs related to the research of my students. That situation has changed today. My earliest students either went into teaching or they went into industry. Those who went into industry did the same things that other chemical engineers did because they had appropriate training and were able to use it.

Daniel Luss: Several faculty members working in this area insist that their students take a large number of courses in the biological sciences. I'm told that it's essential for anyone who wants to do research in this field. I raise the question whether it's the best training for a student who will not be employed in this area. It's clear that the number of available teaching positions in chemical engineering departments for these graduates is declining, as most departments already have young faculty members for biochemical engineering. As a consequence, most of the graduates may have to find industrial employment, and at present the biochemical industry does not hire very many such graduates.

Clark Colton: Whether there is a problem depends on the environment the student is in and the nature of departmental requirements. In our department, all doctoral students must pass a qualifying exam that is based on the core areas of fluid mechanics and transport, thermodynamics, and kinetics and reactor engineering. Although there are no formal course requirements for the Ph.D., in practice virtually all entering students enroll in the associated core graduate subjects. Students are also expected to take a reasonable number of additional graduate subjects in the department. There's a formal requirement for a minor, and it's in this context that most students doing bioengineering research take biology or medical science subjects outside the department. Students who receive this kind of exposure don't lose their chemical engineering identity or capabilities. They can function well in industry. If anything, they enrich the profession.

Edwin Lightfoot: I have two students in this category. The first started in separations, and he's taken a lot of biology, but what he's really working on is NMR techniques. He's going back to the basic quantum mechanics to study NMR techniques. The other student is working on diabetes, and she's taken advanced processing control courses because this is a control problem. Alan Hatton was one of these to a degree, as was Abraham Lenhoff, now at Delaware. These people have done very well in fields that are quite different from where they started. This is a bum rap.

Robert Cohen: I believe this is more than semantics. The problem that Dan brings up is contained in the way he discusses it. If you think of graduate training, the word "train" is what I object to. Graduate education is different from job training, and if you're picking a good problem, there's educational value in that, and the specific field is more or less irrelevant.

Gregory Stephanopoulos: I'm not sure this problem really exists, but even if it does, usually these kinds of problems take care of themselves, in the same way they do in broader disciplines of chemical engineering. If we try to interfere, then we may introduce a lot of factors that will upset the students.

Stuart Cooper: I don't like to come to Dan's defense, but I have an observation. We've never required our Ph.D. candidates at Wisconsin to take a certain number of chemical engineering courses. Recently, for a variety of reasons, we examined the statistics of what our Ph.D.'s were taking on the way through. We concluded that they should be taking more courses in the department, regardless of the area. Now we have the beginnings of a core requirement, which I think will resolve the concern about specialization.

Edwin Lightfoot: The reason that this happened is not the highly specialized nature of research. The reason it happened is the pressure to get money and the tendency of major professors to push their students to spend more time on research and less on course work. That's the real culprit these days.

Robert Brown: Were they taking a sufficient number of courses, but outside the department, or were they taking an insufficient number of courses?

Edwin Lightfoot: There is not a sufficient number of courses inside our department at the upper graduate level.

Stuart Churchill: I'm impressed with our doctoral students who work in this area for Doug Lauffenburger and others. I attend their oral exams, and I understand their language, so they have not lost the ability to communicate with traditional chemical engineers. They still can speak in terms of transport, thermodynamics, and so forth. In this environment, what Dan is worrying about isn't going to happen. The converse is also true. A large fraction of my early doctoral students are now in bioengineering, although I don't take any credit for that. David Hellums, Peter Albrecht, and Irving Miller were all students of mine. They started their careers in something else and then moved into bioengineering. As long as we don't build too high barriers, this is a nonproblem.

SECTION IX
Process Engineering

23

Process Engineering

Arthur W. Westerberg
Engineering Design Research Center
Chemical Engineering Department
Carnegie Mellon University
Pittsburgh, Pennsylvania

I. Introduction

It is an honor to be asked to review the area of process engineering for this centennial celebration of chemical engineering education. This paper provides an overview of the history, current state, and future of some aspects of process engineering. I define process engineering as the application of systems concepts and methodologies to the design and operation of chemical processes. One issue is understanding the behavior of the individual component parts constituting a process; a second is understanding the behavior of a system of integrated parts. In this article, I generally assume an understanding exists of how the individual parts behave and look at the issues arising from combining them into a system. Here, I will discuss only design. I first examine important developments in computer technology that have and will affect design. I propose a classification scheme for design problems and a model for how design is performed. I discuss engineering synthesis—i.e., the automatic generation and selection of design alternatives. A final section traces important developments in analysis methodology—i.e., in the setting up, solving, and optimization of complex models.

ADVANCES IN CHEMICAL ENGINEERING, VOL. 16

A. A Caveat on Coverage

This area is extremely rich with potential topics, each of which would require a separate paper to review in depth. The reader should be aware that, because of page limitations, this paper can cover only one of my favorite topics, analysis, in any depth. To provide perspective, computer technology will also be examined because of its enormous impact on process engineering. I am currently the director of the Engineering Design Research Center at Carnegie Mellon. Based on this experience, a classification scheme is presented to show the diversity of design problems one can face; this is followed by one possible model of the design process itself to suggest how ideas in this area might be organized.

The topic of synthesis will be be covered in the following paper by Douglas [1]. I myself have recently written two papers, each in part a review of process synthesis [2, 3]. Several articles have reviewed synthesis in the last decade, one over 40 journal pages long. The brief summary in this paper will be limited to a few ideas that are not usually discussed but that are thought to be important in process synthesis when heat effects do not dominate the decision making. A more extensive section on analysis completes this work. The paper by Morari [4] will present ideas on plant operation and control.

II. The Impact of Computer Technology

The first computer I worked with had a fast memory of 1024 words in which both the data and program had to reside. The hardware filled an entire room and, when running, could easily have heated the entire building in which it was housed. The computer was vintage mid-1950s and was programmed using assembler code only. It could be used for setting up and solving relatively simple models of single pieces of equipment, like a heat exchanger or a simple column, and had less power than one of today's small hand-held calculators that costs under $100. In the early 1960s, computers such as the IBM 7094 and the CDC 6600, the first really powerful computers, came into being. They were programmed primarily in early versions of Fortran and Cobol. One remembers forecasts that the world could make use of perhaps as many as five CDC 6600 computers.

Today, we see a very different world of computing. Supercomputers abound. The computing center is augmented and in some cases even pushed aside to make way for distributed workstation environments, where everyone has a powerful computer either on his or her desk or in the next room. Everything is networked or rapidly becoming networked; large files are readily transferred from New York to Munich to Trondheim to Tokyo. Students routinely log into supercomputers across the country to run large computations. Electronic mail makes worldwide communications direct and easy, with messages often transferred and responded to in minutes. Three hundred megabyte disk storage devices are available for $3000.

New machine architectures are performing remarkably, provided the software can be written to use the architecture. Hypercube computers available commercially are multidimensional arrays of hundreds of computers, each with the power of a microvax.

Not all computers carry out computations in a traditional way. Neural networks are another form of computer that receive input signals and produce output signals that characterize what was input. These computers can be taught to recognize complex patterns. For example, they can be shown a picture of a person and then can recognize another picture of the same person, even when viewed from a different perspective. They have also been taught to recognize connected speech.

New concepts in software are revolutionizing computing. The Macintosh computer and its easy-to-use icon-based software bring computing to virtually anyone willing to sit at the computer and try. We hear of four-year-old children using computers in ways that many of us could not have done 10 years ago without considerable training.

The UNIX operating system is offering for the first time the possibility of an operating system that is not dependent on the hardware vendor. The battle over operating systems is far from over, however.

A revolution is going on in computer languages, too. An anecdote of the mid-1960s recounts that someone from IBM was asked what would be the most popular language in the year 2000, e.g., Algol, Cobol, or Fortran. He responded, "I do not know how it will look, but it will be called Fortran." Those of us who have seen Fortran evolve in major ways to pick up the characteristics of many of these other languages will find that answer very insightful. However, there are other languages available now that are really different, and that will force more changes on Fortran. Their value is often based on subjective but very appealing arguments. Languages that support object-oriented programming such as Smalltalk, Flavors, or CommonLoops are examples. These languages are not like Fortran, at least not at this time. Among other things, they enforce a style of programming that many suggest reduces the time to program complex systems by factors of about five.

Two features make these languages very interesting. The first is their degree of modularity, with chunks of code, called objects, encapsulating in a distinctive manner the data describing and the methods (procedures) defining the behavior of these objects. The user at the screen is an object; the printer is an object. Communication among objects is carried out by messages through the executive system. For example, one object can send a message to another object representing a vessel and ask for its volume.

The method to compute volume differs depending on the type of object, but many objects can respond to that message. Thus, one can often remove an object in such a programming environment and replace it with another, provided it responds to the same messages without destroying the integrity of the system. The second feature is the power with which these languages can represent information, using the concept of inheritance structures. An object can be a class called column. One can send a message to this object, ask it to create an instance of itself, and call it "B-T splitter." The B-T splitter inherits computational methods from the class called column; it also inherits default values for variables. Asked for the number of stages, the B-T splitter may not yet have a value posted, and the response will be a default from the class that might be a nominal value of 40. The response will be the actual value after it has been computed and posted within the B-T splitter object.

Some computations are not numerical in nature. The early to mid-1970s produced a flurry of computer science and some engineering literature about computers solving problems automatically using complex human reasoning. In the mid-1970s, Minsky [5] published a paper that has had a significant impact on the area of artificial intelligence. He posed the problem of two people having a conversation like the following: (Person A) Tomorrow is Mary's birthday; (Person B) I wonder if Mary would like a kite. We all understand that person B is thinking of giving Mary a kite for her birthday. However, there is nothing in this conversation that would let a computer know this meaning. Minsky argued that the computer would need in memory a number of scenarios about common experiences. One would be about birthdays and that people exchange gifts. Another scenario would be that a kite and objects like it are typical gifts. When the above conversation is examined, the computer would search its memory and retrieve relevant information that could be used to interpret what was implied by person B. This recalled information would be a frame. Thus started a controversy that still exists today as to what a frame is; even Minsky did not define it precisely. An important property of frames is the notion of default information. Minsky argued that when someone mentions a car to you, you are likely to think about your car and not about just any car. If you are then asked about the weight of a car, you would likely estimate the weight based on the specific car you are thinking about. Only when someone particularizes the car by saying it is Jim's car will you move away from the default car. A lot of understanding can come by reasoning using defaults.

Another idea of the late 1960s and early 1970s was to produce a general problem-solving capability suitable for all complex problems. This idea did not succeed. Next came the idea of encoding knowledge about a particular problem domain that could then be applied to solving problems automatically in that domain, the concept behind knowledge-based or expert systems. This idea is being explored extensively by almost every company and university. There are demonstrations that show that it really works. Interestingly, there are many disbelievers. The argument is that every concept promulgated by expert systems is well known and has been "done" in Fortran programs for years. There just was no name given to the concepts. I believe that real issues worthy of research and development exist in this area. These systems offer utility systems called *shells* that make encoding easier and use qualitative information in solving problems. Now, one is not only modeling physical artifacts but also trying to model the process of complex problem solving.

Computer scientists are trying to understand the really difficult issue of automatic learning. Can computer programs be constructed that can learn concepts and in doing so improve their performance when required to solve similar problems a second time? Perhaps the most difficult issue here is how to guess the correct generalizations when one is presented with examples of a concept, so that what is learned is applicable more generally than to just the training examples. Humans guess these generalizations rather well. Not all reasoning is deduction and induction. Complex problem solving seems to rely heavily on abduction, too. Abduction is the process in which one guesses the solution and then proves it using the guess to direct the proof.

There are many other significant unsolved problems. For instance, how can the typical engineer find and use the best tool for the problem at hand? It can take months to find the tool and months to learn to use it. And it may prove to be the wrong tool. Another problem is how to deal with the massive amounts of information that are generated within a particular company. Data logging a process produces more information than anyone cares to think about. How can the process be usefully saved and accessed? Should it be saved? The answer is a resounding yes when a problem occurs in the process.

The future looks very bright in computing. Supercomputers are about to show up on desk tops. Markedly increased speed will come in two ways: (1) new architectures that allow massive parallel processing and (2) new technologies such as submicron line widths in silicon-based chips, which can increase speed by factors of 10^2, and optical computing, where performance factor increases of 10^5 are possible. Understanding of the impact of language and information structuring in problem definition will be substantially improved—a pet topic for this author [6]. Everyone will be linked to a worldwide information network in which just about any piece of information will be available electronically. Means of searching this base will improve to the extent that the system will intelligently guide the user through the maze. Automatic language translation will exist, for example, translating the spoken word so that two participants in a telephone conversation may use different languages. Improved graphics will suitably summarize and present enormous amounts of information to the engineer, some in the form of holograms. Neural network computers coupled with symbolic computers will add new power to problem solving.

It is interesting to speculate whether the future will be limited more by a shortage of people who can determine how to use this power than by the power that will be available. We can predict increases in the size of problems we can solve on the order of 10^8 in the next 20 years, if the past is any indication of the future.

III. Design in Process Engineering

I return to the main theme of this paper—process engineering. As mentioned earlier, process engineering concerns itself with the design and operation of processes. I now look at *design*, which I define as that step during the creation of a new artifact or the modification of an existing artifact at which one gathers together and generates information on all its aspects to plan its fabrication or modification. An artifact can mean anything from a chemical process to a building to a computer code.

Studies in several industries suggest that design activities consume about 10 to 15% of the funds needed to move from the original artifact concept to its final creation. The same studies conclude that one makes decisions in the design step that fix about 80% of the final costs involved. A really bad design decision can almost never be compensated for by those who must fabricate the artifact. Therefore, though a small part of the overall activity, design is the most important. Any manufacturing company that intends to remain competitive in today's marketplace must continually improve its design capability in terms of both people and tools.

To understand the important issues for a design, it is helpful to develop a classification scheme to characterize a design. If the scheme is appropriate, then the methodologies for designing should be related for designs similarly classified. We are using abduction to conjecture the usefulness of this classification scheme (Fig. 1).

Original versus Routine. A major argument typically ensues at any design conference on what is meant by design. Frequently, one set of people means *original* design, where the design team has never designed anything quite like the given artifact before; the other group means *routine* design, where the design concepts and methodologies are well understood by the design team. Many designs fall between these two extremes. A design morphology must classify a design according to this measure because the important issues are very different for these two types of design problems.

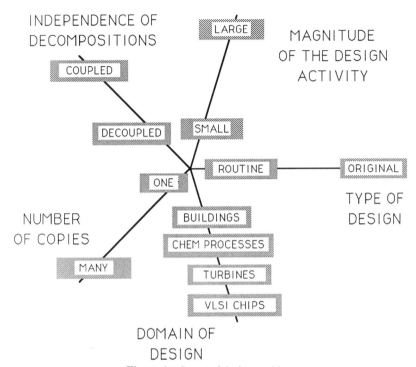

Figure 1. Space of design problems.

For original design, the activity alternates between brainstorming and gathering information, possibly by a geographically dispersed team. This activity generates the concepts on which the design and supporting design calculations will be based.

For routine design, the potential exists to *automate* the design activity completely. An example is designing a house in Japan. The customer looks into a catalogue for the style of house desired—say, a "Frank Lloyd Wright"—and, for that style, enters the number of bedrooms and room size parameters that can be adjusted to customize the design into a form on a computer screen. The computer designs the house; it then generates instructions for the factory that will prefabricate the parts. Produced almost without human intervention in a factory that is creating parts for many different houses simultaneously, the parts for each house arrive at the loading dock just in time to be loaded together onto a truck. When dropped at the construction site, the parts are in the order in which they are to be used. And they fit together. What contractor will survive in that market who refuses to use a similar technology?

Magnitude of Design Activity. There is a significant difference in the magnitude of the design activity needed to design a light switch versus a Boeing 747. For the former, the concept of concurrent engineering can be accomplished by putting a dozen people into a room for a few weeks, with each person representing a different concern such as the design, manufacture, use, or sales of light switches. For the Boeing 747, one can imagine putting the entire team into a single room only at the early stages of design to create the overall design concept. Specialist teams will design the engine, the undercarriage, the wings, and so forth, with each team requiring its own room. Here concurrent engineering requires a very different set of organizational controls. A question is, for example, how to propagate changes from those designing the engine to those designing the wing and vice versa.

One or Many Copies. One typically builds one copy of a chemical process or a major office building. At the other extreme, one will build millions of computer memory chips. The care taken during the design to get all the details right varies dramatically in these two extreme cases. For a process, certain corrections can be taken during fabrication to overcome design flaws—one would like to avoid these, but changes can be made. On the other hand, one could ill afford to design a memory chip that had to be hand-adjusted to function correctly. It has to be absolutely right when designed. The fabrication process for a one-time-only production cycle will seldom be highly automated, though some steps within the whole process may be; the fabrication process for chips is totally automated.

Integration Problems. When a design is accomplished by partitioning it, the parts must then be integrated to form the solution. If the parts are strongly coupled by important overall performance measures or constraints, integration is difficult. A typical grass roots chemical process design is first designed functionally. One then chooses the actual pieces of equipment. These individual decisions to select pieces of equipment are not strongly coupled, which simplifies this aspect of the design activity dramatically. VLSI circuits have a similar strong decoupling, allowing them to be designed functionally first. In contrast, the aerodynamic performance of an airplane is determined by the totality of its shape and weight distribution. This characteristic, which is crucial to the performance of an airplane, cannot be assessed until the component parts are integrated and requires that the design decisions made for the individual subsystems be strongly coupled.

In the retrofit design of a process, we must couple the decision making about function with the reuse of existing equipment. In an example involving the redesign of a distillation sequence to handle 30% more feed, this coupling increased the number of potential sequences from about 140 to over 4.5 million [7].

Improved decomposition of the design activity can strongly affect the number of sequences. Thus, it is a measure of both the inherent characteristics of the artifact to be designed and the current effectiveness of the design methodology available.

The extent of coupling is often difficult to assess for a design. Even for a chemical process, the overall energy integration of the process can complicate the design decision making if one chooses to iterate on the design to find processes that heat integrate well. Safety can strongly couple the selection of equipment to the functional design step. At some later date we may have safety analyses that will allow for better decoupling.

Domain of the Design. A final axis is the domain of the design, i.e., the nature of the artifact being designed, such as a chemical process, computer program, building, VLSI chip, electromechanical device. The knowledge required for reducing the size of design problems comes specifically from the domain of the design. However, the concepts and form for the tools that use this knowledge to support design seem much more affected by the other axes than by this one. This conjecture is the principal hypothesis upon which the Engineering Design Research Center at Carnegie Mellon University is operating.

A. *A Model of the Design Process*

Talukdar and Westerberg [8] proposed that design is the process of moving among and examining a number of different views or *aspects* of the artifact. At the highest level, the aspects are very abstract. At the lower levels, more and more details are built up about the artifact.

Different aspects are needed to translate the design into a set of terms suitable for applying knowledge to it so as to criticize and/or to alter the design. For example, one may create a fault tree for a process to argue about its safety, or a *cost diagram* [9] to look at its capital and operating expenses. Temperature versus enthalpy diagrams help to think about the heat integration of processes.

One uses a synthesis step to move from an abstract representation to a more detailed one, where *synthesis* is the activity of generating design alternatives and selecting the better ones based on incomplete information. The alternatives selected are often modeled in detail (perhaps by moving through several more aspects) and *analyzed* to see if they indeed satisfy the requirements specified in the more abstract representation. Finally, if one is going to allow changes to be made to the more detailed representation, one has to worry about mapping those changes back to the more abstract representation, an activity we can term *abstraction*. This latter activity is very difficult to accomplish in general and there seems to be no formal literature on how to do it. This statement is probably controversial. Typically, one has to guess the changes to make at the more abstract level that will permit attainment of the properties desired at the more detailed level.

The complete design activity can be modeled as a network of aspects with operators among them that allow movement from one to another. Where the operators can be automated, one can create tools; for the others, the designer has to carry out the required step, often in an *ad hoc* manner.

With this model of the design process, a terminology becomes possible for describing different steps in it. One can then imagine formalizing approaches to carrying out design that are more general than the examples used to alert one to the concepts.

IV. Synthesis

To move from an abstract representation to a more refined one is accomplished by a synthesis step. A definition of the synthesis activity is as follows:

the automatic generation of design alternatives and the selection of the better ones based on incomplete information.

Synthesis appears repeatedly throughout a design because of the recursive nature of a design. Synthesis is the activity of *generating* alternatives to increase the net present worth of the company; it is the activity of generating a process flow diagram. In the latter case, this activity may itself have embedded synthesis activities such as the synthesis of separation subsystems or the synthesis of heat exchanger networks. It is also the act of *selecting* the better alternatives based on assessing the value of the design before it can be proved to be among the better ones.

Three issues have been listed as significant in synthesis: representation, evaluation, and search. The representation issue is related to the notion of aspects discussed above. The correct representation often permits one to "see" the solution or how to generate just those competitive alternative solutions to problems that otherwise may seem very difficult to solve. Representation is often based on developing the appropriate insights into the problem. For example, noting that a distillation column degrades heat to accomplish separation work motivates a "cascaded" heat flow representation for a column [10] that is useful in finding how to heat integrate columns with the rest of the process.

An important synthesis problem is that of generating alternative chemical reaction paths by which a desired target product might be made. Thousands of reaction paths can be generated even for small molecules. How can one evaluate each of them? The correct evaluation is to establish the economics for the alternative processes on which each would be based. One simply cannot do this for thousands of alternatives, so simpler evaluation functions are required that allow screening of the poorer alternatives with little computational effort. An example is to look at the thermodynamics of the reactions and reject any in which equilibrium severely limits the production of products from the reactants.

The search space for synthesis is typically enormous. One must search as effectively as possible. Branch and bound methods, hierarchical structuring, constraint propagation, and heuristics to rule out whole sets of possible designs are among the search concepts that are useful. Perhaps one of the most effective is hierarchical structuring, as one then makes high-level decisions that will rule out very large numbers of lower-level alternatives. The following illustrates the potency of this approach. Suppose one has a problem with $m_1 + m_2$ decisions, and one proposes to search by forming a grid of n points in each direction. If m_1 of the decisions can be made first using approximate reasoning and the remaining m_2 are then made, the search space reduces from one of size $n^{m_1 + m_2}$ to one of size $n^{m_1} + n^{m_2}$. Letting n, m_1, and m_2 all equal five, the sizes are 5^{10} (about 10^7) versus $2 \cdot (5^5) = 6250$.

Separation system design offers an example. At the first level of decision making, we could classify the species and the phases involved to decide the type of likely separation methods that might be used. King [11] lists 54 different separation methods. Some are for gas–gas separation, others for gas–liquid, still others for gas–liquid–solid, and so forth. An approach is to make high-level decisions on which type (gas–gas, gas–liquid, etc.) of separation is needed for each of the separation steps required [9]. Then, based on these decisions, lower-level decisions select exactly which method to use to separate species having the same classification.

Constraint propagation also reduces the search space size. This method looks at the problem from every view conceivable and, in each of these views, develops and propagates constraints on the solution space. In a separation system synthesis problem, one might look at the mixture and decide that the species D and E should be separated alone and by using distillation. This constraint dramatically reduces the search space size. Decisions not leading to D and E being isolated into a stream that will then be separated by distillation need not be considered in any enumeration scheme.

In chemical engineering, the design of energy-efficient processes has dominated the synthesis literature. Most well developed is the classical synthesis problem of finding the structure of a heat exchanger network to exchange heat among a number of hot and cold streams within a process. This problem has been significantly aided by the discovery of very useful representations based on plots of temperature versus heat availability and heat need

for a process. These plots allow establishment of minimum utility requirements [12]. Network arguments allow one to determine how many exchangers
are needed. Methods exist for estimating the cost of both the utilities and
the equipment without inventing the actual design. Understanding this
problem has led to synthesis strategies for creating processes that will heat
integrate well, such as the design of multieffect evaporator systems. A recent
paper [13] reviews over 200 articles on this topic alone.

A. Expert Systems

Future synthesis programs will combine both quantitative and qualitative
reasoning about the artifacts being designed. A serious issue will be the
control of these programs so that they can solve the problems opportunis-
tically using every device available to reduce the search space. Lien [14,
15] provides one interesting approach to structuring such an opportunistic
system. Programs should become commercially available in the future for
routine design problems, such as physical property experts or absorption
experts, replacing many of the design handbooks of today.

V. Analysis

The tools for carrying out an analysis step are likely the most well developed as aids for design in virtually every discipline. In analysis, one proposes a model that can describe the behavior of whatever phenomenon is
of interest. Such models can range from how an oil droplet might sit on
top of a pool of water to the production scheduling of an entire company.
They can range in level of detail; the oil droplet could take the power of
a supercomputer to solve, while the company model might exist on a personal computer.

As pointed out in the *Frontiers in Chemical Engineering* report [16],
Chapter 7, the (super)computer of the future will have enough power to allow
ab initio computations that we are learning how to pose to replace laboratory experiments. To perform this *analysis* step, we need to be able to set
up and solve very large scale computations. We might ask how well we can
do this and what the problems are.

To appreciate the problems, imagine having to set up a model comprising 10,000 equations in 11,000 variables. Try doing this without writing
too many equations or too few in terms of the variables involved; then try
to figure out which 1000 variables to fix. Not every set will leave the remaining 10,000 equations in 10,000 variables nonsingular. Then try to give
values to the 1000 variables you have fixed in such a manner that the problem
has a solution, and finally find that solution. Lastly, find all the solutions
that might exist (a new twist added lately).

Not all models are necessarily specified as equations to be solved. Some are specified procedurally, in the form of rules (as in production systems such as OPS-5 or OPS-83) and/or in qualitative terms. These models offer even more interesting problems to solve, which can take us into the domain of expert system concepts. Learning to pose and solve such models effectively is still an area for research.

Since pure equation-based models involving as many as a few hundred thousand equations arise and require solution, ways are needed to ensure that this activity can occur with some degree of success. Perhaps the oldest of such approaches comes with flowsheeting programs.

A. *Steady-State Modeling*

In this section, I examine developments in setting up and solving large-scale, steady-state models for complete chemical processes.

1. Flowsheeting Programs

Today, engineers routinely compute heat and material balances and preliminary sizing and estimations for a traditional chemical process using commercially available flowsheeting programs. These programs permit one to set up a model for an arbitrarily configured process quickly and to solve that model using different degrees of rigor both for the equipment models and for the physical properties required. These models, with the physical property equations included, are often tens of thousands of equations in size.

These programs first came into being in the late 1950s [17]. Flowsheeting systems tied individual unit models together so that entire processes could be modeled. They almost always use the *sequential modular architecture*. Each unit model is represented by a subroutine that is designed to compute the unit output streams given the unit input stream values and enough unit parameters to fix its performance. An example is to compute the vapor and liquid streams out of an isothermal flash unit given the feed stream enthalpy, pressure, composition, and flow rate and the temperature and pressure of the flash unit.

Tying these computations together gives a model with the same recycles that exist in the actual process being modeled. These recycles have to be guessed to start the computations. When their values are computed as outputs of later units, the computed values have to be compared to those guessed. If they are essentially the same, the computation terminates; otherwise, new guesses are needed and the process is iterated.

Early research in flowsheeting involved discovering automatically the better streams to use as guesses (tear streams) by selecting the order in which the unit subroutines should be called. Methods were also published for improving the guessing, which in its simplest form uses the values just computed in what is termed "successive substitution." Often the designer wishes to specify parameter values for intermediate streams which in the sequential modular architecture have to be computed, not specified. Computational controllers compare the computed values for these streams to those specified and force the flowsheeting program to run iteratively while the controller adjusts some input parameters until the desired intermediate-stream behavior is met.

In early attempts at optimization, pattern search based methods put yet another loop around these programs. Fifteen hundred flowsheet simulations to find the optimum were not uncommon; the computation often failed.

By the late 1960s, people had the idea of gathering together the equations for an entire flowsheet model and automatically deriving a solution procedure to solve them. Hardly anyone in industry took this idea seriously, as whatever tests were run on the idea also failed too often. The problems also got too large too fast, particularly if the physical property evaluation equations were included in the set.

The first efforts in academia to derive solution procedures were based on extending the automatic tearing ideas used in solving flowsheets. In tearing, a small subset of the variables are guessed and iterated to solve the entire set of equation. Tearing ideas floundered in the early to mid-1970s. Variables had to be guessed for which no intuition existed, effective solution seemed to require symbolic manipulation of the equations that was much too slow, and finally some neat embedded tearing algorithms—loops within loops—were doomed to failure. This last idea looks really useful when the inner equations are linear. However, all too often the inner loops are singular even though the overall problem is not.

The Newton/sparse matrix methods now used by electrical engineers have become the solution method of choice. Hutchison and his students at Cambridge were among the first chemical engineers to publish this approach, in the early 1970s. They used a quasi-linear model rather than a Newton one, but the ideas were really very similar. (It appears that the COPE flowsheeting system of Exxon was Newton based; it existed in the mid-1960s but slowly evolved into a sequential modular system. One must assume the Newton method failed to compete.)

Presently, several equation-based flowsheeting systems exist; perhaps the best known of these is SPEED-UP, which continues to be developed at Imperial College by Sargent and Perkins and their students. Now commercially available, its strongest attraction in industry seems to be as a dynamic simulator. TISFLO at Dutch State Mines is also often mentioned in the literature. As a complete flowsheeting system, no equation-based approach is yet very popular. One might wonder why.

Equation-based systems still have two characteristics that users find unattractive. First, they are a half order of magnitude slower at solving simple flowsheeting problems; thus they lose when compared in timing studies. Second, they fail to converge more often. This failure can almost certainly be ascribed to the initial guesses used. A large part of the code in the subroutines modeling units in sequential modular systems is used to get a close initial guess. None of the equation-based systems as yet has a comparable capability for making an initial guess. When they do, the two approaches will likely have comparable convergence characteristics—the equation-based approach may even turn out to converge more often.

Finally, the attractiveness of equation-based flowsheeting systems is that one can choose fairly arbitrarily which variables to fix and which to compute. Inputs can be computed in terms of outputs, etc. This attractiveness, however, gives the designer freedom that he or she often cannot use correctly. Deciding which variables to fix is difficult, particularly when there are a thousand of them to select. Since the sequential modular approach has preselected the variables that are fixed, getting the degrees of freedom right for that approach is generally not a serious problem. The equation-solving approach does not preselect—interestingly, its strength and also its weakness. Two forms of aids are possible: default selection to pick most of the variables to be fixed (e.g., the molecular weight for methane should be treated as being fixed), and a set of algorithms that can carry out structural and numerical analyses to aid in getting the degrees of freedom set correctly [18]. These aids are not generally available.

2. Convergence

A major research issue has been the development of approaches to aid convergence of difficult problems or to speed up the solution of easier ones. Making better initial guesses is certainly crucial; it may be the single most important issue in converging stubborn problems. Strangely enough, there is little in the literature about aiding this problem. One should look with a jaundiced eye at many reported convergence results. Seldom does a paper tell of the its author's many false starts that failed to converge until the initial guess became good enough. Slight perturbations in the initial guesses can move one from a problem that refuses to converge to one that marches directly to the solution.

Alternative approaches to setting initial guesses include (1) using default nominal values that reflect the type of the variable—300 K for temperatures, 10,000 Pa for pressures, 0.3 for mole fractions; (2) using *ad hoc* coding to set initial guesses—linearly interpolate the temperatures in a column section; and (3) solving the equations for a computation that is a more natural one first and then switching the degrees of freedom to those desired—solve a column fixing the reflux first, then fix the top product purity and let the reflux be computed with the first solution being the initial guess to the second.

If one has made the best guess possible, then more rugged methods exist for getting to an answer, for example, *continuation methods* [19]. In continuation, a scalar parameter is picked, such as t (think of it as time), that is moved from zero to unity (some versions move to infinity), while the solution moves in a continuous manner from one that is easily found to the one that is difficult to obtain. Thus, one *creeps* up on the solution.

The equation for such a method is as follows.

$$H(x,t) = (1 - t)g(x) + t f(x) = 0 \qquad (1)$$

As t moves from zero to one, $H(x,t)$ moves from the solution of $g(x) = 0$ to that of $f(x) = 0$. $g(x)$ is a set of functions for which the initial solution is easy to find; $f(x)$ consists of the equations that are difficult to solve.

Continuation methods have an interesting degree of freedom: the direction in which t will move. Often, to get to the solution t moves first in the positive direction, then for a time in the negative direction, and then again in the positive direction. Thus, these methods typically change the independent variable for moving from t to s, the distance the continuation path has moved in the $[x,t]$ space.

One continuation method reconstructs exactly the Newton method when t moves in the *positive* direction. Think of the surface that corresponds to summing the squares of the functions one wishes to drive to zero. If the Newton method flounders in a local hole in this surface where the bottom of the hole does not reach down to zero, and thus where the equations do not have a solution, it would be very useful to climb out of the hole by going in the reverse of the Newton direction (i.e., by simply reversing the sign on t), hopefully over the top of a nearby ridge and down the other side into a hole where a solution does exist. A continuation method does just this.

Kuno and Seader [20] show how to use continuation methods to find all the solutions for a set of nonlinear equations. One simply continues past the first instance where t equals unity to find all other instances where it again equals unity. They show how to select a starting point for the search to guarantee that all solutions will be found. No proof exists for their method, but they have tested it extensively without a known failure.

Vickery and Taylor [21] suggest using t as the exponent for the parts of expressions that are giving convergence difficulties. Used in this fashion, t can be termed a *natural continuation* parameter. For example, consider writing the following vapor–liquid equilibrium relationship with the parameter t included as shown.

$$y_i = \left(\frac{\gamma_i}{\phi_i}\right)^t \frac{f_i^o}{P} x^i \qquad (2)$$

When t is zero, the relationship expresses essentially ideal behavior. Moving t to unity slowly introduces the nonideality expressed by liquid activity and the vapor fugacity coefficient. Taylor and others report solving some very difficult problems using this approach.

B. Dynamics

I discuss here the problem of solving models that characterize the dynamic behavior of processes. These models in general comprise a set of ordinary differential equations (ODEs), a set of algebraic equations, and a set of initial conditions from which the solution can be started. One solves by converting the ODEs into a set of approximating algebraic equations that, together with the algebraic equations, can be stepped incrementally through time to trace out the trajectories of the variables. In each step, a set of algebraic equations for the variables at time step $k + 1$ is solved in terms of those at time step k.

1. Stiff Equation Sets

How one forms the approximations for the ODEs is crucial to the performance of this approach. Gear [22] and many others since showed how *implicit* methods convert the ODEs so that the solution method is stable and can therefore be used to solve stiff sets of equations. Implicit methods give algebraic equations that generally must be solved iteratively at each time step, as they usually involve the variables at time step $k + 1$ nonlinearly.

Great debates have raged over what is meant by "stiff." A single equation can be stiff [23]; e.g., consider the following one:

$$\frac{dy}{dt} = -10^6 [y - \sin(t)] + \cos(t), \quad y(0) = 1 \qquad (3)$$

The solution is $y = \sin(t) + \exp(-10^6 t)$. However, any deviation from that solution is blown up by the first term on the right-hand side in the ODE above and will cause numerical problems. Thus, one will have to take very small steps in t to generate what appears to be a slowly moving smooth curve. Stiffness should reflect the size of the step that one has to take versus the size one thinks should be needed when looking at the solution.

2. The Index Problem

Sincovec *et al.* [23] presented a very disquieting example of an initial value problem consisting of two ODEs. They showed that only one of the two state variables involved can be given an independent initial value. It was not hard to see why the problem occurs, but it was evident that such a problem could easily be hidden in a larger example. This work also proved that if an incorrect initial condition is specified and an implicit integration scheme is used, the solution will march directly to a solution that corresponds to one where a legitimate initial condition is used. The initial condition may not be one of interest, however.

Petzold [24] and Gear and Petzold [25] defined the *index* of a mixed set of ODEs and algebraic equations that indicates whether the numerical scheme used to compute the solution might have unexpectedly poor error characteristics. Crudely speaking, the way the degrees of freedom have been specified for a problem can force one or more of the approximation equations to be used *backward*, i.e., to differentiate rather than to integrate. The error estimate is "worse" roughly by h, where h is the integration step size; for example, where one might think the method is first order in step size, the error could in fact be independent of the step size. If more than a single equation is used backward, the error behavior is even worse. One would use the approximation equations backward if one were for computing the control variable trajectory that will give a prescribed state variable behavior. They present a method to compute this index for linear or linearized problems. Gear [26] showed that the index problem could be eliminated by differentiating the equations.

Pantelides [27, 28], in his work with Sargent, defined the index in a manner that exposes its potential to cause problems in initialization as well as in the integration error. They too showed that the index problem can be eliminated by differentiation. Noting that only some of the equations need to be differentiated, they use a method based on the structural properties of the equations to discover these equations. They cite several examples in which the index problem is almost certain to occur in setting up and solving dynamic simulation models, e.g., calculations of flash dynamics and problems in which the trajectory of a state variable is specified.

3. Architectures for Simulators

Implementation of dynamic simulators has led to interesting research issues. For example, many have been implemented in a sequential modular format. To carry out the integration correctly from the point of view of correctly assessing integration errors, each unit model can receive as input a current estimate for the "state" variables (variables x), the unit input stream variables, and any independent input variables specified versus time (variables u). It can then calculate the unit output variables and the right-hand sides (RHSs) for the dynamic equations provided they are written in the following form.

$$\frac{dx}{dt} = f(x,z,u,t) \tag{4}$$

where z are the variables whose values are established by the algebraic equations in each of the unit models.

The executive system can then solve the models in an appropriate sequence to converge the stream variables involved in a process recycle. Once these are converged, the executive can then use the RHSs to integrate simultaneously the differential equations for all the units in this recycle loop.

Brosilow [29] presented an approach to coordinate the solving of unit models in which each integrates its own ODEs internally, using whatever integration method it chooses and using its own step size control. The executive system coordinates this activity. This approach is very appealing, but care is required to ensure that the overall system integration errors are correctly assessed and maintained. The advantage is, of course, that a "quiet" or slowly moving unit will have little computational work to do.

Many current dynamic simulators are equation based. Thus they seem to require that all parts of the simulation move together in the integration. Kuru [30] and Kuru and Westerberg [31] present an approach to take advantage of the modularity of the flowsheet but in an equation-solving environment. A partitioning scheme for the Newton equations that accounts for the modularity of the flowsheet allows the convergence of all the equations but with differing numbers of iterations, e.g., with no iterations for the equations for units that are not moving. Thus, *latency* is moved into the Newton scheme. No approximations are made, so error handling is unaffected.

C. Partial Differential Equations

There is an enormous literature on integration of partial differential equations (PDEs). I have very limited experience in this area and am therefore unwilling to say much here. However, there has to be a strong overlap with the concepts covered above for solving both algebraic and mixed ODE–algebraic models. Typically, PDEs are converted into very large structured sets of approximating algebraic equations in terms of variable values at a number of strategically located grid points, using either finite difference or finite element concepts. These equations relate values of neighboring grid points so that they will approximately satisfy the PDE operator in some optimal sense. In other approaches, they are discretized only in $n - 1$ of the n coordinate directions and thus are converted into a set of ODEs.

One interesting research issue currently receiving much attention is placement of grid points for the discretization. For a single distributed variable such as temperature, one can see how to place the points, i.e., put more points where the variable is changing more rapidly and fewer where it is not. How is this to be done for many distributed but coupled variables that are changing at quite different rates in different parts of the space? This problem is similar to that of slow- and fast-moving units in a dynamic simulation, only here no natural modularity occurs within a flowsheet of interconnected units.

One idea for grid point placement is to guess at a placement, solve, and then, based on the solution, add and delete points where it seems appropriate, solve, regrid, etc. Another is to solve for the variable values and the grid point locations simultaneously. The latter approach can give rise to a very much larger equation set to be solved and to sets of equations that have a high probability of being singular or nearly so. The grid placement is often done to minimize an error criterion that measures the error between the PDE operator and the equations used to discretize it. Often the grid placement has little impact on this error estimate. Then the equations for grid placement become singular. One scheme to correct this problem is to add penalty functions to the error criterion. If that approach is not helpful, a better one might be to remove the degrees of freedom if they are not needed—i.e., remove the equations that place the points and substitute others that space them out in a reasonable fashion. The goal is to leave just enough points whose locations will affect the error and move the others relative to these. Handling nearly singular equation sets for flowsheet modeling is certainly related to this line of thinking.

Another interesting issue is how to handle discontinuities. Often this has been done by increasing the number of grid points. Why not simply remove the constraints that are forcing continuity at the break points and let the temperature or composition profile take on two values at that grid point? Mavaridis *et al.* [32] have shown the utility of this idea. Again, one has to discover where to make these changes as the solution evolves.

Finally, one should be concerned about the index problems for mixed sets of PDEs and algebraic equations. How many problems have been solved where some "derivable" independent equations have been missed?

D. Optimization

I have reviewed optimization in an earlier article [33] in a manner consistent with this presentation. Therefore, this section will only summarize some of the ideas and mention some results that have occurred since that review.

Optimization first became possible for large-scale problems with the development of linear programming codes in the late 1950s. The oil companies quickly adopted this methodology for the scheduling of refineries. With the introduction of mixed-integer linear programming in the 1960s, the same companies started to use these codes to aid in making design decisions. For example, should a new refinery include a fluidized catalytic cracking unit or not? A binary variable that takes on a value of zero or unity only was used to indicate the existence (value of unity) or nonexistence (value of zero) of a part of the proposed design. A modest amount of nonlinear behavior could be added by approximating nonlinear behavior with straight-line segments; the cost is the adding of a large number of binary variables.

Early attempts to add optimization to flowsheeting calculations were not well received. Here, optimization requires the handling of nonlinearities. These attempts used pattern search approaches that proved to be extremely slow and not very reliable. These approaches require only the evaluation of the objective function and constraint violations; i.e., they do not use gradient information. One would regularly see reports of 1000 to 2000 function evaluations (i.e., flowsheet simulations) to carry out an optimization using pattern searches. These approaches often stalled completely on ridges. Industry did not adopt this methodology for solving real problems.

An approach that has enjoyed industrial success is the application of sequential linear programming. The nonlinear problem is locally linearized, with bounds placed on the allowed moves for all the variables so that the local linearization is still appropriate. An iteration is to linearize, solve the corresponding linear program, move to the solution of the linear problem, and converge that solution to the solution of the nonlinear problem. This point is the start of the next iteration. Industry has used this approach successfully for a number of years.

Gradient-based algorithms, such as the generalized reduced gradient and, in the last 10 years, sequential quadratic programming, offer much better performance for optimizing. Experience with the latter suggests that one can often optimize a flowsheet in the time that it takes to solve the flowsheet only two to three times—compare that to the 1000 to 2000 times for pattern search approaches. Most of the commercially available flowsheeting packages have introduced optimization capabilities based on these concepts.

Attempts to use algorithms that require gradients were and often continue to be thwarted by the modeling techniques used within flowsheeting systems. Frequently, the models contain IF statements that are used to switch from one type of model behavior to another depending on the value of the variables. These switches cause unsmooth or even discontinuous behavior, which can destroy gradient-based algorithms. Examples are to switch from a laminar to a turbulent formula to estimate friction factor for flow in a pipe as the Reynolds number passes through 2100 and to switch from a two-phase flash calculation to a single-phase computation as the flash temperature passes below the bubble point or above the dew point. Physical properties packages are notorious for containing such discontinuous or unsmooth behavior, e.g., their complex decision making for handling the roots of (cubic) equation of state models. Even the subtle differences caused by using inner convergence loops that may iterate a different number of times during each calculation can provide unsmooth behavior to the outer optimization algorithm.

These IF statements are really a form of discrete decision making embedded within the model. One possible approach to remove the difficulties it caused is to move the discrete decisions to the outside of the model and the continuous variable optimizer. For example, the friction factor equation can be selected to be the laminar one irrespective of the Reynolds number that is computed later. Constraints can be added to forbid movement outside the laminar region or to forbid movement too far outside the laminar region. If the solution to the well-behaved continuous variable optimization problem (it is solved with few iterations) is on such a constraint boundary, tests can be made to see if crossing the constraint boundary can improve the objective function. If so, the boundary is crossed—i.e., a new value is given to the discrete decision, etc.

Grossmann and his students [34–37] have produced some very exciting new developments in the solving of mixed integer *nonlinear* programs. As indicated above, integer variables can be used to allow discrete changes in a model such as the addition or removal of a unit. This can thus be a tool for design in which one first sets up a superstructure within which are embedded the various alternatives that one wants to consider for a design. Such a tool is very powerful when used for problems with highly coupled decisions.

The approach is as follows. First the designer creates the superstructure for the artifact to be designed, say a flowsheet. Then he or she selects a substructure in that superstructure that is thought to be a good candidate for the optimal one. This selection corresponds to picking values for all the discrete (binary) variables. This flowsheet alternative is optimized using a nonlinear programming code, e.g., one that uses reduced gradients or sequential quadratic approximation concepts. Next, at the optimal solution to this alternative, the flowsheet modeling equations are linearized using an *outer approximation* to the nonlinear equations. An outer approximation is required to ensure that no part of the solution space for the continuous variables is cut off in the linearization. (One cannot always guarantee that such an approximation has been used—more in a moment on this problem.) A constraint known as a *cut* is added to ensure that the previous solution for the discrete decisions is not found again; i.e., the sum of the binary variables set to unity for this solution is set to have an upper bound equal to one less than the number of these variables. This linear approximation with the discrete decisions is solved as a mixed-integer *linear* program (MILP) to find a new set of discrete decisions. These will correspond to a different flowsheet alternative. If one has used a proper outer approximation, the MILP solution is a lower bound on the cost of the flowsheet it finds. This new flowsheet is optimized (a nonlinear program again). Again this flowsheet is linearized and these equations are *added* to the previous linear model; the combined linearization with another cut to eliminate the previous solution is again solved using an MILP code to discover yet another flowsheet alternative. The process terminates when the MILP solution costs more than the best flowsheet alternative found so far. It is not allowed to reuse any previous solution; i.e., the next best solution is too expensive.

Grossmann and his co-workers have shown that this approach converges amazingly fast, often in two to five cycles (i.e., five NLP/MILP cycles). Only a few seconds to minutes of computer time is required for the problems solved, one of which was a complete flowsheet model for the hydrodealkylation of toluene process described in Douglas [9]. The model contained several hundred equations only, so it is a simplification of a rigorous model. Some interesting contributions in the just completed work with Kocis have been (1) a better approach to handle equality constraints, (2) an approach to make improved linearizations for units that are inactive in the initial solution, (3) a linearization scheme for mixers and splitters that guarantees these will be "outer approximated" (their models are nonconvex), and (4) a scheme (admittedly *ad hoc* but apparently often effective) to detect and make adjustments when nonconvexities in the problem have led to incorrect outer approximations [37].

VI. Conclusion

Many new developments are occurring in analysis. The *Frontiers in Chemical Engineering* report [16] indicates that speed of computers is doubling every year, partly from new hardware capabilities and partly from new numerical techniques. There is little evidence to suggest that either will slow for some time. Most exciting will be the use of computers for mixed qualitative–quantitative problems in which human-like reasoning will aid in correctly stating and solving problems.

Acknowledgment

This research has been funded by National Science Foundation Grant CDR-8522616.

References

1. Douglas, J. M., in *Perspectives in Chemical Engineering: Research and Education* (C. K. Colton, ed.), p. 535. Academic Press, San Diego, Calif., 1991 (*Adv. Chem. Eng.* **16**).

2. Westerberg, A. W., in *Recent Developments in Chemical Process and Plant Design* (Y. A. Liu, H. A. McGee, Jr., and W. R. Epperly, eds.), p.127. Wiley, New York, 1987.

3. Westerberg, A. W., *Comput. Chem. Eng.* **13**, 365 (1989).

4. Morari, M., in *Perspectives in Chemical Engineering: Research and Education* (C. K. Colton, ed.), p. 525. Academic Press, San Diego, Calif., 1991 (*Adv. Chem. Eng.* **16**)

5. Minsky, M., in *The Psychology of Computer Vision* (P. Winston, ed.). McGraw-Hill, New York, 1975.

6. Piela, P. C., Epperly, T. G., Westerberg, K. M., and Westerberg, A. W., *Comput. Chem. Eng.* (in press, 1990).

7. Grossmann, I. E., Westerberg, A. W., and Biegler, L. T., in *Foundations of Computed Aided Process Operations* (G. V. Reklaitis and H. D. Spriggs, eds.), p. 403. CACHE/Elsevier, New York, 1987.

8. Talukdar, S., and Westerberg, A. W., "A View of Next Generation Tools for Design." Paper 23a, AIChE Spring Meeting, New Orleans, March 6–10, 1988.

9. Douglas, J. M., *Conceptional Design of Chemical Processes.* McGraw-Hill, New York,1988.

10. Andrecovich, M. J., and Westerberg, A. W., *AICHE J* . **31**, 363 (1984).

11. King, C. J., *Separation Processes,* 2nd Ed. McGraw-Hill, New York, 1980.

12. Hohmann, E. C., "Optimum Networks for Heat Exchange." Ph.D. thesis, Chemical Engineering, University of Southern California, Los Angeles, 1971.

13. Gundersen, T., and Naess, L., *Comput. Chem. Eng.* **12**, 503 (1988).

14. Lien, K. M., "Expert Systems Technology in Synthesis of Distillation Sequences." Ph.D. thesis, University of Trondheim, Norwegian Institute of Technology, Laboratory of Chemical Engineering, Trondheim, Norway, 1988.

15. Lien, K. M., *Comput. Chem. Eng.* **13**, 331 (1989).

16. National Research Council, Committee on Chemical Engineering Frontiers: Research Needs and Opportunities. *Frontiers in Chemical Engineering. Research Needs and Opportunities.* National Academy Press, Washington, D.C., 1988.

17. Piesler, A. H., and Kessler, M. M., *Refining Eng.* **32**, C2 (1960).

18. Barnard, W. L., Benjamin, D. R., Cummings, D. L., Piela, P. C., and Sills, J. L., "Three Issues in the Design of an Equation-Based Process Simulator." Paper 42d, AIChE Spring Meeting, New Orleans, March 6–10, 1988.

19. Weyburn, T. L., and Seader, J. D., *Comput. Chem. Eng.* **11**, 7 (1987).

20. Kuno, M., and Seader, J. D., *Ind. Eng. Chem. Res.* **27**, 1320 (1988).

21. Vickery, D. J., and Taylor, R., *AIChE J.* **32**, 547 (1986).

22. Gear, C. W., *Numerical Initial Value Problems in Ordinary Differential Equations.* Prentice-Hall, Englewood Cliffs, N.J., 1971.

23. Sincovec, R. F., Erisman, A. M., Yip, E. L., and Epton, M. A., *IEEE Trans. Automatic Control* **AC-26**, 139 (1981).

24. Petzold, L. R., *SIAM J. Sci. Stat. Comput.* **3**, 367 (1982).

25. Gear, C. W., and Petzold, L. R., *SIAM J. Numer. Anal.* **21**, 716 (1984).

26. Gear, C. W., *SIAM J. Sci. Stat. Comput.* **9**, 39 (1988).

27. Pantelides, C. C., *SIAM J. Sci. Stat. Comput.* **9**, 213 (1988).

28. Pantelides, C. C., Gritsis, D., Morison, K. R., and Sargent, R. W. H., *Comput. Chem. Eng.* **12**, 449 (1988).

29. Brosilow, C. B., and Liu, Y. C., "Modular Digital Simulation of Chemical Process Dynamics." Paper 26b, AIChE Annual Meeting, Washington, D.C., November, 1983.

30. Kuru, S., "Dynamic Simulation with an Equation Based Flowsheeting System." Ph.D. thesis, Carnegie Mellon University, Pittsburgh, 1981.

31. Kuru, S., and Westerberg, A. W., *Comput. Chem. Eng.* **9**, 175 (1985).

32. Mavaridis, H., Hrymak, A. N., and Vlachopoulos, J., *AIChE J.* **33**, 410 (1987).

33. Westerberg, A. W., in *Foundations of Computer-aided Chemical Process Design* (R. S. H. Mah and W. D. Seider, eds.), p. 149. Engineering Foundation, New York, 1981.

34. Duran, M. A., and Grossmann, I. E., *Math Prog.* **36**, 307 (1986).

35. Duran, M. A., and Grossmann, I. E., *AIChE J.* **32**, 592 (1986).

36. Duran, M. A., and Grossman, I. E., *AIChE J.* **32**, 123 (1986).

37. Kocis, G. R., "A Mixed-Integer Nonlinear Programming Approach to Structural Flowsheet Optimization." Ph.D. thesis, Carnegie Mellon University, Pittsburgh, 1988.

24

Process Control Theory: Reflections on the Past Decade and Goals for the Next

Manfred Morari
Department of Chemical Engineering
California Institute of Technology
Pasadena, California

I. Introduction

In the mid- to late 1970s, interest in process control had dropped to an all-time low. Attendance at conferences had fallen off, and the number of papers presented at these conferences was meager. In 1979, for example, only about 200 people attended the Joint Automatic Control Conference in Denver; a year earlier at the AIChE Annual Meeting, only six papers were presented in the general area of process control.

Now, a decade later, the scenario is dramatically different. Close to 800 attendants flocked to the recent American Control Conference in Atlanta. Roughly 80 papers were submitted for the AIChE Annual Meeting in 1988. What caused this turnaround?

A pivotal event for the surge of interest in process control, we believe, was the publication of three insightful critiques [1–3] that analyzed the gap between academic theory and industrial practice. Bridging this gap became the goal of the "new" generation of process control theory.

Today, industry realizes that excellent payoffs can be achieved from process control projects. As an illustration, Du Pont's process control tech-

ADVANCES IN CHEMICAL ENGINEERING, VOL. 16

nology panel recently reported [4] that by improving process control throughout the corporation, the company could save as much as a half-billion dollars per year.

How can such massive savings be achieved? First of all, most aspects of operations and control are related to the collection, interpretation, communication, and utilization of information. Because of the explosive developments in computers, information processing has undergone revolutionary changes in the last decade. Also, drastic progress has been made in the development of automatic analyzers and intelligent sensors, without which the implementation of advanced feedback control strategies would not be possible. These advances in hardware were accompanied by new software and algorithm breakthroughs.

During the last decade we have seen increasing attention to plant diagnosis with "knowledge-based" approaches moving into the application phase [5]. New techniques based on neural nets [6] may lead to revolutionary changes. The concern for quality has brought about great interest in statistical process control [7]. Many of these concepts have been around for years, however, and breakthrough developments are unlikely.

This paper focuses on two new areas: *model predictive control*, which is rapidly gaining acceptance in the chemical process industry, and *robust control*, a thriving research topic that shows great promise for practical applications. Various aspects of the new technology that can help both to assess the controllability properties of new designs and to assist with the structuring of a control system, i.e., with the selection of manipulated and measured variables and their "pairing," will also be discussed. For details on the new techniques, the reader is referred to two recent monographs [8, 9].

II. The Role of Process Control Theory and Experiment

What should be the role and objective of *good* process control theory?

Controller design is always a matter of compromises and trade-offs. It is well known that if a system is designed to respond very quickly, violent moves in the manipulated variable are necessary, measurement noise is amplified, and the performance degrades drastically as the plant characteristics change—indeed, the control system might become unstable. Thus, rapid response is traded off against smooth, restrained moves of the manipulated variables, damping of measurement noise, and low sensitivity to model error. Theory should provide insight into these trade-offs and make tools available so that the designer can exploit them in a manner appropriate for the specific situation. Theory should not aim at eliminating the engineer from the design process, but should provide him or her with information that makes the decision process more transparent.

Similar to the minimum-work concept in thermodynamics, control theory should provide achievable targets of performance and should do so under practical conditions, e.g., when the manipulated variables are constrained and when the model used for assessing the performance is only an approximate representation of the real system. The targets should be independent of controller complexity. They should allow the designer to assess the merits of simple, empirical control strategies and to judge different process designs in terms of their operability.

Theory should also reveal the advantages and limitations of empirical tools that have a proven success record in industry. Theory can potentially expand the scope of these tools and can play a pedagogical role by uncovering their underlying fundamental principles.

A good control experiment can be designed only with a thorough knowledge of theory. Demonstrating the superiority of an advanced algorithm over a proportional-integral (PI) controller is not proof of its merits, nor is this information very useful. Experiments should be designed to elucidate when, where, and why algorithms fail, not to demonstrate their success under very restricted conditions.

Ed Bristol from Foxboro once stated that "experiments should model the challenges of reality and not reality itself." The reason is that industrial-scale experiments are impossible in a university. Theory should guide what makes a control task challenging and should determine the proper "scale-down" of industrial problems.

III. Control Theory of the 1960s and 1970s

The critiques of the 1970s [1–3] expressed the practitioners' disenchantment with "modern control" theory: The heavy emphasis on optimal control in the 1960s and 1970s was judged to be inappropriate in view of problems as important as the selection of the control structure and the effects of process design on operability. Another point of contention was that optimal control theory was of very limited usefulness in practice: the open-loop switching curves constituting time-optimal strategies are irrelevant because they assume the availability of a perfect model; the popular quadratic objective function is largely fictitious and has no relation to practical performance measures; the order of the optimal controllers is generally too large for them to be implemented; the controllers do not tolerate faults; the sensitivity of performance to modeling errors can be large, and so on.

Progress has been made on the control structure problem. Despite its flaws, optimal control is now used in industry, though in a somewhat modified form and couched in terms of "model predictive control." This new framework is much more intuitive. With this framework, fault tolerance can be accomplished through on-line adaptation. Because of increased computer power, the high order of the controller has become irrelevant.

The new paradigm of *robust optimal control* is well on the way to rendering the linear quadratic Gaussian control obsolete (or at least less important) because it can deal explicitly with model error.

A decade ago, the process control community believed that its problems were unique and that techniques successful in other areas could not be transferred to process control. Since then we have learned that the celebrated tools of the 1960s and 1970s did not find much application in the aerospace field either. Although some aspects of aerospace control are different (for example, the large number of underdamped modes in large space structures have no analog in process control), many problems are actually similar. The models available for aerospace problems are usually better, but then the performance specifications are much tighter than those in process control. Thus, robust control is a very important issue in this area, too. Indeed, large space structures are more "homogeneous" than chemical plants, but there the actuator and/or sensor placement problem is as important as the control structure problem in process control. For design purposes, the time delays that are the trademark of chemical process models can be approximated by rational transfer functions, and the standard tools developed for the design of aerospace control systems can be applied. The many studies on the control of systems described by delay-differential equations are theoretically challenging and interesting but of little practical value.

IV. Model Predictive Control

Model predictive control (MPC) has become widely known as dynamic matrix control (DMC) and model algorithmic control (MAC). A review of the origins of this class of techniques and their theoretical foundations is provided by Garcia *et al.* [10]. Many complex applications were reported at the recent IFAC Workshop [11].

The success of MPC is based on a number of factors. First, the technique requires neither state space models (and Riccati equations) nor transfer matrix models (and spectral factorization techniques) but utilizes the step or impulse response as a simple and intuitive process description. This nonparametric process description allows time delays and complex dynamics to be represented with equal ease. No advanced knowledge of modeling and identification techniques is necessary. Instead of the *observer* or *state estimator* of classic optimal control theory, a model of the process is employed directly in the algorithm to predict the future process outputs.

MPC replaces the fixed-structure explicit control law with on-line optimization, which is a major step forward. This makes it possible to deal easily with (1) systems having unequal numbers of inputs and outputs, (2) constraints on the manipulated or controlled variables, and (3) the failure of actuators. (A stuck valve becomes a constraint that is handled automatically and "optimally" in the algorithm.) The MPC regulatory/supervisory control law can be incorporated conveniently into the higher-level optimization layer. Without modifications or retuning, a variable can be, at different times, a control objective, a manipulated variable, or simply kept constant, depending on the demands of the optimization algorithm.

Operator acceptance of MPC is reported to be very good. Displaying the long-term predicted closed-loop behavior of the process convinces the operator that the input moves, which might appear unusual in the short term, are reasonable. Tuning is accomplished via the prescription of a reference trajectory or the adjustment of a "filter," both of which are related directly to the speed of the closed loop's response.

The successes should not distract one from the shortcomings inherent in the approach. First, the design of MPC controllers for systems like those mentioned previously is not trivial. A complete model describing the effects of all the different process inputs on all the different process outputs must be developed, and all control goals must be incorporated into a single objective function, which is then optimized. To the amazement of academics, this is done by exactly those "weights" that are void of any physical significance and were criticized in the academic approaches of the 1960s and 1970s.

Model predictive control was conceived for multivariable systems with changing objectives and constraints. In simpler situations, a PID controller tuned according to internal model control (IMC) principles [8] can deliver equal performance with much less effort.

The theoretical properties of MPC are not well understood. As Zafiriou [11] showed recently, the stability characteristics of these algorithms can be affected by constraints in very surprising ways.

V. Robust Process Control

As every practicing control engineer knows, a controller that is designed to work well at one operating condition will usually not work as well (if at all) at another. In the same spirit, a controller design based on a process model might not work well if this model is not a "true" representation of the process dynamics. The objective of control system design should be *robust performance*. This means that the controller should be designed so that the closed-loop performance specifications are met despite varying operating conditions, or what we refer to as *model–plant mismatch*. Some typical practical questions related to robust performance are: How large a change in operating conditions is allowed before controller retuning becomes necessary? How large a model–plant mismatch can be tolerated before a control system design becomes *impossible*? Or more generally, what characterizes a "good model" for control system design? This last question seems very basic and at the heart of the design problem. Nevertheless, until recently it has not even been mentioned. Some illustrative examples can be found elsewhere [12].

These issues of model accuracy and model uncertainty were not dealt with at all in the 1960s and 1970s. They are absent from all process control textbooks, an exception being the recent textbook by Seborg *et al.* [13]. The ability of classical control theory to explain the demonstrated effects and difficulties is very limited. In the past 10 years, much progress has been made in the area of robust control. This progress is of obvious practical value. The main results are derived [8] and summarized elsewhere [12, 14].

The systematic *synthesis/design* of controllers that work well despite inaccurate models is about to come into the reach of the industrial designer. The new generation of optimal control techniques (which are referred to as H_∞- and μ-*optimal control*) will be able to tackle the following type of problem: "A two-point composition control system for a distillation column is to be designed and must meet the following requirements. The settling time should be less than 20 minutes for all disturbances and setpoint changes. In the worst case, a disturbance should not be amplified by more than 50% in the transient. A simple model for the column is available. The model is of low order and does not include tray hydraulics and other high-frequency effects that can give rise to effective delays of as much as 1 minute. The rate of valve movement is limited." Within the next few years, software should become available that designs controllers based on specifications such as this. Indeed, software of a somewhat more modest scope is already available [15]. If no existing controller meets all the conflicting criteria, the controller that achieves the best compromise would be proposed.

Such a design procedure is clearly a far cry from the linear quadratic Gaussian techniques in which robustness is obtained in an indirect manner by inventing measurement noise and introducing stochastic processes into an essentially deterministic problem. Nevertheless, the two approaches have amazing mathematical parallels [16].

Despite the optimistic overtones, robust control is not a solved problem. Some difficult theoretical questions remain in the synthesis area. The available software is, at best, experimental; the controller is complex and its structure is not obvious. It generally uses all the measurements and all the manipulated variables in a centralized fashion. On-line tuning is difficult except when the IMC structure is employed [8]. Fault tolerance, that is, continued satisfactory or at least stable performance in the event of an actuator or sensor failure, cannot be guaranteed.

The performance specification and the uncertainty description cannot be stated in quite as simple terms as indicated above. Moreover, the type of uncertainty description needed for the design cannot be readily obtained from standard experiments and identification procedures. Finding an uncertainty description appropriate for controller analysis and design will become a part of the modeling process in the future. Just as with modeling, a number of tools will become available, but, to a certain extent, process and uncertainty modeling will always remain an art, where the engineer must take advantage of both process knowledge and identification techniques.

VI. Control Structure Selection

The optimal robust controller designed with one of the new synthesis techniques is generally not of a form that can be readily implemented. The main benefit of the new synthesis procedure is that it allows the designer to establish performance bounds that can be reached under ideal conditions. In practice, a decentralized (multiloop) control structure is preferred for ease of start-up, bumpless automatic to manual transfer, and fault tolerance in the event of actuator or sensor failures. Indeed, a practical design does not start with controller synthesis but with the selection of the variables that are to be manipulated and measured. It is well known that this choice can have more profound effects on the achievable control performance than the design of the controller itself. This was demonstrated in a distillation example [17, 18] in which a switch from reflux to distillate flow as the manipulated variable removes all robustness problems and makes the controller design trivial.

After the variables are selected, the structure of the controller itself is chosen. This determines which manipulated variables are changed and on which errors the changes are based. Proper control system structuring can have a significant effect on its reliability. In industry, strategies that work well all the time provide more benefits than those that work optimally but are frequently on manual control.

Of course, in model predictive control, the structuring problem is less important. The controller is centralized and reliability is achieved through on-line optimization. There are many cases, however, in which the modeling and design effort necessary for model predictive control is either impossible or cannot be justified economically.

In the 1960s and 1970s, the relative gain array (RGA) [19] was the only systematic tool available for control structure selection. Its simplicity, practical success, and lack of theoretical basis were disturbing to many academics. Since then, the RGA has been largely vindicated: its range of applicability has been defined, its limitations are well understood, and its scope has been extended [8]. For the primary control structuring task, the RGA is still the key tool eliminating potential control structures that cannot be made fault tolerant.

Much progress has been made on the design of decentralized controllers [8]. A methodology has been developed to translate overall performance and robustness specifications into specifications on individual loops. Designing the individual loops according to these specifications guarantees the satisfactory performance of the overall system.

Attempts have been made to systematize the control structure problem on a larger scale ("plant control"). Stephanopoulos [20] provides a good pedagogical summary. From a practical point of view, the choice of feedback and feedforward loops is usually straightforward and can be handled by engineers even with limited experience. Thus, there is little need for defining a systematic procedure. A more difficult task is designing variable structure elements, valve position controllers, and auctioneering and selector systems to ensure that the various operating constraints are not violated and that an economically optimal level of performance is maintained. In the academic approaches, these tasks are dealt with at the supervisory level by using a model. The typical models, however, are usually not accurate enough to be relied on for protection against constraint violation. A more promising problem definition for the structuring of supervisory control is proposed by Bristol [21].

Additional considerations are necessary when the control structure of reactors is to be determined. Reactors are usually highly nonlinear and poorly modeled. Therefore, approaches like those discussed previously are rarely suitable. Exact, offset-free control of all relevant reactor variables is neither possible nor necessary. Usually, keeping the variables in a certain operating range is sufficient. This is often accomplished in an indirect manner—for example, by eliminating as many disturbances as possible at their source. Shinnar [22] has proposed a new philosophical definition of the reactor control problem, a definition that has yet to be put on a quantitative basis and in algorithmic form.

Some of the limitations of the current control theory were demonstrated on the Shell Control Problem [12].

VII. Some Research Topics for the Next Decade

A number of research issues in the areas of robust control, model predictive control, and control structure selection were mentioned previously. Unfortunately, even if all these problems were solved, a practical problem like the Shell Control Problem [23] could still not be tackled in a systematic fashion. All the research topics discussed so far in this paper are related to the lowest (regulatory) control layer. During the past decade, much progress has also been made on the algorithms that make up the highest (optimization) layer. The middle (supervisory) layer, however, is mostly empirical and proprietary and has not attracted attention until recently.

Typical functions of the middle layer are anti-reset windup, variable structure elements, selectors, etc. Although essential for the proper functioning of any practical control system, they have been completely neglected in research circles. We have yet to find an effective multivariable anti-reset windup scheme that works on all our test cases. Can all, or at least most, industrial control problems be solved satisfactorily with some simple loops and minimum–maximum selectors? How can the appropriate logic structure be designed? How should the loops be tuned to work smoothly in conjunction with the logic? How can one detect deteriorating valves and sensors from on-line measurements before these control elements have failed entirely?

Currently, most control researchers begin with the assumption that a model of the process is available in either state space or transfer function form. Historically, this was certainly not always so. The greatest contribution of Ziegler and Nichols is not the famous tuning rule but the "cycling" identification technique: they understood precisely what minimum model information was necessary for tuning and developed a simple, reliable method for obtaining this information. Perhaps the same philosophy should be used to develop the methods needed for designing the supervisory layer. With a complete model available, the design task is trivial. But what is the minimum information necessary (gains, signs of gains, time constants, etc.) for determining the correct structure of the supervisory and optimizing control system for an industrial control problem?

The research breakthroughs of the past decade were possible because of close cooperation between chemical engineers and control engineers from other disciplines, particularly electrical engineering. They were stimulated by continuous interactions with industry. If we are to solve the problems mentioned here, even closer collaboration among the different disciplines will be necessary.

References

1. Foss, A. S., *AIChE J.* **19**, 209 (1973).
2. Kestenbaum, A., Shinnar, R., and Than, F. E., *Ind. Eng. Chem. Proc. Des. Dev.* **15**, 2 (1976).
3. Lee, W., and Weekman, V. W., Jr., *AIChE J.* **22**, 27(1976).
4. National Research Council, Committee on Chemical Engineering Frontiers: Research Needs and Opportunities. *Frontiers in Chemical Engineering. Research Needs and Opportunities.* National Academy Press, Washington, D.C., 1988.

5. Reklaitis, G. V., and Spriggs, H. D., eds., *Foundations of Computer Aided Process Operations*. CACHE-Elsevier, Amsterdam, 1987.

6. Ungar, L. H., and Powell, B., "Fault Diagnosis Using Connectionist Networks." Paper 133e, AIChE Annual Meeting, Washington, D.C., 1988.

7. MacGregor, J. F., *Chem. Eng. Prog.* **84**, 21 (October 1988).

8. Morari, M., and Zafiriou, E., *Robust Process Control*. Prentice-Hall, Englewood Cliffs, N.J., 1989.

9. Prett, D. M., and Garcia, C. E., *Fundamental Process Control*. Butterworths, Stoneham, Mass., 1989.

10. Garcia, C. E., Prett, D. M., and Morari, M., *Automatica* **25**, 335 (1989).

11. McAvoy, T. J., Arkun, Y., and Zafiriou, E., eds., *Model Based Process Control*. Pergamon Press, New York, 1989.

12. Morari, M., *Chem. Eng. Prog.* **84**, 60 (1988).

13. Seborg, D. E. Edgar, T. F. and Mellichamp, D. A. *Process Dynamics and Control*. Wiley, New York, 1989.

14. Morari, M., *Chem. Eng. Res. Des.* **65**, 462 (1987).

15. Lewin, D. R., and Morari, M., *Comput. Chem. Eng.* **12**, 1187 (1988).

16. Doyle, J., Glover, K., Khargonekar, P., and Francis, B., *IEEE Trans. Autom. Control* **34**, 831 (1989).

17. Skogestad, S., "Studies on Robust Control of Distillation Columns." Ph.D. thesis, California Institute of Technology, Pasadena, 1987.

18. Skogestad, S., Morari, M., and Doyle, J. C., *IEEE Trans. Autom. Control* **33**, 1902 (1988).

19. Bristol, E. H., *IEEE Trans. Autom. Control* **AC-11**, 133 (1966).

20. Stephanopoulos, G., *Chemical Process Control. An Introduction to Theory and Practice*. Prentice Hall, Englewood Cliffs, N.J., 1984.

21. Bristol, E. H., *Chem. Eng. Prog.* **11**, 84 (1980).

22. Shinnar, R., *Chem. Eng. Commun.* **9**, 73 (1981).

23. Prett, D. M., and Morari, M., *Shell Process Control Workshop*. Butterworths, Stoneham, Mass., 1987.

25

The Paradigm After Next

James M. Douglas
Department of Chemical Engineering
University of Massachusetts, Amherst
Amherst, Massachusetts

I. Introduction

Westerberg's paper [1] in this symposium presents a comprehensive review of the current status of research in process design that ranges from the process of design, to process synthesis, to process analysis, to the impact of new developments in computing hardware and software on process engineering activities. Thus, his review covers the spectrum from the application of ideas from artificial intelligence (AI) to an examination of the indices of partial differential equations as a way of structuring numerical solution algorithms. He discusses some research issues, and he clearly portrays this area as one of increasing interest and rapid change. To supplement Westerberg's presentation, I would like to present a different view of chemical engineering paradigms, to extend Westerberg's discussion of process synthesis, to describe some additional research opportunities, and to propose some curriculum modifications.

II. The Paradigm After Next

Imperial Chemical Industries in the United Kingdom originated the concept of designing the plant after next. It is becoming clear that the next paradigm in academic chemical engineering will focus on an improved

ADVANCES IN CHEMICAL ENGINEERING, VOL. 16

understanding of molecular-scale phenomena. However, because of my research interest in periodic processing, I expect that the emerging paradigm 20 or 30 years from now will be process engineering, which will include industrial chemistry, industrial biology, industrial materials, etc. That is, no matter who invents the product, chemical engineers will play the major role in developing the process.

Consider the quotation from A. D. Little that is included in Denn's presentation [2], "Chemical engineering ... is not a composite of chemistry and mechanical and civil engineering, but a science of itself, the basis of which is those unit operations which in their proper sequence and coordination constitute a chemical process as conducted on the industrial scale." Thus, Little's interest was not only in obtaining a better understanding of unit operations, i.e., process units, but also in the selection of the units and a better understanding of the interactions between these units in a complete plant, which used to be called industrial chemistry but today is called process engineering.

Denn also states, "Chemical engineering has been viewed as the profession with applications in which physical and chemical rate processes are limiting" and then submits that "this is the unchanging paradigm of chemical engineering." However, the separation units in many petrochemical processes are described by thermodynamic models, and not rate models, so that Denn's assertion seems to be limited. Furthermore, the 1988 Amundson Report [3] states, "Chemical engineering was the first engineering profession to recognize the integral relationship between design and manufacturing, and this recognition has been one of the major reasons for its success." At least in industry, chemical engineers are recognized for their ability to design, build, and operate new processes.

III. A Hierarchy of Paradigms

Of course, in order to understand how to design, build, and operate chemical processes, it is necessary to understand the behavior of the process units, i.e., the unit operations paradigm. However, if we want to develop new types of unit operations, e.g., affinity chromatography or chemical vapor deposition, we must understand the basic conservation principles that apply to that unit, i.e., the transport phenomena paradigm. Similarly, when we build transport models, we often introduce so many parameters that for first designs we must find some way of estimating these parameters based on the structure of the molecules, which will lead us into the molecular phenomena paradigm. As we extend chemical engineering into new application areas, we will need experts in each of these paradigms; each will have a somewhat different culture, since different tools and techniques are involved. A process engineering paradigm can provide a perspective and help to establish research priorities for all of these other activities.

For example, when we consider the design of specialty chemical, polymer, biological, electronic materials, etc. processes, the separation units are usually described by transport-limited models, rather than the thermodynamically limited models encountered in petrochemical processes (flash drums, plate distillations, plate absorbers, extractions, etc.). Thus, from a design perspective, we need to estimate vapor–liquid–solid equilibria, as well as transport coefficients. Similarly, we need to estimate reaction kinetic models for all kinds of reactors, for example, chemical, polymer, biological, and electronic materials reactors, as well as crystallization kinetics, based on the molecular structures of the components present. Furthermore, it will be necessary to estimate constitutive equations for the complex materials we will encounter in new processes.

These estimation procedures will be used to determine whether a proposed new process is profitable and to screen process alternatives. Then the sensitivities of the conceptual design calculations based on these estimates will be used to define the priorities of the experimental development program. With this approach, we can bring new processes to commercialization more rapidly.

IV. A Process Engineering Paradigm

We need to develop new, systematic procedures for inventing new processes, for retrofitting existing processes, and for synthesizing control systems for a wide variety of processes. The types of processes to consider can be classified by their operating characteristics (batch vs. continuous), the phases involved (vapor, liquid, and solid), the product slate (single product, many products at the same time, different products at different times of the year), the product characteristics (pure chemicals, mixtures, distribution specifications such as the molecular weight distribution of polymers or the size distribution of solids), the industrial grouping (petrochemical, refining, polymer, specialty chemicals, biological, pharmaceuticals, electronics, photography, etc.), and the underlying science (organic and inorganic chemistry, biology, electricity and magnetism, optics, etc.). Of course, some processes fit into more than one of these categories, but a classification scheme of this type does help to identify the chemical, physical, and economic features of importance for various types of processes. To give students a better appreciation of design, we must consider examples from all of these categories.

A. Hierarchies of Designs

Experience indicates that less than 1% of the ideas for new designs ever become commercialized. After a chemist, a microbiologist, or a material scientist has developed a new product, the initial design is used to evaluate whether or not the potential process will be profitable, and poor design projects are terminated as quickly as possible. Then the design proceeds through a hierarchy of levels of detail and accuracy, but the purpose of design, at any level of detail, is to generate a cost estimate that can be used to decide whether the project should be terminated or a more detailed design study should be undertaken. The levels discussed in Peters and Timmerhaus [4] are (1) order of magnitude (accuracy within 40%), (2) study estimate (up to 25%), (3) preliminary estimate (within 12%), (4) definitive estimate (within 6%), and (5) detailed estimate (within 3%). Different companies might describe the levels differently and use different accuracy ranges, but the main point is that a hierarchy of levels of detail is considered.

B. A Higher-Level Process Understanding

Instead of focusing on the interactions between process units, as described by Little, a process engineering paradigm will evolve higher-level descriptions of the interactions between process subsystems (reactor subsystems, vapor separation subsystems, liquid separation subsystems, solid separation subsystems, energy management subsystems, etc.). The alternative separation techniques that can be used in one type of subsystem, as well as the interactions of these alternatives with the other subsystems, will be screened as a group. In addition, procedures will be developed to estimate the economic incentive for looking for innovations.

C. "High Value Added" Products

When a new, "first generation" product is developed, there is a significant economic incentive to enter the market as quickly as possible. Thus, the prevailing philosophy is build anything that works but get it into the market. Any new tools that are developed must be sufficiently efficient that they accelerate market entry, rather than slow it down.

However, about 25% of the U.S. market for pharmaceuticals consists of generic drugs. All of these products are made outside the United States, and they are based on U.S. patents that have expired for successful products. Thus, the processing economics of "high value added" materials becomes important after a relatively short time period, and better design tools could make an impact on the industry. Similarly, the Imperial Chemical Industries philosophy of working on the design of "the plant after next" indicates the importance of improving designs that were not carefully considered initially because of the pressure to enter a market as early as possible.

D. Conceptual Design of Batch vs. Continuous Processes

The goal of a conceptual design for a continuous process is to select the process units and the interconnections between these units, identify the dominant design variables and estimate the optimum design conditions, and identify the process alternatives and find the best four or so alternatives. For batch processes, we must also decide which units should be batch and which should be continuous, whether or not some process operations should be carried out in the same process unit or separate units, whether or not parallel units should be used, and how much intermediate storage is required. Thus, batch processes require more decisions to fix the structure of the flowsheet (there are more alternatives to consider). Since there are many situations in which it must be decided whether to develop a batch or a continuous process, both procedures should be present in a general conceptual design code, whereas the current trend is to develop separate codes for batch processes.

E. New Computer-Aided Design Tools

Because of the low success rate for the commercialization of new processes, we will continue to develop new processes by proceeding through a hierarchy of designs, and therefore we will still need shortcut models. To decide whether simple models are applicable in a particular situation, we can develop a perturbation solution around a complex model, so that the simple model is the generating solution. With this approach, we can establish an error criterion that will indicate the validity of the simple model.

Similarly, it often is possible to derive back-of-the-envelope models by applying the scaling concepts from boundary layer theory to the design equations. When these simplified models are used to evaluate the economic trade-offs, it may be possible to derive design heuristics (although many heuristics correspond to large cost discontinuities). In addition to the tools that describe process units, we need to establish a systematic procedure for developing a process flowsheet and for identifying the best process alternative.

V. A Hierarchical Approach to Conceptual Design

As Westerberg mentions in his review [1], hierarchical structuring provides one of the most efficient approaches for deciding among a large number of possible alternatives, because the high-level decisions can be used to discard a very large number of lower-level alternatives. He also notes that "the knowledge required for reducing the size of design problems comes specifically from the domain of the design." Thus, when we attempt to develop a general framework for conceptual design, there will need to be numerous branches in this procedure for the specific application to particular processes. However, we can discern some features of a general framework and identify some of the branch points for particular applications.

A. Products, By-products, and Recycles

There appears to be a natural hierarchical decomposition scheme for processes involving chemical or biological reactions. If the set of reactions is known, the desired production rate is given, and the compositions of the raw material streams are known, then the streams that will leave the process and those that will be recycled to the reactor usually can be identified. Some uncertainties occur for solids, polymer, and biological processes, where separations often are difficult and expensive, so that the isolation of a reactant or a solvent for recycling might not be economically justified. In addition, in some processes, recycle might adversely effect the reactor operation. However, there are relatively few alternatives to consider, and simple economic calculations normally make quick decisions possible. The uncertain decisions can be posed as process alternatives to be evaluated. By keeping careful track of these uncertain decisions, it is possible to develop systematically a list of process alternatives that should be considered.

Once we have decided what components to remove from the process and which should be recycled, we can obtain a quick estimate of the raw materials and by-product flows, as a function of the design variables, for any type of process (including specialties and batch processes) without specifying any physical properties. Normally, we find that the range of the design variables is limited where profitable operation is obtained. By limiting the range of the design variables, we simplify the synthesis/analysis of the separation system. The raw material costs often correspond to between 33 and 85% of the total processing costs, so that these simple calculations provide a significant amount of information.

Normally, it is also a simple task to estimate recycle flows as a function of the design variables. The recycle flows and feed flows provide the information required to conduct reactor synthesis/analysis studies. The cost of the reactor is usually not very important, but the product distribution and the need for heat carriers and/or diluents have a major impact on the synthesis of the separation system.

B. Separation System Specification

The decisions as to which components to remove from the process and which to recycle, along with estimates of the flows, provide the definition of the separations problem. By phase splitting the reactor exit stream, we can usually simplify the separations problem by decomposing it into a vapor recovery system, a solid recovery system, and a liquid separation system. There are numerous alternative types of process units that may appear in vapor, liquid, and solid separation systems, and the types of units normally are highly dependent on the particular type of process under consideration. That is, flash drums are used for vapor–liquid splits for low-viscosity materials, but wiped film evaporators or devolatilizing extruders are needed for high-viscosity polymers. Similarly, biological processes often contain units such as affinity chromatography, and membrane separations, which seldom appear in petrochemical processes.

C. Heat Exchanger Networks

After a separation system has been specified and all of the process flows have been determined as a function of the design variables, then all the information required to synthesize a heat exchanger network is available. The synthesis procedures for heat exchanger networks can then be used to complete a design for one alternative. The process alternatives should then be considered, and normally the best three or four flowsheets should be retained for further examination.

D. Control System Synthesis and Safety

Once a complete flowsheet has been developed, the operability and control of the process can be considered. Moreover, the economic incentive for modifying the flowsheet to improve the control can be considered. Then a (hierarchical) procedure for the synthesis of a control system for the complete plant can be used as an additional tool for screening the process alternatives, and a preliminary hazardous operations study can be initiated. The results of this conceptual design study then provides an estimate of the economic incentive for initiating a more rigorous design study.

VI. Conclusions

There will be major changes in synthesis procedures and computer-aided design tools in the years ahead and eventually a strong focus on process innovation. Concepts from artificial intelligence will play a major role, as will improved algorithms and better hardware. Thus, process synthesis, design, and control will be exciting research areas in the years ahead, and the results will find wide applications in industry.

References

1. Westerberg, A. W., in *Perspectives in Chemical Engineering: Research and Education* (C. K. Colton, ed.), p. 499. Academic Press, San Diego, Calif., 1991 (*Adv. Chem. Eng.* **16**).
2. Denn, M. M., in *Perspectives in Chemical Engineering: Research and Education* (C. K. Colton, ed.), p. 565. Academic Press, San Diego, Calif., 1991 (*Adv. Chem. Eng.* **16**).
3. National Research Council, Committee on Chemical Engineering Frontiers: Research Needs and Opportunities. *Frontiers in Chemical Engineering. Research Needs and Opportunities.* National Academy Press, Washington, D.C., 1988.
4. Peters, M. S., and Timmerhaus, K. D., *Plant Design and Economics for Chemical Engineers*, 3rd Ed. McGraw-Hill, New York, 1980.

26

Symbolic Computing and Artificial Intelligence in Chemical Engineering: A New Challenge

George Stephanopoulos
Laboratory for Intelligent Systems in Process Engineering
Department of Chemical Engineering
Massachusetts Institute of Technology
Cambridge, Massachusetts

I. Computing in Chemical Engineering

The role of computers in chemical engineering research and development has expanded continuously over the last 30 years to a point of pervasive and self-propelling reliance on the machines. In the area of chemical process development and design, computers are used for (1) process simulation and analysis; (2) equipment sizing and costing; (3) optimization; (4) integrated design of energy management systems, reactor networks, and separation sequences; (5) layout of piping networks; and (6) project planning. Chemical process control relies more and more on computers for the implementation of low-level feedback control, on-line parameter estimation, and controller adaptation, as well as for the execution of higher-level tasks such as optimization, planning, and scheduling of plantwide operations. Research in chemical engineering science relies very heavily on numerical simulation of postulated physical and chemical models to reveal behavior that is hard to capture through experimentation or to complement experimental evidence. Thus, very sophisticated codes are generated to solve

ADVANCES IN CHEMICAL ENGINEERING, VOL. 16

numerically integral and differential equations describing the motion of molecules, the behavior of materials, the flow patterns of various systems, the evolution of complex reaction pathways, the performance of separation systems, etc.

Although the variety of computing applications is extensive and growing, the underlying paradigm has been unique and simple, namely, "numerically solve a set of equations." This is a result of the fact that computers have been perceived to be "computational machines" only. Furthermore, the exclusive reliance on numerical computations forces one to use only the knowledge that can be represented by a quantitative scheme, thus limiting the range and utility of the ensuing numerical results. Quite often, the exclusion of available qualitative and approximate-quantitative knowledge is detrimental because the user either overlooks fundamentally sound scientific knowledge or distorts such knowledge to fit it into a quantitative representational scheme. Perhaps needless to say, one should resist the downgrading of reliable quantitative knowledge to inferior qualitative or semiquantitative forms. Instead, one should strive to articulate, represent, and utilize all forms of available knowledge. This new dictum has been imposed by the needs in chemical engineering and made possible by present state-of-the-art advances in computer science and technology.

To better understand the limitations of the paradigm based on numerical computations alone, let us examine a series of representative problems from chemical engineering.

A. Process Design

An advanced computer-aided process engineering environment is composed of a data base management system, which retrieves and directs information among the various facilities such as process simulators, optimizers, process unit sizing and costing routines, estimators of physical or chemical properties, etc., under the guidance of the human designer. But such a computer-aided design system does not know how the design is done and cannot encode and answer questions such as, Where does the design start from? What is to be done next? What simplifications and assumptions should be made for the design to proceed? The design strategy and methodologies for design decision making reside in the expert human designer's mind, and they never become articulated into automatic, computer-implemented procedures. As a result, a computer-aided design system cannot even encode and replay the history of a design.

B. Product and Process Development

Scientists and engineers involved in the design of new products (materials, solvents, pharmaceuticals, specialty chemicals) or the conceptualization of new processes stemming from basic chemical or biochemical reaction schemes do not use computers in their creative tasks because their essential needs are not numerical computations. Thus, fundamental qualitative scientific knowledge or accumulated experiential facts are never formally articulated and represented.

C. Understanding System Behavior

Suppose that we want to describe the behavior of an assumed model for a catalytic reaction and thus investigate the effects of the postulated mechanistic steps. Numerical simulation provides a large body of numerical data, and it is up to the chemist or chemical engineer to interpret the results in terms of the mechanism that generated them. Moreover, the simulation results depend heavily on the assumed values of the inherent parameters and provide only local information. Mathematical analysis, on the other hand, can establish for small problems explicit global properties of the assumed model such as the number of steady-state solutions and their character. It is natural, therefore, that one should attempt to capture all fundamental results from mathematical analysis and formalize them, through proper representations, into a software system that can produce the global descriptive behavior of the assumed catalytic mechanism over various regions of parametric values.

D. Feedback Control

Quantitative models and numerical computations are and will continue to be central in the implementation of model-predictive feedback controllers. Unfortunately, numerical computations alone are weak or not robust in answering questions such as the following: How well is a control system running? Are the disturbances normal? Why is derivative action not needed in a loop? What loops need dead time compensation, and should it be increased or reduced? Have the stability margins of certain loops changed and, if so, how should the controllers be automatically retuned?

E. Monitoring and Diagnosis of Process Operations

Present-day control rooms display thousands of analog or digital process data and hundreds of alarms. During the course of steady-state operations, simple observation of scores of displays is sufficient to confirm the status of the process. But, when the process is in transient operation or crises occur, the dynamic evolution of displayed data can confound even the best operators. Quantitative computations are inadequate to provide a robust "mental model" of what is going on and to carry out routinely tasks such as distinguishing normal from abnormal operating conditions, assessing current process trends and anticipating future operational states, and identifying causes of process trends (e.g., external disturbances, process faults, operator-induced mishandling, and operational degradation due to parametric changes).

F. Planning and Scheduling of Process Operations

The planning of process operations involves specifying an ordered sequence of operations or a partially ordered set of operations that, when carried out, will perturb the state of the chemical plant from some initial state and cause it eventually to attain some prespecified final or goal state. Conceivably, one could formulate this problem as a mixed-integer, nonlinear optimal control problem and solve it numerically, if it were not for the following difficulties: (1) for industrial problems of realistic size, it can be shown that the problem is intractable; (2) nontemporal constraints introduce restrictions on the temporal ordering of process operations; and (3) the objective function cannot be fully articulated *a priori*. Consequently, additional forms of knowledge and symbolic generation and manipulation of primitive operations are essential for the synthesis of operating procedures either *a priori* (i.e., at the process design stage), such as start-up, shutdown, or changeover; or on-line (i.e., during operation), such as response to faults and coordinated plantwide optimization.

It is clear from the previous discussion that present and future needs in chemical engineering cannot be met by the traditional numerical computing paradigm alone. Current developments in computer science and technology allow us to expand significantly the numerical computing paradigm and thus expand the range and scope of tasks that can be carried out by the computer.

II. The Essential Framework of Artificial Intelligence Applications in Chemical Engineering

The use of artificial intelligence (AI) techniques and methodologies and the requisite ways of formulating and solving engineering problems have generated strong emotional reactions among researchers and industrial practitioners, ranging from utopian expectations to an outright dismissal of the whole effort. This diversity of reactions, a typical symptom of a developing but as yet immature field, has propagated confusion: What is new in AI? What can it do that could not be done before? Wherein does the intellectual challenge lie? How is AI distinguishable from routine program development?

A. Making a Mind vs. Modeling the Brain [1]

It is generally accepted that AI is part of computer science and, in the words of Elaine Rich [2], "is the study of how to make computers do things at which, at the moment, people are better." Presently, computers outperform humans in (1) carrying out large-scale numerical computations, (2) storing and efficiently retrieving massive records of detailed data, and (3) efficiently executing repetitive operations. They are currently quite inferior in (1) responding to situations with flexibility, (2) making sense of ambiguous or contradictory messages, (3) recognizing the relative importance of different elements within a situation, and (4) finding similarities despite differences and drawing distinctions despite similarities among various situations. All of these are considered manifestations of human intelligence [3] and therefore subjects of study in the realm of artificial intelligence.

The above definition could be construed as implying that AI is trying to "make computers think exactly like humans," or creating a model of the brain. For engineering work, such interpretation is wrong and obviously sterile. Instead, "making a mind" is a more accurate description of what AI applications in engineering are trying to do. They tackle the same problems as humans with solutions that possess the robustness and flexibility characteristic of human approaches. Such a shift in emphasis has produced excellent examples of computer systems that exploit symbolic processing, novel models to represent all forms of knowledge, and a series of successful problem-solving paradigms, all results of research work in AI.

B. AI and Computer Programming

Rich's definition of AI has another important corollary: the research results should lead to an executable computer program. This requirement places AI squarely in the area of computer science and distinguishes it from operations research, information science, systems theory, mathematical logic, and other fields from which it has been, and still is, drawing ideas and methodologies. Such a requisite computer program, based on a computer language that "is a novel formal medium for expressing ideas about methodology ... [and] ... control[ling] the intellectual complexity" [4], should, ideally, possess provable properties such as tractability, correctness, and completeness. Unfortunately, this is a very hard proposition and, for many applications, impossible to establish.

C. Modeling Knowledge

Looking more closely at the various research advances, one quickly realizes that the practical thrust of AI in engineering applications is to enforce systematic and organized modeling of knowledge, i.e., (1) modeling of physical systems (e.g., at the boolean, qualitative, semiquantitative, or quantitative level), (2) modeling of information processing systems, and (3) modeling of problem-solving paradigms, such as diagnostic, planning, and design paradigms. Without expressive representations of the requisite declarative knowledge (i.e., "what is...") and procedural knowledge (i.e., "how to..."), no computer programs can be written. This preoccupation with all forms of knowledge and their representation distinguishes current efforts from earlier ones in chemical engineering and enables people to deliver what in the past was an "idea." Developing the proper models to represent knowledge and generating programs with, ideally, provable properties are highly challenging propositions with significant intellectual content.

D. *Problem-Solving Paradigms*

Complex codes for numerical computations have enjoyed a significant advantage: the existence of a concise, advanced mathematical background that provides the proof of the computer program's properties, e.g., stability, rate of convergence, and residual errors. Most of the computer programs developed in the field of artificial intelligence are based on problem-solving paradigms that are not as fortunate as their numerical counterparts. However, significant advances in mathematical logic, approximate algebras, and qualitative and semiquantitative calculus are providing the theoretical background for developing algorithms with provable properties.

III. Descriptive Simulation of Physicochemical Systems

To understand the behavior of complex physicochemical systems, researchers rely heavily on the numerical simulation of such systems. Computational methods for integrating systems of differential equations are very well developed, and a variety of powerful general or specific-purpose integration packages are available. But, despite its popularity, numerical simulation alone cannot provide descriptions of global behavior. It is inherently incapable of providing scientific interpretations of the numerical results (this task is delegated to the human) and cannot summarize the results in a physically meaningful manner (simply graphing the results is not sufficient for these purposes).

A growing number of works in computer science address the descriptive simulation of physicochemical systems through novel integration of (1) numerical simulations, (2) symbolic manipulations, and (3) analytic knowledge from mathematics. Typical examples of this attitude are the following:

1. The Piecewise Linear Reasoner (PLR) [5] takes parameterized ordinary differential equations and produces maps with the global description of dynamic systems. Despite its present limitations, PLR is a typical example of a new approach that attempts to endow the computer with large amounts of analytical knowledge (dynamics of nonlinear systems, differential topology, asymptotic analysis, etc.) so that it can complement and expand the capabilities of numerical simulations.

2. The Kineticist's Workbench [6] is a computer program whose purpose is to expand the role that computers play in assisting chemists and chemical engineers

to understand, analyze, and simplify complex reaction mechanisms. It achieves its purpose by integrating numerical simulation techniques with a variety of other algorithms carrying out symbolic tasks. It constructs meaningful descriptions of numerical simulations in terms of dominant episodic phenomena. It possesses rich enough data structures to allow the interpretation of the reaction model's behavior in terms of the underlying assumptions and simplifications and functional expressions used for the individual mechanistic rates.

IV. Formal Description and Analysis of Process Trends

Meaningful description of process trends is essential for the effective monitoring, diagnosis, planning and control of process operations. Current practice, based on either a detailed representation through a set of differential equations or crude time averages, leads to significant losses of information or serious inconsistencies between the various models used for feedback control, adaptive control, diagnosis, and optimal control. Recently, significant research effort has been focused on the systematic and automatic creation of "mental" models for the description of process operations using computers. To achieve this, research has addressed the following two interrelated questions: (1) What is the form of representation needed to model the true process trends, and how can it be used to extract the true scale of the various physicochemical events occurring in a chemical plant? (2) How can the model described above be used to consistently express the modeling needs for control, adaptation, diagnosis, or planning of process operations? Cheung and Stephanopoulos [7, 8] have developed a formal theoretical framework for transforming time records of process variables into meaningful and explicit descriptions of process trends in real time. By providing analytical continuity between quantitative, semiquantitative, and qualitative descriptions, the framework can provide a consistent modeling framework for carrying out control, adaptive control, diagnosis, and planning of process operations. Furthermore, it allows the explicit determination of all events, at all scales of interest, and their logical association to the underlying physicochemical phenomena—something that traditional Fourier transforms fail to provide. Finally, it constitutes the basis for construction of linguistic descriptions by the computer as to what is happening in process operations.

V. Planning of Process Operations

Computer-based process control systems do not know how to plan, schedule, and implement process operations beyond the local confines of single processing units, as required by the evolving needs of computer-integrated manufacturing [9]. Planning of process operations requires the theoretical integration of two distinct and segregated areas of analysis: control theory of continuous systems and temporal logic for control reasoning on discrete events. The work of Cheung and Stephanopoulos [7, 8] provides the bridge between the modeling representations used by these two areas. Work in artificial intelligence has created the theoretical framework for the efficient synthesis of operating procedures. Lakshmanan and Stephanopoulos [10–12] have developed the general scope for nonlinear planning of process operations, which can account for all types of qualitative or semiquantitative engineering constraints imposed on the operations, such as temporal ordering of operations and avoidance of mixing of chemicals. Instead of solving the equivalent mixed-integer, optimal control problem (which incidentally is numerically intractable), Lakshmanan and Stephanopoulos use nonlinear planning ideas from artificial intelligence to propagate the effects of the requisite constraints and thus simplify the ensuing numerical problem of optimal control.

In the area of batch scheduling, the popular academic approach has been the solution of large-scale, mixed-integer linear or nonlinear optimization problems. Given that these problems are numerically intractable, researchers have devised various algorithms for the implicit enumeration of all possible schedules. Unfortunately, very little can be proved about the behavior of these algorithms, especially if the particular scheduling problem cannot fit within the scope of a given algorithm. Work from artificial intelligence is complementing these numerical efforts by providing the following: (1) extensive articulation and activation of knowledge that optimization algorithms cannot accommodate, (2) formal methods for efficient screening of alternative schedules, and (3) automatic learning algorithms for the discovery of rules governing feasible and efficient schedules.

VI. Conceptual Design of Chemical Processes

Various attempts to formalize the conceptual design of chemical processes as mathematical programming problems have yielded limited success with rather narrowly focused problem definitions and quite rigid solution methodologies. Thus, these advances can be viewed as a set of support tools, rather than a theoretical framework for design. In the absence of a general theory of how design is conducted, research in the area of artificial intelligence has addressed the following two distinct but complementary areas of inquiry. The first is axiomatic theory of design, with the objective to establish a theoretically firm ground for the definition of design and thus bring it into the realm of "science," rather than leave it in its present state as "art." Advances in this area have been rather recent, and significant research is currently under way. Second is an engineering science of knowledge-based design, aiming at the development of a rational framework for the formulation, organization, and evaluation of knowledge-based models of how design is done. The central issue to resolve here is the identification and structuring of various forms of knowledge pertinent to the design problem. Two major areas of advancement have resulted from the use of AI-related research: (1) systematic modeling of the process of design, and (2) new, efficient programming styles, which depart from the conventional computer-aided design paradigms and allow the development of large, highly complex computer programs.

Stephanopoulos [13] has been experimenting with a generic model of the design process composed of three facilities—planner, scheduler–advisor, and designer. The planner defines the top-level milestones through which the design is expected to pass. The scheduler–advisor determines the sequence of the design steps to be taken as one attempts to advance the current design to the next milestone. It embodies a theory of how design goals are created, prioritized, and decomposed, and how they interact and are satisfied. As a result, it offers a complete design plan by identifying all the requisite engineering design tasks. The designer simulates the execution of a design step and updates the representation of the artifact that is being designed, which it maintains at all times along with other domain-specific knowledge. It also detects conflicts and, using domain-specific data, prescribes modifications to the design plan. To support this design model, a modeling language has been developed to provide multifaceted representations of the evolving designs [14–16].

Automating the conceptual design of chemical processes offers significant benefits, such as quick evaluation of many alternative chemical production routes; efficient development of economic, operable, and safe processes; intelligent documentation of the design process and its rationale, as well as invaluable accumulation of past experience; and easy verification and modification of past designs.

VII. Epilogue

Artificial intelligence and symbolic computing are expanding the scope of engineering problems in which computers can play an important role. In the previous paragraphs we have discussed only a few of the possibilities. Many more have been left out, such as design of products [17], autonomous process control systems [18], and diagnostic advisors for process operations [19]. The basic message that we tried to convey can be summarized in the following way: Computers offer much more than simple number crunching, and artificial intelligence research leads the way in exploiting this untapped potential.

References

1. Dreyfus, H. L., and Dreyfus, S. E., *Dædalus* **17**, 15 (1988).
2. Rich, E., *Artificial Intelligence*. McGraw-Hill, New York, 1983.
3. Hofstadter, D. R., *Gödel, Escher, Bach: An Eternal Golden Braid*. Vintage Books, New York, 1980.
4. Abelson, H., and Sussman, G. J., *Structure and Interpretation of Computer Programs*. MIT Press, Cambridge, Mass., 1985.
5. Sacks, E. P., "Automatic Qualitative Analysis of Ordinary Differential Equations Using Piece-Wise Linear Approximations." Ph.D. thesis, MIT, Cambridge, Mass., 1988.
6. Eisenberg, M., "Descriptive Simulation: Combining Symbolic and Numerical Methods in the Analysis of Chemical Reaction Mechanisms." Technical Report, Laboratory for Computer Science, MIT, Cambridge, Mass., 1989.
7. Cheung, J.T.-Y., and Stephanopoulos, Geo., *Comput. Chem. Eng.* **14**, 495 (1990).
8. Cheung, J.T.-Y., and Stephanopoulos, Geo., *Comput. Chem. Eng.* **14**, 511 (1990).
9. Stephanopoulos, Geo., in *Foundations of Computer-Aided Process Operations* (G. V. Reklaitis and H. D. Spriggs, eds.). CACHE-Elsevier, New York, 1987.
10. Lakshmanan, R., and Stephanopoulos, Geo., *Comput. Chem. Eng.* **12**, 985 (1988).
11. Lakshmanan, R., and Stephanopoulos, Geo., *Comput. Chem. Eng.* **12**, 1003 (1988).
12. Lakshmanan, R., and Stephanopoulos, Geo., *Comput. Chem. Eng.* **14**, 301 (1989).
13. Stephanopoulos, Geo., in *Foundations of Computer-Aided Process Design '89* (J. J. Siirola, I. Grossmann, and Geo. Stephanopoulos, eds.), p. 21. CACHE-Elsevier, New York, 1989.
14. Stephanopoulos, Geo., Johnston, J., Kriticos, T., Lakshmanan, R., Mavrovouniotis, M. and Siletti, C. A., *Comput. Chem. Eng.* **11**, 655 (1987).
15. Stephanopoulos, Geo., Henning, G., and Leone, H., *Comput. Chem. Eng.* **14**, 813 (1989).
16. Stephanopoulos, Geo., Henning, G., and Leone, H., *Comput. Chem. Eng.* **14**, 847(1990).
17. Joback, K., and Stephanopoulos, Geo., in *Foundations of Computer-Aided Process Design '89* (J. J. Siirola, I. Grossmann, and Geo. Stephanopoulos, eds.), p. 363. CACHE-Elsevier, New York, 1989.
18. Stephanopoulos, Geo., in Prett, D. M., *The Shell Process Control Workshop—II*. Butterworth, Stoneham, Mass., 1988.
19. Calandranis, J., Nunokawa, S., and Stephanopoulos, Geo., *Comput. Chem. Eng.* **86**, 60 (1990).

General Discussion

Process Engineering

Thomas Edgar: Manfred Morari's paper gives a good interpretation of what happened in the 1960s and 1970s. I may be the only person here who was active in control in the 1960s and 1970s and is still involved in the field today. The operative control paradigm of the 1960s and 1970s was totally oriented toward algorithms. Let's develop an algorithm; then if it's successful, let's find a problem that it works on. We invented systems like the linear absorber, which has nothing to do with a real system. The other problem was that we needed an industrial audience to deal with the products of our work, and in the 1960s and 1970s, that didn't exist. I remember going to a conference that Mort Denn chaired back in 1976, where industrial efforts in advanced control were reviewed. All the work on optimal control had been done at that point, and we couldn't find more than 10 people who even used feedforward control, much less optimal control. It was a terrible state of affairs then, and everybody bailed out. John Seinfeld left, Jay Bailey left, as did all the other people I could name. In any event, it was the right thing to do, because they later became successful in other areas.

A number of events have made control a more important discipline today. Manfred didn't mention them, but certainly the change in energy prices in the late 1970s was a big influence, and industry became interested in control because they could actually make money at it. The digital revolution occurred in the 1970s. Before that, we didn't have reliable instruments or reliable equipment to carry out control. Today we do. There was digital control with analog backup in the 1970s. Now we've finally gotten rid of the analog backup, and we have digital control with digital backup. Today, the major emphasis is on quality within the process industries, and that provides a big impetus for process control. We heard about the microelectronics industry today, specifically, as one where quality control is very important.

I'm not sure we've talked about where process control ought to go. Manfred presented mainly what's happening at Caltech. I also subscribe very much to the idea of model-based control. The concept is important, and Manfred deserves much of the credit for introducing it. What's missing from Manfred's interpretation is that we can update nonlinear models simultaneously while doing sophisticated control calculations. The next big jump in our process control capability is advanced modeling, specifically nonlinear models rather than linear models. We've already seen all that will come out of using linear models and control algorithms that can work on them. The next generation of techniques will involve nonlinear model-based control algorithms. This puts a big burden on the control engineer to do the modeling as part of developing the control strategy or at least to work closely with someone who can do the modeling. That other person might be someone in microelectronics, biochemical engineering, or another area. We can no longer sit back and wait for someone to say, here's the model for you to work on. The control people must be involved in the modeling side, which means they must be educated more broadly, not only in control theory but also in simulation, optimization, statistics, and other related areas. We're going to demand more from people working in this area.

Morton Denn: Manfred's historical perspective proves that the first draft of history should always be written while somebody who was there is still alive. A large number of chemical engineers who were involved in control left the area in the early 1970s for primarily one reason—the realization that the crucial problem was to understand enough about the dynamics of chemical systems so that advanced control procedures could be implemented. That's why a substantial shift in interest occurred from control algorithms per se to understanding how to model complex systems. I think that's why most of us made that move.

George Stephanopoulos: Cite a reference where this point was made.

Morton Denn: I've given you oral history right now, George. You can make that the reference if you want. A few years ago, I published a long list of existing models that are absolutely wrong in some of the primary reference sources for chemical engineers. If one were to try to design control systems based on these incorrect models, in particular one that a major company was peddling for about $100,000, then we would be in very serious trouble. I still contend that understanding the dynamics of processes is a major driver in process control.

Manfred Morari: My intended point was that many people talk about modeling. However, I do not know anybody who defines the objective of a model in anything but vague qualitative terms: a system of equations whose solution is representative of the process response, etc. Before the advent of robust control theory in the late 1970s, the model quality necessary for control system design was not understood. All that was available were a few case studies and some sensitivity calculations, which are not very useful when larger, rather than infinitesimal, parameter changes and changes in the model structure have to be considered. The example I showed in my paper demonstrates that in some cases a model that would be judged as terrible by any modeler unfamiliar with robust control theory can actually be completely sufficient for control system design. On the other hand, a seemingly excellent model can be totally inadequate for control system design. There is no evidence in the literature that the researchers who left the control field to work on model development did so with a thorough understanding of the required model accuracy or quality for control system design.

Sheldon Isakoff: I have a quick response to what Mort and Manfred said. I started the first group of process dynamics and control in the Du Pont company. It fell within the chemical engineering research section. We got into process dynamics to learn not only how to design a control system but also how to design a process itself. There are trade-offs that one can make, and it is very useful to get an understanding of the dynamics of the process in the initial process design.

Mark Kramer: I would like to turn the discussion momentarily to educational computing. My point of reference is Project Athena, which was undertaken at MIT as an Institute-wide initiative to bring computers into education. The question posed was, can we, by saturating MIT with computers, revolutionize the way we teach? After five years of the project, there are now over 600 networked UNIX workstations deployed in clusters throughout the campus, all devoted to educational activities.

I see two aspects to computers in education. The first is teaching computer skills, which is straightforward. CACHE Corporation released a set of recommendations in 1985 that outlined what computer skills chemical engineers should have: one or more programming languages and experience in data acquisition, information retrieval from data bases and library systems, word processing, spreadsheets, and graphics processing. The second aspect, use of the computer as an educational tool, is open-ended. Some uses suggested when Project Athena began include a simulator to visualize complex phenomena, such as flows in a fluid mechanics course; a laboratory instrument to collect and analyze data; and a virtual laboratory, a so-called microworld, where the student can interact with this microworld and do manipulations. (An example of this in civil engineering is to see what happens if you change the structure of a bridge.) The fourth use is in a tutorial role, such as question-and-answer programs. Can we use that at a university level in a nontrivial way? More advanced tutorials have the student in command, instead of the computer as the initiator. Finally, computer textbooks—can we use, for example, hypertext as a new way of presenting information? Can we integrate graphics with text in an innovative way?

So far, Project Athena has done the following. It's relieved the computer shortage at MIT. There's more computer usage now among our students. We've also improved our computer skills courses. However, the promise of educational innovation is largely unfulfilled, for several reasons. First, the faculty was ambivalent. The developers assumed that we'd enthusiastically use computers in the classroom. In reality, faculty felt that use of the computer was irrelevant to the theoretical developments they wanted to teach and distracted from the main lessons. Second, there was fear of creating some distance from the real world, especially in terms of dry-labbing experiments. Third, there was student ambivalence, caused by the length of time required to learn and use Athena. Many of our students chose chemical engineering over computer science because they'd prefer not to spend most of their time in front of the computer terminal.

Writing educational software is not easy. Being a good programmer is not sufficient; real insight into the learning process is required. Programs should be interactive, graphical, and sensitive to the level of the user. The effort required to write a good piece of software is akin to writing a textbook. As a colleague said, it might be impossible to excel at writing educational software and at the same time be a research leader in a given field.

These remarks raise several questions. Shall we continue to push the computer aggressively into the instruction of chemical engineering courses? Who will create the educational software—faculty, software companies, or a collaboration of professional consultants with faculty? If we make this aggressive push, how will we deal with the constant evolution of hardware and software systems? These are the challenges to using the computer in education. I am interested in finding out what others have experienced.

Lawrence Evans: I would like to make four points related to the use of computers in the area of design and analysis. First, I agree with Art Westerberg's comment that chemical engineers with specialization in computer-aided engineering are needed to develop and promulgate the decision support tools for industry. I consider the computer-aided engineering software that we're developing largely as tools to help industry make better decisions, whether we're talking about process development, process design, or plant operations. The bottom line is that the use of modeling enables us to make better business decisions. People working in this area need to be educated as chemical engineers, and they also need to interface with computer science and the mathematical decision sciences. The need to be a chemical engineer is primary, though they do need to be able to use the developments in these other areas.

The second point is that academics in this area should focus on long-range, generic problems. This means developing general methods and algorithms for solving problems. It means understanding the structure of the problem. One of the trends we're seeing is expansion of the range of application. Jim Douglas mentioned that the petrochemical plant was the ideal gas of design. I'm not sure that the trend in other applications will be such a natural extension from this ideal gas example. The applications are unique. It's important to solve problems from an industrial context. Tom Edgar pointed out the importance of having an industrial audience for the results of research. It's no accident that virtually all the major research groups now doing work in the area of process systems engineering have some kind of industrial consortium working with them. That's key, and it didn't exist back in the 1960s and 1970s. In seeking industrial collaboration, it is especially important not to oversell what we can do with the computer. Overselling has hurt us before. Also, we should not emphasize developing production software at universities. It's important to develop prototype software that can demonstrate the concepts and techniques, but it's too much of a management effort to develop systems that will be widely used. It'll be picked up by industry if it is useful.

Third, we need to recognize the trend toward the practical application of increasingly rigorous models. We hear a lot about shortcut models and about qualitative models, but I believe the payoff in the future will come from use of more rigorous models. This means that we need a hierarchy of models, and it's important that the simple models should be consistent with and be special limiting cases of the more detailed models, so that when you make a decision or conclusion with a simple model, it will apply to the more rigorous case with the right limiting assumptions. The goal of simple models is to better understand the structure of the problem; it's not good to use a simple model just because a more rigorous model takes too much computer time or is too hard to use. That problem will be solved if we make the more rigorous models easier to use and if we get that 10^8 increase in computer horsepower. My last comment is that we talked a lot about process design and about manufacturing, but there's a whole world out there of product engineering, and we need to look more at computer-aided product engineering.

Robert Brown: There's no disagreement that computation plays a very important role in chemical engineering research and development. Twenty years ago, a number of people, in both chemical engineering and other fields, used what Gary called advanced scaling analysis to solve transport and other problems. The entire community viewed it as an art practiced only by a few, and its relevance to chemical engineering was questioned. Today that's totally changed. Asymptotic analysis is taught in advanced courses on transport processes everywhere. It's not done in the context of an applied math course, but in the context of the analysis of transport systems. How does that relate to computation? People are starting to believe that it's important as a tool and that it's going to become increasingly used in a number of fields.

The supercomputer in the future will not be just a single research tool for a large community. What will be revolutionary is what I call the Crayette. Those are the machines that Art's talking about, at 0.1 megadollars and 10 megaflops. This will radically change how we do things. Just as the PC and the Sun workstation have an effect now, this will be a Cray supercomputer on your desk.

To return to an educational point, if we had believed in 1970 that asymptotic analysis would be important, what would we have done? The point is that there are now context ideas from numerical analysis that should be introduced to our students. I start with our graduate students. How far do we percolate the concepts down? Certain ideas are required for competence in reading literature and making critical decisions and judgments about numerical computations. I'm not suggesting that everyone become a computer scientist, rather that we become literate, in much the same way Clark Colton referred to biology this morning. There are ideas here that focus first on linear algebra and then on nonlinear algebra. The theme reemerges that we live in a nonlinear world. We're starting to grasp and cope with that in many of the ideas about convergence, loss of convergence, and multiplicity. All those concepts are very well presented in the context of numerical analysis where the solution of ODEs, stiffness, and finally PDEs and ideas of convergence are important. These issues are all at the core of all the applications mentioned over the past two days. The question is, do we make the commitment to start presenting these results in a coherent way so that people become literate in them?

George Stephanopoulos: I would like to change topics and discuss the potential role of computer science in chemical engineering. Whenever we use the term programming, we actually mean coding, coding in FORTRAN, Pascal, C, or another language, but coding and debugging of the code are the least interesting and most painful activities in developing a computer program. The cumulative effort our bright students spend coding computer programs is alarming. On the other hand, computer programming is an intellectual problem-solving activity that we never teach our students. This must change. Abelson and Sussman at MIT pioneered an excellent text for teaching students, *The Structure and Interpretation of Computer Programs,* containing all the underlying educational depth of a core subject. There the students learn that a computer program is the formal medium for expressing ideas about methodology and controlling intellectual complexity of a program's declarative or procedural knowledge. In that regard, Abelson and Sussman have demonstrated that computer science plays a central role in engineering education, but no commensurate movement has occurred within chemical engineering curricula, despite the pervasiveness of computers. Computer science is also broadening the horizons of our engineering activities, far beyond the standard equation-solving paradigm we have practiced. Again, where is that taught to chemical engineering students? Information processing is at the core of all recent and future engineering activities. This is a tremendous educational challenge. We cannot respond by assigning a faculty position alone. The whole spectrum of educational activities will eventually use computers in various ways. Are we ready for it? How will we integrate it into our educational efforts? Programming is where the excitement lies. Forget computer coding. Think of how computers can help investigate analytically—not just numerically—physicochemical processes, design polymers or solvents or proteins, explore a variety of chemical production routes, handle vast amounts of knowledge among a variety of scientists and engineers, automate safely and reliably the operation of plants, and interact with humans at a human-like level. The answers will come from chemical engineers with an appreciation of computer science as an educational framework. I expect that computer science will become an integral part of a chemical engineer's education of the same importance as mathematics, physics, chemistry, and biology.

Mark Kramer: I'd like to respond to what I perceive as Manfred's definition of the process control problem, which I think too narrowly defines process control. His emphasis boils down to calculation of a transfer function between the measurements on a process and the valves on a process. Put another way, given some measurements, how do you want to manipulate the valves? The industrial needs in this area go well beyond this, particularly in terms of the gap in the 1970s that he addressed; Alan Foss's article in 1973 and Irv Rinard's article in 1983 pointed out the problems that industry really faces. The real control problems are maintenance, diagnosis problems, managing product and process changeovers, alarm analysis, the design of safety interlock systems, and, in fact, the safety analysis of flowsheets to make sure the process design is complete. There's been very little work in these areas, and I would like to see process control engineers broaden their view and begin to help industry in those areas.

Sheldon Isakoff: I'd like to clarify the data reported by Manfred Morari. The reason for the resurgence of work in process control is that a half-billion dollars a year stake is out there for the Du Pont Company if we can improve our process control technology. What's really important is to understand this stake. A good part of it involves developing new measurement technology. This is a key feature in the control of chemical processes. Chemical engineers have made remarkable contributions to this area in the past. They will also in the future, particularly in composition control or structural control if we're talking about polymers. Another large contribution involves doing the kind of work that lets you understand the process better and therefore results in better models. Putting effort into better process control can provide rewards more rapidly than trying to get into new areas of business based on discoveries and inventions. It is a great field for chemical engineering investigation.

Daniel Luss: Here is a question for Manfred and the other control experts. In control courses, we try to teach the students how to control a process, given a fixed design. Industrial experts claim that if a process creates control problems, one should try to change and modify the design to avoid the control problems. We do a poor job of teaching this approach. How can we introduce this to our students?

Manfred Morari: I completely agree, Dan. In terms of design for control, we really don't have a systematic method. We do have analysis techniques for processes that tell us if control problems are likely to occur. That alone is significant progress. We can start integrating that into the curriculum. It's all a matter of understanding. Five years ago, those methods were around to some extent, but we didn't understand them well enough to teach them to undergraduates. I agree with Mark that there are a number of other areas in process control that are equally important. Finally, Sheldon, about the 500 million figure by Du Pont, this is the only figure that was sent to me when I solicited input from the Du Pont Company. Internally you may have used a range, but what was reported to the outside was only the upper limit.

Anthony Pearson: Mark asked if anybody had any experience to share with him about computer systems in general. I'm certainly not going to share these views now, but if anybody wants to do it over a beer with me, I should say that Schlumberger has probably spent, in the last 10 years, 10^9 dollars on computer systems. It owned a computer manufacturing company called Fairchild, which it has since been pleased to get rid of. It has a computer-aided design company. We have a building in Austin, Texas that was built to hold 600 people who were all to be software engineers—it's now got 100 people. We have actually developed software that we hoped would be useful in meeting the needs of user-friendly knowledge-based systems. In fact, that's still alive because it's been handed over to another computer manufacturing company. The experience at Schlumberger, and I suspect in a lot of other companies as well, frankly might serve to throw cold water on some of the things that have been said today.

James Katzer: I have some quick comments on some of the points made with respect to control. As Mort Denn pointed out, there was some software that was not good. That still exists, and it can be a real problem. Mark Kramer raised the question of fault analysis and safety. These are key issues in control, and they're certainly not being overlooked from an industrial point of view. In many cases, we don't yet have the technology or the understanding to do them well, so we address them as we go along. The dynamics of processes are important, particularly when you make process changes, whose impact you don't fully understand. We need to understand the dynamics better so that we can control the process better. As product requirements and specifications get tighter, we try to tighten our process controls. As we push our controls harder, we're pushing them up against constraints where control is progressively more and more difficult and at times chancy. So we need to better understand the dynamics of processes and have better control hardware. The dynamics are still what we don't understand very well, in real processes.

SECTION X
The Identity of Our Profession

27

The Identity of Our Profession

Morton M. Denn
Department of Chemical Engineering
University of California at Berkeley
Berkeley, California

I. Introduction

We have been hearing a great deal in recent years about the changing nature of chemical engineering. The emphasis on new fields of research has created the appearance of a fragmented profession, comprising specialized research communities with little inclination to interact. New hiring and career patterns in the industries that have traditionally employed chemical engineers, and the emergence of career opportunities in nontraditional industries, have helped to focus attention on the fundamental issue of the very identity of chemical engineering and how that identity may be changing. The National Research Council's report *Frontiers in Chemical Engineering* [1] provides a convenient frame of reference for discussion of this issue because of its emphasis on new directions of the profession.

Chemical engineering, according to *Frontiers in Chemical Engineering*, will be governed by a new paradigm. This new paradigm (which is not clearly defined in the text) is introduced with a table of "enduring" versus "emerging characteristics" and a description of changing social and economic pressures; the paradigm appears to be associated with a change from a focus on macroscale processes to those occurring on a microscale. A paradigm is a profession's intellectual frame of reference; "[it is] what the members of a scientific community, and they alone, share" [2].[1] A new paradigm suggests revolutionary change, with far-reaching implications regarding education research and the practice of the profession.

ADVANCES IN CHEMICAL ENGINEERING, VOL. 16

I do not believe that chemical engineering has been undergoing revolutionary change. Our profession has been experiencing gradual and predictable evolutionary change for four decades and differs little today from chemical engineering as it has been traditionally practiced in North America since World War II. Recognition of this continuity is important if we are to respond effectively to the current pressures on the profession and to meet the societal needs so clearly enunciated in the Frontiers report.

II. What Is the Paradigm?

Chemical engineering traditionally has been viewed as the engineering profession that deals with applications in which physical and chemical rate processes are limiting, and I submit that this is the unchanging paradigm. The introduction of the concept of the unit operation, which has provided a major thread of continuity through seven decades of practice, focused attention on physical rate processes; chemical rate processes were ironically to wait another three decades for introduction into the core of chemical engineering education, research, and practice, until driven by the necessities of World War II.[2] This essential absence of applied chemistry from chemical engineering has been noted in a perceptive essay by Howard Rase [3], and the change is graphically illustrated by comparison of the contents of the *Transactions of the American Institute of Chemical Engineers* from the late 1930s to 1946. In fact, Arthur D. Little himself drew a sharp distinction between chemical engineering and industrial chemistry along a unit operations-based boundary.[3] Chemical engineering as we know it today thus developed during World War II with the integration of industrial chemistry (in the form of the developing speciality of chemical reaction engineering) and unit operations. It is perhaps significant in terms of our current concerns regarding the growing diversity of the profession that the integration of biochemistry into chemical engineering practice and the birth of the specialty of biochemical engineering also began at that time. (Like conventional chemical processing, the industrial roots of modern biochemical engineering can be traced to much earlier dates, at least to the World War I Weizmann process for the large-scale production of acetone.)

The perception that chemical engineering has undergone major changes has caused several institutions to reexamine their undergraduate curricula to ensure that they remain relevant to modern practice. As far as I know, the conclusions reached by my colleagues at Berkeley after a year of study are typical results of such introspection. It was clear to us that the specific content of a number of courses needed modification, generally in the form of examples of applications to new fields of opportunity for chemical engineers. Only a few major structural changes were seen as necessary, however, and none would support the notion of a major change in the paradigm. The separations course was seen as too narrow, and plans are now under way to fashion a course that is philosophically closer to the reaction engineering course in bringing together fundamental concepts and addressing broad issues of separations. Specialized minors, comprising a small number of related courses inside and outside the department, are to be introduced in order to focus students' selections of electives. Finally, an overall plan to integrate computer use throughout the curriculum has been developed. This is all important, but hardly revolutionary.

Chemical engineering practice in the postwar era has been closely tied to advances in the disciplines on which the modern industrial base is built: chemistry, physics, and molecular biology.[4] Whether the major advances in these fields have been "science-pushed" or "technology-pulled" is irrelevant from our professional point of view; any changes in chemical engineering practice have been externally driven by scientific advances. The growth of industrial practice in materials fields at the microscale is a result of advances in chemistry and physics, particularly of the solid state; the traditional skills of the chemical engineer were readily adapted, because they required "only" the addition of the new scientific knowledge. Similarly, biochemical engineering grew with the advances in molecular biology. Chemical engineering research does now consider to an increasing extent phenomena at a "micro" level, using sophisticated instrumentation that was unavailable only a few years ago. These research tools have widened the scope of potential applications, as required by the demands of the modern science-based technologies; they have required no change in the basic disciplinary matrix[5] of transport and reaction rate control.

III. What Are the Implications?

Let us now proceed on the premise that there is a single, identifiable profession of chemical engineering with a common disciplinary matrix, or paradigm: the analysis and design of systems governed by physical and chemical rate processes. Despite interest in specialty areas, chemical engineers share an enduring common culture that provides the opportunity for easy exchange of ideas between subdisciplines and unparalleled opportunities for movement into new problem areas. Why, then, do we appear to be fragmented and uncertain about our future as a profession?

The question was anticipated by Rase in his 1961 essay [3]: "Now it appears that the engineer stands waiting for new developments in science, and as in most waiting games the waiter is the loser. It is because of our hesitation that research in chemical engineering has been relegated to improving the ideas of pure science and doing some of the necessary jobs pure science finds dull or repetitious." This is a proposition that we take hold of our own destiny and define the new scientific directions ourselves, rather than continuing in the reactive mode that has been our tradition. Some of the current discomfort is undoubtedly associated with an attempt to do precisely this, accompanied by a sense that a departure from our traditional role is of necessity a departure from a tradition-bound profession as well.

One strength of chemical engineering as a discipline has been its problem orientation, in which familiarity with physical and chemical rate processes has enabled the chemical engineer to use whatever tools were available to attack a wide range of physicochemical problems. Chemistry and physics (if I may be allowed a broad brush) have been much more technique oriented, with the instrumentation enabling the scientist to seek out problems that *could* be solved. This distinction has always been a profitable one for chemical engineers willing to play the "waiting game," and I disagree with Rase that they have been losers.

Chemical engineers do need to be more aggressive in seeking a leadership role in bringing about advances in technology. Their absence is particularly noteworthy in materials science, where many of the problems are closely related to traditional chemical engineering but are rarely studied by people who identify themselves as chemical engineers. This leadership role can be achieved, however, without abandoning the problem-oriented approach that has been so successful in the past. It will require a continuing awareness of instrumental advances and perhaps collaboration with the specialists whose skills are needed for particular applications. Such collaboration can be synergistic; the engineer has much to offer the scientist and need not feel like a junior partner. True collaborative research is unfortunately unusual in U.S. universities today, as Alan Michaels [4] has noted in an important essay that deserves widespread attention.

Some chemical engineers have taken a leadership role in the science, as well as the technology, in a number of important areas; catalysis, surface science, and polymer and colloid science are among the easiest to identify. With this involvement has come a natural tendency to identify with the area of chemistry or physics in which the advances (and much of the initial excitement) are occurring. This identification has led in many cases to a narrow, outward focus, with accompanying perceptions of uniqueness relative to the core of chemical engineering. The outward focus is reinforced by attendance at specialized scientific meetings and the existence of a specialized literature; the chemical engineering "connection" seems secondary. This tendency is also reinforced by our system of evaluation for tenure in universities and the equivalent evaluation system in major research laboratories. Young chemical engineers perceive the necessity of making a mark within the specialized community with which they interact in research, for these are the peers who will evaluate them for their chemical engineering colleagues after five or six years.

The first generation of chemical engineers moving into an area of basic science undoubtedly retains an engineering problem orientation; it appears to me, however, that quite often this outlook is not passed on to the students, who acquire the viewpoint of the natural scientist. I have found in discussions during the preparation of this article that this proposition is a controversial one. Colleagues have argued that graduate student research can be as science oriented as we may wish; the core graduate educational program in chemical engineering will ensure that the graduate remains a problem-oriented *engineering* scientist, despite working on a dissertation problem that is basic chemistry, physics, or biology. Thus, they argue, we can have all that we might wish if we become more aggressive and seek out the most challenging areas of the natural sciences as they relate to our technological interests. I cannot agree. It has been my observation that many of the best students of science-focused chemical engineering faculty take industrial positions in laboratories where they are indistinguishable from chemists and physicists, and they are effectively lost from the profession.[6] Perhaps the problem is that in some cases the chemical engineering departments that have been most aggressive in seeking new *scientific* directions are those least likely to offer a graduate chemical engineering core.

IV. Some Suggestions

Communication within the chemical engineering profession has been a casualty of the research-driven fragmentation of recent years, and I believe that reestablishment of effective communication is one of our most pressing needs. I have previously commented on the mutual advantages of communication to the specialists and the broader profession, as follows [5]: "some of this research [in new areas] is so strongly based on traditional chemical engineering concepts that it will interest and excite traditional chemical engineers, and help them and the profession to move in new directions. Nor is it to suggest that altruism is the motivation ... ; communication involves the flow of ideas in both directions, and the skills and interests of experienced chemical engineers can often be crucial to the solution of problems in new areas about which they would not otherwise be informed." (Consider how often the effectiveness factor has been rediscovered!)

Our professional societies have been major culprits in creating fragmentation. In my own area of research, the slow response of the AIChE to the growing interest of chemical engineers in polymers was one of the major factors in the formation of still another society, the Polymer Processing Society; most of the North American principals in this new organization are chemical engineers, and the current activities of the society are within the programming responsibilities of the AIChE. The proliferation of specialized societies probably cannot be stopped, but better integration of meetings of these societies with meetings of AIChE and ACS could minimize the damage and enable mutually beneficial interaction with the broad chemical engineering community.[7]

The implementation of the tenure system in universities is another factor in fragmentation. I have already noted the research pressure on young faculty that causes them to isolate themselves from the chemical engineering community. There is an educational loss as well. The young faculty are usually the ones who are most involved with research in new areas, and they are the people who must lead the way in integrating the new areas into the core curriculum. Preparing textual materials that can be used by others is very time-consuming and bears little relation to the time required to prepare one's own lectures. (The closest analogy is probably between writing a computer program for personal use and preparing a "user-friendly" program for distribution. Most of the real work goes into the user-friendly front end.) Young faculty should be encouraged to develop textual material, because it is what the profession needs. Our system of evaluation for tenure needs to be revised to encourage impact on the core profession, both through the preparation of teaching materials intended to broaden the exposure of students in basic chemical engineering courses and through research publications aimed at the entire chemical engineering community.

Finally, it seems obvious that chemical engineering education needs to be built around a core at the graduate as well as the undergraduate level. The core should emphasize the disciplinary matrix: the analysis of chemical and physical rate processes. Whatever new directions are being taken in research, whether oriented toward a traditional engineering outlook or toward basic science, the universality of the profession should be a major focus when these areas are integrated into the course of study.

Notes

1. See also the second edition of Kuhn's *The Structure of Scientific Revolutions* [6]. Kuhn popularized the concept of paradigms for scientific communities, and the term is usually understood in the context of his writings about "scientific revolutions"; I expect that this will be the case for most readers of the *Frontiers* report, although the authors may not intend that to be their meaning. A reading of Kuhn's critics is important to recognize that paradigmatic change need not entail a radical overthrow of established methods and principles; see, for example, Toulmin [7], pp. 96ff.

2. Arthur D. Little's original 1915 concept of "unit operation" did not necessarily exclude chemical rate-determined operations, but it was in practice restricted to physical processes; chemistry entered only in equilibrium-based situations in the standard textbooks. Consider Little's list of examples in his widely quoted letter to the president of MIT: "pulverizing, dyeing, roasting, crystallizing, filtering, evaporation, electrolyzing, and so on." A definition of the profession in 1922 was thus a defensive one, "Chemical engineering ... is not a composite of chemistry and mechanical and civil engineering, but a science of itself, the basis of which is those unit operations which in their proper sequence and coordination constitute a chemical process as conducted on the industrial scale." (Report of the Committee on Chemical Engineering Education of the Ameri-

can Institute of Chemical Engineers, 1922, quoted in [8].) The contrast with the current definition in the AIChE Constitution and Bylaws (Article III) is interesting: "Chemical engineering is the profession in which a knowledge of mathematics, chemistry and other natural sciences gained by study, experience, and practice is applied with judgment to develop economic ways of using materials and energy for the benefit of mankind."

3. "There should always be kept in mind the definite line of demarcation between industrial chemistry, which is concerned with individual processes as entities in themselves, and chemical engineering, which focuses attention upon those unit operations common to many processes and the proper grouping of these unit operations for the production of the desired product as efficiently and cheaply as the ruling conditions permit" [9].

4. This statement seems to be less true of the "systems" aspects of modern chemical engineering, where the expected driver would be the notable advances in information science. Most of the research that is currently being published in chemical engineering journals in this field, however, is concerned with problems that were defined 40 to 50 years ago. The optimization of flowsheets, for example, which has been one area of major interest in recent years, is a straightforward extension of 1930s textbook material, requiring only the availability of more powerful computers and (in some cases) the World War II-driven growth of the mathematical foundations of operations research.

5. This is a term Kuhn [2] would use to replace one of his meanings of "paradigm."

6. I am constantly seeking new reviewers for *AIChE Journal* papers. One of my most important sources of addresses when I have identified potential reviewers is the AIChE membership list. The number of young Ph.D. chemical engineers who are not AIChE members is shockingly large.

7. AIChE's 1987 Conference on Emerging Technologies in Materials is a flawed model; the cosponsorship by 12 other societies was in name only, since they still held their own regular meetings, and the use of a hotel different from that used for the simultaneous AIChE meeting segregated the "materials people" from the rest of the community. Still, it is a better model than the simultaneous 1988 meetings *in different cities* of AIChE, ASME, and the Materials Research Society, all with overlapping materials programming.

References

1. National Research Council, Committee on Chemical Engineering Frontiers: Research Needs and Opportunities. *Frontiers in Chemical Engineering. Research Needs and Opportunities.* National Academy Press, Washington, D.C., 1988.

2. Kuhn, T. S., in *The Structure of Scientific Theories* (F. Suppe, ed.), p. 459. University of Illinois Press, Urbana, 1974. Reprinted in Kuhn, T. S., *The Essential Tension.* University of Chicago Press, Chicago, 1977.

3. Rase, H. F., *The Philosophy and Logic of Chemical Engineering.* Gulf Publishing Co., Houston, 1961.

4. Michaels, A. S., *Chem. Eng. Prog.* **85**, 16 (February 1989).

5. Denn, M. M., *AIChE J.* **33**, 177 (1987).

6. Kuhn, T. S., *The Structure of Scientific Revolutions*, 2nd Ed. University of Chicago Press, Chicago, 1970.
7. Toulmin, S., *Human Understanding: The Collective Use and Evaluation of Concepts.* Princeton University Press, Princeton, 1972.
8. Reynolds, T. S., *Seventy-five Years of Progress—A History of the American Institute of Chemical Engineers 1908–1983.* American Institute of Chemical Engineers, New York, 1983.
9. Little, A. D., preface to Badger, W. L., and McCabe, W. L., *Elements of Chemical Engineering.* McGraw-Hill, New York, 1931.

General Discussion

The Identity of Our Profession

James Wei: Mort Denn has raised some very important points. One that resonates most in my mind is the fragmentation of our profession. What John Prausnitz called a family is in danger of dividing up. Mort also raised questions about meetings and journals. Our ultimate glue probably is still our educational process. Talking about the curriculum seems to be our most important rallying point. The AIChE's number one priority is how to deal with the question of fragmentation. Chemical engineers join different societies, read different journals. How do we keep the family together? I don't believe we can decide what to do in the short time we have remaining, but this is probably one of our most important agenda items.

L. E. Scriven: The critical issue of balancing applications teaching with fundamentals research, of uniting engineering practice with engineering science, was raised in our very first session. Here in our last session it still commands our attention, now in connection with spalling of the profession. I'd like to reiterate my view that most important is the professional orientation that an individual professor adopts. Second to that come the prevailing attitudes in a department and the quality of interactions with chemical engineers and chemical engineering outside academia. By that I mean primarily the industrial world, and not just that part inhabited by established chemical and petroleum processing technologies.

While it is true that some of the best students of fundamentals-oriented professors choose to become applications-oriented scientists in industry, or even knowledge-oriented scientists in universities, my observations don't accord with Mort Denn's that *many* do. Many of the best graduate students

573

who have pursued new, or renewed, scientific directions in chemical engineering are, as I view the record, likely to become leaders of emerging, or rejuvenating, technologies. First to come to my mind are students of Harry Drickamer, perhaps the best scientist among chemical engineers. What's more, there has long been a flux of fine physical scientists into the discipline and the profession. Even though the AIChE is important to both, members surely have to be earned by that organization, not taken for granted.

For instance, from the very beginning there has ben an American Chemical Society alternative, the Division of Industrial and Engineering Chemistry. Its publication has long served chemical engineering well. Those not on the AIChE's membership list are not necessarily separated from the family. But this is not the place for me to try to answer Jim Wei's question, how do we keep the family together?

Clark Colton: Mort, in response to the last issue you raised about what is the core, what did we discuss during the first three sessions here? Did we miss anything?

Gary Leal: I'm certainly not going to define the core, and I'm probably not saying anything new either. I do want to refocus attention on some general issues concerned with education that are an extension of what Mort said and relate to Clark's question. What body of knowledge are we trying to transmit to the students? My starting premise is that there are two classes of information that we try to transmit. One I'd called foundational (Fig. 1), and another I'd call technological or technical applications (Fig. 2). At the top of Fig. 1 is a list of established foundations that we don't have primary responsibility for teaching and shouldn't have unless we're somehow forced by circumstances. Below that are proposed new foundational areas. Biology and materials are subjects which we've heard in various talks are more or less essentials for education. Beyond that are thermodynamics, transport and fluid mechanics, synthesis, design, and control, basically some of the subjects we've heard about today. These are foundations in my opinion. In Fig. 2 is a list of what I call technological applications, in which I've put reaction engineering, processing methods and equipment, environmental topics, biotechnology, and materials. To the right of those, there are some more fundamental topics. These are more specialized foundations that we need for some of the applications areas and are being forced to take on by the the fact that others outside chemical engineering are not often offering relevant course work. For example, for reaction engineering, there are surface science, catalysis, and an understanding of kinetics. In the environmental area, there's a lot of small-particle physics that chemical engineers are heavily involved in because they have to be. In materials, there are topics

CURRICULUM OBJECTIVES

Foundations Taught Within ChE Department

SUBJECTS   QUESTIONS

Taught directly by ChEs ONLY if no
alternative exists

- Established Areas
  Chemistry
  Physics
  Applied Mathmatics

> How do we get others to transmit the correct
> subject matter?
> Who should decide what it is?
> When (or how) do we bring in essential new
> developments, e.g., numerical analysis,
> nonlinear dynamics, etc.?

- Proposed New Areas
  Biology
  Materials Science

> What happens if we add elements of
> these areas to our standard curriculum?

Traditionally taught by ChEs faculty

- Thermodynamics
  (Classsical and Statistical)

- Transport Phenomena /
  Fluid Mechanics

- Principles of Design,
  Syntheiss, and Control

> What to teach?
> What happens when a new area opens to
> teaching–say, non-Newtonian fluid mechanics
> When do such topics cross over from being
> truly foundational to being in the realm of
> technological applications?

Figure 1.

in polymer rheology and polymer science, as well as colloid physics for ceramic materials. In biotechnology, there are areas of applied biology that we need to offer that, as we heard this morning, have been left behind by the biologists.

Now I come back to the critical issues of education. What body of knowledge are we trying to transmit? There are foundations versus technological applications. We can't do it all, and to figure out which or what we want to do, we have to answer the next question. What are the qualities of education that lead to the special characteristics that we perceive in a chemical engineer? One quality that chemical engineers possess that perhaps others don't is an ability or willingness to solve or attempt to solve complex problems with limited data. We heard that this morning in one of the talks. What is the minimum level of preparation that we require to maintain the qualities that got us to our present status? As we become increasingly special-

CURRICULUM OBJECTIVES

Technological Applications

APPLICATION AREA   ADDITIONAL FOUNDATIONS

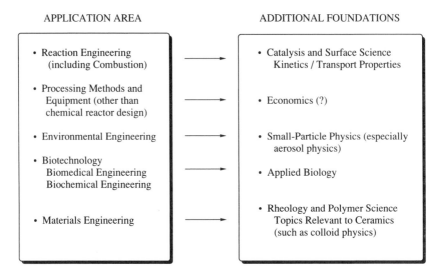

Figure 2.

ized in research, it's not clear in the educational framework what background we need as a profession to remain literate. Next is the objective of functioning in a multidisciplinary environment. What I'm talking about is culture. We can't teach people enough to be cognizant of everything in every area, but somehow we have to impart a way of thinking that enables them to exist in a multidisciplinary world. The second major point goes along with the first one. The foundations become increasingly sophisticated and complex, and from time to time we also want to expand the list of necessary topics—for example, to include something like biology. But, there's obviously a zero-sum game going on here. We can't add additional foundations without subtracting from the foundations we already have, and there's a need for trade-offs and a need to go back and answer the preceding question: what are the qualities that are the essence of what we're trying to transmit to students in chemical engineering? I'm asking questions, not answering them. If we could focus on these issues, our conference could play a helpful role for the profession as a whole, and especially our academic colleagues in other institutions.

Reuel Shinnar: There was one point mentioned here that might require more elaboration. If we want to continue to teach the engineering part of chemical engineering, we need engineers on the faculty. We are slowly becoming a profession that is taught by nonpractitioners. Up till the 1950s, it was expected that engineering professors would have industrial experience and a continuing relationship with industry during their entire career. When we moved from engineering to engineering science, and then moved further to areas of basic science, that need diminished. The lack of industrial experiences and a lack of keen interest in engineering problems are more pronounced in the younger generation. If the profession does not respond to this trend, our future directions may be determined not by any real needs but rather by a lack of understanding of these needs by the academic professor. My favorite analogy is that of the medical school. Its faculty consists of basic scientists, academic clinicians, and hands-on clinicians. We need to maintain a reasonable percentage of "academic clinicians," or, in other words, experienced practitioners who are also academically inclined in our faculty. I realize my comparison is imperfect. Advanced medicine is practiced mainly in the medical school. Engineering is not. Therefore, we need the help of industry. But we also need to find new ways to attract and promote "academic clinicians" for our faculty, i.e., faculty members who practice engineering, are interested in it, and are ready and able to engage in research relevant to advanced engineering design. If we recognize this need, we have to create alternative career pathways for such faculty in our engineering programs.

L. E. Scriven: Here's an observation that perhaps is salutary. The AIChE was founded in 1908. Recently I read transcripts of the deliberations of the Education Committee, one of the first committees set up by AIChE, in 1909, 1910, and 1911. Because I lacked the intervening years, I then skipped to 1921–1922, when the Education Committee finally came to agreement and issued their report, which was tremendously influential. Mort Denn referred to it in his presentation. The points you're raising here are the same points at the forefront of the discussions beginning in 1909—yet then it wasn't at all clear what a chemical engineer should be. But even among those who were trying to lay out curricula in that era, the issues were the same. Only the details change. We're back to Mort Denn's point. Basically, we're engaged in an evolutionary process, and the profession as I perceive it has remained small enough and collegial enough, particularly the university core with the greatest responsibility for the discipline, that these discussions do go on, and through them we have managed collectively to hang together, not four decades, but eight decades.

Morton Denn: There were some changes around World War II in the out-look of the profession. Five years ago the AIChE published their history, which I strongly recommend reading for those of you who haven't done so.

L. E. Scriven: That's a recommendation I'd second. But there's much more to the story than that history tells. Elsewhere in these proceedings I try to sketch how our profession and our discipline emerged and evolved. My view of more recent evolution was printed in *Chemical Engineering Progress* in December 1987.

Robert Armstrong: Mort emphasized the fractionation of our profession into the basic sciences. He did not address the lack of real engineering in our teaching, the lack of synthesis of all these parts. Yesterday Gary Leal talked about that, and Bill Russel talked about our passage into subproblems from which we've never returned. In research and in teaching, we have long lists of topics like Gary's list from a moment ago, all these subjects, but no engineering! Certainly it's important for us to teach these fundamentals, but we have also to teach undergraduates to be engineers, how to take a set of fundamentals and synthesize them into workable solutions for real prob-lems. Mort mentioned that this process makes us engineers. We need to teach how to get back to engineering from a set of fundamentals, which, as we have discussed here, may change from time to time.

Another problem, as Mort observed, is the fractionation of faculty into these basic science areas. We specialize into individual disciplines to the degree that when whoever's assigning courses to teach says, "Who's go-ing to teach engineering?"—there's nobody there to do it. We have some-body to teach fluid mechanics, somebody to teach biology, all the fundamentals. But who's going to teach engineering? We need to hire some-body from industry. That's a serious problem. Again, back to Mort's ideas, we must reward people for teaching synthesis and engineering subjects and doing that type of work. He cited the tenure system as being at fault. Another fault is with our funding system, which would never provide support for that.

Gary Leal: When Mort says we're not in revolution, I would agree. We're in evolution. The only reason that it sometimes feels like revolution is that we lack the vision to respond to the pressure of increasing specialization in certain areas and increasing demands for certain foundational subject matter to be included in our curriculum. If we could go away as a group with some consensus idea, if there is such a thing, of the underlying philo-sophical foundation that must be preserved, that would be the best thing we could do.

Robert Armstrong: I think we handle fundamentals pretty well. What we don't do well is the engineering.

Edwin Lightfoot: I've been sitting so long, I'm just in a bad mood. Mort's paradigm reminds me of an election campaign: it won't cause any trouble but I'm not sure if it does any good. I take issue with two other things. One, we've always been problem oriented. Jim Douglas had this down right in his talk. We're going through some kind of complex limit cycle, where we keep coming back to being problem oriented. From roughly the end of World War II until not too many years ago, we really lost that. I started in 1953 in Wisconsin. Because we needed more tools, we became very tool oriented, and the results have been enormously impressive. I have to agree with Bob Armstrong that in the process, as an academic profession, we're no longer primarily problem oriented. That's a serious issue. I hope that Jim Douglas is right except for his time scale—I hope that we're coming back a little sooner than 20 years. Second, we agree about everything except the essentials: we all agree that we need more math and more chemistry, and maybe more of certain kinds of physics. We don't disagree on the details. We want to modernize our age-old interest in the fundamental sciences, but the one thing you threw out, Gary, was unit operations, which used to be the beginning of the synthesis as well as part of our curriculum. That's where we're in trouble. We've got to get synthesis in very early to the students, right in their first course, and keep it every year. Otherwise, we're going to become nonengineers.

Gary Leal: Is that a separate topic from synthesis in the context of design and control?

Edwin Lightfoot: Yes. For instance, Dale Rudd used to teach a course in process invention to the freshmen, a very successful course. That's the place to tell them what an engineer is, and they loved it.

William Schowalter: I want to relax Bob Armstrong. At the undergraduate level, if we don't teach engineering when we teach thermodynamics, fluid mechanics, and the other so-called core materials, we've lost the whole battle right there. That's where a good share of the engineering belongs.

Sheldon Isakoff: It's helpful, when discussing our identity, to remember that we are engineers and to define what that means. An engineer is one who uses science and mathematics toward some useful end. That end should be of value to either society in general or to the organization to which one belongs. Only then do we modify engineer by the adjective chemical, and I think Mort's definition of a chemical engineer is a good one. It's important to remind the student that the production of something useful is the motivation for all engineers. There are ways to achieve that. Collaboration with industry or industry groups, either actively in research or in other activities, often involving the AICHE, can go a long way in convincing students that chemical engineering is a unique and useful discipline. Focusing on the word engineer is helpful.

Howard Brenner: I want to comment on what Mort Denn said about the revolutionary versus evolutionary aspect of the paradigm. He suggested that the profession is driven by technological and scientific advances, which implies an idealistic approach to problem solving. I feel that it's driven, rather, by economics! While the two are often related in the long run, they are usually unrelated in the short run. What really causes the changes are economic factors. Currently, our profession as a whole is strongly influenced by worldwide competition, the disappearance of heavy industry, and the concomitant growth of high technology—which, as we all know, is a very rapidly changing field. It's simply impossible for our profession to react thoughtfully on the latter time scale.

Chemical engineering faculty are affected by funding patterns, which politically tend to follow these changes. Thus, I suggest that we have very little control over the destiny of our profession except in a very modest sense. Consequently, trying to make long-term predictions about where we will be in the future and the viability of our profession is almost impossible, because most of what happens in the long run is driven by events far outside our profession. It's rather ironic for me, of all people, to be talking about economics, but I believe it constitutes the engine that drives our profession. Without being aware of that important fact, we're in grave danger of totally misunderstanding events affecting us.

James Katzer: As a user of the end product, I support very strongly Mort Denn's assertion that we ought to have a core curriculum. People don't know what they're going to be doing two years or five years after they get out of school, and the better the fundamentals, broadly understood, the better off we are. I'd like to ask about the role of the master's degree in chemical engineering. There is a big difference between a B.S. chemical engineer and an M.S. chemical engineer. The M.S. is much more mature technically, with a much stronger knowledge base. In many activities, we don't necessarily need the research base that much. It is very difficult to find good master's people. It's almost a nonexistent degree if you look broadly. I'd like that question to be put on the table for discussion. For example, an M.S. degree person can move to many locations in the company, from research to refining; a Ph.D. is much more difficult to move into other functions.

Robert Brown: A reply to the comment on the master's student pool. The apparent death of the master's degree is caused by two factors. One is primarily government funding. In a research operation, it's much easier and more efficient to train Ph.D.'s. The external funding for master's students has essentially gone away, but the cost of educating those students is real. If you look at a one-year master's program as just a course work-based master's degree, private schools have no way of supporting that, except in some very special context such as our Practice School Program and Stanford's master's program. Public schools may be a bit different. My viewpoint may again be an oversimplification, but nothing dramatic is going to happen along those lines until industry collectively says they want a master's degree and they're willing to pay for it, in some sense.

Andreas Acrivos: Mort was lamenting that our tenure system forces a young faculty member to specialize and become an expert in a particular field of science. I think that's excellent. Many years ago, when I started out in my academic career, it was well accepted among the general scientific community that yes, there was some fluid mechanics done among chemical engineers, but the work was second-rate. And there was some chemistry done, but it was second-rate, and so on. I don't think we want to be second-rate fluid mechanicians or second-rate physicists or anything else. Those of us who choose fluid mechanics should try to be among the leaders. On the other hand, I feel very strongly that those of us in education should have broad interests as well as our specialization. I'm therefore dismayed by what I see around me, by some of the younger and maybe some of the older faculty members who choose one field to the exclusion of all else. They cannot communicate on any subject except their own, and they're not interested in any other subject. That attitude is passed on to their students. I think that is deplorable. It's getting worse.

John Prausnitz: The tenure system encourages precisely the attitude that you mentioned.

Andreas Acrivos: Once you have achieved tenure, why can't you function as a scholar?

Robert Armstrong: Tenure selects that kind of person, doesn't it?

Robert Brown: It naturally selects against us.

Clark Colton: Is this a selection of the unfittest?

James Bailey: I disagree. My interests often expand to another area of biology rather than another area of chemical engineering. The increased focus within this particular area is not from being parochial or lazy, but just an alternative way of expanding one's interests. It may be a rather natural means for development.

Louis Hegedus: The idea of having core courses and fringe courses seems interesting. The core courses should provide the fundamental tools and techniques, something that has proved to be nonspecific, something that makes a chemical engineer: kinetics, applied math, economics (which in my opinion is not taught sufficiently), thermodynamics, fluid mechanics, transport, material science, separation science, control, and design. Then specific courses, basically the application of the previously listed items to specific areas of technology. Those usually end with the word engineering: biochemical engineering, biomedical engineering, reaction engineering, electrochemical engineering, environmental engineering, polymer science and engineering, electronics processing, combustion, and even catalysis. This structure, as Gary suggested, serves the purpose of giving us the tools, techniques, and identity simultaneously without any of them being in direct conflict with each other.

Gregory Stephanopoulos: I'd like to add the issue of collaboration, which was mentioned in every single area we discussed. Just a few years ago the majority seemed to be very much involved with the single-investigator idea. I find this kind of conversion to be very interesting. I ask, what is the mechanism to achieve this interactive collaboration? Do we expect it will happen by itself?

Edwin Lightfoot: Yes, I think it will. Mort, you gave an example with the seminars. I find that in Wisconsin just walking off to lunch is bad enough, because I'll meet somebody doing something interesting, and a new project is started. In a real university, where most of us come from, the possibilities are enormous.

Morton Denn: Collaborations are not necessarily, and probably not at all, engineering research centers. They're primarily two-body collisions. For example, I'm currently jointly advising Ph.D. students with about four other people.

Gregory Stephanopoulos: Why are we calling for collaboration then? Why is it not happening?

Morton Denn: It is happening.

Gregory Stephanopoulos: Why are we calling for it? There's a need, that's what I've been hearing all these days.

Mark Kramer: There is one very peculiar thing about the tenure system in regard to the collaboration issue. We do have five years at the beginning of our careers, or longer in some cases, where collaboration can hurt your chances of obtaining tenure. One has to establish an independent record of excellence and not a collaborative record. I hate to say it, but one does get trained into a mode of independent research in those early years of one's career. It's ironic if interaction is to be fostered. When we get a new person into the department, interactions are not fostered, quite the opposite.

L. E. Scriven: Collaborative research comes in two kinds. One kind serves to gain access to instruments, techniques, skills, knowledge that the initiator lacks but comes to see the need for in pursuing her —or his—research goal. Sometimes there are multiple initiators and all gain comparable advantage. There may be some synergism and the research could be called truly cooperative. This kind of collaboration can be mandated. It can be managed. It is common in industry, where it is mandated, the initiator being the manager or the manager's manager. There it is found on a grander scale in big development projects, which may have research components. It is not uncommon in academic research. It seems to me to be what the NSF is promoting with engineering research centers.

The second kind is collaboration in pursuit of a mutually conceived research goal, by a mutually selected path. Typically the goal lies in an interdisciplinary area, or has cross-disciplinary scope, or takes multidisciplinary attack. What is required is a synergistic cooperation and more: a genuine partnership, a relationship that has to be budded and grown, a zest for pursuing research together. Joule and Thomson's collaboration is the earliest that springs to my mind. It cannot be mandated. It is too special for ordinary management. It is uncommon in industry, probably less so in university. It can be encouraged by the environment, and at least a few managerial staffs and departmental faculties strive to do that and to take account of the needs of new additions to establish themselves. This kind of collaboration can be marvelously fruitful and satisfying, as many of us know first hand.

But collaborative research tends to be episodic. In the first kind the initiator who has the ability, the drive, and the luck will acquire the knowledge, skills, techniques, and instruments to get to the next goal. Or a newcomer will. Likewise, in the second kind, a newcomer will master the essentials of what the collaborators together supplied and will conceive a new goal. Creativity, innovation, standards, and effective direction in research at a frontier are, in my experience, more likely in a single investigator who comes to integrate the requisite disciplines, and to understand the relevant methods, than in a partnership. These traits also seem far more likely in a real partnership than in a team, whether it's managed by a director or a committee. This is, I think, what is missed by those who have been marketing their brand of collaborative research to engineers in U.S. universities in recent years.

George Stephanopoulos: I'd like to bring to your attention Mort Denn's thinking and ideas about what he calls in his fourth footnote the systems aspects of modern chemical engineering. He has identified the parallel between the problems that were stated 40 years ago and problems that are also being addressed today. But problems of estimating equilibrium states and rates have been haunting chemical engineers for 100 years. That doesn't imply a lack of evolution or a lack of intellectual excitement and content today in those areas. Furthermore, so-called systems engineering has created a series of new problems that were not present in the textbooks 50 years ago. For example, the problems of autonomous process control, product design, interaction between design and operations, structuring corporate-wide production systems, computer-integrated manufacturing, and others are concrete contributions of systems engineering.

Morton Denn: The intent of that note was to draw the contrast with what I think has happened in all other areas of chemical engineering, where modern science has driven the field substantially. The problems have been created by the new sciences as opposed to the systems area, where any reading of the 1930s textbooks defines an awful lot of what is being done, but it's simply being done better.

George Stephanopoulos: In that case, one must recognize the fact that systems engineering is a man-made science. It is not a natural science. Process control theory, process design, operations analysis, and fault diagnosis are not driven by advances in chemistry, biology, or physics. They are driven by advances in a separate set of scientific endeavors, such as applied mathematics, logic, computer science.

James Wei: From my perspective, this meeting has been a great success. We all respect each other, we enjoy each other's company, and we could go on indefinitely. It's been great to take stock and sort out issues and not have an agenda for action. Perhaps we didn't need one. However, I believe that a number of broad themes that concern us have been developed. They include the use of our traditional chemical engineering concepts and tools to conquer new application areas, the need for multidisciplinary teams to solve complex problems instead of doing them alone, the need to learn new sciences and concepts for both new areas and the traditional technologies. In education, we have an old model of a professional four-year B.S. degree, and apparently we want to do a lot more. We have to emphasize fundamentals. We also have a possible model of a professional five- to six-year master's degree with a possibility of not accrediting the four-year B.S. As some of you know, the president of ABET, Russell Jones, also president of the University of Delaware, is calling for that as a possible model. That means we could put more subjects into the curriculum. Finally, how do we keep the family together? This is one of the themes that Mort repeatedly came back to.

* * *

James Wei: The next thing that I want to ask, should we do this again sometime? Every 4 years, 10 years, 100 years? At lunch, the idea of 10 was mentioned several times. Mort Denn suggested that some of the originators be there at the next meeting to keep the history straight. In the future, we should set an agenda for action very high on the list of what we would like to accomplish. The last question is, who will organize it? We need to think about who are the people who want to do it.

Howard Brenner: Let me suggest that you need a context. The context here was the 100th anniversary of chemical engineering education. If there is some emergency or crisis that arises, then such a meeting springs up spontaneously and sets its own agenda. Hence, I don't believe that we need to think seriously about its organization at the present moment.

Clark Colton: Something occurred here, I think for the first time. It won't occur again unless somebody does it. I don't think people in this room representing different parts of our discipline have ever been together before in one room, talking to each other, learning from each other, from the people who are at the leading edge of the field. That doesn't happen these days. Do we want to bring that back again?

James Douglas: It happened a lot 25 years ago, Clark. I think we were very much closer together 25 years ago. People went to hear each other's papers. There's a whole different atmosphere now because of specialization. As we fracture more and more, becoming more and more specialized, it will be essential to bring people back together. I agree with you.

James Wei: Are you saying that it's worth trying to get the family together whether there is an agenda for action or not?

Voices: Yes.

James Douglas: That's the only way you can track the changes and the interactions in the field and where things are going.

James Wei: If we were going to meet again, it has to be of some interest to the people who want to organize it.

L. E. Scriven: Jim, I'd like to make a concrete suggestion. That is, we hereby go on record as favoring some sort of special event at the 100th anniversary of the American Institute of Chemical Engineers, which will be in the year 2008. That's 20 years hence.

Arthur Humphrey: Every other year there's an academic that's president of AIChE. Every four years, that academic president of AIChE could be the foreman, and we might bring the family together. That might be one way of doing it on some kind of a regular basis.

Voice: What's too often?

Voice: Less than five years.

Voices: Ten is a good number.

John Prausnitz: I want to say something that's on my mind; I suspect it's on everybody else's mind also. I want to thank MIT, and in particular Clark Colton, for organizing this conference and choosing such a magnificent place, and most of all for inviting us to participate. I think that all here agree that we haven't really solved any problems, but we are perhaps a little wiser now as we leave, compared to what we were when we came. I wish I had some champagne; water's the best I've got, but happy birthday, MIT.

INDEX

587